For Jeremy
at Chanukah
1996

♡

ORIGINS OF NEUROSCIENCE

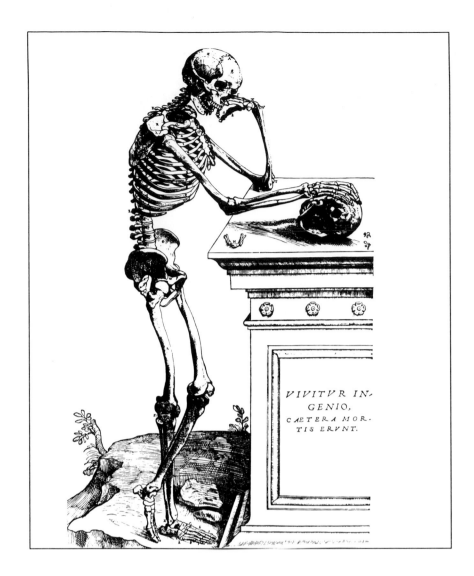

A plate from the first edition of *De humani corporis fabrica (On the Workings of the Human Body)* by Andreas Vesalius (1514–1564). Published in 1543, the Latin text on the base translates as "Man lives through his genius, all the rest is mortal." This drawing also appeared without text in later editions of the *Fabrica*. In addition, the same skeletal figure can be found in Vesalius' shorter work, *Epitome,* also published in 1543. The text on its marble column, however, contains a funereal verse from *Punica* by Silius Italicus (25?–101). The Latin lines on the panel of the tomb in the *Epitome* translate as:

All splendor is dissolved by death, and through
The snow-white limbs steals Stygian hue to spoil
The grace of form.

Vesalius did not identify the artist who made this plate or any of the other plates for the *Fabrica* or *Epitome*. This has led to endless speculation about who should get the credit for giving so much life to a pile of dry bones. One possibility is that the work was done by Jan Stephan van Calcar (1499–1546), who had illustrated an earlier book by Vesalius, his *Tabulae anatomicae* of 1538. Calcar was a Dutch pupil of Titian (c. 1488–1576), the great painter of the Italian Renaissance. Most scholars feel that if this figure did not come from Calcar, it was made by another accomplished artist from the school of Titian.

It has been suggested that the pensive skeleton studying a human skull inspired William Shakespeare (1564–1616) to write Hamlet's famous monologue: "Alas, poor Yorick! I knew him, Horatio: a fellow of infinite jest, of most excellent fancy; he hath borne me on his back a thousand times."

Origins of Neuroscience
A History of Explorations into Brain Function

Stanley Finger

New York   Oxford
OXFORD UNIVERSITY PRESS
1994

Oxford University Press

Oxford New York Toronto
Delhi Bombay Calcutta Madras Karachi
Kuala Lumpur Singapore Hong Kong Tokyo
Nairobi Dar es Salaam Cape Town
Melbourne Auckland Madrid

and associated companies in
Berlin Ibadan

Published by Oxford University Press, Inc.
200 Madison Avenue, New York, New York 10016

Library of Congress Cataloging-in-Publication Data
Finger, Stanley.
Origins of neuroscience : a history of explorations into
brain function / Stanley Finger.
p. cm.
ISBN 0-19-506503-4
1. Neurosciences—History. 2. Neurology—History.
3. Brain—Research—History.
I. Title. [DNLM: 1. Brain—physiology. 2. Neuropsychology—
history. 3. Neurosciences—history. WL 11.1 F497o]
QP353.F55 1994 612.8'09—dc20
DNLM/DLC for Library of Congress 92-48265

9 8 7 6 5 4 3 2 1
Printed in the United States of America
on acid-free paper

To My Wife Wendy
and
My Two Sons Robert and Bradley

Foreword

Not since the late 1960s when Clarke and O'Malley gave us their indispensable history of the brain and spinal cord have we had such an instructive, detailed neurohistorical review of this discipline. But this time, the history includes work on each of the senses and encompasses psychological as well as neuroscientific breakthroughs. In addition to its readability, it offers the readers an unsurpassed opportunity for acquainting themselves with the detailed facts of discovery, opinion, and controversy, function by function, starting in prehistory. Hence it also is a precious source of carefully documented references.

Its organization is natural. From general theories held throughout the ages of how the brain represents the world and guides the body through it, we get to an analysis of the senses, the motor functions, and their failings. Next we ascend from sleep and emotion to the highest functions of intellect and memory, speech and hemisphere dominance, their failings, and medical attempts at restoration.

The author has added the skill and passion of a historian to his time-consuming laboratory and teaching career in biopsychology. Under the influence of Arthur Benton, with whom he spent a summer in 1969, he began to add historical material to his own lectures to stimulate his students. *Origins of Neuroscience* is the result of this longstanding interest in the history of the discipline. To the reader, this book will provide ample and reliable information and guidance in this intricate field.

University of California   Francis Schiller, M.D.
San Francisco

Acknowledgments

The purpose of this book is to convey a feeling for the origins of the modern neurosciences and neurology. Like a tree with many branches and a rich bounty of fruit, the study of the brain can be approached from a variety of orientations, ranging from biochemistry, to gross anatomy, to the overt effects of brain lesions. In this case, a decision was made to emphasize the functions of the brain and how they came to be associated with specific brain parts and systems.

The impetus to write this sort of a book stemmed from two factors, one positive and one negative. The positive factor was the feedback I experienced when I added "neurohistory" to my lectures at Washington University, and to my presentations on recovery of function at scientific meetings. These historical interludes were not only well received, but they led many individuals to ask me where they could obtain more information on the origins of specific brain diseases, functions, systems, and theories.

The negative factor was that individuals interested in the neural sciences did not have a broad text to introduce them to this material. Further, they rarely had courses that discussed the origins of ideas, even theories as basic as cortical localization of function. Some professors told me they would gladly teach a course on the origins of the neurosciences, provided they had a text. The *Zeitgeist* seemed right, and it was for these reasons, coupled with my own insatiable appetite for new sources of stimulation, that I began haunting the libraries, collecting material for this book.

To be sure, a text of this size and scope could not have been written without the encouragement of my family, especially my wife Wendy. She and our two sons, Robert and Bradley, not only gave their blessings to this project, but taught me how to balance my seemingly endless reading and writing with the needs of a close and loving family. To my parents, Harry and Beatrice, I must also express appreciation for demonstrating to me what is truly important, and for endowing me with my inquisitive nature and basic love of learning.

I would also like to express my appreciation to Drs. Aaron Smith and Donald Stein for encouraging me to tackle this project, to Dr. Hans Markowitsch for helping in its formulation, and for all of the assistance and moral support given to me by Joan Bossert at Oxford University Press.

Additionally, I feel that I owe more than a simple debt of gratitude to Dr. Alexander Scriabine of Miles Laboratories, and to Friedel Seuter and Jurg Traber at Bayer AG, the parent company. For the past few years, I have been studying the ability of one of their new drugs to minimize the effects of brain damage. Yet, when I expressed my enthusiasm to write a book on the history of the neural sciences, they were more than willing to assist me on the project. A Faculty Research Grant from Washington University, which helped me to get through my first summer deep in the stacks of university libraries, was also most appreciated.

In terms of technical help, the Interlibrary Loan and Rare Books Library staffs at Washington University could not have been more generous with their time. Thank you Ellen Rubin, Nada Vaughn, Paul Anderson, and Susan Alon for your assistance. And thank you Jan Lazarus, at the National Library of Medicine, for going out of your way to help me obtain most of my photographs.

Several of my students were extremely generous in devoting time to going over drafts of my chapters, looking for inconsistencies and ways to improve the writing. The list is large, but among those students who were truly exceptional in tackling the work with gusto were Christy Wells, Spencer Greene, Grace Knuttinen, Gayle Watters, Jennifer Kolinski, and Lisa Foley.

Saving the best for last, I feel privileged to be able to mention the help of two remarkable scholars who took the time to read each chapter early on and provide me with many helpful suggestions. These true historians of the neurosciences not only added their scholarship to what I had written, but gave encouragement when I most needed it, and steered me to books and articles I might not have found on my own. Having been trained as a biopsychologist, and having worked as a laboratory scientist for more than 25 years, these two men served as my new teachers in the history of the brain sciences, and did so with enthusiasm and a wonderfully infectious love of this subject matter. But more than this, no person could hope for better friends to share a common interest. Thank you Francis Schiller and Arthur Benton, just for being who you are.

Without question, many new discoveries about the origins of the neurosciences will be made in the years ahead. Some of the statements made here will have to be modified as new findings come to light, much as would be the case in any discipline. In this context, I would welcome reprints and correspondence from my readers who come across important facts I might have missed, who wish to express a divergent opinion, or who feel that I should expand the subject matter in a particular way, should there be a future edition of this work. I am anxious to learn more, and hope that you too will be stimulated, and perhaps even enchanted, by the fascinating history of the neurosciences.

St. Louis, Missouri   S. F.
May 1993

Contents

Introduction

We cannot clearly be aware of what we possess till we have the means of knowing what
others possessed before us. We cannot really and honestly rejoice in the advantages of
our own time if we know not how to appreciate the advantages of former periods.
Johann Wolfgang von Goethe, 1810

Johann Wolfgang von Goethe, who was born in 1749 and died in 1832, was a master novelist, dramatist, and Romantic poet. He was also a natural philosopher who searched for unity in nature, and a serious scientist. The epigraph that opens this Introduction was taken from *Zur Farbenlehre,* one of his works on color vision. As can be gleaned from it, Goethe saw history as the proper path for putting scientific achievements into proper perspective. He was convinced that it is only by looking back that one can appreciate how far a discipline or a field has developed, whether the subject is color vision or anything else. He was also convinced that it is only through historical study that one can really appreciate new ideas, see the faults in existing theories, and determine the best path to follow for innovative new research.

Especially in the brain sciences, where new discoveries are being made at a breathtaking rate, it is imperative to look back to appreciate just how far we have come. It is important to distinguish between the truly monumental achievements and the smaller building blocks. It is also fitting and proper that we pay due homage to those earlier philosophers and scientists who were the real pioneering architects of this field of study. More than just an intellectual exercise, a history of neuroscience is essential to put new developments into perspective, to help us to recognize the status or value of our own work (often a humbling experience), and to allow us to see the best paths to follow for future research.

Most books dealing with the origins of the neurosciences have been concerned with the contributions of a central figure, developments within a specific time period, or the evolution of a particular discovery. Representative of the best of the biographies would be Francis Schiller's *Paul Broca: Explorer of the Brain* (1979). In the time period category, one could place *Nineteenth-Century Origins of Neuroscientific Concepts* by Edwin Clarke and L. S. Jacyna (1987) and Mary Brazier's *A History of Neurophysiology in the 19th Century* (1988). Illustrative of the books centered on a specific discovery or phenomenon would be Gordon Shepherd's *Foundations of the Neuron Doctrine* (1991) and Anne Harrington's *Medicine, Mind, and the Double Brain* (1987).

Few twentieth-century authors have set forth to survey the broader or more general history of the neurosciences.

Garrison's History of Neurology, as revised and enlarged by Lawrence McHenry (1969), is organized along a time line and has rightfully become a classic in the field. The present volume is comparable to *Garrison's History of Neurology* in that it too surveys the development of the neurosciences across the millennia. But unlike its predecessor, *Origins of Neuroscience* is organized around specific behavioral functions and describes how they eventually became associated with particular parts of the brain. In short, this volume deals specifically with structure-function relationships in the brain, from antiquity through the opening decades of the twentieth century—those periods that can be called the infancy and the formative years of the basic and applied neural sciences.

This text begins by looking at Neolithic skull openings and at perceived brain function in several of the great cultures of the past. Part I also traces the evolution of the theory of localization of function from antiquity to the twentieth century. It presents the most significant studies and developments that led to broad acceptance of localization theory, and examines what the opponents of localization had to say about this model of brain function, especially when applied to the cerebral cortex.

After the most important general theories and landmark studies are presented, specific functions of the brain are examined in detail. Thus, Part II deals with the history of each of the sensory systems: vision, audition, the skin senses, gustation, and olfaction. In a similar way, advances in understanding the motor systems and several movement disorders (e.g., Huntington's and Parkinson's disease) are presented in Part III.

Part IV looks at the history of sleep, dreaming, and the emotions. In contrast, the next section is concerned with the so-called higher functions of the brain. These chapters not only survey the emergence of the idea that the hemispheres constitute the substrate for intellect, but specifically trace the relationship between intelligence and the association areas, especially the "silent" cortex in the front of the brain. The neurobiology of learning and several disorders of memory are also presented in Part V.

Those functions that do not seem to be equally represented in the right and left hemispheres are among our newest evolutionary achievements. Speech and language, spatial abilities, and voluntary acts such as getting dressed are among the functions now associated with cerebral dominance. The relatively late recognition of cerebral dominance, and the subsequent explosion of research and interest in laterality phenomena, provide the subject matter for Part VI.

This book could easily have concluded with the chapters on cerebral dominance. This, however, would have left several interesting aspects of the history of brain science uncovered. Among the most fascinating integrative topics that would have been left out would have been those dealing with how brain damage has been treated at different times. Thus, much as this book began with a broad look at the development of theories of brain function, it ends with a general survey of how individuals with brain injuries and diseases were cared for from the height of the Egyptian Empire through the first quarter of the twentieth century in the West.

Goethe (1810) seemed to have a fair idea of what he was getting into when he began to write his history of color vision for *Zur Farbenlehre*. In his preface he wrote: "We have a right to expect from one who proposes to give the history of any science, that he inform us how the phenomena of which it treats were gradually known, and what was imagined, conjectured, assumed, or thought respecting them." He then went on to say that "to write a history at all is always a hazardous affair . . . for in such an undertaking, a writer tacitly announces at the outset that he means to place some things in light, others in shade."

Goethe realized that others might see different things in the material he would be describing, and he was willing to recognize that not all readers would agree with his value judgments. To this writer, the statements Goethe made in this context were personally meaningful. As he put it in the Preface to the first edition of *Zur Farbenlehre*:

> The author has, nevertheless, long derived pleasure from the prosecution of his task . . . [but] while the execution is generally accomplished portion by portion, he is compelled to admit that instead of a history he furnishes only materials for one. These materials consist in translations, extracts, original and borrowed comments, hints, and notes; a collection, in short, which, if not answering all that is required, has at least the merit of having been made with earnestness and interest. (1810; 1840 translation; 1967 edition, xxv-xxvi)

There is no need to add more.

REFERENCES

Brazier, M. 1988. *A History of Neurophysiology in the 19th Century*. New York: Raven Press.

Clarke, E., and Jacyna, L. S. 1987. *Nineteenth Century Origins of Neuroscientific Concepts*. Berkeley: University of California Press.

Goethe, J. W. von. 1810. *Zur Farbenlehre*. Weimar. Translated by C. L. Eastlake in 1840 as, "Goethe's Theory of Colours." Frank Cass & Company. 1967.

Harrington, A. 1987. *Medicine, Mind and the Double Brain*. Princeton, NJ: Princeton University Press.

McHenry, L. C., Jr. 1969. *Garrison's History of Neurology*. Springfield, IL: Charles C. Thomas.

Schiller, F. 1979. *Paul Broca: Founder of French Anthropology, Explorer of the Brain*. Berkeley: University of California Press.

Shepherd, G. 1991. *Foundations of the Neuron Doctrine*. New York: Oxford University Press.

PART I
Theories of Brain Function

Chapter 1
The Brain in Antiquity

Ox brains, ready prepared and stripped of most of the cranial parts, are generally on sale in the large cities. If you think more bone than necessary adheres to them, order its removal by the butcher who sells them. . . . When the part is suitably prepared, you will see the dura mater. . . . After the surrounding parts have been examined it is now time to dissect the brain itself. . . . Slice straight cuts on both sides of the mid-line down to the ventricles. . . . Try immediately to examine the membrane that divides right from left ventricle (septum). It has a nature like that of the brain as a whole and is thus easily broken if stretched too vigorously. . . . when you have exposed properly all the parts under discussion, you will observe a third ventricle between the two anterior ventricles with a fourth behind it. You will see the duct on which the pineal gland is mounted passing to the ventricle in the middle.

Galen, second century

The history of modern neurology, neuropsychology, and related disciplines is closely associated with the theory that different parts of the brain serve different functions. This idea, known as localization theory, is consistent with the basic biological principle that structures which look different should have different functions.

In the long history of the brain sciences, it is possible to conceive of the theory of localization as being applied to the whole and then to increasingly smaller parts. At first the question seemed to be, "Why is the brain special, and how is it different from other organs such as the heart?" As more became known about the brain, new questions arose. Some pertained to whether memory and sensation should be localized in the ventricles of the brain or in the neural mass. After this issue was resolved, attention was drawn to functional differences between gross anatomical divisions of the brain, such as the medulla, the cerebellum, and the cerebrum. Surprisingly, it was not until the nineteenth century that scientists seriously entertained the possibility that the cerebral cortex might be divided into distinct parts, each responsible for a different function. The dominant trend since 1861, when Paul Broca (1824–1880) presented the first widely accepted example of cortical localization, has been to try to divide all major brain parts into progressively smaller functional units.

The first four chapters of this book examine the history of localization theory. The present chapter begins by examining findings suggesting that the brain was valued by our early hominid ancestors. It then presents ideas about brain function from ancient Egyptian, Mesopotamian, Chinese, Indian, Greek, and Roman civilizations. The evidence for functional divisions among the major parts of the brain is the major theme of Chapter 2 ("Changing Concepts of Brain Function"), which concentrates on theories of brain function from the time of the Church Fathers through the eighteenth century. Chapter 3 ("The Era of Cortical Localization") primarily addresses the idea of cortical localization of function, while Chapter 4 ("Holism and the Critics of Cortical Localization") presents challenges to rigid cortical localization theory and introduces more holistic ideas about brain function.

These four introductory chapters provide the general background for more detailed discussions of specific brain functions, such as vision, hearing, motor functions, emotion, memory, and speech. These topics are among the many examined in subsequent chapters of this book.

HEAD INJURIES IN EARLY HOMINIDS

Head injuries, capable of causing brain damage, can be found throughout hominid evolution. In some cases, there is even evidence to suggest that early hominids realized damage to the brain could not only be disabling, but could also cause death. For example, among the South African australopithecines, early hominids that walked almost erectly, there is a skull estimated to be 3 million years old with distinct cranial fractures close to each other. Raymond Dart (b. 1893), the eminent anthropologist who examined this *Australopithecus africanus* skull, believed that this individual was clubbed to death from behind by another early hominid. Dart (1949) even showed that the wide condyles of an antelope humerus matched the depressions seen on the fractured skull of the "murdered" australopithecine.

Among 42 baboon skulls found at australopithecine sites in South Africa, 64 percent show evidence of fractures of the cranium, usually in the parietal area (Le Gros Clark, 1967). These baboon skulls also appear to have been hit with club-like weapons, presumably weapons wielded by australopithecines. Such findings further suggest that the brain, or at least the head, was viewed as critical for the vital functions basic to life, even more than a million years ago.

There are many examples of cranial injuries among *Homo erectus,* a more advanced type of hominid. Both Java man (roughly 500,000 to 300,000 years ago) and Peking man (roughly 300,000 to 100,000 years ago) belong to this group. A fair number of the specimens that have been found appear to have been murdered with blows to the head. Four of 11 skulls found on the Solo River displayed head wounds that could have caused death. In Chouk'outien, near Bejing, the cranial remains of about 40 individuals have been recovered, all showing head injuries (Janssens, 1970).

Traumatic head injuries in Neanderthal man (100,000 to 40,000 years ago) have also been well documented. For example, the skull of a Neanderthal specimen found in Shanidar Cave, Iraq, exhibited a healed wound on the top of the cranium and another in the eye region (Soleki, 1957, 1960).

TREPANATION

The assertion that the brain may have been given a special role in higher functions prior to the advent of the great civilizations is based on the fact that skulls with holes deliberately cut or bored in them have been found in a number of Neolithic sites (less than 10,000 years old), and possibly even in some older sites (Janssens, 1970). One such skull is shown in Figure 1.1.

The first trepanned (trephined) skull was found in 1685 by Bernard de Montfauchon (1655–1741) in Cocherel, France, but the significance of his specimen was not appreciated in his lifetime. In 1816, Alexandre François Barbie du Bocage (1797–1834) described a skull found at Nogent-les-Vierges, France. This time it was recognized that the skull was from an individual who survived a craniotomy performed years before he died (Janssens, 1970). The scientific community, however, did not fully appreciate how old this specimen was, and thus again missed the importance of the discovery.

Later in the nineteenth century, prehistoric skulls with manmade openings were found in many European countries, including France, Spain, Portugal, England, Denmark, Sweden, Austria, Poland, Italy, and Russia (Horne, 1894; Lucas-Championnière, 1912). In France alone, hundreds of specimens were discovered. One French site contained 120 skulls, 40 of which had manmade cranial defects (see O'Connor and Walker, 1967).

The holes in the European skulls varied in size from a few centimeters across to almost half the skull. The most common site for trepanation was the parietal region, but many occipital and frontal openings were also found (Walker, 1958). Only a few skulls have been excavated with holes in the temporal bones, presumably because these bones were harder to expose for this sort of surgery.

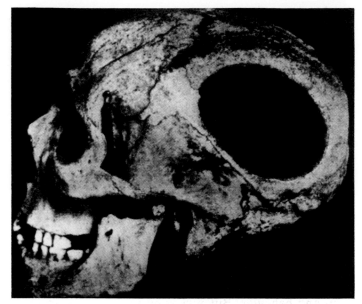

Figure 1.1.   Skull of a man discovered in 1854 in a cavern in Nogent-les-Vierges, near Criel, Oise, France. The skull (No. 343) is from the collection of the Musée de l'Homme, Paris.

The holes seem to have been produced in several ways, varying with the culture, the period, and the location. In the oldest European sites, the bone seemed to have been scraped away with sharp stones (e.g., obsidian or flint) or possibly even shells. In less ancient sites, drills that could be rotated manually (perhaps a pointed stone attached to a wooden shaft) and even primitive metal tools may have been used. The fact that the holes often exhibited smooth margins and clear signs of healing provides convincing evidence that this sort of surgery was conducted on living subjects and was not just a sacrificial or funeral rite.

Skulls with manmade openings were also found in the New World. The first depiction of such a skull appeared in 1839. At first, the opening was thought to be the result of a wound by a war axe. Only later was the nature of the opening appreciated.

Trepanation appeared to be quite common in the Cuzco region of Peru, where more than 10,000 trepanned skulls have been unearthed (estimates of the frequency of trepanation range as high as 47 percent in some areas; Trelles, 1962). In one Paracas Indian necropolis south of Lima, well over 10,000 well-preserved bodies have been found, with about 6 percent showing craniotomies. Many Paracan finds in Peru are about 2,000 years old, but some date back to the first millennia before Christ. Later cultures in this region also practiced craniotomies. This included the Tallan and Mochica cultures, which flourished in the pre-Inca period, and the Inca culture, which was centered in Cuzco and spread across a wide territory at the time of the Spanish conquest in 1532.

Many of the pre-Columbian Peruvian skulls show multiple holes resulting from several successful operations. On the basis of new bone growth, it has been estimated that approximately 63 percent of the individuals subjected to trepanation in ancient Peruvian sites survived at least one operation (survival rates are as high as 84 percent). This fact suggests that the operation probably ceased as soon as the dura came into view and that certain agents were applied to prevent infections (O'Connor and Walker, 1967).

Whether these early Peruvian surgeons used substances with anesthetic properties to help perform their craniotomies is unclear. It is known that the early Peruvians used coco leaves and alcoholic beverages, such as *chicha* and *masato,* to reduce some types of pain (O'Connor and Walker, 1967; Trelles, 1962). Still, the Indians were also taught to endure great pain.

The instruments found in the older Paracan burial sites suggest that the surgery was often accomplished with a piece of sharp obsidian attached to a wooden handle. Nevertheless, some copper and bronze surgical instruments *(tumis)* have been discovered at the newer sites. Some of these newer instruments were found to have elaborate handles and designs, such as the one shown in Figures 1.2 and 1.3. One Paracan skull was even found with a gold plate covering a partially healed hole, but such cases appear to be rare.

The prehistoric practice of opening the skull strongly suggests organized societies with strong beliefs about the brain and behavior. But just why were these operations performed on living people thousands of years ago? This question has led to considerable speculation.

One idea is that trepanation had something to do with tribal rituals and superstitions. Those who survived this surgery may have been treated as if they were endowed with special mental powers. In some New World sites, however, the observations that the skulls came from both sexes and represented a fairly wide range of age groups have been considered indirect evidence against the universality of this possibility.

Some individuals have speculated that these operations may have been conducted to treat headaches, convulsions, and mental disorders. These disorders were likely to have been attributed to demons, and it is conceivable that the holes were made to provide the evil spirits with an easy way out. Some evidence in support of this theory has come from tribes in Africa and the Pacific, where comparable surgeries were performed into the twentieth century as cures for epilepsy, insanity, and headaches (O'Connor and Walker, 1967).

A third theory is that the operations were performed to treat depressed skull fractures. This theory would give trepanation a firm foundation in rational medicine. Although the early European specimens provide little support for this idea, occasional skull injuries from slings and maces were later treated in this way in South America (Majno, 1975).

Figure 1.2. A bronze Peruvian tumi used for trephining the skull. (Courtesy of the Museum für Volkerkunde, Hamburg.)

Figure 1.3. Detail of the handle of the tumi shown in Figure 1.2.

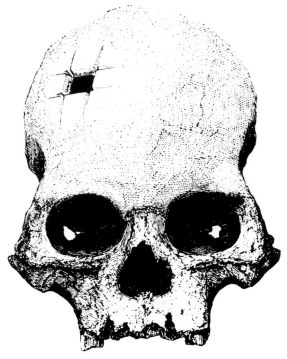

Figure 1.4. A Peruvian skull with a manmade opening. This skull was given to Ephraim George Squier, U.S. Commissioner to Peru, as a gift. Squier presented it to the New York Academy of Medicine and then took it to France for a second opinion. Paul Broca showed it to French scientists and argued that the opening was made deliberately. This individual is believed to have survived the surgery by one or two weeks. (From Squier, 1877, p. 457.)

It is interesting to note that Paul Broca (1874, 1876, 1877a, 1877b), best known for his discoveries relating to cortical localization, was one of the first scientists to try to explain why skulls were surgically opened. Broca's first foray into this area came in 1867. It involved a skull with cross-hatched cuts that came from an Inca cemetery in the valley of Yucay, Peru. The Peruvian skull, shown in Figure 1.4, was brought to him by Ephraim George Squier (1821–1888), an American cultural attaché and amateur anthropologist who had received it as a gift in 1865 from a Mrs. de Romainville of Cuzco (Trelles, 1962). The skull was interpreted (by a Dr. Gardner) as a "case of trephining" before the incredulous members of the New York Academy of Medicine in 1866, but Squier wanted a second opinion. Squier thus sought out Broca, a recognized leader in French anthropology.[1]

This skull left no doubt in Broca's mind that "advanced surgery" was performed in ancient Peru well before the European conquest. He believed that the opening was made while the individual was still alive, although death may have followed within a few days or weeks at most. Broca was perplexed by the fact that there was no sign of fracture on the Peruvian skull. This suggested that the operation might have been performed to relieve a build-up of pressure after a closed head injury, but Broca was not convinced of this and decided it was best to leave the matter open.[2]

After dealing with Squier's specimen, Broca turned his attention to the older trepanned skulls that had been excavated in France. He suggested that Neolithic people were likely to associate various afflictions with supernatural powers. In his mind, this meant that these individuals probably made skull openings to give the demons responsible for convulsions and other problems a means of escape.

Broca mistakenly concluded from his small sample that the operations were performed only on young people. He also surmised this because the young skull was softer and healed more rapidly and because the young were more likely to suffer from epilepsy. He even conducted some experiments to show just how quickly immature skulls could be perforated with a "primitive" glass scraper (Broca, 1877b; Horne, 1894; Janssens, 1970). In fact, Broca found he could scrape a hole in the skull from a deceased 2-year-old child in just 4 minutes, although it took 50 minutes to scrape open the thick skull of an adult (he frequently had to rest his hands from the strain). Some of his craniotomy experiments also involved dogs. Here he showed that it was relatively easy to avoid damaging the dura mater.

Broca (e.g., 1877b) was interested not only in the cranial openings but also in the bone amulets sometimes found with these skulls in European sites. He thought these charms (some chipped out posthumously) were given to the deceased at funeral rites. This ritual may have been conducted to provide the dead with a talisman to drive away evil spirits, to provide evidence of the status of the individual during the past life, or to compensate the dead for bone previously scraped from the skull. If one thing seemed certain to Broca, it was that these amulets provided the first good evidence for a belief in an afterlife.[3]

EGYPTIAN MEDICAL RECORDS

Although certain pre-Columbian cultures depicted craniotomies in their art (e.g., Mochica ceramics), the Egyptians were the first to provide systematic medical records in writing. Of all the known papyruses, the *Edwin Smith Surgical Papyrus* stands out as being the earliest and perhaps most remarkable ancient written record dealing with the effects of head injuries.

Edwin Smith (1822–1906) bought the celebrated papyrus in 1862 in two transactions in the city of Luxor, a part of Thebes. Smith, an Egyptologist, attempted to translate the scroll and seemed to understand much of its contents. But, for unknown reasons he did not publish any of his valuable material. Soon after Smith died, his 4.68-m (15 ft, 3½ in) rolled document was given to the New York Historical Society by his daughter. It was not until 1920, however, that James Breasted (1865–1935), an Egyptologist at the University of Chicago, was asked to translate the papyrus. This was not an easy task, and Breasted was unable to publish his complete translation of the material, with his insights

---

[1]For the history of Broca's life and his impact on science, see Schiller (1979).

[2]Squier included an English translation of Broca's comments in an appendix to his 1877 book, *Peru: Incidents of Travel and Exploration in the Land of the Inca*, which also presented a drawing of the famous skull.

[3]Fragments of bone resembling amulets and powders made from skull bones (especially those from Egyptian mummies) were still being sold in European pharmacies during Broca's lifetime (Schiller, 1979). In particular, Ossa Wormiana (pieces of bone from aberrant skull sutural formations) was widely promoted as an effective treatment for epilepsy.

Figure 1.5. Statue of Imhotep, vizier and physician of King Zoser, a pharaoh of the Third Dynasty (c. twenty–eighth to late twenty–sixth century B.C.). (Courtesy of the Egyptian Museum, Cairo.)

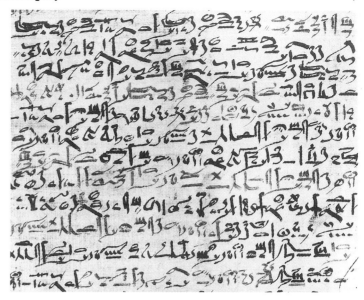

Figure 1.6. A section of the Edwin Smith Surgical Papyrus showing the script and use of two different types of ink. (Courtesy of the University of Chicago Press.)

and commentaries, until 1930. But as he worked on the translation of the papyrus, Breasted realized that scholars who believed that Alcmaeon of Croton (fifth century B.C.) was the first person to write about the brain (whose convolutions were likened to corrugated metal slag in the papyrus he was translating) were in for a rude awakening.

There appeared to be three contributors to the ancient papyrus. The first was the original author, who probably lived near the beginning of the Old Kingdom, the Age of Pyramids. Although Breasted and his contemporaries believed the dates of the Old Kingdom to be roughly from 3000 to 2500 B.C., more contemporary historians have moved its dates forward by about 200 to 300 years, often dating it from about 2780 to about 2200 B.C.

The original author was probably a surgeon who followed the Egyptian armies on some of their campaigns. He was very knowledgeable about wounds caused by spears, swords, and other weapons of war, the injuries described most frequently in the papyrus. He categorized his clinical cases according to body part injured, from the head down, and classified the severity of each as:

1. An ailment which I will treat.
2. An ailment with which I will contend.
3. An ailment not to be treated.

Although the original author is not known with certainty, it has been suggested that he may have been Imhotep, the reputed father of Egyptian medicine depicted in Figure 1.5. Imhotep served King Djozer (Zoser; c. twenty-eighth to late twenty-sixth century B.C.) early in the Third

Dynasty.[4] To honor Imhotep as a physician and healer, temples were later built in Memphis, in nearby Sakkara, and in other locations for the sick and injured to pray and receive medical treatment. Among his many other accomplishments, Imhotep, the royal architect, designed the six-layered "step" pyramid of Sakkara as the eternal resting place for Djoser. This truly spectacular monument is still hailed by architects as the earliest major structure made throughout from stone.

The second contributor came forth several centuries later, probably near the end of the Old Kingdom, and was a most helpful commentator. Some of the original language had become obsolete by his time, and he tried to interpret and explain many of the archaic words in the copy he made of the older document. For example, he wrote that the old phrase "to moor him at his mooring stakes" meant to put the patient on his accustomed diet and not administer any medicine. The analogy was to a boat moored at a dock, one left alone to move up and down in harmony with the waves and tides.

The third contributor appeared around 1650 B.C., during the Twelfth Dynasty. He dipped his pen into red and black ink to make a copy of the annotated version of the text, most likely relying upon a copy of a copy of the original. As can be seen in Figure 1.6, this individual wrote in heiratic, a cursive form of writing that took less time than hieroglyphics, the older and more detailed pictographic form of writing illustrated in Figure 1.7, which shows Imhotep's name in characters. On a case-by-case basis, he copied the history, examination, diagnosis, prognosis, and treatment

---

[4]"Wise Imhotep" achieved great recognition, not only as a skilled physician and the royal architect but as a grand vizier (a minister of state), a scribe, a priest, and an astronomer (Hurry, 1926). His fame and importance seemed to rise with the passage of centuries. At first he was a mere mortal, but soon rose to be a demigod, and then a full deity. The Greeks called Imhotep Imouthes and melded him with their own god, Asklepios. Like the Ancient Egyptians, they believed that gods could evolve out of common men and that the difference between mortals and deities was one of degree. Imhotep, "God of Memphis," was worshiped until the fourth Century A.D.

Figure 1.7. "Imhotep" written using heiroglyphics.

of various injuries sustained in combat. He made his share of errors and corrected some, even employing the first known asterisk to show where some overlooked words should have been placed. And then, for some unknown reason, he stopped writing right in the middle of a word, leaving the bottom portion of the papyrus empty.

We do not know whether this important contributor was a lecturer, a student, or, most likely, a professional scribe hired to do the work. Yet it was his less than perfect copy of the treatise that managed to survive the ravages of time, probably by being buried with its owner in a coffin in an old Thebian cemetery (Breasted, 1930).

Some of the 48 cases presented in the *Edwin Smith Surgical Papyrus* involved individuals who suffered from head injuries. The descriptions revealed that early Egyptian physicians were aware that symptoms of central nervous system injuries could occur far from the locus of the damage. There were examples of head injuries that caused problems in eye-hand coordination and many deficits that occurred on the opposite side of the body from the head injury, such as the one presented in Figure 1.8. There is also evidence for a contrecoup injury, one in which a blow on one side of the head caused the brain to compress and shear against the other side of the skull, resulting in more

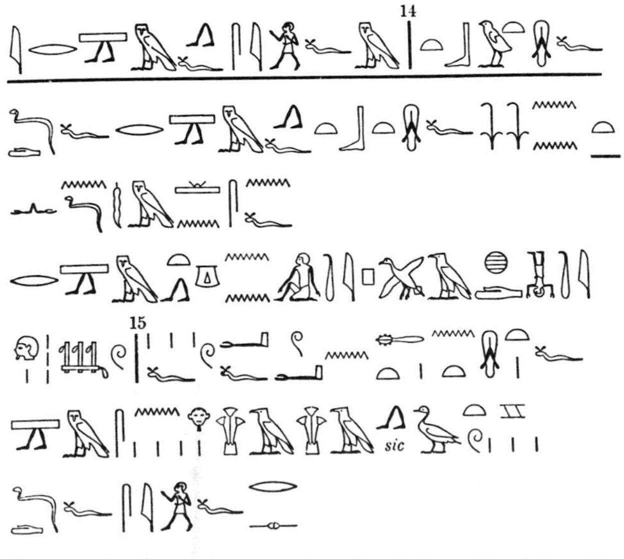

damage and bilateral symptoms (Figure 1.9 shows a head wound from a later era, but one similar to those that may have been encountered by the author of this papyrus.)

The following is a typical head injury case from the papyrus.

Examination: If thou examinest a man having a gaping wound in his head, penetrating to the bone, (and) splitting his skull, thou shouldst palpate his wound. Shouldst thou find something disturbing therein under thy fingers, (and) he shudders exceedingly, while the swelling which is over it protrudes, he discharges blood from both his nostrils (and) from both his ears, he suffers with stiffness in his neck, so that he is unable to look at his two shoulders and his breast, ...

Diagnosis: Thou shouldst say regarding him: "One having a gaping wound in his head, penetrating to the bone, (and) splitting his skull; while he discharges blood from both his nostrils (and) from both his ears, (and) he suffers with stiffness in his neck. An ailment with which I will contend."

Treatment: Now when thou findest that the skull of that man is split, thou shouldst not bind him, (but) moor (him) at his mooring stakes until the period of his injury passes by. His treatment is sitting. Make for him two supports of brick, until thou knowest he has reached a decisive point. Thou shouldst apply grease to his head, (and) soften his neck therewith and both his shoulders. Thou shouldst do likewise for every man whom thou findest having a split skull.

Example of Explanation: As for "Splitting his skull," it means separating shell from shell of his skull, while fragments remain sticking in the flesh of his head, and do not come away.

The original author of the papyrus was an astute observer for his time, but his knowledge of the working of the nervous system was rudimentary at best. Although some of his cases might have suggested otherwise, consciousness and intelligence were viewed as matters of the centrally placed heart during the Old Kingdom, and would remain so in the Land of the Pharaohs for centuries to come. Indeed, the early Egyptians did not distinguish between tendons, blood vessels, and nerves, using the same term, *metu,* meaning "channel," for each. These and other *metu* came together at the heart, and carried blood, air, semen, mucus, food, and waste throughout the body, provided they were not blocked.

Figure 1.8. James Breasted translated the original heiratic of *The Edwin Smith Surgical Papyrus* into both heiroglyphics and English. This shows part of Case 8. To quote: "He walks shuffling with his sole." "As for: 'He walks shuffling with his sole,' he (the surgeon) is speaking about his walking with his sole dragging, so that it is not easy for him to walk, when it (the sole) is feeble and turned over, while the tips of his toes are contracted to the ball of his sole, and they (the toes) walk fumbling the ground. He (the surgeon) says: 'He shuffles,' concerning it." (Breasted, 1930, p. 210; courtesy of the University of Chicago Press.)

Figure 1.9. Mummified head of King Seqenenre (c. 1580 B.C.) showing the head wounds that killed him. (Courtesy of the Cairo Museum, Egypt).

Figure 1.10. Vignette from the *Book of the Dead* of the scribe Ani showing the weighing of the heart. (Courtesy of the British Museum, London. See C. Andrews, *Egyptian Mummies,* Cambridge, MA: Harvard University Press, 1984.)

The papyrus purchased by Edwin Smith was not the only early Egyptian papyrus that dealt with medicine. Another important discovery was the *Ebers Papyrus,* a much longer document known to Edwin Smith (who even may have owned it at one time), and purchased in 1873 by Georg Ebers (1837–1898), an Egyptologist from Leipzig, Germany. This huge papyrus, also discovered in Thebes around 1862, is over 65 feet long and has 108 paragraphs, each containing 20 to 22 lines. It was said to be found between the legs of a mummy (possibly the same mummy associated with the Smith's papyrus) and was written at about the same time. In addition, and perhaps most importantly, it was also recognized as a copy of a far older treatise on medicine, although its ancestry is not quite as old as the *Edwin Smith Surgical Papyrus.*

The *Ebers Papyrus* begins with the line: "Here begins the book on the preparation of medicine for all parts of the body." Indeed, the work contains approximately 900 prescriptions. Some of these prescriptions have been found to contain important medicinal ingredients (see Chapter 28).

ILLNESS AND MYTHOLOGY IN ANCIENT EGYPT

An unsuspecting reader of the *Ebers Papyrus* will probably be surprised to find that 55 prescriptions contain urine or feces, substances thought to spread illnesses, not to cure them, in modern societies. Worse, the excrement often was rubbed over the body of the afflicted. This bizarre practice can be understood only by realizing that Egyptian medicine blended mythology and religion with more rational therapeutic practices. In the case of feces or urine, the logic was primarily to make the body uninhabitable for the demons thought to be responsible for the illness. Especially in the case of disease, the Egyptians believed that the body of a sick individual was one no longer looked upon favorably by the gods, either because of sins or because the good forces had been tricked by evil spirits. Cures therefore required not only medicines but prayers, gifts, charms, and often very elaborate rituals and ceremonies.

As noted, of the various bodily organs, the heart stood well above the brain on the Egyptian ladder of importance.

It, and not the brain, was the seat of the soul, which was why it was so important to keep the 36 channels from the beating heart open. The Egyptians further believed the heart recorded all of the good and evil deeds that one did while alive. At the end of one's life, they thought that the heart was weighed against a feather to see if it were heavy with guilt and evil or free from sin. The weighing ceremony determined whether the deceased would go to heaven or be fed to a crocodile-like mythological figure called the Devourer.

During the judgment, depicted in Figure 1.10, Anubis (the god of mummification, usually shown with a jackal's head) held a scale while Thoth (the god of writing, typically depicted having the head of an ibis) recorded the answers given by the heart to 40 questions. Sentence was passed by an anonymous Great God before the start of the Middle Kingdom (2052 B.C.) or by Osiris, who became supreme god of the dead at that time. It was essential to keep the heart with the body for this ceremony, and for this reason it was left in place prior to the New Kingdom (1567 B.C.).

Although the Egyptians wrote little about their own embalming procedures, Herodotus of Halicarnassus

Figure 1.11. Herodotus (c. 450 B.C.), the Greek historian who traveled to Egypt and other countries where he learned about ancient medical practices. Much of what is known about Egyptian embalming procedures comes from Herodotus. (Courtesy of the Museo Nazionale, Rome.)

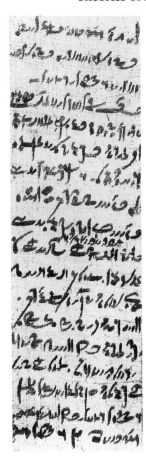

Figure 1.14. Canopic jars from the Twenty-first Dynasty (1000 B.C.). The jars represent the four sons of Horus and were used to house the liver, the lungs, the stomach, and the intestines. The brain was discarded while the heart was considered too important to remove. (Courtesy of the British Museum, London.)

Figure 1.12. Letter from an embalmer in Thebes acknowledging receipt of natron and linen. The embalmer promised to prepare the body and return it on the 72nd day. The letter dates from 270 B.C. (Courtesy of the British Museum, London.)

(c. 490–425 B.C.), the "Father of History," visited Egypt and described embalming procedures still in practice. He also collected information about how embalming was done in earlier periods. Herodotus, shown in Figure 1.11, wrote that the process of embalming took 70 days (see Figures 1.12 and 1.13). It is likely that it actually took less time and 70 days represented the period between death and burial (Spencer, 1982).

The heart was so important in early Egyptian culture that the *paracentetes,* the men who prepared bodies for mummification, sometimes fled in fear of their lives just upon seeing it (Woollam, 1958). In contrast, the brain was treated with complete indifference. Herodotus stated that during the time of the New Kingdom, a common procedure among the wealthy and noble classes was to have most of the brain scooped out through the nostrils and the base of the skull with an iron hook and then simply discarded. Writers after Herodotus have suggested that a small chisel may have been used to pierce the bones of the nose to make this easier. Tissue that could not be scooped out often was rinsed away with chemicals (David, 1978). Although the brain could also be removed through a hole made in the skull or in the eye socket, this procedure was rare (Harris and Weeks, 1973; Leca, 1981). Because the brain was deemed unimportant for the afterlife, it was never saved, and the skull cavity simply was packed with resin-soaked strips of linen (Spencer, 1982).

Even the lungs, liver, stomach, and kidneys received better treatment than the brain. For the wealthy, these organs were either packed into four parcels for replacement back into the body or stored in canopic jars, sometimes decorated to represent the four sons of the mythological figure Horus (the importance of Horus in Egyptian mythology and medicine is explained in Chapter 28). One set of canopic jars is shown in Figure 1.14. The poor, who could not afford the fees for surgical removal and preservation of these organs, simply had the intestines of their dead washed out with cedar oil and the bodies dried in the preservative natrum.

In *The Histories,* Herodotus also stated that the Ancient Egyptian physicians had distinct medical specialties, some dealing with the organs of sensation and the head. In his words:

> The practice of medicine they split up into separate parts, each doctor being responsible for the treatment of only one disease. There are, in consequence, innumerable doctors, some specializing in diseases of the eyes, others of the head, others of the teeth, others of the stomach, and so on; while others, again, deal with the sort of troubles which cannot exactly be localized. (1972 translation, 160)

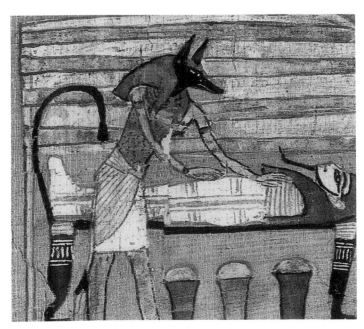

Figure 1.13. Detail from the *Book of the Dead* of the scribe Ani depicting Anubus preparing a mummy within the embalming chamber. Nineteenth Dynasty (c. 1250 B.C.). (Courtesy of the British Museum, London.)

MESOPOTAMIAN MEDICINE

Mesopotamia encompasses the land between the Tigris and Euphrates rivers and between Armenia and the Persian Gulf. Much of the region corresponds to modern Iraq. There is reason to believe that the art of healing developed simultaneously in Mesopotamia and Egypt.

Herodotus thought that the Mesopotamians had no physicians and simply placed their ill or injured in public squares, where passers-by would be obligated to inquire about their problems and to offer sound advice based on personal experience. Modern writers maintain Herodotus erred when he suggested that physicians did not exist in Mesopotamia (see Thorwald, 1963). They point to the fact that Hammurabi (1792–1750 B.C.), the best-known king of Babylon, specifically mentioned the medical profession in many of his laws.

Hammurabi's code dictated the fees to be paid for specific services. The fees were based not only on the type of treatment but on the person who received the help. For example, the code stated that the physician will be paid 10 pieces of silver if he makes a wound and cures a freeman, 5 if the patient is the son of a plebeian, and 2 if he is a slave. The code also listed severe penalties for carelessness and malpractice. One section states that if the patient is treated "with a metal knife for a severe wound and (the physician) has caused the man to die—his hands shall be cut off" (see Walker and O'Connor, 1967, p. 9).

Interestingly, paragraph 278 of Hammurabi's code states that the sale of a slave is null and void if the slave exhibits signs of certain diseases soon after being acquired. One such disease, called *sibtu,* appears to be epilepsy.

The general belief among the Sumerians, the early inhabitants of Babylon, was that demons and other supernatural powers caused disease. Like the Egyptians, they believed their good gods would no longer protect the sinners. This would allow demons and spirits to invade the sinner's body. Consequently, treatments in Mesopotamia also involved prescriptions ensconced in mythology. Prayers and gifts were among the ingredients required to bring the sick back to health (Thompson, 1923, 1924).

Early surgical instruments that could be used to make skull openings have been found in Mesopotamia. Nevertheless, very few skulls with man-made openings have come out of the Middle East. Only in Lachish, a Jewish city (now in Israel), were sawlike holes identified on a collection of skulls. These crania provide evidence of new bone growth, but it is not known if these were the skulls of individuals in an isolated subculture that practiced craniotomies or those of prisoners who came from distant lands where these surgeries had been performed.

In general, much less is known about diseases in the land between the rivers than in Egypt. A major reason is that Mesopotamian soil did not favor preservation of the flesh or even the skeleton.

ANCIENT INDIA

The first written references to medical ideas and practices in Ancient India can be found in the *Atharvaveda* (*veda* means "knowledge"), a work dated after the Aryan ("the Nobles," later known as Sindhu or Hindu) invasion in the middle of the second millennium B.C. Among other things, the *Atharvaveda* provides descriptions of epilepsy, insanity, neuralgia, headaches, and blindness.

Like other ancient peoples, the priests of the Vedic period (c. 2500–800 B.C.) prayed to the gods to help them deal with diseases. The priests relied upon spells, prayers, sacrifices, and rituals. They also performed exorcisms of the mischievous demons. In the *Rigveda* we are told that the sages who accompanied the Aryan tribes also carried bags of medicinal herbs to care for the sick and wounded. They knew how to cauterize wounds, and occasionally created artificial legs and eyes.

The oldest known Indian documents devoted just to medicine are the *samhitas,* a word meaning "collections." These works are compilations of ancient Vedic medicine, largely passed down by oral tradition (Banjeri, 1981; Filliozat, 1964). Indian scholars seem to agree that two of the most important of these documents are the *Charaka Samhita* and the *Susruta Samhita.*

Charaka was a Hindu physician from the court of King Kanishka (either A.D. 78–101 or 120–162). He took on the job of writing down many of the medical ideas that earlier physicians had just listened to and memorized. Charaka opened his text with a statement about the divine origin of medicine. Brahma, the Creator God, transmitted medical knowledge to the mystic sage Arti, who passed it on to other sages. It eventually was given to Atraya, the first physician, who then taught other physicians.

The *Charaka Samhita* contains a large number of lessons. Some are concerned with the management of disorders of the three vital parts, the heart, the head, and the urinary bladder. Of these parts, the heart was viewed as most important. It was the source of power and the center of the body's extensive canal system (70 human blood vessels are listed). *Rasos,* or humors, flowed through these canals to various parts of the body.

As for the head, although it is embedded in a tangle of metaphysical beliefs and notions, there is the following telegraphic listing of some of the symptoms of injury to it:

> If the head is affected stiffness of the carotid regions, facial paralysis, rolling of eye balls, mental confusion, cramps, loss of movement, cough, dyspnoea, lock-jaw, muteness, stammering, ptosis, quivering of cheeks, yawning, salivation, loss of voice, crookedness of face, etc. arise. (Translated by Sharma, 1983, 644)

The *Charaka Samhita* also presents some 500 herbal drugs used by Indian physicians, a list that makes even the extensive Egyptian pharmacies pale by comparison. One of these agents was called moon because it resembled a half moon. It was later identified as *Rauwolfia serpentina,* a sedative that can lower blood pressure and serve as an effective agent for treating headaches, pain from injuries, and anxiety.

The *Susruta Samhita* was probably written between 800 and 400 B.C. Some of the confusion concerning this document may be due to the fact that there are many copies and many versions of it. Some of the passages in this text suggest that certain advances initially attributed to the Greeks may have come from Indian medicine. Of particular interest is the earliest known lesson in dissection:

For these purposes a perfectly preserved body must be used. It should be the body of a person who is not very old and did not die of poison or severe disease. After the intestines have been cleaned, the body must be wrapped in bast (the inner bark of trees), grass, or hemp and placed in a cage (for protection against animals). The cage should be placed in a carefully concealed spot in a river with fairly gentle current, and the body left to soften. After several days the body is removed from the water and with a brush of grassroots, hair, and bamboo it should be brushed off a layer at a time. When this is done, the eye can observe every large or small, outer or inner part of the body, beginning with the skin, as each part is laid bare by the brushing. (Translated in Thorwald, 1963, 202–203)

Susruta taught that surgery should be regarded as the highest, most prestigious of the healing arts. Indeed, surgery was relatively advanced in ancient India, as evidenced by the wide variety of surgical tools that have been found in archeological sites. The *Susruta Samhita* even contains an excellent description of an early cataract operation (see Chapter 5).

It is believed that Indian kings established the first hospitals, following the teachings of Gautama, or Buddha, The Enlightened One (c. 560–480 B.C.). One such hospital was in existence in 427 B.C. in Ceylon (now Sri Lanka). On a rock near Junagarh, one can find several of the edicts of Asoka (274–232 B.C.), an early emperor of the Indian Empire. One reads:

Everywhere King Piyadasi (Asoka) erected two kinds of hospitals, hospitals for people and hospitals for animals. Where there were no healing herbs for people and animals, he ordered that they be brought and planted. (Translated in Thorwald, 1963, 217)

Bishop Basil the Great of Caesarea (330–379) is usually credited with establishing the first "Western" hospitals, but this did not occur until well after many hospitals were built in ancient India and its territories.

ANCIENT CHINA

Ancient Chinese medicine is extremely complex and very much tied to metaphysical notions. The Chinese believed in *Zang Fu,* a general designation of internal organs. There were five main organs, the *Zangs* (heart, lungs, kidneys, liver, and spleen), which were associated with five basic elements (fire, wood, earth, metal, and water). The *Fus* were viewed as auxiliary organs, the main ones of which were involved with digestion. The brain was classified as a "peculiar" *Fu.*

The doctrine of *Zang Fu* appears in the *Huang Di Nei Jing,* named after Huang Di, the legendary first Chinese Emperor. Fragments of this document have been traced to "The Period of Contending States" (475–221 B.C.), but this collection of ancient beliefs, traditions, and healing arts has been revised many times throughout its long history.

Zang Fu looks upon the heart as the most important organ because it houses the spirit and is filled with blood. The heart, along with all the other *Zangs,* is involved in mental activity, but it also hosts happiness. When energy and blood fill the heart, the individual will be vigorous and

smart, but when this is not the case, poor memory, insomnia, and other pathological signs will arise. As for the other *Zangs,* the lungs were associated with sadness, the spleen with consciousness, the liver with anger, and the kidneys with fear.

The brain was believed to be composed of marrow, and in the *Huang Di Nei Jing* it is written that the eyes are connected to the brain. It follows that if evil invades the eyes, it will also invade the brain.

An important part of Chinese medicine concerned *Tao,* the World Spirit. Its contrasting components were Yin and Yang, which were associated with different organs in the body. Yin signified soft, calm, cool, introverted, and feminine, while Yang related to hard, active, warm, outgoing, and masculine. Disease reflected an imbalance between Yin and Yang and could be caused by demons and related mystical agents.

The association of the supposedly affected organ with different elements and with Chinese astrology offered various solutions for curing the different diseases. It is thought that acupuncture arose out of the early notion of driving demons out of the body by wounding and puncturing the flesh with needles. The idea was to make the body an uncomfortable place for the demon, while providing an opening for easy escape. Later, acupuncture was incorporated into more complex frameworks, like the one presented above.

The Chinese pharmacopeias also reflect these complex notions of cosmos and body. Still, as was the case with early Egyptian and Indian medicine, a number of drugs seemed to have good clinical justification. For example, ginseng root, which was frequently recommended, contains substances that intensify vascular tone, increase metabolism, and stimulate the central nervous system.

The Chinese did not perform dissections because ancestor worship forbade the mutilation of the body. This tradition was in part responsible for the endless speculation and philosophy about how disease was related to the cosmos.

The first mention of a surgeon was a man named Hua To, who lived from 190 to 265. Because he had no known predecessors for over 1,000 years, it is thought that he may have been a foreigner, most likely a physician trained in India. Hua To was executed by Tsao Tsao, his prince, but the precise reason is not clear. One version of the story explains that he was killed when he recommended trepanation or possibly acupuncture to his suspicious prince, who suffered from severe headaches. A second version simply holds that he refused to serve the prince properly. Unfortunately, Hua To had his medical library destroyed prior to his scheduled execution (Thorwald, 1963).

THE GREEK ELEMENTS AND THE HIPPOCRATIC REVOLUTION

Greek thought about the body in health and in disease became intimately connected with the theory of fundamental elements. Thales (652–548 B.C.), an Ionian educated by Egyptian priests at Memphis, accepted the Eastern view and that of Homer (c. 700 B.C.), which held that water was the fundamental element of all things.[5] Thales' pupil Anaxi-

---

[5]Thales founded the Ionian school of philosophical thought.

mander (638–547 B.C.), however, adopted another ancient notion, that is, the Mesopotamian idea that all matter was made of water and earth (mud) fused by the sun. A third substance, air, was considered fundamental by Anaximenes (d. 504 B.C.), a pupil of Anaximander. A fourth Ionian, Heraclitus of Ephesus (556–460 B.C.), believed fire was the fundamental element of the cosmos, bringing yet a fourth element into consideration.[6]

Other than the water-earth combination, these philosophers did not package these four "elements" together. The job of combining the elements into one parcel was left to others. In Crotona, Italy, Pythagoras (580–489 B.C.) philosophized that water, earth, air, and fire could account for everything on earth, including the human body. He did not, however, look upon these substances as immutable and distinct, but rather as interchangeable and mixed. For example, fire could be compressed into air, while earth could expand into water. Moreover, Pythagoras thought that these four substances were derived from even more primitive matter. Empedocles (c. 490–430 B.C.), who was intrigued by Pythagorean ideas, is often given the major credit for developing and enunciating the quarternary theory of four basic elements that guided Greek, Roman, and medieval thinking.

At first, each of the four primary substances was associated with one quality, these being moist, dry, heat, and cold. These qualities were really two pairs of opposites: moist versus dry, and heat versus cold. But by the third century B.C., each element was associated with not one but two qualities. Air was associated with heat and moisture, fire with heat and dryness, earth with dryness and cold, and water with cold and moist.

At about the time that the theory of fundamental elements was taking shape, Alcmaeon (c. fifth century), who also taught in Crotona, performed some of the earliest recorded dissections and described the optic nerves (McHenry, 1969). His work led him to propose that the brain was the central organ of sensation and thought. Anaxagoras (500–428 B.C.), who left the Ionian school to found the Athenian school, also proposed that the brain was the organ of mind.

Although these ideas about the brain caused people to rethink long-accepted ideas, they were not uniformly accepted. Many Greek philosophers at this time continued to look upon the heart as the seat of the soul and the organ responsible for many, if not all, mental functions.

Greek thinking during this transition period was still dominated by the idea that gods and demons were responsible for health and illness. This ancient belief remained essentially unchallenged until around 400 B.C. The individual most responsible for exorcising the demons from medicine and calling attention to the role of the brain was Hippocrates (460–370 B.C.).

Hippocrates was the son of a physician and a member of the physicians guild. He was born on the island of Cos, off the Doric coast of what is now Turkey, but studied for a number of years in Athens and traveled extensively to various schools and temples of medicine.

The *Corpus Hippocraticum* is the name given to the col-

lection of treatises written and amended under the name of Hippocrates. It is generally acknowledged that not all of the works in the *Hippocratic corpus* were written by Hippocrates himself, even though his name alone has been attached to the corpus.

The *Hippocratic corpus* contains many references to seizures, paralyses, and other disorders of the nervous system. The belief of the Hippocratic physicians that the brain served as a controlling center for the body is illustrated in the following passage from *On the Sacred Disease,* a work that put seizure disorders into new perspective:

> Men ought to know that from nothing else but the brain come joys, delights, laughter and sports, and sorrows, griefs, despondency, and lamentations. And by this, in an especial manner, we acquire wisdom and knowledge, and see and hear and know what are foul and what are fair, what are bad and what are good, what are sweet, and what are unsavory. . . . And by the same organ we become mad and delirious, and fears and terrors assail us. . . . All these things we endure from the brain, when it is not healthy. . . . In these ways I am of the opinion that the brain exercises the greatest power in the man. This is the interpreter to us of those things which emanate from the air, when the brain happens to be in a sound state. (1952 translation, 159)

Hippocrates and his followers thus rejected the idea that seizures were due to the gods or demons. In fact, they suggested that charlatans and excessively religious people relied on their gods as a means of covering up their own ignorance. Priests knew that they could receive praise if they cured a patient, but could not be held responsible for "the will of the gods" if they failed.

The four physical elements of air, fire, earth, and water, and their respective qualities of dryness, warmth, cold, and moisture, continued to undergo expansion during the long life of Hippocrates. Air, fire, and earth were associated with yellow bile, blood, and phlegm, respectively, early in the *Hippocratic corpus*. Black bile was not mentioned in the earlier writings, but was linked to the element of water and the qualities of cold and moisture by physicians and philosophers soon after this (Mettler, 1947).

Humoral theory provided a new framework for understanding disease states and how to treat them. In short, illnesses were now believed to be due to an imbalance of one or more of the four humors. Physicians thus strove to reestablish the essential balance among the humors (see Chapter 28). To do this, they relied on bleeding, starving, purging, and many other procedures. For example, epilepsy was believed to be caused by too much thick, cold phlegm clogging the vessels of the head. Remedies included bleeding from the lower arm or the thigh, and warming, cauterizing, or opening the skull to thin out the fluids (Marshall, 1967). In general, however, the guiding philosophy seemed to be to let Nature do the healing, and to help with the cautery or the knife only if really needed.

The *iatrós* (physicians) of classical Greece usually left skull fractures (open head injuries) alone, but frequently drilled holes in the skull after closed head injuries. This practice can also be understood in the context of humoral theory. The Greeks believed a blow to the skull could cause blood and other humors to accumulate in the head. These could form harmful dark pus. If there were a fracture, the

---

[6]See Mettler (1947) for a history of the elements and the humors as related to medicine.

opening would provide an exit for the accumulating humors. But if the skull did not break, the physicians performed craniotomies to allow the excessive humors to drain out with the hope of reestablishing the proper balance (Bakay, 1985; Majno, 1975). This practice was probably successful in preventing some deaths that would have been caused by the buildup of intracranial pressure. Thus, the continued use of craniotomies for closed head injuries was sustained, although the logic behind the operation came from a metaphysical theory.

The physicians who wrote the *Hippocratic corpus* were not accomplished or trained in human dissection. In fact, even animal dissection was frowned upon and generally avoided at this time. Systematic human dissections seemed to have begun around 300 B.C. in Alexandria, Egypt. It was here that the Ptolemies established a model Greek city, a truly great center of learning with several magnificent libraries and a museum school.

Herophilus (335–280 B.C.), the "Father of Anatomy," supposedly dissected hundreds of cadavers in Alexandria. Erasistratus (c. 310–250 B.C.), another Alexandrian, also participated in human dissection. These men may even have operated upon live criminals. The condemned were given to them by the first Ptolemies, rulers who were anxious to cooperate with their learned scientists (Clarke and O'Malley, 1968; McHenry, 1969; Woollam, 1958). Although the writings of Herophilus and Erasistratus have been lost, Celsus (25 B.C.–A.D. 50) later said the following about them:

> They hold that Herophilus and Erasistratus did this in the best way by far, when they laid open men whilst alive—criminals received out of prison from the kings—and whilst these were still breathing, observed parts which beforehand nature had concealed. . . . (1935–1938 translation, Vol. 1, 15)

Herophilus, a devoted follower of Hippocrates, was especially interested in the brain and supposedly wrote about some of its parts, including the cerebrum, the cerebellum, and the ventricles (which he associated with the psyche). He attempted to distinguish between the nerves and the tendons, as well as between the sensory and motor nerves. He also described some parts of the eye. Erasistratus, for his part, is best remembered for his comparative studies of the cerebellum and cerebral surface. After looking at these structures across species, he associated the size of the cerebellum with the ability to run fast, and the complexity of the cerebral convolutions with intellect.

It is important to recognize that the concept of experimentation was still in a rudimentary form, even in progressive Alexandria. Scientific "proofs" typically were based on analogies and on ties to other dubious "truths." This was one reason why debates over the precise functions of the heart and the head intensified during the period of Greek cultural supremacy.

THE HEART OR THE HEAD?

Democritus (c. 460–370 B.C.) was an important contemporary of Hippocrates who taught that matter was made up of an infinite number of extremely minute eternal units called "atoms." He maintained that atoms can differ in shape—for example, those in fluids were thought to be round, whereas those in solids were thought to be rough and irregular, features that allowed them to cling together. As will be seen in later chapters, Democritus and his followers applied this theory to sensation. Thus, hooked atoms accounted for sharp acid tastes, while sweet tastes were associated with large round atoms.

Democritus and his contemporary Plato (c. 429–348 B.C.) both believed in a triune soul. One part of the soul was in the head and was associated with intellect. A second part was in the heart and was associated with anger, fear, pride, and courage. The third part of the soul was located in the liver, or at least in the gut, where it functioned in lust, greed, desire, and related lower passions (see Mettler, 1947; Wright, 1925). Democritus believed that the composite soul perished when an individual died, but Plato argued that the intellectual or rational soul was immortal, whereas the souls of the heart and the liver were perishable. Plato expressed these thoughts in his *Timaeus,* a work named after the learned philosopher in the treatise.

The ideas proposed by Democritus and Plato represented a major shift away from the cardiocentric theory championed by Empedocles and others. The new direction was toward a theory linking at least higher functions with the brain. In this respect, the philosophical ideas of Democritus and Plato were in line with the thinking of the Hippocratic physicians.

Still, not everyone was swayed by these new ideas. In particular, Aristotle (384–322 B.C.; shown in Figure 1.15), though a disciple of Plato and the greatest of the Greek natural philosophers, believed that the heart was the seat of intellectual, perceptual, and related functions. One reason for Aristotle's rigid adherence to the cardiocentric position was that the heart felt warm, whereas the brain felt cool. To the Greeks, warmth was more important because it was associated with sensation and distinguished animate from inanimate objects. In addition, Aristotle may have located rational thought in the heart because the beating heart was the first organ that he could observe in the chick embryo.[7] The importance given to the heart by Aristotle must also have been affected by tradition and his appreciation of other cultures. The heart had been recognized as "the acropolis of the body" by the Mesopotamians, Egyptians, Jews, Hindus, Chinese, and most earlier Greeks.

Aristotle, however, did not completely ignore the head. He believed that "the region of the brain" played a role in tempering "the heat and seething" of the heart (see Clarke, 1963). Because heat rises and because the brain is cool, the network of blood vessels blanketing the cool brain seemed well suited to act like a radiator. But what about the large size of the brain, and especially the cerebrum, relative to body size in humans, the most intelligent of all animals? The large size, reasoned Aristotle, was because humans were warmer than other animals. The large, cool human brain with its many surface blood vessels allowed for the most efficient type of cooling and thus assured the most rational of hearts.

It is interesting to reflect upon the fact that the argument over the functions of the heart versus those of the

[7]Aristotle was especially interested in the evidence gleaned from dissection and worked on 49 species, ranging from the sea urchin to the elephant, although probably not on an adult human (see Clarke, 1963).

Figure 1.15. Aristotle (384–322 B.C.) contemplating nature in an illustration from 1791.

brain would remain unresolved for many more centuries. In a reference to emotion and the heart, William Shakespeare (1564–1616) has Portia in *The Merchant of Venice* ask: "Tell me where is fancy bred, / Or in the heart, or in the head?" And even today, shades of this ancient debate seem to carry on as people speak of having a "broken heart," giving "heartfelt" thanks, or memorizing "by heart" (see Chapters 19 and Chapter 23).

GALEN AND THE BRAIN

The later Greek rulers did not share the same strong interest in medicine and science as the first Ptolemies (e.g., Ptolemy I, 367–283 B.C.), the rulers who had played an important role in the cultural development of Alexandria. As the Greek Empire underwent decline, many physicians found the turmoil and changing political climate so intolerable that they left. A number of them settled in Rome, where they competed with the "irregular" practitioners who served the wealthy Roman landlords. Indeed, the foreign physicians were looked upon with such suspicion that they were not even granted citizenship until 46 B.C.

Among the better known Roman writers on medicine was Aurelius Cornelius Celsus. Although he was not formally trained in medicine, he advised the Emperors Tiberius (42 B.C.–A.D. 37) and Caligula (A.D. 12–41). Celsus discussed nervous diseases in his *De medicina,* a work considered to be one of the best accounts of Roman medicine.[8] *De medicina,* as well as related works from this period, show that Roman medicine was largely a continuation of Greek ideas.

[8]Many copies of *De medicina* were made after Celsus died. After lying dormant for hundreds of years, one copy was discovered and printed in 1478 on the Gutenberg Press. This represented the first publication of a medical book on the modern printing press.

Figure 1.16. Galen (130–200), the physician, scientist, and philosopher whose theories and treatments about central nervous system injuries influenced thinking into the post-Renaissance period. In this illustration, Galen encounters a human skeleton whose remarkable construction confirms his belief that there must be a God.

The most influential physician of the Roman Empire was Galen (130–200), who was born in the great Greek city of Pergamon.[9] Shown in Figure 1.16, Galen received most of his training in anatomy in Alexandria, but at the age of 28 he returned to his place of birth where he served as a surgeon to the gladiators. Although he traveled widely in the Mediterranean region, Galen subsequently spent most of his productive adult life in Rome as court physician to four successive emperors.[10]

It is estimated that Galen wrote 500 to 600 treatises in Attic Greek, although many works were lost when his library burned in the fire of 191 in Rome. He idolized both Hippocrates and Aristotle and agreed with them that nothing should be recognized except that which could be experienced through the senses.

Many of Galen's important dissections and experiments were performed on cattle, but he also worked on many other kinds of animals, including sheep, pigs, cats, dogs, weasels, monkeys, apes, and at least one elephant. Galen wrote that the Barbary ape, which can walk properly and run swiftly, most closely resembled man. For this reason, Barbary apes were his favorite subjects even though macaques and other tailed monkeys were more common in the markets, hardier in captivity, and easier to handle. Roman law did not permit human "autopsies" (a word he coined), and there is no evidence that Galen ever performed human dissections, although he had access to two dried human cadavers.

Galen's greatest works were on the nervous system. In his lecture entitled *On the Brain* (pp. 226–237), he provided the directives for a brain dissection (see the quotation at the beginning of this chapter). This lecture was given in the year 177, in Greek, to Roman medical students (see Singer, 1956).

Despite his reverence for Aristotle, Galen did not accept all of his conclusions, especially when they concerned the functions of the brain. In particular, he rejected Aristotle's assertion that the brain simply served to cool the passions of the heart. He thought this was complete nonsense for many reasons, one of which was that Nature would have put the brain closer to the heart if a cooling organ were actually necessary. As stated in *De usu partium (On the Usefulness of the Parts of the Body)*:

> The supposition that the encephalon was formed for the sake of the heat of the heart, to cool it and bring it to a moderate temperament, is utterly absurd, since in that case Nature would not have placed the encephalon so far from the heart. Rather, she either would have entirely surrounded the heart with it, as she has with the lung, or would at least have placed it down in the thorax, and she would not have attached the sources of all the senses to it. (1968 translation, 387)

As far as Galen was concerned, the heart was not the organ of mind.

Galen wrote about the nerves in *De usu partium* as well as in other tracts (see Smith, 1971). He numbered the cra-

nial nerves from anterior to posterior. In addition, he distinguished between the sensory and motor pathways, as did his predecessors in Alexandria, and suggested that the motor nerves went to the cerebellum and the sensory nerves to the cerebrum. The logic behind his early attempt at localization seemed to be the belief that the sensory nerves had to be softer than the motor nerves in order to retain the sensory impressions. Galen found the cerebrum softer than the cerebellum. Hence, the former had to receive the sensory nerves, whereas the latter had to be the source of the motor nerves. In his own words:

> In substance the encephalon is very like the nerves, of which it was meant to be the source, except that it is softer, and this was proper for a part that was to receive all sensations, form all images, and apprehend all ideas. For a substance easily altered is most suitable for such actions and affections, and a softer substance is always more easily altered than one that is harder. This is the reason why the encephalon is softer than the nerves, but since there must be two kinds of nerves, as I have said before, the encephalon itself was also given a twofold nature, that is, the anterior part (the cerebrum) is softer than the remaining hard part (the cerebellum), which is called *encranium* and *parencephalis* by anatomists. Now . . . the posterior part had to be harder, being the source of the hard nerves distributed to the whole body. (1968 translation, 398)

Some of Galen's descriptions involved the autonomic nervous system. He described the sympathetic chain, the rami communicantes, and the autonomic ganglia (see Chapter 20). He viewed the sympathetic trunk chain as one of three branches that made up his "sixth pair" of cranial nerves. And, on a physiological level, he discussed how spirits traveling in the autonomic nervous system could affect one another (i.e., the concept of "sympathy").

Galen did not believe that the convolutions of the brain were associated with intelligence, as had been maintained by Erasistratus.[11] He pointed out that donkeys have exceedingly complex cerebrums, even though they are remarkably stupid animals. He was of the opinion that the temperament of the thinking body ("whatever this body may be") would prove to be a better correlate of intelligence than the convolutions.

One of Galen's most important beliefs was that the vital spirits, produced in the left ventricle of the heart, were carried toward the brain by the carotid arteries. They were then transformed into the highest spirits, animal spirits, in the *rete mirable* ("wonderful net"), a vascular plexus at the base of the brain, or in the ventricles themselves.[12] In *De usu partium*, Galen wrote:

> The plexus called retiform by anatomists is the most wonderful of the bodies located in this region. It circles the gland (pituitary) itself and . . . it is not a simple network but (looks) as if you had taken several fishermen's nets and superimposed them. . . . [Y]ou could not compare this network to any man-made nets, nor has it been formed from any chance material. Nature appropriated the material for this wonderful network. . . . (1968 translation, 430)

---

[9]Pergamon, Galen's birthplace, was known for its sanctuary to the healing god Asclepius and for its library. See Temkin (1973) for a biography of Galen.

[10]The emperors served by Galen were Marcus Aurelius Antoninus (121–180), Lucius Aelius Aurelius Commodus (161–192), Publius Helvius Pertinax (126–193), and Lucius Severus Septimius (146–211).

[11]Both Galen and Erasistratus shared a penchant for deducing structure-function relationships by examining organs across different species (see *De usu partium*, p. 418; also Woollam, 1958).

[12]Galen was not consistent on where the animal spirits were formed.

The animal spirits were stored in the ventricles until needed. When called upon, they were passed through the hollow nerves to force the muscles into action or to mediate sensation. As for the residue products, Galen suggested that the vapors would escape through the sutures of the skull, whereas the liquids, or *pituita* (phlegm), would trickle down to the nose and throat. He declared that no wound of the brain can impair sensation and motion unless it also penetrated to the ventricles, where the spirits resided.

In his *Commentaries on Hippocrates and Plato*, Galen associated the brain substance with a different function. This is because he recognized that a wound of the brain could affect the mind (Pagel, 1958; Spillane, 1981). Thus, he maintained that although the animal spirits formed the "instrument of the soul," the seat of the highest soul and the seat of the intellect had to be the brain itself. He theorized that the animal spirits from the ventricles flowed not only into the nerves but into the brain as well.[13]

Galen considered imagination, cognition, and memory as the basic components of intellect. He acknowledged they could be affected independently, but at least in his known writings he stopped short of actually localizing these faculties in different parts of the brain. As will be seen in Chapter 2, others soon took imagination, cognition, and memory and localized them in separate areas. These early attempts to localize higher functions, however, did not center on the brain substance itself. Rather, it was suggested that these functions could be localized within the cavernous ventricular system.

[13]For the earlier views of Herophilus concerning the ventricles and the soul, see Clarke and O'Malley (1968), McHenry (1969), and Woollam (1958).

REFERENCES

Bakay, L. 1985. *An Early History of Craniotomy.* Springfield, IL: Charles C Thomas.

Banjeri, D. C. 1981. Historical and socio-cultural foundations of health service systems. In G. R. Gupta (Ed.), *The Social and Cultural Context of Medicine in India.* New Delhi: Vikas Publishing House.

Breasted, J. H. 1930. *The Edwin Smith Surgical Papyrus.* Chicago: University of Chicago Press.

Broca, P. 1861. Remarques sur le siège de la faculté du langage articulé; suivies d'une observation d'aphémie (perte de la parole). *Bulletins de la Société Anatomique (Paris), 6,* 330–357, 398–407. Translated as, "Remarks on the seat of the faculty of language followed by an observation of aphemia." In G. von Bonin (1960), *Some Papers on the Cerebral Cortex.* Springfield, IL: Charles C Thomas.

Broca, P. 1874. Sur les trépanations préhistoriques. *Bulletins de la Société d'Anthropologie, 25,* 542–557.

Broca, P. 1876. Trépanations préhistoriques—Crânes trépanés à l'aide d'un éclat de verre. *Bulletins de la Société d'Anthropologie, 11,* 512–513.

Broca, P. 1877a. Sur la trépanation du crâne et les amulettes crâniennes à l'époque néolithique. *Revue d'Anthropologie, 6,* 1–42, 193–225.

Broca, P. 1877b. De la trépanation du crâne practiquée sur un chien vivant par la méthode néolithique. *Bulletins de la Société d'Anthropologie, 12,* 400.

Celsus, A. C. 1935–1938. *De medicina* (3 vols.). (W. G. Spencer, Ed.). Cambridge, MA: Loeb Classical Library.

Clarke, E. 1963. Aristotelian concepts of the form and function of the brain. *Bulletin of the History of Medicine, 37,* 1–14.

Clarke, E., and O'Malley, C. D. 1968. *The Human Brain and Spinal Cord.* Berkeley: University of California Press.

Dart, R. A. 1949. The predatory implemental technique of Australopithecus. *American Journal of Physical Anthropology, 7,* 1–38.

David, R. 1978. *Mysteries of the Mummies: The Story of the Manchester University Investigation.* London: Book Club Associates.

Descartes, R. 1650. *Passions animae.* Amsterdam: L. Elzevir.

Filliozat, J. 1964. *The Classical Doctrine of Indian Medicine.* Dehli: Munshi Ram Manohar Lal, Oriental Publishers.

Galen. 1968. *De usu partium.* Translated by M. T. May as, *On the Usefulness of the Parts of the Body.* Ithaca, NY: Cornell University Press.

Harris, J. E., and Weeks, K. R. 1973. *X-Raying the Pharaohs.* New York: Charles Scribner's Sons.

Herodotus. 1972. *The Histories.* (A. de Sélincourt, Trans., with notes by A. R. Burn). London: Penguin Books.

Hippocrates. 1952. *On the Sacred Disease.* In *Hippocrates and Galen. Great Books of the Western World* (Vol. 10). Chicago: William. Benton. Pp. 154–160.

Horne, J. F. 1894. *Trephining in Its Ancient and Modern Aspect.* London: John Bale and Sons.

Hurry, J. B. 1926. *Imhotep.* Milford, U.K.: Oxford University Press.

Janssens, P. A. 1970. *Paleopathology.* London: Curwen Press.

Leca, A.-P. 1981. *The Egyptian Way of Death.* Garden City, NY: Doubleday.

Le Gros Clark, W. 1967. *Man-apes or Ape-men?* New York: Holt, Rinehart and Winston.

Lucas-Championnière, M. M. 1912. *Les Origines de la Trépanation Décompressive: Trépanation Néolithique, Trépanation Pré-Colombienne, Trépanation des Kabyles, Trépanation Traditionnelle.* Paris: G. Steinheil.

Majno, G. 1975. *The Healing Hand.* Cambridge, MA: Harvard University Press.

Marshall, C. 1967. Surgery of epilepsy and major disorders. In A. E. Walker (Ed.), *A History of Neurological Surgery.* New York: Hafner. Pp. 288–305.

McHenry, L. C., Jr. 1969. *Garrison's History of Neurology.* Springfield, IL: Charles C Thomas.

Mettler, C. C. 1947. *History of Medicine.* Philadelphia PA: Blakiston.

O'Connor, D. C., and Walker, A. E. 1967. Prologue. In A. E. Walker (Ed.), *A History of Neurological Surgery.* New York: Hafner. Pp. 1–22.

Pagel, W. 1958. Medieval and Renaissance contributions to knowledge of the brain and its functions. In M. W. Perrin (Chairman), *The Brain and Its Functions.* Wellcome Foundation. Oxford, U.K.: Blackwell. Pp. 95–114.

Plato. 1952. *Timaeus* (B. Jowett, Trans.). In *Great Books of the Western World: Vol. 7. Plato.* Chicago: Encyclopaedia Britannica. Pp. 442–477.

Schiller, P. 1979. *Paul Broca: Founder of French Anthropology, Explorer of the Brain.* Berkeley: University of California Press.

Sharma, P. 1983. *Caraka-Samhita.* Varanasi: Chaukhambha Orientalia.

Singer, C. 1956. *Galen on Anatomical Procedures.* London: Oxford University Press.

Smith, E. E. 1971. Galen's account of the cranial nerves and the autonomic nervous system. *Clio Medica, 6,* 77–98.

Soleki, R. S. 1957. Shanidar Cave. *Scientific American, 197,* 59–64.

Soleki, R. S. 1960. Three adult Neanderthal skulls from Shanidar Cave, Northern Iraq. *Annual Report of the Smithsonian Institute,* pp. 603–635.

Spencer, A. J. 1982. *Death in Ancient Egypt.* U.K.: Harmondsworth.

Spillane, J. D. 1981. *The Doctrine of the Nerves.* Oxford, U.K.: Oxford University Press.

Squier, E. G. 1877. *Peru: Incidents of Travel and Exploration in the Land of the Incas.* New York: Harper and Brothers.

Temkin, O. 1973. *Galenism. Rise and Decline of a Medical Philosophy.* Ithaca, NY: Cornell University Press.

Thompson, R. C. 1923. *Assyrian Medical Texts.* Milford, U.K.: Oxford University Press.

Thompson, R. C. 1924. Assyrian medical texts. *Proceedings of the Royal Society of Medicine, 17,* 1–34.

Thorwald, J. 1963. *Science and the Secrets of Early Medicine.* New York: Harcourt, Brace and World.

Trelles, J. O. 1962. Cranial trepanation in ancient Peru. *World Neurology, 3,* 538–545.

Walker, A. E. 1958. The dawn of neurosurgery. *Clinical Neurosurger y, 6,* 1–38.

Woollam, D. H. M. 1958. Concepts of the brain and its functions in classical antiquity. In M. W. Perrin (Chairman), *The Brain and Its Functions.* Wellcome Foundation. Oxford, U.K.: Blackwell. Pp. 5–18.

Wright, J. 1925. A medical essay on the *Timaeus. Annals of Medical History, 7,* 117–127.

Chapter 2
Changing Concepts of Brain Function

In saying these things I admit that innumerable tenets of Galen contradict me; such as that the anterior ventricles are the olfactory organs; that these ventricles, gradually increasing in softness, end by constriction into the visual nerves; yet more strangely that these front ventricles emit phlegm to the nose. These and many other things of the kind are, I am convinced, learned rather from copious discussion than from dissection . . .

Andreas Vesalius, 1543

VENTRICULAR LOCALIZATION

In the second century, Galen associated imagination, intellect, and memory with the brain substance in his *Commentaries on Hippocrates and Plato* (see Chapter 1). In *De usu partium,* he stated:

In those commentaries I have given the demonstrations proving that the rational soul is lodged in the encephalon; that this is the part with which we reason; that a very large quantity of psychic pneuma is contained in it; and that this pneuma acquires its own special quality from elaboration in the encephalon. (1968 translation, 432)

Galen contended that these functions could be affected independently by injury and disease, but he stopped short of differentially localizing them in different parts of the encephalon.

The Church Fathers of the fourth and fifth centuries drew upon Galen's ideas. But unlike Galen, they then associated these higher functions with the ventricles of the brain. Moreover, they concluded that the different cavities were responsible for different functions.

One of the earliest advocates of discrete ventricular localization was Nemesius (fl. 390), Bishop of Emesa (now called Homs), a city in Syria. In his most famous work, translated as *On the Nature of Man* (a title first used by Hippocrates), this Byzantine writer blended Galenic medical science with Christian theology. He localized perception in the two lateral ventricles (the anterior ventricles), cognition in the middle ventricle, and memory in the posterior ventricle. Nemesius wrote:

Now if we make this assertion, that the senses have their sources and roots in the front ventricles of the brain, that those of the faculty of intellect are in the middle part of the brain, and that those of the faculty of memory are in the hinder part of the brain, we are bound to offer demonstration that this is how these things work, lest we should appear to credit such an assertion without rational grounds. The most convincing proof is that derived from studying the activities of the various parts of the brain. If the front ventricles have suffered any kind of lesion, the senses are impaired but the faculty of intellect continues as before. It is when the middle of the brain is affected that the mind is deranged, but then the senses are left in possession of their natural functions. . . . If it is the cerebellum that is damaged, only loss of memory follows, while sensation and thought take no harm. (Translated in Telfer, 1955, 341–342)

Nemesius did not describe how he arrived at this "proof" for differential localization within the three ventricles (the two lateral ventricles were treated as a single cavity). He was probably starting from various statements made by Galen, many of whose manuscripts have been lost (Temkin, 1973). It is clear that Galen associated the front of the brain with sensory functions (mainly because it was softer than the cerebellum), while contending that imagination, intellect, and memory could be dissociated. The anterior ventricles, which were associated with perception, also were closest to the peripheral organs of sight, smell, and taste, as well as touch from the face and hands. And philosophically, it probably made much more sense to an early Christian theologian to place the ethereal spirits responsible for the highest functions in hollow cavities of the brain than in the flesh itself.

Although the logic behind ventricular localization is still not fully understood, this early attempt at localization achieved wide acceptance. It was embraced by St. Augustine (354–430) and by Posidonius of Byzantium (fl. 370). Specifically, St. Augustine proposed that recollection belonged to the middle ventricle, motion to the most posterior ventricle, and sensation to the anterior cavities. Posidonius, however, agreed with Nemesius that injury in the front of the head affected only imagination, that damage to the middle ventricle affected understanding, and that injury to the back of the head destroyed memory (see Sudhoff, 1914).

Localization of higher functions in the ventricles continued to be accepted both in Europe and the Middle East for hundreds of years (see Figures 2.1–2.4). For example, the most famous physician of the ancient Persian world, Abu

'Ali al-Husain ibn 'Abdullah ibn Sina (Avicenna; 980–1037), defended ventricular localization. His *Canon* served as a standard reference text from the twelfth to the seventeenth century (Talbot, 1970). In it the following statement can be found:

> One of the animal internal faculties of perception is the faculty of fantasy, i.e., *sensus communis,* located in the forepart of the front ventricle of the brain. It receives all the forms which are imprinted on the five senses and transmitted to it from them. Next is the faculty of representation located in the rear part of the front ventricle of the brain, which preserves what the *sensus communis* has received from the individual five senses even in the absence of the sensed object. . . . Next is the faculty of "sensitive imagination" in relation to the animal soul, and "rational imagination" in relation to the human soul. This faculty is located in the middle ventricle of the brain. . . . Then there is the estimative faculty located in the far end of the middle ventricle of the brain. . . . Next there is the retentive and recollective faculty located in the rear ventricle of the brain. (Translated in Rahman, 1952, 31)

Even during the Renaissance, when old ideas were challenged, many physicians still relied on ventricular localization to explain neurological signs. For example, Antonio Guainerio (d. 1440), an Italian physician, described two old men in his *Opera medica,* published in 1481. One of the two could not speak more than three words, and the other was unable to remember the correct name of anyone. Guainerio believed their disorders, which appeared to be motor and amnestic aphasias, respectively, were due to faulty memory. He postulated that the symptoms were caused by "an excessive accumulation of phlegm in the posterior ventricle so that the 'organ of memory' was impaired" (Benton and Joynt, 1960).

THE MIDDLE AGES

Scholarship and new learning stagnated in the West during the Middle Ages, the "dark" period of about 1,000 years prior to the Renaissance (roughly the fourth to the fourteenth century). In the Middle East, however, natural philosophers such as Avicenna had access to the great Greek and Roman books of science and philosophy (Meyerhof, 1945). The Arabs and the Nestorian Christians venerated, collected, and translated the works of Hippocrates, Aristotle, and Galen. They also based their own medical books on these and other classical sources (see Chapter 5).

The works of Middle Eastern scholars became familiar to Europeans only at the end of the Middle Ages. In fact, Western Europeans discovered much of this material only when they drove south to conquer Moorish Spain, winning Toledo in 1085, Cordoba in 1236, and Seville in 1248. The "new" material soon spread from Spain to France and Italy, and then to other parts of Europe, where it sparked a revival of interest in anatomy, physiology, and medicine.

It is in this context that one finds *The Book of Hunain ibn Is-Hâq on the Structures of the Eye,* translated into Latin near the end of the eleventh century by the medieval Christian medical figure Constantinus Africanus (c. 1020–1087) (see Meyerhof, 1928; also Chapter 5). Since Hunain ibn Is-Hâk (809–873) made no significant departures from Galen

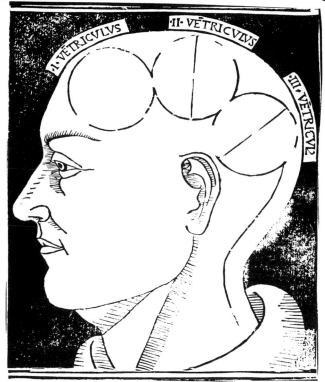

Figure 2.1. Depiction of the ventricles from a 1490 edition of *Philosophia pauperum* by Albertus Magnus, who lived in the thirteenth century.

Figure 2.2. Figure of the ventricles from a 1506 edition of *Philosophia naturalis* by Albertus Magnus.

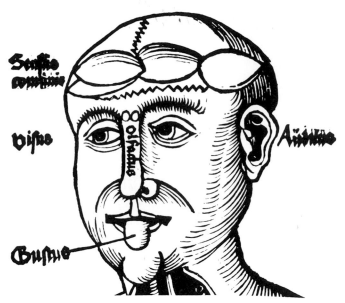

Figure 2.3. A diagram from Johannes Peyligk's (1516) *Compendiosa* showing the ventricles and giving labels to the sensory organs.

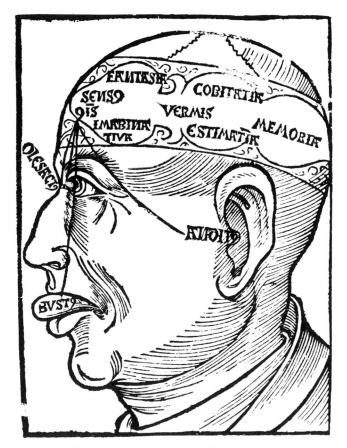

Figure 2.4. The ventricles as depicted by Hieronymous Brunschwig in a 1525 edition of his book from 1497.

on the subject of vision, the West essentially received a new edition of classic Greco-Roman medical science.

In 1347, the Black Death spread throughout Europe, where it devastated families and decimated whole towns.

Estimates are that between one third and one half of the European population perished from the plague. Medicine based on Church dogma and the teachings of Galen failed to halt the loss of human life. In a macabre way, this helped to set the stage for the Renaissance and its changes.

An interesting issue is the extent to which the religious dictate, "the Church abhors the shedding of blood" *(Ecclesia abhorret a sanguine),* was accepted at this time. This dictate has been loosely interpreted to mean that members of the Church could not be surgeons. General edicts to this effect were issued in several European cities in the twelfth and thirteenth centuries (e.g., the Council of Tours in 1163), but there is some question as to what exactly was being demanded and who was most affected. This is because the Church was composed of many levels of clergy, ranging from the Pope to priests and deacons, and down to porters, lectors, and exorcists.

The fact that the Church kept repeating its bans on the drawing of blood suggested that its rules were not rigidly followed. It is also possible that the edicts were viewed as pertaining largely to the higher religious orders and not to all levels of clergy. Thirteenth-century papal records show that if a surgeon were responsible for a woman's death, he might be prevented from continuing to function as a priest, but not necessarily as a monk (Amundsen, 1972).

Because of these edicts, the secular clerks, and perhaps some lower members of the clergy, effectively took surgery out of the hands of the priests and middle-order clergy during this time. By the Renaissance, surgery had passed largely to the barbers in European towns and occasionally to the executioners and bathkeepers.

THE RENAISSANCE

The Renaissance is usually thought of as beginning in Italy in the mid-fourteenth century and ending during the sixteenth century. One of the best-recognized figures of this period was Leonardo da Vinci (1472–1519). Da Vinci dissected over 300 cadavers and made over 1,500 drawings in his lifetime. This was done in secrecy because Pope Boniface VII, who lived in the tenth century, had issued an edict strictly forbidding autopsies—a ruling that was still strongly enforced.

Sometime between 1504 and 1507, da Vinci conducted his famous experiments on the ventricles of cattle to reveal their true structure. His procedure involved injecting molten wax into the cavities. The tubes for injecting the wax were inserted from below, and other tubes were inserted from above to permit the cerebrospinal fluid to be pushed out. After the injected wax hardened, da Vinci cut away the brain. He found that his casts did not come close to corresponding to the drawings that anatomists were using at the time. Nevertheless, he was not mentally ready to reject traditional ventricular ("cell") doctrine. His solution was, in fact, a compromise. It was to produce anatomically correct drawings of the ventricles, while still assigning imagination, cognition, and memory to the different cavities. One such drawing appears as Figure 2.5.

Da Vinci deviated from the generally accepted physiology

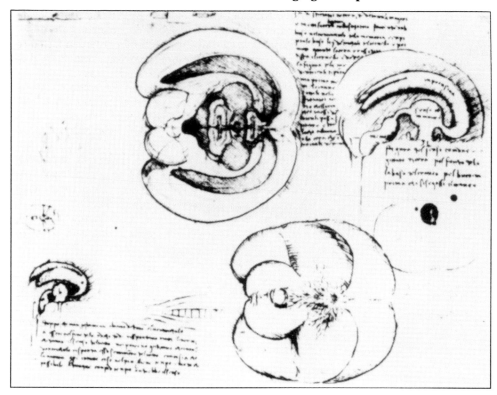

Figure 2.5. Drawings of the brain by Leonardo da Vinci from about 1504.

only when he moved sensation *(sensus communis)* to the middle ventricle from its usual location in the two lateral ventricles (see O'Malley and Saunders, 1952). He did this because he thought most sensory nerves were closer to the middle than to the front of the brain. This change, however, did not affect da Vinci's basic belief in animal spirits traversing the nerves to and from the ventricles.

Others were not as tolerant of older ideas relating to brain function and medicine. A more dramatic break with the past occurred when Paracelsus (Philippus Aureolus Theophrastus Paracelsus Bombastus von Hohenheim, 1493–1541) began his lectures in Basel, Switzerland, by burning the revered works of Avicenna in front of his students (Guthrie, 1940). Although Paracelsus began his own career as an alchemist, he believed that most if not all of the Galenic pharmacy was useless.

At about the time that Paracelsus dramatically burned his bridges with the past, Andreas Vesalius (1514–1564; shown in Figure 2.6), a native of Brussels, published his landmark work *De humani corporis fabrica (On the Workings of the Human Body).* This work, completed in Padua and Venice in 1542 and published in Basel on May 5, 1543, contained seven small books. Of these, Book IV dealt with the nerves and Book VII with the brain. For his art work, Vesalius probably turned to Jan Stephan van Calcar (c. 1499–1546) and/or to other students of Titian (c. 1487–1586), the great painter of the Venetian High Renaissance. Figures 2.7, 2.8, and the cover of this book show three of the *Fabrica* plates. Along with the larger text, Vesalius also published a brief edition under the title *Epitome* (1543).

The *Fabrica* was based on extensive human dissections, conducted largely in Padua, where Vesalius was treated

with honor and provided with ample material for autopsy.[1] With regard to this, Vesalius reflected:

> My study of anatomy would never have succeeded had I, when working at medicine in Paris, been willing that the viscera should be merely shown to me and to my fellow students at one or another public dissection by wholly unskilled barbers, and that in the most superficial way. I had to put my own hand into the business. (Cited in Stirling, 1902, 2)

He wrote that the ventricles of humans were not different in shape from those of other mammals that obviously did not have similar reasoning powers.

> All our contemporaries, so far as I can understand them, deny to apes, dogs, horses, sheep, cattle, and other animals, the main powers of the Reigning Soul—not to speak of other (powers)—and attribute to man alone the faculty of reasoning; and ascribe this faculty in equal degree to all men. And yet we clearly see in dissecting that men do not excel those animals by (possessing) any special cavity (in the brain). Not only is the number (of ventricles) the same, but also all other things (in the brain) are similar, except only in size and in the complete consonance (of the parts) for virtue. (Translated in Singer, 1952, 40)

[1]At the age of 23, Vesalius was appointed professor of surgical anatomy at Padua, the center of the scientific Renaissance (Castiglioni, 1935, 1943). He was instructed by the powerful Senate of the Republic of Venice to conduct public dissections, for which the Senate provided him with a liberal supply of material. The environment was ideal for him to develop his ideas, especially in contrast to Paris or Brussels. He began his human dissections in December 1537. More than 500 students and physicians may have attended some of his demonstrations. Moreover, he was praised as an anatomist and as a teacher in official documents.

Figure 2.6. Andreas Vesalius (1514–1564), a professor of anatomy at Padua, whose dissections and observations rekindled the study of anatomy during the Renaissance. (Archives, Washington University Medical School, St. Louis, MO.)

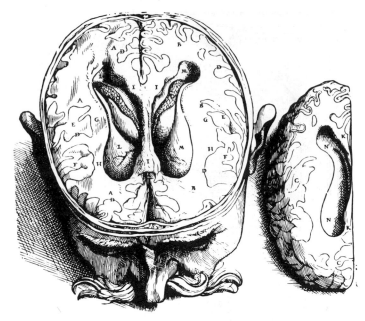

Figure 2.8. A view of the brain from Vesalius' (1543) *De humani corporis fabrica* showing the ventricles.

Vesalius went on to deny the existence of the *rete mirable* in humans, although he did depict it in one figure in a way that matched Galen's descriptions.[2] In retrospect, it seems clear that Galen had described the *rete mirable* of oxen. Although this structure also is present in sheep, cats, goats, and pigs, Galen wrongly assumed that it must therefore exist in humans.

Vesalius was well acquainted with at least three of Galen's anatomical works and stated that he counted some 200 cases in which Galen's anatomy was in error.[3] The message was clear enough. Galen, long idolized by the scientific community, was far from perfect. Vesalius therefore urged his contemporaries to reexamine the structure and function of the brain with an open mind (Rosner, 1974; Singer, 1952; Spillane, 1981). (A realistic portrayal of the ventricles by one of his contemporaries can be seen in Figure 2.9.)

Although Vesalius contributed much to neuroanatomy in the way of new material, his physiology remained more traditional. For example, although he did not think much of ventricular localization, he failed to reject the theory that animal spirits were produced in the ventricles.[4] This can be seen in another passage from his *Fabrica,* where he wrote:

> I ventured to ascribe no more to the ventricles than that they are cavities and spaces in which the inhaled air, added to the vital spirit from the heart, is, by power of the peculiar substance of the brain, transformed into animal spirit.

Figure 2.7. Cover plate to Vesalius' (1543) *De humani corporis fabrica.*

[2]Vesalius was not the first to recognize that the *rete mirable* did not exist in man. In 1518, Jacopo Berengario da Carpi (c. 1460–1530) wrote: "I have dissected many heads, and did not find such a *rete mirable* . . . I have never seen that network" (see Bakay, 1985, p. 19). As noted, Galen thought that the *rete mirable* converted vital spirits to animal spirits.

[3]Galen's three works that clearly influenced Vesalius were *De usu partium, De dissectione nervorum,* and *De anatomicis administrationibus.*

[4]It was Volcher Coiter (1534–1600), not Vesalius, who first cut open the ventricles in living animals and, to everyone's surprise, observed no striking changes in behavior (McHenry, 1969).

Loca tanti proponimus vermiformis, & partium huiufmodi tenuiffimarum,nis qu macmodum ad oculum in confectione difficillime oftendi poffit,ita etiam longe difficillimi fuerit, partes vfqueadeo tenues hic tibi exactiffime demonftrare.

A Fornix, pfallioides,corpus camçratum.
B Principiü vermiformis circa feptü, anteriores vêtriculos diftinguens.
C Conarion glandula.
D Vermiformis lögitudo.
E Glutia.ligamëta et ptes adiacë es vermiformi.
F Via à tertio ventriculo ad quartum.

Figure 2.9. An early realistic representation of the ventricles of the brain. This figure, from *De dissectione partium corporis humani libri tres . . .* by Charles Estienne (Stephanus), appeared in 1545.

past, Vesalius' teacher in Paris, Jacobinus Sylvius (1478–1555), could only remark: "Honest reader, I urge you to pay no attention to a certain ridiculous madman, one utterly lacking in talent who curses and inveighs impiously against his teacher" (see O'Malley, 1964, p. 239). The authoritative Sylvius not only called his student a two-legged ass but requested that "Vesalius be heavily punished and in every way restrained, lest by his pestilent breath he poison the rest of Europe" (see Castiglioni, 1943, p. 775).

WILLIS AND HIS CONTEMPORARIES

In 1664, Thomas Willis (1621–1675; shown in Figure 2.10) published one of the most important books in the history of the brain sciences, his *Cerebri anatome*. At the time it appeared, and for many years afterward, this work was without equal. Although the title implied a book limited to anatomy, Willis, an Oxford physician, also concerned himself with brain function from the first to the last page of his book. As for the illustrations, they were provided by the renowned draftsman and architect Christopher Wren (1632–1723).

Willis proposed that the cerebral gyri controlled memory and the will. Imagination was also seen as a cerebral function. It was specifically delegated to the corpus callosum, which was broadly defined as the white matter of the hemispheres. The corpus striatum also figured prominently in his schema because it was thought to play a role in sensation and movement.

In contrast to the higher functions, the vital and involuntary systems were thought to be mediated largely by the cerebellum. Willis, however, used the term "cerebellum" in

This is presently distributed through the nerves to the organs of sensation and motion, so that these organs, with the help of this spirit . . . perform their functions. (Translated in Singer, 1952, 39)

In addition, Vesalius never challenged Galen's position on the size and complexity of the cerebral convolutions. As noted in Chapter 1, Galen had examined the highly convoluted brain of a donkey and concluded from it and his knowledge of donkey intelligence that the cerebral convolutions could not possibly be related to higher mental functions. As for the parts mediating the higher functions, Vesalius simply lamented that anatomy has its limits and that he could not form an opinion about how the brain regulates imagination, reasoning, and memory.[5]

With respect to his clarion call for human dissection and his belief that it was time to challenge the science of the

[5]Vesalius left Padua in 1544 to become court physician to Charles V (1500–1558). His post was given to Gabriel Fallopius (1523–1562) and then to Fabricius ab Aquapendente (c. 1533–1619), the teacher of William Harvey (1578–1657). The departure of Vesalius from Italy effectively ended his career as an anatomist. He died 20 years later while returning from a pilgrimage to Jerusalem.

Figure 2.10. Thomas Willis (1621–1675), author of *Cerebri anatome* (1664).

Figure 2.11. Franciscus de le Böe (1614–1672), also known as Sylvius, after whom the Sylvian fissure was named. (Archives, Washington University Medical School, St. Louis, MO.)

a way that might have included the pons *(annular protuberance)* and perhaps the colliculi of the midbrain *(orbicular prominences)*. If nothing else, these two brainstem structures were considered processes or appendixes of the cerebellum.

The division of the brain into functional parts, based partly on comparative anatomy, partly on clinical material, and partly on existing theories, stimulated many people to

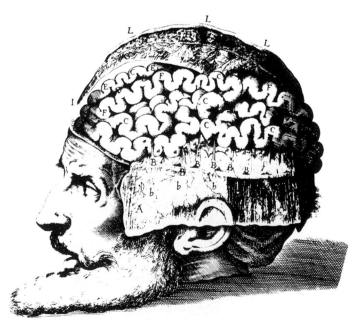

Figure 2.12. A copper plate by Giulio Casserio from a book published in 1627. Clarke and Dewhurst (1972, p. 65) comment that of all the early representations of the brain, "this is one that looks most like coils of small bowel."

accept and follow Willis (see Chapters 15, 19, and 23). Although not all aspects of his thinking would withstand the test of time, Willis, more than any other person in the post-Renaissance period, provided a sound basis and powerful stimulus for looking at the functional contributions of individual brain parts.[6]

Although Willis made the most significant contribution to the study of the brain in the seventeenth century, several other individuals also left their marks. One was Franciscus de le Boë (1614–1672; shown in Figure 2.11), a professor of medicine at Leyden. He adopted the name François Sylvius, leading to some confusion between him and Jacobinus Sylvius, the bitter Galenist who could say nothing good about Vesalius. It was François Sylvius, not Jacobinus Sylvius, who first described the "Sylvian fissure," beginning a trend away from unrealistic drawings of the surface of the brain, such as the one shown in Figure 2.12. In 1663, he wrote:

> The whole surface of the cerebrum is everywhere quite deeply marked by gyri similar to the convolutions of small intestines, and especially by a notable *fissure* or *hiatus,* beginning at the roots of the eyes (optic tracts), passing backwards along the temples above, not farther than the roots of the medulla *(crura cerebri),* and dividing the cerebrum on either hand into an upper, much larger part and a lower, smaller one; with gyri occurring the whole length and depth of the fissure, I may even say with the beginnings of the lesser gyri at the very upper part of the root. (Translated in Baker, 1909, 334)

Another important seventeenth-century figure was Niels Stensen (1638–1686), a Danish physician, geologist, and priest who was better known by his Latin name, Steno. He emphasized the importance of studying the growth of the nervous system and suggested that the brain parts must be organized in a way which would permit them to execute their diverse functions without chaos. This idea was expressed in a lecture *(Discours sur l'Anatomie du Cerveau)* read in 1668 in Paris before an assembly at the house of M. Thévenot. Steno, whose portrait appears in Figure 2.13, declared:

> We need only view a Dissection of that large Mass, the Brain, to have ground to bewail our Ignorance. On the very Surface you see varieties which deserve your admiration: but when you look into its inner Substance you are utterly in the dark, being able to say nothing more than that there are two Substances, one greyish and the other white, which last is continuous with the Nerves distributed all over the body. . . . If this Substance is everywhere Fibrous, as it appears in many places to be, you must own that these Fibres are disposed in the most artful manner; since all the diversity of our Sensations and Motions depend upon them. We admire the contrivance of the Fibres of every Muscle, and ought still more to admire their disposition in the Brain, where an infinite number of them contained in a very small Space, do each execute their particular Offices without confusion or disorder. (Translated in Stirling, 1902, 32)

---

[6]Willis also introduced a number of new words in his text. The most famous of these was "neurology." This word was derived from the Greek word for "sinew" and was used to refer to the doctrine of the nerves. His other contributions to the modern vocabulary included "hemisphere," "lobe," "pyramid," "corpus striatum," and "peduncle."

Figure 2.13. Niels Stensen (Steno, 1638–1686), who argued that scientists must discover the basic functional organization of the brain. (From Sterling, 1902.)

A very Tall and well Set Gentleman, Aged about 24 years, by a Fall from his Horse, had his Skull broken in several places, and being a Person of good Estate, had several Chirugeons to attend him in the course of his Sickness, during which he was divers times Trepan'd, and had several pieces of Skull taken off, which left great Chasms . . . between the remaining parts. Within about three days after his Fall, this Knight . . . was taken with a Dead Palsy on his Right side, which did not equally affect his Arm and his Leg; The use of the latter being sometimes suddenly Restor'd to him in some measure. . . . But his Arm and Head were constantly Paralytical, being wholly depriv'd of Motion; and having so little Sense, that it would sometimes lye under his Body without his Feeling it. . . . And when the Chirugeons were going to close up his Head, as having no more to do; one of them . . . (alleged) that, if they did no more, the Gentleman would lead a Useless and very Melancholy Life; and that he was confident, the Palsey was in some way or other occasion'd by the Fall, which had left something in the Head that they had not yet discover'd. And the Knight himself agreeing to this Man's notion, his Head was further laid open; and at length, under a piece of proud Flesh, they found, with much ado, a Splinter, or rather Flake of Bone, that bore hard upon the dura matter, and was not pull'd out without great Hemorrhage, and at such a stretch of the Parts, as made the Patient think his Brain itself was tearing out. But this Mischief was soon Remedy'd, and his Hurts securely Heal'd up; and he is now a Strong Healthy Man. (1691, 67–73)

CARTESIAN MECHANICS AND THE BRAIN

One of the most influential philosophers of the post-Renaissance period was René Descartes (1596–1650; shown in Figure 2.15). In *De homine,* published posthumously in

Clinical reports at this time were also suggesting that more attention must be paid to the brain and its fundamental organization (Soury, 1899; also see Chapter 25). Robert Boyle (1627–1691; shown in Figure 2.14), who is best known for his studies of the weight and pressure of the atmosphere, included the following case study in his *Experimenta et observationes physicae:*

Figure 2.14. Robert Boyle (1627–1691), the physicist who also described the case of a knight who had a piece of shattered bone successfully removed from his brain. (From Sterling, 1902.)

Figure 2.15. René Descartes (1596–1650), the philosopher who singled out the tiny pineal gland as the logical site where body and mind could interact.

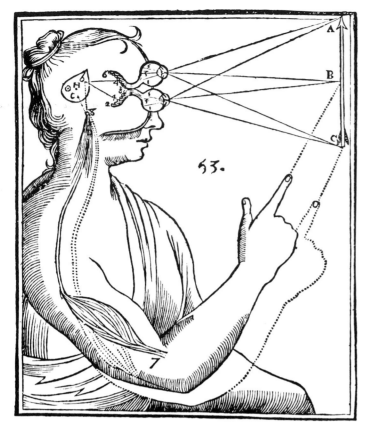

Figure 2.16. A drawing from *De homine* (1662) by René Descartes showing how light enters the eye and forms images on the retina. Hollow nerves from the retina project to the ventricles. The pineal gland (H) then releases the animal spirits into the motor nerves to produce motion.

1662 (the French edition, *L'Homme,* appeared in 1664), Descartes theorized that filaments within each nerve tube operated tiny "valvules" that could control the flow of animal spirits into the nerves. He thought that external stimuli would move the skin, which would pull on these filaments to open valvules in the ventricles. This allowed animal spirits to be released from the ventricular reservoirs into the nerves, which in turn would trigger the muscles to move.[7]

Descartes (1650, 1662) presented this reflexive theory only to account for involuntary behavior. He believed that voluntary behavior demanded the interaction of the rational soul with the automaton. He maintained that this occurred through the pineal gland, a small central body suspended between the anterior ventricles.[8] Small movements by the pineal gland regulated the flow of spirits through the intricate system of pipes and valves, like those depicted in Figure 2.16.

It is often stated that Descartes may have considered the pineal gland simply because it was unitary rather than double. Although this was an important factor, Descartes also selected the pineal gland because it was surrounded by cerebrospinal fluid, which he believed served as a reservoir for the animal spirits.

The solution to the mind-body problem proposed by Descartes was not widely accepted. Among other things, it was argued that the pineal gland was even better developed in wild animals than in humans, even though the brutes were considered soulless automata. Descartes should have expected this. In 1543, Andreas Vesalius had discussed the gross anatomy of the pineal gland when dissecting human brains. In fact, in Book VII of his *Fabrica,* he decried, "I wish that a sheep's brain be at hand, since it shows the gland and the testes (superior colliculi) and nates (inferior colliculi) more distinctly than does the human head."

Even Galen had discussed the pineal gland in his anatomy lesson of A.D. 177, which called for the brain of an ox (see Chapter 1). But perhaps even more interestingly, in the eighth book of his *De usu partium,* Galen mentioned the idea of pineal movements regulating the flow of spirits. He wrote that he evaluated this theory and found it absurd:

> The notion that the pineal body is what regulates the passage of the pneuma is the opinion of those who are ignorant of the action of the vermiform epiphysis *(vermis superior cerebelli)* and who give more than due credit to the gland. Now if the pineal gland were a part of the encephalon itself, as the pylorus is part of the stomach, its favorable location would enable it alternately to open and close the canal because it would move in harmony with the contractions and expansions of the encephalon. Since this gland, however, is by no means part of the encephalon and is attached not to the inside but to the outside of the ventricle, how could it, having no motion of its own, have so great an effect on the canal? . . . Why need I mention how ignorant and stupid these opinions are? (1968 translation, 419–420)

For this and other reasons, opinions varied as to whether the pineal gland could be regarded as the seat of the soul. In this context, a few of Descartes' contemporaries suggested that the honor should be granted to another singular structure, the corpus callosum (see Chapter 26). In retrospect, Descartes' ideas about the brain as a reflexive machine probably had more impact on the development of scientific thought than his specific solution to the mind-body problem.

THE EIGHTEENTH CENTURY

The eighteenth century was a time of continuing advances, especially with regard to clinical descriptions of symptoms and diseases. It was also a time when the first reasonable explanation was provided to account for damage on one side of the brain producing paralysis of the opposite side of the body.

In 1709, Domenico Mistichelli (1675–1715), a professor of medicine in Pisa, described the crossing of the pyramidal tract, albeit in somewhat vague terms. One year later, François Pourfour du Petit (1664–1741), a French military surgeon, observed a patient with a brain abscess and contralateral hemiplegia. Pourfour du Petit became convinced that the animal spirits crossed from one side of the brain to

---

[7]The idea of using mechanics to explain the actions of the brain came to Descartes after he saw some mechanical dolls and new hydraulic machines in Paris. Inventions like these were very popular in France during the seventeenth century, and there was a growing belief that mechanical principles could also explain the actions of biological "machines."

[8]The Latin term *glandula pinealis* translates from *konarion,* a Greek word for the nut of a pine tree common in the Mediterranean area.

the other in the tracts that crossed in the medullary pyramids. He proved his point by severing the pyramidal tracts in some dogs, even though he did not really understand where these tracts originated.[9]

As the eighteenth century progressed, new ideas about the brain were introduced. Animal spirits were replaced by nerve fluid, a substance thought to be produced in the brain and then circulated through the hollow nerves. In the closing decades of the century, the concept of animal electricity also assumed great importance, as will be seen in Chapter 28.

THE DISCOVERY OF THE RESPIRATORY CENTER

During the first half of the 1700s, the medulla was considered little more than an extension of the spinal cord, whereas the cerebellum was believed to be the seat of "animal" functions (Olmstead, 1944a). Some investigators, for example, attributed all vital movements to the cerebellum. This idea was challenged in 1760 by Antoine Charles de Lorry (1726–1783).

Lorry presented two lengthy papers to the *Académie des Sciences* in Paris in which he stressed that dogs could survive without a cerebellum or cerebrum for at least 15 minutes, during which time they had normal pulse and respiration. His observations suggested that the medulla had to be the site of vital functions.

In the opening years of the nineteenth century, Julien-Jean-César Legallois (sometimes written "Le Gallois," 1770–1840), another French physiologist, localized the center for respiration in just one part of the medulla. Legallois conducted experiments in 1806 that left him impressed with the capacity of young, decorticate rabbits to survive without air for as long as 15 minutes. He hoped to find out what parts of the nervous system were essential for life by making sections through the midbrain and medulla after removing the cerebellum.

Legallois found that cutting the medulla at the level of the eighth cranial nerve immediately stopped respiration. He concluded that this had to be the location of the respiratory center. He wrote:

> It is not on the entire brain that respiration depends, but only on a very circumscribed area of the medulla oblongata situated a short distance from the occipital opening and towards the origin of the 8th pair of nerves (the pneumogastric). For if one opens the skull of a young rabbit and extracts the brain in successive portions from front to back by cutting it in slices, one can in this way remove the entire cerebrum, then the entire cerebellum, and finally a part of the medulla without cessation of respiration. But it ceases suddenly when one includes in a slice the origin of the nerves of the 8th pair. (1815, 37)

Legallois' work on the localization of the respiratory center was the first localization within a brain area that was widely accepted. It stimulated Marie-Jean-Pierre Flourens (1794–1867) to conduct his own experiments in which he attempted to confine the critical zone to an even smaller area. At first, Flourens called the whole medulla

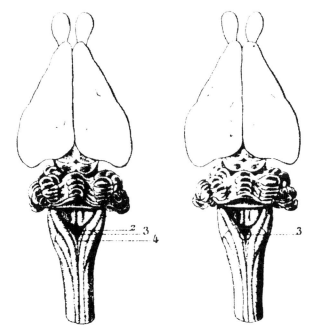

Figure 2.17. Location of the respiratory center in the medulla of young rabbits. This drawing was made by Alfred Vulpian and accompanied a memoir written in 1858 by Marie-Jean-Pierre Flourens, with whom he was working (see Olmstead, 1944).

the *noeud vital,* or "vital node."[10] By 1851, however, he was using this term just to refer to the respiratory area of the medulla, an area he believed was no larger than the head of a pin (Flourens, 1851, 1858, 1859). A drawing from one of his papers appears as Figure 2.17.

THE BELL-MAGENDIE LAW

At about the same time that Legallois was discovering the respiratory center in the medulla, another important discovery about the nervous system was being made. It was beginning to be recognized that the dorsal (posterior) roots of the spinal cord were sensory, whereas the ventral (anterior) roots were motor.

Both Charles Bell (1774–1842), a Scot born in Edinburgh, and François Magendie (1783–1855), a Frenchman, used experimental methods (lesions, stimulation) to study the spinal roots of dogs. Prior to the work of these two men, shown in Figures 2.18 and 2.19, many people believed the spinal nerves were mixed sensory and motor structures and that messages could travel in both directions at once. This view was accepted by Descartes, Albrecht von Haller (1707–1777), and even Bell's teacher, Alexander Monro (1737–1817).

In 1807, Bell, who was teaching in London, wrote a letter to his brother, a professor in Scotland, stating a new position:

> I hinted to you formerly that I was "burning," or on the eve of a grand discovery. I consider the organs of the outward senses as forming a distinct class of nerves from the others.

---

[9]Pourfour du Petit was honored for his discovery by being elected to the Académie des Sciences in 1722 (Bakay, 1985).

[10]The term *noeud vital,* which was favored by Flourens, was actually introduced by the botanist and taxonomist Jean Lamarck (1744–1829) to refer to the collar of a plant (see Olmstead, 1944a). This part, important for the life of the plant, is located between the roots and the stem.

Figure 2.18. Sir Charles Bell (1774–1842), who investigated the functions of the spinal roots and published his findings in a privately printed pamphlet in 1811. (Archives, Washington University Medical School, St. Louis, MO.)

Figure 2.19. François Magendie (1783–1855), the French scientist who made many contributions to anatomy and physiology, including a clear description of the difference between the dorsal and ventral roots of the spinal cord. (Courtesy of the Collège de France.)

I trace them to corresponding parts of the brain totally distinct from the origin of the others.

Less well known but also questioning the notion of mixed spinal roots at this time was an anatomist at the University of Edinburgh. In 1809, Alexander Walker (1779–1852) suggested that the anterior and posterior roots had distinctly different sensory and motor functions (see Brown-Séquard, 1860; Keele, 1957; Walker, 1839). Walker seemed to base his idea on the belief that the *sensorium commune* was located in the cerebrum, anterior to the cerebellum, which mediated motor functions. Walker (1839) wrote the following about the pathway for sensation:

It passes in a general manner to the spinal marrow by the anterior fasciculi of the spinal nerves which are therefore nerves of sensation, and the connections of which within the spinal marrow or brain must be termed their spinal or cerebral terminations; ascends through the anterior columns of the spinal marrow which are therefore its ascending columns; passes forward through the inferior fasciculi of the medulla oblongata, and then through the crura cerebri; extends forwards, outwards and upwards through the corpora striata, and reaches the hemispheres of the cerebrum itself. This precisely is the course of its ascent to the sensorium commune.

Walker went on to argue that the motor pathway went from its posterior origin via the thalamus to the posterior columns of the spinal cord. Although Walker's central physiology was obviously wrong, as was his notion that the posterior roots were motor and the anterior roots were sensory, he did not retreat from his basic position even after the findings of Bell and Magendie achieved acceptance.

In contrast to Walker, Bell wrote the following in 1811:

Next considering that the spinal nerves have a double root, and being of the opinion that the properties of the nerves are derived from their connections with the parts of the brain, I thought that I had an opportunity of putting my opinion to the test of experiment, and of proving at the same time that nerves of different endowments were in the

same cord, and held together by the same sheath. On laying bare the roots of the spinal nerves, I found that I could cut across the posterior fasciculus of nerves, which took its origin from the posterior portion of the spinal marrow without convulsing the muscles of the back; but that on touching the anterior fasciculus with the point of the knife, the muscles of the back were immediately convulsed. Such were my reasons for concluding that the cerebrum and the cerebellum were parts distinct in function, and that every nerve possessing a double function obtained that by having a double root. (1936 reprint, 114)

The scientific community has never been unified in its judgment of Bell or in agreement about who should get credit for the discovery of the law of the spinal roots (Clarke and O'Malley, 1968; McHenry, 1969; Olmstead, 1944b). Although Bell conducted the critical experiment before anyone else, two serious criticisms have been leveled against him.

First, rather than publishing his results in a leading scientific journal or book, Bell described his experiments and ideas in letters to his brother and in his *Idea of a New Anatomy of the Brain; Submitted for the Observations of His Friends.* Only 100 copies of this small privately printed pamphlet were published, and fewer may have been distributed. Few copies of this very rare pamphlet have been discovered and, as far as is known, only one of his "friends" responded to him in writing. If anything, Bell seemed to have intended to present his findings and theories in a series of lectures and not immediately as a book for other scientists.

The second criticism directed at Bell was that while he was sufficiently clear regarding the motor functions of the anterior (ventral) roots, he did not state unequivocally that the posterior roots were sensory. Bell wrote only that one part comes from the cerebrum and the other from the cerebellum. Like Willis and then Walker, Bell considered the cerebellum to be the organ that governed the operations of the viscera essential to life. As for the cerebrum, it was the organ into which "all the nerves from the external organs of the senses enter; and from which all the nerves that are agents of the will pass out." In short, both sensory and voluntary motor functions were associated with the cerebrum, whereas only visceral efferent fibers were firmly associated with the cerebellum.

Bell republished his findings in 1824 after having served the Crown as a military surgeon in its war against France (he was at Waterloo in 1815) and after Magendie's results (see below) had become known. Bell, however, made insertions and subtle changes in his text to make his case for primacy stronger (Brazier, 1959). Surprisingly, Bell did not conduct further experiments on the functions of the two sets of spinal roots but only cited his original data from unconscious animals.

Bell's refusal to subject conscious animals to painful procedures obviously limited his ability to test sensory functions. In fact, Bell and many other scientists in the United Kingdom complained vigorously about "abhorrent" experiments conducted on conscious animals. Magendie, who was claiming that he should get the credit for discovering the law of the spinal roots, was one of their prime targets.[11]

Magendie's famous experiments on the spinal roots were performed on a litter of 8 puppies in 1821. The pups were 6 weeks old and had been given to him as a gift (Keele, 1957). His experiments were conducted at Charenton, near Paris. When Magendie described his findings a year later, he made several statements about the sensory functions of the dorsal roots that were much clearer than those of Bell:

> It is enough for me to announce today as positive, that the anterior and posterior roots of the nerves that arise from the spinal cord have different functions; that the posterior roots appear more specifically related to sensation, and the anterior to movement. (1822; translated in Clarke and O'Malley, 1968, 301)

Magendie was not timid about making his findings public. Nevertheless, he did not exactly state that the posterior roots were sensory and that the anterior roots were motor. He seemed to say only that one was predominantly sensory and the other was predominantly motor. Upon receiving a copy of Bell's pamphlet after he published his first paper on the spinal roots, Magendie coolly announced only that "Mr. Bell was very near discovering the functions of the spinal roots" (Walker, 1839).

These facts led to endless debates about who should be recognized for the discovery that the dorsal roots of the spinal cord are sensory whereas the ventral roots are motor. Because both Bell and Magendie contributed to the discovery, the scientific community eventually honored both men by calling the law of the spinal roots the Bell-Magendie law.

---

[11]Magendie was a native of Bordeaux and is often cited as the founder of the French school of experimental physiology. As a scientist, he showed no reluctance when it came to operating upon fully conscious dogs, including those used to study the functions of the spinal roots. To Bell and many others, this was indisputable evidence of Magendie's unusual cruelty to animals. Interestingly, even after ether began to be employed in operating rooms, Magendie still refused to use anesthetics to reduce pain. (For a history of anesthetics, see Chapter 11; for a history of vivisection, see Schiller, 1967.)

SOME REMARKABLE PREMONITIONS

The discovery of the "law of the spinal roots" probably led some scientists to wonder whether parts of the brain could also be divided into sensory and motor areas. This was an idea that Bell entertained in his 1811 pamphlet. Of course, few people were thinking about cortical localization at this time. But probably unknown to Bell and just about everybody else, a remarkably accurate theory of cortical localization had been postulated in the mid-1700s. The individual who anticipated the birth of modern cortical localization theory was Emanuel Swedenborg (1688–1772; shown in Figure 2.20), a man who established himself in mathematics and mining before he turned to the natural sciences in 1736 (Gibson, 1967; Schwedenberg, 1960).

Swedenborg visited medical centers in France, Italy, and the Netherlands after he decided to study anatomy and medicine. Whether he was skilled in dissection or experimentation is not known, but his writings show he was very well read, especially when it came to anatomy and disorders of the nervous system (Ramstrom, 1910).

In his *Oeconomia regni animalis,* published in Amsterdam in 1740–1741, Swedenborg followed Willis (1664) in concluding that the cerebrum was the source of understanding, thinking, judging, and willing. But Swedenborg went well beyond Willis. He wrote that different functions had to be represented in different anatomical loci at the level of the cortex. Philosophically, he saw this as the only way in which certain functions of the brain would not be confused with each other. It was also the only way to explain certain clinical-pathological phenomena.

Swedenborg called his distinct cerebral territories *sphaerulae* or *cerebellula* and hypothesized that they were separated by fissures and gyri. He localized the motor cor-

Figure 2.20. Emanuel Swedenborg (1688–1772), who concluded that different sensory systems had to be represented in different parts of the cortex and who inferred a separate motor cortex well before the concept of cortical localization came to the fore in the 1800s. (Courtesy of the Royal Swedish Academy of Sciences, Stockholm.)

tex near the front of the brain and seemed to envision a somatotopic arrangement when he stated that the muscles of the extremities are controlled by the upper frontal convolutions, those of the middle part of the body by the middle frontal convolutions, and the head and neck by the lower convolutions. He theorized, however, that the cortex was not responsible for all motor actions. Supposedly, it controlled voluntary movements, whereas more autonomic and habitual movements were mediated by secondary motor centers in the grey substance of the medulla and spinal cord.

Remarkably, Swedenborg also seemed to have inferred the intellectual functions of the frontal lobes. He wrote:

These fibres of the cerebrum proceed from its anterior province, which is divided into lobes—a highest, a middle, and a lowest. . . . These lobes are marked out and encompassed by the carotid artery. . . . If this portion of the cerebrum therefore is wounded, then the internal senses—imagination, memory, thought—suffer; the very will is weakened, and the power of its determination blunted. . . . This is not the case if the injury is in the back part of the cerebrum. (1938 translation, p. 73)

Unfortunately, other than two small fragments of material, Swedenborg's neurophysiological work was not publicly circulated in his lifetime. In fact, it has been maintained that his "premonitions" of what would later be demonstrated experimentally had no impact on his scientific contemporaries or on the history of cortical localization of function (Akert and Hammond, 1962; Schwedenberg, 1960).

The reason is that Swedenborg began to experience mystical visions while he was developing his neurological ideas. These powerful visions led him to abandon the natural sciences for theology in 1744, and he never looked back. As a theologian, Swedenborg attracted many followers. In fact, the Swedenborgian (New Jerusalem) Church was founded in his honor in London in 1772. As for his unpublished neurophysiological manuscripts, these lay buried in the Royal Swedish Archives for years. They were not discovered and translated until after cortical localization was widely accepted (see Swedenborg, 1882–1887, 1938, 1955).

The concept of cortical localization was considered by several other men of science before the eighteenth century ended. Nevertheless, no one developed a theory on the same order as that envisioned by Swedenborg. For example, Jiří ("Georg") Procháska (1749–1820) wrote a widely acclaimed book on the nervous system in which he concluded that different divisions of intellect (generally considered cortical functions) just might occupy separate parts of the brain. To translate Procháska:

It is, therefore, by no means improbable, that each division of the intellect has its allotted organ in the brain, so that there is one for the perceptions, another for the understanding, probably others for the will, and imagination, and memory, which act wonderfully in concert and mutually excite each other to action. (1784, 446)

Such statements from Procháska, however, only hinted at a willingness to accept cortical localization should some real evidence for it emerge (also see Unzer, 1771). This was hardly the sort of writing that confronted people to reevaluate their time-honored ideas.

The real assault would come from Franz Gall (1757–1828) and the phrenologists in the opening decades of the nineteenth century. The history of cortical localization from its birth in Gall's youthful mind to the beginning of the twentieth century is examined in Chapter 3, "The Era of Cortical Localization."

REFERENCES

Akert, K., and Hammond, M. P. 1962. Emanuel Swedenborg (1688–1772) and his contribution to neurology. *Medical History, 6,* 255–266.
Amundsen, D. W. 1972. Medieval cannon law on medical and surgical practice by the clergy. *Bulletin of the History of Medicine, 52,* 22–44.
Augustine. 1955. *Confessions and Enchiridion.* (A. C. Outler, Ed./Trans.). Library of the Christian Classics, Vol. VII. Philadelphia, PA: Westminster Press.
Bakay, L. 1985. *An Early History of Craniotomy.* Springfield, IL: Charles C Thomas.
Baker, F. 1909. The two Sylviuses. An historical study. *Bulletin of the Johns Hopkins Hospital, 20,* 329–339.
Bell, C. 1811. *Idea of a New Anatomy of the Brain; Submitted for the Observations of His Friends.* London: Strahan and Preston. (Reprinted in 1936 in *Medical Classics, 1,* 105–120, and preceded by quotations from some of his letters.)
Benton, A. L., and Joynt, R. 1960. Early descriptions of aphasia. *Archives of Neurology, 3,* 205–221.
Berengario da Carpi, J. 1518. *Tractatus de fractura calve sive cranei.* Bologna: de Benedictis.
Boyle, R. 1691. *Experimenta et observationes physicae.* London: Printed for John Taylor at the Ship and John Wyat at the Rose in S. Paul's Church-Yard.
Brazier, M. 1959. The historical development of neurophysiology. In *Handbook of Physiology: Vol. 1. Neurophysiology.* Baltimore, MD: Waverly Press. Pp. 1–58.
Brown-Séquard, C.-E. 1860. *Course of Lectures on the Physiology and Pathology of the Central Nervous System.* Philadelphia, PA: Collins.
Brunschwig, H. 1525. *The Noble Experyence of the Vertuous Handy Warke of Surgeri . . .* London: P. Treveris.
Castiglioni, A. 1935. The medical school at Padua and the Renaissance of medicine. *Annals of Medical History, 7,* 214–227.
Castiglioni, A. 1943. Andreas Vesalius: Professor at the Medical School of Padua. *Bulletin of the New York Academy of Medicine, 19,* 766–777.
Clarke, E., and Dewhurst, K. 1972. *An Illustrated History of Brain Function.* Berkeley: University of California Press.
Clarke, E., and O'Malley, C. D. 1968. *The Human Brain and Spinal Cord.* Berkeley: University of California Press.
Descartes, R. 1650. *Passions animae.* Amsterdam: L. Elzevir.
Descartes, R. 1662. *De homine figuris et latinitate donatus a Florentio Schuyl.* Leyden: Franciscum Moyardum and Petrum Leffen.
Flourens, M.-J.-P. 1851. Note sur le point vital de la moelle allongée. *Compte Rendu des Séances de L'Académie des Sciences, 33,* 437–439.
Flourens, M.-J.-P. 1858. Nouveaux détails sur le noeud vital. *Compte Rendu des Séances de L'Académie des Sciences, 47,* 803–807.
Flourens, M.-J.-P. 1859. Nouveaux éclaircissements sur le noeud vital. *Compte Rendu des Séances de L'Académie des Sciences, 48,* 1136–1138.
Galen. 1968. *De usu partium.* Translated by M. T. May as, *On the Usefulness of the Parts of the Body.* Ithaca, NY: Cornell University Press.
Gibson, W. C. 1967. The early history of localization in the nervous system. In P. J. Vinken and G. W. Bruyn (Eds.), *Handbook of Clinical Neurology* (Vol. 2). Amsterdam: North Holland Publishing Company. Pp. 4–14.
Guainerio, A. 1481. *Opera medica.* Pavia: Antonius de Carcano.
Guthrie, D. 1940. The history of otology. *Journal of Laryngology and Otology, 55,* 473–494.
Keele, K. D. 1957. *Anatomies of Pain.* Oxford, U.K.: Blackwell.
Legallois, J.-J.-C. 1812. *Expériences sur le Principe de la Vie, Notamment sur Celui des Movemens du Coeur, et sur le Siège de ce Principe.* Paris: Hautel. (English ed., 1815)
Lorry, A. C. de. 1760. Sur les mouvements du cerveau. Second mémoire. Sur les mouvemens contre nature de ce viscère, et sur les organes qui sont le principe de son action. *Mémoires de Mathématique et de Physique, presentés à l'Académie Royale des Sciences, 3,* 344–377.
Magendie, F. 1822. Expériences sur les fonctions des racines des nerfs rachidiens. *Journal de Physiologie Expérimentale et Pathologie, 2,* 276–279.
McHenry, L. C., Jr. 1969. *Garrison's History of Neurology.* Springfield, IL: Charles C Thomas.
Meyerhof, M. 1928. *The Book of the Ten Treatises on the Eye, Ascribed to Hunain ibn Is-Hâq.* Cairo: Government Press.

Meyerhof, M. 1945. Sultan Saladin's physician on the transmission of Greek medicine to the Arabs. *Bulletin of the History of Medicine, 18,* 169–178.

Mistichelli, D. 1709. *Trattato dell'Apoplessia.* Roma: Rossi.

Olmstead, J. M. D. 1944a. Historical note on the Noeud Vital or respiratory center. *Bulletin of the History of Medicine, 16,* 343–350.

Olmstead, J. M. D. 1944b. *François Magendie.* New York: Henry Schuman.

O'Malley, C. D. 1964. *Andreas Vesalius of Brussels, 1514–1564.* Berkeley: University of California Press.

O'Malley, C. D., and Saunders, J. B. de C. M. 1952. *Leonardo da Vinci on the Human Body.* New York: Henry Schuman.

Peyligk, H. 1516. *Compendiosa capitis physici declaratio . . .* Leipzig: Impressit Wolfgangus Monacensis.

Pourfour du Petit, F. 1710. *Lettres d'un Médicin des Hôpitaux du Roy à un Autre Médicin de ses Amis.* Namur: Albert.

Procháska, J. 1784. Adnotationum academicarum. Fasciculus tertius. Prague: W. Gerle. Translated by Thomas Laycock as, *A Dissertation on the Functions of the Nervous System.* In *Unzer and Procháska on the Nervous System.* London: Sydenham Society, 1851.

Rahman, F. 1952. *Avicenna's Psychology.* London: Oxford University Press.

Ramstrom, M. 1910. Emanuel Swedenborg as an anatomist. *British Medical Journal, 2,* 1153–1155.

Rosner, B. 1974. Recovery of function and localization of function in historical perspective. In D. G. Stein, J. J. Rosen, and N. Butters (Eds.), *Plasticity and Recovery of Function in the Central Nervous System.* New York: Academic Press. Pp. 1–29.

Schiller, J. 1967. Claude Bernard and vivisection. *Journal of the History of Medicine and Allied Sciences, 22,* 246–260.

Schwedenberg, T. H. 1960. The Swedenborg manuscripts. *Archives of Neurology, 2,* 407–409.

Singer, C. 1952. *Vesalius on the Human Brain.* London: Oxford University Press.

Soury, J. 1899. *Le Système Nerveux Central: Structure et Fonctions: Histoire Critique des Théories et des Doctrines.* Paris: Carré et Naud.

Spillane, J. D. 1981. *The Doctrine of the Nerves.* Oxford, UK.: Oxford University Press.

Steno, N. 1668. *A Dissertation on the Anatomy of the Brain.* (E. Godfredsen, Trans.). Copenhagen: Nyt Nordisk Forlag Arnold Busck, 1950.

Stirling, W. 1902. *Some Apostles of Physiology.* London: Waterlow and Sons, Ltd.

Sudhoff, W. 1914. Die Lehre von den Hirnventrikeln im textlicher und graphischer Tradition des Altertums und Mittelalters. *Archiv für Geschichte der Medizin, 7,* 149–205.

Swedenborg, E. 1740–1741. *Oeconomia regni animalis.* Amsterdam: Changuion.

Swedenborg, E. 1887. *The Brain, Considered Anatomically, Physiologically, and Philosophically* (2 vols.). (R. L. Tafel, Ed./Trans./Ann.). London: Speirs.

Swedenborg, E. 1938. *Three Transactions on the Cerebrum. A Posthumous Work of Emanuel Swedenborg* (Vol. 1). (A. Acton, Ed./Trans.). Philadelphia, PA: Swedenborg Scientific Association. (Vol. 2, 1940)

Swedenborg, E. 1955. *Psychological Transactions.* Bryn Athyn, PA: Swedenborg Scientific Association.

Sylvius, F. 1663. *Disputationes medicarum . . .* Amstelodami: van den Bergh.

Talbot, J. H. 1970. *A Biographical History of Medicine.* New York: Grune & Stratton.

Telfer, W. 1955. *Cyril of Jerusalem and Nemesius of Emesa.* Philadelphia, PA: The Westminster Press.

Temkin, O. 1973. *Galenism. Rise and Decline of a Medical Philosophy.* Ithaca, NY: Cornell University Press.

Unzer, J. A. 1771. Erste Gründe Physiologie der eigentlichen thierischen Natur thierischer Körper. Leipzig: Weidmanns, Erben und Reich. Translated by Thomas Laycock as, *The Principles of Physiology.* In *Unzer and Procháska on the Nervous System.* London: Sydenham Society, 1851.

Vesalius, A. 1543a. *De humani corporis fabrica.* Basilae: J. Oporini.

Vesalius, A. 1543b. *Epitome.* Basilae: J. Oporini. (Translated from the Latin by L. R. Lind, with a foreword by L. Clendening and anatomical notes by C. W. Asling). New York: Macmillan, 1949.

Walker, A. 1839. *Documents and Dates of Modern Discoveries in the Nervous System.* (Reprinted 1973; Metuchen, NJ: Scarecrow Reprints.)

Willis, T. 1664. *Cerebri anatome: cui accessit nervorum descriptio et usus.* London: J. Martyn and J. Allestry. Tercentenary ed., 1664–1964, *Thomas Willis, The Anatomy of the Brain and Nerves.* Montreal: McGill University Press, 1965.

Chapter 3
The Era of Cortical Localization

> No physiologist . . . can long resist the conviction that different parts of the cerebrum subserve different kinds of mental action. Localization of function is the law of all organization whatever: separateness of duty is universally accompanied with separateness of structure: and it would be marvellous were an exception to exist in the cerebral hemispheres.
>
> Herbert Spencer, 1855

The idea that the cerebral cortex is composed of functionally distinct areas was not given serious consideration by the scientific establishment prior to the nineteenth century. Only Emanuel Swedenborg seemed to be working toward a coherent theory of this sort, but his ideas about cortical localization were not published in his lifetime and do not seem to have affected his contemporaries (Swedenborg, 1887; see also Akert and Hammond, 1962; Gibson, 1967; Schwedenberg, 1960).

As the nineteenth century opened, the idea that major parts of the nervous system could be divided functionally seemed to be in the air (see Chapter 2). Julien-Jean-César Legallois had localized the respiratory center in just one part of the medulla before 1812, when he published this and other findings in book form. Further, Sir Charles Bell seemed to have concluded as early as 1807 that the dorsal and ventral roots of the spinal cord served different functions. Bell (1811) went well beyond Legallois, however, when he suggested that one might have to look at the whole nervous system in a new way.

At this time, Franz Joseph Gall was looking at the cerebral cortex in a radical new way. In the minds of most historians, Gall, more than any other scientist, put the concept of cortical localization of function into play.

The present chapter begins with a look at Gall's ideas. It then turns to some "landmarks" of cortical localization, those few discoveries that changed philosophical orientations more than others. The chapter concludes with a section on the development of neuron doctrine. The reason for considering neuron doctrine here is that the concept of distinct functional areas of the cortex would not have become as established as it did by the end of the nineteenth century if scientists had continued to think of the cerebrum as simply a tangled net of fused cell processes.

GALL AND PHRENOLOGY

Franz Joseph Gall, a physician and an outstanding anatomist, formulated the idea that skull features were an indication of underlying brain development and that the development of specific cortical regions correlated with specific talents or behaviors.

Gall lived at a time when physical features, such as body build and facial characteristics, were associated with temperament and psychological characteristics. For example, Johann Kaspar Lavater (1741–1801) was a Swiss patriot who, in his religious enthusiasm and belief in the interaction of mind and body, looked for traces of the spirit upon the features of the body. It was in this intellectual environment that Gall also assumed "where there is variation in function, there must be variation in controlling structures." He then theorized that the cerebral cortex was composed of different organs whose individual development would be reflected in the bumps and depressions of the overlying skull.

Gall stated that his theory of localization originated when he was only 9 years old. He noted that a classmate with whom he had studied, but who was not as good as he was in composition, had a superior memory for verbal material. This young man also possessed *les yeux à fleur de tête* (cow's eyes). Gall thought that bulging eyes and good verbal memory also appeared together in other students whom he met.

> Although I had no preliminary knowledge, I was seized with the idea that eyes thus formed were the mark of an excellent memory. It was only later on . . . that I said to myself; if memory shows itself by a physical characteristic, why not other faculties? and this gave me the first incentive for all my researches, and was the occasion for all my discoveries. (Gall and Spurzheim, 1810–19, Vol. 4, 68)

Gall postulated that bulging eyes in his fellow students were caused by an underlying brain area devoted to the *Faculty of Attending to and Distinguishing Words, Recollections of Words, or Verbal Memory*. He went on to localize the faculty for memory for words and a distinct faculty for speech in the frontal lobes.

Gall, who settled in Vienna in 1781, began to lecture on his new theory of brain function one year after his graduation from medical school in 1795. In 1802, however, under pressure from the ecclesiastical authorities, the Austrian government issued a decree that put an end to his very popular public lectures. Thus, Gall moved to Paris and, in 1808, began to put his ideas about phrenology on paper. The French Institute (later called the *Académie des Sciences*) officially rejected his theories later that same year. Nevertheless, Gall became friends with some of the finest names in France, made large amounts of money lecturing about phrenology, and lived in considerable luxury (Barker, 1897).

Gall attempted to associate functions with structures primarily by examining the crania of individuals who constituted the extremes of society. These included great writers, poets, and statesmen, as well as lunatics and criminals. When he found men or animals with a special talent, he examined the form of the head for a cranial prominence. To aid him in his correlations, Gall collected skulls and skull casts. In all, he amassed over 300 skulls from individuals with known mental characteristics and over 120 casts of skulls from living persons.

The highly questionable way in which Gall localized faculties can best be illustrated with some examples (see Barker, 1897). "Destructiveness" was localized above the ear for three reasons. First, this is the widest part of the skull in carnivores. Second, a prominence was found here in a student who was "so fond of torturing animals that he became a surgeon." And third, this region was well developed in an apothecary who later became an executioner.

"Acquisitiveness" was located on the upper front part of the squamous suture, largely because this area seemed unduly large among the pickpockets Gall had known. "Ideality" was based on an area found to be large in sculptures of poets and was the part of the head touched or rubbed by poets when writing. As for "Veneration," this was located by the bregma suture when Gall noted that the most fervent worshippers at church services had rather distinct prominences in this location.

Gall believed he could identify and localize 27 faculties in different parts of the human cerebral cortex. He thought that 19 of these faculties could also be demonstrated in animals. Among those unique to humans were wisdom, passion, and sense of satire. Those shared with animals included courage, a sense of space, and a sense of color.

Gall did not argue that he knew the precise anatomical boundaries of each of these innate faculties, nor did he fill in all of the parts of his cortical map. In particular, as can be seen from Figure 3.1, he felt he did not understand much about the ventral surface of the brain. He believed, however, that the hemispheres duplicated each other and that each part of the brain could serve as a complete organ of mind, just as each eye could serve as a complete organ for sight. From this premise, it was held that when unilateral lesions caused symptoms, it was because they upset the functional symmetry between the two sides of the brain.[1]

Gall was assisted on some of his important early works by Johann Spurzheim (1776–1832), a man who met him in Vienna in 1800, but who did not begin to work with him until 1804. Spurzheim, who traveled extensively, was more

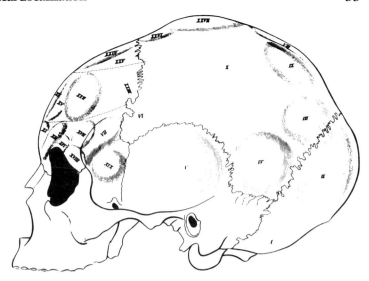

Figure 3.1. One of Franz Gall's plates using numbers to show the locations of the faculties of the mind.

than a devoted follower of the master. He developed some of his own ideas as well. For example, Spurzheim (1825, 1832) increased the number of organs of mind and distinguished between intellectual faculties and those pertaining to feelings. In contrast to Gall, he maintained there were no "bad" faculties (Carlson, 1959). One of Spurzheim's maps of the brain appears as Figure 3.2.

It is sometimes thought that Gall and Spurzheim had absolutely no interest in clinical material. This notion most likely stems from their stated convictions that (1) these

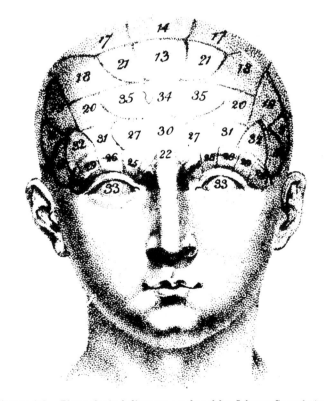

Figure 3.2. Phrenological diagram produced by Johann Spurzheim in 1825.

---

[1]This idea was also associated with the popular theories of Marie-François-Xavier Bichat (1771–1802) (see Chapter 25).

"accidents of nature" could not be duplicated from patient to patient; (2) such mutilations are associated with a host of secondary effects; (3) clinical subjects often are difficult to test reliably; and (4) only short survival times exist after severe brain injuries.

Nevertheless, Gall and Spurzheim did present a number of very interesting clinical cases in their writings.[2] For example, Gall discussed two men who could not remember the names of friends or family members after being injured above the eye by the point of a sword. In addition, Baron Dominique Jean Larrey (1776–1842), Napoleon Bonaparte's (1769–1821) most prominent battle surgeon, sent him the case of a man who had a left anterior lobe injury resulting from a fencing foil. Although this man, Edouard de Rampan, exhibited a loss of memory for words and a paralysis on the right side of the body, he retained a good memory for images. Gall also described a language disorder following a stroke in one patient. Still, it seems clear that Gall and his followers selected only those clinical cases that supported phrenological theory and did not hesitate to discard or explain away the contradictions.

Some physicians attempted to apply the new ideas of Gall and Spurzheim to clinical medicine. Three prominent physicians, John C. Warren (1778–1856), John Bell (1796–1872), and Charles Caldwell (1772–1853), heard Gall speak in Paris and returned to the United States in 1822 to spread the word (Freemon, 1991). Warren developed a research program in Boston to study phrenology, whereas Bell and Caldwell founded the Central Phrenological Society in Philadelphia in 1822.

In one of his lectures, Bell stated that Spurzheim had mentioned seeing 30 women guilty of infanticide, of which 26 had defective development of the inferoposterior organ of "philoprogenitiveness," or love for children. The implication was that their crimes resulted from a physically defective brain. Bell (1822) also cited the case of a mother who treated her children with excessive indulgence. She was asked if she had been fond of her children, upon which she remarked, "Oh sir, what comfort would we poor people have but for our children." As one might expect, her skull revealed that the organ of philoprogenitiveness was very highly developed.

In 1824, Caldwell wrote *Elements of Phrenology*, the first American textbook on phrenology, and in 1827 he published an expanded second edition of this book. Nevertheless, interest in his Central Phrenological Society was already waning, and by 1828, the year Gall died, the Society had ceased to exist.

Spurzheim, now the leading phrenologist in the world, went to the United States to revive interest in the theory (Freemon, 1991; Walsh, 1972). He drew large audiences in Boston for his lecture series aimed at the general public. He also attracted considerable attention with a lecture series aimed at physicians. Unfortunately, Spurzheim died from typhoid fever before all of his talks were completed (Walsh, 1972). Like Gall, Spurzheim had requested that his skull be preserved. Consequently, it was removed, studied in detail, and then turned over to the Harvard Medical School. It is shown in Figure 3.3.

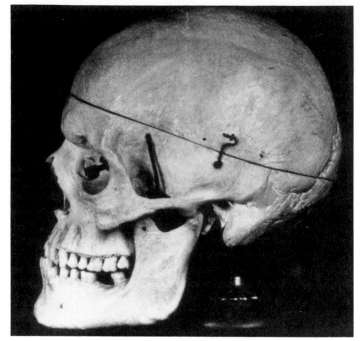

Figure 3.3. The skull of Johann Spurzheim (1776–1832). (Courtesy of the Warren Museum, Harvard Medical School, Cambridge, MA.)

George Combe (1788–1858), who at first thought the system of phrenology was absurd, became an admirer of Spurzheim in 1816 after watching him dissect a brain (Walsh, 1971). In 1820, Combe founded the Edinburgh Phrenological Society. The *Phrenological Journal and Miscellany,* which was largely under his editorship, followed three years later.[3] Combe also distinguished himself with his books on phrenology. His *Constitution of Man,* which first appeared in 1827, sold more than 50,000 copies between 1835 and 1838 and well over 100,000 copies before the presses were stopped. At this time, only the Bible and John Bunyan's (1628–1688) *Pilgrim's Progress* (1678) surpassed it in popularity as a book to be found on the shelves in English-speaking homes (Young, 1985).

To say the least, there were also more extreme forms of phrenology and physiognomy than those advocated by Gall, Spurzheim, and Combe. Figure 3.4 shows one of the most complex schemas.

THE REACTION AGAINST PHRENOLOGY

Many individuals in the United States opposed phrenology, even during the years of its greatest popularity. John P. Harrison (1796–1849) argued that the borders of Gall's proposed cerebral organs could not be delineated, making it impossible to correlate a specific ability with the size of a faculty. Harrison (1825) measured a collection of human skulls and showed that it was not even possible to predict brain size from the volume of the intracranial space. Similar sentiments were expressed by Thomas Sewall (1786–1845) in 1839 and by others.

[2]For synopses of some of Gall's clinical cases, see Head (1926), Stookey (1954), and Young (1970).

[3]See Fulton (1927) for a brief early history of the phrenological societies and their journals.

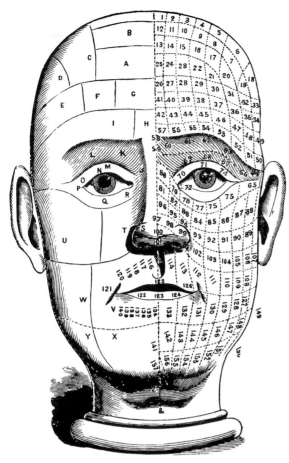

Figure 3.4. A map of the head and face by I. W. Redfield. This represented an extreme form of physiognomy. (From Wells, 1866; reprinted 1894.)

Although impressed by his anatomical skills, Flourens tried to portray Gall as a madman—a human driven beyond reason to build up a sizable collection of skulls—a man to be feared. In his book, *Phrenology Examined,* he included the following passage from a letter written in 1802 to Georges Cuvier (1769–1832) to enhance this perception:

"At one time," says Charles Villers, "every body in Vienna was trembling for his head, and fearing that after his death it would be put in requisition to enrich Dr. Gall's cabinet. He announced his impatience as to the skulls of extraordinary persons—such as were distinguished by certain great qualities or by great talents— which was still greater cause for the general terror. Too many people were led to suppose themselves the object's of the doctor's regards, and imagined their heads to be especially longed for by him, as a specimen of the utmost importance to the success of his experiments. Some very curious stories are told on this point. Old M. Denis, the Emperor's librarian, inserted a special clause in his will, intended to save his cranium from M. Gall's scalpel." (1846, 102)

In his *Psychologie Comparée,* Flourens, intending to shatter the scientific reputation of Gall's second in command, told the following story about Spurzheim:

The famous physiologist, Magendie, preserved with veneration the brain of Laplace. Spurzheim had the very natural wish to see the brain of a great man. To test the science of phrenology, Mr. Magendie showed him, instead of the brain of Laplace, that of an imbecile. Spurzheim, who had already worked up his enthusiasm, admired the brain of the imbecile as he would have admired that of Laplace. (1864, 234)

Flourens believed in laboratory science and emphasized the use of animals to study the effects of ablation and

To the extent that Gall and Spurzheim generalized from limited case material to large populations and dealt poorly with cases that did not support their ideas, some of the light jabs and heavier punches were clearly justified. Those disciples who took more extremist positions than did their mentors added fuel to the fire.

Combe made an American tour in 1838, after Gall and Spurzheim had died. But by now, scientific interest in phrenology had largely abated in the United States, and the new leader of phrenology found his lectures in support of the doctrine virtually ignored by leading American doctors. Moreover, the *American Journal of Phrenology,* which was founded in the same year, flopped completely as a magazine for scientists and physicians. Many U.S. medical books and journals of this period were either ceasing to mention phrenology, were noting it only in passing, or were setting it up for criticism.

The reaction against phrenology was stronger in Europe, especially in France, where Gall, a foreigner, was perceived as an unacceptable threat to the French scientific establishment. Marie-Jean-Pierre Flourens, shown in Figure 3.5, was one of Gall's most caustic and outspoken critics. He was a respected French scientist who had published many papers on comparative anatomy, anesthesiology, embryology, and physiology. He was the recipient of a seat in the *Académie des Sciences* and a member of the Legion of Honor.

Figure 3.5. Marie-Jean-Pierre Flourens (1794–1867), the leading French physiologist who conducted many experiments on the brains of animals. Flourens vigorously opposed the conclusions drawn by the phrenologists.

Figure 3.6. A late-nineteenth-century cartoon by Cocking ridiculing phrenology.

stimulation of various brain parts.[4] He described his ablation method in detail in a volume entitled *Recherches Expérimentales sur les Propriétés et les Fonctions du Système Nerveux dans les Animaux Vertébrés.* This book, published in 1824 and revised and expanded in 1842, included a wealth of behavioral observations.[5]

From his many experiments on a variety of animals, Flourens concluded that the cerebellum was responsible for coordinated movement and that the medulla played a role in basic functions necessary for life. At the same time, he argued vociferously that the cerebral cortex could not be divided into functional units. Flourens was convinced that, in contrast to what the phrenologists were claiming, the cortex functioned as a whole and that all parts of the cortex were responsible for intelligence, the will, and perception. As stated in the 1824 edition of his monograph:

> All sensations, all perceptions, and all volition occupy concurrently the same seat in these organs. The faculty of sensation, perception, and volition is then essentially one faculty.

Flourens witnessed recovery of function in a number of his experiments, especially those conducted on birds. He thought that when one cortical function returned, all returned. This, of course, was entirely consistent with his belief in cortical "equipotentiality."

---

[4]Flourens was not the first to use the lesion ("experimental") method. For example, in 1673, Joseph DuVerney (1648–1730) made brain lesions in pigeons (McHenry, 1969). In addition, Luigi Rolando (1773–1831) used the ablation method at least as early as 1809 in his laboratory. Flourens, however, had considerably more influence than his contemporaries in popularizing the lesion method.

[5]Flourens' behavioral assessments were unsophisticated, as exemplified by his description of a hen who had the equivalent of her cerebral hemispheres removed:

> I brought nourishment under her nose, I put her beak into grain, I put grain into her beak, I plunged her into water, I placed her on a shock of corn. She did not smell, she did not eat anything, she did not drink anything, she remained immobile on the shock of corn, and she certainly would have died of hunger if I had not returned to the old process of making her eat myself. Twenty times, in lieu of grain, I put sand into her beak; and she ate this as she would have eaten grain (1842, p. 57).

The failure of Flourens to find differences following lesions in different parts of the cortex might in part be explained by the fact that much of his work was done on hens, ducks, pigeons, and frogs, vertebrates displaying less cortical dependency than higher mammals. Nevertheless, he also worked on mice, cats, and dogs, which would suggest that he relied too much on the use of very young animals and/or that he might have been too quick to generalize from lower forms to higher forms.

It is also conceivable that Flourens might have been "blinded" by his enthusiasm to show that the phrenologists simply had to be wrong about cortical localization. Thus, although he successfully promoted the use of methodologies that should have suggested functional differences among cortical areas, his strong bias against cortical divisions could have reduced the chances of his finding any functional specificity.

In this context, it is possible to think of Gall as the visionary who had the right idea but the wrong method, and of Flourens as a laboratory scientist with the better method but the wrong theory. Each individual had something important to contribute to the study of cortical function, although their primary contributions did not overlap.

Initially, Flourens and his followers were able to withstand the challenge from the phrenologists, who believed in cortical localization. But with the passage of time, the basic idea of a functionally divisible cerebral cortex would grow from the seeds sown by Gall and his followers (Young, 1968, 1970). By the 1870s, cortical localization would achieve wide acceptance, but it would not be a localization based on skull features. This would be in disrepute (see Figure 3.6). Rather, it would be a localization based largely on brain damage and brain stimulation, the two techniques accepted and promoted by Flourens.

SPEECH AND THE FRONTAL LOBE

The first cortical localization that became widely accepted linked fluent, articulate speech to the frontal cortex. The issue of cortical localization of function was being debated in the French learned societies when Paul Broca presented his famous clinical case showing this in 1861. Although many other scientists had previously presented data both for and against the theory, earlier contributions failed to have the impact of Broca's case study (see Chapter 25).

For example, Jean-Baptiste Bouillaud (1796–1881), a student of Magendie and a respected figure in French science, had long emphasized the use of clinical examinations and autopsy material to support the case for localization of function. In 1825, after examining data from a large number of cases, an approach that clearly distinguished him from his predecessors, he wrote:

> In the brain there are several special organs, each one of which has certain definite movements depending upon it. In particular, the movements of speech are regulated by a special cerebral centre, distinct and independent. Loss of speech depends sometimes on loss of memory for words, sometimes of want of the muscular movements of which speech is composed. . . . Loss of speech does not necessitate want of movements of the tongue, considered as an organ of prehension, mastication, deglutination of food; nor does it necessitate loss of taste. The nerves animating the muscles, which combine in the production of speech arise from the anterior lobes or at any rate possess the necessary communications with them. (Translated in Head, 1926, 13–14)

Bouillaud cited Claude-François Lallemand (1790–1854) and Léon Louis Rostan (1790–1866) to bolster his case. In 1820 and 1823, respectively, these men had also reported loss of speech after anterior lobe damage. Bouillaud then argued that lesions elsewhere did not affect fluent speech (see Head, 1926; Ombredane, 1951; Stookey, 1963).

Unfortunately, Bouillaud had two negative factors working against him. First, he had been a founding member of the *Société Phrénologique*. Although he soon questioned Gall's cranioscopy procedures and recommended the use of brain-injured patients to study structure-function relationships, the scientific community remained overly cautious about anything or anyone associated in any way with Gall or phrenology.

Second, Bouillaud's descriptions of the causal lesions, like those of his predecessors and contemporaries, were less than satisfactory. As stated by Adolf Kussmaul (1822–1902) in 1878 (p. 721):

> In order to understand how superficially localizations are recorded even in the writings of the best of the older observers, let one cast a glance at the principle works of such men as Bouillaud, Lallemand and Rostan, to whom, notwithstanding, the pathology of the brain owes so much. They are content to indicate the affected lobe with the general designation, "in front," "behind," "in the middle," etc., and with equally indefinite appreciation of the extent of the lesion. One seldom learns whether the cortex or the white medulla, corpus striatum, centrum semiovale, etc., was the seat of the lesion.

It should also be noted that not everyone at this time was finding impairments in speech after frontal lobe lesions. For instance, Gabriel Andral (1797–1846) discussed 37 cases of frontal lobe lesions collected at the Charité hospital between 1820 and 1831 and found speech to be seriously impaired in only 21 of these cases. Andral (1840) did not say anything about the side of damage, but

he did mention that he also had 14 cases of speech loss who had sustained brain damage posterior to the frontal lobes. These findings left him cautious at best about accepting localization.

Even more disconcerting was the response from Jean Cruveilhier (1791–1874) to Bouillaud's assertions. Cruveilhier stated:

> If it is pathologically demonstrated that a lesion of the anterior lobes is constantly accompanied by a corresponding alteration of speech; on the other hand, if lesions of all parts of the brain other than the anterior lobes never entail speech alterations, the question is solved, and I immediately become a phrenologist. . . . In fact, the loss of the faculty of sound articulation is not always the result of a lesion of the anterior lobes of the brain; moreover, I am able to prove that the loss of the faculty of sound articulation can accompany a lesion of any other part of the brain. (Translated in Riese, 1977, 59)

Thus, rather than being a time of calm, this was a time of controversy and debate, especially in the halls of the French scientific academies. On the whole, the conservatives were on Flourens' side, while the liberals were the localizers (Head, 1926). It was in this far from tranquil environment, with men like Simon Alexandre Ernest Aubertin (1825–1893) arguing for localization and Pierre Gratiolet (1815–1865) taking the opposing position, that an enfeebled Monsieur Leborgne was transferred to Broca's surgical ward. This dying man had been hospitalized for 21 years, and he now showed epilepsy, loss of speech, and right hemiplegia. Broca, shown in Figure 3.7, took it upon himself to examine Leborgne as a test of Bouillaud's and

Figure 3.7. Paul Broca (1824–1880), whose association between aphasia and damage to the frontal cortex in M. Leborgne ("Tan"), in 1861, became the first cortical localization that was widely accepted. (Courtesy of the Académie de Médecine, Paris.)

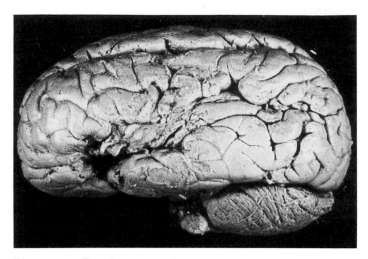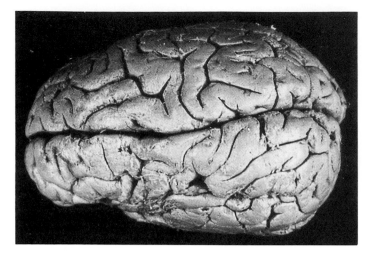

Figure 3.8. Two photographs of the brain of Leborgne ("Tan"), Paul Broca's celebrated case. (From the Musée Dupuytren. Courtesy of Assistance Publique, Hôpitaux de Paris.)

Aubertin's contention that speech loss will always be associated with a large lesion of the anterior lobes.

Six days after Broca saw him, Leborgne died. His brain, which proved to be in poor condition due to infarctions, was removed and presented to the *Société d'Anthropologie* the next day (Figure 3.8 shows the brain). Broca issued a weak statement about localization at this time, but at a decidedly more animated meeting of the *Société d'Anatomie* later that year, he boldly proclaimed that his case study of Leborgne placed him in firm agreement with the localizationists. Broca was especially careful to emphasize that his localization of a faculty for articulate language in the frontal lobe differed from that proposed by Gall, although both he and the phrenologists believed that articulate speech was an anterior lobe function.

Broca's "conversion" was a true landmark event, and many people who were vacillating about localization accepted the localizationist doctrine after Broca took his stand. As for Monsieur Leborgne, because he had often uttered a word that sounded like "tan," Broca's first and most famous case became known as Tan to the scientific community.

Just why did the case of Tan have so much impact relative to the material presented by Bouillaud, Aubertin, and others who favored localization? The answer to this question probably rests on four important factors that converged in Broca's report (Sondhaus and Finger, 1988). First, Broca's paper contained detail: detail in the case history, detail in its emphasis on articulate speech as opposed to any defect in speech, and detail in trying to find a more circumscribed locus to account for the symptoms. Second, the findings did not support the site of the faculty proposed by the phrenologists. Third, there was a willingness on the part of the scientific community to listen to these ideas at this time; the *Zeitgeist* was right. And fourth, it was Paul Broca, the highly respected scientist, physician, and esteemed head of academic societies, who was now willing to lead the fight for cortical localization of function.[6]

[6]Schiller (1979) provides an excellent description of how Broca was esteemed by many other French scientists. In general, he was perceived as a thorough, independent thinker of great energy and as a fair-minded person with great leadership ability.

It is worth noting that two more years would pass before Broca would note that the lesion affecting speech was usually on the left side. But even in his 1863 statements, Broca seemed uncomfortable about dealing with the issue of dominance. Perhaps it was because the two sides of the brain looked so similar to him and because, as a scientist, he felt it best always to proceed cautiously. Hence, it was not until 1865 that he addressed this issue in a more direct and meaningful way (see Chapter 26). Further, while Broca might at first have believed that he was observing an invariant principle when he argued that the center for articulate language resided in the third left frontal convolution, he soon found himself discussing and trying to account for exceptions to this "rule" (Berker, Berker, and Smith, 1986; Finger and Wolf, 1988).

EXPERIMENTAL CONFIRMATION OF A MOTOR CORTEX

If Broca's 1861 report can be regarded as the most important clinical paper in the history of cortical localization, the discovery of the dog's cortical motor area nine years later must be regarded as the most significant laboratory discovery. This is because Eduard Hitzig (1838–1907; shown in Figure 3.9) and Gustav Fritsch (1838–1927), the two men who conducted the experiment, proved that cortical localization was not restricted to a single function. Their work overthrew Flourens' doctrine that the cortex was not directly involved with the control of movement. Further, they showed that much could be learned about cortical function by conducting experiments on laboratory animals.

Fritsch and Hitzig (1870) performed their famous experiment on dogs not in a laboratory, but on a dressing table in a bedroom of Hitzig's house in Berlin. Prior to this, Hitzig (1867, 1869), who had been a physician to the Berlin garrison hospital, developed an apparatus to generate therapeutic electrical currents and tested it on his patients. When currents applied to the back of a patient's head reliably elicited eye movements, he thought it worthwhile to investigate this finding further.

Although Hitzig also obtained some encouraging results

Figure 3.9. Edouard Hitzig (1838–1907), codiscoverer of the motor cortex in 1870 with Gustav Fritsch. (From L'Académie Nationale de Médecine, Paris.)

with rabbits, his efforts between 1867 and 1869 did not prove there was a motor cortex. There was still the nagging possibility that the cerebellum was responding to the spread of the stimulation. Additionally, currents might also have spread to the basal ganglia. In fact, as Fritsch and Hitzig later pointed out, many others had tried to elicit movements by stimulating the cortex. Almost uniformly, earlier findings were either negative, questionable, or unappreciated for what they really represented (see Chapter 14).

For example, Luigi Rolando (1809), who was not mentioned by Fritsch and Hitzig, was one of the first to use galvanic current to stimulate the cortex in animals. He associated the strongest muscle contractions with stimulation of the cerebellum, but still obtained vigorous movements with electrodes on the cerebrum. However, because he used very strong stimuli, the possible spread of current made his findings very difficult to assess.

Fritsch and Hitzig, like nearly everyone else, seemed unaware of the observations that Robert Todd (1809–1859) had published in 1849. Todd's main interest was to stimulate the brainstem, but while inserting his electrodes through the cortex, he passed some current and observed involuntary twitching. Unfortunately, Todd did not appreciate what this meant (Jefferson, 1953).

It was in this scientific environment that Hitzig, a psychiatrist, invited Fritsch, an anatomist, to join him in an experiment on dogs. They decided to use low levels of electrical stimulation, just enough to be felt when applied to the tip of the tongue. They found distinctive cortical sites that gave muscular responses corresponding to forepaw (extension and flexion), hindpaw, face, and neck on the

opposite side. This is shown in Figure 3.10. The investigators concluded:

> A part of the convexity of the hemisphere of the brain is motor, another part is not motor. The motor part, in general, is more in front, the non-motor part more behind. By electrical stimulation of the motor part, one obtains combined muscular contractions of the opposite side of the body. These muscle contractions can be localized on certain very narrowly delimited groups by using very weak currents. . . . The possibility to stimulate narrowly delimited groups of muscles is restricted to very small foci which we shall call centers. (1960 translation, 81)

In the same paper, Fritsch and Hitzig went on to describe how they used a scalpel handle to ablate the forepaw area unilaterally in two dogs. Although this did not abolish all movement from the contralateral forepaw, motor performance was impaired and abnormal postures were observed. In contrast, sensation appeared to be normal. This convinced Fritsch and Hitzig that they had, in fact, discovered a discrete motor zone with a somatotopic organization. For Hitzig in particular, cortical localization satisfied his compulsive need for order in science as well as in other facets of life (Pauly, 1983).

An interesting question is why many of the predecessors of Fritsch and Hitzig did not obtain movements or find this degree of specificity in their experiments. After all, many more established scientists had attempted to stimulate the cortex electrically before 1870. The answer seemed to reside at least partly in the fact that the motor cortex is difficult to expose, especially in dogs. Evidently, earlier

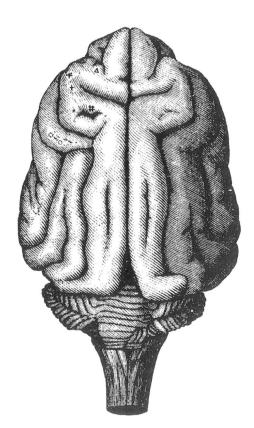

Figure 3.10. Drawing of the brain of a dog from Fritsch and Hitzig (1870). This figure shows the areas (triangle, cross, hatch marks) from which movements of the opposite side of the body were evoked with electrical stimulation.

Figure 3.11. Roberts Bartholow (1831–1904), the American physician who electrically stimulated the brain of Mary Rafferty in 1874 and showed that the human cerebral cortex was electrically excitable.

experimenters exposed the brain over nonmotor regions and, after failing to elicit movements, did not proceed much further. They may have assumed that Flourens was correct in his assertion that functions represented in one part would also be found in all other loci.

Had some of these investigators been more familiar with John Hughlings Jackson's (1835–1911) early publications on epilepsy and hemiplegia, the discovery of the motor cortex might have come a few years earlier. Although Jackson (e.g., 1863) did not attempt to pinpoint the location of the motor zone, the experiments of Fritsch and Hitzig essentially confirmed Jackson's growing conviction that there had to be a somatotopically organized cortical region involved with movement (see Chapter 14).

Fritsch and Hitzig concluded their paper in a way that encouraged other researchers to try to confirm their findings and to search for other cortical areas. They maintained it was indeed worthwhile to search for areas concerned with sensation, and even for regions involved with intelligence. Fritsch and Hitzig (1870, p. 96) wrote that "certainly some psychological functions, and perhaps all of them, in order to enter matter or originate from it, need circumscribed centers of the cortex."

Although little was heard of Fritsch after 1870, Hitzig continued his experiments and published additional studies on dogs and monkeys (1874a, 1874b). He also described human case material, including a soldier wounded during the Franco-Prussian War in the Battle of Sedan (1870) in northeastern France. The soldier developed a cortical abscess, and his symptoms included localized seizures and aphasia.

The idea of cortical excitability was confirmed in humans in 1874 by Roberts Bartholow (1831–1904), a professor of medicine in Cincinnati, Ohio. Bartholow, shown in Figure 3.11, stimulated the dura and then the cortex just behind the Rolandic fissure of a dying, feeble-minded girl. He found that Mary Rafferty moved her limbs and felt tingling sensations on the opposite side of her body when he stimulated her Rolandic cortex.

Bartholow stated that he did his experiments with the girl's approval. Further, her skull was ulcerated so badly that her pulsating brain could be seen, and incisions had already been made to allow pus to escape. Although he felt he could introduce needles into Mary Rafferty's cortex "without material injury," he was severely criticized in the United States and Europe for conducting these physiological experiments on a human being (Walker, 1957).

FERRIER'S EXPERIMENTS

The man who emerged as the leader of the new "localizers" was the British experimentalist Sir David Ferrier (1843–1928), whose portrait is shown in Figure 3.12. He replicated the motor cortex experiments of Fritsch and Hitzig with even finer currents and produced more detailed maps of the monkey brain. Ferrier described behaviors as intricate as the twitch of an eyelid, the flick of an ear, and the movement of one digit.

Ferrier's first publication on this subject appeared in 1873. The work was conducted at the West Riding Lunatic Asylum, where he had been invited by the superintendent, James Crichton-Browne (1840–1938), to pursue the work of Fritsch and Hitzig (see Spillane, 1981).

By 1875, Ferrier was describing many different parts of the monkey motor cortex. In 1881, Ferrier even took a monkey with a unilateral lesion of the left motor cortex to the Seventh International Medical Congress in London. Over 120,000 people were invited to this huge Victorian spectacle, including "all the crowned heads of Europe." When the hemiplegic monkey, which had been operated upon seven months earlier, limped into the demonstration room, Jean-Martin Charcot (1825–1893) remarked, "It is a patient!"

This animal was killed after the session, and a committee, which included William Gowers (1845–1915), Edward Albert Schäfer (1850–1935), and John Newport Langley

Figure 3.12. David Ferrier (1843–1928), one of the leaders of the localizationist movement in the last quarter of the nineteenth century. (Courtesy of Yale Medical Historical Library, New Haven, CT.)

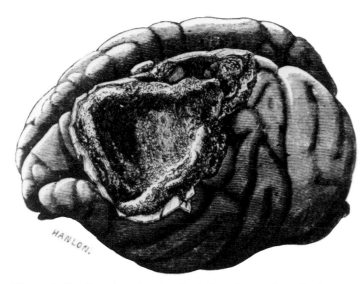

Figure 3.13. Drawing showing the lesion of a monkey who became hemiplegic after cortical surgery. The monkey was presented at the International Medical Congress in 1881 by David Ferrier.

(1852–1925), examined the brain (Klein, Langley, and Schäfer, 1883).[7] They reported that the lesion, shown in Figure 3.13, was indeed placed in those areas that Ferrier and his associate, Gerald Yeo (1845–1909), said they had destroyed.[8]

In his continuing studies on the motor cortex, Ferrier showed that fish, birds, and amphibians either gave no responses or nondiscrete responses to electrical stimulation. The observations on these lower vertebrates helped to explain why Flourens did not find the motor cortex.

Ferrier, who was associated with the National Hospital and the Medical School of King's College after he left West Riding, also worked on the superior temporal gyrus. He found that monkeys perked up the opposite ear and turned both head and eyes to the opposite side when this region was stimulated. This suggested a role for this area in audition. An ablation experiment confirming this hypothesis soon followed (see Chapter 9).

But while Ferrier appeared to be accurate in this association and in his localization of a center for smell in the uncinate region of the temporal lobe, his localization of taste in the temporal lobe was off the mark (see Chapter 12). In addition, he was very slow to recognize the importance of the occipital lobe in vision (see Chapter 6).

Many of Ferrier's findings were summarized in his book *The Functions of the Brain,* published in 1876. He not only presented his own data in this book but went on to show how they related to some of John Hughlings Jackson's clinical cases. The work had immense impact on the scientific community. One outcome was that it opened the "modern" era of neurosurgery. Neurosurgeons now turned to "functional maps" of the brain for guidance (Gibson, 1962; Jef-

---

[7]After 1918, Schäfer changed his name to Sharpey-Schafer to honor one of his teachers (see Sherrington, 1935).

[8]The unfortunate postscript to this story was that after the Congress was over, Ferrier was taken to the Bow Street Police Court under the Cruelty to Animals Act for operating on the animals without a government license and in a way that caused them pain. The trial was followed by scientists, sympathizers, and antivivisectionists all over the world. It culminated with Ferrier being acquitted because the surgeries were actually performed under general anesthetics by Yeo, who did hold a government license (Spillane, 1981).

ferson, 1960; Macewen, 1879; Trotter, 1934; Walker, 1967a, 1967b; also see Chapter 29). New surgical cases, in turn, offered further confirmation of the localization idea.

ELECTROPHYSIOLOGICAL RECORDINGS

In addition to the findings obtained from neurosurgical patients, three other parallel developments supported the type of localization that Ferrier and many other experimentalists and clinicians were advocating. These were (1) electrophysiological recordings from the brain, (2) cytoarctitectonic studies of the cortex, and (3) the growing belief that neurons are independent entities.

Richard Caton (1842–1926; shown in Figure 3.14) was probably the first to record the spontaneous electrical activity of the brain. Caton, who had been a medical student with Ferrier in Edinburgh, worked at the Liverpool Royal Infirmary School of Medicine. His research was based on Ferrier's descriptions of the effects of stimulation and ablation of discrete areas of the cortex.

Caton's first report was a presentation before the British Medical Association in Edinburgh in July 1875 and was summarized in the *British Medical Journal* later that year. Caton told his audience that the electrical changes taking place in the brain varied in location with the specific peripheral stimuli he was using:

> In every brain hitherto examined, the galvanometer has indicated the existence of electric currents. The external surface of the grey matter is usually positive in relation to the surface of a section through it. Feeble currents of varying direction pass through the multiplier when the electrodes are placed on two points of the external surface of the skull. The electric currents of the grey matter appear to have a relation to its functions. When any part of the grey matter is in a state of functional activity, its electric current usually exhibits negative variation. For example, on the areas shown by Dr. Ferrier to be related to rotation of

Figure 3.14. Richard Caton (1842–1926), a pioneer in the study of the electrical activity of the brain. Caton recorded action potentials from the brains of rabbits and monkeys in 1875.

the head and to mastication, negative variation of the current was observed to occur whenever those two acts respectively were performed. Impressions through the senses were found to influence the currents of certain areas; e.g., the currents of that part of the rabbit's brain which Dr. Ferrier has shown to be related to movements of the eyelids, were found to be markedly influenced by stimulation of the opposite retina by light. (1875, 278)

Caton published another short report in 1877. In 1887, he even traveled to the United States to present electrophysiological evidence for cortical localization at the Ninth International Medical Congress. He described how he applied small electrodes directly to the brain surface in animals, which were then allowed to eat, drink, and move about on a small table. He had no success in his attempts to find parts of the brain responsive to sounds or odors, but he told his audience that light was a very effective stimulus for modifying regional cortical activity.

Although Caton's (1875, 1877, 1887, 1891) electrophysiological discoveries on rabbits and monkeys preceded those that were announced by other nineteenth-century pioneers in electrophysiology, most seemed ignorant of what Caton had done. In fact, there were a number of claims to being the first to record evoked potentials from the brain (Brazier, 1959, 1988; Cohen, 1959; also see Chapter 6).

Among others, Ernst Fleischl von Marxov of Vienna (1846–1892) and Adolf Beck (1863–1942) of Krakow thought they were the first to demonstrate cortical localization with electrophysiological methods. The former expressed this in a sealed letter (dated 1883 and published in 1890), whereas Beck based this on his thesis material, which was presented in 1890.

In 1891, Caton wrote a letter in German to the editor of the *Centralblatt für Physiologie* to dispute these claims. Included was a summary of his work from 1875, which had been published in the *British Medical Journal*. Caton stated:

> In the year 1875 I gave a presentation before the Physiological Section of the British Medical Association in which electrical currents of the brain in warm-blooded animals were demonstrated and, in addition, their undoubted relationship with regard to function was established. May I be permitted to draw your attention to the following publication (Brit. Med. J., 1875, 2, 178). . . . In the transactions of the 9th Medical Congress in Washington there is yet another published communication (Vol. 3, 246) under the title "Researches on electrical phenomena of cerebral grey matter." It is by no means my intention to detract from these learned physiologists, nevertheless I myself have made these observations, . . . I have published them, so I think it must be conceded that I am already an earlier discoverer. (Translated by Brazier, 1988, 207)

Electrophysiological recordings became much more fashionable after Hans Berger (1873–1941) published his electroencephalograph (EEG) work on humans in 1929. Unlike almost everyone else, Berger cited Caton's valuable contribution to the field. In 1929, he wrote:

> Caton had already (1874) published experiments on the brains of dogs and apes in which bare unipolar electrodes were placed either on the surface of both hemispheres or one electrode on the cerebral cortex and the other on the surface of the skull. The currents were measured by a sensitive galvanometer. There were found distinct variations in current, which increased during sleep and with the onset of death strengthened, and after death became weaker and then completely disappeared. Caton could show that strong current variations resulted in brain from light shone into the eyes, and he speaks already of the conjecture that under the circumstances these cortical currents could be applied to localisation within the cortex of the brain. (Translated by Cohen, 1959, 258)

Berger's acknowledgment of Caton's discoveries appeared four years after Caton had died. A few years earlier, in its obituary columns, the *Lancet* did not even mention Caton's contribution to electrophysiology, while the *British Medical Journal* noted only that he did some work on the localization of movements in the 1870s. Most people seemed to have remembered Caton better as the gentleman who became Lord Mayor of Liverpool in 1907.

CYTOARCHITECTONICS

Early in the twentieth century, several anatomists, including Oscar Vogt (1870–1950), Cécile Vogt (1875–1962), Grafton Elliot Smith (1871–1937), Alfred Walter Campbell (1868–1937), and Korbinian Brodmann (1868–1918), attempted to study the fine anatomy of the cortex. They reported that the structure and cellular composition of some areas clearly differed from those of other areas (Brodmann, 1909; Campbell, 1905; Vogt and Vogt, 1919). Brodmann, for instance, recognized 52 discrete cortical areas (see Figure 3.15), whereas the Vogts thought there might

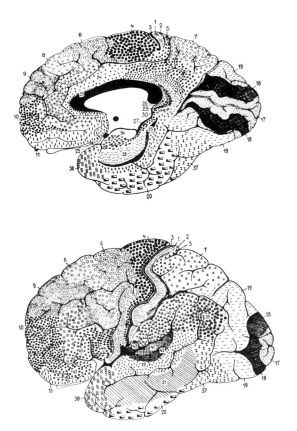

Figure 3.15. Korbinian Brodmann's 1909 map of the human brain showing his numbering system.

be over 200 areas, each with its own distinguishing cytoarchitectonic pattern.

The criteria used included absolute thickness of the cortex, number of horizontal laminations, relative thickness of the laminae, arrangements of the cells, presence or proportions of specific cell types, variations in cell shape, rate of change in cell size within and between layers, degree of intermingling of cell types, cell density, and staining affinities. There was no real agreement among the various investigators as to how these criteria should be weighted. Because of this and their small samples, the maps that were produced differed from each other.[9]

In spite of the variability among experimenters and across specimens, the fact that regional differences in cytoarchitecture were found at the cortical level clearly suggested to these histologists that there must also be some corresponding functional specificity. Brodmann (1909) may have stated this best when he postulated:

> The specific histological differentiation of the cortical areas proves irrefutably their specific functional differentiation—for it rests as we have seen on the division of labor—the large number of specially built structural regions points to a spatial separation of many functions and from the sharp delineation of some fields there follows finally the sharply delimited localization of the physiological processes which correspond to it. (1960 translation, 217)

In general, this was an idea that attracted many supporters.

"NEURON DOCTRINE" AND LOCALIZATION

Another development which had a major effect on the emergence of localization theory was the discovery that neurons are independent units and not entities fused to each other. This discovery involved many scientists and spanned the greater part of the nineteenth century.[10] It began with improved microscopy, resulting largely from the achromatic objectives developed in 1826. These newer devices allowed scientists to obtain high magnifications without blurred images and thus permitted them to see nucleated cells more clearly.

The new microscopists included Christian Gottfried Ehrenberg (1795–1876), Gabriel Valentin (1813–1883), and Jan Purkyně (pronounced "Poorkyne" and spelled Purkinje in German and English; 1787–1869), who is shown in Figure 3.16. Each of these investigators recognized nerve cell bodies and some of their "branches" (Shepherd, 1991). Purkyně's (1837) early account of cerebellar cells led to the "Purkinje cells" of the cerebellum being named in his honor (see Figure 3.17).

In 1839, Theodor Schwann (1810–1882; shown in Figure 3.18)—who had just described the myelin sheath as a fat-like secondary deposit on the neurilemma—proposed that

[9]In 1946, Lashley and Clark wrote a lengthy paper discussing the variability in the maps. They concluded that some differences were due to experimenter biases, but that primate subjects also differed from each other much more than had been recognized, even within a single species. On the basis of their own findings, they argued that one should not place much faith in cytoarchitechtonic maps based on single subjects and that maps constructed by averaging data made just as little sense (see Chapter 4).

[10]The history of the neuron doctrine was traced by Shepherd (1991) and Van der Loos (1967), among others.

Figure 3.16. Jan Evangelista Purkyně (1787–1869), one of the first anatomists to draw neurons in the brain.

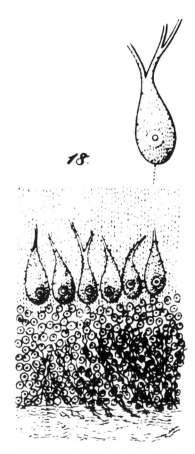

Figure 3.17. Cerebellar cells drawn by Jan Purkyně for the Congress of Physicians and Scientists in Prague in 1837 and published in 1838. These cells, the first identified nerve cells, would be called Purkinje cells in his honor.

Figure 3.18. Theodor Schwann (1810–1882), who described the myelin sheath and then, in 1839, proposed that the entire body was made of individual cells.

the entire body inside and out was made of individual cells. Schwann's "cell theory" was immediately accepted for all organs with the singular exception of the nervous system. It was not possible to determine with the methods available whether the branches from the nerve cells formed a continuous net or whether they retained some independence. It was not even clear if the branches were derived from the cell bodies. Valentin (1836), for example, was

unable to find connections between nerve cells and the fibers he saw under his microscope.

The nerve cell with its uninterrupted processes was described by Otto Friedrich Karl Deiters (1834–1863) in a work that was completed by Max Schultze (1825–1874) in 1865, two years after Deiters died of typhoid fever. This work portrayed the cell body with a single chief "axis cylinder" and a number of smaller "protoplasmic processes" (see Figure 3.19). The latter would become known as "dendrites," a term coined by William His (1831–1904) in 1889.

William His, an embryologist, was a student of Robert Remak (1815–1865; shown in Figure 3.20), whose own embryological studies had already indicated that the axon was not only continuous with the nerve cell body but that it arose from it (Remak 1836, 1837, 1838). In 1849, Rudolph Albert von Kölliker (1817–1905), another Remak student, also argued that the axis cylinder must be a process of the nerve cell.

Kölliker was among the mid-nineteenth century investigators who recognized that the methods available still left much to be desired. Largely for this reason, not everybody was willing to accept his conclusions—or those of Deiters, His, or Remak, who had seen many of the same things in their preparations. In particular, little was known about the relationships among the nerve fibers, axis cylinders, and dendrites. Was the nervous system one huge reticulum, or was it made of anatomically distinct units or systems? Were the processes of different cells fused by anastomosis, or did they retain some sort of independence?

Deiters (1865) and Kölliker (1867) denied that axons might fuse, but thought the dendrites might form a net (see Figure 3.21). In 1872, Joseph von Gerlach (1820–1896), the discoverer of the popular carmine and gold chloride staining methods, agreed. He wrote: "I have been able to show with the gold method the continuity of the network with

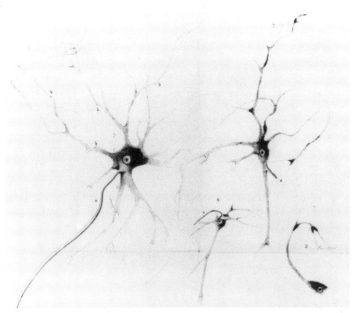

Figure 3.19. Nerve cells from the spinal cord with axons and dendrites. This illustration appeared in Otto Friedrich Karl Deiters' posthumous work of 1865.

Figure 3.20. Robert Remak (1815–1865), who in the 1830s proposed that the axon came from the nerve cell body.

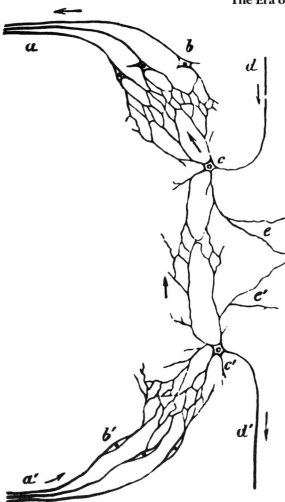

Figure 3.21. Rudolph Albert von Kölliker's 1867 portrayal of spinal nerve branches forming fused nets.

Figure 3.22. Camillo Golgi (1843–1926), who published his first paper on the silver nitrate staining method in 1873. Golgi's procedure allowed scientists to see elements in the nervous system in a new way and contributed to the birth of the neuron doctrine.

the protoplasmic prolongations (dendrites) of the nerve cells." Gerlach also asserted that the axonal processes formed complex nets.

The need for more conclusive histological techniques was satisfied when a new stain was discovered, one that allowed anatomists to examine elements in the nervous system with considerably more precision. Camillo Golgi (1843–1926; shown in Figure 3.22) was a resident physician at a home for incurables (Casa degli Incurabili) in the small town of Abbiategrasso, near Milan, Italy, when he developed his silver nitrate staining method (Da Fano, 1926; Viets, 1926). The discovery took place in a kitchen at the hospital, which he had converted into a laboratory. There was minimal instrumentation, and much of the work took place in the evening by candlelight. Although Golgi never explained how the idea came to him, he mentioned that he first hardened his specimens with potassium dichromate and ammonia and then immersed them in a solution of silver nitrate. His procedure did not stain all nerve cells, but for those occasional cells that stained black against the yellow background, all major morphological features seemed to be present.

In 1873, Golgi published the first brief but "adequate" picture of *la reazione nera* (the black reaction), which showed the whole nerve cell, including its cell body, axon,

and branching dendrites. This was followed by many additional studies by Golgi, who was honored with a university position at Pavia in 1875. One of his earlier drawings appears as Figure 3.23 and a later one as Figure 3.24.

In his initial report and over the ensuing years, Golgi assigned a nutritive role to the dendrites (see Golgi, 1903). As for the axons, he firmly believed his studies showed that the axis cylinders formed a dense intertwining, fused reticulum of the type described by Gerlach. This led him to oppose the growing concept of cerebral localization and to take a decidedly holistic approach to brain function (see Chapter 4). At most, he was willing to talk about broad territories, into which incoming fibers were distributed, but not the finely demarcated zones that Ferrier, Hitzig, and others were postulating (see Golgi, 1906).

In 1886, His broke with the past when he concluded "transmission of a stimulus without direct continuity is possible" in the sensory organs. He thought the evidence gleaned from the study of neural development favored contiguity, not continuity. This new idea was also entertained by August Forel (1848–1931), whose important paper on the subject appeared in 1887. Forel, shown in Figure 3.25, came to this conclusion in part because he had seen highly selective atrophy of nerve cells after some fibers had been destroyed. In his autobiography, Forel wrote about the then-prevailing theory of anastomosis and the idea of close contacts as opposed to fusion:

During a holiday . . . I worked out quite a different idea, relating to the anatomy of the brain. . . . At that time the problem was still quite obscure. We spoke of anastomoses between ganglion-cells of the nervous system, without really knowing how such connections between the elements of cells which are originally quite independent of one another could be effected. But recently the Italian anatomist, Professor Golgi, had invented a new method of colouring, by which he

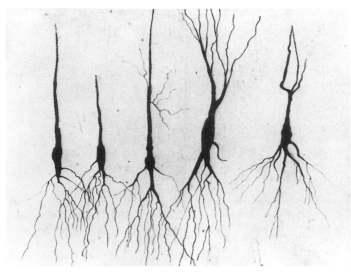

Figure 3.23. An early drawing by Camillo Golgi showing nerve cells stained black (his *reazione nera*) (see Zanobio, 1975).

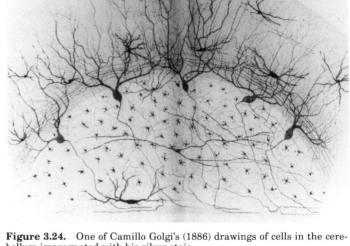

Figure 3.24. One of Camillo Golgi's (1886) drawings of cells in the cerebellum impregnated with his silver stain.

was able to show that the so-called protoplasmic processes of the ganglion-cells . . . are not connected to anything. However, this author was so completely ensnared by the old notion of anastomoses that on showing that the so-called nerve-processes of the ganglion-cells undergo ramification, he now assumed a network of anastomoses for these ramifications. . . . In my laboratory we succeeded in preparing specimens by Golgi's method, and we saw the blind terminations of the protoplasmic processes, but not the network of anastomoses connecting the nerve-processes.

In the next paragraph, Forel added:

It was as though the scales had fallen from my eyes. I asked myself the question: "But why do we always look for anastomoses? Could not the mere intimate contact of the protoplasmic processes of the nerve-cells effect the functional connection of nervous conduction just as well as absolute continuity? . . . All the data supported the theory of simple contact. (1937, pp. 162–163)

The theory of anastomosis was also opposed by Fridtjof Nansen (1861–1930) of Norway, whose doctoral dissertation on Golgi-stained neurons in invertebrates was published in 1887. Nansen's statements included the following: "A direct combination between the ganglion cells, by direct anastomosis of the protoplasmic process does not exist" and "The branches of the nervous processes do not anastomose."[11]

The new ideas of His, Forel, and Nansen had a significant impact. Still, nobody played a greater role in promoting "cell doctrine" than Santiago Ramón y Cajal (1852–1934), a young, free-thinking Spaniard who began his work with silver stains in 1887, soon after a visit to Madrid, where he saw sections stained with the Golgi technique.[12]

From the beginning, Ramón y Cajal, whose photograph is shown in Figure 3.26, knew that the "capricious and uncertain character" of the Golgi method represented a sig-

nificant obstacle that had to be overcome. He set forth to make better slides by staining more intensely, by cutting thicker sections, and by studying neurons prior to myelination (the Golgi stain does not stain myelinated fibers well). After doing this, he began to publish a series of papers describing his findings, mainly in a journal he initiated and paid for at great personal expense: his *Revista de Histologia Normal y Patologica (Trimestral Review of Normal and Pathological Histology).*

Figure 3.25. The idea that axonal or dendritic fusion (anastomosis) is not necessary for communication among nerve cells was presented in 1887 by August Forel (1848–1931), shown here late in his life.

---

[11]Nansen later achieved far greater fame as a polar explorer on the *Fram* and then for his involvement with World War I refugees, for which he received the Nobel Peace Prize in 1922.

[12]At the time, Ramón y Cajal held a position in Valencia, but later that same year he moved to Barcelona (Ramón y Cajal, 1937).

Figure 3.26. Santiago Ramón y Cajal (1852–1934), winner of the 1906 Nobel Prize for his extensive contributions to neuroanatomy.

Ramón y Cajal's first paper on the Golgi stain was on the bird cerebellum, and it appeared in the *Revista* in 1888. He acknowledged that he found the nerve fibers to be very intricate, but stated in the strongest terms that he could find no evidence for either axons or dendrites undergoing anastomosis and forming nets. He called each nervous element "an absolutely autonomous canton." He went on to study the retina, the olfactory bulb, the cerebral cortex, the spinal cord, and the brainstem. Still, he could find

Figure 3.27. Wilhelm von Waldeyer (1836–1921), who in 1891 wrote a highly-influential review favoring neuron doctrine. Waldeyer strongly supported Santiago Ramón y Cajal and also coined the word "neuron."

no evidence for the reticular theories of Gerlach and Golgi. Reviewing his many findings in 1889, he argued that nerve cells were independent elements.

Ramón y Cajal recognized that Spain stood in relative isolation from the scientific centers of Europe. He knew that sending copies of his journal from Spain to Germany and other European centers was not enough, since few people were able to read Spanish. Thus, he decided to have his papers translated into German and joined the German Anatomical Society. In 1889, with his precious Zeiss microscope and his slides, he set forth to Berlin to present his material at a scientific meeting (Ramón y Cajal, 1937).

It was at this meeting that Ramón y Cajal met Kölliker. The elder German scientist was so impressed with the young Spaniard that he entertained him and introduced him to many of the other leading scientists. Over the next year, Kölliker, who was universally admired for his calm yet exceptional judgment, confirmed many of Ramón y Cajal's observations, abandoned his own dendritic network theory, and taught himself Spanish in order to read Ramón y Cajal's original manuscripts and to translate them into German. Kölliker would give the "axon" its name in 1896.

One of the other scientists who Ramón y Cajal met at the 1889 meetings was Wilhelm von Waldeyer (1836–1921), a man who would write a highly influential review of the evidence in favor of the neuron doctrine two years later. In his paper, Waldeyer (1891), shown in Figure 3.27, wrote that nerve cells terminate freely with end arborizations and that the "neuron" is the anatomical and physiological unit of the nervous system. The word "neuron" was born in this way.

In 1890, Ramón y Cajal demonstrated how axons grew from growth cones, thus more firmly establishing the viability of the neuron doctrine. But there was still one nagging question—If neurons do not fuse to one another, how exactly do they communicate? Do they just touch, or is it something else?

The solution to this problem came largely from the work of Sir Charles Scott Sherrington (1857–1952), who recognized the gaps between neurons and muscles and between one neuron and another. In 1897, Sherrington called these junctions synapses (Foster, 1897). Many years later, Sherrington, whose photograph appears as Figure 3.28, wrote a letter to John Fulton (1899–1960) explaining his choice of terms:

> You enquire about the introduction of the term "synapse"; it happened thus. M. Foster had asked me to get on with the Nervous System part (Part iii) of a new edition of his "Text of Physiol." for him. I had begun it, and had not got far with it before I felt the need of some name to call the junction between nerve-cell and nerve-cell (because the place of junction now entered physiology as carrying functional importance). I wrote him of my difficulty, and my wish to introduce a specific name. I suggested using "syndesm." . . . He consulted his Trinity friend Verrall, the Euripidean scholar, about it, and Verrall suggested "synapse": (from the Greek "clasp") and as that yields a better adjectival form, it was adopted for the book. (Cited in Fulton, 1938, 55)

Although Sherrington developed the concept of the synapse, he was not the first to propose chemical trans-

Figure 3.28. Sir Charles Scott Sherrington (1857–1952), the Nobel Prize-winning neurophysiologist who helped to establish the concept of the synapse (Courtesy of the Royal Society of Medicine, London.)

mission. In the 1870s, Emil du Bois-Reymond (1818–1896) had hypothesized that the transmission of the excitatory process from nerves to effector cells could take place either electrically via currents or chemically using excitatory substances liberated by nerve endings (see du Bois-Reymond, 1848–84, 1874). As for the transmitters themselves, these began to be discovered early in the twentieth century (for reviews, see Brooks, 1959; Eccles, 1959; also see Chapter 20).

Sherrington, Ramón y Cajal, and Golgi eventually were each awarded the Nobel Prize for their work on various aspects of the neuron doctrine. Yet when Golgi gave his Nobel address in December 1906, he still spoke in favor of axons fusing by anastomosis and forming large neural nets. Ramón y Cajal (1906) could not understand how Golgi could be so blind to the evidence favoring the neuron doctrine. Here was the man who provided the scientific community with the procedure to see the nervous system in a revolutionary way, steadfastly refusing to accept the new doctrine. To quote Ramón y Cajal from his own Nobel address:

> True, it would be very convenient and very economical from the point of view of analytical effort if all the nerve centres were made up of a continuous intermediary network between the motor nerves and the sensitive and sensory nerves. Unfortunately, nature seems unaware of our intellectual need for convenience and unity, and very often takes delight in complication and diversity. . . . The irresistible suggestion of the reticular complex, of which I have spoken to you (and the form of which changes every five or six years) has led several physiologists and zoologists to object to the doctrine of the propagation of nerve currents by contact or at a distance. All their allegations are based on the findings by incomplete methods showing far less than those which have served to build the imposing edifice of the neuronal conception. . . . I will only say that in spite of the pains I have taken to perceive the supposed intercellular anastomoses in preparations made with diverse coloration processes . . . I have never succeeded in finding any definite ones . . . (1906, pp. 240–242)

Because Golgi maintained that axons were fused together, he also opposed the widespread belief in localization of function. Localization demanded a certain independence of pathways and centers, and nerve net theory was more in line with holistic interpretations of brain function.

Although Ramón y Cajal thought Golgi's views on these matters were ridiculous, the fact was that Golgi was not the only scientist still opposing cortical localization. Although not necessarily for the same reasons, a number of other prominent scientists were not willing to accept the type of localization proposed by the Hitzigs and Ferriers of the scientific world. The critics of localization and their opposing views are the subject of the next chapter.

REFERENCES

Akert, K., and Hammond, M. P. 1962. Emanuel Swedenborg (1688–1772) and his contribution to neurology. *Medical History, 6,* 255–266.

Andral, G. 1840. *Clinique Médicale.* (4th ed.). Paris: Fortin, Massonet Cie.

Barker, L. F. 1897. The phrenology of Gall and Flechsig's doctrine of association centres in the cerebrum. *Bulletin of the Johns Hopkins Hospital, 8,* 7–14.

Bartholow, R. 1874. Experimental investigations into the functions of the human brain. *American Journal of Medical Science, 67,* 305–313.

Beck, A. 1891. Oznaczenie Lokalizacyi z mózgu i Rdzeniu za Pomoca Zjawisk Elektrycznych. Presented October 20, 1890. *Polska Akademia Umiejetnosci, Ser, 2, 1,* 186–232.

Bell, C. 1811. *Idea of a New Anatomy of the Brain; Submitted for the Observations of His Friends.* London: Strahan and Preston. (Reprinted in 1936 in *Medical Classics, 1,* 105–120, and preceded by quotations from some of his letters.)

Bell, J. 1822. On phrenology, or the study of the intellectual and moral nature of man. *Philadelphia Journal of Medical and Physical Sciences, 4,* 72–113.

Berger, H. 1929. Über das Elektrenkephalogramm des Menschen. *Archiv für Psychiatrie und Nervenkrankheiten, 87,* 527–580.

Berker, E. A., Berker, A. H., and Smith, A. 1986. Translation of Broca's 1865 report: Localization of speech in the third frontal convolution. *Archives of Neurology, 43,* 1065–1072.

Bichat, M.-F.-X. 1805. *Recherches Physiologiques sur la Vie et la Mort* (3rd ed.). Paris: Brosson/Gabon. Translated by F. Gold as, *Physiological Researches on Life and Death.* Boston, MA: Richardson and Lord, 1827. Reprinted in D. N. Robinson (Ed.), *Significant Contributions to the History of Psychology, Ser. E, Physiological Psychology: Vol. II, X. Bichat, J. G. Spurzheim, P. Flourens.* Washington DC.: University Publications of America, 1978. Pp. i–334.

Bouillaud, J.-B. 1825. Recherches cliniques propres à démontrer que la perte de la parole correspond à la lésion des lobules antérieurs du cerveau et à confirmer l'opinion de M. Gall, sur le siège de l'organ du langage articulé. *Archives Génerale de Médecine (Paris), 8,* 25–45.

Brazier, M. 1959. The historical development of neurophysiology. In (Ed.), *Handbook of Physiology: Vol. 1. Neurophysiology.* Baltimore, MD: Waverly Press. Pp. 1–58.

Brazier, M. 1988. *A History of Neurophysiology in the 19th Century.* New York: Raven Press.

Broca, P. 1861. Remarques sur le siège de la faculté du langage articulé; suivies d'une observation d'aphémie (perte de la parole). *Bulletins de la Société Anatomique (Paris), 6,* 330–357, 398–407. In G. von Bonin, *Some Papers on the Cerebral Cortex.*. Translated as, "Remarks on the seat of the faculty of articulate language, followed by an observation of aphemia." Springfield, IL: Charles C Thomas, 1960. Pp. 49–72.

Broca, P. 1863. Localisation des fonctions cérébrales. Siège du langage articulé. *Bulletins de la Société d'Anthropologie (Paris), 4,* 200–203.

Broca, P. 1865. Sur le siège de la faculté du langage articulé. *Bulletins de la Société d'Anthropologie (Paris), 6,* 337–393.

Brodmann, K., 1909. *Vergleichende Lokalisationslehre der Grosshirnrinde in ihren Prinzipien dargestellt auf Grund des Zellenbaues.* Leipzig: J. A. Barth. In G. von Bonin, *Some Papers on the Cerebral Cortex.*. Translated as, *"On the Comparative Localization of the Cortex."* Springfield, IL: Charles C Thomas, 1960. Pp. 201–230.

Brooks, C. McC. 1959. Discovery of the function of chemical mediators in the transmission of excitation and inhibition to effector tissues. In C. McC. Brooks and P. F. Cranefield (Eds.), *The Historical Development of Physiological Thought.* New York: Hafner. Pp. 171–181.

Brown-Séquard, C.-E. 1860. *Course of Lectures on the Physiology and Pathology of the Central Nervous System.* Philadelphia, PA: Collins.

Caldwell, C. 1824. *Elements of Phrenology.* Lexington, KY: T. T. Skillman.

Caldwell, C. 1827. *Elements of Phrenology* (2nd ed.). Lexington, KY: A. G. Meriweather.

Campbell, A. W. 1905. *Histological Studies on the Localisation of Cerebral Function.* Cambridge, U.K.: Cambridge University Press.

Carlson, E. T. 1959. The influence of phrenology on early American psychiatric thought. *American Journal of Psychiatry, 115,* 535–538.

Caton, R. 1875. The electric currents of the brain. *British Medical Journal, 2,* 278.

Caton, R. 1877. Interim report on investigation of the electric currents of the brain. *British Medical Journal, 1, Suppl. L,* 62.

Caton, R. 1887. Researches on electrical phenomena of cerebral grey matter. *Transactions of the Ninth International Medical Congress, 3,* 246–249.

Caton, R. 1891. Die Ströme des Centralnervensystems. *Centralblatt für Physiologie, 4,* 758–786.

Clarke, E., and O'Malley, C. D. 1968. *The Human Brain and Spinal Cord.* Berkeley: University of California Press.

Cohen of Birkenhead, Lord. 1959. Richard Caton (1842–1926); pioneer electrophysiologist. *Proceedings of the Royal Society of Medicine, 52,* 645–651.

Combe, G. 1827. *The Constitution of Man, Considered in Relation to External Objects* (5th Amer. ed.) Edinburgh: Neill.

Da Fano, C. 1926. Camillo Golgi, 1843–1926. *Journal of Pathological Bacteriology, 29,* 500–514.

Deiters, O. 1865. In M. Schultze (Ed.), *Untersuchungen über Gehirn und Rückenmark des Menschen und der Säugethiere.* Braunschweig: Wieweg.

du Bois-Reymond, E. 1848–1884. *Untersuchungen über thierische Elektricität.* Berlin: G. E. Reimer.

du Bois-Reymond, E. 1874. Experimentalkritik der Entladungshypothese über die Wirkung von Nerf auf Muskel. In *Gesammelte Abhandlungen zur allgemeinen Muskel-und Nervenphysik* (Vol. II). 1877. P. 698.

Eccles, J. C. 1959. The development of ideas on the synapse. In C. McC. Brooks and P. F. Cranefield (Eds.), *The Historical Development of Physiological Thought.* New York: Hafner. Pp. 39–66.

Ehrenberg, C. G. 1833. Nothwendigkeit einer feineren mechanischen Zerlegung des Gehirns und der Nerven vor der chemischen, dargestellt aus Beobachtungen von C. G. Ehrenberg. *Annalen der Physik und Chemie, 28,* 449–465, 471–473.

Ferrier, D. 1873. Experimental researches in cerebral physiology and pathology. *West Riding Lunatic Asylum Medical Report, 3,* 30–96.

Ferrier, D. 1875. Experiments on the brain of monkeys. *Philosophical Transactions, 165,* 433–488.

Ferrier, D. 1876. *The Functions of the Brain.* London: Smith, Elder and Company.

Finger, S., and Wolf, C. 1988. The "Kennard effect" before Kennard: The early history of age and brain lesion effects. *Archives of Neurology, 45,* 1136–1142.

Fleischl von Marxov, E. 1890. Mittheilung betreffend die Physiologie der Hirnrinde. *Centralblatt für Physiologie, 4,* 538.

Flourens, M.-J.-P. 1842. *Recherches Expérimentales sur les Propriétés et les Fonctions du Système Nerveux dans les Animaux Vertébrés* (2nd ed.). Paris: J. B. Ballière. (First ed., 1824)

Flourens, M.-J.-P. 1846. *Phrenology Examined* (2nd ed.). (C. de L. Meigs, Trans.). Philadelphia, PA: Hogan and Thompson. Reprinted in D. N. Robinson (Ed.), *Significant Contributions to the History of Psychology, Ser. E, Physiological Psychology: Vol. II, X. Bichat, J. G. Spurzheim, P. Flourens.* Washington, DC: University Publications of America. Pp. iv-144.

Flourens, M.-J.-P. 1864. *Psychologie Comparée* (2nd ed.). Paris: Garnier Frères.

Forel, A. H. 1887. Einige hirnanatomische Betrachtungen und Ergebnisse. *Archiv für Psychiatrie, 18,* 162–198.

Forel, A. H. 1937. *Out of My Life and Work.* New York: Norton.

Foster, M. 1897. *A Text-Book of Physiology* (7th ed.). London: Macmillan.

Freemon, F. R. 1991. American medicine and phrenology. Paper presented at Neurohistory Conference, Ft. Myers, FL.

Fritsch, G., and Hitzig, E. 1870. Über die elektrische Erregbarkeit des Grosshirns. *Archiv für Anatomie und Physiologie,* 300–332. In G. von Bonin, *Some Papers on the Cerebral Cortex.* Translated as, "On the electrical excitability of the cerebrum." Springfield, IL: Charles C Thomas, 1960. Pp. 73–96.

Fulton, J. 1927. The early phrenological societies and their journals. *Boston Medical and Surgical Journal, 196,* 398–400.

Fulton, J. 1938. *Physiology of the Nervous System.* London: Oxford University Press.

Gall, F. J., and Spurzheim, J. 1810–1819. *Anatomie et Physiologie du Système Nerveux en Général, et du Cerveau en Particulier.* Paris: F. Schoell. (Gall was the sole author of the first two volumes of the four in this series.)

Gerlach, J. von. 1872. Über die struktur der grauen Substanz des menschlichen Grosshirns. *Zentralblatt für die medizinischen Wissenschaften, 10,* 273–275.

Gibson, W. C. 1962. Pioneers in localization of function in the brain. *Journal of the American Medical Association, 180,* 944–951.

Gibson, W. C. 1967. The early history of localization in the nervous system.

In P. J. Vinken and G. W. Bruyn (Eds.), *Handbook of Clinical Neurology* (Vol. 2). Amsterdam: North Holland Publishing Company. Pp. 4–14.

Golgi, C. 1873. Sulla struttura della grigia del cervello. *Gazetta Medica Italiani Lombardi, 6,* 244–246. Translated by M. Santini as, "On the structure of the gray matter of the brain," in M. Santini (Ed.), *Golgi Centennial Symposium, Proceedings.* New York: Raven Press, 1975. Pp. 647–650.

Golgi, C. 1886. *Sulla Fina Anatomia degli Organi Centrali del Sistema Nervoso.* Milano: Hoepli.

Golgi, C. 1903. *Opera Omnia.* Milano: Hoepli. (Golgi's collected works, 1868–1902)

Golgi, C. 1906. The neuron doctrine—theory and facts. In *Nobel Lectures Physiology or Medicine 1901–1921.* New York: Elsevier, 1967. Pp. 189–217.

Harrison, J. P. 1825. Observations on Gall and Spurzheim's theory. *Philadelphia Journal of Medical and Physical Sciences, 11,* 233–249.

Head, H. 1926. *Aphasia and Kindred Disorders of Speech.* New York: Macmillan.

His, W. 1886. Zur Geschichte des menschlichen Rückenmarkes und der Nervenwurzeln. *Sächsische Akademie der Wissenschaften zu Leipzig. Math-Phys. Cl., 13,* 147–209, 477–513.

His, W. 1889. Die Neuroblasten und deren Entstehung im embryonalen Marke. *Sächsischen Akademie der Wissenschaften zu Leipzig. Math-Phys. Cl., 15,* 313–372.

Hitzig, E. 1867. Über die Anwendung unpolarisirbarer Elektroden in Elektrotherapie. *Berliner klinische Wochenschrift, 4,* 404–406.

Hitzig, E. 1869. Verhandlungen ärzlichen Gesellschaften. *Berliner klinische Wochenschrift, 6,* 420.

Hitzig, E. 1874a. Über die Resultate der elektrischen Untersuchung der Hirnrinde eines Affen. *Berliner klinische Wochenschrift, 11,* 65-67.

Hitzig, E. 1874b. *Untersuchungen über das Gehirn.* Berlin: Hirschwald.

Jackson, J. H. 1863. Convulsive spasms of the right hand and arm preceding epileptic seizures. *Medical Times and Gazette, 1,* 110–111.

Jefferson, G. 1953. The prodromes to cortical localization. *Journal of Neurology, Neurosurgery and Psychiatry, 16,* 59–72.

Jefferson, G. 1960. Sir William Macewen's contributions to neurosurgery and its sequels. In *Selected Papers of Sir Geoffrey Jefferson.* Springfield, IL: Charles C Thomas, 1960. Pp. 132–149. Reprinted from Glasgow University Publications, LXXXI, 1950.

Klein, Langley, and Schäfer. 1883. On the cortical areas removed from the brain of a dog, and from the brain of a monkey. *Journal of Physiology, 4,* 231–247.

Kölliker, A. 1849. Neurologische Bemerkungen. *Zeitschrift für wissenschaftliche Zoologie, 1,* 135–163.

Kölliker, A. 1867. *Handbuch der Gewebelehre des Menschen* (5th ed.). Leipzig: Engelmann.

Kölliker, A. 1896. *Handbuch der Gewebelehre des Menschen* (6th ed.). Leipzig: Engelmann.

Kussmaul, A. 1878. Disturbances of speech: An attempt in the pathology of speech. In H. V. Ziemssen (Ed.), *Cyclopedia of the Practice of Medicine* (Vol. 14). New York: Wood.

Lallemand, C-F. 1820. *Recherches Anatomico-Pathologiques sur l'Encéphale et ses Dépendances* (Vol. 1). Paris: Baudouin Frères.

Lashley, K. S., and Clark, G. 1946. The cytoarchitecture of the cerebral cortex of Ateles: A critical examination with architectonic studies. *Journal of Comparative Neurology, 85,* 223–305.

Legallois, J.-J.-C. 1812. *Expériences sur le Principe de la Vie, Notamment sur Celui des Mouvements du Coeur, et sur le Siège de ce Principe.* Paris: Hautel.

Macewen, W. 1879. Tumour of the dura mater removed during life in a person affected with epilepsy. *Glasgow Medical Journal, 12,* 210-213.

McHenry, L. C., Jr. 1969. *Garrison's History of Neurology.* Springfield, IL: Charles C Thomas.

Nansen, F. 1887. *The Structure and Combination of the Histological Elements of the Central Nervous System.* Bergen: Bergens Muset Aarsberetning.

Ombredane, A. 1951. *L'Aphasie et l'Elaboration de la Pensée Explicite.* Paris: Presses Universitaires de France.

Pauly, P. J. 1983. The political structure of the brain: Cerebral localization in Bismarkian Germany. *International Journal of Neuroscience, 21,* 145–150.

Purkyně, J. E. 1838. *Bericht über die Versammlung deutcher Naturforscher und Ärzte in Prag im September, 1837.* Prague. Part 3, Section 5: A. Anatomische Verhandlungen. Pp. 177–180.

Ramón y Cajal, S. 1888. Estructura de los centros nerviosos de las aves. *Revista de Histologia Normal y Patologica, 1,* 1–10.

Ramón y Cajal, S. 1889. Conexíon general de los elementos nerviosos. *La Medicina Práctica, 2,* 341–346.

Ramón y Cajal, S. 1890. A quelle époque apparaissent les expansions des cellules nerveuses de la moëlle épinière du poulet? *Anatomischer Anzeiger, 5,* 631–639.

Ramón y Cajal, S. 1906. The structure and connexions of neurons. In *Nobel Lectures Physiology or Medicine 1901-1921.* New York. Elsevier, 1967. Pp. 220–253.

Ramón y Cajal, S. 1937. *Recollections of My Life.* (E. H. Craigie, Trans., with assistance of J. Cano). Philadelphia, PA: American Philosophical Society.

Remak, R. 1836. Vorläufige Mitteilung mikroscopischer Beobachtungen über den innern Bau der Cerebrospinalnerven und über die Entwicklung ihrer Formelemente. *Archiv für Anatomie und Physiologie,* 145–161.

Remak, R. 1837. Weitere mikroscopische Beobachtungen über die Primitivfasern des Nervensystems der Wirbelthiere. *Neue Notizen, 3,* 35–40.

Remak, R. 1838. *Observationes anatomicae et microscopicae de systematis nervosi structura.* Berlin: G. E. Reimer.

Riese, W. 1977. Discussions about cerebral localization in the learned societies of the 19th century. In K. Hoops, Y. Lebrun, and E. Buyssens (Eds.), *Selected Papers on the History of Aphasia: Vol. 7. Neurolinguistics.* Amsterdam: Swets and Zeitling.

Rolando, L. 1809. *Saggio sopra la Vera Struttura del Cervello dell'uomo e degli Animali e sopra le Funzioni del Sistema Nervoso.* Sassari: Stamperìa Da S. S. R. M. Privilegiata.

Rostan, L. L. 1823. *Recherches sur une Maladie Encore Peu Connue qui a Reçu le Nom du Ramolissement du Cerveau* (2nd ed.). Paris: Bechet.

Schiller, F. 1979. *Paul Broca: Founder of French Anthropology, Explorer of the Brain.* Berkeley: University of California Press.

Schwann, T. 1839. *Mikroscopische Untersuchungen über die Übereinstimmung in der Struktur und dem Wachsthum der Thiere und Pflanzen.* Berlin: G. E. Reimer.

Schwedenberg, T. H. 1960. The Swedenborg manuscripts. *Archives of Neurology, 2,* 407–409.

Sewall, T. 1839. *Examination of Phrenology.* Boston, MA: D. S. King.

Shepherd, G. 1991. *Foundations of the Neuron Doctrine.* New York: Oxford University Press.

Sherrington, C. S. 1935. Sir Edward Sharpey-Schafer, 1850–1935. *Quarterly Journal of Experimental Physiology, 25,* 99–104.

Sondhaus, E., and Finger, S. 1988. Aphasia and the C.N.S. from Imhotep to Broca. *Neuropsychology, 2,* 87–110.

Spencer, H. 1855. *The Principles of Psychology.* London: Longmans.

Spillane, J. D. 1981. *The Doctrine of the Nerves.* Oxford, U.K.: Oxford University Press.

Spurzheim, J. 1825. *Phrenology, or, the Doctrine of the Mind; and the Relations Between Its Manifestations and the Body.* London: C. Knight.

Spurzheim, G. 1832. *Outlines of Phrenology.* Boston, MA: Marsh, Capen and Lyon.

Stookey, B. 1954. A note on the early history of cerebral localization. *Bulletin of the New York Academy of Medicine, 30,* 559–578.

Stookey, B. 1963. Jean-Baptiste Bouillaud and Ernest Aubertin: Early studies on cerebral localization and the speech center. *Journal of the American Medical Association, 184,* 1024–1029.

Swedenborg, E. 1887. *The Brain, Considered Anatomically, Physiologically, and Philosophically* (2 vols.). (R. L. Tafel, Trans./Ed./Ann.). London: Speirs.

Todd, R. B. 1849. On the pathology and treatment of convulsive diseases. *London Medical Gazette, 8,* 661–671, 724–729, 766–772, 815–822, 837–846.

Trotter, W. 1934. A landmark in modern neurology. *Lancet, 2,* 1207-1210.

Valentin, G. C. 1836. Über den Verlauf und die letzten Ende der Nerven. *Nova Acta Physico-medica Academiae Caesareae Leopoldinp-Carolinae Naturae Curiosorum, 18,* 51–240.

Van der Loos, H., 1967. The history of the neuron. In H. Hydén, *The Neuron.* Amsterdam: Elsevier. Pp. 1–47.

Viets, H. R. 1926. Camillo Golgi, 1843–1926. *Archives of Neurology and Psychiatry, 15,* 623–627.

Vogt, O., and Vogt, C. 1919. Allgemeine Ergebnisse unserer Hirnforschung. *Journal für Psychologie und Neurologie, 25,* 273–462.

Waldeyer-Hartz, W. von. 1891. Über einige neuere Forschungen im Gebiete der Anatomie des Centralnervensystems. *Deutsche medizinische Wochenschrift, 17,* 1213–1218, 1244–1246, 1267–1269, 1287–1289, 1331–1332, 1352–1356.

Walker, A. E. 1957. Stimulation and ablation: Their role in the history of cerebral physiology. *Journal of Neurophysiology, 20,* 435–449.

Walker, A. E. 1967a. Sir William Macewen. In A. E. Walker (Ed.), *A History of Neurological Surgery.* New York: Hafner. Pp. 178–179.

Walker, A. E. 1967b. Prologue. In A. E. Walker (Ed.), *A History of Neurological Surgery.* New York: Hafner. Pp. 1–22.

Walsh, A. A. 1971. George Combe: A portrait of a heretofore generally unknown behaviorist. *Journal of the History of Behavioral Sciences, 7,* 269–278.

Walsh, A. A. 1972. The American tour of Dr. Spurzheim. *Journal of the History of Medicine and Allied Sciences, 27,* 187–205.

Wells, S. R. 1866. *New Physiognomy, or Signs of Character, as Manifested through Temperament and External Forms.* New York: Fowler and Wells. (Reprinted 1894)

Young, R. M. 1968. The functions of the brain: Gall to Ferrier (1808–1886). *Isis, 59,* 251–268.

Young, R. M. 1970. *Mind, Brain and Adaptation in the 19th Century.* Oxford, U.K.: Clarendon Press.

Young, R. M. 1985. *Darwin's Metaphor.* Ch. 3: The role of psychology in the nineteenth-century evolutionary debate. Cambridge, U.K.: Cambridge University Press. Pp. 56–78.

Zanobio, B. 1975. Golgian memorabilia at the Museum for the History of the University of Padua. In M. Santini (Ed.), *Golgi Centennial Symposium Proceedings.* New York: Raven Press. Pp. 650–655.

Chapter 4
Holism and the Critics of Cortical Localization

> I ought however to emphasize the fact that anatomical localization is in no way denied or abridged. It is only the mechanism of function and of the symptoms that is concerned. The conventional cortical centers as at present conceived are symptom producing centers and not centers of functions as conventionally ascribed to them.
>
> Morton Prince, 1910

The idea of cortical localization was very slow to develop, especially before 1861, when Paul Broca presented his famous case of Tan and sided with the localizationists (see Chapter 3). The strong bias against cortical localization up to Broca's time might seem somewhat surprising given that a theory of localization of higher functions was broadly embraced from the fourth century through the Renaissance. The schema accepted by the Church Fathers and their followers, however, did not involve the cortex or even the brain substance. Rather, it concerned the ventricles, and it most often held that the two lateral ventricles (treated as a single cavity) served as the common receptacle for sensory information; that the middle ventricle mediated fantasy, ideation, and cognition; and that the posterior ventricle was responsible for memory (see Chapter 2).

After Thomas Willis published his *Cerebri anatome* in 1664, most scientists became more willing to accept the idea that different functions might be mediated by the cerebellum, cerebrum, corpus striatum, and medulla (see Chapter 2). This signaled a major change in thinking about the brain itself. But while cases of brain injury should have suggested even finer divisions than this, serious attempts to divide the cerebral cortex or major brainstem structures into smaller functional units did not materialize.

Albrecht von Haller was one of the strongest opponents of localization in the post-Willis period.[1] In fact, Haller (1759–66) wrote that he was not convinced of any specialization above the peripheral nerve. He maintained that no nerve has its own private territory in the spinal cord, in the brainstem, or in the cortex. He even rejected the ideas that the corpus striatum was the seat of sensation and the source of movement, that the cerebellum played a role in vital functions, and that imagination, memory, and other higher functions resided in the cerebral hemispheres.

Haller was able to attract some followers, but most eighteenth-century scientists still agreed with Willis that the cerebral cortex played a special role when it came to mediating higher functions. These individuals also followed

---

[1]Haller completed his medical degree in Leyden in 1727.

Willis in looking upon the cerebral cortex as a single functional organ. The cortex simply was not a structure made up of separate centers, each specialized for different higher functions.

The idea that the cortex might contain various specialized parts remained dormant throughout the eighteenth century. Although the work of Emanuel Swedenborg might be cited as an exception to this statement, his ideas about cortical localization were not published in his lifetime and appear to have had no influence on his contemporaries (see Chapter 3).

Finally, as the nineteenth century opened, the phrenologists began to present their theories of cortical function. For the first time, the belief in the equipotentiality of the cortex received a serious challenge.

This chapter looks at the case against localization at the cortical level in the nineteenth century and into the opening decades of the twentieth century. It begins with the reaction against the phrenologists, but emphasizes the literature following Paul Broca's 1861 announcement of a cortical center for articulate language. It examines the thoughts of several eminent physiologists and physicians who questioned the growing concept of cortical specialization and the theories of those scientists who were willing to accept localization for some (but not all) "cortical" functions.

This chapter also considers Gestalt and related "holistic" approaches to brain function. The individuals representing these schools of thought typically did not deny some specialization at the cortical level. Nevertheless, they stressed that the cerebral cortex—and indeed the brain itself—must be looked upon as a single working unit rather than as a collection of discrete independent centers.

Some of the methodological and logical issues relating to the interpretation of brain lesion material are considered as these different orientations are contrasted with classical localizationist doctrine. As will be seen, one of the most important issues raised by the opponents of cortical localization theory is the idea that localization of symptoms should not be equated with localization of function.

CORTICAL EQUIPOTENTIALITY AND THE CHALLENGE FROM PHRENOLOGY

The theory of cortical equipotentiality received its first serious challenge from Franz Joseph Gall, who was soon joined by Johann Spurzheim (see Chapter 3). While some individuals were swayed by the logic of these phrenologists and impressed with their emphasis on skull measurements, many others responded to the challenge with sharply worded reactions, typically rejecting outright their concepts of cortical localization and questionable methods.

Marie-Jean-Pierre Flourens, a professor of comparative anatomy in Paris, led the fight against phrenology (see Chapter 3). He was critical of both the methods and the theories of Gall and his followers. Flourens (1842, 1846) conducted many ablation and stimulation experiments on birds, rodents, and other small animals. His findings led him to conclude that discrete cortical faculties, ranging from speech to spatial vision, could not be localized in separate cortical regions. As far as he was concerned, when any one of the cortical functions was lost, all were lost, and when one returned, all returned.

Although his beliefs about the cerebral cortex placed him squarely in Haller's camp, Flourens differed from his predecessor at Göttingen by assigning very different functions to the cerebrum, the cerebellum, and the medulla. Indeed, Flourens (1842, 1851, 1858, 1859) even attempted to limit the medullary center for respiration to a smaller area than that defined by Legallois in 1812 (see Chapter 2). The fight against localization, as Flourens saw it, was something that pertained only to the cerebral cortex; to him, the cortical theories proposed by the phrenologists were nothing short of scientific heresy.

A number of studies involving hydrocephalics with extensive cortical damage appeared in the literature at this time. Some of the observations made on these individuals were used to argue against the propositions of the phrenologists, whose cortical localizations included such lofty functions as love of children, acquisitiveness, and "ideality." These case studies also led at least a few investigators to question whether the cortex even played an essential role in certain more basic functions, including perception, as had been theorized by Flourens.

One such paper was published by John Harrison (1825), a physician at the University of Louisville, in Kentucky. Harrison, who was critical of phrenology, described his own and others' observations in his report. For example, he discussed the case of a young hydrocephalic child "who appeared to have possessed, during life, the full enjoyment of all its senses" (taken from Delafield, 1822). Yet, "no part of the brain remained, except a small mass at the base of the skull, which appeared to be the pons varolii." Harrison also mentioned another child who retained her senses and many of her other faculties up to the time of her death. He wrote, "the brain was nearly all absorbed," and what was left was "not larger than a hen's egg" and rested on the base of the cranium.[2]

Although these cases raised some interesting questions, almost all scientists still believed that the cerebral cortex

functioned in perception, ideation, the will, and memory, as suggested by Flourens. They also agreed with Flourens that the cortex was equipotential.

The bias against cortical localization was so strong between 1825 and 1861 that mounting clinical evidence suggesting that fluent speech was most likely to be affected by lesions in the anterior lobes received a decidedly cold reception. In particular, the early efforts of Jean-Baptiste Bouillaud to localize speech on the basis of pathology succeeded in converting only a small number of scientists to his side.[3,4]

After Broca underwent what Bouillaud called his 1861 "conversion" to the doctrine of localization, and especially after 1870, when Gustav Fritsch and Eduard Hitzig described the motor cortex in the dog, the situation changed dramatically. Individuals who continued to defend cortical equipotentiality now found themselves on the outside looking in. And, with every passing year, as the new defenders of the faith fortified their localizationist armories, the assaults by the opponents of localization tended to meet with increased resistance.

CHALLENGES IN THE POST-BROCA ERA

Charles-Edouard Brown-Séquard (1817–1894) was one of the individuals highly critical of any sort of hard-wired conception of cortical localization in the period following Broca's (1861, 1865) reports. In 1877, he wrote a paper that dealt with the supposed localization of speech, which by this time was one of the most agreed-upon faculties in the localizationists' arsenal. Brown-Séquard argued that children can recover from cortical injuries in Broca's or Wernicke's area and that the evidence linking the third left frontal convolution with speech in adults was far from impressive.

Brown-Séquard cited three types of "evidence" to support his position. First—noting that Broca (1865) localized speech in the left cerebral cortex in right-handed people— he mentioned finding 42 cases of aphasia after damage to the right side of the brain. He emphasized that he knew many of these individuals were right-handed and said he could not believe that there were more than a few left-handed people in the group.

Second, he presented evidence to show that lesions outside Broca's region in either of the frontal lobes can affect speech. He wrote:

> If we followed the mode of proceeding of the localizers, these facts would lead us to say that, in certain individuals, the centre of speech is in the posterior lobe—that in certain other individuals it is in the very front part of the anterior

---

[2]To say the least, the behavioral findings were often overstated in these studies.

[3]One reason for this was that localization in any form at all remained associated with the stigma of phrenology. Worse still, Bouillaud had studied phrenology. The fact that considerable variability in lesions and symptoms existed across subjects also served as a limiting factor. These factors are discussed in more detail in Chapters 3 and 24.

[4]Although positions did not shift immediately, the books, articles, and presentations by Bouillaud (1825a, 1825b, 1830, 1839–40, 1848) and his followers, including Simon Alexandre Ernest Aubertin, had the positive effect of keeping the concept of cortical localization in the foreground. They paved the way for Paul Broca to present his first and most famous aphasia case ("Tan") in 1861, the case study that placed him firmly on the side of the early localizationists.

lobe—that in other individuals, according to a most excellent case of a friend of mine, it is in the top part of the middle lobe, very near the longitudinal sinus—that in other individuals it exists in the corpus callosum on the left side (there are two such cases)—that it also exists in the optic thalamus, and in the pons Varolii. (1877, 214)

The third line of evidence involved adult cases with damage to Broca's area who did not exhibit speech defects. Brown-Séquard stated that the number of such cases was not small and that he had no doubt that the "pretended organs of speech on the left side of the brain in right-handed people have been fully and completely destroyed" (p. 218). He then cited a number of cases of bilateral destruction of the two anterior lobes without any alteration of speech.

From these premises, Brown-Séquard concluded that cortical localization, despite its rapid growth in popularity, had to be rejected in its present form. Instead, he proposed that cells responsible for speech and related faculties were spread throughout the cerebrum. Specifically,

> each function of the brain is carried on by special organs, but that those organs, instead of being composed of cells forming a cluster or mass in one part, are composed of scattered cells diffused in many parts of the brain, in communication, of course, one with the other by fibres, and forming a whole by this union of fibres, but still so diffused that a great many parts of the brain—I would not want to be bold enough to say all parts—contain the elements endowed with each of the various functions that we know to exist in the brain. (1877, 212)

In the same year as Brown-Séquard was presenting his arguments, Eugène Dupuy published his highly critical review of cortical localization. He too focused on experiments contradicting prevailing localization theories. He cited many experiments that did not result in the expected effects or that were associated rapid recovery of the symptomatology. For instance, Dupuy discussed how his own cats and dogs with motor cortex lesions could still use their "would-be paralyzed limbs," not only for basic movements but also for purposeful voluntary movements. He specifically mentioned one dog with damage to the motor cortex that was still able to give its paw or stand up on command. This represented a direct challenge to the theories presented by Eduard Hitzig and David Ferrier, both of whom advocated that the motor cortex was essential for voluntary movements (see Chapter 14).

NERVE NETS AND HOLISTIC FUNCTION

The case against cortical localization was argued not only by some physicians and experimental physiologists but also by some histologists, especially those who believed the nervous system was comprised of massive fused nets of dendrites and/or axons (see Chapter 3; also Shepherd, 1991). They believed that localization of function in specific areas was not consonant with the idea that huge structural territories were formed when axons and/or dendrites fused together by anastomosis.

Neural reticulum theory achieved considerable popularity in the second and third quarters of the nineteenth century, a time when histological procedures were not very precise. It was championed by Joseph von Gerlach, who developed the carmine and then the gold chloride staining methods. In 1872, Gerlach stated he was convinced that both dendritic and axonal processes formed nets by anastomosis.

Camillo Golgi, who developed the revolutionary silver nitrate staining method, published his first report using the new stain in 1873. In it, and in his later papers, he also argued for fused neural nets (see Golgi, 1903, for his early collected papers). Golgi differed from Gerlach in that he did not think the dendrites fused, but the two scientists were in agreement when it came to axonal processes and anastomosis.

Because Golgi was vigorously opposed to the segregation of units in the nervous system, he argued for a holistic approach toward brain function. In the 1880s, when cortical localization was viewed by most of his colleagues as a fact of life, Golgi wrote the following about the brain:

> The concept of so-called *location of the cerebral functions*, should it be insisted on . . . in a rigorous sense, would not be in perfect harmony with the anatomical data. . . . It being demonstrated, for example, that a nervous fibre is in relation with extensive groups of gangliar cells, and that the gangliar elements of entire provinces, and also of various neighboring provinces, are conjoined by means of a diffuse network, to the formation of which all the various categories of cells and nervous fibres of these provinces contribute, it is naturally difficult to understand a rigorous functional localization, as many would have it. (1883, 395)

Golgi concluded that there were so many connections among the cells in the nervous system that it was impossible to formulate a law pertaining to transmission in individual cells or groups of cells. He argued that the concept of sharply circumscribed centers simply disregarded "minute anatomical researches" and took investigators away from how the whole organ must function. He felt that at most one might be able to talk about very large territories gradually merging into other broad territories and about specificity in the peripheral nerves, but certainly not much more. It appeared to Golgi that the "laws" formulated by Fritsch and Hitzig (1870) and Ferrier (1876) were inconsistent with the anatomical record.

Golgi steadfastly refused to budge from this position. Upon receiving the Nobel Prize, he told his audience:

> I found myself unable to follow current opinion, because I was confronted by one concrete anatomical fact; this was the existence of the formation which I have called the diffuse nerve network. I attached much more importance to this network, which I did not hesitate to call a nerve organ, because the very manner in which it is composed clearly indicated its significance to me. In fact, although in various ways and to varying extent, every nerve element of the central nervous system contributes to its formation. (1906, 193)[5]

---

[5]Near the end of his Nobel address, Golgi (p. 216) stated:

As for the specific function of the central nervous system, I have, on several occasions, contested that it was correlated with a specificity of organization of the nerve centres, and I have come round to the idea that specific function is not associated with the characteristics of the organization of centres, but rather with the specificity of peripheral organs destined to receive and transmit impulses, or again, with the particular organization of peripheral organs which must receive the central stimuli.

By this time, the evidence against nerve nets had become overwhelming, and neuron doctrine with all of its implications had become widely accepted. In particular, Santiago Ramón y Cajal (1906), who began to work with Golgi's silver stain in 1887 and who shared the Nobel Prize with Golgi in 1906, was baffled by Golgi's inability to abandon the older nerve net theory. Golgi was convinced, however, that the case against anastomosis had not been proved and that the literature on recovery of function demanded a holistic orientation.

THE BRAIN ACCORDING TO GOLTZ

Friedrich Goltz (1834–1902) was appointed professor of physiology at the new University of Strassburg in 1872, as a part of Chancellor Otto von Bismarck's (1815–1898) campaign to "Germanify" Alsace after the Franco-Prussian War (1871–1872). Goltz had studied anatomy with Hermann Helmholtz (1821–1894), but he soon rejected physiological reductionism and any reliance on complex instrumentation. Instead, he turned to relatively simple experimentation and extensive observation.

Goltz was certain that intellect could not be confined to discrete parts of the cerebrum. He based his conviction on the behavioral changes he observed in dogs after a variety of large cortical lesions, usually made imprecisely with a water jet. The resulting dementias did not seem to be related to destruction of any one cortical area. Instead, they seemed to be a function of the overall amount of cerebral cortex destroyed. Goltz (1881a) stated this as follows:

> In general, one may affirm that the degree of dementia is proportionate to the spatial extent of the lesion. After removal of both posterior quadrants the dementia appears to be more striking than after removal of the anterior. But the surface of the former is appreciably greater than that of the latter. An animal with the posterior half of one hemisphere and the anterior half of the other destroyed approaches the mean between animals with the anterior and those with the posterior quadrants destroyed. The dementia after removal of three quadrants is severe and most severe of all after destruction of the four quadrants. (Translated in Lashley, 1963, 4)

Goltz (1881a, 1881b, 1888) attempted to define the nature of the dementia and concluded it was not due to a defect of any specific type of sensation or cognition. If anything, he thought the failures of memory, judgment, and integration were more likely to be due to a general defect in attention, and this, he believed, was a function of the whole cerebrum.

Goltz attracted widespread attention by exhibiting his "decorticate" dogs at various scientific meetings throughout Europe. For example, he took two of his dogs with large cortical lesions to the 1881 International Medical Congress in London (see Chapter 3) and to the 1889 Physiological Congress in Basel.

The Medical Congress of 1881 was especially noteworthy. Goltz began his talk by taking the lectern, opening his suitcase, and producing the damaged skull of a dog with its remnant of brain. He told his audience the dog survived four major operations on the cerebrum and that Ferrier's

somatosensory and motor areas had been removed. The dog had been kept alive for a year after its last surgery. It acted "idiotic," Goltz said, but it was not robbed of its sensation and was not paralyzed. Goltz believed his dog was proof that Ferrier's theories about cortical localization of sensory and motor functions had to be wrong. As he put it: "A fruit can look extremely tempting and nevertheless be wormy at the core. It is not difficult to detect the wormy core in all the hypotheses of cerebral localization."

Goltz then exhibited a live dog, one that had undergone five operations, largely on the parietal and occipital lobes. He showed the audience that this dog was not paralyzed and that it could run, jump, see, hear, smell, and feel.

This dog, along with the two monkeys that Ferrier and Yeo exhibited at the Congress (one deaf and the other hemiplegic), were sacrificed soon after Goltz and Ferrier made their presentations. Their brains were examined by William Gowers and other scientists who formed a panel of internationally acclaimed experts (Klein, Langley, and Schäfer, 1883; also see Chapter 3).

The select committee was very impressed by the lesions in the predicted locations in the deaf and hemiplegic monkeys presented by Ferrier and Yeo. They were considerably less impressed, however, with the lesions found in the dog operated upon by Goltz. As can be seen in Figure 4.1, the

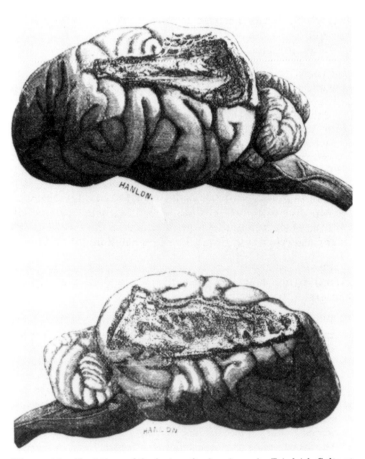

Figure 4.1. Depictions of the lesion of a dog shown by Friedrich Goltz at the International Medical Congress in 1881. Goltz emphasized that this dog was not paralyzed by the lesion and that its senses were intact. The behavioral findings were not contested, but upon sacrifice, the lesions were found to be less extensive than Goltz had indicated.

damage was considerably less extensive than Goltz had maintained. In particular, the frontal lobes—including the motor areas—were intact, and the visual areas showed considerable sparing. The committee felt that the remaining parts of critical areas could have mediated sensory and motor functions. They also recognized that the lower phylogenetic status of the dog could also have contributed to its ability to retain its functions. Ferrier and the defenders of localization savored their victory.

Throughout his life, Goltz remained firm in his belief that intellect could not be localized in discrete cortical areas (e.g., in the frontal association cortex). Yet he appeared to become less certain about vision, motor functions, touch, and related functions. As the years passed, he felt compelled to admit that the four cortical lobes were not perfectly equivalent for sensory and motor functions.

Goltz recognized that anterior lesions were most likely to affect motor functions and that posterior lesions were most likely to affect vision. But he also emphasized that the deficits following lesions in these areas (1) typically are not absolute, (2) often are accompanied by various other changes (e.g., in affect), and (3) usually are followed by at least some recovery.

These findings were more indicative of some type of mixed specificity than they were of sharply defined, unique functional loci. The data suggested a slow blending of areas with maximum cooperation among the parts more than Hitzig's or Ferrier's conceptions of cortical areas with strict law-and-order boundaries.

To quote Goltz (1888):

> Even if I was convinced that the assumption of small circumscribed centers did not correspond with the facts, that did not mean that the idea of Flourens, that the substance of the brain was equivalent everywhere, was correct. I tried to find out what consequences the amputation of single lobes of the brain had, and found that the destruction of the anterior lobes led to entirely different disturbances from the destruction of the occipital lobes. . . . The assumption of small circumscribed centers I consider today more nonsensical than ever. But to assume that I should hold back my own discoveries on consequences of destruction of whole lobes of the brain because this would be a kind of localization, that is simply funny. (1960 translation, 129–130)

In his last important report on "decorticate" dogs, Goltz (1892) again addressed the concept of well-defined cortical centers. The rhetoric was uniquely Goltz:

> When the present report is published an attempt will undoubtedly be made to deny the obvious shipwreck of the doctrine of the small, outlined centers, by inventing auxiliary hypotheses. The sad shambles of the wreck by the old theory will be patched up so that it may become seaworthy again. "Be fertile and multiply," the centers will be told, and the host of the outlined cerebral centers which is already increasing daily will be fortified by as many reserve centers in the midbrain and hindbrain, ready to go into battle when the upper ones in the cerebrum meet with an accident." (Translated in Clarke and O'Malley, 1968, 564)

In the end, while Goltz was willing to accept at least a rudimentary form of cortical localization for some func-

tions, he was never willing to accept the extreme, sharply defined type of localization that was favored by many of his map-making contemporaries.

LOCALIZATION FOR SOME BUT NOT ALL FUNCTIONS

The views expressed by Goltz on higher functions were shared in part by Hermann Munk (1839–1912). Munk agreed that higher intellectual functions could not be confined to small parcels of cortical tissue. But when it came to sensory and motor functions, Munk (1890) placed himself much more firmly in the localizationist camp. Unlike Goltz, he hypothesized that there were distinct cortical sensory areas (spheres) that contained the images and ideas associated with different types of sensation (e.g., vision, hearing). But he also thought that the interconnections between these specific sensory areas could account for intellectual functions without recourse to the concept of association areas.

Jacques Loeb (1859–1924), who had worked as Goltz's assistant in Strassburg, also felt it was a mistake to believe that intellect could be housed in parcels of association cortex. Loeb (1902), shown in Figure 4.2, ablated the association areas in various combinations in his dogs and reported that his animals were not affected by this damage. In addition, Loeb cited several clinical cases to support his belief that intellect and associative memory were diffuse functions of the whole cortex.

One case involved a person who was no longer able to recognize numbers over 3. Loeb thought it was inconceivable that the parts of the brain containing the memory images for 1, 2, and 3 could be preserved, whereas those for all other numbers could be lost. The more likely explanation, he reasoned, was that the numbers 1, 2, and 3 had

Figure 4.2. Jacques Loeb (1859–1925), who was among the turn-of-the-century scientists who questioned the idea of localization of higher cognitive functions.

many more associations throughout the cerebrum and thus were more resistant to injury of just one part of the brain. Loeb wrote: "In processes of association the cerebral hemispheres act as a whole, and not as a mosaic of a number of independent parts. . . . Association processes occur everywhere in the hemispheres" (1902, pp. 262, 275).

Although Loeb dismissed the idea that intelligence was a summation of diverse functions coming together in the association areas, he was willing to accept some specialization at the cortical level. For example, Loeb (1885, 1886) recognized that damage to the occipital cortex affected vision. But, by using his new method of double simultaneous stimulation (see Benton, 1956), he concluded that this was more of a weakness or inattention for the contralateral visual field (hemiamblyopia) than an absolute blindness.

Thus, the arguments against cortical localization underwent a fundamental change as the twentieth century opened. Now, just about everyone was willing to accept some sort of cortical localization for sensory and motor functions. The debate essentially narrowed down to how sharply these cortical areas could be defined and whether specific engrams (e.g., visual memory traces) or basic sensory functions were affected by cortical damage.

In contrast, the situation concerning memory and intellect was still an active battleground for the opponents and defenders of localization. On the one hand, many scientists believed intellectual functions and associative memory were governed by specific "association" areas, and in particular by the frontal association cortex (see Chapter 22). But, on the other hand, a fairly large contingent of scientists felt just as strongly that most if not all "executive" functions could only be the result of the cerebrum acting as a whole.

MATTERS OF LOGIC

To a number of individuals in the scientific community, many of the debates about what was localized and where this compartmentalization took place stemmed from how willing the combatants were to equate localization of symptoms with localization of function. The point raised was that symptoms were observable and were the sort of thing that all should be able to agree on. In contrast, localized functions were simply inferences or hypotheses derived primarily from symptoms.

Toward the end of the nineteenth century, John Hughlings Jackson warned his colleagues that localization of symptoms and localization of function were not identical. Jackson realized that focal brain damage allowed one to localize particular symptoms, at least in a general or relative way. Thus, there was no argument from Jackson about fluent propositional speech being more likely to be affected by damage to the left hemisphere than the right hemisphere. This was observable; cases could be counted, and once Broca (1865) pointed it out, the correlation was easy to see. Jackson had difficulty, however, with the belief that observable symptoms specified the precise locations of circumscribed special centers for the affected functions. He felt this was a major theoretical jump, and he was not fully prepared for the leap.

Jackson was bothered by the growing number of scien-

tists who were forgetting that localizing behavioral functions into small areas of cortex was not factual but theoretical—just one way of looking at the world. He reasoned that it was entirely possible that some symptoms could be due to secondary effects of the damage on other structures, as noted in the concept of "release" that he used to explain "positive" symptoms (e.g., increased activity) after brain injury (Jackson, 1881, 1884, 1894). He was also far from convinced that lost behavioral functions, such as impaired geographical sense or the inability to perform a string of motor acts on command, always represented fundamental deficits (he classified many different symptoms under his broad heading of "imperception"). Additionally, he seemed to recognize that there was always a possibility that damage to a structure involved in a given function might not seem to cause deficits if one failed to ask precisely the right question (e.g., his distinction between automatic and voluntary movement).

There were scientists before Jackson who also realized that interpreting brain lesion effects was no easy matter. For instance, Gabriel Andral (1833) recognized that brain lesions can have both proximal and distal consequences. He maintained that this feature made it hard to state with confidence that the site of the damage is the source of a specific symptom, much less a specific function. Andral made the following comment:

> On the one hand, (the brain) is a great whole composed of a number of parts, each of which performs a special function; while on the other hand, these different parts are intimately connected with each other, in such a manner that they are mutually, jointly and severally responsible, if I can express it thus. Hence it follows that at the place where you discover a lesion there does not always reside the direct cause of the effects which are produced. (1833, 734)

Andral and Jackson were followed by a number of individuals who appreciated and adopted their cautious approaches. One was Morton Prince (1854–1929), an American who spoke about three distinct types of localization: (1) anatomical localization, based on where the nerve tracts terminated; (2) localization of symptoms, based on behavior after brain lesions; and (3) localization of function, the inference derived from the response to brain damage and, in some cases, anatomy.

Prince gave a notable Presidential Address on localization to the American Neurological Association in 1910. Early in his presentation he cited the work of Pierre Marie (1853–1940) on Broca's speech area. Marie (1906) claimed he found symptoms of Broca's aphasia in only about half of his patients with lesions of the third left frontal convolution. He also claimed Broca's aphasia can often be encountered in patients whose lesions spare the third frontal convolution of the left hemisphere. Prince (1910, p. 339) called Marie's studies "the first rude blow to the traditional localization of function in the cortex" and added that "our former naive conception of aphasia can never be revived." As far as he was concerned, the popular diagrams of the speech circuits found in many textbooks belonged "to the scrap heaps of the phantasies of science."

Prince went on to argue that the cortex had some specialization, although it had to function more as a dynamic whole in higher "psychical" processes. For higher functions, damage to any of a number of parts could affect performance. Prince also questioned whether the apex of the occipital cortex was concerned with the function of associative visual memories, as many of his colleagues were claiming, or was simply involved in seeing, which he viewed as more reasonable.

Like Jackson, Prince believed that the real problem was that symptoms were too readily being equated with functions:

> The present doctrine of cerebral localization regarded as a mapping of the brain into areas within which lesions give rise to particular groups of symptoms is one of the triumphs of neurology which cannot be valued too highly. Regarded as a localization of the psycho-physiological functions represented by these symptoms within narrowly circumscribed areas it is in large part naive to a degree which will excite the smiles of future neurologists. (1910, 340)

Prince felt it was a terrible mistake to forget that the brain had to function as a whole. This was a position he shared with the great French physiologist Claude Bernard (1838–1878). In the middle of the nineteenth century, before Broca (1861) even presented the case of "Tan," Bernard (1855–56) had written:

> If it is possible to dissect all the parts of the body, to isolate them in order to study them in their structure, form and connections it is not the same in life, where all parts cooperate at the same time in a common aim. An organ does not live on its own, one could often say it did not exist anatomically, as the boundary established is sometimes purely arbitrary. What lives, what exists, is the whole, and if one studies all the parts of any mechanism separately, one does not know the way they work. In the same way, anatomically, we take the organism apart, but we cannot grasp the whole. This whole can only be seen when the organs are in motion. (Translated in Spillane, 1981, 258)

HOLISM AND THE GESTALT MOVEMENT

The idea that one must not lose site of the whole was a major theme in Gestalt psychology and biology. The German word *Gestalt* literally means "form" or "configuration." The Gestalt movement began as a reaction against reductionism, reflexology, and various machine models of the mind. In neurology, it emphasized dynamic, changing relationships among the parts of the brain, and it offered solutions to a number of nagging problems, including understanding the complex phenomena of perception and recovery from brain damage.

The founder of the Gestalt movement was an Austrian, Christian von Ehrenfels (1859–1932). He reasoned that one cannot appreciate music by analyzing individual notes, or the concept of "squareness" by looking at each line separately. In fact, Ehrenfels (1890) maintained, form quality *(Gestaltqualität)* is something over and above the elements; the whole is different from the sum of the parts (Boring, 1950). With this as his theme, Ehrenfels (1890) withdrew from the ideas of Wilhelm Wundt (1832–1920) and the Structuralists, who looked upon the whole as the simple sum of the parts, and launched the Gestalt movement.

While World War I (1914–1918) was raging, Ehrenfels saw his call for *Ganzheit* (oneness or wholeness) as something applicable not just to perception but to his culture. In his mind, it was a necessary defense against racial mixing and ultimate chaos (see Harrington, in press).

Others, such as Baron Jacob von Uexküll (1864–1944), also used the Gestalt banner as a call for oneness or unity on both cultural and biological levels. Uexküll (1920), a holistic biologist who stressed the interaction of the organism with its perceived environment, ardently maintained that Germany must be handled as if it were a living organism. Germany too was a unified whole that had to be protected from foreign bodies and racial impurity.

Fortunately, not all German-speaking biological and behavioral scientists who embraced the concept of Gestalt were willing to apply this sort of thinking to culture and politics. A number of prominent individuals fled from the situation in Germany to the United States. Among these men were Wolfgang Köhler (1887–1967), Kurt Koffka (1886–1941), Max Wertheimer (1880–1943), and Kurt Goldstein (1878–1965), all major figures in the Gestalt movement.[6]

Koffka, Köhler, and Wertheimer used Gestalt principles of figure and ground relationships and organization to explain complex perceptual and memory functions (see Chapter 24; also Koffka, 1935; Köhler, 1929). Kurt Goldstein, however, was considerably more neurologically oriented. His major work, *Der Aufbau des Organismus,* appeared in 1935 and was translated into English as *The Organism* four years later. His major theme was that one must consider each and every biological event in the context of a whole organism striving to adapt and endure.

Goldstein did not forcefully argue against the study of the parts of the brain, yet he clearly worked on the assumption that reductionism by itself was not sufficient for understanding his patients. He maintained that the focus of the brain damage should be viewed as the figure, while the remaining brain should be perceived as the ground. Both the figure and the ground assumed equal importance when it came to understanding why his patients exhibited certain symptoms.

Goldstein admired the ideas of John Hughlings Jackson and was similarly careful to distinguish between localization of function and localization of symptoms. He approached symptoms as responses given by a struggling organism to challenging environmental demands. Attempting to localize lost functions in discrete parcels of cortex was not nearly as meaningful to him as trying to understand why his patients became more "concrete" or why they often showed "catastrophic" reactions after brain injuries (see Frommer and Smith, 1988).

THE HOLISM OF MONAKOW AND HEAD

The idea that the brain functions as a dynamic organizing entity was a key feature of Goldstein's neurology and of the Gestalt theories of Koffka, Köhler, and Wertheimer. It was also an important aspect of the neurology of Constantin

---

[6]For a history of the German academic community between 1890 and 1933, see Ringer (1969).

von Monakow (1853–1930) and Henry Head (1861–1940), two non-German scientists who embraced Gestalt concepts in their own special ways.

Monakow, a professor of neurology at the University of Zürich, began his scientific career doing classical neuroanatomy. At the close of the nineteenth century, his approach became considerably more holistic, and his emphasis shifted to the functional relationships among brain regions. In particular, Monakow stressed that brain lesions can have both proximal and distal effects. This meant that it was impossible to damage any one part of the brain without affecting other parts. In 1905, Monakow began to develop his theory of *diaschisis* as a way of accounting for the distal effects of brain lesions and then wrote extensively on this subject before World War I.

Monakow (1911, 1914) did not really oppose the general concept of cortical localization of function. Like Munk and Loeb, he felt there was some sort of localization for sensory and motor functions. Nevertheless, the evidence was insufficient to convince him that the association areas had any special intellectual functions. Instead, he reasoned that intellect probably required the coordination of many diverse parts of the brain. It was not something that could be the domain of just one part of the cortex.

Similarly, Monakow (1911) concluded that associative memory was not confined to discrete cortical areas. He wrote:

> The general conception that in the visual sphere we have centers for the optic perception and images, in the auditory sphere that for the perception of tones, and in Broca's and Wernicke's centers the memory images for word sounds, the center for intelligence in the frontal lobe . . . is to be declined with certainty; in the future, it will probably generally be called naive, and it was called so by a few intelligent researchers since the beginning of investigations of localization. (1960 translation, 232)

Monakow was mentally devastated by the useless loss of life in World War I and slowly withdrew from the scientific mainstream to ponder ethics, morals, and a higher unity. He became especially interested in the concept of *horme,* a hypothetical, self-actualizing force that brought individual processes together into a moral and functional whole.[7]

In contrast to Monakow, Henry Head, who served with the British in World War I, never blended his scientific holism with notions of morality or spiritual "oneness." Head (1918, 1920) looked upon the cortex as a mosaic composed of foci of integration. He argued that a brain lesion which affects behavior does so not just by affecting specific parts but by throwing the entire system into disorder. He reasoned that the damaged brain must be looked upon as a whole new system, and not simply as the old system minus one or more discrete parts. To quote Head:

> So far as the loss of function or negative manifestations are concerned, this response does not reveal the elements out of

which the original form of behaviour was composed. . . . It is a new condition, the consequence of a fresh readjustment of the organism as a whole to the factors at work at the particular functional level disturbed by the local lesions. (1920, 498)

Head continued to emphasize the whole and how new schemas might be formed after brain lesions. He reasoned that although these new schemas may be somewhat like old ones, they may no longer be accurate representations of reality. For this reason, they are likely to be associated with aberrant behaviors. Especially in this regard, Head's ideas can be compared to those of Kurt Goldstein, who also looked upon the damaged brain as a new entity.

FRANZ AND THE AMERICAN SCHOOL

During the first two decades of the twentieth century, Shepherd Ivory Franz (1874–1933) was the leading American scientist who fought against compartmentalization of associative and intellectual functions. Franz, whose photograph appears as Figure 4.3, also had his doubts about discrete, well-localized centers for sensory and motor functions.

With regard to memory, Franz (1912, p. 328) wrote that comparable deficits have been associated with diverse cortical lesions:

> We have no facts which at present will enable us to locate the mental processes in the brain any better than they were located fifty years ago. . . . I at least am unwilling to stand with the histological localizationists on the ground of a special mental process for special cerebral areas or for special cerebral cell groups.

Franz based this partly on data he collected while work-

Figure 4.3. Shepherd Ivory Franz (1874–1933), one of the Americans who questioned rigid cortical localization, especially for higher functions. Franz studied both laboratory animals and human patients with brain damage. Karl Lashley conducted several experiments with Franz in the second decade of the twentieth century and shared many of his dynamic views about the brain.

---

[7]Monakow, depressed and broken by the war, retreated to his mountain cabin, where he wrote not only about the *horme* but produced a series of papers on good and evil, the causes of war, and the emotions (e.g., Monakow, 1916). He believed it was only by training one's core biological instincts that mankind could kindle a spiritual rebirth and a return to an enduring morality. He died in Switzerland at the age of 77 while working on a manuscript to be entitled *The Value of Life.* See Yakovlev (1953) for a brief sketch of Monakow's scientific life.

ing with animals in his laboratory and partly from observing neurological patients in hospitals. His work with cats and monkeys sustaining frontal lobe lesions showed that even though these animals occasionally had difficulty remembering acts learned just prior to damage, they could still relearn the problems (Franz, 1902, 1906a, 1907). In addition, when monkeys were trained beyond a reasonable criterion for mastery before sustaining frontal lesions, they did not show severe deficits. Franz (1912, 1913) was convinced that data like these indicated that the frontal lobes were not the exclusive seat for memory and intellectual functions.

Franz was impressed by the role motivation played in enabling brain-damaged patients to overcome even long-standing deficits. He felt this was something that had been ignored for too long by therapists, as well as by the champions of localization. For example, he described a man who had walked with a cane for 19 years as a result of a paralysis. During a baseball game in which this man successfully hit the pitched ball, the excitement was so great that he forgot to use his cane as he ran to first base. After he reached base safely, he called for his cane again, saying he could not walk without it (see Garrett, 1951).

Another of his cases involved a soldier who had been hemiplegic for nine months as a result of a war wound. This man also used a cane. One day he was told by a very stern examiner that he had to put his cane aside and walk without assistance to his chair. The man insisted he could not walk the required 12 feet. When ordered to do so by his superior, he protested, but then crossed the examination room without assistance.

Franz argued that cases like these showed that the circuits needed to perform "lost" acts were not destroyed in these patients. This formed the philosophical basis of his foray into the field of "re-education" (Franz, 1906b, 1924, 1961; Franz, Scheetz, and Wilson, 1915). Findings from his therapeutic programs strengthened Franz's conviction that the type of rigid functional localization that many of his colleagues were postulating simply could not be correct.

Franz made some of his strongest statements against rigid functional localization in 1911 and 1917.[8] In the latter paper, he first described a number of individuals who either did not show the expected deficits or who exhibited good recovery. He then reiterated that a behavioral defect after a lesion does not necessarily mean that the damaged part is solely concerned with that function or that the person has permanently lost his capacity to deal with that function. He added that he had no qualms with those who favored anatomical (histological) localization at the cortical level, but he was prepared to do battle with those who equated this with functional localization. In his words:

This means that the motor cortex is not necessary for the execution of voluntary movement, that the visual cortex is not necessary for vision, that the tactile cortex is not neces-

sary for tactile discrimination, and that the speech cortex is not necessary for speech functions. When these facts are admitted, as they must be admitted, the whole structure of cerebral organology breaks down. The histological localization of function which has been in vogue takes its true place as a histological differentiation of an anatomical nature, without the functional implications which have been assumed. (1917, 140)

Franz ended his attack on the extreme localizationists by hypothesizing that there was a safety factor in the nervous system. He argued that an individual may live with only one half of a kidney and questioned why the same would not be true for the brain.

Franz (e.g., 1921, 1923) made comparable statements in a few of his later papers, effectively stating that he stood somewhere between those who believed in strict localization of function (e.g., Broca, Hitzig) and those who took equipotentiality to the opposite extreme. Thus, he admitted to some specialization at the cortical level, while also accepting considerable overlap in the functional capabilities of different anatomical areas. He maintained that even if there were more specialization, the brain must still function as a whole.

LASHLEY'S EXPERIMENTS AND THEORIES

Upon receiving his Ph.D. from Johns Hopkins University, Karl Lashley (1890–1958) went to see Franz, who at the time worked at the Government Hospital for the Insane in nearby Washington, DC.[9] The two men had met previously, but they now decided to collaborate on several research projects. They agreed to use rats in their experiments; it was proposed that Franz would do the surgeries and histology in Washington, while Lashley would conduct the behavioral testing in Baltimore.

Franz and Lashley started their investigations in 1916 and reported the results of their first two experiments a year later. They found that rats previously trained on a brightness discrimination showed no appreciable loss of this "simple" habit when large parts of their frontal lobes were damaged (Franz and Lashley, 1917). The importance of this finding derived from the fact that the frontal association region was thought by many of their contemporaries to be essential for normal learning and intellect (see Chapter 22). In addition, the posterior part of the frontal region housed the motor cortex, believed by some to be a storehouse for the engrams necessary for performing learned motor patterns (see Chapter 14).

In their second publication, Lashley and Franz (1917) reported that the capacity for simple maze learning may remain intact after limited damage to any part of the cortex. This finding in particular led Lashley to agree with Franz that intelligence did not depend on any one cortical

[8]In 1911, Franz (p. 118) wrote:

Finally it may be said that the recent work on the histological differentiation of the cerebral areas is most important for a definite understanding of functional localization, but it does not appear, either from the experimental or the clinical investigations, that the brain is a simple organ with simple but numerous functions, nor does it appear that efforts to bring about a revival of the doctrines of Spurzheim, even though it be based upon these anatomical studies, will help materially to advance the knowledge of this "vehicle of mind."

[9]Lashley did not begin his training in psychology or physiology. Nevertheless, while studying genetics at Johns Hopkins University, he fell under the influence of John B. Watson (1878–1958), the leading proponent of Behaviorism and the President of the American Psychological Association (Bruce, 1986). Lashley conducted a number of studies with Watson, who clearly stimulated him to study learning and memory. But, unlike Watson, who seemed content to examine normally developing children and healthy adults, Lashley decided to approach learning and memory by assessing the effects of brain damage in laboratory animals.

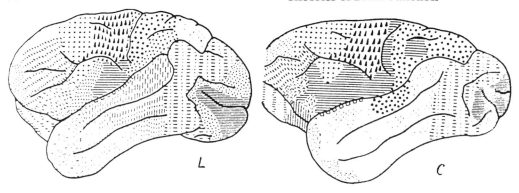

Figure 4.4. Cytoarchitectonic maps of two brains from Ateles monkeys. Lashley and Clark (1946) drew these maps independently. They stressed differences in experimenter criteria, and the variability across brains even within a species, to account for the fact that their maps differed in so many ways. (Courtesy of the *Journal of Comparative Neurology.*)

area. Instead, when decrements in intellectual capability were observed, they seemed to be a function of lesion size, not lesion locus.

Lashley summarized these and many other findings in *Brain Mechanisms and Intelligence,* a book that appeared in 1929.[10] The recurring theme of the monograph was that there was not the slightest shred of evidence to suggest that maze habits were made of independent associational elements. Instead, and in accordance with prevailing Gestalt theory, Lashley argued that "higher-level integrations" were a function of the dynamic organization of the entire cerebral system.

Lashley discussed two "laws" in this context. One was the law of equipotentiality, and the other was the law of mass function (mass action). "Equipotentiality," a term used by Lashley and Franz in 1917, was defined in 1929 as follows:

> The term "equipotentiality" I have used to designate the apparent capacity of any intact part of a functional area to carry out, with or without reduction in efficiency, the functions which are lost by destruction of the whole. This capacity varies from one area to another and with the character of the functions involved. It probably holds only for the association areas and for functions more complex than simple sensory or motor co-ordination. (1963 reprinting, 24)

In the next paragraph, Lashley defined mass action:

> I have already given evidence, which is augmented in the present study, that the equipotentiality is not absolute but is subject to a law of mass action whereby the efficiency of performance of an entire complex function may be reduced in proportion to the extent of brain injury within an area whose parts are not more specialized for one component of function than for another.

Lashley's definitions make it clear that he was willing to accept some sort of cortical specialization for sensory and motor functions, although not for intellectual functions. But how much? And in what form? His answer was based on experimentation.

Lashley taught rats a simple brightness habit, one that required them to choose between a light and a dark alley. After the rats were trained, he damaged the posterior cortex and then retested his subjects. These rats showed a complete loss of the brightness habit, whereas rats with other cortical areas destroyed performed well. But while this argued for visual localization, he added that the rats with posterior cortical lesions still relearned. In fact, they relearned in the same number of trials as they required for original learning. Lashley concluded that the occipital region was important but not critical for vision. Supposedly, other parts of the cortex had the capacity to take over quickly and effectively.[11]

Lashley also conducted anatomical studies to challenge mainstream ideas about cortical localization. In particular, he questioned the anatomical uniformity of the various cortical areas across individuals, a feature implied by the cytoarchitectonic maps of Korbinian Brodmann and others (see Chapter 3).

In an attempt to demonstrate just how different brains could be, and how misleading standardized maps could be, Lashley examined a brain from an Ateles monkey and made a cytoarchitectonic map. He gave an associate a second Ateles brain and had him do the same. The two drawings appear in Figure 4.4. The investigators wrote that the differences between their drawings were such "as to cast doubt upon the reliability of both and to rule out either as a satisfactory guide to experimental work" (Lashley and Clark, 1946, p. 231).

The two investigators then compared their two specimens. They found that the differences in their drawings were not just a function of differences in experimenter criteria. There were instances in which an area clearly distinguishable in one brain could not even be found in the other. At the end of their paper, they wrote the following about anatomical variability and the value of cytoarchitectonic maps for localization of function:

> In planning experimental work on the cortex, it is desirable to have some guide to probable functional units in order that the work should not be undertaken wholly at random. For this purpose, however, the "ideal" architectonic chart is nearly worthless, because individual variation is too great

---

[10]An interesting review of Lashley's book was written in 1931 by James Papez (1883–1958). Papez seemed less than enthusiastic about Lashley's theories.

[11]These and other findings that did not fit with classic localization theory were presented by Lashley in one of his most cited publications, *In Search of the Engram,* published in 1950.

to make the chart significant for a single specimen, because the areal subdivisions are in large part anatomically meaningless, and because they are misleading as to the presumptive functional divisions of the cortex. . . . The charting of areas in terms of poorly defined and variable characters, in the hope that future physiologic studies may some time reveal their significance, has contributed nothing to knowledge of cerebral organization and gives no promise of better achievement in the future. (1946, 298–299)

Overall, Lashley's conclusions about localization were not really revolutionary. His ideas about intellect were somewhat like those reached by Flourens (e.g., 1842). But, unlike Flourens, who did not believe in any specialization at the cortical level, Lashley accepted the general concepts of specialized sensory and motor areas of the cortex. In this regard, Lashley's general conceptualization of cortical function was more like that proposed by Loeb, who accepted the idea of cortical territories for specific sensory and motor functions, but who concluded that associative functions could not be localized in specific regions.

CONCLUSIONS

After Lashley's death, the evidence in favor of cortical specialization continued to grow, even for some of the so-called intellectual or executive functions. In part, this reflected improved methodologies for investigating the brain and better techniques for measuring behavior. In this sense, the war was won by those individuals who favored at least some form of cortical localization. Nevertheless, many debates continued to rage over exactly what was being localized, especially within the association areas.

As for holism, this is a concept that has continuously waxed and waned in popularity over the years. Yet the importance of looking at the workings of the whole brain, whether before or after injury, never had to be viewed as a direct challenge to the localizationist position. To be sure, there were clear differences in emphasis. The orientation chosen by Ferrier and Hitzig was reductionistic, and that favored by Goltz and Goldstein was not. Localization theory was aimed more at answering questions about the fundamental biological organization of the brain. Holism appealed more to individuals who were trying to understand why brain-damaged patients might show both positive as well as negative symptoms, and why maladaptive behaviors might vary with contextual and situational conditions.

But even beyond these differences in emphasis, there was ample room and a definite need for both approaches. The importance of the holistic message was that investigators studying the individual parts of the brain must not forget that the brain had to function as a finely integrated unit, and not simply as a collection of autonomous or independent parts. At the same time, the localizationist idea that the brain is composed of many specialized parts is something that could no longer be ignored.

Unfortunately, the localizationist and holistic positions were not always seen in this sort of a noncompetitive way. This fact will become more apparent in subsequent chapters of this book.

REFERENCES

Andral, G. 1833. *Clinique Médicale . . .* (Vol. 4, 2nd ed.). Paris: Librairie de Deville Cavellin.

Benton, A. L. 1956. Jacques Loeb and the method of double stimulation. *Journal of the History of Medicine and Allied Sciences, 11,* 47–53.

Bernard, C. 1855–1856. *Leçons de Physiologie Expérimentale Appliqué à la Médecine.* Paris: J. B. Ballière.

Boring, E. G. 1950. *A History of Experimental Psychology.* New York: Appleton-Century-Crofts.

Bouillaud, J.-B. 1825a. Recherches cliniques propres à démontrer que la perte de la parole correspond à la lésion des lobules antérieurs du cerveau et à confirmer l'opinion de M. Gall sur le siège de l'organe du langage articulé. *Archives Générale de Médecine (Paris), 8,* 25–45.

Bouillaud, J.-B. 1825b. *Traité Clinique et Physiologique de l'Encéphalite ou Inflammation du Cerveau.* Paris: J. B. Ballière.

Bouillaud, J.-B. 1830. Recherches expérimentales sur les fonctions du cerveau (lobes cérébraux) en général, et sur celles de sa portion antérieure en particulier. *Journal Hebdomadaire de Médecine (Paris), 6,* 527–570.

Bouillaud, J.-B. 1839–1840. Exposition de nouveaux faits à l'appui de l'opinion qui localise dans les lobules antérieurs du cerveau le principe législateur de la parole; examen préliminaire des objections dont cette opinion a été sujet. *Bulletin de l'Académie Royale de Médecine (Paris), 4,* 282–328, 333–349, 353–369.

Bouillaud, J.-B. 1848. *Recherches Cliniques Propres à Démontrer que le Sens du Langage Articulé et le Principe Coordinateur des Mouvements de la Parole Résident dans les Lobules Antérieurs du Cerveau.* Paris: J. B. Ballière.

Broca, P. 1861. Remarques sur le siège de la faculté du langage articulé; suivies d'une observation d'aphémie (perte de la parole). *Bulletins de la Société Anatomique (Paris), 6,* 330–357, 398–407. In G. von Bonin, *Some Papers on the Cerebral Cortex.* Translated as, "Remarks on the seat of the faculty of articulate language, followed by an observation of aphemia." Springfield, IL: Charles C Thomas, 1960. Pp. 49–72.

Broca, P. 1865. Sur le siège de la faculté du langage articulé. *Bulletins de la Société d'Anthropologie, 6,* 337–393. Translated by E. A. Berker, A. H. Berker, and A. Smith as, "Localization of speech in the third left frontal convolution." *Archives of Neurology, 43,* 1065–1072, 1986.

Brodmann, K., 1909. *Vergleichende Lokalisationslehre der Grosshirnrinde in ihren Prinzipien dargestellt auf Grund des Zellenbaues.* Leipzig: J. A. Barth.

Brown-Séquard, C.-E. 1877. Aphasia as an effect of brain-disease. *Dublin Journal of Medical Science, 63,* 209–225.

Bruce, D. 1986. Lashley's shift from bacteriology to neuropsychology, 1910–1917, and the influence of Jennings, Watson, and Franz. *Journal of the History of Behavioral Sciences, 22,* 27–44.

Clarke, E., and O'Malley, C. D. 1968. *The Human Brain and Spinal Cord.* Berkeley: University of California Press.

Delafield, E. 1822. Remarkable case of hydrocephalus in which the operation of tapping was performed. *American Medical Recorder, 4,* 448–453.

Dupuy, E. 1877. Prevailing theories concerning the physiology and pathology of the brain. *Medical Times and Gazette, 2,* 11–13, 32–34, 84–87, 356–358, 474–475, 488–490.

Ehrenfels, C. von. 1890. Über Gestaltqualitäten. *Vierteljahrschrift für wissenschaftliche Philosophie, 14,* 249–292.

Ferrier, D. 1876. *The Functions of the Brain.* London: Smith, Elder and Company.

Flourens, M.-J.-P. 1842. *Recherches Expérimentales sur les Propriétés et les Fonctions du Système Nerveux dans les Animaux Vertébrés* (2nd ed.). Paris: J. B. Ballière. (First ed., 1824)

Flourens, M.-J.-P. 1846. *Phrenology Examined* (2nd ed.). (C. de L. Meigs, Trans.). Philadelphia, PA: Hogan and Thompson. Reprinted in D. N. Robinson (Ed.), *Significant Contributions to the History of Psychology, Ser. E, Physiological Psychology: Vol. II, X. Bichat, J. G. Spurzheim, P. Flourens.* Washington, DC: University Publications of America. Pp. iv-144.

Flourens, M.-J.-P. 1851. Note sur le point vital de la moelle allongée. *Compte Rendu des Séances de L'Académie des Sciences, 33,* 437–439.

Flourens, M.-J.-P. 1858. Nouveaux détails sur le noeud vital. *Compte Rendu des Séances de L'Académie des Sciences, 47,* 803–807.

Flourens, M.-J.-P. 1859. Nouveaux éclaircissements sur le noeud vital. *Compte Rendu des Séances de L'Académie des Sciences, 48,* 1136–1138.

Franz, S. I. 1902. On the functions of the cerebrum: I. The frontal lobes in relation to the production and retention of simple sensory-motor habits. *American Journal of Physiology, 8,* 1–22.

Franz, S. I. 1906a. Observations on the functions of the association areas (cerebrum) in monkeys. *Journal of the American Medical Association, 47,* 1464–1467.

Franz, S. I. 1906b. The reeducation of an aphasic. *Journal of Philosophy, Psychology and Scientific Method, 2,* 589–597.

Franz, S. I. 1907. On the functions of the cerebrum: The frontal lobes. *Archives of Psychology, 2,* 1–64.

Franz, S. I. 1911. The functions of the cerebrum. *Psychological Bulletin, 8,* 111–119.

Franz, S. I. 1912. New phrenology. *Science, 35,* 321–328.

Franz, S. I. 1913. The functions of the cerebrum. *Psychological Bulletin, 10,* 125–138.

Franz, S. I. 1917. Cerebral adaptation vs. cerebral organology. *Psychological Bulletin, 14,* 137–140.

Franz, S. I. 1921. Cerebral-mental relations. *Psychological Review, 28,* 81–95.

Franz, S. I. 1923. Conceptions of cerebral functions. *Psychological Review, 30,* 438–446.

Franz, S. I. 1924. Studies in re-education: The aphasias. *Journal of Comparative Psychology, 4,* 349–429.

Franz, S. I. 1961. Shepherd Ivory Franz. In C. Murchison (Ed.), *A History of Psychology in Autobiography.* New York: Russell and Russell. Pp. 89–113.

Franz, S. I., and Lashley, K. S. 1917. The retention of habits by the rat after destruction of the frontal portion of the cerebrum. *Psychobiology, 1,* 3–18.

Franz, S. I., Scheetz, M. E., and Wilson, A. A. 1915. The possibility of recovery of motor function in long-standing hemiplegia. *Journal of the American Medical Association, 65,* 2150–2154.

Fritsch, G., and Hitzig, E. 1870. Über die elektrische Erregbarkeit des Grosshirns. *Archiv für Anatomie und Physiologie,* 300–332. In G. von Bonin, *Some Papers on the Cerebral Cortex.* Translated as, "On the electrical excitability of the cerebrum." Springfield, IL: Charles C Thomas, 1960. Pp. 73–96.

Frommer, G. P., and Smith, A. 1988. Kurt Goldstein and recovery of function. In S. Finger, T. E. LeVere, C. R. Almli, and D. G. Stein (Eds.), *Brain Injury and Recovery: Theoretical and Controversial Issues.* New York: Plenum Press. Pp. 71–88.

Gall, F., and Spurzheim, J. 1810–1819. *Anatomie et Physiologie du Système Nerveux en Général, et du Cerveau en Particulier.* Paris: F. Schoell. (Gall was the sole author of the first two volumes of the four in this series.)

Garrett, H. E. 1951. *Great Experiments in Psychology* (3rd ed.). New York: Appleton-Century-Crofts.

Gerlach, J. von. 1872. Über die struktur der grauen Substanz des menschlichen Grosshirns. *Zentralblatt für die medizinischen Wissenschaften, 10,* 273–275.

Goldstein, K. 1934. *Der Aufbau des Organismus.* The Hague: Martinus Nijhoff.

Goldstein, K. 1939. *The Organism.* New York: American Book Company.

Goldstein, K. 1942. *Aftereffects of Brain Injuries in War: Their Evaluation and Treatment.* London: Heinemann.

Golgi, C. 1873. Sulla struttura della grigia del cervello. *Gazetta Medica Italiani Lombardi, 6,* 244–246. Translated by M. Santini as, "On the structure of the gray matter of the brain," in M. Santini (Ed.), *Golgi Centennial Symposium, Proceedings.* New York: Raven Press, 1975. Pp. 647–650.

Golgi, C. 1883. Continuation of the study of the minute anatomy of the central organs of the nervous system. Ch. II. *Alienist and Neurologist, 4,* 383–416.

Golgi, C. 1903. *Opera Omnia.* Milano: Hoepli. (Golgi's collected works, 1868–1902)

Golgi, C. 1906. The neuron doctrine—theory and facts. In *Nobel Lectures Physiology or Medicine 1901–1921.* New York: Elsevier, 1967. Pp. 189–217.

Goltz, F. 1881a. Über die Verrichtungen des Grosshirns. *Archiv für die gesamte Neurologie und Psychiatrie, 26,* 1–49.

Goltz, F. 1881b. *Über die Verrichtungen des Grosshirns, Gesammelte Abhandlungen.* Bonn.

Goltz, F. 1888. Über die Verrichtungen des Grosshirns. *Pflüger's Archiv für die gesammte Physiologie, 42,* 419–467. In G. von Bonin, *Some Papers on the Cerebral Cortex.* Translated as, "On the functions of the hemispheres." Springfield, IL: Charles C Thomas, 1960. Pp. 118–158.

Goltz, F. 1892. Der Hund ohne Grosshirn. Siebente Abhandlung über die Verrichtungen des Grosshirn. *Archiv für die gessamte Physiologie, 51,* 570–614.

Haller, A. von. 1759–1766. *Elementa physiologie.* Lausannae: Marci-Michael, Bosquet et Sociorum.

Harrington, A. In press. A feeling for the whole. In R. Porter and M. Teich (Eds.), *The Fin-de-Siècle and Its Legacy.* London: Cambridge University Press.

Harrison, J. P. 1825. Observations on Gall and Spurzheim's theory. *Philadelphia Journal of Medical and Physical Sciences, 11,* 233–249.

Head, H. 1918. Some principles of neurology. *Brain, 41,* 344–354.

Head, H. 1920. *Studies in Neurology.* London: Hodder and Stroughton.

Jackson, J. H. 1881. Remarks on dissolution of the nervous system as exemplified by certain post-epileptic conditions. *Medical Press and Circular,* April, 329–332ff.

Jackson, J. H. 1884. Evolution and dissolution of the nervous system. *British Medical Journal, 1,* 591, 660, 703. (Also in *Popular Science Monthly, 25,* 171–180, 1884)

Jackson, J. H. 1894. The factors of the insanities. *Medical Press and Circular,* June, 615–619.

Klein, Langley, and Schäfer. 1883. On the cortical areas removed from the brain of a dog, and from the brain of a monkey. *Journal of Physiology, 4,* 231–247.

Koffka, K. 1935. *Principles of Gestalt Psychology.* New York: Harcourt, Brace.

Köhler, W. 1929. *Gestalt Psychology.* London: Bell and Sons.

Korsakoff, S. S. 1887. Disturbance of psychic functions in alcoholic paralysis and its relation to the disturbance of the psychic sphere in multiple neuritis of non-alcoholic origin. *Vestnik Psichiatrii, 4,* Fasc. 2.

Lashley, K. S. 1929. *Brain Mechanisms and Intelligence.* Chicago: University of Chicago Press. (Reprinted 1963; New York: Dover Publications)

Lashley, K. S. 1950. In search of the engram. *Symposia of the Society for Experimental Biology, 4,* 454–482.

Lashley, K. S., and Clark, G. 1946. The cytoarchitecture of the cerebral cortex of Ateles: A critical examination with architectonic studies. *Journal of Comparative Neurology, 85,* 223–305.

Lashley, K. S., and Franz, S. I. 1917. The effects of cerebral destruction upon habit-formation and retention in the albino rat. *Psychobiology, 1,* 71–139.

Legallois, J. J. C. 1812. *Expériences sur le Principe de la Vie, Notamment sur Celui des Movemens du Coeur, et sur le Siège de ce Principe.* Paris: Hautel.

Loeb, J. 1885. Die elementaren Störungen Funktionen nach oberflächlicher, umschriebener Verletzung des Grosshirns. *Pflüger's Archiv für die gesammte Physiologie, 37,* 51–56.

Loeb, J. 1886. Beiträge zur Physiologie des Grosshirns. *Pflüger's Archiv für die gesammte Physiologie, 39,* 265–346.

Loeb, J. 1902. *Comparative Physiology of the Brain and Comparative Psychology.* New York: G. P. Putnam's Sons.

Marie, P. 1906. La troisième circonvolution frontale gauche ne joue aucun role spécial dans la fonction du langage. *Semaine Médicale, 26,* 241–247. In M. F. Cole and M. Cole (Eds./ Trans.). *Pierre Marie's Papers on Speech Disorders.* New York: Hafner, 1971. Pp. 51–71.

Marie, P., and Moutier, F. 1906. New case of cortical lesion of the foot of the third left frontal in a right-handed man without language disorders. *Bulletins et Mémoires de la Société Médicale des Hôpitaux de Paris, Ser 3, 23,* 1295–1298. In M. F. Cole and M. Cole (Eds./Trans.). *Pierre Marie's Papers on Speech Disorders.* New York: Hafner, 1971. Pp. 141–143.

Monakow, C. von. 1897. *Gehirnpathologie.* Wien: Holder.

Monakow, C. von. 1911. Lokalisation der Hirnfunktionen. *Journal für Psychologie und Neurologie, 17,* 185–200. In G. von Bonin, *Some Papers on the Cerebral Cortex.* Translated as, "Localization of brain functions." Springfield, IL: Charles C Thomas, 1960. Pp. 231–250.

Monakow, C. von. 1914. *Die Lokalisation im Grosshirn und der Abbau der Funktion durch Kortikale Herde.* Wiesbaden: J. F. Bergmann. (G. Harris, Trans./Excerpter). In K. H. Pribram (Ed.), *Mood, States and Mind.* London: Penguin Books, 1969. Pp. 27–37.

Monakow, C. von. 1916. Gefühl, Gesittung und Gehirn. *Arbeiten Hirnanatomie Institut Zürich, 10,* 115–213.

Munk, H. 1890. *Über die Funktionen der Grosshirnrinde.* Berlin: A. Hirschwald.

Papez, J. W. 1931. Book review of K. S. Lashley's *Brain Mechanisms and Intelligence. American Journal of Psychology, 43,* 527–529.

Prince, M. 1910. Cerebral localization from the point of view of function and symptoms. *Journal of Nervous and Mental Disease, 37,* 337–354.

Ramón y Cajal, S. 1906. The structure and connexions of neurons. In *Nobel Lectures Physiology or Medicine 1901–1921.* New York: Elsevier, 1967. Pp. 220–253.

Ringer, F. K. 1969. *The Decline of the German Mandarins. The German Academic Community, 1890–1933.* Cambridge, MA: Harvard University Press.

Shepherd, G. 1991. *Foundations of the Neuron Doctrine.* New York: Oxford University Press.

Spillane, J. D. 1981. *The Doctrine of the Nerves.* Oxford, U.K.: Oxford University Press.

Uexküll, J. von. 1920. *Staatsbiologie* (2nd ed.). Special issue of *Deutsche Rundschau.* Berlin: Bebrüder Poetel.

Willis, T. 1664. *Cerebri anatome: cui accessit nervorum descriptio et usus.* London: J. Martyn and J. Allestry. Tercentenary ed., 1664–1964, *Thomas Willis, The Anatomy of the Brain and Nerves.* Montreal: McGill University Press, 1965.

Yakovlev, P. I. 1953. Constantin von Monakow (1853–1930). In W. Haymaker and F. Schiller (Eds.), *The Founders of Neurology.* Springfield, IL: Charles C Thomas. Pp. 336–340.

PART II
Sensory Systems

Chapter 5
Vision: From Antiquity through the Renaissance

Of the coats of the eye the first, which is visible, is called "horn-like" (cornea); the second, "grape-like" and "chorioid." It is called "grape-like" because the part lying under the cornea is like a grape in its external smoothness and internal roughness; it is called "chorioid" because the part under the white, being full of veins, resembles the chorion that encloses the foetus. The third coat encloses a vitreous humor; because of its fineness its ancient name was "arachnoid" (i.e., like a spider's web). But since Herophilus likens it to a drawn net, some call it net-like (retiform; cf. retina), others call it glass-like (vitreous) from its humor.

Rufus of Ephesus, second century

VISUAL DISTURBANCES IN ANCIENT EGYPT

The Ancient Egyptians suffered from a high incidence of eye problems, and their records contain a plethora of descriptions of eye diseases and statements about how they were treated. Among the diseases encountered in the "land of the blind" were conjunctivitis, "barleycorn" (probably cataract and glaucoma), and leucoma (a white scarring on the cornea). They also suffered from trachoma, an infectious disease caused by a microorganism, which might have left 50 percent of its victims blind.[1]

The Egyptians divided the body into 36 parts, as set out in their *Book of the Dead,* and each part was thought to be controlled by a different deity who answered to Thoth. The deity of the eyes was Hathor. They believed eye diseases were due to malevolent spirits, and thus many of the prayers, incantations, and divinations that were said were intended to force these demons to leave the body. This was often accomplished by appealing to strong gods, such as Re (Gordon, 1939).

Besides asking their gods for help, the Ancient Egyptians also used a variety of agents to treat eye problems. Medicine chests of rich and powerful Egyptians were filled with eye ointments to help protect them against the winds and the sun and to help fight eye diseases. The *Ebers Papyrus,* copied before 1500 B.C. from considerably older sources, contains a special supplement on how to treat diseases of the eye. The papyrus shows that copper, sulphur, alum, fish gall, and beef liver were often made into ointments for application to the eyes. These agents all possess anti-infectious properties. The Ancient Egyptian eye doctors, however, also used many questionable substances. These included urine (sometimes used for eyewash) and feces, although it has been suggested that such "waste" may be rich in antibiotic substances (see Thorwald, 1963).

[1]Even during the Napoleonic Wars (1798–1815), trachoma continued to plague armies that fought in Egypt. Further, soldiers returning from Egypt unknowingly carried trachoma back to their homelands and caused epidemics of blindness in other parts of the world (Feibel, 1983).

A letter written by a painter (Poi) to his son, who lived at the time of Ramses II (1304–1237 B.C.), describes the loss of vision, the despair this caused, and how it was treated. It read:

Do not leave me. I live in despair. . . . I live in darkness. My god Amon has left me. Bring me honey for my eyes, and fat . . . and genuine eye-paint, as soon as possible. Am I not your father? I wish to see, but my eyes are deserting me. (Translated in Thorwald, 1963, 82)

Egyptian mythology contains clues about how Egyptian eye specialists may have tested vision. After Horus suffered an eye injury, the god Re placed him in front of a light-colored wall on which a small black line and a larger black pig were painted. Re asked Horus if he were able to see the line, upon which Horus replied that he could not see it. Re then tested vision for the painted pig and found Horus could make out the bigger figure. Thus, Re concluded that Horus still had partial sight in his injured eye.

ASSYRO-BABYLONIAN OPHTHALMOLOGY

After Hammurabi conquered Babylonia during the eighteenth century B.C., he formulated his famous code of laws, basing it largely on older Sumerian codes (see Chapter 1). The code was engraved on a large upright stone, which was set in front of one of the major temples, where it could be seen by the public. Hammurabi's code contained a number of sections relating to the eye. Some specified the penalties for causing the loss of an eye. These penalties varied with the nature of the crime and the economic status of the parties involved. One of the most severe penalties called for the destruction of the eye of the accused, a penalty consonant with the phrase that most people associate with Hammurabi's laws: "An eye for an eye, a tooth for a tooth." In other instances, the payment was as small as one third of a small silver coin *(maneh).*

Hammurabi's code also specified the fees that could be charged for treating eye diseases and injuries, as well as the penalties for "medical malpractice." Some of these sections read:

215. The doctor who treats and cures a gentleman's wound, or has operated on the eye with a copper lancet, shall charge 10 shekels of silver.

216. If the patient be a poor man, the doctor shall charge 5 shekels of silver.

217. If the patient be a servant, his master shall pay the doctor 2 shekels of silver.

218. If a doctor operates on a wound with a copper lancet, and the patient dies, or on the eye of a gentleman, who loses his eye in consequence, his hands shall be cut off.

Like the Egyptians, the Babylonians thought visual problems were due to demons. Even injuries could be tied to animistic concepts, especially the idea that wounds occurred when the gods deserted the body because of some sin. Further, it was thought that an afflicted person could cause a demon to enter the body of another individual. Thus, a man with a cataract, a squint, or an eye infection was deemed capable of causing the demon to enter another's eye. This was the original "evil eye" (Krause, 1934).

Because the Babylonians were highly susceptible to eye diseases, they considered the demons responsible for visual problems to be especially powerful. Thus, therapy for eye diseases among the late Assyrians usually involved prayers, rituals, and charms to lure out the demons (Thompson, 1923, 1924). This is evident from the clay tablets that were found by Sir Austin Henry Layard (1817–1894) at Nineveh, in what is now Iraq. A surprising number of the clay tablets from the Royal Library of Assurbanipal, dated from the seventh century B.C., were found to deal with disorders of sight.

The prayers and pleas of the people from the region of the Tigris and Euphrates rivers may have been relatively simple at first, but records show that as time went by the texts became increasingly complex and associated with very elaborate rituals. For example, the water ritual of the goddess Ea at Eridu was based on the ancient notion of water as a purifying force. This evolved into a special set of prayers and acts begging Ea to cure blindness.

In addition, a number of drugs, plants, and minerals were used for treating eye diseases, as in Egypt. For obvious reasons, rubbing an eye with an onion or drinking a beer with onion juice was recommended for dry eyes (Thompson, 1923, 1924). The sources of some other treatments are not as well understood. For example, one called for disemboweling a yellow frog, mixing its gall with curd, and applying it to the eyes!

The following was given as a prescription for night blindness:

If a man's eyes suffer from *Sinlurmâ* thou shalt thread *makut* of the liver of an ass and flesh of its neck on a cord and put it on his neck, prepare a water pot; on the morrow thou shalt spread a cloth in the sun, prepare a censer of pine-gum; then thou shalt let this man stand behind the cloth in the sun. A priest shall take seven rounds of bread; he whose eyes are sick shall take seven rounds of bread: then the priest shall say to the sick man, "Receive, O clear of the eye": the sick man shall say to the priest, "Receive, O dim of the eye." Thou shall chop up the *makut* of the liver. . . . Thou shall mix curd and the best oil together, apply to his eyes. (Cited in Krause, 1934, 53)

ANCIENT INDIA

Susruta, the author of the first known surgical treatise from India, is mentioned in hymns from the Vedic period of about 1000 B.C. (see Chapter 1). This would suggest that Susruta must have lived at around this time or somewhat earlier, although it is doubtful that he lived before 1500 B.C. (Dutt, 1938). Susruta was an enlightened physician who allowed his students to perform dissections. He played a significant role in opening the golden age of Hindu surgery and medicine (Bidyadhar, 1940).

The *Susruta Samhita,* which was written in classical Sanskrit, describes the anatomy, physiology, and treatment of many eye diseases, including glaucoma and night blindness. Susruta advocated eating liver and applying ointments with liver extracts for night blindness. In modern terms, he was treating this disorder appropriately by giving the patient vitamin A. Although Susruta had an extensive "pharmacy," he also relied on other agents and "tools" for treating visual problems. These included leeches, the cautery, and magnets for removing pieces of metal from the eye.

The claim has been made that Susruta was the first known "surgeon" to perform the "couching" operation for a cataract (Bidyadhar, 1939, 1940, 1941; Dutt, 1938). He used a special tool called the *Jabamukhi Salâkâ,* which looked like a curved needle, to loosen the lens and push the cataract out of the field of vision. Following the surgery, the eye was soaked with warm butter and bandaged. Susruta had considerable success with the couching operation. He stated: "The patient at once regains clear eyesight, as the sun shines after the sky is clear of the clouds" (see Dutt, 1938, p. 5). Still, being fully aware of the complications and shortcomings of couching, he mentioned that not everyone with cataracts should be operated upon.

Many famous Greek philosophers and scientists visited India and the Middle East, where physicians performed Susruta's operation for cataracts. It has been suggested that these travelers brought the couching technique, along with many of Susruta's other ideas about medical treatment, to the West (Bidyadhar, 1940). Among the more notable of these early Greek travelers were Pythagoras, Empedocles, and Democritus.

GREEK ANATOMY AND METAPHYSICAL THEORIES OF VISION

The Ancient Greeks made a number of anatomical discoveries in their studies of the eye. For example, in the fifth century B.C., Alcmaeon, who was associated with the ancient medical school in Crotona, may have been the first scientist to dissect and study the optic nerves. Nevertheless, he erroneously believed that the nerves mediating vision were hollow canals.

Hippocrates and his followers studied the anatomy of the eye. The Hippocratic writers described three "tunics" of the eye: (1) the sclera (with the cornea), (2) the uveal tunic (iris, ciliary body, and choroid), and (3) the retina (about which very little was known).

Herophilus, who had a deep reverence for Hippocrates, presented a more advanced scientific discussion of the eye. He mentioned many parts of the eye, including the vitreous ("glasslike") humor and the optic nerves. His descriptions, which appeared during the most advanced period of Greek science (the Hellenistic Period; 323–212 B.C.), were based largely on dissections of animals and humans (see Chapter 1). In contrast, Aristotle, who also lived in the fourth century B.C., described only those parts of the eye easily seen in a living person.

The theoretical notions that accompanied the anatomical advances differed from one another but still shared certain common features. Alcmaeon hypothesized that sight required "fire" (a term that included not only flame but red heat and light) in a watery medium, and he compared the eye to a shining lantern. His evidence was that a blow to the eye produced a flash within it. We learn of this from Theophrastus (c. 371–288 B.C.), who wrote the following in his essay *"On the Senses"*:

> (Alcmaeon states that) eyes see through the water round about. And the eye obviously has fire within, for when one is struck (this fire) flashes out. Vision is due to the gleaming—that is to say, the transparent—character of that which (in the eye) reflects to the object; and sight is the more perfect, the greater the purity of this substance. All the senses are connected in some way with the brain; consequently they are incapable of action if (the brain) is disturbed or shifts its position. (1917 translation, 543–544)

Alcmaeon believed vision had something to do with emanations (beams, rays). In various forms, this idea was adopted by many later Greek philosophers.

Empedocles, who like Alcmaeon lived in the fifth century B.C., was very clear about his belief in emanations from an object to the eye. Aristotle stated in *De sensu* that Empedocles also believed in an outward-flowing fire. Nevertheless, Empedocles' acceptance of a second emanation from the eye to the object and the possible union of the two emanations have been matters of protracted debate.[2] The diverse interpretations of Empedocles' theory stem from the fact that he only made a passing reference to the image of a lantern while describing the structure and composition of the eye. His simile suggests that fire may leave the eye, but he really said nothing about the way in which the outward fire might contribute to the act of seeing.

Empedocles agreed with Alcmaeon that "visual fire" surrounded by water served an important function in the interior of the eye (O'Brien, 1970; Siegel, 1959). In particular, dark emanations were thought to enter the larger watery pores of the eyes, whereas the smaller fiery pores received bright effluences. Empedocles thought good vision resulted when there was a balance between fire and water. He maintained that too much fire created dazzling effects that made vision problematic. In contrast, too much moisture dimmed vision.

Like other Ancient Greeks, Empedocles believed white and black were colors and not just shades of light intensity.

Thus, he postulated that white objects were seen through the fiery pores in the eye, whereas black objects were seen through the watery pores. He also thought that species differences in the locations and ratios of the two types of pores allowed some animals to see better during the day and others to see better at night (see Siegel, 1959).

Democritus expanded many of the ideas set forth by Empedocles. He hypothesized that objects emitted images *(eidola)* into the air, where they made imprints comparable to a seal pushed into wax (see Siegel, 1959). Theophrastus described Democritus' theory as follows:

> Vision he explains by the reflection (in the eye) of which he gives a unique account. For the reflection does not arise immediately in the pupil. On the contrary, the air between the eye and the object of sight is compressed by the object and the visual organ, and thus becomes imprinted . . . this imprinted air, because it is solid and is of a hue contrasting (with the pupil), is reflected in the eyes which are moist. A dense substance does not receive (this reflection), but what is moist gives it admission. (1917 translation, 544)

Theophrastus believed the whole idea of imprints in the air was extravagant and argued that this theory should not be taken too seriously. Nevertheless, the theory did persist.[3]

Democritus also wrote about color perception. He stated that there were four basic colors (black, white, red, and yellow), which corresponded to the four basic elements, and that they differed in atomic structure. Unfortunately, Democritus' two most important books about vision have been lost, and only their titles are known today (Siegel, 1959). These are (1) *About the Mind, Perception and Colors* and (2) *About the Different Arrangements and the Concepts of Forms.*

Some of these early metaphysical ideas appear in Plato's *Timaeus,* a work named after its main character, a genial, mathematically oriented philosopher who engaged Socrates (c. 470–399 B.C.) in a lengthy dialogue. Timaeus held that a nonburning visual fire issued from the eye and united with rays of "outer light" (i.e., from the sun) to form a single, compacted homogeneous body. This body stretched from the eye into visual space. He explained that there also were emanations from visible objects. In his schema, the coalesced emanations formed by daylight and visual fire transmitted features of the object back to the eye. Timaeus explained:

> And of the organs they first contrived the eyes to give light, and the principle according to which they were inserted was as follows: So much of fire as would not burn, but gave a gentle light, they formed into a substance akin to the light of every-day life; and the pure fire which is within us and related thereto they made flow through the eyes in a stream smooth and dense, compressing the whole eye; and especially the centre part, so that it kept out everything of a coarser nature, and allowed to pass only this pure element. When light of day surrounds the stream of vision, then like falls upon like, and they coalesce, and one body is formed by natural affinity in the line of vision; wherever the light that falls from within meets with an external object. And the whole stream of vision, being similarly affected in virtue of similarity, diffuses the motions of what

---

[2]For interpretations of Empedocles' theory of vision, see Lindberg (1976) and O'Brien (1970).

[3]Remnants of this ancient theory can even be found in the writings of René Descartes (1638, 1644), who postulated a visible object will provide a thrust in the air that is essential for seeing the object.

it touches or what touches it over the whole body, until they reach the soul, causing that perception which we call sight. (1952 translation, 454)

Timaeus went on to say that vision ceases at night because the fire from the eye now faces a dissimilar element in space. He added that closing one's eyes causes sleep and dreaming by preventing the internal fire of the eyes from leaving. Timaeus also spoke about how different-sized particles in motion produced different color sensations by interacting with the fire and water of the eye.

Plato's theory of vision, especially the hypothesis of two emanations described by Timaeus, is thought of as being highly original by those historians who believe that Empedocles did not previously postulate the actual union of rays from the object with emanations from the eye (O'Brien, 1970).

A somewhat different approach was taken by Epicurus (c. 341–270 B.C.), who hypothesized that particles or films given off by the visible object entered the eye:

> For particles are continually streaming off from the surface of bodies, though no diminution of the bodies is observed, because other particles take their place. . . . We must also consider that it is by the entrance of something coming from external objects that we see their shapes and think of them. For external things would not stamp on us their own nature of colour and form through the medium of the air which is between them and us, or by means of rays of light or currents of any sort going from us to them, so well as by the entrance into our eyes or minds, to whichever their size is suitable, of films coming from the things themselves. (From Letter to Herodotus; see Diogenes Laertius, 1925, Vol. 2, 577–579)

Aristotle stressed the importance of an invisible, stationary, transparent medium for transmitting the visual message, but in general did not present ideas that represented a radical break from the metaphysical visual theories of the past. His guiding theoretical notion was the Ionian doctrine, "like is only known by like." In *De sensu*, Aristotle wrote the following about internal and external fires in the context of this doctrine:

> Now, as vision would be impossible without light (between the object and the eye), so also would it be impossible if there were no light inside (the eye). There must, therefore, be some translucent medium within the eye, and, as this is not air, it must be water. . . . It is a matter of experience that soldiers wounded in battle by a sword slash on the temple, so inflicted as to sever the passages (inward from) the eye, feel a sudden onset of darkness, as if a lamp had gone out; because what is called the pupil, i.e., the translucent, which is a sort of lamp for the interior of the eye, is then cut off (from its connection with the soul). (1908a translation, 438b2–16)

Aristotle, like other Greek natural philosophers, also showed an interest in visual illusions. In *De somniis (On Dreams)*, he described both positive and negative afterimages.

> When we have looked steadily for a long while at one colour, e.g., white or green, that to which we next transfer our gaze appears to be of the same colour. Again if, after having looked at the sun or some other brilliant object, we close the eyes, then, if we watch carefully, it appears . . .

first in its own colour; then it changes to crimson, next to purple, until it becomes black and disappears. (1908b translation, 459)

ROMAN THEORIES OF VISION

Early Roman theories of vision did not represent marked advances over those of the Greeks, primarily because early Roman medicine was largely performed by visiting Greeks (see Chapter 1). Indeed, it was considered degrading for Roman citizens to even practice medicine at this time (Chance, 1934).

For example, Lucretius (c. 98–55 B.C.) accepted Epicurus' notion that films or images *(simulacra)* leaving visible objects entered the eye to produce sensations. As stated in his *De rerum natura:*

> Amongst visible things many throw off bodies, sometimes loosely diffused abroad, as wood throws off smoke and fire heat; sometimes more close-knit and condensed, as often when cicalas (i.e., cicadas) drop their thin coats in summer, and when calves at birth throw off the caul from their outermost surface, and also when the slippery serpent casts off his vesture among the thorns. (1937 translation, 251–253)

The anatomical contributions of Herophilus were also accepted by Aurelius Cornelius Celsus. In his classic *De medicina,* from the first century, Celsus described a number of visual disorders, including ophthalmoplegia, night blindness, and cataracts. He wrote that night blindness could be cured by bleeding and by eating roasted goat liver. He accepted the Alexandrian notion that cataracts were due to the coagulation of the humor between the iris and the lens and described how to push the thickened humor out of the line of vision. Celsus introduced the reclining method for the couching operation and suggested other variations to increase the ease of the surgery. He wrote:

> Now either from disease or from a blow, a humour forms underneath the two tunics in what I have stated to be an empty space; and this as it gradually hardens is an obstacle to the visual power within. . . . Then he is to be seated opposite the surgeon in a light room, facing the light, while the surgeon sits on a slightly higher seat; the assistant from behind holds the head so that the patient does not move. . . . Thereupon a needle is to be taken pointed enough to penetrate, yet not too fine; and this is to be inserted straight through the two outer tunics . . . in such a way that no vein is wounded. . . . When the spot is reached, the needle is to be sloped against the suffusion itself and should gently rotate there and little by little guide it below the region of the pupil; when the cataract has passed below the pupil it is pressed upon more firmly that it may settle below. . . . After this the needle is drawn straight out; and soft wool soaked in white of egg is to be put on, and above this something to check inflammation; and then bandages. (1935–1938 translation, Vol. 3, 349–353)

Rufus of Ephesus practiced medicine in Rome a century later. The quotation at the opening of this chapter explaining the names of many parts of the eye came from Rufus' treatise *On the Names of the Parts of the Body*. This work contained a compilation of material from older Greek sources along with one of the first systematic descriptions of the nervous system. This included a description of the optic nerves ("optic channels"). Rufus wrote that these nerves emerged from the base of the brain and divided into two branches that entered the eye obliquely.

Galen dismissed the idea that an object sends a part of itself to the eye. He thought such a theory could not be true because it would not allow a person to appreciate the real size of an object. A mountain, for example, would have to shrink dramatically to enter the pupil.

Instead, Galen proposed that the ventricles released pneuma *(spiritus animalis)* through the optic nerves to the eyes. The pneuma supposedly emerged and altered the surrounding air, which was illuminated by sunlight, to endow that air with special powers. The pneuma in this schema extended only far enough from the eye to transform the air that then made contact with the object. The transformed air conveying the qualities of the image (e.g., color, brightness, shape) then returned along the same path back to the eye (see Lindberg, 1976).

Good descriptions of many parts of the eye were provided in Galen's work, *On the Eyes and Their Accessory Organs,* which was a part of his *De usu partium.* His descriptions included the conjunctiva, cornea, iris, crystalline lens, choroid, sclera, and the aqueous and vitreous humors (Daremberg, 1854). He also described a number of diseases of the eye, including glaucoma, which he considered the most common cause of blindness.[4]

Galen accepted Herophilus' comparison of the retina with a fishing net. This comparison probably came about because the retina was found to collapse around the lens, like a net around a fish, when it was dissected carelessly. The Ancient Greek word for "retina" was *amphiblestroëidés,* which means "netlike."[5]

Among other things, Galen distinguished between hard nerves, which he thought were motor, and soft nerves, which he assumed were sensory. He contended the eye was supplied with both varieties. He listed the optic nerves as the first of the cranial pairs and contended these were the only nerves with "a clear perceptible pore, whence some anatomists have called them canals, not nerves." He assumed that this feature, previously mentioned by Alcmaeon, allowed the *pneuma* (spirits) to communicate between higher centers and the eye.

Galen maintained that spirits traveling down the optic nerve emerged at the retinal tunic and then made their way to the centrally located crystalline lens. It was here, in the lens and in the watery anterior chamber of the eye, that the visual spirits met the "outer light" entering the eye. His belief was that the spirits then returned along the same route with the newly acquired visual information.

In Galen's mind, the crystalline lens was the "receptive organ" of the eye, while the retina was an organ of nutrition (along with the choroid because both had numerous blood vessels) and a conveyor of visual spirits to and from the lens. Galen emphasized the similarities between the retina and the brain and maintained that the retina and optic nerves were really displaced parts of the brain.[6] This

anatomical unity fit well with the prevailing belief of spirits traveling easily from the brain to the optic nerve and then to the retina.

Galen believed that the optic nerves only seemed to cross in the chiasma, a structure which was given its name because its X-like shape resembled the Greek letter chi. Instead, he contended that each optic nerve actually remained on its own side but that the two nerves communicated at the chiasma. The convergence of the two channels allowed visual spirits associated with each eye to interact so that a single binocular image would be perceived (Siegel, 1970). This joining, Galen said, was suggested to him by a god during a dream. As for binocular vision, this had the advantage of allowing the viewer to see a larger area than would be possible with one eye alone. He added that if one eye were lost, all the pneuma would go to the other eye to compensate for the loss.

Exactly where did the optic nerves terminate centrally? Galen suggested that they went to the sides of the lateral ventricles. He called these cavities *thalami,* a term that translates as bedchamber, and one that probably came from the Egyptian word *thalam,* meaning the inner sanctum of a temple. He wrote that the place of termination was unknown to earlier anatomists.

The doctrine of the spirits, which dominated thinking at the time of Galen, was highly complex and often contradictory (Temkin, 1951–52; also see Chapter 1). Galen called those animal spirits that traveled through the hollow optic nerves "visual spirits."

ARABIC CULTURES

In the West, advances were slow in coming from the height of the Roman Empire until the fourteenth century, when the Renaissance began in Italy. Even animal dissections were discontinued during the Dark Ages, both in Europe and the Middle East. Still, in contrast to the West, the ancient treatises of Hippocrates, Aristotle, Galen, and others were not only collected and preserved but also venerated in Arabic-speaking countries.

Middle Eastern scholars used these texts as the essential works in their medical schools. Thus, while the study of intellectual subjects declined in Europe, physicians in places like Iraq and Egypt were able to maintain strong classical traditions (Hirschberg, 1905). In fact, by the end of the ninth century, Middle Eastern scholars were in possession of almost all of the classical manuscripts that were of interest to them (Meyerhof, 1945). Places of learning, such as Baghdad's *Bait al-Hikma (School of Wisdom),* housed large collections of the classics, and the business of translating manuscripts from the original Greek flourished.

The earliest known Middle Eastern text on the eye was written by Hunain ibn Is-Hâq (809–873), a distinguished Nestorian Christian.[7] Also known as Johannitius in the West, Hunain ibn Is-Hâq served as a court physician and directed the *Bait al-Hikma* library in Baghdad. He was a prolific translator of Greek, whose skills in understanding

---

[4]See Laughlin (1934) for a history of glaucoma.

[5]The Greek word *amphiblestroëidés* was translated into Arabic by Avicenna in the eleventh century. It was from this source that the Latin word "retina" was derived by Gerard of Cremona (c. 1114–1187), a scholar well known for translating Arabic works into Latin during the Middle Ages (see Polyak, 1941).

[6]Polyak (1957) stated that this idea originated with Aristotle.

[7]Nestorius (d. 451) and his "heretic" followers who formed the Church of the East accepted Christ as a God-inspired man rather than as a God-made man.

the spirit of different languages far exceeded those of earlier translators. In addition to translating from Greek into Arabic for his Muslim patrons, Johannitius also translated the classics from Greek into Syriac for other prominent Nestorians (Meyerhof, 1928b).

Johannitius not only translated but also wrote more than 100 books. His most famous book on the eye was *The Book of Hunain ibn Is-Hâq on the Structures of the Eye, Its Diseases, and Their Treatment, According to the Teachings of Hippocrates and Galen, in Ten Treatises*. The first nine parts of this book came from treatises written over a 30-year period. The tenth treatise came later.[8]

The visual pathways and concepts in the *Ten Treatises* were essentially those spelled out by Galen (see English translation by Meyerhof, 1928a). Thus, in the hands of Johannitius, the crystalline lens was still the "photoreceptive" organ, and the pneuma still carried impressions to the "posterior sides of the anterior ventricles of the brain" via the hollow optic nerves, which were briefly united at the chiasma. Johannitius also accepted Galen's idea that air activated by the visual pneuma of the eye in daylight ("it strikes it as it were with the shock of a collision") was essential for vision.

In addition to sections on Galenic anatomy and physiology, the *Ten Treatises* contains sections on the diagnosis, causes, and treatment of various eye disorders, ranging from itch and ulcers to tumors. The recommended treatments involved plant, mineral, and animal remedies and showed the use of an extensive pharmacopeia. Surgical treatments, including couching for cataracts, were also mentioned.

The *Ten Treatises* contains the earliest known diagrams of the anatomy of the eye. Like other diagrams from Islamic countries, these drawings were more schematic than realistic in character. This style was adopted because Islamic law prohibited the representation of the human form or its parts. One early drawing appears as Figure 5.1.

Another important work of this period was a short book written by an Egyptian known as Ammar (c. 1000). It was entitled *Choice of Eye Diseases* and included his description of a new operation for removing cataracts. Ammar wanted to avoid the problem of cataracts migrating back into the field of vision, a problem often encountered with traditional couching. To reduce the possibility of cataract migration, surgeons often required their patients to lie on their backs for seven days. But not all patients were willing to do this. Ammar thus proposed extracting the cataract:

> Then I constructed the hollow needle, but I did not operate with it on anybody at all, before I came to Tiberias. There came a man for an operation who told me: Do as you like with me, only I cannot lie on my back. Then I operated on him with the hollow needle and extracted the cataract; and he saw immediately and did not need to lie, but slept as he liked. Only I bandaged his eye for seven days. With this needle nobody preceded me. I have done many operations with it in Egypt. (Translated by Hirschberg, 1905, 1128–1129)

Early in the eleventh century, Alhazen wrote the best-known Arabic treatise on physiological optics (Lindberg,

Figure 5.1. Schematic of the eye from Hunain ibn Is-Hâq. This picture, thought to be the oldest known detailed diagram of the eye, came from an Arabic manuscript dated 1197, which was a copy of a treatise from 1003. The latter was based on Hunain ibn Is-Hâq's original work, written around 860 (see Meyerhof, 1928b).

1967). Although he was born in Basra, Iraq, he spent most of his scientific life in Cairo.[9]

The works of two great mathematicians, Euclid (fl. 300 B.C.) and Claudius Ptolemaeus (fl. second century), formed the basis for Alhazen's contributions to the science of vision. But unlike those who came before him, Alhazen might have been the first to understand the principles underlying the formation of inverted and reversed images, such as those seen on the wall behind the aperture of a *camera obscura* (pinhole camera). As stated in a translation from a Latin edition of his *Perspectiva*:

> The evidence that lights and colors are not intermingled in air or in transparent bodies is that when a number of candles are in one place, (although) in various and distinct positions, and all are opposite an aperture that passes through to a dark place and in the dark place opposite the aperture is a wall or an opaque body, the lights of those candles appear on the (opaque) body or the wall distinctly according to the number of candles; and each of them appears opposite one candle along a (straight) line passing through the aperture. If one candle is covered, only the light opposite (that) one candle is extinguished; and if the cover is removed, the light returns. . . . Therefore, lights are not intermingled in air, but each of them is extended along straight lines. (Risner, 1572, Bk. 1, 17; translated in Lindberg, 1968, 154)

Alhazen did not believe rays were emitted from the eye, but he did accept the idea that rays from the object seen entered the eye. Specifically, in his *Kitab al-manazir* he proposed that light and color radiated forth in straight lines in every direction from each point of an object. To prevent confusion, he also postulated that each part of the center of the eye (the glacial humor) received a converging perpendicular ray from only one part of the object. These perpendicular rays were thought to be stronger than the rays of light striking at oblique lines (e.g., a sword penetrates a surface more easily when it strikes perpendicularly). Thus, Alhazen gave visual theory a mathematical basis while discarding the old

---

[8]In addition, an eleventh treatise written by Johannitius was annexed to the 10 in one of the few known Arabic copies.

[9]Alhazen was summoned to Cairo by Fatimid Khalif, al-Hakim (996–1021), a powerful ruler who thought Alhazen knew how to regulate the flow of the Nile.

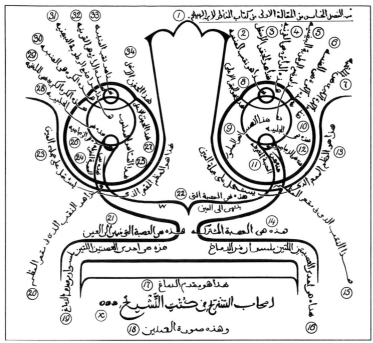

Figure 5.2. Schematic of the visual system from *Kitab al-manazir* by Alhazen, dated 1083. (The numbering is contemporary; see Polyak, 1941. Courtesy of the University of Chicago Press.)

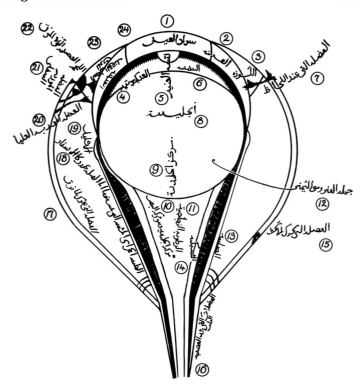

Figure 5.3. Diagram of the eye from *Tanqih al-manazir,* by Kamal al-Din Abu'l-Hasan al-Farisi, which appeared in 1316. This figure may have been based on an earlier diagram by Alhazen. (The numbering is contemporary; see Polyak, 1941. Courtesy of the University of Chicago Press.)

extramission theory of sight. One of his diagrams can be seen in Figure 5.2 (see also Figure 5.3).

Avicenna, who was perhaps the most influential of the Islamic natural philosophers, considered sight as one of his five "external senses," along with touch, taste, hearing, and smell. Avicenna showed strong Aristotelian sympathies. The following thoughts, which could have been written more than 1,000 years earlier, were expressed in his *The Canon of Medicine,* which first appeared in 1025.

> Sight is a faculty located in the optic nerve which discerns images projected on the crystalline humor, whether of figures or solid bodies, through the medium of a translucent substance which extends from it to the surface of reflecting bodies. (1930 translation, 135)

From this passage it can be seen that the doctrines of the classical authorities were held in such high esteem in the Middle East that both their advances and their errors tended to be accepted. In particular, the following ancient ideas did not change significantly during the Middle Ages: (1) that the optic nerves were hollow, (2) that the lens was located in the middle of the eye, (3) that rays emanated from objects, and (4) that visual spirits or pneuma conveyed messages centrally.

One of the few real advances was the emergence of the belief that perhaps the optic nerves did cross. This was entertained by Avicenna and by Rhazes (Abu Bakr Muhammad ibn Zakariyya al-Razi; c. 865–925). The latter is thought to have written a number of books on optics, although not one has been found (see Lindberg, 1976).

Perhaps the most controversial question revolving around Arabic visual science in the Middle Ages is whether Averroes (Abu-l-Walid Muhammad ibn Rushd; 1126–1198) really suggested that the retina, and not the lens, played

the critical role in photoreception. Averroes lived in Cordoba, Spain. The idea that he argued for a receptive role for the retina was suggested by the Austrian ophthalmologist Vincenz Fukala (1847–1911) in 1901. Fukala cited passages from Averroes' major medical book, *Colliget,* to make the case that Averroes recognized retinal sensitivity (see also Polyak, 1941, 1957). Others who have examined this work, however, have argued that despite some "superficial ambiguity" in the passages, there are only a few hints of a radical departure from the traditional ideas of Galen (Eastwood, 1969; Lindberg, 1975).

A stronger case for Averroes' departure from the past has been made by scholars who have referred to passages in his *Epitome of parva naturalia,* a more philosophical work not cited by Fukala. Averroes wrote:

> The innermost of the coats of the eye (i.e., the retina) must necessarily receive the light from the humors of the eye, just as the humors receive the light from the air. However, inasmuch as the perceptive faculty resides in the region of this coat of the eye, in the part which is connected with the cranium and not in the part facing the air, these coats, that is to say, the curtains of the eye, therefore protect the faculty of the sense by virtue of the fact that they are situated in the middle between the faculty and the air. (1961 translation, 6)

It has been claimed that in passages like this one Averroes was maintaining that (1) forms change from a corporeal transmission to a spiritual one in the retina and (2) the lens, which receives the forms of visual things, is not sensitive and cannot be the organ of perception. Some have argued that even if his wording does not signify a modern understanding of retinal function, he at least came close to suggesting a revised view of how the eye may work (e.g., Koebling, 1972).

Nevertheless, the case has also been made that in order to appreciate the meaning of the critical passages in the *Epitome,* one must look at other statements made in this work and in the *Colliget,* where nearly identical passages can be found. Such an analysis has led at least one historian to conclude that Averroes never intended to argue that the retina was the primary sensitive structure for sight (Lindberg, 1976). If this is true, the views presented by Averroes are very much like those of Galen, who wrote the following about the retina in Book 10 of *De usu partium:*

> Its principal and greatest usefulness, that for the sake of which it was brought down from above, is to perceive the alterations of the crystalline humor and in addition to convey and transmit nutriment to the vitreous humor. (1968 translation, 463)

But when Galen continued, he made it perfectly clear that it was the lens which was the principle organ for vision:

> The crystalline humor itself is the principle instrument of vision, a fact clearly proved by what physicians call cataracts, which lie between the crystalline humor and the cornea and interfere with vision until they are couched.

RENAISSANCE OPTICS AND PHYSIOLOGY

The Arabic writings, which closely reflected the thoughts of Greek and Roman naturalists, became familiar to the Europeans at the end of the Middle Ages. Specifically, the Europeans became acquainted with the Arabic (and Sephardic Jewish) material as they drove south to conquer Moorish Spain. The invaders won Toledo in 1085, Cordoba in 1236, and Seville in 1248. The "new" material soon spread from Spain to France and Italy and afterward to other parts of Europe.

It is in this context that one finds *The Book of Hunain ibn Is-Hâq on the Structures of the Eye* translated into Latin by Constantinus Africanus, a Tunisian who converted to Christianity. Constantine was associated with the medical school at Salerno in the eleventh century and translated many Arabic works into Latin. He presented *The Book of Hunain ibn Is-Hâq on the Structures of the Eye* as *Constantini Africani liber de oculis.* The indication was that it, as well as several other treatises from the Middle East, were his own (Meyerhof, 1928b). The important point is not so much the plagiarism but the fact that Westerners were essentially treated to classic Greco-Roman science when they discovered these works.

The earliest Western books on optics were mostly compilations and syntheses of existing knowledge about vision, based largely on Alhazen's works. Alhazen's *Kitab al-manazir* was translated by an unknown person into *De aspectibus* late in the twelfth century. (Lindberg, 1967, 1976). This work, written in Latin, had a major impact on the writings of the Pole Witelo (c. 1220–1275) and on two Englishmen, John Pecham (d. 1292) and Roger Bacon (c. 1220–1292). Bacon's optical works, dispatched to the papal court around 1268, came first. It is very likely that they were read by both Witelo, who may not have known Bacon personally, and by Pecham, who knew

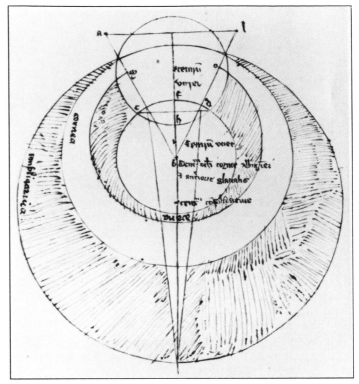

Figure 5.4. Diagram of the eye from a fifteenth-century manuscript based on Roger Bacon's thirteenth-century *Perspectiva.*

Bacon at Oxford. It is also possible that Pecham became acquainted with Witelo's *Perspectiva* when he was writing his own *Perspectiva communis* in the 1270s (Lindberg, 1971). These three authors continued to present Arabic (Greco-Roman) optical ideas in a largely unadulterated form. Subsequent generations of writers followed their lead.

Even the eye diagrams that appeared in the Western books, such as those shown in Figures 5.4, 5.5, and 5.6, were nothing more than adaptations of those found in Arabic texts. The optic nerves continued to be treated as hollow tubes, and only the question of the crossing or merging of the optic nerves caused disagreement. Dissections still were not attempted to verify assertions or to clear up these confusions.

A renewed interest in human and animal dissection and experimentation emerged in the fifteenth century with Leonardo da Vinci, and was especially stimulated by the demonstrations and writings of Andreas Vesalius in the mid-sixteenth century (see Chapter 1). Vesalius (1543) broke with all of his predecessors by not depicting the optic nerves as hollow in Book VII of his *De humani corporis fabrica.* His break with Galenic tradition was based on the fact that he simply could not observe a hole in the middle of the optic nerves in the animal specimens he had examined or in the optic nerves of humans who had just been executed. One of his woodcuts appears as Figure 5.7.

Skeptical of most previous teachings, Vesalius also considered the receptive role of the lens as questionable at best and hinted that the retina was the most important part of the eye for photoreception. Still, Vesalius could not make a complete break from the past. He continued to place the lens in the center of the eye and did not show the eye chambers in correct proportion. In addition, he still

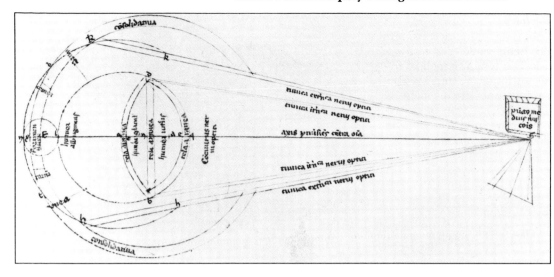

Figure 5.5. Fourteenth-century diagram of the eye from Vitello's thirteenth-century *Perspectivam.*

agreed with Galen that the crossing in the chiasma was illusory and that each optic nerve went from the eye to the same side of the brain. Vesalius, however, did not mention the place of termination.

Perhaps the first Renaissance scientist to recognize correctly that the optic nerves terminated in the posterior thalamus, and not in the lateral ventricles, was Bartolomeo Eustachio (Eustachius, 1524–1574). Although the plates for his *Tabulae anatomicae* were made in 1552, they were not published until 1714, leaving his work relatively unknown for over 150 years. Constanzo Varolio (Varolius, 1543–1575), who also recognized the thalamic termination of the optic nerves, published his plates in 1573 and is often given the credit for correcting this Galenic error.

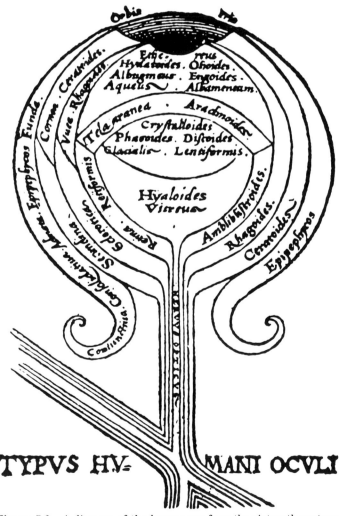

Figure 5.6. A diagram of the human eye from the sixteenth century showing the optic nerves as hollow channels (from Ryff, 1541.)

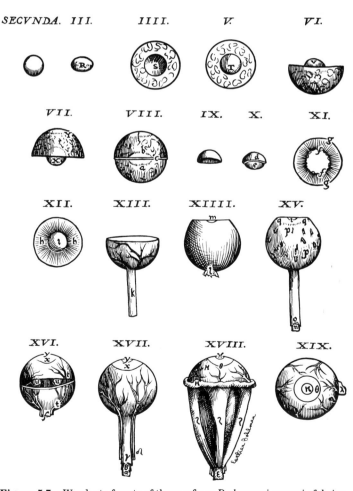

Figure 5.7. Woodcut of parts of the eye from *De humani corporis fabrica* by Andreas Vesalius (1543).

In 1583, Felix Platter (1536–1614), an anatomist at Basel who cited Varolio's contribution, made a clear and strong statement about how the lens served only to focus the rays of light converging on it. The lens was called *perspicillum nervi visorii,* which meant "the looking through of the optic nerve." This was a significant turning point in visual science. Thus, although Platter did not really formulate a theory of the retinal image or recognize the inverted nature of the image on the retina, his work focused new attention on the possible role of the retina in photosensitivity (Lindberg, 1976).

Platter also differed from others before him by testing his ideas experimentally. For example, he cut the suspensory ligament of the lens, by which pneuma were supposed to pass from the lens to the retina. When he found vision could still take place, he concluded that he had proved that the time-honored lens theory of photoreception must be incorrect. Platter also recognized that the lens was not located in the center of the eye. In accordance with this observation, he moved it to its accepted anterior location.

The idea that the lens merely focused the light on the retina had posed a problem for Leonardo da Vinci, who followed Alhazen in comparing the eye to a camera obscura.[10] Leonardo was disturbed by the fact that the image seemed to be inverted. It was inconceivable to him that the image carried by the optic nerve could be anything but upright. One solution he entertained was that there was an inversion in front of the lens and a second inversion within the lens. Another of his solutions had the inverted image from the back of the eye reflected back upon the lens in a way that made it erect once again.[11] Leonardo's numerous diagrams and notes showed, however, that he was never really satisfied with these ideas.

Leonardo's interest in vision was largely that of an artist interested in perspective. Perhaps for this reason, his drawings still showed the eye sending messages to the ventricles. In his earlier work, the target was the anterior of the three ventricles shown on his drawings. This location was consistent with the ancient idea that the first ventricle received all sensory impressions. After he injected wax into the ventricles of an ox, and after it became known that most sensory nerves arise near the middle ventricle, he showed the optic tract ending in the middle cavity, behind the lateral ventricles (see Chapter 1).

Johannes Kepler (1571–1630), shown in Figure 5.8, recognized that the aperture of the eye must produce the same "errors" of vision as would be seen in the closed chamber of the camera obscura. Operating in a medieval perspectivist tradition, Kepler (1604, 1611) thought the image was not "caught" by the lens of the eye but "painted" on the retina, the sensitive element. Thus, after considerable consternation, he reasoned that the retinal image had to be inverted and reversed (left-right). He called this optical image a *pictura.*

Figure 5.8. Johannes Kepler (1571–1630), the outstanding astronomer and mathematician, who was also a pioneer in understanding the nature of the ocular image on the retina. (From Mach, 1925.)

Kepler (1604) broke with tradition by suggesting that the problem of the retinal inversion did not belong to the realm of optics. He knew very little about the pathways of vision and simply reasoned that the inverted image probably was corrected by the "activity of the Soul":

I say that vision occurs when the image of the whole hemisphere of the world that is before the eye . . . is fixed on the reddish white concave surface of the retina. How the image or picture is composed by the visual spirits that reside in the retina and the nerve, and whether it is made to appear before the soul or the tribunal of the visual faculty by a spirit within the hollows of the brain, or whether the visual faculty, like a magistrate sent by the soul, goes forth from the administrative chamber of the brain into the optic nerve and the retina to meet this image, as though descending to a lower court—this I leave to be disputed by the physicists. For the armament of opticians does not take them beyond this first opaque wall encountered within the eye. (Translated in Lindberg, 1967, 203)

The reality of Kepler's inverted retinal image was demonstrated experimentally by Christopher Scheiner (1575–1650), a Jesuit friar. Using animals and then humans, he removed part of the opaque layer from the back of the eyeball, leaving only the semitransparent retina. Scheiner (1619) thus was able to see and describe the inverted retinal image.[12]

---

[10]See Lindberg (1968, 1970) for a history of the pinhole image.

[11]See Ferrero (1952) and Keele (1955) for reproductions of Leonardo's eye diagrams.

[12]The same experiment seemed to have been performed by Julius Caesar Arantius (1530–1589) in 1595, but Arantius failed to understand its meaning (see Polyak, 1957).

REFERENCES

Alhazen. *Kitab al-manazir* (see Risner, 1572).

Arantius, J. C. 1595. *De humano foetu . . .* Venetiis: Apud Bartholomeum Carampellum.

Aristotle. 1908a. *De sensu.* (J. E. Beare, Trans.). In W. D. Ross (Ed.), *The Oxford Translation of Aristotle.* Oxford, U.K.: Clarendon Press.

Aristotle. 1908b. *De somniis.* (J. E. Beare, Trans.). In W. D. Ross (Ed.), *The Oxford Translation of Aristotle.* Oxford, U.K.: Clarendon Press.

Averroes (Ibn Rush'd). 1560. *Colliget.* Venetiis: Apud Cominum de Tridino.

Averroes. 1961. *Epitome of parva naturalia.* (H. Blumberg, Trans.). Cambridge, MA: Medieval Academy of America.

Avicenna. 1930. *The Canon of Medicine.* (O. C. Gruner, Trans.). London: Luzac & Company. (Originally printed 1025)

Bacon, R. 1614. *Perspectiva.* (I. Combach, Ed.). Francofurti: Antonij Hommij.

Bidyadhar, N. K. 1939. Susruta and his ophthalmic operations. *Archives of Ophthalmology, 22,* 550–574.

Bidyadhar, N. K. 1940. Evolution of cataract surgery from the pre-historic age up to the present time. *Antiseptic, 37,* 339–346.

Bidyadhar, N. K. 1941. Principles of Susrutian ocular therapy interpreted in the light of modern ophthalmic science. *Archives of Ophthalmology, 25,* 582–628.

Celsus, A. C. 1935–1938. *De medicina* (3 vols). (W. G. Spencer, Ed.). Cambridge, MA: Loeb Classical Library.

Chance, B. 1934. A view into the ophthalmology of Galen. *American Journal of Ophthalmology, 17,* 718–721.

Cohen, M. R., and Drabkin, I. E., 1948. *A Sourcebook in Greek Science.* New York: McGraw-Hill.

Daremberg, C. 1854. *Oeuvres Anatomiques, Physiologiques et Médicales de Galien.* Paris: J. B. Ballière.

Descartes, R. 1638. La Dioptrique. In R. Descartes, *Discours de la Méthode.* Leyde: J. Maire.

Descartes, R. 1644. *Principia philosophiae.* Amstelodami: Ludovicum Elzevirium.

Diogenes Laertius. 1925. *Lives of Eminent Philosophers.* (R. D. Hicks, Trans.). London: Loeb Classical Library.

Dutt, K. C. 1938. Cataract operations in the prehistoric age. *Archives of Ophthalmology, 20,* 1–15.

Eastwood, B. S. 1969. Averroes' view of the retina—A reappraisal. *Journal of the History of Medicine and Allied Sciences, 24,* 77–82.

Eustachi, B. 1714. *Tabulae anatomicae . . .* Roma: F. Gonzagae.

Feibel, R. M. 1983. John Vetch and the Egyptian ophthalmia. *Survey of Ophthalmology, 28,* 128–134.

Ferrero, N. 1952. Leonardo da Vinci: Of the eye. *American Journal of Ophthalmology, 35,* 507–521.

Fukula, V. 1901. Historischer Beitrag zur Augenheilkunde. *Archiv für Augenheilkunde, 42,* 203–208.

Galen. 1968. *De usu partium.* Translated by M. T. May as, *On the Usefulness of the Parts of the Body.* Ithaca, NY: Cornell University Press.

Gordon, B. L. 1939. Oculists and occultists. Demonology and the eye. *Archives of Ophthalmology, 22,* 25–65.

Hirschberg, J. 1905. Arabian ophthalmology. *Journal of the American Medical Association, 45,* 1127–1131.

Keele, K. D. 1955. Leonardo da Vinci on vision. *Proceedings of the Royal Society of Medicine, 48,* 384–390.

Kepler, J. 1604. *Ad vitellionem paralipomena . . .* Francofurti: C. Marnium & Haeredes.

Kepler, J. 1611. *Dioptrice.* Augsburg: Augustae Vindelicorum, Typis D. Franci.

Koebling, H. M. 1972. Averroes' concepts of ocular function—Another view. *Journal of the History of Medicine and Allied Sciences, 27,* 207–213.

Krause, A. C. 1934. Assyro-Babylonian ophthalmology. *Annals of Medical History, 6,* 42–55.

Laughlin, R. C. 1934. Glaucoma. *Bulletin of the Institute of the History of Medicine: The Johns Hopkins University, 2,* 141–163.

Lindberg, D. C. 1967. Alhazen's theory of vision and its reception in the West. *Isis, 58,* 321–341.

Lindberg, D. C. 1968. The theory of pinhole images from antiquity to the thirteenth century. *Archive for the History of Exact Sciences, 5,* 154–176.

Lindberg, D. C. 1970. The theory of pinhole images in the fourteenth century. *Archive for the History of Exact Sciences, 6,* 299–325.

Lindberg, D. C. 1971. Lines of influence in thirteenth-century optics: Bacon, Witelo, and Pecham. *Speculum, 46,* 66–83.

Lindberg, D. C. 1975. Did Averroes discover retinal sensitivity? *Bulletin of the History of Medicine, 49,* 273–278.

Lindberg, D. C. 1976. *Theories of Vision from Al-Kindi to Kepler.* Chicago: University of Chicago Press.

Lucretius Carus, T. 1937. *De rerum natura.* (W. H. D. Rouse, Trans.). London: Loeb Classical Library.

Mach, E. 1925. *The Principles of Physical Optics.* London: Methuen.

Meyerhof, M. 1928a. New light on Hunain ibn Ishâq and his period. *Isis, 8,* 685–724.

Meyerhof, M. 1928b. *The Book of the Ten Treatises on the Eye, Ascribed to Hunain ibn Is-Hâq.* Cairo: Government Press.

Meyerhof, M. 1945. Sultan Saladin's physician on the transmission of Greek medicine to the Arabs. *Bulletin of the History of Medicine, 18,* 169–178.

O'Brien, D. 1970. The effect of a simile: Empedocles' theories of seeing and breathing. *Journal of Hellenistic Studies, 90,* 140–179.

Plato. 1952. *Timaeus.* (B. Jowett, Trans.). In *Great Books of the Western World: Vol. 7. Plato.* Chicago: Encyclopaedia Britannica. Pp. 442–477.

Platter, F. 1603. *Felicis Plateri Archiatri et Profess. Basil. De corporis humani structura et usu libri . . .* Basileae: Apud Ludovicum König. (First ed., 1583)

Polyak, S. L. 1941. *The Retina.* Chicago: University of Chicago Press.

Polyak, S. L. 1957. *The Vertebrate Visual System.* Chicago: University of Chicago Press.

Risner, F. 1572. *Opticae thesaurus Alhazeni Arabis libri septem, nunc primum editi a Federico Risnero.* Basel.

Rufus of Ephesus. 1948. On the names of the parts of the body. In M. R. Cohen and I. E. Drabkin (Eds.), *A Sourcebook in Greek Science.* New York: McGraw-Hill. P. 476.

Ryff, W. H. 1541. *Das aller fürtrefflichsten geschöpfes . . . wahrhaftige Beschreibung der Anatomie.* Strassburg: Balthasar Beck.

Scheiner, C. 1619. *Oculus hoc est: fundamentum opticum.* Innsbrück: Agricola.

Siegel, R. E. 1959. Theories of vision and color perception of Empedocles and Democritus: Some similarities to the modern approach. *Bulletin of the History of Medicine, 33,* 145–159.

Siegel, R. E. 1970. Principles and contradictions of Galen's doctrine of vision. *Zeitschrift für Wissenschaftsgeschichte, 54,* 261–276.

Singer, C. 1952. *Vesalius on the Human Brain.* London: Oxford University Press.

Temkin, O. 1951–1952. On Galen's pneumatology. *Gesnerus, 8,* 180–189.

Theophrastus. 1917. On the senses. In G. M. Stratton (Ed.), *Theophrastus and Greek Physiological Psychology before Aristotle.* New York: Macmillan.

Thompson, R. C. 1923. *Assyrian Medical Texts.* Milford, U.K.: Oxford University Press.

Thompson, R. C. 1924. Assyrian medical texts. *Proceedings of the Royal Society of Medicine, 17,* 1–34.

Thorwald, J. 1963. *Science and the Secrets of Early Medicine.* New York: Harcourt, Brace and World.

Varolius, C. 1573. *De nervis opticis . . .* Patavij: Apud Paulum & Antonium Meiettos Fratres.

Vesalius, A. 1543. *De humani corporis fabrica.* Libri septem. Basilae: J. Oporini.

Vitello (Witelo). Ca. 1270–1278. *Vitellionis mathematici doctissimi . . . Perspectivam vocant . . .* Norimbergae: Apud Io, Petreium.

Chapter 6
Post-Renaissance Visual Anatomy and Physiology

If one has extirpated in a dog both sides of the cortex in the place A1 . . . only in the optic sense a curious anomaly presents itself. Completely free and unhindered, the dog moves around in the room and in the garden without ever bumping into an object . . . (But) the dog has become *Seelenblind,* i.e., he has lost the optical pictures which he possessed . . . so that he does not know or recognize anything he sees.

Hermann Munk, 1881

THE DISCOVERY OF THE BLIND SPOT

In the seventeenth century, l'Abbé Edme Mariotte (1620–1684) discovered the blind spot in the eye through careful study. He determined where the optic nerve entered the eye and set about to test the sensitivity of this point. Mariotte's discovery of the blind spot was communicated in an "Epistle" to a M. Pecquet, which was then sent to the Royal Society in London. Mariotte (1668, p. 668) wrote:

I fastn'd on an obscure Wall, about the hight of my Eye, a small round paper, to serve me for a fixt point of Vision; and I fastned such an other on the side thereof towards my right hand, at the distance of about 2. foot; but somewhat

lower than the first, to the end that it might strike the *Optick Nerve* of my Right Eye, whilst I kept my Left shut. Then I plac'd my self over against the First paper, and drew back little and little, keeping my Right Eye fixt, and very steddy upon the same; and being about 10. foot distant, the second paper totally disappear'd.

This discovery was deemed so important that it was demonstrated to the king. Nevertheless, Mariotte erroneously concluded that the choroid and not the retina was the sensitive layer of the eye.

Mariotte's phenomenon was easy to confirm, but his interpretation was subject to criticism. In a response published in the same issue of the *Philosophical Transactions of the Royal Society,* the case was made for the importance of the retina, and it was suggested that the presence of thick blood vessels at the point of entry of the optic nerve was the reason for the absence of vision in one part of the eye. Mariotte refused to budge from his original position, however, and responded negatively to the retinal hypothesis in a follow-up communication published two years later.

LEEUWENHOEK'S MICROSCOPY OF THE RETINA

One of the best-known scientific figures of the seventeenth century was Antony van Leeuwenhoek (1632–1723), a linen draper from Delft.[1] Leeuwenhoek, who is shown in Figure 6.1, learned to grind his own lenses and developed the simple microscope.[2] In fact, he had hundreds of microscopes, each consisting of a small biconvex lens placed in a socket between two brass plates with a small hole below the lens. The object to be examined was set at a reasonable

Figure 6.1. Antony van Leeuwenhoek (1632–1723), the Dutch scientist who developed the simple microscope and used it to study the eye.

[1]Leeuwenhoek also had his name spelled "Leewenhook" (1674) and "Leewenhoeck" (1675, 1684) at the end of his letters to the Royal Society.

[2]Leeuwenhoek was not the first to build a microscope. This honor goes to Robert Hooke (1635–1703), an English physicist who constructed an instrument with a compound lens and an eyepiece. Hooke used his instrument to observe the cells in a slice of cork and described his findings in 1665. Leeuwenhoek was probably the first to build a microscope and to use it to describe the structure of animal tissues.

distance from the lens, and the focal length was adjusted by screws.

Leeuwenhoek began to use his newly developed microscope to study the eye early in his career. His first letter describing his findings to members of the Royal Society in London appeared in 1674. The Dutch scientist stated that he very much wanted to know if the optic nerves were hollow. This had been suggested by the Greeks and Romans and was accepted by his immediate predecessors as essential for vision. Indeed, it was through the hollow nerve that the spirits carried the visual information to the higher centers (see Chapter 5).

Leeuwenhoek worked on a variety of animals, but could not find the "hollowness" that Galen had seen on a "clear sun-shiny day." Leeuwenhoek wrote:

> I communicated these Observations to *Doctor Schravesande,* and shew'd him the Crystallin humor; and he mentioning, that some Anatomists affirm'd the *Optic* Nerve to be hol'ow, and that themselves had seen that hollowness, through which they would have the Animal spirits, that convey the visible species, represented in the eye, pass into the Brain; I thereupon concluded with my self, that, if there were such a cavity visible in that Nerve, that it might also be seen by me, especially since, if it be so, it must be pretty bigg, and the body of it pretty stiff, or else, the circumjacent parts would press together. And in order to this discovery, I sollicitously view'd three Optic Nerves of Cows; but I could find no hollowness in them; I only took notice, that they were made up of many filimentous particles, of a very soft substance, as if they only consisted of the corpuscles of the Brain joined together, the threads were so very soft and loose: They were composed and conjoined of globuls, and wound about again with particles consisting of other transparent globuls. (1674, 179–180)

Leeuwenhoek's (1674, 1675) observations led him to theorize that visible objects set the soft globules at the proximal end of the optic nerve into motion, "similar to the motion imparted to water by touching its surface" and that "like motions" were conveyed to other globules until they were communicated to the brain. In fact, Leeuwenhoek directly compared the eye and its nerve to a glass of water.

> I represent to my self a tall Beer-glass full of Water: This Glass I imagine to be one of the Filiments of the *Optick Nerve,* and the Water in the Glass to be the Globuls of which the Filiments of that Nerve are made up, and then, the Water in the Glass being touch'd on its Surface with the Finger, that to this Contact did resemble the Action of a visible Object upon the Eye, whereby the outermost Globuls of the Fibres in the Optick Nerve next to the Eye are touched. This Contact of the Water made by the Finger cannot be said to touch and move only on the Surface of the Water, but we must also grant, that all the Water in the Glass comes to suffer, and to be more pressed by it, than it was before the Finger touched the Water, and that also all the Parts of the Water are moved thereby. This Motion then of the Water, said to be made by the Contact of the Finger, I imagine to be like the Motion of a visible Object made upon the soft Globuls, that lie at the End of the Optick Nerve of the Eye, which outermost Globuls do communicate the like Motion to the other Globuls so as to convey it to the Brain. (1675, 379)

In retrospect, it is likely that Leeuwenhoek may have been the first to see the rods and cones of the retina. But whether his globules were fat cells, degeneration products, or optical artifacts has never been clear (Clarke and O'Malley, 1968). Unfortunately, Leeuwenhoek's work did not lead to an immediate flood of additional microscopic studies of the retina by other investigators. The *Zeitgeist* was not yet ripe for his observations to be treated as anything more than perhaps some interesting secrets of nature.

THE PROJECTIONS OF THE OPTIC NERVE IN THE SEVENTEENTH CENTURY

A number of seventeenth-century scientists attempted to follow the optic nerve into the brain. For example, Thomas Willis (1664) traced the optic nerves to the *optic thalami* (a term that included both the brainstem and striate bodies at this time). He believed this represented the highest level of the visual system, whose physiology he still linked with the production and distribution of visual spirits.

In 1684, Raymond Vieussens (1641–1716) proposed that the fibers of the optic nerve continued to the cerebral cortex. In contrast to this and to the neuroanatomy of Willis, René Descartes (1686) returned the optic nerve terminals to the ventricles. He also hypothesized that images from the two eyes were unified in the pineal gland, where the rational soul and the body act upon one another (see Chapter 2).

Descartes' metaphysical ideas about the visual system were entirely speculative and hopelessly inadequate. Yet this seventeenth-century philosopher may have been the first to propose that the optic nerve fibers, as well as their sources and points of termination, are organized topographically.

SOME EIGHTEENTH-CENTURY ADVANCES

Samuel Thomas von Sömmerring (1755–1830), shown in Figure 6.2, saw the macula lutea, the lemon-yellow part of the retina, in January 1791.[3] He drew it (see Figure 6.3) and called it the *limbus luteus.* He also observed the fovea centralis, a small round depression in the macula lutea. Because Sömmerring believed that this depression was a "puncture-like" hole, he gave it the name *foramen centrale.* He thought this hole was Mariotte's (1668) blind spot. This error would remain uncorrected until the 1830s.

The fovea and macula were first seen in the eye of a person who had drowned a few hours earlier. Sömmerring then observed these retinal structures in at least 50 other people. He even noted that the macula changed color with age. It was pale yellow in children, a rich yellow in adolescents, and pale yellow again in old age.

These observations remained essentially unknown for several years. Finally, in 1795, Sömmering submitted a paper with his "unexpected discoveries" to the Royal Society of Sciences. A summary of his material was published later that year, and figures appeared in 1801.[4]

---

[3]Sömmerring, an outstanding anatomist, was affectionately called Herr Neurologue.

[4]Sömmerring did not know that the macula lutea was also seen by Francesco Buzzi (1751–1805), an Italian scientist who only briefly mentioned his observations in 1782 (see Mazzolini, 1991).

Figure 6.2. Samuel Thomas von Sömmerring (1755–1830), who described the macula region of the retina. (From Kilian, 1828.)

The projections from the eye were also of interest to scientists at this time. This led to a better understanding of the crossing of the nerves from the eye at the optic chiasma and a shift from the Galenic notion that optic nerve fibers came together but stayed on the same side of the eye they served. The credit for this discovery is usually given to Sir Isaac Newton (1642–1727), who concluded that nerve fibers from the same (homonymous) sides of the two eyes

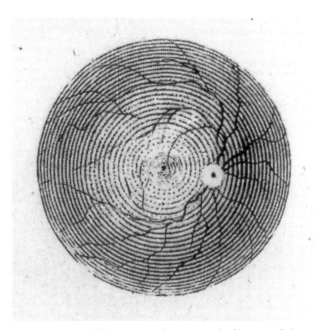

Figure 6.3. Samuel Thomas von Sömmerring's diagram of the retina showing the macula lutea that he had observed in 1791. (This figure formed one part of Plate 5 from Sömmerring, 1801.)

became unified in the optic chiasma. He also provided a drawing showing how the optic nerve fibers from the nasal side of each retina crossed at the chiasma, whereas those from the temporal side proceeded straight back (Newton, 1704).

Newton's remarkable insight was based on intuition, not on experimentation or dissection (see Sweeney, 1984). He realized a partial crossing of the optic nerves was the most logical explanation for unified binocular vision. In 1755, his proposed partial crossing of the optic fibers in the chiasma was confirmed with anatomical methods by Johann Gottfried Zinn (1727–1759).

At a more central level, the geniculate bodies were noted by Giovanni Santorini (1681–1737) in 1724. Santorini, an illustrator and careful investigator, used the term *corpus geniculatum laterale* to describe where the optic tract ended (Meyer, 1971). The geniculates were subsequently observed by a number of other eighteenth-century anatomists. It was not until 1822, however, that Karl Friedrich Burdach (1776–1847) referred to the lateral geniculate nucleus as the *corpus geniculatum externum* to distinguish it from the *corpus geniculatum internum,* or the medial geniculate body.

Some of the advances from the eighteenth century related to a better understanding of visual perceptual phenomena.[5] In his *Primae lineae physiologiae,* Albrecht von Haller wrote that perception could be affected by learning and experience and that understanding the image on the retina was just a beginning. Haller (1786) stated this as follows in the English edition of his book:

> But the mind not only receives a representation of the image of the object by the eye, impressed on the retina, and then transferred to the common sensory or seat of the soul; but she learns or adds many things from mere experience, which the eye itself does not really see, and other things the mind confers or interprets to be different from what they appear to her by the eye. (1966 translation, 29–30)

THE DISCOVERY OF THE STRIPE OF GENNARI

In 1776, a young medical student in Parma, Italy, was dissecting a frozen human cadaver when he discovered a broad white line running through the calcerine cortex in the occipital lobe (see Figure 6.4). As can be seen in his statement from 1782, Francesco Gennari (1750–1797) was unaware of the significance of this discovery or of the lamination of the cortex:

> This line I first saw the 2nd of February 1776 and having often followed it, it seemed to run finally to the ultimate substance of the hippocampus. For the rest I am ignorant to what purpose that substance may be formed, like so many others of which the use is still hidden. . . . I exhort the anatomists in every way to enquire into this substance. (Translated in Gibson, 1962, 946)

[5]A number of striking demonstrations pertaining to the sense of sight were made in the eighteenth century. In one case, the experimenter discharged a Leyden jar between the head and a leg of a young man who had been blind due to congenital cataracts. The electrical spark led to a phosphene that was perceived as a blaze of light (Le Ray, 1755, cited by Le Grand, 1975).

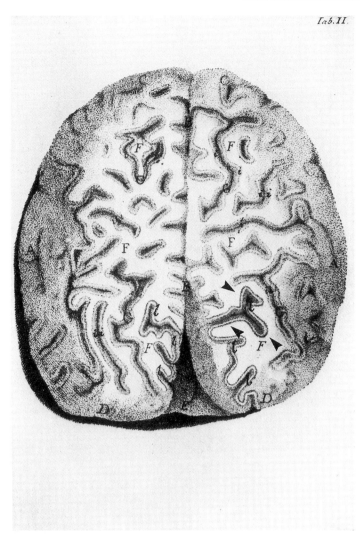

Tab. II

Figure 6.4. Francesco Gennari's (1792) figure of the human brain showing the *lineola albidior* (white line) in the mesial part of the occipital lobe.

In his book on cerebral structure, Gennari (1782) called this line *lineola albidior*.[6] The same conspicuous white line was described in 1786 by the French anatomist Félix Vicq d'Azyr (1748–1794), who was unaware of Gennari's observations.[7] In the final years of the eighteenth century, the striped line was also seen by Sömmerring and, in 1828, by Luigi Rolando.

In 1840, Jules-Gabriel-François Baillarger (1806–1891), a French psychiatrist interested in the morphological correlates of mental illness, recognized that a fainter white line that ran through the entire cerebral cortex was continuous with the bolder stripe of Gennari.[8] Nevertheless, neither Baillarger nor any of his contemporaries had even the

[6]Other than this book, Gennari apparently wrote just two other medical works; one was a case report and the other was a paper that seems to have been lost (Fulton, 1937).

[7]Vicq d'Azyr saw the line in 1781, if not earlier. This French anatomist should also be remembered for naming the central portion of the optic nerve the optic tract.

[8]Baillarger also showed that the cortex was made of layers of cells. He made his discoveries by pressing thin slices of human brain between two glass plates that he then held up to the light for inspection with the naked eye. Robert Remak published his important studies on the six cortical cell layers in 1844.

vaguest idea that the thicker band in the occipital lobe would serve as a marker for the visual cortex.

The thick white band that ran through the occipital cortex was finally given its name in 1888 when Heinrich Obersteiner (1847–1922) referred to the band as the "stripe of Gennari" to honor the Italian investigator who had discovered it 112 years earlier (McHenry, 1969, 1983). Later, it would be recognized that this stripe is sharper in humans and apes than in lower primates, that its boundaries are even less clear in ungulates and carnivora, and that its location varies with the development of the association areas (Smith, 1904a; also see below).

RODS, CONES, AND DUPLICITY THEORY

In 1838, when Johannes Müller (1801–1858) wrote about the functions of the retina, he did not know about the rods and the cones. He knew only that the retina was constructed of a layer of cylindrical "papillae" that looked like "rods" placed close together. This information was extracted from the work of Gottfried Reinhold Treviranus (1776–1837), a German comparative anatomist.

Treviranus (1828, 1835–38) corrected the error of earlier microscopists who saw only globules. His thin "cylinders," shown in Figure 6.5, were seen at 300X magnification in the retinas of birds, mammals, amphibians, and fish. He associated these structures and their terminal papillae with the optic nerve and with the reception of light. Nevertheless, Treviranus mistakenly believed these papillae faced into the vitreous chamber rather than away from it. In his mind, the optic nerve had to branch behind the receptors. He even drew the retina as he believed it should be rather than as it really was.

The belief that the rods faced the vitreous chamber was accepted by Johannes Müller (1837) and many others. The individual who first recognized the inverted nature of the retinal receptors was Friedrich Heinrich Bidder (1810–1894). In 1839, Bidder observed that the tips of the rods extended into the choroid layer. Nevertheless, he was perplexed by the reversed position of the receptors. Hence, he guessed that the rods acted like mirrors or reflectors which strengthened the image by passing the light twice through the optic nerve fibers.

A better understanding of the structure of the retina came after 1840, the year in which Adolph Hannover (1814–1894) began to use chromic acid to harden tissue, in order to slice thinner sections. Heinrich Müller (1820–1864) further improved Hannover's method. He then described the receptors as facing away from the vitreous chamber and identified and numbered the principle layers of the retina. Other microscopists added that the optic nerve fibers were continuous with the retinal ganglion cells (e.g., Bowman, 1849; Corti, 1850).

In 1852, better fixation techniques permitted Kölliker to describe two distinct types of endings (rods and cones) in the retina. This important discovery was confirmed by Heinrich Müller (1856, 1857), who associated the rods with a purplish color, yet did not realize the different functions served by the two types of receptors.

The distinction between the rods and the cones commanded more attention after Max Schultze, shown in Figure

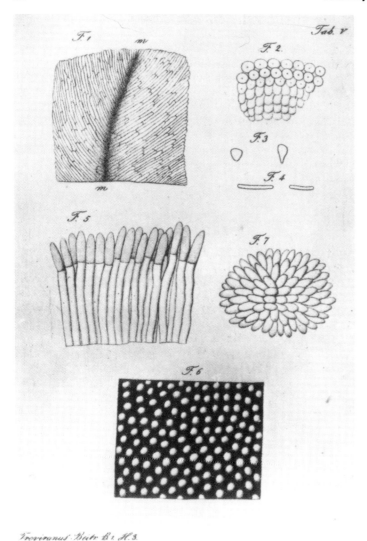

Figure 6.6. Max Schultze (1825–1874), who differentiated the rods from the cones in his studies of the retina and also localized the receptors for olfaction.

Figure 6.5. A series of figures from the frog, tadpole, and snake retinas by Gottfried Reinhold Treviranus (1837), a German scientist who conducted extensive microscopic research on the eye. This figure shows the rods, called papillae by Treviranus, in different orientations.

6.6, published a series of studies on comparative retinal anatomy. These studies were conducted in Bonn and appeared in 1866 and 1867 in Schultze's new journal, *Archiv für mikroskopische Anatomie*. Schultze provided good descriptions of the inner and outer segments of the photoreceptors and other elements of the retina, including the bipolar cells. He also included detailed descriptions of all 10 layers of the retina. Some of his plates are shown as Figures 6.7, 6.8, and 6.9.

Schultze (1866, 1867a, 1867b, 1872) made the case that the two different kinds of photoreceptors were associated with different functions. The rods were thought responsible for colorless night (scotopic) vision, whereas the cones were correlated with daylight (photopic) vision. The rods were thought to be the more primitive organ of vision and the cones, the more differentiated.

Schultze was aided in his thinking by a book written by Hermann Rudolf Aubert (1826–1892) in 1865. Aubert noted that color discrimination and visual acuity diminished as one went from the center of the retina toward the periphery. Schultze was able to show that the cones were concentrated near the center of the human retina, whereas the rods predominated in the periphery. Thus, by combining his anatomy with Aubert's behavioral observations, Schultze came to the inescapable conclusion that the cones alone were responsible for color vision and high-resolution spatial vision.

Schultze (e.g., 1872/1873) felt his hypothesis was strengthened by his own work on the proportions of rods and cones in the retinas of animals who were active either during the day or at night. He noted that the retinas of nocturnal animals, such as bats, owls, moles, and hedgehogs, contained few if any cones. He also found few rods in animals who were active only during broad daylight and who went into the ground or roosted at night (e.g., lizards, certain birds, chameleons, and snakes). Because human vision is colorless at night, it was a relatively simple step to assign achromatic vision in dim light to the rods. In was in this way that *Duplizitätstheorie* (duplicity theory), a term introduced by Johannes von Kries (1853–1928) in the 1890s, was born (Parsons, 1924; Tschermak, 1902).

Two additional early contributors to duplicity theory were Henri Parinaud (1844–1905) and Augustin Charpentier (1852–1916), both of whom seemed unaware of Schultze's papers when their own treatises were written.[9] Parinaud

[9]Charpentier's name should also be mentioned in another context. In 1886, he reported an unusual illusion of movement when a small dimly illuminated object was observed in a dark room. The movement of the object was influenced by such things as a change in attention and a loud sound. The moving object usually showed an angular velocity of 2 to 3 degrees per second. Charpentier noted that this illusion was not due to unconscious movements of the eyes and that it could be partially inhibited by having other points of light present. His illusion of movement subsequently became known as the autokinetic effect (Exner, 1896).

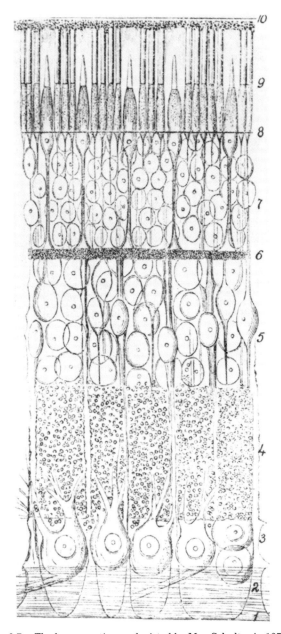

Figure 6.7. The human retina as depicted by Max Schultze in 1872.

BOLL'S DISCOVERY OF RHODOPSIN

In 1876, Franz Christian Boll (1849–1879) detached a frog retina from the choroid. When he held it up to the light he observed that the reddish-purple color, noticed by Heinrich Müller some two decades earlier, now changed to yellow. Boll called the "unbleached" pigment *Sehpurpur,* or "visual purple," a term that is still used, along with its synonym, "rhodopsin." Boll (1876, 1877a, 1877b) found that visual purple existed only where there were rods. He noted that bleached visual purple could regenerate and speculated that this depended on substances derived from the epithelium. He went on to theorize that chemical changes in the pigment were at least partly responsible for phenomena of brightness adaptation (e.g., the Purkyně shift).

Boll's discoveries and ideas were extended by Willy Kühne (1837–1900) and Carl Anton Ewald (1845–1915) in 1877 and 1878. Working together, these scientists proved the existence of rhodopsin in the human retina by dissecting the eye of a criminal who had just been executed in the dark. Rhodopsin was absent from the fovea centralis, which contains only cones. Kühne also showed that certain salts could liberate visual purple from retinas macerated in solution, making it possible to study the photopigment

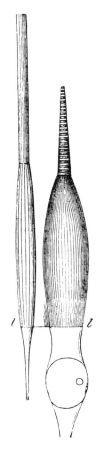

Figure 6.8. A rod (left) and a cone (right) from the human retina showing the inner and outer segments, as depicted by Max Schultze. In his caption, Schultze (1873, p. 253) wrote: "The outer segment of the cone is broken up into discs, which, however, are still adherent to one another." He said nothing about the outer segment of the rod.

(1884), an ophthalmologist associated with Charcot's neurological service at the Salpêtrière, noted that night blindness did not diminish foveal sensitivity and that the fovea was not subject to dark adaptation. Thus, he postulated *deux rétines* (two retinas), one for daylight (cones) and the other for dim light (rods). As far as Parinaud was concerned, night blindness (hemeralopia) was due to a defect in the rods.

Charpentier (1877) had also concluded that differences in vision in central and peripheral parts of the retina suggested the existence of two independent classes of receptors. He distinguished between *vision nette,* which he thought mediated color and accurate spatial information, and *perception lumineuse brute,* which mediated a diffuse sense of light. The latter was associated with the rods.

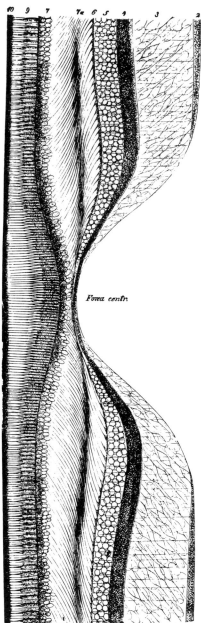

Fig. 361. Diagrammatic section through the macula lutea and fovea centralis of the retina of Man. Magnified 110 diameters. 2, optic-nerve fibres; 3, ganglion cells; 4, internal granulated layer; 5, internal granule layer; 6, external granulated layer; 7a, external fibrous layer; 7, external granule layer; 9, rods and cones; 10, pigment layer.

Figure 6.9. A diagram of the macula and fovea by Max Schultze. This figure appeared in Schultz's chapter, "The Eye: The Retina," published in 1873.

in vitro. In addition, he isolated a second photopigment (porphyropsin, or visual violet) from the fish retina.

By the end of the nineteenth century, investigators were busy studying "optograms," or fixed retinal images, on animals and humans who had just been killed. The logic and formulas were relatively simple:

> The bleaching is limited to the part of the retina on which the light falls, and produces an optogram, or image of an external object, both in the living and the recently killed animal. The eye should be kept from the light for ten minutes, and then exposed for about three minutes in a room lighted by one window, so that an image of the window falls on the retina; the eyeball should then be extirpated by the light of a sodium flame, and the retina laid in a 4 per cent solution of alum. The eye of a recently killed animal will then show a well-defined colourless optogram. (Rivers, 1900, 1045)

The study of optograms caught the imagination of the public as well as the scientific community. Some scientifically minded sleuths even hoped that the optograms of murder victims might betray their killers! This possibility, however, proved more lucrative to the writers of "thrillers" than to the criminologists.

The absorption coefficients of visual purple, and psychophysical luminosity functions obtained with humans, firmly linked rod-mediated vision with dark adaptation (König, 1894). It was soon accepted that the great increase in sensitivity in the dark-adapted eye was due to the increase in rhodopsin. The fact that adaptation to brightness was much more rapid also fit well with the finding that rhodopsin can be bleached much more rapidly than it can be regenerated.

The isolation of the cone pigments proved to be much more difficult. This was not accomplished satisfactorily in the nineteenth century, or even in the first half of the twentieth century. The difficulty may seem surprising since Schultze (1866) showed that fowl have predominantly cone eyes and a cone/rod ratio 1,000 times that of humans. Nevertheless, better techniques had to be developed to identify the pigments that would eventually become known as erythrolabe ("red catching"), chromolabe ("green catching"), and cyanolabe ("blue catching") (see Rushton, 1962).

FURTHER STUDIES OF RETINAL STRUCTURE

Advances in fixation and staining techniques made late in the nineteenth and early in the twentieth centuries led to increased knowledge of the retina. The silver nitrate method, discovered by Camillo Golgi in 1873, played an important role in these discoveries (see Chapter 3). Until the time of Golgi and Ramón y Cajal, it was assumed that the retinal receptors were endings of the optic nerve or directly connected to the optic nerve. But using a modified Golgi stain, Ramón y Cajal (1893, 1909–11) showed conclusively that there were not only independent receptors and retinal ganglion cells but also bipolar cells between them.[10] One of his drawings appears as Figure 6.10.

The horizontal cells, which run between the receptors and the bipolar cells in the retina, and the amacrine cells, which are located between the bipolar cells and the ganglion cells, were not a part of Ramón y Cajal's original scheme. These discoveries came later, as did theories about their inhibitory and modifying roles in retinal perception (see Ramón y Cajal, 1923).

Among the issues debated late in the nineteenth century was whether each eye communicated directly with the other. In fact, the idea of a retino-retinal pathway was entertained not only in the nineteenth century but also into the opening years of the twentieth century (e.g., Arnold, 1838–42; Parsons, 1902). As the twentieth century advanced, however, hard evidence for a direct pathway between the two eyes failed to materialize (Polyak, 1957).

THE OPTIC CHIASMA

During the Renaissance, it was generally believed that the optic nerves did not cross in the chiasma. After Newton

---

[10]The number of some cells in the human retina was soon approximated. It was guessed that there were about 120 million rods, 7 million cones, and 800,000 retinal ganglion cells. It was shown that better visual acuity in the central retina was due to fewer receptors converging onto higher-order cells than was true for peripheral receptors.

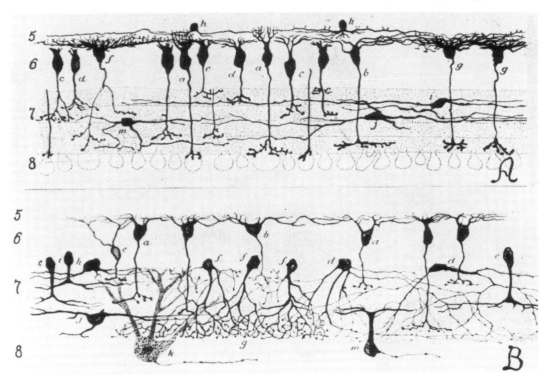

Figure 6.10. The bull retina as shown by Santiago Ramón y Cajal in 1892. The upper part of the figure was based on the Golgi stain and the lower part on the use of methylene blue.

(1704) postulated a partial crossing and Zinn (1755) showed experimentally that there was a crossing, the prevailing idea changed. But instead of accepting the idea that the crossing was only partial in organisms with eyes in the front of the head, most scientists, perhaps unaware of Newton's diagram and logic, incorrectly assumed all of the optic nerve fibers crossed in every species.

François Magendie, for example, who worked largely on birds (where the crossing is total), scoffed at the idea of incomplete decussation. Magendie, who was not above biting sarcasm, wrote:

> With respect to its crossing with that of the opposite side, no doubt can reasonably exist; the fact that I have reported I consider demonstrative. M. Pouillet, in his *Treatise on Physics*, does not agree in this opinion. He believes that it may be true, perhaps, with regard to animals, but not in man, and that Wollaston had only spoken of the latter. To this I reply, that, with respect to the anatomical arrangements here referred to, man does not differ from the mammiferi. I will add, that, having occasion to make my objections, in England, to that profound philosopher, whom the intellectual world has so many reasons to deplore, he did not appear to doubt that if the section of the decussation, over the *sella turcica,* produces blindness, it may be concluded that the crossing is total, and not partial. I do not think that he insisted upon my conjecture after the publication of my experiments. (1844, 65)

Magendie would not have made his broad generalization if he had looked more carefully at the emerging literature on homonymous hemianopia in humans.[11]

The first good description of the loss of vision in the same half of the visual field for both eyes was reported by Giovanni Battista Morgagni (1682–1771) in 1719. It involved an individual with a unilateral brain lesion. Four years after

Morgagni's case appeared, homonymous hemianopia was correctly attributed to a partial decussation of the optic nerve fibers at the chiasma (Vater and Heinicke, 1723).

John Taylor (1703–1772), the most famous English itinerant oculist and surgeon of the eighteenth century, came to a similar conclusion about homonymous hemianopia when evaluating his clinical cases.[12] Taylor and his diagram of the visual system are shown in Figures 6.11 and 6.12, respectively.

William Hyde Wollaston (1766–1828), who was mentioned in the scathing quotation from Magendie, was a renowned English chemist, physicist, and student of optics. He not only studied homonymous hemianopia in several of his acquaintances but suffered from transient attacks of the disorder himself (Chance, 1922). The fact that he saw only half of each field with each eye during these attacks led him to agree that there was only "semidecussation of the optic nerves." Wollaston began to present his ideas in 1824, just a few years before he recognized he was dying from a brain tumor.

Clinical findings such as these continued to appear and, together with better anatomical studies, eventually showed to everyone's satisfaction that the crossing in the chiasma was only partial, at least in humans.

BRAINSTEM TERMINATIONS OF THE RETINAL PROJECTIONS

The nature of the crossing in the chiasma was not the only problem facing nineteenth-century investigators of the visual system. Another question that had to be addressed

---

[11]"Hemianopia" ("hemianopsia") was called "hemiopia" in the earlier literature (see Hirschberg, 1875).

[12]Taylor claimed to have treated over 80,000 patients with cataracts, glaucoma, and various other eye diseases during his 30 years of practice. Many of "Chevalier Taylor's" (1727, 1750) statements about his successes in treating eye disorders, however, were clearly extravagant. He was considered a quack by many of his contemporaries (Pusey, 1903). He even managed to elicit the wrath of Samuel Johnson (1709–1784), who called him the most ignorant man he had ever known.

Figure 6.11. John Taylor (1703–1772) was one of the first to suggest that the optic nerves from the same places on each retina come together centrally to bring about a single visual experience. Taylor's eccentric style and personality led many of his learned contemporaries to consider him an ignorant charlatan. (From Taylor, 1750.)

was where the optic nerve fibers terminated (a reference map from 1920 is shown as Figure 6.13). While Magendie was arguing that the optic nerves were completely crossed, little was said about the visual system above the optic chiasma. In fact, if occasional statements were made about higher structures, the bias was against the localization of visual functions in their own specific neural territories.

This bias may have reached an extreme earlier in the eighteenth century with the Swiss naturalist Albrecht von Haller. He maintained that no nerve has its own exclusive territory in the brain and that structures, such as the superior colliculi or the cerebral hemispheres, should not be assigned functions different from those of the brain as a whole. Yet in striking contrast to this theory of central equipotentiality, Haller recognized the specialization inherent in the peripheral and cranial nerves.

The opposing position came from the phrenologists, who believed that different parts of the brain, including the cortex, functioned in different ways (see Chapter 3). This opinion was expressed by Franz Joseph Gall and Johann Spurzheim, the two leaders of the phrenological school. Gall and Spurzheim, however, were not just theorists but also outstanding anatomists. In 1809, they noted that both the external (lateral) geniculate nucleus and the superior colliculus in the midbrain atrophied following damage to the optic nerve. This observation led them to suggest that the lateral geniculate and superior colliculus were the important brainstem nuclei for vision.

Karl Friedrich Burdach (1822) believed the superior colliculus sent some fibers to the lateral geniculate. He also found the lateral geniculate nucleus to be made of gray and white bands, which he described as interlacing "like the tongues of a flame." Burdach's findings were forgotten until the laminated structure of the lateral geniculate nucleus

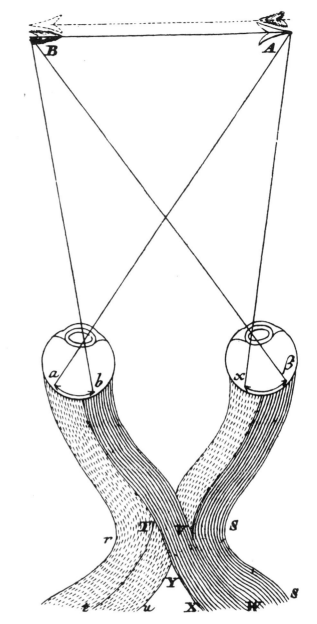

Figure 6.12. John Taylor's (1750) diagram showing the partial crossing of the optic nerves at the chiasma.

was "rediscovered" by Theodor Meynert (1833–1892) and Jakob Henle (1809–1885) later in the century. Meynert (1870) thought the lateral geniculate body was one of three loci of optic nerve termination, the other two being the medial geniculate body and the pulvinar. He believed that each of these areas sent projections to the cortex.

Neither Meynert nor Henle understood that each eye projected to different layers of the lateral geniculate and that the projections from the two eyes did not come together at this level. The discrete terminations of ipsilateral and contralateral optic tracts in the lateral geniculate nuclei were only later recognized by Mieczyslaw Minkowski (1884–?). Minkowski (1912, 1913, 1920) spoke of three "external" geniculate layers in the cat and six in the monkey and described how they were affected by enucleation (eye removal). Nevertheless, his belief that the projections from

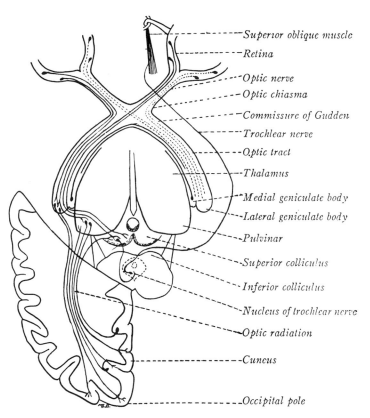

Superior oblique muscle

Retina

Optic nerve

Optic chiasma

Commissure of Gudden

Trochlear nerve

Optic tract

Thalamus

Medial geniculate body

Lateral geniculate body

Pulvinar

Superior colliculus

Inferior colliculus

Nucleus of trochlear nerve

Optic radiation

Cuneus

Occipital pole

Figure 6.13. Diagram of the major visual pathways and terminations as depicted in Ranson's *The Anatomy of the Nervous System* in 1920.

the two eyes could interact in the lateral geniculate would not be substantiated.

Minkowski (1913) also studied the effects of lesions of the striate cortex on the lateral geniculate body. His results indicated that the lateral geniculate was the only subcortical structure to show extensive anatomical changes after striate lesions. He further determined that the anterior part of the striate cortex was connected with the anterior part of this thalamic nucleus, while the posterior part of the cortex was associated with the posterior part of the lateral geniculate. This work suggested a strong linkage between the different parts of the retina, the lateral geniculate, and the striate cortex.

Like the lateral geniculate body, the superior colliculus also began to be studied in the nineteenth century. Henri Parinaud (1883, 1886) reported that paralysis of vertical eye movements ("upgaze") and problems with convergence followed tectal damage (usually tumors) in humans. In contrast, reflexes to light remained intact. This complex has been called Parinaud's syndrome (Wilkins and Brody, 1972).

Most other nineteenth-century findings about the superior colliculus were essentially negative. The better lesion experiments conducted at the turn of the century showed that superior colliculus lesions did not cause blindness in animals (Ferrier and Turner, 1898, 1901). In addition, animals with collicular damage still constricted their pupils to light, in accordance with Parinaud's findings in humans (Bechterew, 1883, Ferrier and Turner, 1898, 1901; Levinsohn, 1904).

EARLY HINTS OF AN OCCIPITAL CORTICAL REGION FOR VISION

One of the earliest hints of a cortical role in vision was provided by Herman Boerhaave (1668–1738). He described a beggar in Paris who used his calvarium, which had been removed from the rest of his skull, to collect alms. In return for a coin, this man even allowed people to touch his brain, which occasionally elicited visual sensations.

> Upon gently pressing the dura mater with one's finger, he suddenly perceived, as if it were, a thousand sparks before his eyes, and upon pressing a little more forcibly his eyes lost all their sight; by pressing the hand still stronger on the dura mater, he fell down in a deep sleep, which was attended with all the symptoms of a slight apoplexy, merely by this pressure with the hand, which was no sooner removed but he as gradually recovered from the symptoms as they were brought on, the apoplectic symptoms first vanishing, then the lethargy, and lastly the blindness, all his senses recovering their former perfection. (Translated in Gibson, 1962, 945–946)

During the first half of the nineteenth century, some clinicians began to argue that vision not only involved the cortex but that sight demanded its own cortical locus. One such person was Jean-Baptiste Bouillaud. Although trained in phrenology, he emphasized the use of clinical examinations and autopsy material for studying localization of function (see Chapter 3). Bouillaud (1830) believed all sensory systems had to have their own distinct cortical centers. Nevertheless, he was unable to localize any cortical sensory centers in his studies.

The work of Bartolomeo Panizza (1785–1867) brought more attention to the possibility that cortical lesions may cause blindness. Working in Pavia, and admittedly influenced by the phrenologists, Panizza became convinced that the visual projection involved the thalamus and the posterior cortex. Panizza (1855, 1856) examined pathological human case material and also used data from lesion experiments with dogs to argue that bilateral posterior cortical lesions could cause blindness. He further maintained that this faculty could be affected independently of other faculties.

In one of his clinical cases, Panizza claimed that a hemorrhage of the left thalamus caused "blindness" in the right eye, as well as paralysis and loss of sensation on the right side of the body. Panizza seemed to have made the same mistake as many of his predecessors when he inferred blindness of the opposite eye after unilateral thalamic and cortical lesions; it seems fairly certain he was observing homonymous hemianopia (see Polyak, 1957). In a related case, which involved a person with a lesion of the thalamus and posterior cortex, he described the fiber tract that originated in the posterior thalamus and terminated in the posterior superior gyri of the cortex.

Unfortunately, Panizza's work was ignored (Benton, 1990). Three reasons for this are that (1) he published in a provincial periodical; (2) he was not well known; and (3) his works appeared in Italian, as opposed to German, English, or French. A fourth reason may well have been that during the 1850s, the visual center was still believed to be located in the optic thalamus. To argue that it could be

Figure 6.14. Albrecht von Gräfe (1820–1870), known for his systematic testing of the visual fields, described homonymous hemianopia and believed that it was often associated with cerebral lesions. (From Michaelis, 1877.)

localized in a small piece of cerebral cortex may have seemed foolish. Even worse, Panizza's ideas may have conjured up images of a mortally wounded, but still not buried, phrenology.

Only in 1880, after Hermann Munk's initial lesion studies on the occipital cortex had been published, was attention called to Panizza's studies. His contribution was recognized by Augusto Tamburini (1848–1919), another Italian with an interest in vision and the brain.

In contrast to the isolated, relatively unknown Panizza, Albrecht von Graefe (1828–1870), shown in Figure 6.14, was an established leader in the field of clinical ophthalmology and a pioneer in eye surgery (Michaelis, 1877; Perera, 1935). In 1852, he established one of the most respected eye clinics in Europe. Two years later, when he was only 26 years old, he founded the *Archiv für Ophthalmologie.*

Well known for his systematic testing of the visual fields, Graefe (1856) not only described homonymous hemianopia, but also realized that this sort of visual disturbance could be caused by any lesion above the optic chiasma in humans. Notably, he often found the defect associated with cerebral lesions.

After Graefe's work, many other clinical cases that seemed to suggest an occipital cortex localization for vision appeared in the literature (see Polyak, 1957, for references). One of the better ones, reported in 1863, involved an elderly individual who lost his sight for objects on the right side of midline. The disorder was found to be associated with a softening of the medial part of the left occipital lobe and the posterior part of the left thalamus (Chaillou, 1863).

Even before the 1880s, when Panizza's studies became known, there developed a belief among the anatomists that some visual projections might eventually go to the cortex (see Polyak, 1957). In 1810, Gall and Spurzheim hinted at a projection from the lateral geniculate nucleus to the posterior part of the cerebral cortex. Further, the French anthropologist and anatomist Pierre Gratiolet (1854) expressed the belief that some fibers of the optic tract which headed toward the thalamus continued on to the occipital and parietal cortices. After studying dissected fixed specimens, Gratiolet wrote that the visual radiation spread out like a fan after leaving the geniculate nuclei of the thalamus.

Gratiolet's idea that the posterior cortex received visual projections represented both a challenge to, and a confirmation of, the views of Marie-Jean-Pierre Flourens (1824, 1842). Flourens conducted the first laboratory studies suggesting that the cerebral cortex played a role in visual perception. Working mostly with pigeons, hens, and other birds, he noted that damage to one cerebral hemisphere seemed to make these animals blind in the opposite eye. This finding implied that vision was dependent on more than just subcortical structures. It also fit well with Gratiolet's notion of a sensory pathway to the cortex.

In contrast, Gratiolet's suggestion that only one part of each hemisphere received the visual input was anything but consistent with Flourens' idea of cortical equipotentiality. Since Flourens associated the right eye with the left cerebral cortex and the left eye with the right cerebral cortex, one could argue that this French experimentalist, who so adamantly opposed cortical localization, was, in fact, implying some specialization for perception at the cortical level. Flourens stated, however, that there was no specialization for vision or any other function within a hemisphere.

A later anatomist who maintained that there was a restricted visual radiation to the cortex was Theodor Meynert (1870). He agreed with Gratiolet that the occipital cortex played an important role in vision and, strictly on the basis of anatomy, singled out the calcarine sulcus (fissure) as the most important part of the occipital lobe. Meynert suggested that in addition to the occipital lobe, almost the whole temporal lobe may be involved in vision.

FERRIER'S STUDIES WITH MONKEYS

In the early 1870s, David Ferrier observed that electrical stimulation of the angular gyrus of the posterior parietal lobe in monkeys caused eye movements toward the opposite side. In contrast, he was unable to obtain eye movements when he stimulated the occipital lobe. When Ferrier destroyed one angular gyrus, he reported that his animals showed temporary blindness in the opposite eye. Bilateral lesions resulted in a loss of vision in both eyes. Ferrier also stated that the greater portion of both occipital lobes may be removed without noticeable impairments in vision. Armed with these experimental results, he concluded that vision was localized in the angular gyrus in primates (see Ferrier, 1876).

Ferrier (1878) further supported his contentions by describing some clinical cases. Although he associated lesions of the angular gyrus with four cases of blindness, each case was of questionable validity (see Starr, 1884a). Ferrier also cited some 13 clinical cases to show that lesions of the occipital lobe did not always produce blindness. Eleven of these cases were recorded before 1868, and even Ferrier himself later questioned the reliability of this material.

Ferrier softened his position after he and Gerald Yeo made lesions of the angular gyrus and occipital cortex in another group of monkeys and after he looked more carefully at additional clinical material. Ferrier (1881) now found that even bilateral angular gyrus lesions did not cause permanent or complete blindness. It was only when he also destroyed the occipital lobes that the monkeys seemed totally blind.

Ferrier (1881) also noticed that his animals showed a partial loss of vision in both eyes when the angular gyrus and the occipital lobe were damaged on just one side of the cortex. This loss affected the visual field on the side opposite the lesion. The homonymous hemianopia, however, did not appear to be permanent. Further, Ferrier thought the contralateral eye was more severely affected than the ipsilateral eye.

After obtaining this new information, Ferrier announced that the visual system was only partially crossed in primates and that vision was mediated by both the angular gyrus and the occipital lobe. He did not state, however, that the angular gyrus and the occipital cortex were functionally equivalent. He still looked upon the angular gyrus as more important because he thought it was concerned with central ("macular") vision.[13] In contrast, he felt that when the occipital lobes alone were damaged, even with very large bilateral lesions, the animals could still behave as if they had normal vision.

Why Ferrier did not obtain severe visual deficits after large lesions of both occipital lobes in monkeys is not clear. Quite possibly the most important parts of the occipital cortex were spared in some of his cases. In contrast, the blindness observed following damage to the angular gyrus was attributed to two important anatomical facts. First, his deep lesions of the angular gyrus could have cut the visual radiation from the thalamus to the occipital cortex (Starr, 1884a). And second, damage to the midcerebral artery could have affected the visual radiations, even if they were spared by the primary lesions (Starr, 1884b).

Ferrier's (1876, 1886) insistence on the importance of the angular gyrus was questioned by a number of vocal critics. Some of these investigators pressed for just an occipital cortex localization for vision, whereas others simply argued against any specific cortical localization for vision. Friedrich Leopold Goltz, for instance, strongly opposed any sort of rigid cortical localization (see Chapter 4).

MUNK'S "DISCOVERY" OF THE VISUAL CORTEX

Hermann Munk, a professor of physiology in Berlin, conducted lesion experiments on animals with more refined techniques than those used by Goltz. Munk, shown in Figure 6.15, announced the results of his first series of experiments on the occipital cortex in 1878. His initial findings supportive of an occipital cortical localization for vision were met with considerable doubt, in part because his data differed from those reported by David Ferrier. Thus, Munk felt it necessary to repeat his experiments.

Figure 6.15. Hermann Munk (1839–1912) of Berlin. Munk did some of the pioneering lesion experiments that helped to establish the occipital cortex as a center for vision. (From Rothmann, 1909.)

In 1881, he presented the results of another series of investigations, in which he confirmed his earlier findings. He went on to summarize his case for the occipital cortex before the Physiological Society of Berlin in 1883. The British journal *Nature* presented his findings later that same year (see Anonymous, 1883).

Munk stated he had studied the effects of brain lesions on both dogs and monkeys, with the majority of his experiments conducted on dogs. Unlike Ferrier and Goltz, who came under criticism for using short survival periods, Munk kept his animals alive for up to five years after injury, allowing considerable time to study their recovery.

He found that when the posterior part of the occipital cortex was partially damaged, and when the initial inflammation subsided, his dogs acted as if they were unable to remember the meanings of various objects they had previously been shown. He called this *Seelenblindheit,* or "psychic blindness," a type of agnosia. Munk (1881) wrote the following about such a dog:

(He) remains completely cold when looking at people which he used to greet with joy; he remains just as cold in the presence of dogs with which he formerly used to play. As hungry and thirsty as he may be . . . he does not look as formerly to those places in the room where he used to find his food and even if one puts his food and water right in his way he frequently goes around them without paying any attention to them. . . . Finger and fire approaching the eye do not make him blink. The sight of the whip which used to drive him regularly into the corner does not frighten him any more in the least. (1960 translation, 97–98)

Psychic blindness was not considered blindness in the true sense of the word because the dogs were able to use their eyes to avoid obstacles and to navigate successfully. Importantly, the dogs were not "intellectually blunted," as shown by the fact that they could recognize the meanings of objects when they were allowed to use their other senses. Rather, psychic blindness was interpreted as the loss of memory images, or optical impressions, that were stored in the occipital lobe of the brain.

---

[13]This theory was also supported by others (e.g., Ewens, 1893).

The dogs usually recovered from this state in four to six weeks. Munk interpreted this recovery as similar to the new learning seen in a young animal. Although this learning seemed to take less time the second time through, it appeared to Munk as if the visual slate had been wiped clean and a new memory bank had been established. Munk hypothesized that it was through relearning that the lost visual memories were replaced.

Munk (1879a, 1879b, 1881) believed that the posterodorsal area of the occipital lobe was the part of the cortex most responsible for sharpest vision in the dog, and he associated this area with the macula. He hypothesized that other parts of the occipital cortex also had to contribute to the sense of sight and that the whole region was organized in an orderly manner, much like the receptors of the retina. This can be seen in Figure 6.16. Munk thought that areas near the macular region of the cortex were the ones responsible for relearning the meanings of visual images during recovery from psychic blindness. This process involved learning to use healthy parts of the eye to compensate for the blinded macular region. Because these more peripheral retinal regions were not associated with sharp vision, the compensatory process was thought to have its limits.

The idea that these neighboring areas were involved in

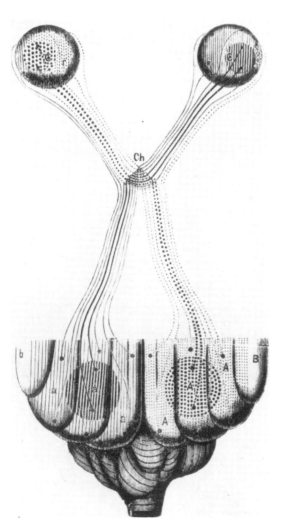

Figure 6.16. One of Hermann Munk's (1879b) drawings of the projections from the two eyes to the dog's occipital cortex. The circles in the center of each visual cortex represent the macular area.

vision seemed even more certain to Munk after he made even larger occipital cortex lesions in his dogs. These lesions resulted in true blindness, or *Rindenblindheit* ("cortical blindness"), a condition very different from the psychic blindness state associated with smaller lesions in the macular area of the cortex.

Munk's findings stood in opposition to those of Ferrier, especially after Munk claimed to have demonstrated that only the occipital area, and not the angular gyrus, was critical for vision in monkeys. Munk did, however, give the angular gyrus a secondary role in vision. He postulated it might function in eye movements and in blinking reflexes, but not in sight per se (Munk and Obregia, 1890).

At first, Munk thought that the cortical blindness caused by a large unilateral lesion was limited to the opposite eye (see Munk, 1881). This was the same error that had been made by Flourens when generalizing from birds and by Panizza and Ferrier, who worked with humans and primates. Munk later corrected his error but did not consider the dog, which he favored in his experiments, and the monkey, which Ferrier favored, directly comparable. Munk noted that more of the optic tract decussated in the optic chiasma of the dog than in the monkey. For this reason, the hemispheres were more involved with the opposite eye in the dog than in primates, where the optic nerve fibers showed a more even split at the chiasma.

THE EXPERIMENTALIST RESPONSE TO MUNK'S IDEAS

In general, Munk's basic findings were accepted by most experimentalists. These included the Italian investigators Luigi Luciani (1840–1919) and Augusto Tamburini, who conducted their own lesion experiments. They assessed vision by throwing small pieces of meat (taking care to avoid sounds) to their dogs (Luciani, 1884; Luciani and Tamburini, 1879). The dogs were tested with one eye closed, with both eyes opened, or with both closed. The investigators began by throwing the meat directly in front of the dog's nose and subsequently threw scraps to the left or right of the dog's midline.

Luciani and Tamburini stated that visual disturbances can follow severe damage to any part of the cortex, but that only occipital and parietal cortex lesions give lasting deficits. Thus, they concluded that the visual sphere was concentrated in the occipitoparietal zone of the hemisphere. Luciani (1884, p. 150) added:

> The experimental results obtained on monkeys are in accordance with this mode of interpreting the facts which relate to the visual sphere of the dog. My new experiments confirm on the one hand the capital importance of the occipital lobes as centres of vision, as already demonstrated by Munk, whilst on the other they confirm the view, that these centres spread by irradiation into the parietal lobes, especially into the angular gyri, and their fold towards the parieto-occipital fissure, as was shown by Tamburini and myself, and by Ferrier and Yeo. My new researches prove in addition, that the isolated destruction of the angular gyri produce slight, and rather transitory, visual disturbances; and that the effects of destruction of the occipital lobes are aggravated then considerably, but not rendered permanent by a posterior decortication of the angular gyri.

In addition, Luciani and Tamburini agreed with Munk that the visual deficit caused by a unilateral posterior cortical lesion was one that affected the homologous fields of the two eyes. To quote Luciani (1884, p. 150) again:

My recent experiments confirm the view that the fibres of the optic nerve of the dog undergo in the chiasma an incomplete decussation, as was seen by Tamburini and myself, and that the extirpation of one of the occipital lobes produces a bilateral homonymous hemiopia, as Munk was the first to prove clearly, though more extensive in the eye of the opposite side.

Nevertheless, these workers did not agree with Munk on all points. In particular, they felt that absolute (cortical) blindness may not appear or may be only transient, even after large posterior cortical lesions in dogs. They emphasized that the more meaningful deficit was the psychic blindness, characterized by weak vision and a loss of the meaning of things seen. Thus, Luciani and Tamburini did not look upon the cortical visual center as the seat of basic visual sensations but instead emphasized that its true function was to "elaborate psychically the visual sensations," which take place at lower levels (see Luciani, 1884).

Munk also had a supporter in Edward Schäfer, who conducted his own experiments on vision in monkeys at University College, London. In contrast to Ferrier, Schäfer found he could elicit contralateral movements of the eyes when he stimulated the occipital lobe (e.g., Brown and Schäfer, 1888; Schäfer, 1888a, 1888b, 1888c, 1888d, 1900). Schäfer also noted he could elicit eye movements with stimulation of adjacent structures, including the posterior limb of the angular gyrus and the medial temporal gyrus. This led to his initial conclusion that electrical excitation of the occipital lobes and possibly adjoining areas produced subjective visual sensations. This is illustrated in Figure 6.17.

When Schäfer made lesions of the occipital and angular gyri, he realized that the contribution of the occipital lobes was more significant than that of the angular gyrus. He stated this in his presentation to the Royal Society in 1887. The paper was published one year later, and in it Schäfer (1888c, p. 1) maintained:

The visual area of the cerebral cortex in the monkey, so far as it is determinable from the results of extirpation and of electrical excitation, comprises the whole of the occipital lobe. It perhaps includes a part or the whole of the angular gyrus, although the results of extirpation would appear to show, that the part played by the latter convolution is comparatively small. Removal of both occipital lobes produces total and permanent blindness, whereas destruction of the cortex of both angular gyri is not followed by any appreciable permanent defect of vision.

Ferrier (1888) questioned Schäfer's findings, mentioning that he had made extensive occipital cortical lesions in his monkeys, only to observe good evidence of vision just two hours after surgery. Nevertheless, Schäfer argued that Ferrier's lesions were incomplete and that the monkeys probably quickly learned to compensate for small blind spots in their visual fields, making it impossible to detect smaller losses of vision. Indeed, Schäfer also found evidence for the return of good vision after "incomplete" lesions in some of his own experiments on the occipital

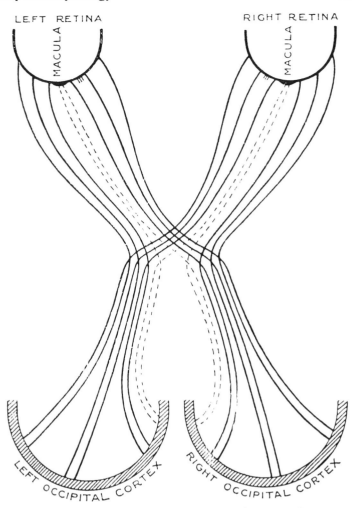

Figure 6.17. Edward Albert Schäfer's diagram from 1900 showing the "probable relationships" between the different areas of the retina and the occipital cortex.

lobes (see Brown and Schäfer, 1888; Schäfer, 1888d). In fact, he later stated that to produce complete blindness the lesions may have to extend "somewhat in advance of what is generally taken to be the limit of the lobe, on the inner and lower surface" (Schäfer, 1900, p. 753).

Like Luciani and Tamburini, Schäfer (1888c) did not fully accept everything Munk had to say. He questioned Munk's finding, based largely on dogs, that central vision was associated within an area near the middle of the convexity (the "apex") of both occipital lobes. Schäfer (1888c) found that bilateral lesions in the middle of the convexity were associated with only small effects in his monkeys. In contrast, he wrote, "there was a marked defect of central vision produced by bilateral injury of both mesial surfaces" (p. 6). Thus, in opposition to the dog, he correctly localized central vision in monkeys and humans in the medial part of the occipital lobe, in the neighborhood of the calcarine fissure.

Munk found less support and stronger opposition from Constantin von Monakow. This Swiss scientist questioned the "stability" of the retinal projection to the striate cortex and proposed a more dynamic view of cortical function. Monakow thought the macula could be represented throughout a large part of the posterior cortex.

Another individual who challenged Munk's concepts was Jacques Loeb. Loeb applied his new method of double simultaneous stimulation to dogs who had unilateral occipital cortex ablations. He found his visual stimuli were still appreciated when they were presented separately to either the right or the left of his dogs. In contrast, the stimulus contralateral to the cortical lesion tended to be ignored when both stimuli were presented together.

Loeb described this phenomenon in 1884 and discussed it in more detail in 1885 and 1886.[14] To Loeb, the visual deficit seemed more like a weakness or inattention for the opposite visual field (hemiamblyopia) than a true loss of vision.

Along with such questioning came additional experiments. These experiments placed the ipsilateral and contralateral homonymous parts of the retina in the same occipital cortical locus, rather than side by side as Munk had initially assumed.

MYELINATION AND THE VISUAL AREA

The experimental findings provided by Munk and Schäfer fit well with what was being learned from the study of myelogenesis. Beginning in the 1870s and continuing for more than 50 years, Paul Emil Flechsig (1847–1929), a professor of psychiatry at Leipzig, worked on the process of myelogenesis of different tracts in the central nervous system (Schröder, 1931). He noted that different systems became myelinated at different times; that cortical sensory and motor areas "matured" before "association centers" (Assoziationszentren); and that the association areas, which he believed formed the foundation for man's superiority, took up more of the cortex as one ascended the phylogenetic scale (e.g., Flechsig, 1895, 1896).

Flechsig thought there were 40 distinct cortical areas at first, but later he combined some of these areas to reduce their number to 36 in humans (in dogs the number was only 20). Numbers 1–10 were "primordial zones" whose myelination could be seen with the naked eye in a full-term infant; areas 11–31 were "intermediate zones" of moderate development; and areas 31–36 were "terminal zones" whose myelination did not begin until weeks after birth (Flechsig, 1901).

Flechsig began to investigate the visual system in 1874. He found that the visual radiation projected to the calcarine fissure in the area characterized by the stripe of Gennari. His area No. 4 corresponded to Brodmann Area 17 and to the region of cortex called *area striata* by Grafton Elliot Smith, the Australian who studied the morphology and evolution of the human and infrahuman primate occipital cortex (see Figure 6.18; Brodmann, 1909; Smith, 1904a, 1904b, 1904c). Flechsig (1895) also recognized that these were the first of several fiber systems projecting to the occipital lobe to become myelinated. Consequently, he called this projection, which appeared in human babies at the age of one month, the *primäre Sehstrahlung,* or "primary visual radiation."

At the end of the first postnatal month, radiating fibers, which Flechsig thought might be collaterals of the primary

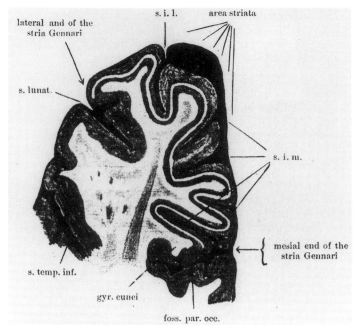

Figure 6.18. Horizontal section of a human brain from Smith (1904a) showing the stripe of Gennari.

projection, spread over a much larger territory of the posterior cortex. These areas were considered the "marginal zone of the visual sphere." Flechsig (1895, 1896) believed this secondary zone was concerned with higher visual functions than the primary zone.

At first, Flechsig maintained that the visual radiation came from three sources: the superior colliculus, the pulvinar, and the lateral geniculate nucleus. As his work continued, however, Flechsig (1896) broke away from this view and stated that the projections to the visual cortex came only from the lateral geniculate nucleus. The course of these fibers was described in detail, including the sharp turn in the temporal lobe that became known as Flechsig's temporal knee or Flechsig's detour (see Polyak, 1957). These observations, which extended the earlier findings of Gratiolet and Meynert, put the concept of cortical centers for vision on a sturdier foundation.

CLINICAL CONFIRMATION OF THE OCCIPITAL LOCALIZATION

The basic findings of Munk and Flechsig were also supported by clinical investigators who were looking more carefully at their cases as the nineteenth century drew to a close. One such individual was Hermann Wilbrand (1851–1935), an ophthalmologist from Hamburg. Wilbrand (1881, 1887, 1890) studied many clinical cases and concluded vision was in fact localized in the occipital cortex. He postulated that the visual cortex was comprised of mini units, each made of ipsilateral and contralateral cells from homologous spots on the retina.

Another supporter was the American physician Moses Allen Starr (1854–1932). Starr (1884a, 1884b, 1884c) reviewed a large number of autopsy reports of individuals who suffered from homonymous hemianopia and confirmed that the occipital lobes were responsible for vision. He also

---

[14]For a history of the method of double simultaneous stimulation, see Benton (1956).

Figure 6.19. Salomon Henschen (1847–1930), the Swedish neurologist who believed in cerebral localization and used extensive clinical material to help localize both the visual and the auditory cortices in man. (From Grote, 1925.)

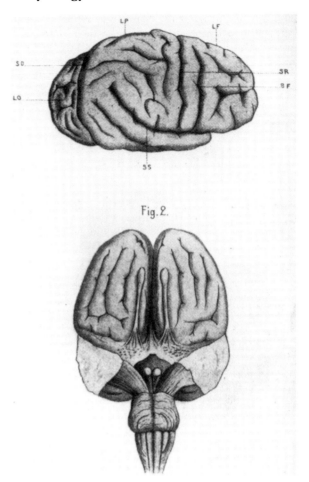

Figure 6.20. The brain of a 14-month-old child who was born with all parts of the visual system absent up to the lateral geniculate nuclei, and with underdeveloped occipital lobes. This case was presented as strong evidence for vision being localized in the occipital cortex. (From Giovanardi, 1881.)

showed that in the absence of damage to other structures, the homonymous hemianopias seen after cortical, thalamic, and lower brainstem lesions were indistinguishable from one another. He agreed with Ludwig Mauthner (1840–1894), the Austrian ophthalmologist, who stated that there was not a single authenticated case of a lesion of one hemisphere that produced blindness in just the opposite eye (Mauthner, 1881).

Support for an occipital cortex localization also came from Salomon E. Henschen (1847–1930; shown in Figure 6.19), a leading Swedish scientist and the man whom Flechsig (1901) considered "the most careful worker in the field." Henschen reviewed all the recorded cases of blindness and hemianopia resulting from cerebral lesions and, around 1890, began his own systematic studies on the visual pathways. By 1893, he reported that over 160 such cases were on record.[15]

The material Henschen collected led him to localize the human visual cortex in the area around the calcarine fissure (Henschen, 1893, 1894, 1897, 1903, 1910). He wrote:

> Firstly, a lesion on the medial surface causes hemianopsia only if the cortex of the calcarine fissure, or the fibres derived from it, are affected. Secondly, a lesion limited to the calcarine cortex can induce a complete hemianopsia, and of this the most instructive case is, I believe, one now published in my book, which includes all guarantees for an exact conclusion. It was stationary, uncomplicated, and the clinical examination, as well as the *post-mortem,* was accurate. The lesion was limited to the cortex in the depth of the calcarine fissure, and the hemianopsia was complete and absolute. (1893, p. 177)

Henschen argued that lesions of the angular gyrus of the posterior parietal lobe gave rise to visual defects only if they damaged "the optic radiation of Gratiolet," that is, visual fibers en route to the occipital lobe. Furthermore, as

far as he was concerned, the visual cortex was a replica of the retina itself (i.e., "the cortical retina") and was very stable, in spite of what Constantin von Monakow (e.g., 1900) was saying about "mobile retinas." Still, Henschen was incorrect about the organization of the visual cortex. Among other things, he localized the human foveal region anteriorly and put the retinal periphery in the back of the calcarine cortex (e.g., Henschen, 1893). This error, however, would soon be corrected by others (e.g., Lenz, 1909).

At the turn of the century, there were many other clinical investigators who provided data in agreement with these general ideas (e.g., Haab, 1882; Mott, 1907). A more unusual study involved a baby who was born without eyes and showed no calcarine region (Giovanardi, 1881). The drawing accompanying this report is shown in Figure 6.20. The conclusion was undeniable: The primary visual cortex was in fact synonymous with the broad white stripe first seen in 1776 by Francesco Gennari.

THE BIRTH OF VISUAL ELECTROPHYSIOLOGY

As noted in Chapter 3, cortical electroencephalography began in 1875 with Richard Caton's short report to the

[15]Seguin had only found some 40 cases when he reviewed the literature in 1886.

British Medical Association. With regard to vision, Caton (1875, p. 278) stated:

> Impressions through the senses were found to influence the currents of certain areas; e.g., the currents of that part of the rabbit's brain which Dr. Ferrier has shown to be related to movements of the eyelids, were found to be markedly influenced by stimulation of the opposite retina by light.

Caton presented another report on his experiments with sensory stimulation in the *British Medical Journal* in 1877. He wrote that by this time he had studied some 40 animals and felt confident that visual stimuli evoked changes in the posterior part of the cortex.

> A point was found on the posterior and lateral part of the hemisphere in which, in three rabbits out of seven experimented on, variation of the current was seen to occur whenever a bright light was thrown upon the retina. (1877, 62)

In a report before the Ninth International Medical Congress 10 years later, Caton described some newer experiments with flashes of light as stimuli. Once again, he described electrical changes to light in the posterior region of the cortex.

Adolf Beck, shown in Figure 6.21, began to develop his method for localizing functions in the brain in 1888, when he was still a student in Poland (Brazier, 1988). He also attempted to record brain potentials in response to visual stimuli. Using a burning magnesium ribbon as his visual stimulus, and without the benefit of an amplifier, Beck (1890, 1891) observed that bright light altered the electri-

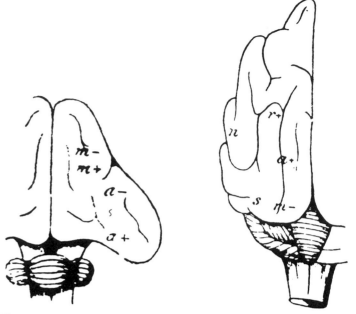

Figure 6.22. Two diagrams drawn by Adolf Beck for his doctoral thesis, published in 1891. Light affected the electrical activity of the rabbit brain (left) in places (a–) and (a+) and in the dog brain (right) in (m–) and (m+). (Courtesy of Mary Brazier.)

cal activity in the occipital lobe of his rabbits and curarized dogs.[16] Two of his diagrams appear in Figure 6.22.

THE PARASTRIATE AND PERISTRIATE AREAS

By the beginning of the twentieth century, it seemed clear that the most rapid growth of the occipital lobe was anterior to Brodmann Area 17. This suggested that the striate cortex was probably "pushed" into its mesial position in primates by the emergence of phylogenetically newer areas. These areas included the parastriate and peristriate cortices of the occipital lobe as well as the posterior parietal lobe. As stated by Grafton Elliot Smith:

> The area striata is situated partially on the mesial and partly on the lateral aspect of the hemisphere and the exact proportion of the pars medialis to the pars lateralis varies in different Apes and in Man. The great distinctive feature of the area striata (when contrasted with most other regions of the neopallium) is the fact that it undergoes little if any increase in size in the series consisting of old world Apes and Man: and, as the neopallial region immediately in front of the lateral part of the area striata undergoes a progressive expansion, the stria-bearing cortex gradually becomes pushed on to the mesial surface. In many human brains the area striata becomes wholly relegated to the mesial surface: but in other cases more than a quarter of it may still retain its primitive situation on the lateral surface. In the smallest Cercopithecidae, in which the great

Figure 6.21. Adolf Beck (1863–1942), the Polish neurophysiologist who recorded cortical evoked potentials to light and other stimuli at the end of the nineteenth century. (Courtesy of Mary Brazier.)

---

[16]Beck's work succeeded in attracting considerably more attention than Caton's earlier reports, but it also led to a debate over priority for these discoveries (see Brazier, 1988). Beck appeared to be unaware of Caton's findings when he claimed priority. The debate that ensued between Beck and others eventually led Richard Caton to republish his 1875 report verbatim, in the form of a letter dated 1891. This story is told in Chapter 3.

parietal expansion has not yet become specially pronounced, the pars lateralis may include as much as one half of the stria-bearing cortex. (1904b, 439)

The idea that the striate, parastriate, and peristriate cortices might have different functions was also beginning to attract serious attention. Henry Hun (1854–1924), a clinical professor in Albany, New York, hypothesized:

On the median surface of the occipital lobe take place those actions which are associated with simple visual sensations. On the convex surface of the occipital lobe take place those actions which are associated with complete visual perception and recognition. (1887, 150)

SOME UNANSWERED QUESTIONS

Despite these advances, there were many unanswered questions. The callosal connections between the visual areas remained a bit of a mystery. This was because the corpus callosum, present across more anterior parts of the occipital lobe, appeared to be absent between the striate areas (Area 17) of each hemisphere.

Furthermore, some behavioral phenomena seemed to remain a good step ahead of physiological explanation. For example, illusions and the ways in which previous learning and early experience could affect perception were easily demonstrated but very poorly understood.

On a physiological level, practically nothing was known in the opening decades of the twentieth century about how pattern vision might be coded in the optic nerve, much less in the brainstem or the cortex. In addition, color vision presented some very special challenges. Many advances had been made in understanding color vision after the Renaissance, and theories of color vision abounded as the twentieth century began. But which theories made sense? What was really known about the cones and congenital color blindness? And was there a separate cortical area for seeing and understanding colors? The answers to these questions are examined in the next chapter, which focuses specifically on color vision.

REFERENCES

Anonymous. 1883. Notes of the Berlin Physiological Society. *Nature, 28,* 431–432. (Report of H. Munk's presentation of July 27, 1883)

Arnold, F. 1838–1842. *Tabulae anatomicae* (4 vols.). Turici: Orell, Fuessli et Soc.

Aubert, H. 1865. *Physiologie der Netzhaut.* Breslau: E. Morgenstern.

Baillarger, F. 1840. Recherches sur la structure de la couche corticale des circonvolutions du cerveau. *Mémoires de l'Académie de Médecine de Paris, 8,* 487–493.

Bechterew, W. 1883. Ueber den Verlauf der die Pupille verengenden Nervenfasern im Gehirn und über die Localisation eines Centrums für die Iris und Contraction der Augenmuskeln. *Archiv für die gesammte Physiologie, 31,* 60–87.

Beck, A. 1890. Die Bestimmung der Localisation der Gehirn- und Rückenmarkfunctionen vermittelst der elektrischen Erscheinungen. *Centralblatt für Physiologie, 4,* 473–476.

Beck, A. 1891. Oznaczenie Lokalizacyi z mózgu i Rdzeniu za Pomoca Zjawisk Elektrycznych. Presented October 20, 1890. *Polska Akademia Umiejetnosci, Ser, 2, 1,* 186–232.

Benton, A. L. 1956. Jacques Loeb and the method of double stimulation. *Journal of the History of Medicine and Allied Sciences, 11,* 47–53.

Benton, A. L. 1990. The fate of some neuropsychological concepts: An historical inquiry. In E. Goldberg (Ed.), *Contemporary Neuropsychology and the Legacy of Luria.* Hillsdale, NJ: Lawrence Erlbaum. Pp. 171–179.

Bidder, F. 1839. Zur Anatomie der Retina, insbesondere zur Würdigung der stabförmigen Körper in derselben. *Archiv für Anatomie, Physiologie und wissenschaftliche Medizin,* 371–385.

Boll, F. 1876. Zur Anatomie und Physiologie der retina. *Monatsberichte. Königliche Preussichen Akadamie der Wissenschaften zu Berlin,* 783–787.

Boll, F. 1877a. Zur Anatomie und Physiologie der Retina. *Archiv für Physiologie,* 4–37.

Boll, F. 1877b. Zur Physiologie des Sehens und Farbenempfindung, *Monatsberichte der Königliche Preussichen Akadamie der Wissenschaften zu Berlin,* 2–7.

Bouillaud, J.-B. 1830. Recherches expérimentales sur les fonctions du cerveau. *Journal de Physiologie Expérimentale et Pathologique, 10,* 36–98.

Bowman, W. 1849. *Lectures on the Parts Concerned in the Operations on the Eye, and on the Structure of the Retina, Delivered at the Royal London Ophthalmic Hospital, Moorfields, June, 1847. To which are added, a Paper on the Vitreous Humor; and also a Few Cases of Ophthalmic Disease.* London: Longman, Brown, Green and Longmans.

Brazier, M. 1988. *A History of Neurophysiology in the 19th Century.* New York: Raven Press.

Brodmann, K. 1909. *Vergleichende Lokalisationslehre der Grosshirnrinde in ihren Prinzipien dargestellt auf Grund des Zellenbaues.* Leipzig: J. A. Barth.

Brown, S., and Schäfer, E. A. 1888. An investigation into the functions of the occipital and temporal lobes of the monkey's brain. *Philosophical Transactions of the Royal Society of London (Biology), 179,* 303–327. (Paper read December 15, 1887)

Burdach, K. F. 1819, 1822, 1826. *Von Baue und Leben des Gehirns* (3 vols.). Leipzig: Dyk'schen Buchhandlung.

Buzzi, F. 1782. *Nuove sperienze fatte sull'occhio umano.* Milano: Opuscoli Scelti Sulle Scienze e Sulle Arti.

Caton, R. 1875. The electric currents of the brain. *British Medical Journal, 2,* 278.

Caton, R. 1877. Interim report on investigation of the electric currents of the brain. *British Medical Journal, 1, Suppl. L,* 62.

Caton, R. 1887. Researches on electrical phenomena of cerebral grey matter. *Transactions of the Ninth International Medical Congress, 3,* 246–249.

Caton, R. 1891. Die Ströme des Centralnervensystems. *Centralblatt für Physiologie, 4,* 758–786.

Chaillou, F. H. 1863. Observations sur ramollissement multiple du cerveau et du cervelet. *Bulletins de la Société d'Anatomie (Paris), 2d. Ser., 8,* 70–73.

Chance, B. 1922. Some English worthies of science of interest to ophthalmologists. *Annals of Medical History, 4,* 251–270.

Charpentier, A. 1877. De la vision avec les diverses parties de la rétine. *Archives de Physiologie, Ser. 2, 4,* 894–945.

Clarke, E., and O'Malley, C. D. 1968. *The Human Brain and Spinal Cord.* Berkeley: University of California Press.

Corti, A. 1850. Beitrag zur Anatomie der Retina. *Archiv für Anatomie, Physiologie und wissenschaftliche Medizin,* 273–275.

Descartes, R. 1686. *Tractatus de homine, et de formatione foetus.* Amstelodami: Blaviana.

Ewens, G. F. W. 1893. A theory of cortical visual representation. *Brain, 16,* 475–491.

Exner, S. 1896. Über autokinetische Empfindung. *Zeitschrift für Psychologie und Physiologie der Sinnesorgane, 12,* 313–330.

Ferrier, D. 1876. *The Functions of the Brain.* London: Smith, Elder and Company.

Ferrier, D. 1878. *Localisation of Cerebral Disease.* London: Smith, Elder and Company.

Ferrier, D. 1881. Cerebral ambylopia and hemiopia. *Brain, 3,* 456–477.

Ferrier, D. 1886. *The Functions of the Brain* (2nd ed.). New York: G. P. Putnam's Sons.

Ferrier, D. 1888. Schäfer on the temporal and occipital lobes. *Brain, 11,* 7–30.

Ferrier, D., and Turner, W. A. 1898. An experimental research upon cerebro-cortical afferent and efferent tracts. *Proceedings of the Royal Society of London, 62,* 1–3.

Ferrier, D., and Turner, W. A. 1901. Experimental lesion of the corpora quadrigemina in monkeys. *Brain, 24,* 27–46.

Flechsig, P. E. 1895. Weitere Mittheilungen über die Sinnes- und Associationscentren des menschlichen Gehirns. *Neurologisches Centralblatt, 14,* 1118–1124, 1177–1179.

Flechsig, P. E. 1896. *Gehirn und Steele.* Leipzig: Veit.

Flechsig, P. E. 1901. Developmental (myelogenetic) localisation of the cerebral cortex in the human subject. *Lancet, ii,* 1027–1029.

Flourens, M.-J.-P. 1824. *Recherches Expérimentales sur les Propriétés et les Fonctions du Système Nerveux dans les Animaux Vértebrés.* Paris: J. B. Ballière.

Flourens, M.-J.-P. 1842. *Recherches Expérimentales sur les Propriétés et les Fonctions du Système Nerveux dans les Animaux Vertébrés* (2nd ed.). Paris: J. B. Ballière.

Fulton, J. F. 1937. A note on Francesco Gennari and the early history of cytoarchitectural studies of the cerebral cortex. *Bulletin of the Institute for the History of Medicine, 5,* 895–913.

Gall, F., and Spurzheim, J. 1809. *Recherches sur le Système Nerveux en Général, et sur Celui du Cerveau en Particulier* . . . Paris: F. Schoell, H. Nicolle.

Gall, F., and Spurzheim, J. 1810–1819. *Anatomie et Physiologie du Système Nerveux en Général, et du Cerveau en Particulier.* Paris: F. Schoell. (Gall was the single author of the first two volumes of the four in this series.)

Gennari, F. 1782. *De peculiari structura cerebri, nonnolisque ejus morbis.* Parmae: Ex Reg. Typog.

Gibson, W. C. 1962. Pioneers in localization of function in the brain. *Journal of the American Medical Association, 180,* 944–951.

Giovanardi, E. 1881. Intorno ad un caso di anoftalmia doppia congenita (mancanza dei nervi ottici, atrofia dei lobi occipitali). *Revista Sperimentale di Freniatria e Medicina Legale delle Alienazioni Mentali, 7,* 244–250.

Golgi, C. 1873. Sulla struttura della grigia del cervello. *Gazetta Medica Italiani Lombardi, 6,* 244–246. Translated by M. Santini as, "On the structure of the gray matter of the brain," in M. Santini (Ed.), *Golgi Centennial Symposium, Proceedings.* New York: Raven Press, 1975. Pp. 647–650.

Goltz, F. 1881. *Ueber die Verrichtungen des Grosshirns.* Bonn: E. Strauss.

Graefe, A. von. 1856. Ueber die Untersuchung des Gesichtsfeldes bei amblyopischen Affectionen. *Archiv für Ophthalmologie, 2,* 258–298.

Gratiolet, P. 1854. Note sur les expansions des racines cérébrales du nerf optique et sur leur terminaison dans une région déterminée de l'écorce des hémisphères. *Comptes Rendus Hebdomadaires des Séances de l'Académie des Sciences de Paris, 29,* 274–278.

Grote, L. R. 1925. *Die Medizin der Gegenwart in Selbstdarstellung* (Vol. 5). Hamburg: Meiner Verlag.

Haab, O. 1882. Ueber Cortex-Hemianopie. *Klinische Monatsblätter für Augenheilkunde, 20,* 141–153.

Haller, A. von. 1786. *First Lines of Physiology.* Translated from the Latin. (Reprinted 1966; New York: Johnson Reprint Company)

Hannover, A. 1840. Die Chromsäure ein vorzügliches Mittel bei mikroskopischen Untersuchungen. *Archiv für Anatomie, Physiologie und wissenschaftliche Medizin,* 548–558.

Helmholtz, H. von. 1856, 1860, 1866. *Handbuch der physiologischen Optik,* 3 vols. Hamburg: Voss. (Edition 3 translated by Southall as, *Handbook of Physiological Optics;* New York: Dover Publications. 1962)

Henschen, S. E. 1893. On the visual path and centre. *Brain, 16,* 170–180.

Henschen, S. E. 1894. Sur les centres optiques cérébraux. *Revue Générale d'Ophthalmologie, 13,* 337–352.

Henschen, S. E. 1897. Ueber Lokalisation innerhalb des äusseren Kniehöckers. *Neurologisches Centralblatt, 16,* 923–924.

Henschen, S. E. 1903. La projection de la rétine sur la corticalité calcarine. *La Semaine Médicale, 23,* 125–127.

Henschen, S. E. 1910. Zentrale Sehstörungen. In M. Lewandowsky (Ed.), *Handbuch der Neurologie* (Vol. 2). Berlin: Springer. Pp. 891–918.

Hirschberg, J. 1875. Zur Semidecussation der Sehnervenfasern im Chiasma des Menschen. *Virchow's Archiv für pathologische Anatomie und Physiologie, 65,* 116.

Hooke, R. 1665. *Micrographia: Or some Physiological Descriptions of Minute Bodies Made by Magnifying Glasses.* London: J. Martyn and J. Allestry.

Hun, H. 1887. A clinical study of cerebral localization, illustrated by seven cases. *American Journal of Medical Sciences, 93,* 140–168.

Kilian, H. F. 1828. *Die Universtäten Deutschlands in medicinisch-naturwissenschaftlicher Himsicht* . . . Heidelberg: Karl Groos.

Kölliker, A. von. 1852a. Zur Anatomie und Physiologie der Retina. *Verhandlungen physiks-medizin Gesellschaft, Würtzburg, 3,* 316–336.

Kölliker, A. von. 1852b. *Mikroscopische Anatomie, II.* Leipzig: Engelmann.

König, A. 1894. Ueber den menschlichen Sehpurpur und seine Bedeutung für das Sehen. *Sitzungsberichte. Preussische Akademie der Wissenschaften,* 577–598.

Kries, J. von. 1894. Ueber den Einfluss der Adaptation auf Licht- und Farbenempfindung und über die Funktion der Stäbchen. *Berichte naturforschen de Gesellschaft, Freiburg im Breisgau, 9,* 61–70.

Kries, J. von. 1895. Ueber die Funktion der Netzhautstäbechen. *Zeitschrift für Psychologie, 8,* 81–123.

Kühne, W. 1878. Über den Sehpurpur. *Untersuchungen physiologisches Institut, Universität Heidelberg, 1,* 15–103.

Kühne, W., and Ewald, C. A. 1877–1878. Untersuchungen über den Sehpurpur. *Untersuchungen physiologisches Institut, Universität Heidelberg, 1,* 139–218, 248–290, 370–455.

Leeuwenhoek, A. van. 1674. More observations from Mr. Leewenhook, in a letter of Sept. 7. 1674. sent to the publisher. *Philosophical Transactions of the Royal Society of London, 9,* 178–182

Leeuwenhoek, A. van. 1675. Microscopical observations of Mr. Leewenhoeck, concerning the optic nerve, communicated to the publisher in Dutch, and made by him in English. *Philosophical Transactions of the Royal Society of London, 10,* 378–380.

Leeuwenhoek, A. van. 1684. A letter from Mr. Anthony Leewenhoeck, Fellow of the Royal Society, dat. Apr. 14. 1684. containing observations about the crystallin humor of the eye, etc. *Philosophical Transactions of the Royal Society of London, 14,* 780–789.

Le Grand, Y. 1975. History of research on seeing. In E. C. Carterette and

M. P. Friedman (Eds.), *Handbook of Perception: Vol. 5. Seeing.* New York: Academic Press. Pp. 3–23.

Lenz, G. 1909. Zur Pathologie der cerebralen Sehbahn unter besonderer Berücksichtigung ihrer Ergebnisse für die Anatomie und Physiologie. *Archiv für Ophthalmologie, 72,* 1–85, 197–273.

Levinsohn, G. 1904. Beiträge zur Physiologie des Pupillarreflexes. *Archiv für Ophthalmologie, 59,* 191–200, 436–458.

Loeb, J. 1884. Die Sehstörungen nach Verletzung der Grosshirnrinde. *Archive für die gesammte Physiologie, 34,* 67–172.

Loeb, J. 1885. Die elementaren Störungen Functionen nach oberflächlicher, umschriebener Verletzung des Grosshirns. *Pflüger's Archiv für die gesammte Physiologie, 37,* 51–56.

Loeb, J. 1886. Beiträge zur Physiologie des Grosshirns. *Pflüger's Archiv für die gesammte Physiologie, 39,* 265–346.

Luciani, L. 1884. On the sensorial localisations in the cerebral cortex. *Brain, 7,* 145–160.

Luciani, L., and Tamburini, A. 1879. *Sulle Funzioni del Cervello.* Regio-Emilia: S. Calderini. Abstracted by A. Rabagliati in *Brain,* 1879, *2,* 234–250.

Magendie, F. 1844. *An Elementary Treatise on Human Physiology* (5th ed.). (J. Revere, Trans). New York: Harper and Brothers.

Mariotte, L'Abbé. 1668. A new discovery touching vision. *Philosophical Transactions of the Royal Society of London, 3,* 668–671.

Mariotte, L'Abbé. 1670. The answer of Monsieur Mariotte to Monsieur Pecquet, concerning the principal organ of vision; where occurr divers considerable experiments. *Philosophical Transactions of the Royal Society of London, 5,* 1023–1042.

Mauthner, L. 1881. *Gehirn und Auge.* Wiesbaden: J. B. Bergmann.

Mazzolini, R. G. 1991. Schemes and models of the thinking machine (1662–1762). In P. Corsi (Ed.), *The Enchanted Loom.* New York: Oxford University Press. Pp. 68–143.

McHenry, L. C., Jr. 1969. *Garrison's History of Neurology.* Springfield, IL: Charles C Thomas.

McHenry, L. C., Jr. 1983. Classics in neurology. *Neurology, 33,* 765.

Meyer, A. 1971. *Historical Aspects of Cerebral Anatomy.* London: Oxford University Press.

Meynert, T. 1870. Beiträge zur Kenntniss der centralen Projection der Sinnesoberflächen. *Sitzungberichte der Kaiserlichten Akademie der Wissenschaften, Wien. Mathematisch-Naturwissenschaftliche Classe, 60,* 547–562.

Michaelis, E. 1877. *Albrecht von Graefe. Sein Leben und Wirken.* Berlin: G. F. Reimer.

Minkowski, M. 1912. Experimentelle Untersuchungen über die Beziehungen des Grosshirns zum Corpus geniculatum externum. *Neurologisches Centralblatt, 31,* 1470–1472.

Minkowski, M. 1913. Experimentelle Untersuchungen über die Beziehungen des Grosshirnrinde und der Netzhaut zu den primären optischen Zentren, besonders zum Corpus geniculatum externum. *Arbeiten. Hirnanatomisches Institut, Zürich, 7,* 255–362.

Minkowski, M. 1920. Über den Verlauf, die Endigung und die zentrale Repräsentation von gekreuzten und ungekreuzten Sehnervenfasern bei einigen Säugetieren und beim Menschen. *Schweizer Archiv für Neurologie und Psychiatrie, 6,* 268–303.

Monakow, C. von. 1900. Pathologische und anatomische Mittheilungen über die optischen Centren des Menschen. *Neurologisches Centralblatt, 19,* 680–681.

Morgagni, G. B. 1719. *Adversaria anatomica omnia.* Patavii: Excudebat Josephus Cominus.

Mott, F. W. 1907. The progressive evolution of the structure and functions of the visual cortex in mammalia. *Archives of Neurology, 3,* 1–48.

Müller, H. 1856. Observations sur la structure de la rétine de certains animaux. *Comptes Rendus de l'Académie des Sciences de Paris, 43,* 743–745.

Müller, H. 1857. Anatomish-physiologische Untersuchungen über die Retina bei Menschen und Wirbeltierin. *Zeitschrift für wissenschaftliche Zoologie, 8,* 1–124.

Müller, J. 1837. Jahresbericht über die Fortschritte der anatomisch-physiologischen Wissenschaften im Jahre 1836. *Archiv für Anatomie, Physiologie und wissenschaftliche Medizin,* i–cxxxxiii.

Müller, J. 1838. *Handbuch der Physiologie des Menschen für Vorlesungen.* Coblenz: J. Hölscher.

Munk, H. 1878. Weitere Mittheilungen zur Physiologie der Grosshirnrinde. *Verhandlungen der Physiologischen Gesellschaft zu Berlin,* 162–178.

Munk, H. 1879a. Weiteres zur Physiologie der Sehsphäre der Grosshirnrinde. *Verhandlungen der Physiologischen Gesellschaft zu Berlin,* 581–594.

Munk, H. 1879b. Physiologie der Sehsphäre der Grosshirnrinde. *Centralblatt für praktische Augenheilkunde, 3,* 255–266.

Munk, H. 1881. *Über die Funktionen der Grosshirnrinde.* 3te Mitteilung, pp. 28–53. Berlin: A. Hirschwald. In G. von Bonin, *Some Papers on the Cerebral Cortex.* Translated as, "On the functions of the cortex." Springfield, IL: Charles C Thomas, 1960. Pp. 97–117.

Munk, H., and Obregia. 1890. Sehsphäre und Augenbewegungen. *Sitzungberichte der Königlichen Preussischen Akademie der Wissenschaften zu Berlin, 1,* 53–74.

Newton, I. 1704. *Opticks*. London: Smith and Walford. (Reprinted 1952; New York: Dover Publications)

Obersteiner, H. 1888. *Anleitung beim Stadium des Baues der nervösen Centralorgane im gesunden und kranken Zustände*. Leipzig: Toeplitz.

Panizza, B. 1855. Osservazioni sul nervo ottico. *Instituto Lombardo di Scienze e Lettere, Milan. Giornale di Scienze e Lettere, 7*, 237–252.

Panizza, B. 1856. Osservazioni sul nervo ottico. *Memoria, Istituto Lombardo di Scienze, Lettere e Arte, 5*, 375–390.

Parinaud, H. 1883. Paralysie des mouvements associés des yeux. *Archives de Neurologie, 5*, 145–172.

Parinaud, H. 1884. De l'intensité lumineuse des couleurs spectrales; influence de l'adaptation rétinienne. *Comptes Rendus de l'Académie des Sciences de Paris, 99*, 937–939.

Parinaud, H. 1886. Paralysis of the movement of convergence of the eyes. *Brain, 9*, 330–341.

Parsons, J. H. 1902. Degenerations following lesions of the retina in monkeys. *Brain, 25*, 257–269.

Parsons, J. H. 1924. *An Introduction to the Study of Colour Vision* (2nd ed.). London: Cambridge University Press.

Pecquet. 1668. The answer to M. Mariotte. *Philosophical Transactions of the Royal Society of London, 5*, 669–671.

Perera, C. A. 1935. Albrecht von Graefe, founder of modern ophthalmology. *Archives of Ophthalmology, 14*, 742–773.

Polyak, S. L. 1957. *The Vertebrate Visual System*. Chicago: University of Chicago Press.

Pusey, B. 1903. An old-time quack eye doctor. *Journal of the American Medical Association, 41*, 1142–1144.

Ramón y Cajal, S. 1893. La rétine des vertébrés. *La Cellule, 9*, 119–255.

Ramón y Cajal, S. 1909–1911. *Histologie du Système Nerveux de l'Homme et des Vertébrés*. Paris: A. Maloine.

Ramón y Cajal, S. 1923. *Recuerdos de Mi Vida* (3rd ed.). Madrid: J. Pueyo.

Ranson, S. W. 1920. *The Anatomy of the Nervous System*. Philadelphia, PA: W. B. Saunders.

Remak, R. 1844. Neurologische Erläuterungen. *Archiv für Anatomie, Physiologie und wissenschaftliche Medizin*, 463–472.

Rivers, W. H. R. 1900. Vision. In E. A. Schäfer (Ed.), *Text-Book of Physiology* (Vol. 2). Edinburgh: Young J. Pentland. Pp. 1026–1148.

Rolando, L. 1828. *Saggio Sopra la Vera Struttura del Cervello e Sopra le Funzoni del Sistema Nervoso*. Torino: Pietro Marietti.

Rothmann, M. 1909. Hermann Munk zum 70. Geburtstag. *Deutsche medicinische Wochenschrift, 35*, 258–259.

Rushton, W. A. H. 1962. *Visual Pigments in Man*. (Sherrington Lecture VI). Springfield, IL: Charles C Thomas.

Santorini, G. 1724. *Observationes anatomicae*. Venetiis: Apud Jo. Baptistam Recurti.

Schäfer, E. A. 1888a. Experiments on special sense localisations in the cortex cerebri of the monkey. *Brain, 10*, 362–380.

Schäfer, E. A. 1888b. On electrical excitation of the occipital lobe and adjacent parts of the monkey's brain. *Proceedings of the Royal Society of London, 43*, 408–410.

Schäfer, E. A. 1888c. Experiments on the electrical excitation of the visual area of the cerebral cortex in the monkey. *Brain, 11*, 1–6.

Schäfer, E. A. 1888d. On the functions of the temporal and occipital lobes: a reply to Dr. Ferrier. *Brain, 11*, 145–166.

Schäfer, E. A. 1900. The cerebral cortex. In E. A. Schäfer (Ed.), *Textbook of Physiology*. Edinburgh: Young J. Pentland. Pp. 697–782.

Schröder, P. 1930. Paul Flechsig. *Archiv für Psychiatrie und Nervenkrankheiten, 91*, 1–8.

Schultze, M. 1866. Zur Anatomie und Physiologie der Retina. *Archiv für mikroskopische Anatomie, 2*, 175–286.

Schultze, M. 1867a. Ueber Stäbchen und Zapfen der Retina. *Archiv für mikroskopische Anatomie, 3*, 215–247.

Schultze, M. 1867b. Bemerkungen über Bau und Entwickelung der Retina. *Archiv für mikroskopische Anatomie, 3*, 371–382.

Schultze, M. 1872. Sehorgan. I. Die Retina. In S. Stricker (Ed.), *Handbuch der Lehre von den Geweben des Menschen und der Thiere*. Leipzig: Engelmann. Pp. 977–1034. Translated as, "The eye: The retina," in S. Stricker (Ed.), *Manual of Human and Comparative Histology*. London: New Sydenham Society, 1873. Pp. 218–298.

Smith, G. E. 1904a. Studies in the morphology of the human brain with special reference to that of the Egyptians. No. 1. The occipital region. *Records of the Egyptian Government School of Medicine* (Vol. 2). Cairo: National Printing Department. Pp. 125–173.

Smith, G. E. 1904b. The morphology of the occipital region of the cerebral hemisphere in man and the apes. *Anatomischer Anzeiger, 24*, 436–451.

Smith, G. E. 1904c. The fossa parieto-occipitalis. *Journal of Anatomy, 38*, 164–169.

Sömmerring, S. T. von. 1795. Paper summarized under *Göttingische Unzeigen von gelehrten Sachen: Göttingen. Göttingische Akademie der Wissenschaften, 140*, 1401–1402.

Sömmerring, S. T. von. 1801. *Icones oculi humani*. Francofurti ad Moenum: Varrentrapp und Wenner.

Starr, M. A. 1884a. The visual area in the brain determined by a study of hemianopsia. *American Journal of Medical Sciences, 87*, 65–83.

Starr, M. A. 1884b. Cortical lesions of the brain. A collection and analysis of the American cases of localized cerebral disease. *American Journal of Medical Sciences, 88*, 114–141.

Starr, M. A. 1884c. Cortical lesions of the brain. A collection and analysis of the American cases of localized cerebral disease. *American Journal of Medical Sciences, 87*, 366–391.

Sweeney, P. J. 1984. Isaac Newton and the optic chiasm. *Neurology, 34*, 309.

Sterling, W. 1902. *Some Apostles of Physiology*. London: Waterlow and Sons, Ltd.

Tamburini, A. 1880. Rivendicazione al Panizza della scoperta del centro visivo corticale. *Revista Sperimentale di Freniatria e Medicina Legale, 6*, 153–154.

Taylor, J. 1727. *An Account of the Mechanism of the Eye* . . . Norwich: H. Cross-grove.

Taylor, J. 1750. *Mechanismus* . . . Frankfurt am Maym: Bey Stoks seel. erben und Schilling.

Treviranus, G. R. 1828. *Beiträge zur Anatomie und Physiologie der Sinneswerkzeuge des Menschen und der Thiere*. Bremen: Heyse.

Treviranus, G. R. 1835–1838. *Beiträge zur Aufklärung der Erscheinungen und Gesetze des organischen Lebens*. Bremen: Heyse.

Tschermak, A. 1902. Die Hell-Dunkeladaptation des Auges und die Funktion der Stäbchen und Zapfen. *Ergebnisse der Physiologie, 1*, 694–800.

Vater, A., and Heinicke, J. C. 1723. *Dissertatio inauguralis medica qua visus vitia duo rarissima alterum duplicati, alterum dimidiati physiologice et pathologice considerata*, . . . Wittenbergae: Literis Viduae Gerdesiae.

Vicq d'Azyr, F. 1781. Sur la structure de cerveau, de cervelet, de la moelle alongée, de la moelle épinière; et sur l'origine des nerfs de l'homme et des animaux. *Histoire de l'Académie Royale des Sciences*, 495–622.

Vicq d'Azyr, F. 1786. *Traité d'Anatomie et de Physiologie*. Paris: F. A. Didot.

Vieussens, R. de. 1684. *Neurographia universalis*. Lyons: J. Certe.

Wilbrand, H. 1881. *Ueber Hemianopsie und ihr Verhältnis zur topischen Diagnose der Gehirnkrankheiten*. Berlin: A. Hirschwald.

Wilbrand, H. 1887. *Die Seelenblindheit als Herderscheinung und ihre Beziehungen zur homonymen Hemianopsie*. Wiesbaden: J. F. Bergmann.

Wilbrand, H. 1890. *Die hemianopischen Gesichtsfeld-Formen und das optische Wahrnehmungszentrum*. Wiesbaden: J. F. Bergmann.

Wilkins, R. H., and Brody, I. A. 1972. Parinaud's syndrome. *Archives of Neurology, 26*, 91–93.

Willis, T. 1664. *Cerebri anatome: cui accessit nervorum descriptio et usus*. London: J. Martyn and J. Allestry.

Wollaston, W. H. 1824. On semi-decussation of optic nerves. *Philosophical Transactions of the Royal Society of London, 114*, 222–231.

Zinn, J. G. 1755. *Descriptio anatomica oculi humani iconibus illustrata* . . . Gottingae: B. A. Vandenhoeck.

Chapter 7
Color Vision

As it is almost impossible to conceive of each sensitive point of the retina to contain an infinite number of particles, each capable of vibrating in perfect unison with every possible undulation, it becomes necessary to suppose the number limited, for instance, to the three principle colours, red, yellow, and blue.

Thomas Young, 1802

Several very speculative early ideas relating to color vision were briefly presented in Chapter 5. These were mostly Greek metaphysical theories, which included three basic thoughts. The first was that white and black are colors, and not just shades of light intensity. This was a notion held by Empedocles, who lived in the fifth century B.C. The second important idea maintained that there are a finite number of basic colors. The third suggested that different colors have something to do with atoms or particles differing in size, shape, or motion.

The idea that there are color primaries is usually associated with Democritus, who stated that the four basic colors were black, white, red, and yellow. He thought that mixtures of these four "primaries" could produce all of the known colors. Democritus experimented with painters' dyes. His colors most likely involved pulverized marble (or shells) for white, charred bone or soot for black, a mercury compound (vermilion) for red, and the earth color, ocher, for yellow. Democritus believed his four primaries corresponded to the four basic elements and that each color was made up of atoms differing in spatial arrangement.

Plato, who accepted much of what Democritus had to say, wrote that different-sized particles produced different motions, which in turn led to the sensations of different colors. In varying forms, ideas like these were accepted by the Roman and Arabic writers and appear in Latin texts into the Renaissance.

The emphasis of this chapter is on advances after this period, beginning with the discoveries of Sir Isaac Newton and ending with color theories that dominated thinking at the beginning of the twentieth century.

NEWTON'S THEORY OF LIGHT

In 1665, Robert Hooke suggested that perceived differences in color were due to the different angles at which the pulses of light struck the retina.[1] He further theorized that

red and blue were the two basic colors and that all other colors were intermediates between these two.

Hooke's theory of color was rejected by Sir Isaac Newton, whose many accomplishments included the invention of calculus (with Gottfried Wilhelm Leibniz, 1646–1716), the formulation of a theory of planetary motion, the discovery of the three laws of mechanics (Newton's laws), and the formulation of the law of gravity.

The story often told is that Newton, who worked at Cambridge University, bought a prism at a fair to study the refraction of light.[2] He observed that the prism split the light into different spectral colors. This led him to conclude that the lenses in conventional telescopes had to introduce serious chromatic errors. Newton first applied his knowledge to the construction of a better device, the reflecting telescope. Then, in 1671–1672, he presented his theory of color to the Royal Society of London.

In this report he wrote that rays of light differ in degrees of *refrangibility* (the theorized angle of refractibility) and correspond to different colors (the most refrangible rays are deep violet and the least refrangible are red). He maintained that the rays of primary colors may mix to create compound colors (e.g., yellow and blue mix to create green). And, in a radical departure from the past, he added that white is not a basic color at all but a combination of all of the spectral colors.

Newton was astonished by the opposition to his color theory by Hooke, Mariotte, Christiaan Huygens (1629–1695; shown in Figure 7.1), and many other natural philosophers and experimentalists. The surprise came largely because he thought he had already presented solid arguments that adequately addressed their objections to his ideas (Sepper, 1988). In fact, it is usually stated that Newton's initial communication resulted in so much criticism that it was not until 1704 that he finally gave a full account of his propositions, theorems, and experiments. This took place in his *Opticks*.

Newton suffered from two periods of severe emotional and mental disturbances, and it is possible that they also

---

[1]Hooke served as an assistant to Robert Boyle and was appointed Curator of Experiments by the newly founded Royal Society. Hooke also played a role in the development of the microscope (see note 2, Chapter 6).

[2]For a brief biography of Newton that emphasizes his work on vision, see Chance (1922).

Figure 7.1. Christiaan Huygens (1629–1695), discoverer of the basic laws of refraction and a proponent of the wave theory of light. (From Huygens, 1724.)

effectively delayed the presentation of *Opticks*. His first "breakdown" began in 1676 and resulted in his intellectual isolation for six years; the second, which has been better documented, started in 1693 and again lasted a few years. In both cases, Newton showed a tremor (hatter's shakes in the United Kingdom, Danbury shakes in the United States) that affected his handwriting and created severe insomnia, loss of appetite, delusions of persecution, problems with memory, mental confusion, and a decline in personal relationships. These symptoms fit the profile of mercury poisoning, a condition described more than a century earlier by Paracelsus among the miners of Idria and by others in Newton's own time (Johnson and Wolbarsht, 1979; Klawans, 1991; Spargo and Pounds, 1979).

Support for the hypothesis that Newton suffered from mercury poisoning comes from records which show Newton excessively exposed himself to mercury, especially in connection with his studies on alchemy (1676–79) and optics (1692–93). Newton not only inhaled mercury vapors but rubbed, handled (e.g., to make "quick-silver'd" surfaces), and tasted mercury, in association with these experiments. He even joked that his hair turned prematurely gray (around age 30) from his work with quicksilver.[3]

When *Opticks* finally did appear, Newton showed precisely how each color had its own specific refractive properties. His critical experiment involved passing white light through two prisms (one with a large refracting angle) that split the light into its spectral colors, each with its own angle of refractibility (Sepper, 1988). After demonstrating white light was composed of rays corresponding to different colors, he showed that multiple refractions could reconstitute the white beam. Newton thus proved that white light is a mixture of all the colors and that gray is only a less luminous white.

[3]Direct evidence for the mercury poisoning hypothesis emerged when some of Newton's hair was assayed (Spargo and Pounds, 1979).

Newton went on to formulate the laws of color mixture. He concluded that the color of an object was due to selective absorption and reflection. He even showed that rapid succession of colors produces the same effects as simultaneous mixture.[4]

But whereas Newton was correct in most of his ideas, he erred when he hypothesized that because there are seven musical tones of an octave, there should also be seven basic colors that follow the same laws. His conception of retinal and higher processes was also based on theory more than facts. In his search for mathematical precision, he hypothesized that the optic nerve transmitted the specific vibrations of his seven primary colors to the sensorium. Fortunately, this physiological theorizing did not seem to detract from Newton's contributions to the physics of color and light.

COLOR VISION IN THE EIGHTEENTH CENTURY

The theory of trichromatic color vision, usually thought of as a nineteenth-century development, had clear antecedents in the eighteenth century. For example, Mikhail Vasil'evich Lomonosow (1711–1765) used the framework of the particle theory of light to suggest three kinds of "ether particles." The large ones, he believed, corresponded to red, the medium ones to yellow, and the small ones to blue (Lomonosow, 1756; see Weale, 1957). As stated in the *Monthly Review* from 1759, which abstracted Lomonosow's theory from the Latin:

> This is an attempt to establish a new theory of light and colours. Mr. Lomonosow supposes light to consist of a subtle matter, agitated by peripheral vibrations and gyrations, on which latter species of motion depend our sensations of colours. He conjectures the particles of light to be spherical, and to be of three different sizes, so adapted to each other that the smallest may be included in the interstices of the largest: the gyrations of the largest spheres, producing the sensations of red, those of the middle size, yellow, and of the smallest blue. On the whole, the piece is ingenuous enough, but being purely hypothetical, we shall not trouble our readers with the conclusions our academic draws from premises so chimerical.

Shortly after Lomonosow's theory appeared, a different trichromatic color theory was put forth by a mysterious individual who went by the name G. Palmer (Walls, 1956). Palmer published his first paper in London in 1777. This paper was presented in the form of a dialogue between Palmer and a man named Johnson, who was made to sound like Samuel Johnson himself. This may have been done to confer prestige upon the work.

Palmer made a case for three receptors and based it on a series of experiments conducted with prisms. Some of the "principles" that Palmer laid down in accordance with his belief in the wave theory of light were:

> (1) Each ray of light is compounded of three other rays only; one of these rays is analogous to the yellow, one to red, and the other to blue.

[4]Newton relied heavily upon the scientific approach advocated by Francis Bacon (1561–1626). Bacon's type of science emphasized method, experimentation, and observation. It contrasted with Cartesian approaches, which were decidedly more mechanical, metaphysical, and speculative.

(2) A white surface, by reflecting all these rays, shews an absolute want of colours.

(3) A surface painted of the three colouring principles . . . by absorbing the three rays of light . . . shews an absolute want of light, or a perfect blackness. (1777, 5)

Some of his other principles related to vision. Two examples are:

(1) The superficies of the retina is compounded of particles of three different kinds, analogous to the three rays of light; and each of these particles is moved by his own ray.

(2) The motion of these particles by discomposed rays, whether by coloured bodies, or by prismatic refractions, produces the sensation of colours. (1777, 6)

Palmer followed this paper with a private communication and then with a pamphlet in French, dated 1786 (Weale, 1957).

EARLY DESCRIPTIONS OF COLOR BLINDNESS

The interest in color theory stimulated by Newton's discoveries was enhanced by John Dalton's (1766–1844) work on color blindness. Dalton was a highly respected chemist and physicist who taught pharmaceutical chemistry at the Pine Street School of Medicine in Manchester and who founded the atomic theory of chemistry. In 1794, after being elected to the Manchester Literary and Philosophical Society, he published a paper entitled *Extraordinary Facts Relating to the Vision of Colours: With Observations*. In this paper, Dalton, shown in Figure 7.2, told about his own color defect and described color defects in several other people.

Dalton explained that while observing a pink geranium in candle-light he discovered he could not distinguish between pink and blue.

Figure 7.2. John Dalton (1766–1844), the chemist who described his own case of color blindness in 1794. (From Sterling, 1902.)

With respect to colours that were *white, yellow,* or *green,* I readily assented to the appropriate term. *Blue, purple, pink,* and *crimson* appeared rather less discriminable; being, according to my idea, all referable to *blue.* I have often seriously asked a person whether a flower was *blue* or *pink,* but was generally considered to be in jest. (1794, 29)

Dalton stated that while most people could distinguish six colors in the spectrum when they used a glass prism, he could distinguish only two or three (yellow, blue, and purple). He lamented:

My yellow comprehends the *red, orange, yellow,* and *green* of others; and my *blue* and *purple* coincide with theirs. That part of the image which others call *red* appears to me little more than a shade, or defect of light; after that the orange, yellow and green seem *one* colour, which descends pretty uniformly from an intense to a rare yellow, making what I should call different shades of yellow. (p. 31)

In addition to collecting data on his brother, Dalton collected information on 28 other people who seemed to show visual defects similar to his own. The fact that this type of color blindness was more likely to affect males than females was also noted. To quote Dalton: "It is remarkable that I have not heard of one female subject to this peculiarity" (p. 40).

Dalton thought his type of color blindness might be due to abnormal coloring in one of the humors in the eye. He postulated that the most likely possibility was some blue coloring in the vitreous humor. An examination of Dalton's own eyes after death failed to support this theory, but the interesting nature of Dalton's discoveries led to the term "Daltonism" being applied to his type of color blindness.

Although he drew considerable attention to this sort of color blindness, Dalton was not the first to describe it. The same disorder was described in 1777 by Joseph Huddart (1741–1816). In fact, Huddart's case was cited by Dalton in his 1794 paper. Huddart presented his material in the form of a letter that appeared in the *Philosophical Transactions of the Royal Society.* It involved a shoemaker named Harris who seemed to confuse red and green. Harris had two brothers who were similarly affected, but he stated that his parents and other siblings did not appear to be color blind.

Huddart wrote that Harris

had reason to believe that other persons saw something in objects which he could not see; that their language seemed to mark qualities with confidence and precision, which he could only guess at with hesitation, and frequently with error. His first supposition of this arose when he was about four years old. Having by accident found in the street a child's stocking, he carried it to a neighbouring house to inquire for the owner: he observed the people called it a *red* stocking, though he did not understand why they gave it that denomination, as he himself thought it completely described by being called *a stocking.* . . . He observed also that, when young, other children could discern cherries on a tree by some pretended difference of colour, though he could only distinguish them from the leaves by their difference of size and shape. He observed also, that by means of this difference of colour, they could see the cherries at a greater distance than he could, though he could see other objects at as great a distance as they; that is, where the sight was not assisted by colour. (1777, 261–262)

Michael Lort (1725–1790) published another case of color blindness one year later in the *Philosophical Transactions of the Royal Society*. Lort also cited the Harris case and presented a personal letter from an Englishman named J. Scott, who appeared to have the same defect as Harris. Scott complained:

> I do not know any green in the world; a pink colour and a pale blue are alike, I do not know one from the other. A full red and a full green the same, I have often thought them a good match; but yellows (light, dark, middle) and all degrees of blue, except those very pale, commonly called sky, I know perfectly well. . . . I married my daughter to a genteel, worthy man a few years ago; the day before the marriage he came to my house, dressed in a new suit of fine cloth cloaths. I was much displeased that he should come (as I supposed) in black: said, He should go back to change his colour. But my daughter said, No, no; the colour is very genteel; that it was my eyes that deceived me. He was a gentleman of the law, in a fine rich claret-coloured dress, which is as much a black to my eyes as any black that ever was dyed. (1778, 612–613)

Scott attributed his problem to "a family failing." His father, his sister, his sister's two sons, and his mother's brother all had the same visual problem. Scott went on to say that he had no other trouble with his eyesight and hoped that the men of this "learned Society can fetch out the cause of this very extraordinary infirmity, and find a method for an amendment" (p. 614).

Dalton's case was followed by a number of other reports of color blindness (e.g., Goethe, 1810). Notably, the people examined by August Seebeck (1805–1849) led him to divide his cases into two categories. Seebeck (1837) separated "deuteranopes," whose color system is reduced to blue and yellow but who have a normal spectrum length, from "protanopes," who also see yellow and blue but who have a shortened spectrum at the red end.[5]

In contrast to red-green color blindness, the existence of a congenital blue-yellow color blindness, which is much rarer, was not established in the nineteenth century (Maxwell, 1872). Some early cases of "tritanopia" seemed to be attributed to jaundice or sclerosis of the lens and unusually dense pigment in the macula (Collins, 1925).

YOUNG'S TRICHROMATIC THEORY

Thomas Young (1773–1829), a physician by training, was considered a genius from early childhood. He claimed he could read fluently at the age of 2, and that he had read the Bible twice before the age of 4 (Chance, 1922). By his late teens, he was familiar with 14 languages and showed many outstanding abilities. His interests included music, hydrodynamics, the study of tides, tables of mortality, theory of structures, and vision.

Young's first paper on vision concerned accommodation in the eye. He showed accommodation was not due to changes in the cornea, the humors of the eye, or the length

Figure 7.3. Thomas Young (1773–1829), best known for his trichromatic theory of color vision. (Front plate from Peacock, 1855).

of the eye. Instead, Young, shown in Figure 7.3, proved experimentally that the eye's ability to accommodate was due to a change in the curvature of the lens. This work, published in 1793, preceded his research on color vision and led to his election to the Royal Society when he was only 21 years old.

Young's theory of color vision was presented in a speech before the Royal Society in 1801 and published in 1802, the year he was appointed a professor of natural and experimental philosophy at the Royal Institute. His theory was developed as an alternative to what he viewed as an implausible idea. He simply could not accept the notion that each individual retinal element could respond with an infinite number of different vibrations to white light (light that contained all of the frequencies of the spectrum).

Young postulated that there were several types of distinct retinal elements, each best tuned to a specific color and each with its own connection to the central nervous system. Young's clearest statement about this included the segment at the opening of this chapter. To quote Young:

> As it is almost impossible to conceive of each sensitive point of the retina to contain an infinite number of particles, each capable of vibrating in perfect unison with every possible undulation, it becomes necessary to suppose the number limited, for instance, to the three principle colours, red, yellow, and blue, of which the undulations are related in magnitude nearly as the numbers 8, 7, and 6; and that each of the particles is capable of being put in motion more or less forcibly by undulations differing less or more from a perfect unison; for instance, the undulations of green light being nearly in the ratio 6–1/2, will affect equally the particles in unison with yellow and blue, and produce the same effect as light composed of those two species; and each sensitive filament of the nerve may consist of three portions, one for each principle color.

Soon after giving the 1801 *Bakerian Lecture* to the Royal Society, Young modified his initial ideas about the retinal resonators to accommodate the suggestion that red, green, and violet (and not red, yellow, and blue) were the

---

[5]The terms "protanopes" and "deuteranopes" were not used when Seebeck presented his material. These and related words were coined near the beginning of the twentieth century by Johannes von Kries (see Collins, 1925).

most important spectral colors (see Peacock, 1855). The numbers 8, 7, and 6 thus changed to 7, 6, and 5, which were in rough proportion to the wavelengths of the corresponding colors.

Young described the experimental support behind his new choices in *A Course of Lectures on Natural Philosophy and the Mechanical Arts* (Lecture 37, "On Physical Optics") published in 1807. Among other things, he suggested that Dalton's color blindness was probably due to the absence of the retinal resonator for red.

The seeds of the theory of specific nerve energies can be found in Thomas Young's work. This theory, which would blossom within the next few decades, stressed the importance of stimulating specific receptors and nerves for perception (Müller, 1826, 1838). Unfortunately, it is believed that Young never realized the full significance or the implications of his theory. Still, it was Young who may have been the first to explain three primary colors on the basis of visual physiology as opposed to the physics of light. As he put it, trichromatic theory is based "not in the nature of light, but in the constitution of man" (Maxwell, 1872, p. 260).

HELMHOLTZ'S MODIFICATIONS OF THE YOUNG THEORY

Young's theory attracted scant attention until 1852, when Hermann von Helmholtz, a disciple of Johannes Müller and already the inventor of the ophthalmoscope, brought Young's trichromatic theory to the fore.[6] This was long after Young had died and after the doctrine of specific nerve energies had been accepted.

Helmholtz was an ardent experimentalist who conducted many color-mixing experiments. These led him to accept Young's trichromatic theory of color, which he then proceeded to develop as a theory of three specific nerve energies. Helmholtz made the following statement about Young:

The Theory of Colours, with all its marvelous and complicated relations was a riddle . . . until I, at last, discovered that a wonderfully simple solution had been presented at the beginning of the century and had been in print ever since for everyone to read who chose. This solution was found and published by the same Thomas Young who first showed the right method of arriving at the interpretation of Egyptian hieroglyphics. He was one of the most acute men who ever lived, but had the misfortune to be too far in advance of his contemporaries. They looked on him with astonishment, but could not follow his bold speculations, and thus a mass of his most important thoughts remained buried and forgotten in the Transactions of the Royal Society, until a later generation by slow degrees arrived at the rediscovery of his discoveries, and came to appreciate the force of his arguments and the accuracy of his conclusions. (Cited in Chance, 1922, 268)

Helmholtz provided data to support and quantify Young's theory. As a result of his efforts and modifications, the trichromatic theory became known as the Young-Helmholtz theory of color vision.

A basic tenet of the new version of Young's trichromatic theory was that the red, green, and violet "fibers" were all affected by different wavelengths of light, although in different degrees. Thus, Helmholtz (1860) envisioned three different mathematical curves of excitation, each one with a different peak, but each overlapping the others. As Helmholtz stated in 1866:

1. The eye is provided with three distinct sets of nervous fibers. Stimulation of the first excites the sensation of red, stimulation of the second the sensation of green, and stimulation of the third the sensation of violet.

2. Objective homogeneous light excites these three kinds of fibers in various degrees, depending on its wavelength. The red-sensitive fibers are stimulated most by light of longest wavelength, and the violet-sensitive fibers by light of shortest wavelength. But this does not mean that each color of the spectrum does not stimulate all three kinds of fibers, some feebly and others strongly; on the contrary, in order to explain a series of phenomena, it is necessary to assume that that is exactly what does happen. (Translated in MacAdam, 1970, 97)

Helmholtz's work on vision and optics was presented in his *Handbuch der physiologischen Optik (Handbook of Physiological Optics),* a monumental work that appeared in three volumes (1856, 1860, 1866) and included extensive literature review. Other scientists soon attempted to modify the curves derived by Helmholtz and to work out more precisely the mathematics of the relationships of the primaries. Establishing such relationships often involved studying people with protanopia and deuteranopia (see Figure 7.4; König, 1903). These studies tended to concentrate on the fovea, especially after this region was demonstrated to be blue-blind by Arthur König (1856–1901) near the end of the century (e.g., König, 1894). This meant that the foveas of protanopes and deuteranopes contained only a single type of functional cone, an ideal situation for assessing the properties of that cone (Maxwell, 1872).

One criticism leveled at the theory was that the three proposed cone varieties seemed to look alike. That is, different red, green, and blue cones could not be identified by histologists. The failure to find distinct markers for each of the three cones led to the idea that perhaps each cone had all three retinal pigments. It was conceivable that as colored light affected them, different degeneration products might be liberated, each of which could affect the retinal neurons in a somewhat different way. Only well into the twentieth century was it proved that there really were three different cones in the normal human eye.

THE COLOR SCIENCE OF GOETHE

The visual science embraced by Newton, Young, and Helmholtz was different from a second approach emerging in German-speaking countries at this time. This was a period characterized by idealism in Germany and Austria, one in which poetry and literature were valued highly. To some individuals, personal experience became just as important, if not more important, than experimentation.

One of these men was Johann Wolfgang von Goethe (1749–1832). At the time of his foray into the visual sciences, this young poet, playwright, and novelist had

---

[6]Helmholtz invented the ophthalmoscope in 1850 and called it the "eye-mirror."

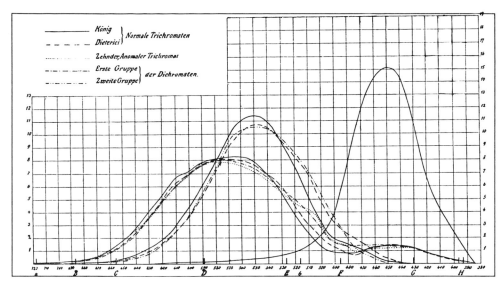

Figure 7.4. The Helmholtz-König-Dieterici distribution curves for normal trichromatic vision. The three basic curves (solid lines) were obtained from normal individuals and were presumed to represent the sensitivity functions of the three different retinal cones. The curves of individuals with various forms of defective color vision were compared to these standard curves to learn more about cone functions. (From Ladd-Franklin, 1924.)

already captured the German imagination with two great literary works, *Götz von Berlichingen mit der eisernen Hand. Ein Schauspiel (Götz von Berlichingen with the Iron Hand: A Play)* in 1773, and *Die Leiden des jungen Werthers (The Sorrows of Young Werther)* in 1774.

Goethe had always shown an intense interest in botany, anatomy, and other sciences and he felt well prepared to ascertain the rules that governed the artistic use of color. This venture was stimulated by a trip to Italy (1786–1788), where he noted that artists were able to enunciate rules for all of the elements of painting except color.

Goethe began to conduct his own optical experiments around 1790 (Sepper, 1988). Unlike his predecessors, however, he wished to study the phenomena of color in all of their manifestations. In a way, he hoped to discover principles pertaining to the unity of nature.

In his demand for a comprehensive science of color, Goethe opposed the thrust of the Newtonians, who attempted to approach color on a purely physical basis. Goethe recognized that understanding the physics of light was important, but emphasized this was but one approach among many needed to achieve a comprehensive understanding of color. Indeed, he argued, many perceptual effects, such as the effects of physical pressure on the eye, afterimages, and adaptation and contrast phenomena, could not be understood just by knowing the physics of light. No, the science of color could not be thought of as a simple appendage of physics. Natural things must be studied in context, and one must understand the nature and conditions responsible for them. The best approach was a broad effort involving chemists, physicists, philosophers, historians, physiologists, and people with other interests and orientations toward color.

Goethe's first publication on color was his *Beiträge zur Optik (Contribution Towards Optics)*, which appeared in two volumes in 1791 and 1792. This work dealt primarily with experiments using prisms. It also contained an implicit critique of Newton's theory, although Goethe's critics maintained it did not refute Newton's ideas. The book was written in a popular style intended to appeal to the reader's sense of beauty and experience. It was Goethe's intention to keep adding to the *Beiträge,* but the series ended with just two parts.

This effort was followed by *Zur Farbenlehre (Concerning Color Theory),* a work Goethe began to write in 1793 but one that did not appear until 1810. The three volumes of *Zur Farbenlehre* incorporated and expanded the phenomenological philosophy of the *Beiträge.* It was composed of four parts: (1) a plan for a color doctrine with a large collection of cases, (2) a critique of Newton's theory, (3) a part on history starting with the ancients, and (4) a fragmentary fourth part that contained a supplemental essay on the chemical effects of colored light.

In *Zur Farbenlehre,* Goethe distinguished among the physical, physiological, and chemical aspects of color. His observations with a man who was color blind had convinced him more than ever that color had to have a subjective contribution.

Goethe believed *Zur Farbenlehre* was his most important scientific achievement. Yet he misjudged the way in which his audience would react to it, especially to his initial premise that Newton's theory of white light and colors was wrong. He based his refutation of Newton's theory on a simple prism experiment involving borders. He observed that the central portion of a broad white figure on a dark background remained white when seen through a prism. In contrast, a narrow black stripe (e.g., window bars) on a light gray background (sky) was broken into colors by the prism.

In short, Goethe believed his observations with prisms and light-dark boundaries contradicted Newton's relatively simple ideas. Newton's simplistic theory could not account for all of the relevant perceptual phenomena. In this conviction, Goethe was uncompromising. Overlooked was the fact that Goethe's statements about edges and boundaries could not explain the appearance of a rainbow or, even worse, the fact that the "warm" colors (e.g., red, orange, and yellow) were the "wrong side."

Although Goethe had expressed himself in a relatively reserved way in the *Beiträge,* he launched a fierce, direct assault on Newton in *Zur Farbenlehre.* Indeed, in his famous letter to the Royal Society of London, Newton seemed to discount the possibility that lighting and external influences could affect the vivid and intense colors he witnessed. As far as Goethe was concerned, Newton had not addressed the science of perceived color. Instead, Newton's science was only a science of optics or rays.

The phenomena observed by Goethe through his prism appear to have been known to Newton. Goethe simply misunderstood what Newton was saying even though he had intensely studied Newton's optical writings (Sepper, 1988). A number of Goethe's friends tried to get him to correct himself and to change his caustic and insulting language toward Newton and his disciples. Unfortunately, urging him to change his mind or to be more diplomatic fell on deaf ears. Goethe continued to use terms such as "word-rubbish" *(Wortkram)*, "incompetent," and "fairy tale" to describe Newton's ideas and accomplishments. The fact that Goethe proposed an orientation that relied little on mathematics and physics in an environment which was showing a steadily growing respect for these sciences also did not help him.

Goethe's *Zur Farbenlehre,* however, involved much more than a scathing attack on Newton. It included 920 statements about the psychology and sociology of color, dealing with such subjects as the relationships between colors and feelings. Some examples from this category ("Effect of Color with Reference to Moral Associations") are:

769. This impression of warmth may be experienced in a very lively manner if we look at a landscape though a yellow glass, particularly on a grey winter's day. The eye is gladdened, the heart is expanded and cheered, a glow seems at once to breathe towards us.

783. Rooms which are hung with pure blue, appear in some degree larger, but at the same time empty and cold.

793. History relates many instances of the jealousy of sovereigns with regard to the quality of red. Surrounding accompaniments of this colour have always a grave and magnificent effect.

Among those statements made later in Goethe's long list was one dealing with color contrast and another dealing with nationalities and color preferences. These are:

831. The colours of the active side placed next to black gain in energy, those of the passive side lose. The active conjoined with white and brightness lose in strength, the passive gain in cheerfulness. Red and green with black appear dark and grave; with white they appear gay.

838. Colours, as connected with particular frames of mind, are again a consequence of peculiar character and circumstances. Lively nations, the French for instance, love intense colors, especially on the active side; sedate nations like the English and Germans, wear straw-coloured or leather-coloured yellow accompanied with dark blue. Nations aiming at dignity in appearance, the Spaniards and Italians for instance, suffer the red colour of their mantles to incline to the passive side.

Goethe's *Naturphilosophie,* with its attempt to put the human spirit in touch with nature, advanced in a direction that seemed too subjective and too speculative to many scientists who preferred to embrace *Physik* in a more rigorous sense. In addition, Goethe's needlessly disparaging and vicious attacks on Newton succeeded in isolating him from many of the individuals in the scientific community he had hoped to impress. The lines were drawn, but lost somewhere in the heat of the battle was an understanding that Newton's and Goethe's approaches could complement each other because they were emphasizing color science at two different levels.

PURKYNĚ AND THE GOETHE TRADITION

Jan Purkyně (Purkinje; see Chapter 3), who was born in what would later become Czechoslovakia, was one of the foremost scientists of the nineteenth century. Although he is often regarded as a physiologist, his explorations also concerned anatomy, psychology, botany, embryology, pharmacology, philosophy, and physics (Hykes, 1936; Studnicka, 1936).

Purkyně, shown in Figure 7.5, studied the subjective aspects of vision, often using himself as his subject. This approach began with his thesis, *Beiträge zur Kenntniss des Sehens in subjektiver Hinsicht (Contributions to the Knowledge of Vision from the Subjective Point of View),* which was published in 1819. In this work he showed an interest in the effects of electrically stimulating the eye and in placing pressure on the eyeball. He also studied afterimages, memory for images, and changes in images as the eye underwent accommodation. In 1822, Goethe viewed Purkyně's approach as consonant with his own phenomenological orientation. It is said that he may have helped Purkyně obtain his prestigious appointment at the University of Breslau, the Prussian university that would soon be called "the cradle of histology." Purkyně paid Goethe a visit later in 1822, and the two men became good personal friends.

Figure 7.5. Jan Evangelista Purkyně (1787–1869)—usually spelled "Purkinje" in German and English—the Czech scientist who maintained a strong interest in color perception. Another engraving of Purkyně can be found in Chapter 3. (From Sterling, 1902.)

Purkyně made hundreds of observations pertaining to visual perception. The one for which he remains best known concerned the relative brightness of different colors with changes in illumination. He observed that red and blue flowers, which appeared equally bright with ordinary illumination, differed from each other at low levels of illumination (when they appeared as shades of gray), the blue flower now being brighter. This phenomenon, which Purkyně (1825) did not think was particularly important, has since become known in English as the Purkinje shift or the Purkinje effect. The basis of the effect lies in the fact that the rods, which are responsible for vision in dim light, are more sensitive to short wavelengths than long wavelengths (see Chapter 6).

HERING AND HIS OPPONENT PROCESS THEORY

In 1870, Ewald Hering (1834–1918) replaced Purkyně in Prague and remained there until 1895, when he went to the Leipzig Institute of Physiology. In Prague, Hering began to develop his theory of the light sense—a theory based on four fundamental colors: red, yellow, green, and blue. The idea of four fundamental colors, as opposed to the Young-Helmholtz trichromatic theory, however, did not originate with Hering. It can be found in the writings of Leonardo da Vinci and Goethe (1810), among others.

Hering's theory rested in part on the idea that yellow appeared to be just as much a primary as red or green; that is, it did not appear subjectively as a greenish-red color. As he and his followers seemed to put it, Helmholtz was psychically color blind—he was incapable of appreciating yellow! The theory also rested on Hering's belief that Helmholtz was unable to give a good physiological explanation for color-contrast phenomena. Some of the questions Hering faced were how to account for contrast, how to explain black and white, how to understand the striking subjective phenomena of afterimages, and how to explain certain types of color blindness.

In 1874, Hering took on these problems by postulating a "natural system of color sensations" (see Hering, 1878, 1964). He maintained that red and green were opposites and yellow and blue were opposites. He called this *Gegenfarben,* or "the theory of opponent colors." Hering stressed that there could be no yellow-blues or red-greens.

Hering believed yellow and blue were the first of the fundamental colors because half of the spectrum appears yellowish and half bluish. Hence, he theorized that red and green evolved later because they alone might disappear in color blindness or in peripheral vision, whereas yellow and blue are rarely lacking unless red and green are also affected (see Parsons, 1924, for a review of Hering's ideas).

Hering further theorized that white-black was the oldest of the three systems, partly because it was the most stable. His approach to black was perhaps the most novel of his various ideas. The generally accepted view among physiologists was that black was the absence of retinal stimulation. Hering argued this could not be the case since there was a considerable degree of brightness remaining with the total exclusion of light from the eye. He called this "mean grey" or "middle grey." True black, he maintained, occurs only under conditions of simultaneous or successive contrast; black

demands external stimulation, not its absence. Thus, to experience a true black sensation, one must look at a dark spot on a stark white background, and so on.

Hering (1878) then postulated that there were three substances which could change in either of two directions (assimilation and dissimilation). Assimilation (anabolism), he maintained, was responsible for white, yellow, and red. Dissimilation (catabolism) was responsible for black, blue, and green. Hering went on to propose an elaborate system of valences to account for the effects of different colors, such as a light orange or a deep blue-green.

Dissimilation and assimilation were not believed to be like the simple bleaching and regeneration of rhodopsin, but instead were thought to be events in neural tissue. Hering assumed that light entering the eye was absorbed by a photosensitive material. The energy released by this absorption served as the stimulus for the neural response.

Hering provided many types of observations to account for his theory of opposites and reciprocal physiological interactions. The model fit well with what was known about afterimages, color blindness, and the evolution of color vision. For example, when the eye has been adapted to a bright light that is then removed, the catabolic change in the opposite direction could be cited to account for the complementary afterimage. Similarly, contrast was explained by supposing that changes stimulate opposite kinds of events in adjoining areas—the effects being most marked at the border of the area of stimulation.

The explanations of contrast and related phenomena provided by Hering rested on his belief that the visual system could no longer be regarded as a mosaic of independent elements. He effectively demonstrated that a simple visual stimulus treated in isolation from other stimuli in the field could not account for perceptual phenomena. The message was that one must also take neighboring stimuli into account, as well as stimuli previously experienced. This idea would soon be implanted into Gestalt theories of vision (Boring, 1942).

One of the difficulties faced by Hering's theory was the problem of how to account for the two varieties of red-green blindness. Both protanopes and deuteranopes see yellow and blue, just as the opponent-process theory would predict, but they differ in other characteristics, such as maximum brightness for yellows. Hering thought such differences could be due to variations in the amount of yellow pigmentation in the macula and lens. The problem with his explanation was that one would expect to find differing amounts of pigmentation and gradual changes in sensitivity as one looked across cases. The evidence indicated, however, that protanopia and deuteranopia were not points of a continuum but clearly dichotomous disorders. As a result, even some supporters of Hering's general ideas did not feel confident about the theory (Collins, 1925; Rivers, 1900; Southall, 1924).

Another serious problem Hering faced was that his behavioral observations were far ahead of solid physiological evidence for the interactions he postulated. This was especially true for his ideas relating to assimilation and dissimilation of the sensory substance. Even the seemingly simple idea that physiological excitation of one neuron could inhibit another had not yet been demonstrated electrophysiologically.

For these reasons in particular, Hering's theory was rejected by many of his contemporaries. Although Ernst Mach (1838–1916), the brilliant Austrian physicist and positivist philosopher, could be counted among his true supporters, the fact is that Hering's most important concept, that of opponent processes, would achieve much greater recognition after his death than during his lifetime.

MODIFICATIONS BY LADD-FRANKLIN AND DONDERS

The ideas of Hering and Helmholtz were combined into a single theory by Christine Ladd-Franklin (1847–1930). Like Hering, Ladd-Franklin (1892, 1893, 1924, 1929) also approached the phenomenon of color from an evolutionary perspective. She theorized that the visual apparatus responded only to the grays (from white to black) in the earliest stage of its development. This corresponded to the hypothesized decomposition of a molecule that set free a chemical substance which stimulated the primitive retinal elements. This molecule was thought to consist of a firm core with loosely attached atoms around it. The loose atoms could be torn off by vibrations from the visual stimulus and, for the rods, the molecule "goes to pieces all at once under the influence of light of any kind."

In the second stage of its evolution, the undeveloped "mother substance" became differentiated so that only a part of it would undergo cleavage when stimulated by selective parts of the visible spectrum. The presumption was that this advance corresponded to the evolution of the cones. In this evolutionary stage, one part of the molecule was "shaken to pieces" by light in the longer part of the spectrum and the other by light in the shorter part. These parts of the original chemical substance gave rise to yellow or blue sensations, respectively.

The yellow-producing part of the color molecule was further differentiated in the third stage of evolution. This was associated with red and green sensations. Ladd-Franklin (1924, pp. 464–465) wrote:

It is assumed that there is a light-sensitive substance in the rods which gives off, under the influence of light, a reaction-product which is the basis of the primitive sensation of whiteness. In the cones, in the next higher stage of development of the colour-sense (the yellow and blue vision . . .) this same light-sensitive substance has become, by a simple molecular rearrangement, more specific in its response to light, and in such a way that the two ends of the spectrum act separately to produce nerve-excitant substances which, however, when they are produced both at once, unite chemically to form the "white" nerve-excitant out of which they were developed. In the third and final stage, the "yellow" nerve-excitant has again undergone a development in the direction of greater specificity, and red and green vision are acquired.

Normal color vision thus emerged with the evolution of selective receptors for blue, red, and green, as had been postulated by Helmholtz. Nevertheless, Ladd-Franklin considered the Young-Helmholtz trichromatic theory important only for understanding the initial photochemical processes that eventually cause sensation.

The Ladd-Franklin theory regarded color blindness as an atavistic condition. In total color blindness, the cleavage of the primitive molecule into parts did not take place. In red-green blindness, the gray molecule was cleaved into yellow and blue, but there was no further differentiation.

The theory did not treat black as a true "light sensation" but rather as a constant permanent background sensation capable of blending with colors, especially those that are faint.

In fairness to Ladd-Franklin, she, like Hering, did not claim to know much about the hypothetical molecules that supposedly underwent decomposition to light. In fact, Ladd-Franklin (1893, p. 489) stated:

They are not so much intended as real molecules, but rather as diagrammatic molecules,—as molecules, that is, which are possessed of such properties as would enable them to carry out the logical requirements of a light-sensation theory. I make no claim to having hit upon *the* process which goes on in the photo-chemical substance, but merely to having described *a* process which might with perfect plausibility *result from* the action of ether-waves upon the retina, and *from which would result* all the facts of light-sensation. More than this no hypothesis, in the present state of our knowledge, can hope to do.

The idea of the decomposition of a primary molecule, which is central to this theory, did not originate with Ladd-Franklin. Specifically, Frans Cornelius Donders (1818–1889), a leading Dutch scientist who had a strong interest in vision, suggested such a model in 1881.[7] Donders, who worked in Utrecht, proposed that white corresponds to complete molecules, that colors correspond to partial dissociation of complex molecules, and that the partial dissociation gives rise to molecules whose secondary changes form the basis of complementary perceptual events.

Ladd-Franklin acknowledged that she came up with her ideas while she was trying to prove Donders' theory better than the models proposed by Hering or Helmholtz. She concluded, however, that Donders did not do an adequate job explaining why there are no red-greens or yellow-blues. In addition, she differed from him when she suggested that gray must represent the equilibrium condition.

In opposition to more contemporary models of color vision, Donders theorized that while a fourfold process might be valid for the periphery, a threefold scheme might be operative centrally. He postulated that red and green could combine centrally to account for yellow and that the perception of white could also be due to central processing. Hence, he felt it was possible to accept both the theory of three fundamental colors, as well as an opponent-process type of theory, because these two seemingly alternative models described processing at different levels (i.e., "zones") of the system.

Contemporary texts suggest that Donders was correct in his belief that two fundamentally different mechanisms might be playing roles in encoding color. But whereas Donders thought there were four basic colors in the retina and three centrally, the prevailing view which emerged in the twentieth century was that he had the solution to the problem of color coding reversed. That is, three cone receptors

---

[7]See Pfeiffer (1936) for a biography of Donders as a physiologist and ophthalmologist.

and two more centrally located pairs of opponent process cells would be found to provide a better model for understanding how color information is processed (DeVallois and Jacobs, 1968, 1984).

PHRENOLOGY, THE CEREBRAL CORTEX, AND COLOR

As noted thoughout this chapter, most scientists attributed color blindness to a defect in the eye itself. Nevertheless, this was not a universally accepted belief. Some individuals thought color blindness might represent a defect of the brain, most likely one involving the cerebral cortex. This notion emerged with the phrenologists, who believed perception of color was mediated by a specific cortical organ (see Chapter 3). Color-blind individuals, it was suggested, may suffer from hereditary underdevelopment of the cerebral organ responsible for the appreciation of color or from disease or injury to that organ.

In 1824, two papers supporting this idea were presented before the Central Phrenological Society in Philadelphia. The abstract of one read as follows:

> Mr. Robert Tucker . . . at the age of seventeen discovered that he was unable to distinguish several of the primitive colours from each other. Yellow is the only one known by him to a certainty. "A difference in the shade of green he can distinguish, though not the green colour itself from the orange." . . . "In comparing," says Dr. B., "Mr. Robert Tucker's cranium with casts," and with the plates in Dr. Spurzheim's book, I was forcibly struck with the flatness of his os frontis, at the place of the orbitary ridge, where the organ of colouring is said to be situated—and this flatness, it is well known, indicates a small development of the organ. (Butter, 1824, 198)

The other paper was presented by George Combe, one of the three leading phrenologists in the world (Walsh, 1971). Combe's abstract was only a paragraph long, but the message was clear. It read:

> Mrs. Milne's grandfather, on the mother's side, had a deficiency in the power of perceiving colours, but could distinguish forms and distance easily. In himself and two brothers the imperfection appeared in a decided manner—while in his sisters, four in number, no trace of it is to be found, as they distinguish colours easily. He rather excels in distinguishing forms and proportions. By a cast of Mr. Milne's head it appears that the organ of colouring is decidedly deficient. At one of the meetings of the Society in Edinburgh, he and Mr. Gibson, landscape painter, were placed behind a screen at the door, and the head of each covered with a handkerchief, when more than half a dozen of gentlemen, who were strangers to both, named them respectively and correctly, by feeling their heads at the part indicated. (1824, 199)

Even as late as 1844, the psychiatrist Pliny Earle (1809–1892) discussed color blindness in his own family in the context of phrenology.

Some phrenologists entertained the idea that one might be able to treat this disorder by stimulating the cortical organ for color. This optimistic approach was expressed in the context of deficiencies in general when one individual wrote:

> When, however, the faculties become torpid and weak from long continued overaction, or from want of mental stimuli, it may, and often does become an object of attention, to arouse them to greater activity, by supplying such physical and mental exercise and amusements, as are adapted to secure this effect. (Buttolph, 1849, 135)

The idea of exercising the faculty for color by presenting intense red and green stimuli to color-blind subjects was attempted. These efforts, as would be expected, were not successful. Indeed, most phrenologists thought their models of color blindness afforded little hope for therapy (Freemon, 1991).

CORTICAL COLOR BLINDNESS

The idea that a person may lose the ability to see colors after a brain injury which does not cause total blindness has been looked upon with skepticism since the late 1800s, when the first "serious" cases of "acquired achromatopsia" appeared in the scientific literature. There were probably many reasons why the scientific community was hesitant to accept a brain disorder characterized by perception limited to shades of gray. One was that the early case descriptions were fragmentary and incomplete. In addition, a distinction between color vision and other kinds of vision based on neurological material suggested at least two separate visual centers in the cortex. The idea of multiple visual areas with different functions was diametrically opposed to the prevailing view held at the end of the nineteenth century and into the twentieth century, which maintained that there was but a single "cortical retina." As stated by one late-nineteenth-century investigator:

> It would appear necessary, if there is a separate area for the centre for colour, that there should be one for each individually and separately. It is much more reasonable on these grounds to suppose the colour loss to be due to a lessening of function, which, if more pronounced would cause a loss for light. (Ewens, 1893, 490)

One of the first good cases of acquired color blindness involved a man who worked as a printer in Germany (Steffan, 1881). Because the use of color was basic to his profession, it was believed that this man had a good eye for colors before he suffered the stroke which left him without color vision but with good visual acuity. The color defect abated only to the extent that he could make out some intense colors four years later. This case was considered so "clean" that the author felt it proved there had to be a separate center for color vision in the brain. Unfortunately, the brain was not autopsied.

A second good case of acquired achromatopsia was described by Nathan E. Brill (1860–1925) one year later. His patient also suffered a stroke and showed severe color confusions without a loss of visual acuity or a noticeable scotoma. In this instance, however, a post-mortem was conducted, and the lesion was found to involve the calcarine fissure and the lingual gyrus. Brill found this sufficient to conclude that there were two visual areas in the cortex, one responsible for sight in general (whose destruction causes hemianopia) and the other, situated next to the first, which was responsible for color perception.

These cases were followed by others.[8] For example, Henry R. Swanzy (1843–1913) described the case of a clergyman who lost color perception in his left visual field when he was 77 years old. Swanzy argued for the existence of a separate color center at the cortical level. He stated:

> The first three cases of chromatic hemianopia which I have quoted, and the one I now bring under the notice of the Society, all clearly dependent upon a cerebral lesion and without retinal or optic nerve changes, go to prove that the centre for colour perception is situated in the brain and not in the eye or peripheral parts of the optic nerve, and, moreover, that in the brain is a separate centre distinct from that for the form sense and for ordinary light perception. (1883, 188)

More often than not, the color center was thought to lie in front of the center for seeing, which was placed at the back of the occipital cortex. Nevertheless, spatially discrete areas were not always postulated. For example, Hermann Wilbrand (1884) hypothesized that the visual projections first go to a light center, then to a form center, and finally to a color center. In Wilbrand's model, all three centers were located in the primary visual cortex, but in different layers.

Perhaps the most significant of these early cases was one reported in 1888 by a Frenchman named Verrey. His patient showed a complete hemiachromatopsia. After careful examination of the autopsy material, Verrey stated that the critical region for color vision involved the lingual and fusiform gyri, structures below the calcarine sulcus. Like Wilbrand, he postulated three cortical visual centers in the occipital lobe. Unlike Wilbrand, however, he was convinced these were lateral to each other, not layered in a single visual area. Verrey's most dorsal center was responsible for light, his middle center for form, and his ventral center for color. As he put it, "the centre for the colour sense will be found in the most inferior part of the occipital lobe, probably in the posterior part of the lingual and fusiform convolutions" (translated in Mackay and Dunlop, 1899, p. 511). Because the form center was adjacent to the color center, he thought it likely that some problems in form perception would be found in conjunction with achromatopsia.

Nevertheless, by the end of the 1880s the number of achromatopsia cases of was still small, and the individual reports were still open to criticism for lacking autopsy material and/or because the clinical examinations were deemed unsatisfactory (MacKay, 1888). As additional cases of hemiachromatopsia and bilateral achromatopsia continued to appear in the clinical literature, some skeptics converted to the idea of a separate center for color at the cortical level. In one such report, which dealt with complete achromatopsia, the authors wrote:

> With Henschen's further assertion that the centres for luminous and for colour perception coincide we would fain agree, for one of us has hitherto seen no cause to alter the opinion expressed in 1888 that the existence of a special colour centre was "not proven." But in this particular case the total loss of colour sense is associated with a bilateral lesion of the fusiform convolution so well defined and symmetrical, that it becomes difficult to avoid the conclusion that the grey matter of that convolution is probably concerned in the perception of colours. (MacKay and Dunlop, 1899, 510)

[8]For a thorough review of the history of acquired achromatopsia, see Zeki (1990). Critchley (1965) also reviewed some of this literature.

The autopsy reports on patients with achromatopsia suggested that the most likely location for a color center was on the fusiform and lingual gyri of the ventral occipital lobe (e.g., Lenz, 1921). Nevertheless, because most of the lesions were associated with scotomas, those opposed to the idea of a separate cortical center for color perception seriously doubted whether the lesions really involved anything more than destruction of the primary areas for seeing or the projections to those areas. Gordon Holmes (1876–1966), for example, wrote: "My observations, which extend over a very large number of cases, consequently tend to show that an isolated loss or dissociation of colour vision is not produced by cerebral lesions" (1918, p. 380).

Typically, the scotomas involved the lower representation of the retina (upper visual field) on the inferior bank of the calcarine fissure, just above the lingual and fusiform gyri. They never involved the upper representation of the retina alone.

In addition, the achromatopsia was occasionally associated with prosopagnosia, the inability to recognize faces or comparable objects within a group (Meadows, 1974a, 1874b; also see Chapter 27). In fact, in some early reports the primary consideration was the prosopagnosia, with the achromatopsia being mentioned only briefly, as exemplified by an 1892 case study by the famous neurologist Joseph-Jules Dejerine (1849–1917). Nevertheless, in accordance with the notion that centers for basic vision, form, and color might be separate, not all cases of prosopagnosia exhibited a loss of color vision.

If anything, the idea of a specific color vision area ventral to the striate region was hindered by early cytoarchitectural studies. Although Korbinian Brodmann delineated the striate (Area 17), peristriate (Area 18), and parastriate (Area 19) areas, Areas 18 and 19 sliced right across the fusiform and lingual gyri where the best evidence for a separate color center existed.

The issues concerning central achromatopsia were not resolved in the first half of the twentieth century. Some newer clinical reports, however, have confirmed many of the findings reported in earlier studies and have utilized better techniques for identifying the locus of the damage. Subtle anatomical differences within the fusiform-lingual region have also been identified. Together, these newer clinical and anatomical findings suggest that those early investigators who looked upon acquired achromatopsia as a distinct neurological disorder may have been correct after all.[9]

[9]See Zeki (1990) for newer anatomical and physiological support for a distinct area, called V4, responsive to colors in the fusiform and lingual cortex. For a newer clinical report on acquired achromatopsia, see Damasio et al (1980).

REFERENCES

Boring, E. G. 1942. *Sensation and Perception in the History of Experimental Psychology.* New York: Appleton-Century Crofts.

Brill, N. E. 1882. A case of destructive lesion in the cuneus, accompanied by color-blindness. *American Journal of Neurology and Psychiatry, 1,* 356–368.

Brodmann, K., 1909. *Vergleichende Lokalisationslehre der Grosshirninde in ihren Prinzipien dargestellt auf Grund des Zellenbaues.* Leipzig: J. A. Barth.

Butter, D. 1824. Remarks on the faculty of perceiving colors. *Philadelphia Journal of Medical and Physical Sciences, 8,* 198.

Buttolph, H. A. 1849. The relation between phrenology and insanity. *American Journal of Insanity, 6,* 127–136.

Chance, B. 1922. Some English worthies of science of interest to ophthalmologists. *Annals of Medical History, 4,* 251–270.

Collins, M. 1925. *Colour-Blindness.* New York: Harcourt, Brace.

Combe, G. 1824. Case of deficiency in the power of perceiving and distinguishing colours, accompanied with a small development of the organ, in Mr. James Milne, brass founder in Edinburgh. *Philadelphia Journal of Medical and Physical Sciences, 8,* 199.

Critchley, M. 1965. Acquired anomalies of colour perception of central origin. *Brain, 88,* 711–724.

Dalton, J. 1794. Extraordinary facts relating to the vision of colours: with observations. *Memoirs and Proceedings of the Manchester Literary and Philosophical Society, 5,* 28–45.

Damasio, A., Yamada, T., Damasio, H., Corbett, J., and McKee, J. 1980. Central achromatopsia: Behavioral, anatomic, and physiologic aspects. *Neurology, 30,* 1064–1071.

Dejerine, J.-1892. Contribution à l'étude anatomo-physiologique et clinique des différents variétés de cécité verbale. *Comptes Rendus Hebdomidaire des Séances de Mémoires de la Société de Biologie.* Paris: Masson.

DeVallois, R. L., and Jacobs, G. H. 1968. Primate color vision. *Science, 162,* 533–540.

DeVallois, R. L., and Jacobs, G. H. 1984. Neural mechanisms of color vision. In I. Darien-Smith (Ed.), *Handbook of Physiology: Section 1. The Nervous System. VIII. Sensory processes.* Baltimore, MD: Waverly Press. Pp. 425–455.

Donders, C. 1881. Ueber Farbensysteme. *Archiv für Ophthalmologie, 27,* 155–223.

Earle, P. 1844. *Memoires.* Boston, MA: Damreil and Upham.

Ewens, G. F. W. 1893. A theory of cortical visual representation. *Brain, 16,* 475–491.

Freemon, F. R. 1991. American medicine and phrenology. Paper presented at Neurohistory Conference, Ft. Myers, FL.

Goethe, J. W. von. 1773. *Götz von Berlichingen mit der eisernen Hand. Ein Schauspiel.* Frankfurt am Main: Selbstverlag.

Goethe, J. W. von. 1774. *Die Leiden des jungen Werthers.* Leipzig: Weygandsche Buchhandlung.

Goethe, J. W. von. 1791–1792. *Beiträge zur Optik.* (2 vols.). Berlin: A. Weimar.

Goethe, J. W. von. 1810. *Zur Farbenlehre.* Weimar. Translated by C. L. Eastlake as, "Goethe's theory of colours." Frank Cass & Company.

Helmholtz, H. von. 1856, 1860, 1866. *Handbuch der physiologischen Optik* (3 vols.). Hamburg: Voss. Vol. 3 translated by J. P. C. Southall as, *Handbook of Physiological Optics.* New York: Dover Publications, 1962.

Hering, E. 1878. *Zur Lehre vom Lichtsinne.* Wien: C. Gerold's Sohn.

Hering, E. 1964. *Outline of a Theory of the Light Sense.* (L. M. Hurvich and D. Jameson, Trans.). Cambridge, MA: Harvard University Press.

Holmes, G. 1918. Disturbances of vision by cerebral lesions. *British Journal of Ophthalmology, 2,* 353–384.

Hooke, R. 1665. *Micrographia: or some Physiological Descriptions of Minute Bodies Made by Magnifying Glasses.* London: J. Martyn and J. Allestry.

Huddart, J. 1777. An account of a person who could not distinguish colours. *Philosophical Transactions of the Royal Society, 67,* 260–265.

Hykes, O. V. 1936. Jan Evangelists Purkyně (Purkinje). His life and his work. *Osiris, 2,* 464–471.

Johnson, I. W., and Wolbarsht, M. L. 1979. Mercury poisoning: A probable cause of Isaac Newton's physical and mental ills. *Notes and Records of the Royal Society of London, 34,* 1–9.

Klawans, H. L. 1991. Newton's madness. In H. L. Klawans (Ed.), *Newton's Madness.* New York: HarperCollins. Pp. 30–39.

König, A. 1894. Ueber den menschlichen Sehpurpur und seine Bedeutung für das Sehen. *Sitzungsberichte. Preussische Akademie der Wissenschaften,* 577–598.

König, A. 1903. *Gesammelte Abhandlungen zur physiologischen Optik.* Leipzig: J. A. Barth.

Ladd-Franklin, C. 1892. Eine neue Theorie der Lichtempfindungen. *Zeitschrift für Psychologie, 4,* 211–221.

Ladd-Franklin, C. 1893. On theories of light-sensation. *Mind, 2, Ser. 2,* 473–489.

Ladd-Franklin, C. 1924. The nature of the colour sensations. In J. P. C. Southall (Ed.), *Helmholtz's Treatise on Physiological Optics: Vol. 2. The Sensations of Vision.* Rochester, NY: Optical Society of America. Pp. 455–468.

Ladd-Franklin, C. 1929. *Colour and Colour Theories.* New York: Harcourt, Brace.

Lenz, G. 1921. Zwei Sektionsfälle doppelseitiger zentraler Farbenhemianopsie. *Zeitschrift für die gesamte Neurologie und Psychiatrie, 71,* 135–186.

Lomonosow, M. V. 1756. *Oratio de origine lucis sistens novam theoriam colorum.* Petropoli: Academiae Scientiarum.

Lort, M. 1778. An account of a remarkable imperfection of sight. Letter from J. Scott to the Rev. Mr. Whisson, Trinity College, Cambridge. *Philosophical Transactions of the Royal Society, 68,* 611–614.

MacAdam, D. L. 1970. *Sources of Color Science.* Cambridge, MA: MIT Press.

MacKay, G. 1888. A discussion on a contribution to the study of hemianopsia, with special reference to acquired colour-blindness. *British Medical Journal, 2,* 1033–1037.

MacKay, G., and Dunlop, J. C. 1899. The cerebral lesions in a case of acquired colour-blindness. *Scottish Medical and Surgical Journal, 5,* 503–512.

Maxwell, J. C. 1872. On colour vision. *Proceedings of the Royal Institution of Great Britain, 6,* 260–271.

Meadows, J. C. 1974a. Disturbed perception of colours associated with localized cerebral lesions. *Brain, 97,* 615–632.

Meadows, J. C. 1974b. The anatomical basis of prosopagnosia. *Journal of Neurology, Neurosurgery, and Psychiatry, 37,* 489–501.

Müller, J. 1826. *Zur vergleichenden Physiologie des Gesichtssinnes des Menschen und der Thiere.* Leipzig: C. Cnobloch.

Müller, J. 1838. *Handbuch der Physiologie des Menschen für Vorlesungen.* Coblenz: J. Hölscher.

Newton, I. 1671–1672. New theory about light and colours. *Philosophical Transactions of the Royal Society of London, 80,* 3075–3087.

Newton, I. 1704. *Opticks.* London: Smith and Walford. (Reprinted 1952; New York: Dover Publications)

Palmer, G. 1777. *Theory of Colours and Vision.* London: Printed for S. Leacroft at the Globe, Charing-Cross.

Parsons, J. H. 1924. *An Introduction to the Study of Colour Vision* (2nd ed.). London: Cambridge University Press.

Peacock, G. 1855. *Life of Thomas Young.* London: J. Murray.

Pfeiffer, R. 1936. Frans Cornelis Donders, Dutch physiologist and ophthalmologist. *Bulletin of the New York Academy of Medicine, 12,* 566–581.

Purkyně, J. E. 1819. *Beiträge zur Kenntniss des Sehens in subjektiver Hinsicht.* Prague: J. G. Calve.

Purkyně, J. E. 1825. *Neue Beiträge zur Kenntniss des Sehens in subjektiver Hinsicht.* Berlin: G. Reimer.

Rivers, W. H. R. 1900. Vision. In E. A. Schäfer (Ed.), *Text-Book of Physiology* (Vol. 2). Edinburgh: Young J. Pentland. Pp. 1026–1148.

Seebeck, A. 1837. Ueber den bei manchen Personen vorkommenden Mangel an Farbensinn. *Annalen der Physik und Chemie, 42,* 177–233.

Sepper, D. L. 1988. *Goethe Contra Newton.* Cambridge, U.K.: Cambridge University Press.

Southall, J. P. C. 1924. *Helmholtz's Treatise on Physiological Optics: Vol. 2. The Sensations of Vision.* Rochester, NY: Optical Society of America.

Spargo, P. E., and Pounds, C. A. 1979. Newton's "derangement of the intellect": A new light on an old problem. *Notes and Records of the Royal Society of London, 34,* 11–32.

Steffan. 1881. Beitrag zur Pathologie des Farbensinnes. *Archiv für klinische und experimentelle Ophthalmologie, 27,* 1–24.

Sterling, W. 1902. *Some Apostles of Physiology.* London: Waterlow and Sons.

Studnicka, F. K. 1936. J. E. Purkinje's "physiology" and his services to science. *Osiris, 2,* 472–483.

Swanzy, H. R. 1883. Case of hemiachromatopsia. *Transactions of the Ophthalmological Society of the United Kingdom, 3,* 185–189.

Verrey, D. 1888. Hémiachromatopsie droite absolue.—Conservation partielle de la perception lumineuse and des formes.—Ancien kyste hémorrhagique de la partie inférieure du lobe occipital gauche. *Archives d'Ophthalmologie, 8,* 289–301.

Walls, G. L. 1956. The G. Palmer story (or what's it like, sometimes to be a scientist). *Journal of the History of Medicine and Allied Sciences, 11,* 66–96.

Walsh, A. A. 1971. George Combe: A portrait of a heretofore generally unknown behaviorist. *Journal of the History of Behavioral Sciences, 7,* 269–278.

Weale, R. A. 1957. Trichromatic ideas in the seventeenth and eighteenth centuries. *Nature, 179,* 648–651.

Wilbrand, H. 1884. *Ophthalmometrische Beiträge zur Diagnostik der Gehirn-Krankheiten.* Wiesbaden: J. F. Bergmann.

Young, T. 1793. Observations on vision. *Philosophical Transactions of the Royal Society, 83,* 169–181.

Young, T. 1802. On the theory of light and colours. *Philosophical Transactions of the Royal Society, 92,* 20–71.

Young, T. 1807. *A Course of Lectures on Natural Philosophy and the Mechanical Arts.* London: Printed for J. Johnson.

Zeki, S. 1990. A century of cerebral achromatopsia. *Brain, 113,* 1721–1777.

Chapter 8
The Ear and Theories of Hearing

In short, this lamina is not only capable of receiving the aerial vibrations, but its structure makes it seem likely that it is capable of responding to all their various characteristics; for since it is larger at the beginning of the first turn than at the extremity of the last one (where it ends at a point), and the other parts diminish proportionately in size, we may declare that the larger parts can be vibrated by themselves alone, and are capable of being agitated only slowly and hence respond to the grave tones; and that on the contrary the narrower portions on being agitated are more rapid in their movements and respond to the acute tones. . . . The result is that by virtue of the different vibrations of the spiral lamina the spirits of the nerve . . . receive different impressions which represent in the brain the various characteristics of tones.

Joseph Duverney, 1683

GREEK ACOUSTICS AND THE IMPLANTED AIR THEORY

The Ancient Greeks associated sound with motion. Their analytical approach stressed the importance of air and other media in transmitting sound. The basic belief was that "like is perceived by like." This belief was embraced in the fourth century by Empedocles, who assumed there was an internal air in the ear, by which we perceive the moving air that constitutes sound, as well as a light in the eye, by which we perceive external light.

Only some 400 lines remain of the entire writings of Empedocles. As related by Theophrastus, the Greek philosopher who went to Athens to study under Aristotle:

He (Empedocles) says that hearing results from sounds within (the head), whenever the air, set in motion by a voice, resounds within. For the organ of hearing, which he calls a "fleshy off-shoot," acts as the "bell" of a trumpet, ringing with sounds like (those it received). When set in motion (this organ) drives the air against the solid particles and produces there a sound. (1948 translation, 547)

Empedocles and his contemporaries (e.g., Alcmaeon) had some knowledge about the nature of sound. Indeed, Pythagoras had introduced mathematics to acoustics and to the musical scale in the sixth century B.C. It is said that Pythagoras began his studies of acoustics when he passed a blacksmith's shop and heard the sound of a hammer on an anvil (Robinson, 1941). He went on to demonstrate experimentally the relationship between perceived pitch and the length of a vibrating string. Pythagoras also investigated the role played by the thickness and tension of the string.

It also seems clear that these Greek scientists had some basic knowledge about the ear. Empedocles described the auditory labyrinth, and it was recognized that the tympanic cavity, located behind the eardrum, was filled with air. From this premise it was assumed that it was the middle ear which housed the implanted air necessary for the perception of sounds. The implanted air, however, was not thought to be identical to the air we breathe. It was a purer substance, a highly refined permanent component of the middle ear.

Plato, who discussed many aspects of sensation and perception in his *Timaeus,* accepted this theory. He assumed the purified air was implanted in the middle ear in utero. His knowledge of the sensory pathways was even more speculative and metaphysical. The brain, which formed a part of the tripartite soul (with the liver and the heart), was treated as an intermediary for sensation. Although it was the source of "divine reason," it also had the role of conveying sensory impulses to the blood vessels, which took the impulses to the heart and the liver. Thus, we are told:

We may in general assume sound to be a blow which passes through the ears, and is transmitted by means of the air, the brain, and the blood, to the soul, and that hearing is the vibration of this blow, which begins in the head and ends in the region of the liver. (1952 translation, 465)

Aristotle also believed air was set in motion by sound, and he too accepted the implanted air theory. In his treatise *On the Soul,* he presented perhaps the clearest brief statement of the *aër ingenitus* or *aër implantatus* theory:

Now there is air naturally connected with the organ of hearing. And because this organ is in air, motion of the external air produces motion of the air within the organ. (1948 translation, 547)

The implanted air theory was also mentioned in *De audibilibus,* but whether Aristotle wrote this work is not certain. It has been suggested that it may have been written by Strato (d. c. 270 B.C.) or another person in the Lyceum (Clagett, 1955).

Unlike Plato, and in contrast to the theoretical positions

108

of Alcmaeon and Democritus, Aristotle denied the brain a direct role in sensation, emotion, and intellect. Instead, he believed these were functions of a single dominant organ, the heart—the "Acropolis of the body." In this regard, Aristotle showed a conservative adherence to the views that dominated other early civilizations, such as that of the Ancient Egyptians (see Chapter 1).

ROMAN AUDITORY ANATOMY AND PHYSIOLOGY

Before the middle of the first century A.D., Celsus described the external auditory meatus, the middle ear, and the inner ear. Celsus also discussed the reconstruction of the external ear in cases of injury and the means for opening constricted auditory canals. He accomplished the latter by having a patient lie on a table with the affected ear touching the surface. He then struck the table with a hammer, hoping to shake out the foreign body.

In the second century, Rufus of Ephesus gave the anatomists the terms "pinna," "helix," and "concha," while Galen taught that the external ear was largely ornamental in humans (Robinson, 1941). Galen also recognized that the small (2.7 cm), canal-like pathway of the external ear (external auditory meatus) served to protect the receptive tissue situated further in the ear. He did not, however, describe the tympanic membrane, most likely because it was no longer visible in his old, poorly preserved specimens (Wever, 1949).

Galen also encountered two nerves in his explorations of the ear. One was traced to the muscles of the face and the other, a smaller nerve, to the inner ear (the "labyrinth"). Galen classified these two nerves as the fifth cranial pair. Most likely, he was describing the facial (VII) and auditory (VIII) nerves, respectively.

In Galen's time, it was believed that sensory nerves were soft, making them more suitable for registering sensory impressions, whereas the motor nerves were hard. In accordance with this belief, Galen taught that the nerves going to the inner ear were soft, but not as soft as the optic nerves or the brain (the cerebrum being the softest of all nervous tissue), and those going to the muscles were hard. He and his contemporaries believed the nerves were activated by psychic pneuma, or animal spirits (see Chapters 1 and 5).

The Romans also showed a clear interest in acoustics and auditory prostheses. Among the more notable accomplishments, the Roman architect Vitruvius, in the first century A.D., proposed a wave theory to account for sound. He made the analogy between the propagation of sound in air and the spread of ripples on water. Anicius Boethius (c. 480–524), a later Roman scholar and statesman with interests in music and the arts, also favored a wave theory and took into account the reflections of sound waves (Lindsay, 1973).

As for prosthetics, ear trumpets were used to help the hearing impaired in Ancient Rome. It has been claimed that Archigenes of Apamea, who lived in the second century A.D., was the first to use these prosthetic devices (Robinson, 1941). Galen, who lived one century earlier, however, mentioned the use of ear trumpets. In fact, it is very likely such aids were used well before Galen's time (Mettler, 1947).

THE RENAISSANCE

Two of the middle ear bones, the malleus and the incus, were mentioned by Jacopo Berengario da Carpi in his *Isagogae,* published in 1514. He wrote: "On the tympanic membrane within the tympanic cavity are two small ossicles." In 1536, Niccolò Massa (1489–1569) described the ossicles as mobile and connected with the tympanic membrane in a way that allowed them to move with the membrane. Andreas Vesalius was the first to illustrate the malleus and the incus and to show their correct association with the tympanic membrane. Vesalius gave these two ossicles their common names in his *De humani corporis fabrica,* published in 1543 (Chapter 2). He did not, however, describe the third ossicle, the stapes.

Very soon after this, in 1546, Giovanni Filippo Ingrassia (1510–1580) provided the first accurate description of the stapes. This Sicilian anatomist further described the oval window, upon which the stapes ended, and the round window (cited by Schelhammer, 1684). Ingrassia also became known for his investigations of bone conduction.

Gabriel Fallopius, a pupil of Vesalius, made a number of additional contributions to the anatomy of the ear. Fallopius (1561), shown in Figure 8.1, provided excellent descriptions of the three ossicles of the "tympanum," his middle ear. In *Observationes anatomicae,* he also described their articulations, the oblique position of the tympanic membrane, and the chorda tympani branch of the facial nerve (VII), which passes through the middle ear.

Fallopius (1561) divided the inner ear into two parts. He named the vestibule with the semicircular canals the "labyrinth" and the snail shell division the "cochlea." He

Figure 8.1. A number of important contributions to the anatomy of the ear were made by Gabriel Falloppius (1523–1562). These included excellent descriptions of the three ossicles, the tympanic membrane and the chorda tympani. Falloppius (1561) also divided the inner ear into the "labyrinth" and the "cochlea."

Figure 8.2. Bartholomeo Eustachio (1524–1574), whose contributions include descriptions of the Eustachian tube, the tensor tympani and stapedius muscles of the middle ear, and the membranous spiral lamina of the cochlea.

recognized that the latter began at the round window and had three turns. He also noted the bony spiral lamina of the cochlea and remarked that the form and size of the inner ear showed little change from the time of birth.

In this same general period, Bartholomeo Eustachio (1564) and Costanzo Varolio (1591) described the tensor tympani and stapedius muscles of the middle ear. These tiny structures were not recognized as muscles by others who had seen them previously. Eustachio and Varolio both correctly suggested that the small muscles served a protective function.

Eustachio, who is shown in Figure 8.2, recognized the membranous spiral lamina in addition to the bony lamina of the cochlea. Nevertheless, he still thought the cochlea housed implanted air. Eustachio also gave a detailed description of the passage connecting the middle ear with the pharynx. The Eustachian tube, however, was not "discovered" by him. Alcmaeon, in the fifth century B.C., knew that there was a *tuba auditiva* between the middle ear and the throat.

Volcher Coiter (Koyter, 1534–1600) worked for some time with Eustachio, whom he considered his good friend. Coiter's book about the organ of hearing, *De auditus instrumento* (1572), traced the vibrations constituting sound from the external ear to the cochlea.[1] He wrote that the sound was collected by the pinna of the external ear, reinforced by the external auditory canal, and transmitted to the tympanic membrane. This moved the auditory ossicles of the middle ear, which in turn moved the oval window of the cochlea.

Coiter (1572) opposed the idea of "pure" implanted air in the tympanic cavity because he recognized that the Eustachian tube opened during swallowing, allowing ordinary air to enter the middle ear. Coiter correctly reasoned

that sound was picked up in the inner ear by the branches of the auditory nerve. Nevertheless, he believed the sound was intensified and propagated in the cochlea by its *aër ingenitus*.

Fabricius ab Aquapendente, a pupil of Fallopius and his successor at Padua, wrote a manuscript in 1600 that not only described the structures of the ear but also discussed why the ears are located on the sides of the head and why the pinna faces forward. Fabricius ab Aquapendente, whose portrait appears in Figure 8.3, also gave advice on how to remove foreign bodies from the ear. He favored a small hook, but cautioned that great care must be taken to avoid damaging the tympanic membrane (Guthrie, 1940b). His talented pupil Giulio Casserio (c. 1545–1616) investigated the structure of the ear in a great number of domestic animals and provided many beautiful and exact illustrations of the ear in his works (Casserio, 1601).

The credit for the first clinical text on hearing is usually given to Girolamo Mercuriale (Mercurialis; 1530–1606), a practitioner with an enormous clinical practice who wrote *De oculorum et aurium affectibus praelectiones* late in the sixteenth century (1591; see Guthrie, 1940a). In order to treat disorders of the ears, Mercurialis utilized a wide variety of medicaments, including honey, seawater, ant eggs, oil of mustard, and extracts from many different plants.

As might be expected, a number of new educational techniques for working with the deaf emerged during the Renaissance. Pedro Ponce de León (1520–1584), for example, utilized a system in which his pupils first learned to write the names of objects pointed out to them. The motor patterns associated with speaking the appropriate word for the object were then studied. Attempts to say the word flu-

Figure 8.3. Fabricius ab Aquapendente, who described the structures of the ear and discussed why the ears are located on the sides of the head and why the pinna faces forward. Fabricius ab Aquapendente also suggested ways to remove foreign bodies from the ear.

---

[1]Coiter's book has been called the earliest textbook of otology (Guthrie, 1940a).

ently followed. Ponce de León's original treatise has been lost, but an account of his ideas was given in 1620 by Juan Pablo Bonet (1579–1633).

In addition, Girolamo Cardano (1501–1576), the Italian physician and mathematician who had used raised letters to communicate with the blind, described a formal method for teaching deaf mutes to communicate with signs (see Cardano, 1663). This idea led to a number of related signing systems. One devised by Johann Conrad Amman (1663–1730) achieved considerable popularity (Amman, 1692, 1700).

THE SEVENTEENTH CENTURY

The clinical significance of bone conduction was recognized in the opening years of the seventeenth century by Girolamo Capivaccio (1523–1589). Capivaccio (1589) placed one end of a rod between the deaf person's teeth and the other end on the vibrating string of a zither. If the patient heard the sound, he diagnosed the deafness as a disorder of the drum membrane. But if the deaf individual did not hear the sound, the disorder was attributed to a problem of the labyrinth.

This clinical finding was verified by Felix Platter (1625), but it remained for Günther Christoph Schelhammer (1649–1716) to show that these sounds were, in fact, transmitted by bone and not through the Eustachian tube. He demonstrated this by placing a vibrating rod in the patient's mouth under alternating conditions where it did and then did not touch the teeth. Schelhammer reported his results in 1684.[2]

The wave theory of air conduction, proposed in the first century by Vitruvius and adopted in the sixth century by Boethius, was further expanded by Galileo Galilei (1564–1642) in the seventeenth century. Galileo, shown in Figure 8.4, was interested in resonance, especially as it applied to narrow halls, domed towers, pendulums, and musical instruments. In 1638, he wrote that "waves are produced by the vibrating of a sonorous body, which spread through the air, bringing to the tympanum of the ear a stimulus which the mind interprets as sound."

Galileo was followed by Kaspar Bauhin (1560–1624), the anatomist, botanist, and professor of medicine at the University of Basel, who formulated the first of a long line of resonance theories. Bauhin (1605) proposed that the dimensions and shapes of the various cavities of the ear worked as selective resonators. The large, roomy spaces were responsive to the deep sounds, whereas the small, narrow cavities responded to the "acute" sounds.

Although Bauhin's theory set the framework for later resonance theories, it was limited by the lack of knowledge about the inner ear. Although he did not rule out the possibility that the cochlea and labyrinthine organs may also contribute to sound reception, Bauhin thought the tympanic space, in which the implanted air was located, played the major role in the receptive process.

The increasing interest in the physics of sound, and in resonance in particular, led to a better appreciation of the role of the pinna and the external auditory meatus in fun-

Figure 8.4.   An engraving of Galileo Galilei (1564–1642) by Samuel Sartain after H. Wyatt. Galileo was interested in resonance and believed the ears were stimulated by waves from a vibrating body, which spread through the air.

neling the sound inward. It also suggested that the middle ear played an important role in amplifying the sound and in improving reception of the signal.

In England, Thomas Willis wrote that the external ear gathers together the particles constituting sounds and directs them inward to a thin membrane by way of twisting passages. The sound shakes this membrane like a drum, and this delivers the impression to the sonorous particles in the cavity beyond it. Willis (1672) knew about the three ossicles of the middle ear and the tensor tympani, which he thought controlled the tension of the membrane. He stated that his predecessors believed implanted air resided in this cavity and added that he did not consider the implanted air theory unlikely. Although he did not discuss the role of the ossicles, Willis went on to say that the sound was then transmitted to the cochlea (he actually spoke of two cochleas). Willis also traced the course of the eighth cranial nerve, but said nothing about how high and low tones were discriminated.

Later in the seventeenth century, major anatomical and theoretical advances were made by Joseph Duverney, a respected professor of anatomy and surgery in Paris and the King's medical adviser.[3] Duverney prefaced his *Traité de l'Organe de l'Ouïe* by stating: "Although I do not pretend that this work is entirely perfect, I hope, at least, that the

---

[2]See Kelley (1937) for a history of bone conduction.

[3]Duverney's work on the ear appeared in French in 1683 and was one of the first medical works to be written in the native language of the author. A Latin edition appeared in 1684. Although Duverney stated that he wished to write on all of the senses, none of his other works on the other sensory systems have been found (see Teed, 1936).

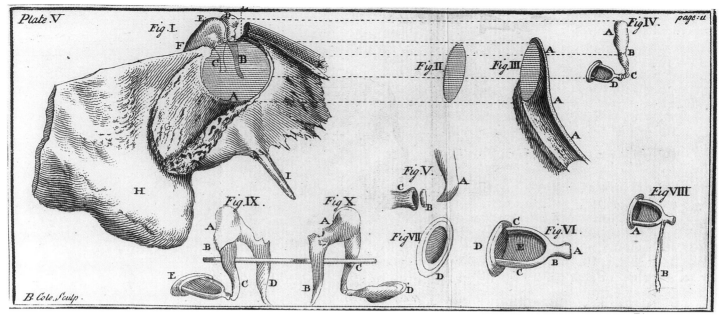

Figure 8.5. An illustration from Joseph Guichard Duverney's *Traité de l'Organe de l'Ouïe,* originally published in 1683. This plate shows the bones of the middle ear.

reader may here find something which has not been already described."

Duverney went on to describe the tensor tympani and stapedius muscles in detail (plus a third "muscle" that he called the external muscle of the malleus, but which was probably the anterior ligament). Figure 8.5 shows some of these structures. Duverney also divided the spiral cochlea with its two and one-half turns into two parts, the bony lamina and the membranous lamina. This is depicted in Figure 8.6. The basilar membrane was also portrayed more accurately than before. Duverney presented this membrane as a thinner and darker structure than the bony lamina.

On a functional level, Duverney made his share of incorrect assumptions. One was that sound was in part propagated by air through the middle ear. A second was that the bony lamina contained the receptors for hearing (see the quotation about the bony lamina and resonance at the opening of this chapter). Third, he believed low tones were detected at the base of the cochlea and high tones at its apex. A fourth erroneous assumption was that the inner

ear was filled with air. And a fifth was that the vestibule and semicircular canals were also involved with human hearing. The last assumption was based on the premise that fish and birds can hear even though they lack cochleas. Actually, Duverney reasoned the cochlea must be the organ of hearing in humans, but that the vestibule and semicircular canals can amplify the sound. One by one these errors were corrected by later scientists.

One year after Duverney presented his ideas, Claude Perrault (1613–1688; shown in Figure 8.7) gave a noteworthy description of the comparative anatomy of the ear and presented a theory in which he emphasized the importance of the implanted air in the cochlea. Perrault's (1684) ideas were attacked almost immediately by Günther Schelhammer (1684). Schelhammer argued strongly that waves in the air must be stopped and transferred to a new medium if perception is to take place. The venerable *aër ingenitus* theory was also attacked in 1686 by Johannes Bohn (1640–1718), a Leipzig physiologist. Still, the implanted air idea would not die. Even after fluid was found in the cochlea, some scien-

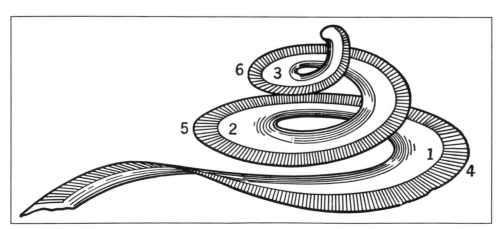

Figure 8.6. Joseph Guichard Duverney's (1683) drawing of the spiral organ of the cochlea. The white part of the ribbon (1, 2, 3) is the bony lamina and the striated part (4, 5, 6) is the membranous lamina. The numbering is contemporary.

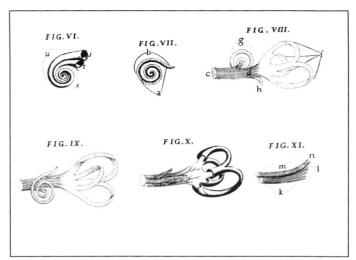

Figure 8.8. A drawing by Antonio Maria Valsalva (1704) showing the cochlea, the semicircular canals, and their nerves.

Figure 8.7. Claude Perrault (1613–1688) gave a noteworthy description of the comparative anatomy of the ear in 1684. Nevertheless, he continued to emphasize the importance of implanted air in the cochlea.

tists continued to maintain that air must still reside in other parts of the cochlea.

The seventeenth century was also a period in which a number of significant clinical observations were made. In this context, Théophile Bonet's (1620–1689) work on the pathology of the ear deserves mention. In *Sepulchretum*, Bonet (1679) presented protocols of some 3,000 autopsies. Included were cases of deafness associated with bilateral absence of the stapes, deafness associated with small ossicles, and hearing losses resulting from brain tumors.

THE EIGHTEENTH CENTURY

The importance of the membranous lamina (as opposed to the bony lamina) in hearing was recognized in the opening decades of the eighteenth century. This significant development can be traced to Antonio Maria Valsalva (1666–1738), who dissected over 1,000 human heads before publishing his *De aure humana tractus* in 1704. Valsalva recognized that the membranous lamina, which he called the "zona cochlea," grew wider as it ascended the cochlea (also see Valsalva, 1717). Effectively correcting Duverney's (1683) model of tonal localization, he maintained that low tones were detected at the apex of the cochlea and high tones at the base.

Although it is doubtful that Valsalva was able to trace nerve endings to the membranous lamina, he wrote that the zona cochlea was composed of the finest terminations of the auditory nerve. Among his other contributions were (1) the use of the terms "scala vestibuli" and "scala tympani" instead of the "upper" and "lower" galleries of the cochlea, (2) the division of the ear into external, middle, and inner parts, and (3) the recognition of perilymph within the inner ear (also see Cassebohm, 1735). Nevertheless, Valsalva still accepted the

implanted air theory and the idea that the semicircular canals served an acoustic function. Figure 8.8 shows one of his drawings of the cochlea and semicircular canals. Another plate from this period is shown as Figure 8.9.

When his *Primae lineae physiologiae* appeared in the mid-eighteenth century, Albrecht von Haller felt unsure of the nature of the terminations of the auditory nerve. As stated in the English translation of the Latin text:

> Do the transverse nervous filaments pass out from the nucleus of the cochlea, all the way successively shorter through the spiral plates? Is it the organ of hearing? These are questions, which we are yet hardly able to resolve from anatomy; though this seems repugnant to the course which we observe nature takes in brute animals, in birds, and in fishes, who hear very exquisitely without any cochlea. However this may be in the human body, it is there probable, that the spiral plate, spread full of nerves, is agitated with tremors from the oscillations of the membrane of the tympanum, by which the air in the cavity of the tympanum is agitated, so as to press the membrane of the round fenestra, which again agitates the air contained in the cochlea. (1786; 1966 reprint, 286–287)

Although Haller clearly overstated the role of air conduction in the middle ear, in his next paragraph he theorized that the receptive area had to be composed of an infinite number of cords of various lengths. But like Duverney, as opposed to Valsalva, Haller (p. 287) associated "the longest cords in the basis of the cochlea with grave sounds; and the shortest cords nearer the tip or apex, with sharper sounds."[4]

The best ideas of Duverney, Haller, and Valsalva were eventually incorporated into a more solid theory of hearing by Domenico Cotugno (1736–1822; shown in Figure 8.10), a professor of anatomy and surgery at Naples. Cotugno

---

[4]Haller drew his ideas largely from *Praelectiones academicae*, written in 1740 by Herman Boerhaave, his teacher and a leading physician of the age. Boerhaave, in turn, credited the idea of an infinite number of cords to Claude Perrault. Yet Perrault did not write about an infinite number of cords and erroneously assumed that the resonators were on the bony labyrinth.

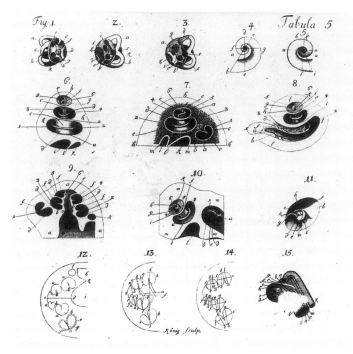

Figure 8.10. Domenico Cotugno (1736–1822), the individual who showed that all parts of the cochlea were filled with fluids. Cotugno proposed that the auditory receptive surface was on the membranous lamina, that this lamina grew wider as it ascended the cochlea, and that it was made of a large number of cords.

Figure 8.9. Illustration of the cochlea by J. Zaccharias Petsche. This figure appeared in a treatise on the ear by Johann Friedrich Cassebohm (1734).

relied on the use of fresh cadavers for his anatomical studies. In 1761, he proposed that the auditory receptive surface was on the membranous lamina of the cochlea, that this lamina grew wider as it ascended the cochlea, and that it was made of a large number of cords.

Cotugno showed that all parts of the cochlea were filled with fluids.[5] His work included descriptions of the cochlear and vestibular aqueducts and recognition of the fact that the perilymphatic space communicated with the subarachnoid space at the base of the brain.

Cotugno's efforts represented the death knell for the resilient implanted air theory. This theory had been maintained into the eighteenth century in part because it was believed liquids could not transmit sound vibrations. Cotugno's new theory of hearing held that the stapes set the labyrinthine fluid into vibration, which in turn stimulated the fibers of the membrane.

Although little was heard of the implanted air theory after this time, Cotugno's other ideas were not immune to criticism. In fact, serious questions were raised even before the eighteenth century ended. The major doubts related to the properties of the resonators. Could they move singly? Did they have sufficient tension to be independent of one another? Were they long enough to vibrate suitably? Would there not be confusion from harmonics if the strings vibrated in segments in addition to vibrating as a whole?

NINETEENTH-CENTURY PHYSIOLOGY

Charles Bell (1812) discussed hearing in some detail in his *Anatomy of the Human Body*. He cited the anatomical facts

as they were known at the time, but maintained that the entire labyrinth of the inner ear (shown in Figure 8.11 from Sömmerring) was the organ for hearing. Because the cochlea was very highly developed in horses and humans, Bell suggested that it may function specifically in the finer aspects of audition, including musical perception.

Bell also commented on the round window of the cochlea. In the first decade of the nineteenth century it was known that the stapes pressed directly upon the oval window, but it was still believed that air conducted sounds to the round window. Bell pointed out that because the cochlea was a closed chamber filled with fluid, the stapes could not push the oval window in unless the elastic round window could move out. Bell realized that if the bone and air conduction methods were equivalent, the two modes of transmission would block each other, since both oval and round membranes could not be pressed in at the same time. Thus, he looked upon movements of the stapes as being important, whereas air pressure changes on the round window were considered insignificant.

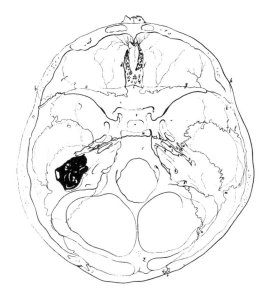

Figure 8.11. Samuel Thomas Sömmerring's 1806 figure of the position of the inner ear in the human cranium.

[5]Cotugno acknowledged that Valsalva and Morgagni had known about perilymph. He was unaware that Johann Theodor Pyl (1749–1794) had discovered a second fluid in the inner ear, the endolymph, in 1742.

After conducting research on the transmission of sound from one medium to another, Johannes Müller (1838) also concluded that conduction via the ossicles was much more effective than air conduction. But unlike Bell, Müller believed that stimulation of the cochlea by waves of air could still be significant. In addition, he mistakenly thought the ossicles were immobile and that sounds passed through them as they would through any other solid body (see Boring, 1942).

In 1824, Félix Savart (1791–1841) showed that the tympanic membrane vibrated when sound impinged upon it. He did this by placing sand on the membrane and watching it move to sounds. This procedure was an extension of the one used by Ernst Chladni (1756–1827), who had earlier described the effects of sound by putting sand on metal plates ("Chladni figures"; 1787).

In the same year, Flourens published his findings suggesting that the semicircular canals did not function in hearing. Flourens argued that the semicircular canals played a role in reflexive orientation. Although other theories about the semicircular canals would be forthcoming, they were not closely associated with hearing after this time.

These advances allowed Ernst Weber (1795–1878) to write the following synopsis in 1851:

> The transverse vibrations of the tympanic membrane set up by the vibrating particles of the air, are communicated by this lever-system to the other lever arm connected with the stapes, which is attached by its base in the oval window and consequent to the transverse vibrations of the tympanic membrane moves to and fro. Such a movement would, however, be impossible if the incompressing fluid filling the osseous labyrinth had no way of escape. For this purpose a second opening in the labyrinth, the round window, closed by a delicate membrane exists. . . . The movement of this membrane may even be observed with the naked eye, if the stapes is moved in the oval window. In this movement of the stapes, which is communicated from the oval to the round window, the whole labyrinth fluid necessarily participates. The vibrations of the fluid set in motion suspended parts of the membranous labyrinth and the nerves embedded in this structure. The spiral lamina of the cochlea is stretched in the labyrinthine fluid between the oval and round window, and the vibrations in passing to the round window have to traverse the membrane of the spiral lamina and thus encounter the nerve-terminals embedded in the lamina. (Translated in Kostelijk, 1950, 28)

A few years later, the first experiments actually showing that the ossicles vibrated to sound stimuli were made by Adam Politzer (1835–1920), whose portrait appears as Figure 8.12. In 1864, Politzer reported that he had fixed tiny hairs on the malleus and incus and had recorded vibrations on a rotating cylinder drum. The accuracy of this technique, which put a considerable load on the system, was questioned at first, but later studies using luminescent paint and strobe lights tended to confirm Politzer's findings (see Kostelijk, 1950).

It is interesting to note that the accepted instrument for measuring hearing in Politzer's time was not the tuning fork, which had been discovered by John Shore (fl. 1715) in 1711, but the pocket watch (see Kelley, 1937). The distance away that the ticks could be heard served as the unit of measurement. Because of differences in tone and intensity, the "watch test," although useful for clinical examinations, could not be standardized. Recognizing this, Politzer (1913)

Figure 8.12. Adam Politzer (1835–1920), who demonstrated that the ossicles vibrated to sound stimuli.

constructed the "acoumeter." This instrument made a noise like a watch tick, could be heard at known distances by normal subjects, and had a feature that allowed the practitioner to compare air and bone conduction.[6]

NINETEENTH-CENTURY ANATOMICAL ADVANCES

The study of the fine structure of the inner ear was made possible by many advances in microscopy in the first half of the nineteenth century. In particular, the development of the compound microscope around 1830 allowed investigators to study the cochlea in more detail.

In 1851, two of the most significant anatomical discoveries in the field of audition were made. First, Ernst Reissner (1824–1878), of Riga, Latvia, discovered the membrane in the cochlea that now bears his name. This finding allowed the cochlea to be divided into three scalae (media, tympani, and vestibularis). Second, Alphonso Giocomo Corti (1822–1876) published his extensive studies of the cochlea. Corti examined some 200 specimens from humans, cattle, pigs, sheep, cats, rabbits, moles, and mice. He discovered the hair cells, the rods ("arches") of Corti, and the tectorial membrane ("Corti's membrane"), which rested on top of these structures.

Corti's samples were not in the best condition, and his preparations did not allow him to appreciate depth. Hence, his descriptions of the rods and hair cells required considerable improvement. In particular, Corti saw the basilar membrane greatly flattened and the cells as spherical. These shapes were due to the escape of the perilymph and to the specimens drying and shrinking on flat plates. Corti's 1851 drawing of the basilar membrane is shown as Figure 8.13.

In 1858, Max Schultze traced the nerve fibers to the base of the inner and outer hair cell regions and showed that the rods of Corti were not nervous structures. A much

[6]Sometime between 1796 and 1801, Giovanni Venturi (1746–1822) hypothesized that differences in sound intensity between the two ears were basic for localizing sound in space. To test this hypothesis, Venturi played brief notes from some flutes and examined normal people and individuals who were deaf in one ear to see if they could locate sounds in space. He showed that people with one deaf ear could not do this (see Gulick, 1971).

Figure 8.13. Alphonso Corti's (1851) drawing of the sensory apparatus of the ear.

better description of Corti's rods was provided by Deiters in 1860. He observed the rods in their correct form as arches and saw how the hair cells were related to each other and their supporting cells.

Gustaf Magnus Retzius (1842–1919), who published two important volumes on the anatomy of the auditory system between 1881 and 1884, observed the nerves ending on the hair cells (Retzius, 1892, 1893). The "modern" drawings of the cochlea, made by Retzius in Sweden and by Louis Antoine Ranvier (1835–1922) in France, soon appeared in many textbooks on anatomy and physiology. One of Ranvier's plates is shown as Figure 8.14.

The finer morphology of the hair cells became even better understood in the beginning of the twentieth century (see Held, 1904, 1926). Still, the exact innervation of the hair cells remained uncertain. It was not until 1937 that Rafael Lorente de Nó (b. 1902) showed that each inner hair cell is innervated by one or two nerve fibers and that each nerve fiber branches to innervate only a few hair cells.

HELMHOLTZ AND HIS RESONANCE THEORY

Weber's ideas (1848, 1851), as well as Corti's anatomical work, stimulated Hermann von Helmholtz at the University of Bonn to propose his own theory of hearing. Helmholtz was also inspired by recent discoveries in acoustics, including Georg Simon Ohm's (1787–1854) "acoustical law," presented in 1843 This law held that complex sounds can be broken into simpler sinusoidal components (each associated with a different pitch) by the ear in the manner suggested by Jean-Baptiste Fourier's (1768–1830) mathematical theorem for periodic motion.[7]

The ideas of Johannes Müller, who was Helmholtz's teacher and associate, also played a prominent role in the development of Helmholtz's theory. Müller's "doctrine of specific nerve energies" first appeared in 1826 and was presented in a more elaborate form in 1838. This law held that each nerve can produce only its own specific kind of sensation, no matter how it is stimulated.[8] Although Müller had applied this concept to only the five general sensory systems, it was readily expanded by his followers to account for finer units of sensation (Natanson, 1844; also see Chapter 10). In 1844, for example, Alfred Wilhelm Volkmann (1800–1877) wrote that the law could be applied to every pitch which could be heard.

Helmholtz presented his resonance theory of hearing at a public lecture on music and science in 1857 and continued to formulate as well as expand his ideas over the coming years. His fully developed theory was presented in 1863 in *Die Lehre von den Tonempfindungen,* a monumental work that was translated in 1885 as *On the Sensations of Tone.*

Helmholtz theorized that the cochlea was made up of a collection of spatially tuned analyzers, each with a separate fiber. This was to account for the approximately 5,000 pitches he believed could be distinguished by the human ear. He thought the outer rods of Corti, being under tension and resting on the basilar membrane, were favorably

[7]Fourier's major paper on this subject appeared in 1822. His first paper on periodic motion, dated 1807, can be found in a book by Grattan-Guinness (1972) that covers all of Fourier's contributions.

[8]A similar theory can be found in Charles Bell's *Idea of a New Anatomy of the Brain* (1811).

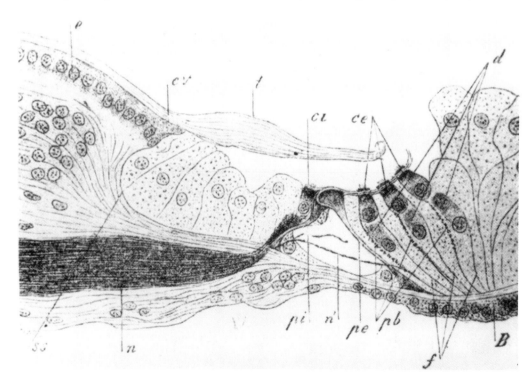

Figure 8.14. Louis Ranvier's 1875 drawing of the organ of Corti.

suited for movement and functioned as the resonators. Helmholtz postulated that high-frequency sounds affected resonators at the base of the cochlea and that low-frequency sounds affected resonators at the apex of the cochlea. He based this on what he knew about the anatomy of the rods and the basilar membrane and on his belief that rods in different loci varied in tension or stiffness. He further maintained that each resonator had its own nerve fiber to deliver a specific message to the brain.

Helmholtz's choice of resonators was criticized by several influential anatomists. In two papers published in 1863, Victor Hensen (1835–1924) pointed out that the rods (arches) of Corti do not show enough of a variation in length to account for the range of audible sounds. In contrast, he showed the basilar membrane is a 30-mm-long trapezoidal strip made of transverse fibers that vary in width from 0.04 mm at the base to 0.50 mm at the tip of the cochlea (for measurements by other scientists, see Kostelijk, 1950). In Hensen's opinion, the transverse fibers of the basilar membrane, and not the rods of Corti, were the more likely resonators.

Hensen went on to suggest that the changes produced in different groups of hair cells by movements of the basilar membrane could account for frequency discrimination. This idea was based in part on his studies of hair cell resonance in crustaceans. It was also based on recent studies showing that the auditory nerve ended on the hair cells and not on the rods of Corti (Schultz, 1858).

These findings and the comparative studies published by Carl Hasse (1841–1922) in 1867 forced Helmholtz to revise his theory in 1869. The revision appeared in subsequent editions of *Die Lehre von den Tonempfindungen*. Specifically, he left the rods of Corti and turned toward the transverse fibers of the basilar membrane as the most likely resonators. Nevertheless, he still maintained that the rods of Corti were affected by the movements of the basilar membrane, and consequently affected the hair cells.

The revised resonance theory was still not free from criticisms. Four objections to it were (1) that the size of the organ of Corti was too small to resonate to such large waves (Can one really compare a piece of tissue 1/10 mm in length with a string 1 m in length?); (2) that the idea of stretched, tense strings in the cochlea was absurd; (3) that the number of resonators was probably much more limited than Helmholtz believed (the eye had only three receptors, and most people believed the skin had just four at this time); and (4) that the resonators could not be finely tuned because of damping (see Wever, 1949).

Helmholtz did not help his cause when he argued that whereas the cochlea was sensitive to tones, other inner ear structures were sensitive to noises. Although he retreated somewhat from this position by 1877, he still maintained that the cilia of the ampulla and macula were likely mediators of squeaking, hissing, chirping, and crackling noises.

Helmholtz attempted to defend his theory against criticisms in a number of ways. For instance, in dealing with the evidence that the transverse fibers of the basilar membrane were not separate, he suggested that the longitudinal tension of the basilar membrane was too weak relative to the transverse (radial) tension to be significant. As for the comment that the resonators might not be as specific as he thought, Helmholtz seemed increasingly open to the possibility that there might be a best tone, as well as less favored tones, that could affect each resonator.

PLACE THEORIES AFTER HELMHOLTZ

As the anatomical and physiological demands of Helmholtz's highly specific resonance theory looked as if they could not be met, alternative or revised resonance theories came to the fore. Many of the variations on the theme emphasized waves of stimulation and placed more reliance on the role of central neural mechanisms for decoding the message (Wever, 1949, 1962).

Carl Hasse (1867) dismissed the idea of stretched strings in the cochlea. But rather than proposing a radically different theory from that of Helmholtz, he hypothesized that differences in the tectorial membrane made it the more reasonable candidate for resonance. Hasse favored the tectorial membrane because it was near the hair cells that had to be stimulated and was easily excited by movements. He later questioned his own hypothesis and eventually swung back to the view that the basilar membrane was the primary resonator.

Augustus D. Waller (1856–1922) briefly suggested a pattern theory for basilar membrane stimulation in 1891. This idea was more fully developed in 1899 by Julius R. Ewald (1855–1921), a professor of physiology at the University of Strassburg. Ewald theorized that the basilar membrane was affected by broad vibratory patterns ("acoustic images") which were set up by standing waves. He maintained that the patterns of the waves differed with the frequencies of the sounds.

Ewald (1899a, 1899b) based his ideas on extensive tests with tuning forks and a mechanical model called the "camera acoustica." He stimulated a thin rubber membrane stretched over a wedge in a block of wood and photographed the light reflecting from it. The camera acoustica represented an advance over earlier models of the inner ear because it avoided distortions resulting from sand and other substances placed on the membranes.

Ewald thought that high tones were likely to produce patterns which would affect the basal region of the basilar membrane, whereas low tones would involve the whole membrane. The ability of low tones to affect larger areas, he postulated, accounted for their expansive quality. Ewald hypothesized that the analysis of complex sounds was affected by practice. This was a function of the brain and it varied significantly across individuals.

A related theory was proposed by C. Herbert Hurst (1856–1898) in 1895. The basic assumption of his "traveling bulge" theory was that sounds produced waves which moved up and down the basilar membrane. He postulated that these waves met at certain places, the intersections varying with spectral frequency. Hurst presumed that cochlear fluid, concentrated at these intersections, moved the tectorial membrane and maximally stimulated the hair cells. He maintained that the greater the time interval between waves, the lower the intersection on the basilar membrane and the lower the perceived pitch. This assumption was contradicted by the evidence for high tones being localized near the basal end of the cochlea and served as a serious drawback to his theory.

Figure 8.15. Georg von Békésy (1889–1972), winner of the Nobel Prize for Physiology and Medicine in 1961 for his research on the ear and hearing.

Hurst's theory was followed by a number of other wave theories, including one presented by Henry Jackson Watt (1879–1925) in 1914 and again in 1917. The most important of these theories, however, was the traveling wave theory that began to be developed by Georg von Békésy (1889–1972) in 1928 when he lived in Budapest. Békésy, shown in Figure 8.15, later emigrated to the United States.

Békésy constructed a number of models of the inner ear. One was a brass and glass "cochlea" (box) with round and oval rubber "windows." This model was divided down the middle (except at the "helicotrema") with a rubber "basilar membrane" and was filled with fluid in which coal dust and gold particles were suspended. When Békésy stimulated the "oval window" with a short metal shaft (his "stapes"), he observed wavelike movements of the particles sweeping toward the helicotrema. But he also found that the waves typically peaked before reaching the end of his artificial basilar membrane. By trying stimuli with different acoustic frequencies, he observed that the place of maximum displacement of the basilar membrane changed, with higher-frequency sounds producing waves that peaked closer to the oval window than lower-frequency sounds.

Békésy confirmed the accuracy of his traveling wave theory with experimental work on preserved human and animal cochleas. He decalcified bone, added particles of coal or metal to the fluids of the cochlea for better observation, and observed cochlear movements to stimulation under stroboscopic illumination.[9] His findings strongly suggested that the part of the basilar membrane maximally stimulated allowed one to distinguish between different high-frequency and middle-frequency sounds.

[9]For a summary of the work that won him the Nobel Prize, see Békésy (1960). For the statement of Békésy's accomplishments made before the King of Sweden, see Bernhard (1962).

PATHOLOGY AND PLACE THEORIES OF HEARING

The place theories proposed by Helmholtz and his followers were subjected to a number of tests. One involved attempting to destroy selected parts of the cochlea surgically to see if there would be selective hearing losses. This approach began in the 1880s with an experiment which used dogs to show that removal of the apical turn of the cochlea resulted in the loss of low-tone perception (Baginsky 1883). This finding was later confirmed by many other experimenters (Held and Kleinknecht, 1927; Marx, 1909; Munk, 1881; Pavlov, 1927).

A second method of assessing the theory involved studying the cochleas of humans who showed selective hearing losses (see Guild et al, 1931). The term "stimulation deafness" was applied to those cases where exposure to intense noises caused the deafness (Kemp, 1935). The literature from the late 1800s reveals that experienced locomotive engineers, boilermakers, and smiths often had a hard time hearing sounds in their working environments.

In one study, 25 boilermakers were examined and showed themselves to be especially hard of hearing for the high tones (e.g., Habermann, 1890). When one of the men died, a post-mortem examination was carried out and his organ of Corti was found to show degeneration near the base of the cochlea. This finding was confirmed and extended in a follow-up study of 107 cases of occupational deafness, which included 5 cases with histology (Habermann, 1906).

A more detailed psychophysical study involved boilermakers differing in number of years on the job. This investigation showed that an auditory deficit may start off being relatively specific, but with continued loud sound exposure can grow to affect hearing for a greater number of frequencies (Rodger, 1923; also see Crowe, Guild, and Polvogt, 1934). Figure 8.16 presents a diagram showing the wide cochlear lesions in cases of selective stimulation deafness.

The first extensive study on animals exposed to intense sounds was presented by Karl Wittmaack (1876–?) in 1907. He exposed guinea pigs to sounds, such as bells and whistles, and then examined their ears. Daily exposure to a loud whistle blown directly into the ear led to degeneration in the basal end of the cochlea. The damage was proportional to the extent of the exposure, and for subjects exposed to this sound for many months, there was no trace of the organ of Corti in the basal region.

Guinea pigs were exposed to different tones from whistles and sirens in two follow-up experiments that included behavioral measures (Marx, 1909; Yoshii, 1909). The data provided some evidence for the place theory of hearing. Still, the hearing losses extended beyond the specific frequencies of exposure, and the lesions encompassed more of the basilar membrane than expected.

Later, more sophisticated experiments with high-frequency sounds on guinea pigs, cats, mice, pigeons, dogs, and infrahuman primates produced essentially the same results. For example, Hallowell Davis (1896–1992) and his coworkers exposed cats and guinea pigs to 600, 800, and 2,400–2,500-Hz sounds at different intensities for several weeks. They then recorded auditory nerve potentials and looked for pathological changes in the ear (Davis et al, 1935a, 1935b).

The animals exposed to the low-frequency sounds did not

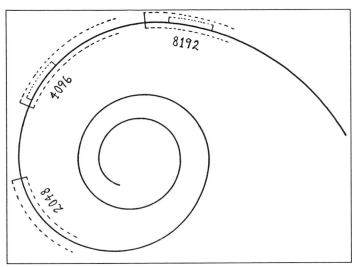

Figure 8.16. A diagram of the cochlea showing the parts usually damaged in humans cases with deafness for high pitches. (From Crowe, Guild, and Polvogt, 1934, with the permission of the Johns Hopkins Press.)

show major effects, but the animals exposed to the high-frequency sound at 95–107 db for 15 hours a day over 40 days exhibited severe deficits. Audiograms showed the loss of auditory sensitivity was fairly wide (700–1,700 Hz), peaking at about 1,200 Hz, where the guinea pig has its most sensitive hearing, and not at the 2,400 Hz-exposure frequency. Histological examination of the inner ear revealed a corresponding loss of external hair cells and a rupturing of Reissner's membrane in the parts of the cochlea known to be associated with 700–1,700-Hz sounds.

In general, both human and animal stimulation deafness studies with high- and middle-frequency sounds provided some support for the place principle of hearing, at least to those willing to consider a "zone of reception" instead of a single point of reception. Exposure to low-frequency sounds, however, did not appear to be nearly as effective in causing audiometric deficits or histological damage.

Another test of the place principle involved organisms born with abnormal organs of Corti, such as albino cats (Howe and Guild, 1933). In one case, a cat was found with a fairly normal apex of the cochlea, but no hair cells at the base (Lurie, Davis, and Derbyshire, 1934). This animal's audiogram showed only a moderate loss of sensitivity for sounds below 250 Hz, but a steep decline and an absolute loss of sensitivity to frequencies above 3,500 Hz. Figures 8.17 and 8.18 illustrate this work.

Overall, pathological material provided welcome support for the place principle of hearing. But because even pure tones appeared to be associated with a spread of activity along the basilar membrane, it soon became evident that some additional principle had to be added to the place model to explain the ability of the organism to detect very small changes in frequency. Some proposals related to the point of maximal stimulation and the role of contrast (Békésy, 1930; Wilkinson and Gray, 1924). In addition, although the evidence supported a place principle for high and middle tones, the data for low-frequency tones suggested that a different principle might be at work at the low end of the spectrum.

FREQUENCY THEORIES

In 1865, Heinrich Aldof Rinne (1819–1868) launched an attack against existing place theories of hearing. He wrote that if each nerve fiber were responsible for a different sound, more had to be said about the synthesis and integration taking place at higher levels. He believed the law of parsimony would be better served if complex sounds were not broken into simpler sounds only to be reconstructed.

Rinne's opposition to the place theories was shared by Friedrich E. R. Voltolini (1819–1889). In 1885, he proposed that every hair cell is stimulated by every audible sound and that having many hair cells activated increases the acuity of the system.

By far the best known of the "frequency theories" of hearing was that proposed by William Rutherford (1839–1899), a professor of physiology at the University of Edinburgh. His work on rabbits led him to the same conclusions as Voltolini—that is, every sound stimulates every hair cell, and the larger the number of hair cells involved, the greater the sensitivity and accuracy of the transmission. Rutherford presented his theory in 1886. He stressed that it was the brain's job to interpret the vibratory patterns of stimulation and that this could be affected by practice.

Rutherford's theory became known as the telephone theory because he made the direct analogy between the auditory apparatus and the telephone, invented by Alexander Graham Bell (1847–1922) just 10 years earlier. The following passage appeared as a part of a summary of Rutherford's lecture:

> Some five years ago it struck the lecturer that the case of the telephone might throw light on the difficulties regarding the sense of hearing. When sound-waves fall on the plate of one of the telephones it vibrates. The vibrations of the iron near the magnet affect the magnetism, and so induce in the wire currents of electricity whose frequency and amplitude correspond to those of the vibrations of the

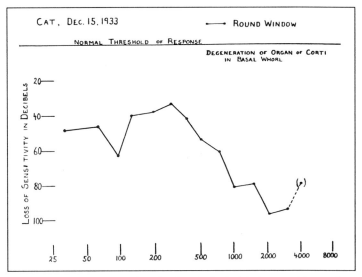

Figure 8.17. Audiogram (hearing loss in decibels) of a cat with complete degeneration of the organ of Corti near its base. Note the moderate loss of sensitivity for tones below 250 Hz and the very severe hearing loss for tones above 1500 Hz. (From Lurie et al., 1934, with permission of the *Annual of Otology, Rhinology and Laryngology.*)

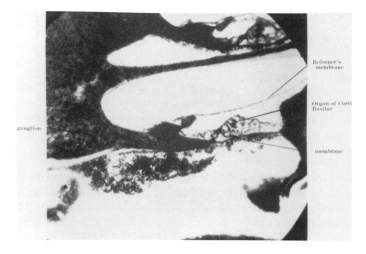

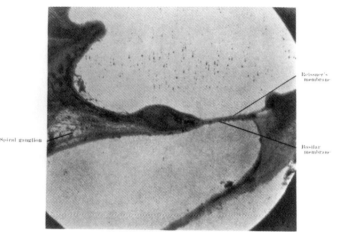

Figure 8.18. Photograph showing the absence of the organ of Corti on one part of the basilar membrane from one ear of an albino cat. This animal showed severe hearing losses for tones above 1500 Hz (From Lurie, Davis, and Derbyshire, 1934, with the permission of the *Annuals of Otology, Rhinology and Laryngology.*)

iron plate induced by the sound. The currents travel to the second telephone and induce oscillations of its magnetism, which in turn causes its iron plate to vibrate and produce sounds similar to those communicated to the first telephone. There is no analysis of the sound waves. (1886, 167)

The article continued:

The theory which the lecturer arrived at . . . might be called the telephone theory of the sense of hearing—The theory that the cochlea does not act on the principle of sympathetic vibration, but that the hairs of all its auditory cells vibrate to every tone just as the drum of the ear does; that there is no analysis of complex vibrations in the cochlea or elsewhere in the peripheral mechanism of the ear; that the hair cells transform sound vibrations into nerve-vibrations similar in frequency and amplitude to the sound vibrations; that simple and complex vibrations of nerve energy arrive in the sensory cells of the brain, and there produce, not sound again of course, but the sensations of sound, the nature of which depends not upon the stimulation of different sensory cells, but on the frequency, amplitude, and form of the vibrations coming into the cells, probably through all the fibres of the auditory nerve.

A related idea appeared in the closing years of the nineteenth century when Max Friedrich Meyer (1873–?), a

research assistant at the University of Berlin, began to develop his "frequency analytic" theory of hearing. Meyer tried to account for sound patterns by a geometric analysis of the waves that he believed formed in the inner ear. Meyer's theory is best understood by thinking about a leather seat. The cushion first stretches when one sits upon it, but then stabilizes and offers maximum resistance. Meyer believed this is how the basilar membrane must respond to the pressure caused by movements of the stapes upon the oval window. When the membrane is relaxed, it imposes little resistance to small displacements, but when the slack is taken up, the resistance increases markedly, and movement practically ceases.

Meyer's ideas were outlined in 1896 and presented formally in 1898. He continued to develop his theory during his long tenure at the University of Missouri (e.g., Meyer, 1928). Meyer's theory resembled Rutherford's in that he believed the nerves would be excited at frequencies corresponding to those of the stimulus. He theorized that the number of hair cells involved would correlate with the loudness of the sound (also see Kuile, 1900).

Although these theories were imaginative, many scientists wondered whether they rested on a solid scientific foundation. The basic problem with each was that nerves were required to fire up to 20,000 times per second in humans and even higher in some animals. In fact, when Rutherford's theory was presented in 1886, it was believed that the upper limit for human hearing might be two to three times as high as this.

Rutherford (1886) recognized the problem before he presented his theory and tried to prepare for this criticism by conducting nerve stimulation experiments on rabbits. He reported only a dismal maximal response rate of 352 muscle contractions per second. Yet because his experiment involved only nerves to the muscles, he did not see this result undermining his theory of auditory nerve conduction.

Rutherford (1898) later referred to electrophysiological experiments showing that action potentials lasted about 0.0007 sec. Thus, he calculated that frequencies of 1,400 impulses per second could be expected. Without any real explanation, he added that frequencies of more than 10 times this number might still be possible.

A different sort of problem facing the frequency theories, even during the late 1800s, was that mounting evidence from pathology favored some kind of a place theory (e.g., Baginsky, 1883; Moos and Steinbrügge, 1881; Stepanow, 1886). At the very least, the accumulating data confirmed that the basal part of the cochlea was most responsive to high-frequency sounds. Rutherford acknowledged this, but emphasized that the evidence indicated low tones were not localized in any small region of the cochlea. The problem was that he did not restrict his telephone theory to just the low tones.

AUDITORY ELECTROPHYSIOLOGY AND THEORIES OF HEARING

Attempts to record electrical responses from the auditory nerve were made late in the nineteenth century, but the results of early experiments that varied sound frequencies were far from conclusive (Beauregard and Dupuy, 1896). Even the results of experiments conducted in the 1920s did not provide unequivocal evidence for either place or telephone theories (see Forbes, Miller, and O'Connor, 1927).

In 1930, Ernest Glen Wever (b. 1902) and Charles W. Bray (b. 1904) inserted a large electrode into the auditory nerve of a cat. The electrical potentials were amplified and carried to a telephone receiver in an adjacent room, which was isolated from the test area. Wever and Bray found that 100–5,000-Hz sounds made in the room with the cat were reproduced in their recordings. They felt this "microphonic" was biological because it decreased when they reduced blood to the head, increased again when the blood supply was restored, and ceased when the cat died. They attributed it to the auditory nerve.

These findings appeared to support the frequency theory of hearing because the sounds did not undergo a spectral breakdown of the type predicted by place theorists. The problem was that nerve-firing frequencies of several thousand cycles per second were well outside the refractory periods thought possible for neurons. The evidence at this time suggested neurons required at least 0.001 sec to recover from one action potential before firing again. Although some investigators seemed willing to grant somewhat shorter refractory periods to neurons in the auditory system, the frequencies seen in the "microphonic" seemed beyond the realm of possibility.

In an attempt to account for their findings, Wever and Bray (1930) proposed that different fibers could respond to alternate waves of sounds if the frequencies were too high to be followed by individual neurons. For example, one neuron might fire to waves 1, 4, 7, 10, and so on, whereas another might follow waves 2, 5, 8, 11, and so on. In this way a higher-frequency stimulus could be followed by a platoon of neurons. This idea became known as the volley theory.

Whether the mechanism proposed by Wever and Bray could account for more than low- or middle-frequency sounds was questioned. There were numerous attempts to determine whether they had, in fact, recorded their potentials from neurons. The physiologist Edgar D. Adrian (1889–1977), who was soon to be awarded the Nobel Prize, at first believed that the source of the "microphonic action of the cochlea" was not neural (Adrian, 1931).[10] Later, Adrian changed his mind and accepted the idea that nerve cells or nerve fibers were responsible for the microphonics (Adrian, Bronk, and Phillips, 1931).

In 1932, Leon J. Saul (b. 1901) and Hallowell Davis concluded that Wever, Bray, and Adrian had rejected the correct explanation (also see Davis, 1935; Davis et al, 1934; Derbyshire and Davis, 1935). Saul and Davis used a shielded electrode to record the electrical activity of the auditory nerve and the cochlea independently. Their findings with cats strongly suggested that Wever and Bray had seen the summation of cochlear tissue and auditory nerve potentials.

Davis and his coworkers went on to demonstrate that the cochlear microphonic could follow frequencies at least a hundred times greater than the frequency that even the most sensitive nerve could track. They also showed that the microphonic exhibited more fidelity, more resistance to

Figure 8.19. Hallowell Davis (1896–1992) receiving the National Medal of Science in 1976 for his research in audition from President Gerald R. Ford (b. 1913). (Courtesy of the Central Institute for the Deaf.)

anesthetics or loss of blood supply, and a shorter latency than the nerve response. The microphonic also seemed practically immune to fatigue and showed no evidence of acting in an all-or-none fashion like a neuron.

But could the auditory nerve still follow frequencies above the refractory period of a single cell? The coaxial electrode allowed Davis (shown in Figure 8.19) and his associates to test this hypothesis. Their findings appear in Figure 8.20. The authors wrote:

> If action currents are led from the auditory nerve by suitable coaxial electrodes which exclude the direct microphonic response of the cochlea and are measured on a cathode ray oscillograph, it is found that their size is roughly constant in response to maximal auditory stimulation at frequencies of 700 and below. The frequency of the action current waves corresponds exactly to that of the stimulating sound. Between 700 and 900, and quite sharply in any given preparation, the size of the action currents falls to approximately one half of this value. A second, though less

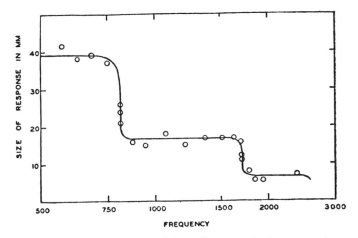

Figure 8.20. An illustration of the volley principle. Stevens and Davis (1934, p. 9) wrote: "The initial size of the action potential in the auditory nerve as a function of frequency. The sudden drops in the curve are due to the fact that above a certain frequency (800–1,000 cycles) the individual nerves fibers cannot follow the frequency of the stimulus, but must alternately respond to every other vibration. At twice this critical frequency each fiber responds to every third vibration." (Reproduced with permission of the American Institute of Physics.)

[10]Lord Adrian shared the 1932 Nobel Prize in medicine with Sir Charles Sherrington. Both men were honored for their work in neurophysiology.

spectacular drop, is usually seen at about 1,700. The frequency of the stimulating sound may still be detected in the response up to 2,800 per second but no higher. (Davis, Forbes, and Derbyshire, 1933, 522)

The first reduction was explained by individual nerves volleying to every second sound wave. The second reduction, at 1,700 Hz, was thought to be due to three groups of auditory nerve cells, each following every third sine wave. Davis and his coworkers concluded that the volley principle was no longer applicable above 2,800 Hz. While frequency of firing could explain tonal perception up to this level, confirming the theory of Rutherford, the place principle provided the best explanation for high-frequency sound perception, in accordance with the theory of Helmholtz.[11]

[11]See Derbyshire and Davis (1935) for slighxtly higher cutoffs for the volley principle.

REFERENCES

Adrian, E. D. 1931. The microphonic action of the cochlea: An interpretation of Wever and Bray's experiments. *Journal of Physiology, 71,* xxviii–xxix.

Adrian, E. D., Bronk, D. W., and Phillips, G. 1931. The nervous origin of the Wever and Bray effect. *Journal of Physiology, 73,* 2–3.

Amman, J. K. 1692. *Surdus loquens, seu methodus, qua qui surdus natus est loqui discere possit.* Amstelaedami: Apud Henricum Wetstenium.

Amman, J. K. 1700. *Dissertatio de loquela.* Amstelaedami: Joannem Wolters.

Aristotle. 1913. *De audibilibus.* (T. Loveday and E. S. Forster, Trans.). Oxford, U.K.: Clarendon Press.

Aristotle. 1948. *On the Soul.* II. 8. (R. D. Hicks, Trans.). In M. R. Cohen and I. E. Drabkin (Eds.), *A Sourcebook in Greek Science.* New York: McGraw-Hill. P. 547.

Baginsky, B. 1883. Die Function der Gehörschnecke. *Virchow's Archiv für die pathologische Anatomie und Physiologie, 94,* 65–85.

Bauhin, K. 1605. *Theatrum anatomicum.* Francofurti at Moenum: Matthaei Beckeri.

Beauregard, H., and Dupuy, E. 1896. Sur la variation électrique (courant d'action) déterminée dans le nerf acoustique par le son. *Archives de Laryngologie, de Rhinologie et des Maladies des Premières Voies Respiratoires et Digestives, 9,* 383–386.

Békésy, G. von. 1928. Zur Theorie des Hörens; Die Schwingungsform der Basilarmembran. *Physikalische Zeitschrift, 29,* 793–810.

Békésy, G. von. 1930. Sue la théorie de l'audition. *Année de Psychologie, 31,* 63–96.

Békésy, G. von. 1960. *Experiments in Hearing.* (E. G. Wever, Trans.). New York: McGraw-Hill.

Bell, C. 1811. *Idea of a New Anatomy of the Brain; Submitted for the Observations of His Friends.* London: Strahan & Preson. Reprinted in *Journal of Anatomy and Physiology,* 1869, *3,* 147–182.

Bell, C. 1812. *Anatomy of the Human Body.* New York: Collins and Company.

Berengario da Carpi, J. 1514. *Isagogae breves perlucide ac uberime in anatomiam humani corporis a communi medicorum academia usitatam.* Bologna: Hectoris.

Bernhard, C. G. 1962. The Nobel Prize for Physiology and Medicine, 1961. *Journal of the Acoustical Society of America, 34,* i–ii.

Boerhaave, H. 1740. *Praelectiones academicae. . . .* Göttingen: A. Vandenhoeck.

Bohn, J. 1686. *Circulus anatomico-physiologicus.* Leipzig: J. F. Gleditsch.

Bonet, J. P. 1620. *Reduction de las Letras, Arte para Ensenar á Ablar los Mudos.* Madrid: F. Abarca de Angulo.

Bonet, T. 1679. *Sepulchretum.* Geneva: Sumtibus L. Chouët.

Boring, E. G. 1942. *Sensation and Perception in the History of Experimental Psychology.* New York: Appleton-Century-Crofts.

Capivaccio, G. 1589. *Opera omnia.* Francofurti: E. Paltheniana, curante J. Rhodio.

Cardano, G. 1663. *Opera omnia tam hactenus excusa hic tamen aucta et emendata quam nunquam alias visa ac primum ex auctoris ipsius autographo eruta Cura Caroli Sponii* (10 vols.). Lugduni.

Cassebohm, J. F. 1734. *Tractus quintus anatomicus de aure humana.* Halae Magdeburgicae: Sumtibus Orphanatrophei.

Casserio, G. 1601. *De vocis auditusque organis historia anatomica.* Ferrariae: V. Baldinus.

Celsus. 1948. *On Medicine.* VIII. 1.5–6. (W. G. Spencer, Trans.). In M. R. Cohen and I. E. Drabkin (Eds.), *A Sourcebook in Greek Science.* New York: McGraw-Hill. Pp. 475–476.

Chladni, E. F. F. 1787. *Entdeckungen über die Theorie des Klanges.* Leipzig: Weidmanns erber und Reich.

Clagett, M. 1955. *Greek Science in Antiquity.* New York: Abelard-Schuman.

Coiter, V. 1572. *De auditus instrumento.* Noribergae: Theodorici Gerlatzeni.

Corti, A. 1851. Recherches sur l'organe de l'ouïe des mammifères. *Zeitschrift für wissenschaftlicte Zoologie, 3,* 109–169.

Cotugno, D. 1761. *De aquaeductibus auris humanae internae.* Neapoli: Simoniana.

Crowe, S. J., Guild, S. R., and Polvogt, L. M. 1934. Observations on the pathology of high-tone deafness. *Bulletin of the Johns Hopkins Hospital, 54,* 315–379.

Davis, H. 1935. The electrical phenomena of the cochlea and the auditory nerve. *Journal of the Acoustical Society of America, 6,* 205–215.

Davis, H., Derbyshire, A. J., Kemp, E. H., Lurie, M. H., and Upton, M. 1935a. Experimental stimulation deafness. *Science, 81,* 101–102.

Davis, H., Derbyshire, A. J., Kemp, E. H., Lurie, M. H., and Upton, M. 1935b. Functional and histological changes in the cochlea of the guinea-pig resulting from prolonged stimulation. *Journal of General Psychology, 12,* 251–278.

Davis, H., Derbyshire, A. J., Lurie, M. H., and Saul, L. J. 1934. The electric response of the cochlea. *American Journal of Physiology, 107,* 311–332.

Davis, H., Forbes, A., and Derbyshire, A. J. 1933. The recovery period of the auditory nerve and its significance for the theory of hearing. *Science, 78,* 522.

Deiters, O. 1860. *Untersuchungen über die Lamina spiralis membranacea.* Bonn: Henry und Cohen.

Derbyshire, A. J., and Davis, H. 1935. The action potentials of the auditory nerve. *American Journal of Physiology, 113,* 476–504.

Duverney, J. 1683. *Traité de l'Organe de l'Ouïe.* Paris: E. Michallet. (Second ed. translated by J. Marshall as, *A Treatise on the Ear.* London: Printed for John Whifton at Mr. Boyle's-Head, 1748.

Eustachi, B. 1564. *Opuscula anatomica.* Venetiis: Vincentius Luchinus.

Ewald, J. R. 1899a. *Eine neue Hörtheorie.* Bonn: E. Strauss.

Ewald, J. R. 1899b. Zur Physiologie des Labyrinthes. VI. Eine neue Hörtheorie. *Archiv für die gesammte Physiologie, 76,* 147–188.

Fabricius, Hieronymus ab Aquapendente, H. 1600. *De visione, voce, auditu.* Venetiis: F. Bolzettam.

Fallopius, G. 1561. *Observationes anatomicae.* Venetiis: Petrum Mannam.

Flourens, M.-J.-P. 1824. *Recherches Expérimentales sur les Propriétés et les Fonctions du Système Nerveux dans les Animaux Vertébrés.* Paris: J. B. Ballière.

Forbes, A., Miller, R. H., and O'Connor, J. 1927. Electric responses to acoustic stimuli in the decerebrate animal. *American Journal of Physiology, 1927, 80,* 363–380.

Fourier, J.-B. 1822. *Théorie Analytique de la Chaleur.* Breslau: G. Koebner.

Galen. 1968. *De usu partium.* Translated by M. T. May, as *On the Usefulness of the Parts of the Body.* Ithaca, NY: Cornell University Press.

Galilei, G. 1638. *Dialogues Concerning Two New Sciences.* (H. Crew and A. DeSalvio, Trans.). Evanston, IL: Northwestern University Press, 1939.

Grattan-Guinness, I. 1972. *Joseph Fourier, 1768–1830.* Cambridge, MA: MIT Press.

Guild, S. R., Crowe, S. J., Bunch, C. C., and Polvogt, L. M. 1931. Correlations of differences in the density of innervation of the organ of Corti with differences in the acuity of hearing, including evidence as to the location in the human cochlea of the receptors for certain tones. *Acta Oto-Laryngologica, 15,* 269–308.

Gulick, W. L. 1971. *Hearing: Physiology and Psychophysics.* New York: Oxford University Press.

Guthrie, D. 1940a. Early text-books of otology. *Journal of Laryngology and Otology, 55,* 109–112.

Guthrie, D. 1940b. The history of otology. *Journal of Laryngology and Otology, 55,* 473–494.

Habermann, J. 1890. Ueber die Schwerhörigkeit der Kesselschmiede. *Archiv für Ohrenheilkunde, 30,* 1–25.

Habermann, J. 1906. Beiträge zur Lehre von der professionellen Schwerhörigkeit. *Archiv für Ohrenheilkunde, 69,* 106–130.

Haller, A. von. 1786. *First Lines of Physiology.* Translated from the Latin. (Reprinted 1966; New York: Johnson Reprint Company)

Hasse, C. 1867. Die Schnecke der Vögel. *Zeitschrift für wissenschaftlicte Zoologie, 17,* 56–104.

Held, H. 1904. Untersuchungen über den feineren Bau des Ohrlabyrinthes der Wirbeltiere, I. *Berichte über die König Sächsischen Gesellschaft der Wissenschaften zu Leipzig. Math-Phys. Cl., 28,* 1–74.

Held, H. 1926. Die Cochlea der Säuger und der Vögel, ihre Entwicklung und ihr Bau. In A. Bethe (Ed.), *Handbuch der normalen und pathologischen Physiologie, 11, Receptionsorgane I.* Berlin: Springer. Pp. 467–534.

Held, H., and Kleinknecht, F. 1927. Die lokale Entspannung der Basilarmembran und ihre Hörlücken. *Pflüger's Archiv für die gesammte Physiologie, 216,* 1–31.

Helmholtz, H. L. F. 1863. *Die Lehre von den Tonempfindungen als physiologische Grundlage für die Theorie der Musik.* Braunschweig: Vieweg.

Helmholtz, H. L. F. 1885. *On the Sensations of Tone.* (A. J. Ellis, Trans.). London: Longmans, Green.

Hensen, V. 1863a. Studien über das Gehörorgan der Decapoden. *Zeitschrift für wissenschaftliche Zoologie, 13,* 319–412.

Hensen, V. 1863b. Zur Morphologie der Schnecke des Menschen und der Säugethiere. *Zeitschrift für wissenschaftliche Zoologie, 13,* 481–512.

Howe, H. A., and Guild, S. R. 1933. Absence of the organ of Corti and its possible relation to electric auditory nerve responses. *Anatomical Record, 55,* 20–21.

Hurst, C. H. 1895. A new theory of hearing. *Transactions of the Liverpool Biological Society, 9,* 321–353.

Kelley, N. H. 1937. Historical aspects of bone conduction. *Laryngoscope, 47,* 102–109.

Kemp, E. H. 1935. A critical review of experiments on the problem of stimulation deafness. *Psychological Bulletin, 32,* 325–342.

Kostelijk, P. J. 1950. *Theories of Hearing.* Leiden: University Pers Leiden.

Kuile, E. H. ter. 1900. Die Uebertragung der Energie von der Grundmembran auf die Haarzellen. *Pflüger's Archiv für die gesammte Physiologie, 79,* 146–157.

Lindsay, R. B. 1973. *Acoustics: Historical and Philosophical Development.* Stroudsburg, PA: Dowden, Hutchinson and Ross.

Lorente de Nó, R. 1937. The sensory endings in the cochlea. *Laryngoscope, 47,* 373–377.

Lurie, M. H., Davis, H., and Derbyshire, A. J. 1934. The electrical activity of the cochlea in certain pathological conditions. *Annals of Otology, Rhinology and Laryngology, 43,* 321–343.

Marx, H. 1909. Untersuchungen über experimentelle Schädigungen des Gehörorganes. A. Ueber mechanische Zerstörungen des Gehörorganes. *Zeitschrift für Ohrenheilkunde, 59,* 192–210.

Massa, N. 1536. *Liber introductorius anatomiae.* Venice.

Mercurialis, H. 1584. *De oculorum et aurium affectibus praelectiones.* Francofurdi: Ionnem Wechelum.

Mettler, C. C. 1947. *History of Medicine.* Philadelphia, PA: Blakiston.

Meyer, M. F. 1898. Zur Theorie der Differenztöne und der Gehörsempfindungen überhaupt. *Zeitschrift für Psychologie, 16,* 1–34.

Meyer, M. F. 1928. The hydraulic principles governing the function of the cochlea. *Journal of General Psychology, 1,* 239–265.

Moos, S., and Steinbrügge, H. 1881. Ueber Nervenatrophie in der ersten Schneckenwindung. *Zeitschrift für Ohrenheilkunde, 10,* 1–15.

Müller, J. 1826. *Zur vergleichenden Physiologie des Gesichtssinnes des Menschen und der Thiere.* Leipzig: C. Cnobloch.

Müller, J. 1838. *Handbuch der Physiologie des Menschen für Vorlesungen.* Coblenz: J. Hölscher.

Munk, H. 1881. *Über die Funktionen der Grosshirnrinde.* Berlin: Hirschwald.

Natanson, L. N. 1844. Analyse der Functionen des Nervensystems. *Archiv für Physiologie und Heilkunde 3,* 515–535.

Ohm, G. S. 1843. Ueber die Definition des Tones, nebst daran geknüpfter Theorie der Sirene und ähnlicher tonbildender Vorrichtungen. *Annalen der Physik und Chimie, 135,* 497–565.

Pavlov, I. P. 1927. *Conditioned Reflexes.* (G. V. Anrep, Trans.). London: Oxford University Press.

Perrault, C. 1684. *Mémoire pour Servir à l'Historie Naturelle des Animaux.* Paris.

Plato. 1952. *Timaeus.* (B. Jowett, Trans.). In *Great Books of the Western World: Vol. 7. Plato.* Chicago: Encyclopaedia Britannica. Pp. 442–477.

Platter, F. 1625. *Praxeos medicae tomi tres.* Basileae: L. Regis, J. Schroeteri.

Politzer, A. 1864. Untersuchungen über Schallfortpflanzung und Schalleitung im Gehörorgane. *Archiv für Ohrenheilkunde, 1,* 59–73.

Politzer, A. 1913. *Geschichte der Ohrenheilkunde, 1850–1911.* Stuttgart: Ferdinand Enke.

Pyl, J. T. 1742. *Dissertatio medica de auditu in genere et de illo qui fit per os in specie.* Griefswald.

Ranvier, L. 1875. *Traité Technique d'Histologie.* Paris: F. Savey.

Reissner, E. 1851. *De auris internae formatione.* Dorpati (Livonorum): H. Laakmann.

Retzius, G. 1881–1884. *Das Gehörorgan der Wirbeltiere: Vol. 1. Gehörorgan der Fische und Amphibien; II, Das Gehörorgan der Reptilien, der Vögel, und der Säugetiere.* Stockholm: Samson and Wallin.

Retzius, G. M. 1892. Die Endigungsweise des Gehörnerven. *Biologische Untersuchungen, 3,* 29–36.

Retzius, G. M. 1893. Weiters über die Endigungsweise des Gehörnerven. *Biologische Untersuchungen, 5,* 35–38.

Rinne, H. A. 1865. Beitrag zur Physiologie des menschlichen Ohres. *Zeitschrift für rationelle Medizin, 24,* 12–64.

Robinson, V. 1941. Chronology of otology. *Bulletin of the History of Medicine, 10,* 199–208.

Rodger, T. R. 1923. The pathological effects of excessive sounds on the cochlea apparatus, considered in relation to the theories of sound perception. *Journal of Laryngology and Otology, 38,* 66–71.

Rutherford, W. 1886. Physiological Notice: A new theory of hearing. *Journal of Anatomy and Physiology, 21,* 166–168.

Rutherford, W. 1898. Tone sensation with reference to the function of the cochlea. *Lancet, 2,* 389–394.

Saul, L. J., and Davis, H. 1932. Action currents in the central nervous system. 1. Action currents of the auditory tracts. *Archives of Neurology and Psychiatry, 28,* 1104–1116.

Savart, F. 1824. Recherches sur les usages de la membrane du tympan et de l'oreille externe. *Annales de Chimie et de Physique, Ser. 2, 26,* 5–39.

Schelhammer, G. C. 1684. *De auditu.* Lugduni Batavorum: Apud Petrum de Graaf.

Schultz, M. 1858. Ueber die Endigungsweise des Hörnerven im Labyrinth. *Archive für Anatomie und Physiologie,* 343–381.

Stepanow, E. M. 1886. Zur Frage über die Function der Cochlea. *Monatschrift für Ohrenheilkunde, 20,* 116–124.

Stevens, S. S., and Davis, H. 1936. Psychophysical acoustics: Pitch and loudness. *Journal of the Acoustical Society of America, 8,* 1–15.

Teed, R. W. 1936. The otology of Duverney. *Annals of Medical History, 8,* 453–455.

Theophrastus. 1948. *On the Senses. 25,* 9. (G. M. Stratton, Trans.). In M. R. Cohen and I. E. Drabkin (Eds.), *A Sourcebook in Greek Science.* New York: McGraw-Hill. Pp. 546–547.

Valsalva, A. M. 1704. *De aure humana tractus. . . .* Bononiae: C. Pissarii.

Valsalva, A. M. 1717. *De aure humana tractus . . .* Trajeti ad Rhenum: Guilielmum vande Water.

Varolio, C. 1591. *De nervis opticis, Nonnullisque aliis praeter communem opinionem in humano capite observatis.* Francofurti, Ioannem, Welchelum and Petrum: Fischerum.

Vesalius, A. 1543. *De humani corporis fabrica.* Libri septum. Basilae: J. Oporinus.

Volkmann, A. W. 1844. Nervenphysiologie. In R. Wagner (Ed.), *Handwörterbuch der Physiologie.* II. Braunschweig: F. Vieweg. Pp. 521–526.

Voltolini, R. 1885. Einiges Anatomische aus der Gehörschnecke und über die Function derselben resp. des Gehörorganes. *Virchow's Archiv für die pathologische Anatomie und Physiologie, 100,* 27–41.

Waller, A. D. 1891. A possible part played by the membrana basilaris in auditory excitation. *Journal of Physiology, 12,* xlix–l.

Watt, H. J. 1914. Psychological analysis and theory of hearing. *British Journal of Psychology, 7,* 1–43.

Watt, H. J. 1917. *The Psychology of Sound.* Cambridge, U.K.: Cambridge University Press.

Weber, E. H., 1848. Über die Umständ durch welche man geleitet wird manche Empfindungen au aüssere Objecte zu beziehen. *Berichte über die Verhandlungen der Königlichen Sächsischen Gesellschaft der Wissenschaft (Math.-Phys. Kl.), 2,* 226–237.

Weber, E. H., 1851. Ueber den Mechanismus des menschlichen Gehörgans. *Berichte über die Verhandlungen der Königlichen Sächsischen Gesellschaft der Wissenschaften (Math.-Phys. Kl.), 3,* 29–31.

Wever, E. G. 1949. *Theory of Hearing.* New York: John Wiley & Sons.

Wever, E. G. 1962. Development of traveling-wave theories. *Journal of the Acoustical Society of America, 34,* 1319–1324.

Wever, E. G., and Bray, C. W. 1930. The nature of acoustical response: The relationship between sound frequency and frequency of impulses in the auditory nerve. *Journal of Experimental Psychology, 13,* 373–387.

Wilkinson, G., and Gray, A. A. 1924. *The Mechanism of the Cochlea.* London: Macmillan.

Willis, T. 1672. *De anima brutorum . . .* Londini: Prostant Apud Guilielm. Wells & Robertum Scott.

Wittmaack, K. 1907. Ueber Schädigung des Gehörs durch Schalleinwirkung. *Zeitschrift für Ohrenheilkunde, 54,* 37–80.

Yoshii, U. 1909. Experimentelle Untersuchungen über die Schädigung des Gehörorganes durch Schalleinwirkung. *Zeitschrift für Ohrenheilkunde, 58,* 201–251.

Chapter 9
Audition and the Central Nervous System

> With regard to hearing, Munk was of the opinion that the whole of the temporal region is its sensory area. Ferrier and Yeo reaffirm the conclusion that destruction of the first temporal convolution on both sides causes complete and permanent loss of hearing. . . . On the other hand the statement that the temporal lobe is auditory, is formally contradicted by Schäfer, who found no impairment in the hearing of monkeys for weeks after complete destruction of that region. We must therefore suspend our opinion with regard to the cortical locus of hearing.
>
> Augustus D. Waller, 1891

The history of the anatomy of the ear and the parallel development of theories of hearing were examined in Chapter 8. Although the present chapter also deals with the auditory system, its emphasis is on the role of structures in the central nervous system. These include the auditory pathways, their brainstem nuclei, and the auditory cortex. Figure 9.1 presents a diagram from 1920 showing these structures.

THE COCHLEAR NUCLEI

Anatomical studies of auditory nerve (VIII) terminations were conducted in the second half of the nineteenth cen-

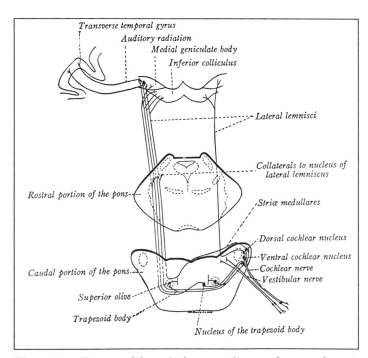

Figure 9.1. Diagram of the major human auditory pathways and terminations as depicted in Ranson's *Anatomy of the Nervous System*, 1920.

tury by some of the same individuals who had studied the cochlea. Among these scientists was Otto Friedrich Karl Deiters, a talented anatomist who lived to be only 29 years of age. Deiters' (1865) major work on the subject was an uncorrected and unfinished book, which was prefaced, edited, and finished by his highly regarded chief, Max Schultze.

Vladimir Bekhterev (1885, 1887) added to Deiters' work by distinguishing between the anterior and posterior roots of the eighth nerve. He called the former *ramus vestibularis* and the latter *ramus cochlearis*. He associated the vestibular root with "Deiters' nucleus."[1] His work led to a better understanding of the fact that only some components of the eighth nerve have acoustic functions.

Efforts to shed more light on the terminations of the eighth nerve in the cochlear nuclei came from Sigmund Freud (1856–1939) in 1886, Hans Held (1866–?) in 1891 and 1893, Rudolph Albert von Kölliker in 1896, and Santiago Ramón y Cajal in 1896. The investigations of these individuals added to the growing knowledge about the anatomy of the medullary nuclei involved in hearing. These studies also laid the foundation for many twentieth-century investigations, including Ramón y Cajal's (e.g., 1909–11) later work on the auditory projections.

In the opening years of the twentieth century, Arthur van Gehuchten (1861–1914) was able to conclude with some confidence that all fibers from the cochlear branch of the eighth cranial nerve terminate in the dorsal and ventral cochlear nuclei (see van Gehuchten, 1906).

In the early 1930s, Rafael Lorente de Nó used the Golgi stain on the brains of mice, rats, cats, and monkeys to show "the remarkable arrangement" of the auditory projection as it entered the cochlear nuclei. Lorente de Nó (1933a, 1933b), shown in Figure 9.2, described how the spiral shape of the cochlea was maintained in the composition of the auditory nerve. His illustrations, such as the one presented as Figure 9.3, revealed in much greater detail how each auditory nerve fiber sends one branch to the dor-

[1]See Schiller (1974) for the history of Deiters' nucleus.

Figure 9.2. Rafael Lorente de Nó (b. 1902), who studied the central connections of the auditory pathways. (Courtesy of the Central Institute for the Deaf.)

sal cochlea nucleus and one to the ventral cochlear nucleus on the ipsilateral side of the brain. Lorente de Nó estimated that the number of cells in the cochlear nuclei is twice as large as the number of incoming auditory nerve fibers. He concluded that single incoming cells could make hundreds and perhaps even thousands of synapses.

Lorente de Nó's studies showed that the cochlear nuclei were indeed complex structures. He pointed to at least 13 distinct histological regions and wrote that "each cochlea fibre establishes contacts with cells of every one of these

thirteen regions" (1933b, p. 329). In 1937, he categorized the incoming fibers into those with localized endings (radial fibers) and those with extensive synaptic connections (spiral fibers).

Stephen Poliak (also spelled "Poljak" and "Polyak"; 1889–1955), using the Weigert staining method on bats, also found a very orderly and systematic arrangement of the fibers from the ear to the cochlear nuclei (Poljak, 1926, 1927). Nevertheless, many later scientists and authors of secondary sources simplified the complex schemes of Lorente de Nó and Poliak, content to distinguish only between the dorsal and ventral (or anterior and posterior in the older literature) cochlear nuclei of the medulla.

THE SUPERIOR OLIVARY COMPLEX

Scientists soon found that the projections from the cochlear nuclei went to a number of loci, but these projections were not easy to follow. The cochlear nucleus efferents appeared to take at least three routes across the midline to the opposite side of the brain (see Papez, 1930; Ramón y Cajal, 1896, 1909–11). The largest was found to leave the ventral part of the ventral cochlear nucleus and to traverse the midline via the trapezoid body (ventral stria). A second pathway, the intermediate stria, originated in the dorsal part of the ventral cochlear nucleus (Held, 1893). And a third pathway, the dorsal stria described by Constantin von Monakow (1890, 1891), originated in the dorsal cochlear nucleus.

The precise terminations of these pathways were matters of debate, some of which seemed to be due to species differences. It was clear, however, that most cochlear nucleus fibers terminated in the contralateral superior olivary complex. Investigators typically included the superior olive proper, the accessory olive, the preolivary nuclei, and the trapezoid nuclei in this nuclear complex (see Barnes, Magoun, and Ranson, 1943; Papez, 1930). The nucleus of the lateral lemniscus was occasionally considered a rostral extension of this complex.

THE INFERIOR COLLICULUS: A REFLEX CENTER?

Many of the cells leaving the superior olivary complex proper were found to project via the lateral lemniscus to the inferior colliculus. This was observed by Held (1891, 1892, 1893) with Weigert and Golgi stains applied to cat, rabbit, chimpanzee, and human brains. Held added that ascending fibers might also synapse in the nucleus of the lateral lemniscus and that there could be a small projection to the superior colliculus. Kölliker (1896), using similar stains on cat and human brains, came to the same conclusions.

Ramón y Cajal (1909–11) thought the lateral lemniscus received some fibers from the contralateral cochlear nuclei and that the lateral lemniscus projected largely and directly to the medial geniculate body. He maintained that the inferior colliculus received mostly collaterals of axons headed toward the medial geniculate body in the thalamus. His findings with Golgi preparations of the mouse brain led him to theorize that the superior olivary complex and the inferior colliculus essentially serve as reflex centers.

The idea that the inferior colliculus might be a reflex center was also entertained before 1909. Early electrical

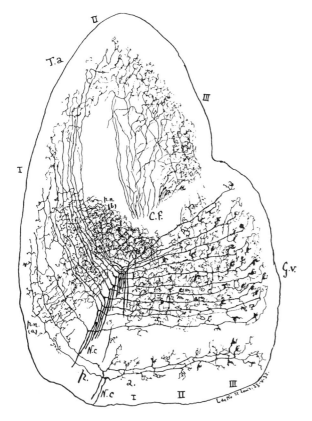

Figure 9.3. Drawing by Lorente de Nó (1933) showing the endings of the auditory nerve (VIII) branching into the dorsal and ventral cochlear nuclei. (Reproduced with the permission of *Laryngoscope*.)

stimulation experiments by Prus (1899) on the inferior colliculus of a variety of animals seemed to elicit many different reflexive responses. These included walking movements, eye rotation, deviations of the head and body to the opposite side, changes in blood pressure, slowing of respiration, and most important, erection of the ears. On the basis of these findings, and some hints from the work of David Ferrier (1876a), there was fairly widespread acceptance of the theory that the colliculi functioned primarily in reflexive acts.

This idea was further strengthened when Vladimir Bekhterev (1909) reported that unilateral removal of the inferior colliculus led to a transient diminution of reflexive movements of the contralateral ear to sounds. Moreover, lesions of the inferior colliculi did not cause deafness in monkeys (Ferrier and Turner, 1901).

The major problem with the electrical stimulation findings was simply that later experimenters were unable to confirm them.[2] In fact, there was every reason to believe earlier workers were seeing the effects of the spread of current in the region of the superior colliculus to neighboring nuclei and tracts (Spiegel and Kakeshita, 1926). Newer anatomical findings were also more compatible with the growing belief that the inferior colliculus lacked the necessary connections to lower brainstem structures that would be required of a reflex center.

By the mid-1930s the role of the inferior colliculus as a reflex center had few supporters. If anything, scientists were now viewing it as an important "relay" station in the auditory projection system (Rasmussen, 1946).

THE MEDIAL GENICULATE BODY

The medial geniculate body was distinguished from the lateral geniculate body by Karl Burdach in 1822. He called the medial geniculate the *corpus geniculatum internum* and the lateral geniculate body the *corpus geniculatum externum*.

Later in the nineteenth century, studies of fibers ascending from the inferior colliculus and the superior olivary complex showed that the medial geniculate was the thalamic nucleus most concerned with audition. This conclusion was entirely consistent with "top-down" studies on the source of the afferents to those parts of the temporal cortex thought to be most important for hearing.

For example, Paul Flechsig (1896), the pioneer histologist and professor of psychiatry at Leipzig, reported that the fibers from the medial geniculate body to the temporal lobe were already showing myelination in the 2-month-old child. He saw this occurring almost exclusively in the superior temporal gyrus, which was his area No. 5. Flechsig (1901) looked upon these projections as the only sensory projections concerned directly with hearing.

In 1895, Constantin von Monakow agreed that the medial geniculate body projected most densely to the ipsilateral temporal lobe. He arrived at this conclusion by removing a large part of the known cortical auditory area from cats and, after a six-month period, examining the degeneration in the thalamus.

Anatomical studies on primates, including the Marchi degeneration experiments conducted on monkeys by Poliak (1932), showed that small regions of the medical geniculate body degenerated after small circumscribed lesions of the "auditory cortex" (see also Rundles and Papez, 1938; Walker, 1937).

The "bottom-up" strategy of destroying parts of the medial geniculate body and then looking for cortical changes was appreciably slower in emerging, but consistent with the conclusions drawn from the aforementioned studies (Woollard and Harpman, 1939).

FERRIER AND THE DISCOVERY OF THE AUDITORY CORTEX

Prior to the nineteenth century, some individuals believed the cerebellum played a role in the perception of sounds. For example, Constanzo Varolio localized the sense of hearing in the cerebellum in 1591. In the next century, Thomas Willis (1664) associated the cerebellum with the perception of music. He believed the perception of music had something to do with the softness of the cerebellum. He wrote that audible sounds pass through the cerebellum and leave behind traces in people who are endowed with a musical ear. But in those with harder cerebellums, traces are not produced, and there is no "Musical faculty" (see Neuburger, 1981).

Interestingly, the relation between the cerebellum and music persisted well after the seventeenth century had ended. As late as 1838, Antoine Louis Dugès (1797–1838) continued to hold an opinion like that of Thomas Willis.

The discovery of the auditory cortex is usually attributed to David Ferrier, who, in the mid-1870s, used faradic current to map the cortex. Ferrier's method was viewed as an improvement over galvanic current, the type of current used by Gustav Fritsch and Eduard Hitzig when they discovered the motor cortex in the dog just a few years earlier (1870).

Ferrier (1876a) wrote that when he electrically stimulated the upper part of the temporal lobe, his monkeys lifted the opposite ear, turned their heads and eyes to the opposite side, and dilated their pupils. He believed these were auditory and not somatic responses because they closely resembled the startle responses made by monkeys to sudden sounds. Although Ferrier seemed unaware of the degree to which the primary auditory cortex was lodged in the depths of the Sylvian fissure, his stimulation studies suggested that the auditory receptive region occupied the upper two thirds of the superior temporal gyrus ("superior temporo-sphenoidal convolution").

Ferrier knew that the findings from his stimulation experiments had to be validated by ablation. In 1876, he ablated the superior temporal gyrus in five monkeys from a group of 25 given various lesions (Ferrier, 1876b). Two of these five monkeys had unilateral lesions, and the remaining three had bilateral lesions. He then examined the animals to see if they would turn when he called them, show startle responses to sound, and so on. He especially wanted to know if they would still be able to show responses like those elicited with faradic stimulation before surgery.

Ferrier found that the three animals with bilateral lesions were unresponsive to sound. He also found that the two monkeys with unilateral lesions were deaf in one ear—

---

[2]See Woollard and Harpman (1940) for a summary of this literature.

something he "proved" by plugging the ear ipsilateral to the lesion with cotton before attempting to test his animals. He took these findings as strong confirmation of the hypothesis that the superior temporal gyrus played a critical role in audition.

Ferrier's findings were subjected to many questions and criticisms. The issue was not just whether the superior temporal gyrus played a role in hearing but whether these lesions should have caused total deafness. Primarily because his surgeries were not performed under aseptic conditions, Ferrier tried to test his animals and sacrifice them for histology before they could develop serious infections. Thus, it remained possible that some hearing would have returned to his animals had they been granted recovery periods of a few months.

Ferrier was sensitive to this criticism and conducted a second series of experiments under aseptic conditions with Gerald Yeo. One monkey had lesions of the middle temporal gyrus and, as expected, the damage did not affect hearing. Another monkey, known as "Monkey F," had bilateral lesions of the superior temporal gyri. Ferrier wrote that this monkey remained deaf for the 13 months it was studied.

Monkey F was seen by Ferrier as his real test case. In fact, this monkey was considered so important that he took it to the 1881 International Medical Congress in London (see Chapters 3 and 4). There, before an astonished audience, he demonstrated that it did not even startle to the sound of a gun (Ferrier and Yeo, 1885).

THE GREAT DEBATE

In contrast to Ferrier's findings (see Figure 9.4), others reported some recovery of auditory function after large lesions of the auditory cortex. For example, Luigi Luciani and Augusto Tamburini found considerable recovery in the lesion experiments they reported in 1879 (also see Luciani, 1884). The Italian investigators used dogs and tested hearing by throwing pieces of meat on a plate, varying both the size of the piece and the force of the throw. They felt that

Figure 9.5. Luigi Luciani (1842–1919), an Italian investigator who conducted many lesion studies on the sensory cortical areas of dogs and monkeys.

loud, sudden noises and even calls were more suitable for demonstrating profound hearing losses, but that this method was the better one for detecting more subtle hearing changes.

Luciani concluded that each ear projected to both sides of the brain, but that the opposite side received the more significant projections. The whole temporal lobe seemed to be important for hearing. Luciani, shown in Figure 9.5, made the following statements about deafness after large temporal lobe lesions:

> The effects of more or less extensive unilateral extirpations within the limits of the auditory sphere last, according to our recent results, for a longer or shorter time according to the amount of cortical matter removed, but always disappear after some time, or at least cannot be discovered by the means at our disposal. The symptoms consist of a more or less profound blunting of the acuity of hearing, but which never reaches complete deafness, and gradually fades off. An incomplete psychical deafness only subsists, by which the animal seems not to appreciate the meaning of sounds, noises, or calls, though it appears to hear them. The effects of bilateral extirpations are always more serious: at first the auditory troubles may amount to absolute deafness, but soon pass into a condition of obtuseness of hearing; this also becomes attenuated and leaves behind it only more or less evident signs (according to the extent of the injuries) of psychical deafness. (1884, 156)

Because these experiments involved dogs, Ferrier never really felt they contradicted his own findings with monkeys. He reasoned there were likely to be species differences in auditory recovery—differences comparable to those noted after motor cortex lesions (see Chapter 14).

More serious problems arose, however, when Edward Albert Schäfer, shown in Figure 9.6, began to report the results of his experiments on monkeys. Schäfer's first experiments were conducted with Victor Horsley (1857–1916). They were presented orally before the Royal Society in February 1887 and published a year later as an experimental report with little in the way of commentary or discussion (Horsley and Schäfer, 1888). Later that year they appeared as part of a much broader, important theoretical paper (Schäfer, 1888a).

Schäfer (1888a, p. 373) described one monkey with "almost complete" bilateral lesions of both temporal lobes

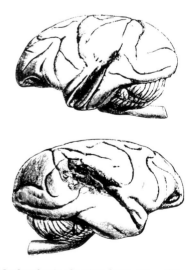

Figure 9.4. Monkey brain showing destruction of the superior temporosphenoidal convolution on both sides. This lesion was thought to cause complete deafness by David Ferrier (1886).

Figure 9.6. Edward Albert Schäfer (Sharpey-Schäfer 1850–1935). In the closing years of the nineteenth century, David Ferrier and Schäfer had several major disagreements, one over the effects of lesions of the superior temporal gyrus. (From the *Quarterly Journal of Experimental Physiology*, 1935. Reproduced with the permission of the editors.)

which shows the size of these lesions). Their behavioral and anatomical findings led them to conclude that Ferrier had not given his monkeys adequate survival time for recovery, with the possible exception of Monkey F. Brown and Schäfer then suggested that Monkey F may have been deaf going into surgery!

Ferrier was enraged by Schäfer's verbal and written attacks. In April of the same year, he countered with a paper in which he presented all of the evidence he knew of that linked hearing with the superior temporal gyrus. He included his own work with electrical stimulation and ablation techniques, as well as two cases of cortical deafness in the clinical literature.

One of the cited cases was described by John Cargyll Shaw (1845–1900) in a full report in *Archives of Medicine* (1882a) and in a short report in *Brain* (1882b). It involved a 34-year-old woman who became deaf and blind after "complete atrophy of the angular gyri and superior temporo-sphenoidal convolutions of both hemispheres, almost exactly symmetrical" (1882b, p. 430).

The second case cited by Ferrier involved an "absolutely deaf" woman described by Carl Wernicke (1848–1904) and Carl Friedländer (1847–1887) in 1883. This patient also appeared to suffer successive strokes that affected the superior temporal convolutions of both hemispheres.

and six additional cases, in which "we have more or less completely destroyed the superior temporal gyrus on both sides." None of these monkeys seemed to suffer a hearing loss. Immediately after surgery, they responded to sounds such as the smacking of the lips and the crumpling of newspapers. Months later, hearing did not deteriorate. Schäfer (1888b) argued that this showed the temporal lobes were not critically involved in hearing in primates.

Sanger Brown (1852–1928) and Schäfer conducted a second, more detailed analysis of monkeys with various lesions, including some with bilateral temporal lobe lesions. This study was read before the Royal Society in December 1887 and was published in 1888. As with Ferrier's second study, all of the animals were operated upon under antiseptic conditions.

Although some of the animals with large lesions of both temporal lobes seemed to have trouble understanding sights and sounds, there was still no evidence for "stone deafness," nor even for changes in auditory sensitivity. Brown and Schäfer wrote:

> Animals with both superior temporal gyri completely destroyed give evidence of the possession of acute powers of hearing; they turn at the slightest rustle, look up at the smallest noise, even immediately after the operation and when still drowsy from the prolonged influence of the anaesthetic, and follow with the head and eyes the direction of footsteps along a corridor outside the room in which they are confined. Some of these animals we have had under observation for several months; they have been seen and tested by many people, and the absence of the gyrus has been attested by *post-mortem* examination. (1888, 324–325)

Brown and Schäfer maintained that they made lesions which were larger and more complete than those of Ferrier and Yeo (1885), who claimed the superior temporal gyrus was the exclusive cortical auditory center (see Figure 9.7,

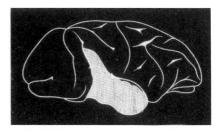

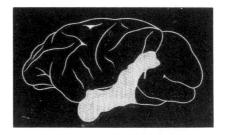

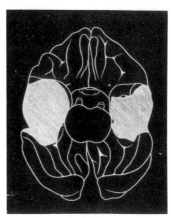

Figure 9.7. Brain of a monkey with almost complete removal of both temporal lobes. Edward Albert Schäfer (1888) wrote that this animal could hear even slight sounds soon after its operation.

Ferrier's reply also included an excerpt from a letter of March 19, 1885, sent to him by Schäfer. In it, Schäfer suggested that the temporal lobe was in fact critical for hearing. It read:

> We have some monkeys now which you ought to see. The whole of the temporo-sphenoidal lobe, including the hippocampus, has been removed in them, and we cannot detect any paresis of sensation (except auditory).

Ferrier then suggested that Schäfer's monkeys must have had incomplete lesions relative to his own Monkey F. He wrote:

> But Professor Schäfer's figures (Figs. 5 and 6) of the brains of these monkeys must be regarded more as a sort of pictorial representation of his own idea . . . rather than of the reality. Having myself examined both preparations, I most emphatically deny the accuracy of Professor Schäfer's diagrams. In neither one nor the other were the superior temporal gyri completely destroyed:—considerable portions remaining absolutely intact in both hemispheres, and notably so in the brain which Fig. 6 professes to represent. And considering how extensive lesions may be in any given centre—witness the visual centre—without total annihilation of its functions, nothing short of absolute destruction, primarily or secondarily, is likely to abolish all vestiges of the sense to which it is related. (1888, 10–11)

Ferrier included his own notes on Monkey F in his publication. These notes provided a few indications that Monkey F might have occasionally reacted to loud sounds, although Ferrier stressed that these were probably nothing more than chance effects.

> July 7th. Have been making numerous experiments, also Dr. Yeo, as to whether the animal hears the noise of a percussion-cap exploded near. It is difficult to judge. Occasionally it seems to start coincidentally with the sound, and yet at other times gives not the slightest sign, while the others invariably start.
>
> July 27th. . . . Loud whistling, knocking, ringing electric bells in the neighbourhood failed to elicit any sign of attention. Occasionally, as before, there was an appearance of a start, or a look around, apparently conditioned by the noise, but repeated experimentation showed that this was pure coincidence. (1888, 16)

Schäfer (1888b) seemed as enraged by Ferrier's 1888 article as Ferrier was by Schäfer's paper. He issued a cutting reply to Ferrier three months later in *Brain,* the same journal in which Ferrier's response to him had appeared. Schäfer retracted his initial notion that Ferrier's celebrated monkey was totally deaf even before surgery and now cited Ferrier's just published notes as proof that Monkey F could, in fact, hear after surgery. He wrote that Ferrier was an "observer unskilled in the examination of cerebral lesions" (p. 147). He again emphasized that his monkeys' lesions were more complete than those made by his adversary and that the lesions in the clinical cases cited by Ferrier affected much more than the superior temporal gyrus. Schäfer specifically noted that the Wernicke and Friedländer case study involved a syphilitic woman with epilepsy who showed bilateral seizures. He stressed her "general stupidity" and "aberrant intellect," saying in just so many words that her hearing was untestable.

Schäfer continued his barrage upon Ferrier in his *Textbook of Physiology,* published in 1900. But perhaps a bit weary from the prolonged battle, and aware of the growing human and animal literature, he retreated somewhat from his original hardline position. Schäfer now maintained that "the function of sound perception is not entirely localised to the superior temporal gyrus, although the focal point for it is probably there" (p. 761). When making his case for a broader auditory sphere, Schäfer cited his own experiments, as well as the work of Luciani and Munk on dogs.

In the context of Schäfer's citing of Munk's work, it is worth noting that Munk's findings had much in common with those of Luciani. Munk (1881) localized the most important area for hearing in the posterolateral temporal lobe, but then suggested the whole temporal lobe, including the *cornu ammonis,* might be a part of the "auditory sphere."

Munk also thought absolute deafness *(Rindentaubheit)* could be seen in dogs with temporal lobe lesions. But if the lesions did not involve the whole "auditory sphere," this condition seemed to disappear within a few weeks. In contrast, if the lesion involved only the center of the auditory sphere, rather than being totally deaf for a time, such dogs exhibit "psychic deafness" *(Seelentaubheit).* He defined this as a condition in which the dog can hear, but one in which it cannot understand the meaning of sounds—that is, a disorder analogous to his condition of psychic blindness (see Chapter 6). Munk viewed psychical deafness as a specific memory problem, one due to the loss of stored auditory images.

THREE COMPETING THEORIES

With Ferrier, Schäfer, Luciani and Tamburini, and Munk, three possible outcomes had been placed on the table. They were: (1) lesions of the superior temporal gyrus cause complete deafness; (2) lesions of this part of the cortex, or its homologue, have no effects, or only transient effects, on hearing; and (3) superior temporal gyrus lesions (or damage to its homologue) can affect hearing, but produce something short of total deafness.

Along with these behavioral possibilities were three physiological possibilities. For primates, these were: (1) the superior temporal gyrus alone houses the cortical organ for hearing; (2) it does not house the cortical organ for hearing; and (3) it is a part of a larger functional area concerned with hearing.

Late-nineteenth-century investigators were left wondering who was right and who was wrong. The distinguished American psychologist William James (1842–1910), among others, was absolutely baffled by the debates, although James wrote in his *Principles of Psychology* (1890) that he had less reason to doubt Schäfer's observations. Augustus Waller (1891), after contrasting the findings of Munk, Ferrier, and Schäfer, did not even attempt to guess who was correct (see Waller's quotation at the beginning of this chapter).

AUDITORY AURAS

The war that erupted over the superior temporal cortex and the concept of cortical deafness primarily involved monkeys. By the end of the nineteenth century, however, the field of combat had shifted from the laboratory to the hospital and the clinic. One reason was that laboratory scientists, following Munk's example, turned more to dogs and cats for their experiments. Ferrier agreed with Luciani and Munk that these animals do not necessarily show permanent cortical deafness. The real unanswered question had to do with primates, and especially with humans.

Some light was shed on the special role of the temporal lobes in human hearing in studies involving temporal lobe epileptics. Auditory auras were noted in a number cases by William Gowers (1885), who described them in his book *Epilepsy*. He mentioned six individuals who exhibited a loss of hearing during their seizures. In addition, he presented 21 cases in which there was an auditory sensation preceding the seizure.

In four of these cases, there was a "crash," which Gowers noted was a type of aura mentioned in classical times by Aretaeus of Cappadocia (second or third century A.D.). Gowers had a fifth patient who described "thunder," a noise that could also be thought of as a crashing sound. In the remaining cases, the descriptions of the auras included "sound of a drum," "whiz," "hissing," "ringing," a "rushing sound," and a "whistle." These sounds were considered to be the lowest types of auditory sensations.

Gowers also mentioned three patients with higher auras, sensations that involved words or music. Interestingly, these patients never seemed to be able to make out the tune. He thought these more elaborate sensations were associated with milder forms of epilepsy. One patient, who had heard a piece of music before a seemingly slight attack, experienced only the sound of a bell before a more severe seizure.

With regard to localization, Gowers stated:

> I am aware of no published post-mortem of a case in which an auditory aura preceded the fits, but in a case of Dr. Hughlings Jackson's, of cerebral tumor . . . a sound of bells ringing, referred to the left ear, preceded left-sided fits in an early period of the disease. After death we found a tumour outside the opposite optic thalamus. It was extensive, but had evidently commenced between the thalamus and the upper temporal-sphenoidal convolution. (1885, 57)

Thus, the study of auditory auras in the late nineteenth century was thought to support the role of the superior temporal cortex in hearing.

CORTICAL DEAFNESS IN HUMANS

Before the nineteenth century ended, Charles K. Mills (1845–1931) presented the case of a person who had been deaf for 30 years, a man "whose brain showed marked atrophy of both superior temporal convolutions" (1891, pp. 467–468). Mills also provided a complete description of a woman whose successive strokes severely damaged her superior temporal gyri and left her totally deaf. These findings, like those of Shaw (1882a, 1882b) and Wernicke and Friedländer (1883), were viewed as highly supportive of Ferrier's findings with monkeys.

More studies on cortical deafness in humans appeared early in the twentieth century. For example, Frederick W. Mott (1853–1926) described a deaf woman's brain that showed "complete bilateral destruction of the auditory cortical centres" (Mott, 1907, p. 313). The damage affected the posterior third of the first temporal gyrus, the posterior part of the second temporal gyrus, and the transverse gyri of Heschl.[3]

By 1918, Salomon Henschen was able to cite and discuss 16 cases of cortical deafness, each with an accompanying postmortem examination. Even more studies appeared in the 1920s (e.g., Misch, 1928). A typical case was described as follows:

> A right-handed woman of 62, the subject of mitral stenosis, suddenly developed . . . a sensory aphasia and paraphasia. Twelve days later she presented symptoms of another cerebral attack. Upon admission to hospital the patient was found to be completely deaf. Examination of the ears showed no abnormality of the membranes. According to the patient's sister her hearing had been quite acute prior to her illness, and Dr. A. Y. Cockrane, who saw the patient on several occasions between her first and second attacks was satisfied that she was not deaf during this period. The diagnosis of embolic softenings in both temporal lobes was suggested. . . . The post-mortem examination showed an extensive softening in the left hemisphere, involving a large part of the two upper temporal convolutions, and on the right side a more localized softening involving the supramarginal gyrus as well as the upper and posterior part of the left superior temporal convolution. (Bramwell, 1927, 579)

Case reports such as these seemed to provide good support for Ferrier's concept of cortical deafness in primates. They also suggested that the superior temporal area was the most important part of the cortex for hearing in humans, although the possibility remained open that auditory functions were not confined to this one gyrus.

As might be expected, the critics were ready to point to the limitations of these clinical studies. In their minds, the so-called evidence might be looked upon as tentative at best. The more significant concerns were: (1) there was little in the way of convincing evidence for the absence of subcortical or peripheral lesions, or problems such as presbycusis, in these cases; (2) there usually were no premorbid measures for comparison, which would control for such things as the effects of noise exposure or the use of ototoxic drugs; (3) the naturally occurring lesions were not confined to just the two superior temporal gyri; (4) there were rarely controls for inattention and "testability"; (5) autopsy material was rarely available; and (6) survival times were often too short for results to be meaningful.

Regarding the last point, Adolf Meyer (1866–1950), in his discussion of the Bramwell case, stated that "in the present case there was evidently a general deafness, but it may have been only a temporary symptom which might in part have cleared up if the patient had lived longer" (1927, p. 580). The importance of one of Schäfer's main criticisms of Ferrier's experiments was not lost on the clinicians.

To a certain extent, the growing criticisms of the cortical deafness literature by scientists demanding more experimental rigor opened the door for the resurrection of Schäfer's original hypothesis, which held that the temporal lobes might not be essential for hearing. The idea reemerged with no less a personage than Walter Dandy (1886–1946), the American neurosurgeon. Dandy (1930, 1933) shocked the establishment by stating that the cerebral hemispheres

---

[3]This area was named after Richard L. Heschl (1824–1881), who described it in 1878.

may not be critical for hearing (also see Squires, 1935).

Dandy based his extremist position largely upon observations of individuals who underwent radical surgeries for the removal of cerebral tumors. These included some cases of hemispherectomies. Audiometric testing was conducted on one of his right-hemispherectomy cases. The audiograms of the right and left ear were found to be approximately equal with no gaps in the patterns (Bunch, 1928). Dandy, never one to shy from a fight, argued that he never encountered deafness from a tumor or any other type of lesion in either cerebral hemisphere. In his mind, there was no proof that a center for hearing existed in either cerebral hemisphere!

MORE CONTEMPORARY LESION STUDIES

A number of studies were conducted after this time to try to resolve the question of cortical deafness in primates. A few involved monkeys, and some involved human cases. Relatively recent studies on infrahuman primates showed that bilateral lesions of the superior temporal gyrus can result in total deafness for weeks, but that this is often followed by some recovery (Heffner and Heffner, 1986). The recovery appears best for low-frequency sounds, but these monkeys never seem to reach normal levels of performance and continue to show marked hearing losses. That is, they do not show protracted cortical deafness, but neither do they show complete recovery.

More contemporary human case studies have also revealed few lasting defects in frequency and intensity discrimination, and thus also contradict the idea that "stone deafness" must appear after large superior temporal cortical lesions. Nevertheless, the patients in these studies have been found to suffer from persisting deficits in the perception of sound duration, and in recognizing auditory sequence patterns (including speech), thereby indicating that these lesions still have serious long-term effects on some aspects of hearing (Jerger, Lovering, and Wertz, 1973; Jerger, et al, 1969; Lhermitte et al, 1971; Rosati et al, 1982).

If anything, investigators seem to have concluded that Ferrier was correct when he pointed to the importance of the superior part of the temporal lobe for hearing in primates. But most scientists have also concluded he was wrong in asserting that lesions in this area cause absolute deafness in primates and that this is the only part of the cerebral cortex significantly involved in hearing. Ferrier's belief that the cortex is responsible for sensation and that lesions of the appropriate cortical area should lead to total abolition of sensation may have biased him away from recognizing that even complete superior temporal cortical lesions can result in something less severe than total deafness.

MULTIPLE CORTICAL AREAS
AND THEIR ORGANIZATION

Richard Caton (1875), who presented the first description in the scientific literature on spontaneous electrical activity in the brain, provided a report of his experiments with sensory stimulation in 1877 (see Chapter 3). This report, based on over 40 rabbits, cats, and monkeys, showed that visual stimuli evoked regional changes in the electrical activity of the hemispheres. Caton added, however, that intense sounds, such as those made by a loud bell, were ineffective in producing comparable changes.

In Poland, Adolf Beck also attempted to record electrical changes in the cortex to loud sounds (see Brazier, 1988). In his thesis, he placed electrodes on the surface of the brain and stimulated the sensory receptors of his animals. In one of two experiments on dogs, he found that a shout led to small changes in cortical activity, provided the electrode was positioned in the temporal region. Beck (1890) noted it was more difficult to record from the auditory system than from the visual system because the electrodes had to be placed on the inner surface of the temporal lobe.

The first shred of evidence for some sort of "tonotopic" organization at the level of the cortex came not from Beck's electrical recording studies but from some of the ablation studies previously described. Hermann Munk, testing dogs with the anterior part of the auditory area destroyed, observed that their deficits were greatest for high tones. He added that destruction of the posterior part of this area was associated with deficits when lower-pitched sounds were presented (cited in Schäfer, 1900, p. 763).

Munk's assertions were supported and extended by Larionow (1899), who tested the ability of dogs to hear different musical notes after small ablations in the posterior part of the suprasylvian fissure of the temporal lobe. On the basis of his findings, he divided the dog's auditory cortex into three zones, one for high, one for middle, and one for low tones. This is illustrated in Figure 9.8.

In addition to these studies with laboratory animals, Salomon Henschen (1918), after showing that the macula of the human retina projects to just one part of the visual cortex, postulated that there was a comparable orderly arrangement for sound frequencies at the cortical level. He felt this was well supported by his clinical cases.

The first detailed studies on the electrophysiology of the auditory cortex did not appear until the early 1940s (Woolsey and Waltz, 1942). By stimulating different parts of the cat cochlea with small electrodes, and by recording from different parts of the temporal lobe, the specificity of the connections from the cochlea to the auditory cortex was

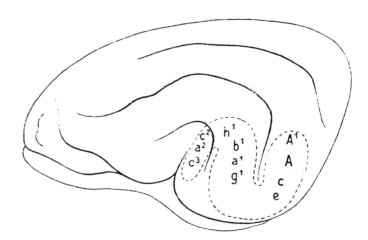

Figure 9.8. A "functional" map of the cortical auditory area. The map was based on the ability of dogs to hear different musical notes after small cortical ablations. (From Larinow, 1899.)

observed. Not only was one tonotopically organized region found at the cortical level with the base of the cochlea represented anteriorly, but a second tonal map was found ventral to the first with the sound spectrum in reverse order. These two responsive regions became known as AI and AII (Woolsey and Fairman, 1946).

At about the same time, electrophysiological studies with strychnine revealed a previously "silent" area in the posterior ectosylvian gyrus of the cat (Ades, 1943). This third area was thought to be dependent upon the integrity of the primary auditory cortex and, for this reason, was presented as a true auditory association area.

The finding of multiple auditory areas in the cat cortex was soon extended to the dog (Tunturi, 1944). Two responsive cortical areas were found. The dorsal part of the middle ectosylvian gyrus was organized with high frequencies located anteriorly and low frequencies posteriorly. The ventral area showed an inverse arrangement, with frequencies ranging from 100 to 16,000 Hz. In addition, a third auditory area, which overlapped the second somatosensory area in the anterior ectosylvian gyrus, was identified in the dog a year later (Tunturi, 1945).

The period following the publication of these studies was one marked by the discovery of other cortical areas thought to be involved in audition, as well as by attempts to understand more fully the contributions of the primary and secondary cortical areas.[4] The electrophysiological findings now indicated that Munk (1881), Luciani and Tamburini (1879), and Schäfer (1900) were all probably correct when they asserted that the superior (primates) and/or middle part (dogs, cats) of the temporal lobe may be the cortical area most involved with audition, but that the auditory sphere most certainly involved other cortical areas as well.

The tonotopic organization of auditory brainstem structures also became clear at about this time. In 1939, small localized lesions were combined with behavioral testing procedures to study the tonotopic organization of the medial geniculate body. Different loci in the medial geniculate body for 4,000-, 2,000-, 1,000-, 500-, and 250–125-Hz tones were identified. The authors wrote the following about the organization of the auditory system as a whole:

These facts clearly indicate that impulses concerned with each frequency of the audible range enter and traverse the medial geniculate bodies by separate pathways. In no other way can we account for the differential loss of hearing that results after the interruption of limited fractions of the acoustic pathway. This being true, some form of "place" or "resonance" theory is indicated, since it is apparent that each frequency has already, upon reaching the medial geniculate, been routed through its own group of fibers. It is to be presumed that this localization continues to the primary acoustic cortex, in view of the intimate connection between specific areas in the medial geniculate and specific areas in the temporal cortex. . . . We can say with assurance that whatever obscure and circuitous paths are followed by impulses from the cochlea . . . they reach that level by way of an orderly spatial projection of the organ of Corti on successively higher levels. (Ades, Mettler, and Cutler, 1939, 21)

---

[4]See Woolsey (1960) for a review of this literature.

REFERENCES

Ades, H. W. 1943. A secondary acoustic area in the cerebral cortex of the cat. *Journal of Neurophysiology, 6,* 59–63.

Ades, H. W., Mettler, F. A., and Culler, E. A. 1939. Effect of lesions in the medial geniculate bodies upon hearing in the cat. *American Journal of Physiology, 125,* 15–23.

Barnes, W. T., Magoun, H. W., and Ranson, S. W. 1943. The ascending auditory pathway in the brainstem of the monkey. *Journal of Comparative Neurology, 79,* 129–152.

Bauhin, K. 1605. *Theatrum anatomicum.* Francofurti at Moenum: Matthaei Beckeri.

Beck, A. 1890. Die Bestimmung der Localisation der Gehirn- und Rückenmarkfunctionen vermittelst der elektrischen Erscheinungen. *Centralblatt für Physiologie, 4,* 473–476.

Beck, A. 1891. Oznaczenie Lokalizacyi z mózgu i Rdzeniu za Pomoca Zjawisk Elektrycznych. Presented October 20, 1890. *Polska Akademia Umiejetnosci, Ser. 2, 1,* 186–232.

Bekhterev, V. M. 1885. Ueber die innere Abtheilung des Strickkörpers und den achten Hirnnerven. *Neurologisches Centralblatt, 3,* 145–147.

Bekhterev, V. M. 1887. Zur Frage über den Ursprung des Hörnerven und über die physiologische Bedeutung des N. vestibularis. *Neurologisches Centralblatt, 6,* 193–198.

Bekhterev, V. M. 1909. *Die Funktionen der Nervenzentra.* Jena: G. Fischer.

Bramwell, E. 1927. A case of cortical deafness. *Brain, 50,* 579–580.

Brazier, M. 1988. *A History of Neurophysiology in the 19th Century.* New York: Raven Press.

Brown, S., and Schäfer, E. A. 1888. An investigation into the functions of the occipital and temporal lobes of the monkey's brain. *Philosophical Transactions of the Royal Society of London (Biology), 179,* 303–327. (Paper read December 15, 1887)

Bunch, C. C. 1928. Auditory acuity after removal of the entire right cerebral hemisphere. *Journal of the American Medical Association, 90,* 2102.

Burdach, K. F. 1822. *Von Baue und Leben des Gehirns.* Leipzig: Dyk'schen Buchhandlung.

Caton, R. 1875. The electric currents of the brain. *British Medical Journal, 2,* 278.

Caton, R. 1877. Interim report on investigation of the electric currents of the brain. *British Medical Journal, 1, Suppl. L,* 62.

Dandy, W. 1930. Changes in our conceptions of localization of certain functions in the brain. *American Journal of Physiology, 93,* 643.

Dandy, W. E. 1933. Physiological studies following extirpation of the right cerebral hemisphere in man. *Bulletin of the Johns Hopkins Hospital, 53,* 31–51.

Deiters, O. 1865. In M. Schultze (Ed.), *Untersuchungen über Gehirn und Rückenmark des Menschen und der Säugethiere.* Braunschweig: Wieweg.

Dugès, A. L. 1838. *Traité de Physiologie Comparée de l'Homme et des Animaux* (Vol. 1). Montpellier: L. Castel.

Ferrier, D. 1876a. *The Functions of the Brain.* London: Smith, Elder and Company.

Ferrier, D. 1876b. Experiments on the brain of monkeys: Second series (Croonian Lecture). *Philosophical Transactions of the Royal Society of London, 165,* 433–488.

Ferrier, D. 1886. *The Functions of the Brain* 2nd ed. New York: G. P. Putnam's Sons.

Ferrier, D. 1888. Schäfer on the temporal and occipital lobes. *Brain, 11,* 7–30.

Ferrier, D., and Turner, W. A. 1901. Experimental lesion of the corpora quadrigemina in monkeys. *Brain, 24,* 27–46.

Ferrier, D., and Yeo, G. 1885. A record of experiments on the effects of lesions of different regions of the cerebral hemispheres. *Philosophical Transactions of the Royal Society of London (Biology), 175,* 479–564.

Flechsig, P. E. 1896. *Localisation der geistigen Vorgänge* . . . Leipzig: Veit and Comp.

Flechsig, P. E. 1901. Developmental (myelogenetic) localisation of the cerebral cortex in the human subject. *Lancet, ii,* 1027–1029.

Freud, S. 1886. Ueber den Ursprung des N. acusticus. *Monatsschrift für Ohrenheilkunde, 20,* 245–251.

Fritsch, G., and Hitzig, E. 1870. Über die elektrische Erregbarkeit des Grosshirns. *Archiv für Anatomie und Physiologie,* 300–332. In G. von Bonin, *Some Papers on the Cerebral Cortex.* Translated as, "On the electrical excitability of the cerebrum." Springfield, IL: Charles C Thomas, 1960. Pp. 73–96.

Gehuchten, A. van. 1906. *Anatomie du Système Nerveux de l'Homme.* Louvain: A. Uystpruyst-Dieudonné.

Gowers, W. R. 1885. *Epilepsy.* London: William Wood and Company. (Reprinted 1964; New York: Dover Publications)

Heffner, H. E., and Heffner, R. S. 1986. Hearing loss in Japanese macaques following bilateral auditory cortex lesions. *Journal of Neurophysiology, 55,* 256–271.

Held, H. 1891. Die centralen Bahnen des Nervens acusticus bei der Katze. *Archiv für Anatomie und Physiologie,* 271–288.

Held, H. 1892. Die Endigungsweise der sensiblen Nerven im Gehirn. *Archiv für Anatomie und Physiologie*, 33–39.

Held, H. 1893. Die centrale Gehörleitung. *Archiv für Anatomie und Physiologie*, 201–248.

Henschen, S. E. 1918. Über die Hörsphäre. *Journal für Psychologie und Neurologie, 22,* 319–474.

Horsley, V., and Schäfer, E. A. 1888. A record of experiments upon the functions of the cerebral cortex. *Philosophical Transactions of the Royal Society of London (Biology), 179,* 1–45. (Read Feb. 1887.)

James, W. 1890. *Principles of Psychology.* New York: Henry Holt.

Jerger, J., Lovering, L., and Wertz, M. 1973. Auditory disorder following bilateral temporal lobe insult: Report of a case. *Journal of Speech and Hearing Disorders, 37,* 523–553.

Jerger, J., Weikers, N. J., Scharbrough, F. W., III, and Jerger, S. 1969. Bilateral lesions of the temporal lobe: A case study. *Acta Oto-Laryngologica, Suppl., 258,* 1–51.

Kölliker, A. 1896. *Handbuch der Gewebelehre des Menschen* (Vol. 2). Leipzig: W. Engelmann.

Larionow, W. E. 1899. Ueber die musikalischen Centren des Gehirns. *Archiv für die gesammte Physiologie, 76,* 608–625.

Lhermitte, F., Chain, F., Escourolle, R., Ducarne, B., Pillon, B., and Chedru, F. 1971. Etude des troubles perceptifs auditifs dans les lésions temporales bilatérales. A propos de trois observations dont deux anatomo-cliniques. *Revue Neurologique, 124,* 329–351.

Lorente de Nó, R. 1933a. Anatomy of the eighth nerve. The central projection of the nerve endings of the internal ear. *Laryngoscope, 43,* 1–38.

Lorente de Nó, R. 1933b. Anatomy of the eighth nerve. III. General plan of structure of the primary cochlear nuclei. *Laryngoscope, 43,* 327–350.

Lorente de Nó, R. 1937. The sensory endings in the cochlea. *Laryngoscope, 47,* 373–377.

Luciani, L. 1884. On the sensorial localisations in the cortex cerebri. *Brain, 7,* 145–160.

Luciani, L., and Tamburini, A. 1879. *Sulle Funzioni del Cervello.* Regio-Emilia: S. Calderini. Abstracted by A Rabagliati in *Brain,* 1879, *2,* 234–250.

Lurie, M. H., Davis, H., and Derbyshire, A. J. 1934. The electrical activity of the cochlea in certain pathological conditions. *Annals of Otology, Rhinology and Laryngology, 43,* 321–343.

Meyer, A. 1927. Discussion of "A case of cortical deafness," by E. Bramwell. *Brain, 50,* 579–580.

Mills, C. K. 1891. On the localization of the auditory centre. *Brain, 14,* 465–472.

Misch, W. 1928. Über corticale Taubheit. *Zeitschrift für die gesamte Neurologie und Psychiatrie, 115,* 567–573.

Monakow, C. von 1890. Ueber Striae acusticae und untere Schleife. *Archiv für Psychiatrie und Nervenkrankheiten, 22,* 1–26.

Monakow, C. von. 1891. Striae acusticae und untere Schleife. *Archiv für Psychiatrie und Nervenkrankheiten, 22,* 1–26.

Monakow, C. von. 1895. Experimentelle und pathologisch-anatomische Untersuchungen über die Haubenregion, den Sehhügel und die Regio subthalamica, nebst Beiträgen zur Kenntniss früh erworbener Gross- und Kleinhirndefecte. *Archiv für Psychiatrie und Nervenkrankheiten, 27,* 1–128.

Mott, F. W. 1907. Bilateral lesion of the auditory cortical centre: Complete deafness and aphasia. *British Medical Journal, 2,* 310–315.

Munk, H. 1881. *Über die Funktionen der Grosshirnrinde.* Berlin: Hirschwald.

Neuburger, M. 1981. *The Historical Development of Experimental Brain and Spinal Cord Physiology Before Flourens.* (Translated and edited with additional material by E. Clarke). Baltimore, MD: Johns Hopkins University Press.

Papez, J. W. 1930. Superior olivary nucleus: Its fiber connections. *Archives of Neurology and Psychiatry, 24,* 1–20.

Poliak, S. 1932. *The Main Afferent Fiber Systems of the Cerebral Cortex in Primates* (Vol. 2). Berkeley: University of California Publications in Anatomy.

Poljak, S. 1926. The connections of the acoustic nerve. *Journal of Anatomy, 60,* 465–469.

Poljak, S. 1927. Ueber den allgemeinen Bauplan des Gehörsystems und über seine Bedeutung für Physiologie, für die Klinik und für die Psychologie. *Zeitschrift für die gesamte Neurologie und Psychiatrie, 110,* 1–49.

Prus, J. 1899. Bemerkungen zu dem Aufsatze des Herrn Docenten Dr. Bernheimer: Die Beziehungen der vorderen Vierhügel zu den Augenbewegungen. *Weiner klinische Wochenschrift, 112,* 1311–1312.

Ramón y Cajal, S. 1896. *Beiträge zum Studium der Medulla oblongata, des Kleinhirns und Ursprungs der Hirnnerven.* Leipzig: J. A. Barth.

Ramón y Cajal, S. 1909–1911. *Histologie du Système Nerveux de l'Homme et des Vertébrés.* Paris: A. Maloine.

Ranson, S. W. 1920. *The Anatomy of the Nervous System.* Philadelphia: W. B. Saunders.

Rasmussen, G. L. 1946. The olivary peduncle and other fiber projections of the superior olivary complex. *Journal of Comparative Neurology, 84,* 141–219.

Rosati, G., De Bastiani, P., Paolino, E., Prosser, S., Arslan, E., and Artioli, M. 1982. Clinical and audiological findings in a case of auditory agnosia. *Journal of Neurosurgery, 227,* 21–27.

Rundles, R. W., and Papez, J. W. 1938. Fiber and cellular degeneration following temporal lobectomy in the monkey. *Journal of Comparative Neurology, 68,* 267–296.

Schäfer, E. A. 1888a. Experiments on special sense localisations in the cortex cerebri of the monkey. *Brain, 10,* 362–380.

Schäfer, E. A. 1888b. On the functions of the temporal and occipital lobes: A reply to Dr. Ferrier. *Brain, 11,* 145–166.

Schäfer, E. A. 1900. The cerebral cortex. In E. A. Schäfer (Ed.), *Text-book of Physiology.* Edinburgh: Young J. Pentland. Pp. 697–782.

Schiller, F. 1974. The intriguing nucleus of Deiters. *Bulletin of the History of Medicine, 48,* 276–286.

Shaw, J. C. 1882a. Note on a case of localized cerebral atrophy. *Archives of Medicine, 7,* 77–86.

Shaw, J. C. 1882b. Blindness and deafness with bilateral cerebral lesion. *Brain, 5,* 430.

Spiegel, E. A., and Kakeshita, T. 1926. Experimentalstudien am Nervensystem. II. Zur zentral Lokalisation cochlearer Reflexe. *Archiv für die gesammte Physiologie, 212,* 769–780.

Squires, P. C. 1935. The problem of auditory bilateral cortical representation, with special reference to Dandy's findings. *Journal of General Psychology, 12,* 182–193.

Tunturi, A. R. 1944. Audio frequency localization in the acoustic cortex of the dog. *American Journal of Physiology, 141,* 397–403.

Tunturi, A. R. 1945. Further afferent connections to the acoustic cortex of the dog. *American Journal of Physiology, 144,* 389–394.

Varolio, C. 1591. *De nervis opticis, Nonnullisque aliis praeter communem opinionem in humano capite observatis.* Francofurti, Ioannem, Welchelum et Petrum: Fischerum.

Walker, A. E. 1937. The projection of the medial geniculate body to the cerebral cortex in the macaque monkey. *Journal of Anatomy, 71,* 319–331.

Waller, A. D. 1891. *Human Physiology.* London: Longmans, Green.

Wernicke, C., and Friedländer, C. 1883. Ein Fall von Taubheit in Folge von doppelseitiger Laesion des Schläfelappens. *Fortschritte der Medicin, 1,* 177–185.

Willis, T. 1664. *Cerebri anatome: cui accessit nervorum descriptio et usus.* London: J. Martyn and J. Allestry.

Woollard, H. H., and Harpman, J. A. 1939. The cortical projection of the medial geniculate body. *Journal of Neurology and Psychiatry, 2, New Ser.,* 35–44.

Woollard, H. H., and Harpman, J. A. 1940. The connexions of the inferior colliculus and of the dorsal nucleus of the lateral lemniscus. *Journal of Anatomy, 74,* 441–458.

Woolsey, C. N. 1960. Organization of cortical auditory system: A review and a synthesis. In G. L. Rasmussen and W. F. Windle (Eds.), *Neural Mechanisms of the Auditory and Vestibular Systems.* Springfield, IL: Charles C Thomas. Pp. 165–180.

Woolsey, C. N., and Fairman, D. 1946. Contralateral, ipsilateral, and bilateral representation of cutaneous receptors in somatic areas I and II of the cerebral cortex of pig, sheep, and other mammals. *Surgery, 19,* 684–702.

Woolsey, C. N., and Waltz, E. M. 1942. Topical projection of nerve fibers from local regions of the cochlea to the cerebral cortex of the cat. *Johns Hopkins Hospital Bulletin, 71,* 315–344.

Chapter 10
The Cutaneous Senses

> If an animal is to survive, its body must have tactile sensation . . . it is clear that without touch it is impossible for an animal to exist. . . . The loss of this one sense alone must bring about the death of an animal . . . for it is the only one which is indispensably necessary to what is an animal.
>
> Aristotle, fourth century B.C.

CUTANEOUS SENSATION: ONE SENSE OR MANY?

The idea that there may be five senses can be traced to the Ancient Greeks, who spoke of vision, hearing, taste, smell, and touch. Questions arose about the touch category, however, and even Aristotle, who is often associated with the notion that there are five senses, appeared to wonder whether touch really represented one sense or many.

Aristotle believed the skin senses required direct contact for stimulation. In his *Treatise on the Principles of Life,* he assigned several paired qualities to "touch," including hard-soft, hot-cold, and rough-smooth.

Pain represented a particularly difficult problem. Aristotle, like many others in classical times, put pain with pleasure and classified the two not as senses but as passions of the soul. (Because pain was often treated separately from the five senses, and has always been the subject of considerable attention, Chapter 11 is devoted to the history of pain.)

Galen, like many of his contemporaries in Ancient Rome, theorized that light stimulation of the skin resulted in sensations of touch or temperature, and that violent irritations produced pain. But unlike Aristotle, he argued that the nerves carried the messages to the head, not the heart. He also thought that qualities such as dry, moist, hard, and soft involved judgments based on experience and were not just attributable to peripheral nerve stimulation (Siegel, 1970).

Early attempts to analyze touch and define its subsystems spanned both continents and time (Dallenbach, 1939). These efforts involved Avicenna, Averroes, Albertus Magnus (1193–1280), Aegidus (1247–1316), Geronimo Cardano, Francis Bacon, and Immanuel Kant (1694–1778), among others.

Some individuals added senses to the traditional five. Charles Bell, for example, included a sense of movement. Erasmus Darwin (1731–1802) added several senses and maintained, among other things, that nature provided nerves for the perception of heat. He based this on a patient who had lost tactile sensitivity but who retained normal appreciation of warmth. In general, there has been considerable disagreement on how to classify the various sensations

that one can feel through the skin, not only when Bell and Darwin lived but into the twentieth century.

WEBER'S PSYCHOPHYSICAL STUDIES

Efforts to study the skin senses were accentuated by the contributions of Ernst Weber. Born and educated in Wittenberg, Weber was the leading physiologist in Leipzig when he published a monograph in 1834 that examined touch. A more influential and expanded analysis of his original Latin text appeared in Rudolph Wagner's (1805–1864) *Handwörterbuch der Physiologie,* published in 1846. Weber's chapter in the *Handwörterbuch* was entitled *"Der Tastsinn und das Gemeingefühl"* ("Touch and Common Sensibility"). He distinguished between touch due to receptors in the skin and common sensitivity, which was associated with a wide variety of often vague experiences, including fatigue, hunger, general well-being, and pain.[1]

Weber developed a number of tests for studying the cutaneous sensation. The best known was his two-point test, which asked subjects whether one point or two points of a compass were being touched to the skin (Weber, 1834). Weber thought the skin was composed of a mosaic of sensory circles (although he drew them as hexagons), each supplied by a nerve. He believed the ability to discriminate was largely dependent on whether the compass points were placed in the same, in adjacent, or in nonadjacent sensory circles.

Weber also devised measures of tactual space (see Sinclair, 1967). One of these asked subjects to localize the point on the skin that the experimenter had just touched. This was accomplished by touching the skin of a blindfolded subject with a blunt point dipped in charcoal. The subject could not use vision and had to touch the same spot with another charcoal pointer. The distance between the two points was then measured.

Weber found that the error in localization, like the two-point threshold, varied with body region. Soon after this,

[1]See Schiller (1984) for a history of common sensory sensibility.

Karl von Vierordt (1818–1884) proposed his "law of mobility," which held that sensitivity increases as one goes from the midline toward the periphery (e.g., fingertips and toes). This phenomenon was later confirmed by others (e.g., Kottenkamp and Ullrich, 1870).

Scientists soon demonstrated that many factors could affect the results of tactile psychophysical experiments like these. In this regard, the age of the subject appeared to be an important variable (Czermak, 1855). Thresholds were also found to vary with attention and fatigue (Griesbach, 1895; Vannod, 1896). The latter observation led to Weber's two-point test being given to schoolchildren as a practical test of fatigue.

In this context, William McDougall's (1871–1938) data, which showed that Murray Island "savages" had lower two-point thresholds than Englishmen, were interpreted as being due to the fact that the Murray Islanders did not fatigue as fast as Englishmen. Later, it was pointed out that the Murray Islanders desire to do well for the testers probably led them to use different criteria for reporting two-point sensations (see Boring, 1942, for a review of this literature).

Weber is perhaps best known for his work on discriminating changes in pressure. He believed that a just noticeable increase in the pressure of a weight was a constant fraction of the weights of the stimuli. He wrote that if 1 g had to be added to a 20-g weight for it to be perceived as noticeably heavier, 2 g had to be added to a 40-g weight, and so on. This idea was also applied to other sensory modalities. It soon became known as Weber's law, although as a "law" it did not hold up perfectly, especially with weak stimuli.

Weber believed the sense of touch *(Tastsinn)* was restricted to the skin and included temperature, pressure, and position (locality). He argued that these components did not function independently and demonstrated how they could interact by placing a cold silver coin (a Thaler) on the forehead of a volunteer. Subjectively, the cold coin was perceived to be twice as heavy as a "neutral" Thaler.

THE LAW OF SPECIFIC NERVE ENERGIES

Johannes Müller, who was from Coblenz but worked in Bonn and Berlin, emphasized in 1826, and again in 1838, that there was some sort of specificity in the nerves serving the different sensations (Brazier, 1957; Keele, 1957; Riese, 1970). Actually, Müller, shown in Figure 10.1, developed an idea that Charles Bell had stated very clearly in 1811. Bell wrote:

> The nerves of sense, the nerves of motion, and the vital nerves, are distinct throughout their whole course, though they seem sometimes united in one bundle. . . . It is admitted that neither bodies nor the images of bodies enter the brain. It is indeed impossible to believe that colour can be conveyed along a nerve, or that a vibration . . . can be retained in the brain; but we can conceive and have reason to believe, that an impression is made upon the outward senses when we see, hear or taste. In this inquiry it is most essential to observe that while each organ of sense is provided with a capacity for receiving certain changes to be played upon it, as it were, yet each is utterly incapable of

Figure 10.1. Johannes Müller (1808–1858), whose *Handbuch der Physiologie* (1838) became a standard reference in the filed and who was well known for his law of specific nerve energies.

receiving the impression destined for another organ of sensation. (1936 reprint, 107–108)

Prior to Bell's time, some scientists thought the nerves might be transmitting messages to a common center in the brain and that movement involved the same and not different nerves. In his 1811 pamphlet, Bell described this as follows:

> It is imagined that impressions, thus differing in kind, are carried along the nerves to the sensorium, and presented to the mind; and that the mind, by the same nerves which receive sensation, sends out the mandate of the will to the moving parts of the body. (1936 reprint, 106)

Müller presented his ideas about "specific energies" and "specific irritability" of nerves to explain why similar sensations occasionally occurred when qualitatively different stimuli were applied to the same nerves (e.g, light and pressure on the eye both elicit visual sensations) and why nerves are not sensitive to all stimuli (light does not affect the cutaneous nerves). Müller hypothesized that different sensations were largely due to different centers of nerve termination, but used "specific irritability" to suggest that nerves were also specialized to respond differently.

In the 1840 edition of his *Handbuch der Physiologie*, which became the standard reference in the field, Müller described his 10 laws of how a sense is related to its anatomical system. In it, he wrote:

> Sensation consists in the sensorium receiving through the medium of its nerves, as a result of the action of an external cause, a knowledge of certain qualities or conditions not of external bodies but of the nerves of sense themselves. . . . The immediate objects of our senses are merely particular states induced in the nerves, and felt as sensations either by the nerves themselves or by the sensorium. (Translated in Hardy, Wolff, and Goodell, 1967, 4)

Interestingly, Müller, who was a prolific writer (over 200 papers), did not apply his concepts of specificity directly to the question of the cutaneous senses. In contrast, his follow-ers quickly accepted the notion that there must be a sepa-rate set of nerve fibers for each sensory quality, as well as different receptors and brain areas. The emerging belief was that there were several specific nerve endings or fibers in the skin, which were associated with the different types of cutaneous sensation. Carl Wilhelm Hermann Nothnagel (1841–1905), for example, argued that there must be specific nerve fibers to account for differences between warm and cold (also see Natanson, 1844).

Hermann von Helmholtz, a disciple of Müller, is often credited with developing the concept of sensory modalities (Boring, 1942). The skin senses could be divided into several modalities, each a class of sensations along a qualitative continuum. By the end of the nineteenth century, touch, warm, cold, and pain were being treated as modalities of somesthesis by many people—that is, as subsystems that had the potential to interact with each other as Weber had shown.

Among his other contributions, Helmholtz, shown in Figure 10.2, also paved the way for electrophysiological studies of somatic sensation when he measured the veloc-ity of the nervous impulse. He conducted his cerebrated experiment on frog nerves, using the pendulum myograph he invented in 1850, while a professor of physiology at Königsberg.[2]

[2]The measurement of the nerve impulse was something Johanne Müller and many others at this time thought was impossible to accom-plish. The feeling was that the event took place too rapidly to be measured accurately (Leake, 1970).

Figure 10.3. Magnus Blix (1849–1904), the Swedish investigator who pioneered the study of the punctate sensitivity of the skin in the 1880s. (Courtesy of the Uppsala Universitetsbibliotek.)

SENSORY SPOTS ON THE SKIN

The idea of specific cutaneous subsystems gained ground in the 1880s. At that time, three investigators, Magnus Blix (1849–1904) from Sweden, Alfred Goldscheider (1858–1935) from Germany, and Henry Donaldson (1857–1938) from the United States, described spots on the skin associated with different sensations.

Blix (1882/1884), shown in Figure 10.3, used low-inten-sity faradic current, delivered to the skin through a small pin. He first reported that warm, cold, touch, and pain spots did not overlap. He then built a device that allowed him to pass warm or cold water through a metal tube shaped like a cone which could be touched to the skin. Blix found warm and cold spots and observed that the warm spots could not be activated by cold stimuli, nor could the cold spots be activated by warm stimuli. This finding led him to believe that warm and cold receptors had distinctly different endings in the skin, as predicted by the law of specific nerve energies.

Goldscheider (1884) was also interested in testing the theory of specific nerve energies. He used cork for touch, ether on brushes or in capillary tubes for cold, heated brass cylinders for warm, and needles for pain. He conducted his initial experiments with no knowledge of Blix's work and became aware of it only when he was about to publish his own findings. Like Blix, Goldscheider found warm spots to be rarer than cold spots. Probably because he used more intense thermal stimuli, Goldscheider also reported more temperature spots than Blix.

While conducting an experiment with G. Stanley Hall (1844–1924), an American pioneer in experimental psy-chology, Donaldson (1885), shown in Figure 10.4, con-cluded that there might be temperature-sensitive spots on

Figure 10.2. Hermann von Helmholtz (1821–1894) studied the speed of nerve conduction and sensory psychophysics. Helmholtz also discussed the concept of sensory modalities in the skin senses.

Figure 10.4. Henry Herbert Donaldson (1857–1938), one of the first scientists to recognize the punctate sensitivity of the skin, tried to associate sensitive spots with specific types of receptors.

the skin. He wrote that he came to notice the punctate sensitivity of the skin in the following way:

> The sensations of motion as derived from the skin were being studied by means of a metal point which was slowly drawn over the surface. When the motion of this point, which was controlled by a suitable apparatus, was very slow, it often happened that it seemed to stand still for a time or even be lost, when suddenly a sharp sensation of cold, distinctly localized, would recall its presence and position. This occurred so often that I find in my protocol for April 18th, 1884, the note: "Point always felt as cold." . . . This fact arrested my attention, and in connection with other work I made several maps of these cold-spots on different parts of the body. When the experiments had reached this point, an important paper by Magnus Blix came into my hands. (1885, 399)

Donaldson went on to confirm many of the observations of Blix and Goldscheider.

SPECIFIC END ORGANS FOR CUTANEOUS SENSATION

A number of specialized end organs in the skin were observed well before attempts were made to correlate these endings with sensation. Vater-Pacini corpuscles were identified by Abraham Vater (1684–1751) in 1741, only to be rediscovered in the nerves of the fingers by Filippo Pacini (1812–1883) in 1835. In addition, Meissner's corpuscles were observed in 1852, Krause end bulbs in 1859, and Merkel discs in 1875. Ruffini endings, however, were not described until 1894.[3]

Donaldson (1885) and Goldscheider (1886a, 1886b) went on to excise cold and warm spots in an attempt to associate specific sensations with receptors beneath the skin. They had little success and reported only that there were many

[3]Descriptions and some of the history of the cutaneous end organs can be found in Boring (1942) and Sinclair (1967).

nerve endings. Donaldson stated his objectives and findings as follows:

> If then the organs are in the skin, it should be possible to cut them out and examine them histologically. A cold and a heat spot were localized on my own skin, and then cut out for me by Dr. Councilman. The bits of skin were . . . cut into serial sections. No difference could be made out between the spot at which cold had been felt and that at which heat had been observed. There were numerous nerves beneath these spots, but they were almost as numerous as in neighbouring parts. (1885, 408–409)

This did not seem to discourage Max von Frey (1852–1932; Figure 10.5). He developed his own theory of specific skin sense receptors in the 1890s. His theory, which achieved surprising popularity, held that Ruffini end organs were responsible for warmth, Krause end bulbs for cold, Meissner corpuscles for touch in glabrous skin, and free nerve endings for pain (Frey, 1895, 1896). In the hairy skin, Frey suggested that the touch receptor was made of nerve endings which wrapped around hair follicles. He further believed that these end organs were associated with specific nerve fibers which went to different loci in the central nervous system.

Frey's associations were not based on solid anatomical foundations. Rather, as his critics have pointed out, he simply linked the last unassigned organ to the last available sense to account for warm, and associated pain with free nerve endings only because both unencapsulated endings and pain spots were so numerous.

Because Frey's theory was based on inadequate anatomical evidence, it was questioned immediately after it was introduced. For example, histologists pointed out that there were many intermediate types of receptors in the skin and that the categories chosen by the German investigator were not discrete. The fact that some anatomists thought they had found many more than just these end organs also did not bode well for the theory. Alexander Dogiel (1852–1922), for

Figure 10.5. Max von Frey (1852–1932), who proposed that specific receptors in the skin were responsible for pain, cold, warmth, and touch sensations.

instance, spoke of 14 varieties, whereas others (e.g., Botezat) described as many as 34 discrete types (see Sinclair, 1967). In addition, although H. Strughold (b. 1898) mapped the cold sensitivity of the conjunctiva and mouth cavity and reported in 1925 that this showed a fair association with Krause end bulbs, others failed to associate Krause end bulbs or any other encapsulated end organ with cold (Dallenbach, 1927; Gilbert, 1929; Häggqvist, 1914; Pendleton, 1928).

Still more problems arose when some of these investigators reported that encapsulated end organs were in fact rare in the hairy skin (Dallenbach, 1927; Gilbert, 1929; Pendleton, 1928). Most previous studies had examined only the palms, soles, eyes, lips, and other hairless parts.

Subsequent experiments showed that skin containing free nerve endings could be cut without pain being experienced; that Ruffini cylinders may be both pressure and heat sensitive; and that some encapsulated end organs could be found in places not known to be sensitive (Keele, 1957; Sinclair, 1967).

Nevertheless, Frey did generate considerable interest in the skin senses. In addition, he invented some interesting ways of measuring sensory thresholds. For instance, he devised a technique for mechanically presenting stimuli of different weights to the skin. He also used hairs of different lengths and diameters. The latter were pressed on the skin until they bent. The "von Frey hairs" were calibrated by pressing them on a balance so that the force needed to bend each one was known. These stimuli were readily accepted by investigators in many different countries who wished to quantify their work on the skin senses.

PERIPHERAL NERVE

The idea of specificity, readily applied to receptors in the skin, was also applied to the peripheral nerve. One line of research which suggested that cold, warmth, touch, and pain employed different peripheral nerves used nerve blocks.

Cold blocks were attempted by Weber in 1847, and compression was used by Edme Felix Alfred Vulpian (1826–1887) in the 1850s (Bastien and Vulpian, 1855). In 1886, chemical anesthetics were tried by Goldscheider (1886a, 1886b). In one notable experiment, a tourniquet was used to block the nerves, and it was noted that cold and touch were the first sensations to be lost, followed by heat, superficial pain, and finally deep pain (Herzen, 1886). Upon release of the block, the sensations returned in the reverse order. Together, the nerve block data were viewed as consonant with Müller's doctrines of specificity.

Perhaps the most influential work on nerve blocks began in the 1920s, when Joseph Erlanger (1874–1965) and Herbert Gasser (1888–1963), used local anesthetics to correlate the sequence of sensory loss with nerve fiber size. This research was conducted at Washington University in St. Louis. Gasser and Erlanger's (1929) pioneering use of the cathode ray oscilloscope and three-stage amplifier to demonstrate compound action potentials enabled them to associate specific components of the compound nerve response with fiber size and sensation. Touch was associated with large fibers, pain with small ones, and temperature with intermediate fibers. In 1944, Erlanger and Gasser, whose photographs are shown as Figures 10.6 and 10.7, won the Nobel

Figure 10.6. Joseph Erlanger (1874–1965), who shared the Nobel Prize with Herbert Gasser in 1944 research on the electrophysiological properties of peripheral nerves. This photograph was taken on the occasion of Erlanger's 84th birthday. (Courtesy of the Washington University Archives, St. Louis, MO.)

Prize for their many "discoveries relating to the highly differentiated functions of single nerve fibers."

This is not to say there were no attempts to associate sensation with fiber size before this time. For example, Stephen Ranson (1880–1942) cut the small fibers in the medial division of the cat's posterior spinal root while sparing the large fibers. He found that this procedure eliminated pain reflexes (Ranson and Billingsley, 1916). Ranson (1931) later demonstrated that cutting large fibers while sparing unmyelinated and lightly myelinated fibers did not have this effect. The growing literature on nerve blocks was consonant with these findings.

Among the most cited papers to appear early in the twentieth century were those of Henry Head, a neurologist (shown in Figure 10.8), and William Rivers (1864–1922), a psychologist. They studied the return of sensation following section of a cutaneous nerve. This work began in 1903, and Head, who was associated with the London Hospital, served as the subject in the investigations, even cutting the superficial ramus of his own radial nerve (Head and Rivers, 1908; Head, Rivers, and Sherren, 1905). Head volunteered in part because he was impatient with subjects who would become tired during the lengthy test sessions. He even gave up smoking and restricted his drinking to prepare for the experiments (Denny-Brown, 1970).

The time needed for different sensations to return after the superficial ramus of the radial nerve was cut was not the same. This led Head to the conclusion that in addition to a system serving deeper sensibility, there were two sets of peripheral fibers that had regenerated at different rates. The one that mediated extremes of temperature or pain was called protopathic, and it recovered first. The system capable of mediating light touch and fine localization of a stimulus was called epicritic, and its recovery was slower (Head, 1920; Head and Rivers, 1908; Head, Rivers, and Sherren, 1905).

The sensory mechanism in the peripheral nerves is thus found to consist of three systems:—

Figure 10.7. Herbert Gasser (1888–1963), co-winner of the 1944 Nobel Prize with Joseph Erlanger. (Courtesy of the Washington University Archives, St. Louis, MO.)

Figure 10.8. Sir Henry Head (1861–1940), who conducted many studies on the dermatomes, the changes that occur after peripheral nerves are cut, and the effects of brain lesions on somesthesis. Head also contributed significantly to the study of aphasia.

(I) Deep sensibility, capable of answering to pressure and to the movement of parts, and even capable of producing pain under the influence of excessive pressure, or when the joint is injured. The fibres, subserving this form of sensation, run mainly with the motor nerves, and are not destroyed by division of all the sensory nerves to the skin.

(II) Protopathic sensibility, capable of responding to painful cutaneous stimuli, and to the extremes of heat and cold. This is the great reflex system, producing a rapid widely diffused response, unaccompanied by any definite appreciation of the locality of the spot stimulated.

(III) Epicritic sensibility, by which we gain the power of cutaneous localization, of the discrimination of two points, and of the finer grades of temperature, called cool and warm. (Head, Rivers, and Sherren, 1905, 111)

This new way of looking at cutaneous sensation generated considerable research, but others who tried to do comparable experiments came away with different findings. In particular, the sharp boundaries reported by Head and Rivers were rarely seen, and some experimenters argued that all forms of sensibility tended to reappear gradually and normalize together.

Some replications of Head's work showed that when perception for moderate temperatures was not very good, stronger temperatures felt weak or moderate (e.g., Trotter and Davies, 1909, 1913). It was also found that even when two-point discrimination and pain perception seemed very poor, they could still be elicited from the skin with stronger stimuli. This suggested quantitative rather than qualitative differences in the course of recovery.

Edwin Boring's (1886–1968) study from 1916 is especially interesting because his background was in rigorous experimental psychology and because he followed Head's example by training himself for 13 months to be an ideal observer. Boring had a cutaneous nerve cut in his own arm and studied the return of sensation over a two-year period. He found that the return of sensation was gradual. It went from anesthesia to hypoesthesia to normal. He also found that pain, cold, and touch seemed to reach control levels in about the same amount of time (warm lagged a bit). Boring

was forced to conclude that Head's work could not be supported by his findings.

Head's data were also inconsistent with studies of World War I soldiers with peripheral nerve injuries (Cobb, 1919; Frazier and Silbert, 1920; Stookey, 1916). For example, one doctor in the U.S. Army studied over 1,000 cases of peripheral nerve lesions. He stated that he believed Head's pattern of return of sensitivity was not due to differential reinnervation by distinct subsystems but instead was due to the actions of adjacent nerves which were not cut (Pollock, 1919).

Another investigator argued that the terms "protopathic" and "epicritic" should be dropped from the vocabulary. Instead, he suggested, better descriptions should be given of the stimuli (Stookey, 1916). A third individual wrote that he accepted the epicritic-protopathic dissociation after superficial analyses of his 540 cases, but that as his procedures became more sensitive, areas of dissociation could no longer be recognized (Cobb, 1919).

THE PERIPHERAL NERVE AND ILLUSORY SENSATIONS

A number of interesting illusions that seemed at least partly related to the peripheral nerve were reported at the turn of the century. One such phenomenon was the experiencing of cold sensations when a hot stimulus was applied to the skin. This was noted with stimuli in the 45–50-degree Celsus range and was named "paradoxical cold" by Frey (1895). The belief was that the cold fibers could be activated by either extreme of temperature.

"Paradoxical warmth" was not found, or was noted much less frequently, when cold stimuli were used. Still, it was also the subject of some reports (Alrutz, 1897; Goldscheider, 1912; Rubin, 1912).

Another "illusion" that attracted considerable attention involved sensations which seemed to come from limbs that had been amputated. These sensations were known as "phantoms," and most investigators thought that they had something to do with the nerve endings in the stump.

Because individuals with amputated limbs often complained about the phantoms being painful, the topic of phantom limbs is discussed in Chapter 11 on pain.

DERMATOMES

The region of skin supplied by a dorsal root, commonly called a dermatome, was studied by Charles Sherrington in the 1890s (e.g., 1898). He cut dorsal roots above and below the one he wished to study and then mapped the remaining "island of sensibility." This corresponded to the territory of the middle, intact root. Sherrington used monkeys and first tried pinching the skin and recording reflexes. He later recognized that the territories of the spinal roots were smaller for pain and heat than they were for touch.

The standard maps of the dermatomes in humans also owe much to the efforts of Henry Head. He began to study the dermatomes as a part of his doctoral thesis, which was submitted to Cambridge University in 1892. He examined patients with "shingles" *(herpes zoster),* a disease that affects the dorsal root ganglia and produces skin rashes confined to the territory of affected spinal roots (Head and Campbell, 1900). See Figures 10.9 and 10.10.

A third individual whose work on the dermatomes should be acknowledged is Otfrid Foerster (1873–1941; Wartenberg, 1970). Foerster, shown in Figure 10.11, published an important paper on this subject in 1933. He combined Sherrington's method of residual sensibility with vasodilation and electrical stimulation to map out nearly all of the dermatomes of the human body. He spent over 30 years on these projects and confirmed Sherrington's observation that the tactile dermatomes were larger than those for pain and temperature stimuli. Foerster, however, described more root overlap than Head had reported. In fact, he noted that cutting just one root did not seem to result in any loss of sensibility.

SPINAL CORD AND BRAINSTEM PROJECTIONS

The Roman physician Galen, who attributed paresthesia in a patient's hand to an injury of the cervical cord, also conducted experiments on the spinal cord in animals. He anticipated some important nineteenth-century findings when he wrote the following in his *De locis affectibus:*

> You have been taught that a transverse incision of the entire cord deprives all parts of the body below it of sensa-

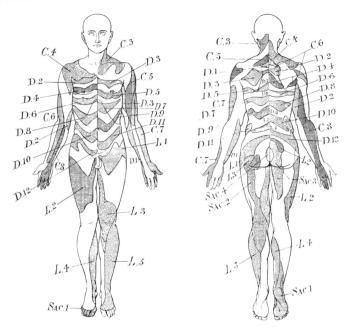

Figure 10.10. Head and Campbell produced this composite map of the dermatomes in 1900.

tion and motion, seeing that the cord derives the faculty of sensation and of voluntary motion from the brain. You have seen further in our dissection that transverse hemisections, which do not cut deeper than the centre of the cord, do not paralyze all the inferior parts of the body but only those directly underneath the incision, the right when the right side of the cord has been cut and vice versa. (1917 translation, 367)

Figure 10.11. Otfrid Foerster (1873–1941), who studied the somatosensory and motor systems, developed the technique of rhizotomy and produced early maps of the dermatomes. (From Kolle, *Grosse Nervenärzte,* 1970. Reproduced with the permission of Georg Thieme Verlag.)

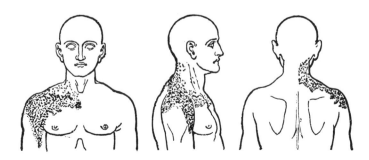

Figure 10.9. Head and Campbell (1900) compared the rashes in individual cases of herpes zoster, like the one shown above, to map the dermatomes in humans.

Nevertheless, just prior to this observation, Galen erroneously stated that "dissection has taught us that for all the parts of an animal below the neck which are capable of voluntary motion, the corresponding motor nerves arise from the *dorsal* part of the spinal cord."

A better understanding of the spinal tracts had to wait for the experiments of Charles Bell in 1811 and François Magendie in 1822. Both men agreed that the dorsal (posterior) roots transmitted sensation, whereas the ventral (anterior) roots were motor in function (see Chapter 2).

Bell thought all sensory information passed up the posterior half of the spinal cord to terminate in the brain, but he knew little about the central projections. François Achelle Longet (1811–1871) and many others also thought the dorsal columns conveyed all sensation (see Chapter 11). This was the prevailing opinion until the middle of the nineteenth century.

The idea that at least some sensory pathways may cross in the spinal cord and the belief that touch may utilize different pathways than pain came largely from the laboratory experiments conducted in the 1840s through the 1860s by Charles-Edouard Brown-Séquard and Moritz Schiff (1823–1896). As will be seen in Chapter 11 on pain, these important concepts were also based on the clinical observations of Brown-Séquard and William Gowers.

As the nineteenth century ended, there was a flourish of anatomical work on the central projections of the major spinal cord pathways with the exception of the spinocervical tract, which would not be described adequately for another half decade. This work was enhanced by Augustus Waller's histological work on anterograde and retrograde degeneration after axonal injury.

These efforts dispelled the belief that the cerebellum mediated somatic sensation, as had been thought by some scientists at the Salpêtrière (e.g., Delaye, Foville). Instead, the findings supported Luigi Rolando's contention that the medulla was involved in cutaneous sensation and the opinion of Antoine Serres (1776–1868), who spoke of more diffuse brainstem projections and a wider field of influence.

A number of investigators showed that, at least in the cat, some of the long spinothalamic projections terminated in thalamic nuclei, such as the pulvinar, lateral posterior nucleus; suprageniculate nucleus; and medial geniculate body (Goldstein, 1910; Papez, 1929; Wallenberg, 1900). At about the same time, it became clear that the projections from the dorsal column nuclei of the medulla crossed to the opposite side of the brain and formed the medial lemniscus, which terminated in the ventrolateral nuclei (ventralis posterolateralis for nonfacial projections) of the thalamus (Probst, 1900).

These findings were consistent with the myelination studies of Paul Flechsig (see Chapters 6 and 21). He noted that the postcentral gyrus received significant inputs from the ventrolateral nuclei of the thalamus. He viewed these nuclei as key stations between the posterior roots entering the spinal cord and the sensory cortex. Constantin von Monakow also contributed significantly to the anatomical study of these pathways.

During the opening decades of the twentieth century, there was a strong belief that the posterior white columns, at first thought to be responsible for all sensation, were more implicated in the finer aspects of touch. Henry Head and Gordon Holmes (1911–12), in a classic paper on the

subject, stated that impulses for posture also were conveyed via the dorsal columns to the dorsal column nuclei en route to the thalamus and parietal cortex. A related idea was that simple contact sensibility ascended in at least two spinal pathways, one of which crossed in the cord (see Chapter 11).

SOMATOSENSORY CORTEX: LESIONS IN LABORATORY ANIMALS

Fewer lesion experiments were carried out on the postcentral gyrus than on the motor cortex in the late nineteenth century (see Chapter 14). In 1875, David Ferrier wrote that tactile sensibility was impaired not by lesions of the parietal lobes but by lesions of the temporal lobes (hippocampal and temporo-sphenoidal regions). In his book *The Functions of the Brain,* he reviewed his ablation experiments on monkeys, including those conducted with Gerald Yeo. Ferrier wrote:

> I have effected destruction of the hippocampal region, more or less completely, both by the method of penetration by means of a wire cautery . . . and by cutting the outer aspect of the inferior temporo-sphenoidal region inwards towards the hippocampal gyrus, so as to destroy or detach it. . . . Sight and hearing were found to be unimpaired, and the intelligence quick and active as before. But cutaneous stimulation by pricking, pinching, or pungent heat sufficient to cause lively manifestations of sensation when applied to the right side of the body, failed in general to elicit any reaction whatever on the left side, whether hand or foot. (1886, 27–329)

Ferrier went on to say that this was confirmed in a second series of experiments on 10 monkeys. He added that in none of his experiments was the analgesia absolute, and in none was the whole hippocampal region destroyed. He then described how considerable recovery was seen in animals surviving for a few weeks after this surgery. This led him to conclude that the sensory center had to involve more than just the hippocampal gyrus.

The gyrus fornicatus was added to Ferrier's definition of the center for cutaneous sensation after further experiments on monkeys showed that lesions of this structure made the deficits after hippocampal gyrus lesions even more severe (Ferrier, 1886; also see Schäfer, 1900).

In contrast to Ferrier, Hermann Munk developed the idea that the area of the monkey's brain between the arcuate fissure and the intraparietal fissure, encompassing both frontal and parietal regions, was the *Fühlsphäre,* a broadly defined sensorimotor sphere (McHenry, 1969). Munk viewed this whole area as electrically excitable and thought that different body parts were associated with different individual regions. Munk (1892) held that total removal of the *Fühlsphäre* caused loss of touch and pressure sensations as well as paralysis. He thought that cutaneous deficits seen after lesions outside this zone were due to unavoidable injury to this area. Munk did not, however, produce and analyze lesions restricted to just the postcentral gyrus.

Frederick Mott (1894), working with monkeys, confirmed many of Munk's observations and supported his general theory. Like Munk, Mott did not write about lesions restricted to just the postcentral gyrus. Due partially to the studies of Munk and of Mott, the prevailing view at the end of the nineteenth century was that sensory and motor areas were at least somewhat co-extensive.

A major problem was that neither the boundaries of the so-called sensory-motor (or kinesthetic) region, nor the lesions made in the early experiments, were described in adequate detail by many authors. The "Rolandic region," for example, was often mentioned, but this sometimes meant the precentral gyrus, sometimes the postcentral gyrus, and sometimes both areas (see Russel and Horsley, 1906). The result, as might be expected, was considerable confusion. In fact, when Edward A. Schäfer (1900) published his *Text-book of Physiology,* he stated in his chapter on the cortex that it could well be that somatosensory functions do not even involve the cerebral cortex. Still, he acknowledged Mott (1894, 1895) might have been correct when he asserted that somatosensory fibers branch out to a large region of the cortex.

To say the least, experimenters differed tremendously in technique and orientation. Many had little idea how to test for sensory functions in laboratory animals. Yet it was clear, at least from the better-controlled experiments of Albert Sidney Frankau Grünbaum (1869–1921; known as A. S. F. Leyton after 1915) and Charles Sherrington, that large ablations of the postcentral gyrus in monkeys and apes did not cause notable paralysis (Grünbaum and Sherrington, 1901, 1903). This led some individuals to begin to assert that the postcentral region alone was the seat of cutaneous sensory representation (see Ruch, 1935; Russel and Horsley, 1906).

Motivated by the observations of Grünbaum and Sherrington, others began to study the effects of lesions of the postcentral gyrus (Lewandowsky and Simons, 1909; Rothmann, 1914; Vogt, 1906). Like their predecessors, these investigators concentrated on motor functions. They found these lesions did not result in paralyses in their monkeys, although the animals tended to use the affected limbs less.

Oskar Minkowski (1858–23) used monkeys in his experiments, but unlike most earlier investigators, his work concentrated more on the sensory effects of parietal lobe lesions. Minkowski (1917–18) observed that light touch was clearly impaired, whereas pain was essentially unaffected by damage to the parietal region of the brain. He also noted abnormal movements and postures. Because he found that motor cortex lesions resulted in more transient tactile deficits, he correctly concluded that cutaneous discriminative ability was chiefly, but not exclusively, represented in the postcentral gyrus.

Johannes Dusser de Barenne's (1885–1940) studies involving "strychninization" of the cortex represented another way of studying cortical function. He found that applying strychnine to the cortex could lead to positive sensory signs such as paresthesias, hyperalgesias, and biting and scratching body parts, as if they had been stimulated (Dusser de Barenne, 1924, 1932, 1933). The sensory overreaction occurred immediately after he activated the parietal cortex or the motor-premotor area. Like Minkowski, Dusser de Barenne concluded that the precentral cortex has some sensory functions, but that the parietal cortex is the major projection area for cutaneous sensation. (Dusser de Barenne and one of his cortical maps are shown as Figures 10.12 and 10.13, respectively.)

Figure 10.12. Johannes Dusser de Barenne (1885–1940), best known for his use of strychnine to assess the functional properties of different parts of the cortex.

During the 1930s, it was reported that monkeys given postcentral gyrus lesions showed a loss of position sense, placing and hopping reflexes, and spontaneous limb movements (Ferraro and Barrera, 1935). These behaviors were also affected by lesions of the dorsal white columns of the spinal cord and the dorsal column nuclei in the medulla. The deficits seen after lesions lower in the system, however, typically were more severe than the deficits found after cortical damage. Still, these structures appeared to be part of a single system, as also suggested by the known anatomy.

At about the same time, T. C. Ruch and John Fulton ablated the somatosensory and motor cortices of some mon-

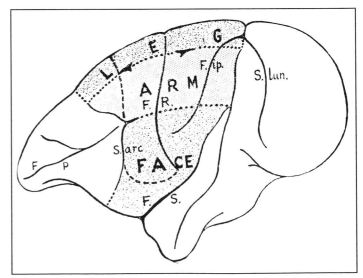

Figure 10.13. Johannes Gregorius Dusser de Barenne's (1933) diagram of the sensory cortex of the macaque monkey as revealed by the method of focal strychninization. (From *Archives of Neurology and Psychiatry,* 1933. Courtesy of the American Medical Association.)

keys. They reported the animals could still discriminate small differences in weight. But when these investigators also ablated the posterior parietal lobe, this ability decreased significantly.

Ruch and his colleagues went on to study the effects of partial and complete parietal lobe lesions on roughness and weight discrimination in other monkeys and chimpanzees (Ruch, 1935; Ruch and Fulton, 1936; Ruch, Fulton, and Kasdon, 1937; Ruch, Kasdon, and Fulton, 1940). They found that monkeys with unilateral parietal lobectomies showed deficits in weight discrimination soon after their lesions, but could recover to preoperative levels. Chimpanzees with similar lesions also showed deficits on the roughness and weight discriminations. If the posterior parietal lobe were also ablated, the recovery was very slow, and the animals failed to reach preoperative levels, even after a few years.

Ruch's data further suggested that both precentral and postcentral cortices are involved in sensation, at least in the monkey. Ruch (1935) thought cats and dogs showed even more sensory-motor overlap, whereas the motor cortex has less of a sensory role in humans. He hypothesized that the pre-Rolandic area was a primitive sensory area whose role varied with the position of the organism on the phylogenetic scale.

The discovery of a second somatosensory cortex, just above the Sylvian fissure, did not occur until 1940. The second area corresponded to Brodmann's Area 43 and was first discovered by Edgar D. Adrian during his electrophysiological studies. The second somatosensory area was soon confirmed by others (Woolsey, Marshall, and Bard, 1942). It was not until 1947 that the effects of lesions in the two somatosensory areas were compared. This was done on dogs in an experiment that looked at conditioned leg movements to tactile stimuli (Allen, 1947). The results of this and later studies have suggested that the second somatosensory area is not just a backup for the first, although the specific functions of each have been debated.

LESIONS INVOLVING THE POSTCENTRAL GYRUS IN HUMANS

The fact that lesions in the region of the Rolandic fissure can cause sensory disturbances in human patients was recognized in the 1880s by Moses Allen Starr (1884a, 1884b). In a review of a large series of cases, it was stated that in 75 percent of the cases of somatosensory disturbances, the lesion was posterior to the fissure of Rolando, most often in the postcentral gyrus or in the adjacent parietal region. Starr added:

If any distinction is to be made between motor and sensory areas in the central region it seems to be clear that the relative position of the motor and sensory fibres in the internal capsule is preserved in their distribution in the cortex— the sensory areas lying posterior to the motor. Cases have already been cited in which a lesion of the parietal lobules alone has produced partial anaesthesia of the opposite side of the body. It is, therefore, probable that the sensory areas extend further backward than the motor areas, and that they include the parietal lobes in their extent to some degree. (1884a, 137)

Henri Verger (1873–1930) and Joseph Dejerine were among the first to demonstrate convincingly that damage to the human parietal lobe may result in relatively minor disturbances of pain and temperature sensitivity, but clear deficits in tactile localization, postural sensitivity, and fine tactile discrimination (see Dejerine and Mouzon, 1914; Verger, 1900; also the review by Critchley, 1953). Nevertheless, occasional patients showing marked analgesia and/or a blunting of the temperature sense have also been reported (e.g., Russel and Horsley, 1906).

Among the most cited papers dealing with the effects of parietal lobe lesions on somesthesis were those published by Henry Head with Gordon Holmes (1911–12). They noted that simple touch, pain, and temperature thresholds were rarely affected by lesions confined to the parietal cortex, at least after shock and the general effects of trauma had dissipated. In contrast, significant difficulties with differentiation, comparisons, synthesis, and localization (atopognosis) were found. That is, recognition of objects, two-point discrimination, stimulus localization, weight discrimination, and texture appreciation were often affected.

The ability to appreciate passive movement of the limb was also often poor. Normally, movements of about 2 degrees could be detected, but on the affected side, the movements had to be in the range of 10–20 degrees to be appreciated. Head and Holmes wrote the following about posture and the ability to appreciate passive movements:

Inability to recognize the position of the affected part in space is the most frequent sensory defect produced by lesions of the cerebral cortex. In some cases, this and the allied faculty of recognizing passive movement may be the only discoverable abnormalities. Whenever sensation is disturbed at all, these two forms of spatial recognition will certainly be affected. (1911–12, 157)

Head and Holmes realized that getting reliable data from this patient population was difficult. They found that individuals with parietal cortex damage were much more likely than those with subcortical or peripheral lesions to give irregular, unpredictable, and incongruous responses. One example would be saying "two points" when only one point of a compass was applied to the skin. Another would be saying "touched" on catch trials not associated with the application of any stimulus.

Lesions at lower levels of the nervous system produce, on the whole, forms of altered sensibility which react with remarkable constancy to graduated stimuli. A touch of definite intensity, if sufficiently strong to evoke a sensation, will do so in a large proportion of instances; while some less intense stimulus will cause no sensation of any kind. . . . But the characteristic change produced by a cortical lesion consists in a want of constancy and uniformity of response to the same tactile stimulus. Increasing the stimulus does not necessarily improve the patient's answers and in many cases no threshold can be obtained. (1911–12, 147)

After reviewing extensive data on soldiers wounded in World War I, Head (1918) published another important study in which he confirmed that the deficits often included poor limb coordination, impaired tactile sensitivity as measured with von Frey hairs, and difficulties discriminating weights on the side of the body contralateral to the lesion. He again concluded that the parietal lobe was largely involved with comparing stimuli to each other and not with "primitive" aspects of touch, temperature, or pain. This fit well with his earlier notion of an epicritic system.

Otfrid Foerster (1916), Head's counterpart in Germany during World War I, stressed the following points about cortical damage and somesthesis:

1. The boundaries of loss are gradual and not sharp.
2. The midline structures may show fewer deficits.
3. Some body regions may show complete sparing with unilateral lesions.
4. Distal parts of limbs may be more affected than proximal parts.
5. The face and extremities are more affected than the trunk.
6. The arm is more susceptible to deficits than the leg.
7. The front of the body is more affected than the back.

One problem with the research conducted on soldiers at this time was that the degree of subcortical involvement was often hard to determine. As a result, how a given subject might be classified could vary considerably across studies. In this context, Gordon Holmes worked with gunshot cases in which he felt certain that the damage was limited to the cerebral cortex. He agreed with Head's assertion that, after the initial shock has worn off, unilateral lesions of the somatosensory cortex result only in contralateral deficits. Holmes (1927) also found that peripheral areas of abnormal sensitivity were never very sharply defined. He added that he felt fairly sure the precentral gyrus had no sensory functions.

> The question of whether the precentral cortex has sensory functions has been a matter of much debate. Most of the older physiologists refused to recognize separate localization of sensory and motor functions, and many clinicians . . . have insisted that sensation may be disturbed by lesions limited to the precentral gyrus. But there has lately been a tendency to limit the sensory zone to the parietal lobe. My own observations on gunshot wounds and other lesions of the brain have led me to the conclusion that the sensory area lies entirely behind the central fissure, and that the precentral gyrus has no sensory functions. (1927, 424)

Later investigators seemed to agree with Head and Holmes that the somatosensory cortex was involved in fine comparisons and spatial relationships (e.g., Evans, 1935; Stopford, 1930). This included localization of the stimulus, two-point discrimination, and position sense, and excluded the detection of pain or the presence of temperature.[4]

---

[4]The postural defect was much better defined in the period after World War I. It was found to involve a loss of passive movement of the joint, failure to recognize the spatial position of the limb without sight, and not knowing the direction of movement. Problems in the kinesthetic sphere resulted in "postural drift" after parietal lobe lesions. This phenomenon was examined by having subjects hold their hands at 90-degree angles from their bodies while not looking. The typical finding was that the hand contralateral to the lesion drifted down well before the healthy limb. Tests were even developed that involved moving folds of the patient's skin in different directions (Tshlenow, 1928). In general, these defects proved to be very sensitive indicators of parietal lobe damage.

ASTEREOGNOSIS

Five patients who could not identify objects with their hands were described in an 1844 report by Friedrich A. Puchelt (1784–1856). One patient showed preserved sensitivity for the various tactile modalities—she described the temperature, hardness, and size of the object—but could not identify the object she was handling and feeling. Puchelt considered this to be a case of "partial paralysis" (the term "paralysis" was not restricted to motor functions at the time) and wrote that its cause would not be found in the limb or in the spinal cord but in the brain.

Puchelt's report, which might have been the first to describe astereognosis adequately, seemed to generate little interest when it appeared (Benton, 1990). Hardly any attention was drawn to astereognosis until the 1880s, when H. Hoffmann (1819–1891) described 16 patients with hemiplegia who could not identify objects with their affected hands. Hoffmann (1883, 1885), a Strassburg neurologist, called this recognition problem a "defect of stereognosis"—his new word combining *stereos,* meaning "solid form," with *gnosis,* Greek for "knowledge." Soon the disorder was being called astereognosis.[5]

An excellent description of astereognosis appeared in 1898.

> B.C. was 24 years old when he presented himself to Dr. Burr for treatment. When he was about 10 years old he was accidently struck on the side of the head by an axe handle with such force that he was thrown into a river, on the bank of which he had been standing. Examination of the head showed that he had a simple depressed fracture of the right parietal bone over the motor area. He remained in a state of alternating coma and delirium for about three weeks. On recovering he found himself partially paralyzed on the left side of the body and face, and completely anesthetic on the same side. The palsy and anesthesia entirely passed away in a few months, sensation returning before motion. He was supposed to have recovered completely, until, on putting his left hand into his coat pocket for the first time after his illness, he discovered that he could not tell what he had in his grasp, though he had preserved the sense of touch. (Burr, 1898, 37)

Some debate emerged at the turn of the century over whether the term "astereognosis" should be used descriptively when there were also more elementary sensory losses or whether basic sensations must be largely intact for such a term to have meaning. In addition, most investigators wanted to feel confident that motor problems were not responsible for the failure to identify objects. In the patient just described, there seemed to be no basic sensory or motor losses that could account for his failure to recognize objects by touch.

Agreement was found over the point that although lesions outside the parietal region can occasionally give rise to defects in stereognosis, the disorder was most likely to

---

[5]Many other terms were used to denote astereognosis over the years. Among the words used synonymously with it were *Tastlahmung* (tactile paralysis), "tactile blindness," "tactile" or "psychic anaesthesia," "tactile asymboly," "tactile aphasia," "tactile agnosia," and "tactile apraxia."

accompany posterior parietal lobe damage. Investigators also seemed to agree that the higher-order defect was somewhat like word blindness or word deafness (e.g., Williamson, 1897).

STIMULATION AND EPILEPSY OF THE HUMAN PARIETAL CORTEX

In a number of monkey studies, David Ferrier (1875, 1876, 1886) electrically stimulated the region behind the Rolandic fissure. He found only pupil and eye movements, and perhaps some head turning. This conveyed only vague information about this area and, as has been noted, Ferrier opted to localize somesthesis in the temporal lobes. A better understanding of the effects of activating this region required verbal human subjects.

The initial study on electrical stimulation of the human cortex was conducted by Roberts Bartholow in 1874 (see Chapter 3). His feeble-minded patient seemed to experience tingling sensations when he stimulated behind the Rolandic fissure. Better studies, largely involving severe epileptics, came a number of years later. Using weak currents, investigators typically found that most somatosensory responses occurred when the postcentral gyrus close to the Rolandic fissure was stimulated (Cushing, 1909; Foerster, 1936; Valkenburgh, 1914). The usual responses included numbness, tingling, and feelings of electricity, all localized on the contralateral side.

For example, Harvey Cushing (1869–1939; shown in Figure 10.14), one of the pioneers in the surgical use of electrical stimulation, studied two epileptic men. They told him they experienced sensations on the opposite side of the body, especially in their arms, hands, and fingers, when he stimulated the postcentral gyrus (see Figure 10.15). The experiences were said to be just like the auras that preceded their epileptic attacks. The sensations were described as numbness, "funny feelings," touch, stroking, and some vague and indescribable feelings suggestive of warmth. The subjects also reported dryness, stiffness, and temperature sensations from the tongue region. In general, these findings were con-

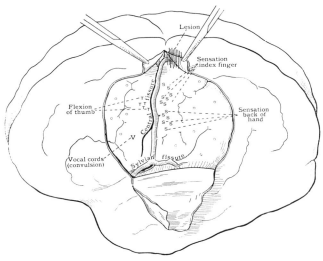

Figure 10.15. Diagram from Harvey Cushing (1909) showing distribution of sensory responses from the postcentral gyrus and motor responses from the precentral gyrus in a conscious human patient who underwent electrical stimulation of the brain.

firmed a number of years later in the celebrated investigations of Wilder Penfield (1891–1976) on epileptic patients (Penfield and Boldrey, 1937; Penfield and Gage, 1933).

For many centuries, the odd sensations accompanying an epileptic attack were believed to commence in the periphery of the body. To quote Gowers:

> The word "aura" was first used by Pelops, the master of Galen, who was struck by the phenomena with which many attacks begin—a sensation, commencing in the hand or the foot, apparently ascending to the head. The sensation having been described to him as a "cold vapor," he suggested that it might really be such, passing up the vessels, then believed to convey air. . . . This notion of peripheral origin of the aura was maintained until recent times. (1885, 33)

Studies like those of Cushing and Penfield proved that stimulation of the cerebral cortex could trigger specific types of epileptic auras, and added to the growing body of knowledge about cortical function in health and in disease.

REFERENCES

Adrian, E. D. 1940. Double representation of the feet in the sensory cortex of the cat. *Journal of Physiology, 98,* 16–18.

Allen, W. F. 1947. Effect of partial and complete destruction of the tactile cerebral cortex on correct conditioned differential foreleg responses from cutaneous stimulation. *American Journal of Physiology, 151,* 325–337.

Alrutz, S. 1897. On the temperature senses. *Mind, 6,* 445–448.

Bartholow, R. 1874. Experimental investigations into the functions of the human brain. *American Journal of Medical Science, 67,* 305–313.

Bastien, J. B., and Vulpian, A. 1855. Mémoire sur les effets de la compression des nerfs. *Gazette Médicale de Paris,* 794.

Bell, C. 1811. *Idea of a New Anatomy of the Brain; Submitted for the Observations of His Friends.* London: Strahan & Preson. (Reprinted in *Medical Classics, 1936, 1,* 105–120.)

Benton, A. L. 1990. The fate of some neuropsychological concepts: An historical inquiry. In E. Goldberg (Ed.), *Contemporary Neuropsychology and the Legacy of Luria.* Hillsdale, NJ: Lawrence Erlbaum. Pp. 171–179.

Blix, M. 1884. Experimentelle Beiträge zur Lösung der Frage über die specifische Energie der Hautnerven. *Zeitschrift für Biologie, 20,* 141–156. (Translation of paper read in Sweden in 1882)

Boring, E. G. 1916. Cutaneous sensation after nerve division. *Quarterly Journal of Experimental Physiology, 10,* 1–95.

Figure 10.14. Harvey Cushing (1869–1939), the American neurosurgeon, with a young patient in 1928.

Boring, E. G. 1942. *Sensation and Perception in the History of Experimental Psychology.* New York: Appleton-Century-Crofts.

Brazier, M. A. B. 1957. Rise of neurophysiology in the nineteenth century. *Journal of Neurophysiology, 20,* 212–226.

Burr, C. W. 1898. A case of psychic amnesia. *Journal of Nervous and Mental Disease, 25,* 37–38.

Cobb, S. 1919. Cutaneous sensibility in cases of peripheral nerve injury. Epicritic and protopathic hypotheses of Head untenable. *Archives of Neurology and Psychiatry, 2,* 505–517.

Critchley, M. 1953. *The Parietal Lobes.* Baltimore, MD: Williams & Wilkins.

Cushing, C. 1909. A note upon the faradic stimulation of the post-central gyrus in conscious patients. *Brain, 32,* 44–53.

Czermak, J. H. 1855. Beiträge der Physiologie des Tastsinnes. *Sitzungsberichte der Kaiserlichen Akademie der Wissenschaften. Math-Phys. Cl., 15,* 466–521.

Dallenbach, K. M. 1927. The temperature spots and end organs. *American Journal of Psychology, 39,* 402–427.

Dallenbach, K. M. 1939. Pain: History and present status. *American Journal of Psychology, 52,* 331–347.

Darwin, E. 1796. *Zoonomia.* New York: T & J. Swords.

Dejerine, J., and Mouzon, J. 1914. Deux cas de syndrome sensitif cortical. *Revue Neurologique, 28,* 388–392.

Denny-Brown, W. 1970. Henry Head, 1861–1940. In W. Haymaker and F. Schiller (Eds.), *The Founders of Neurology.* Springfield, IL: Charles C Thomas. Pp. 449–452.

Donaldson, H. H. 1885. On the temperature sense. *Mind, 10,* 399–416.

Dusser de Barenne, J. G. 1924. Experimental researches on localization in the cerebral cortex on the monkey (Macacus). *Proceedings of the Royal Society of London, Ser. B, 96,* 272–291.

Dusser de Barenne, J. G. 1932. Some aspects of the problem of "corticalization" of function and of functional localization in the cerebral cortex. *Proceedings of the Association for Research in Nervous and Mental Disease, 13,* 85–106.

Dusser de Barenne, J. G. 1933. "Corticalization" of function and functional localization in the cerebral cortex. *Archives of Neurology and Psychiatry, 30,* 884–901.

Evans, J. P. 1935. A study of the sensory deficits resulting from excision of cerebral substance in humans. *Proceedings of the Association for Research in Nervous and Mental Disease, 15,* 331–370.

Ferraro, A., and Barrera, S. E. 1935. Summary of clinical and anatomical findings following lesions in the dorsal column system of Macacus rhesus monkeys. In Association for Research in Nervous and Mental Disease, *Sensation: Mechanisms and Disturbances* (Vol. 15). Baltimore, MD: Williams & Wilkins. Pp. 371–395.

Ferrier, D. 1875. Experiments on the brains of monkeys. *Philosophical Transactions of the Royal Society of London, Pt. 2, 165,* 433–488.

Ferrier, D. 1876. *The Functions of the Brain.* London: Smith, Elder and Company.

Ferrier, D. 1886. *The Functions of the Brain.* New York: G. P. Putnam's Sons. (Reprinted 1978; Washington, DC: University Publications of America)

Flechsig, P. 1896. *Ueber die Lokalization der geistigen Vorgänge insbesondere der Sinnesempfindungen des Menschen.* Leipzig: Veit.

Foerster, O. 1916. Die Topik der Sensibilitätsstörungen bei Unterbrechung der sensibilen Leitungsbahnen. *Neurologisches Centralblatt, 35,* 807–810.

Foerster, O. 1933. The dermatomes in man. *Brain, 56,* 1–39.

Foerster, O. 1936. The motor cortex in man in the light of Hughlings Jackson's doctrines. *Brain, 59,* 135–159.

Frazier, C. H., and Silbert, S. 1920. Observations in five hundred cases of injuries of the peripheral nerves at U.S.A. Hospital No. 11. *Surgery, Gynecology and Obstetrics, 30,* 50–63.

Frey, M. von. 1895. Beiträge sur Sinnesphysiologie der Haut. *Sächsischen Akademie der Wissenschaften zu Leipzig. Math-Phys. Cl., 47,* 166–184.

Frey, M. von. 1896. Untersuchungen über die Sinnesfunctionen der menschlichen Haut: Druckempfindung und Schmerz. *Abhandlungen der Sächsischen Akademie der Wissenschaften zu Leipzig. Math-Phys. Cl., 23,* 175–266.

Galen. 1917. Experimental section and hemi-section of the spinal cord. Translation of a passage from *De locis affectibus. Annals of Medical History, 1,* 367.

Gasser, H. S., and Erlanger, J. 1929. The role of fiber size in the establishment of a nerve block by pressure or cocaine. *American Journal of Physiology, 88,* 581–591.

Gilbert, R. W. 1929. Dermal sensitivity and the differentiated nerve-terminations of the human skin. *Journal of General Psychology, 2,* 445–459.

Goldscheider, A. 1884. Die specifische Energie der Temperaturnerven. *Monatschrift für Praktische Dermatolologie, 3,* 198–208.

Goldscheider, A. 1886a. Zur Dualität des Temperatursinns. *Archiv für die gesammte Physiologie, 39,* 96–120.

Goldscheider, A. 1886b. Die Wirkungen des Kokains und anderer Anaesthetica auf die Sinnesnerven der Haut. *Monatschrift für praktische Dermatologie, 5,* 49–67.

Goldscheider, A. 1912. Ueber die Empfindung der Hitze. *Zeitschrift für klinische Medizin, 75,* 1–14.

Goldstein, K. 1910. Über sie aufsteigende Degeneration nach Querschnittsunterbrechung des Rückenmarks (Tractus spinocerebellaris posterior, Tractus spino-olivaris, Tractus spino-thalamicus). *Neurologisches Centralblatt, 29,* 898–911.

Gowers, W. R. 1885. *Epilepsy.* London: William Wood and Company. (Reprinted 1964; New York: Dover Publications)

Griesbach, H. 1895. Ueber Beziehungen zwishen geistiger Ermüdung und Empfindungsvermögen der Haut. *Archiv für Hygiene, 24,* 124–212.

Grünbaum, A. F. S., and Sherrington, C. S. 1901. Observations on the physiology of the cerebral cortex of some of the higher apes. *Proceedings of the Royal Society, 69,* 206–209.

Grünbaum, A. F. S., and Sherrington, C. S. 1903. Observations on the cerebral cortex of the anthropoid apes. *Proceedings of the Royal Society, 72,* 152–155.

Häggqvist, G. von. 1914. Histophysiologische Studien über die Temperatursinne der Haut des Menschen. *Anatomischer Anzeiger, 45,* 46–63.

Hardy, J. D., Wolff, H. G., and Goodell, H. 1967. *Pain Sensations and Reactions.* New York: Hafner. (Reprint of 1952 edition)

Head, H. 1918. Sensation and the cerebral cortex. *Brain, 41,* 57–253.

Head, H. 1920. *Studies in Neurology.* London: Oxford University Press.

Head, H., and Campbell, A. W. 1900. The pathology of herpes zoster and its bearing on sensory localization. *Brain, 23,* 353–523.

Head, H., and Holmes, G. 1911–1912. Sensory disturbances from cerebral lesions. *Brain, 34,* 102–254.

Head, H., and Rivers, W. H. R. 1908. A human experiment in nerve division. *Brain, 31,* 323–450.

Head, H., Rivers, W. H. R., and Sherren, J. 1905. The afferent nervous system from a new aspect. *Brain, 28,* 99–115.

Hering, E. 1877. Grundzüge einer Theorie des Temperatursinns. *Sitzungberichte der Kaiserlichten Akademie der Wissenschaften, Wien. Math-Nat. Kl., 75,* 101–135.

Herzen, A. 1886. Über die Spaltung des Temperatursinnes in zwei gesonderte Sinne. *Archiv für die gesammte Physiologie, 38,* 93–103.

Hoffmann, H. 1883. *Stereognostiche Versuche.* Dissertation, Strassburg.

Hoffmann, H. 1885. Stereognostiche Versuche, angestellt zur Ermittelung der Elemente des Gefühlssinnes, aus denen die Vorstellungen der Körper im Raume gebildet werden. *Deutsches Archiv für klinische Medizin, 36,* 398–426.

Holmes, G. 1927. Disorders of sensation produced by cortical lesions. *Brain, 50,* 413–427.

Keele, K. D. 1957. *Anatomies of Pain.* Oxford, U.K.: Blackwell.

Kolle, K. 1970. *Grosse Nervenärzte.* Stuttgart, Germany: Georg Thieme Verlag.

Kottenkamp, R., and Ullrich, H. 1870. Versuche über den Raumsinn der Haut der oberen Extremität. *Zeitschrift für Biologie, 6,* 37–52.

Leake, C. D. 1970. Hermann von Helmholtz, 1821–1894. In W. Haymaker and F. Schiller (Eds.), *The Founders of Neurology.* Springfield, IL: Charles C Thomas. Pp. 225–229.

Lewandowsky, M., and Simons, A. 1909. Zur Physiologie der vorderen und der hinteren Zentralwindung. *Pflüger's Archiv für die gesammte Physiologie, 129,* 240–254.

Magendie, F. 1822 Experiences sur les fonctiones des racines des nerfs rachidiens. *Journal de Physiologie Expérimentale et Pathologie, 2,* 276–279.

McHenry, L. C., Jr., 1969. *Garrison's History of Neurology.* Springfield, IL: Charles C Thomas.

Minkowski, M. 1917–1918. Etude sur la physiologie des circonvolutions rolandiques et pariétales. *Schweizer Archiv für Neurologie und Psychiatrie, 1,* 389–459.

Mott, F. W. 1894. The sensory functions of the central convolutions of the cerebral cortex. *Journal of Physiology, 15,* 464–487.

Mott, F. W. 1895. Experimental inquiry upon the afferent tracts of the central nervous system of the monkey. *Brain, 18,* 1–20.

Müller, J. 1826. *Zur vergleichenden Physiologie des Gesichtssinnes des Menschen und der Thiere.* Leipzig: C. Cnobloch.

Müller, J. 1838. *Handbuch der Physiologie des Menschen für Vorlesungen.* Coblenz: J. Hölscher.

Müller, J. 1840. *Handbuch der Physiologie des Menschen für Vorlesungen* (2nd ed.). Coblenz: J. Hölscher.

Munk, H. 1892. Über die Fühlsphären der Grosshirnrinde. *Sitzungberichte der Kaiserlichen Akademie der Wissenschaft, 101,* 679–723.

Natanson, L. N. 1844. Analyse der Functionen des Nervensystems. *Archiv für Physiology und Heilkunde, 3,* 515–535.

Nothnagel, H. 1867. Beiträge zur Physiologie des Temperatursinns. *Deutsches Archiv für klinische Medizin, 2,* 284–299.

Papez, J. W. 1929. *Comparative Neurology.* New York: Crowell.

Pendleton, C. R. 1928. The cold receptor. *American Journal of Psychology, 40,* 353–371.

Penfield, W., and Boldrey, E. 1937. Somatic motor and sensory representation in the cerebral cortex of man as studied by electrical stimulation. *Brain, 60,* 389–443.

Penfield, W., and Gage, L. 1933. Cerebral localization of epileptic manifestations. *Archives of Neurology and Psychiatry, 30,* 709–727.

Pollock, L. J. 1919. Overlap of so-called protopathic sensibility as seen in peripheral nerve lesions. *Archives of Neurology and Psychiatry, 2,* 667–700.

Probst, M. 1900. Experimentelle Untersuchungen über die Schleifenendigung die Haubenbahnen, das dorsale Längsbündel und die hintere Commissur. *Archiv für Psychiatrie und Nervenkrankheit, 33,* 1–57.

Puchelt, F. 1844. Über parielle Empfindungslähmung. *Heidelberg Medizinische Annalen, 10,* 485–495.

Ranson, S. W. 1931. Cutaneous sensory fibers and sensory conduction. *Archives of Neurology and Psychiatry, 26,* 1122–1144.

Ranson, S. W., and Billingsley, P. R. 1916. The conduction of painful afferent impulses in the spinal nerves. *American Journal of Physiology, 40,* 571–589.

Riese, W. 1970. Johannes Müller, 1801–1858. In W. Haymaker and F. Schiller (Eds.), *The Founders of Neurology* (2nd ed.). Springfield, IL: Charles C Thomas. Pp. 243–247.

Riley, W. H. 1894. A study of the temperature sense. *Journal of Nervous and Mental Disease, 21,* 549–566.

Rothmann, M. 1914. Über die Grenzen der Extremitätenregion der Grosshirnrinde. *Monatsschrift für Psychiatrie und Neurologie, 36,* 319–341.

Rubin, E. 1912. Beobachtungen über Temperaturempfindungen. *Zeitschrift für Sinnesphysiologie, 46,* 388–393.

Ruch, T. C. 1935. Cortical localization and somatic sensibility. In Association for Research in Nervous and Mental Disease, *Sensation: Mechanisms and Disturbances* (Vol. 15). Baltimore, MD: Williams & Wilkins. Pp. 289–330.

Ruch, T. C., and Fulton, J. F. 1935. Cortical localization of somatic sensibility: the effect of precentral, postcentral, and posterior parietal lesions upon the performance of monkeys trained to discriminate weights. *Proceedings of the Association for Research in Nervous and Mental Disease, 15,* 289–330.

Ruch, T. C., and Fulton, J. F. 1936. Somatic sensory function of the cerebral cortex in monkey and chimpanzee. *American Journal of Physiology, 116,* 134–135.

Ruch, T. C., Fulton, J. F., and Kasdon, S. 1937. Further experiments on the somatosensory functions of the cerebral cortex in monkey and chimpanzee. *American Journal of Physiology, 119,* 394–395.

Ruch, T. C., Kasdon, S., and Fulton, J. F. 1940. Late recovery of sensory discriminability after parietal lesions in the chimpanzee. *American Journal of Physiology, 129,* 453.

Russel, C. R., and Horsley, V. 1906. Note on the apparent re-representation in the cerebral cortex of the type of sensory representation as it exists in the spinal cord. *Brain, 29,* 137–152.

Schäfer, E. A. 1900. The cerebral cortex. In E. A. Schäfer (Ed.), *Text-book of Physiology.* Edinburgh: Young J. Pentland. Pp. 697–782.

Schiller, F. 1984. Coenesthesis. *Bulletin of the History of Medicine, 58,* 496–515.

Sherrington, C. S. 1898. Experiments in examination of the peripheral distribution of the fibres of the posterior roots of some spinal nerves. *Philosophical Transactions, B, 190,* 45–186.

Siegel, R. E. 1970. *Galen on Sense Perception.* Basel: S. Karger.

Sinclair, D. 1967. *Cutaneous Sensation.* London: Oxford University Press.

Starr, M. A. 1884a. Cortical lesions of the brain. A collection and analysis of the American cases of localized cerebral disease. *American Journal of Medical Sciences, 88,* 114–141.

Starr, M. A. 1884b. Cortical lesions of the brain. A collection and analysis of the American cases of localized cerebral disease. *American Journal of Medical Sciences, 87,* 366–391.

Stookey, B. 1916. Gunshot wounds of peripheral nerves. *Surgery, Gynecology and Obstetrics, 23,* 639–656.

Stopford, J. S. B. 1930. *Sensation and the Sensory Pathway.* London: Longmans, Green.

Strughold, H. 1925a. Die Schwellen des Kältesinnes am Auge, bestimmt mit Reizen von kleiner Fläche und geringer Wärmekapazität. *Zeitschrift für Biologie, 83,* 201–206.

Strughold, H. 1925b. Die Topographie des Kältesinnes in der Mundhöhle. *Zeitschrift für Biologie, 83,* 515–534.

Strughold, H., and Karbe, M. 1925a. Die Topographie des Kältesinnes auf Cornea und Conjunctiva, ein Beitrag zur Frage nach den specifischen Empfängern desselben. *Zeitschrift für Biologie, 83,* 189–200.

Strughold, H., and Karbe, M. 1925b. Die Dichte der Kältpunkte im Lidspaltenbereiche des Auges. *Zeitschrift für Biologie, 83,* 207–212.

Trotter, W. B., and Davies, H. M. 1909. Experimental studies in the innervation of the skin. *Journal of Physiology, 38,* 134–246.

Trotter, W. B., and Davies, H. M. 1913. The peculiarities of sensibility found in cutaneous areas supplied by regenerating nerves. *Journal für Psychologie und Neurologie, 20,* 102–150.

Tshlenow, L. G. 1928. Sur la cinesthésie cutanée. *Revue Neurologique, 2,* 506–508.

Valkenburgh, C. T. van. 1914. Zur fokalen Lokalisation der Sensibilität in der Grosshirnrinde des Menschen. *Zeitschrift für die gesamte Neurologie und Psychiatrie, 24,* 294–312.

Vannod, T. 1896. La fatigue intellectuelle et son influence sur la sensibilité cutanée. *Revue Médicale de la Suisse Romande, 16,* 712–751.

Verger, H. 1900. Sur les troubles de la sensibilité générale consécutifs aux lésions des hémisphères cérébraux chez l'homme. *Archives Générales de Médecine, 6,* 641–713.

Vierordt, K. 1869. Ueber die Ursachen der verschiedenen Entwickelung des Ortsinnes der Haut. *Archiv für die gesammte Physiologie, 2,* 297–306.

Vierordt, K. 1870. Die Abhängigkeit der Ausbildung des Raumsinnes der Haut von den Beweglichkeit der Körpertheile. *Zeitschrift für Biologie, 6,* 53–72.

Vogt, O. 1906. Über strukturelle Hirncentra mit besonderer Berücksichtigung der strukturellen Felder des Cortex pallii. *Anatomischer Anzeiger, 29,* 74–114.

Wallenberg, A. 1900. Secundäre sensible Bahnen im Gehirnstamme des Kaninchens; ihre gegenseitige Lage und ihre Bedeutung für den Aufbau des Thalamus. *Anatomischer Anzeiger, 18,* 81–105.

Wartenberg, R. 1970. Otfrid Foerster, 1873–1941. In W. Haymaker and F. Schiller (Eds.), *The Founders of Neurology.* Springfield, IL: Charles C Thomas. Pp. 555–559.

Weber, E. H. 1834. *De pulsu, resorptione, auditu at tactu: annotationes anatomicae et physiologicae.* Lipsiae: C. F. Koehler.

Weber, E. H. 1846. Temperatursinn. In R. Wagner (Ed.), *Handwörterbuch der Physiologiez* (Vol. 3). Braunschweig: F. Vieweg. Pp. 481–588.

Williamson, R. T. 1897. On "touch paralysis," or the inability to recognize the nature of objects by tactile impressions. *British Medical Journal, 2,* 787–788.

Woolsey, C. N., Marshall, W. D., and Bard, P. 1942. Representation of cutaneous tactile sensibility in the cerebral cortex of the monkey as indicated by evoked potentials. *Bulletin of the Johns Hopkins Hospital, 70,* 399–441.

Chapter 11
Pain

It is evidently impossible to transmit the impression of pain by teaching, since it is only known to those who have experienced it. Moreover, we are ignorant of each type of pain before we have felt it.

Galen, second century

PAIN AS PENALTY

In early civilizations, pain related to disease was believed to be caused by supernatural forces as a penalty for sins. Both the Greek word *poine,* and the Latin *poena,* words from which the term "pain" is derived, mean "penalty" or "punishment."

In Ancient Egypt, Seth and Sekhmet were the two gods most associated with painful diseases, although 36 different gods were associated with the care of specific body parts. Demons were thought to enter via the left (sinister) nostril or the left ear and to cause pain and suffering by attacking the blood vessels or the heart. The pains caused by these harmful demons were treated with magic, prayer, and rituals, as well as with drugs (Keele, 1957). The basic idea was to entice or force the intruding demon to leave.

The belief that painful diseases were the result of sin also characterized early Mesopotamian medicine (see Chapter 1). The notion that diseases could have "natural causes" simply did not exist in Ancient Babylon, where charms, amulets, and prayers were used to prevent painful objects and malevolent spirits from even entering the victim's body.

ATOMS, HUMORS, AND PAIN

Two theories that the early Greeks shared with the Egyptians were the idea that pain resulting from disease was caused by the supernatural and that the heart controlled sensation and emotion. These ancient ideas were challenged in fifth century B.C. by the Hippocratic physicians. These men turned from the supernatural to associate the four elements and their affiliated humors with disease and physical pain. In the same century, the theory that the brain played a major role in sensation began to be taught by Alcmaeon in the learning center of Crotona, Italy.

One of the most important figures of this period was Democritus. He believed that everything was composed of atoms, and he reasoned that these atoms could undergo changes in arrangement, shape, and position in space. He taught that "shed images" (emanations) from objects stimulated the different receptors of the body and affected the small atoms of the soul. The intrusion of irregularly shaped, hooked atoms into the vessels of the body was viewed as the cause of pain. Because the soul atoms were distributed throughout the body, the pain was believed to be experienced in the part of the body stimulated, instead of in one central location (e.g., the heart or the head). This change from the older demonic intrusion idea was accepted and modified by many others.[1]

Plato, who accepted this general thesis, substituted sharp triangular atoms in motion for the irregularly shaped, hooked atoms of Democritus. In particular, he thought fire was made of the most mobile and sharp minute triangles. Plato's schema centered on the idea that atoms which violently and "irrationally" intruded into the veins affected the soul and caused pain. A return to the natural condition, if intense, was posited as the basis of pleasurable feelings. Pain and pleasure were viewed as "affections" common to the entire body.

Although Plato associated the brain with reason and ideation, he showed no awareness of the role of the central nervous system in mediating pain. Instead, he thought that pain, whether caused by peripheral stimulation or emotional experience, was associated with the mortal soul. In *Timaeus,* Plato explained that this soul, which was also responsible for lust and other passions, was housed in the liver and the more noble heart, far away from the rational and immortal soul of the cerebrum. Plato taught that excessive pain or pleasure could affect the ability of the immortal soul to act rationally, even though the two souls were physically separated.

Aristotle believed both pain and pleasure were conveyed by waves or ripples in the blood vessels to the heart, his

---

[1]Democritus' theory was accepted not just by the Greeks but later by many Romans. In his *De rerum natura,* Lucretius included the following line: "For every shape which gratifies the senses has been formed not without a smoothness in its elements; but, on the other hand, whatever is painful and harsh has been produced not without some roughness of matter."

"Acropolis of the body." He maintained that areas rich in blood had increased sensitivity to these impressions and reasoned that excessive stimulation, or extreme sensitivity, could cause pain. Thus, among other things, Aristotle thought pain could result if the blood ran too hot, or if the heart were too soft.[2]

Aristotle followed Plato when he chose not to include pain as a quality of the sense of touch or as a separate sense, but rather as a passion of the soul. But like Plato, he recognized that touch and pain were closely associated and that touch could become painful under some conditions. Thus, in *De partibus animalium*, Aristotle wrote: "All animals have one sense at least, that of touch, and whatever has this sense has the capacity for pleasure and pain" (see Keele, 1962, p. 17).

In Aristotle's mind, humans had to be superior to other animals because humans are most sensitive to touch and because touch is the root of the higher feelings of pain and pleasure, as well as the desires. Plants, he reasoned, do not feel touch or experience pain, whereas lowly animals have a poor sense of touch and can be aware of only a diffuse sense of pain. Aristotle warned, however, that excessively intense stimulation could destroy animal life.

During the Roman period, Galen theorized that the senses were subserved by soft nerves that allowed external objects to make impressions on them. Unlike Aristotle, he maintained that the brain, which was even softer, received these sensations and played a role in perception, imagination, and reasoning. Galen thought that physical pain involved intense violent irritations of the nerves and viewed pain as the lowest form of sensation. Because intense stimulation of the skin, the eye, and the ear could cause pain, he did not think nerves specialized exclusively for pain could exist.

Galen accepted the earlier Greek idea that diseases cause pain by disrupting the balance among the humors. In fact, he interpreted his own extremely painful intestinal colic in this way. Usually it was the influx of toxic black bile or acrid yellow bile that was associated with painful disease, but imbalances in other humors could cause problems as well. Galen and other physicians in Rome often used cooling to shift the balance of the humors in patients suffering from pain (Siegel, 1970).

Galen's belief in the humors led him to an explanation for referred pain. He reasoned that movements of the humors and transmission through the nerves allowed pain in one organ of the body (*protopatheia*) to be felt in another organ (*deuteropatheia*). He called this "sympathy" and was especially impressed by how pain could spread from the stomach to the heart (for a discussion of sympathy, see Chapter 20).

From his clinical and personal experience, Galen believed it was easy to distinguish patients with real pain from those who were faking. He stated that patients who are really distressed will be willing to submit to any treatment or remedy so long as the individual believes it has a chance of reducing the suffering. In contrast, malingerers and individuals who greatly exaggerate mild pain may fling themselves from one position to another, but will then make every excuse to avoid bitter medicines, fasts, or other interventions that can hurt or be uncomfortable.

MULTIPLE THEORIES OF PAIN

The classification of pain continued to divide physicians and philosophers through the Middle Ages into the Renaissance and even into relatively modern times. Many theories and ideas from classical antiquity remained, some of which were staunchly defended. For instance, late in the eighteenth century, Erasmus Darwin still championed an "intensive theory" of pain, the theory advocated by Plato in *Timaeus*. Darwin (1796) chose not to include pain as a special sense since "whenever the sensorial motions are stronger than usual . . . a great excess of light . . . of pressure or distension . . . of heat . . . of cold produces pain" (p. 90). The same general idea was entertained by Wilhelm Erb (1840–1921) and Edward Bradford Titchener (1867–1927) in the next century, although Titchener later switched to a sensory theory of pain.

In 1846, Ernst Weber argued that pain should be classified with common sensibility or feeling (*Gemeingefühl*). This grouping, which involved "the ability to perceive our own state of sensation," also included hunger, thirst, dizziness, sickness, itch, sting, tickle, nausea, and other vague organic perceptions.[3]

Even the traditional notion that pain should be classified with pleasure under the emotions continued to have some eighteenth- and nineteenth-century adherents, especially among philosophically oriented theorists (see Dallenbach, 1939). For example, in 1749, David Hartley (1705–1757) defined pain as "pleasure carried beyond a due limit." He also postulated that violent vibrations were produced in the peripheral nerves and the brain by painful stimuli (see Chapter 23).

The observation that pain could come from many sources (internal, external, or mental) kept alive the classical belief that pain was different from the traditional five senses. The separation was also maintained because pain did not appear to have just one specific stimulus, but instead seemed to be the result of any form of very intensive stimulation. The belief that pain lacked its own specific end organ, like an eye, an ear, the nose, or the tongue, was also consistent with the view that pain should not be equated with vision, hearing, smell, or taste.

The theory that pain should, in fact, be considered as an independent sense, or at least as a sensory modality of the skin senses, was suggested by Avicenna, the leading Persian physician-philosopher of the eleventh century. Acceptance of Avicenna's position, however, was surprisingly slow in coming. Lengthy papers specifically aimed at discrediting the sensory theory of pain continued to emerge through the beginning of the twentieth century. One frequently cited article appeared in 1894 and was entitled "Are There Special Nerves for Pain?" Henry Rutgers Marshall (1852–1927), the author, faulted the methods used to "prove" pain was a sense and questioned why nerves and tracts for pleasure have not been discovered if pain pathways really existed.

---

[2]This sort of thinking probably gave rise to the expression "softhearted," a term used to describe a person who tended to be overly sensitive and easily hurt.

[3]See Schiller (1984) for a history of *coenesthesis*.

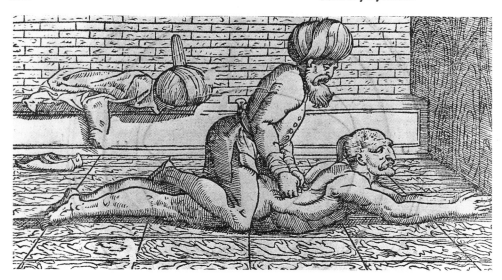

Figure 11.1. Woodcut showing Avicenna (980–1037) massaging a man's back. Avicenna described many different kinds of pain, but relied heavily upon Galenic theory when it came to accounting for (and treating) pain. (From *Avicennae arabum medicorum principis;* Venice, 1585.)

LIMITING THE DEFINITION OF PAIN

Developments in neurology and anatomy strengthened the case for pain as a modality of cutaneous sensation (see below). But this meant additional care had to be exercised to distinguish "pain the sensation" from vague feelings of discomfort, emotional hurt, and suffering. To some extent, the changes that took place in thinking about pain can be appreciated from the lists of different types of pain from different periods.

The earliest lists typically included discomfort and emotional hurt as well as the painful effects of physical injury and disease. These can be found in the ancient Indian document the *Nirvana Sutra*. The pains listed in the *Nirvana Sutra* are (1) birth pains, (2) pains of age, (3) pains of disease, (4) pain of death, (5) pain of parting with loved ones or things, (6) pain of meeting with something or someone disliked, (7) pain of not reaching a goal, and (8) pain of the five elements, or pain produced by the body (Gruner, 1930).

In the eleventh century, Avicenna, who is depicted in Figure 11.1, listed 15 types of pain. He included physical discomfort in *The Canon of Medicine*, but left emotional hurt off his list. In alphabetical order, Avicenna mentioned the following types of pain: (1) boring, (2) compressing, (3) corrosive, (4) dull, (5) fatigue-pain, (6) heavy pain, (7) incisive, (8) irritant, (9) itching, (10) pricking, (11) relaxing, (12) stabbing, (13) tearing, (14) tension, and (15) throbbing. Each type was defined. For example, the pain of compression was described as that pain "produced by a fluid or gas, when it is confined in too small a space in a member, and so compresses or squeezes the tissues" (Gruner translation, 1930, p. 249). Avicenna followed his intellectual predecessors in Greece and Rome by relating each of these types of pain to changes in the humors.

In 1920, Henry Head made only a terse statement on how pain should be defined. He did not feel compelled to distinguish between pain as a sensation and pain meaning emotional upset. Head's real concern was just to clarify the difference between true sensory pain and mild feelings of discomfort.

> Pain is a distinct sensory quality equivalent to heat and cold, and in its intensity can be roughly graded according to the force expended in stimulation. Discomfort on the other

hand is that feeling tone which is directly opposed to pleasure. It may accompany sensations not in themselves painful, as for instance that produced by tickling the sole of the foot. (1920, 665)

Head's separation of pain from discomfort, and his classification of pain as a sensory modality, helped to establish pain as a sensation. A number of important developments late in the nineteenth century paved the way for Head's particular orientation. One involved the discovery of different types of sensory spots on the skin. Recognition of some specialization in the peripheral and spinal pathways for the skin senses was also important. Additionally, brain lesions seemed to suggest that pain was just as much a sense as warmth or cold.

These and other significant discoveries about pain will now be described. Information about the receptors is presented first, followed by discussions of the peripheral nerve, spinal cord, brainstem, and cerebral cortex.

SENSORY SPOTS AND RECEPTORS FOR PAIN

Magnus Blix, Alfred Goldscheider, and Henry Donaldson all felt comfortable with the concept of pressure spots, warm spots, and cold spots (see Chapter 10). The same could not be said about how they felt about pain spots. Because pain spots were so numerous, they and many other investigators felt less certain they were discrete. Goldscheider (1891), for example, eventually concluded that pain might not even have its own receptor (also see Marshall, 1894; Strong, 1895).

Although Goldscheider accepted the intensive theory of pain, which held that pain was the result of strong stimuli acting on other types of receptors, he postulated that pain could still have its own pathways and centers. Specifically, he theorized that tactile stimuli could activate a special pain pathway in the spinal gray matter if the stimuli were of sufficiently high intensity.[4] In this respect, Goldscheider was

---

[4]Wilhelm Wundt, among others, accepted the idea that there were two spinal paths sharing the same peripheral receptors. He thought that with moderate impulses the message of touch, warmth, or cold was mediated by a pathway in the white matter of the spinal cord. With very intense stimulation, the impulses overflowed into the spinal gray matter and activated a second, high-threshold pathway, one that conveyed the sensation of pain.

still viewing pain as a sensation and not as an affect or an emotion.

More than anyone else, Max von Frey was responsible for the idea that pain should be treated as a separate modality with its own distinct architecture. His position was clearly different from that of Goldscheider, who thought the same peripheral apparatus might serve both pain and touch. Specifically, Frey attempted to associate pain with free nerve endings in the skin and the other modalities of cutaneous sensations with other receptors.

Frey found an ardent supporter of the pain issue in Friedrich Kiesow (1858–1940), who went to Leipzig in 1891 to study the new discipline of psychology with Wilhelm Wundt. While working on his first independent research project in Leipzig, Kiesow found places in the mouth sensitive to touch but not to pain. Kiesow, who continued his research at the University of Torino, later wrote:

Thus, I made the determination that the mucous membrane of the cheeks as well as the back parts of the mouth, including the back half of the tongue, are but slightly sensitive to pain; that the lower half of the uvula is completely insensitive to pain stimuli; and that there is a place in about the middle of the mucous membrane of the cheeks, which with suitable mechanical and electrical stimulation . . . is sensitive to touch but not to pain. (1928, 207)

Kiesow did not seem to realize the significance of his discovery when he first made it. But after collecting additional data on sensitive spots and describing pathological material, Kiesow concluded

that the condition of arousal of a superficial sensation of pain must be different from that of the sensation of touch; that no unitary nervous mechanism can be assumed for the two sensory qualities; but that, instead, the possibility of their arousal is to be sought in the existence of separate nervous mechanisms. (1928, 211)

In his autobiography, Kiesow (1930) wrote that he had collaborated with Frey on a number of projects and that his mentor had verified these findings under different stimulus conditions.[5]

DOUBLE PAIN

The idea that two types of pain may be elicited with some stimuli, the first being a sharp pain and the second a burning or aching pain, was recognized by Goldscheider (1891). Some questions arose about whether this phenomenon was really a pain-pain sequence or actually touch followed by pain (see Frey, 1894; Rosenbach, 1884). Nevertheless, most investigators at the turn of the century accepted the idea that a pricking pain can be followed about one second later by a burning pain.

A number of theories were presented to account for the double pain phenomenon. One theory maintained the two pains reflected two different pathways (Gad and Goldscheider, 1892). A second theory held that the end organs in different layers of skin could account for the delay and differ-

ent sensations (Rosenbach, 1884). A third hypothesis was that the first response was due to rapid stimulation of the nerves, whereas the second came from the slow liberation of a chemical which stimulated the receptors (Frey, 1922a, 1922b; Thunberg, 1902).

Better acceptance of the theory that the sharp, initial pain sensation is mediated by myelinated fibers, whereas the dull, slow throbbing pain was mediated by unmyelinated fibers had to wait for electrophysiological studies of the peripheral nerve (Zotterman, 1933, 1939). As for chemical mediation of pain, good evidence for the role of histamine in pain sensitivity did not emerge until the late 1930s (Rosenthal and Minard, 1939).

PHANTOM LIMBS

Those sensations that seem as if they are coming from an amputated part of the body are called phantoms. Very often amputees feel as if a whole missing limb is still present (hence the term "phantom limb"). Frequently the phantom limbs are painful (Stone, 1950).

An individual suffering from phantom limb was described in 1551 by Ambroise Paré (1510–1590), a French pioneer in surgery and prosthetics. His patient complained the limb felt as if it still existed. René Descartes (1644) also mentioned a case of phantom limb. Almost two hundred years later, phantoms were again described, this time by François Magendie.

The surgeon Silas Weir Mitchell (1829–1914; shown in Figure 11.2) was stationed in Philadelphia during the 1861–1865 American Civil War (Haymaker, 1970; Spillane, 1981). Especially after the Battle of Gettysburg (July 1–3, 1863), Philadelphia's South Street and Turner's Lane hospitals filled with wounded soldiers for Mitchell and others to treat.

Figure 11.2. Silas Weir Mitchell (1829–1914). Mitchell's early studies involved soldiers wounded during the American Civil War (1861–1865). Some his work involved treating "phantom" limbs and causalgia.

---

[5]In general, Kiesow liked to use pin pricking to test for pain. Frey preferred to test for pain with mild electrical current.

Because South Street Hospital was so involved with the care of amputees, the soldiers simply called it "Stump Hospital" (Middleton, 1966).

Mitchell (1871, 1872) noted that 86 of 90 amputees that he examined in his service developed "sensory ghosts" soon after the loss of a limb. He wrote:

> It has been known to surgeons that when a limb has been cut off the sufferer does not lose the consciousness of its existence. This has been found to be true in nearly every such case. Only about five per cent of the men who have suffered amputation never have any feeling of the part as being still present. Of the rest, there are a few who in time come to forget the missing member, while the remainder seem to retain a sense of its existence so vivid as to be more definite and intrusive than is that of its truly living fellow member. (1871, 565)

Mitchell (1871) emphasized that the phantoms typically were incomplete and that the hallucinated limb usually felt shorter than the real limb. Many of them were painful. He found that painful phantoms could be triggered by a variety of activities, including yawning, touch, or even changes in the wind.[6] Mitchell also noted that wearing an artificial limb can influence the phantom and even elicit phantom sensations in soldiers whose "ghost" sensations had previously disappeared.

Many scientists in the nineteenth century, including Mitchell, believed that phantoms were due to irritation of the nerves in the cut stumps. Nevertheless, Mitchell pointed out that it was extremely difficult to treat painful phantoms successfully. He tried to cauterize the nerves, used drugs, and even attempted acupuncture. Fustrated, he wrote the following:

> The stump . . . is liable to the most horrible neuralgias and to certain curious spasmodic maladies. In fact, a stump is rarely a perfectly comfortable portion of the body, and for years is apt to be tender, easily hurt, and liable at any time to pain. The form of neurologic torture to which stumps are liable arises from inflamed or hardened conditions of the divided nerves, and very often obliges the sufferer to submit to a second amputation. (1871, 564–565)

CAUSALGIA

Mitchell's wartime colleagues included George Reed Morehouse (1829–1905) and William Williams Keen (1837–1932). In 1864, as the Civil War was drawing to a close, these three U.S. Army surgeons, known affectionately as the Turner's Lane group, wrote *Gunshot Wounds and Other Injuries of Nerves,* one of the classics of American medicine. This small book, the first devoted just to peripheral nerve injuries, provided many excellent descriptions of gunshot injuries and accompanying pain states. One type of pain was particularly well defined. This was the burning pain of causalgia. Mitchell, Morehouse, and Keen said the following about it.

> It is a form of suffering as yet undescribed, and so frequent and terrible as to demand from us the fullest description. . . .

It was described as "burning," "mustard red hot" or as a "red-hot file rasping the skin." . . . In these parts, it is to be found . . . on the palm of the hand or palmar face of the fingers and on the dorsum of the foot. . . . The part itself is not alone subject to an intense burning sensation, but becomes exquisitely hyperaesthetic, so that a touch or a tap of the finger increases the pain. Exposure to the air is avoided by the patient with a care which seems absurd, and most of the cases keep the hand constantly wet, finding relief in the moisture rather than in the coolness of the application. Two of these sufferers carried a bottle of water and a sponge, and never permitted the part to become dry for a moment. (1864, 101–103)

Mitchell and his colleagues went on to say that the pain made the soldiers irritable and gave them an anxious yet weary expression. The pain, which was found to be associated with a warmer temperature of the affected part, interfered with sleep. Just about any stimulus, from intense sounds (e.g., a military band) to the rustling of a newspaper, had the potential to make the pain worse. In fact, the pain state often became so unbearable that soldiers asked to have their affected limbs amputated.

The Turner's Lane group hypothesized that the irritation of the nerve at the wound site was the cause of the pain, perhaps acting by changing the circulation and nutrition of the affected parts. They tried to treat causalgia by blistering the affected parts or by injecting morphine.[7]

After he wrote his 1864 monograph, Mitchell named the spontaneous, burning type of pain that follows peripheral nerve injuries "causalgia." The term was suggested by his friend Robley Dunglison (1798–1869), a professor at the Jefferson Medical College in Philadelphia. Dunglison combined the Greek words for "heat" and "pain" to form his new word. It first appeared in an article written by Mitchell in 1867 in an obscure journal, *United States Sanitary Commission Memoirs.*

Mitchell, Morehouse, and Keen have been recognized for being first to present the entire syndrome of causalgia (Richards, 1967). Other reports, however, suggested that they were not the first to see this painful state (Denmark, 1813; Hamilton, 1838). One involved a soldier who was wounded earlier in the same century during the storming of Badajoz, Spain, in the Peninsular Wars.[8] The burning pain was so intense that the soldier's arm had to be amputated (Denmark, 1813).

Many more reports on causalgia appeared in the literature in the years after Mitchell, Morehouse, and Keen published *Gunshot Wounds and Other Injuries of Nerves.* In general, the belief that this pain state was due to acute injuries (high-velocity missiles) which fail to sever the nerves completely was confirmed. Less clear was why causalgia appeared to affect only some 5 percent of those individuals who experienced this type of acute nerve injury (Richards, 1967).

Most of the early observations about causalgia proved to be accurate. Among those beliefs that have been questioned are (1) that causalgia typically appears well after the time of

---

[6]Mitchell wrote an especially interesting paper on pain and the weather in 1877.

[7]The more modern techniques of interrupting the appropriate sympathetic nerves surgically or by injecting local anesthetics were unknown to Mitchell and his colleagues (Barnes, 1953; Slessor, 1948).

[8]These wars were a part of the Napoleonic campaigns in Iberia. They lasted from 1808 to 1814. The battles that took place during 1811–1812 in Badajoz, Spain, were especially fierce.

injury; (2) that it is always associated with "nutritional" changes (hot glossy skin) in the affected parts; and (3) that it always disappears with sufficient time. For example, Weir Mitchell's son, John Kearsley Mitchell (1869–1927), followed 15 of his father's patients for years and in 1895 reported that several of them still experienced severe pain.

Mitchell's work during the Civil War helped to establish his outstanding reputation as a neurologist. After the war, he resumed his general practice in Philadelphia and spent time at the Philadelphia Orthopaedic Hospital and Infirmary for Nervous Diseases. He seemed to have no interest in visible academic appointments. In fact, he was "Professor Mitchell" for only five minutes of his life:

> He was a teacher of those who taught, although he was never a professor in any medical school except for five minutes during the time that he was a trustee at the University of Pennsylvania. At a board meeting he was asked to retire, and during his absence he was elected Professor of Physiology. He immediately resigned after being informed of his appointment. (Mumey, 1934, 47)

Mitchell published well over 150 papers and a number of important medical books during the postwar period (e.g., Mitchell, 1897). He was elected president of the American Neurological Association when it was formed in 1875, but he declined this honor. His books of poetry, short stories, and novels established his fame outside medicine and afforded him a literary popularity that equaled his reputation as a neurologist.

When Mitchell died in 1914, Keen eulogized:

> Never have I known so original, suggestive, and fertile a mind. I often called him a "yeasty" man. His mind was ever fermenting, speculating, alert, and over-flowing with ideas. With these he leavened the minds of his fellows and set their ideas fermenting. He was always desirous of putting everything to the test of experiment, and never satisfied until he had exhausted all possibilities. Almost every research opened a vista of other and more interesting problems to be solved. (1914, 347)

SPINAL PATHWAYS: EARLY NINETEENTH-CENTURY STUDIES

Both Charles Bell (1811) and François Longet believed the dorsal (posterior) columns conveyed pain and other sensations from the skin through the spinal cord to the brain (see Chapter 10). This view was based partly on the observations that stimulation of the dorsal columns seemed to give responses indicative of pain and that this did not happen when other parts of the spinal cord were stimulated.

New knowledge about the spinal pathways mediating pain came from many studies conducted by Charles-Edouard Brown-Séquard. Also contributing to the picture, although not as prominently, was Moritz Schiff.

Brown-Séquard and Schiff made select spinal cord lesions in laboratory animals. The atmosphere for their important studies was set by earlier experimenters who also made lesions of the spinal cord.[9] Most of their prede-

---

[9]For reviews of these early experiments, see Bezold (1858), Brown-Séquard (1860), and Chauveau (1857).

Figure 11.3.   Charles-Edouard Brown-Séquard (1817–1894). Based on work started in 1846, Brown-Séquard suggested that the dorsal columns were not the only pathways involved in somesthesis and that at least some important sensory projections crossed in the spinal cord.

cessors, however, provided little detailed information about their methods or results. Another drawback was that few tested for sensations from the opposite side of the body. Nevertheless, some of the findings concerning the posterior columns were intriguing.

For example, Michele Foderà (1793–1848) found that sensation seemed to be preserved in rabbits given lesions of the dorsal columns. Further, Izaak van Deen (1804–1869), in 1841, concluded from his own animal experiments that the posterior columns seemed to share sensory functions with the spinal gray matter. Other early investigators (e.g., Bellingeri, Schoeps, Rolando, and Calmeil) also appeared to find that sectioning the posterior columns did not always lead to major changes in sensitivity.

Despite these findings, when Brown-Séquard and Schiff began their experiments on the spinal cord, it was still widely believed that the dorsal columns alone conveyed all cutaneous sensations.

BROWN-SÉQUARD'S EXPERIMENTS ON THE SPINAL CORD

Brown-Séquard is shown in Figure 11.3. His interest in the spinal pathways began with his dissertation on the physiology of the spinal cord, *Recherches et Expériences sur la Physiologie de la Moelle Epinière*, which was published in 1846. He cut the dorsal columns, the lateral columns, or both, in frogs, eels, birds, and mammals. His dorsal column lesions seemed to result in a small disruption of movement, but created no noticeable effects on sensation. His anterolateral lesions, while impairing motor functions, also did not appear to affect sensation. When both regions were destroyed on one side, he described diminished movement and, more important, stated, "I have found the parts situated behind the section and on the same side to be seemingly as sensitive as those on the sound side" (translated by Olmstead, 1946, p. 17). Thus, his thesis indirectly suggested that at least some sensory fibers crossed within the spinal cord.

Although Brown-Séquard's initial work had virtually no effect on the doctrines of Bell or Longet, he conducted additional lesion studies on rabbits, dogs, and guinea pigs. In some cases, the dogs were blindfolded and allowed to lie down quietly before one of the toes was touched. To test for temperature, he applied ice or heat to the toes of the relaxed dog. Pinching and pricking the skin, galvanization, and cauterization were among the ways in which he tested for pain.

The findings obtained between 1849 and 1851 provided Brown-Séquard with the additional ammunition he needed to fight the idea that all sensory transmission through the cord utilized the ipsilateral dorsal columns. He stated that even if some sensory impressions are conveyed by the posterior columns, the majority are not. He added that the most important sensory pathways crossed in the cord and, if anything, damage will cause hypersensitivity on the ipsilateral side of the spinal cord.

These findings demanded a marked change in thinking. Prior to 1849, it was generally accepted that there was no crossing of sensory fibers in the spinal cord and that if any pathways crossed, this occurred in the brain itself. Although Brown-Séquard agreed that the motor pathways crossed in the brain, his research indicated this was not completely true for the skin senses.

Oddly, Brown-Séquard's experiments did not have a major impact on established doctrine until 1855 (see Olmstead, 1946, 1970). This was when Paul Broca, already a scientist of considerable stature, summarized his colleague's findings to the *Société de Biologie*. Brown-Séquard's reputation then grew rapidly. He went on to publish an important collection of lectures in 1860 and then a set of papers in which a number of his human cases with spinal cord lesions were described along with material from other observers (Brown-Séquard, 1868a, 1868b).

Brown-Séquard continued to emphasize that paralysis, loss of muscle sense, hyperasthesia to touch and painful stimuli, and conservation of sensation to cold and warmth, all appeared on the same side as the spinal cord lesion. On the opposite side, he found conservation of voluntary movement and an intact muscle sense, but a loss or diminution of pain, warmth, cold, and touch.

Brown-Séquard's laboratory and human studies led him to the conclusion that the most important sensory pathways had to ascend in the central gray matter on the opposite side of the spinal cord. He recognized his position went against another view which was just beginning to gain some support, one which held that the opposite anterior or lateral white columns conveyed sensation to the brain. Some investigators (Calmeil, Nonat) argued for these other areas because sectioning of the entire spinal cord except for the anterolateral region did not eliminate sensation. Brown-Séquard (1860) responded that the lesions sparing sensation must also have spared at least some of the central gray matter.

Brown-Séquard's pioneering laboratory animal work did

not really allow him to conclude with assurance that the "conductors" for pain, warmth, cold, and touch ascended in different parts of the spinal cord. But largely from clinical cases he and others had collected (1860, 1869), he confidently argued:

> The various sensitive impressions—of touch, of pain, of temperature, of muscular contractions, etc.—are transmitted by conductors which are quite distinct from one another, and so much so that the conductors of painful impressions, for instance, are not more able to convey other kinds of impressions than to transmit the orders of the will to the muscles. (1860, 39)

In this context, Brown-Séquard described patients who showed severely impaired responsiveness to noxious and temperature stimuli, but who retained tactile awareness. He also described others who lost the heat sense but not touch and still others with normal touch and hypersensitivity to pain.

SCHIFF'S TWO PROJECTION SYSTEMS

In the mid-1800s, Moritz Schiff, a student of both François Longet and François Magendie, examined more than 100 dogs and cats with various spinal cord lesions (Schiller and Haymaker, 1970). Schiff, whose portrait appears as Figure 11.4, started with the bias that the posterior columns mediated all sensory input from the body to the brain. But in 1854, before a commission of the *Académie Française*, which included Magendie, Flourens, and Serres, he reported pain could not be aroused after he cut through the spinal gray

Figure 11.4. Moritz Schiff (1823–1896). In 1849, Schiff argued that temperature and pain projections crossed in the spinal cord and did not ascend in the dorsal columns.

matter and was not affected by lesions of the posterior columns (Dallenbach, 1939; Keele, 1957; McHenry, 1969). These findings were published in 1858 and were in accord with Brown-Séquard's conclusion that the dorsal columns were not the only spinal pathways for sensation.

Schiff went on to hypothesize that the dorsal columns may contribute to awareness of movement and to touch, but not to the appreciation of pain or temperature. In contrast, it was his belief that another system crossed over soon after entering the spinal cord and was more likely to be involved with perception of pain. He called this system the "aesthsodic" system, and the dorsal column system the "kinaesthsodic system."

Like Brown-Séquard, Schiff thought the crossed pathway ascended in the spinal gray matter. His reasoning was based partly on his belief that intense stimulation of any skin sense modality could cause pain. From this premise, it seemed logical to suggest that there was a spillover from specific tracts into the gray matter of the spinal cord.

In his 1864 lectures, Edme Félix Alfred Vulpian, a student of Flourens, ridiculed Schiff's suggestions concerning separate fibers for touch and pain. In contrast, Brown-Séquard, who by this time was convinced he had seen such a dissociation in his human cases, sided with Schiff and maintained that touch, tickle, pain, and temperature utilized distinct parts of the spinal cord. Brown-Séquard stated this very clearly in his 1860 lecture series and again in 1868 and 1869.

After reconsidering the issue, Vulpian acknowledged that Brown-Séquard was probably correct. In 1868, he wrote that the conductors of the impressions of touch, temperature, and pain are absolutely distinct from one another and that each of these types of conductors occupies a distinct part of the spinal cord. This represented an important victory for both Brown-Séquard and Schiff.[10]

THE CASES OF GOWERS AND SPILLER

William Gowers, who appears in Figure 11.5, became more comfortable with the notion that the anterolateral quadrant of the spinal cord was responsible for pain and temperature information after he studied a student who had attempted suicide by shooting himself. Although he experienced no blunting of tactile sensitivity, this student could neither move his right arm nor feel pain in his left limbs. After he died, an autopsy revealed that a piece of bone had severely bruised the anterolateral right half of his spinal cord. Gowers (1878, 1886) took this as good evidence for the pain pathway crossing in the spinal cord and for temperature and pain utilizing pathways different than touch.

Another important clinical case cited by Gowers (1886) involved an individual who was stabbed in the back. The knife severed the posterior white columns on both sides and the anterolateral column on the right side. Pain below the level of the lesion was lost only on the opposite side, whereas touch seemed to be affected on both sides.

It is interesting to note that in 1879 Gowers referred to the ascending tract which seemed to carry pain and temperature information as the "anterolateral tract," although he did not discuss its possible role at this time. A few years later, he wrote that all sensations may pass through the dorsal columns, but suggested that pain may utilize the lateral tract, at least in animals. Only in 1886, after encountering the two aforementioned cases, did he finally associate projections in the anterolateral cord with pain sensation in humans (see Illingworth, 1982).

Gowers' change of heart about the anterolateral part of the spinal cord might have been influenced by an earlier report by another Englishman. In 1859, William Ogle (1827–1912), a physician to St. George's Hospital, described seven neurological cases. His material led him to conclude that the posterior columns were not the only pathway for sensations and that some sensory fibers must cross upon entering the cord (this report was also cited by Brown-Séquard).

It should also be noted that Gowers was not alone in his slow acceptance of a crossed, anterolateral projection system for pain and temperature. Ludwig Edinger (1855–1918), who contributed much to the anatomy of the crossed tract and who traced it to where it seemed to join the medial lemniscus, chose not to associate it with pain in 1889 and 1890. In addition, Vladimir Bekhterev (1899) also traced the crossed tract and at first chose not to associate it with pain.

The idea that there was a crossed tract responsible for pain and temperature sensations was strengthened considerably in 1905 when William Gibson Spiller (1863–1940;

Figure 11.5. Sir William Gowers (1845–1915). In the last quarter of the nineteenth century, Gowers presented a number of clinical cases that led to a better understanding of the pain pathways in the spinal cord. His textbook *Diseases of the Nervous System* became a classic in neurology.

---

[10]Surprisingly, Schiff's name did not figure prominently in discourses given by scientists during the 1850s and 1860s (Bezold, 1858; Chauveau, 1857). Although this could reflect the fact that Schiff's material was presented only orally in 1854 and not published until 1858, even Brown-Séquard (1860) hardly mentioned Schiff's name in his lengthy book of lectures. Late-nineteenth-century scientists had mixed opinions about Schiff's importance, although all praised the continuous efforts of Brown-Séquard to define the pathways of sensation in the spinal cord.

Figure 11.6. William Gibson Spiller (1863–1940). In 1905, Spiller associated loss of pain with tubercular lesions of the anterolateral columns of the spinal cord in one of his patients. He suggested cutting the anterior columns to alleviate severe pain.

shown in Figure 11.6) presented a patient with a tubercular lesion of the anterolateral quadrant of the lower thoracic cord (see Figure 11.7). The influential Spiller, who worked in Philadelphia, was considered "the most distinguished American clinical neurologist of his time" (Ornstein, 1970). His 23-year-old male patient demonstrated severely impaired pain and temperature sensitivity on

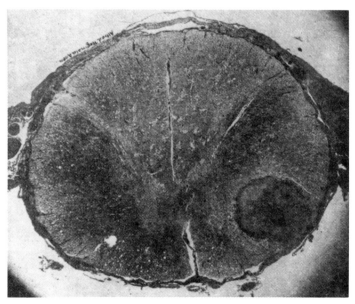

Figure 11.7. A cross section of the spinal cord (dorsal columns on top) showing the tuberculoma in the anterolateral column that disrupted the contralateral appreciation of pain in Spiller's patient. (From *Journal of Nervous and Mental Disease*, 1905, 32, p. 320)

body parts opposite and below the anterolateral spinal cord lesion. Tactile and motor abnormalities, in contrast, were minimal or nonexistent on that side. At the time, however, Spiller believed it was the spinocerebellar tract that was responsible for pain and temperature. One year later, Henry Head examined some cases of syringomyelia and hemisections of the spinal cord and verified Spiller's (1905) clinical findings (Head and Thompson, 1906).

Histological studies in the first decade of the twentieth century showed that some anterolateral projections went directly to the thalamus (Collier and Buzzard, 1903; Goldstein, 1910; Quensel and Kohnstamm, 1907). It was also recognized that other projections from the anterolateral spinal cord terminated more diffusely in the midbrain and other parts of the lower brainstem (Bruce, 1898; May, 1906; Mott, 1892; Russell, 1898; Stewart, 1901; Thiele and Horsley, 1901; Tooth, 1892). These histological studies, when combined with the findings of Gowers and Spiller, showed that the spinothalamic pathways could be differentiated from the nearby spinocerebellar tracts with which they had previously been confused both structurally and functionally (Walker, 1940).

ATTEMPTS TO TREAT PAIN BY CUTTING NERVES AND TRACTS

Neurosurgical attempts to treat pain by cutting the peripheral nerves have a long history. Paré stated in his commentaries that King Charles IX of France (1550–1574) suffered from intense pain and contractions in one arm due to a nerve injury suffered while he was being bled for smallpox. Although the King recovered in a few months, Paré wrote that had the King failed to recover, oil would have been put on his wound, or the troublesome nerve would have been severed completely (see White and Sweet, 1969). This statement suggests there was little hesitation to treat pain with surgery in the mid-1500s.

Toward the end of the seventeenth century, Georges Mareschal (1658–1736), the personal surgeon of Louis XIV of France (1638–1715), cut branches of the trigeminal nerve to treat tic douloureux (see Leriche, 1940). The first book on the surgical relief of pain was written in 1873 by Jean-Joseph Létiévant (1830–?), who cited the work of Mareschal.

The modern era of surgery for pain may have begun with the work of Victor Horsley, who treated trigeminal neuralgia by destroying the trigeminal ganglion (Horsley, Taylor, and Colman, 1891). This procedure had been tried by others before Horsley (Abbe, 1889; Bennett, 1889). More important, it became popular and was improved after Horsley's initial report (Frazier, 1928; Krause, 1893).

The idea that the anterolateral columns of the spinal cord or the more central projections of this pathway could be cut to treat pain was understandably later in coming. This approach, which might have been suggested by Brown-Séquard's or Schiff's experiments, was stimulated by Spiller's (1905) observations on the patient with tubercular lesions of the anterolateral cord. It was also based on some newer work on laboratory animals.

In 1910, Artur Schüller (1874–1957), an Austrian, cut the anterolateral columns in some monkeys. He called his operation *Chordotomie.* In his report, Schüller suggested that cutting the anterolateral columns may be more effective than cutting the posterior roots of the spinal cord in patients suffering from severe pain.

One year after Schüller's paper and six years after Spiller's tubercular case appeared, Spiller convinced Edward Martin (1859–1938) to cut the anterolateral quadrants in one of his patients. The patient subsequently reported a great reduction in his pain and suffering. Spiller and Martin wrote:

> On January 22 there was great relief of pain on both lower limbs. The patient appeared very grateful for the relief from suffering and received only 1/6 grain of morphine on the day following the operation and a similar amount two days later, and this was given for the pain caused by the operation. The intern on the surgical service, who had the opportunity of observing the man constantly, believed that he was not suffering pain in the lower limbs. Sensation in its various forms objectively tested did not seem to be much impaired above the hip joints. Pain was felt occasionally in the lower limbs during the three weeks following the operation, but the man was positive that it was less than before the operation. (1912, 1490)

The Spiller and Martin paper appeared in the *Journal of the American Medical Association.* The very next article in this journal described how the anterolateral pathways were cut bilaterally in two dogs at Spiller's suggestion. The authors reported that pain (pinprick and crushing) and temperature sensations were abolished or severely blunted by this surgery, whereas touch and pressure were not affected (Cadwalder and Sweet, 1912).

These findings raised serious questions about Frederick Mott's (1895) contention that monkeys with anterolateral column lesions respond normally to heat and pinprick stimulation. Instead, they were more in line with the reports of Max Lewandowsky (1876–1918) and others who claimed that anterolateral cord lesions in the region of the dorsal spinocerebellar tract could produce analgesia in animals (cited in Cadwalder and Sweet, 1912).

These observations were supported by clinical reports on metastatic cancer and tabetic patients (Beer, 1913; Foerster, 1913; Foerster and Gagel, 1932). In general, warm and cold temperature discrimination deficits accompanied their pain losses. These studies also helped to demonstrate the segmental arrangement of fibers within the spinal columns.

Not until 1938 was an attempt was made to cut the spinothalamic tract in the brainstem to relieve pain.[11] Achille Mario Dogliotti (1897–1966), working in Catania, Italy, transected the long pain pathway near the pons-midbrain border in four cancer patients, one of whom died. Two of the others showed an initial loss of pain on one side of the body.

> The postoperative pains disappeared immediately in two patients with a complete secondary hemianalgesia and were greatly reduced in the other patient. A week later the patients had become uneasy, only complaining of some

light unpleasant and undetermined feelings (paresthesias) diffused to the half of the body opposite to the side operated on. No more or very little morphine was used during the months of remaining life. (1938, 145)

The literature following Dogliotti's report seemed to indicate that pain, cold, and warmth were usually affected together by lower brainstem lesions (see Head and Holmes, 1911–12). There were, however, some occasional reports of loss of pain and heat without cold, loss of temperature without pain, and loss of pain without changes in temperature thresholds (Head and Holmes, 1911–12; Kutner and Kramer, 1907; Mann, 1892; Monakow, 1895).

Although the idea of doing cordotomies or related surgical interventions at first seemed worthwhile, it was gradually recognized that a different kind of pain, which was at least as distressing as the original pain for which the surgery was performed, could emerge with the passage of time. This, coupled with the surgical risk, the development of better drugs to treat pain, and often incomplete relief of the original pain, tended to dampen the initial enthusiasm for this sort of surgery.

THALAMIC SYNDROME

A finding that began to receive considerable attention early in the twentieth century was that some brain lesions may result in enhanced responsiveness to pain. Joseph-Jules Dejerine and Gustave Roussy (1874–1948), both born in Switzerland and practicing in France, examined this phenomenon (see Figures 11.8 and 11.9 for portraits of these men). Dejerine, who had been Vulpian's most distinguished pupil, and Roussy, who had been Dejerine's intern, defined

Figure 11.8. Jules-Joseph Dejerine (1849–1917), the neurologist who made many contributions to neurology, including important early studies on *le syndrome thalamique.*

---

[11] This was also the year in which Sjöqvist published the first paper on sectioning the spinal tract of the trigeminal nerve to treat intractable facial pain ("Sjöqvist's operation").

Figure 11.9. Gustave Roussy (1874–1948), Jules-Joseph Dejerine's intern, who studied and defined the thalamic syndrome in his thesis.

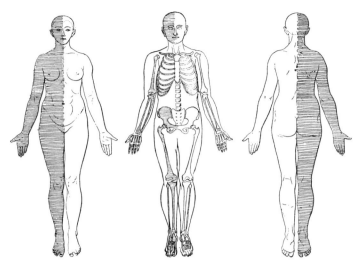

Figure 11.10. Diagram showing thalamic syndrome affecting the right side of the body (shaded) in a 52-year-old woman. This woman suffered from thalamic syndrome for six years. (From Dejerine, 1914.)

their *syndrome thalamique* in a landmark paper published in 1906, one year before Roussy submitted the material for his doctoral dissertation (Guillain, 1970; Zabriske, 1970).

As originally defined by Dejerine and Roussy (1906), thalamic syndrome included the following characteristics:

1. Pain on one side of the body described as aching, burning, crushing or gnawing.

2. Paroxysms of hypersensitivity to cutaneous stimulation, which can prevent an adequate sensory examination.

3. Slightly elevated sensory thresholds.

4. Referred pain over large areas.

5. Evoked pain much longer in duration than the stimulus.

6. Evocation of very disagreeable sensations by touch or temperature stimuli.

7. Absence of associated motor deficits.

Figure 11.10 presents a diagram from Dejerine (1914) showing the parts of the body affected by thalamic syndrome in one of his cases.

In 1911–12, Head and Holmes associated thalamic syndrome with occlusion of the thalamogeniculate artery.[12] They pointed out it was rare to see this syndrome in cases of slow-growing thalamic tumors. Head and Holmes also emphasized there may be blunted appreciation of the sensory characteristics of a stimulus, even though stimuli could hurt more or feel more uncomfortable. After presenting one of their many cases, they wrote:

> If a pin is lightly dragged across the face or trunk from the right to the left she exhibits intense discomfort when it

passes the middle line; she not only calls out that it hurts her more, but her face becomes contorted with pain. She insists that, although the stimulus is more painful, it is "less plain" and "less sharp" than over normal parts; the prick is less distinct but it hurts her more.

> This "hyperalgesia," or over-reaction, would seem to point to a lowered threshold to the prick of a pin. But measured stimulation with both the spring and the weight algesimeters shows that, if anything, the threshold is a little raised; she never responds with certainty over the abnormal (left) half of the body to a stimulus which can evoke a sensation of pricking on similar normal parts to the right of the middle line. And yet, if a measured stimulus of the same strength is applied to similar parts, more pain is evoked over the affected than over the normal half of the body; the same stimulus, provided it is sufficiently strong to cause pain, produces a more uncomfortable sensation on the abnormal side. (Head and Holmes, 1911–12, 128)

Head and Holmes found that just gently scraping such a patient's palm could result in screaming and hand withdrawal. Some of the male patients became so sensitive they even had to stop shaving.

Head (1920) theorized thalamic syndrome occurred because partial destruction of the rapidly conducting epicritic system left the protopathic system unchecked (Chapter 10). In his 1921 Croonian Lecture, Head also discussed the idea that newer structures were responsible for exercising control over older structures. The title of his lecture was "Release of Function in the Nervous System." He told his audience that the idea was not really his own, acknowledging that John Hughlings Jackson's writings had stimulated his interest in the subject. Jackson's classic paper on release of function, "Evolution and Dissolution of the Nervous System," was published in 1884.[13]

[12]More recent studies have suggested that features of thalamic syndrome may be observed after acute lesions which may spare the thalamus, although this situation occurs much less frequently. The commonalty appears to be the interruption of fast pain pathways to higher centers—that is, lesions which leave the slower pain fibers uninhibited (Waltz and Ehni, 1966).

[13]Jackson based his idea on the work of Charles Darwin (1809–1882) and Herbert Spencer (1820–1903); see Chapter 19.

GATE CONTROL THEORY

The idea that large nerve fibers can inhibit the small fibers associated with pain is a central premise of gate control theory (Melzack and Wall, 1965). The concept of gate control was developed in the 1920s by Otfrid Foerster, who in turn attributed it to Brown-Séquard. Foerster (1927, pp. 79–80) wrote the following about one system inhibiting another under normal circumstances and the effects of removing that inhibition on pain:

> The cause of hyperpathia, in my opinion, must be sought in the fact that the more highly differentiated, phylogenetically younger systems of perception exert a moderating influence on the phylogenetically age-old systems; loss of this normal inhibition causes an abnormal increase in excitability of the pain system, of course assuming its own integrity. As especially Brouwer has shown, the long posterior column fibers represent a phylogenetically very young afferent system. . . . We must assume that in the posterolateral tract, possibly close to the so-called marginal zone of the gray matter, there runs a corticofugal pathway which normally exerts an inhibiting influence on the excitability of the cellular elements associated with the pain system in the posterior horns; its abolition leads to an increased excitability of these posterior horn elements. Even Brown-Séquard has already adduced such an explanation for the hyperalgesia on the side of the lesion in spinal hemisection. (Translated by Schiller, 1990, 42)

Foerster also gave some credit to Henry Head, to a little-known Finnish psychiatrist named Harald Fabritius (1877–1946), and to a number of other scientists for their ideas on inhibition.

CEREBRAL CORTEX

The role of the cerebral cortex in pain has never been clear and, in fact, was one of the reasons pain was treated separately from the other senses by some investigators (e.g., Marshall, 1894). Harvey Cushing (1909), for example, who was one of the first to try to stimulate the postcentral gyrus in conscious human patients, did not elicit any responses indicative of pain, although numbness and touch were encountered with some frequency. It is worth noting that Wilder Penfield, who confirmed Cushing's findings a number of years later, also found pain to be a very infrequent response to cortical stimulation (see Penfield and Boldrey, 1937).

As for cortical lesions and pain, it can be reiterated that most authors seemed to agree that pain is the skin sense least affected by cortical lesions (see Chapter 10). Gordon Holmes (1927), for instance, stated he never found raised pain thresholds that persisted long after focal cortical lesions. When pain was affected, Holmes thought it was most likely due to lesions that also involved subcortical sites. Whether Holmes was correct in his clinical assertions or whether the finer aspects of pain discrimination can be affected by cortical lesions is still debated.

EARLY ANESTHETICS

The use of pharmacological agents to treat pain and suffering appears to be as old as written history. For example, the power of henbane to treat pain was described on an Ancient Babylonian clay tablet found at the city of Nippur. This tablet is estimated to be over 4,000 years old (Keys, 1963). The juice of the poppy has also been in use for several thousand years (Brau, 1968; Scott, 1969). In Greece, Xenophon (c. 430 to after 355 B.C.) mentioned the medical use of opium, and his contemporary Socrates (c. 470–399 B.C.), was certainly aware of its sedative powers (Davison, 1965). Galen considered opium foremost among the antidotes to treat pain and the strongest of the agents that can dull sensitivity.

The root of the mandragora, a flowering plant in the nightshade family, was another pharmacological agent used to treat pain at this time. Pedanius Dioscorides (fl. c. A.D. 50), a Greek physician, described the properties of mandrake root, along with those of some 600 medicinal plants, in *De materia medica*. This early pharmacological book achieved considerable popularity and remained in use for some 16 centuries.

In the same century, Pliny the Elder (c. 23–79), who also considered mandrake root a dangerous poison, described how to gather and put small amounts of the root in wine for use as a narcotic prior to "cuttings and puncturings, lest they should be felt." Dioscorides administered mandrake root boiled in wine prior to surgery. Even Galen and Celsus spoke of the root's medicinal properties.

The Greeks and Romans also relied on other plants to deaden pain and knew they could also produce some degree of surgical anesthesia by compressing the carotid artery. The effects of arterial compression, however, were short-lived, and this forced surgeons to hurry through their operations (Walker, 1951).

A number of topical pain-killing agents were also known to the Greeks and Romans. For example, both Dioscorides and Pliny mentioned that local anesthesia was produced when the stone memphitis was powdered and mixed with vinegar and then placed on the skin (see Morton, 1859). Because the stone was made of carbonate, the acetic acid from the vinegar caused it to release carbon dioxide, which dulled the sensations from the area. It was described as making the area insensible to pain for cutting or cauterization.

In South America, the chewing of coca leaves to affect sensation has been known for centuries (Keys, 1963; Walker, 1967). Cocaine, however, was not isolated from coca leaves until 1860. This was accomplished by Albert Niemann (1831–1917). In 1884, William Halsted (1852–1922), working in New York, injected cocaine to block nerve trunks.[14]

Attempts at gaseous anesthetics can be traced back to at least the ninth century, at which time a recipe for a soporific sponge *(sponga somnifera)* appeared in the Bamberg *Antidotarium*. Similar recipes were found in a *Monte*

---

[14]Novocaine itself was not synthesized until 1904.

Carlo Codex of the same period and in the *Antidotarium* of Nicholas of Salerno in the twelfth century.

The record shows that Ugo, or Hugh of Lucca (c. 1250), developed an anesthetic sponge. It was saturated with opium and mandrake and dried in the sun until needed. The surgeon was instructed to soak the sponge in hot water and then hold it under the patient's nose during surgery. Hugh himself left no known written records, but Theodorico Borogognoni (1205–1296), a surgeon and possibly his son, described Hugh's method of obtaining anesthesia.[15]

Theodorico used the soaked sponge technique for amputations. A related mode of anesthesia was used by Guy de Chauliac (c. 1298–1368). Nevertheless, these sponges did not achieve wide acceptance. One reason might be that poor control over the ingredients could have caused many deaths (Keys, 1963; Walsh, 1919). Another idea is that the sponges were not particularly effective.[16] It is also possible that some surgical pain was considered beneficial.

During the Renaissance, most techniques used to deaden pain were fairly similar to those used before this time. Mandrake root, cooling, opium, and alcoholic beverages were still popular. Some surgeons, including Ambroise Paré, also attempted to deaden nerves with compression (Davison, 1965; Keys, 1963).

THE SURGICAL USE OF NITROUS OXIDE

Although many of the aforementioned agents and techniques were of some help in controlling pain, they were hopelessly inadequate for surgeries that required an extended period of time. For many operations, patients either had to be tied down or held down by strong attendants. Even the pain of having a tooth extracted was unbearable for many people. In fact, surgeons were judged largely on speed (i.e., the ability to do amputations in less than a minute or two) and the ability to work on moving targets. Major operations involving body parts like the chest were virtually impossible before the advent of inhalation anesthesia in the mid-nineteenth century.

The story behind modern gaseous anesthetics is often traced to Sir Humphry Davy (1778–1829), who was invited to work on therapeutic inhalations by Thomas Beddoes (1754–1805). Pneumatic medicine was a fad in England in the 1790s, and Beddoes was a leader in this movement (Duncam,

1947). Davy was only 20 years old when he noticed that breathing nitrous oxide, a gas discovered by Joseph Priestley (1733–1804) in 1772, made him insensible. He conducted a number of additional tests on this gas, which seemed highly pleasurable in lower doses, and suggested it might be useful in surgery.

Davy's suggestion seemed to have no immediate effect on the surgical community. Nevertheless, laughing gas parties became popular among the students, and itinerant chemists began to give public demonstrations with the gas. The entertainment and commercial value of the gas was recognized by "Professor" Gardner Quincy Colton (1814–1898), who demonstrated it to a large audience in Hartford, Connecticut, in the winter of 1844. A subject who was under the exhilarating influence of the gas injured himself by running into some furniture on the stage, but seemed strangely unaware of any pain. Horace Wells (1815–1848), a young dentist with a practice in Hartford, noticed this and thought that perhaps the gas could be used to perform operations without pain. He then volunteered to join the inhalers on the stage, much to his wife's chagrin.

The next day, Wells decided to have his own decayed molar extracted under nitrous oxide. "Professor" Colton, the itinerant chemist who had given the public demonstration, administered the gas while another dentist performed the extraction. Wells did not feel any pain and was so impressed that he began to use the gas with some of his own dental patients.

Wells wanted to make his discovery public and convinced Dr. John C. Warren (1842–1927), a leading American surgeon, to allow him to demonstrate his new anesthetic at Massachusetts General Hospital. A man who was going to have a tooth extracted was chosen for the demonstration and was put under the anesthetic in January 1845. Unfortunately, the gas bag was withdrawn too soon. The man cried out as his tooth was pulled, although he later testified that the pain was less than usual for this operation. Nevertheless, as Wells put it, "several expressed their opinion that this was a humbug affair" (see Pullen, 1979, p. 88). Wells left the hospital in disgrace, not realizing that at least some of the doctors, including Henry Bigelow (1818–1890), thought he had proved his case.[17]

Wells shared his experience with a number of doctors and dentists after returning to Hartford and, by the summer of 1845, nitrous oxide was being used for dental surgery in several cities in Connecticut and Massachusetts (Bigelow, 1846). By 1847, Wells was serving as a general anesthetist, administering nitrous oxide for much more difficult operations, such as removing a thigh.

THE SURGICAL USE OF ETHER

The story behind the early use of ether in surgery is closely related to that of nitrous oxide. In the sixteenth century, Paracelsus mixed sulfuric acid with alcohol and distilled his mixture. In doing so, he formed "sweet vitriol." He noted that chickens taking this substance fell asleep, but awak-

---

[15]The following statement, attributed to Theodorico Borogognoni, is found in a book published in 1498:

> The preparation of a scent for performing surgical operations, according to Master Hugo. It is made of thus: take of opium and the juice of unripe mulberry, of hyoscyamus, of the juice of the hemlock, of the juice of the leaves of the mandragora, of the juice of the woody ivy, of the juice of the forest mulberry, of the seeds of lettuce, of the seed of the burdock, which has large and round apples, and of the water hemlock, each one ounce; mix the whole of these together in a brazen vessel, and then in it place a new sponge, and let the whole boil, as long as sun on the dog days, till it (the sponge) consumes it all, and let it be boiled away in it. As often as there is need of it, place this same sponge into warm water for one hour, and let it be applied to the nostrils till he who is to be operated on has fallen asleep; and in this state let the operation be performed. When this is finished, in order to rouse him, place another dipped in vinegar, frequently to his nose, or let the juice of the roots of fenigreek be squirted into his nostrils. Presently he awakens. (Translated in Morton, 1859, 5)

[16]Tests made with similar ingredients have led some contemporary investigators to conclude that these devices were no more effective than placebos (O'Connor and Walker, 1967).

[17]Bigelow achieved considerable fame for his involvement with the Phineas Gage "crowbar" case. This case is discussed in Chapters 19 and 22.

Figure 11.11. Crawford W. Long (1815–1878), a pioneer in the use of ether for general anesthesia. Long worked in Georgia during the 1840s.

ened unharmed. Paracelsus wrote that sweet vitriol "quiets all suffering without any harm, and relieves all pain, and quenches all fevers, and prevents complications in all illnesses" (translated in Keys, 1963, p. 9). Sweet vitriol, or "ether," was also described by Valerius Cordus (1515–1544), who was working with Paracelsus around 1540. Whether it was known before the sixteenth century is unclear, but it is a distinct possibility (Morton, 1859).

Ether was found to produce effects somewhat like those of nitrous oxide early in the 1800s, and "ether frolics" also became popular entertainment (Davison, 1965). William E. Clarke, a medical student in Rochester, New York, held ether parties and, in January 1842, also tried to use ether for a tooth extraction. Unfortunately, Clarke, who may have been the first to employ ether in surgery, did not publish his results.

In Georgia, Crawford W. Long (1815–1878), a young physician who attended his share of ether frolics, heard a story about how a young slave at an ether party given in 1839 in South Carolina was temporarily rendered unconscious. Long, shown later in life in Figure 11.11, noticed that one of the people at a party he now attended had two small tumors on his neck (Davison, 1965). He suggested to the man that he might want to have the tumors removed while made unconscious by ether. The man (James Venable) gave his consent. On March 30, 1842, Long saturated a towel with ether and had the man breathe. Long successfully removed one tumor without pain and the second with slight pain (Long, 1849; Young, 1942). He later wrote:

He gave no evidence of suffering during the operation, and assured me, after it was over, that he did not experience the slightest degree of pain from its performance. . . . The second operation I performed upon a patient was on the 6th June, 1842, and was on the same person, for the removal of another small tumor. . . . The patient was insensible to pain

during the operation, until the last attachment of the cyst was separated, when he exhibited signs of slight suffering, but asserted, after the operation was over, that the sensation of pain was so slight as scarcely to be perceived. (1849, 708)

Long subsequently used ether to conduct other operations, including amputation of toes and fingers. He waited until 1849 before he wrote an article on his experiences with sulphuric ether. Long's manuscript included the following affidavit from his first patient, the man with the neck tumors:

I commenced inhaling the ether before the operation was commenced, and continued it until the operation was over. I did not feel the slightest pain from the operation, and could not believe the tumour was removed until it was shown to me. (Venable, 1849; see Long, 1849, 709)

The patient explained that the slight pain he felt when the second tumor was removed occurred because he stopped inhaling the ether from the towel before the operation was completed.

Long wrote that his application of ether in 1842 would not have gone unnoticed had he lived in a major city in which physicians assisted each other and shared their experiences. He stated that he felt hesitant to present his findings in 1842 because he wanted to ensure they were not due to "the imagination, or owing to any peculiar insensibility to pain" (Long, 1849, p. 710). In addition, Long complained of his laborious practice, which infringed upon the time he needed for making his findings public.

In 1846, before Crawford Long's findings became known, William T. G. Morton (1819–1868) began to experiment with chloric and sulfuric ethers in his Boston dental practice. Morton took the idea from his friend Charles T. Jackson (1805–1880), who told him he had tried inhaling ether to treat his own sore throat and had become unconscious in the process. Jackson later stated that when he woke up, he immediately knew he had found a way to reduce pain during surgery. Jackson wrote a manual on etherization in 1861, but Morton (1859) later said that whereas Jackson knew of ether's ability to cause sleep, he did not anticipate the effects of ether on pain.

At the time, both Jackson and Morton were friends of Horace Wells, who had visited them early in 1845 when he had gone to Boston to demonstrate nitrous oxide anesthesia. Wells informed both men of his experiences with nitrous oxide, and Morton even assisted Wells with his seemingly ill-fated demonstration at Massachusetts General Hospital.

After some successes with his dental patients, Morton persuaded John C. Warren to try ether for removal of a tumor. Warren was the surgeon who had been present in the operating arena of Massachusetts General Hospital early in 1845 when Horace Wells tried to show the benefits of nitrous oxide in surgery.

Morton arrived late on October 16, 1846 because he decided to add valves to his apparatus. Some members of the audience believed he would not show up at all. They then saw him administer the ether and tell Warren that his patient was ready for surgery. The doubting audience next saw Warren remove a tumor from the man's lower jaw without pain. The impact on everyone was tremendous. Warren faced his audience and dramatically announced, "Gentlemen, this is no humbug" (Pullen, 1979).

Over the next few weeks, Morton used ether to perform a number of more difficult operations, including a leg amputation. The news of his successes spread rapidly, and the operating arena was soon being called the "ether dome." Oliver Wendell Holmes (1809–1894), at Harvard University, wrote to Morton proposing the use of the words "anaesthesia" and "anesthetic" to describe the loss of sensitivity and the drugs that induce this state, respectively. These terms, derived from the Greek, quickly caught on.

Credit for the discovery of gaseous anesthetics, however, has remained a matter of debate. The reason is that the discovery involved a number of different men, controversies over the success of their efforts, and trials not always made public soon after they were run (e.g., Colton, 1896; Morton, 1859; Pullen, 1979). The fights for fame, honor, and fortune among Horace Wells, William Morton, Crawford Long, and Charles Jackson left some of these men bitter and in financial ruin.[18] Yet one thing could not be challenged. It was that the discovery of inhalation anesthetics in the 1840s signaled the end to those 60-second "slapdash" operations on frightened "moving targets" that had characterized surgery up to this time.[19]

---

[18]Major cash prizes from governments and medical societies were common at this time. The competition was often extremely intense, and scientists often used their own financial resources to publicize and lobby for these prizes. Horace Wells went insane and committed suicide; only Crawford Long seemed to live a contented life in the aftermath of these discoveries (Young, 1942).

[19]There remained a belief that the different races were not equally sensitive to pain. In 1864, Carl Christoph Vogt (1817–1895) expressed the opinion that the Negro stood far below the Caucasian when it came to acuteness of the senses, even though some of Vogt's contemporaries clearly felt the senses might be even more highly developed among the "savages" in the wild. Vogt (1864, p. 188) quoted an 1863 paper by Franz Pruner (Pruner-Bey; 1808–1882), a member of Paul Broca's *Société d'Anthropologie* and a physician to the viceroy of Egypt, as follows:

> The most remarkable phenomenon relates to the Negroes apparent insensibility to pain. We have never seen the least spontaneous expression of pain. . . . Mishaps, or bad treatment will draw from the Negress, the child and even the adult Negro, an abundant flow of tears, but physical pain never.

REFERENCES

Abbe, R. 1889. A contribution to the surgery of the spine. *Medical Record, 35,* 149–152.

Avicenna. 1930. *The Canon of Medicine.* (O. C. Gruner, Trans.). London: Luzak & Company. (Originally printed 1025.)

Barnes, R. 1953. The role of sympathectomy in the treatment of causalgia. *Journal of Bone and Joint Surgery, 35B,* 172–180.

Beer, E. 1913. The relief of intractable and persistent pain due to metastases pressing on nerve plexuses by section of the opposite anterolateral column of the spinal cord above the entrance of the involved nerves. *Journal of the American Medical Association, 60,* 267–269.

Bekhterev, V. M. 1899. *Les Voies de Conduction du Cerveau et le la Möelle.* Paris: Octave, Doin.

Bell, C. 1811. *Idea of a New Anatomy of the Brain; Submitted for the Observations of His Friends.* London: Strahan & Preson. (Reprinted in *Medical Classics,* 1936, *1,* 105–120.)

Bennett, W. H. 1889. A case in which acute spasmodic pain in the lower extremity was completely relieved by sub-dural division of the posterior roots of certain spinal nerves, all other treatment having proved useless. Death from sudden collapse and cerebral haemorrhage on the twelfth day after the operation, at the commencement of apparent convalescence. *Medico-Chirurgical Transactions, 72,* 329–348.

Bezold, A. von 1858. Ueber die gekreuzten Wirkungen des Rückenmarkes. *Zeitschrift für wissenschaftliche Zoologie, 9,* 307–364.

Bigelow, H. J. 1846. Insensibility during surgical operations produced by inhalation. *Boston Medical and Surgical Journal, 35,* 309–317.

Brau, J. L. 1968. *Histoire de la Drogue.* Paris: Tchou.

Brown-Séquard, C.-E. 1846. *Recherches et Expériences sur la Physiologie de la Moelle Epinière.* Paris: Rignoux.

Brown-Séquard, C.-E. 1860. *Course of Lectures on the Anatomy and Physiology of the Central Nervous System.* Philadelphia, PA: Collins.

Brown-Séquard, C.-E. 1868. Nouvelle recherches sur le trajet des diverses espèces de conducteurs d'impressions sensitives dans la moelle épinière. *Archives de Physiologie Normale et Pathologique, 1,* 610–621, 716–724.

Brown-Séquard, C.-E. 1869. Nouvelle recherches sur le trajet des diverses espèces de conducteurs d'impressions sensitives dans la moelle épinière (troisième partie). *Archives de Physiologie Normale et Pathologique, 2,* 236–250.

Bruce, A. 1898. Note on the upper terminations of the direct cerebellar and ascending antero-lateral tracts. *Brain, 21,* 374–387.

Cadwalder, W. B., and Sweet, J. E. 1912. Experimental work on the function of the anterolateral columns of the spinal cord. *Journal of the American Medical Association, 58,* 1490–1493.

Chauveau, A. 1857. Nouvelle étude expérimentale des propriétés de la moelle épinière. *L'Union Médicale, 11,* 250ff.

Collier, J., and Buzzard, E. F. 1903. The degeneration resulting from lesions of the posterior roots and from transverse lesions of the spinal cord in man. A study of twenty cases. *Brain, 26,* 559–591.

Colton, G. C. 1896. *A True History of the Discovery of Anaesthesia.* New York: A. G. Sherwood & Company.

Cushing, H. 1909. A note upon the faradic stimulation of the postcentral gyrus in conscious patients. *Brain, 32,* 44–53.

Dallenbach, K. M. 1939. Pain: History and present status. *American Journal of Psychology, 52,* 331–347.

Darwin, E. 1796. *Zoonomia.* New York: T. & J. Swords.

Davison, M. H. A. 1965. *The Evolution of Anesthesia.* Baltimore, MD: Williams & Wilkins.

Deen, I. van. 1841. *Traités et Découvertes sur la Physiologie de la Moelle Epinière;* Leyden: S. & J. Luchtman.

Dejerine, J.-J. 1914. *Sémiologie des Affections du Système Nerveux.* Paris: Masson et Cie.

Dejerine, J.-J., and Roussy, G. 1906. Le syndrome thalamique. *Revue Neurologique, 12,* 521–532.

Denmark, A. 1813. An example of symptoms resembling tic douloureux produced by a wound in the radial nerve. *Medico-Chirurgical Transactions, 4,* 48–52.

Descartes, R. 1644. *The Philosophical Works: Vol. 1. Principles of Philosophy.* (E. Haldane and G. R. T. Ross, Trans.). New York: Dover Publications.

Dogliotti, A. M. 1938. First surgical sections in man of the lemniscus lateralis (pain and temperature path) at the brain stem, for the treatment of diffused rebellious pain. *Current Researches in Anesthesia and Analgesia, 17,* 143–145.

Duncam, B. M. 1947. *The Development of Inhalation Anesthesia.* London: Oxford University Press.

Edinger, L. 1889. Verleichend-entwicklungsgeschichtliche und anatanomische Studien im Bereiche des Centralnervensystems. II. Ueber die Fortsetzung der hinteren Rückenmarkswurzen zum Gehirn. *Anatomischer Anzeiger, 4,* 121–128.

Edinger, L. 1890. Einiges vom Verlauf der Gefühlsbahnen im centralen Nervensysteme. *Deutsche Medicinische Wochenschrift, 26,* 421–426.

Foerster, O. 1913. Vorderseitenstrangdurchschneidung im Rückenmark zur Beseitgung von Schmerzen. *Berliner klinische Wochenschrift, 50,* 1499.

Foerster, O. 1927. *Die Leitungsbahnen des Schmerzgefühls und die chirurgische Behandlung der Schmerzzustände.* Berlin: Urban and Schwartzenberg.

Foerster, O., and Gagel, O. 1932. Die Vorderseitenstrangdurchschneidung beim Menchen. Eine Klinische-patho-physiologisch-anatomische Studie. *Zeitschrift für die gesamte Neurologie und Psychiatrie, 138,* 1–29.

Frazier, C. H. 1928. Operation for the radical cure of trigeminal neuralgia. *Archives of Neurology and Psychiatry, 13,* 378–384.

Frey, M. von. 1894. Beiträge zur Physiologie des Schmerzsinns. *Sächsischen Akademie der Wissenschaften zu Leipzig. Math-Phys. Cl., 46,* 185–196, 283–296.

Frey, M. von. 1922a. Versuche über schmerzerregende Reize. *Zeitschrift für Biologie, 76,* 1–24.

Frey, M. von. 1922b. Verspätete Schmerzempfindungen. *Zeitschrift für die gesamte Neurologie und Psychiatrie, 79,* 324–333.

Gad, J., and Goldscheider, A. 1892. Über die Summation von Hautreizen. *Zeitschrift für klinische Medizin, 20,* 339–373.

Goldscheider, A. 1891. Ueber die Summation von Hautreizen (nach gemeinshaftlich mit Hr. Gad angestellen Versuchen). *Archiv für Physiologie, 15,* 164–169.

Goldstein, K. 1910. Ueber die aufsteigende Degeneration nach Querschnittsunterbrechung des Rückenmarks (Tractus spino-cerebellaris posterior, Tractus spino-olivaris, Tractus spino-thalamicus). *Neurologisches Centralblatt, 29,* 898–911.

Gowers, W. R. 1878. A case of unilateral gunshot injury to the spinal cord. *Transactions of the Clinical Society of London, 11,* 24–32.

Gowers, W. R. 1886. *Manual of Diseases of the Nervous System.* London: Blakiston.

Gruner, O. C. 1930. *A Treatise on the Canon of Medicine of Avicenna.* London: Luzac & Company.

Guillain, G. 1970. Gustave Roussy, 1874–1948. In W. Haymaker and F. Schiller (Eds.), *The Founders of Neurology* (2nd ed.). Springfield, IL: Charles C Thomas. Pp. 510–512.

Hamilton, J. 1838. On some effects resulting from wounds of the nerves. *Dublin Journal of Medical Science, 13,* 38–55.

Hartley, D. 1834. *Observations of Man, His Fame, His Duty, His Expectations* (6th Ed.). London: Thomas Tegg. (First published 1749)

Haymaker, W. 1970. Weir Mitchell, 1829–1914. In W. Haymaker and F. Schiller (Eds.), *The Founders of Neurology.* Springfield, IL: Charles C Thomas. Pp. 479–484.

Head, H. 1920. *Studies in Neurology.* London: Oxford University Press.

Head, H. 1921. Release of function in the nervous system. *Proceedings of the Royal Society, B, 92,* 184–208.

Head, H., and Holmes, G. 1911–1912. Sensory disturbances from cerebral lesions. *Brain, 34,* 102–254.

Head, H., and Thompson, T. 1906. The grouping of afferent impulses within the spinal cord. *Brain, 29,* 537–741.

Hippocrates. *Aphorisms.* (W. H. S. Jones, Trans.). Cambridge, MA: Harvard University Press, 1959.

Holmes, G. 1927. Disorders of sensation produced by cortical lesions. *Brain, 50,* 413–427.

Horsley, V., Taylor, J., and Colman, W. S. 1891. Remarks on the various surgical procedures devised for the relief or cure of trigeminal neuralgia (tic douloureux). *British Medical Journal, 2,* 1139–1143, 1191–1193, 1249–1252.

Illingworth, R. 1982. Historical concepts of spinal cord pathways subserving pain. In F. C. Rose and W. F. Bynum (Eds.), *Historical Aspects of the Neurosciences.* New York: Raven Press. Pp. 209–212.

Jackson, C. T. 1861. *A Manual of Etherization. . . .* Boston, MA: J. B. Mansfield.

Jackson, J. H. 1884. Evolution and dissolution of the nervous system. *British Medical Journal, i,* 591ff. Reprinted in J. Taylor (Ed.), *Selected Writings of John Hughlings Jackson* (Vol. 2). New York: Basic Books, 1958. Pp. 45–75.

Keele, K. D. 1957. *Anatomies of Pain.* Oxford, U.K.: Blackwell.

Keele, K. D. 1962. Some historical concepts of pain. In C. A. Keele and R. Smith (Eds.), *The Assessment of Pain in Man and Animals.* London, E. & S. Livingstone Ltd. Pp 12–27.

Keen, W. W. 1914. Tribute to S. Weir Mitchell, 1829–1914. *Transactions of the College of Physicians, Philadelphia, 3rd Ser.,* 345–349.

Keys, T. E. 1963. *The History of Surgical Anesthesia.* Huntington, NY: Robert E. Krieger Publishing Company.

Kiesow, F. 1928. The problem of the condition of arousal of the pure sensation of cutaneous pain. *Journal of General Psychology, 1,* 199–212.

Kiesow, F. 1930. F. Kiesow. In C. Murchison (Ed.), *A History of Psychology in Autobiography.* Worcester, MA: Clark University Press.

Krause, F. 1893. The question of priority in devising a method for the performance of intracranial neurectomy of the fifth nerve. *Annual of Surgery, 18,* 362–364.

Kutner, R., and Kramer, F. 1907. Sensibilitätsstörungen bei acuten und chronischen Bulbärerkrankungen. *Archiv für Psychiatrie und Nervenkrankheiten, 42,* 1002-1060.

Leriche, R. 1940. *La Chirurgie de la Douleur* (2nd ed.). Paris: Masson et Cie.

Létiévant, J.-J. E. 1873. *Traité des Sections Nerveuses: Physiologie Pathologie, Indications, Procédés Opératoires.* Paris: J. B. Ballière.

Long, C. 1849. An account of the first use of sulphuric ether by inhalation as an anaesthetic in surgical operations. *Southern Medical and Surgical Journal, 5,* New Ser., 705–713.

Lucretius Carus, T. 1937. *De rerum natura.* (W. H. D. Rouse, Trans.). London: Loeb Classical Library.

Mann, L. 1892. Casuistischer Beitrage zur Lehre vom central entstehenden Schmerze. *Berliner klinische Wochenschrift, 29,* 244–245.

Marshall, H. R. 1894. Are there special nerves for pain? *Journal of Nervous and Mental Disease, 21,* 71-84.

May, W. P. 1906. The afferent path. *Brain, 29,* 742–803.

McHenry, L. C., Jr. 1969. *Garrison's History of Neurology.* Springfield, IL: Charles C Thomas.

Melzack, R., and Wall, P. D. 1965. Pain mechanisms: A new theory. *Science, 150,* 971–979.

Middleton, W. S. 1966. Turner's Lane Hospital. *Bulletin of the History of Medicine, 40,* 14–42.

Mitchell, J. K. 1895. *Remote Consequences of Injuries of Nerves and Their Treatment.* Philadelphia, PA: Lea.

Mitchell, S. W. 1871. Phantom limbs. *Lippincott's Magazine, 8,* 563–569.

Mitchell, S. W. 1872. *Injuries of Nerves, and Their Consequences.* Philadelphia, PA: Lippincott.

Mitchell, S. W. 1877. The relations of pain to weather, being a study of the natural history of a case of traumatic neuralgia. *Journal of Nervous and Mental Disease, 73,* 370–373.

Mitchell, S. W. 1897. *Clinical Lessons on Nervous Diseases.* Philadelphia, PA: Lea.

Mitchell, S. W., Morehouse, G. R., and Keen, W. W. 1864. *Gunshot Wounds and Other Injuries of Nerves.* Philadelphia, PA: Lippincott.

Monakow, C. von. 1895. Experimentelle und pathologische-anatomische Untersuchungen über die Haubenregion, den Sehhügel und die Regio subthalamica; nebst Beiträgen zur Kenntniss früh erworbener Gross- und Kleinhirn Defecte. *Archiv für Psychiatrie und Nervenkrankheiten, 27,* 1–128.

Morton, W. T. G. 1859. *Sulphuric Ether.* Washington, DC: U.S. House of Representatives.

Mott, F. W. 1892. Ascending degenerations resulting from lesions of the spinal cord in monkeys. *Brain, 15,* 215–229.

Mott, F. W. 1895. Experimental inquiry upon the afferent tracts of the central nervous system of the monkey. *Brain, 18,* 1–20.

Mumey, N. 1938. *S. Weir Mitchell. The Versatile Physician (1829–1914). Sketch of His Life and His Literary Contributions.* Denver, CO: Range Press.

O'Connor, D. C., and Walker, A. E. 1967. Prologue. In A. E. Walker (Ed.), *A History of Neurological Surgery.* New York: Hafner. Pp. 1–22.

Ogle, J. W. 1859. Series of cases (with observations) illustrating the views recently put forward by Dr. Brown-Séquard, as regards certain points connected with the physiology of the nervous system. *British and Foreign Medico-Chirurgical Review, 24,* 363–373.

Olmstead, J. M. D. 1946. *Charles-Edouard Brown-Séquard.* Baltimore, MD: Johns Hopkins University Press.

Olmstead, J. M. D. 1970. Edouard Brown-Séquard (1817–1894). In W. Haymaker and F. Schiller (Eds.), *The Founders of Neurology.* Springfield, IL: Charles C Thomas.

Ornstein, A. M. 1970. William Gibson Spiller, 1863–1940. In W. Haymaker and F. Schiller (Eds.), *The Founders of Neurology.* Springfield, IL: Charles C Thomas. Pp. 520–524.

Paré, A. 1551. *La Méthode de Curer les Combustiones Principalement Faictes par la Pouldre à Canon.* Paris: Jean de Brie.

Penfield, W., and Boldrey, E. 1937. Somatic motor and sensory representation in the cerebral cortex of man as studied by electrical stimulation. *Brain, 60,* 389–443.

Plato. 1952. *Timaeus.* (B. Jowett, Trans.). In *Great Books of the Western World: Vol. 7. Plato.* Chicago: Encyclopaedia Britannica. Pp. 442–477.

Pruner-Bey, F. 1863. *De la Chevelure comme Charactéristique des Races Humaines d'après des Recherches Microscopiques.* Paris. (Reprinted from *Mémoires de la Société d'Anthropologie de Paris)*

Quensel, F., and Kohnstamm, O. 1907. Praparate mit aktiven Zelldegenerationen nach Hirnstammverletzung bei Kaninchen. *Neurologisches Centralblatt, 26,* 1138–1139.

Richards, R. L. 1967. Causalgia: A centennial review. *Archives of Neurology, 16,* 339–350.

Rosenbach, O. 1884. Über die unter physiologischen Verhältnissen zu beobachtende Verlangsamung der Leitung von Schmerzempfindungen bei Anwendung von thermischen Reizen. *Deutsches Medizinische Wochenschrift, 10,* 338–341.

Russell, J. S. R. 1898. Contributions to the study of some of the afferent and efferent tracts in the spinal cord. *Brain, 21,* 145–179.

Schiff, M. 1858. *Lehrbuch der Physiologie des Menschen. Muskel-und Nervenphysiologie.* Lahr: Schauenburg.

Schiller, F. 1984. Coenesthesis. *Bulletin of the History of Medicine, 58,* 496–515.

Schiller, F. 1990. The history of algology, algotherapy, and the role of inhibition. *History and Philosophy of Life Sciences, 12,* 27–50.

Schiller, F., and Haymaker, W. 1970. Moritz Schiff, 1823–1896. In W. Haymaker and F. Schiller (Eds.), *The Founders of Neurology.* Springfield, IL: Charles C Thomas. Pp. 258–264.

Schüller, A. 1910. Über operative Durchtrennung der Rückenmarkestrange (Chordotomie). *Wien Medizinische Wochenschrift, 60,* 2291–2296.

Scott, J. M. 1969. *The White Poppy.* London: Heinemann.

Siegel, R. E. 1970. *Galen on Sense Perception.* Basel: S. Karger.

Sjöqvist, O. 1938. Studies on pain conduction in trigeminal nerve: Contributions to surgical treatment of facial pain. *Acta Psychiatrica (Kobenhaven), Suppl. 17,* 1–139.

Slessor, A. J. 1948. Causalgia: A review of 22 cases. *Edinburgh Medical Journal, 55,* 563–571.

Spillane, J. D. 1981. *The Doctrine of the Nerves.* Oxford, U.K.: Oxford University Press.

Spiller, W. G. 1905. The location within the spinal cord of fibers for temperature and pain sensations. *Journal of Nervous and Mental Disease, 32,* 318–320.

Spiller, W. G., and Martin, E. 1912. The treatment of persistent pain of organic origin in the lower part of the body by division of the anterolateral columns of the spinal cord. *Journal of the American Medical Association, 58,* 1489–1490.

Stewart, P. 1901. Degeneration following a traumatic lesion of the spinal cord, with an account of a tract in the cervical region. *Brain, 24,* 222–237.

Stone, T. T. 1950. Phantom limb pain and central pain. *Archives of Neurology and Psychiatry, 63,* 739–748.

Strong, C. A. 1895. The psychology of pain. *Psychological Review, 2,* 329–347.

Thiele, F. H., and Horsley, V. 1901. A study of the degeneration observed in the central nervous system in a case of fracture dislocation of the spine. *Brain, 24,* 519–531.

Thunberg, T. 1902. Untersuchungen über die bei einer momentanen Hautreizung auftretended zwei stechenden Empfindungen. *Skandinavisches Archiv für Physiologie, 12,* 394–442.

Tooth, H. G. 1892. On the destination of the antero-lateral ascending tract. *Brain, 15,* 397–402.

Vogt, C. C. 1864. *Lectures on Man: His Place in Creation, and in the History of the Earth.* London: Longman, Green, Longman, and Roberts.

Walker, A. E. 1940. The spinothalamic tract in man. *Archives of Neurology and Psychiatry, 43,* 284–298.

Walker, A. E. 1951. *A History of Neurological Surgery.* New York: Hafner.

Walsh, J. J. 1919. Two chapters in the history of laryngology and rhinology. *Annals of Medical History, 2,* 23–33.

Waltz, T. A., and Ehni, G. 1966. The thalamic syndrome and its mechanism. *Journal of Neurosurgery, 24,* 735–742.

Weber, E. H. 1846. Der Tastsinn und das Gemeingefühl. In R. Wagner (Ed.), *Handwörterbuch der Physiologie* (Vol. 3). Braunschweig: F. Vieweg. Pp. 481–588.

White, J. C., and Sweet, W. H. 1969. *Pain and the Neurosurgeon.* Springfield, IL: Charles C Thomas.

Young, H. H. 1942. In commemoration of the one hundredth anniversary of the first application of ether anesthesia. *Bulletin of the History of Medicine, 12,* 191–225.

Zabriske, E. G. 1970. Joseph-Jules Dejerine, 1849–1917. In W. Haymaker and F. Schiller (Eds.), *The Founders of Neurology.* Springfield, IL: Charles C Thomas. Pp. 426–430.

Zotterman, Y. 1933. Studies in the peripheral nervous mechanism of pain. *Acta Medica Scandinavica, 80,* 185–242.

Zotterman, Y. 1939. The nervous mechanism of touch and pain. *Acta Psychiatrica et Neurologica, 14,* 91–97.

Chapter 12
Gustation

Among the most important facts in the physiology of taste is the observation that none of the four qualities, sweet, acid, salt, bitter, can be subdivided into components. Each taste is an elementary quality that exhibits only qualitative differences; it is not possible to pass gradually from one taste to another, as in the colours of the solar spectrum and the tones of the musical scale; as yet we have no rational criterion for arranging the four qualities in a given order in the series.

Luigi Luciani, 1917

EARLY CONCEPTIONS OF TASTE

A number of early Greek philosophers and naturalists discussed the sense of taste and the mechanisms thought to underlie it. One of the earliest was Alcmaeon, who lived in the fifth century B.C. He hypothesized that the tongue allowed taste particles to enter the sensorium through small pores on its surface.

It is with the tongue we discover tastes. For this being warm and soft dissolves the sapid particles by its heat, while by the porousness and delicacy of its structure it admits them into its substance and transmits them to the sensorium. (Translated in Keele, 1957, 16)

Alcmaeon thought hollow tubes were essential for the passage of the elements to the sensorium and that the sensorium was located in the brain—a position that hardly obtained unanimous agreement among the Greek philosophers.

Democritus, who was a contemporary and friend of Hippocrates, thought that sensation in general had much to do with the shape of small particles. As stated by Theophrastus:

Democritus investigating each taste with its characteristic figure makes the sweet that which is round and large in its atoms; the astringently sour that which is large in its atoms, but rough, angular, and not spherical; the acid, as the name imports, that which is sharp in its bodily shape, angular, and curving, thin, and not spherical; the pungent that which is spherical, thin, angular, and curving; the saline, that of which the atoms are angular, and large, and crooked and isosceles; the bitter, that which is spherical, smooth, scalene, and small. The succulent is that which is thin, spherical, and small. (1906 translation, 164)

Thus, Democritus may have been the first to propose a stereochemical theory for the chemical senses (taste and smell). With regard to the sensorium, he, like Alcmaeon, pointed to the brain.

A variation on this theme was accepted by Plato, who believed the sensation of taste was mediated by the heart, leaving the brain free for loftier functions. Plato wrote in *Timaeus* that small bodies entered the veins of the tongue and that different tastes had something to do with the shape of these bodies and how they affected the small veins that went to the heart.

Aristotle placed his emphasis in a somewhat different direction and was among the first to try to list the basic tastes. In *De Anima,* he included sweet, bitter, sour, and salty on his list; these were the four "primaries" that would be accepted by most taste researchers in the twentieth century. Aristotle's list, however, did not end with these four tastes. It went on to include astringent, pungent, and harsh (see Bartoshuk, 1978).

Aristotle believed taste involved the transfer of qualities from the object to the tongue. The qualities were then carried by the blood to the heart, which was responsible for sensory experience. The brain, in contrast, was not considered the organ of perception by Aristotle or his faithful followers (see Chapter 1).

As Roman civilization ascended, the tendency was simply to accept the basic tastes proposed by Aristotle. Galen, for instance, who viewed taste as a more reliable sensory system than smell, did not modify Aristotle's list appreciably. But, at the same time, he indicated that the number of the gustatory qualities was uncertain and that there was nothing special about Aristotle's magic number seven.[1]

Galen also accepted Aristotle's idea that the elements (corpuscles) which stimulated the gustatory system were coarser than those that stimulated the olfactory system. In contrast, he did not accept Aristotle's notion that qualities from the object were transferred to the tongue and then carried to the heart (see Chapter 1).

Galen emphasized that only a moist tongue could function normally. Thus, saliva was vital for taste. He thought the saliva came from the humors in the blood vessels and

---

[1]Galen's follower, Nemesius of Emesa (fl. 390), added oily to Aristotle's sweet, sour, bitter, salty, harsh, pungent, and astringent.

Figure 12.1. Jean Fernel (1496–1558) devised a taste classification system based upon Aristotle's ideas.

that the salivary glands served as intermediaries in the process of keeping the tongue moist.[2]

Galen made a number of other contributions in the domain of gustation. In particular, he attempted to identify the nerves innervating the tongue. In modern terms, he singled out (1) a combination of nerves (his number VII), which included the glossopharyngeal nerve (IX); (2) the lingual branch of the trigeminal nerve (V); and (3) the hypoglossal nerve (XII). Galen thought the glossopharyngeal nerve was gustatory, an idea accepted during his lifetime and afterward. In contrast, he mistakenly believed the lingual branch of the trigeminal nerve was gustatory. As for the hypoglossal nerve, it was thought to be important for movements of the tongue, not for taste. Galen wrote that movements of the tongue were abolished when it was damaged, whereas taste was unaffected (Siegel, 1970).

Avicenna, the eleventh-century Persian physician-philosopher, accepted the idea that substances must be soluble to be tasted. He wrote that "the sense of taste detects soluble nutriments in those objects which come in contact with the tongue, discriminating between sweet, bitter, sharp, sour, etc." (p. 135). Interestingly, he considered smell "a very delicate kind of taste."

The first major work dealing solely with taste was *De sapore dulci et amaro,* by Laurentius Gryllus (1484–1560). This largely philosophical treatise was published in Prague

in 1566. It appeared only a few years before Jean Fernel (1497–1558; shown in Figure 12.1) published his *De naturali parte médicine* (1542), the first Western book on human physiology since Galen's time (McHenry, 1969; Sherrington, 1946). Fernel, who introduced the term "physiology," was also interested in the gustatory system. He took Aristotle's basic classification scheme and added fatty and insipid to the ancient list of seven tastes. The latter was described as "probably not a flavor, but the absence of flavor."

This period of early, largely philosophical ideas about taste was followed by the emergence and very rapid growth of physiology and psychophysics in the nineteenth century. This work would put the study of the sense of taste on considerably more solid scientific ground.

BASIC TASTES

As noted, there were a number of early attempts to list the basic tastes. These included the efforts of Aristotle in classical times, the attempts of Avicenna in the Middle Ages, and the work of Jean Fernel in the sixteenth century (also see Bravo, 1591). All included sour, sweet, salty, and bitter, but then went on to add other primaries to their list. This pattern continued with Carolus Linnaeus (Linné, 1707–1778), the Swedish botanist and naturalist, who included the following in his 1752 classification: sweet, acid, bitter, saline, astringent, sharp, viscous, fatty, insipid, aqueous, and nauseous.

In his *Primae lineae physiologiae,* which first appeared in 1747 and was translated into English in 1786, the Swiss anatomist and physiologist Albrecht von Haller (shown in Figure 12.2) also classifed tastes as sweet, bitter, sour, salty. Like Linnaeus, Haller did not end his classification

Figure 12.2. Albrecht von Haller (1708–1777), the eighteenth-century scientist who described different types of papillae on the tongue, tried to classify tastes and attempted to explain species differences in taste sensitivity.

[2]Galen and other physicians in Rome associated changes in the color of the tongue with disease. Using the doctrine of the humors as a basis, they believed yellow signified too much yellow bile, dark colors indicated too much black bile, and pale colors suggested too much whitish phlegm. They noted that the color of the tongue also seemed to correlate with the color and smell of the urine, which was also important for diagnostics.

with these four tastes. His additions were rough, urinous, spirituous, aromatic, acrid, putrid, and insipid.

Some investigators in the first quarter of the nineteenth century succeeded in reducing their lists to five tastes (e.g., Horn, 1825). Tastes other than these were seen not as "pure" tastes but as mixtures of primary tastes and cutaneous stimuli, or blends of the primaries. In particular, it seemed absurd to discuss Fernel's "insipid" as a primary; this was like giving the name of a color to something invisible. Interestingly, each of these new listings still included sweet, acid (sour), bitter, and salty, the four qualities that would later be accepted as the only primaries.

The chapter on taste written by Maximilian Ritter von Vintschgau (1832–1902), which appeared in *Hermann's Handbuch der Physiologie der Sinnesorgane* in 1880, restricted the list to just sour, sweet, salty, and bitter. Vintschgau believed the sensitivity of each was independent of the others, but also chose these four because he strove for parsimony.

These primaries, however, were not unanimously accepted at this time. Some investigators wanted an even shorter list and suggested there were only two basic taste categories, sweet and bitter. To some extent, this came close to Aristotle's belief that sweet and bitter were the extreme qualities between which the other taste qualities could be placed (Bartoshuk, 1978). Gabriel Valentin (1853), in accord with this position, put sour, salty, sharp, burning, and cool together with touch, leaving sweet and bitter as tastes. This classification was challenged by other investigators. Among other things, they showed the entire mouth was sensitive to touch, but not to sour (Stitch, 1857).

There were also a number of individuals late in the nineteenth century who insisted that four basic tastes were too few. Wilhelm Wundt took sour, sweet, bitter, and salty and added alkaline and metallic to his 1880 and 1887 classifications. He is shown in Figure 12.3.

Figure 12.3. Wilhelm Wundt (1832–1920), the "father of modern psychology," who believed that the primary tastes included alkaline and metallic as well as sweet, sour, bitter, and salty.

Figure 12.4. Hjalmar Öhrwall (Öhrvall; 1851–1929), the Swedish investigator who conducted many important studies on the psychophysics and physiology of the gustatory system. (Courtesy of the Uppsala Universitetsbibliotek.)

Edward Bradford Titchener, Wundt's most famous student, similarly began with the four primaries, but emphasized a fifth taste, vapid. Titchener (1915) considered vapid a salt-sweet taste and the only instance of a mixture resulting in a new taste or real blend. He stressed that taste owed much to the senses of smell and touch in real life. His observations on the blunting of taste when one catches a cold can be found in *A Beginner's Psychology*. Titchener (1915, p. 49) almost certainly minimized the importance of the gustatory system when he added that we "taste" entirely by smell or touch.

Hjalmar Öhrwall (also spelled Öhrvall; 1851–1929), who appears in Figure 12.4, supported the idea of four basic tastes and argued that alkaline and metallic, originally favored by Wundt, were really taste-touch fusions (Öhrwall, 1891; also see Herlitzka, 1908). Perhaps as a result of Öhrwall's stance, Wundt soon began to question alkaline and metallic. Similarly, in 1900, John Haycraft (1857–1922), who had earlier cited astringent as a possible "true taste," now deduced that perhaps touch had something to do with it.

The case for only four primaries was strengthened when some topical anesthetics were found to affect the four basic tastes at different times and/or in different ways (Fontana, 1902; Kiesow, 1894b; Ponzo, 1909). Topical applications of cocaine affect general feeling and pain first, then bitter, sweet, salty, acid, and touch, respectively (Shore, 1892). In addition, the leaves and the acid extracted from the plant *Gymnema sylvestre* and the berries from "miracle fruit" (*Synsepalum dulcificum*) affect some taste qualities more than others (Hooper, 1887; Kiesow, 1894b; Shore, 1892;

Skramlik, 1926). The different reaction times to these substances and their regional variations on the tongue were also viewed as consistent with the concept of four taste primaries.

Nevertheless, the data obtained with these agents did not really prove there were four basic independent tastes.

> We can best explain the action of *Gymnema* by supposing that the nerve fibres or nerve endings capable of being stimulated by pure sweet and bitter substances are different from those which are excited only by acid and salt. The selective action of cocaine, not only on the nerve endings concerned with taste, but on others associated with more general sensory impressions, points also to the multiplicity of the kinds of endings of sensory nerves in the tongue. The more powerful action of cocaine on bitter taste than on sweet, and of gymnema on sweet taste than on bitter, may also be an indication that the nerve fibres or nerve endings concerned with these tastes are also distinct. The selective actions of drugs between these two tastes is however much less marked and the constant grouping of these together gives some support to the view that they may be due to different molecular activities of the same end-organs. The phenomena of "after-tastes" recorded by some observers, between salt and bitter sensations, points in the same direction. When bitter taste is absent at the tip of the tongue, sweet taste is very weak or absent also. (Shore, 1892, 216)

In addition, it is important to note that there did not appear to be any clear cases of individuals deficient in just one of these primaries (Moncrieff, 1944). The analogy was drawn to vision. If different types of color blindness could be produced by specific problems in red, green, or blue retinal receptors, just where were the analogous cases of inability to taste salt but not the other three "primaries," and so on?

CHEMISTRY OF THE PRIMARIES

Once scientists began to speak about just four primary tastes, the push was on to identify the chemical bases of the primaries. Early workers thought this would not be a difficult task, since it seemed very likely that sweet would be associated with sugars, bitter with alkaloids, sour with acids (free hydrogen ions), and salty with anions, like those characterizing the organic salts (Parker, 1922). After some studies were conducted, however, it became clear that these were at best only rough approximations. As expressed by Luigi Luciani (1917, p. 145):

> At first sight it does not seem difficult to discover a series of correlations between the chemical composition of bodies and their taste. Almost all acids have a sour taste, many salts a salty taste, a large number of alkaloids are bitter, many carbohydrates are sweet. It might reasonably be supposed that some of the properties which enable us to classify these substances into definite or distinct chemical groups represent the cause or constitute the origin of their respective flavours. But this correlation is more apparent than real, because not all compounds chemically known as acids give an acid taste, not all salts a salt taste, not all alkaloids are bitter, nor are all sugars sweet. There are bodies of a very different constitution from the sugars which arouse a sensation of sweetness, such as the glucosides, saccharine, chloroform, and certain mineral salts, as lead acetate and salts of beryllium. On the other hand, it is interesting to note that one sugar, d-mannose, has a bitter

taste, while many other mineral and organic substances of varying chemical composition which do not belong to the alkaloids have the same bitter taste. Lastly, there are compounds of closely allied chemical composition, with different tastes. Such are the two asparagines, of which the dextro-rotatory is sweet, and the laevo tasteless.

The complexity of the issue can also be appreciated from a statement made about the sweet taste, which appeared in a monograph on the chemical senses published in 1922. The author wrote:

> The sweet taste is excited by the diatomic and polyatomic alcohols of the aliphatic series, by the aldehydes and ketones derived from these alcohols, and especially by the hexoses whose polymerization products, the disaccharides and polysaccharides, are in this respect particularly important. Besides these carbohydrates other organic compounds, such as chloroform, dextro-asparagine, and saccharine, have sweet tastes. Among inorganic substances neutral acetate of lead, often called sugar of lead, and the salts of glucinum are known to be sweet. Solutions of the alkalis, if they are of appropriate dilution, are said likewise to excite this taste. (Parker, 1922, 142)

THE TONGUE AS THE ORGAN FOR TASTE

In *Primae lineae physiologiae,* Albrecht von Haller described various types of papillae and noted how they looked like reasonable end organs for taste. He started with Avicenna's premise that a substance must be soluble in saliva to be tasted. He then expressed the belief that healthy substances taste good and that substances which taste disagreeable are never healthy. Humans, Haller emphasized, not only learn this rule but can communicate it to others.[3] He theorized that humans therefore have less of a need for well-developed papillae than other animals.

> But brute animals, who have not like ourselves the advantage of learning from each other by instruction, have the faculty of distinguishing flavours more accurately, by which they are admonished to abstain cautiously from poisonous or unhealthy food; and therefore it is that herbivorous cattle, to which a great diversity of noxious plants is offered amongst their food, are furnished with such large and long papillae of so elegant a structure in the tongue, which are not so necessary to man. (1786 translation; 1966 reprint, 265)

Early in the nineteenth century, Charles Bell and Johannes Müller agreed with Haller that the papillae are the gustatory organs. Bell wrote the following in 1811:

> There are four kinds of Papillae on the tongue, but with two of those only we have to do at present. Of these, the Papillae of one kind form the seat of the sense of taste; the other Papillae (more numerous and smaller) resemble the extremities of the nerves in the common skin, and are the organs of touch in the tongue. When I take a sharp steel point, and touch one of these Papillae, I feel the sharpness. The sense of touch informs me of the shape of the instrument. When I touch a Papilla of taste, I have no sensation similar to the former. I do not know that a point touches the tongue, but I

---

[3]Haller may have taken his ideas from Thomas Willis. Similar notions appear in Willis' *Cerebri anatome* of 1664. (See Willis' ideas about the sense of smell in Chapter 13.)

am sensible of a metallic taste, and the sensation passes backward on the tongue. (1936 reprint, 108)

But are the gustatory organs limited to the tongue? Haller (1786) went into a lengthy discussion of the different types of papillae on the tongue and then presented the idea that the papillae were elevated to perform the office of taste. He began his section on taste by stating that "the power of taste is exercised by the tongue chiefly," but added that with penetrating sapid bodies "the palate, root of the tongue, uvula, and likewise the oesophagus, are affected with the taste" (1966 reprint, p. 258).

The belief that the tongue could not be the only organ having taste receptors was also adopted by Claude Le Cat (1700–1768), who wrote *A Physical Essay on the Senses*, printed in London in 1750. Le Cat described two children. One was born without a tongue, and the other lost it because of a gangrenous infection caused by smallpox. Still, both were able to taste different substances. Thus, Le Cat maintained that taste sensitivity was not restricted to the tongue, even in humans.

The conclusions of Haller and Le Cat would prove to be consistent with the later discovery of taste buds on the dorsal surface of the tongue, anterior and posterior epiglottis, inner surface of the arytenoid process of the larynx, soft palate above the uvula, anterior pillar of the fauces, and posterior wall of the pharynx. In contrast, taste buds were found to be absent on the lips, gums, cheeks, inferior surface of the tongue, hard palate, uvula, and tonsils—that is, on those mouth parts where people lack a sense of taste.

DIFFERENTIAL SENSITIVITY OF THE TONGUE

A number of early nineteenth-century workers noted that the tongue surface was not uniformly sensitive to different taste qualities (e.g., Horn, 1825). This finding was more convincingly demonstrated in the 1890s (Shore, 1892). The tip of the tongue was found to be most sensitive to sweet, whereas the edges were most sensitive to sour, and the base of the tongue was most sensitive to bitter. In contrast, the mid-dorsum of the tongue was found to be insensitive to all tastes, even with strong solutions. As for salty tastes, the tongue surface was found to be moderately sensitive to it in all places but the mid-dorsum. These conclusions were based on studies that typically used quinine for bitter, sodium chloride for saline, sulfuric acid for sour, and glycerine for sweet.

These findings were quantified early in the twentieth century (Hänig, 1901). They were also found to be consonant with Friedrich Kiesow's (1894a) reaction time data. Kiesow studied the tip of the tongue and showed that salt had the shortest reaction time, followed by sugar, acid, and bitter. Together, the differential sensitivity of the tongue surface seemed to suggest that there were not fewer than four primaries for taste (Boring, 1942).

SPECIFICITY OF THE PAPILLAE

Investigators at the end of the nineteenth century did not just study the distribution of the papillae and the sensitivity of the different parts of the tongue. Some were already beginning to ask whether individual papillae were specific to different tastes. Hjalmar Öhrwall was one of the most important scien-

tists investigating this problem. He was a student of Magnus Blix, a fellow Swede whose efforts to map the punctate sensitivity of the skin are described in Chapter 10.

Öhrwall tried to study the gustatory system with the same approach Blix was employing for studying the sensory spots on the skin. That is, he attempted to test the sensitivity of individual papillae by applying chemical solutions to them, one at a time, with a small brush. He used strong solutions (40 percent sugar, 5 percent tartaric acid, 20 percent NaCl, 2 percent HCl) and told his subjects which solution would be applied in advance of each trial, a procedure Luciani (1917) and others would later criticize.

In his publication from 1891, Öhrwall reported on 125 papillae. After dropping the results obtained with sodium chloride because the sensations elicited were not distinct enough, Öhrwall stated that 98 papillae were sensitive to his solutions. But whereas some papillae were sensitive to just one of the primaries, most were sensitive to more than one. For example, of the 91 sensitive to acid, only 12 were exclusively sensitive to acid. And whereas 79 were sensitive to sweet, only 3 papillae were sensitive to just sweet. In fact, no papilla responded only to bitter, although 71 responded to bitter and another taste solution.

Öhrwall concluded that the more basic unit of function had to be the taste bud and that while a papilla may contain a number of different types of taste buds, the taste buds themselves were probably specific to the primaries.

TASTE BUDS

Taste buds were discovered by Gustav Schwalbe (1844–1916) and Otto Christian Lovén (1835–1904) in 1867. Their independent anatomical studies were conducted on the tongues of rabbits and hares and were soon confirmed in man, horse, dog, pig, squirrel, guinea pig, bat, mouse, and various marsupials (e.g., Lovén, 1868; Schwalbe, 1868; Tuckerman, 1888).

Christian Lovén, whose portrait appears as Figure 12.5,

Figure 12.5. Christian Lovén (1835–1904) of the Karolinska Institute in Stockholm. Lovén discovered taste buds on the tongue in 1867. (Courtesy of the Royal Swedish Academy of Sciences.)

was a professor at the Karolinska Institute in Stockholm, and Gustav Schwalbe was a pupil of Max Schultze, the leading histologist in Germany. Schwalbe called the structures he observed in the papillae walls and ditches *Schmeckbecher,* meaning "taste cups." Lovén referred to them as *Geschmacks knospen* and *Geschmackszwiebeln,* meaning "taste bulbs."[4]

While the location and specialized appearance of the buds left little doubt that they were the end organs for taste, investigators soon found themselves disagreeing over how many taste buds were affiliated with each papilla. The range extended from under 15 to over 2,500, depending upon the type of papilla, location, species, and age of the subject (Parker, 1922). The average seemed to be about 250 buds per papilla in humans. Luciani (1917), for example, said there were about 400 buds in each circumvallate papilla and that each bud was innervated by multiple nerve endings. In the foliate papillae, the average appeared to be somewhere between 50 and 100, with large individual differences.

This finding created a real problem, which early experimenters could not overcome. On the one hand, they were anxious to study individual taste buds and their responsiveness to different stimuli. But on the other hand, the proximity of one bud to another precluded this kind of research. Further, in most taste buds there appeared to be between 2 and 12 separate taste cells.

By the end of the nineteenth century, it was accepted that not all papillae were associated with taste buds. Although all of the circumvallate papillae and most of the fungiform papillae seemed to be endowed with gustatory capacity, the same could not be said for the filiform papillae (Öhrwall, 1891).[5]

At this time it was also well accepted that each taste bud was made up of modified epithelial cells. Some investigators classified the epithelial cells as either supporting cells or gustatory receptor cells. The receptor cells were more slender and had "taste hairlets" at one end that formed a small brush which reached the surface of the taste bud. They also tended to be found in the interior of the taste bulb (Englemann, 1873). The neurons innervating the receptors were unmyelinated in the bud itself (e.g., Tuckerman, 1888). Figure 12.6 shows an early drawing of some taste buds.

Some investigators raised the possibility that the gustatory cells and the supporting cells might be different developmental stages of a single type of cell. While this would prove to be true, the issue was not resolved in the first quarter of the twentieth century (Parker, 1922).

DEVELOPMENT AND DEGENERATION OF THE PAPILLAE AND TASTE BUDS

One of the first studies on the development of the taste buds was conducted by Ludimar Hermann (1838–1914) in 1884. Although his study involved dogs, others soon followed with research on humans (e.g., Tuckerman, 1888, 1890). When the tongues of 10-week-old human embryos

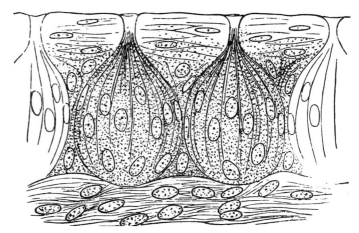

Figure 12.6. The "gustatory bulbs" of the rabbit. This drawing appeared in an early chapter on the histology of the gustatory system (Englemann, 1873).

were examined, it was noted that the gustatory papillae not only were undeveloped but showed no hint of their eventual topography. Four weeks later, some papillae could be differentiated from the surrounding epithelium. In particular, the circumvallate papillae were present in their adult locations and showed some bulbs, although their trenches were still undeveloped. The fungiform papillae also were in various stages of growth and could be seen in abundance at the sides and back of the tongue. In the 5-month-old fetus, the lateral folds still were not seen on the tongue, but by 7 months these folds were well defined. One early depiction of immature human taste buds appears as Figure 12.7.

The number and loci of the taste buds were observed to change with development. In general, investigators noted a much wider distribution of taste buds in young children and fetuses than in older children and adults (see Luciani, 1917; Parker, 1922). For example, the mid-dorsum of the tongue, which does not have taste buds in adults, was found to have taste buds in young children. There also appeared to be a major loss of buds beginning around age 45, which correlated with a blunting of sensitivity at this stage of life.

The finding that the integrity of the taste buds depended upon the innervation of the peripheral nerve was suggested by developmental studies on the taste buds of dogs and humans (Hermann, 1884; Marchand, 1902). It was found that the growth of the nerve into appropriate loci preceded

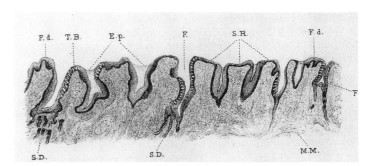

Figure 12.7. Transverse section through the foliate papillae of a 4-month-old infant. T. B. signifies "taste-bulbs." Other abbreviations: Fd = folds; S. R. = secondary ridges; Ep = epithelium; F = furrow; F = invaginated furrow; S. D. = sereous ducts; Mm = mucous membrane. (From Tuckerman, 1888.)

---

[4]A number of other descriptive terms were also used to describe the taste buds in the literature following their discovery (see Titchener, 1915; Tuckerman, 1888).

[5]Although taste buds were associated with some types of papillae on the tongue, they appeared to be buried in the epithelium of the throat.

the development of taste buds. The importance of the nerves was also shown by cutting the gustatory nerves on one side of the tongue in animals (Vintschgau and Hönigschmied, 1877). This resulted in the loss of taste buds and taste sensations from the papillae previously served by these nerves.

At first, it was thought that when their neural innervation was lost, taste cells turned back into ordinary epithelial cells (Vintschgau and Hönigschmied, 1877). Louis Ranvier, who worked on the rabbit, however, could not find any evidence for taste cells undergoing metamorphosis into ordinary epithelial cells. In his 1889 paper, he suggested that taste cells simply degenerate. Others confirmed that taste buds and taste cells do, in fact, degenerate when their nerves are cut, whereas new receptor cells form after the nerves regenerate (e.g., Olmstead, 1920, 1921).

VIBRATION AS A MECHANISM OF RECEPTOR ACTIVATION

A number of theories were proposed to explain how the taste receptors can be activated. One of the more interesting ones was expounded by John Haycraft, who believed receptors in different sensory systems followed similar rules and were activated in comparable ways. Haycraft (1887, p. 163) suggested that the taste receptors, just like receptors for hearing, vision, and olfaction, were activated by vibration.

> Just as . . . stimulating the eye produces the same colour sensations; just as certain strings of definite length and consistency vibrate in a similar way, and produce the same sound sensation—so in like manner, similar sapid compounds (containing similar elements, or the same compound radicals) vibrate in a similar way, and produce the same taste.

Haycraft turned to the periodic table of the elements to make the point that substances of very high or very low molecular weight were likely to be tasteless. He suggested that molecular weight was related to vibration, and hence may set the limits for taste.[6]

CRANIAL NERVES

In 1818, more than 1,600 years after Galen attempted to trace the cranial nerves involved with taste, Carlo Francesco Bellingeri (1789–1848) argued that the chorda tympani branch of the facial nerve (VII) was the nerve mediating taste from the anterior part of the tongue. This nerve passes through the middle ear cavity to enter the brain.[7] The fact that loss of taste from the front of the tongue may accompany middle ear disease was known in the nineteenth century.

Nevertheless, some nineteenth-century scientists still thought the trigeminal nerve was involved with taste—an idea whose origin can be traced back to Galen. These investigators included Moritz Schiff, Wilhelm Erb, William Gowers, and William Turner (1832–1916). At first, this

idea seemed to be supported by some cases of "trigeminal paralysis" caused by tumors. The patients cited exhibited a loss of tactile and taste sensitivity on one side of the tongue (e.g., Bishop, 1833).

In contrast, Fedor Krause (1857–1937) and many others held that taste afferents from the tongue did not involve the trigeminal nerve. Case studies and lesion experiments showing a dissociation between touch and taste were cited to support this position. The situation was confusing, to say the least.

The anatomy of the first-order gustatory fibers remained unclear through the time of World War I. This can be gleaned from Luciani's (1917, p. 136) erroneous comment that "although the lingual nerve ramifies in the anterior part of the tongue, and the glossopharyngeal in the base, the gustatory fibres of the lingual probably come from the glossopharyngeal."

In fact, when the literature on taste was reviewed by Walter Börnstein (1890–?) in 1940, he still felt compelled to write that "whether the trigeminal nerve carries gustatory fibers is still a controversial subject" (see 1940a).

Nevertheless, by the time of Börnstein's review, the weight of the evidence clearly went against a direct role of the trigeminal in taste.[8] Two pertinent reasons for rejecting the trigeminal hypothesis had appeared. First, the experimentalists were able to show that sectioning the trigeminal did not lead to degeneration of taste buds. Second, the taste deficits noted after "clean" trigeminal lesions in humans were likely to be very short-lived, if they even appeared at all. This was observed by Harvey Cushing in 1904. His study involved 13 patients with the Gasserian ganglion sectioned.

The notion that taste is mediated by projections from cranial nerves VII (facial) from the front of the tongue, IX (glossopharyngeal) from the back of the tongue, and X (vagus) from the throat, only slowly gained acceptance in the twentieth century.

CENTRAL PROJECTIONS

Even slower to appear in textbooks of neurology and physiology, as well as in specialty monographs, were statements about the diverse peripheral projections coming together to form the solitary tract to the solitary nucleus of the medulla (see Parker, 1922). Investigators in the first quarter of the twentieth century also did not seem to recognize that taste projections from the solitary nucleus terminated around the dorsomedial tip of nucleus ventralis posteromedialis (arcuate nucleus) of the contralateral thalamus. Nor did they know that the afferents then projected to or near the mouth region of the somatosensory cortex on the postcentral gyrus.

The role of the thalamic arcuate nucleus in taste was suggested by Adler in 1934. He studied an individual who had progressively lost gustatory sensitivity from one half of the tongue and cutaneous sensitivity from one side of the face. This person was found to have a glioblastoma of the

---

[6]Haycraft also applied the vibration theory to olfaction. A lengthier discussion of this idea appears in Chapter 13.

[7]The course of the chorda tympani through the ear made it a very accessible target for electrical recordings in the next century (Pfaffman, 1959).

[8]The role of the trigeminal in mediating heat sensations from peppers and other "hot" foods was not the debated issue. Arguments centered around true tastes and not related sensations that might affect the descriptions of some foods.

third ventricle, which involved the arcuate nucleus. This case led Adler to conclude that taste was represented in the medial part of the arcuate nucleus.

The arcuate nucleus was also suggested by anatomical and clinical studies of the facial region of the somatosensory cortex (Walker, 1938). When this cortical region was damaged, there was extensive degeneration in this part of the thalamus along with changes in taste (see next sections). It was not until 1944, however, that animals were subjected to lesions of the thalamic arcuate nucleus and tested. When given a choice between water and quinine, monkeys with these lesions were less discriminating than other monkeys (Patton, Ruch, and Walker, 1944).

CORTICAL LESIONS: LABORATORY ANIMAL EXPERIMENTS

In general, taste deficits following cortical lesions are not very striking. This is one reason so little has been written about cortical lesions and taste. For instance, Dusser de Barenne (1933) noted that decorticate cats and dogs, as well as anencephalic children, still grimaced at bitter foods. This fact suggested a role for the brainstem in basic taste responsiveness.

David Ferrier was of the opinion that taste was represented in or near the uncus at the base of the temporal lobe, close to his center for smell. He based this idea partly on a series of experiments involving monkeys with various cortical lesions. He watched to see if the monkeys ate food after surgery and how they responded to acetic acid placed on their tongues. In his Croonian Lecture of 1875, Ferrier (p. 471) made the following statement:

> As to the sense of taste, the positive indications are less distinct than those of smell or hearing. Yet the phenomena occasionally observed on stimulation of the lower part of the middle temporo-sphenoidal convolution, viz. movements of the lips and cheek- pouches, may be taken in connexion with lesions affecting this region, and accompanied by loss of reaction to stimuli of taste, to afford evidence of no weak character for the localization of taste in or near this region.

Ferrier's results were not supported by experiments on monkeys conducted by other investigators (Brown and Schäfer, 1888). In addition, his experiments were criticized for not being well controlled and on the grounds that some animals lived for only a few days and could have been showing more general effects of their surgery. Further, some of Ferrier's figures showed that the "temporo-sphenoidal" lesions sometimes involved the parietal region (Börnstein, 1940a).

There were conflicting suggestions about damage to the "masticatory cortex" affecting taste preferences in animals. Some investigators reported that lesions in the precentral region affected taste in rabbits and other animals, whereas others failed to confirm this finding (Bremer, 1923; Schtscherbak, 1892).

The idea that the postcentral gyrus played a role in taste lacked strong support until 1928, when Börnstein published an important paper on the cortical representation for taste. Börnstein then continued to work on the gustatory system and published several other papers on the role of the forebrain in taste, including a series of studies in 1940. He tested monkeys with bilateral lesions of the sensory and/or motor face areas on a two-bottle preference test. One bottle was filled with water and the other with a quinine solution. The quinine concentration was started at a very low level and slowly increased until a monkey rejected it. It was assumed this took place when the solution began to taste bitter.

Börnstein found that damage to the postcentral gyrus face area resulted in deficits on this test, but that recovery was rapid. In contrast, when the lesions also included the mouth area in the precentral gyrus, the deficits were much more severe. These findings led him to argue that the primary taste areas were not located on the temporal lobe but in or near the facial region of the sensory and possibly the motor cortex. Later work on animals seemed to indicate that the most important part of the taste area lay buried in the operculum, just below the facial regions (Patton, 1950).

GUSTATORY CORTEX: HUMAN CLINICAL DATA

The locus of the cortical representation for taste was also addressed by studying human epileptic patients who experienced gustatory auras. William Gowers wrote that true gustatory auras were very rare and professed he had come across only one such case by 1885. This patient described the sensation as a taste between sour and bitter and localized it on the back of the tongue.

Later studies showed that some epileptic patients may grimace, shake their heads, and rub their mouths. A typical response is that the taste is bad but difficult to describe. The simple qualities of sour, sweet, salty, and bitter were rarely used to describe the taste auras.

The locus of these disturbances, however, was not clear until the middle of the twentieth century. As noted, the consensus was that taste, like smell, was localized in the temporal lobe. Only later did attention shift to the Sylvian fissure above the insula (see Börnstein, 1940a; Penfield and Jasper, 1954). The major problems were that taste hallucinations were rare, and autopsies were not always possible. Further, when post-mortems were performed, the lesions were not always confined to one lobe.

Nevertheless, the data from the epileptic population caused some people to think that Ferrier's association of taste with the temporal lobe might not be correct. The question raised was: Why should olfactory auras be reported much more frequently than gustatory auras if both sensory systems are mediated by closely related parts of the temporal lobe? The slowly growing answer to this question was that the temporal lobe was not playing a major role in gustation and that the temporal tumors and epilepsy associated with gustatory hallucinations were affecting the cortex above the Sylvian fissure.

For example, Harvey Cushing described 59 cases of temporal lobe tumors, 10 of which were analyzed in detail in 1921. Four were associated with olfactory hallucinations and only two were associated with taste sensations. In one of these cases, there was a question as to whether taste (versus smell) was really being reported when the patient described smelling and tasting roasted peanuts. In the other, the patient described peppermint, which could involve more than taste, while showing paresthesias of the teeth and fingers, signs that suggested the "temporal lobe" tumor might be affecting the parietal lobe.

Figure 12.8. Wilder Penfield (1891–1976) conducted many studies involving electrical stimulation of the brain in conscious human patients undergoing surgery for epilepsy.

Also aiding in the inquiry were newer studies of conscious human patients undergoing electrical stimulation of the brain. Wilder Penfield, shown in Figure 12.8, elicited gustatory responses in very few of his cases, but found that most of the active sites for taste were in or near the tongue area of the postcentral gyrus (Penfield and Boldrey, 1937).

Prior to World War II, there was a paucity of well-conducted studies on humans having acute brain lesions and taste defects against which the data from epileptics and tumor victims could be compared. In 1940, however, Börnstein described his cases with gunshot wounds. He showed that the cortical lesions having the greatest effects on gustation involved the base of the parietal lobe (the operculum; Brodmann Area 43). He based this on systematic testing of his patients.

Börnstein (1940a, 1940b) felt that localizing taste in the lower parietal region made sense since it was adjacent to the masticatory area, which was associated with taste deficits in rabbits. He also knew that taste and cutaneous afferents ran together through the brainstem and reasoned there was no reason to expect them to split apart at the cortex. In addition, he recognized that the opercular region was an area which could easily be affected by temporal lobe tumors growing below it. Moreover, this was the area most clearly associated with taste auras in Penfield's electrical stimulation studies.

TASTE PSYCHOPHYSICS

The demand to learn more about gustatory psychophysics paralleled the demand to learn more about the anatomy and physiology of taste. Many psychophysical studies have been conducted on this system since Albrecht von Haller wrote the following in *Primae lineae physiologiae* in the middle of the eighteenth century:

> But the nature or disposition of the covering with which the papillae are clothed, together with that of the juices, and of the aliments lodged in the stomach, have a considerable share in determining the sense of taste; insomuch, that the same flavour does not equally please or affect the organ in all ages alike, nor in persons of all temperatures; nor even in one and the same person at different times, who shall be differently accustomed in health or variously diseased. (1786 translation; 1966 reprint, 264)

Ernst Weber, for example, noted that putting very warm or cold solutions on the tongue affected sweet and then expanded this finding to include other tastes (see Luciani, 1917). Weber's discovery, that temperature is an important parameter in taste stimulation experiments, was the subject of many later studies. Some investigators showed that the tongue was virtually insensitive to taste outside the range of 0–50° centigrade.[9] If anything, the reaction time data proved even more sensitive than the threshold data to small temperature differences (Chinaglia, 1916).

As noted previously, one of the major debates at the end of the nineteenth century was whether the so-called basic tastes should be treated as qualities within one sensory system or as distinct systems, albeit with much in common. The two primary figures in this debate were Kiesow and Öhrwall.

Kiesow went to Leipzig in 1891 to study with Wilhelm Wundt, where he began his research career by studying the effects of *Gymnema sylvestre* and cocaine on the gustatory system. He claimed that taste contrast and compensation (defined as changes in one quality that could be produced by experiencing a second quality) existed for the four basic tastes. For example, he argued that when he applied sucrose to one edge of the tongue and salt to the opposite edge, his subjects reported a blended sensation. Kiesow (1894b, 1896, 1930) thus followed Johannes Müller, who described cross-adaptation early in the nineteenth century. Müller pointed out that sweet things take away from the flavor of wine, whereas a piece of cheese can increase a wine's taste.

Öhrwall, in contrast, did not agree that interactive phenomena could be demonstrated in the gustatory system in the same way they could be demonstrated for color mixing. He stated that when two or more basic tastes are mixed, the mixture tastes like the individual parts and not like a real blend, unless a new chemical compound is formed. In lemonade, for example, the taste is one of sweet and sour, and not an intermediate blend with the loss of these two basic tastes. Armed with many such examples, Öhrwall wrote a paper in 1901 that was extremely critical of Kiesow's ideas.

A number of studies from other laboratories on cross-adaptation appeared at this time. Some data seemed to suggest Kiesow and Öhrwall might both have been right.

> If a sapid substance be applied at one region of the tongue, other parts of the organ are rendered more susceptible to other stimuli. Thus, if one border of the tongue be rubbed with salt, the other border will give a reaction with sugar out of all proportion to the strength of the stimulus employed.

[9]Many psychophysical findings, including the importance of temperature on taste, have been reviewed by Pfaffman (e.g., 1959).

Salt and acid likewise give simultaneous contrast, but it is questionable whether any facts can be brought forward to show that bitter sensations are affected by other stimuli simultaneously applied. (Haycraft, 1900, 1245)

The issue of interactions continued to be debated by later investigators (Hahn, 1934; McBurney, 1978; Parker, 1922). In general, Kiesow's idea of four "qualities" capable of blending achieved wider acceptance than Öhrwall's notion of four distinct modalities that retain their individual identities when mixed (Bartoshuk, 1978).

HENNING'S TASTE TETRAHEDRON

In 1916, Hans Henning (1885–?) proposed that tastes could best be understood as a tetrahedron with the four primaries making up the four corners of the structure. The idea was that substances which tasted like two of the primaries would be positioned on the edge of the tetrahedron between the two relevant corners. Substances tasting like three of the primaries would appear on the face of the tetrahedron in a locus consistent with the three basic tastes making it up. Henning believed the tetrahedron had to be hollow because no substance could produce all four taste sensations.

Henning's model generated considerable interest but attracted few loyal followers. In addition, the utility of the model proposed by Henning was questioned in a number of theoretical reviews (e.g., Skramlik, 1926).

CONTROL OVER THE STIMULUS AND TESTING PROCEDURES

The specific findings described in this chapter and the debates they generated were not independent of the procedures used to study taste. In a thoughtful discussion that bears directly on how difficult it is to compare the procedures used by different investigators, Luigi Luciani (1917) made the point that many experimenters did not control for olfactory cues or hold cutaneous sensations constant when they tested for taste. His suggestion was to block the nostrils, to use solutions of a constant temperature, and to test for cutaneous sensations, such as pain or irritation, by also applying solutions to parts of the tongue insensitive to taste. Although these precautions were rarely taken, it is interesting to note that when olfactory and cutaneous stimuli were controlled, held constant, or eliminated, a number of substances at first thought to be basic taste qualities were eliminated from the lists of primaries. These included aromatic, alcoholic, nauseous, oily, sharp, astringent, and dry.

Luciani also mentioned that the part of the tongue stimulated should be taken into account because of regional variations in sensitivity to different substances. This, of course, also related to the size of the area stimulated. Another point made was that tongue movements, which could diffuse the stimulus and alter its characteristics, should be minimized. He added that the mouth must be rinsed between stimuli, and time must be permitted to overcome previous sensations. In the latter domain, Luciani suggested at least five minutes between stimuli and catch trials involving just distilled water as a proce-

dural check. Lastly, Luciani recommended not telling the subject anything about the substance to be presented. Öhrwall (1891), it will be recalled, told his subjects about the stimuli to be applied to individual papillae.

ELECTRICAL TASTES

In 1752, Johann Sulzer (1720–1779) put his tongue into a "sandwich" made by placing a piece of zinc on one side of it and copper on the other. Sulzer experienced a taste sensation when the two pieces of metal were joined at one end. Unaware of what Sulzer had found, Alessandro Volta (1745–1827) repeated this experiment in 1792. He also described tastes with different metals applied to the tongue. In fact, when Volta put one metal strip on his tongue and another near his eye, he experienced both an unpleasant taste and light sensations.

Volta thought the tastes experienced were due to the passage of electrical currents. In fact, he was able to get the same effects with instruments that produced and passed current. He believed the current directly stimulated the end organs for taste. In contrast, his contemporary, Alexander von Humboldt (1769–1859), argued that the currents affected chemicals produced by the tongue, which in turn caused taste sensations (cited in Luciani, 1917).

The idea of electrical tastes became a popular subject after this, and many new things were discovered about these tastes. For example, one investigator late in the eighteenth century kept the current running through the tongue for a long time. He reported the sensation at the anode changed from sour to bitter, and then to alkaline (Ritter, cited in Parker, 1922). At about the same time, it was noted that alternating current was even more effective for eliciting taste sensations than direct current (Gertz, 1919). The work with alternating current also showed that electrical tastes were not totally dependent upon electrolysis or its by-products, an idea which had been favored by some early investigators. To say the least, most investigators found these effects fascinating.

REFERENCES

Adler, A. 1934. Zur Topic des Verlaufes der Geschmackssinnfasern und anderer afferenter Bahnen im Thalamus. *Zeitschrift für gesamte Neurologie und Psychiatrie, 149,* 208–220.

Aristotle. 1907. *De anima.* (R. D. Hicks, Trans.). Cambridge, U.K.: Cambridge University Press.

Bartoshuk, L. M. 1978. History of taste research. In E. C. Carterette and M. P. Friedman, *Handbook of Perception: Vol. 6. Taste and Smelling.* New York: Academic Press. Pp. 3–18.

Bell, C. 1811. *Idea of a New Anatomy of the Brain; Submitted for the Observations of His Friends.* London: Strahan & Preson. (Reprinted in *Medical Classics,* 1936, *1,* 105–120.)

Bishop, J. 1833. Observations on the physiology of the nerves of sensation illustrated by a case of paralysis of the fifth pair. *Proceedings of the Royal Society of London, 3,* 205–206.

Boring, E. 1942. *Sensation and Perception in the History of Experimental Psychology.* New York: Appleton-Century-Crofts.

Börnstein, W. S. 1928. Beobachtungen an einem Gehirnverletzen. *Monatsschrift für Psychiatrie und Neurologie, 67,* 216–222.

Börnstein, W. S. 1940a. Cortical representation of taste in man and monkey. I. Functional and anatomical relations of taste, olfaction, and somatic sensibility. *Yale Journal of Biology and Medicine, 12,* 719–736.

Börnstein, W. S. 1940b. Cortical representation of taste in man and monkey. II. The localization of the cortical taste area in man and a method of measuring impairment of taste in man. *Yale Journal of Biology and Medicine, 13,* 133–156.

Börnstein, W. S. 1940c. The cortical taste area in monkeys and a semiquantitative method of testing taste in monkeys. *American Journal of Physiology, 29,* 314.

Bravo, J. 1591. *De saporum et odorum differentiis, causis et effectionibus.* Venetiis: Apud Joan Baptisam Ciottum.

Bremer, F. 1923. Physiologie nerveuse de la mastication chez le chat et le lapin. Réflexes de la mastication. Réponses masticatrices corticales et centre cortical du goût. *Archives Internationales de Physiologie, 21,* 208–252.

Brown, S., and Schäfer, E. A. 1888. An investigation into the functions of the occipital and temporal lobes of the monkey's brain. *Philosophical Transactions of the Royal Society of London (Biology), 179,* 303–327. (Paper read December 15, 1887)

Chinaglia, L. 1916. Richerche intorno all'influenze esercitata dalla temperatura sulla sensibilità gustativa. *Revista di Psicologia, 11,* 196–226.

Cushing, H. 1904. The taste fibers and their independence of the N. trigeminus. Deductions from thirteen cases of Gasserian ganglion extirpations. *Johns Hopkins Hospital Bulletin, 14,* 71–78.

Cushing, H. 1921. Distortions of the visual fields in cases of brain tumour. The field defects produced by temporal lobe lesions. *Brain, 44,* 341–396.

Dusser de Barenne, J. G. 1933. "Corticalization" of function and functional localization in the cerebral cortex. *Archives of Neurology and Psychiatry, 30,* 884–901.

Engelmann, T. W. 1873, The organs of taste. Ch. XXXXIII in S. Stricker (Ed.), *Manual of Human and Comparative Histology* (Vol. 3). (H. Power, Trans.). London: New Sydenham Society. Pp. 1–26.

Fernel, J. 1542. *De naturali parte Médicine.* Paris: Simon de Colines.

Ferrier, D. 1875. Experiments on the brain of monkeys. *Philosophical Transactions of the Royal Society of London, 165,* 433–488.

Ferrier, D. 1886. *The Functions of the Brain* (2nd ed.). New York: G. P. Putnam's Sons.

Fontana, A. 1902. Ueber die Wirkung des Eucain B auf die Geschmacksorgane. *Zeitschrift für Psychologie und Physiologie der Sinnesorgane, 28,* 253–260.

Gertz, H. 1919. Une expérience critique relative à la théorie du goût électrique. *Acta Oto-laryngologica, 1,* 551–556.

Gowers, W. R. 1885. *Epilepsy.* William Woods and Company. (Reprinted 1964; New York: Dover Publications)

Gruner, O. C. 1930. *A Treatise on the Canon of Medicine of Avicenna Incorporating a Translation of the First Book.* London: Luzac & Company.

Gryllus, L. 1566. *De sapore dulci et amaro.* Prague: Georgium Melantrichum ab Aventino.

Hahn, H. 1934. Die Adaptation des Geschmackssinnes. *Zeitschrift für Sinnesphysiologie, 65,* 105–145.

Haller, A. von. 1786. *First Lines of Physiology.* Translated from the Latin. (Reprinted 1966; New York: Johnson Reprint Company)

Hänig, D. P. 1901. Zur Psychophysik des Geschmackssinnes. *Philosophische Studien, 17,* 576–623.

Haycraft, J. B. 1887. The nature of the objective cause of sensation. II. Taste. *Brain, 10,* 147–163.

Haycraft, J. B. 1900. The sense of taste. In E. A. Schäfer (Ed.), *Text-book of Physiology.* Edinburgh: Young J. Pentland. Pp. 126–159.

Henning, H. 1916. Die Qualitätsreihe des Geschmacks. *Zeitschrift für Psychologie, 74,* 203–219.

Herlitzka, A. 1908. Sul "sapore metallico," sulla sensazione astringente e sul sapore dei sali. *Archivo di Fisiologia, 5,* 217–242.

Hermann, F. 1884. Beitrag zur Entwicklungsgeschichte des Geschmacksorgans beim Kanichen. *Archiv für Microscopische Anatomie, 24,* 216–229.

Hooper, D. 1887. An examination of the leaves of *Gymnema sylvestre. Nature, 35,* 565–567.

Horn, W. 1825. *Ueber den Geschmacksinn des Menschen.* Heidelberg: Kal Groos.

Keele, K. D. 1957. *Anatomies of Pain.* Oxford, U.K.: Blackwell.

Kiesow, F. 1894a. Beiträge zur physiologischen Psychologie des Geschmackssinnes. *Philosophische Studien, 10,* 329–368, 523–561.

Kiesow, F. 1894b. Ueber die Wirkung des Cocain und des Gymnemasäure auf die Schleimhaut der Zunge und des Mundraums. *Philosophische Studien, 9,* 510–527.

Kiesow, F. 1896. Beiträge zur physiologischen Psychologie des Geschmackssinnes. *Philosophische Studien, 12,* 255–278.

Kiesow, F. 1930. F. Kiesow. In C. Murchison (Ed.), *A History of Psychology in Autobiography.* Worcester, MA: Clark University Press. Pp. 163–190.

Le Cat, M. 1750. *A Physical Essay on the Senses.* London: Printed for R. Griffiths at the Dunciad in St. Paul's Church-yard.

Linnaeus, K. von. 1752. Odores medicamentorum. *Amoenitates Academicae, 3,* 183–201.

Lovén, C. 1868. Beiträge zur Kenntniss vom Bau der Geschmackwärzchen der Zunge. *Archiv für Mikroscopische Anatomie, 4,* 96–110.

Luciani, L. 1917. *Human Physiology.* London: Macmillan. (F. Welby, Trans.).

Marchand, M. L. 1902. Dévelopment des papilles gustatives chez le foetus humain. *Comptes Rendus de la Société de Biologie, 54,* 910–912.

McBurney, D. H. 1978. Psychological dimensions and perceptual analysis of taste. In E. C. Carterette and M. P. Friedman (Eds.), *Handbook of Perception: VIA. Tasting and Smelling.* New York: Academic Press. Pp. 125–155.

McHenry, L. C., Jr. 1969. *Garrison's History of Neurology.* Springfield, IL: Charles C Thomas.

Moncrieff, R. W. 1944. *The Chemical Senses.* New York: John Wiley & Sons.

Müller, J. 1838. *Handbuch der Physiologie des Menschen.* Coblenz: Verlag J. Hölscher.

Öhrwall, H. 1891. Untersuchungen über den Geschmacksinn. *Skandinavisches Archiv für Physiologie, 2,* 1–69.

Öhrwall, H. 1901. Die Modalitäts—und Qualitätsbegriffe in der Sinnesphysiologie und deren Bedeutung. *Skandinavisches Archiv für Physiologie, 11,* 245–272.

Olmstead, J. M. D. 1920. The nerve as a formative influence in the development of taste-buds. *Journal of Comparative Neurology, 31,* 465–468.

Olmstead, J. M. D. 1921. Effects of cutting the lingual nerve of the dog. *Journal of Comparative Neurology, 33,* 149–154.

Parker, G. H. 1922. *Smell, Taste, and Allied Senses in the Vertebrates.* Philadelphia, PA: Lippincott.

Patton, H. D. 1950. Physiology of smell and taste. *Annual Review of Physiology, 12,* 469–484.

Patton, H. D., Ruch, T. C., and Walker, A. E. 1944. Experimental hypogeusia from Horsley-Clarke lesions of the thalamus in *Macaca mulatta. Journal of Neurophysiology, 7,* 171–184.

Penfield, W., and Boldrey, E. 1937. Somatic motor and sensory representation in the cerebral cortex of man as studied by electrical stimulation. *Brain, 60,* 389–443.

Penfield, W., and Jasper, H. 1954. *Epilepsy and the Functional Anatomy of the Human Brain.* Boston, MA: Little, Brown.

Pfaffman, C. 1941. Gustatory afferent impulses. *Journal of Cellular and Comparative Physiology, 11,* 245–272.

Pfaffman, C. 1959. The sense of taste. In J. Field, H. W. Magoun, and V. E. Hall (Eds.), *Handbook of Physiology: Vol. 1. Neurophysiology.* Washington, DC: American Physiological Society. Pp. 507–533.

Plato. 1952. *Timaeus* (B. Jowett, Trans.). In *Great Books of the Western World: Vol. 7. Plato.* Chicago: Encyclopaedia Britannica. Pp. 442–477.

Ponzo, M. 1909. Über die Wirkung des Stovainsauf die Organe des Geschmacks, der Hautempfindungen, des Geruchs und des Gehörs, nebst einigen weiteren Beobachtungen über die Wirkung des Kokains, des Alpins und der Karbolsäure im Gebiete der Emfindungen. *Pflüger's Archiv für die gesammte Physiologie, 14,* 385–436.

Ranvier, L. 1889. *Traité Technique d'Histologie.* Paris: F. Savy.

Schtscherbak, A. 1892. Zur Frage über die Localisation der Geschmackscentren in der Hirnrinde. *Zentralblatt für Physiologie, 5,* 289–298.

Schwalbe, G. 1867. Ueber das Epithel der Papillae vallatae. *Archiv für Mikroscopische Anatomie, 3,* 504–508.

Schwalbe, G. 1868. Ueber die Geschmacksorgane der Säugethiere und des Menschen. *Archiv für Mikroscopische Anatomie, 4,* 154–187.

Sherrington, C. S. 1946. *The Endeavor of Jean Fernel.* Cambridge, U.K.: Cambridge University Press.

Shore, L. E. 1892. A contribution to our knowledge of taste. *Journal of Physiology, 13,* 191–217.

Siegel, R. E. 1970. *Galen on Sense Perception.* New York: S. Karger.

Skramlik, E. von. 1926. *Handbuch der Physiologie der Niederen Sinne.* Leipzig: Georg Thieme.

Stitch, A. 1857. Ueber die Schmeckbarkeit der Gase. *Annalen Charité-Krankenhauses, Berlin, 8,* 105–115.

Telfer, W. 1955. *Cyril of Jerusalem and Nemesius of Emesa.* Philadelphia, PA: Westminster Press.

Theophrastus. *On the Causes of Plants.* (J. I. Beare, Trans.). In *Greek Theories of Elementary Cognition from Alcmaeon to Aristotle.* Oxford, U.K.: Clarendon Press, 1906.

Titchener, E. B. 1915. *A Beginner's Psychology.* New York: Macmillan.

Tuckerman, F. 1888. The anatomy of the papilla foliata of the human infant. *Journal of Anatomy and Physiology, 22,* 499–501.

Tuckerman, F. 1890. Further observations on the development of the taste-organs of man. *Journal of Anatomy and Physiology, 24,* 130–131.

Valentin, G. 1853. *A Textbook of Physiology.* London: Henry Renshaw.

Vintschgau, M. von. 1880. Physiologie des Geschmackssinns und des Geruchssinns. In L. Hermann (Ed.), *Handbuch der Physiologie der Sinnesorgane, Zweiter Theil.* Pp. 145–286.

Vintschgau, M. von, and Hönigschmeid, J. 1877. Versuche über die Reactionszeit einer Geschmacksempfindung. *Archiv für die gesammte Physiologie, 14,* 529–592.

Walker, A. E. 1938. *The Primate Thalamus.* Chicago: University of Chicago Press.

Willis, T. 1664. *Cerebri anatome: cui accessit nervorum descriptio et usus.* London: J. Martyn and J. Allestry. Tercentenary ed., 1664–1964, *Thomas Willis, The Anatomy of the Brain and Nerves.* Montreal: McGill University Press, 1965.

Wundt, W. 1880. *Grundzüge der physiologischen Psychologie* (Auf. 2). Leipzig: Engelmann.

Wundt, W. 1887. *Grundzüge der physiologischen Psychologie* (Auf. 3). Leipzig: Engelmann.

Chapter 13
Olfaction

In the case of the sense of taste, it is probable that there are only four sensations—those called sweet, sour, bitter, and salt; and these four sensations are universally recognized and have definite and well understood names. Were it possible in the same way to group bodies that produce similar or allied smells, mankind would long ago have introduced a well-understood smell nomenclature. This has not been done, and odours are called by the names of bodies which give them out, and of these there are an infinite number.

John Haycraft, 1900

EARLY HISTORY

The idea that unpleasant odors can be evil, powerful, or dangerous has a long history. In fact, in some languages, words such as "wizard" or "witch" were derived from terms meaning "to smell evil" (Summers, 1926). In addition, feces and other substances with repulsive odors were used routinely to drive away demons and harmful spirits in Egyptian and other early cultures (see Chapter 1).

In the fifth century B.C., in Ancient Greece, Alcmaeon associated the brain with sensation and proposed that smell was produced by the aspiration of odorous particles through the nose to the brain, much like sight resulted from the passage of fire through the eye (see Chapter 5).

Less than a century later, Plato wrote the following in *Timaeus,* about the four elements and olfaction:

No element is so proportioned to have any smell. The veins about the nose are too narrow to admit earth and water, and too wide to detain fire and air; and for this reason, no one ever perceives the smell of any of them; but smells . . . are perceptible only in the intermediate state when water is changing into air and air into water. . . . Wherefore the varieties of smell have no name, and they have not many, or definite or simple kinds; but they are distinguished only as painful and pleasant, the one sort irritating and disturbing the whole cavity . . . the other having a soothing influence, and restoring this same region to an agreeable and natural condition. (1952 translation, 464–465)

Plato was followed by Aristotle, who believed the smell receptors were located inside the nose and who made analogies between smell and taste (Beare, 1906). Aristotle thought olfactory sensations were less definite than those of taste, but mentioned some odors that corresponded to his taste qualities. These were sweet, sour, pungent, harsh, and succulent. He added fetid, which was compared to bitter tastes. Aristotle also commented that humans possess a sense of smell inferior to that of many animals.

Theophrastus, a student of Aristotle, wrote the first known treatise on odor. He discussed odor as a distance and a contact sense. He believed contact occurred when eating food. Theophrastus supposedly described seven odor qualities, but records of them have not been found.

The Ancient Greeks also addressed the question of why some odors were pleasant and others repulsive or unpleasant. Democritus and Epicurus—who both held that all matter is made up of atoms—believed pleasant smells were caused by smooth, round particles, whereas harsh or unpleasant smells were due to hooked atoms, which tore open passages from the areas stimulated. In Ancient Rome, Lucretius continued to accept these ideas.

Galen, the "Prince of Physicians," had great difficulty classifying odors and wound up saying only that smells were either agreeable or disagreeable, a position much like that of Plato. Galen thought odors traveled as very fine particles (but larger than those required for hearing or seeing) to the olfactory areas in the nose. These particles supposedly went unchanged through holes in the ethmoid bone. Tubular nerves then transported them to the hollow olfactory bulbs and to the lateral ventricles of the brain. In *The Organ of Smell,* a part of his *Opera omnia,* Galen wrote:

The nose has a septum in the middle and hence two important channels, these being apparent, one in each nostril, but one must be aware that they are divided into two higher up. One of these divisions goes backward to the mouth and the other straight, as it is directed from the beginning, mounting upwards to the brain itself. The brain beneath it, has two outgrowths elongated and hollow, having their beginnings from the anterior ventricles coming down to that part of the cranium where the nose begins. Here is the seat of the sieve like bones . . . and the dura mater overspreading these bones is pierced with minute holes. (1924 translation, 2)

Galen adhered to a pneumatic theory, one in which the brain expanded to bring fresh air with new odorants into the olfactory bulbs and ventricles and contracted to expel the stale air. He explained:

Inquiring why the sensation of odors arises in those inhaling, and that at no other time, if indeed its origin is in the channels of the nose, this one thing seems evident to me, that not at the external surface of the tunic of the nose, but very deep in the organ the sense of smell is located, certain pores extending up to its site, not open continually, but having a thin membrane as a covering, and these being opened at the time of respiration by the impetus of the pneuma, thus the object perceived comes to the differentiating part. (1924 Translation, 6)

Galen saw the brain's pulsation during trepanation and after head injuries. Yet he did not seem to associate expansion and contraction of the pneumatic brain with the heart or the pulse.

Galen and his followers came to the conclusion that the loss of smell could have three causes: (1) obstruction of the nasal passages (e.g., from too much mucous coming from the ventricles of the brain), (2) blockage of pores in the ethmoid bone, and (3) diseases of the anterior (lateral) ventricles.[1]

After his death, Galen's influence was still being felt. Near the end of the fourth century, Nemesius wrote that odors could be fragrant, foul, or "in between" and that smells reach the surface of the front ventricles of the brain (Telfer, 1955; also see Chapter 2).

Six hundred years later, Avicenna described the olfactory system. He considered smell to be a very delicate type of taste. In his *Canon of Medicine,* completed in 1025, he stated:

> Smell is a faculty located in a protuberance situated in the fore part of the brain, and resembling a nipple of the female breast, which apprehends what the air inhaled brings to it of odours mingled with the vapours wafted by air currents, or impressed upon it by diffusion from the odorific body. (1930 translation, 135)

FROM THE RENAISSANCE TO HALLER

During the Renaissance, Andreas Vesalius concluded that the central seat of olfaction was probably the brain itself, and not the ventricles. In particular, he pointed to the mammillary bodies. This represented a major change in thinking (Cain, 1978).

In 1587, Joachim (Johannes) Camerarius the Younger (1534–1598) wrote a thesis at Marburg comparing Galen's pneumatic views of olfaction with the newer ideas of Vesalius. Camerarius also classified odors as heavy, acute, and fragrant. He discussed the relation between smell and taste and questioned whether odors could be smelled in water.

The work of Camerarius set the stage for Conrad Victor Schneider (1614–1680) to reject and finally destroy the idea that the nose was just a hollow tube to and from the ventricles (Kenneth, 1928). In a work published in 1655, Schneider argued that the nasal mucosa, and not the ventricles or the brain, was the source of nasal secretions.[2] Schneider also cited the Bolognese anatomist Eustachio

---

[1]Like other physicians of his day, Galen was interested not only in diseases that affected the sense of smell but in the physician's use of his sense of smell to diagnose disease. He routinely used body odors and the smell of urine to diagnose stomach ailments, ulcers, tooth decay, and many other diseases.

[2]Although Schneider rejected Galen's pneumatic theory, he did not know that the nasal secretions came specifically from the racemose glands (Wright, 1898).

Rudio (d. 1611), who claimed to know a youth without smell. Post-mortem examination of the youth revealed no olfactory nerve (see Luciani, 1917).

To those individuals who still believed in the pre-Schneiderian notion that odorants could enter the brain, the growing use of smoking tobacco was a major source of concern. Jean le Royer Sieur de Prade (pseudonym, Edme Baillard; 1648–1677) wrote a treatise in 1668 to calm the populace on this matter. In *Discours sur le Tabac, où il est Traité Particulièrement du Tabac en Poudre* he maintained that tobacco and related substances do not enter the brain. Many other writers, such as Cornelius van Bontekoe (1640–1685), also wrote discourses on tobacco and related "toxic" substances.

Albrecht von Haller discussed smell in his *Primae lineae physiologiae,* which first appeared in 1747 (English edition, 1786), and in his eight-volume *Elementa physiologiae* (1757–66). Haller wondered why "man's curiosity about the particles that produce smell has been so minimal." He believed all bodies released their effluvia into the air, especially if rubbed or heated, and that these served as the stimulus for olfaction.

Haller maintained that it was difficult to categorize odors, yet he still classified them into (1) sweet (ambrosial), (2) foul (stenches), and (3) intermediate categories (such as roasted coffee). He wrote that smell was most important in allowing organisms to discover and avoid poisonous foods before they are tasted, a theory comparable to his ideas on the importance of the taste receptors (see Chapter 12). His contention that humans do not need smell as much as wild animals, which lack man's powers of reasoning and language, was similar to the case Thomas Willis made in 1664 and 1672. In 1664, Willis had written:

> The smelling Nerves, which have within the Skull their mammillary Processes depending on them, are much greater in an Ox, Goat, and in Cattle, and such like beasts that live on herbage, than in flesh-eating Animals; to wit, because in those there seems to be more need of the sense of smelling to be more exquisite for the knowing the virtues of the manifold herbs. Also these nerves are larger in all Brutes than in Man: the reason on which is, because they discern things only by these senses, and especially their food by the smell; but Man learns many things by education or nurture and discourse, and is rather led by the taste and sight, than by the smell in chusing his aliments. (1965 reprint, 138)

PUTREFACTION, AROMATICS, AND OLFACTION

Until the second half of the eighteenth century, air was viewed as an element and passive carrier of foreign particles that could affect the health of an organism. People had learned to avoid air infected with putrid exhalations. They thought stale air could hasten the putrefaction of the body by causing changes in the balance of the four humors. Thus, certain smells in pre-Pasteurian France (Louis Pasteur: 1822–1895), as well as in many other places, were closely associated with emotion and anxiety.

The release of foul-smelling organic substances, such as those that might occur with land excavation, could cause a panic. The smell of a corpse was even believed capable of causing death. It is said that a student in anatomy, who was given a liver from a putrefied corpse, fainted, was carried home, and died within 70 hours; another, the famous Fourcroy, was affected with burning eruptions on the skin; the other two . . . languished for a long time; the latter never recovered. . . . Chambon (Dean of the Paris Faculty of Medicine) completed his lesson amid stewards drenching their handkerchiefs in aromatic water, and probably owed his health to that cerebral overexcitement which brought on a copious sweating during the night, after a few outbreaks of fever. (Londe, 1838; see Corbin, 1986, 30–31)

People became afraid of crowding together in barracks, churches, and especially in prisons and aboard ships. Indeed, Roger Bacon, who lived in the thirteenth century, previously had called the smell of the plague most dangerous. He added that the odor of a prison was the second most feared smell of all. In contrast, many people believed that air coming from the body of a child was sweet and beneficial.

Racial differences related to odor were also thought to exist. It was presumed that Eskimos, Hottentots, and Black Africans emitted stronger odors than Western Europeans. This thinking was consonant with the idea that these races were evolutionarily closer to wild animals than Caucasians (also see Chapter 21).

Before the end of the eighteenth century, aromatics were used to "revitalize" bad air. In addition, perfumes were used not only because they made a person smell more attractive but because they were thought to prevent infection.

Traditionally, in a period of epidemic people tried to protect themselves by covering themselves with aromatics. Papon summarized these old practices in 1800: a sponge soaked in vinegar or a lemon studded with cloves or an odoriferous ball will be carried in the hand and sniffed from time to time. For people who are not in a position to afford odoriferous balls or perfume-pans, the best authors recommend sachets of rue, melissa, marjoram, mint, sage, rosemary, orange blossom, basil, thyme, serpolet, lavender, bay leaves, orange and lemon bark, and quince rind; they recommend that these always be present in apartments at the time of plague. Buc'hoz recommended sniffing red carnations and sprinkling clothing with pulverized angelica. (Corbin, 1986, 63–64)

Even during the reign of Charles II (1630–1685), King of England, Scotland, and Ireland, it was thought that sweet herbs could effectively fight off the plague (the smell of death). Rosemary, another effective agent, commanded high prices at the time. In contrast, the growing smell of industrial pollutants (e.g., from burning coal) did not cause any of the anxiety associated with putrefaction.

Not until late in the eighteenth century did Félix Vicq d'Azyr denounce the presumed effectiveness of aromatic fumigations. This followed acceptance of the fact that air was not an element but a mixture of gases, and helped pave the way for the use of substances that would destroy the source of the odor rather than simply mask the smell. More effective chemical management of stenches was soon accomplished with limewater, sulfuric acid, nitric acid, and related substances.[3]

CHANGING NINETEENTH-CENTURY ORIENTATIONS TO OLFACTION

In 1821, Hippolyte Cloquet (1787–1840) wrote a 758-page book on olfaction that must be considered in the context of both scientific developments in olfaction and the social history of odors. His *Osphrésiologie* (*osphre* is the Greek word for "smell") became the basic source on olfaction and continued to be cited frequently through the nineteenth century. Cloquet classified odors in a systematic way, described pathological material, talked about uses for scents, attempted to discuss the physiology of smell, and wrote about individual differences. His book on smell described some of the few known scientific discoveries, such as Robert Boyle's seventeenth-century observation that a substance loses weight as it emits its odor. Nevertheless, these discoveries were combined with intuitions and social philosophy, and Cloquet made many statements that were not based on facts.

Johannes Müller, in contrast, devoted only 7 pages to olfaction in his more scientific handbook, which was published in 1838. Further, Rudolf Wagner's (1844) *Handwörterbuch der Physiologie* contained a chapter on olfaction (by Bidder) just 11 pages long. These works, however, represented a decidedly more scientific orientation toward olfaction and a fairly clear break with the mixed naturalistic, sociological, and philosophical approach that characterized most earlier efforts.

By 1880, many scientific findings had become available on the anatomy, physiology, and phenomenology of olfaction. Consequently, the chapters were now longer and more detailed. For example, Ludimar Hermann's (1880) *Handbuch der Physiologie der Sinnesorgane* contained a 61-page section on olfaction, written by Maximilian Ritter von Vintschgau. In 1900, John Haycraft's 113-page chapter on olfaction appeared in Edward Schäfer's *Text-book of Physiology*. Then came the landmark *Physiologie des Geruchs*, written by the Dutch physiologist Hendrik Zwaardemaker (1857–1930) in 1895. This 324-page book stimulated a great amount of new work in the science of olfaction.[4]

OLFACTORY PRIMARIES

Much of the long history on olfaction centered on the search for "primary" odors, from which all complex odors could be derived. Linnaeus, the Swedish scientist shown in Figure 13.1, who devoted most of his life to botany, provided one of the first classifications of smells that appeared to have had some impact. His list of seven odors included aromatic, fragrant, ambrosial (musky), alliaceous (garlic), hircine (goaty), repulsive, and nauseous (see Linnaeus, 1756). At about the same time, however, Albrecht von Haller was writing that odors could be either sweet/ambrosiac, stench, or intermediaries between these extremes (e.g., Haller, 1786).

Neither Linnaeus nor Haller based his system on controlled scientific tests involving the identification of mixed odors. Nor were their lists based on chemistry. Instead, like most scientists at the time, their ideas came largely

---

[3]For a scholarly review and discussion of the social history of olfaction, including the use of perfumes, see Corbin (1986).

[4]These were the major works of the period, but not the only influential reviews. Among the others was a good review by Passy (1895).

Figure 13.1. Carolus Linnaeus (Linné, 1707–1748), the noted Swedish architect, naturalist, and physician who devised a system for the classification of odors.

from daily experience and philosophy. Not until the end of the eighteenth century did lists begin to be based on chemical structure (see Lorry, 1788).

Surprisingly, Zwaardemaker (1895) went back to Linnaeus, Haller, and other early investigators for his primaries. He took the original classification of seven odors proposed by Linnaeus and added empyreumatic (from Haller) and ethereal (from Lorry) (see Table 13.1). Zwaardemaker also presented subclasses within each of the primaries and provided common examples of each of the basic odors (e.g., empyreumatic = tobacco smoke or roasted coffee; ethereal = perfume or beeswax).

Testing for cross-adaptation emerged as the most popular procedure for identifying primary odors.[5] It involved

Table 13.1. The Nine Classes of Odors Presented by Zwaardemaker in 1895

1. Odores aetherei (ätherische Gerüche). Lorry.
2. Odores aromatici (aromatische Gerüche). Linné.
3. Odores fragrantes (balsamische Gerüche). Linné.
4. Odores ambrosiacei (Amber-Moschus-Gerüche). Linné.
5. Odores alliacei (Allyl-Cacodyl-Gerüche). Linné.
6. Odores empyreumatici (benzliche Gerüche).
7. Odores hircini. (Caprylgerüche). Linné.
8. Odores tetri (widerliche Gerüche). Linné.
9. Odores nausei (Erbrechen erregende oder ekelhafte Gerüche). Linné.

[5]The early literature on adaptation and cross-adaptation had been reviewed by Luciani (1917) and Woodworth (1938).

presenting one odorant and, after the subject adapted to it, seeing if this odorant affected perception of a different odorant. The idea was that if substance *A* affected perception of substance *B*, the two were related and should be put into the same category. If they did not interact, they should be placed into different categories.

For example, in one experiment the nose was "fatigued" to ammonium sulfide (see Haycraft, 1900). Testing with hydrogen sulfide and vapors of chlorine or bromine revealed that these odorants were now more difficult to detect (same category), whereas ethereal odors were unaffected (different categories). Vanilla and cumarin were used in other studies. These two substances were placed in the same category by Zwaardemaker in 1895. Nagel (1897), however, had a subject adapt to vanilla and then presented either vanilla or cumarin. In this experiment, only the cumarin was detected. Thus, it was concluded that these two substances should not have been put into the same category.

Hans Henning (1885–?) emerged as Zwaardemaker's chief adversary. His alternative classification was based on six classes of odors: (1) fragrant, (2) ethereal (fruity), (3) resinous, (4) spicy, (5) putrid, and (6) burned. He represented these odorants on a hollow prism (see below).

Although the primaries given by different investigators often overlapped, the lack of agreement among scientists led Edward Bradford Titchener, a student of Wilhelm Wundt and a founder of American psychology, to conclude:

> Smell . . . has more sensations than we can count or name; more sensations, probably, than all the rest of our senses put together. . . . When we add that odors can freely blend or combine to give new scents, you will understand that the number of smell sensations is enormous. (1915, 49)

The same sentiments had been echoed by John Haycraft in 1900. As can be seen in the quotation at the opening of this chapter, Haycraft felt that not only were there no established primaries but that this void was reflected in the absence of a special smell nomenclature.

THE ADEQUATE STIMULUS

A question asked by early scientists was whether the smell receptors had to be stimulated by particles in air. This issue had been addressed by Camerarius in 1587, who believed odorants could be smelled only in air. The hypothesis was tested again and again in the nineteenth century (Kenneth, 1928). For example, in 1827 Caspar Tourtual (1802–1865) put solutions into the nose and found they could not be "smelled."

Ernst Weber (1847) also experimented with odorous substances administered as liquids. He put a 10 percent solution of cologne in his nose, tilted his head back, and tried to smell the dilute cologne. He perceived no odor and also concluded that liquids could not activate the olfactory receptors (see Murphy, 1929). In retrospect, cologne might not have been the best stimulus to use because it could have anesthetized the olfactory receptors. Nevertheless, the finding that air was the essential normal medium for smell was confirmed and accepted (Haycraft, 1900; Nagel, 1894; Zwaardemaker, 1895).

This is not to say that smell could not be produced in unusual ways. Cloquet stated that Guillaume Dupuytren (1777–1835), the famous French surgeon, injected odorous fluids into the vein of a dog. Presumably because the olfactory area is heavily vascularized, the dog then went around sniffing as if it smelled something in the air. The same experiment was later performed on humans, and the results were comparable (see Bednar and Langfelder, 1930).

In 1882, William Ramsay (1852–1916) hypothesized that a substance had to have a particular molecular weight to be smelled under ordinary conditions. The threshold found was about 15 times the weight of hydrogen, or roughly the weight for ethane. Marsh gas, with a molecular weight of 8, could not be smelled and was compared to ultraviolet light. In contrast, propane, which weighed 22 times as much as hydrogen, was easily detected. In general, elements in Groups V, VI, and VII of the periodic table were associated with odors. Extremely heavy molecules, like very light molecules, were odorless.

Haycraft (1889) noted that substances close to each other within a chemical group had related but not identical smells. A good example comes from Group VII, where chlorine, bromine, and iodine have different smells but also share common features.

Ramsay and Haycraft believed molecular weight was associated with vibratory patterns and that the olfactory receptors were sensitive to these vibrations. As expressed by Ramsay:

> There is a probability that our sense of smell is excited by vibrations of a lower period than those which give rise to the sense of light or heat. These vibrations are conveyed by gaseous molecules to the surface network of nerves in the nasal cavity. The difference of smells is caused by the rate and the nature of such vibrations; just as difference in tone of musical sounds depends on the rate and on the nature of the vibration, the nature being influenced by the number and pitch of the harmonies. (1882, 188)

Both Ramsay and Haycraft hypothesized that all of the sensory systems functioned according to the same principles.[6] Thus, the endings of the olfactory nerves were tuned to vibrations in a manner comparable to the hair cells of the organ of Corti in the ear. Nevertheless, these investigators were forced to concede that "we know nothing as to how it is that ether vibrations stimulate the cones of the retina, still less can we guess at the actions of vibrating atoms and molecules of ordinary matter on the sensitive end-organs of the nose" (Haycraft, 1889, p. 177).

By the second decade of the twentieth century, at least one prominent scientist stated that the vibratory theory of olfaction had no basis in fact and should not command further attention (Luciani, 1917).

OLFACTORY RECEPTORS

It was recognized early on that the nose was the organ for smell, but the question remained open as to just where in the nose the receptors were located. The earliest idea was that the receptors were located throughout the lining of the nose. Haller still advocated this view in the eighteenth century. In the English edition of his *Primae lineae physiologiae,* one finds:

> The sense of smelling is performed by means of a soft pulpy membrane, full of pores and small vessels, which lines the whole internal cavity of the nostrils, being thicker upon the septum and principle convolutions, but thinner in the sinuses. Within this membrane are distributed abundance of soft nerves throughout the middle of its fabric, from the first pair, which descend through the holes of the os cribiform into the septum narium; but in such a manner, that it is very difficult to trace them to their extremities in the septum. (1786, 265–266)

In the mid-nineteenth century, the "whole nose" theory was challenged by anatomical work that showed the receptive region was confined to a small patch of tissue in the upper alcoves of the air passages. The investigators who made this discovery also tried to identify the various types of cells in the receptor area.

In 1855, Conrad Eckhard (1822–1905) described two distinct classes of cells in the nasal epithelium of the frog. One was a cylindrical cell and the other was a fusiform cell. Although Eckhard suspected the two were functionally different, he was unsure which was the receptor for olfaction. In 1876, working with amphibia, he described a third type of cell, the basal cell.

It was Max Schultze who firmly localized mammalian olfactory receptors high in the nasal cavity. In 1856, he described a small area in the superior turbinated bone where there resided special cells with hairlike processes. Because this distinctive area was yellowish in humans, sheep, and cattle, it was given the name *locus luteus.*[7]

After first reporting on olfactory and columnar epithelial cells (believed to be special supporting cells), Schultze (1862) described Eckhard's three types of cells in mammalian specimens. Schultze argued that the nucleated cells were the sense cells. Although he was unable to see how the receptors were connected to the olfactory bulb, he felt sure that time would prove him correct. Nevertheless, he wrongly assumed that the olfactory receptors in mammals do not have hairs or fine cilia.

With the advent of better histological stains, more was learned about the olfactory receptive region (Babuchin, 1873). This work involved the use of gold chloride and later methylen-blue stains. Later, Golgi stains became available to trace the olfactory nerve fibers from the sensory surface (Parker, 1922; Read, 1908; also see Chapter 3).

Scientists soon accepted the idea that there were supporting cells and bipolar receptor cells, with spindle-shaped cell bodies and cilialike hairs, in the olfactory mucosa. Some 6–12 very delicate cilia were associated with each olfactory receptor cell in humans (Brunn, 1892). The very thin axons of the olfactory cells still proved extremely difficult to trace individually. Together they formed the olfactory bundle, which passed through the cribiform plate en route to the olfactory bulb.

---

[6]The theory of vibrations has strong roots in Greek philosophy and in the later works of Thomas Hobbes (1588–1679), David Hartley, and Herbert Spencer. In medicine, vibratory theory was discussed by William Ogle in 1870.

[7]The area comparable to the *locus luteus* of humans and cattle was found to be considerably more brownish in rabbits, dogs, and guinea pigs. Later histological work showed that the distinct coloration in the receptor area is due to discrete pigment granules in the supporting cells and in the acinar cells of Bowman's gland.

Even with these advances, there were still arguments about the nature of the olfactory endings in the epithelium. In particular, some investigators thought there were olfactory buds, analogous to those seen on the tongue for taste (Blaue, 1884; Disse, 1896a; but also see Kamon, 1904). Another mistake involved confusing the endings of the trigeminal nerve with the olfactory nerve endings in the nose. Many researchers saw trigeminal nerve endings at the border of the respiratory region in man, but not all recognized these endings as distinct from the olfactory nerve endings (see Brunn, 1892; Disse 1896b).

In general, the studies from the turn of the century suggested the olfactory system had certain "primitive" anatomical features.

> In the position of the olfactory nerve cell we apparently have a primitive condition. This is the only case in vertebrates where the nerve cells are within the epithelium, as are those for the tactile sense in many invertebrate forms. In other organs of sense there is a gradual recession of the ganglion cell until, in the ganglion of the dorsal root of the spinal cord, the central nervous system is approximated. The branches for the tactile sense end freely either in special organs (tactile corpuscles) or in the free end-knots within the epithelium, but do not reach the surface of it; while the branches of the nerves of other sense organs end freely *among* special cells, but do not reach the free surface of the epithelium. In the olfactory region the olfactory hairs are above the free surface of the epithelium and in direct contact with the air. (Read, 1904, 43)

In addition, there was a growing belief that the olfactory cells may be primitive in another way; namely, these cells may be lacking physiological specificity. The inability to agree on olfactory primaries suggested this. To quote Titchener:

> It is highly-probable that the sensory cells of smell are the seat of only a few chemical processes, by whose combination all the wealth of odors is created, just as the cone-cells of the retina are the seat of those three reversible processes (black-white, blue-yellow, red-green) whose combination endows us with the variety of daylight vision. We have yet, however, no such definite grounds for hypothesis as we have in the case of sight; we cannot even guess what these processes are, or how many of them are taking place in the smell-membrane. (1915, 63)

AIR CURRENTS AND RECEPTOR ACTIVATION

The olfactory epithelium seemed to be out of the pathway of the main air currents that accompanied breathing (see Figure 13.2). Consequently, people began to wonder how odorants could get to such a seemingly inaccessible place.

In one experiment, the investigator cut the head of a cadaver lengthwise (Paulsen, 1882). As shown in Figure 13.3, he put squares of red litmus paper in various parts of the nasal cavities and fitted the head back together. He then pumped ammonia through the nostrils. Only the litmus paper well below the area where the olfactory receptors were located turned blue from the ammoniated air. From this observation, it was concluded that if air were to reach the olfactory area, it would have to be by smaller currents caused by sniffing or breathing or by eventual diffusion from lower regions of the cavity.

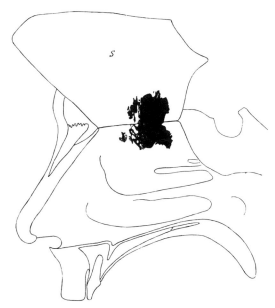

Figure 13.2. Diagram of the nasal cavities of a 40-year-old man. The darkened zone signifies the distribution of the olfactory nerve. (From Brunn, 1892.)

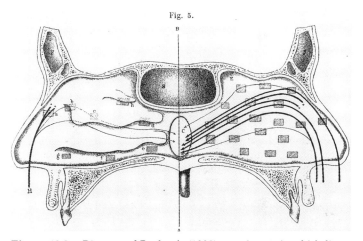

Figure 13.3. Diagram of Paulsen's (1882) experiment, in which litmus paper was used to line the air passages of a cadaver while ammonia was pumped through the passages. (From Zwaardemaker, 1895.)

This type of experiment was repeated with additional human cadavers. In one investigation, the mucous membrane was stained black, and the experimenter used a bellows to blow light-colored tobacco smoke through the nostrils (Franke, 1893). A variation was also tried by Zwaardemaker (1895), who used a plaster model of the nasal cavities of a horse. He placed a glass plate against one of the split halves, and the soot of a petrol lamp was aspirated through the nose and followed visually. The findings, shown in Figure 13.4, again revealed that the olfactory area was not in the direct path of the major air currents. This was confirmed on a living person by aspirating fine magnesia powder into the nose of a subject, and then viewing the mucous membrane with a rhinoscope to see where the powder went (Kayser, 1890).

These studies led to the conclusion that the olfactory receptors had to be exquisitely sensitive to minute stimulation. Although this could come about by diffusion, this was a slow process. More likely than not, smell also depended

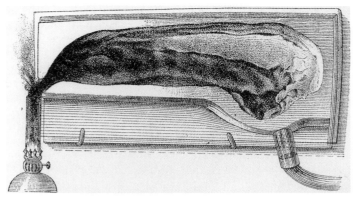

Figure 13.4. Results of an experiment in which Zwaardemaker aspirated the soot of a lamp through a model of the nasal cavities of a horse. The intention was to see if the particles would reach the olfactory receptor area on the roof of the nasal cavity. (From Zwaardemaker, 1895.)

upon the air being moved out of the main pathway for breathing. This could be accomplished by swallowing, during which the soft palate is raised and air is driven toward the olfactory region.

William Ogle (1870) emphasized that sniffing also closed the usual respiratory pathway and forced air into the olfactory area. He pointed to individuals with facial palsy who lost the use of muscles on one side of the face and were now unable to sniff or smell effectively on the affected side. Ogle wrote:

> It is plain how facial palsy must interfere with voluntary olfaction. . . . It interferes by preventing the active dilation of the nostrils . . . (and) by preventing the lateral compression which is required to close the respiratory channel. At the same time it will not materially interfere with the perception of flavours, for there is no hindrance to the passage of odours from the mouth upwards through the posterior nares. (1870, 270)

Many treatments for clogged nasal passages, due to colds and other illnesses, work as "odorivectors." Agents such as eucalyptus, ammonia, and methylol produce a rapid flow of mucous, which softens the thicker mucous obstructing the olfactory channels, thus permitting the passages to be used more effectively. Centuries earlier, Robert Boyle had written that spirits of *sal armoniack* opened nasal passages blocked by colds (Moncrieff, 1944).

Work on the nasal pathways also led to the conclusion that lighter molecules might be reaching the upper portion of the mucosa, whereas heavier molecules might reach only lower parts. Thus, different mosaics or patterns of stimulation were envisioned for different olfactory stimuli (Zwaardemaker, 1930).

OLFACTORY NERVE AND BULB

Interestingly, a number of people, including François Magendie (e.g., 1824), claimed the trigeminal nerve as one nerve, or as *the* nerve, of olfaction. Magendie came to this conclusion after he cut the olfactory nerve in dogs and found his animals still responded to ether and ammonia. Moreover, after he blindfolded a dog, it still found meat and cheese, but left the meat alone if it were covered with tobacco.

Magendie's critics (e.g., Valentin, Schiff, and Prévost) pointed out that the substances he used typically were irritants and consequently could have acted on the trigeminal nerve. They argued that had he used mild, non-noxious substances, he would not have reported "olfactory" discrimination after cutting the olfactory nerve. Magendie, not one to give in, responded that not all of his substances were irritants and pointed to the use of lavender oil. Even so, Magendie's position encountered strong opposition from the start and did not attract many followers.

Early evidence for the importance of the olfactory nerve in smell came from a number of clinical reports of anosmic ("without smell") people who had lost their olfactory nerves (e.g., Eschricht, 1825). Nevertheless, here too there were occasional contradictions in the literature. One involved Claude Bernard, who, while working with Magendie, examined the brain of a woman who he thought showed a complete absence of both olfactory nerves. Bernard (1858) inquired among friends and family as to whether the woman retained a sense of smell. He was told she enjoyed flowers, hated pipe smoke, and was a good cook.

Cases of trigeminal "paralysis" without loss of smell also appeared in the literature at this time. One concerned a person with a tumor that affected the trigeminal nerve (Bishop, 1833). The patient was left sensitive to odors, but insensitive to touch and irritants, like ammonia and snuff, on the side of the tumor. Magendie's position was further weakened by additional cases of otherwise anosmic people who could still respond to fumes from ammonia or ether (e.g., Dugès, 1838).

In 1839, Gabriel Valentin, a pupil of Johannes Purkyně, showed that a rabbit with cut olfactory nerves would not sniff a dead rabbit, whereas other rabbits would sniff the dead animal. Similarly, Moritz Schiff (1859) cut the olfactory nerves in four of five puppies and noted that only those with cut olfactory nerves could not find the mother's nipples (see Zwaardemaker, 1895). All of the pups still sneezed and wheezed when ammonia or ether was presented.

Some debates about the olfactory nerve (cranial nerve I) concerned whether it should be classified as a peripheral nerve or a tract. Rudolph Albert von Kölliker initially believed the first cranial nerve had a central origin, but then reversed his position to one favoring a peripheral origin (see review by Bedford, 1904). The issue seems to have been settled early in the twentieth century when it was noted that "at no stage have structures been observed originating in the brain that may be considered to take part in the formation of the olfactory nerve" (Bedford, 1904, p. 408).

The olfactory nerve axons enter the olfactory bulb from the anterior and ventral positions and fan out over the surface of the bulb. The olfactory bulb is very large in monotremes (egg-laying mammals) and insectivores, of moderate size in most mammals, and relatively small in most primates including man.

The general structure of the olfactory bulb was worked out when the silver stain was developed. In 1875, Camillo Golgi applied his new stain to the olfactory bulb and identified the endings of the olfactory fibers, the granule cells, and the large mitral cells, along with their arborizations within each glomerulus. Soon after, Santiago Ramón y Cajal began to describe his findings on the olfactory bulb. Arthur van Gehuchten and Kölliker (1896) added more detail to the anatomical descriptions provided by Golgi (1875) and Ramón y Cajal (1890, 1894).

The first electrical recordings in the olfactory bulb sampled multiple units. The fibers were found to be very thin and fragile, precluding single unit recordings for many years. When Adrian finally recorded from single units in the 1950s, he reported that the evidence for unit specificity was not very good.

Stimulation of the olfactory bulb elicited olfactory sensations in conscious human subjects undergoing surgery for epilepsy. This observation was made by Wilder Penfield and his associates (e.g., Penfield and Erickson, 1941). Unpleasant or strong olfactory sensations, exemplified by burning leather, were reported most frequently.

CENTRAL PROJECTIONS

The early anatomical studies of the projections from the olfactory bulb suggested there were three distinct pathways (stria) from the glomeruli. These were the lateral, medial, and intermediate striae. The basic schema held that the lateral olfactory tract went to the temporal lobe (pyriform cortex and the amygdala), whereas the medial olfactory tract projected to the septum. The intermediate tract was thought to cross in the anterior commissure to the opposite olfactory bulb. There was also a fairly widespread belief that the olfactory bulb projected directly to the hippocampus and entorhinal cortex.[8]

As the twentieth century progressed, the trend was to reduce the three systems to two. The lateral olfactory tract, occasionally referred to simply as the olfactory tract, was found to be made up largely of mitral cells, as had been suggested by Ramón y Cajal (1911). This tract projected back without crossing to end primarily in the temporal prepyriform cortex and in the corticomedial nuclei of the amygdala.

Projections in the anterior limb of the anterior commissure, the second system, terminated in the nucleus of the stria terminalis, in the central amygdaloid nucleus, and in the deeper layers of the opposite olfactory bulb. The major input to this system was the tufted cells of the olfactory bulb.

With more careful analyses, it became clear that there were probably no direct connections from the olfactory bulb to the septum, to the hippocampus, or to the entorhinal area (see Chapter 20). Nevertheless, these structures, part of the so-called rhinencephalon, may receive higher-order projections from some of the regions already mentioned.

CASE STUDIES OF THE LIMBIC LOBE AND THE CEREBRAL CORTEX

The position generally accepted in the late 1800s, and later refined, was that the ventral temporal lobe and/or its subcortical structures played an important role in olfaction. This idea was supported by the clinical work of William Ogle and John Hughlings Jackson.

Ogle (1870) called attention to the fact that many people with anosmia caused by strokes show lesions in or near the ventral region of the temporal lobe. Later clinical reports have supported this general finding, but, as might be expected, the human clinical literature has not been consis-

tent. For example, Salomon E. Henschen (1919) looked at patients with abscesses and tumors of the temporal lobes and noted that even if the uncus, the hippocampal gyrus, and Ammon's horn were destroyed on one side, smell might not be impaired. Still, with considerably more refined testing procedures, it was found that unilateral lesions of the temporal lobes usually did affect olfaction on at least one side (Elsberg, 1935).

As for Jackson, he presented a paper in 1889 dealing with olfactory auras and epilepsy. This particular report did not include autopsy material, but it was important for associating smell with intellectual auras ("dreamy states") in epileptics. In general, the sensations were described as horrible. Smells like those of phosphorus or rotten eggs occasionally led to nausea in these patients.

Jackson wrote up his clinical findings in greater detail in 1890, with Charles Beevor (1854–1908) presenting the anatomy. One case involved a 53-year-old epileptic who was admitted to Queen's Square Hospital, London. The subject experienced terrible smell sensations ("burning dirty stuff," "nasty dreadful smell") that made her feel as if she were suffocating. These auras were associated with "dreamy states" and complex visual imagery. The patient died, and a sarcoma the "size of a tangerine orange" was found in the right hemisphere. It involved the hippocampal lobule and the amygdala but spared the uncus.

In his discussion of this case, Jackson cited related cases in which olfactory sensations were experienced and in which the olfactory bulb and tract were spared, as was found in his case. One patient experienced "a dreadful disagreeable smell" prior to motor seizures and had a tumor that involved half of the temporo-sphenoid lobe. In another case, a woman suddenly perceived disagreeable odors "of a foetid character," like burning rags or a match, which felt as though they were choking her (from Hamilton, 1882). In this case, in which the woman had epilepsy for 10 years following a fall, there was shrinkage of the right temporo-sphenoid lobe.

Jackson and Beevor (1890) noted that their fellow countryman William Gowers (1885) had found that olfactory auras were associated with a sense of suffocation in two cases and with visual hallucinations in two cases. Actually, Gowers found olfactory auras to be quite rare. He noted sensory auras in only 119 cases from his sample of 1,000 epileptic cases, and in only 7 did olfactory auras precede the fit. The sensations were usually unpleasant and were compared to "rotten eggs," "Tonquin beans," "sulfur," odors from "the pit of the stomach," and so on.

Jackson and Beevor also believed their observations were in accord with David Ferrier's statements about olfaction in *The Functions of the Brain*. Ferrier (1886, p. 320) wrote that disorders of smell "are evidently related to lesions of the hippocampal lobule and the neighbouring regions." Ferrier stated that electrical stimulation of the hippocampal lobule resulted in a particular torsion of the lip and nostril on the same side. He argued that this was just how a normal monkey would respond to a strong or disagreeable odor.[9]

---

[8]Reviews of the central anatomy of the olfactory system can be found in Allison (1953a, 1953b).

[9]Olfactory auras were noted by the Ancient Greeks and Romans. Galen, for example, wrote about olfactory hallucinations and stated that they constituted signs of disease. But unlike Jackson, Ferrier, and Penfield, Galen and his contemporaries had no idea that olfactory auras would prove to be a good indicator of seizure disorders in the temporal lobe.

The studies of olfactory auras by Wilder Penfield were largely in agreement with what Jackson had said. In a review of many human cases studied in Montreal, it was found that "the sensation seems to be always disagreeable, a sharp or smoky or disgusting odor" (Penfield and Erickson, 1941). Penfield also held that olfactory auras were not common except in cases of brain tumors in or near the temporal lobes.

Penfield's work on electrical stimulation of the brain in conscious human subjects suffering from epilepsy revealed that olfactory sensations occasionally could be evoked from the uncus. It was uncertain, however, as to whether the uncus itself was responsible for the sensations or whether they were due to stimulation of the amygdala just below the cortical surface (Penfield and Jasper, 1954).

FOREBRAIN LESIONS IN LABORATORY ANIMALS

In the 1870s, David Ferrier began to study the sense of smell in monkeys with various cortical lesions. He used fumes of acetic acid to test for olfaction, while at the same time acknowledging that this substance can also irritate the nerves of the trigeminal in the nose. He also observed how a monkey recently operated upon would sniff its food before eating it. Ferrier (1886, p. 318) wrote that "when a piece of apple was again offered, the animal seized it, and held it to its nostrils repeatedly, as if it had difficulty in making out the nature of the object by its sense of smell."

Although Ferrier's (1875, 1886) methods lacked precision, he concluded that most cortical lesions did not affect the sense of smell, whereas those in or near the uncinate convolution impaired smell from the nostril on the side of the lesion.

> The effects of irritation of this region are very consistent and characteristic, and are of the same nature as direct irritation of the nostril or the olfactory bulb itself. Destruction of this region causes abolition or diminution of reaction to smell on the same side of the lesion. Taken together, these facts establish the localization of the sense of smell in the subiculum, or tip of the temporo-sphenoidal lobe. (1875, 471)

Thus, Ferrier localized the sense of smell in the same general area (the hippocampal lobule) in which he erroneously believed taste was localized. Yet, in his favor, Ferrier was considerably more confident about his data on olfaction than he was about the results of his tests for taste, where his methods of testing were even more questionable. Figure 13.5 shows one of his diagrams of the olfactory region of the cortex.

Ferrier's olfactory findings were confirmed some years later when Gorschkow (1901), also using "crude" testing procedures, found that large ablations of the pyriform lobe severely affected smell in the ipsilateral nostril of his dogs. Dogs with unilateral lesions showed milder contralateral effects.

In 1933, Dusser de Barenne noted normal smell reactions in cats with neocortex extirpated, leaving only the phylogenetically older cortex intact. He felt this was consistent with the findings reported in the small but growing human clinical literature on cortical lesions and olfactory deficits.

Lesion studies involving rats conditioned to make olfactory discriminations typically failed to identify small corti-

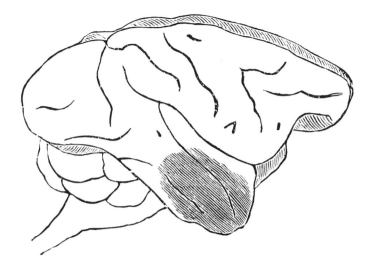

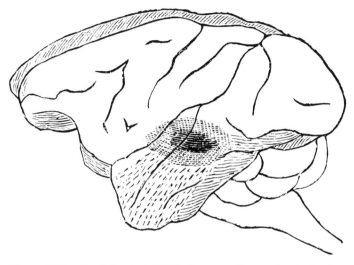

Figure 13.5. David Ferrier's (1886) diagram of the monkey brain showing a lesion that caused a loss of taste and smell. Ferrier localized both functions in the temporal lobes.

cal or limbic areas in which focal lesions produce reliable deficits. It should be noted, however, that stimuli well above threshold levels were used in the earlier conditioning experiments. In one such study, rats sustained lesions of the pyriform lobe, other cortical areas, the hippocampus, the amygdala, or the fornix (Swann, 1934, 1935). They were then taught to paw through one of two piles of wood shavings (anise versus creosote). No significant lesion effects appeared on this task. Rats with lesions of the olfactory tracts, however, performed poorly.

More negative findings were obtained by experimenters who tested rats with lesions of the anterior thalamus and septum (Brown and Ghiselli, 1938; Lashley and Sperry, 1943). Clear olfactory deficits also failed to emerge in conditioning studies with dogs that sustained forebrain lesions (Allen, 1940, 1941).

THE ANOSMIAS

William Ogle (1870) described a number of individuals who were hit on the back of the head and subsequently developed anosmias. He made the point that blows to the back of the head were more likely to cause anosmias than any other head injuries because they may rupture the olfactory nerves.

> It is very easy to understand how a blow, which is not sufficiently violent to do serious mischief to the anterior brain generally, may still suffice to tear the olfactory nerves, owing to their very small size, and, still more, owing to their excessive softness. In only one recorded case of loss of smell from a blow on the head, have I found mention of the exact part struck. There also, as in these cases, the blow was on the occiput. (1870, 266)

Ogle wrote that his subjects with anosmia complained about a loss of taste. Although he believed that the gustatory and olfactory pathways were separate, he recognized that flavor depended to a large degree on smell interacting with taste. He noted that individuals with unilateral facial palsy lose most of the sense of smell on the paralyzed side of the face and complain about food not tasting the same, even though the gustatory system remains functional.

Ogle made two other interesting points in his 1870 paper. One was that there might be an association between anosmia and aphasia, possibly because lesions of the temporal lobes could affect both smell and speech. His second point was that olfactory acuity may be related to skin pigmentation. Here he discussed differences in olfaction and pigmentation across species, with development, and as related to race.

Ogle also discussed mutations. He cited examples of white sheep, horses, pigs, and rhinoceroses that presumably eat plants and foods which pigmented members of the species ostensibly avoid. He hypothesized that the degree to which cells in the olfactory epithelium are pigmented may explain behavioral differences in foraging and olfactory acuity. His idea was that darker colors can absorb more odorants than light or white pigments.

> Mr Darwin and other writers . . . suppose that both coloured and white animals alike eat the poisonous food; but that, owing to some hidden constitutional peculiarity correlated with their colour, the white animals are poisoned by a plant which is innocuous to the black or coloured. . . . In the absence of clear proof to the contrary I cannot but think it more likely that the white animals suffer because they eat poisonous plants, while the black animals escape because they avoid them; and I think I have shown fair grounds for suspecting that the explanation for this difference of habits is to be found in differences of acuteness of smell, depending on the presence or absence of abundant pigment in the olfactory region. (1870, 282).

Ogle's theory attracted some supporters who also noted that albino animals seemed more susceptible to poisoning. For example, it was said that white pigs eat the poisonous paint root *(Lachnanths tinctoria),* which pigmented pigs avoid, and that in parts of Italy only black sheep are raised because white ones are often poisoned by the curly-leaved St. John's wort *(Hypericum cristum)* (see Allison, 1953a, 1953b).

Ogle's ideas about albinism linked some anosmias to genetics. Inherited defects in olfaction in nonalbinos was the subject of a short paper that appeared in 1918. The author described a man born in Kiev, Russia, who was unable to distinguish odors unless they were irritants capable of stimulating the trigeminal nerve (Glaser, 1918). The family was studied, and it was found that many other individuals also showed the defect, which was associated with dental abnormalities, an unusually large thumb, and excessive sexuality. The subject had one brother just like him and one who was a bit better at smell, as well as two normal sisters. Their mother could not detect odors, nor could the mother's father and some cousins. The subject himself wrote (p. 648) that "after making inquiries among other people I know from my former place of residence, I came to the conclusion that locality inbreeds this defect so that quite a number are affected with it." Pedigree analysis suggested the inherited anosmia could have been caused by a sex-linked recessive gene.

As might be expected, scientists have shown more interest in cases of "specific anosmias," since the belief was that these would shed more light on the olfactory primaries. This field has been characterized by contradictions and inconsistent findings. In the same issue of *Science* in which the aforementioned pedigree analysis appeared, there was another short paper that went beyond a simple anecdotal approach to this subject (Blakeslee, 1918). The writer was involved with classifying verbena flowers and noted that one had an extremely pleasing odor. He mentioned this to his assistant, but the man could not smell the flower presented to him. Instead, he selected another flower for its special fragrance, one the author was not able to smell! The two men then tried to rank the verbena plants by fragrance and repeatedly came out with opposite lists and different descriptive terms.

To see whether this odd anosmia would be found in other members of the population, the two men tested other people. Given a variety of plants to choose from, their subjects either chose the one the author had appreciated or the one his assistant could smell, and not others from the group. Each subject could not understand how anyone could choose the other verbena plant, as a result of being "smell blind" to the other. There were no gender differences, but the ratio was two people smelling the author's plant, for every one person who could smell only the plant selected by his assistant. Clearly, genetics could affect olfaction.

ZWAARDEMAKER AND OLFACTORY PSYCHOPHYSICS

Hendrik Zwaardemaker began his career at a military hospital in Amsterdam. From the beginning, he believed that smell should be analogous to vision. That is, it should obey the same laws of mixture and adaptation. The question Zwaardemaker faced in the mid-1880s, while at the College of Veterinary Medicine in Utrecht, was how to get enough control over the stimulus to test his hypothesis.

> As I thought over this question, I took in my hand a glass tube and a piece of rubber, with which to join the tubes, and, thoughtlessly slipping one over the other, I suddenly had the inspiration that we, in that manner, had at hand a method of measuring the odorous sense, and, through that, the sensory value of any given odor. The laboratory assistant was immediately called . . . and in a quarter of an hour, the "olfactometer" was born.
> A few days after the "invention" of the olfactometer, I walked with Donders from his house to the laboratory. Passing along one of the quiet canals that cross the Dutch town, I took the olfactometer out of my pocket and handed it to Donders. He stopped, looked at the little instrument,

tried it with slight astonishment, and, returning it to me, said with his warm voice, "That is beautiful, for it is simple," and happily I could reply, "It is useful too, for we can now mix odors, as you mix colors in your double-slit spectroscope, using odor equations instead of color equations," and then we talked along on the analogies between the two senses. (1930, 493–494)

In this way, in 1887, the first olfactometer was constructed to study olfactory perception and to assess less than complete defects in the sense of smell. This instrument is shown in Figure 13.6. In 1889, a drawing and description of the olfactometer with the outer tube made of rubber appeared in *Lancet*. Zwaardemaker reported that the rubber tubing had been impregnated with an odorant, such as garlic, and that he had measured how much of the tubing had to be exposed for an odor to be detected or identified. The number of centimeters of exposure required to detect an odor was called an "olfactie."[10]

Zwaardemaker described his techniques and instruments and brought his findings together in *Die Physiologie des Geruchs*. In this monograph, which appeared in 1895, he showed several olfactometers (see Figures 13.7 and 13.8) and presented absolute threshold, difference threshold, reaction time, adaptation, and cross-adaptation data. In general, he observed very rapid adaptation and plotted curves to show this.[11] He also noted that adaptation to one odor sometimes blunted sensitivity to others. For example, adaptation to ammonium sulfide affected sensitivity to hydrogen sulfide, bromine, and chlorine. In contrast, he found some odors were antagonistic and competed with each other. Occasionally, one would dominate the other before the situation would reverse.

Zwaardemaker was unable to recognize any simple rules that would predict or explain his diverse findings. Thus, he had to conclude that the effects of mixing odorants were not comparable to the effects of mixing colors (where one could no longer perceive the wavelengths that went into the mixture). Therefore, although he listed nine qualities and gave them olfactie values, he cautioned against taking them or his equations too literally.

Zwaardemaker continued to publish books and papers on olfaction after *Die Physiologie des Geruchs* appeared.[12] His second major book on smell, *L'Odorat*, appeared in 1925. In addition, in 1927 he described his *camera inodorata*, which was a large box with glass walls into which the subject's head was placed. A mercury lamp and a ventilation system reduced unwanted odors in the box and permitted the subject to work under tightly-controlled conditions.

[10]Zwaardemaker was not the first investigator to try to get absolute thresholds for smell. In the 1880s, some experimenters used a room into which a weighted quantity of odorant was released and mixed with a fan to obtain thresholds (Fischer and Penzolt; see review by Troland, 1930). The subject's task was to enter the room and detect the odor. By varying the dilutions of the odorant, they measured the minimum amount of odorant that could be detected. For the chemical, mercaptan, the threshold was found to be one part chemical to approximately 50 billion parts air. The actual weight in 50 cc of air was calculated to be 1/46 million of a milligram.

[11]Titchener's comment on the rapid adaptation of the olfactory system is worth quoting. In *A Beginner's Psychology* (1915, p. 51) he wrote: "Sight and hearing are better suited than smell to our everyday needs; for smells very soon fade out and disappear; indeed, if they did not, the work of garbage collectors or of medical students in the dissecting room would be permanently disagreeable."

[12]Of Zwaardemaker's approximately 300 lifetime publications, about 50 dealt with olfaction. An autobiographical essay of his career as an olfactory researcher was published in 1930, the year in which he died. For one of his obituaries, see Noyons (1931).

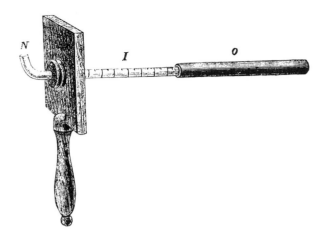

Figure 13.6. Hendrik Zwaardemaker's olfactometer as it first appeared in the *British Medical Journal* in 1888.

Figure 13.7. A modification of Zwaardemaker's original olfactometer. This instrument utilized a porcelain cylinder over a glass barrel. (From Zwaardemaker, 1895.)

Figure 13.8. A double olfactometer for presenting the same or different odors to the two nares simultaneously. (From Zwaardemaker, 1895.)

A MYRIAD OF TECHNIQUES OF OLFACTOMETRY

In the preceding section, two basic techniques of olfactometry were mentioned: the portable olfactometer and a larger apparatus that fit over the subject's head. Other techniques were tried, and the diversity of methods used for studying olfaction was almost as great as the number of laboratories involved in the study of smell at this time.[13]

One of the first techniques of olfactometry, and a procedure quite different from those described above, was described in the mid-1800s by Valentin. He sealed a measured amount of an odorant in a glass tube and put the tube into a larger tube that was then sealed. When the smaller tube was broken, a known amount of odorant filled the larger tube. By using larger and larger tubes, Valentin was able to vary the dilution of the odorant.

A variation involved opening a stoppered bottle filled with odorant. But rather than vary the amount of odorant in the bottle, the subject was positioned at different distances from the bottle. This procedure was of some use for comparing the sensitivity of different subjects to the same odor. This technique, however, was more questionable for comparing sensitivity to different odorants, mainly because substances differ in rates of diffusion.

Another procedure related to Valentin's technique was to evaporate varying amounts of odorant in a flask of ethyl alcohol and then have the subject smell the container. Hans Henning also used flasks, but in his case the methodology was different. He evaporated odorant in one flask, which in turn was connected to any number of additional flasks to make smaller and smaller concentrations.

Another variation put filter paper into a given concentration of odorous liquid. The paper was hung in a wide glass tube, and a pump was used to blow air past the paper into the nostrils.

The procedures described above, including those used by Zwaardemaker, had many limitations (see Wenzel, 1948). For each technique, one or more of the following problems existed: (1) lack of control over the size of the inhalation; (2) temperature inconsistencies; (3) using a medium that has its own odors; (4) lack of control over the number of molecules actually in the air; (5) loss of substance before it is presented and subsequent changes in dilution; (6) reuse of the same container; and (7) absorption by the container.

These problems were addressed in part with the development of the "blast injection" in the 1930s (Elsberg and Levy, 1935–36). The experimenters pumped air into a 500-cc bottle with odorant and then released a pinch clamp on the tube leading to a nosepiece to send a jet of air into the subject's nostrils. The experimenters could put varying amounts of water in the bottom of the flask to control the amount of air in the container and were able to apply a mercury manometer to measure the pressure of the "blast."

This technique achieved some control over the force (pressure) of the stimulus and precluded sniffing by the subject, variables known to be important (Elsberg, Brewer, and Levy, 1935–36). Still, the number of molecules reaching the receptor surface remained unknown, and with a hand-operated clamp there were variations in stimulus onset and offset characteristics. The stream injection, which used a compressed-air tank to provide a continuous

"stream" of stimulus, helped address the latter problem, but as with all earlier techniques, the question of the number of molecules reaching the nasal receptors in different subjects was a matter of speculation.

Needless to say, the same criticisms and many more pertained to the early olfactory experiments on laboratory animals. In 1870, Ogle wrote: "We possess then no gauge by which to test directly the acuteness of smell in dumb animals; and we must content ourselves with the less direct plan of observing how far they perform unerringly such actions as imply a keen scent" (p. 285). He agreed with Haller that actions such as the avoidance of poisonous foods reflected the development and integrity of the olfactory system.

The early lesion studies did not involve quantitative assessments on the amount of olfactory stimulus released into the testing environment, and absolutely nothing at all was known about how much reached the dog's or rat's olfactory epithelium. The experimenters also did not attempt to use odorants near threshold levels. This may have been one reason for the sparsity of brain lesion effects in studies involving laboratory animals given focal lesions.

HENNING'S PRISM

In 1916, Hans Henning published a lengthy book on olfaction, which represented the largest work on the subject since Cloquet's 758-page book appeared in 1821. Henning was not kind to earlier researchers in this work. In a review of the book, published in the same year, the author remarked that Henning's "attitude toward earlier or contemporary work is minutely critical and, in the main, destructive" (Gamble, 1916, p. 135). Many felt Henning was particularly unfair to Zwaardemaker.

The book included a number of articles Henning had just published and described olfactory stimulation of one nostril (monorhinic stimulation), both nostrils with the same substance (dirhinic stimulation), and the two nostrils with different substances (dichorhinic stimulation). It also described differences in findings with the subject knowing what the smell was to be beforehand ("object smell" or *Gegenstandsgeruch*), as contrasted with "true smell quality" *(Gegenheitsgeruch),* which was tested when the subject was not informed about the upcoming stimulus.

The book included Henning's smell "prism," which was based on six principle odors. This is shown in Figure 13.9. Henning's basic idea was that complex odors could be shown on the face, edges, and sides of the prism in appropriate positions between the corners occupied by the primaries (flowery, fruity, resinous, spicy, burnt, and foul). He used 15 subjects for his psychophysical experiments, but most of his conclusions were based on findings with two well-trained observers.

Henning (1916) tried to fit organic chemistry to his prism. For example, "fragrant" was believed to be due to the orthodistribution (adjacent corners) of the benzene ring, and "burnt" was attributed to the smooth heterocyclic ring with nitrogen at one corner instead of carbon.

Although the prism stimulated interest from laboratories around the world, it was granted little support (Cain, 1978). Zwaardemaker (1925), for one, complained he could not find a place for garlic and certain other odors on the prism and hence viewed it as useless. In general, others

---

[13]For reviews of some of the methods used, see Henning (1916), Skramlik (1926), and Wenzel (1948).

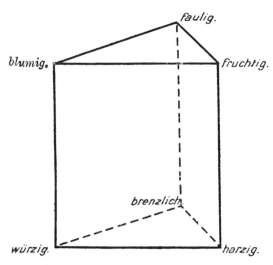

Figure 13.9. Henning's smell prism as shown in *Zeitschrift für Psychologie*, 1916. From top (clockwise): foul, fruity, resinous, burnt, spicy, flowery.

agreed that Henning's configuration was a gross approximation at best, lacking in essential research and detail, and requiring much more experimental confirmation (Boring, 1942; Gamble, 1916). As stated by one reviewer, "The clarity of Henning's scheme is at once its most attractive and its most suspicious feature" (Parker, 1922, p. 76).

A DEGENERATING SENSORY SYSTEM?

Is the olfaction system a "degenerating" system in primates, and especially in humans? This question was debated intensely in the past. Indeed, the distinction between anosmatic (animals with a poor sense of smell) and osmatic (animals with a highly developed sense of smell) species was the subject of an important paper by Paul Broca in 1878 (see Chapter 20).[14]

An argument in favor of degeneration of the olfactory system in man was presented by Haycraft (1900) in his chapter on the sense of smell in Schäfer's authoritative *Text-Book of Physiology*. Haycraft stated that he believed the olfactory apparatus had undergone "involution" in man, monkeys, and some whales. The sensory surface seemed small, and "it is probable, therefore, that the sense of smell was more acute in ancestral forms, and that it played a relatively greater part in the psychology of man's ancestry, a condition which we find to-day among the macrosmatic mammalia" (p. 1248).

Edward Titchener (1915), in contrast, defended the opposite position—that humans do not display a good sense of smell because of system disuse, rather than because the system had actually degenerated with evolution. He viewed smell as a "ground sense "— as a sense that humans naturally would rely on less when they assumed an upright position which raised the nose well above the ground. Titchener added that "the very fact that odors have no settled system of names, like cold or pain, red or blue, shows that they have not been utilized in human life" (p. 51). In support of the

idea that humans still have an acute sense of smell that can be trained, Titchener described the hunting patterns of the Botocudos of Brazil and other aboriginal tribes from Malasia. These hunters, he noted, learn to be very effective in tracking game, and they accomplish this largely by smell.

A position close to that of Titchener can be found in the 1917 textbook by Luigi Luciani. Although he stated that the olfactory sense seemed to be more highly developed among the "savage races" than in "civilized man," he added that behavioral differences probably were due to use and olfactory "education." Luciani mentioned Alexander von Humboldt's observation that Peruvian Indians can follow scents like hunting dogs. He added that because of their training, pharmacists can recognize drugs, physicians can diagnose diseases, and blind and deaf individuals can distinguish people from each other.

As more was learned about the extent to which the olfactory capabilities of humans could be trained, the less likely it seemed that there were major innate differences in olfactory acuity across the family of man or that humans were that much worse in their olfactory abilities than many animals. But why people have such a hard time identifying odors by name has remained a mystery.

REFERENCES

Allen, W. F. 1940. Effect of ablating the frontal lobes, hippocampi, and occipito-parieto-temporal (excepting pyriform areas) lobes on positive and negative olfactory conditioned responses. *American Journal of Physiology, 128,* 754–771.

Allen, W. F. 1941. Effect of ablating the pyriform-amygdaloid areas and hippocampi on positive and negative olfactory conditioned reflexes and on conditioned olfactory discrimination. *American Journal of Physiology, 132,* 81–92.

Allison, A. C. 1953a. The morphology of the olfactory system in the vertebrates. *Cambridge Philosophical Society Biological Reviews, 28,* 195–244.

Allison, A. C. 1953b. The structure of the olfactory bulb and its relationship to the olfactory pathways in the rabbit and the rat. *Journal of Comparative Neurology, 98,* 309–353.

Aristotle. ca. 330 B.C. *De sensu.* In G. R. T. Ross (Ed./Trans.). *Aristotle: De sensu and De memoria.* Cambridge, U.K.: Cambridge University Press, 1906.

Avicenna. 1025. *The Canon of Medicine.* (O. C. Gruner, Trans.). London: Luzac & Company, 1930.

Babuchin. 1873, The olfactory organ. Chapter XXXV in S. Stricker (Ed.), *Manual of Human and Comparative Histology* (Vol. 3). (H. Power, Trans.). London: New Sydenham Society. Pp. 201–217.

Beare, J. 1906. *Greek Theories of Elementary Cognition from Alcmaeon to Aristotle.* Oxford, U.K.: Clarendon Press.

Bedford, E. A. 1904. The early history of the olfactory nerve in swine. *Journal of Comparative Neurology, 14,* 390–410.

Bednar, M., and Langfelder, D. 1930. Über das intravenöse (Hämatogene) Riechen. *Monatsschrift für Ohrenheilkunde Laryngo-Rhinologie, 64,* 1133–1139.

Bell, C. 1811. *Idea of a New Anatomy of the Brain.* London: Strahan and Preston. (Reprinted in *Medical Classics,* 1936, *1,* 105–120.)

Bernard, C. 1858. *Leçons sur la Physiologie et la Pathologie du Système Nerveux* (Vol. 2). Paris: J. B. Ballière.

Bidder, F. 1844. Riechen. In R. Wagner (Ed.), *Handwörterbuch der Physiologie* (Vol. 3). Pp. 916–926.

Bishop, J. 1833. Observations on the physiology of the nerves of sensation illustrated by a case of paralysis of the fifth pair. *Proceedings of the Royal Society of London, 3,* 205–206.

Blakeslee, A. F. 1918. Unlike reactions of different individuals to fragrance in verbena flowers. *Science, 48,* 298–299.

Blaue, J. 1884. Untersuchungen über den Bau der Nasenchleimhaut bei Fischen und Amphibien, namentlich über Endknospen des Nervus olfactorius. *Archiv für Anatomie und Entwickelungsgeschichte, 8,* 231–309.

Bontekoe, C. van. 1684. *Korte verhandeling van's menschen leven. . . .* Gravenhage: Pieter Hagen.

Boring, E. G. 1942. *Sensation and Perception in the History of Experimental Psychology.* New York: Appleton-Century-Crofts.

Broca, P. 1878. Anatomie comparée des circonvolutions cérébrales. Le grand lobe limbique et la scissure limbique dans la série des mammifères. *Revue d'Anthropologie, Ser. 2, 1,* 385–498.

[14]The synonyms "microsmatic" and "macrosmatic" were introduced by William Turner (1890–91).

Brown, C. E., and Ghiselli, E. E. 1938. Subcortical mechanisms in learning. IV. Olfactory discrimination. *Journal of Comparative Neurology, 26,* 109–120.

Brunn, A. von. 1892. Beiträge zur mikroscopischen Anatomie der menschlichen Nasenhöhle. *Archiv für Mikroscopische Anatomie, 39,* 632–651.

Cain, W. S. 1978. History of research on smell. In E. C. Carterette and M. P. Friedman (Eds.), *Handbook of Perception: Vol. VIA. Tasting and Smelling.* New York: Academic Press. Pp. 197–229.

Cloquet, H. 1821. *Osphrésiologie ou Traité des Odeurs, du Sens et des Organes de l'Olfaction* (Ed. 2). Paris: Chez Méquignon-Marvis.

Corbin, A. 1986. *The Foul and the Fragrant.* Cambridge, MA: Harvard University Press.

Disse, J. 1896a. Ueber Epithelknospen in der Regio olfactoria der Säuger. *Anatomische Hefte, 6,* 21–58.

Disse, J. 1896b. Ueber die erste Entwickelung des Riechnerven. *Sitzungsberichte der Gesellschaft zu Beförderung der gesamten Naturwissenschaften in Marburg, 7,* 77–91.

Dugès, A. 1838. *Traité de Physiologie* (Vol.1). Montpellier: L. Castel.

Dusser de Barenne, J. G. 1933. "Corticalization" of function and functional localization in the cerebral cortex. *Archives of Neurology and Psychiatry, 80,* 884–901.

Eckhard, C. 1855. Ueber die Endigungsweise des Geruchsnerven. *Beiträge zur Anatomie und Physiologie, 1,* 77–84.

Elsberg, C. A. 1935. The sense of smell. XIII. Olfactory fatigue. *Bulletin of the Neurological Institute of New York, 4,* 479–495.

Elsberg, C. A., Brewer, E. D., and Levy, I. 1935–1936. The sense of smell. V. The relative importance of volume and pressure of the impulse for the sensation of smell and the nature of the olfactory process. *Bulletin of the Neurological Institute of New York, 4,* 264–269.

Elsberg, C. A., and Levy, I. 1935–1936. The sense of smell. I. A new and simple method of quantitative olfactometry. *Bulletin of the Neurological Institute of New York, 4,* 5–19.

Eschricht, D. F. 1825. *De functionis nervorum faciei et olfactus organi.* Hafniae. Typis Schlesinger.

Ferrier, D. 1875. Experiments on the brain of monkeys. *Philosophical Transactions of the Royal Society of London, 165,* 433–488.

Ferrier, D. 1876. *The Functions of the Brain.* London: Smith, Elder and Company.

Ferrier, D. 1886. *The Functions of the Brain.* New York: G. P. Putnam's Sons. (Reprinted 1978; Washington, DC: University Publications of America)

Franke, G. 1893. Experimentelle Untersuchungen über Luftdruck, Luftbewegung und Luftwechsel in der Nase und ihren Nasenhöhle. *Archiv für Laryngologie und Rhinologie, 1,* 230–249.

Galen. 1924. The Organ of Smell (J. Wright, Trans.). *Laryngoscope, 34,* 1–11.

Gamble, E. A. McC. 1916. Taste and smell. *Psychological Bulletin, 13,* 134–137.

Gehuchten, A. van, and Martin, I. 1891. Le bulbe olfatif chez quelques mammifères. *La Cellule, 7,* 205–237.

Glaser, O. 1918. Hereditary deficiencies in the sense of smell. *Science, 48,* 647–648.

Golgi, C. 1875. *Sulla Fina Struttura dei Bulbi Olfactorii.* Regio-Emilia.

Gorschkow, L. 1901. Ueber Geschmacks- und Geruchszentren in der Hirnrinde. *Neurologisches Centralblatt, 20,* 1902.

Gowers, W. R. 1885. *Epilepsy.* London: William Wood and Company. (Reprinted 1964; New York: Dover Publications)

Haller, A. von. 1757–1766. *Elementa physiologie.* Lausannae: Marci-Michael, Bousquet & Sociorum.

Haller, A. von. 1786. *First Lines of Physiology.* Translated from the Latin. (Reprinted 1966; New York: Johnson Reprint Company)

Hamilton, A. M. 1882. On cortical sensory discharging lesions (sensory epilepsy). *New York Medical Journal, 35,* 575–584.

Haycraft, J. B. 1889. The objective cause of sensation. Part III. The sense of smell. *Brain, 11,* 166–178.

Haycraft, J. B. 1900. The sense of smell. In E. A. Schäfer (Ed.), *Text-Book of Physiology* (Vol. 2). Edinburgh: Young J. Pentland. Pp. 1146–1258

Henning, H. 1916. *Der Geruch.* Leipzig: Barth.

Henschen, S. E. 1919. Ueber die Geruchs- und Geschmackszentern. *Monatsschrift für Psychiatrie und Neurologie, 45,* 121–165.

Jackson, J. H. 1889. On a particular variety of epilepsy ("intellectual aura"), one case with symptoms of organic brain disease. *Brain, 11,* 179–207.

Jackson, J. H., and Beevor, C. E. 1890. Case of tumour of the right tempero-sphenoidal lobe, bearing on the localisation of the sense of smell and on the interpretation of a particular variety of epilepsy. *Brain, 12,* 346–357.

Kamon, K. 1904. Ueber die "Geruchsknospen." *Archiv für Mikroscopische Anatomie, 64,* 653–664.

Kayser, R 1890. Ueber den Weg dur Atmungsluft durch die Nase. *Zeitschrift für Ohrenheilkunde, 20,* 96–109.

Kenneth, J. H. 1928. A note on a forgotten sixteenth century disputation on smell. *Journal of Laryngology and Otology, 43,* 103–104.

Kölliker, A. von. 1896. *Handbuch der Gewebelehre des Menschen.* Leipzig: Englemann.

Lashley, K. S., and Sperry, R. W. 1943. Olfactory discrimination after

destruction of the anterior thalamic nuclei. *American Journal of Physiology, 139,* 446–450.

Linnaeus (Linné), C. 1756. *Amoenitates academicae, III.* Lugduni Batavorum: Apud Cornelium Haak.

Lorry, D. 1788. Observations sur les parties volitiles et odorantes. *Histoire de la Société Royale de Médecine, Paris, Année 1784–1785,* 306–308.

Luciani, L. 1917. *Human Physiology.* (F. Welby, Trans.). London: Macmillan.

Magendie, F. 1824. Le nerf olfactif est-il l'organe de l'odorat? *Journal de Physiologie et Expérimentale et Pathologique, 4,* 169–176.

Moncrieff, R. W. 1944. *The Chemical Senses.* New York: John Wiley & Sons.

Müller, J. 1838. *Handbuch der Physiologie des Menschen.* Coblenz: Verlag J. Hölscher.

Murphy, G. 1929. *Historical Introduction to Modern Psychology.* New York: Harcourt, Brace.

Nagel, W. 1894. Verleichend physiologische und anatomische Untersuchungen über den Geruchs- und Geschmachssin. *Bibliotheca Zoologica, 7,* Heft 18.

Nagel, W. 1897. Ueber Mischgerüche und die Komponentengliederung des Geruchssinnes. *Zeitschrift für Psychology und Physiologie des Sinnesorgans, 15,* 82–101.

Noyons, A. K. M., 1931. Hendrik Zwaardemaker: 1857–1930. *American Journal of Psychology, 43,* 525–526.

Ogle, W. 1870. Anosmia; or, cases illustrating the physiology and pathology of the sense of smell. *Medico-Chirurgical Transactions, 53,* 263–290.

Parker, G. H. 1922. *Smell, Taste and Allied Senses in the Vertebrates.* Philadelphia, PA: Lippincott.

Passy, J. 1895. Revue générale sur les sensations olfactives. *Année Psychologie, 2,* 363–410.

Patton, H. D. 1950. The physiology of taste and smell. *Annual Review of Physiology, 12,* 469–484.

Paulsen, E. 1882. Experimentelle Untersuchungen über die Strömung der Luft in der Nasenhöhle. *Sitzungsbericht der Kaiserlichen Akademie der Wissenschaften, 85,* 348–373.

Penfield, W., and Erickson, T. C. 1941. *Epilepsy and Cerebral Localization.* Springfield, IL: Charles C Thomas.

Penfield, W., and Jasper, H. 1954. *Epilepsy and the Functional Anatomy of the Human Brain.* Boston, MA: Little Brown.

Plato. 1952. *Timaeus.* (B. Jowett, Trans.). In *Great Books of the Western World: Vol. 7. Plato.* Chicago: Encyclopaedia Britannica. Pp. 442–477.

Prade, J. le R. S. de 1668. *Discours sur le Tabac, où il est Traité Particulièrement du Tabac en Poudre.* Paris. Impr. de M. le Prest aux Dépens de l'Auteur.

Ramón y Cajal, S. 1894. La fine structure des centres nerveux. (Croonian Lecture). *Proceedings of the Royal Society of London, Ser. B, 55,* 444–467.

Ramón y Cajal, S. 1911. *Histologie du Système Nerveux de l'Homme et des Vertébrés.* Paris: A. Maloine.

Ramsay, W. 1882. On smell. *Nature, 26,* 187–189.

Read, E. A. 1908. A contribution to the knowledge of the olfactory apparatus in dog, cat, and man. *American Journal of Anatomy, 8,* 17–47.

Schneider, C. 1655. *Liber de osse cribiforme et sensu de organe odoratus et morbus ad utrumque spectantibus, de coryza, hemorrhagia narium, polypo, sternutatione, admissione odoratus.* Wittenberg.

Schultze, M. 1856. Über die Endigungsweise des Geruchsnerven und die Epithelialgebilde der Nasenschleimhaut. *Monatsberichte der Königlichen Preussiche Akadamie des Wissenschaften zu Berlin,* 504–514.

Schultze, M. 1862. Untersuchungen über den bau der Nasenschleimhaut, namentlich die Structur und Endigungsweise der Geruchsnerven beim Menschen und bei den Wirbeltieren. *Abhandlungen der Naturforschenden Gesellschaft zu Halle, 7,* 1–100.

Skramlik, E. von. 1926. *Handbuch der Physiologie der niederen Sinne. Band 1. Die Physiologie des Geruchs- und Geschmackssinnes.* Leipzig: Tieme.

Summers, M. 1926. *The History of Witchcraft and Demonology.* New York: Knopf.

Swann, H. G. 1934. The function of the brain in olfaction: II. The results of destruction of olfactory and other nervous structures upon the discrimination of odors. *Journal of Comparative Neurology, 59,* 175–201.

Swann, H. G. 1935. The function of the brain in olfaction: The effects of large cortical lesions on olfactory discrimination. *American Journal of Physiology, 111,* 257–262.

Telfer, W. 1955. *Cyril of Jerusalem and Nemesius of Emesa.* Philadelphia, PA: Westminster Press.

Theophrastus. ca. 320 B.C. *Concerning Odors.* In A. Hort (Ed./Trans.), *Theophrastus: Enquiry into Plants and Minor Works on Odours and Weather Signs* (Vol. 2). London: Heinemann, 1916. Pp. 325–189.

Titchener, E. B. 1915. *A Beginner's Psychology.* New York: Macmillan.

Tourtual, C. T. 1827. *Die Sinne des Menschen in den wechselseitigen Beziehungen ihres psychischen und organischen Lebens.* Münster: F. Regensberg.

Troland, L. T. 1930. *The Principles of Psychophysiology. Vol. II: Sensation.* New York: Van Nostrand.

Turner, W. 1890–1891. The convolutions of the brain: A study in comparative anatomy. *Journal of Anatomy and Physiology, 15,* 105–153.

Valentin, G. 1839. *De Functionibus Nervorum Cerebralium et Nervi Sympathici*. Bernae et Sangalli Helvetiorum: Huber.

Vintschgau, M. von. 1880. Physiologie des Geschmacksinns und des Geruchssinnes. In L. Hermann (Ed.), *Handbuch der Physiologie, Zweiter Theil*. Leipzig: F. C. W. Vogel. Pp. 145–286.

Weber, E. H. 1847. Ueber den Einfluss der Erwärmung und Erkältung der Nerven auf ihr Leitungsvermögen. *Archiv für Anatomie und Physiologie*, 342–356.

Wenzel, B. M. 1948. Techniques in olfactometry: A critical review of the last one hundred years. *Psychological Bulletin, 45*, 231–247.

Willis, T. 1664. *Cerebri anatome: cui accessit nervorum descriptio et usus*. London: J. Martyn and J. Allestry. Tercentenary ed., 1664–1964, *Thomas Willis, The Anatomy of the Brain and Nerves*. Montreal: McGill University Press, 1965.

Willis, T. 1672. *De anima brutorum*. . . . London: Typis E. F. Impensis Ric. Davis, Oxon.

Wright, J. 1898. *The Nose and Throat in Medical History*. St. Louis, MO: Matthews.

Woodworth, R. S. 1938. *Experimental Psychology*. New York: Henry Holt.

Zwaardemaker, H. 1888. Die Bestimmung der Geruchsschärfe. *Berliner klinische Wochenschrift, 25*, 950–951.

Zwaardemaker, H. 1889. On measurement of the sense of smell in clinical examination. *Lancet, 1*, 1300–1302.

Zwaardemaker, H. 1895. *Die Physiologie des Geruchs*. Leipzig: Engelmann.

Zwaardemaker, H. 1925. *L'Odorat*. Paris: Doin.

Zwaardemaker, H. 1927. The sense of smell. *Acta Oto-laryngologica, 11*, 3–15.

Zwaardemaker, H. 1930. An intellectual history of a physiologist with psychological aspirations. In C. Murchison (Ed.), *A History of Psychology in Autobiography* (Vol. 1). Pp. 491–516.

Part III
Motor Functions

Chapter 14
The Pyramidal System and the Motor Cortex

> The map of the human motor cortex to be found nowadays in every textbook of neurology is nothing else but a copy of the picture engraved in Hughlings Jackson's brain.
> Otfrid Foerster, 1936a

EARLY OBSERVATIONS

The relationship between the brain and the musculature has a long written history. This is attested to by the fact that a number of cases of brain damage with impaired motor functioning appear in the *Edwin Smith Surgical Papyrus,* which is based on records well over 4,000 years old (see Chapter 1). In Case 20, for example, it is stated that "if thou puttest thy fingers on the mouth of the wound . . . he shudder exceedingly" (translated by Breasted, 1930). This statement has been interpreted as showing that the Egyptians recognized that direct stimulation of the brain can cause a person to have a motor seizure.

There are several references to disturbed motor function following head injuries in the writing of Hippocrates and his followers, dating from the fourth century B.C. Some of the most intriguing of these involved head wounds associated with contralateral hemiplegias and seizures. It was written that "if the wound be situated on the left side [of the head], the convulsion attacks the right side of the body" (see Courville, 1946, p. 16). The Hippocratic writers specifically warned against making incisions in the brain because this can result in convulsions on the opposite side of the body. Nevertheless, the role of the nerves was not understood when Hippocrates lived, and it was not even certain that the brain alone, much less the cortex, was the seat of the mind and in direct control of the motor functions.

Within a few hundred years after the death of Hippocrates, the contralateral effects of brain damage were more clearly attributed to the crossing of the nerves from higher brain centers. Yet it still was not known where the critical nerves originated or where the crossing took place. Aretaeus the Cappadocian, who lived near the end of the second century A.D., wrote the following about the two sides of the brain and the control of motor function:

> If, therefore, the commencement of the affection is below the head, such as the membrane of the spinal marrow, the parts which are homonymous and connected with it are paralyzed; the right on the right side and the left on the left side. But if the head be primarily affected on the right side,

the left side of the body will be paralyzed. If on the left side, the right. The cause of this is the interchange in the origins of the nerves, for they do not pass along on the same side, the right on the right side, until their terminations; but each of them passes over to the other side from that of its origin, decussating each other in the form of the letter **X.** (1856 translation, 306)

During the golden years of Arabic culture, contralateral paralyses were usually explained by citing the classical scholars of Greece and Rome. The old tradition was adhered to by Rhazes, who accepted the decussation of the nerves responsible for movement and who compiled an encyclopedia of medicine around A.D. 900. But in contrast to Rhazes, Avicenna, a century later, had great difficulty accepting the very notion of contralateral paralysis.

Guglielmo da Saliceto (1210–1280), a surgeon of the school of Bologna, was perhaps the first to suggest that voluntary movements originated in the "brain," whereas "natural and necessary" movements originated in the cerebellum (Giannitrapani, 1967). He contradicted the authority of Avicenna by accepting and describing contralateral paralysis. And, like others before him, he supported his statements with citations from Greek and Latin manuscripts, preserved by the monks in his region.

MOVEMENT AND THE CEREBRAL HEMISPHERES BEFORE THE NINETEENTH CENTURY

In 1709, Domencio Mistichelli, a professor of medicine in Pisa, described the medullary crossing and provided the first published illustration of the decussation in the pyramids (see Figure 14.1).[1,2] He reasoned that the nerve crossing in the pyramids was responsible for the hemiplegia following wounds of the opposite side of the head. He also argued

---

[1]Although Eustachio had illustrated the pyramids in 1552, the work of this Italian investigator was not published until 1714.

[2]The name pyramids was given to this structure by Thomas Willis (1664). He considered them reservoirs for the animal spirits.

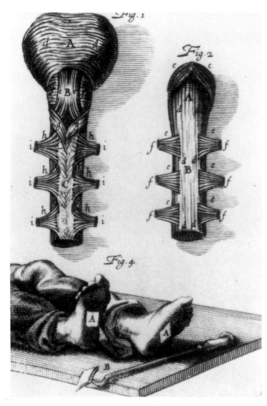

Figure 14.1. Domencio Mistichelli's (1709) figure showing the decussation of the pyramids and the outward rotation of a paralyzed limb.

against the then popular notion of nervous fluids. Nevertheless, he erroneously maintained the nerves originated in the dura mater and suggested that paralysis of the leg could be cured by the application of a hot cautery to the foot.

At almost the same time, François Pourfour du Petit, a French military surgeon and physician in the King's Hospitals, observed soldiers who were paralyzed on one side of the body after injuries believed to be confined to the cerebral hemispheres. He found the damage contralateral to the paralysis and deduced, as had others before him, that the right side of the brain must control the left side of the body, whereas the left side of the brain must control the right side of the body. In 1710, one year after Mistichelli described the pyramids, Pourfour du Petit presented his own pictures of the fibers from higher centers crossing in the pyramids.

Pourfour du Petit also made lesions in dogs—often with a small knife—in the parietal region. When the lesions included the corpus striatum, they produced contralateral hemiplegia. Yet even when the lesions were thought to be limited to the cortex, there was weakness in the extremities.[3]

Pourfour du Petit, like other early-eighteenth-century physicians, believed that animal spirits in the nerves were responsible for the contractions of the muscles. He thought that these spirits moved from the cortex through the striatum and basal ganglia and across the pyramids *(corps cannellez)* to the muscles. He broke with classical tradition when he ascribed a direct role in movement to the cerebral cortex.[4]

[3]These findings were verified in 1721 by Pietro Paolo Molinelli (1702–1764), who subjected dogs to hemispherectomies (see Vedrani, 1920).

[4]Pourfour du Petit's esteemed teacher, Joseph Duverney, accepted the crossing of the major motor pathway in the pyramids and also presented diagrams in 1761 to show this.

In 1724, Giovanni Domenico Santorini put the medulla in water, macerated it, and gently pulled it apart. His experiment demonstrated convincingly that the fibers crossed in the pyramids. Santorini, however, adhered to the belief that movements were due to fluids in the nerves.

Among the clinical reports from the mid-eighteenth century on the motor pathways and structures, one by Joseph Lambert Baader (1723–1773), dated 1762, stands out.[5] Baader, a professor of materia medica, botany, and chemistry at Freiburg im Breisgau, described a patient who displayed convulsions of the left arm and complained about an abscess on the right side of his head. Baader thought this showed that the "elements and action of the brain are subject to decussation, so that sensation and mobility of one side of the body are under the dependence of the opposite cerebral hemisphere."

Baader then observed a second patient whose symptoms led him to conclude that "the region of the brain which is located beneath the parietal bone commands the motility and sensitivity of the upper extremity of the opposite side." On the basis of these early localizations, Baader felt it was possible to

> know and predict, for the greatest benefit of practitioners, what part of the brain gave this or that extremity sensation or movement; as a result, knowing which member is involved, one can determine what part of the brain is ill, and inversely, being given a certain lesion in the brain, predict which member must be affected. (Translated in Jeannerod, 1985, 52)

The percentage of fibers from the forebrain that crossed in the medulla was not known at this time. Giovanni Battista Morgagni (1740), the renowned Italian pathologist, still believed that most fibers to the muscles were not crossed. Even so, he was not one to dispute the clinical evidence for contralateral hemiparesis and proposed that a small bundle of fibers crossed in the medulla and was responsible for the contralateral paralysis. Morgagni wrote:

> When the right hemisphere has been injured, why is the left side of the body which obtains fewer fibers from that hemisphere affected, while the right side which receives many more fibers from that side of the brain is not affected? It comes to this, as indicated above, that there are nerves which originate before all those decussations and to which all those decussations do not pertain. (Translated by Giannitrapani, 1967, 360)

Morgagni assumed that cases of hemiplegia on the same side as a forebrain lesion could be accounted for by the uncrossed projection system.

Perhaps the most remarkable eighteenth-century "premonition" about the motor cortex came from the Swedish religious philosopher Emanuel Swedenborg, who wrote several volumes on the brain between 1738 and 1744 (see Swedenborg, 1882–87). Most of his anatomy was taken from other authors, but his theory was not. Swedenborg localized the motor cortex in the front of the brain and stated that the muscles of the extremities are controlled by the upper frontal convolutions, those of the middle part of the body by the middle frontal convolutions, and the head and neck by the lower convolutions. Unfortunately, as was

[5]Discussions of Baader's important work may be found in Soury (1899, p. 473) and Jeannerod (1985, p. 52).

noted in Chapter 2, Swedenborg's works on the nervous system were not circulated during his lifetime, and their impact appeared to be minimal on emerging scientific thought (Akert and Hammond, 1962).

THE CEREBRAL HEMISPHERES IN THE FIRST HALF OF THE NINETEENTH CENTURY

In the first half of the nineteenth century, Marie-Jean-Pierre Flourens conducted a number of decerebration experiments. Most were on birds, but he also operated on mammals. Flourens (1824) found that the decorticated birds sat as if they were asleep and would not even feed themselves when brought close to starvation. When a bird was tossed in the air, however, it flew, and when food was put into its beak, it swallowed it. Other animals also remained immobile, but responded when adequately stimulated. Largely on the basis of these observations, Flourens concluded that the cerebral cortex did not house the circuitry critical for movements of the musculature. Instead, he maintained, it was the seat of the will and also played an important role in perception and intelligence.

These ideas were in opposition to the beliefs of the phrenologists (Chapter 3). Compartmentalization or localization of function at the level of the cortex was basic to phrenological doctrine, even though the placements of the faculties of mind differed in the schemas of different phrenologists. Among other things, many phrenologists, including Franz Gall himself, attempted to localize speech—considered at least in part a motor function—in the frontal cortex.

Jean-Baptiste Bouillaud, a professor of clinical medicine at the Charité in Paris, studied phrenology early in his career. Although he retained a deep-seated belief in localization at the cortical level, he rejected Gall's questionable phrenological methods. Bouillaud suggested that the better way to determine the cortical function was to correlate behavioral deficits with pathology.

Bouillaud was especially interested in localizing the cortical movement centers. In particular, he noted that lesions of the anterior lobes were most likely to affect the ability to speak (see Chapters 3 and 25). He also noticed that patients who lost voluntary movements on one side of the body could still respond with automatic and instinctive movements. For example, one paralyzed patient still withdrew his leg to pinching.

Thus, in 1825, Bouillaud proposed that the cortex controls those movements subject to the will and intelligence. He also concluded there was not just one movement center but many. Thus, the center for the will to move the muscle groups necessary for speech was not localized in the same place as the will to move an arm.

The position taken by Bouillaud obviously represented a challenge to the beliefs of Flourens and his loyal followers. Nevertheless, to say that Bouillaud believed that the cortex had a motor function whereas Flourens completely dissociated the cortex from movement would be oversimplifying the issue. Both men believed the cortex played a role in the will to do something; neither thought of the cortex as a pure motor center. The fundamental difference between these two adversaries was that only Bouillaud believed that the cortex could be subdivided into discrete functional units, each best suited to conduct a particular function; only Bouillaud was driven by the belief that there had to

Figure 14.2. John Hughlings Jackson (1835–1911), who inferred the presence of a somatotopically organized motor cortex on the basis of epileptic seizures and other clinical evidence.

be discrete cortical centers to "will" and to control the movements of different parts of the body.

JACKSON AND THE MOTOR CORTEX CONCEPT

John Hughlings Jackson, shown in Figure 14.2, is often cited for his brilliant clinical observations and deductions that suggested the existence of a motor cortex before such a structure could be demonstrated experimentally. Eduard Hitzig, who "codiscovered" the motor cortex in the dog in 1870 (see next section), gave Jackson due credit for his remarkable insights and inferences. Hitzig (1900) stated that his physiological experiments only confirmed the conclusions Jackson had reached by clinical observations and deduction (also see the quotation from Foerster at the beginning of this chapter).

Jackson began to show an interest in the nervous system in 1858 and started to write about epilepsy in 1861 when he joined the staff of the *Medical Times and Gazette* as a medical reporter.[6,7] Jackson's (1861) first article in this periodical was about syphilis. Here he discussed incomplete, unilateral "epileptiform convulsions" without loss of consciousness, the disorder that would later become known as Jacksonian epilepsy.

Jacksonian seizures must have been seen during classical times (Temkin, 1972). It was not until 1827, however, that L. F. Bravais tried to establish these seizures as a distinct variety of disease (Greenblatt, 1865). Four years later, Richard

[6]See Critchley (1960) and Taylor (1925) for biographies of Jackson. For more focused articles dealing with the major influences on Jackson's early thinking, see Greenblatt (1965, 1977).

[7]Johnathan Hutchinson (1828–1913), with whom Jackson lived at the time and who was to become his lifelong friend, was probably responsible for this important appointment. He had himself been a reporter for the *Medical Times and Gazette* since 1855.

Bright (1789–1858) of Guy's Hospital combined clinical and anatomical approaches to study these unilateral convulsions, which Robert Todd (1849) referred to with the adjective "epileptiform." Nevertheless, most physicians and laboratory scientists were not especially interested in "non-genuine" (unilateral) seizures at this time.

In 1863, Jackson received an appointment to the National Hospital for the Paralyzed and Epileptic. In his *Suggestions for Studying Diseases of the Nervous System on Professor Owen's Vertebral Theory* (1863a), he again wrote about unilateral seizures associated with cerebral syphilis. At the time, it was largely believed that the corpus striatum was the uppermost part of the motor tract and was where the centers for movements of the muscles of the limbs were probably located. Another popular theory was that epilepsy was caused by disturbances in the lower brainstem (Foerster, 1936a). Jackson (1863b, 1863c, 1863d) was coming to the realization, however, that the hemispheres were often affected in unilateral convulsive patients.

> In very many cases of epilepsy, and especially in syphilitic epilepsy, the convulsions are limited to one side of the body; and, as autopsies of patients who have died after syphilitic epilepsy appear to show, the cause is obvious organic disease on the side of the brain, opposite to the side of the body convulsed, frequently on the surface of the hemisphere. (Jackson, 1863b, 110)

Jackson's developing ideas were not radically different from those of Robert Bentley Todd, who had made the following statements a few years earlier:

> Such phenomena . . . as a convulsive affection of the whole of one side of the body, evidently brought on by the deposition of tubercular matter on the surface of the cerebral hemisphere on the opposite side, and its consequent irritation, show that convulsive movements may be excited by a superficial lesion of the hemispheric lobes. Hence we must not deny to these lobes a certain power of exciting motion, either directly or indirectly, through their influence upon other ganglia of the brain . . . From all these facts, then, I infer that a disturbed state of the hemispheric lobes may undoubtedly give rise to so much of the phenomena of the epileptic paroxysm as refers to the affection of consciousness and sensibility, and that it may, *in some degree* at least, contribute to the development of the convulsions. (1849, 820)

But unlike Todd, Jackson was intrigued by the "march" of the seizure over the body parts, such as from the hand to the arm to the face (see the photographic sequence in Figure 14.3). Yet he still was not ready to discuss cortical organization or unequivocally to localize movements in the cortex.

Jackson's emerging thoughts on the role of the cerebral cortex in movements were further influenced by his observations of hemiplegia. By 1865, he had seen some 500 individuals with unilateral paralysis. He was struck both by the general absence of sensory deficits and by the degree to which some parts of the body, such as the trunk, were spared from paralyses that severely affected the limbs.

Jackson recognized that unilateral seizures were most likely to affect those parts of the body which were also most affected in hemiplegia. He considered epilepsy to be

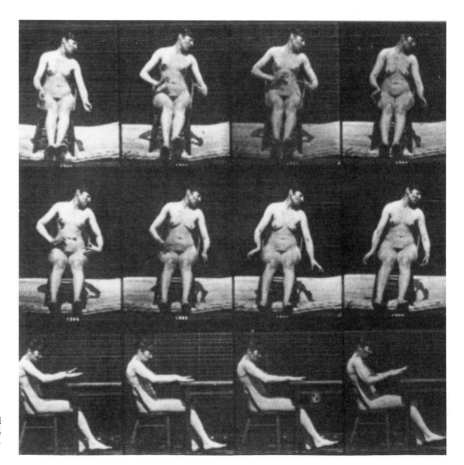

Figure 14.3. A patient with an induced seizure. The demonstration was made in 1884 by Francis X. Dercum for the book *Animal Locomotion* by Eadweard Muybridge (1887).

the "mobile counterpart of hemiplegia." Together the two disorders afforded him a "double method" for comparing the effects of discharging (or "over-functioning") and destroying lesions.[8]

Jackson's more mature ideas about seizures, hemiplegia, and localization appeared in his papers after 1865. Here one finds additional descriptions of the spread of convulsions up a limb, intensifying as they go, leaving muscles relaxed in their wakes. The spread of these nonsimultaneous spasms was found to be fairly predictable. Those starting in the foot spread up the leg and then to the upper arm before the hand. Those starting in the hand went up the arm to the face and the hip and then down the leg on the same side to the toes. Consciousness was rarely lost with this type of seizure unless the convulsion was severe enough to cross the midline and affect the whole body. But even under these circumstances, the muscles of the face, hands, and toes were found to be the most vulnerable. Among other things, Jackson also noted that only convulsions involving the right side of the face seemed to affect speech directly (also see Jackson, 1863d).

Jackson continued to build his case for cortical involvement in motor function. Yet at the same time, he had difficulty abandoning the older theory that epileptic convulsions were caused by discharges in the corpus striatum. The striatal position played a significant role in his writings on epilepsy until 1868, although he had previously acknowledged that the cerebral cortex could be involved.

Jackson's bias clearly shifted from the striatum to the cortex around 1870 (see Temkin, 1972, pp. 333–334). He now entertained three related ideas: (1) the cortex is partly to blame for severe convulsions; (2) cortical fits could trigger secondary convulsions in the corpus striatum; and (3) the corpus striatum may be the site of discharge in smaller convulsions. He maintained that both the corpus striatum and many cortical convolutions were supplied by the middle cerebral artery and that both were made of gray matter whose instability could account for the unilateral convulsions.

Jackson's shift in emphasis gave new meaning to the role of the cortex. He theorized that the cortex contained "impressions" and hypothesized that there were more impressions from the arm in one place and more from the leg in another locus. A degree of somatotopic organization was the only way to account for the progression of the seizure across successive muscle groups in cases of spreading epilepsy, as well as for epilepsy just affecting one part of the body. Jackson thus believed in differing degrees of representation for different functions in given areas. As he put it, he was neither a strict localizer nor a "universalizer." Consistent with this orientation, his discussions centered on the logical organization of the motor representation at the level of the cortex, rather than on precise anatomical loci for these functions.

Jackson proposed that the hand, face, and foot, or the three "leading" parts (i.e., those capable of the greatest number of specialized voluntary movements), had to have the greatest representation at the level of the cortex. This notion was consistent with his belief that greater differentiation of function implies larger representation in the brain. He added, however, that the more automatic, primi-

tive, and less skilled the movement, the more likely it was to be mediated not by the cortex but by the corpus striatum and the cerebellum. He presumed the cortex contained more intricate impressions of motions than the corpus striatum, the generally accepted site of the motor apparatus (see Chapter 15). He also firmly believed that the higher nervous centers could only be viewed as evolutionary outgrowths of lower centers.

FRITSCH AND HITZIG'S DISCOVERY

In 1870, Gustav Theodor Fritsch and Eduard Hitzig finally provided unequivocal experimental confirmation of a "motor" cortex in the frontal lobes of the dog (see Chapter 3). Fritsch, an anatomist, and Hitzig, a psychiatrist, followed up on some preliminary findings by Hitzig (1867, 1869) involving electrical stimulation of the back of the human head and some work on rabbits. Working in Berlin, they found electrical stimulation of limited parts of the dog's frontal cortex could elicit discrete movements of the face or limbs on the opposite side of the body.

This finding contrasted with the earlier literature suggesting the cerebral cortex was unexcitable, or that when electrical stimulation caused movements, it was due to the spread of current to the basal ganglia or cerebellum. The nonexcitability of the cortex had been a primary belief of Flourens (1824), Magendie (1839), Schiff (1858–59), and just about everyone else.

Fritsch and Hitzig even quoted from François Longet and Julius Ludwig Budge (1811–1884) on the nonexcitability of the cortex. Both wrote negatively about the idea of a motor cortex. Longet, for example, had written:

> On dogs and rabbits, on some young goats, we have irritated the white substance of the cerebral lobes with a scalpel; we cauterized with potassium, azotic acid, etc., we applied galvanic current in all directions, without successfully setting into play the involuntary muscle contractions to develop convulsive tremors; same negative result, in placing the same agents on the gray or cortical substance. (1842, 644)

Budge was even more dogmatic when he wrote:

> If one can conclude from the present standpoint of science that there are no motor fibers in a nervous part in which after excitation no contractions occur, one can say with the greatest certainty there is not one fiber in the hemisphere of the brain which goes to voluntary muscles. Not a single observer saw movements of such muscles after stimulation of the central parts. (1842, 84)

According to Fritsch and Hitzig (1870), in the past there had been only one experiment even suggestive of a cortical motor zone. They stated that an investigator had noticed fine movements of the extremities when slicing through the anterior lobes of the brain. They did not give the investigator's name, the source, or any further details after making this statement.

This, however, was overstating the case against the excitability of the cortex. There was a significant exception. In 1849, Robert Todd had applied galvanic stimulation to cortical and subcortical sites of rabbits. He obtained motor responses from both sites, but the responses were different from one another. To quote Todd:

---

[8]Jackson's double dissociation method is discussed in Foerster (1936a), Temkin (1972), and Walsh (1961)

I then tried the corpora quadrigemina and the mesocephale. Having passed fine bradawls into the cranium in such a direction as I had previously satisfied myself would lead to this organ, I subjected it to the influence of the machine: general convulsions were produced, of a character essentially different from those which resulted from stimulating the spinal cord or the medulla oblongata. They were combined movements of alternate contraction and relaxation, flexion and extension, affecting the muscles of all the limbs, of the trunk, and of the eyes, which rolled about just as in epilepsy.

On inserting the awls into the hemispheric lobes, still different effects were produced by the application of the machine. I could observe nothing like true convulsions; but slight convulsive twitchings of the muscles of the face took place, which were no more than what would be caused by the stimulus of galvanism acting upon the nerves of the face. (1849, 821)

Unfortunately, Todd failed to see the significance of his own findings and did not use his data to argue for a cortical motor region (Lyons, 1982).

Fritsch and Hitzig pointed out that other experimenters probably never stimulated the frontal motor areas. They thought this was because it was much more difficult to expose and stimulate this part of the brain than the cortex located more posteriorly or dorsal to it. Upon failing to elicit discrete motor responses from the occipital cortex or the posterior parietal areas, most workers probably concluded Flourens was correct and simply did not bother to study other parts of the cortex.

Fritsch and Hitzig confirmed their stimulation findings by damaging the electrically excitable cortex. They found that removal of small pieces of cortex from the critical area affected motion. They described two dogs who had some motor cortex removed (with the handle of a scalpel) on one side of the brain. The lesions did not produce a paralysis like that seen in humans with severe cortical injuries, but the animals still showed clear deficits.

In running the animals used the right forepaw wrongly, sometimes more inward and sometimes more outward than the other ones, and frequently glided out with this paw, never with the other ones, so that they fell to the ground . . . the anterior paw is put down with the dorsal instead of the ventral side without the dog noticing it. In sitting on the hind legs with both front paws standing on the ground, the right front paw gradually glides away toward the outside until the dog lies on the right side. (1960 translation, 95)

Fritsch and Hitzig also noticed that dogs with unilateral lesions of the motor area failed to move their contralateral paws to more natural locations when they were put in awkward positions. In contrast, the same dogs quickly shifted their limbs to comfortable positions when the same test was conducted on their ipsilateral limbs.

Soon after completing his landmark experiment, Hitzig left Berlin to serve in the Franco-Prussian War of 1870–1871. After his service in Nancy, France, he returned to Germany to study some of the defects of movement that followed bilateral lesions in his animals. Again, the dogs used their limbs actively, but in "unsuitable" ways. For example, when placed at the edge of a table they stepped blindly into the air even though the sense of sight had not been destroyed.

These unusual behaviors led Fritsch and Hitzig (1870), and later Hitzig (e.g., 1874, 1900) alone, to avoid speaking of a purely motor function for these areas. Instead, they described the loss of the ability to excite the motions necessary for carrying out the will. Yet, unlike Flourens, who associated the whole cortex with the will to perform actions, they felt they had shown that one cortical area in particular was critical for voluntary movement production. In this regard, the position they took was decidedly closer to that of Bouillaud.

Fritsch and Hitzig emphasized that they often saw some recovery of function from the motor cortex lesions in their dogs. This suggested there was more than one center involved with the control of the musculature. Developmental and phylogenetic studies led Hitzig to the same conclusion as Hughlings Jackson—that is, there had to be at least two basic motor systems in the brain. Hitzig wrote:

We have known for a long while that animals of lower order without a cerebrum or with only a rudimentary cerebrum are capable of very complicated forms of motion that are excited in a reflex way and that are mediated by *special lines of communication* with the intercentral and centrifugal course of which we are but imperfectly acquainted. In higher animals there is associated with this system a second motor system that stands in manifold relation to it and that serves for *conscious arbitrary motion*. This second system starts in the motor centres to which I have referred and takes its course in the pyramid tract, which constantly develops itself more and more with the higher development of the animal. (1900, 577)

FERRIER AND THE MOTOR CORTEX

Fritsch and Hitzig's basic findings were confirmed by Edouard Fournié (1833–1886). In 1873, he put corrosive substances (chromic acid, zinc chloride) on the hemispheres and into the brains of some animals.

The findings of the two scientists from Berlin were also expanded by David Ferrier (1876) in Great Britain. Ferrier attempted to see if the findings on dogs would generalize to macaque monkeys and other species. He applied faradic current through an induction coil to produce contractions of long duration. This allowed him to examine movements, which he believed more closely resembled intentional movements than those produced by galvanic current. Whereas galvanic current produced only brief muscular contractions when the circuit was opened or closed, faradic current allowed him to see monkeys that seemed to be trying to walk, grab at objects, scratch, and so on.

Ferrier reported that the excitable cortex of the monkey was very large. It included not only the usual motor area but much of the parietal lobe and part of the temporal lobe. He also showed a very large excitable cortex in other mammals, including the dog, the jackal, the cat, the rabbit, the guinea pig, and the rat. Figure 14.4 shows one of Ferrier's cortical diagrams.

Ferrier noted, however, that "the mere fact that movements result from stimulation of a given part of the hemisphere does not necessarily imply that the same is a motor centre in the proper sense of the term" (1886, p. 249). He thought the movements he obtained from some of the areas considered questionable by Hitzig (1874) were, in fact, more "expressive of sensation." To cite but one example, Ferrier pointed to stimulation of the temporo-sphenoidal

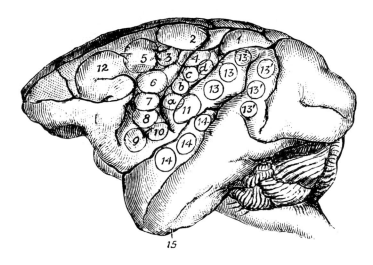

Figure 14.4. David Ferrier's (1886) diagram of the left hemisphere of the monkey brain showing the regions that gave rise to movements (circles with numbers) when electrically stimulated.

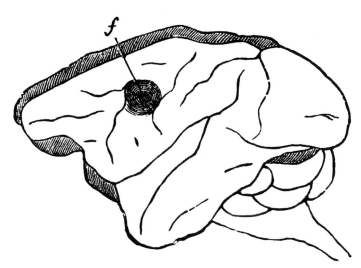

Figure 14.5. David Ferrier's (1886) diagram of the left hemisphere of the monkey brain showing the locus of the ablation that resulted in a paralysis of the biceps on the right side.

convolution. Stimulation of this area gave rise to movements of the ears, the head, and the eyes. These findings were more consistent with this region's being a center for hearing than for its housing a true motor center.

Ferrier also made lesions of the precentral region. He noted that the lesions which led to degeneration in the internal capsule and the motor region of the spinal cord were associated with paralyses and weakness of the muscles (paresis) in monkeys and other animals. In one case, a monkey was given a relatively small motor cortex lesion. The paralysis this animal developed was restricted to the opposite hand and forearm. This deficit remained for the two months the animal was kept alive. In contrast, the monkey always showed normal sensitivity to touch and to noxious stimuli, such as pinching or touching with a hot wire. This monkey's brain is presented in Figure 14.5.

Another case involved a monkey with a lesion that encompassed almost all of the excitable cortex of the left hemisphere. This animal displayed a right hemiplegia. In 1881, some eight months after the surgery, the monkey was brought to the International Medical Congress in London where it attracted considerable attention as a model of "incurable" hemiplegia in man (Ferrier and Yeo, 1885; also see Chapters 3 and 4). Independent anatomical study showed degeneration in the pyramidal tracts extending as far as the lumbar spinal cord (Ferrier, 1886).

Ferrier was consistent in his belief that the paralyses following large ablations of the motor cortex were permanent in primates, just as he firmly believed lesions of the superior temporal gyri led to permanent deafness in monkeys (see Chapter 9). The postulated permanence of the deficit in primates, however, was questioned early on by Luigi Luciani and Arturo Tamburini. They kept their own monkeys alive for a number of months and concluded that the only lasting changes were in hand preference (Luciani and Tamburini, 1878; also see Rabagliati, 1879). They also stated that the recovery was even more rapid and complete when they worked with animals lower on the phylogenetic scale. Consequently, they suggested Ferrier probably would have seen the same thing if he had kept his mon-

keys alive for more than just a few days. Ferrier, of course, went on to do this when he presented his monkey model of "incurable" hemiplegia at the 1881 medical meetings in London (see Chapters 3 and 4). The result was a clear difference of opinion on the recovery issue.

ELECTROPHYSIOLOGY AND THE BOUNDARIES OF THE MOTOR CORTEX

In the 1880s, Victor Horsley began to work on the boundaries and characteristics of the "so-called" motor cortex. In 1885, he and Charles Beevor observed that electrical stimulation of the postcentral region was not always effective in producing movements and that the movements obtained from this area were often feeble (see review by Horsley, 1909). These and other findings suggested that the main representation of the motor cortex had to be in front of the Rolandic fissure.

In experiments conducted over many years, Horsley and others showed that the ability to elicit movements from the postcentral gyrus decreased as one ascended the phylogenetic scale. Horsley (1909, p. 126) even added that "in man I have never elicited a motor response by stimulating with a minimal current the post-central gyrus."

Nevertheless, Horsley did not believe the precentral gyrus was purely motor. Instead, he believed the precentral gyrus functioned in a wide range of both sensory and motor functions. These included "slight tactility, topognosis, muscular sense, arthric sense, stereognosis, pain, (and) movement" (Horsley, 1909, p. 132). Further, because he observed recovery after small lesions in the precentral gyrus, he also concluded that "the pre-central gyrus is not in man the only out-going or motor centre for the voluntary movements of the upper limb" (p. 131).

To most workers in the field at this time, the "definitive" topographical maps were those published by Grünbaum and Sherrington (1902, 1903). These investigators studied the great apes and reported that stimuli which could evoke responses from the precentral gyrus failed to evoke

Figure 14.6. Fedor Krause (1857–1937), the Berlin physician who first mapped the human motor cortex.

ton's experiments were conducted so carefully, and presented with such detail, that it did not seem difficult to generalize from apes to humans.

Although Roberts Bartholow had stimulated the human cortex electrically in a dying feeble-minded girl in 1874, the first person to attempt to map the human motor cortex in a systematic way was Fedor Krause, who is shown in Figure 14.6. Krause spent most of his career in Berlin, where he worked with patients who occasionally underwent surgery for removal of cortical scars. These patients were lightly anesthetized, and in some instances he was able to stimulate the motor cortex with faradic current while his assistants kept records of patient movements.

Krause (1911) presented figures of his results with electrical stimulation along with good verbal descriptions. Figure 14.7 shows one of his diagrams. He wrote:

> All the foci belong to the anterior central (precentral) convolution. It shows that those for the lower extremity are arranged on the upper surface of the brain near the longitudinal sinus, and it can be inferred from animals that they extend on or to the medial surface of the hemisphere; approximately the upper quarter of the central convolution is claimed by the lower extremity. About one half of the middle portion responds with contractions in the contralateral upper extremity from the shoulder down to the fingers. Finally, in the lower quarter the areas for the facial and masticatory muscles are added; here also should be found those for the muscles of the larynx, the platysma, and the tongue. (Translated in Clarke and O'Malley, 1968, 524)

From this statement it can be surmised that Krause's maps of the human brain may have been supplemented by the growing knowledge about the motor cortex in animals and in particular by the experiments of Grünbaum and

responses from the postcentral gyrus. They also showed that ablations of the hand area of the precentral gyrus resulted in paresis, which did not occur when lesions were made in the postcental gyrus. Grünbaum and Sherring-

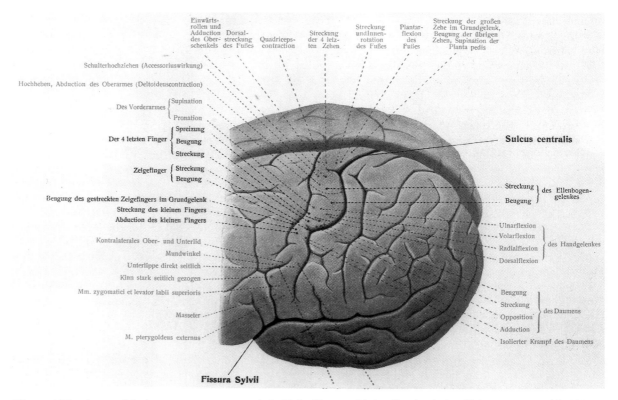

Figure 14.7. A map of the human motor cortex made by Fedor Krause with faradic stimulation. This map appeared in 1911.

Sherrington (1902, 1903). The papers of the latter investigators were, in fact, cited, and a full-page detailed drawing of their results also appeared in Krause's (1911) celebrated book.

THE CORTICOSPINAL TRACT

Through the first half of the nineteenth century, the major pathway for the control of movement was presumed to arise in the corpus striatum and to cross in the pyramids. This view first began to be challenged by Ludwig Türck (1849, 1850, 1853), who, around 1850, published a number of important papers on the anatomy of the pyramidal pathway. Türck based his conclusions largely on secondary degeneration in the pyramidal and "capsular" tracts of the spinal cord after cortical lesions. Still, with the "motor cortex" not yet discovered, his ideas were slow to gain adherents. In his famous paper of 1861, for example, Paul Broca wrote that a lesion of the corpus striatum was "the only one to look for as the cause of the (hemiplegic) paralysis."

The experiments of Fritsch and Hitzig (1870) generated not only greater acceptance of the cortical origin of the pyramidal tract, but a desire to learn much more about the details of the motor pathway from the cortex. This necessitated electrically stimulating the motor cortex to define the boundaries of the areas controlling different parts of the body and then removing small portions of this area. After periods of about 10 to 40 days, the animals were sacrificed. When their brains were removed, sections were cut and stained to study the degeneration.[9]

In some studies, monkeys were used, and it was seen that the great majority of the cortical fibers crossed in the pyramids of the medulla to the lateral columns of the opposite side (Mellus, 1894, 1899). This proved there was a direct pyramidal tract from the cortex in infrahuman animals, still an emotionally charged issue at the turn of the century.

That a direct pyramidal tract existed in man was not as controversial at this time. This was largely because Paul Flechsig had provided strong anatomical evidence for a corticospinal tract in humans. It was also because Jackson, Ferrier, and many others presumed the motor cortex was largely concerned with voluntary control of skilled movement. This was a feature more characteristic of human beings than of phylogenetically lower animals (see below).

The relationship between these findings and cortical cytoarchitecture was the subject of additional studies. In 1874, Vladimir Alexandrovich Betz (1834–1894), a Russian histologist, described giant pyramidal cells in the motor region of dogs, monkeys, apes, and humans. These "Betz cells" were located in the fifth cell layer of the paracentral lobule on the medial surface of the hemisphere, the area that would become known as Brodmann Area 4 (Brodmann, 1909).

Four years later, Bevan Lewis (1878) confirmed that this region contained giant pyramidal cells and suggested these cells were especially concerned with motion. Figure 14.8 shows one of his drawings. Because these cells did not correspond perfectly with the foci of representation obtained with electrical stimulation, and because there were parts of the motor cortex that showed a sparsity of

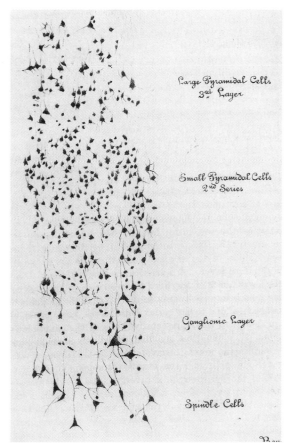

Figure 14.8. A drawing by W. Bevan Lewis (1878) showing the layers of the neocortex.

these cells, the studies of Lewis and Betz attracted relatively little attention at first.

The situation changed when Sherrington showed that destruction of the region rich in Betz cells was associated with degeneration of the pyramidal tract in dogs. Still, questions remained. One question was whether some pyramidal fibers also came from non-Betz cells or from other cortical areas, especially as one went down the phylogenetic scale. For example, Constantin von Monakow argued that a small portion of the corticospinal tract also came from the parietal lobe and the frontal lobe anterior to the precentral convolution, two areas in which Betz cells were not found (see Levin and Bradford, 1938).

Additional studies showed that large "pyramidal" cells, but not the giant cells of Betz, existed outside Brodmann Area 4 and that these cells did, in fact, contribute to the corticospinal tract, at least in monkeys. In macaques, for example, only about 80 percent of the corticospinal tract originates in Area 4, with the remaining 20 percent coming from other frontal and parietal regions with large pyramidal cells (see Levin and Bradford, 1938).

VOLUNTARY ACTION AND THE MOTOR CORTEX

As already noted, John Hughlings Jackson believed that the motor region was responsible for voluntary motor activities and that lower centers played a role in the more automatic activities. He felt this explained why voluntary ges-

[9]Much of this work involved the Marchi stain.

tures could be markedly impaired after cortical lesions, whereas more automatic gestures may remain intact.

David Ferrier (1878), who admired Jackson, accepted his distinction between voluntary and involuntary activities. Ferrier showed the voluntary-involuntary distinction also accounted for the observation that motor functions were less affected by lesions of the cortex in organisms lower on the phylogenetic scale, whereas they can be very much affected by lesions in the Rolandic region in apes and especially in humans. The basic idea was that humans rely more on conscious, voluntary movements than do apes; that apes do this more than monkeys; that monkeys use more skilled voluntary movements than dogs; and that dogs use the motor cortex more than lower, more reflex-bound organisms, such as frogs.

Ferrier was convinced that the less a movement was affected by development and learning, the less it would be affected by cortical damage. To quote:

> The centres or regions in question are strictly motor, but motor in a special or limited sense, viz. psychomotor, or concerned in the execution of consciously discriminated or volitional movements proper. The differences observable as to the effects of destruction of the cortical centres in different animals depend on the degree or extent in which conscious discrimination, as distinct from automaticity or mere reflex action, is concerned in the ordinary modes of activity manifested by the animals operated on. . . . The more machine-like or automatic the movements are at birth, the less the disturbance created by the destruction of the centres concerned in volitional action. . . . Where voluntary control is speedily acquired, or automaticity inherited or rapidly established, as in the rabbit and dog, the centres of voluntary motor acquisition may be removed without completely or permanently interfering with the powers of locomotion. (1886, 362–363)

To quote Ferrier from an earlier publication:

> I ventured to predict that, even in the case of animals whose motor powers did not seem permanently to suffer from destructive lesions of the cortical motor centres, those movements must be paralyzed which involved conscious discrimination, and were not automatically organized. This has been amply verified by Goltz's experiments on dogs. Goltz found that, though a dog's paw is not permanently paralyzed as an organ of locomotion by destruction of the cortex, yet it remains permanently paralyzed for all those actions in which it is employed as a hand. (1878, 443–444)

Neither Ferrier nor Jackson believed one part of the brain could vicariously take over the functions of a different area after an injury. Thus, in humans, where voluntary actions play such a critical role in movements, relatively little recovery would be expected after destruction of the motor cortex. At best, one would find an imperfect compensation resulting from the use of remaining structures.

In contrast, for those organisms whose behavior patterns are decidedly more automatic, the recovery following cortical damage might seem impressive, at least until the right questions are asked. For example, within days after sustaining bilateral cortical lesions, a dog may show recovery of the use of its limbs for locomotion. But, noted Ferrier, the same dog will probably not be capable of extending a paw to a verbal command, will be unable to steady a large bone that it wishes to carry, and will not extend a limb successfully to get a piece of food just out of the reach of its teeth.

It is interesting to note that Luciani and Tamburini (1878), who disagreed with Ferrier about the permanence of cortical lesion effects, agreed with him on the basic issue of cortical control of voluntary motion.

> It cannot be doubted, after the experiments on monkeys, both of excitation and destruction of the centres, that the movements excited by the cortical centres have the characters of purely voluntary movements. We are therefore in complete accord with those who consider the cortical centres as centres of voluntary motion. We express this fact by calling the cortical centres not simply motor, but psycho-motor, that is, motor centres which are in direct relation with psychic centres, and by means of which the psyche, or the conscious activity of the cerebrum, expresses with different movements certain forms of excitements, or modifications which take place in it and have their origin in it. (Translated by Rabagliati, 1879, 541)

The difference between the Italian investigators and Ferrier on the voluntary-involuntary issue was that the former believed the basal ganglia also possessed some control over voluntary movement in animals. Luciani and Tamburini (1878) reasoned this had to be true, since even large cortical lesions in the motor areas in animals did not lead to a complete loss of voluntary movement. They felt that the role played by the basal ganglia in "psycho-motor" functions was significant in dogs, cats, and rabbits, less marked in monkeys, and nonexistent in humans. As they put it: "We find that the basilar centres in man are quite devoid of psycho-motor function" (p. 541).

CUTANEOUS SENSATION AND THE MOTOR CORTEX

David Ferrier (1886) was of the opinion that damage to the motor cortex in higher organisms had no effect on the sense of touch, on temperature, or on pain. This position, however, was not shared by all investigators and, in some cases, precisely the opposite conclusion was drawn. Moritz Schiff (1883), for example, thought that the loss of movement following lesions of the motor centers was due to loss of tactile sensitivity.

David Ferrier (1886) thought Schiff's tests and conclusions were ridiculous. He also questioned the supposed loss of tactile sensitivity that Hermann Munk (1892) reported after damage to the motor zone. Munk had claimed that after lesions of the cortical arm and leg area, tactile sensibility of the opposite extremities was permanently lost and that neither tactile reflexes nor eye or head movements could be produced.

Ferrier had no quarrel with the point that an animal may not react strongly to cutaneous stimulation after a cortical lesion. Indeed, he noted the same thing could be seen in humans with hemiplegia. But he made the case that the diminished responsiveness to tactile stimulation was not adequate to prove sensation was directly affected. Instead, he argued that the basis of this sluggishness resided in the motor defect which prevented vigorous responding.

> Cutaneous reflexes are less readily excitable on the paralyzed side, though the patients testify to the fact that they feel and localise the tactile stimulus as readily and acutely on the one

side as the other. It is easy by devising methods calculated to evoke signs of true sensation, as distinct from merely reflex reaction, in the lower animals, to prove beyond all question that in them also sensation is entirely unaffected by lesions in the cortical motor zone. In monkeys attentive to all their surroundings, it is as a rule only necessary to gently touch the paralyzed side to elicit the most indubitable signs of attention. (Ferrier, 1886, 375)

Ferrier believed that although most of his own work was on monkeys and apes, the same conclusions could be drawn for the dog. He reasoned that all mammalian brains must be organized in a similar manner. Anything else would be "incomprehensible."

Hitzig, to say the least, was surprised by Ferrier's conclusions about the absence of sensory deficits after these lesions. He responded:

It can under no circumstances be in any way doubted that any lesions at all in the motor part of the gyrus sigmoides of the dog are always and without exception combined with disturbances which affect the sensibility of the whole limb, that is to say, of the skin, of the muscles, of the joints, etc, and that these disturbances can in no case be an illusion caused by purely motor disturbances, or, more correctly expressed, by a "motor defect." In this reference to a matter of fact I agree with Schiff, Goltz, Munk and the others, that is to say, with authors whose views I in other respects most decidedly oppose. . . . Since the dog then undoubtedly shows disturbances of his sensory functions as a consequence of those lesions, it appears to me to be *a priori* impossible, or to speak with Ferrier, it would be "incomprehensible" if such disturbances were not to be observed in the ape. (1900, 565)

Edward Schäfer, who disagreed with Ferrier on just about everything else, sided with him after finding that ablations of the motor zones in his monkeys did not seem to affect sensitivity to light touch. He stated:

It cannot therefore be the case that the motor paralysis which is produced by a lesion in the Rolandic area is due to a sensory disturbance. And it also follows that tactile sensibility is not localized in the same part of the cortex from which voluntary motor impulses directly emanate. It is quite another question whether the motor cortex directly receives any branches from the sensory (afferent) nerve-tracts. That it must receive afferent fibres from the several sensory regions of the brain, and amongst these from the region concerned with the perception of cutaneous and muscular sensations, wherever this may be, is certain. The secondary effects . . . which result from section of these fibres in removing the motor cortex may explain some of the cases in man in which motor paralyses *are* accompanied by sensory disturbances (1898, 311–312)

Nevertheless, just a few years earlier, another of Ferrier's countrymen, Frederick Mott (1894), stated that his experiments on monkeys supported Munk's view, which held that touch and pressure sensations were perceived in the so-called motor cortex.

THE MOTOR CORTEX AND THE KINESTHETIC SENSE

Hitzig, who clearly had trouble with the concept of a purely cortical motor center, proposed that the deficits exhibited by his animals were due to the loss of the muscular sense and

Figure 14.9. Henry Charlton Bastian (1837–1915), the British neurologist who concluded that the so-called motor cortex really functioned as a "kinaesthetic centre." Bastian also made significant contributions to the study of aphasia. (Courtesy of the Bibliothèque de l'Académie National de Médecine, Paris.)

the subsequent "defective consciousness of the condition of their limbs" (e.g., Hitzig, 1900, p. 562). Ferrier (1886, p. 380), however, felt that "the existence of the so-called muscular sense does not confer mobility on the limb." He maintained that a paralyzed limb can still have its muscular sense and that the loss of muscular sense does not paralyze a limb.

It is interesting to note that the proposed role of the muscle sense, which was debated so intensely by Ferrier and Hitzig, made its appearance even before Fritsch and Hitzig conducted their landmark experiment in 1870. The person who first began to develop the muscle sense theory, and who then defended it with gusto, was Henry Charlton Bastian (1837–1915), shown in Figure 14.9.[10]

Bastian believed a considerable amount of sensory information was provided by the muscles and that this information was used by the brain to coordinate motor acts. He further thought this information did not have to register in consciousness. In an article from 1869 (p. 463), he stated:

There is evidence to show that the brain is assisted in the execution of voluntary movements by guiding impressions of some kind which, whilst they differ from the impressions producible by means of the ordinary cutaneous and deep sensibility, may differ still further from these, owing to the fact of their not being revealed in consciousness.

Bastian introduced the term "kinaesthetic" in 1880 and then continued to work on his theory. But where was the

[10]Bastian, who worked in London, is sometimes remembered as one of the people who stubbornly defended the concept of spontaneous generation, even after the classic work of Louis Pasteur and John Tyndall (1820–1893) on germ-free air had become widely accepted (Jones, 1972). In neurology, he made many significant contributions, especially on the role of the cortex in language functions (see Chapter 25).

center that would receive these unconscious impressions and then use them to initiate voluntary movement? Although Bastian (1887) discussed the possibility that such movement could be mediated by the cerebellum, in 1895 and again in 1909 he pointed directly to the cerebral cortex. He specifically implicated the region close to or in the area known to be concerned with voluntary motor activity—that is, the "motor cortex."

In this regard, Bastian may have been influenced by his friend Herbert Spencer (1870–72), who theorized that previously experienced sensory impressions must be "revived" before the muscles could be governed to reproduce a specific movement. Through sensory-motor arcs, new movements could be assessed and then modified to be made even more effective.

Seen in this way, only the lower motor neurons, or final common path, was purely "motor." Stated somewhat differently, what Bastian had really proposed was that the motor cortex actually served as a sensory center, and specifically as a center for kinesthetic imagery. He believed a lesion in the Rolandic region could cause a "paralysis," but not by damaging a motor structure per se. Rather, it was a paralysis due to the inability to recall appropriate movement patterns.[11]

Bastian's concept of a cortical kinesthetic center was not given a warm reception by the scientific elite. Only some of his contemporaries accepted his specific theory of ideal motor recall and its localization. The major supporter of his theory was Hermann Munk (1881), who spoke of the motor area as being one part of a *Fühlsphäre,* an area involved in storing the sensory memories necessary for movements (see Chapter 10). Like Bastian, Munk argued that there was no such thing as a purely motor area of the cortex. Instead, the so-called motor cortex "consisted of organs at the service of a special memory, the memory of movements, and constitutes a sort of cerebral register that stores the information acquired by the sensory organs" (translated in Jeannerod, 1985, p. 63). And like Bastian, he also believed that the paralysis which followed destruction of the motor cortex was nothing more than a "pseudo-paralysis"—a deficit due to the loss of prior impressions.

Victor Horsley was also a member of the minority that thought the kinesthetic memory theory had its merits. He wrote:

> From the list of sensory defects I observed in 1886 to follow removal of the so-called motor cortex, it was clear that we had before us precisely the kinds of anaesthesia which should be exhibited in accordance with Munk's views, that is, those sensory functions were disordered which are directly connected with the accurate evolution of a movement, and constitute his "memories of movements." (1909, 128)

By this time, Bastian's term "kinaesthetic sense" was already being replaced by Sherrington's (1906) new word, "proprioception." To Bastian's credit, however, there was growing acceptance of his basic idea that sensory and

---

[11]The favored tests for the kinesthetic sense at this time involved passive movement, position sense, and coordination. These tests required patients to touch a part of their body with their eyes closed, as well as related tasks (Slinger and Horsley, 1906; Wesphal, 1882; also see Chapter 10).

motor functions had to be integrated somewhere in the central nervous system.

THE FRONTAL EYE FIELDS

Eduard Hitzig (1904) noted that electrical stimulation of a region in front of the precentral gyrus in the dog caused combined movements of the head and eyes. He believed this was due to excitation of the organ of coordination and that the cortex contained a separate center for isolated movements of the eyes.

This area had also been studied in two monkeys by David Ferrier (1886). He thought that stimulation of parts of the posterior superior and middle frontal convolutions affected eye movements. Monkeys electrically stimulated here opened their eyes, dilated their pupils, and moved their heads and eyes to the opposite side. But Ferrier was uncertain about what this meant at the time.

Other experimenters whose work suggested the existence of separate frontal eyefields were Victor Horsley and Edward Sharpey-Schäfer, who worked together, and Vladimir Bekhterev (1857–1927). In experiments from 1889, these three scientists found that stimulation of an area in front of the prefrontal gyrus caused movements of the head and eyes to the opposite side. The separate eyefield area was also identified in apes by Horsley and Beevor a year later.

The findings of the electrical stimulation experiments were in accord with the effects of lesions of the area anterior to the motor strip. As shown by Eduard Hitzig, frontal lobe damage anterior to the classical motor areas often caused disturbances of vision in the opposite eye in dogs and monkeys. Although Hermann Munk questioned this finding, subsequent observations left little doubt about Hitzig being correct. As noted by Leonardo Bianchi (1848–1927) with his monkeys: "If some solid piece of food, like an apple or chestnut, is thrown at some distance from it, the animal runs after it for a certain distance and then stops, either as the result of deficient attention or because it loses sight of the object owing to defect of the attentive accommodation" (1922, p. 196).

In 1931, Otfrid Foerster wrote that stimulation of this area in humans caused the eyes to move to the opposite side and occasionally upward. Damage to the area, in contrast, caused a temporary paralysis of movement of the eyes to the opposite side.

By this point in time, there was consensus about the frontal cortex containing separate centers for eye movements.

MOTOR VERSUS PREMOTOR CORTEX FUNCTIONS

During the 1930s, a number of studies were conducted on infra-human primates and on human patients to compare the functional contributions of Brodmann Area 4, associated with the precentral gyrus, with those of Area 6, located just in front of it. John Fulton and his colleagues at Yale University, including Margaret Kennard (1899–1976), Carlyle F. Jacobsen (1902–1974), and Paul Bucy (b. 1904), conducted many experiments to compare these two areas

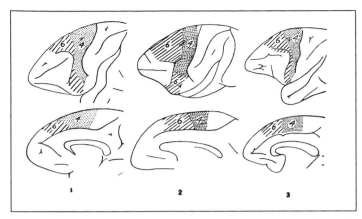

Figure 14.10. The motor (Area 4) and premotor (Area 6) areas laterally and medially in Old World monkeys. Paul Bucy (1935) based this figure on the work of other investigators (Brodmann, the Vogts, Mauss) who did not always identify the specific type of monkey used. Bucy concluded that most likely (1) and (3) were based on rhesus monkeys whereas (2) could have been a baboon. (From *Journal of Comparative Neurology* with the permission of Wiley-Liss.)

in monkeys and apes.[12] The work of Otfrid Foerster (1931, 1936a, 1936b) on human patients also led to a better understanding of the different functions of these two "motor" areas.

The lesion studies published in the 1930s were based on the observation that Areas 4 and 6 differ in cytoarchitecture and on the age-old presumption that where there is a difference in structure, there must be a difference in function. Research efforts were also stimulated by the work of Oskar and Cécile Vogt (1919), who were the first to demonstrate the excitability of Brodmann Area 6 in detail.

Area 6, which is found in front of Area 4, differs from Area 4 cytoarchitectonically because it lacks giant Betz cells. It also sends a much smaller percentage of fibers to the pyramidal tract. Most of its efferents go to Area 4 and to brainstem structures such as the substantia nigra and the red nucleus (Fulton, 1937; Levin, 1936). In part because Area 6 sends many projections to Area 4, the term "motor cortex" was maintained for Area 4, and "premotor cortex" was given to Area 6. These two areas are illustrated in Figure 14.10.

Although the efferent systems associated with Areas 4 and 6 have been called pyramidal and extrapyramidal, respectively, these terms have been used in different ways, leading to considerable confusion (see Bucy, 1936). Some individuals have defined the pyramidal system as the projection system from Area 4 alone; others limited it to just the Betz cell contribution; and still others included pyramidal cell projections from other cortical areas in their definitions. The one thing that seemed clear, however, was that the overlap of the projections from the motor and premotor areas decreased as one ascended the phylogenetic scale.

The Vogts (1919) and others found that stimulation of Area 6 can give rise to discrete movements of the skeletal muscles on the opposite side of the body (e.g., Bucy, 1934;

Foerster, 1931). In fact, the movements associated with Area 6 stimulation were found to be as fine as those seen with Area 4 stimulation. Nevertheless, thresholds of excitability differed between Areas 4 and 6. More current was needed to elicit skeletal responses when Area 6 was stimulated. Further, because Area 6 feeds into Area 4, damage to Area 6 did not affect the Area 4 response to electrical stimulation, although damage to Area 4 severely impaired movements to stimulation of Area 6.

In the first quarter of the twentieth century, it was generally assumed that damage to the pyramidal system was the cause of spastic paralysis. This assertion began to be challenged in 1929 when John Fulton first observed that lesions confined to Area 4 were not associated with spasticity in primates (see Fulton, 1937). Instead, damage to Area 4 resulted in a flaccid paralytic condition. This finding was confirmed in subsequent experiments, in which it was shown that severing the pyramidal tracts in the medulla was also associated with a flaccid paralysis.

In contrast, damage limited to Area 6 was found to be associated with spasticity. The degree of spasticity appeared to increase as one went up the phylogenetic scale, whereas the extent of the recovery seemed to be greater as one descended the phylogenetic scale. These observations led Fulton (1937, p. 1021) to write, "It is clear from the evidence now available that the extrapyramidal rather than the pyramidal projections from the cortex are those which, when interrupted, give rise to the spastic state."

When Area 6 was damaged in conjunction with Area 4, the affected extremities still became spastic. This was probably why spasticity was associated with "motor cortex" damage in many earlier studies. Following bilateral destruction of both Area 4 and Area 6, primates have great difficulty walking, standing, feeding, and caring for themselves, even months after injury. In addition to these differences, Fulton and his associates showed that the motor and premotor cortices also differ in several other significant ways. These findings were widely accepted by the scientific community.

REFERENCES

Akert, K., and Hammond, M. P. 1962. Emanuel Swedenborg (1688–1772) and his contribution to neurology. *Medical History, 6,* 255–266.

Aretaeus, the Cappadocian. On epilepsy; on paralysis. In F. Adams (Ed.), *The Extant Works of Aretaeus, the Cappadocian.* London: Sydenham Society, 1856.

Avicenna. 1930. *A Treatise on the Canon of Medicine.* (O. C. Gruner, Trans.). London: Luzac.

Bartholow, R. 1874. Experimental investigations into the functions of the human brain. *American Journal of Medical Science, 67,* 305–313.

Bastian, H. C. 1869. On the muscular sense and the physiology of thinking. *British Medical Journal, 1,* 394–396, 437–439, 461–463, 509–512.

Bastian, H. C. 1880. *The Brain as an Organ of Mind.* New York: Appleton.

Bastian, H. C. 1887. The "muscle sense": Its nature and cortical localization. *Brain, 10,* 1–137.

Bastian, H. C. 1895. Note on the relations of sensory impressions and sensory centres to voluntary movements. *Proceedings of the Royal Society, 58,* 89–98.

Bastian, H. C. 1909. The functions of the kinaesthetic areas of the brain. *Brain, 32,* 328–341.

Betz, W. 1874. Anatomischer Nachweis zweier Gehirncentra. *Centralblad für die medizinische Wissenschaft, 12,* 578–580, 595–599.

Bianchi, L. 1922. *The Mechanism of the Brain.* Edinburgh: E. & S. Livingstone. Translated into English by J. H. MacDonald from *La Meccanica del Cervello e la Funzione dei Lobi Frontali.* Torino: Bocca, 1920.

Bieber, I., and Fulton, J. F. 1933. The relation of forced grasping and grop-

---

[12]The papers by Bucy and Fulton comparing motor and premotor cortices in infrahuman primates include Bucy (1933a, 1933b, 1934, 1935, 1936); Bucy and Fulton (1933); Fulton (1937); Fulton, Jacobsen, and Kennard (1932); Fulton and Kennard (1934); Kennard and Fulton (1933); and Welch and Kennard (1944). Bucy (1942) also presented human clinical material.

ing to the righting reflexes. *American Journal of Physiology, 105,* 439–466.

Bieber, I., and Fulton, J. F. 1938. Relation of the cerebral cortex to the grasp reflex and to postural and righting reflexes. *Archives of Neurology and Psychiatry, 39,* 433–454.

Bouillaud, J.-B. 1825. *Traité Clinique et Physiologique de l'Encéphalite ou Inflammation du Cerveau.* Paris: J. B. Ballière.

Bravais, L. F. 1827. *Recherches sur les Symptômes et le Traitement de l'Epilepsie Hémiplégique.* Thèse 118. Paris.

Breasted, J. H. 1930. *The Edwin Smith Surgical Papyrus.* Chicago: University of Chicago Press.

Bright, R. 1831. *Reports of Medical Cases Selected with a View of Illustrating the Symptoms and Cure of Diseases by Reference to Morbid Anatomy. Diseases of the Brain and Nervous System.* London: Longman.

Broca, P. 1861. Remarques sur le siège de la faculté du langage articulé; suivies d'une observation d'aphémie (perte de la parole). *Bulletins de la Société Anatomique (Paris), 6,* 330–357, 398–407. In G. von Bonin, *Some Papers on the Cerebral Cortex.* Translated as, "Remarks on the seat of the faculty of articulate language, followed by an observation of aphemia." Springfield, IL: Charles C Thomas, 1960. Pp. 49–72.

Brodmann, K. 1909. *Vergleichende Lokalisations-lehre der Grosshirnrinde.* Leipzig: J. A. Barth.

Bucy, P. C. 1933a. Electrical excitability and cyto-architecture of the premotor cortex in the monkey. *Archives of Neurology and Psychiatry, 30,* 1205–1224.

Bucy, P. C. 1933b. Representation of the ipsilateral extremities in the cerebral cortex. *Science, 78,* 418.

Bucy, P. C. 1934. The relation of the premotor cortex to motor activity. *Journal of Nervous and Mental Disease, 79,* 621–630.

Bucy, P. C. 1935. A comparative cytoarchitectonic study of the motor and premotor areas in the primate cortex. *Journal of Comparative Neurology, 62,* 293–331.

Bucy, P. C. 1936. Areas 4 and 6 of the cerebral cortex and their projection systems. *Archives of Neurology and Psychiatry, 35,* 1396–1400.

Bucy, P. C. 1942. Cortical extirpation in the treatment of involuntary movements. *Research Publications of the Association for Research in Nervous and Mental Disease, 21,* 551–595.

Bucy, P. C., and Fulton, J. F. 1933. Ipsilateral representation in the motor and premotor cortex of monkeys. *Brain, 56,* 318–342.

Budge, J. L. 1842. *Untersuchungen über das Nervensystem.* Frankfurt am Main: Jäger.

Carville, C., and Duret, H. 1875. Sur le fonction des hémisphères cérébraux. *Archiv der Physiologie, 7,* 352–490.

Clarke, E., and O'Malley, C. D. 1968. *The Human Brain and Spinal Cord.* Berkeley: University of California Press.

Courville, C. B. 1946. The ancestry of neuropathology, Hippocrates and *De vulneribus capitis. Bulletin of the Los Angeles Neurological Society, 11,* 1–19.

Critchley, M. 1960. Hughlings Jackson, the man; and the early days of the National Hospital. *Proceedings of the Royal Society of Medicine, 53,* 613–618.

Duverney, J. G. 1761. *Oeuvres Anatomiques.* Paris: Jombert.

Eckhard, K. 1867. *Experimentalphysiologie des Nervensystems.* Giessen: E. Roth.

Eustachi, B. 1714. *Tabulae anatomicae.* Roma: F. Gonzaga.

Ferrier, D. 1876. *The Functions of the Brain.* London: Smith, Elder and Company.

Ferrier, D. 1878. The Goulstonian lectures on localization of cerebral diseases. *British Medical Journal, 1,* 399–402, 443–447.

Ferrier, D. 1886. *The Functions of the Brain.* New York: G. P. Putnam's Sons. (Reprinted 1978; Washington, DC: University Publications of America)

Ferrier, D., and Yeo, G. 1885. A record of experiments on the effects of lesions of different regions of the cerebral hemispheres. *Philosophical Transactions of the Royal Society of London (Biology), 175,* 479–564.

Flourens, M.-J.-P. 1824. *Recherches Expérimentales sur les Propriétés et les Fonctions du Système Nerveux dans les Animaux Vértebrés.* Paris: J. B. Ballière.

Foerster, O. 1931. The cerebral cortex in man. *Lancet, 2,* 309–312.

Foerster, O. 1936a. The motor cortex in man in the light of Hughlings Jackson's doctrines. *Brain, 59,* 135–159.

Foerster, O. 1936b. Motorische Felder und Bahnen. In O. Bumke and O. Foerster (Eds.), *Handbuch der Neurologie* (Band 6). Berlin: Springer. Pp. 1–357.

Fournié, E. 1873. *Recherches Expérimentales sur la Fonctionnement du Cerveau.* Paris: A. Delahaye.

Fritsch, G., and Hitzig, E. 1870. Über die elektrische Erregbarkeit des Grosshirns. *Archiv für Anatomie und Physiologie,* 300–332. In G. von Bonin, *Some Papers on the Cerebral Cortex.* Translated as, "On the electrical excitability of the cerebrum." Springfield, IL: Charles C Thomas, 1960. Pp. 73–96.

Fulton, J. F. 1937. Spasticity and the frontal lobes. A review. *New England Journal of Medicine, 217,* 1017–1024.

Fulton, J. F., Jacobsen, C. F., and Kennard, M. A. 1932. A note concerning the relation of the frontal lobes to posture and forced grasping in monkeys. *Brain, 55,* 524–536.

Fulton, J. F., and Kennard, M. A. 1934. A study of flaccid and spastic paralyses produced by lesions of the cerebral cortex in primates. *Research Publications—Association for Nervous and Mental Disease, 13,* 158–210.

Giannitrapani, D. 1967. Developing concepts of lateralization of cerebral functions. *Cortex, 3,* 355–370.

Greenblatt, S. H. 1965. The major influences on the early life and work of John Hughlings Jackson. *Bulletin of the History of Medicine, 39,* 346–376.

Greenblatt, S. H. 1977. The development of Hughlings Jackson's approach to diseases of the nervous system 1863–1866: Unilateral seizures, hemiplegia and aphasia. *Bulletin of the History of Medicine, 51,* 412–430.

Grünbaum, A. S. F., and Sherrington, C. S. 1902. Observations on the physiology of the cerebral cortex in some of the higher apes (preliminary communication). *Proceedings of the Royal Society, 69,* 206–209.

Grünbaum, A. S. F., and Sherrington, C. S. 1903. Observations on the physiology of the cerebral cortex in anthropoid apes. *Proceedings of the Royal Society, 72,* 152–155.

Hippocrates. 1886. *The Genuine Works of Hippocrates.* New York: Wood.

Hitzig, E. 1874. *Untersuchungen über das Gehirn.* Berlin: A. Hirschwald.

Hitzig, E. 1867. Über die Anwendung unpolarisirbarer Elektroden in Elektrotherapie. *Berliner klinische Wochenschrift, 4,* 404–406.

Hitzig, E. 1869. Verhandlungen ärzlichen Gesellschaften. *Berliner klinische Wochenschrift, 6,* 420.

Hitzig, E. 1900. Hughlings Jackson and the cortical motor centres in the light of physiological research. *Brain, 23,* 545–581.

Hitzig, E. 1904. *Untersuchungen über das Gehirn.* Berlin: A. Hirschwald.

Horsley, V. 1909. The function of the so-called motor area of the brain. *British Medical Journal, 2,* 125–132.

Horsley, V., and Beevor, C. E. 1890. A record of the results obtained by electrical excitation of the so-called motor cortex in the orang-outang (Simia satyrus). *Philosophical Transactions of the Royal Society, B, 181,* 129–158.

Horsley, V. A. H., and Sharpey-Schäfer, E. A. 1889. A record of experiments upon the functions of the cerebral cortex. *Philosophical Transactions of the Royal Society, B, 179,* 1–45.

Jackson, J. H. 1861. Syphilitic affections of the nervous system. *Medical Times and Gazette, 1,* 648–652; *2,* 59–60, 83–85, 133–135, 456, 502–503.

Jackson, J. H. 1863a. *Suggestions for Studying Diseases of the Nervous System on Professor Owen's Vertebral Theory.* London: H. K. Lewis.

Jackson, J. H. 1863b. Convulsive spasms of the right hand and arm preceding epileptic seizures. *Medical Times and Gazette, 1,* 110–111.

Jackson, J. H. 1863c. Epileptiform convulsions (unilateral) after injury to the head. *Medical Times and Gazette, 2,* 65–66.

Jackson, J. H. 1863d. Epileptiform seizures—Aura from the thumb—Attacks of coloured vision. *Medical Times and Gazette, 1,* 589.

Jackson, J. H. 1866. A physician's notes on ophthalmology. Cases of disease of the nervous system in which there were defects of sight, sight and hearing. *Royal London Ophthalmic Hospital Reports, 5,* 252–306.

Jackson, J. H. 1870. A study of convulsions. *Transactions of the St. Andrew's Medical Graduates Association, 3,* 162–204.

Jeannerod, M. 1985. *The Brain Machine.* Cambridge, MA: Harvard University Press.

Jones, E. G. 1972. The development of the "muscle sense" concept during the nineteenth century and the work of H. Charlton Bastian. *Journal of the History of Medicine and Allied Sciences, 27,* 298–311.

Kennard, M. A., and Fulton, J. F. 1933. The localizing significance of spasticity, reflex grasping and the signs of Babinski and Rossolimo. *Brain, 56,* 213–225.

Krause, F. 1911. *Chirurgie des Gehirns und Rückenmarkes nach eigenen Erfahrungen.* Berlin: Urban and Schwartzenberg.

Levin, P. M. 1936. The efferent fibers of the frontal lobe of the monkey, Macaca mulatta. *Journal of Comparative Neurology, 63,* 369–419.

Levin, P. M., and Bradford, F. K. 1938. The exact origin of the corticospinal tract in the monkey. *Journal of Comparative Neurology, 68,* 411–422.

Lewis, B. 1878. On the comparative structure of the cortex cerebri. *Brain, 1,* 79–96.

Longet, F. A. 1842. *Anatomie et Physiologie du Système Nerveux de l'Homme et des Animaux Vértebrés.* Paris: Masson et Cie.

Luciani, L., and Tamburini, A. 1878. *Richerche Sperimentali sui Centri Psico-Motori Corticali.* Reggio-Emilia: Tip. di S. Calderini e Figlio.

Lyons, J. B. 1982. The neurology of Robert Bentley Todd. In F. C. Rose and W. F. Bynum (Eds.), *Historical Aspects of the Neurosciences.* New York: Raven Press. Pp. 137–150.

Magendie, F. 1839. *Leçons sur les Fonctions et les Maladies du Système Nerveux.* Paris: Ebrard.

Malacarne, M. V. G. 1776. *Nuova Esposizione della Struttura del Cervelletto Umano.* Turin: G. Brolio.

Mellus, E. L. 1894. Preliminary note on bilateral degeneration in the spinal cord of monkeys (*Macacus sinicus*) following unilateral lesion of the cerebral cortex. *Proceedings of the Royal Society of London, 55,* 208–210. (Communicated by V. Horsley)

Mellus, E. L. 1899. Motor paths in the brain and cord of the monkey. *Journal of Nervous and Mental Disease, 26*, 197–209.

Mistichelli, D. 1709. *Trattado dell'Apoplessia.* Roma: A. de Rossi alla Piazza di Ceri.

Morgagni, J. B. 1740. *Dissertationes Anatomicas.* Venetiis: Pitteri.

Mott, F. W. 1894. The sensory motor functions of the central convolutions of the cerebral cortex. *Journal of Physiology, 15*, 465–487.

Mott, F. W., and Sherrington, C. S. 1895. Experiments upon the influence of sensory nerves upon movement and nutrition of the limbs (preliminary communication). *Proceedings of the Royal Society, 57*, 481–488.

Munk, H. 1881. *Über die Funktionen der Grosshirnrinde.* Berlin: A. Hirschwald.

Munk, H. 1892. Über die Fühlsphären der Grosshirnrinde. *Sitzungsberichte der Kaiserlichen Akademie der Wissenschaften, 101*, 679–723.

Muybridge, E. 1887. *Animal Locomotion.* Philadelphia, PA: Lippincott.

Pourfour du Petit, F. 1710. *Lettres d'un Médecin des Hôpitaux du Roi, à un Autre Médecin de Ses Amis.* Naumur: Charles Gerhard Albert.

Rabagliati, A. 1879. Luciani and Tamburini on the functions of the brain. *Brain, 1*, 529–544.

Rhazes. 1486. *Liber elhavi seu totum continentus bubikir zacharie errasis filii.* Traductio ex Arabice in Latinum per Mag. Ferragium. Bescia: Britannicus.

Santorini, J. D. 1724. *Observationes anatomicae.* Venetiis: Recurti.

Schäfer, E. A. 1898. On the alleged sensory functions of the motor cortex cerebri. *Journal of Physiology, 22*, 310–314.

Schiff, M. 1858–1859. *Lehrbuch der Physiologie des Menschen* (Vol. 1). Lahr: Schauenburg.

Schiff, M. 1883. Ueber die Erregbarkeit des Rückenmarks. *Pflüger's Archiv für Physiologie, 30*, 199–275.

Sherrington, C. S. 1906. *The Integrative Action of the Nervous System.* New Haven, CT: Yale University Press.

Slinger, R. T., and Horsley, V. 1906. Upon the orientation of points in space by the muscular, arthrodial, and tactile sense of the upper limbs in normal individuals and in blind persons. *Brain, 29*, 1–28.

Soury, J. A. 1899. *Le Système Nerveux Central; Structure et Fonctions; Histoire Critique des Théories et des Doctrines.* Paris: Carré et Naud.

Spencer, H. 1870–1872. *Principles of Psychology* (2nd ed.). London: Williams and Norgate.

Swedenborg, E. 1882–1887. *The Brain, Considered Anatomically, Physiologically, and Philosophically.* (R. L. Tafel, Trans./Ed./ Annotator). London: Speirs.

Taylor, J. (Ed.). 1925. Biographical memoir. In *Neurological Fragments by John Hughlings Jackson.* London: Oxford University Press.

Temkin, O. 1972. *The Falling Sickness.* Baltimore, MD: Johns Hopkins University Press.

Todd, R. B. 1849. On the pathology and treatment of convulsive diseases. *London Medical Gazette, 8*, 661–671, 724–729, 766–772, 815–822, 837–846.

Türck, L. 1849. Mikroskopischer Befund des Rückenmarkes eines paraplegischen Weibes. *Zeitschrift für die Gesellschaft der Ärzte zu Wien, 1*, 173–176.

Türck, L. 1850. Über ein bisher unbekanntes Verhalten des Rückenmarkes bei Hemiplegie. *Zeitschrift für die Gesellschaft der Ärzte zu Wien, 1*, 6–8.

Türck, L. 1853. Über sekundäre Erkrankung einzelner Rückenmarkstränge und ihrer Fortsetzungen zum Gehirne. *Zeitschrift für die Gesellschaft der Ärzte zu Wien, 2*, 289–317.

Vedrani, A. 1920. L'innervazione controlaterale. *L'Illustrazione Medica Italiana, 2*, 81–83.

Vogt, C., and Vogt, O. 1919. Allgemeinere Ergebnisse unserer Hirnforschung. *Journal für Psychologie und Neurologie, 25*, 279–462.

Walsh, F. M. R. 1961. Contributions of John Hughlings Jackson to neurology. *Archives of Neurology, 5*, 119–131.

Welch, W. K., and Kennard, M. A. 1944. Relation of cerebral cortex to spasticity and flaccidity. *Journal of Neurophysiology, 7*, 255–268.

Wesphal, C. F. O. 1882. Zur Lokalisation der Hemianopsie und des Muskelgefühls beim Menschen. *Charité-Annalen, 7*, 466–489.

Willis, T. 1664. *Cerebri anatome: cui accessit nervorum descriptio et usus.* London: J. Flesher.

Chapter 15
The Cerebellum and the Corpus Striatum

> the Cerebel is a peculiar Fountain of animal Spirits designed for some works, wholly distinct from the Brain. Within the Brain, Imagination, Memory, Discourse, and other more superior Acts of the animal function are performed. . . . But the office of the Cerebel . . . seems to be for the animal Spirits to supply some nerves; by which involuntary actions . . . are performed.
>
> Thomas Willis, 1664

This chapter is divided into two parts. The first examines the history of the cerebellum, a structure described and differentiated from the cerebrum by the Ancient Greeks. The second is devoted to the caudate nucleus, the putamen, and the globus pallidus, three nuclei that make up the corpus striatum.

The Cerebellum

THE CEREBELLUM THROUGH THE RENAISSANCE

The cerebellum was differentiated from the cerebrum by Aristotle in the fourth century B.C. Erasistratus, who lived a century later, is said to have taught that deer, hare, and other fast-running animals have more intricately folded cerebellums than less active animals. Although this may have been the first attempt to associate a function with the cerebellum, it proved off the mark. As Erasistratus' critics pointed out, animals not noted for their fleetness of foot, such as two-toed sloths and oxen, have cerebellums just as convoluted as the speedy deer.

During the Roman period, the cerebellum was described in more detail by Galen, who thought it was a source of motor nerves and perhaps the origin of the spinal cord. Galen believed the vermis of the cerebellum acted as a valve that regulated the flow of animal spirits through the ventricular system (see Clarke and O'Malley, 1968).

The anatomy of the cerebellum was better defined during the Renaissance. Charles Estienne (Stephanus, 1503–1564) was probably the first to illustrate the true convolutional patterns of the cerebellum as well as those of the cerebral cortex. His copper plates were published in 1539. Not long after this, Constanzo Varolius, who was the first to popularize the method of dissecting the brain from below, noted the connections between the pons and the cerebellum. Varolius (1573), who is shown in Figure 15.1, gave the pons its name because it reminded him of a bridge (pons) spanning a canal. His plate showing the pons appears as Figure 15.2. He followed this by publishing one of the earliest texts on neu-

roanatomy. But, like Galen, Varolius (1591) believed some cranial nerves originated in the cerebellum. For this reason, he associated the cerebellum with hearing and taste.

Guglielmo da Saliceto, who lived in the thirteenth century, was one of the first to suggest that "natural and necessary" movements originated from the cerebellum (Giannitrapani, 1967). As noted in Chapter 14, this Italian surgeon also proposed that voluntary movements have their origins in the cerebral hemispheres.

Nevertheless, these ideas were not immediately accepted. Instead, the cerebellum was thought to house memory by most investigators, at least prior to the seventeenth century

Figure 15.1. Constanzo Varolio (1543–1575), the Italian physician-anatomist who gave the pons its name. Varolio was honored by having his own name attached to the pons ("pons Varolii").

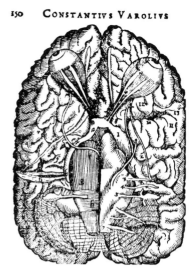

150   CONSTANTIVS VAROLIVS

Figure 15.2. A late-sixteenth-century woodcut by Constanzo Varolio of the underside of the brain. The pons is labeled *h* and is described as *Processus transversalis cerebri, qui dicitur pons.*

and the contributions of Thomas Willis. This belief was largely based on the association of the cerebellum with the fourth ventricle, the usual seat of memory in classical "cell doctrine" (see Chapters 2 and 23). Among others, Johann Vesling (1598–1649) and Nicolaas Tulp (1593–1674) considered the cerebellum the seat of memory functions (Neuberger, 1981).

The association between the cerebellum and memory was also mentioned by Andreas Vesalius (1543) in his *Fabrica,* although he did not seem to have as much faith in it. With regard to the anatomy of the cerebellum, Vesalius stated that the brain has customarily been divided into an anterior part, called the cerebrum, and a posterior part, called the cerebellum, and then into right and left halves. In Book VII of the *Fabrica,* he described the external features of the cerebellum and noted the cerebellum was only one tenth or one eleventh the size of the cerebrum.

THOMAS WILLIS AND HIS INFLUENCE

In 1664, in a manner somewhat reminiscent of Guglielmo da Saliceto four centuries earlier, Thomas Willis proposed that the organ of thought was the cerebrum, whereas the cerebellum (which he at times defined broadly to include its processes and appendages, in particular, the pons and colliculi) was responsible for involuntary motor functions. The full passage from which the short quotation at the beginning of this chapter was taken stated this as follows:

> The Cerebel is a peculiar Fountain of animal Spirits designed for some works, wholly distinct from the Brain. Within the Brain, Imagination, Memory, Discourse, and other more superior Acts of the animal function are performed; besides, the animal spirits flow also from it into the nervous stock, by which all the spontaneous motions, to wit, of which we are knowing and will, are performed. But the office of the Cerebel seems to be for the animal Spirits to supply some Nerves; by which involuntary actions (such as are the beating of the Heart, easie Respiration, the Concoction of the Aliment, the protrusion of Chyle, and many

others) which are made after a constant manner unknown to us, or whether we will or no, are performed. As often as we go about voluntary motion we seem as it were to perceive within us the Spirits residing in the fore-part of the Head to be stirred up to action, or an influx. But the Spirits inhabiting the Cerebel perform unperceivedly and silently their works of Nature without our knowledge or care. (1965 reprint, 111)

Willis, a professor at Oxford, probably came to this conclusion on the basis of three factors. First, an alternative to the old idea that the posterior ventricle functioned in memory was one which associated the posterior ventricle with movement. Saint Augustine (354–430), for example, accepted this association, as did a number of Willis' contemporaries.

Second, Willis relied upon pathology and clinical observations. Even though the lesions he examined were not "clean" and probably affected other structures (e.g., the medulla), his empirical orientation was a significant improvement over purely philosophic conjecture.

Third, his theory was based on observable differences in neuroanatomy. Here he noted that whereas the cerebral hemispheres had an irregular surface, the cerebellum was more uniformly shaped. From this he deduced that the animal spirits in the cerebellum responded in a more automatic manner. Similarly, comparative studies revealed that the cerebellum was very similar in all animals. This further justified the conclusion that only those functions common to all living organisms (e.g., the heartbeat and breathing) were under its jurisdiction.

Willis' *Cerebri anatome* was considered the most complete and accurate account of the nervous system at the time of its publication (McHenry, 1969). It stimulated considerable research, and his ideas about the cerebellum and involuntary "vital" bodily functions were soon subjected to experimental verification.

For example, Théophile Bonet, who practiced medicine during the seventeenth century in Geneva, demonstrated that injury to the cerebellum could be associated with circulatory and respiratory arrest and death, as predicted by the Willis theory (Irons, 1942). (Bonet is shown in Figure 15.3, and the cover page of his most famous work appears as Figure 15.4.) Supportive data (literally death after cerebellar lesions) were also presented by Claude Perrault in Paris and Raymond Vieussens in Montpellier (see Figures 15.5 and 15.6; Clarke and O'Malley, 1968; Neuburger, 1981).

COMPETING IDEAS IN THE POST-WILLIS PERIOD

To say the least, not all experimenters agreed with Willis that the cerebellum was responsible for involuntary motions basic to life. Joseph Duverney, Pierre Chirac (1650–1732), François Pourfour du Petit, Antonio Valsalva, and Giovanni Morgagni all showed that animals could survive lesions of the cerebellum, often with breathing not disturbed.[1]

In addition, both Antoine Charles Lorry and Justus Arnemann (1763–1806) argued that the earlier association between lesions of the cerebellum and death was due to excessive bleeding and/or to damage that affected other

[1]Morgagni discussed this in 1740. For additional references and commentary, see Neuburger (1981).

Figure 15.3. Théophile Bonet (1620–1689), who showed that injury to the cerebellum led to circulatory and respiratory arrest, in accord with the theory of Thomas Willis (1664). This representation of Bonet appeared in the 1700 edition of his *Sepulchretum*.

Figure 15.4. The title page of Théophile Bonet's *Sepulchretum* (2nd ed., 1700).

Figure 15.5. Raymond Vieussens (1641–1716), professor of anatomy at Montpellier, France. Vieussens made a number of discoveries in his explorations of the cerebellum and basal ganglia. (From Sterling, 1902.)

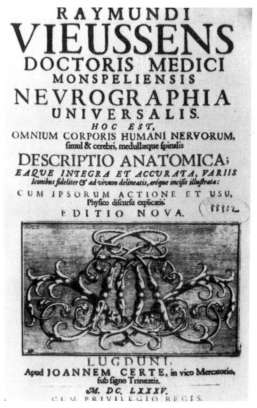

Figure 15.6. Title page of Raymond Vieussens' *Neurographia universalis* (1665).

parts of the brainstem. In particular, the medulla was singled out in this context.[2]

A further blow to the Willis theory was administered by Albrecht von Haller and his followers, including Johann Georg Zimmermann (1728–1795) and Johann Zinn. These investigators demonstrated that the cerebellum did not directly influence cardiac or respiratory activity and that cerebellar damage was no more fatal than damage to the cerebrum. They claimed that when cerebellar operations were done with proper care, animals could survive for long periods of time (see Haller, 1762).

As a result of experiments like these, different theories of cerebellar function abounded, especially through the eighteenth century and into the opening decades of the nineteenth century:

1. Willis' basic idea that the cerebellum was responsible for the involuntary movements vital to life was still supported by some scientists. In the ranks of this camp were Jacob Ackermann (1765–1815) and Adam Eschenmayer (1768–1852).

2. Michele Malacarne (1744–1816), who published an entire book on the cerebellum in 1776, linked it with intellect.[3] Specifically, he hypothesized that the number of laminae making up the cerebellum reflected the intellectual capacity of the organism.

3. Some investigators saw the cerebellum as a common center for sensation (sensorum commune). Achille Louis Foville (1799–1878) and Philippe Pinel (1745–1826) represented this view.

4. Alexander Walker and Johann Grohmann (1769–1847) associated the cerebellum with the will.

5. Phillip Franz von Walther (1782–1849) saw it as the seat of animal instincts.

6. Franz Joseph Gall hypothesized that the cerebellum played an important role in sexual excitation. This was based on a boyhood observation, in which he associated thick necks, like those found in a bull or a stallion, with large cerebellums and sexual prowess.

The latter theory aroused great controversy, but survived for a fairly long time because cerebellar injuries seemed to be associated with testicular atrophy. Some cases showing this were seen by Baron Larrey, surgeon to Napoleon Bonaparte's armies. In one instance, a young soldier was struck on the back of the neck while he was serving in Egypt. After his return to France, it was found that his genital organs were reduced to the size of an infant's; he could no longer achieve an erection; and he had lost his sexual appetite.

Gall's theory and Larrey's reports were later criticized by William A. Hammond (1828–1900). In 1869, Hammond pointed out that many animals with small cerebellums (e.g., kangaroos and monkeys) are extremely sexually active. He

also noted that castrated animals do not show atrophy of the cerebellum. Hammond thought injury to the medulla and spinal cord might have been even more responsible than cerebellar damage for the genital atrophy observed in cases like Larrey's.

Nevertheless, some evidence to suggest Gall might not have been entirely mistaken came from cases of hereditary cerebellar ataxia. This disorder was found to be associated with testicular atrophy and loss of sexual prowess (Holmes, 1907). Notably, hereditary cerebellar ataxia was also associated with poor coordination, a symptom that became even more closely linked with cerebellar dysfunction.

ROLANDO, FLOURENS, AND MAGENDIE

The path out of this speculative nightmare was paved by Luigi Rolando, who worked in Sassari, Sardinia, before becoming a professor of anatomy in Turin in 1814. Working with animals, Rolando stimulated different parts of brain with galvanic current. He found the most violent convulsions occurred when his electrodes were placed on the cerebellum. He also noted that when this organ was removed from pigs, sheep, birds, and reptiles, the animals acted as if they were paralyzed. Smaller lesions resulted in uncertain and hesitant movements and led to swaying and loss of coordinated motor control. When the lesions were on only one side of the cerebellum, the effects were ipsilateral to the damage.

In 1809, Rolando described one of his experiments as follows:

> By means of a very thin saw and a small trephine I took off, in scales, a large part of the skull of a kid. . . . Later I ablated some thin layers of the cerebellar lobules, and . . . did not observe any change in movement. After five minutes, I took from the median lobe, and laterally, some deeper layers, about two "linee" thick. The animal then began shaking, every once in a while it would kneel, if slightly hit it would fall, sometimes on one side sometimes on the other. It would try to eat, although it could hardly manage because of contractions and tremor of the muscles of its head and neck. (Translated in McHenry, 1969, 210)

Rolando's lesions and behavioral observations were admittedly crude. But as a result of seeing "weakness," paralysis, and instability after his cerebellar lesions, he assigned the function of muscle movement to this structure, and distinguished this from the stupor that followed large cortical lesions.

These findings, and his theory that the cerebellum functioned as a whole, were described in a book he engraved, printed, and bound himself in 1809 (Figure 15.7 shows the cover page of this work). His *Saggio*, however, had a very limited printing and was not widely disseminated. The limited availability of Rolando's findings proved to be significant because his experiments on the cerebellum remained largely unknown to the wider scientific community for a number of years.

It was during this time that Marie-Jean-Pierre Flourens began to publish his own "new" experiments on the cerebellum. This touched off a debate over primacy. Rolando's writings, unknown to Flourens when he began his work,

---

[2]Lorry's (1760) experiments involved inserting needles into the brains of some pigeons. He found that needles pushed to the cerebellum caused his bird to sway. Although the significance of this effect escaped him, this probably was one of the first experimental demonstrations of the association between cerebellar damage and problems with coordination.

[3]Malacarne's book emphasized the anatomy of the cerebellum. This work stimulated Johann Christian Reil (1759–1813) to conduct an important series of anatomical studies on the cerebellum in the first decade of the 1800s. Prior to Malacarne, Raymond Vieussens (1684) described the dentate nucleus (rhomboid bodies) of the cerebellum.

Figure 15.7. The title page of Luigi Rolando's *Saggio* (1809), the classic in which Rolando described his ablation experiments on the cerebellums of different animals.

were only later translated into French. It was then realized that the two men had been doing some of the same things.

Although Flourens (1824) did, in fact, repeat some of Rolando's basic experiments, he showed more precision in his procedures. Like Rolando, he did not see changes in vital or intellectual functions after cerebellar lesions, but he did observe changes in motor function. Further, Flourens agreed with Rolando that the cerebellum was equipotential. In this context, Flourens stressed that it possessed a remarkable capacity for recovery from small lesions.

Nevertheless, Flourens' ideas differed from Rolando's in one very important way. The former hypothesized that the cerebellum was the special organ for motor coordination, rather than a center for the production of movements (a function Flourens assigned to the spinal cord and medulla). In particular, he stressed that lesions of the cerebellum caused irregular, uncertain, voluntary movements, which reminded him of the swagger of a drunk. According to Flourens, the animals were not in any way paralyzed, as Rolando had thought.

Flourens (1824, p. 36) wrote the following about a pigeon with cerebellar damage:

> I removed the cerebellum from a pigeon slice by slice. During the removal of the first slices, only slight weakness and lack of coordination of movement appeared. With the middle slices, an almost universal restlessness was manifested though there was no sign of convulsions; the animal performed sudden and disordered movement; it heard and saw. On removal of the last slices, . . . far from remaining calm and steady as in pigeons deprived of the cerebral lobes, it became vainly and almost continually agitated, but

it never moved in a steady and firm manner. (Translated in Clarke and O'Malley, 1968, 657)

Flourens found a similar loss of muscular coordination when he repeated his experiments on a dog. Even though vision and intellect seemed to be unimpaired, the animal had difficulty turning, fell repeatedly, and bumped into objects. The same sort of result was obtained with a pig. Flourens (1823) wrote that progressive slicing caused the pig to "stagger in a drunken fashion; its feet moved crudely and clumsily; normal motion was impaired; and when it fell over its efforts to rise were very awkward." With the removal of additional layers of the cerebellum "the animal was completely unable to stand erect or walk; it lay on its belly or on its side, often moving its feet as if running or walking, making many fruitless attempts to rise but only tumbling over again when it managed to succeed."

Antoine Serres (1823), who worked on dogs and horses, also noted that cerebellar lesions resulted in abnormal movements and postures. After unilateral lesions, his animals bent, rolled, and showed severe disturbances of equilibrium, without corresponding changes in respiration or sensation.

These experiments were important in finally laying to rest at least four old ideas: (1) that the cerebellum was the organ of sensation, (2) that it played a role in intellectual functions, (3) that it was the center for memory, and (4) that it was responsible for the vital functions essential to life.

While Flourens was publishing his findings, François Magendie (1823) postulated there were two opposing motor forces in the brain. He thought one was located in the cerebellum and the other in the corpora striata. He postulated the former was responsible for forward motion, whereas the latter was responsible for backward motion. Magendie reached this remarkable conclusion when he observed uncontrollable backward movements after the cerebellum was ablated. In fact, one of his ducks persisted in swimming backward for about eight days. In his mind, with the elimination of the cerebellar center for forward motion, and with the preservation of the striatal center for backward motion, the organism could only go backward.

In 1824, Magendie wrote that he had also observed rotatory movements after unilateral cerebellar lesions. This led him to conclude that the cerebellum was also responsible for the maintenance of equilibrium. To a fair extent, Magendie shared this idea with Jean-Baptiste Bouillaud, who had studied with him. Bouillaud (1827) noted that cerebellar patients usually had difficulty with posture, stance, and locomotion.

Although Rolando, Flourens, and Magendie each associated the cerebellum with some aspect of motor function, there still were some nagging cases exhibiting just the contrary. One of these involved a child of 11 years of age who appeared to have a congenital absence of the cerebellum. Jean Cruveilhier (1829–42) described the case and wrote that only speech and intellect were defective. This observation suggested to him that cerebellar lesions were not associated with any specific motor signs.

In addition, William Hammond (1869) concluded that cerebellar lesions caused vertigo, which was responsible for the drunken walk and other motor symptoms. In Hammond's opinion, those lesions that affected motor function directly were not confined to the cerebellum.

Still, of the various ideas proposed in the first half of the nineteenth century, those of Flourens seemed most influential and best supported. For example, his belief that the cerebellum acted as a center for the coordination of voluntary movements, and his notion that animals can show good recovery after subtotal cerebellar lesions, received support from John Call Dalton (1825–1889) at Columbia University. Dalton repeated some of the same experiments Flourens had performed on the pigeon (see Figure 15.8). He wrote:

I have always found that, as soon as any considerable portion of the cerebellum has been injured, the animal shows an irregularity in its gait, in its posture, and in the movements of its head, neck, and wings; and this irregularity increases in proportion to the quantity of nervous mass which is removed, so that when nearly the whole of the cerebellum has been taken away, the animal can neither stand, walk, nor fly, and is only capable of making confused and ineffectual struggles.

Dalton went on:

This is not the effect of debility or of partial paralysis; for, at the time when the condition of the animal is most helpless, his muscular contractions are often very vigorous and even violent. It is just such an effect as would be produced by the loss or diminution of co-ordinating power. When the injury has been moderate in extent, so that the pigeon can still stand and walk, though imperfectly, there is often a very close and ludicrous resemblance to the effects of intoxication—the movements being still quite natural in force and rapidity, but their harmony and certainty being lost. (1861, 84–85)

Dalton also found that lost functions were often restored within a few weeks. Even after he removed two thirds of the cerebellum, his pigeons regained the power of muscular coordination, as well as the ability to compete for food and

make vocalizations. Nevertheless, the birds were not perfectly normal. For instance, his "recovered" birds still had difficulty landing when flying from a high perch to the floor.

The experiments of Rolando, Flourens, and Magendie stimulated interest in the microscopic features of the cerebellum. The "Purkinje cell" was seen in 1837 by the Czech scientist Jan Purkyně. His classic description of these "flasked-shaped ganglion bodies" appeared one year later.[4]

Later in the nineteenth century, Camillo Golgi and Santiago Ramón y Cajal, using the newly developed Golgi stain, contributed much to the growing body of knowledge concerning the fine structure of the cerebellum.

THE CEREBELLUM EARLY IN THE TWENTIETH CENTURY

The belief that the cerebellum is essential for the smoothness and effectiveness of movements was well accepted as the twentieth century began. Nevertheless, scientists still accounted for this behavioral finding by alluding to different underlying mechanisms. Charles Sherrington, for example, considered the cerebellum the head ganglion of the proprioceptive system. He viewed the cerebellum as a structure that received impulses from joints, tendons, muscles, and labyrinths. He believed that it regulated muscle tone.

On the clinical side, advances were made in describing the nature and range of deficits that followed cerebellar injury. Joseph Babinski (1857–1932) wrote about the role of the cerebellum in equilibrium. He defined this as the ability to maintain an upright posture without leaning or swaying to either side. Babinski (1902) coined the word *adiadococinésie* to describe the loss of the ability to carry out rapid, successive voluntary movements after cerebellar damage.

In 1904, Gordon Holmes and T. Granger Stewart (1837–1900) published an important paper on 40 patients at the National Hospital with tumors that directly or indirectly affected the cerebellum. Among the symptoms commonly encountered in the British patient population were vertigo; nystagmus; loss of power and muscle tone on the ipsilateral limbs; abnormal postures; and a drunken, staggering, or reeling gait. Some degree of ataxia and Babinski's adiadococinesia were also noted. The ataxic defect, Holmes and Stewart (1904) stated, "consists in a deficient control of the various component muscular contractions by which a complex act is executed" (p. 534). Two pages later, they added:

We have carefully investigated in a number of our cases the sign described by Babinski as pathognomonic of cerebellar disease, to which he has given the name "diadococinesia." It consists in the inability of the patient to accurately perform rapid alternate movements with the homolateral limb, although the individual movements are easily possible. . . . It seems to us that this inability to rapidly repeat movements is dependent upon the defective cooperation of the muscles and their antagonists, due in part, at least, to the diminution of the reflex muscle tone described, and to the consequent failure of the antagonists to control the primary movement owing to their lack of tone.

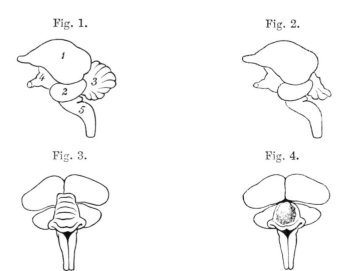

Fig. 1.   Fig. 2.

Fig. 3.   Fig. 4.

Figure 15.8. Lateral (Figs. 1, 2) and posterior views (Figs. 3, 4) of the pigeon brain. Figures 1 and 3 show the normal cerebellum, and Figures 2 and 4 show the cerebellar lesions that led John C. Dalton to conclude that Marie-Jean-Pierre Flourens was correct in theorizing the cerebellum served as a center for coordinated voluntary movement. Dalton (1861) reported excellent recovery from this lesion.

[4]For a history of the Purkinje cell, see Clarke and Jacyna (1987). Purkyně's role in the history of neuron doctrine appears in Chapter 3.

Sensation was found to be unimpaired, and tendon reflexes were variable in these patients. When the cerebellar tumors were small, the fine tremor was not always present.[5] This tremor, which increases with voluntary movement, was more likely to accompany larger, bilateral cerebellar tumors and to be displayed soon after cerebellar tumor surgery, especially by unsupported ipsilateral limbs.

In 1917, Holmes turned his attention to gunshot wounds of the cerebellum. He noted that when a person with a unilateral cerebellar lesion attempted to grasp the experimenter's two hands simultaneously, the power of the affected limb (ipsilateral to the lesion) was reduced. In addition, there was slowness of movement, an irregular and intermittent grasp, and rapid fatigue of the hand from grasping. Further, Holmes saw that when a patient with a cerebellar injury tried to reach for an object, only the contralateral limb was able to do so easily and successfully.

> He raises his right (ipsilateral) hand from the bed less promptly, and as soon as the limb is unsupported, it sways unnecessarily and aimlessly at its more proximal joints, and during the movement his finger deviates from the direct course by which it could most easily reach its aim. Further, he rarely succeeds in touching the object at once, but usually brings his index to one side or the other side of it, and projects it too far or more rarely stops the movement too soon. Errors consequently occur in both the direction and in the range of its movement. (1917, 477)

Holmes noted that many of the soldiers with cerebellar damage displayed tremors when voluntary muscular contractions were called for, although not necessarily when the muscles were at rest and fully supported.

The investigations conducted by Gordon Holmes (1907, 1917, 1922) on patients with cerebellar tumors and on soldiers wounded in World War I (1914–1918) produced findings largely in agreement with those of Luigi Luciani, who worked in Italy with dogs and monkeys. Luciani found he could keep his animals alive for at least one year after total extirpations of the cerebellum. This was important because it allowed him to study them more meticulously than had any of his predecessors. In 1891, Luciani wrote that the cerebellar deficit had three components: (1) atonia (a loss of muscle resistance or tension), (2) asthenia (diminished energy of the muscles), and (3) astasia (a defect in the stability of muscular contractions).

Rudolf Magnus (1873–1927) added to the growing body of knowledge about the cerebellum when he showed that the cerebellum and labyrinth organs were not directly connected to each other. With his Dutch associates, Magnus demonstrated that postural reflex abnormalities, sometimes thought to be symptoms of cerebellar damage, were in fact due to extracerebellar injuries. In his Croonian Lecture, he stated:

> Experiments have proved that all postural reflexes discussed in this lecture are present and perfectly undisturbed after total extirpation of the cerebellum. . . . Unfortunately, we know not of a single function or reflex positively connected with the cerebellum, in such a way that it is absent after cerebellar extirpation, and present after ablation of other parts of the brain, as long as the cerebellum remains uninjured. (1925, 352–353)

By this point in time, most investigators also had come to the conclusion that the cerebellum was not equipotential, as had been claimed by Rolando and Flourens. This doctrine actually began to fall with the experiments of Max Solly Löwenthal (1867–1960) and Victor Horsley, scientists who mapped the "excitable" areas of the cerebellum in dogs and cats in 1897. Louis Bolk (1866–1930) also contributed to the change in thinking when he concluded, on the basis of embryology and anatomy, that each muscular movement had to be represented in a definable area of the cerebellar cortex. Bolk (1906) stated:

> The anterior (central) lobe contains the coordination center for the group of muscles of the head (eyes, tongue, muscles of mastication, facial muscles) and in addition of the larynx and pharynx; in the lobulus simplex is the coordination center of the cervical muscles; the upper portion of the lobulus medianus posterior contains the unpaired coordination center for the left and right extremities; in each of the lobuli ansiformis and paramediani is one of the paired centers for the two extremities; and in the remaining part of the cerebellum we find the coordinating centers for the muscles of the trunk. (Translated in Clarke and O'Malley, 1968, 697)

These and related advances represented a marked change in thinking.[6] Not many years earlier, William Gowers had written:

> The function of the hemispheres of the cerebellum is still mysterious. They lessen in size as we descend the scale of animals, until they disappear in birds, in which the whole cerebellum corresponds to the middle lobe of man. They are connected chiefly with those parts of the cortex of the cerebrum which chiefly subserve psychical processes. With these parts, moreover, the cerebellar hemispheres have this in common, that simple loss of substance causes no definite and recognizable loss of any function of the brain. The loss can apparently be compensated by other parts. Hence it seems possible that the old theory may be correct which assumes that the cerebellar hemispheres are in some way connected with psychical processes. (1888, 495)

Much of what Gowers had to say has withstood the test of time, but in the case of the cerebellum, he was less than accurate.

The Corpus Striatum

DEFINING THE CORPUS STRIATUM

By the end of the nineteenth century, the cerebellum and the motor cortex of the cerebrum were not the only structures firmly associated with motor functions. A number of forebrain nuclei were also associated with movement. Foremost among these structures were the nuclei of the corpus

---

[5]The tremor was noted by earlier investigators, including Rolando in 1809. Luciani (1891) and Ferrier (1886) also noted it in their experiments on animals.

[6]The idea of localization within the cerebellum was confirmed and expanded by Olaf Larsell (1886–1964) after World War I (see Larsell, 1937).

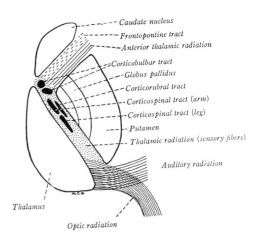

Figure 15.9. Diagram of the human basal ganglia as depicted in Ranson's *The Anatomy of the Nervous System* in 1920.

striatum, shown in a textbook drawing from 1920 as Figure 15.9.

The term "corpus striatum" has been used in the literature to describe the lenticular and caudate nuclei, along with the internal capsule located between them. The origin of the term comes from the fact that the caudate nucleus and the dorsal part of the lenticular nucleus are connected by slender, gray bridges across the internal capsule. These bridges give the area a striped, or striated, appearance.

The corpus striatum was illustrated by Andreas Vesalius in 1543. His pictures, such as Figure 15.10, showed the caudate and lenticular nuclei separated by the internal capsule, and the division of the lenticular nucleus into its two distinct parts, the globus pallidus and the putamen. Nevertheless, Vesalius did not give these structures their names, nor did he discuss their functions. His intent seemed only to distinguish "gray" *(fulva sive subpallida)* from white matter (Schiller, 1967).

It was the terminology of Thomas Willis in the seventeenth century that eventually led to the common term "corpus striatum." Willis (1664) presented illustrations of the corpus striatum (see Figure 15.11) and wrote:

These bodies, if they should be dissected along the middle, appear marked, with medullar streaks, as it were rays or beams; which sort of chamferings or streaks have a double aspect or tendency; to wit, some descend from the top of this body, as if they were tracts from the Brain into the oblong Marrow; and others ascend from the lower part, and meet the aforesaid, as if they were paths of Spirits from the oblong marrow into the Brain. And it is worth observation, that in the whole head besides there is no part found chamfered or streaked after the like manner. (1965 reprint, 102)

The lenticular or lentiform nucleus was named after Willis' *prominentiae lentiformes.* Karl Burdach was the first to use the names "globus pallidus" *(blasser Klumpen)* and "putamen" to describe the two divisions of this nucleus, which he envisioned as a core and a shell (Meyer, 1966; Schiller, 1967). The globus pallidus ("pale clump"), which is medial to the putamen, was given its name because of its

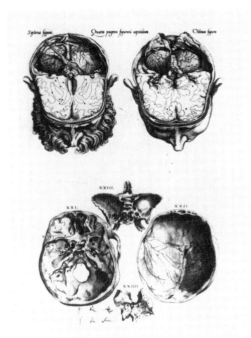

Figure 15.10. A figure from Andreas Vesalius' *Fabrica* showing the basal ganglia and the internal capsule (upper left), as well as the superior surface of the cerebellum (upper right).

pale appearance, due to the many myelinated fibers that traversed it. The putamen, or *Schale,* seemed to form a "shell" or crescent around it.[7,8]

EARLY IDEAS ABOUT THE CORPUS STRIATUM

Thomas Willis (1664) found degeneration of the corpus striatum in patients who had suffered severe paralyses. He also noted the absence of striations in newborn animals. He wrote:

For as often as I have opened the bodies of those who dyed of a long Palsie, and most grevious resolution of the Nerves, I have always found these bodies less firm than others in the Brain, discoloured like filth or dirt, and many chamferings obliterated. Further, in Whelps newly littered, that want their fight, and hardly perform the other faculties of motion and sense, these streaks or chamferings, being scarcely wholly formed, appear only rude. (1965 reprint, 102)

[7]Burdach called the caudate nucleus *Streifenhügel,* meaning "striatal hillock." He also named the red nucleus *(roter Kern)* and the claustrum. These two structures were discovered earlier by Félix Vicq d'Azyr (1805, 1813). Burdach, perhaps the greatest originator of anatomo-medical terms, also introduced the words "biology" and "morphology" into the scientific nomenclature (Meyer, 1970). His monumental work, *Vom Baue und Leben des Gehirns* (1819–26), contained many of these discoveries, names, and descriptions.
[8]The caudate and putamen are now known to be composed primarily of small- and medium-sized cells that connect these nuclei to each other and to the globus pallidus. Nevertheless, these nuclei also include larger cells with longer connections (e.g., to the substantia nigra). The caudate and putamen constitute the newer part of the striate complex. Together the two nuclei make up the "neostriatum." Because they alone are sometimes referred to as the striatum, the terminology in the literature on the corpus striatum can be confusing.
The globus pallidus, the main efferent center of the striatum, is made up of a greater percentage of large cells and is phylogenetically older than the caudate and the putamen. It is considered paleostriatal. It sends its fibers to the thalamus, hypothalamus, subthalamic nucleus, reticular formation, and (at least indirectly) to the red nucleus.

Figure 15.11. The brainstem of a sheep with the cerebral hemispheres removed. The left corpus striatum (A) is bisected to show its striped appearance. This figure was drawn by Sir Christopher Wren to illustrate Thomas Willis' *Cerebri anatome* (1664).

Willis thus associated these nuclei with movement and thought they contained channels for the flow of the spirits controlling the muscles. Specifically, he believed the animal spirits collected in the corpus striatum and initiated voluntary movement via the medulla (see Meyer and Hierons, 1964).

Willis (1664) further believed the corpora striata were a meeting place for both sensory and motor spirits. He expressed these thoughts as follows:

> This part is the common Sensory . . . and so causes the perception of every senses . . . when from hence they are passed further into the Brain . . . and further, these bodies as they receive the forces of all the Senses, so also the first instincts of spontaneous local motions. . . . For here, as in a most famous Mart, the animal Spirits, preparing for the performance of the thing willed, are directed into appropriate nerves. (1965 reprint, 102)

This theory was embraced by many scientists in the seventeenth and eighteenth centuries. Emanuel Swedenborg (see Chapter 2), for example, also maintained that the corpus striatum was important for movement as well as sensation. He wrote: "The royal road of the sensations of the body to the soul is through the corpora striata and all determinations of the will also descend by that road" (translated in Wilson, 1914, p. 430).

During the 1750s and 1760s in Italy, Leopoldo Caldani (1725–1813), who was influenced by Albrecht von Haller's concept of irritability, conducted a series of animal experiments to study the sensitivity of the brain. Caldani used a rod dyed red to probe the brain substance in an attempt to define the sensitive parts of the brain. He found the medulla to be quite sensitive, but concluded that the corpus striatum was the most irritable structure when it came to producing convulsions (see Neuburger, 1981, pp. 139–142).

On the basis of his lesion experiments on dogs, lambs, and goats, Caldani went on to associate the corpus striatum with voluntary movement. He noticed that by injuring or partially removing the striatum on one side, the animals became hemiplegic and arched themselves toward the side of the lesion. He considered the possibility that the nerves controlling the muscles even decussated in the corpus striatum. In 1786, Caldani wrote:

> I am satisfied with having provided proof, by means of experiments and (clinical) observations, that large or small lesions of the corpora striata are followed by paralysis, more or less severe and more or less extensive, of the opposite side of the body and that therefore the crossing of the medullary fibers ought to be placed in those bodies more than anywhere else. (Translated in Neuburger, 1981, 141)

For more than 200 years after Willis (1664) introduced the idea, the striatum remained the "fountainhead of motor power" and the source of crossed hemiplegia (Schiller, 1967). The sensory role that Willis had also assigned to these nuclei, however, slowly moved to the nearby thalamus.

THE CORPUS STRIATUM THROUGH THE FIRST HALF OF THE NINETEENTH CENTURY

Around 1820, Marie-Jean-Pierre Flourens, who conducted many stimulation experiments on the corpus striatum, came to the conclusion that electrical stimulation of this area was not associated with movements (see Flourens, 1842). In contrast, many other investigators did associate this region with movements. Nevertheless, the associations made were often based on limited data from poor preparations. As a result, the conclusions reached were sometimes imaginative and fanciful.

For example, in 1827 Antoine Serres reported that movements of the posterior limbs followed mechanical stimulation of the corpus striatum in his dogs. To the extent that motor functions were associated with the corpus striatum, this aligned him more with Burdach and Willis than with Flourens. Nevertheless, Serres added that movements of the anterior limbs followed irritation of the thalamus (see Meyer, 1971).

Just a few years later, François Magendie (1823, 1839) reported that destroying the corpus striatum caused his animals to leap forward. He thought the propulsion was like an irresistible impulse; the animals behaved as though they were hearing the barking of hunting dogs behind them. This observation was the opposite of what he had observed after cerebellar lesions. Together these findings led Magendie to conclude that the striatum controlled backward movements under normal conditions. Theoretically, his animals moved forward because he had destroyed the center for moving backward. As might be expected, these findings could not be verified (Longet, 1842).

The work of Louis Sebastian Saucerotte (1741–1814) should also be noted in this context.[9] In 1819, he reported that he had conducted a series of studies on dogs, which

---

[9]The forename Nicholas is sometimes given for Saucerotte, but this is incorrect according to Neuburger (1981).

involved mechanical stimulation of the cortex and subcortical structures. His experiments were very crude and often simultaneously involved the cortex, the corpus striatum, and the thalamus. Yet he did not shy from speculation. Saucerotte postulated that the forebrain exercised its influence on the posterior extremities and that the hindbrain played a role in controlling the anterior limbs. Fortunately, this idea also did not seem to have a major effect on scientific thinking (see Neuburger, 1981).

During the next two decades, the debates over the possible role of the corpus striatum in movement became more intense. In 1840, Julius Budge hypothesized that the corpus striatum controlled the internal organs, in particular those organs involved with digestion. This idea was developed when he accidently found that stimulating the corpus striatum in a recently killed dog increased the motions of the intestines and the stomach.

Budge's (1840) work did not shake the steadily growing conviction that the corpus striatum was involved in some way with control over the skeletal muscles. At the time Budge presented his theory, many scientists maintained that the corpus striatum was probably the point of origin of the motor tract, which coursed through the medullary pyramids en route to the spinal cord. This idea was suggested by many stimulation experiments on animals. Hemorrhages in this area were also often associated with paralyses in humans.

Nevertheless, the issue remained open because not all experimenters were able to induce movements reliably when they stimulated the striatum (e.g., Schiff, 1858; van Deen, 1860). Even Robert Bentley Todd (1849) wrote that mechanical irritation of the corpora striata did not produce convulsions. Although he also stated that damaging the striatum did not disturb motion, Todd still entertained the idea that the corpus striatum might have something to do with the will and voluntary movement.

THE IMPACT OF THE DISCOVERY OF THE MOTOR CORTEX

After the discovery of the motor cortex in 1870 by Gustav Theodor Fritsch and Eduard Hitzig, scientists had to question whether the corpus striatum served as the point of origin of the pyramidal tract, as had been previously believed (see Chapter 14). David Ferrier (1876, 1886), for one, accepted the newer idea that the pyramidal tract passed directly from the motor cortex through the pyramids of the medulla. Ferrier came to the conclusion that the ganglia of the corpus striatum were terminal centers similar to the cortex itself. Projections from these ganglia, he maintained, formed two divisions. One went to the substantia nigra and the pons; the other, a larger division, was more closely linked to cerebellar function and went to the red nucleus and olivary bodies.

Ferrier, like most other scientists in the final quarter of the nineteenth century, wondered whether earlier findings concerning the functions of the corpus striatum were due to lesions that infringed upon the internal capsule or to current that spread to this structure and affected the corticospinal tract. He conducted a series of stimulation and lesion experiments on his own in an attempt to come to a better understanding of the functions of these mysterious structures. These studies involved monkeys, cats, dogs, jackals, and rabbits. He found that current applied to the surface of the lenticular nucleus elicited muscular contractions on the opposite side of the body. This caused the head to be drawn to the side, the body to arch, the face to go into spasmodic contractions, and the limbs to become rigid. He, like other investigators at this time, did not see any specificity in the response (e.g., Carville and Duret, 1875).

Ferrier went to great lengths to argue that the contractions he elicited were not due to the spread of current to the thalamus or to the internal capsule. This was essential because Flourens, Todd, Schiff, and van Deen had claimed that the corpora striata were not electrically excitable.

In his articles and books, Ferrier (e.g., 1886) described the work of Carl Nothnagel (1873), but felt the difficulties of exact localization were too great to allow Nothnagel's results to be accepted without question. Nothnagel had injected chromic acid into the caudate nucleus on one side and found it produced violent forward movements in his rabbits. When he did the same thing to the lenticular nucleus, the rabbits bent toward the side stimulated, also implicating this structure in motor function, although in a different way. Nothnagel further reported that bilateral lesions of the lenticular nucleus made his animals immobile, much like rabbits deprived of their cerebral hemispheres.

In contrast, Ferrier's own lesion experiments on the basal forebrain were associated with somewhat milder movement disorders. He concluded from his experiments that the corpus striatum was responsible for a more primitive form of movement than the cerebral cortex. Ferrier wrote:

> It appears from these facts that the corpora striata proper are centres of innervation of the same movements as are differentiated in the cortical motor centres, but of a lower grade of specialization. The innervation of the limbs in all that relates to their employment as instruments of consciously discriminated acts is dependent on the cortical centres, while for all other purposes involving mere strength or automatism, primary or secondary, the corpora striata with the lower ganglia are sufficient. In man almost every movement has to be laboriously acquired by conscious effort through the agency of the cortical centres, and continues to involve the activity of these centres to a greater or less extent thoughout. Hence the destruction of the corpora striata adds little if anything to the completeness of the paralysis which results from destruction of the cortical motor centres alone. (1886, 419)

Ferrier's conclusion had a major impact on nineteenth- and twentieth-century thinking and was very much in accord with the thoughts of his intellectual mentor, John Hughlings Jackson. Nevertheless, some of Ferrier's results and conclusions were questioned. At issue was whether the movements Ferrier obtained with electrical stimulation still could have been due to currents affecting the motor pathways descending from the cerebral cortex through the lenticular nucleus. It was in this context that some experimenters first removed the motor cortex and then tried to stimulate the corpus striatum. Some individuals found that these subcortical structures were no longer electrically excitable after the cortex had been ablated (e.g., Minor, 1889).[10]

---

[10]See Wilson (1914) for a discussion of this literature and references to the work of Ziehen, Stieda, and Bechterev, who attempted such experiments.

Another matter of debate was whether the striatum possessed any "psycho-motor" functions at all, or whether it was strictly involved with nonvoluntary and more primitive types of movements. Luigi Luciani and Arturo Tamburini (1878), basing their conclusions on the degree and permanence of the paralyses found after cortical lesions, concluded that the functions of the striatum varied with the species. They wrote:

> This function of the basal ganglia is not equally developed in different animals. Thus, following Dr. Maragliano, we find that the basilar centres in man are quite devoid of psycho-motor function, which is entirely confined to the cortical. Descending from man to the monkey, and thence to dogs, cats, and rabbits, we find a progressive development of the functions of the basal ganglia, inversely proportional to the intensity of the paralytic effects which follow destruction of the cortical centres. (Translated by Rabagliati, 1879, 541)

As the nineteenth century ended, the exact functions of the striatum still remained unclear.

INFRAHUMAN PRIMATE STUDIES IN THE FIRST HALF OF THE TWENTIETH CENTURY

In 1914, S. A. Kinnier Wilson (1878–1937) presented the results of his extensive experiments on 25 monkeys subjected to electrical stimulation and then electrolytic lesions in different parts of the corpus striatum. These animals were operated upon by Victor Horsley using the new Horsley-Clarke stereotaxic instrument. Wilson found that when the internal capsule was spared by the electrolytic lesions, stimulation of the lenticular nucleus did not elicit movements. He also found that small lenticular nucleus lesions did not give rise to motor impairments or tremor. Larger unilateral lesions, which involved the greater part of the lenticular nucleus, did not cause paralysis either, although some monkeys showed a preference for the ipsilateral limb. In contrast to David Ferrier, Wilson wrote:

> Whatever functions the corpus striatum once possessed there is no experimental evidence in apes to show that it exercises any motor function comparable to that of the motor cortex. There is no evidence to suggest that it is a centre for so-called automatic movements. It is electrically inexcitable, and comparatively large unilateral lesions do not give rise to any unmistakable motor phenomena. (1914, 482–483)

Nevertheless, Wilson continued to maintain that the corpus striatum functioned in movement. He cited anatomical facts, including the results of his 1912 study on lenticular degeneration, to suggest that the corpus striatum probably exercises a steadying influence on the lower motor neurons ("the final common path").

Wilson elaborated upon these ideas in 1924, in a paper in which he classified the corpus striatum as a part of "the old motor system." As far as he was concerned, the corpus striatum was concerned with muscle tone and steadiness in humans and not with originating movements.

> The relation of the corpus striatum to the rest of the old motor system is one of tone control, and of steadiness of innervation. Remove its influence by disease, and cerebello-mesencephalospinal motor mechanisms come into overaction in spite of the normal activity of the pyramidal system; tonic postures become overemphasized; in fact, a universal

muscular rigidity appears, distinct from the selective rigidity of corticospinal disease; or, alternatively, incessant involuntary movements develop, over which the control of the master system, the corticospinal, is at best fleeting and imperfect. (1924, 400)

Wilson's experimental studies on primates were followed by studies by Margaret Kennard (1940, 1944; Kennard and Fulton, 1941). In general, she found that small lesions of the caudate nucleus, the putamen, or the globus pallidus did not result in paralyses or motor disturbances in primates, thus confirming Wilson's observations. After large bilateral lesions of the striatum, however, Kennard found that tremors and other motor disturbances appeared during voluntary movements (also see Mettler and Mettler, 1942).

Kennard thought one of her most significant findings was that the striatal lesions greatly augmented the effects of premotor cortex damage. Combined Area 6 and striatal lesions were associated with a marked tremor, one that was greatest during complex voluntary movements, such as those needed for climbing or using the hands. The combined lesion effects were more severe in chimpanzees than in monkeys; she observed chorea as well as tremor after these lesions in her apes.

The data obtained by Kennard and Wilson led them to theorize that the corpus striatum played an important role as one part of a larger system responsible for smooth, regulated motor functioning. Although Wilson (1914, 1924) shied from the word "inhibition," Kennard and her contemporaries did not. Thus, by the mid-twentieth century it was commonly maintained that the striatum probably functioned at least in part as "an inhibitory mechanism, subject to cortical control and forming a significant link in the inhibitory path from the cortex to the motor neurone" (Mettler and Mettler, 1942, p. 250).

Clinical data seemed to support their general conclusions. In fact, so much was learned about the basal ganglia and involuntary movement disorders in people by this time that Chapter 16 is devoted entirely to these developments.

REFERENCES

Babinski, J. F. F. 1902. Sur le rôle du cervolet dans les actes volitionnels nécessitant une succession rapide de mouvements (diadococinésie). *Revue Neurologique, 10,* 1013–1015.

Bolk, L. 1906. *Das Cerebellum der Säugethiere.* Haarlem: Bohn.

Boneti, T. 1700. *Sepulchretum sive anatomia practica.* Lugduni: Cramer & Perachon.

Bouillaud, J.-B 1827. Recherches cliniques tendant à réfuter l'opinion de M. Gall sur le fonction du cervelet . . . *Archives Générale de Médecine (Paris), 15,* 225–247.

Brodmann, K. 1909. *Vergleichende Lokalisations-lehre der Grosshirnrinde.* Leipzig: J. A. Barth.

Budge, J. L. 1840. On the dependence of the intestines' motions on the central organs of the nervous system. *London Medical Gazette, 1,* 957–959.

Burdach, K. F. 1819–1826. *Von Baue und Leben des Gehirns* (3 vols.). Leipzig: Dyk'schen Buchhandlung.

Caldani, L. M. A. 1786. Ezperienze ed osservazioni del Signor Leopoldo M. A. Caldani dirette a determinare qua sia il luogo principale del cervello, in cui, piu'che altrove, le fibre midollari dello stesso viscere s'incrocicchiano (17 March, 1871). *Sagga Scientifica e Letterari dell'Acadamia de Padova, 1,* 1–15.

Carville, C., and Duret, H. 1875. Sur les fonctions des hémisphères cérébraux. *Archives de Physiologie, 2,* 352–490.

Clarke, E., and Jacyna, L. S. 1987. *Nineteenth Century Origins of Neuroscientific Concepts.* Berkeley: University of California Press.

Clarke, E., and O'Malley, C. D. 1968. *The Human Brain and Spinal Cord.* Berkeley: University of California Press.

Cruveilhier, J. 1829–1842. *L'Anatomie Pathologique du Corps Humain. . . .* (2 vols.). Paris: J. B. Ballière.

Dalton, J. C. 1861. On the cerebellum, as a center of coordination of the voluntary movements. *American Journal of Medical Science, 41,* 83–88.

Estienne, C. 1546. *La Dissection des Parties du Corps Humain Diversée en Trois Livres.* Paris: Simon de Colines.

Ferrier, D. 1876. *The Functions of the Brain.* London: Smith, Elder and Company.

Ferrier, D. 1886. *The Functions of the Brain.* New York: G. P. Putnam's Sons. (Reprinted 1978; Washington, DC: University Publications of America)

Flourens, M.-J.-P. 1823. Recherches physiques sur les propriétés et les fonctions du système nerveux dans les animaux vertébrés. *Archives Génerale de Médecine (Paris), 2,* 321–370.

Flourens, M.-J.-P. 1824. *Recherches Expérimentales sur les Propriétés et les Fonctions du Système Nerveux dans les Animaux Vertébrés.* Paris: J. B. Ballière.

Flourens, M.-J.-P. 1842. *Recherches Expérimentales sur les Propriétés et les Fonctions du Système Nerveux dans les Animaux Vertébrés* (2nd ed). Paris: J. B. Ballière.

Giannitrapani, D. 1967. Developing concepts of lateralization of cerebral functions. *Cortex, 3,* 355–370.

Gowers, W. R. 1888. *A Manual of Diseases of the Nervous System.* Philadelphia, PA: Blakiston.

Haller, A. von. 1762. *Mémoires sur la Nature Sensible et Irritable, des Parties du Corps Animal.* Lausannae: F. Grasset.

Hammond, W. A. 1869. The physiology and pathology of the cerebellum. *Quarterly Journal of Psychological Medicine and Medical Jurisprudence, 3,* 209–244.

Holmes, G. 1907. A form of familial degeneration of the cerebellum. *Brain, 30,* 466–489.

Holmes, G. 1917. The symptoms of acute cerebellar injuries due to gunshot injuries. *Brain, 40,* 461–535.

Holmes, G. 1922. On clinical symptoms of cerebellar disease and their interpretation. *Lancet, 1,* 1177–1182.

Holmes, G., and Stewart, T. G. 1904. Symptomatology of cerebellar tumours: A study of forty cases. *Brain, 27,* 522–591.

Irons, E. E. 1942. Théophile Bonet, 1620–1689; his influence on the science and practice of medicine. *Bulletin of the History of Medicine, 12,* 623–665.

Kennard, M. A. 1940. Functions of the basal ganglia in monkeys. *Transactions of the American Neurological Association, 66,* 131–140.

Kennard, M. A. 1944. Experimental analysis of the functions of the basal ganglia in monkeys and chimpanzees. *Journal of Neurophysiology, 7,* 127–148.

Kennard, M. A., and Fulton, J. F. 1941. Experimental tremor and chorea in monkey and chimpanzee. *Transactions of the American Neurological Association, 67,* 126–130.

Larsell, O. 1937. The cerebellum: A review and interpretation. *Archives of Neurology and Psychiatry, 38,* 580–607.

Longet, F. A. 1842. *Anatomie et Physiologie du Système Nerveux de l'Homme et des Animaux Vertébrés.* Paris: Masson et Cie.

Lorry, C.-A. 1760. Sur les mouvemens du cerveau et de la dure-mère. *Mémoires de Mathématique et de Physique, 3,* 277–313, 344–377.

Löwenthal, M., and Horsley, V. 1897. On the relations between the cerebellar and other centres (namely cerebral and spinal) with special reference to the action of antagonistic muscles (preliminary account). *Proceedings of the Royal Society, 61,* 20–25.

Luciani, L. 1891. *Il Cervelletto. Nuovi Studi de Fisiologia Normale e Patologica.* Florence: Le Monnier.

Luciani, L., and Tamburini, A. 1878. *Richerche Sperimentali sui Centri Psico-Motori Corticali.* Reggio-Emilia: Tip. di S. Calderini e Figlio.

Magendie, F. 1823. Note sur les fonctions des corps striés et des tubercules quadrijumeaux. *Journal de Physiologie et Expérimentale Pathologie, 3,* 376–381.

Magendie, F. 1824. Mémoire sur les fonctions de quelques parties du système nerveux. *Journal de Physiologie et Expérimentale Pathologie, 4,* 399–407.

Magendie, F. 1839. *Leçons sur les Fonctions et les Maladies du Système Nerveux.* Paris: Ebrard.

Magnus, R. 1925. Animal posture (Croonian Lecture). *Proceedings of the Royal Society, 98B,* 339–353.

Malacarne, M. V. G. 1776. *Nuova Esposizione della Struttura del Cervelletto Umano.* Turino: G. Brolio.

McHenry, L. C., Jr. 1969. *Garrison's History of Neurology.* Springfield, IL: Charles C Thomas.

Mella, H. 1924. The experimental production of basal ganglion symptomatology in Macacus rhesus. *Archives of Neurology and Psychiatry, 11,* 405–417.

Mettler, F. A., and Mettler, C. C. 1942. The effects of striatal injury. *Brain, 65,* 242–255.

Meyer, A. 1966. Karl Friedrich Burdach on Thomas Willis. *Journal of the Neurological Sciences, 3,* 109–116.

Meyer, A. 1970. Karl Friedrich Burdach and his place in the history of neuroanatomy. *Journal of Neurology, Neurosurgery and Psychiatry, 33,* 553–561.

Meyer, A. 1971. *Historical Aspects of Cerebral Anatomy.* London: Oxford University Press.

Meyer, A., and Hierons, R. 1964. A note on Thomas Willis' views on the corpus striatum and the internal capsule. *Journal of the Neurological Sciences, 1,* 547–554.

Minor, L. 1889. *Signification et Rôle du Corps Strié.* Dissertation, Moskow.

Morgagni, J. B. 1740. *Dissertationes anatomicas.* Venetiis: Pitteri.

Neuburger, M. 1981. *The Historical Development of Experimental Brain and Spinal Cord Physiology before Flourens.* (Translated and edited with additional material by E. Clarke). Baltimore, MD: Johns Hopkins University Press.

Nothnagel, H. 1873. Experimentelle Untersuchungen über die Funktionen des Gehirns. *Archiv für pathologische Anatomie und Physiologie und für klinische Medizin, 57,* 184–214.

Rabagliati, A. 1879. Luciani and Tamburini on the functions of the brain. *Brain, 1,* 529–544.

Rolando, L. 1809. *Saggio sopra la Vera Struttura del Cervello dell'uomo e degli Animali e sopra le Funzioni del Sistema Nervoso.* Sassari: Stamperìa Da S. S. R. M. Privilegiata.

Schiff, M. 1858. *Lehrbuch der Physiologie des Menschen. Muskel-und Nervenphysiologie.* Lahr: Schauenburg.

Schiller, F. 1967. The vicissitudes of the basal ganglia (further landmarks in cerebral nomenclature). *Bulletin of the History of Medicine, 41,* 515–538.

Serres, E. R. A. 1823. Suite des recherches sur les maladies organiques du cervelet. *Journal de Physiologie et Expérimenta le Pathologie, 3,* 114–153.

Serres, E. R. A. 1827. *Anatomie Comparée du Cerveau.* Paris: Gabon.

Todd, R. B. 1849. On the pathology and treatment of convulsive diseases. *London Medical Gazette, 8,* 661–671, 724–729, 766–772, 815–822, 837–846.

van Deen, I. 1860. Ueber die Unempfindlichkeit der Cerebrospinalcentra für elektrische Reize. In *Untersuchungen zur Naturlehre des Menschen und der Thiere* (Band 7). Pp. 380–392. Frankfort am Main: Moleschott.

Varolius, C. 1573. *Anatomia sive de resolutione corporis humani.* Patavii: P. et A. Meiettos.

Varolius, C. 1591. *De nervis opticis, nonnullisque aliis praeter communem opinionem in humano capite observatis.* Frankfurt, Ioannem Welchelum & Petrum: Fischerum.

Vesalius, A. 1543. *De humani corporis fabrica.* Book VII. Basel: J. Oporinus.

Vicq d'Azyr, F. 1805. *Oeuvres.* (J. L. Moreau, Ed.). Paris: L. Duprat-Duverger.

Vicq d'Azyr, F. 1813. *Planches pour le Traité d'Anatomie du Cerveau.* Paris: Duprat-Duverger.

Vieussens, R. de. 1684. *Neurographia universalis.* Lyon: J. Certe.

Wesphal, C. F. O. 1882. Zur Lokalisation der Hemianopsie und des Muskelgefühls beim Menschen. *Charité-Annalen, 7,* 466–489.

Willis, T. 1664. *Cerebri anatome: cui accessit nervorum descriptio et usus.* London: J. Martyn and J. Allestry. Tercentenary ed., 1664–1964, *Thomas Willis, The Anatomy of the Brain and Nerves.* Montreal: McGill University Press, 1965.

Wilson, S. A. K. 1912. Progressive lenticular degeneration: A familial nervous disease associated with cirrhosis of the liver. *Brain, 34,* 295–509.

Wilson, S. A. K. 1914. An experimental research into the anatomy and physiology of the corpus striatum. *Brain, 36,* 427–492.

Wilson, S. A. K. 1924. The old motor system and the new. *Archives of Neurology and Psychiatry, 11,* 385–404.

Chapter 16
Some Movement Disorders

> Over fifty years ago, in riding with my father on his professional rounds, I saw my first case of "that disorder," which was the way in which the natives always referred to the dreaded disease. I recall it as vividly as though it had occurred but yesterday. It made a most enduring impression upon my boyish mind, an impression every detail of which I recall today, an impression which was the very first impulse to my choosing chorea as my virgin contribution to medical lore.
>
> George Huntington, 1910

ERGOT POISONING

During the Middle Ages, a number of "movement" disorders that affected groups of people were witnessed. One type, referred to as the dancing manias, was typically associated with religious gatherings where the multitudes would congregate and then dance. The frenzied dancers often experienced delusions and hallucinations of a religious nature and sometimes fell to the ground completely delirious and exhausted. This "hysteria" was seen in Aix-la-Chapelle, France, in 1374 and spread rapidly throughout Europe during the time of the plague (Barbeau, 1958).

The term "chorea," which was the Greek word for "dance," often was applied to these episodes of delirium. In France, it was called *danse de Saint-Guy* and in Germany Saint John's dance or Saint Vitus' dance *(chorea Sancti Viti* in Latin). The latter term derived from the fact that a magistrate in Strassburg ordered the sufferers of the dancing mania to visit the Chapel of Saint Vitus to be cured. The term "Saint Vitus' dance," however, became very widely used and was soon applied to all forms of chorea, including movement disorders not psychogenic in nature (de Jong, 1937).

Another type of movement disorder that affected Europeans before, during, and after the Renaissance had a definite organic basis. This disorder could also affect large groups of people simultaneously, and was caused by ergot *(Claviceps purpurea),* a fungus that grows on grain. The ergot destroys the ovaries of the grain, and the kernel is replaced by a brownish-violet, protruding horn-shaped mass. This is shown in Figure 16.1. Rye is most vulnerable to this fungus, especially during wet weather.

Ergot poisoning may have been encountered more than 2,500 years ago; a "noxious pustule in the ear of grain" was described on an Assyrian tablet from about 600 B.C. (Goodman and Gilman, 1970). The Greeks and Romans were not very fond of rye, but still recognized the "smut" of rye. It remained for the Teutons, who cultivated rye in conquered areas, to popularize its consumption in Europe. It is esti-

Figure 16.1. Rye and wheat (far right) infected with ergot *(Claviceps purpurea).*

mated that some 132 outbreaks of ergotism occurred in Europe by 1789 (Dotz, 1980; Haller, 1981).

The epidemics of ergotism that erupted during the Middle Ages involved both the convulsive and gangrenous forms. The gangrenous variety was seen near the Rhine in 857 and was most common in Germany. The convulsive variety was first adequately described in the eleventh century and was more common in France. Both were associated with severe vasospasm, gastrointestinal distress, and impaired mental function.

In gangrenous ergotism, the effects were greatest on the

blood vessels, reducing the blood supply to the extremities. The person afflicted with this type of ergotism experienced intense burning pains, referred to as "fires."

In the convulsive (spasmodic) variety, the central nervous system was most affected, giving rise to severe contractions of the limbs and fingers, along with tingling and paralyses, leading to staggering, awkward movements. Convulsive ergotism was thus described as the "fire which twisted the people" and as the "creeping sickness." This type of ergotism, which generally lasted one to two months, was also associated with hallucinations and a host of other symptoms ranging from headache and insomnia to hunger and diarrhea (Haller, 1981).

The victims of ergotism appealed to the Virgin Mary and to various saints for help. The saint best known in this regard was Saint Anthony of Egypt (c. 250–355), shown in Figure 16.2, who gave up all his worldly goods to help spread Christianity and to live a life of prayer and contemplation. Long after Saint Anthony died, his remains were taken to Alexandria, and afterward to Constantinople. In 1070, Geilin II, the Count of Dauphiné, received permission to bring the saint's remains to Dauphiné for burial (Dotz, 1980).[1]

Soon after the arrival of Saint Anthony's remains in France, many of the faithful began to go to the Church of Motie-au-Bois, where the remains were kept, to pray for their salvation during outbreaks of ergotism. One of the richest noblemen of the Dauphiné region, Gaston, and his afflicted son Gérin, were among the pilgrims to this church in 1090. Gaston swore before the tomb of Saint Anthony that if Gérin recovered from the sacred fire *(ignis sacer)* that brought him such pain, he would donate his wealth to the church to help other victims of ergotism. Gérin did recover, and true to his word, Gaston established the order of Saint Anthony, a brotherhood of monks who would care for the victims of ergotism. In addition, in 1093 he built these monks a hospital with flame-red walls. All of these things may have contributed to ergotism being called Saint Anthony's fire (Bove, 1970; Dotz, 1980; Tanner, 1987).

Whether ergot poisoning played a role in some of the uncontrollable "dancing" and trancelike states associated with religious gatherings in the New World is not certain. This possibility, however, is suggested by records dating from 1801 to 1803. At that time, 20,000 to 30,000 people attended a series of religious revival meetings on the Kentucky frontier. Many showed convulsive dancing and trancelike behaviors and complained of feeling heaviness on the chest. Those afflicted also experienced prickly sensations, severe convulsions, hallucinations, tremors, twitches, and jerky movements. These symptoms resembled those of ergot poisoning more closely than the psychogenic variety of chorea.[2]

It has also been suggested that ergot poisoning may have played a role in the witchcraft craze in the New World, especially in Salem in 1692 (Caporael, 1976; Matossian, 1982). Rye was an established crop in New England by this time,

Figure 16.2. Sixteenth-century woodcut by an unknown artist showing a peasant who lost his right foot and left arm in flames appealing to Saint Anthony for help.

and many of the "witches" showed convulsions; complained of being pricked, burned, pinched, or bitten; and had visions, often involving fire.[3,4]

Ergot also has had its benefits (Dotz, 1980; Haller, 1981). In 1582, Adam Lonitzer (1528–1586) wrote that midwives used low doses of ergot to stimulate uterine contraction (Goodman and Gilman, 1970). Its use in childbirth was also mentioned by Kaspar Bauhin, Rudolph Camerarius, and John Ray (1627–1705). Women also relied on ergot for abortions. Nevertheless, control over the drug was difficult, and many pregnant women died while taking it. A passage from the sacred book of the Parsees (400–300 B.C.) states: "Among the evil things created by Angro Maynes are noxious grasses that cause pregnant women to drop the womb and die in childbed" (Goodman and Gilman, 1970). This would suggest that the difficulties involved in administering ergot were recognized well before the Renaissance.[5]

---

[1]This was just before the Turks won the Battle of Manzikert (1071), which disrupted pilgrimages to Jerusalem and led to the first Crusade in 1095.

[2]It has been argued that the rye harvested in 1801 from Kentucky was infected with the fungus (Massey and Massey, 1984; Matossian, 1978).

[3]In the twentieth century, the potent hallucinogen lysergic acid diethylamide (LSD) was isolated from ergot.

[4]Epidemics of ergot poisoning have continued into the twentieth century. A major epidemic erupted in 1951 when a baker in Pont Saint Esprit, France, made bread with contaminated, illegal flour in an attempt to avoid the taxes on inspected flour. Over 200 people were infected, and four died (Gabbai, Lisbonne, and Pourquier, 1951).

[5]Ergot was added to the *U.S. Pharmacopeia* for obstetric purposes in 1820. It was recognized later in the nineteenth century as useful for treating headaches. The purified amino acid derivative of ergot, ergotamine, was isolated in 1918. This drug has been used to treat migraine headaches and in obstetrics.

SYDENHAM'S CHOREA

Paracelsus, a sixteenth-century nonconformist with a disdain for classical medical practices, was one of the first to try to distinguish among the different forms of chorea. The person whose name became synonymous with one specific form of chorea, however, lived a century later. He was Thomas Sydenham (1624–1689).[6]

In 1686, Sydenham published his last book, *Schedula monitoria de novae febris ingressa*. Chapter XVI, entitled "On Saint Vitus's Dance," greatly restricted the broad use of the term *"chorea Sancti Viti"* to one specific movement disorder. Sydenham wrote the following about Saint Vitus' Dance:

> This is a kind of convulsion, which attacks boys and girls from the tenth year to the time of puberty. It first shows itself by limping or unsteadiness in one of the legs, which the patient *drags*. The hand cannot be steady for a moment. It passes from one position to another by a convulsive movement, however much the patient may strive to the contrary. Before he can raise a cup to his lips, he makes as many gesticulations as a mountebank; since he does not move it in a straight line, but has his hand drawn aside by spasms, until by some good fortune he brings it at last to his mouth. He then gulps it off at once, so suddenly and so greedily as to look as if he were trying to amuse the lookers-on. (1848a translation; 1979 reprint, Vol. 2, 257–258)

Sydenham's work was quickly followed by a number of other treatises on movement disorders. In his honor, this particular form of chorea began to be called Sydenham's chorea.

The major motor symptoms of Sydenham's chorea are poor coordination, spontaneous tics, and muscular weakness. The tics and twitching are the first symptoms to appear, usually starting in the hands or face. As the twitching increases, it can interfere with purposeful movements. Planned actions can be markedly disrupted at the peak of the disorder.

The spasmodic movements are irregular in frequency and severity. A person may move his head slightly and sway from one side to another. As the movements become more considerable, the limbs may be thrown around with enough force to disrupt standing or sitting. Patients with this form of chorea have been known to break bones and to throw themselves out of bed. Once asleep, however, the writhing almost invariably ceases.

Speech is affected by the involuntary movements of the muscles of the lips and face. These patients cannot maintain tongue protrusion, and consequently words and phrases become explosive, jerky, and indistinct. In addition, there may be irritability, restlessness, and emotional swings, although the senses and intellect typically remain normal throughout the disease (Aron, Freeman, and Carter, 1965).

Sydenham's chorea, or chorea minor, is a disease of young people. The great majority of cases occur between the ages of 5 and 20, and even early tabulations showed it was two or three times more common in females than in

males (Gowers, 1893; Huntington, 1872). The disorder usually lasts from six weeks to six months, although in rare cases it may persist for a few years. It has a tendency to recur, often within two years of the first attack. Nevertheless, the vast majority of cases show complete recovery from the movement disorder. These facts, when combined with the symptoms mentioned above, have made this disorder relatively easy to diagnose.

Thomas Sydenham thought this form of chorea was due to "some humour falling on the nerves," and he added that "such irritation causes the spasm . . . hence the treatment is first to bleed and purge, and then to restore the strength" (1979 reprint, p. 199). The specific treatment he recommended called for bleeding from the arm, cathartics (a special concoction and paregoric), other potions, and a special plaster applied to the soles of the foot. He thought the same treatment might cure epilepsy in adults.

The treatment changed over the ensuing years. In the late-eighteenth century, for example, Sydenham's chorea was treated primarily with rest and a reduction of those factors that could cause emotional or physical excitement. Treatments also involved drugs such as arsenic, which were given to stimulate the appetite. In the nineteenth century, patients were occasionally treated with galvanism and magnets, typically to no avail, and with sedatives and zinc, which may have helped a little.

The fact that many patients who have Sydenham's chorea also suffer from rheumatism was noticed early in the nineteenth century, although it has been suggested that Sydenham himself may have recognized this. In 1810, Etienne Michel Bouteille (1732–1816) pointed to the frequency of joint pains in people suffering from Sydenham's chorea. He noticed that in four of his cases arthritic rheumatism preceded the chorea, whereas in another case it followed it. In 1831, Richard Bright (1789–1858) even classified chorea under the heading of rheumatism. Germain Sée (1818–1896) wrote that rheumatism was important in the etiology of this form of chorea. In 1851, he stated that for every two rheumatic children, at least one was choreic.

In 1840, Jean-Baptiste Bouillaud pointed to a relationship between rheumatism and heart disease (see Thiebaut, 1968). Others also noticed a relationship between chorea and heart disease (see Barbeau, 1958; Greenfield and Wolfsohn, 1922). In 1866, Henri Roger (1809–1891) brought these observations together when he postulated that chorea, rheumatoid arthritis, and endocarditis or valvular heart disorders share a common source.

The observation that cardiac problems can be found in approximately half of the choreic cases suggested that the condition of the blood or the blood vessels was the cause of Sydenham's chorea. Specifically, scientists attempted to implicate small embolisms, thromboses, and "the morbid condition of the blood" (Greenfield and Wolfsohn, 1922; Straton, 1885).

The exact cause of Sydenham's chorea was unknown until the end of the nineteenth century. The possibility that Sydenham's chorea could be a communicable disease due to a coccus received increasing recognition at this time (see Straton, 1885). Finally, a diplococcus was isolated from the cerebrospinal and pericardial fluids of a child who died from chorea with rheumatic pericarditis (Wesphal, Wassermann, and Malkoff, 1899). It was soon demonstrated that an intra-

---

[6]Sydenham also provided the world with detailed descriptions of many other diseases, ranging from gout to chicken pox to scarlet fever.

venous injection of the coccus from a fatal case could produce endocarditis and chorea in rabbits and that the coccus could make its way into the brain. The causal agent was called *diplococcus rheumaticus* (Poynton and Paine, 1913).

But what part of the central nervous system was most affected in Sydenham's chorea? This question was not easily answered because there were very few post-mortem studies on these patients, owing to the fact that Sydenham's chorea usually had a favorable outcome. The absence of fresh autopsy material led to considerable speculation.

One of the more popular theories in the nineteenth century was that the chorea was due to cerebellar insult.

> The most probable theory, and one which I believe is most generally accepted at the present day is, that the disease depends upon some *functional* derangement in the cerebellum. Modern physiologists pretty generally agree upon the opinion first advanced by Flourens, that the function of the cerebellum is to direct, govern and coördinate the movements of the muscular system. This being the case, then, the irregular ungoverned movements of the muscles in chorea would most decidedly and emphatically point to the cerebellum as the seat of the difficulty. (Huntington, 1872, 318)

After the motor cortex was discovered in 1870, another popular belief was that this structure was most affected by the disorder. As stated by William Gowers (1893, p. 614), one of the most authoritative voices at the time: "It is doubtful whether there are at present any facts to justify us in going beyond the motor cortex in our search for the primary disturbance."

Opinions changed once again after it was discovered that the pathophysiology of Sydenham's chorea involved inflammation of both the cortex and the basal ganglia (especially the caudate nucleus and the putamen), with less damage in areas such as the cerebellum, the thalamus, and the mesencephalon (e.g., Delcourt and Sand, 1908; Greenfield and Wolfsohn, 1922; Marie and Trétiakoff, 1920).[7]

PARKINSON'S DISEASE

Another disorder associated with involuntary movements and changes in the basal forebrain is Parkinson's disease. The clinical signs of this disorder usually appear in the second half of life. Affecting men and women equally, it is characterized by tremor and muscular rigidity.

The characteristic tremor usually starts locally (often in one hand) and spreads slowly.

> The head is bent forward, and the expression of the face is anxious and fixed, unchanged by any play of emotion. The arms are slightly flexed at all joints from muscular rigidity, and (the hands especially) are in constant rhythmical movement, which continues when the limbs are at rest so far as the will is concerned. The tremor is usually more

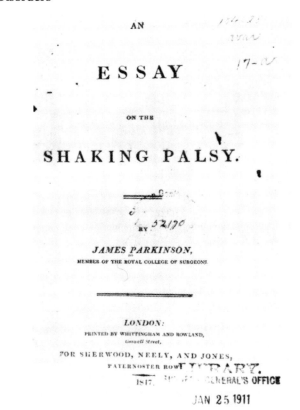

Figure 16.3. Cover page of James Parkinson's (1817) *An Essay on the Shaking Palsy.*

marked on one side than the other. Voluntary movements are performed slowly, and with little power. The patient often walks with short quick steps, leaning forward as if about to run. (Gowers, 1893, 638)

The individual for whom Parkinson's disease is named, James Parkinson (1755–1824), was born in the London suburb of Shoreditch. He was the son of an apothecary and surgeon, and he apprenticed to his father and later took over his practice. By 1817, when his *An Essay on the Shaking Palsy* appeared (see Figure 16.3), James Parkinson had firmly established himself as a leading paleontologist, an author of medical books for laymen, the writer of a chemical handbook, and an ardent supporter of liberal medical and political causes (under the pseudonym Old Hubert).[8]

Although Parkinson's name is well known today, it seemed to have been rescued from near oblivion by an American, Leonard Rowntree (1883–1959), who was vacationing in England early in the twentieth century (Schiller, 1970). Rowntree (1912) stimulated renewed interest in Parkinson's contribution to the disease that now bears his name with an article that began, "English born and bred, an English physician and scientist, forgotten by the English and by the world at large—such is the fate of James Parkinson" (p. 33).

Indeed, when Germain Sée (1850) differentiated between Sydenham's chorea and Parkinson's shaking palsy, he cited Parkinson as "Patterson" without a single objection from the

---

[7]In the newer literature, one find's Sydenham's chorea discussed as a complication of a streptococcal infection that can be treated with penicillin. Opinions about the sites of the central nervous system lesions and the presumed roles of the basal ganglia and the cortex have not changed much since the first quarter of the twentieth century (for more recent anatomical studies, see Greenfield, 1963; Hallervorden, 1957; Thiebaut, 1968).

[8]For biographical information on Parkinson, see McHenry (1968), McMenemey (1955), Morris (1955), Tyler and Tyler (1986), and Yahr (1978).

Paris Academy of Medicine (Schiller, 1970). To this it can be added that there are no known portraits of Parkinson nor a published obituary. However, one of his associates wrote:

> Mr. Parkinson was rather below the middle stature, with an energetic, intelligent, and pleasing expression of countenance and of mild and courteous manners; readily imparting information either on his favorite science or on professional subjects. (Mantell, cited in Rowntree, 1912, 45)

Parkinson's *An Essay on the Shaking Palsy* was based on only six cases, just three of which were examined in detail. Of the other three, two were met "within the street" and one "was only seen at a distance." In his short (66 pages) essay, the 62-year-old Parkinson (1817) described the disorder he had seen as:

> Involuntary tremulous motion, with lessened muscular power, in parts not in action, and even when supported; with a propensity to bend the trunk forward, and to pass from a walking to a running pace: the senses and intellects being uninjured. (1989 reprint, 152)

He stated the disease progressed so slowly that his patients rarely recollected when their first symptoms, usually weakness and proneness to trembling in the hands, began to appear. Parkinson described how the symptoms then grew to incapacitate these patients. He argued that *paralysis agitans,* the Latin name for his shaking palsy, should be considered a disease entity.

Tremor with some paralysis was well recognized prior to this time. One can even find a good description of it in William Shakespeare's *Henry VI,* which was first produced for the 1590–1591 theater season. One of the characters, Say, shook, quivered, and nodded his head in a manner characteristic of Parkinson's patients. In one scene, Dick asks Say, "Why dost thou quiver, man?" "The palsy and not fear provokes me," answered Say.

Very little was known about the underlying pathology of tremor when James Parkinson lived. There had been little interest in even classifying shaking beyond tremors of voluntary motion and tremors of rest. This simple distinction can be traced back to Galen, who saw both types of tremors and who called the coarser tremor, which could be seen at rest, *palmos.* Galen's contribution was acknowledged by Parkinson.

Parkinson also cited François Boissier de Sauvages de la Croix (1706–1767), who described different types of *scelotyrbe,* one of which he called *scelotyrbe festinans* ("running disturbance of the limbs"). When patients with *scelotyrbe festinans* try to walk, there is resistance and sometimes tripping, followed by short moves and running with hasty steps. Sauvages de la Croix (1763) thought this was due to diminished flexibility in the muscle fibers. He postulated that these individuals try to overcome the resistance with more exertion, which causes the running. He admitted he had come across only two such cases, but already was convinced that *scelotyrbe festinans* was a disorder of advanced years. He did not, however, correlate the abnormal gait and the running with the mass of other symptoms that make up the Parkinson symptom complex. Nor was he the only early writer to describe *scelotyrbe festinans,* as Parkinson himself noted.

Parkinson included *scelotyrbe festinans* as just one of the symptoms of the shaking palsy in his *Essay* and commented that this symptom may not appear until 10 to 12 years after the onset of the tremor. Nevertheless, he did not discuss the rigidity that Sauvages de la Croix emphasized in the context of *scelotyrbe festinans*—a surprising fact, especially since the rigidity is usually considered the most disabling aspect of the disease.[9]

Later writers, such as Johannes von Oppolzer (1808–1871), emphasized the permanent state of muscle contraction and the stiffness exhibited by patients with *paralysis agitans.* In 1911, Harold N. Moyer (1858–1923) observed that the resistance of the extremities to stretch did not proceed smoothly but rather in a jerky or intermittent manner. He likened this interrupted muscular action to that of a cogwheel, giving rise to the term "cogwheel rigidity." Moyer stated he never saw this in any other disease and felt it was an excellent early diagnostic sign of Parkinson's disease.

Parkinson clearly delineated almost all of the other major symptoms of the disease that now bears his name. The typical age at the time of onset was seen as being over 50 years, and the shaking disorder was recognized as progressive. He noted that his patients had a hard time initiating activities and walked with short steps and few swinging arm movements. He also saw that they tended to drool excessively (ptyalism), especially while eating.[10]

Parkinson was not able to study the brains of his patients, but believed this disorder was not due to pathology of the peripheral nerves. Instead, he suggested it represented a problem with the central nervous system. He speculated that the site might be in the cervical cord, with possible expansion into the medulla oblongata as the disease progressed. But this was only a guess, and he expressed the hope that future medical men would have the opportunity to conduct autopsies to determine the causal lesion.

Parkinson's concept of *paralysis agitans* as a complex of symptoms forming a distinct disease entity appeared to be ignored by all but a few medical men in the period immediately following the publication of his pamphlet. But those few who read his work gave his ideas a warm reception (Morris, 1989). For example, it was highly praised by John Cooke (1756–1838) in his scholarly work *A Treatise on Nervous Disorders.* Published in two volumes in 1820 and 1824 (Vol. 2 dealt with the palsies), Cooke's text was considered "the first separate work in neurology, . . . the most significant single contribution to neurology for its time" (McHenry, 1969, p. 271; also see Spillane, 1981). Parkinson's contribution was also applauded by Dr. John Elliotson (1791–1868), who delivered a clinical lecture at Saint Thomas' Hospital in which he stated:

> The best account of this disease which I have seen is one given by a general practitioner, now deceased, by the name of Parkinson, a highly respectable man, who wrote an essay upon the subject in 1817, from which I have derived nearly all I know upon the complaint. . . . It does not appear as if this disease of which I am at present speaking, was well characterized or distinguished before Mr. Parkinson wrote on the subject. (1830, 120)

---

[9]Possibly the omission was because muscle tone was not really examined early in the 1800s (Schiller, 1986a). Another hypothesis is that the descriptions given of the other symptoms may have been sufficient to signify the presence of rigidity.

[10]Johann Sagar (1732–1813) was not cited by Parkinson, although he had described ptyalism in the context of *scelotyrbe festinans* in 1776. Sagar wrote that "in Vienna I saw a man above the age of fifty who was running involuntarily, being also incapable of changing direction so as to avoid obstacles; in addition he suffered from ptyalism" (translated by Schiller, 1986b, p. 1583).

The first new case report of Parkinson's disease appeared in 1833. The author, Adolphe Toulmouche (1798–?), described a 76-year-old man who showed tremors, difficulty walking, and loss of balance. Toulmouche did not cite Parkinson's *Essay*. Other reports slowly followed his (e.g., Elliotson, 1839).

A major reason for the lack of recognition of Parkinson's contribution was that his 1817 monograph appeared in only a single, small edition. Only a few copies have been found (Greenfield, 1956). But despite its slow start, Marshall Hall (1790–1857) succeeded in incorporating Parkinson's term *paralysis agitans* into the medical language in 1841 (McHenry, 1969). And in 1882, Thomas Buzzard (1831–1919) wrote that Parkinson's *Essay* "presents so graphic and admirable a description of the disease that comparatively little has been left for subsequent observers to add to his account" (p. 329).

The term "Parkinson's disease" *(la maladie de Parkinson)* was coined by Jean-Martin Charcot, the leading neurologist at the Salpêtrière in Paris, who also contributed much to the accumulating knowledge about this disorder. (Figures 16.4 and 16.5 show a portrait of Charcot and a print of the Salpêtrière, respectively.) In 1861 and 1862, Charcot and Félix Vulpian gave a detailed account of what was known about "this strange affection" resembling chorea. Charcot also included sections on this disease in his *Lectures on the Diseases of the Nervous System* (1877).

In general, Charcot accepted Parkinson's classical description of *paralysis agitans*. He criticized him and most of his successors only for overlooking rigidity as one of the key features of the disease. In this regard, Charcot spoke highly of Armand Trousseau's descriptions. Trousseau had noted that one of his patients "sat down with some difficulty as if his trunk and legs were stiff." Charcot thought this rigidity caused the slight inclination of the body when standing and the bent head that is difficult to turn to either side (see Figure 16.6, from Gowers, 1893). He also discussed the

Figure 16.5. The Salpêtrière, the huge Paris hospital that received its name because Louis XIII stored his gunpowder here. This hospital, which housed 5,000 to 8,000 welfare cases, was where Jean-Martin Charcot and many other leading French neurologists conducted their studies. (Courtesy of Charles C Thomas, Springfield, IL.)

laboriousness and slowness in the execution of the movements (bradykinesia) and pointed out that it was difficult for a Parkinson patient to get out of a chair. He emphasized that this could not be attributed to the tremor.

In 1877, Charcot discussed the positions of the hands and the tremor when writing. He noted the thicker and more irregular lines made by these patients. Three years later, he spoke about the unusual shrinkage in the size of the handwriting (micrographia).

Charcot also focused considerable attention on differentiating the type of tremor displayed by Parkinson's patients from that displayed by patients with multiple sclerosis, with which it was often confused (e.g., Hall, 1841). He illustrated the different types of tremors to his students by having patients hold large feathers from women's hats. Charcot agreed with Parkinson that, unlike the tremor of multiple sclerosis, the tremor of *paralysis agitans* is present at rest.

Figure 16.4. Jean-Martin Charcot (1825–1893), the eclectic French neurologist who long maintained an interest in movement disorders.

Figure 16.6. Drawing from Gowers (1893) showing postures associated with Parkinson's disease.

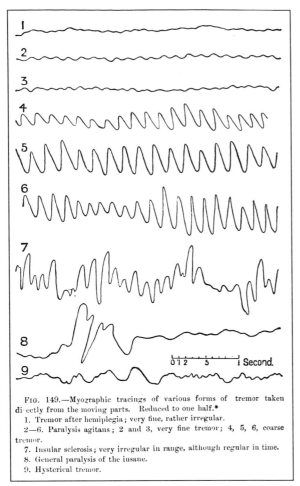

FIG. 149.—Myographic tracings of various forms of tremor taken directly from the moving parts. Reduced to one half.*
1. Tremor after hemiplegia; very fine, rather irregular.
2–6. Paralysis agitans; 2 and 3, very fine tremor; 4, 5, 6, coarse tremor.
7. Insular sclerosis; very irregular in range, although regular in time.
8. General paralysis of the insane.
9. Hysterical tremor.

Figure 16.7. William Gowers' (1893) recordings of the tremors associated with different diseases. Lines 2–6 show the fine (2, 3) and coarse (4, 5, 6) tremors of Parkinson's disease.

The first clear tracings of the tremor patterns were published by William Gowers and Frederick Peterson (1859–1938) (Fine, Soria, and Paroski, 1990). In 1886, Gowers attached a metal rod to the trembling part of the body and placed the stylus end of the rod on a rotating drum covered with smoky paper. One of his figures was published in his highly influential *A Manual of Diseases of the Nervous System* (1888, 1893). This is shown in Figure 16.7.

Gowers noticed that as the disease progressed, the tremor of *paralysis agitans* decreased in frequency, but increased in range (size). He believed that the tremor, which was most noticeable in the extremities, was caused by an alternating contraction in opposing muscles. He also noticed it ceased during sleep and that:

> The great characteristic of the tremor of *paralysis agitans* is, as Parkinson pointed out, that it continues during rest. The hands go on moving when they are resting on the patient's knee, and the legs when he is sitting. A voluntary movement may stop the tremor for a few seconds, sometimes for many, but it recommences and accompanies the movement. Hence the patient's handwriting reveals his disease; the letters may be fairly formed, but every line is a zigzag. (1893, 642)

Peterson (1889) used a sphygmograph, originally designed to record the pulse, to make records of the tremors. Attached

to a body part, it was connected through a system of gears and levers to a pen that recorded on rolling paper. Like Gowers, he found the frequency of the tremor was 4–7 Hz in Parkinson patients (as compared to 8–11 Hz in chronic alcoholism) and that voluntary movements could temporarily arrest the tremors in many individuals.

In addition to studying the tremor, late-nineteenth-century investigators began to analyze more complex movements in greater detail (see Gowers, 1893). These included the running movements and the tipping forward (propulsion) once walking started. When jerked backward, these patients exhibited an unavoidable tendency to walk backward (retropulsion) while still stooped forward (Charcot, 1877). Monotonous speech and delays in starting sentences were also noted, as was a tendency to show festination in speech—that is, words being uttered rapidly and often with some confusion once a sentence began. Another feature that received some attention was the piping character of the voice (Buzzard, 1882).

Following initial attempts to study the motor aspects of Parkinson's disease in more detail, investigators began to pay more attention to associated nonmotor disturbances. Most authors at the end of the nineteenth century or early in the twentieth century, including Kinnier Wilson, believed the major mental symptoms were depression (e.g., loss of interest and crying) and irritability. Many physicians thought these were secondary reactions to knowing that the disease was incurable. It was typically stated that any symptoms beyond these were due to arteriosclerosis or other medical conditions (Patrick and Levy, 1922). Nevertheless, the possibility that one could see "mental weakness" or loss of memory and intellect late in the disease was not denied (Gowers, 1893).

The basis of the disorder remained a mystery for many years after Parkinson first described it (Rose, 1989). Hermann Oppenheim (1858–1919) in 1902, and Charcot before this time, considered it a neurosis. This term, however, was used only to signify a disorder that "possesses no proper lesion" (Charcot, 1877). Charcot attributed it to excessive damp, cold, and emotion. John Hughlings Jackson thought the tremor was a release phenomenon due to cerebral damage. In remarks published in *The Medical Examiner* in 1877 and again in 1878, Jackson stated he believed that cerebral lesions led to the unopposed influence of the cerebellum.

> (1) In health the cerebellar influx is fully antagonized; in (2) the early stages of *paralysis agitans* it is intermittently antagonized—the movement constituting each single tremor occurring betwixt the cerebral impulses; in (3) late stages it is not antagonized at all, and there is such a stream of cerebellar impulses that rigidity occurs. (Jackson, 1878, 267)

Other contributors to the pathology of Parkinson's disease proved to be more on the mark. In 1871, Theodor Meynert, shown in Figure 16.8, presented a case of general paresis with severe tremor on one side of the body, shrinkage of the corpus striatum and lenticular nucleus on the opposite side, and no notable cortical atrophy. The tremor was not that of *paralysis agitans,* since it ceased during rest, but Meynert suggested that Parkinson's disease, as well as Sydenham's chorea, might be the result of defective functioning of the basal ganglia.

In 1912, S. A. Kinnier Wilson wrote that damage to the

corpus striatum, and in particular to the lenticular nucleus, removed the steadying influence the striatum normally exerted on the corticospinal paths, thereby causing hypertonus. Wilson termed the pathway involved in the development of rigidity the "extrapyramidal pathway."

Another contributor to the pathology of the disorder was Edouard Brissaud (1852–1909). In 1894, Brissaud made the bold suggestion that *paralysis agitans* may be due to a vascular lesion in the substantia nigra, a structure containing a pigmented group of neurons. After examining a 38-year-old patient with a tuberculoma in the region of the substantia nigra, Brissaud (1895) concluded that "the locus niger might well be its anatomical substratum." This patient had been described by Paul Oscar Blocq (1860–1896) and Georges Marinesco (1864–1938) in 1893. Brissaud's idea, however, was not taken very seriously by his contemporaries (Greenfield, 1956).

It was not until 1919, when Constantin Trétiakoff studied the substantia nigra in patients with *paralysis agitans,* that the importance of the *locus niger de Soemmerring* began to be appreciated. Trétiakoff found nigral lesions or inflammations in his idiopathic Parkinson's disease cases.[11] He also saw nigral lesions in his cases showing Parkinson's symptoms in association with *encephalitis lethargica,* the sleeping sickness that became epidemic in Europe in 1915.[12] The mortality rate from the encephalitis disorder was about 40 percent, and about half of the survivors developed Parkinson's symptoms within five years (Rose, 1989).

Edward Farquhar Buzzard (1871–1945) published one of the first clear descriptions of a clinical encephalitis case showing the features of *paralysis agitans*. This appeared in the 1918 issue of *Lancet.*

> April 1st he lay in bed in a state of generalized rigidity, presenting every appearance of a case of advanced *paralysis agitans* including the "mask-like" facies and the characteristic attitude of the limbs. He was so weak and rigid that he could not turn in bed, but he could carry out movements with every limb slowly and feebly. (1918, 616)

In 1921, Trétiakoff commented that the lesions of the substantia nigra can precede the symptoms of Parkinson's disease by weeks or months. He added that he never saw a case of longstanding, severe damage to the substantia nigra without Parkinsonism.

Nevertheless, the debate over the causal lesion was not over. In particular, Trétiakoff was unable to convince Oskar or Cécile Vogt (1920, 1923) of the importance of the substantia nigra (also see Bielschowsky, 1922). In addition, some scientists suggested that idiopathic and postencephalitic Parkinsonism could have somewhat different lesions (see Greenfield and Bosanquet, 1953). For example, one author wrote:

> We may surmise then that in idiopathic *paralysis agitans* a lesion of the globus pallidus, or possibly of the subthalamic structures with which it is connected, is the usual pathological basis of the disease; whereas post-encephalitic Parkin-

[11]The inflammations are often called *corps de Lewy*, "hyaline inclusions of the Lewy type," or "Lewy bodies."
[12]Previous epidemics of *encephalitis lethargica* occurred in Europe in 1580, in London in 1674 (Sydenham's *febris comatosa*), in Germany in 1712, and in Italy in 1889 (where it was called *nona*).

Figure 16.8.   Theodor Meynert (1833–1892), who in 1871 suggested that Parkinson's disease may be the result of defective functioning of the basal ganglia.

sonism is usually due to degeneration of the substantia nigra. (McAlpine, 1923, 268)

For the most part, however, studies after 1925 confirmed the strong association between lesions of the substantia nigra (e.g., shrinkage of the pigmented cells, chromatolysis) and Parkinson's disease (e.g., Foix and Nicolesco, 1925; Hassler, 1938). Furthermore, after it was demonstrated that axons of cells in the substantia nigra ended almost entirely in the lenticular nucleus of the corpus striatum, the picture became clearer, and it was believed that the rigidity and tremor were due to the loss of inhibition from this circuit (Freeman, 1925).

These studies put to rest the idea that Parkinson's disease was a "functional" disorder, a position still advocated by some writers early in the twentieth century (see Wilson, 1924). Nevertheless, years would pass before Parkinson's disease in humans would be associated with depleted dopamine production in the subcortical motor circuitry (Ehringer and Hornykiewicz, 1960; Hornykiewicz, 1963).

Prior to the advent of levodopa, which can pass through the blood-brain barrier and increase brain dopamine concentrations, it was generally believed that the symptoms of Parkinson's disease could not be treated very effectively. Parkinson (1817) thought internal remedies were useless, but added that early treatment with physical therapy, liniments, and bleeding might be able to slow the progress of the disease. Others tried physical therapy and also concluded that while exercise might have transient effects, it could not stop the course of the disease or serve as a cure (Swift, 1916).

Charcot noticed that patients with *paralysis agitans* often showed some relief of their symptoms during long journeys in a train or carriage. He thought bouncing and vibration were beneficial and hence constructed a trembling armchair *(trépidant)* to use with his patients at the Salpêtrière. Charcot also used an apparatus that suspended his patients in the air with a harness (see Charcot, 1887–88; Goetz, 1987). This is shown in Figure 16.9. Gowers (1893), however, believed that neither the shaking chair nor electrical stimulation, another technique used by Charcot, could improve the condition of these patients (see Chapter 29).

Surgery to ablate either the motor cortex, the premotor cortex, or both, was also tried. In 1937, cortical surgery—introduced by Victor Horsley for athetosis in 1890—was first attempted on Parkinson patients. Three problems with it were: (1) it was associated with a high mortality rate (10–17 percent); (2) there was residual paresis; and (3) many patients were not relieved of their involuntary movements (Cooper, 1961; Oliver, 1953).

Sectioning the pyramidal tract as it entered the spinal cord was also attempted (Putnam, 1938). Here too, the operation was considered worthwhile by only a minority of physicians. Whereas there was often some relief of tremor, there was also a loss of muscle power and no improvement in the rigidity or the incapacitation (Oliver, 1953).

Surgery of the basal ganglia began a bit later. It involved removing the head of the caudate nucleus and making incisions in the globus pallidus (Meyers, 1940). These procedures were also associated with unacceptably high mortality, convulsions, and inadequate relief. Although attempts were made to modify these surgeries, they too were abandoned, especially as advancements continued to be made in pharmacology.

As noted, drugs such as arsenic and opium, as well as a variety of sedatives, were used at the end of the nineteenth century. These agents gave mixed results and were slowly replaced by drugs belonging to the atropine group, such as hyoscine hydrobromide (scopolamine) and stramonium (Oliver, 1953). These, in turn, were followed by agents with antispasmodic properties, such as artane trihexyphenidyl (artane) and caramiphen hydrochloride (parpanit). Levodopa became the drug of choice only in the late 1960s.[13]

HUNTINGTON'S CHOREA

Huntington's disease is named after George Huntington (1850–1916), who, in 1872, provided a good clinical description of this progressive disease.[14] Huntington studied several generations of people in East Hampton (originally called Maidstone), New York. East Hampton was settled in 1649 by English immigrants who had been in America for a number of years. The Huntingtons came to the New World in 1633 from Norwich on the "Elizabeth Bonaventura." Simon Huntington, the father of the family, died

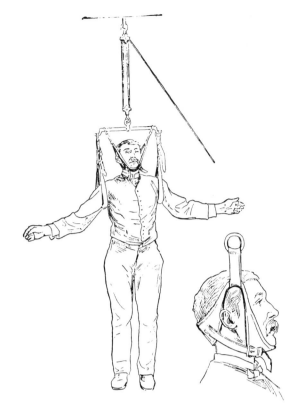

Figure 16.9.   Charcot's stretching apparatus, used to treat Parkinson's disease, ataxias, and multiple sclerosis. This device was the subject of one of Charcot's famous "Tuesday lessons" (January 15, 1889).

during the trip, but was survived by his wife and children, who probably landed at Saybrook, Connecticut. George Huntington's grandfather moved to East Hampton in 1797 and became the leading physician in the area. His son George, the father of the George Huntington who wrote the classic paper on hereditary chorea, was also a physician. The Huntingtons themselves never suffered from the hereditary chorea.

George Huntington graduated from the College of Physicians and Surgeons of Columbia University in 1871. He then returned to Long Island to assist in the family practice. This gave him the opportunity to observe patients with hereditary chorea—patients whose families had been followed for years by his father and grandfather, both of whom provided him with written records. Indeed, Huntington's grandfather had written that by 1797 the disease was already established in this area.

Later in 1871, George Huntington moved to Pomeroy, Ohio. He first presented his report on the choreas to the Meigs and Mason Academy of Medicine at Middleport, Ohio, on February, 15, 1872. Later that year, his observations were published in the *Medical and Surgical Reporter* in a paper entitled "On Chorea." This brief report mostly concerned Sydenham's chorea; only at the end of the manuscript did Huntington describe the "hereditary" chorea. This was restricted to seven short paragraphs, the first three of which read:

> And now I wish to draw your attention more particularly to a form of the disease which exists, so far as I know, almost exclusively on the east end of Long Island. It is peculiar in itself and seems to obey certain fixed laws. In the first place,

---

[13]Initial attempts to use levodopa, however, did not take place without difficulty. In the early 1960s, some notable successes were reported (Birkmeyer and Hornykiewicz, 1961, 1962), but these were not always confirmed (Fehling, 1966; McGeer et al, 1961; McGeer and Zeldowicz, 1964). Many patients showed severe side effects, including temporary symptoms of psychosis, and adjusting the drug dosage often proved so difficult and unpredictable that many patients preferred not to be treated after an initial encounter with L-DOPA (Sacks, 1983). This drug was administered more effectively to a greater percentage of patients by the mid-1960s, although side effects and patient variability still represented very real problems.

[14]See de Jong (1970) and Winfield (1908) for biographies of Huntington.

let me remark that chorea, as it is commonly known to the profession, and a description of which I have already given, is of exceeding rare occurrence there. I do not remember a single instance occurring in my father's practice, and I have often heard him say that it was a rare disease and seldom met with by him.

The *hereditary* chorea, as I shall call it, is confined to certain and fortunately a *few* families, and has been transmitted to them, an heirloom from generations away back in the dim past. It is spoken of by those in whose veins the seeds of the disease are known to exist, with a kind of horror, and not at all alluded to except through dire necessity, when it is mentioned as *"that disorder."* It is attended generally by all the symptoms of common chorea, only in aggravated degree, hardly ever manifesting itself until *adult* or *middle* life, and then coming on gradually but surely, increasing by degrees, and often occupying years in its development, until the hapless sufferer is but a quivering wreck of his former self.

It is as common and is indeed, I believe, *more* common among *men* than women, while I am not aware that season or complexion has any influence in the matter. There are three marked peculiarities of this disease: 1. Its hereditary nature. 2. A tendency to insanity and suicide. 3. Its manifesting itself as a grave disease only in adult life. (1872, 320)

In his remaining four paragraphs, Huntington went into more detail on these three features of the disorder, noting in particular that the disease never skips a generation to manifest itself in another and that "no treatment seems to be of any avail." The original paper bears the penciled corrections made by his father, whose advice he sought before sending his manuscript to press.

In 1892, William Osler (1849–1919), one of the most respected physicians of the period, told the Philadelphia Neurological Society: "In the whole range of descriptive nosology there is not, to my knowledge, an instance in which a disease has been so accurately and fully delineated in so few words" (see Osler, 1893, p. 97).[15]

Being a modest man, Huntington downplayed the significance of his observations. He considered the disease to be of interest, but "merely as a medical curiosity." Nevertheless, his paper stimulated great scientific interest in the disorder and in the concept of hereditary diseases. Soon there were many descriptions of others afflicted with the disease and attempts to trace family trees (Jelliffe, 1908; Maltsberger, 1961; Vessie, 1932). By 1908, the disease was so well known that a special issue of *Neurographs* was devoted just to Huntington's chorea.

George Huntington made no other scientific contributions and did not even decide to practice neurology. Nevertheless, he later presented his recollections of how he became interested in the disease that now bore his name. This was in a speech to the members of the New York Neurological Society in 1909. His thoughts were published in a journal article a year later as "Recollections of Huntington's Chorea as I Saw It at East Hampton, Long Island, during My Boyhood." Huntington recalled that:

Driving with my father through a wooded road leading from East Hampton to Amagansett, we suddenly came

upon two women, mother and daughter, both tall, thin, almost cadaverous, both bowing, twisting, grimacing. I stared in wonderment, almost in fear. What could it mean? My father paused to speak with them and we passed on. Then my Gamaliel-like instruction began: my medical education had its inception. From this point on my interest in the disease has never wholly ceased. (1910, 255–256)

It seems certain that Huntington's disease was present in England in 1630 when King Charles I (1600–1649) allowed the Church of England to increase its persecution of nonbelievers and nonconformists. Witch-hunting became a sacred duty of the populace, and at significant risk were adult women whose actions and appearances had changed. These odd behaviors were thought to be signs of a covenant with the devil.

It is thought that these religious and superstitious actions forced some families with Huntington's genes to flee England to the Americas. The most notable crossing was that of John Winthrop's (1588–1649) fleet of 11 ships that sailed from Yarmouth, England, to Salem, Massachusetts, in 1630. Yet in spite of extensive genealogical work, the names of those few individuals who brought the Huntington's genes to the New World has been a matter of considerable debate.[16] As Huntington (1910) had himself recognized, this was in part due to failures in keeping records of the maiden name of the mother; to questions abounding about illegitimate children; and to the fact that early sufferers of Huntington's chorea may have been classified simply as catatonic, demented, or manic.

Those individuals who carried Huntington's chorea across the Atlantic typically faced persecutions and terrors in the New World not very different from those they were fleeing. Not only were they still not free from suspicion of witchcraft but their weaving and jerking motions were viewed in the colonies as an inherited curse because their forefathers had blasphemously imitated Christ's suffering during the crucifixion (see Lyon, 1863). At least seven "witches" in New England were known to come from families that had Huntington's disease, and at least one was hanged after being convicted of witchcraft. This was Ellin, the wife of Nickolas Haste, who was executed in 1653. The "Groton Witch," whose genealogy was also traced to one of the original carriers of the disease to the New World, was convicted in 1671 but then pardoned.[17]

Huntington's chorea was not just a disease of the English. The disease affected pockets of individuals in other parts of Europe, some of whom also sailed to the New World for religious freedom. For example, after the 1598 Edict of Nantes was revoked in 1685, some Huguenots fled France to Halifax to escape the escalating religious persecution. It seems certain that some members of this group also brought the hereditary chorea with them. Other affected individuals appear to have come to the New World from Ireland, Germany, Norway, and southern Europe (Bruyn and Went, 1986).

A strong case can be made to show that George Huntington,

---

[15]Osler was so intrigued with movement disorders that he published an entire book on chorea one year later.

[16]Many versions of the transmission of Huntington's disease to the New World abound. Records are traced and discussed in Barbeau (1958), Bruyn and Went (1986), Caro and Haines (1975), Caro and Obst (1977), Critchley (1934, 1964, 1984), Hattie (1909), Klawans (1989), Maltsberger (1961), and Vessie (1932).

[17]The witch craze, which began in Massachusetts in 1645, ended there in 1692 (Potts, 1920).

whose contribution is universally acknowledged, was not the first to describe this hereditary disease. Chronic, hereditary chorea was previously described in a number of reports, possibly beginning in 1793 with a municipal death report from Belgium (see Bruyn, 1968; Bruyn and Went, 1986; de Jong, 1937; Gowers, 1893; Maltsberger, 1961). In 1832, John Elliotson wrote that he had seen hereditary adult chorea, which he associated with dementia. In contrast to Sydenham's chorea, he doubted whether the adult chorea could be cured. To quote Elliotson (p. 163):

> When it occurs in adults, it is frequently connected with paralysis or idiotism, and will perhaps never be cured. It is very rare for you to remove the affection if it occurs in adults, or if it occurs in a local form. It will sometimes take place in one arm only, or in the head, or some of the muscles of the face, so that the person makes faces continually. In cases of this description I have never seen the affection cured. It then appears to arise for the most part from something in the original constitution of the body, for I have often seen it hereditary. But the cases which occur in females, particularly in childhood and at the beginning of the adult period, may be almost every one of them cured.

One of the better early descriptions of the hereditary chorea may have been that of the Reverend Dr. Charles O. Waters (1816–1892) of New York. In a letter dated May 5, 1841, he described a type of hereditary chorea seen in southeastern New York that gradually resulted in dementia. This letter was cited in Robley Dunglison's *The Practice of Medicine*, which was first published in 1842 and again in 1848. Waters wrote that this disorder rarely occurred before adulthood and never ceased spontaneously. It was called the magrums by the common people, a word probably derived from the Dutch for "restlessness" or "nervousness."

Another early report came from Dr. Irving W. Lyon (1840–1896), a physician at Bellevue Hospital in New York. Lyon's paper, entitled "Chronic Hereditary Chorea," appeared in the *American Medical Times* nine years before Huntington's paper was published. As can be seen in his statements about Case 1, his emphasis was on the hereditary nature of this form of chorea:

> Mr. A. . . . has a well marked chorea, which is quite general; so that he is constantly, when awake, making irregular movements with the upper and lower extremities, facial muscles, and more or less with those of the body. This condition has existed for many years, but seems not to interfere materially with his general health, the vegetative functions being well performed. Mr. A. has two brothers and three sisters; the two brothers have themselves never had choreal symptoms, but one of them has two children in whom well defined chorea has existed for many years; of the three sisters, two have had the chorea for most of their lives, being now past middle age.
>
> The progenitors of Mr. A., on the male side, were perfectly free from chorea, but not so on the maternal side; his mother had well developed choreal manifestations from early life, which continued till her decease; she also had a brother who died during adult life from the severity of the disease; but to go still further, both the grandfather and great-grandfather of Mr. A., on the maternal side, had the same disorder which we find in their children. (1863, 289)

In 1859, Johann Christian Lund reported on hereditary chorea in Norway. He traced the pedigrees of two families for four generations and not only noticed the inheritance of

this form of chorea but found it affected both men and women. He also recognized its late onset and its association with marked mental changes.

> As mentioned in the previous report, St. Vitus' dance appears to occur as an hereditary disease in Stersdalen. It is commonly called *Rykka* (i.e., "jerking disease"), sometimes also *Arvesygen* (i.e., the hereditary disease"). It occurs as a rule between the ages of 50 and 60; generally starting with less conspicuous phenomena, which sometimes only slowly progress and do not become very marked, so that the patient's normal activities are not particularly hindered; but more often after a few years they increase to a considerable degree so that every kind of work becomes impossible for them, and they even have great difficulty in bringing food to the mouth by means of several detours. Their whole body, particularly, however, the head, the arms and the upper part of the body are in constant, severe jerking and flinging motion, except during sleep, when they usually are quiet. Some of the severely affected patients have in the last period of their life become *fatui* (i.e., psychotic or demented). (Translated by Refsum, 1982, 426)

Lund's description, which was written in Norwegian, was part of a report published in 1863 for the Ministry of the Interior. It was followed by another statement on the hereditary chorea in 1868, also in an obscure Norwegian publication.

Depictions of the key features of the disorder have remained essentially the same as those described by these individuals, especially by Huntington in 1872. Major early symptoms of the disease are restlessness, facial grimacing, and movements of the fingers. Voluntary gaze is affected early in the course of the disorder, and constant rotation of the eyes may be observed. Subtle "piano playing" movements of the fingers are another early sign. As the disease progresses, the chorea spreads through the body and affects voluntary actions such as walking. The victim may lurch from side to side with unequal steps and have difficulty with balance. Hands may twist and wave up and down, and speech becomes dysarthric at this time.

Personality changes may appear early in the course of the disease, the most frequent early "mental" symptoms being depression, attentional impairments, and outbursts of irritability and anger. As the disease progresses, cognitive and memory impairments manifest themselves. These later symptoms usually occur about two years after the initial personality changes are seen. More recent events become the most difficult to recall, and judgment and insight, which may be reasonably intact in the early stages of the disease, fade as the dementia progresses (Diefendorf, 1908).

The disorder is caused by an autosomal dominant gene, and the penetrance of symptoms in Huntington's disease is 100 percent. Even during Huntington's lifetime, family trees were traced back to the sixteenth century (Tilney, 1908). The frequency of the gene mutation which gives rise to Huntington's disease is so rare that investigators believe there may be no known examples of new spontaneous mutations (see Klawans, 1989; Myers and Martin, 1982).

Huntington's disease generally shows itself between the ages of 35 and 42, but there is variability in the time of onset. It is generally thought that the earlier the disease appears, the more rapidly it will run its course. The average survival time is about 17 years, but some patients have endured this progressive disorder for as long as 30 years (Myers and Martin, 1982). Although various drugs, diets, exercise programs, and other therapies have been tried, even in the nineteenth century it was felt that nothing could be done to aid these patients (Elliotson, 1832; King, 1885).

The changes in the brain that underlie this disease were not well understood in the nineteenth century. Cortical changes were suspected by some investigators (e.g., Fisher, 1890; Gowers, 1893). There was also some evidence from autopsy material for the possible involvement of the basal ganglia (e.g., Anton, 1896; Jelgersma, 1909).

In 1911, Alois Alzheimer (1864–1915) described the extensive loss of neurons in the caudate and putamen nuclei that accompanied Huntington's disease. The changes in the cells that he saw with the newer Nissl stain had not been fully appreciated when the older Weigert method of staining myelin was used, and were missed in unstained material (see Lewy, 1942). Alzheimer also recognized diffuse cortical cell loss in Huntington's disease, but emphasized it was the striatal damage that caused the motor disturbances.

In 1914, Pierre Marie and Jean Lhermitte (1877–1959) wrote that the lesions affected the frontoparietal cortex more than other cortical areas and that the caudate and putamen also underwent degeneration. These findings were upheld by later investigators.

> Gross measurements of the corpus striatum in Huntington's chorea showed this organ to be, on an average, less than half as large as in the control cases. All parts of the striatum appeared to take part in this reduction in size. . . . Only a small part of the loss of weight in the forebrain in Huntington's chorea can be accounted for by the smallness of the corpus striatum, for in control brains of average size, the combined weights of the two corpora striata was found to be only about 25 Gm.; . . . the total loss from this source would hardly exceed 15 Gm., while the average loss of substance in the cerebrum is more than ten times this amount. (Dunlap, 1927, 894–896)

The theory that evolved from this work was that the cortical losses were those responsible for the cognitive impairments of the disorder (i.e., the dementia), whereas the striatal losses primarily accounted for the choreic movements (Hunt, 1917).[18,19]

ATHETOSIS

William Hammond, who served as Surgeon General and president of the American Neurological Association and is shown in Figure 16.10, was the first to distinguish athetosis from other choreic disorders.[20] His important contribution appeared in 1871 in his celebrated book *A Treatise on the Diseases of the Nervous System.*

[18]This idea was further supported by research on animals. In 1924, for example, Hugo Mella (1888–1969) published his studies on choreic monkeys with lesions of the basal ganglia.
[19]The association between Huntington's chorea and GABA, dopamine, and other neurotransmitters was not adequately recognized in the first half of the twentieth century.
[20]Hammond served briefly as the U.S. Surgeon General under Abraham Lincoln (1809–1865). Partly as a result of his arrogance and pompous personality, Brigadier General Hammond, who took the post in 1862, was court-martialed in 1863. He was charged with "conduct to the prejudice of military discipline" and "unbecoming of an officer and a gentleman" and was forced out of his post. Without any money, and in debt, Hammond moved to New York where he opened a private practice. Although his court-martial was eventually reversed by an act of Congress, Hammond did not return to the military (Klawans, 1982; Spillane, 1981).

Figure 16.10. William Hammond (1828–1900), Surgeon General of the United States, who contributed to the study of athetosis, myriachit, and a number of other movement disorders.

The disorder Hammond called athetosis was characterized by grotesque involuntary movements of the fingers and toes and by abnormal postures. He noted that the sufferers of this disorder exhibited an inability to retain the fingers and toes in any position in which they were placed. In fact, he derived the word "athetosis" from the Greek term meaning "without fixed position."

Hammond described two cases of athetosis, only one of which was actually seen by him. The second was described to Hammond by an Ohio physician. This patient's records included photographs that showed the abnormal postures of the hands. After discussing these two cases, Hammond predicted that the responsible lesion would be found in the striatum. He did not, however, have post-mortem data at the time.

It was well known that flaccid hemiplegia was often followed by spasticity and a condition in which the fingers would open and close as they went from a state of overextension to one of overflexion. These odd movements were mentioned by Jean-Baptiste Cazauvielh (fl. 1825) in 1827 in his description of cortical hemiatrophy. By Hammond's time, they were called spasmoparalysis or spastic contracture. It is also possible that these were the movements alluded to by Parkinson (1817) when he quoted Linnaeus on a disorder involving continuous distortion of the limbs without mental deterioration ("hieranosis," or *Morbus Sacer*).

The importance of Hammond's contribution was not recognized immediately. Silas Weir Mitchell (1874), his close friend and countryman, still called these posthemiplegic movements hemichorea. And William Gowers (1876), who performed the first autopsy of an athetosis case—and erroneously thought the causal lesion was in the thalamus—called the disorder "mobile spasm." In fact, Hammond's separation of athetosis from other disorders was even criticized in some circles (e.g., in 1879, Charcot said athetosis was only a variety of posthemiplegic chorea).

Nevertheless, with the passage of time, Hammond received more recognition for his descriptions of athetosis and for his terminology. Regarding the latter, because jerky movements may accompany the slower changes, the term "choreo-athetosis" was preferred over "athetosis" by some writers who otherwise accepted much of what Hammond had to say.

The abnormal movements that characterize athetosis were studied by Otfrid Foerster in 1921 and by S. A. Kinnier Wilson in 1925. The bizarre, cramplike, spasmodic movements were complex and irregular in sequence. Foerster compared them with climbing movements, and Wilson looked upon them as purposeless caricatures.[21] The involuntary movements were found to increase when the patient's attention was drawn to them and in periods of increased emotion. When this occurred, it became even more difficult for the patients to execute smooth, voluntary movements. The disorder affected speech as well, making it jerky, shrill, explosive, and tremulous. In more extreme cases, patients can have great difficulty even caring for themselves (Richter, 1923). As with many other movement disorders, the movements decreased significantly during sleep (Berger, 1903).

It was found that athetosis could result from many different kinds of trauma. In children, it is often the result of birth injuries and neonatal asphyxia and is typically accompanied by mental impairment and convulsive seizures. Measles, whooping cough, and diphtheria can also cause athetosis in children. In adults, cerebrovascular accidents are the most frequent cause, but athetosis can also be the result of carbon monoxide, barbiturate, or manganese poisoning (Spiegel and Baird, 1968).

Lesion studies at first suggested that the neural mechanism for the production of choreo-athetosis was in the precentral area (Wilson, 1925). It was thought that small lesions here might be stimulating remaining motor efferents (Brissaud, 1895). In part for this reason, Victor Horsley attempted to treat this disorder by removing pieces of the motor cortex.

> The pathology of athetosis is as yet obscure. I have, however, always regarded it as a form of cortical discharge. Quite recently my colleague, Dr. Beevor, asked me to operate in a hopeless case, and one in which he, having detected a successive invasion of segments by the movement commencing in the thumb, was led to conclude that the affection was one of cortical origin. I therefore removed the focus for the representation of the movements of the thumb, with the effect that they were arrested for about a fortnight. They however returned as the cortex around resumed its functional activity, and it is therefore evident that the whole representation of a part must be removed, a course which the paralyzed state of the limb fully warrants. This operation, in fact, offers the only means of relieving the conditions of spasm. (1890, 1291)

Horsley's (1890, 1909) surgical findings led others to attempt the removal of limited parts of the motor and/or premotor cortex as a treatment for athetosis (Bucy, 1942; Bucy and Buchanan, 1932; Bucy and Case, 1939). In one case,

which involved a young girl, Area 6a of the premotor cortex was exposed and electrically stimulated. This elicited athetoid-like movements. When the arm region was removed, her athetosis ceased and did not return during the ensuing nine-year observation period (Bucy and Buchanan, 1932). In other studies, the findings were not as encouraging. This prompted more radical operations, some of which involved removing the arm representations in both Area 6 and Area 4 (Bucy, 1942).

The opposing position, the idea that the movements might be generated in the corpus striatum, also had its share of supporters at this time (e.g., Foerster, 1921; Vogt, 1912). As noted, William Hammond (1871) had predicted that the striatum was involved. Hammond's original case finally came to autopsy in 1890. As Hammond had predicted, scar tissue was found in the corpus striatum. Although the lesion destroyed the inner parts of the globus pallidus, it also damaged the internal capsule and parts of the thalamus.

After this finding became known, other investigators also began to associate athetosis with damage to the basal ganglia (see Lewy, 1942). In 1896, Gabriel Anton (1858–1933) found lesions in the putamen of a 9-year-old boy with athetosis and postulated that this lesion upset the balance in his motor systems. Anton thought the "basal ganglia tracts" specifically served to inhibit such movements when they were healthy.

The most important confirmation of a striatal lesion in this disorder took place in 1911. In that year, Cécile Vogt published an important paper on *athétose double,* which was based on a patient seen by Hermann Oppenheim (see Freund and Vogt, 1911; Oppenheim and Vogt, 1911; Vogt, 1911). This 23-year-old woman showed a particular type of bilateral atrophy of the striatum, a feature called *status marmoratus* or *état marbré*. It was characterized by a marbled or mottled appearance of the striatum, which was also reduced in size. This was especially noticeable in the putamen. *Status marmoratus* was also seen in a second patient and appeared to be due to areas of hypermyelination and areas of demyelination, as well as to a loss of cells.[22]

Experimental confirmation of the importance of the striatum in athetosis occurred when Hugo Mella (1924) exposed monkeys to manganese. Mella found this produced athetoid movements in the animals. Upon autopsy, their brains showed severe striatal damage with considerably milder cortical changes. Autopsies on athetosis patients further showed the extreme sensitivity of the basal ganglia, and the globus pallidus in particular, to carbon monoxide poisoning (Poelchen, 1882).

Although many questions about athetosis remained unanswered, by the 1930s most investigators agreed with Hammond (1871) that athetoid movements were at least partly due to damage to the extrapyramidal system.

TOURETTE'S SYNDROME

Tourette's syndrome is a movement disorder characterized by tics, twitches, grimaces, and unusual noises. It is also associated with coprolalia (use of obscenities), echolalia

---

[21]In 1931, Ernst Herz (b. 1900) studied these strange athetotic movements by using motion picture analysis.

[22]See Spiegel and Baird (1968) for a review of the early literature on *status marmoratus*.

(repeating words and phrases), and echopraxia (imitating actions).

This distinctive disorder was named after Georges Gilles de la Tourette (1857–1904), who, like James Parkinson and George Huntington, came from a family of doctors (Guilly, 1982). In 1884, Gilles de la Tourette was a house physician at the Salpêtrière in Charcot's department. He became Charcot's registrar in 1886 and was described as the master's most enthusiastic supporter and disciple.[23]

In 1884, two years before he defended his thesis, Gilles de la Tourette published a paper that examined the compulsive movement disorders known as *faisait un saut du Maine, latah de Malaisie,* and *myriachit de Sibérie.* These three movement disorders served as the door that led Gilles de la Tourette into the study of Tourette's syndrome.

The "jumping Frenchmen" of Maine were described in 1880 by George M. Beard (1839–1883) of New York, who is shown in Figure 16.11.[24] He wrote that the "jumpers" were people who responded immediately to commands, even when told to strike someone, and that "it was dangerous to startle them in any way when they had an axe or knife in their hand" (p. 188). The jumpers (over 50 were observed) were both echolalic and echopraxic, but generally modest, quiet, and in good health. Coprolalia was not mentioned, but barking accompanied the jumping response to sudden noises. Beard maintained this disorder was hereditary, that it showed itself in young children, and that it rarely affected women. He believed it was analogous to "mental or psychical hysteria . . . a temporary trance induced by reflex irritation, and the emotion of fear," and could not be cured ("once a jumper, always a jumper").

Latah, the second disorder that attracted the attention of Gilles de la Tourette, had been described by J. A. O'Brien in 1883. Malaysians with latah showed an impulse to strike out at objects and to shout involuntary obscenities. Like the "jumpers," they were also echolalic and echopraxic.

Miryachit (*myriachit* in French), the third disorder, had been the subject of a report by William Hammond. This condition was seen not by him but by officers of an American ship sent to Siberia. The steward on the ship had shown some very aberrant behaviors. Hammond (1884, p. 191) wrote:

It seemed that he was afflicted with a peculiar mental or nervous disease, which forced him to imitate everything suddenly presented to his senses. Thus, when the captain slapped the paddle-box suddenly in the presence of the steward, the latter instantly gave it a similar thump; or if any noise were made suddenly, he seemed compelled against his will to imitate it instantly, and with remarkable

Figure 16.11. George Beard (1839–1883), who described the "jumping Frenchmen of Maine" in 1880, but who became better known for his concept of neurasthenia.

accuracy. To annoy him, some of the passengers imitated pig grunting, or called out absurd names; others clapped their hands and shouted, jumped, or threw their hats on the deck suddenly, and the poor steward, suddenly startled, would echo them all precisely and sometimes several consecutively. . . . In speaking of the steward's disorder, the captain of the general staff stated that it was not uncommon in Siberia; that he had seen a number of cases of it, and that it was commonest among the Yakutsk, where the winter cold is extreme. Both sexes were subject to it, but men much less than women. It was known to the Russians by the name of *Miryachit.*

Hammond thought miryachit was identical to the jumping disorder seen by Beard.

Gilles de la Tourette (1884) concluded that jumping, latah, and miryachit were identical disorders. He ended the body of his paper by mentioning he had seen a case of a seemingly related movement disorder when in the service of Professor Charcot. This patient showed hyperexcitability, echolalia, and tics. He then added a footnote stating he just had the good fortune to encounter a second such subject. He told his readers that these two cases, and three others belonging to Professor Charcot, would form the basis of a follow-up report.

Charcot, for his part, discussed this paper in one of his Tuesday lectures. Although Charcot's lecture was not published in French, it was summarized in Italian soon afterward (Melotti, 1885). More important, Gilles de la Tourette's second paper, as promised, appeared in 1885. He maintained that the unusual movement disorder which he was now attempting to document was previously placed under the general category of chorea, that it had little recognition in the past, and that earlier descriptions were lacking in essential details.

Tourette's syndrome can best be appreciated by looking

---

[23]Gilles de la Tourette was an individual of great talent, subject to overexcitement and extraordinary activity. He wrote books, research articles, communications, and historical papers in psychiatry and neurology, but was well versed and interested in an even wider range of subjects. Late in his life, a young paranoid woman who was confined to a mental hospital shot him three times while he was in a consulting room. One bullet hit him in the head, and although the bullet was easily removed, it was said he never fully recovered from his injury (Stevens, 1971). Tourette suffered from episodes of depression and mania in his last years and died insane, most likely from syphilis (Guilly, 1982).

[24]The jumping Frenchmen of Maine were also seen in New Hampshire and Canada. George Beard (1839–1883), who described these individuals, was the psychiatrist who also introduced in 1869 the controversial concept of "neurasthenia," or functional weakness of the nervous system (see Rosenberg, 1962).

directly at his case of Madame de Dampierre. This woman
was described as Observation 10 by Jean Itard (1775–1838;
see Figure 16.12) in 1825 and later in life she was seen by
Charcot himself. She was Case 1 in Gilles de la Tourette's
1885 paper.

> Soon it became clear that the movements were indeed invol-
> untary and convulsive in nature. The movements involved
> the shoulders, the neck, and the face, and resulted in contor-
> tions and extraordinary grimaces. . . . Her examination
> showed spasmodic contractions that were continual or were
> separated only by momentary intervals of time. . . . In the
> midst of an interesting conversation, all of a sudden without
> being able to prevent it, she interrupts what she is saying or
> what she is listening to with horrible screams and with
> words that are even more extraordinary than her screams.
> All of this contrasts deplorably with her distinguished man-
> ners and background. These words are for the most part
> offensive curse words and obscene sayings. These are no less
> embarrassing for her than for those who have to listen, the
> expressions being so crude that an unfavorable opinion of the
> woman is almost inevitable. (Translated by Goetz and
> Klawans, 1982, 2–3)

Gilles de la Tourette pointed out that although the disease
was more likely to affect men than women, this case was
rather typical of the cases he had either seen or learned
about from other sources. For example, the range for onset
was 6 to 16 years in his sample, with five individuals (including Madame de Dampierre) showing initial symptoms at
ages 7 to 8. In addition, her tics and verbal utterances were
very much in line with those shown by the other patients.

Gilles de la Tourette wrote that twitches and uncoordinated movements were the first indications of the disease.
These generally start in the face or upper extremities, with
one arm usually being affected before the other. The arms
may shake, the fingers extend and flex, and the shoulders
flinch. The face may grimace, the eyes squint, the tongue
project from the mouth, and the teeth grind. The head as a
whole may swing from side to side or go forward and backward. After a while, the lower extremities become involved.
These contractions can affect an entire limb and can lead to
jumping or propelling movements or stamping of the feet.

He was amazed by how abruptly the movements could
appear. He also noted that the limited movements often
recurred within a few minutes, whereas the more global
movements were separated by longer intervals (from 15
minutes to hours), with considerable variability across
patients. He also commented that whereas physical or mental discomfort could aggravate the movements, they stopped
during sleep, which he thought was particularly sound.

Gilles de la Tourette stressed that the disease can be
limited to these motor symptoms for years, and possibly
indefinitely. Yet he maintained that other signs were even
more characteristic of the disease than just isolated tics
and movements. Here he spoke of the noises the patients
made during the height of some of their movements. These
were often sounds like "ouh," "ah," and "hem."

Another unusual aspect of the disease was the tendency
for the patients to be echolalic and echopraxic. The patient
might be listening to a conversation only to find himself
repeating some of the words, aware of the situation, but
being unable to control this effectively. This included repeating words out loud that were just seen in print.

Although Gilles de la Tourette considered the tics and gri-

Figure 16.12. Jean Itard (1775–1838), the Frenchman who described a
young woman with Tourette's syndrome in 1825. The same woman was
seen by Jean-Martin Charcot many years later and was described by
Gilles de la Tourette in his 1885 paper (Case 1).

maces the first category of symptoms, and placed the grunting sounds and echolalia in the second category, he emphasized that there was also a third category of symptoms,
which was a still better indicator of the disorder. It was in
this category that he placed the coprolalia that accompanied
the movements in many of the patients. For instance,
Madame de Dampierre, a woman of good background, occasionally shouted obscene words and phrases like *merde*
("shit") and *foutu cochon* ("dirty pig") while making involuntary jerking movements. These verbal obscenities were
never accompanied by obscene gestures.

The spontaneous cursing that accompanied the involuntary muscular movements is usually the last symptom to
appear. While it was hailed as the unique feature or unmistakable signature of Tourette's disease, Gilles de la Tourette
saw it in only five of the nine patients he described in 1885.
Thus, he wrote that this sign may not be universal and left
the door open for a change of conclusion based on further
study.[25]

Gilles de la Tourette's paper from 1885 considered not
just the symptoms characteristic of the disorder but those
functions not affected by it. Among the important faculties
that were not directly impaired by the disease were intelligence and mental status. He emphasized that these patients
were often highly intelligent, that they were conscious of
their state, and that their acts were not those of the insane.
He also found no evidence for epilepsy in his population. In
addition, sensory perception and overall physical health
were normal, as was life span.

Gilles de la Tourette found evidence of neurological

---

[25]The original percentage of patients showing coprolalia seems to have
been upheld. Shapiro and Shapiro (1982) stated that involuntary cursing
was seen in only about 60 percent of their patients and later disappeared
in about 33 percent of the cases.

problems in the family members of some of his cases and surmised that the disorder was hereditary. He maintained it could be seen in a diversity of countries and climates and that social or professional background was of no importance in its etiology.

Near the end of his description, he went on to say that there may be spontaneous periods of remission, but that the symptoms never disappear completely. Whereas the patients may consider themselves cured during such periods, the features of the disorder always reappear. Madame de Dampierre, for example, suffered from this disorder from childhood until the age of 80, but supposedly experienced an 18- to 20-month period of remission when she went to Switzerland as a young woman to be treated by an expert in nervous diseases. Her treatment, he noted, consisted largely of soothing milk baths!

Gilles de la Tourette (1885) was pessimistic about the effectiveness of this and other treatments.

> The treatment for this singular condition is still to be discovered. By reading our observations, one sees clearly that curative attempts have thus far been unsuccessful; all central nervous system sedatives have failed. The only treatment that has seemed to help the symptoms and has been associated with periods of remission has been isolation combined with various treatments: iron preparations, hydrotherapy, etc. . . . Perhaps these treatments can slow the natural progression of the disease, especially in patients who are treated early. In any case, we do not feel that we have a definite treatment for this condition. (Translated by Goetz and Klawans, 1982, 14)

Charcot, who corrected Gilles de la Tourette's overemphasis of echolalia and echopraxia in the disorder, honored his devoted student by naming the disorder after him soon after his paper appeared (see Melotti, 1885).

Parkinson and Huntington were not the first to describe the disorders named after them. The same can be said for Tourette's syndrome, although in this case, if anything, Gilles de la Tourette may have been overly liberal in seeing the disorder in some of the papers, books, and reports of his predecessors.

Perhaps the first report of an individual with Tourette's syndrome can be seen in *Malleus maleficarum,* or the *Witch's Hammer,* a book that appeared in 1489 (Shapiro and Shapiro, 1982). This work by Jakob Sprenger (1436–1495) and Hans Kraemer (d. 1508) described a priest with both motor and phonetic tics. It was published at the time of the Inquisition and was concerned with identifying witches and then either successfully exorcising them to free their bodies of the devil or burning them at the stake.

Another possible case is that of a French nobleman in the court of Louis XIV of France (1638–1715). The Prince de Condé reportedly stuffed objects into his mouth to suppress his involuntary barking sounds (Stevens, 1971).

It was already stated that Itard (1825) had examined Madame de Dampierre (Gilles de la Tourette's Case 1 in 1885) and had published a case report on her in 1825. Gilles de la Tourette (1885) mentioned Itard's name and added that this patient was also described in two other reports in the 1850s (Roth, 1850; Sandras, 1851). These three reports concerned themselves with the distinctions between this case and classic chorea. None of the authors tried to link the case of Madame de Dampierre with other cases.

Figure 16.13. Armand Trousseau (1801–1867) as a young man. Trousseau provided an excellent early description of Tourette's syndrome and also described other disorders of movement, including Parkinson's disease. A picture of Trousseau later in life appears in Chapter 26.

In 1861, Armand Trousseau (shown in Figure 16.13) provided a particularly good description of the syndrome in Volume 2 of *Clinique Médicale de l'Hôtel Dieu de Paris.* Gilles de la Tourette (1885) quoted Trousseau as saying:

> These tics may be in some cases accompanied by a cry, which is a more or less abrupt and very characteristic sound. In this respect, I remember an encounter with one of my old school colleagues. Although he was behind me in the street and I had not seen him for 20 years, I knew it was he immediately by the barking sound that I had so often heard when we had been together studying. This shout, this yelp, this bursting noise—true forms of laryngeal or diaphragmatic chorea—can constitute the entire tic disorder. Besides the strange sounds, there is also a singular tendency to repeat the same word or expression, and the person may even loudly utter words that he would prefer to keep silent. These tics are often hereditary. (Translated by Goetz and Klawans, 1982, 2)

Sixteen papers on Tourette's syndrome, written by other authors, appeared within 15 years of Gilles de la Tourette's (1885) most important publication.[26]

"Neuropathic heredity" remained widely cited into the 1900s as the etiology of Tourette's syndrome. There was also a trend, which increased in the twentieth century, to ascribe the disease to psychological factors and to consider it hysterical, as an obsessive-compulsive disorder or as a narcissistic or aggressive problem. Those afflicted with the disorder were often called higher degenerates in this literature. Only

[26]Tourette's disease was thought to be rare before the 1960s, and a survey found that only 50 cases of this disorder appeared in the literature between 1884 and 1965 (Guilly, 1982). The disorder received considerably more attention after this time. The newer literature shows quite clearly that earlier estimates of the incidence of the disease were on the low side (Shapiro and Shapiro, 1982).

slowly did this disorder begin to be treated as an organic brain dysfunction, one associated with a neurotransmitter imbalance.[27]

THE STRANGE CASE OF SAMUEL JOHNSON

It is difficult to leave the topic of Tourette's syndrome without mentioning the case of Samuel Johnson. This eighteenth-century figure, best known for his dictionary, letters, plays, biographies, and brilliant conversation, was considered one of the greatest men, if not the greatest man, of his era (Murray, 1982). Yet Johnson suffered from odd gesticulations, endless posturings, and unusual movements of his arms and legs. He also made strange noises, described as ruminating sounds, half whistles, and the blowing of his breath "like a whale." These personal characteristics led to Johnson's being ridiculed by those who did not know him well and prevented him from securing schoolmaster and assistant headmaster jobs when he was in his twenties.

Johnson's problems seem to have emerged in childhood. Even at the age of 7, his appearance was described as being "little better than that of an idiot" (see McHenry, 1967). He was well aware of his disability throughout his life and had a great aversion to gesticulating in company. Thus, he tried, with fair success, to suppress his movements when in social settings.

As described by Fanny Burney (Madame d'Arblay, 1752–1840), the female novelist who knew Johnson very well:

> His mouth is almost constantly opening and shutting as if he were chewing. He has a strange method of frequently twirling his fingers and twisting his hands. His body is in continual agitation seesawing up and down; his feet are never a moment quiet; and in short his whole person is in perpetual motion. (Cited in Gibbs, 1960, 1243)

Johnson wrote about many things, including his other illnesses, but he rarely mentioned his movement disorder, which shocked people who met the eminent man for the first time (Murray, 1979, 1982). One exception occurred when his hand, which was moving back and forth at a dinner table, knocked the shoe off a woman who jokingly put her foot in front of his moving hand. He told the people at the table that he should not be rebuked for this act because "the motion was involuntary, and the action not intentionally rude" (Hawkins, 1961).

Johnson had a difficult birth and was not expected to live. This alone might suggest perinatal anoxia and athetoid cerebral palsy. Nevertheless, his movements were not the typical athetoid movements, which cannot be controlled and can interfere with all voluntary motion. As noted, Johnson could, and would, consciously suppress his tics for at least short periods of time while in social settings (McHenry, 1967).

Along with many of his contemporaries, James Boswell (1740–1795), Johnson's friend and noted biographer, thought the great man suffered from Saint Vitus' dance. But the fact that Sydenham's chorea does not last a lifetime makes this diagnosis unlikely. In addition, it is also unlikely that Johnson had epilepsy, another possible cause of aberrant motor behaviors.

It is interesting to note that some of Johnson's contemporaries believed his movement disorder was not due to organic disease. Sir Joshua Reynolds (1723–1792), the renowned artist who painted Johnson's portrait, attributed the aberrant actions to bad habits and obsessive thoughts that he believed Johnson was trying to suppress. Sometimes cited in this context is a story in which a young girl asked him why he made "such strange gestures." "From bad habit," he answered, "Do you, my dear, take care to guard against bad habits?" (Boswell, 1791). Nevertheless, this quotation, given as a lighthearted answer to a child's question, is directly at variance with the aforementioned quotation in which Johnson asked for understanding from his friends because his abnormal movements were involuntary.

Because Johnson showed involuntary movements, odd complex motor acts, strange vocalizations, and unusual compulsive behaviors, it has been suggested that he suffered from Tourette's syndrome (Murray, 1979, 1982). The fact that each of the other diagnoses seems to have a flaw has also been used to argue the case that Johnson may have suffered from this disorder.

The evidence is obviously incomplete, and the case is not yet proved. Yet the curious possibility remains that one possible sufferer of Tourette's syndrome may not only have achieved fame but may have risen above the weight of this disorder to become one of the greatest intellects the world has ever known.

REFERENCES

Alzheimer, A. 1911. Huntingtonische chorea und die choreatischen Bewegungen uberthaupt. *Zeitschrift für die gesamte Neurologie und Psychiatrie, 3,* 891–892.

Anton, G. 1896. Ueber die Betheiligung der grossen basalen Gehirnganglien bei Bewegungsstörungen und insbesondere bei der Chorea. *Jahrbuch für Psychiatrie, 14,* 141–181.

Aron, A. M., Freeman, J. M., and Carter, S. 1965. The natural history of Sydenham's chorea. *American Journal of Medicine, 38,* 93–95.

Barbeau, A. 1958. The understanding of involuntary movements: An historical approach. *Journal of Nervous and Mental Disease, 127,* 469–489.

Beard, G. M. 1880. Experiments with the "jumpers" or "jumping Frenchmen" of Maine. *Journal of Nervous and Mental Disease, 7,* 487–490.

Berger, A. 1903. Zur Kenntnis der Athetose. *Jahrbuch für Psychiatrie, 23,* 214–233.

Bielschowsky, M. 1922. Weitere Bemerkungen zur normalen und pathologischen Histologie des striären Systems. *Journal für Psychologie und Neurologie, 27,* 233–288.

Birkmeyer, W., and Hornykiewicz, O. 1961. Der L-3, 4-dioxyphenylalanine (= L-DOPA)—Effekt bei der Parkinson-Akinese. *Wiener klinische Wochenschrift, 73,* 787–788.

Birkmeyer, W., and Hornykiewicz, O. 1962. Der L-Dioxyphenylalanine (= L-DOPA)—Effekt bei Parkinson-syndrom des Menschen zur pathologenese und behandlung der Parkinson-Akinese. *Archiv für Psychiatrie, 203,* 560–574.

Blocq, P., and Marinesco, G. 1893. Sur un cas de tremblement Parkinsonien hémiplégique. *Revue Neurologique, 2,* 265.

Boswell, J. 1791. *Life of Johnson.* London: H. Baldwin for C. Dilly.

Bouteille, E. M. 1810. *Traité de la Chorée, ou Danse de St. Guy.* Paris: Vincard.

Bove, F. J. 1970. *The Story of Ergot.* New York: S. Karger.

Bright, R. 1831. *Reports of Medical Cases Selected with a View of Illustrating the Symptoms and Cure of Diseases by Reference to Morbid Anatomy.* London: Longman.

Brissaud, E. 1895. *Leçons sur les Maladies Nerveuses.* Paris: Masson.

Bruyn, G. W. 1968. Huntington's chorea: Historical, clinical and laboratory synopsis. Ch. 13 in P. J. Vinken and G. W. Bruyn (Eds.), *Handbook of Clinical Neurology* (Vol. 6). Amsterdam: North Holland Publishing Company. Pp. 298–378.

---

[27]During the 1960s, Tourette's syndrome was associated with high levels of dopamine. Subsequently, it has been treated by drugs that decrease the production or uptake of dopamine in the brain.

Bruyn, G. W., and Went, L. N. 1986. Huntington's chorea. In P. J. Vinken, G. W. Bruyn, and H. L. Klawans (Eds.), *Handbook of Clinical Neurology: Vol. 5. Extrapyramidal Disorders*. Amsterdam: Elsevier. Pp. 267–313.

Bucy, P. C. 1942. Cortical extirpation in the treatment of involuntary movements. *Research Publications—Association for Research in Nervous and Mental Disease, 21*, 551–595.

Bucy, P. C., and Buchanan, D. N. 1932. Athetosis. *Brain, 55*, 479–492.

Bucy, P. C., and Case, J. T. 1939. Tremor: Physiological mechanism and abolition by surgical means. *Archives of Neurology and Psychiatry, 41*, 721–746.

Bucy, P. C., and Case, J. T. 1942. Cortical extirpation in the treatment of involuntary movements. *Diseases of the Basal Ganglia. Proceedings of the Association for Research in Nervous and Mental Disease, 21*, 551–595.

Buzzard, E. F. 1918. Toxic encephalitis. *Lancet, 1*, 616.

Buzzard, T. 1882. *Clinical Lectures on Diseases of the Nervous System*. London: Churchill.

Caporael, L. R. 1976. Ergotism: The satan loosed in Salem? *Science, 192*, 21–26.

Caro, A., and Haines, S. 1975. The history of Huntington's chorea. *Medical History, 7*, 91–95.

Caro, A., and Obst, D. 1977. *A genetic problem in East Anglia Huntington's chorea*. Thesis, University of East Anglia.

Cazauvielh, J.-B. 1827. Recherches sur l'agénésie cérébrale et la paralysie congénitale. *Archives Générales de Medécine, 14*, 347–366.

Charcot, J.-M. 1877. *Lectures on the Diseases of the Nervous System*. (G. Sigerson, Trans.). London: New Sydenham Society. Pp. 129–156.

Charcot, J.-M. 1880. *Leçons sur les Maladies du Système Nerveux Faites à la Salpêtrière* (4th ed.). Paris: Dalahaye and Lecrosnier.

Charcot, J.-M. 1887–1888. *Leçons du Mardi à la Salpêtrière*. Paris: Bureau du Progrès Médical.

Charcot, J.-M., and Vulpian, E. F. A. 1861–1862. De la paralyse agitante (A propos d'un cas tiré de la Clinique du Professeur Oppolzer). *Gazette Hebdomadaire, 8*, 765–767, 816–820; *9*, 54–59.

Cooke, J. 1824. *A Treatise on Nervous Disorders* (Vol. 2). London: Longman, Hurst, Rees, Orme and Brown.

Cooper, I. S. 1961. *Parkinsonism: Its Medical and Surgical Therapy*. Springfield, IL: Charles C Thomas.

Critchley, M. 1934. Huntington's chorea and East Anglia. *Journal of State Medicine, 42*, 575–587.

Critchley, M. 1964. Huntington's chorea: Historical and geographical considerations. In M. Critchley (Ed.), *The Black Hole and Other Essays*. London: Pitman Medical Publishing Company. Pp. 210–219.

Critchley, M. 1984. The history of Huntington's chorea. *Psychological Medicine, 14*, 725–727.

De Jong, R. N. 1937. George Huntington and his relationship to the earlier descriptions of chronic hereditary chorea. *Annals of Medical History, 9, New Ser.*, 201–210.

De Jong, R. N. 1970. George Huntington (1850–1916). In W. Haymaker and F. Schiller (Eds.), *The Founders of Neurology*. Springfield, IL: Charles C Thomas. Pp. 453–456.

Delcourt, A., and Sand, H. 1908. Un cas de chorée de Sydenham terminé par la mort. *Archives de Médecine des Enfants, 11*, 826–836.

Diefendorf, A. R. 1908. Mental symptoms of Huntington's chorea. *Neurographs, 1*, 128–136.

Dotz, W. 1980. St. Anthony's fire. *American Journal of Dermatology, 2*, 249–253.

Dunglison, R. 1842. *The Practice of Medicine, or a Treatise on Special Pathology and Therapeutics*. Philadelphia, PA: Lea and Blanchard.

Dunglison, R. 1848. *The Practice of Medicine, or a Treatise on Special Pathology and Therapeutics* (3rd ed). Philadelphia, PA: Lea and Blanchard.

Dunlap, C. B. 1927. Pathologic changes in Huntington's chorea. *Archives of Neurology and Psychiatry, 18*, 867–943.

Ehringer, H., and Hornykiewicz, O. 1960. Verteilung von Noradrenalin und Dopamin (3-hydroxytyramin) im Gehirn des Menschen und ihr Verhalten bis Erkrankungen des extrapyramidalen Systems. *Klinische Wochenschrift, 38*, 1236–1240.

Elliotson, J. 1830. Clinical lecture given at St. Thomas' Hospital, October 11, 1830. Reprinted in *Lancet, 2*, 119–123.

Elliotson, J. 1832. Hypochondriasis—Haemoptysis: Clinical lecture delivered at St. Thomas's Hospital. *Lancet, 1*, 161–167.

Elliotson, J. 1839. *Principles and Practice of Medicine*. London: J. Butler.

Fehling, C. 1966. Treatment of Parkinson's disease with L-DOPA: Double blind study. *Acta Neurologica Scandinavica, 42*, 367–372.

Fine, E. J., Soria, E. D., and Paroski, M. W. 1990. Tremor studies in 1886–1889. *Archives of Neurology, 47*, 337–340.

Fisher, E. D. 1890. Remarks on the pathology of chorea. *Journal of Nervous and Mental Disease, 15*, 221–224.

Foerster, O. 1921. Zur Analyse und Pathophysiologie der Striären Bewegungsstörungen. *Zeitschrift für die gesamte Neurologie und Psychiatrie, 73*, 1–164.

Foix, C., and Nicolesco, J. 1925. *Les Noyaux Gris Centraux et la Région Mésencéphalo-Sous-Optique*. Paris: Masson.

Freeman, W. 1925. The pathology of paralysis agitans. *Annals of Clinical Medicine, 4*, 106–116.

Freund, C. S., and Vogt, C. 1911. Ein neuer Fall von Etat marbré des Corpus striatum. *Journal für Psychologie und Neurologie, 18*, 489–500.

Gabbai, Lisbonne, and Pourquier. 1951. Ergot poisoning at Pont St. Esprit. *British Medical Journal, 27*, 650–651.

Gibbs, L. 1960. *The Diary of Fanny Burney*. London: Dent.

Gilles de la Tourette, G. 1884. Jumping, latah, myriachit. *Archives de Neurologie, 8*, 68–84.

Gilles de la Tourette, G. 1885. Etude sur une affection nerveuse caractérisée par de l'incoordination motrice accompagnée d'écholalie et de copralalie (jumping latah, myriachit). *Archives de Neurologie, 9*, 19–42, 158–200.

Goetz, C. G. 1987. *Charcot the Clinician: The Tuesday Lectures*. New York: Raven Press.

Goetz, C. G., and Klawans, H. L. 1982. Gilles de la Tourette on Tourette syndrome. In A. J. Friedhoff and T. N. Chase (Eds.), *Gilles de la Tourette Syndrome*. New York: Raven Press. Pp. 1–16.

Goodman, L. S., and Gilman, A. 1970. *The Pharmacological Basis of Therapeutics* (4th ed.). New York: Macmillan.

Gowers, W. R. 1876. On athetosis and post-hemiplegic disorders of movement. *Medical and Chirurgical Society Transactions, 59*, 271–326.

Gowers, W. R. 1888. *A Manual of Diseases of the Nervous System*. Philadelphia, PA: Blakiston.

Gowers, W. R. 1893. *A Manual of Diseases of the Nervous System* (2nd ed., Vol. 2). Philadelphia, PA: Blakiston.

Greenfield, J. G. 1955. The pathology of Parkinson's disease. In M. Critchley (Ed.), *James Parkinson (1755–1824)*. London: Macmillan. Pp. 219–243.

Greenfield, J. G. 1956. Historical landmarks in the pathology of involuntary movements. *Journal of Neuropathology and Experimental Neurology, 15*, 5–11.

Greenfield, J. G. 1963. Rheumatic chorea. In W. Blackwood, W. H. McMenemey, A. Meyer, R. M. Norman, and D. S. Russell (Eds.), *Greenfield's Neuropathology*. London: Edward Arnold. Pp. 117–118.

Greenfield, J. G., and Bosanquet, F. D. 1953. The brain-stem lesions in Parkinsonism. *Journal of Neurology, Neurosurgery and Psychiatry, 16*, 213–226.

Greenfield, J. G., and Wolfsohn, J. M. 1922. The pathology of Sydenham's chorea. *Lancet, 2*, 603–606.

Guilly, P. 1982. Gilles de la Tourette. In F. C. Rose and W. F. Bynum (Eds.), *Historical Aspects of the Neurosciences*. New York: Raven Press. Pp. 397–413.

Hall, M. 1841. *On Diseases and Derangements of the Nervous System*. London: J. B. Ballière.

Haller, J. S. 1981. Smut's dark poison: Ergot in history and medicine. *Transactions and Studies of the College of Physicians of Philadelphia, 3*, 62–79.

Hallervorden, J. 1957. Chorea minor, Chorea gravidarum, senile Chorea und andere Choreaformen. In O. Lubarsch, F. Henke, and R. Roessle (Eds.), *Handbuch der speziellen Pathologie, Anatomie und Histologie* (Vol. 13, Part 1A). Berlin: Springer. Pp. 823–826.

Hammond, W. A. 1871. *A Treatise on the Diseases of the Nervous System*. New York: Appleton.

Hammond, W. A. 1884. Miryachit, a newly described disease of the nervous system, and its analogues. *New York Medical Journal, 39*, 191–192.

Hassler, R. 1938. Zur Pathologie der Paralysis agitans und des postencephalitischen Parkinsonismus. *Journal für Psychologie und Neurologie, 48*, 1–55, 387–476.

Hattie, W. H. 1909. Huntington's chorea. *American Journal of Insanity, 66*, 123–128.

Hawkins, J. 1961. *The Life of Samuel Johnson, L.L.D.* New York: Macmillan.

Herz, E. 1931. Die anyostatischen Unruheerscheinungen. Klinisch-kinematographische Analyse ihrer Kennzeichen und Begleiterscheinungen. *Journal für Psychologie und Neurologie, 43*, 3–181.

Hornykiewicz, O. 1963. Die topische Lokalisation und das Verhalten von Noradrenalin und Dopamin (3-Hydroxytyramin) in der Substantia Nigra des Normalen und Parkinsonkranken Menschen. *Wiener klinische Wochenschrift, 75*, 309–312.

Horsley, V. 1890. Surgery of the central nervous system. *British Medical Journal, 2*, 1286–1292.

Horsley, V. 1909. The function of the so-called "motor" area of the brain. *British Medical Journal, 21*, 125–132.

Hunt, J. R. 1917. Progressive atrophy of the globus pallidus. *Brain, 40*, 58–148.

Huntington, G. 1872. On chorea. *Medical and Surgical Reporter, 26*, 317–321.

Huntington, G. 1910. Recollections of Huntington's chorea as I saw it at East Hampton, Long Island, during my boyhood. *Journal of Nervous and Mental Disease, 37*, 255–257.

Itard, J. M. G. 1825. Mémoire sur quelques fonctions involontaires des appareils de la locomotion, de la préhension et de la voix. *Archives Générales de Medécine, 8*, 385–407.

Jackson, J. H. 1877. Remarks on rigidity of hemiplegia. *Medical Examiner, 2*, 271–272.

Jackson, J. H. 1878. On cerebral paresis or paralysis with cerebellar tremor or rigidity. *Medical Examiner, 3*, 266–267.

Jackson, J. H. 1899. On certain relations of the cerebrum and cerebellum

(On rigidity of hemiplegia and on paralysis agitans). *Brain, 22,* 621–630.

Jelgersma. 1909. Neue anatomischer Befunde bei Paralysis agitans und bei chronischer Chorea. *Neurologisches Zentralblatt, 27,* 995.

Jelliffe, S. E. 1908. A contribution to the history of Huntington's chorea—A preliminary report. *Neurographs, 1,* 116–124.

King, C. 1885. Hereditary chorea. *New York Medical Journal, 41,* 468–470.

Klawans, H. 1982. The court-martial of William A. Hammond. In H. Klawans (Ed.), *The Medicine of History.* New York: Raven Press. Pp. 107–126.

Klawans, H. 1989. Spontaneous generation. In H. Klawans, *Tuscanini's Fumble and Other Clinical Tales.* New York: Bantam Books. Pp. 93–112.

Lewy, F. H. 1942. Historical introduction: The basal ganglia and their diseases. *Research Publications—Association for Research in Nervous and Mental Disease, 21,* 1–20.

Lonitzer, A. 1582. *Kreuterbuch, künstliche conterfeytunge der baüme, Stauden, hecken, kreuter, getreyde, gewürtze.* Frankfort: G. Egenolff.

Lund, J. C. 1859. "Chorea Sancti Viti i Saetesdalen." Beretning om sundhetsstilstanden og medicinalforholdene i. *Norge,* 137.

Lund, J. C. 1863. In *Beretning om Sundhedstilstanden og Medicinalforholdene i Norge i 1860.* Christiania: Trykt i Det Steenske Bogtrykkeri.

Lund, J. C. 1868. In *Beretning om Sundhedstilstanden og Medicinalforholdene i Norge i 1868.* Christiania: Trykt i Det Steenske Bogtrykkeri.

Lyon, I. W. 1863. Chronic hereditary chorea. *American Medical Times, 7,* 289–290.

Maltsberger, J. T. 1961. Even unto the twelfth generation—Huntington's chorea. *Journal of the History of Medicine and Allied Sciences, 1,* 1–17.

Marie, P., and Lhermitte, J. 1914. Les lésions de la chorée chronique progressive. *Annales de Médecine, 1,* 18–47.

Marie, P., and Trétiakoff, C. 1920. Examen histologique de centres nerveux dans un cas de chorée aiguë de Sydenham. *Revue Neurologique, 27,* 428–438.

Massey, J. M., and Massey, E. W. 1984. Ergot, the "jerks," and revivals. *Clinical Neuropharmacology, 7,* 99–105.

Matossian, M. K. 1978. Religious revivals and ergotism in America. *Clinical Medicine, 16,* 185–192.

Matossian, M. K. 1982. Ergot and the Salem witchcraft affair. *American Scientist, 70,* 355–357.

McAlpine, D. 1923. The pathology of the Parkinsonian syndrome following encephalitis lethargica. *Brain, 46,* 255–281.

McGeer, P. L., Boulding, J. E., Gibson, W. C., and Foulkes, R. G. 1961. Drug induced extrapyramidal reactions treatment with diphenhydramine hydrochloride and dihydroxyphenylalanine. *Journal of the American Medical Association, 177,* 665–670.

McGeer, P. L., and Zeldowicz, L. R. 1964. Administration of dihydroxyphenylalanine to Parkinsonian patients. *Canadian Medical Association Journal, 90,* 463–466.

McHenry, L. C., Jr. 1967. Samuel Johnson's tics and gesticulations. *Journal of the History of Medicine and Allied Sciences, 22,* 152–168.

McHenry, L. C., Jr. 1968. James Parkinson. *Journal of the Oklahoma State Medical Association, 51,* 521–523.

McHenry, L. C., Jr. 1969. *Garrison's History of Neurology.* Springfield, IL: Charles C Thomas.

McMenemey, W. H. 1955. The pathology of Parkinson's disease. In M. Critchley (Ed.), *James Parkinson (1755–1824).* London: Macmillan.

Mella, H. 1924. The experimental production of basal ganglion symptomatology in Macacus rhesus. *Archives of Neurology and Psychiatry, 11,* 405–417.

Melotti, G. 1885. Lavori Originali e Lezioni: Ospedale della Salpêtrière, J. M. Charcot. Intorno ad alcuni casi di tic convulsivo con coprolalia ed ecolalia. *La Riforma Medica, No. 184, 2; No. 185, 2; No. 186, 2.*

Meyers, R. 1940. A surgical procedure for postencephalitic tremor, with notes on the physiology of the premotor fibers. *Archives of Neurology and Psychiatry, 44,* 455–459.

Meynert, T. 1871. Beiträge zur Differential-Diagnose des paralytischen Irrsinns. *Wiener medizinische Presse, 12,* 646–648.

Mitchell, S. W. 1874. Post-paralytic chorea. *American Journal of Medical Science, 68,* 342–352.

Morris, A. D. 1989. *James Parkinson: His Life and Times.* Boston, MA: Birkhäuser.

Moyer, H. N. 1911. A new diagnostic sign in paralysis agitans: The cogwheel resistance of the extremities. *Journal of the American Medical Association, 57,* 2125.

Murray, T. J. 1979. Dr. Samuel Johnson's movement disorder. *British Medical Journal, 1,* 1610–1614.

Murray, T. J. 1982. Doctor Samuel Johnson's abnormal movements. In A. J. Friedhoff and T. N. Chase (Eds.), *Gilles de la Tourette Syndrome.* New York: Raven Press. Pp. 25–30.

Myers, R. H., and Martin, J. B. 1982. Huntington's disease. *Seminars in Neurology, 2,* 365–372.

O'Brien, J. A. 1883. Latah. *Journal of the British Asiatic Society, 1,* 381–429.

Oliver, L. C. 1953. *Parkinson's Disease and Its Surgical Treatment.* London: H. K. Lewis and Company.

Oppenheim, H. 1902. Zur Symptomatologie der Paralysis agitans. *Journal für Psychologie und Neurologie, 1,* 134–139.

Oppenheim, H., and Vogt, C. 1911. Nature et localisation de la paralysie pseudobulbaire congénitale et infantile. *Journal für Psychologie und Neurologie, Suppl. 18,* 293–308.

Osler, W. 1893. Remarks on the varieties of chronic chorea, and a report upon two families of the hereditary form, with one autopsy. *Journal of Nervous and Mental Disease, 18,* 97–111.

Osler, W. 1894. *On Chorea and Choreiform Affections.* Philadelphia, PA: Blakiston.

Parkinson, J. 1817. *An Essay on the Shaking Palsy.* London: Whittingham and Rowland, for Sherwood, Neely, and Jones. Reprinted 1989 in A. D. Morris, *James Parkinson, His Life and Times.* Boston, MA: Birkhäuser. Pp. 152–175.

Patrick, H. T., and Levy, D. M. 1922. Parkinson's disease: A clinical study of one hundred and forty-six cases. *Archives of Neurology and Psychiatry, 7,* 711–720.

Peterson, F. 1889. A contribution to the study of muscular tremor. *Journal of Nervous and Mental Disease, 16,* 99–112.

Poelchen. 1882. Gehirnerweichung nach Vergiftung mit Kohlendunst. *Berliner klinische Wochenschrift, 19,* 396–399.

Potts, C. S. 1920. An account of the witch craze in Salem, with reference to some modern witch crazes. *Archives of Neurology and Psychiatry, 3,* 465–484.

Poynton, F. J., and Paine, A. 1913. *Researches on Rheumatism.* London: Churchill.

Prescott, O. 1813. *A Dissertation on the Natural History and Medical Effects of the Secale Cornutam, or Ergot.* Boston, MA: Cummings and Hilliard.

Putnam, T. J. 1938. Relief from unilateral paralysis agitans by section of the pyramidal tract. *Archives of Neurology and Psychiatry, 40,* 1049–1050.

Refsum, S. 1982. Some aspects of the history of neurology in Norway. In F. C. Rose and W. F. Bynum (Eds.), *Historical Aspects of the Neurosciences.* New York: Raven Press. Pp. 415–433.

Richter, H. 1923. Beiträge zur Klinik and pathologischen Anatomie der extrapyramidalen Bewegungsstörungen. *Archiv für Psychiatrie und Nervenkrankheiten, 67,* 226–294.

Roger, H. 1866. Recherches cliniques sur la chorée, sur le rheumatisme et sur les maladies du coeur chez les enfants. *Archives Générales de Médecine, 2,* 641–665.

Rose, F. C. 1989. Parkinsonism since Parkinson. In A. D. Morris (Ed.), *James Parkinson, His Life and Times.* Boston, MA: Birkhäuser. Pp. 176–187.

Rosenberg, C. E. 1962. The place of George M. Beard in nineteenth century psychiatry. *Bulletin of the History of Medicine, 36,* 245–259.

Roth, D. C. 1850. *Histoire de la Musculation Irrésistible ou de la Chorée Anormale.* Paris: Germer-Ballière.

Rowntree, L. G. 1912. James Parkinson. *Bulletin of the Johns Hopkins Hospital, 23,* 33–45.

Sacks, O. 1983. *Awakenings.* New York: E. P. Dutton.

Sagar, J. B. M. 1776. *Systema morborum symptomaticum.* Vienna: Joannis Pauli Kraus.

Sandras, C. M. S. 1851. *Traité Pratique des Maladies Nerveuses.* Paris: Germer-Ballière.

Sauvages de la Croix, F. B. de. 1763. *Nosologia methodica.* Amstelodami: Sumptibus Fratrum de Tournes.

Schiller, F. 1970. James Parkinson (1755–1828). In W. Haymaker and F. Schiller, *The Founders of Neurology* (2nd ed.). Springfield, IL.: Charles C Thomas. Pp. 496–499.

Schiller, F. 1986a. Parkinsonian rigidity: The first hundred-and-one years (1817–1918). *History and Philosophy of the Life Sciences, 8,* 221–236.

Schiller, F. 1986b. Rigidity and drooling in Parkinson's disease. *Neurology, 36,* 1583.

Sée, G. 1850. *De la Chorée.* Paris: J. B. Ballière.

Shapiro, A. K., and Shapiro, E. 1982. Tourette Syndrome: History and present status. In A. J. Friedhoff and T. N. Chase, *Gilles de la Tourette Syndrome.* New York: Raven Press. Pp. 17–23.

Spatz, H. 1927. Physiologie und Pathologie der Stammganglien. In H. Bethe (Ed.), *Handbuch der normalen und pathologischen Physiologie* (Vol. 10). Berlin: Springer. Pp. 318–417.

Spiegel, E. A., and Baird, H. W. 1968. Athetotic syndromes. In P. J. Vinken and G. W. Bruyn (Eds.), *Handbook of Clinical Neurology: Vol. 6. Diseases of the Basal Ganglia.* Amsterdam: North Holland Publishing Company. Pp. 440–475.

Spillane, J. D. 1981. *The Doctrine of the Nerves.* Oxford, U.K.: Oxford University Press.

Sprenger, J., and Kraemer, H. 1489. *Malleus maleficarum.* (M. Summers, Trans.). 1948. London: Pushkin Press, 1948.

Stevens, H. 1971. Gilles de la Tourette and his syndrome by serendipity. *American Journal of Psychiatry, 128,* 489–492.

Straton, C. R. 1885. The prechoreic stages of chorea. *British Medical Journal, 2,* 437–438.

Swift, W. B. 1916. A new treatment for paralysis agitans. *Journal of the American Medical Association, 67,* 1834–1835.

Sydenham, T. 1848a. On St. Vitus's dance. Chapter XVI of *Processus Integri*. (R. G. Latham, Trans.). In *The Entire Works of Thomas Sydenham* (Vol. 2). London: Sydenham Society. (Reprinted 1979; Birmingham, AL: The Classics in Medicine Society. Pp. 257–259)

Sydenham, T. 1848b. On the appearance of a new fever. Chapter I of *Schedula monitoria de novae febris ingressu*. (R. G. Latham, Trans., Latin ed. 1686). In *The Entire Works of Thomas Sydenham* (Vol. 2). London: Sydenham Society. (Reprinted 1979; Birmingham, AL: The Classics in Medicine Society. Pp. 190–213)

Tanner, J. R. 1987. St. Anthony's fire, then and now: A case report and historical review. *Canadian Journal of Surgery, 30,* 292–293.

Thiebaut, F. 1968. Sydenham's chorea. In P. J. Vinken and G. W. Bruyn (Eds.), *Handbook of Clinical Neurology: Vol. 6. Diseases of the Basal Ganglia*. Amsterdam: North Holland Publishing Company. Pp. 409–434.

Tilney, F. 1908. A family in which the choreic strain may be traced back to colonial Connecticut. *Neurographs, 1,* 124–127.

Todd, R. B. 1843. Clinical lectures on cases of diseases of the nervous system. The pathology and treatment of chorea. *Lancet, 2,* 463–466.

Toulmouche, A. 1833. Observations de quelques fonctions involontaires des appareils de la locomotion et de la préhension. *Mémoires de l'Académie de Médecine, 2,* 368–398.

Trétiakoff, C. 1919. *Contribution à l'Etude de l'Anatomie Pathologique du Locus Niger*. Thèse de Paris, No. 293.

Trétiakoff, C. 1921. Syndrome parkinsonien par état lacunaire du corps strié (Discussion on pathology of Parkinson's disease). *Revue Neurologique, 37,* 592–593.

Trousseau, A. 1861. *Clinique Médicale de l'Hôtel-Dieu de Paris*. Vol. 2. Paris: J. B. Ballière.

Trousseau, A. 1867–1871. *Lectures on Clinical Medicine Delivered at the Hotel Dieu*. (V. Bazue, Trans.). Philadelphia, PA: Lindsay and Blakiston.

Tyler, K. L., and Tyler, H. R. 1986. The secret life of James Parkinson (1755–1824): The writings of Old Hubert. *Neurology, 36,* 222–224.

Vessie, P. R. 1932. On the transmission of Huntington's chorea for 300 years—the Bures family group. *Journal of Nervous and Mental Disease, 76,* 533–570.

Vogt, C. 1911. Quelques considérations générales à propos du syndrome du corps strié. *Journal für Psychologie und Neurologie, 18,* 479–488.

Vogt, C., and Vogt, O. 1920. Zur Lehre der Erkrankungen des striären Systems. *Journal für Psychologie und Neurologie, 25,* 627–646. Abstracted by Winkelman in *Archives of Neurology and Psychiatry, 1923, 10,* 563–583.

Wesphal, P., Wassermann, and Malkoff. 1899. Ueber den infectiösen Charakter und den Zusammenhang von acutem Gelenkrheumatismus und Chorea. *Berliner klinische Wochenschrift, 36,* 638–640.

Wilson, S. A. K. 1912. Progressive lenticular degeneration: A familiar nervous disease associated with cirrhosis of the liver. *Brain, 34,* 295–509 (Part of an Edinburgh M.D. thesis dated 1911)

Wilson, S. A. K. 1924. The old motor system and the new. *Archives of Neurology and Psychiatry, 11,* 385–404.

Wilson, S. A. K. 1925. Disorders of motility and muscle-tone with especial reference to the corpus striatum. *Lancet, 2,* 1–10, 53–62, 169–178, 215–219, 268–276.

Winfield, J. M. 1908. A biographical sketch of George Huntington, M.D. *Neurographs, 1,* 89–95.

Yahr, M. D. 1978. A physician for all seasons: James Parkinson, 1755–1824. *Archives of Neurology, 35,* 185–188.

Part IV
Sleep and Emotion

Chapter 17
The Process of Sleep

SWEET PLEASING SLEEP! of all the Pow'rs the best!
O Peace of Mind, Repairer of Decay,
Whose Balms renew the Limbs to Labours of the Day;
Care shuns thy soft Approach, and sullen flies away.
Ovid, first decade

Two very closely related topics are sleep and dreaming. This chapter looks at the nature of sleep, and Chapter 18 examines theories and research on dreaming. The present chapter begins with some general ideas about sleep from ancient civilizations and works into the nineteenth century, when experimenters began to study sleep more scientifically. Much of the discussion here focuses on the major theories of sleep proposed during the late nineteenth and early twentieth centuries. Two sleep disorders, African sleeping sickness and narcolepsy, are also examined.

GRECO-ROMAN THEORIES OF SLEEP

The early Greek myths described Nyx as the goddess of the night. She was often represented as a young woman in a star-studded garment holding an inverted torch and driving a quadriga (a chariot pulled by four horses abreast). Nyx (Night) was a descendant of Chaos (Infinite), who lived in the remote regions of the sunset. She had many children, including the goddesses of Destiny and Fate (Moros, Ker). Nyx also had twin sons, Hypnos (Somnus to the Romans) and Thanatos. Thanatos (Death) was associated with destruction, and Hypnos (Sleep) with quiet, gentle sleep.

Thus, Sleep was "Death's half-brother" and one of the "sons of gloomy Night" to Hesiod, the Greek epic poet who lived around 800 B.C. and who claimed to be inspired by the muses when he wrote his *Theogony.* These mythological relationships can also be found in the *Iliad,* in which Homer (c. 700 B.C.) called Sleep and Death "two twins of winged race." The same idea would later appear in the *Aeneid,* in which the Roman poet Virgil (70–19 B.C.) wrote that sleep was a "kinsman to death" (Book 6, line 273).[1]

After the Homeric period and with the advent of Hippocratic medicine, the Greeks attempted to account for sleep with more naturalistic explanations. The theories that achieved the greatest popularity revolved around the four elements—fire, air, water, and earth—and their associated humors (see Chapter 1). In this context, Empedocles maintained that sleep was due to a slight cooling of the blood, or a separation of fire from the other three elements.

A different idea was proposed by Alcmaeon, who, like Empedocles, lived in the fifth century B.C. He proposed that sleep was caused by the withdrawal of blood from the brain, whereas waking followed a return of blood to the brain. He theorized that if the blood remained drained from the brain, the result would be sleep's half brother, death.

The idea that sleep constitutes a time for the body to heal or restore itself can also be found in many Greek writings from this period. Sophocles (c. 497–406 B.C.), for example, wrote the following in *Philoctetes:*

Sleep, to whom pain is an unknown thing,
Sleep, with whom no griefs dwell,
Come, breathing thy balmy spell,
Come wafter of peace to our lives, O king!
Over these eyes still, still hold thou
Thy shimmering veil over-drooping them now!
Come, come, with healing under thy wing.

In the third century B.C., Aristotle included two sections on sleep, *De somno et vigilia* and *De somniis,* in his *Parva naturalia.* He wrote that sleep and wakefulness reflected the activity of the faculty of sense perception, which was located in the heart. He thought sleep was due to the inhibition of sense perception and pointed to eating as a major cause of sleep. Specifically, Aristotle hypothesized that when one ingests food, fumes enter the veins and are pushed up to the brain by the heat of the body. When enough vapors collect in the cool brain, he reasoned, they descend en masse to the torso. This cools the heart and reduces its ability to function effectively. The result is a blunting of the senses and sleep.

Like the Greeks, the Romans saw a close relationship between sleep and death. Lucretius, a follower of Empedocles, wrote that the soul broke into parts and that some parts flew away during sleep, whereas others stayed with the body. Sleep differed from death in that the entire soul left the body at the time of death. These thoughts were expressed in Lucretius' *De rerum natura* as follows:

---

[1]Many additional examples of the close relation between sleep and death can be found in classical and early Western poetry and literature (see Gayley, 1911; Phillips, 1924).

When the divided Soul flies, part abroad,
And part, oppress'd with an unusual Lord,
Retiring backward, closely lurks within,
Then sleep comes on, and Slumbers then begin.
For then the Limbs grow weak, soft Rest does seize
On all the Nerves: they lie dissolved in ease,
For since Senses rise from the Mind alone,
And all the Sense is lost, as Sleep comes on;
Since heavy Sleep can stop, dull Rest control
The sense, it must divide, and break the Soul:
Some Parts must fly away, but some must keep
Their Seats within; else 'twould be Death, not Sleep.

The Romans accepted the idea, previously expressed by Sophocles, that sleep was important for healing and restoring the body. This can be seen in the short passage at the beginning of this chapter from Ovid's *Metamorphoses*.

SLEEP IN THE "PRESCIENTIFIC ERA"

Many of the aforementioned ideas were accepted by the early Christians. In the New Testament Book of Ephesians, one can find a relationship between sleep and death: "Awake, O sleeper, and arise from the dead, and Christ shall give you light." At this time, sleep was looked upon as rewarding. For example, in Ecclesiastes 5:12 one finds: "Sweet is the sleep of a laborer, whether he eats little or much; but the surfeit of the rich will not let him sleep." Nevertheless, too much sleep was considered bad, or at least counterproductive. In Proverbs 20:13 one is told, "Love not sleep, lest you come to poverty."

These various ideas were maintained through the Renaissance. At that time, there were also various opinions about how much sleep one should have and whether sleep was as basic to health as eating and breathing. Paracelsus (1493–1541), for example, maintained that sleep was a mechanical operation, just like breathing. He recommended six hours of sleep a night and stressed that one should sleep neither too much nor too little.

Opinions through the eighteenth century were based largely on conjecture and personal experience. There was little in the way of formal experimentation on sleep prior to the middle of the nineteenth century. In 1846, however, Edward Binns (d. 1851) published *The Anatomy of Sleep*. Although his book was filled with many opinions and anecdotes about sleep, it also contained some threads of knowledge about the physiology of sleep.

Binn's work has been called "the capstone of the prescientific era" of sleep research (Webb, 1973). The experimental study of sleep was beginning to emerge. Among the questions scientists were starting to ask were: Is the depth of sleep constant throughout the night? What are the effects of sleep deprivation? And can people wake up at set or predicted times?

MEASURING THE DEPTH OF SLEEP

One of the first things researchers interested in sleep attempted to measure was the depth of sleep throughout the night. In 1862, the first widely accepted figures on changes in the depth of sleep appeared. The experiment was conducted by Kohlschütter and involved one subject who was studied for eight nights. This subject showed a marked decrease in sensitivity to sound at the end of the first hour of sleep (i.e., deep sleep). There was much more sensitivity to noise at the end of the third hour (i.e., lighter sleep) and a slow but gradual increase in sensitivity (lowered threshold) for the remainder of the night. Kohlschütter's curve of the depth of sleep is shown in Figure 17.1.

Investigators after Kohlschütter showed that depth of sleep could be measured not only by responsiveness to sounds but also by using other forms of graded stimuli. For example, graded electrical stimuli and differential pressure on the skin were used to study depth of sleep (see de Sanctis and Neyroz, 1902; Howell, 1897).

The "Kohlschütter curve" was favorably accepted by the scientific community. Only 67 years later would it be realized that Kohlschütter decided not to connect all of his data points because he wanted to produce a curve that looked smooth, orderly, and regular. In particular, he was distressed by fluctuations after the first two hours of sleep. In 1929, Swan found that the peak for deep sleep remained about where it had been placed by Kohlschütter in 1862 (one to two hours after the onset of sleep). He realized,

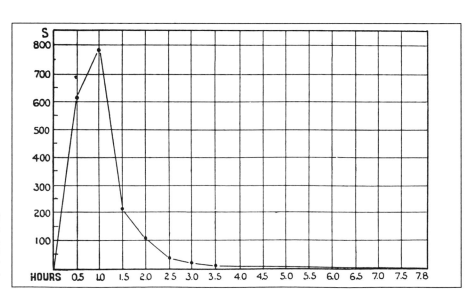

Figure 17.1. Kohlschütter's curve of the depth of sleep as measured by the distance an object had to be dropped to produce a sound to awaken a sleeper. The test was made at 30-minute intervals and was repeated over a number of days. The curve was "normalized" to make it appear more lawful and regular. (From Donaldson, 1895.)

however, that there was considerable irregularity in the depth of sleep throughout the rest of the night and into the morning hours. Swan did not know that some of these fluctuations reflected regular intervals of dreaming.

SLEEP DEPRIVATION IN HUMANS

Nineteenth-century scientists exhibited at least as great an interest in the effects of sleep deprivation. Their interest might have been spawned by personal experiences or might have been stimulated by descriptions of forced sleeplessness as an instrument of torture and, in at least some parts of the world, as a painful means of execution. For example, the following story appeared in the *Louisville Semi-Monthly Medical News* in 1859:

A Chinese merchant had been convicted of murdering his wife, and was sentenced to die by being deprived of sleep. This painful mode of death was carried into effect under the following circumstances: The condemned was placed in prison under the care of three of the police guard, who relieved each other every alternate hour, and who prevented the prisoner from falling asleep night or day. He thus lived nineteen days without enjoying any sleep. At the commencement of the eighth day his sufferings were so intense that he implored the authorities to grant him the blessed opportunity of being strangled, guillotined, burned to death, drowned, garotted, shot, quartered, blown up with gunpowder, or put to death in any conceivable way their humanity or ferocity could invent. This will give a slight idea of the horrors of death from want of sleep.[2]

The first "controlled" sleep-deprivation study conducted on human patients took place at the University of Iowa in the 1890s (Patrick and Gilbert, 1896). The experimenters kept three healthy young men awake for 90 hours, during which time they were given various psychological and performance tests. There were few pathological signs on the first night, but on the second night the urge to sleep became very strong, and one subject fell asleep every time he was left idle. All three showed slower reaction times, impaired motor performance, and memory problems, especially as the experiment progressed. Notably, one subject suffered from visual hallucinations and other gross disturbances in perception after only the second night of sleeplessness. The authors described his bizarre thought process as follows:

The subject complained that the floor was covered with a greasy-looking, molecular layer of rapidly moving or oscillating particles. Often this layer was a foot above the floor and parallel with it and caused the subject trouble in walking, as he would try to step up on it. Later the air was full of these dancing particles which developed into swarms of little bodies like gnats, but colored red, purple or black. The subject would climb upon a chair to brush them from about the gas jet or stealthily try to touch an imaginary fly on the table with his finger. These phenomena did not move with movements of the eye and appeared to be true hallucinations . . . they entirely disappeared after sleep. (Patrick and Gilbert, 1896, 470–471)

The other two men did not show severe perceptual abnormalities, and all recovered completely after sleeping soundly at the end of the experiment. The period of sleep that followed the deprivation was longer than normal and seemed to be deepest at the end of the second hour. Nevertheless, it represented only a fraction of the amount of sleep lost during the experiment. These findings would be confirmed in later experiments.

SLEEP DEPRIVATION IN ANIMALS

Research on laboratory animals allowed experimenters to push sleep deprivation to the limits and to study the effects this had on the brain itself. Marie de Manacéïne (Manasiena; d. 1903) started to deprive animals of sleep in a controlled manner in 1894. Three years later, she wrote:

I have found that by experimenting on 10 puppies that the complete deprivation of sleep for four or five days (96 to 120 hours) causes irreparable lesions in the organism, and in spite of every care the subjects of these experiments could not be saved. Complete absence of sleep during this period is fatal to puppies in spite of the food taken during this time, and the younger the puppy the more quickly he succumbed. (1897, 66–67)

In a related study, three dogs were kept awake by continuously walking them (Tarozzi, 1899). The animals died 9, 13, and 17 days after the beginning of sleep deprivation. Although they did not show cerebral hemorrhages like the dogs described by de Manacéïne, there was diffuse chromatolysis and neuropathology, especially in the frontal lobes (Daddi, 1898).

Henri Piéron (1881–1964) and his associates also studied sleep-deprived dogs, but did not deprive them to the point where they died. They found that sleep-deprived dogs became hyperirritable, showed muscular weakness, and exhibited degenerative changes in the nervous system (e.g., Legendre 1911; Legendre and Piéron, 1908; Piéron, 1913). The changes in the brain included cell shrinkage, vacuolization of cytoplasm, loss of Nissl granules, and displacement of cell nuclei. These changes were most noticeable in the prefrontal cortex and least noticeable in the occipital cortex and the lower brainstem. Behaviorally, the animals seemed completely restored after a single long sleep session, a finding consistent with that seen in sleep-deprived people. No pathological changes could be found in the brains of the recovered dogs.

A major problem with the dog studies was that they required keeping animals awake by continuous activity. Because activity and sleep deprivation were experimentally confounded, critics of these investigations questioned their validity. Were the observed effects due to sleep deprivation itself, or could they have been due to stress or fatigue from sustained activity? The two factors could not be disentangled very easily, and the experiments had to be interpreted with great care.

THE ABILITY TO WAKE AT EXPECTED TIMES

Another interesting question asked by nineteenth- and early-twentieth-century investigators was whether one could control the time of waking by some sort of internal clock

---

[2]This story was retold by Forbes Winslow (1810–1874) in a monograph published in 1860 (p. 604). The quotation is from Winslow's book.

(Tart, 1965). George T. Ladd (1842–1921) began to address this issue in 1892, as did Mary Calkins (1863–1930) in 1893. In general, studies showed that subjects could readily wake from sleep within 10 to 15 minutes of desired times on a high percentage of trials (Bond, 1929; Frobenius, 1927; Hall, 1927). Interestingly, these subjects had no idea how this was accomplished.

A later study of this phenomenon divided women into two groups, those who thought they could wake at a specified time and those who believed they could not (Omwake and Loranz, 1933). In a two-week test period, the women who thought they could do well on the waking task averaged 10 percent waking exactly on time, 30 percent within 15 minutes of the selected time, and 49 percent within 30 minutes of the set time. These women slept considerably less soundly than women in the comparison group, whose ability to wake at a set time was much poorer.

A PLETHORA OF SLEEP THEORIES

The rapid growth of the sciences in general, and findings dealing with sleep in particular, led investigators during the "scientific period of sleep research" to propose and evaluate a number of different sleep theories. A common element running through most of the theories was the idea that sleep gave the body time to repair itself. Espoused in Greco-Roman times and during the Renaissance, this was hardly a new idea. It was even incorporated into Shakespeare's plays, such as when Macbeth bemoaned:

> . . . the innocent sleep,
> Sleep that knits up the ravell'd sleeve of care,
> The death of each day's life, sore labour's bath,
> Balm of hurt minds, great nature's second course,
> Chief nourisher in life's feast.
> (*Macbeth,* c. 1606, Act II, scene II)

The twin notions of rest and repair can be found in the works of Arthur E. Durham (1833–1895), who wrote:

> Considered physiologically, I believe sleep may be correctly regarded as the period of the brain's repose. Every living structure passes through alternating conditions of repose and activity: when active, the tissue is consumed; when at rest, the tissue is nourished, and the waste repaired. Temporary inactivity, we have reason to believe, is a condition essential to perfect repair. (Durham, 1860, 150)

It can also be found in the writings of other scientists from this era. To cite but one more example:

> The state of general repose which accompanies sleep is of especial value to the organism in allowing the nutrition of the nervous tissue to go on at a greater rate than its destructive metamorphosis. (Hammond, 1865, 89)

For the most part, the individual theories differed not in their emphasis on the need for repair but in the trigger for the sleep essential for that repair. The more widely accepted theories tended to emphasize (1) blood flow and cerebral anemia, (2) combustion, (3) autointoxication, and (4) deafferentation.

Each of these theories was based at least in part on trends and new discoveries in the life sciences, as well as in physics and chemistry. For example, theories of cerebral

Figure 17.2. Johann Friedrich Blumenbach (1752–1840). Blumenbach examined a young man with an opening in his skull. He observed that his brain seemed to press upon his skull when he was awake and seemed to shrink when he was asleep. Blumenbach concluded that sleep was caused by a diminished flow of blood to the brain.

congestion and anemia grew in popularity as more was learned about the vascular system and the oxygen needs of the brain. In a similar way, deafferentation theories became popular when cell mobility was envisioned during development and when the concept of inhibition began to attract considerable attention.

BLOOD FLOW AND ANEMIA THEORIES

In the mid-eighteenth century, Albrecht von Haller maintained in his *Elementa physiologiae* that sleep was due to pressure in the skull which caused a slowing of the spirits in the blood vessels of the brain. This theory was generated in part by the observation that pressure from an intracranial tumor could cause sleepiness. The idea that children sleep more than adults because a child's growing brain puts more pressure on the skull was also cited as a fact consistent with this theory.[3]

A somewhat different notion attributed sleep to the retreat of blood from the skull. The roots of this theory can be traced to Alcmaeon in the fifth century B.C. In 1795, Johann Friedrich Blumenbach (1752–1840; shown in Figure 17.2) examined a man with a skull fracture caused by a fall. This individual had an opening in the front of his skull that allowed Blumenbach to see his brain. When the man was awake, his brain seemed to be near the skull, but when he went to sleep, the space between his brain and skull deepened. From this observation, and from the study of hibernating animals, Blumenbach concluded that the primary cause of sleep was the diminished flow of oxygenated blood to the brain.

---

[3]Ideas like these were also suggested by Marshall Hall in the first half of the nineteenth century.

Blumenbach's observations were supported by others who studied patients with skull wounds and malformations that exposed the brain (see Hammond, 1865, 1869). One case involved a French woman whose brain was depressed below her skull when she was in very deep sleep. While she was dreaming, the brain elevated slightly, and during wakefulness it protruded through the skull.[4]

Angelo Mosso (1846–1910) of Turin was one of the first scientists to measure systematically some of the changes in blood flow. In 1881, he wrote about people with different kinds of brain injuries. He used plethysmographic tracings to show a relaxation of the blood vessels of the limbs and a decrease in the blood supply to the brain during sleep, as well as the opposite changes with increasing wakefulness.

Some scientists even compressed their own carotid arteries to produce sleep by depriving the brain of its blood supply. One individual who did this wrote:

> While preparing a lecture on the mode of operation of narcotic medicines, I thought of trying the effect of compressing the carotid arteries on the functions of the brain. I requested a friend to make the first experiment on my own person. He compressed the vessels at the upper part of the neck, with the effect of causing immediate deep sleep. This experiment has been frequently repeated on myself with success, and I have made several cautious but successful trials on others. . . . The mind dreams with much activity, and a few seconds appear as hours, from the number and rapid succession of thoughts passing through the brain. . . . The period of profound sleep, in my experiments, has seldom exceeded fifteen seconds, and never half a minute. (Fleming, 1855, 529–530)

The anemia theory was also popularized by experiments on laboratory animals (Foster, 1901; Hammond, 1865). Arthur E. Durham, who cemented a watch glass over an opening made in the skull of a dog, wrote:

> After the effects of the chloroform passed off, the animal (dog) sank into a comparatively natural and healthy sleep. Corresponding changes took place in the appearance of the brain: its surface became pale, and sank down rather below the level of the bone; the veins were no longer distended; a few small vessels, containing blood of arterial hue, could be distinctly seen; and many which had before appeared congested, and full of dark blood, could scarcely be distinguished. After a time the animal was roused; a blush seemed to start over the surface of the brain, which again rose into the opening through the bone. As the animal was more and more excited, the pia mater became more and more injected, and the brain substance more and more turgid with blood; the surface was of a bright red colour; innumerable vessels, unseen while sleep continued, were now everywhere visible, and the blood seemed to be coursing through them very rapidly. (1860, 154)

Durham believed that blood leaving the brain was directed to the alimentary canal (to aid in digestive processes) and to the excretory organs. He hypothesized that this was why people feel tired after eating a large meal. In addition, Durham stated that variations in the amount of blood in the vessels of the brain were accompanied by corresponding variations in the amount of cerebrospinal fluid in the ventricles and the subarachnoid spaces. Durham's observations were confirmed in later studies on young dogs, which showed that their brains became pale and their blood pressure dropped every time they fell asleep (Tarchanoff, 1894).

These experimental observations were consistent with reports involving drugs and diseases associated with sleepiness. In one case, a boy whose skull was broken when he was kicked by a horse underwent surgery to remove pieces of bone (Brown, 1860). It was found that ether and chloroform depressed the blood supply to his brain and gave it a pale appearance. When the anesthetics began to wear off, as they did a number of times during his surgery, the brain presented "a florid and injected appearance" and exhibited a "heaving and bulging." Eight months after the successful operation, it was noted that one could still see an increase in the pulsations of the boy's brain when he was excited.

Studies of healthy children, whose brains could be seen through their thin skulls, showed comparable changes with sleep and wakefulness (Hammond, 1865). Nevertheless, as the nineteenth century drew to a close, there was a growing belief that decreased circulation to the brain was only one factor that correlated with sleep. Further, it was unclear whether it was a primary or a secondary factor. William Howell (1860–1945), whose portrait is shown as Figure 17.3, felt uncertain. Moreover, as improved ways of measuring blood flow to the brain became available in the twentieth century, some scientists observed cerebral hyperemia, instead of anemia, upon falling asleep (e.g., Shepard, 1914). After reviewing this aspect of the sleep literature, Nathaniel Kleitman (b. 1895), a leading sleep researcher in the twentieth century, concluded that cerebral anemia probably was not the primary cause, and certainly not the only cause, of sleep (Kleitman, 1939, 1963).

Figure 17.3. William Howell (1860–1945). Late in the nineteenth century, Howell proposed that sleep could have multiple interacting causes. He looked upon cerebral anemia and blood flow as being especially important factors.

---

[4]This was the case of Dendy (1821). It was cited by Hammond, (1865), who also presented a comparable case.

CHEMICAL THEORIES

Some early chemical theories of sleep concentrated on the brain's need for oxygen and some on the idea that the oxygen is converted into carbon dioxide (carbonic acid gas), believed to be sleep inducing.

The importance of oxygen was suggested by Jacob Fidelis Ackermann (1765–1815), a German physiologist, who wrote that the oxygen in the air gives off an "ether of life," which reaches the brain where it is extracted and stored. This ether can enter the nerves and muscles to produce "animal motion." Sleep occurs when the body is temporarily depleted of this ether, and it is at this time that the body replenishes itself (cited in Borbély, 1986).

In the first half of the nineteenth century, Alexander von Humboldt and a number of his contemporaries also proposed that sleep was due to a lack of oxygen. As stated in one article:

> The less rapidly that the heart beats, the less rapidly can the blood be aërated, and the oxygen-bearing fluid can be supplied to the brain. The slower that the lungs act, the slower must the oxygen enter the system to supply the diminished circulation. And as the brain in sleep is not in a state in which it can change, from a deficiency in the supply of oxygen, the consequence is . . . that the manifestations of the mind are prevented, and it becomes no longer apparent to the external world. . . . Any thing which removes oxygen from the blood will in the same manner cause sleep. . . . The tendency to sleep in different animals is in inverse proportion to the amount of oxygen consumed by them, and to the amount of carbonic acid produced. (Playfair, 1844, 26–27)

In 1875, Eduard Friedrich Wilhelm Pflüger (1829–1900) wrote that the conditions accompanying the formation of carbon dioxide from oxygen are such that the atoms oscillate most violently at the moment of carbon dioxide production, just as happens with an explosive blast. These "explosions" were thought to cause additional explosions in other cells. Pflüger, who is shown in Figure 17.4, held that the cortex is more active than any other part of the body and requires the most oxygen. As the available oxygen to the brain undergoes depletion, cerebral activity slows and sleep results.

The idea that the carbon dioxide replacing the oxygen might itself be an active sedative or a toxin soon followed. An early advocate of autointoxication was Arthur E. Durham (1860). He felt the brain's need for oxygen was the reason behind the increased blood flow during wakefulness. Durham theorized that when the end products of oxidation reached a certain level, they interfered with the brain's activity. He thought oxidation increased the acidity of the brain and singled out the build-up of carbonic acid gas (carbon dioxide) as a likely factor that could interfere with processes needed for the brain's continued activity. Durham wrote:

> It is at any rate certain that carbonic acid is among the substances first produced. Thus, by the very functional activity of the brain, are generated compounds which interfere with the mutual reaction of oxygen and tissue, and thereby prevent over-exhaustion, and indirectly tend to induce, at the right moment, that state of repose which is essential to repair. (1860, 162)

Heinrich Obersteiner (1872, p. 176) agreed that sleep was due to "the accumulation of acid products in the brain, just as in the case of muscular fatigue." Like Durham (1860), he

Figure 17.4. Eduard Pflüger (1829–1910), who related sleep and wakefulness to the brain's need for oxygen.

based his conclusion on data showing that brain acidity increased with exposure to air or with fatigue. He also assumed sleep occurred involuntarily when the acidity level passed a certain threshold. But unlike Durham, Obersteiner suggested that lactic acid, not carbon dioxide, was the most likely trigger for sleep.

Thierry Wilhelm Preyer (1841–1877) agreed with Obersteiner that lactic acid could play a primary role in producing sleep. Preyer explained:

> During an active condition of mind or body certain substances are brought into existence which are not found (or, at all events, very sparingly) during a state of rest—lactic acid, for example, and creatine. These latter substances may accumulate in the blood, and as they have a great affinity for oxygen they appropriate a principle required for active exertion. The first stage of this accumulation characterizes fatigue; the second stage gives rise to sleep; the third stage, when oxidation has been completed, is followed by awakening. (1877, 547)

Preyer attempted to produce fatigue in animals by giving them lactates, but his results were inconclusive.

Autointoxication theories received some experimental support in a report published in 1894. The paper was written by Charles-Jacques Bouchard (1837–1915), who collected human urine at different times of the day, neutralized and filtered it, and then injected it into the veins of rabbits. He found that daytime urine was much more likely to cause sleep (and death) than nighttime urine. The latter, in contrast, had more convulsive properties and could produce awakening. His data suggested that one toxin builds up during the day and is broken down or eliminated more efficiently during sleep, whereas another builds up during sleep and is broken down during wakefulness.

These findings and general ideas were incorporated into many more specific sleep theories. In Brussels, Leo Errera (1858–1905) wrote:

Work in the organism is indissolubly bound up with a chemical breaking down. Among the products which result, the leucomaines figure. Borne along by the blood, they are without doubt retained by the cerebral centres, and as many of them have a fatiguing and narcotic action, they must at length produce fatigue and sleep. During activity more leucomaines are formed by this breaking down than oxidation can destroy. But during sleep they are destroyed and carried away. Their oxidation products, having no special affinity for the protoplasm of the grey substance, are washed away by the blood stream. The nerve cell is thus cleansed, and a slight stimulus suffices to produce awakening. (1895; translated by de Manacéïne, 1897, 49–50)

Early in the twentieth century, Henri Piéron and R. Legendre tried to inject blood from sleeping dogs into awake dogs to test the idea that a sleep-causing substance may build up in the blood. They were not successful in inducing sleep when the blood was injected into a vein or into the brain itself. More encouraging results were obtained when they injected cerebrospinal fluid from fatigued dogs into awake animals. The results with cerebrospinal fluid were stated as follows:

The . . . cerebro-spinal fluid, of an animal exhausted by loss of sleep, if injected under these conditions into a normal animal, produces in the latter in about half an hour an imperative need of sleep. The animal so injected is benumbed little by little, its eyelids blink, its limbs relax, its eyes close, it loses all attention, and it responds but feebly to strong stimulation. Its brain presents the characteristic lesions of insomnia. The injections, under the same conditions, of liquids from a normal animal have no effect at all. You, therefore, may conclude that it is possible to transmit the absolute need of sleep from an exhausted animal to a normal one, and also that the liquids of exhausted animals have a property or contain a substance capable of producing sleep. (Legendre, 1911, 602)

These findings also suggested that sleep-causing agents built up during the active part of the daily cycle. Because Piéron and his associates failed to detect any changes in the carbon dioxide content of the blood of their sleep-deprived dogs, they dismissed the idea that the toxin in question was carbon dioxide (Legendre and Piéron, 1910).

Piéron hypothesized that the unidentified, sleep-inducing "hypnotoxin" probably worked "by eliciting an inhibitory reflex according to the Brown-Séquard conception" (Legendre and Piéron, 1912). Charles-Edouard Brown-Séquard first proposed an inhibitory reflex theory of sleep in 1889. He thought the center for the reflex had to be in the brainstem. Piéron more fully developed the hypothesis of a hypnotoxin in a book entitled *Le Problème Physiologique du Sommeil*, which appeared in 1913.

Additional experimental support for Piéron's hypnotoxin theory came during the 1930s with more experiments involving cerebrospinal fluid (e.g., Ivy and Schnedorf, 1937; Schnedorf and Ivy, 1939). In addition, brain tissue taken from sleeping rabbits and cats and injected intravenously or suboccipitally into waking animals was found to put them to sleep, as did extract from a hibernating hamster's brain (Kroll, 1933). In fact, the hamster brain extract caused cats to sleep from 2 to 35 days and was accompanied by a sharp drop in body temperature and blood sugar. Extracts from other hamster organs and from the brains of nonhibernating hamsters were ineffective.

Nevertheless, the hypnotoxin idea and the early experiments of Piéron and Legendre were subjected to questioning. One criticism held the very act of injecting or tapping cerebrospinal fluid might have been so stressful that it could have fatigued an awake dog or aroused a sleeping dog. Later experimenters showed that just injecting cerebrospinal fluid from alert dogs raised temperatures and made recipient dogs sleepy. In fact, even withdrawing and replacing cerebrospinal fluid from the same dogs had this effect (Ivy and Schnedorf, 1937; Schnedorf and Ivy, 1939). In support of the hypnotoxin theory, however, these dogs did not become as sleepy as those dogs that received comparable volume injections from fatigued dogs.

The sleep patterns of fused twins sharing a common circulatory system were also used to argue against the hypnotoxin idea. One of the earliest reports of such a case was published in 1836 by Etienne Geoffroy Saint-Hilaire (1772–1844). He noted that while one of the joined twins nursed, the other often slept.[5] In addition, observations on "Siamese twins" and cross-circulation experiments on animals showed that one twin might sleep while the other remained awake (Neri, 1934).

There was, however, a logical flaw with the criticisms stemming from these observations. Scientists simply presumed that the underlying nervous system of each member of a joined pair was physiologically equivalent. This was a risky assumption, especially with joined twins, where one may be much healthier than the other. Thus, circulating toxins could have been present, but they could also have differentially affected the members of each pair because of differences in their health and physiological status.

Arguments for and against sleep toxins also involved experiments which seemed to indicate that prolonged sleep deprivation could result in death (e.g., Marie de Manacéïne, 1894). Advocates of the theory maintained that death would be the expected result if a poison continued to build up with sleeplessness. Opponents of the toxin theory responded that sleep-deprived organisms do not show steadily increasing sleepiness with the passage of time. Instead, there is a wavelike descending curve of greater tiredness at night, with a decrease in sleepiness during subsequent daylight hours. Again, there were two ways of interpreting the experimental findings.

Later in the twentieth century, better evidence would be presented to suggest that sleep-promoting factors do, in fact, build up in the body with prolonged wakefulness.

DEAFFERENTATION THEORIES

The "deafferentation" hypothesis holds that sleep occurs when highest centers in the brain are deprived of sensory input. This theory, mentioned briefly in the previous section when Brown-Séquard's hypothesis was discussed, stems from early observations that some brain lesions produce sleeplike states. For example, in 1809 Luigi Rolando removed the cerebral hemispheres from some ducks, hawks, hens, and mammals. He noted they acted largely as if they were asleep. He described a decorticate raven as follows:

Although (the animal) was standing, it remained nonetheless so drowsy, that it never opened its eyes, unless a very

[5]The Saint-Hilaire experiment was cited by Kleitman (1963, p. 354). For a more recent report on twins joined cranially, see Lahmeyer, (1988).

strong noise was made. It raised and flexed its head, or placed it under its wing, as (the intact animal) was accustomed to do when really asleep. (Translated in Moruzzi, 1964, 21)

Rolando's findings were confirmed by Marie-Jean-Pierre Flourens in 1822. Flourens made bilateral ablations of the cerebral hemispheres in a pigeon and wrote:

Just imagine an animal which has been condemned to be permanently asleep, one that has been devoid even of the ability to dream during sleep; this was more or less the situation of the pigeon in which I had ablated the cerebral hemispheres. . . . (It) never made spontaneous movements and its behavior was one of an animal asleep or drowsy; when it was stimulated during this state of lethargy, its behavior was one of an aroused animal (Translated in Moruzzi, 1964, 21)

In 1846, Jan Purkyně theorized that a continuous flow of information through the afferent systems to the cortex was necessary for normal wakefulness and that blocking the inputs to the cortex caused sleep. Purkyně proposed that the sensory tracts were blocked in the corona radiata and that sleep was due to a hyperemia of the basal nuclei.

But how might this "turning off" be accomplished? One idea was that the neurons were capable of a kind of amoebic movement, which allowed them to make or break contact with other cells by extending or retracting their axons or dendrites. This idea was proposed by Hermann Rabl-Rückhard (1839–1905) in 1890. It attracted some supporters, including Mathias Duval (1844–1907), who discussed it in two papers in 1895, and Jacques Raphaël Lépine (1840–1919), who wrote about it in 1894 and 1895. The theory was stimulated by Santiago Ramón y Cajal's discovery of the growth cone in 1890.

Ramón y Cajal also pondered the idea, but concluded that the kind of neural "amoeboidism" proposed by Rabl-Rückhard and Lépine was not consistent with the known facts. Among other things, neural movements were too slow. He thought, however, that glial cells were capable of greater and more rapid mobility. Thus, rather than abandon the concept entirely, he theorized that glial cells might be able to extend their "pseudopodia" among the nerve cells in the molecular layer of the cortex to block the input of nervous impulses. Theoretically, reasoned Ramón y Cajal (1895), this could control the spread of currents and lead to sleep, whereas retraction of the glial pseudopodia would lead to renewed wakefulness.

Like the neuron version of the theory, the glial theory languished from a lack of good experimental support and never attracted many followers (DeFelipe and Jones, 1988). It was assailed by Rudolph Albert von Kölliker (1896) as a flight of imagination without any experimental basis. Kölliker, well known for his sense of fairness and clarity of thought, was in other respects one of Ramón y Cajal's greatest admirers (see Chapter 3; also Shepherd, 1991).

Some nineteenth-century case studies were cited as support for the deafferentation hypothesis. One case involved a patient who had a paralysis of the third cranial nerve and displayed extreme lethargy. An autopsy revealed a lesion in the midbrain and the thalamus (Gayet, 1875). Fifteen years after the case was reported, Ludwig Mauthner, an ophthalmologist in Vienna, pointed out that paralyses

of the motor nerves to the eye and pathological sleep tended to occur together in a number of diseases (e.g, in Wernicke's *polio-encephalitis haemorrhagica* and Gerlier's *vertige paralysant*). Mauthner (1890) reasoned that damage in the gray matter surrounding the aqueduct of Sylvius in the midbrain could account for both symptoms. From this, Mauthner deduced that normal sleep could be due to "fatigue" of the cells in the gray matter of the midbrain near the aqueduct. This fatigue could cause a functional break in the sensory pathways between the brainstem and the cerebral cortex, effectively deafferenting the cortex.

The general idea that sleep may involve some sort of a deafferentation mechanism attracted considerable attention after this time. William Gowers (1907), for example, wrote that "we know that sleep is a state of rest, during which the highest centers, subserving consciousness, are functionally separated from the lower centers, motor and sensory" (p. 105).

Mauthner's hypothesis was given even more recognition in the 1920s, several years after the European outbreak of *encephalitis lethargica* (see Chapter 16; also Economo, 1923, 1929a, 1929b; Spiller, 1926). In 1923, Constantin von Economo (1876–1931; shown in Figure 17.5), who studied patients afflicted with viral encephalitis, theorized:

The great apparatus that controls periodic sleep is nervous in part only; some of the most important links in the nervous chain lie in the 'tween brain and mid-brain, especially in the posterior wall of the third ventricle and in the adjoining grey matter of the interpeduncular region, the aqueduct and tegmentum. How this region regulates sleep we do not know; it may be by altering the cerebral blood supply; it may be that it influences the sum of cerebral functions concerned in consciousness; or it may be that hormonal influences activated or inhibited by these centres produce effects upon the whole body including the hemispheres. In any case it must be accepted as a physiological fact that a centre exists here from which sleep may be influenced primarily; the fact that other important vegetative functions of the organism are also under the influence of this region makes our conception very plausible. (Translated in Adie, 1926, 300–301)

Mauthner's basic theory was further supported by Stephen Ranson's work on laboratory animals. He reported that lesions between the third cranial nerve and the mammilary bodies made his cats lethargic, and that while they spent long periods of time sleeping, they were not comatose and could be aroused (Ranson and Ingram, 1932). Ranson later made hypothalamic lesions in a large number of monkeys, but chose to interpret his effects as being due to a decrease in the activity of a waking center, rather than to damage to a sleep center.

The theory of a so-called sleep center is based on a misinterpretation of facts. . . . We have shown that the hypothalamus is the center for the integration of emotional expression and that when it is active there is excitement. It is therefore a center for waking, and when it is thrown out of function the withdrawal of this source of excitation permits quiet and relaxation, which favor the onset of sleep. (1939, 23)

The deafferentation hypothesis also had its fair share of critics, including Ivan Pavlov (also spelled Pawlow; 1849–

Figure 17.5. Constantin von Economo (1876–1931), who studied patients with viral encephalitis and concluded sleep might be caused by damage to the periventricular gray matter of the brainstem.

Figure 17.6. Frédéric Bremer (1892–1959), the Belgian scientist who conducted many studies on sleep in *cerveau isolé* and *encéphale isolé* brainstem preparations.

1936).[6] Nevertheless, the theory flourished, especially after Frédéric Bremer (1892–1959; see Figure 17.6), a Belgian neurophysiologist, presented his evidence in support of it in the 1930s. Using cats, Bremer (1935 1937) either cut through the upper brainstem between the superior and inferior colliculi or below the oculomotor nucleus and found that this made the animals act as if they were in a permanent state of sleep. Their pupils were constricted, their eyes diverged, and their EEGs showed deep sleep activity.[7] Bremer called this the *cerveau isolé* preparation, and it seemed to show that sleep involved isolating the brainstem from higher parts of the nervous system.

Bremer then took other cats and made cuts much lower in the brainstem. He called this the *encéphale isolé,* or the whole brain isolation. This preparation was associated with wakefulness, dilated pupils, and EEGs unlike those of deep sleep. Bremer concluded that the difference between the two preparations was that in the *encéphale isolé* condition most of the brainstem afferents still reached the cerebral cortex.

At about the same time Bremer was conducting his experiments, Walter Hess (1881–1973), of the University of Zurich, showed that stimulation of the gray matter surrounding the third ventricle of the brainstem caused animals to go to sleep (Hess, 1929, 1939). The animals could then be aroused normally.

The concept of a "reticular activation system" figured prominently in deafferentation theories of sleep after this time. In particular, Guiseppe Moruzzi (1910–1986) and Horace Magoun (1907–1991) emphasized the role of the brainstem reticular formation in sleep and wakefulness, and wrote a number of important books and papers on the subject (e.g., Moruzzi, 1964; Moruzzi and Magoun, 1949). The reticular afferent formation was also linked to attention and learning at this time.

EVOLUTIONARY THEORIES

A number of researchers took basic ideas from other sleep theories, such as the need to eliminate toxins causing fatigue, and worked them into models based upon the different sleep patterns of higher and lower organisms and on sleep patterns in subjects of different ages.

For example, it was reasoned that very primitive organisms, such as protozoa, probably have only a very rudimentary form of consciousness (Foster, 1901). As one ascends into multicellular organisms there is better evidence for consciousness, but one also finds that the capacity to respond to stimuli decreases as an organism becomes fatigued. This

[6]Pavlov, the Russian physiologist and Nobel Prize winner, found animals tended to fall asleep when they were classically conditioned to withhold responses for more than a few seconds in monotonous environments (Pavlov, 1923a, 1923b, 1923c, 1927; Pawlow, 1923). He thought some higher brain mechanisms were becoming exhausted from inhibiting responses so vigorously and that this inhibition caused small groups of cells to fall asleep. From this idea he postulated that "inhibition is partial sleep, or sleep distributed in localized parts, forced into narrow limits; true sleep is a diffused and continuous inhibition of the whole of the hemispheres" (Pawlow, 1923, p. 605). Pavlov found the idea of specialized sleep centers crude and contradictory to his own theory that sleep exists in every cell. Pavlov's theory, however, did not withstand the test of time.

[7]The first human EEGs were recorded in 1929 by Hans Berger, the head of neurology at a county hospital in Jena, Germany (Chapter 3). It was not until the mid-1930s, however, that Alfred Loomis (1887–1975) and his associates showed that EEG patterns changed dramatically over the course of a night's sleep. Loomis and his coworkers (1935, 1937) classified the patterns of electrical activity they observed into five types. Their categories, from lightest to deepest sleep, were called *A, B, C, D,* and *E.* These stages corresponded to: *(A)* persistence of the alpha rhythm; *(B)* suppression of the voltage and disappearance of 9–11-second alpha; *(C)* emergence of spindle waves; *(D)* spindles interspersed with high-voltage slow waves; and *(E)* diffuse high-voltage slow waves. This type of a classification of the stages of sleep, with the addition of the REM period of dream sleep, achieved considerable popularity after this time.

tendency to fatigue is especially noticeable in very young organisms, which show alternating periods of sleep and wakefulness throughout the day. The theory holds that more mature higher organisms have successfully adapted to the problem of maintaining alert consciousness for longer periods of time by consolidating their naps into one sleep period and by sleeping in comparatively secure places at a relatively safe time.

To a certain extent, the theory that Nathaniel Kleitman (1939) developed was based on these evolutionary and ontogenetic ideas. Kleitman observed that organisms with very large cortical lesions and very young organisms showed extensive sleep, but still woke up to take care of their most basic functions. He noted this was also seen by Friedrich Goltz (1892), who observed that decorticate dogs slept more often than usual. These animals woke up when hungry, only to quickly fall asleep again after eating (also see Rothmann, 1923). Comparable phenomena of waking only when soiled, hungry, or thirsty also characterized anencephalic children (e.g., Edinger and Fischer, 1913).

Kleitman (1939) called this primitive behavioral pattern "wakefulness of necessity" and maintained it was dependent upon "the hypothalamus, extending perhaps into the mesencephalon and thalamus." He postulated that these subcortical regions produce sleep by blocking inputs to the cortex. Thus, without a cortex, alternating sleep and wakefulness should still be seen, but only to execute functions basic to survival itself. With the development of the cerebral cortex, organisms achieved greater control over their sleep patterns. Kleitman called this type of wakefulness, which characterizes healthy adult humans, "wakefulness of choice."

AFRICAN SLEEPING SICKNESS

The literature on sleep disorders is voluminous, and the early journals contain many examples of unusual types of sleep in subjects not suffering from terminal diseases (Gimson, 1863). For some sleep disorders, the literature, occasionally with medico-legal opinions, can be traced back many hundreds of years. For example, sleepwalking was discussed in 1313 at the Council of Vienna where it was decreed that "the act of a sleeper is not of itself a sin" (Roper, 1989; Walker, 1968).

A discussion of all the major diseases of sleep is beyond the scope of this chapter. Nevertheless, because some sleep disorders have contributed significantly to an understanding of the basic mechanisms of sleep, and have attracted great attention for their clinical features, the history of two deserve special attention. They are African sleeping sickness and narcolepsy.

African sleeping sickness—also known as hypnosia (Forbes, 1894) or African lethargy (Gowers, 1888)—was a popular topic of discussion in the nineteenth century when European countries were sending explorers and settlers to Africa. The disorder was described in the opening years of the nineteenth century by Thomas Winterbottom (1766–1859), an Englishman who lived in Africa for four years. Winterbottom's classic description of African sleeping sickness appeared in a book entitled *An Account of the Native Africans in the Neighborhood of Sierra Leone*. To quote:

> The Africans are very subject to a species of lethargy, which they are much afraid of, as it proves fatal in every instance.

> . . . This disease is frequent in the Foola country, and it is said to be much more common in the interior parts of the country than upon the sea coast. . . . At the commencement of the disease, the patient has commonly a ravenous appetite, eating twice the quantity of food he was accustomed to take when in health and becoming very fat. When the disease has continued some time, the appetite declines, and the patient gradually wastes away.

Winterbottom added:

> The disposition to sleep is so strong, as scarcely to leave a sufficient respite for the taking of food; even the repeated application of a whip, a remedy which has been frequently used, is hardly sufficient to keep the poor wretch awake. The repeated application of blisters and of setons has been employed by the European surgeons without avail, as the disease, under every mode of treatment, usually proves fatal within three or four months. The natives are totally at a loss to what this complaint ought to be attributed. (1803, 29)

Later explorers frequently encountered sleeping sickness on the West Coast of Africa, in Niger, in the Antilles, and in the region stretching from Senegal to the Congo. The illness was believed to affect only black-skinned people: "Europeans living in the same localities are exempt" (Gowers, 1888, p. 1336); "It has never been known to affect any but the negro race." (Forbes, 1894, p. 1185).

Many treatments were tried to aid the victims of sleeping sickness, but few were successful. For example, it was suggested that applying galvanism to the spine might stimulate the nervous system and rouse the sleeping victim from apathy (Forbes, 1894). To put this into perspective, the use of electricity was not just suggested for cases of African sleeping sickness; galvanism was used to treat almost every kind of nervous and mental disease as the nineteenth century came to a close (see Chapter 29).

Nineteenth-century authors placed the mortality figure between 80 and 100 percent of those afflicted with African sleeping sickness. The cause of the disease was still unknown when Gowers wrote his *A Manual of Diseases of the Nervous System* in 1888. In contrast, descriptions of its effects, largely in accord with what Winterbottom had written in 1803, were plentiful by this time.

Individuals with sleeping sickness showed an enlargement of the lymph nodes. In addition, early post-mortem examinations seemed to reveal that the cerebral hemispheres of these individuals became pale and unusually firm and that the membranes covering the brain were inflamed (Forbes, 1894).

Until the twentieth century, it was not generally accepted that the disease was caused by a trypanosome (a flagellate protozoan), which colonized in the choroid plexus and surrounding brain areas. As one writer in the last decade of the nineteenth century put it:

> The causation of this curious disease may be said to be wrapped in the deepest obscurity. . . . The three most acceptable theories are: (1) That it may be due to a septic condition of the blood, borne out by swelling of glands, etc.; (2) it is due, as Dr. Manson was the first to point out, to the presence of filaria in the blood; (3) or it may be (though this is an assumption on my part) a neurosis, eventually affecting the neurotrophic system and causing ultimate emaciation and death: this is somewhat substantiated by appearances and lesions sometimes found post-mortem in the brain and its membranes. (Forbes, 1894, 1185)

The association between the disease and the trypanosome was almost made in 1890. Dr. Stephen MacKenzie (1853–1925) studied a 22-year-old native of the Congo and observed that threadlike parasitic worms were always present in this man's blood. He concluded, however, that the association was probably "accidental," in part because filariae did not seem to be present in another supposed case of African sleeping sickness. Only later would the role of the trypanosome and its carrier, the tsetse fly, be understood (also see Manson, 1892).

NARCOLEPSY

In 1888, William Gowers wrote that the term "narcolepsy" had been used in a variety of ways. In general, he noted, it seemed best used to describe a repetitive condition, one in which individuals fall into sleep for short periods of time during the day, usually for a few minutes. Patients can be aroused from this condition, which Gowers considered "the result of some peculiarity of the nervous system."

A typical patient may show many such attacks each day, with each preceded by drowsiness. These sleep attacks can increase in frequency with deficient sleep and decrease during periods of activity. Although the sleep attacks can be held off for some time, they cannot be put off indefinitely, and narcoleptics trying to fight off sleepiness tend to feel uncomfortable. Their sleep episodes are characterized by intensely vivid dreams, but their memory of the dreams tends to lack details.

Gowers (1888) made the point that narcolepsy could be distinguished from trance states by the brevity of the sleep attacks and their recurrent nature in narcoleptics. He also noted that narcolepsy was more likely to be confused with minor attacks of epilepsy, but stated it differs from seizure disorders because narcoleptics look as if they are in normal sleep and because petit mal seizures tend to be shorter in duration.

Nineteenth-century journals provided many good descriptions of narcolepsy. One involved a man from Dublin who fell asleep over his soup (Graves, 1851). Another concerned a maid who was said to have fallen asleep while announcing a visitor and bringing in a tray holding cups of coffee. Marie de Manacéïne (1894), who described the latter case, also mentioned a nurse who nearly killed a baby by dropping the infant during one of her sleep attacks. It was stated that both the maid and the nurse showed very low pulse rates and that "every excitement caused, instead of an acceleration, a retardation of the heartbeats, the pulse falling to 35 in the woman and 40 in the girl."

The term "narcolepsy" was itself introduced by Edouard Gélineau (1828–1906) in 1880. He derived it from words meaning "somnolence" and "to seize" and defined it as an ailment characterized by a compelling need to sleep for short durations at close intervals. Gélineau (1880a, 1880b) considered it a rare functional condition characterized by a need for short-duration sleep.

Gélineau described a merchant whose attacks began when he was 36 years old. Each attack lasted from one to five minutes, and the individual exhibited as many as 200 attacks in a single day. The narcoleptic attacks resembled normal sleep, but were brought on by emotion, exertion, and even changes in the weather. This patient also exhibited a loss of muscle tone during the attacks.

Gélineau's definition of narcolepsy was rather open ended, and some investigators spoke of "the narcolepsies" whereas others, such as Gowers (1907), attempted to restrict use of the term to cases showing very short sleep times—usually on the order of minutes—on a background of wakefulness (see also Adie, 1926; Passouant, 1976; Wilson, 1928). This restrictive definition excluded diseases associated with long periods of sleep or progressive deterioration.

The trend early in the twentieth century was to divide narcolepsy into two basic types (Löwenfeld, 1902). The first was the narcolepsy described by Gélineau, in which short attacks of sleepiness occurred with cataplexy, a loss of physical power without loss of consciousness. The two main symptoms of this type of narcolepsy, short sleep attacks and cataplexy, were thought to be due to a single process, such as inhibition. Emotion often triggered these attacks.

The second, more questionable, category was made up of cases of short sleep attacks without muscular "powerlessness" or "tonelessness." These cases were rarer and harder to diagnose than Gélineau's variety. As stated by W. J. Adie (1886–1935):

> To my mind cataplexy in narcolepsy is as characteristic of the disease as the sleep attacks themselves. Given a case with sleep attacks alone the diagnosis is difficult, for sleep attacks indistinguishable from those of true narcolepsy occur as a symptom in many dissimilar diseases; but if definite cataplectic attacks are present as well the diagnosis is made certain, for the combination is seen in no other condition whatsoever. (1926, 279)

One interesting feature of the early study of narcolepsy was the predominance of male victims. Adie (1926) found 17 cases of men, but only three of women. The usual age of onset seemed to range from the early teens to about 40, with a mean in the low twenties.

Nineteenth- and early-twentieth-century investigators were unable to agree upon the cause or lesion responsible for narcolepsy. Tumors and vascular disorders were suggested, as well as brain abscesses, seizure disorders, endocrine abnormalities, alcoholism, and degenerative conditions (Parkes, 1982; Wilson, 1928).

Attention seemed most concentrated on the periventricular gray *(Höhlengrau)* of the brainstem (Mauthner's area), following some studies that seemed to show lesions in this area in patients with possible viral encephalitis and other disorders that prolonged sleep (Economo, 1919; Gayet, 1875). As noted by S. A. Kinnier Wilson, "tumours situated in relation to the third ventricle (walls, floor, anterior and posterior extremities) together form a far higher percentage of abnormal sleep cases than any other group, and the association must be of significance" (1928, p. 98).

In 1877, F. C. O. Wesphal (1833–1890) presented the case of a man with narcolepsy and pointed out that the patient's mother experienced similar attacks. This relationship suggested a genetic contribution to narcolepsy, but strong supporting evidence did not come until the mid-twentieth century.[8] Adie (1926, p. 267), for instance, wrote that "as a rule there is nothing in the family history of these patients to indicate that they belong to a neuropathic or degenerate stock."

---

[8]For references on the genetics of narcolepsy, see Parkes (1982).

Regarding treatment, Gélineau attempted to help his patients by giving them a number of drugs (e.g., strychnine and amyl nitrate). He seemed to have had his best success with caffeine. Later in the same decade, Gowers (1888, p. 1336) recommended "an active life with a change in scene" and agreed that "no drugs did so much good as a combination of caffeine and nitroglycerine, which almost entirely arrested the attacks." Gowers (1907) continued to recommend stimulant drug treatments, as well as activity, into the twentieth century.

Interestingly, the use of stimulants to treat medical disorders can be traced to ancient times. The herb *ma huang* (*Ephedra vulgaris*) was used in China as far back as 5,000 years ago as a stimulant for respiratory diseases (Kornetsky, 1969). Ephedrine, which was isolated from *ma huang* early in the twentieth century, was found to be too toxic for clinical use at that time. For this reason, a synthetic substitute for ephedrine was sought. When amphetamine was synthesized in response to this demand in the 1920s, it was found that its clinical effects included insomnia. The drug was given the name benzedrine (Smith, Kline and French Laboratories), and it began to be used to treat narcolepsy in the mid-1930s (Prinzmetal and Bloomberg, 1935).

REFERENCES

Adie, W. J. 1926. Idiopathic narcolepsy: A disease *sui generis;* with remarks on the mechanism of sleep. *Brain, 49,* 257–306.

Aristotle. 1908. *Parva naturalia.* (J. I. Beare, Trans.). Oxford, U.K.: Clarendon Press.

Berger, H. 1929. Über das Elektroenkephalogramm des Menschen. *Archiv für Psychiatrie und Nervenkrankheiten, 87,* 527–570.

Binns, E. 1846. *The Anatomy of Sleep: or The Art of Procuring Sound and Refreshing Slumber at Will.* London: Churchill.

Blumenbach, J. F. 1795. *Anfangsgründe der Physiologie.* Translated by J. Elliotson as, *The Elements of Physiology* (4th ed.). London: Longman, 1828.

Bond, N. B. 1929. The psychology of waking. *Journal of Social and Abnormal Psychology, 24,* 226–230.

Borbély, A. 1986. *Secrets of Sleep.* (D. Schneider, Trans.). New York: Basic Books.

Bouchard, C.-J. 1894. *Lectures on Auto-Intoxication in Disease, or Self-Poisoning of the Individual.* (T. Oliver, Trans.). Philadelphia, PA: F. A. Davis Company.

Bremer, F. 1935. Cerveau "isolé" et physiologie du sommeil. *Comptes Rendus Hebdomadaires des Séances et Mémoires de la Société de Biologie, 118,* 1235–1341.

Bremer, F. 1937. L'activité cérébrale au cours du sommeil et de la narcose. Contribution à l'étude du mécanisme du sommeil. *Bulletin de l'Académie Royale de Médecine de Belgique, 4,* 68–86.

Brown, B. 1860. Case of extensive compound fracture of the cranium. Severe laceration and destruction of a portion of the brain, followed by fungus cerebri, and terminating in recovery. *American Journal of Medical Sciences, 40,* 399–403.

Brown-Séquard, C.-E. 1889. Le sommeil normal, comme le sommeil hypnotique, est le résultat d'une inhibition de l'activité intellectuelle. *Archives de Physiologie Normale et Pathologique, 1,* 333–335.

Calkins, M. 1893. Statistics of dreams. *American Journal of Psychology, 5,* 311–343.

Daddi, L. 1898. Sulle alterazioni degli elementi del sistema nervoso centrale nell'insonnia sperimentale. *Rivista di Patologia di Nervosa e Mentale, 3,* 1–12.

DeFelipe, J., and Jones, E. G. 1988. *Cajal on the Cerebral Cortex.* New York: Oxford University Press.

De Sanctis, S., and Neyroz, U. 1902. Experimental investigations concerning the depth of sleep. *Psychological Review, 9,* 254–182.

Donaldson, H. H. 1895. *The Growth of the Brain.* London: Walter Scott, Ltd.

Durham, A. E. 1860. The physiology of sleep. *Guy's Hospital Gazette, 3rd Ser., 6,* 149–173.

Duval, M. 1895a. Hypothèse sur la physiologie des centres nerveux; théorie histologique du sommeil. *Comptes Rendus Hebdomadaires des Séances et Mémoires de la Société de Biologie, 47,* 74–77.

Duval, M. 1895b. Remarques apropos de la communication de M. Lépine. *Comptes Rendus Hebdomadaires des Séances et Mémoires de la Société de Biologie, 47,* 86–87.

Economo, C. von. 1919. Grippe-encephalitis und Encephalitis lethargica. *Wiener klinische Wochenschrift, 32,* 393–396.

Economo, C. von. 1923. Encephalitis lethargica. *Wiener medizinische Woschenschrift, 73,* 777–782, 835–838, 1113–1117, 1243–1249, 1334–1338.

Economo, C. von. 1929a. *Die Encephalitis lethargica, ihre Nachkrankheiten und ihre Behandlung.* Berlin: Urban and Schwartzenberg.

Economo, C. von. 1929b. Schlaftheorie. *Ergebnisse der Physiologie, 28,* 312–339.

Edinger, L., and Fischer, B. 1913. Ein Mensch ohne Grosshirn. *Archiv für die gesammte Physiologie, 152,* 535–561.

Errera, L. 1895. *Sur le Mécanisme du Sommeil.* Bruxelles.

Fleming, A. 1855. Note on the induction of sleep and anesthesia by compression of the carotids. *British and Foreign Medical and Chirurgical Review, 15,* 529–530.

Flourens, M.-J.-P. 1824. Détermination du rôle qui jouent les diverses parties du système nerveux dans les mouvements dits volontaires, ou de locomotion et de préhension. *Mémoires Lus à l'Académie Royale des Sciences de l'Institut le 31 Mars et 27 Avril 1822.* Published in *Recherches Experimentales sur les Propriétés et les Fonctions du Système Nerveux dans les Animaux Vertébrés.* Paris: Crévot.

Forbes, C. 1894. The "sleeping sickness of West Africa." *Lancet, 1,* 1185–1186.

Foster, H. H. 1901. The necessity for a new standpoint in sleep theories. *American Journal of Psychology, 12,* 145–177.

Frobenius, K. 1927. Über die zeitliche Orientierung im Schlaf und einige Aufwachphänomene. *Zeitschrift für die gesamte Psychologie, 103,* 100–110.

Gayet, M. 1875. Affection encéphalique (encéphalite diffuse probable) localisée aux étages supérieurs des pédoncules cérébraux at aux couches optiques, ainsi qu'au plancher du quatrième ventricule at aux parois latérales du troisième. *Archives de Physiologie Normale et Pathologique, Sér. 2, 2,* 341–351.

Gayley, C. M. 1911. *Classic Myths in English Literature and in Art.* Boston, MA: Ginn and Company.

Gélineau, J. 1880a. De la narcolepsie. *Gazette des Hôpitaux Civils et Militaires, 53,* 626–628.

Gélineau, J. 1880b. De la narcolepsie. *Gazette des Hôpitaux Civils et Militaires, 54,* 635–637.

Gimson, W. G. 1863. Case of prolonged and profound sleep, occurring at intervals during twenty years. *British Medical Journal, 1,* 616.

Goltz, F. 1892. Der Hund ohne Grosshirn. *Archiv für die gesammte Physiologie, 51,* 570–614.

Gowers, W. R. 1888. *A Manual of Diseases of the Nervous System.* Philadelphia, PA: Blakiston.

Gowers, W. R. 1907. *Border-land of Epilepsy.* Philadelphia, PA: Blakiston.

Graves, R. J. 1851. Observations on the nature and treatment of various diseases. *Dublin Journal of Medical Sciences, 11,* 1–20.

Hall, W. W. 1927. The time sense. *Journal of Mental Science, 73,* 421–428.

Haller, A. von. 1769. *Elementa physiologiae corporis humani. Vol. V: Sensus externi interni.* Lausannae: Sumptibus Franciscis Grasset et Sociorum.

Hammond, W. A. 1865. On sleep and insomnia. *New York Medical Journal, 1,* 89–101.

Hammond, W. A. 1869. *Sleep and Its Derangements.* Philadelphia, PA: Lippincott.

Hess, W. R. 1929. Hirnreizversuche über den Mechanismus des Schlafes. *Archiv für Psychiatrie und Nervenkrankheiten, 86,* 287–292.

Hess, W. R. 1939. Beziehungen zwischen Winterschlaf und Aussentemperatur beim Siebenschlaefer. *Zeitschrift für vergleichende Physiologie, 26,* 529–536.

Howell, W. H. 1897. A contribution to the theory of sleep, based upon plethysmographic experiments. *Journal of Experimental Medicine, 2,* 313–345.

Ivy, A. C., and Schnedorf, J. G. 1937. On the hypnotoxin theory of sleep. *American Journal of Physiology, 119,* 342.

Kleitman, N. 1939. *Sleep and Wakefulness.* Chicago: University of Chicago Press.

Kleitman, N. 1963. *Sleep and Wakefulness* (2nd ed.). Chicago: University of Chicago Press.

Kölliker, A. 1896. *Handbuch der Gewebelehre des Menschen* (6th ed.). Leipzig: Engelmann.

Kohlschütter, E. 1862. Messungen der Festigkeit des Schlafes. *Zeitschrift für rationelle Medizin, 17,* 210–253.

Kornetsky, C. 1969. The pharmacology of the amphetamines. *Seminars in Psychiatry, 1,* 227–235.

Kroll, F. W. 1933. Ueber das Vorkommen von übertragbaren schlaferzeugenden Stoffen im Hirn Schlafender Tiere. *Zeitschrift für die gesamte Neurologie und Psychiatrie, 146,* 208–218.

Ladd, G. T. 1892. Contribution to the psychology of visual dreams. *Mind, 1,* New Ser., 299–304.

Lahmeyer, H. W. 1988. Sleep in craniopagus twins. *Sleep, 11,* 301–306.

Legendre, R. 1911. The physiology of sleep. *Reports of the Smithsonian Institute, 12,* 587–602.

Legendre, R., and Piéron, H. 1908. Distribution des altérations cellulaires du système nerveux dans l'insonnie expérimentale. *Comptes Rendus*

Hebdomadaires des Séances et Mémoires de la Société de Biologie, 64, 1102–1104.

Legendre, R., and Piéron, H. 1910. La théorie de l'autonarcose carbonique comme cause du sommeil et les données expérimentales. *Comptes Rendus Hebdomadaires des Séances et Mémoires de la Société de Biologie, 68,* 1014–1015.

Legendre, R., and Piéron, H. 1912. De la propriété hypnotoxique des humeurs développée au cours d'une veille prolongée. *Comptes Rendus Hebdomadaires des Séances et Mémoires de la Société de Biologie, 72,* 210–212.

Lépine, R. 1894. Cas d'hystérie a forme particulière. *Revue de Médecine, 14,* 727–728.

Lépine, R. 1895. Théorie mécanique de la paralysie hystérique, du somnambulisme, du sommeil naturel et de la distraction. *Comptes Rendus Hebdomadaires des Séances et Mémoires de la Société de Biologie, 47,* 85–86.

Löwenfeld, L. 1902. Über Narkolepsie. *Münchener medizische Wochenschrift, 49,* 1041–1045.

Loomis, A. L., Harvey, E. N., and Hobart, G. A. 1935. Potential rhythms of the cerebral cortex during sleep. *Science, 81,* 597–598.

Loomis, A. L., Harvey, E. N., and Hobart, G. A. 1937. Cerebral states during sleep as studied by human brain potentials. *Journal of Experimental Psychology, 21,* 127–144.

Lucretius Carus, T. 1937. *De rerum natura.* (W. H. D. Rouse, Trans.). London: Loeb Classical Library.

MacKenzie, S. 1890. Report of the Clinical Society of London. *Lancet, 1,* 1100–1101.

Manacéïne, M. de. 1894. Quelques observations expérimentales sur l'influence de l'insomnie absolue. *Archivio Italiano di Biologia, 21,* 322–325.

Manacéïne, M. de. 1897. *Sleep: Its Physiology, Pathology, Hygiene and Psychology.* London: Walter Scott. (French ed., 1894)

Manson. P. 1892. The treatment of filaria *Sanguinis hominis. Lancet, 2,* 765–766.

Mauthner, L. 1890. Zur Pathologie und Physiologie des Schlafes, nebst Bemerkungen über die "Nona." *Wiener medizinische Woschenschrift, 40,* 961–964, 1001–1004, 1049–1052, 1093–1094, 1144–1146, 1185–1188.

Moruzzi, G. 1964. The historical development of the deafferentation hypothesis of sleep. *Proceedings of the American Philosophical Society, 108,* 19–28.

Moruzzi, G., and Magoun, H. W. 1949. Brain stem reticular formation and activation of the EEG. *EEG and Clinical Neurophysiology, 1,* 455–476.

Mosso, A. 1881. *Über den Kreislauf des Blutes im menschlichen Gehirn.* Leipzig: Viet.

Neri, V. 1934. Studi sperimentali sul meccanismo del sonno. *Cervello, 13,* 301.

Obersteiner, H. 1872. Cause of sleep. *Journal of Anatomy and Physiology, 7,* 176.

Omwake, K. T., and Loranz, M. 1933. Study of ability to wake at a specified time. *Journal of Applied Psychology, 17,* 468–474.

Parkes, D. 1982. History of narcolepsy. In F. C. Rose and W. F. Bynum (Eds.), *Historical Aspects of the Neurosciences.* New York: Raven Press. Pp. 151–163.

Passouant, P. 1976. The history of narcolepsy. In C. Guilleminault, W. C. Dement, and P. Passouant (Eds.), *Advances in Sleep Research* (Vol. 3). New York: Spectrum Publications. Pp. 3–14.

Patrick, G. T. W., and Gilbert, J. A. 1896. On the effects of loss of sleep. *Psychological Review, 3,* 469–483.

Pavlov, I. P. 1923a. Die normale Taetigkeit und allgemeine Konstitution der Grosshirnrinde. *Scandinavisches Archiv für Physiologie, 44,* 32–41.

Pavlov, I. P. 1923b. "Innere Hemmung" der bedingten Reize und der Schlaf—ein und derselbe Prozess. *Scandinavisches Archiv für Physiologie, 44,* 42–58.

Pavlov, I. P. 1923c. On the identity of inhibition—as a constant factor in the waking state—with hypnosis and sleep. *Quarterly Journal of Experimental Physiology, 17, Suppl.,* 39–43.

Pavlov, I. P. 1927. *Conditioned Reflexes. An Investigation of the Physiological Activity of the Cerebral Cortex.* New York: Oxford University Press.

Pawlow, I. P. 1923. The identity of inhibition with sleep and hypnosis. *Scientific Monthly, 17,* 603–608.

Pflüger, E. 1875. Theorie des Schlafes. *Archiv für die gesammte Physiologie, 10,* 468–478

Phillips, C. A. 1924. *An Anthology of Sleep.* London: Guy Chapman.

Piéron, H. 1913. *Le Problème Physiologique du Sommeil.* Paris: Masson.

Playfair, L. 1844. On the functions of oxygen in the production of natural phenomena in the animal economy. I. Sleep, and some of its concomitant phenomena. *Northern Journal of Medicine, 1,* 24–34.

Preyer, W. 1877. On the cause of sleep. *American Journal of Medical Sciences, 74,* 546–547.

Prinzmetal, M., and Bloomberg, W. 1935. The use of benzedrine for the treatment of narcolepsy. *Journal of the American Medical Association, 105,* 2051–2054.

Purkinje, J. 1846. Wachen, Schlaf, Traum und verwandte Zustände. In R. von Wagner (Ed.), *Handwörterbuch der Physiologie mit Rücksicht auf physiologische Pathologie herausgeg* (Band III, Abth. 2). Braunschweig, Vieweg und Sohn. Pp. 412–480.

Rabl-Rückhard, H. 1890. Sind die Ganglienzellen amöboid? Eine Hypothese zur Mechanik psychischer Vorgägne. *Neurologisches Centralblatt, 9,* 199.

Ramón y Cajal, S. 1890. A quelle époque apparaissent les expansions des cellules nerveuses de la moëlle épinière du poulet? *Anatomischer Anzeiger, 5,* 631–639.

Ramón y Cajal, S. 1895. Algunas conjeturas sobre el mechanismo anatómico de la ideación, asociación y atención. *Revista de Medicina y Cirugia Practica* (Madrid: Moya), 1–14. Reprinted in German as, "Einige Hypothesen über den anatomischen Mechanismus der Ideenbildung, der Association und der Aufmerksamkeit." *Archiv für Anatomie und Physiologie,* 367–378.

Ranson, S. W. 1939. Somnolence caused by hypothalamic lesions in the monkey. *Archives of Neurology and Psychiatry, 41,* 1–23.

Ranson, S. W., and Ingram, W. R. 1932, Catalepsy caused by lesions between the mammilary bodies and third nerve in the cat. *American Journal of Physiology, 101,* 690–696.

Rolando, L. 1809. *Saggio Sopra la Vera Struttura del Cervello dell'Uomo e Degli Animali e Sopra la Funzioni del Sistema Nérvoso.* Sassari: Stamperie di S. S. R. M. privilegiata.

Roper, P. 1904. Bulimia while sleepwalking, a rebuttal for sane automatism? *Lancet, 2,* 796.

Rothmann, H. 1923. Zusammenfassender Bericht über den Rothmannschen grosshirnlosen Hund nach klinischer und anatomischer Untersuchung. *Zeitschrift für die gesamte Neurologie und Psychiatrie, 87,* 247–313.

Schnedorf, J. G., and Ivy, A. C. 1939. An examination of the hypnotoxin theory of sleep. *American Journal of Physiology, 125,* 491–505.

Shakespeare, W. 1961. *Macbeth.* In *The Works of William Shakespeare Gathered into One Volume.* New York: Oxford University Press. Pp. 858–884.

Shepard, J. F. 1914. *The Circulation and Sleep.* New York: Macmillan.

Shepherd, G. 1991. *Foundations of the Neuron Doctrine.* New York: Oxford University Press.

Spiller, W. 1926. Narcolepsy occasionally a post-encephalitic syndrome. *Journal of the American Medical Association, 86,* 673–674.

Swan, T. H. 1929. A note on Kohlschütter's curve of the "depth of sleep." *Psychological Bulletin, 26,* 607–610.

Tarchanoff, J. 1894. Quelques observations sur le sommeil normal. *Archivio Italiano di Biologia, 21,* 318–321.

Tarozzi, G. 1899. Sull'influenza dell'insonnio sperimentale sul ricambio materiale. *Rivista di Patologia di Nervosa e Mentale, 4,* 1–23.

Tart, C. T. 1965. Toward the experimental control of dreaming: A review of the literature. *Psychological Bulletin, 64,* 81–91.

Walker, N. 1968. *Crime and Insanity in England.* Edinburgh: University of Edinburgh Press.

Webb, W. B. 1973. Sleep research past and present. In W. B. Webb (Ed.), *Sleep: An Active Process.* Glenview, IL: Scott, Foresman. Pp. 1–10.

Wesphal, C. 1877. Zwei Krankheitsfälle. II. *Archiv für Psychiatrie und Nervenkrankheiten, 7,* 631–635.

Wilson, S. A. K. 1928. The narcolepsies. *Brain, 51,* 63–109.

Winslow, F. 1860. *On Obscure Diseases of the Brain.* Philadelphia, PA: Blanehard & Lea.

Winterbottom, T. M. 1803. *An Account of the Native Africans in the Neighborhood of Sierra Leone.* London: Hatchard and Mawman.

Chapter 18
The Nature of Dreaming

On the road between Tralleians and Nysa . . . is a cave that lies above the sacred precinct, by nature wonderful; for they say that those who are diseased and give heed to the cures prescribed by these gods resort thither and live in the village near the cave among experienced priests, who, on their behalf, sleep in the cave and through dreams prescribe cures. These are also the men who invoke the healing power of the gods. And they often bring the sick into the cave and leave them there, to remain in quiet, like animals in their lurking-holes, without food for many days. And sometimes the sick give heed to their own dreams, but still they use those of other men, or priests, to initiate them into the mysteries and to counsel them.

Strabo, first century A.D.[1]

ANCIENT EGYPT

The earliest written records pertaining to dreaming are those of the Ancient Egyptians. One "dream book" has been ascribed to Pharaoh Merikare, who lived around 2070 B.C. The *Chester Beatty Papyrus,* dated approximately 1250 B.C., contains the records of some 200 dreams *(omina)* and their interpretation according to the priests of the god Horus. This papyrus, which is in the British Museum, is known to be based on centuries of tradition. Unfortunately, the records compiled by the priests of another important god, Set, have been lost. Thus, the extent to which the interpretations of the different gods overlap is not known.

Records, such as the *Chester Beatty Papyrus,* show that the Ancient Egyptians paid considerable attention to the interpretation of dreams. They believed dreams were a way of seeing into the future, as well as a way for their gods to communicate with the people in order to warn them of dangers, answer questions, and order acts of piety.

In 1818, a huge stela was found at the base of the Sphinx (see Figure 18.1). The inscription on the stone revealed that Prince Thutmes (Thutmos) IV dreamed a great god (Hormakhu, Khepra, Ra, Toom) spoke to him "just as a father speaks to his son." The god told him:

The kingdom shall be given to thee. . . . The earth shall be thine in its length and in its breadth as far as the light of the eye of the lord of Alls shines. . . . My countenance is gracious towards thee, and my heart clings to thee; I will give the best of all things. The sand of the district in which I have my existence has covered me up. Promise me that thou will do what my heart wish; then will I acknowledge that thou art my son, that thou art my helper. (Translated by Brugsch-Bey, 1947, 49–50)

Thutmes IV cleared the sand and freed the body of the Sphinx, as he had promised the god in his dream. He erected the huge stela in front of the freed Sphinx soon after he took the throne, around 1450 B.C.

Figure 18.1. The Sphinx at Giza showing the stone stela at its base telling how Thutmes IV listened to the god Hormokhu in a dream. Thutmes IV cleared the sand covering the great Sphinx, as requested by Hormokhu, and was granted prosperity. Thutmes IV lived approximately 1450 B.C.

The Ancient Egyptians used rituals, fasting, incantations, potions, and ointments to help in contacting their gods. They drew magical pictures, put messages into the mouth of a dead black cat, and even went to special places, especially if they were sick or infertile, to have important

[1]Strabo was born in about 63 B.C. and died in A.D. 24. This passage was translated by Jones (1947, pp. 60–61).

dreams. These places were called incubation sites, and many were special temples, such as the one in Memphis dedicated to Imhotep (see Chapter 1). Other better-known incubation sites were the Temple of Thoth at Khimunu, the Temple of Isis at Philare, and the Temple of Hathor at Sinai. It is thought that the Egyptians learned dream incubation from the Mesopotamians and modified it to suit their culture, much as they had acquired and then modified astrology from the Mesopotamians (MacKenzie, 1965).

The Egyptians relied on trained priests at their temples to help them interpret their dreams. These "Masters of the Secret Things" *(pa-hery-tep)* typically gave special instructions for inducing dreams. They were experts in dream symbolism and often recognized contrary meanings in dreams. For example, they associated long life with dreams of death.

Animism seemed to emerge slowly in Egypt, but around 1200 B.C. one finds references to the soul, or *ba,* leaving the body not only after death but during dreams. During this time, the *ba* was thought to roam in the world of the spirits. This sort of animistic concept can be found in the *Thebian Book of the Dead.*

EARLY EASTERN, BIBLICAL, AND SPIRITUAL APPROACHES TO DREAMING

The idea that sleep or dreaming could release the spirit from the body is also found in many other early cultures. A release of the soul during sleep was accepted by the Pythagoreans in Ancient Greece, the Gnostics after the time of Christ, the Renaissance occultists, the Theosophists, and later, various mystics. In some instances, this release also allowed the body to be entered by other spirits.

The concept of sleep as a release of the spirit may be seen in the philosophy of the Chinese philosopher Chuang Tzu, who died in approximately 300 B.C. He delighted in nature and serenity and wrote: "Everything is one; during sleep the soul, undistracted, is absorbed into this unity; when awake, distracted, it sees the different beings" (Tracol, 1982). He also wondered about the true reality. Was it dreaming or was it waking? Chuang Tzu wrote:

> Once upon a time, I, Chuang Tzu, dreamt I was a butterfly fluttering hither and thither, to all intents and purposes a butterfly. I was conscious only of my fancies as a butterfly, and was unconscious of my individuality as a man. Suddenly I awaked, and there I lay, myself again. Now I do not know whether I was a man dreaming I was a butterfly, or whether I am now a butterfly dreaming I am a man. (Translated in Giles, 1947, 60)

Concepts of spiritual unity and animism are also found in Ancient India (Tracol, 1982, p. 7). There are many references to these beliefs in the *Upanishads.* In addition, it is stated that the father of Gautama Buddha, who lived in the sixth century B.C., had a dream foretelling the birth of his son. Buddha himself interpreted the world on the basis of his dreams.

The idea that dreams can be prophetic is found in the Book of Genesis. For example, Joseph angered his brothers with stories from his dreams:

> And Joseph dreamed a dream. . . . For, behold, we were binding sheaves in the field, and lo, my sheaf arose, and also stood upright; and, behold, your sheaves stood round

about, and made obeisance to my sheaf. And his brethren said to him, "Shalt thou indeed reign over us? Or shalt thou indeed have dominion over us?" And they hated him yet the more for his dreams. (Genesis 37: 5–11)

Later, Joseph went in search of his brothers and found them in Dothan. They considered slaying him to "see what shall become of his dreams," but decided to sell him as a slave. Joseph eventually was taken to Egypt where he was called upon to assist the Pharaoh, who had a dream he could not interpret. Joseph told the Pharaoh his dream meant that seven good years would come, but they would be followed by seven years of famine.

> And Pharaoh said unto Joseph, I have dreamed a dream, and there is none can interpret it. . . . And Joseph said unto Pharaoh, The dream of Pharaoh is one: God hath shewed Pharaoh what he is about to do. The seven good kine are seven years and the seven good ears are seven years. . . . And let them gather all the food of those good years that come, and lay up corn . . . against the seven years of famine. (Genesis 41:15–36)

The importance of communicating with God through dreams, which characterized the *Old Testament* (see Figure 18.2), can also be seen in Saul's (c. 1020–1000 B.C.) lament:

> I am sore distressed; for the Philistines war against me, and God is departed from me, and answereth me no more, neither by prophets, nor by dreams. (I Samuel 28:15)

Unlike the Egyptians, however, the Jews did not use magic to invoke divine dreams, although they often did turn to incubation sites to pray for special messages from God.

Figure 18.2. A picture from the Lambeth Bible (twelfth century) illustrating Jacob's dream of a ladder connecting heaven and earth.

There are also passages in the New Testament suggesting that dreams were viewed as a window for looking into the future and as a way for God to communicate with people (see Bigelow, 1896, 1905). In the Christmas story, for instance, Joseph first dreamed of taking Mary as his wife, then of fleeing to Egypt, later of returning home in safety, and finally of the coming of the Magi.

Many of the same orientations toward dreaming are found in the Koran, the first half of which was heavily based on dreams. Mohammed the Prophet (c. 570–632) attached great importance to dreams. He based his mission on a dream in which the Archangel Gabriel guided him through the depths of hell and then through the seven divine spheres until they reached the light of Allah. Mohammed asked his disciples about their dreams every morning and told them his dreams. Early converts to Mohammedism believed in incubation, the use of skilled interpreters, divine dreams, and rituals to protect themselves from false or evil dreams. In addition, the personality characteristics of the dreamer were taken into account when dreams were interpreted.

Spiritual approaches to dreaming may also be observed in other religions, with numerous examples to be found in the Americas. The Iroquois, who lived in present New York State, considered dreaming to be the language of the soul and the means by which it expressed itself. These extremely religious people piously attended to every dream and felt obligated to follow all they had witnessed in their dreams.

The Maricopa Indians of Colorado believed the soul of the dreamer searched for spirits to teach the tribe cures. They relied on dream specialists, who underwent considerable training. The use of dream specialists was a feature the Maricopa shared with the great civilizations of Central and South America. The dream interpreters were called *teopexqui* by the Aztecs and *cocome* by the Mayans.

GREEK MYTHOLOGY AND MEDICINE

The pre-Homeric Greeks believed the cosmos was made of concentric circles of time and space and that Zeus ruled the known, rational world in the center. As one left the known world, barbarous and strange lands were encountered. After this, there were the mystical lands and Okeanos, the frontier of the real world, which was located before the Realm of the Dead. Because dreams did not "live" in the real world, the pre-Homeric Greeks believed that mortals were "visited" by dreams.

Morpheus, who was looked upon as the god of dreams, was one of the three sons of Hypnos, who was one of the sons of Nyx, or Night (see Chapter 17).[2] The Ancient Greeks believed that the abode of dreams had two gates; one was made of horn and the other of beautiful ivory. The Gate of Horn was associated with truthful dreams, but the Gate of Ivory was associated with false dreams and flattery (Gayley, 1911).

The Greeks, like the Egyptians, visited incubation sites and relied on special priests to help them interpret their dreams. The most famous of their sites was a temple in Epidaurus (c. 1100 B.C.), named after Asclepius, the Greek

god of medicine. More than 300 incubation sites were devoted to this "son of Apollo and the nymph Coronis."

In the post-Homeric period, the Greeks began to account for dreaming with less recourse to the supernatural. Although Democritus and many others still believed dreams came from outside the body and entered the sleeper through pores to bring encoded messages, Hippocrates and his followers maintained that dreaming and "untimely wanderings" were the products of the activities of the brain and the blood.[3] This view is illustrated in the following passage from *On the Sacred Disease:*

> He calls out and screams at night when the brain is suddenly heated. The bilious endure this. . . . But when the man happens to see a frightful dream and is in fear as if awake, then his face is in a greater glow, and the eyes are red and the patient is in fear. And the understanding mediates doing some mischief, and thus it is affected in sleep. But if, when awakened, he returns to himself, and the blood is again distributed along the veins, it ceases. (1952 translation, 159)

Aristotle wrote about dreaming in various sections of his *Parva naturalia*. In one part, *De somno et vigilia*, he pondered "whether the truth is that sleepers always dream but not always remember their dreams" (453b). But in another part, *De somniis*, he wrote:

> There are persons who in their whole lives have never had a dream, while others dream when considerably advanced in years, having never dreamed before. The cause of their not having dreams appears to be somewhat like that which operates in the case of infants, and (that which operates) immediately after meals. It is intelligible enough that no dream-presentation should occur to persons whose natural constitution is such that in them copious evaporation is borne upwards, which, when borne back downwards, causes a large quantity of motion. (1908 translation, 462a)

Aristotle maintained that personal opinions did not shape dreams because opinions occur only after an event. He thought, however, that dreams can have predictive value, but only as a warning of disease (see below). Here he differed from Homer, Virgil, and Ovid (43 B.C.–A.D.18), authors of the *Iliad*, the *Aeneid*, and *Metamorphoses*, respectively. In each of these Greek and Roman masterpieces, important prophetic dreams are described.

DREAMING IN ANCIENT ROME

Many Romans adopted a decidedly more common sense attitude toward dreaming. Petronius Arbiter (d. A.D. 66), the reputed author of the *Satyricon*, was also Nero Claudius Caesar's (A.D. 37–68) personal adviser. As such, he had to listen to many of the emperor's frightful dreams of monsters and distortions. He comforted Nero by telling him: "It is neither the gods nor divine commandments that send the dreams down from the heavens, but each of us makes them for himself" (MacKenzie, 1965, pp. 60–61). Similarly, Cicero (104–43 B.C.) challenged the idea of the divine nature of

---

[2]The other two sons of Hypnos were Icelus and Phantasus. They appeared in dreams as animals and as inanimate objects, respectively.

[3]In the Roman period, Lucretius still wrote that dreams were caused by material images that detached themselves from visible objects, floated through the air, and stimulated the organs of perception.

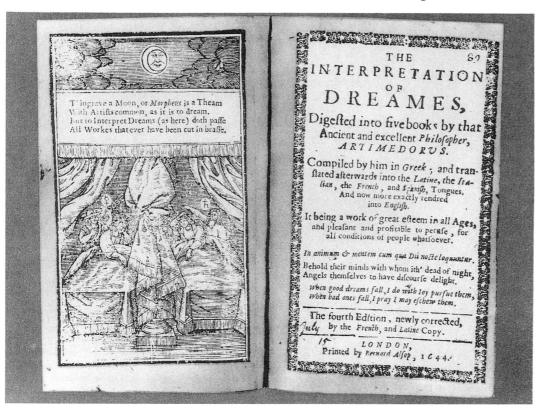

Figure 18.3. Cover page of the English edition (1644) of the popular dream book by the ancient writer Artemidorus (c. A.D. 140).

dreams simply by questioning why the gods would choose to communicate with people in such an unreliable way.

Like Aristotle, Galen felt that dreams offered clues about the status of the humors. He believed that this information enabled the astute physician to predict or uncover hidden diseases. In his short treatise *Diagnosis from Dreams,* Galen wrote:

> To dream indicates an abnormal condition of the body. To see fire in sleep means this condition is due to yellow bile, to see smoke or fog or profound darkness, black bile, to see violent rain means it is due to an overabundance of cold humidity, to see snow and ice, an overabundance of cold phlegm. . . . We conclude that the wrestler who dreams that he is standing in a cistern of blood which he barely overtops is affected with an overabundance of blood and requires blood letting. (Translated in Walsh, 1934, 7)

Interestingly, Galen was at least somewhat influenced by a dream to enter medicine. Specifically, he was 16 years old when his father dreamed that the gods wanted his son to take up this profession. As an obedient only child, Galen followed the suggested path and never married in order to devote himself to a proper medical career.

During the Roman period there were sustained efforts to study and categorize dreams. The individual best known for systematic efforts to classify and interpret dreams was Galen's contemporary, Artemidorus of Ephesus (c. 140).[4] Inspired by Apollo, who appeared before him in a dream, Artemidorus devoted his life to traveling through the civilized world to compile all of the known literature on dreaming. This collection included earlier Egyptian, Assyrian, and

Greek dream records. "I have done no other by day and night but meditate and spend my spirit in the judgment and interpretation of dreams."

Artemidorus presented his observations and theories in his multi-volume *Oneirocritica.* He believed that dreams were given to mortals for instructional purposes and that spirits roamed during sleep.

One dream discussed in *Oneirocritica* involved the word "scalped." Even in Rome, this was a slang expression for "cheated." Artemidorus wrote:

> To have one's scalp scratched signifies a cancellation of interest in the case of a debtor, but for others it means a loss through those by whom they have been scratched, if indeed they have dreamed of having their scalp scratched by others. For we say that a man has been scalped if he has suffered a loss and been deceived by another person. (Translated in MacKenzie, 1965, 56)

Artemidorus emphasized that in order to interpret dreams correctly, the interpreter had to know about the history and personality of the dreamer (rich, poor, temperament, etc.). He thought knowing about the individual characteristics of a dreamer was essential and that there were no set rules for interpreting dreams. His dream book was translated into many languages and continued to be used centuries after his death (see Figure 18.3).

FROM CHURCH DOCTRINE THROUGH THE EIGHTEENTH CENTURY

Some early Christian writers opposed the view held by Aristotle and other Greeks and Romans about dreams—namely,

---

[4]Artemidorus' name is sometimes given as "Artemidorus of Daldis," to honor his mother's birthplace.

the belief that dreams were neither divine nor prophetic. A few of these men viewed dreams as a way of communicating with God, but thought that dreams had to be interpreted on an individual basis using one's own experiences.

Saint John Chrysostom (347–407) accepted the ancient idea that dreams conveyed symbolic messages. He maintained, however, that individuals were not responsible for what they did or felt in their dreams and that dreams should not be thought of as disgraceful. In contrast, Saint Thomas Aquinas, who lived nine centuries later, viewed dream prophecy as an impossibility in his own era. He felt that claims of prophecy amounted to nothing more than people deliberately making something happen after having dreamed about the event.

The question of prophecy and messages in dreams continued to generate heated debate through the Renaissance. Thomas Mancinius (1550–1620), who lived in Ravenna, believed that God used dreams to convey messages to his followers. In 1591, Mancinius wrote an important manuscript on this subject, in which he quoted many biblical examples to make his points. Many others, however, strongly disagreed with his position.

DREAM FREQUENCY

An early belief appeared to be that not everyone dreamed. Aristotle pondered this idea, and Pliny the Elder and Plutarch (c. 46–119) described men who supposedly never dreamed. John Locke (1632–1704), in *An Essay Concerning Human Understanding,* wrote that he knew a scholar who did not dream until he was in his mid-twenties and had then suffered a life-threatening fever. Locke believed that most individuals do not dream for a good part of their lives.

In 1769, Albrecht von Haller devoted 36 pages to sleep and dreaming in *Elementa physiologie.* He also discussed this subject in *Primae lineae physiologiae.* Haller postulated that dreaming was increased by large amounts of hard, indigestible food, uncomfortable postures, and strong, violent ideas. He argued against the ideas that there is perpetual dreaming when we sleep and that the sleeping mind is never at rest. In fact, in his *First Lines of Physiology* (1786) he stated that sleep without dreams is more refreshing and that dreaming is due to stimulation which "interrupts the perfect rest of the sensorium."

In 1888, when dream research became fashionable, Joseph Jastrow (1863–1944) asked 183 people if they dreamed. He found that 1 percent said "no" and 43 percent replied "seldom." Only 8 percent of the subjects in his study said they dreamed every night. These numbers seemed at least somewhat supportive of Locke's contention.

In contrast, Mary Calkins, who had studied psychology with William James, conducted an extensive study of dreaming using herself and another person as subjects. An alarm clock was used to disrupt sleep at different hours throughout the night. At such times, notes were written about whether the subjects were dreaming and, if so, about dream content. Calkins concluded that people who are convinced they do not dream are probably only showing poor recall. She wrote that her basic finding

> vigorously disputes the assertions of people who report that they never dream; and this on strictly empirical grounds. For

I have had more than one instance of waking without the faintest memory of having dreamed and of discovering by my side the night record of one dream or of several. (1930, 32)

An interesting twist on this subject was that at least one objective measure of dreaming appeared to have been found in those mid-nineteenth-century observations on people who had breaks in their skulls that exposed their brains (see Chapter 17). In one study, it was observed that the brain became elevated when a woman whose skull was open was dreaming. The brain even protruded through the skull opening when the woman was having a particularly vivid and stimulating dream (Caldwell; cited by Durham, 1860, p. 153).

This was also found to be true for a man with a large skull opening.

> Standing by his bedside one evening, just after he had gone to sleep, I observed the scalp slightly rise from the chasm in which it was deeply depressed. I was sure he was going to awake, but he did not, and very soon he became restless and agitated, while continuing to sleep. Presently he began to talk, and it was evident that he was dreaming. In a few minutes the scalp sank down to its ordinary level when he was asleep, and he became quiet. I called his wife's attention to the circumstance, and desired her to observe this condition thereafter when he slept. She subsequently informed me that she could always tell when he was dreaming from the appearance of the scalp. (Hammond, 1869, 145)

Unfortunately, scant attention was paid to what these observations were revealing about the frequency and duration of dreaming. As a result, myths about dreaming persisted, even among scientifically oriented investigators. Marie de Manacéïne, for example, wrote that dream frequency increased with old age and was positively correlated with intellectual development.

> In persons occupied with intellectual work (professors, writers, scientific men), and in good health, the number of nights in the month without dreams oscillated between three and ten; any source of exhaustion would increase this number even up to twenty. On the other hand, in subjects belonging to the lower classes of society (cooks, servant girls, dressmakers, coachmen, soldiers, etc.) dreams were generally absent from eight to twenty-five nights during the month, varying with the individual. In short, dreams increase with the variety and activity of the intellectual life. (1897, 273)

STIMULATION AND DREAM CONTENT

In 1895, Edward B. Titchener summarized the scientific literature on dreaming. He wrote that most dreaming is visual, although "auditory dreams, especially those in which the auditory ideas are verbal, probably stand high in the order of frequency" (p. 506). He added that dreams involving the skin senses were not unusual. Titchener agreed with his mentor, Wilhelm Wundt, that gustatory and olfactory dreams were rare.

Mary Calkins came to the same general conclusions in 1893, but added two more facts. The first was that most dreams occurred during morning sleep. The second went against the folklore and science fiction notion that dreams were instantaneous events. As Calkins put it, "It is by no means so certain as some of us think that the train of thought and imagery in dreams is swifter than that of the waking life" (p. 339).

One question that arose from the analyses of the frequency of dreaming and types of dreams concerned the extent to which dream content could be influenced by peripheral stimulation and suggestion. The first notable experimental work on this subject may have been conducted by Louis Maury (1817–1892). After he fell asleep, Maury (1861) had an assistant tickle and pinch him, open perfume bottles near his nose, make a ringing sound near his ears, drop water on him, and so on. He found that these stimuli tended to be incorporated into his dreams. For example, he dreamed he was at sea and that his ship had blown up when a burning match was held close to his nose and the wind blew through his open window. Maury (1878) was awakened one night by a piece of board that fell on his neck. He reported having dreamed about a series of events that culminated in his being guillotined at just that moment! Unfortunately, this dream helped to prolong the erroneous belief that all dreams were always instantaneous, stimulus-bound events.

The literature following Maury's work confirmed that both internal stimuli (stomach cramps, pains) and external events (thunder, breezes, cold temperatures) frequently tend to be incorporated into the content of dreams.[5] In general, studies involving peripheral stimulation showed that tactile stimuli were more likely to be incorporated into dreams than stimuli in other modalities.

In particular, taste dreams were singled out for study in a number of stimulation and suggestion experiments because they seemed to occur so rarely under ordinary conditions. Titchener (1895) reported that he had a taste dream induced by "autosuggestion." He also mentioned that another person, after reading a report on taste dreams, set out to verify this and also had a taste dream.

One problem with many studies at this time was that they did not address the question of how many naturally occurring stimuli were *not* being incorporated into dreams. Nevertheless, there were some early experiments in which external stimuli were presented to sleeping subjects in a more highly controlled manner (De Sanctis and Neyroz, 1902). In one study, crushed cloves were put on the tongues of 20 women before they went to bed (Monroe, 1899). The women reported 254 dreams, of which 17 involved taste and 8 involved smell. These statistics were well above the numbers expected by chance.

DREAMING AMONG THE BLIND

Some nineteenth-century investigators set out to study dreaming among the blind. In what may well be the first serious study of this sort, G. Heermann (1838) examined 101 blind subjects and wrote that they dreamed less than sighted people. He found that none of his 14 subjects who had lost sight before age 5 still had visual dreams. In contrast, all 35 of his subjects who became blind after age 7

had visual dreams. This was true even for subjects who had gone for more than 50 years without sight. Autopsies on a few of the blind subjects showed that the optic nerves degenerated. Heermann therefore concluded that visual dreaming in the blind was dependent upon the cerebrum and not on the optic nerves.

Joseph Jastrow's (1888) investigation of dreaming in the blind confirmed much that Heermann had to say. Jastrow, shown in Figure 18.4, also found that the congenitally blind and those blinded in early childhood lacked visual images in their dreams. Auditory and motor stimuli filled their dreams, but they also showed tactile, taste, and other sensory images at higher levels than those that characterized control subjects. Typical was a dream by a boy who imagined he was in a raging battle. He heard the thunder of the cannons, but did not see flashing when they were fired.

A somewhat different finding emerged among subjects who became blind between the ages of 5 and 7. A few of these individuals reported visual imagery in their dreams. In contrast, all 20 people who became blind after age 7 had visual dreams. Thus, Jastrow's cutoff for visual dreams was identical to that reported by Heermann (also see Wheeler, 1920).

Working from the unpublished notes of G. Stanley Hall, Jastrow also described one subject who was born both deaf and blind (the case of Laura Bridgman). Her dreams lacked both visual and auditory images and were instead filled with tactile and motor images. She reported that people "talked" to her with their fingers, and she responded in kind. To quote:

Her sleep seemed almost never undisturbed by dreams. Again and again she would suddenly talk a few words or letters with her fingers, too rapidly and too imperfectly to be intelligible (just as other people utter incoherent words

Figure 18.4. Joseph Jastrow (1863–1944). In 1888, Jastrow published an important study on the nature of dreaming in the blind.

---

[5]For additional references to the early literature on how physical stimuli can be incorporated into dreams, see Nelson (1888), Ramsey (1953), and Tart (1965).

and articulate sounds in sleep), but apparently never making a sentence. (Hall, cited in Jastrow, 1888, 25)

DREAMING AND CEREBRAL DOMINANCE

Schroeder van der Kolk (1797–1862) hypothesized that the hemisphere of the brain lying closer to the pillow was the more active hemisphere during sleep (see Huppert, 1869, 1872). Soon after this, Delaunay (1882) had subjects sleep on either the right or left side. He thought this would send more blood to the hemisphere on the bottom, causing it to become more active than the other. Delaunay found that subjects lying on the right side had dreams that were more threatening, more illogical, and filled with more sensory images. The dreams of the left hemisphere, at least those that arose when the subject was lying on the left side, were considered more intelligent and less absurd and involved a greater use of language.

A different sort of experiment on the contributions of the two hemispheres was conducted by Marie de Manacéïne (1894, 1897). She simply tickled the sleeper's face. When tickled on the right side, the sleeper always brushed with the left hand. This was the case even for sleepers lying on the left side. She also found that left-handed people brushed with the right hand, even when it was the left hand that was free. On the basis of these observations, she concluded that the more active hemisphere during the day was the least active and most restive hemisphere during sleep.

In general, the view that emerged at the turn of the century was that the right hemisphere was more visual and imaginative, at least in right-handed people. In addition to its supposed roles in hypnosis, hysteria, hallucinations, and delusions, it seemed likely to many investigators that the right hemisphere also played the dominant role in dreaming and the unconscious.[6]

DREAMING, THE UNCONSCIOUS, AND PSYCHOANALYSIS

The unconscious has been called a revolt against romanticism, against introspection and observation, and against the Enlightenment's "rule of reason" (Schiller, 1985). In 1845, Ernst von Feuchtersleben (1806–1849), a Viennese professor, wrote about the unconscious serving as the driving force in dreams. (See Figure 18.5 for Francisco José Goya y Lucientes' [1746–1828] famous depiction of a terrifying dream.) He proposed that obscure ideas could attain particular prominence in dreams and that dreams revealed the nature of the mind. Feuchtersleben blended the Greco-Roman notion that dreaming can serve as an indicator of the health of the body with ideas later associated with psychoanalysis. He maintained:

(Dreaming) as the precursor and accompaniment of diseases, deserves continued investigation, not because it is to be considered as a spiritual divination, but because, as the unconscious language of the *coenesthesis,* and the *senso-*

rium commune, it often very clearly shows, to those who comprehend its meaning, the state of the patient, though he himself is not aware of this; and the interpretation of dreams deserves the attention and study of the physician, if not of anyone else. (Translated in Schiller, 1985, 905)

Feuchtersleben's book, which was translated as *Principles of Psychological Medicine* in 1847, was followed by two other important books on the unconscious. One was Carl Gustav Carus' (1789–1869) *Psyche, A Developmental History of the Soul,* which appeared in German in 1846. The other was Karl Eduard Hartmann's (1842–1906) *Philosophy of the Unconscious,* which was first published in German in 1868.

Sigmund Freud did not seem to be aware of Feuchtersleben's theories and emphasis on unconscious mental activity, although both men worked in Vienna. Freud's own work, *Die Traumdeutung (The Interpretation of Dreams),* appeared in 1899. In Freud's book, which has always been regarded as one of his greatest achievements, the theory espoused was that the manifest part of the dream, or the part remembered, was not as important as its latent content, or the symbolic meaning of the dream. (Figure 18.6 shows another book on interpreting dreams from earlier in the nineteenth century.)

Freud thought the unconscious mind produced thematic material too traumatic and emotional for the rational, conscious mind to accept. For these reasons, important themes that represented unconscious wishes had to emerge through symbols. The deep-seated desires of the individual were fulfilled by the dream, which was indeed a major road to the unconscious for Freud and other psychoanalysts.

Figure 18.5. Etching from 1799 by Francisco José de Goya y Lucientes (1746–1828). The work was called *The Dream of Reason Produces Monsters.*

---

[6]For references on cerebral dominance and visual imagery, see Harrington (1985, 1987).

Figure 18.6. A popular nineteenth-century dream book from Glasgow, Scotland.

EYE MOVEMENTS AND DREAMING

A number of eighteenth-century investigators, including François Boissier de Sauvages de la Croix, noticed movements of the eyelids during sleep (see Schiller, 1984). In 1868, Wilhelm Griesinger (1817–1868) not only observed twitching eyelids in sleeping humans and animals, and recognized that they corresponded to movements of the body, but postulated that they were in some way related to dreaming.

George T. Ladd wrote the following about eye movements and dreaming in the closing years of the nineteenth century:

I am inclined to think that on closing the eyes for sleep the eyeballs are . . . turned upward and inward. This position is probably most favourable to the disappearance from consciousness of all disturbing visual images. Perhaps in deep and dreamless sleep this position of the eyeballs is maintained unchanged. But I am inclined to believe that, in somewhat vivid visual dreams, the eyeballs move gently in their sockets, taking various positions induced by the retinal phantasms, as they control the dreams. As we look down the street of a strange city, for example, in a dream

we focus our eyes somewhat as we should do in making the same observation when awake. (1892, 304)

Marie de Manacéïne (1897) added that since the pupils dilated with emotion during wakefulness, it was not unreasonable to think that they should also change with emotions during dreaming.

E. Jacobson, the author of a popular book that appeared in 1938, claimed not only that eye movements accompanied dreaming but that they could be measured electrically and recorded photographically. Jacobson would have been the first to do these things, but records to support his claims have not be found.

In fact, eye movements during sleep were not of great scientific interest until 1953. In that year, Eugene Aserinski (b. 1921) and Nathaniel Kleitman wrote that eye movements were associated with EEG patterns and dreaming. In a series of controlled experiments, they showed that when people were awakened during periods of rapid eye movements and low-voltage, fast EEG activity, they were much more likely to say they had been dreaming than at other times. The researchers also noticed that there were several periods of rapid eye movement (REM) sleep throughout the night, each lasting approximately 10 to 20 minutes.

Aserinski and Kleitman found that so-called nondreamers showed eye movements just like those of the dreamers in their experiment. In a way, dream research had run a full circle back to Aristotle, who wrote in *De somno et vigilia* that he thought it possible sleepers always dreamed, whereas nondreamers simply had very poor dream recall.

REFERENCES

Aristotle. 1908. *Parva naturalia.* (J. I. Beare, Trans.). Oxford, U.K.: Clarendon Press.
Aserinski, E., and Kleitman, N. 1953. Regularly occurring periods of eye motility, and concomitant phenomena during sleep. *Science, 118,* 273–274.
Bigelow, J. 1896. *The Mystery of Sleep.* New York: Harper and Brothers.
Bigelow, J. 1905. *The Mystery of Sleep* (3rd ed.). New York: Harper and Brothers.
Brugsch-Bey, H. 1947. The dream of King Thutmes IV. In R. L. Woods (Ed.), *The World of Dreams: An Anthology.* New York: Random House. P. 60.
Calkins, M. 1893. Statistics of dreams. *American Journal of Psychology, 5,* 311–343.
Calkins, M. W. 1930. Mary Whiton Calkins. In C. Murchison (Ed.), *A History of Psychology in Autobiography* (Vol. 1). New York: Russell and Russell. Pp. 31–37.
Carus, C. G. 1846. *Psyche: Zur Entwicklungsgeschichte der Seele.* Pforzheim: Flammer und Hoffmann.
Delaunay, G. 1882. Sur deux nouveaux procédés d'investigation psychologique. *Lancette Française: Gazette des Hôpitaux, 55,* 22–23.
De Sanctis, S., and Neyroz, U. 1902. Experimental investigations concerning the depth of sleep. *Psychological Review, 9,* 254–182.
Durham, A. E. 1860. The physiology of sleep. *Guy's Hospital Gazette, 3rd Ser., 6,* 149–173.
Feuchtersleben, E. von. 1847. *The Principles of Psychological Medicine.* (H. E. Lloyd, Trans.). London: Sydenham Society.
Freud, S. 1900. *The Interpretation of Dreams.* (English ed., 1956). New York: Basic Books
Gayley, C. M. 1911. *Classic Myths in English Literature and in Art.* Boston, MA: Ginn and Company.
Giles, H. A. 1947. A dream dilemma (translation of Chuang Tzu). In R. L. Woods, *The World of Dreams: An Anthology.* New York: Random House. P. 60.
Griesinger, W. 1868. Berliner medicinisch-psychologische Gesellschaft. X. *Archiv für Psychiatrie und Nervenkrankheiten, 1,* 200–204.
Haller, A. von. 1769. *Elementa physiologiae corporis humani. Tomus V: Sensus externii internii.* Lausanne: Sumptibus Franciscis Grasset et Sociorum.

Haller, A. von. 1786. *First Lines of Physiology.* Translated from the Latin. (Reprinted 1966; New York: Johnson Reprint Company).

Hammond, W. A. 1869. *Sleep and Its Derangements.* Philadelphia, PA: Lippincott.

Harrington, A. 1985. Nineteenth-century ideas on hemisphere differences and "duality of mind." *Behavioral and Brain Sciences, 8,* 617–660.

Harrington, A. 1987. *Medicine, Mind and the Double Brain.* Princeton, NJ: Princeton University Press.

Hartmann, E. 1931. *Philosophy of the Unconscious* (English ed.). New York: Harcourt, Brace. (German ed., 1868)

Heermann, G. 1838. Beobachtungen und Betrachtungen über die Träume der Blinden, ein Beitrag zur Physiologie und Psychologie der Sinne. *Monatsschrift für Medicin, 3,* 116–180.

Hippocrates. 1952. *On the Sacred Disease.* In *Hippocrates and Galen. Great Books of the Western World* (Vol. 10). Chicago: William Benton. Pp. 154–160.

Hippocrates. 1959. *On Dreams.* In W. H. S. Jones (Trans.), *Hippocrates* (Vol. 4). Cambridge, MA: Harvard University Press. Pp. 422–427.

Huppert, M. 1869. Doppelwahrnehmung und Doppeldenken. Eine psychologische Studie. *Allgemeine Zeitschrift für Psychiatrie, 26,* 529–550.

Huppert, M. 1872. Ueber das Vorkommen von Doppelvorstellungen, eine formale Elementarstörung. *Archiv für Psychiatrie und Nervenkrankheiten, 3,* 66–110.

Jacobson, E. 1938. *You Can Sleep Well. The ABC of Restful Sleep for the Average Person.* New York: Whittlesey House.

Jastrow, J. 1888. Dreams of the blind. *New Princeton Review, 5,* 18–34.

Jones, H. A. 1947. Cures through dreams (Translation of Strabo's *Geography*). In R. L. Woods (Ed.), *The World of Dreams: An Anthology.* New York: Random House. P. 60.

Ladd, G. T. 1892. Contribution to the psychology of visual dreams. *Mind, 1,* New Ser., 299–304.

Lucretius Carus, T. 1937. *De rerum natura.* (W. H. D. Rouse, Trans.). London: Loeb Classical Library.

MacKenzie, N. 1965. *Dreams and Dreaming.* New York: Vanguard Press.

Manacéïne, M. de. 1894. Quelques observations expérimentales sur l'influence de l'insomnie absolue. *Archivio Italiano di Biologia, 21,* 322–325.

Manacéïne, M. de. 1897. *Sleep: Its Physiology, Pathology, Hygiene and Psychology.* London: Walter Scott. (French ed., 1894)

Mancinius, C. 1591. *De somniis, de synesi per somnia.* Ferrara, Italy: Baldinus.

Maury, A. 1861. *Le Sommeil et les Rêves; Etudes Psychologiques sur ces Phénomènes et les Divers Etats qui s'y Rattachent.* Paris: Didier et Cie.

Maury, A. 1878. *Le Sommeil et les Rêves; Etudes Psychologiques sur ces Phénomènes et les Divers Etats qui s'y Rattachent (4th ed.).* Paris: Didier et Cie.

Monroe, W. S. 1899. A study of taste dreams. *American Journal of Psychology, 10,* 245–246.

Nelson, J. 1888. A study of dreams. *American Journal of Psychology, 1,* 367–400.

Ramsey, G. V. 1953. Studies of dreaming. *Psychological Bulletin, 50,* 432–455.

Sauvages de la Croix, F. B. de. 1768. *Nosologia methodica sistens morborum classes juxta Sydenhami mentem et botanicorum ordinem.* Amsterdam: de Tournes.

Schiller, F. 1984. Historical note on sleep and eye movements. *Sleep, 7,* 199–201.

Schiller, F. 1985. The inveterate paradox of dreaming. *Archives of Neurology, 42,* 903–906.

Tart, C. T. 1965. Toward the experimental control of dreaming: A review of the literature. *Psychological Bulletin, 64,* 81–91.

Titchener, E. B. 1895. Minor studies from the psychological laboratory of Cornell University. *American Journal of Psychology, 6,* 505–509.

Tracol, H. 1982. Why sleepest Thou, O Lord? *Parabola, 7,* 6–9.

Walsh, J. 1934. Galen's writings and the influences inspiring them. *Annals of Medical History, 6,* New Ser., 1–14.

Wheeler, R., H. 1920. Visual phenomena in the dreams of a blind subject. *Psychological Review, 27,* 315–322.

Chapter 19
Theories of Emotion from Democritus to William James

The equilibrium or balance, so to speak, between his (Gage's) intellectual faculties and animal propensities seems to have been destroyed. He is fitful, irreverent, indulging at times in the grossest profanity . . . impatient of restraint or advice when it conflicts with his desires. . . . A child in his intellectual capacity and manifestations, he has the animal passions of a strong man.

John Harlow, 1868

This chapter examines a number of theories of emotion, beginning almost 2,500 years ago with those of the Ancient Greeks and ending at the close of the nineteenth century with the theories of William James and Carl G. Lange (1834–1900). Among the topics considered are the notions of François-Xavier Bichat, and the revolutionary ideas of Charles Bell and Guillaume-Benjamin-Amand Duchenne de Boulogne (1806–1870) on the muscles of emotion. Charles Darwin's highly influential theories about how the emotions evolved and are controlled are also examined in this chapter.

Because the literature on the emotions is so large, Chapter 20 covers material dealing with the role of the autonomic nervous system in emotion, the development of the limbic system concept, and the history of "psychosurgical" interventions for so-called emotional disorders.

THE CLASSICAL PERIOD

The word "emotion" has been associated with a number of terms in the literature from Ancient Greece and Rome. These terms include "feelings," "desires," "appetites," and "passions." Such words were sometimes used interchangeably, but often with very specific meanings intended. As a result, early statements about the emotions can be confusing, even in the works of the best ancient writers.

Democritus, a contemporary of Hippocrates in the fifth and fourth centuries B.C., distributed the functions of the soul to three bodily organs: the liver, the heart, and the head. He associated the head with reason, matched the liver with desire, and the heart with anger.

Plato, who lived at about the same time, also believed that the head, the liver, and the heart housed the centers for the rational, the appetitive, and "the spirited" aspects of the soul. Plato wrote that pleasure was the fulfillment of a need. In contrast, he coupled pain with the presence of a need state. Overall, Plato was more concerned with the intellectual processes than with pleasure and pain or with other feelings and emotions.

Aristotle, who followed Plato, closely tied the emotions to the "passions" in his ethical writings, including *Nichomachean Ethics,* a work named after his son. Aristotle wrote that the passions included "appetite, anger, fear, confidence, envy, joy, love, hatred, longing, emulation, pity, and in general the feelings that are accompanied by pleasure or pain." He felt animals can have passions and emphasized that humans have much in common with those animals that display courage or timidity, gentleness or fierceness, mildness or cross tempers, and so on.

Aristotle maintained that humans and animals differed in one important way when it came to the passions. The brutes show their passions instinctively and automatically, whereas human beings may be able to influence their emotions by intention and design. Indeed, he reasoned, only humans have virtue and vice. In Aristotle's schema, emotion took place when the intellect was involved, whereas the passions were more instinctive, being associated with the lower bodily processes.

Aristotle taught that the passions were attended by pleasure and pain and led to actions aimed at increasing pleasurable feelings. He also taught that emotional acts, especially the virtues of temperance and justice, could reveal the true character of a person. As a naturalist, he added that passions can be distracting and that perception can suffer as feelings increase in intensity. He also noted that the passions do not operate uniformly. Rather, they can have different effects on different types of men. In parts of *Parva naturalia,* he wrote that the coward suffers most when excited by fear, whereas the amorous person is most affected by amorous desire. His point was that the passions can intensify those characteristics of the individual which are already in place and can affect the minds of even the best men (see Robinson, 1989).

Like Plato, Aristotle emphasized the roles of the internal organs in governing emotional expression. In "On the Motion of Animals," Aristotle made this statement about the heart in relation to expressions indicative of emotion:

And it is not hard to see that a small change occurring at the center makes great and numerous changes at the circumference, just as by shifting the rudder a hair's breadth you get a wide deviation at the prow. And further, when by reason of heat or cold or some kindred affection a change is set up in the region of the heart, or even in an imperceptibly small part of the heart, it produces a vast difference in the periphery of the body—blushing, let us say, or turning white, goose-skin and shivers and their opposites. (1966 translation, 8)

By the time of Zeno (356–264 B.C.) and the Stoics, a trend to list the major passions and then subdivide then into smaller units was firmly in place. Zeno himself recognized four basic passions: grief, fear, desire, and pleasure. His subdivisions included pity, emulation, envy, jealousy, pain, sorrow, perturbation, anguish, and confusion.

Plato's tripartite conception of the soul endured through the decline of the Greek empire and into Roman times. Galen adopted it and attributed desire to the liver, temper and vitality to the heart, and intellect to the brain. Galen believed that the balance among these three parts determined the character of the individual. Further, he defined different temperaments on the basis of the humoral theory of the Hippocratic school. His well-known fourfold classification of temperament included choleric (heat and dryness), sanguine (heat and moisture), melancholic (cold and dryness), and phlegmatic (cold and moisture).

Saint Thomas Aquinas (Tommaso d'Aquino, 1225–1274), like Nemesius some eight centuries earlier, stands out for attempting to blend Greco-Roman philosophy and natural science with Christian dogma (Telfer, 1955). Aquinas agreed with Aristotle that the emotions were likely to disrupt pure thought and had to be controlled by reason. And, from a theological perspective, he also considered different forms of the emotions. For example, in the case of "love," he distinguished between love based on sensuous desire *(amor)*, love with reason *(dilectio)*, and love in the highest sense of the word—that is, a love of God *(caritas)*.

THEORIES OF EMOTION IN THE POST-RENAISSANCE PERIOD

The drive to classify the emotions into basic types continued unabated into and through the Renaissance (Ruckmick, 1936). In his seventeenth-century work *Les Passions de l'Ame (The Passions of the Soul)*, René Descartes (1650) reduced emotional life to six basic states. These were wonder, love, hate, desire, joy, and sadness. All others were seen as variants or "species" of these six basic emotions, which were explained in terms of motions of the spirits in the brain, blood, and vital organs.

At this time, the brain was given a more definitive role in controlling the expression of the emotions. Thomas Willis, whose *Cerebri anatome* of 1664 represented the most complete and modern account of the nervous system before the eighteenth century, attempted to localize central control of the viscera (heartbeat, breathing, and so forth) in his broadly-defined cerebellum.

Willis actually postulated that the spirits generated in the cerebellum governed involuntary motion, whereas those produced in the cerebrum were responsible for willed action. Jerking the head in the direction of the cerebellum, wrote Willis, resulted in irregularities in heartbeat and respiration. He also noted that severe injuries of the cerebellum led to an instantaneous loss of the heartbeat.

Willis had many followers, including Hermann Boer-haave, who first published his *Institutiones medicae* in 1708. Boerhaave took a more mechanical orientation toward the working of the nervous system. Still, he agreed with Willis that the involuntary nerves arose from the cerebellum.

BICHAT'S THEORIES

The theories of François-Xavier Bichat, which came forth as the eighteenth century drew to a close, played a major role in how most nineteenth-century scientists were to view the emotions.[1] Bichat (1805) thought life was divisible into two parts. One was *la vie organique* (the organic life). This was the life of the heart, the intestines, and the other inner organs. Of these organs, the heart was considered the center of organic life.

The organs regulating organic life were mostly asymmetrical and worked continuously. *La vie organique* was associated with the passions and the metabolic functions of the body. Bichat theorized that this life was regulated through the *système des ganglions* (the ganglionic nervous system), a collection of small independent "brains" in the chest cavity (see Chapter 20 for the history of the autonomic nervous system).

In contrast, *la vie animale* (the animal life) was put on a decidedly higher plane. Animal life, Bichat wrote, involved symmetrical, harmonious organs, such as the eyes, ears, and limbs. It included habit and memory and was ruled by the will and the intellect. This was the function of the brain itself, but it could not exist without the heart, the center of organic life.[2]

Bichat's (1805) concept of two independent systems for two types of lives led some very imaginative clinical investigators to associate a wide variety of diseases with pathology of the system governing organic life and its ganglionic nervous system (Ackerknecht, 1974). In 1823, Johann G. Lobstein (1777–1835) ascribed mania, depression, colic, asthma, arthritis, migraine, insomnia, toothache, and malaria to disorders of the organic system. In 1863, two other investigators, Albert Eulenburg (1840–1917) and Paul Guttmann (1834–1893), followed his lead. Their authoritative list of diseases of the system responsible for organic life included glaucoma, optic nerve paralysis, progressive muscular atrophy, angina, epilepsy, locomotor ataxia, diabetes, Addison's disease, and Grave's disease.

To a fair extent, nineteenth-century science can be envisioned as a battle over whether to return control over the apparatus of the emotions to the brain. Before turning to this development, however, the ideas of three men who, in addition to Bichat, had a marked influence on emerging nineteenth- and twentieth-century conceptions about the emotions must be mentioned. In chronological order, these men are Charles Bell and Duchenne de Boulogne, both of whom studied the anatomical basis of facial expressions, and Charles Darwin, who set forth to explain the adaptive value of emotional displays.

---

[1]Bichat had been an army surgeon before teaching anatomy and physiology in Paris.
[2]Bichat probably would have written much more on his two independent nervous systems had he not died in 1802 at the young age of 31. In particular, it is felt that he would have turned to the microscope. Bichat wholly neglected the microscope during his short life, even though others, including Marcello Malpighi (1628–1694), had already begun to advance the cause of science with their new optical instruments (see Spillane, 1981).

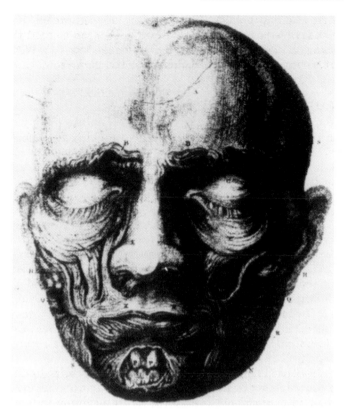

Figure 19.1. The muscles of facial expression as revealed by "careful dissection." This figure appeared in Sir Charles Bell's *Essays on the Anatomy of Expression in Painting,* dated 1806.

THE MUSCLES OF FACIAL EXPRESSION

Facial expressions began to be studied in earnest by anatomists at the beginning of the nineteenth century. The scientific study of the facial muscles was especially stimulated by Sir Charles Bell's *Essays on the Anatomy of Expression in Painting.* This book appeared in 1806 and was based on meticulous dissection and observation. It included figures of the facial muscles after the skin and fat had been removed (see Figure 19.1). Bell, an accomplished artist, theorized that facial expressions were limited in scope by the relationships of the facial muscles to one another. He also postulated that humans had been created with certain muscles specifically to allow them to express their feelings.

This theory was further elaborated in Bell's clinical lecture entitled "Partial Paralysis of the Face." With regard to the portio dura of the facial nerve, he stated:

And as this portio dura takes its circuitous course, for the purpose of associating parts necessary to the act of respiration, for the same reason it must be the nerve of expression; since the self-same parts are the organs of expression that are the organs of respiration. . . . you see that with the destruction of this nerve, the expression of laughing is gone from the side of the face. You will, perhaps, take it on my authority, that crying would be all on one side of the face too. . . . Now these are the extremes of expression; and all the intermediate gradations which are the signs of emotion, are frequently lost. (1827, 750)

Bell's work on facial anatomy was noted by Duchenne de Boulogne. After studying medicine in Paris, Duchenne returned to his home town of Boulogne-sur-Mer in 1831 to

set up a clinical practice. He remained there for a few years, but after his wife died he returned to Paris to spend most of his time at the Salpêtrière.[3]

Duchenne (1850, 1862a, 1862b, 1867) maintained a longstanding interest in electricity (faradism) as a means of stimulating the skeletal muscles. He was especially interested in the facial muscles of expression. In 1850, he began his work on faradic stimulation of these muscles with an induction coil. In 1862, he published the second edition of his *Mécanisme de la Physionomie Humaine,* a scholarly volume now accompanied by a separate atlas of 84 original photographs.

Duchenne accomplished many firsts, especially in his *Mécanisme de la Physionomie Humaine.* He was the first to combine clinical photography with faradism, the first to illustrate a medical publication with original photographs, and the first to record facial expressions as they related to systematic stimulation of the facial muscles (Cuthbertson, 1978, 1985; Cuthbertson and Hueston, 1979; Hueston and Cuthbertson, 1978).[4]

Duchenne used models for some of the photographs in his *Mécanisme de la Physionomie Humaine,* but apologized for the amateurishness of some of his items. One problem was that his films lacked light sensitivity, so that he was often forced to take pictures in front of windows with people staring into the sunlight. A second problem was that the subject's eyes and nose could not be put into simultaneous focus with the lenses available between 1855 and 1857. Although the depth of field problem was solved by 1862, Duchenne wanted to include his original photographs in his books because he considered some of this material unique.

On a theoretical level, Duchenne postulated that there was a specific facial muscle for each emotion. He painstakingly applied localized faradic stimulation to the various facial muscles and studied the results, many of which he photographed. He found some muscles highly expressive of emotion, others somewhat expressive, and still others inexpressive. In addition, some muscles were found to be complementary to others. Duchenne classified those facial muscles expressive of the emotions in part on the basis of the specific emotions that the muscle movements seemed to convey. He called the *Zygomaticus major* the "muscle of joy." Certain other muscles were termed "muscles of disdain and doubt," and some were called "muscles of crying."

In the supplementary "aesthetic" section of the *Méchanisme de la Physionomie Humaine,* Duchenne applied his findings to paintings and sculpture. He examined the emotions in several paintings and explained them on the basis of his new theory of human facial expression. Occasionally he noted that artists depicted two muscles fixed in positions which could not possibly be occurring simultaneously. As a result, he chided some artists for not examining emotional states more carefully.

Of course, not everybody agreed with Duchenne that facial expression depends on the activity of single muscles (see Landis, 1929). Nevertheless, he motivated scientists to

[3]Surprisingly, Duchenne never held an official teaching or hospital position, even after he had achieved considerable fame (Cuthbertson, 1979; Guilly, 1936). His expressed goal was to be a "searcher," not to be "riveted" to a hospital ward (Spillane, 1981).

[4]Before Duchenne started to use photographs in his medical books, photographs had been used only for making engravings for medical illustration. Nonmedical books, however, were already including photographs.

Figure 19.2. Charles Darwin (1809–1882), whose ideas about evolution and natural selection affected thinking in the natural sciences and led to new ways of viewing behavior and the emotions in man and lower animals.

think about the emotions in a more scientific way. As for the future of his applied photography, the advantages it offered attracted many scientists to adopt his methods. For example, Charles Darwin (1872, p. 147) later wrote, "It is easy to observe infants whilst screaming; but I have found photographs made by the instantaneous process the best means for observation, as allowing more deliberation."[5]

DARWIN ON EMOTION

The young naturalist Charles Darwin (shown later in life in Figure 19.2) returned to England in October 1836 after having served for five years as a naturalist aboard the "Beagle."[6] Within three years, Darwin won election to the London Geological Society (1836) and the Royal Society (1839). Before the decade closed, he also published his celebrated book *A Journal of the Voyage of the Beagle* (1839), an authoritative yet beautiful work that established him as a leading writer and naturalist.

As Darwin pondered his voyage, an evolutionary theory based on natural selection took shape in his mind. But while he jotted down some notes about his beliefs, he remained silent in public. This was largely because he recognized that his thinking represented a direct challenge to

[5]In 1860, Jean-Martin Charcot, Duchenne's close personal friend, established the first clinical photography department in a hospital. Appropriately, this was at the Salpêtrière, where Duchenne spent many hours pursuing his science.

[6]The "Beagle" was commanded by Robert FitzRoy (1805–1865), the brilliant but moody navigator and meteorologist who remained a rigid creationist throughout his life (Clark, 1984; Marks, 1991). Darwin, who as a young man thought he was destined for the clergy, became more agnostic as his research progressed and went out of his way to remain friendly toward FitzRoy and most of the others who did not share his views.

the literal interpretation of Creation as presented in the Bible, the view accepted by many of his colleagues as well as his wife, Emma. Finally, in 1856, he decided to write a book to be entitled *Natural Selection*. The plan was for this book to be some 3,000 pages in length.

Two years later, Alfred Russel Wallace (1823–1913) completed a manuscript describing his own theory of evolution. The manuscript was sent from Asia with an accompanying letter to Charles Lyell (1797–1875) and Joseph Hooker (1817–1911), two of the scientists who had been urging Darwin to publish his thoughts. These men, who were Darwin's close friends, arranged for both Darwin's abstracts and Wallace's paper to be presented jointly at the July 1858 meetings of the Linnean Society. With Wallace's work already on the table, Darwin worked feverishly on a much shorter book to be called *On the Origin of the Species by Means of Natural Selection, or the Preservation of Favoured Races in the Struggle for Life.* Darwin's landmark work of "just" 490 pages was published at the end of 1859. Its impact was tremendous, and it rapidly went through many editions.[7]

In *On the Origin of the Species*, Darwin discreetly side-stepped how his theories of evolution and natural selection pertained to human emotions. But in 1871, he took aim at this subject in *The Descent of Man and Selections in Relation to Sex.* This book was followed one year later by *The Expression of the Emotions in Man and Animals.*

In *The Descent of Man*, Darwin argued that humans came from animal-like ancestors. He took the assumption of man's antiquity largely for granted and simply recommended that his readers consult certain "admirable treatises" on man's early appearance, including some books written by his good friend, the geologist Sir Charles Lyell. From this premise, he concluded that humans had to share some emotions with other higher mammals. For example, Darwin, who had dogs of his own, felt that a dog may experience jealousy when its master devotes attention to another dog. He wrote that his own experiences convinced him that dogs can experience shame and pride, as well as something akin to a sense of humor, such as when playing with an object about to be returned to its master. The difference between humans and animals with regard to the basic emotions was one of degree, not one of kind.

Darwin did not treat the emotions as extensively as he had hoped in *The Descent of Man* because of the book's large size. In fact, he had so much more to tell his readers that he decided to expand the appendix in order to publish another book just on the emotions. *The Expression of the Emotions in Man and Animals* was the finished product.

Darwin opened *The Expression of the Emotions in Man and Animals* by stating that most of the earlier treatises on emotion were of little help to him. But he did single out two individuals for special praise, one of whom was Sir Charles Bell. Darwin (p. 2) wrote that Bell's work on emotion, which first appeared in 1806, "may with justice be said, not only to have laid the foundation of the subject as a branch of science, but to have built up a noble structure." He noted, however, that Bell did not try to explain why cer-

[7]For a series of essays on how Darwin's ideas were received by other scientists, the public, and the clergy, see Young (1965). Irving Stone's popular book *The Origin* also provides many good quotations from individuals who read or were told about Darwin's *On the Origin of the Species*, as well as an excellent description of Darwin's own temperament and values.

tain expressive acts, such as smiling or crying, were triggered by different emotional states.

The second person praised by Darwin was Duchenne, who allowed Darwin to include some of his photographs on the facial muscles of expression in his book (see Figures 19.3 and 19.4). Darwin thought Duchenne exaggerated the importance of single muscles in expression, but was gentle in his criticism of Duchenne's theory and applauded the Frenchman's empirical studies involving electricity and his splendid photographic records. As for his own interest in recording the emotions, Darwin stated that this began in 1838.

Darwin now took more time to argue that the basic emotions can characterize a species just as much as bones or teeth. From this premise, he asserted that expressive acts, such as the disappointed look of the chimpanzee in Figure 19.5, must be the result of adaptive actions that have had value for the survival of the species. He stated that human behavior, in particular, should be analyzed from an evolutionary perspective, especially as shaped by natural selection. But in the absence of a knowledge of genetics, he, like Herbert Spencer, who coined the term "survival of the fittest," speculated freely about mental inheritance, the effects of use and disuse, and the transmission of acquired characteristics.

For most expressive acts, Darwin began, the purpose of the behavior pattern or habitual action is not hard to discern because it still is, or has been, of obvious "service" to the species. For example, staring with wide eyes in surprise may have come about because dilating the pupils allowed the organism to see things more clearly. Snarling and showing one's teeth when in rage could have emerged from the act of actual biting and the ability of these acts to scare or disarm opponents in a fight. The sneer could have come from movements of the nose when smelling something irritable, and so on. Obvious service, at least in the past, called "The Principle of Serviceable Associated Habits," was Darwin's first principle to account for emotional displays.

Still, not all emotional acts fit this description. Thus, Darwin proposed that some expressive behaviors could have been stamped in because they successfully signaled the opposite of easily recognized expressive gestures. Darwin reasoned that if an aggressive animal raises its fur and shows its teeth while assuming a defensive posture, it would make sense to think that an animal that wants to be playful would do the opposite. For a dog, this might entail rolling over, wagging the tail, letting the lips hang loosely, and crouching. This was his second principle, "The Principle of Antithesis."

Darwin's third principle was concerned with stimuli "over-activating" the nervous system. Here he borrowed heavily from Spencer, who suggested that an excess of "nerve-force" could result in a general disturbance of organic function, independent of will or habit. Such a state might be manifested in trembling, alterations in heart rate, contortions of the body, and grimacing. Darwin illustrated this state in his discussion of rage:

> Under this powerful emotion the action of the heart is much accelerated, or it may be much disturbed. The face reddens, or it becomes purple from the impeded return of the blood, or may turn deadly pale. The respiration is laboured, the chest heaves, and the dilated nostrils quiver. The whole body often trembles. The voice is affected. The teeth are clenched or

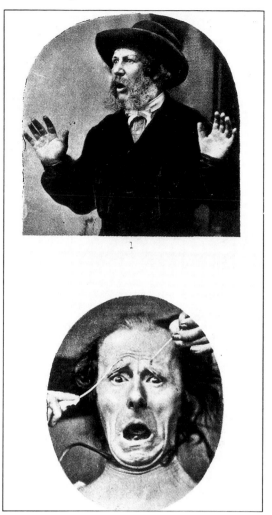

Figure 19.3. Above: An actor showing fear and terror. Below: An attempt to elicit the same emotions with galvanism. These two photographs appeared in Charles Darwin's (1872) *The Expression of the Emotions in Man and Animals.* The lower figure was given to Darwin by Duchenne de Boulogne.

ground together, and the muscular system is commonly stimulated to violent, almost frantic action. . . . All these signs of rage are probably in large part, and some of them appear to be wholly, due to the direct action of the excited sensorium. (1872, 74)

Darwin gave his third principle of emotional expression the cumbersome title "The Principle of Actions Due to the Constitution of the Nervous System, Independently from the First of the Will, and Independently to a Certain Extent of Habit."

Darwin (1872) emphasized that some human emotional expressions no longer have obvious survival value. For this reason, emotional posturing must be understood by the important role that it might have played in the past. This idea went hand in hand with one of Darwin's most important observations, namely, that emotions tend to be expressed in the same way across the human group. A smile, a sneer, or crying with one's head down can convey a message that does not know racial or group boundaries. This, of course, would be expected if all humans shared a common ancestor. In Darwin's words:

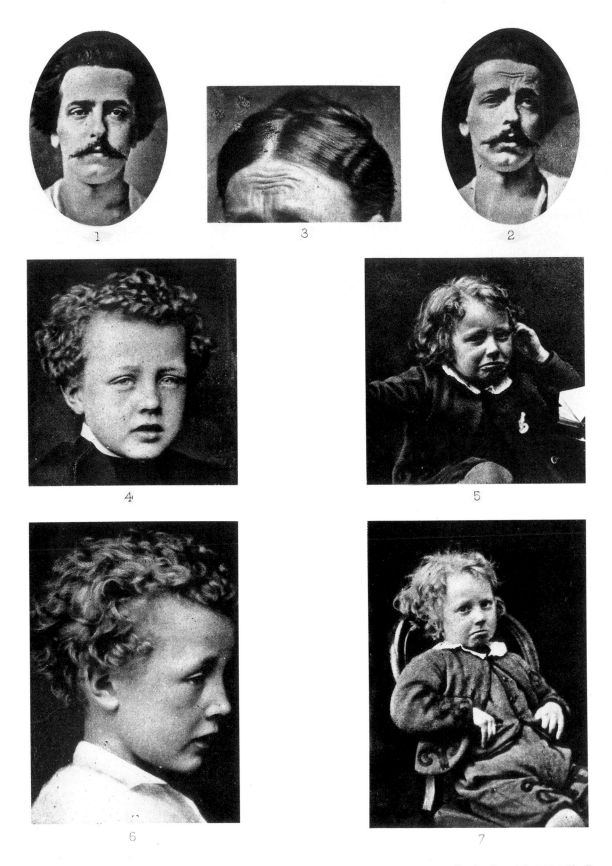

Figure 19.4. A series of photographs showing dejection and despair. These photographs appeared in Charles Darwin's (1872) *The Expression of the Emotions in Man and Animals.* Figures 1 and 2 came from Duchenne de Boulogne and depict a good actor only simulating grief.

Figure 19.5. A drawing of a "disappointed and sulky" chimpanzee from Charles Darwin's (1872) *The Expression of the Emotions in Man and Animals.*

With mankind some expressions, such as the bristling of the hair under the influence of extreme terror, or the uncovering of the teeth under that of furious rage, can hardly be understood, except on the belief that man once existed in a much lower and animal-like condition. The community of certain expressions in distinct though allied species, as in the movements of the same facial muscles during laughter by man and by various monkeys, is rendered somewhat more intelligible, if we believe in their descent from a common progenitor. (1872, 12)

Darwin kept detailed records on the development of his first child, his son "Doddy" (William), who was born in 1840. After reading a paper on the acquisition of language in children, he decided to write "A Biographical Sketch of an Infant." This synopsis of his son's development appeared in the journal *Mind* some 37 years after Doddy was born. It provided many descriptions of his son's emotional development, including the observation that some acts, such as screaming and startling in response to a loud sound, can be seen during the first few days of life. Darwin noted, however, that not all of the basic emotions appeared this early. It was 10 weeks before anger began to appear. This was first seen when the milk given to Doddy was not warmed sufficiently. At 4 months of age, Doddy showed "violent passion" when he dropped an object with which he was playing. Darwin concluded that the most important human emotions are not learned, but that some of these innate emotions require practice for perfection. Here he cited laughing, which can be controlled and consciously employed as a means of communication.

Overall, Darwin's work had a marked influence on how scientists approached the emotions in subsequent decades. He asked some of the most important questions about the origins of emotional expressions and shed light on the similarities between humans and other animals. And although Darwin did not associate the emotions with specific neural circuits—which is somewhat surprising since localization was one of the most exciting developments at this time—others were quick to grasp how his theory of evolution could apply to the organization of the nervous system.

DARWIN'S IMPACT: PASSION VERSUS REASON

In the last paragraph of *The Descent of Man,* Darwin (1871) wrote that the human mind still contained "the indelible stamp of his lowly origin." To the extent that basic emotional expressions were considered primitive acts, humans could differ from the animals only by their ability to control these passions—that is, to suppress the beast within. This notion, which can be traced back to Aristotle, was now seen as one of the triumphs of evolution and a glory of civilization. But it was also viewed as a reversible process. In particular, disorders of higher brain structures could open the cage, permitting the beast to run amok, unchecked and uncontrolled by reason, creating a danger to the individual and to others.

This sort of reasoning made perfect sense to many Victorians. In particular, the application of Darwinian theory to behavior was embraced by John Hughlings Jackson. He maintained a deep-seated belief that the evolution of the nervous system involved the gradual building of more complex structures on more primitive structures. The newer structures were envisioned as having increased complexity, specialization, and integration and as having more numerous interconnections. Jackson reasoned that the earlier evolutionary achievements remained intact as the brain continued to evolve. Thus, rational thinking represented a new, specialized achievement that could be tied to the rapidly developing cerebral cortex, whereas more automatic acts, such as the expression of rage, represented more primitive functions with their circuits housed at lower levels.

During the 1880s and 1890s, Jackson (1881, 1884, 1887–1888, 1894) theorized that insanity was due to a loss of higher control over lower centers. The same could occur after injuries to higher centers, epileptic seizures, or alcoholic intoxication.[8] He realized that the newest parts were most vulnerable to insult and used the term "dissolution" to signify a loss of higher control or a change taking place opposite that encountered in evolution.

These ideas can be traced to Darwin, Spencer, and Bell. The belief that any "release" from higher control could make even the most civilized human being act more like his or her primitive ancestors was clearly "Darwinian." As for the term "dissolution," this was borrowed from Herbert Spencer.[9] Further, the idea that the highest centers were those most vulnerable to insult can be traced to Charles Bell, whose writings must have been familiar to Jackson. In 1823, Bell wrote:

It is a fact familiar to physiologists, that when debility arises from affection of the brain, the influence is greatest on those muscles which are, in their natural condition, most under the command of the will. We may perceive this in the progressive states of debility of the drunkard, when successively the muscles of the tongue, the eyes, the face, the limbs, become unmanageable; and, under the same circumstances, the muscles which have a double office, as those of the chest, lose their voluntary motions, and retain

---

[8]The idea of lower center release from cortical control was further developed by Henry Head in 1918.

[9]Jackson was influenced in many ways by Spencer (Greenblatt, 1965; Riese, 1956). In a footnote to his paper, "Remarks on Evolution and Dissolution of the Nervous System" (1877–1878, p. 31), Jackson expressed his gratitude to Spencer by writing: "I should consider it a great calamity, were any crudities of mine imputed to a man to whom I feel profoundly indebted."

their involuntary motions, the force of the arms is gone long before the action of breathing is affected. (182)

Jackson's schema had a profound influence on the thinking of Sigmund Freud, who acknowledged an intellectual debt to Britain's foremost neurologist. Although Freud was trained in anatomy, he soon became more interested in function than in structure. In psychoanalytic theory, the id is that part of the self that contains the animal instincts—that is, the raw passions and drives. The ego, in turn, has the job of controlling these passions and drives and preventing them from breaking into consciousness where they could affect civilized behavior. In Freud's (1920, 1927) mind, this was comparable to a man trying to ride a strong, passion-driven horse. Seen in this light, Freud essentially took the principles of Jacksonian neurology, which were based on the ideas of Darwin, Spencer, and Bell, and incorporated them into his own psychodynamic schema.

But where was the evidence to suggest that primitive emotions were controlled by the newer parts of the nervous system, as Darwin had implied and Jackson had written? The field of comparative neurology provided fertile ground for such assertions. Developmental studies, such as Darwin's work on his son, also suggested that the emergence of higher brain structures was associated with increasing control over the passions. Still, when it came to hard data, scientists looked increasingly to the records of individuals with well-defined injuries of higher brain structures, and particularly to the growing literature on diseases and injuries of the most recently evolved part of the cerebrum, the prefrontal lobes. This was the part of the brain that most distinguished modern humans from their apish ancestors and the part last to develop. In the minds of many investigators, this was the part of the brain that had to mediate the highest mental functions. This was the "keeper of the beast" within.

THE FRONTAL LOBES AND EMOTION IN HUMANS

Along with Franz Joseph Gall and Johann Spurzheim, the two leading phrenologists, both Emil Huschke (1854) and Paul Broca singled out the frontal lobes as the seat of intellectual functions and suggested that emotion, being a lower function, must reside elsewhere. Broca (1861, p. 338) wrote: "The most noble cerebral faculties . . . have their seat in the frontal convolutions, whereas the temporal, parietal and occipital lobe convolutions are appropriate for the feelings, penchants and passions."

But was this, in fact, the case? The literature by the mid-1800s was already strongly suggesting that both the intellect and the emotions could be affected by frontal insult. The most striking clinical study demonstrating this involved Phineas Gage (1823–1860), the subject of the famous American "crowbar" case that had been described by Henry Bigelow (Figure 19.6) and John Harlow (1819–1907).[10]

Gage, aged 25, worked as a foreman for the Rutland and Burlington Railroad in New England. He achieved immortality when he accidently dropped his tamping iron on a rock while looking over his shoulder. This set off a spark that ignited some blasting powder. The result was that the huge

Figure 19.6. Henry Bigelow (1818–1890), one of the physicians who wrote about Phineas Gage. Bigelow was also involved with the early use of gaseous anesthetics (see Chapter 11).

tamping iron—more than a meter in length and weighing roughly 6 kg—was shot point first through the left side of his head (Bigelow, 1850; Harlow, 1848, 1849, 1868).

> (It) entered the cranium, passing through the anterior left lobe of the cerebrum, and made its exit in the medial line, at the junction of the coronal and sagittal sutures, lacerating the longitudinal sinus, fracturing the parietal and frontal bones extensively, breaking up considerable portions of the brain, and protruding the globe of the left eye from its socket, by nearly half its diameter. (Harlow, 1848, 20)

The huge iron was found some distance away, covered with blood and pieces of brain. It is shown next to Gage's skull in Figure 19.7.

Gage was put on a cart and taken to his hotel, badly hurt but still able to get out and walk upstairs with assistance from his men. Fragments of broken skull were taken off his brain, and his wound was dressed. His mind seemed surprisingly clear, and he said he expected to return to work in a few days. The hemorrhaging ceased two days later, but Gage's mind now seemed affected, and it was concluded that he could not survive the damage. Over the next few days, he did not speak unless aroused, and then only with great difficulty. Gage, who became semicomatose for more than two weeks, also developed a "fungus" on his exposed brain that had to be removed surgically. Miraculously, one month after the injury, it looked as if he would live after all.

Gage returned to his home in New Hampshire, where he continued to recover. After sufficient time to recuperate, he tried to win back his job with the railroad. But because of his personality changes, he was turned down. As a result, he worked in a livery stable and then, in 1852, he left for Chile to help establish a coach line. During this time, he continued to carry the tamping iron that had been shot through his

---

[10]Henry Bigelow also played a role in the history of gaseous anesthetics (see Chapter 11).

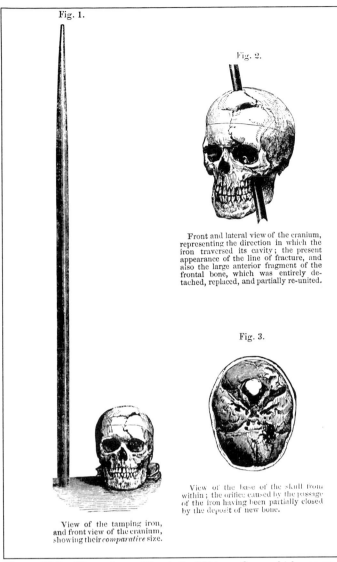

Figure 19.7. Three views of the skull of Phineas Gage, which was penetrated by the railroad tamping iron shown at the left. Gage surprised the authorities by living for 12½ years after this accident. His accident and recovery were described in a series of papers by John M. Harlow and Henry Bigelow. This figure came from Harlow's 1868 report in the *Bulletin of the Massachusetts Medical Society.*

brain. After eight years in South America, he sailed back to California, a dying man suffering from epilepsy. He passed away in 1860.[11]

An autopsy was not performed on Gage's brain, but Gage's mother gave her son's skull and the tamping iron to John Harlow after the body was exhumed from a cemetery in San Francisco. In 1868, Harlow published a full history of Phineas Gage's accident and partial recovery. Harlow later gave the skull and the iron to the museum of the Medical Department at Harvard University.

Harlow (1868) wrote that although Gage showed remarkable recovery, he exhibited changes in his intellect, personality, and emotional stability. He became surprisingly childish, impulsive, and capricious after his accident and now used uncharacteristic profanity (see the quotation at the

beginning of this chapter). In Harlow's (1868) words, Gage was "no longer Gage."

The idea that frontal cortex lesions could affect emotional faculties was confirmed in many clinical case studies soon after this. In 1884, the American neurologist Moses Allen Starr, shown in Figure 19.8, published two papers in which he attempted to review the U.S. literature on patients with localized brain damage confirmed by autopsy. Starr (1884a, 1884b) stated that tumors, abscesses, and even bullet wounds of the frontal lobes may not give rise to notable neurological symptoms. Nevertheless, he emphasized that not all cases of frontal lobe damage were free of symptoms. Here he added that some cases show deficiencies in self-control, lack of self-restraint, and undue excitability. These changes result in a change of character, which he attributed to a loss of inhibition and good judgment.

The idea that temperament could change after damage to the front of the brain was also maintained by Henry Hun, another American. Hun (1887, p. 165) noted that large tumors of the prefrontal areas may not give rise to many noteworthy symptoms until late in the patient's history. In an advanced stage, however, these tumors may cause changes in the character and disposition of the patient.

In Switzerland, Dr. Leonore Welt (1888) described a 37-year-old craftsman who damaged his frontal lobes when he fell from the fourth floor of a house. Before his fall, he had been a relaxed man of good humor. Afterward he became critical, aggressive, and annoying. He blamed other patients and the hospital staff for his problems, played mean tricks on them, and unjustly criticized the way in which he was treated (also see Müller, 1902). These disruptive behaviors were exhibited for about a month, after which there was some improvement. The workman died some months later from an infection, and an autopsy revealed a destruction of

Figure 19.8. Moses Allen Starr (1854–1932), the American neurologist who conducted a number of lengthy analyses of the effects of damage to different lobes of the human brain.

---

[11]Gage's death certificate says 1860. Harlow (1868) erroneously gave the date as 1861.

the right inferior frontal gyrus, as well as damage to the gyrus rectus on both sides.

Welt (1888) added other descriptions of character changes in her patients with frontal cortical lesions. She also provided detailed tables on roughly 50 cases from the literature, some of which dated back to 1819. By comparing those lesions that affected temperament with those that did not, Welt concluded that changes in temperament were most likely to follow damage to the right, medial orbital surface of the frontal lobes.[12]

At the same time, Moritz Jastrowitz (1839–1912) noted that whereas some of his patients with frontal tumors became depressed, others acted elated and euphoric—becoming loquacious, using obscene language, and making many puns. Jastrowitz (1888) considered this odd elevation in mood, which he called *Moria* (stupidity), to be a pathognomonic sign of damage to the frontal lobes.

Hermann Oppenheim observed this behavior in a few of his own patients. But because his patients seemed more sarcastic and biting in their joking, as opposed to being just silly or stupid, Oppenheim (1889, 1890, 1891) used the term *Witzelsucht*, from the German word *Witz* (meaning "wit" or "joke") for this condition. Upon autopsy, Oppenheim seemed surprised to find that his cases of "addiction to joking" seemed to confirm Jastrowitz's frontal lobe localization (Benton, 1991). Four of his patients had tumors of the right frontal lobe, three of which invaded the medial and basal region.

After the observations of Jastrowitz and Oppenheim became better known, the euphoric state with its silly puns and overtalkativeness became a recognized symptom of frontal lobe tumors.[13] Nevertheless, *Witzelsucht* was not something displayed by all frontal lobe cases, nor was this symptom first seen in the late 1800s. Two reports in the early-nineteenth-century German literature, one by Johann F. M. Heyfelder (1798–1869) from 1839 and one by Carl Friedrich Mohr (1806–1879) from 1840, described tumor patients who had become silly and childish. In 1881, M. Bernhardt (1844–1915) noted the same behaviors in some of his patients with frontal lobe tumors.

Paul Flechsig, noted for his theory of cortical myelogenesis, was also intrigued by the emotional changes that may follow prefrontal cortex damage. Flechsig (1905) wrote:

> In the face of these and other facts it seems to me that we have to assume a different idea about the prefrontal defects. The most constant of them concern the emotional reactions to changes in the state of the own person, that is to say the I. The emotional basis of the consciousness of the self, the collaboration of those feelings which form the nucleus of the I and the association of external impressions with this complex seems to me . . . mainly a function of the prefrontal region, and I am here, as far as I can tell, mainly repeating excellent Italian investigators, such as Bianchi, Sergi and Tamburini, who partly experimentally, partly pathologically, have maintained similar ideas for quite a while. (1960 translation 195)

To a certain extent, some of the debates that emerged about the role of the prefrontal cortex in emotional states might have been less heated had more attention been drawn to the fact that symptoms may change with time, with the type of damage, and with the side of insult (see Chapters 26 and 27 for a discussion of the differential effects of right and left hemispheric lesions).

For example, in 1934, a woman with a large, slowly growing tumor centered ventromedially in the anterior frontal lobes was described in a case report by Roy Glen Spurling (1894–?). During the first few months after her tumor was removed, this patient showed many classical "frontal lobe signs," including impaired emotion and altered personality. The following was written about her during this period:

> She was definitely exhilarated and euphoric; there was a youthful lustre in her eyes; she was over-active, playful, facetious, played in a childish manner on the floor with the children, teased them very much and laughed at trivial things that even the children would not laugh at; she tried to dictate to her husband and family. There was an increase in her sexual demands . . . she was extravagant with money. (Ackerly, 1935, 725)

In contrast, the woman was described by her family and neighbors as being "again her old self" within six months after the operation, and her status did not deteriorate in a two-year follow-up (Ackerly, 1935; Spurling, 1934).

By the 1940s, it seemed clearer that extremely aberrant emotional behaviors might diminish considerably after excision of frontal lobe tumors.[14] These findings led some investigators to suggest that some of the changes in personality and emotion seen in tumor cases prior to surgery may be due to the irritative and compressing actions of the tumors on surrounding and neighboring brain areas. Nevertheless, this pattern was not seen in all patients. For instance, there were lasting changes in personality and emotional make-up in the case described by Richard M. Brickner (1896–1959) in 1932 and again in 1936.[15]

THE FRONTAL LOBES AND EMOTION IN ANIMALS

The findings on humans with frontal lobe damage were very intriguing, but much was also being learned about the emotions by studying the effects of damaging the frontal association cortex in laboratory animals, particularly in higher primates. Leonardo Bianchi, who conducted extensive research on the frontal lobes of dogs and monkeys, consistently noted certain emotional changes in his animals after bilateral lesions in these areas. For example, Bianchi wrote the following about one monkey whose major symptoms after surgery included listlessness and abnormal fearfulness:

> She takes no interest in her neighbors and their actions, does not respond to calls, and manifests a quite unusual feeling of terror. She is even afraid of the other cynocephalus, whom

---

[12]An interesting earlier case was described in 1865 by Knörlein. It involved a young soldier who exhibited bizarre social behavior after he sustained damage to the basal part of the prefrontal cortex.

[13]Among the books and papers dealing with this euphoric state are Bernhardt and Borchardt (1909), German and Fox (1934), Rylander (1940, 1941, 1948), and Schuster (1902).

[14]There is a fairly large literature on symptomatology diminishing after frontal lobe surgery to remove tumors (see German and Fox, 1934; Hebb, 1939, 1945; Hebb and Penfield, 1940; Lidz, 1939; Penfield and Evans, 1934, 1935; Stookey, Scharf, and Teitelbaum, 1941).

[15]For a more complete discussion of Brickner's case, and for the contrasting idea that "clean" frontal lobe lesions in humans may produce little in the way of symptomatology, see Chapter 22.

she previously used to hug with evident pleasure. She has lost her previous interests, e.g., in the puppy, of whom she is even afraid if placed near her. (1895, 509)

This monkey no longer showed jealousy when attention was given to the other monkeys, and she no longer showed signs of self-defense or revolt when threatened.

A second monkey with frontal lobe lesions also seemed to have lost her sociality. Bianchi (1895) wrote that she no longer showed affection toward the people who cared for her or toward the other monkeys with whom she had previously played.

Unlike David Ferrier, Eduard Hitzig, or Hermann Munk, who played down the emotional changes they might have witnessed after prefrontal cortex damage, Bianchi (1895, 1920, 1922) viewed these symptoms as so important that he integrated them into his theory of prefrontal lobe function. Specifically, he theorized that the prefrontal areas of the cortex are responsible for the "psychical tone" of the subject. These areas supposedly accomplish this by integrating impressions with emotional states. To quote Bianchi again:

Fear is an immediate result of psychical disaggregation, from defective sense of personality, and unbalanced perception and judgment. As the oyster closes its valves on the approach of a cloud, thus the operated monkey trembles at the simulated hostility of its keeper, ceases to read on his face the smile of kindness, does not improve by past experience, nor perceive the means of escape at hand. Courage rests upon the treble basis of self-conscious force, rapid perception of the enemy's powers . . . and the influence of certain feelings; our animals show an absence of all these characteristics. . . . On the one hand, their affective nature, friendliness and sociability is impaired; on the other, their avidity becomes reckless and insatiable. . . . These symptoms depend . . . upon a dissolution of the psychical personality. (1895, 522)

In 1922, Bianchi went to great lengths to explain that the frontal lobes were not responsible for the basic biological drives of reproduction, temperature regulation, sleeping, and so on, which he called the "fundamental emotions." Bianchi singled out the social sentiments (friendship, social understanding, true love, social obligation, etc.) as those emotions most clearly dependent upon the prefrontal lobes. He wrote that these sentiments, or spiritual aspects of mind, require the greatest synthesis of intellect, perception, and emotion in consciousness and an appreciation of the environment and the feelings of other members of the social group.

"Sociality," Bianchi argued, has had a long and slow evolution—one that parallels the emergence of the prefrontal lobes. Thus, he reasoned, it is best developed in humans, although rudiments of sociality can be found in monkeys and perhaps other mammals, such as dogs. It is lost after severe damage to the frontal lobes in humans and animals and may be rudimentary in individuals whose frontal lobes fail to develop (e.g., idiots and imbeciles).

In every case of mutilation of the frontal lobes in monkeys, what was found to be suppressed was, without exception, that sentimentality or feeling for others that we designate as *sociality*. A similar condition exists in human beings who have suffered severe injury to both frontal lobes and in those whose frontal brains are imperfectly developed. The majority of the insane exhibit the same feature. Sociality is

thus a psychic manifestation which comes with evolution. It varies in different individuals and races and becomes suppressed by mental disease and severe frontal lesions. (Bianchi, 1922, 289)

THE JAMES-LANGE THEORY

In 1884, William James, who, like Jackson, admired the ideas of Charles Darwin, published a controversial paper entitled "What Is an Emotion?" James proposed that the nervous system was predisposed to respond in certain ways to particular features of the environment. He argued that sensation immediately triggered certain bodily changes, which he classified as "object-simply-apprehended." The subsequent conscious experience of these bodily responses was "object-emotionally-felt." Thus, James envisioned a circuit running from the sense organs to the brain for sensation, then to the viscera and muscles for action, and finally back to the cortex for perceptual awareness.

This sort of thinking turned conventional wisdom upside down. Traditionally, it was thought that awareness of emotional situations triggered sweating, shaking, and changes in heart rate and respiration. James (1884) argued that not only was this wrong but it was contrary to what was really happening. The emphasis had to be on the awareness of the backflow from the viscera. In his words:

Common sense says, we lose our fortune, are sorry and weep; we meet a bear, are frightened and run; we are insulted by a rival, are angry and strike. The hypothesis here to be defended says that this order of sequence is incorrect . . . and that the more rational statement is that we feel sorry because we cry, angry because we strike, afraid because we tremble, and not that we cry, strike, or tremble, because we are sorry, angry, or fearful as the case may be. Without the body states following on the perception, the latter would be purely cognitive in form, pale, colourless, destitute of emotional warmth. We might then see the bear . . . but we could not actually *feel* afraid. (1969 reprint, 247–248)

A theory related to the one proposed by William James (1884) appeared in a small pamphlet written by Carl G. Lange in 1885.[16] This work was translated from the original Danish into German in 1887 (see James and Lange, 1922, for the English translation of the German version). Like James, Lange limited his theory to the simple emotions (called the "coarse emotions" by James) of fright, joy, anger, and sorrow. Love, hate, scorn, admiration, and the like were more complicated higher-level passions to which his theory was not meant to apply.

Lange examined the physiological changes that accompanied each of his four basic emotions. He then asked whether vascular changes played a primary role in emotion and whether other bodily changes that can contribute to the feeling of an emotion were actually secondary to these vascular changes. Lange wrote:

---

[16]James and Lange both taught at the institutions that granted them their degrees. James graduated from Harvard University Medical School in 1872 and remained there. Lange was born in Zealand, Denmark, and studied medicine at the University of Copenhagen. He served as a professor of pathological anatomy at that institution from 1877 to 1900.

Is it possible that vasomotor disturbances, varied dilation of the blood vessels, and consequent excess of blood, in the separate organs, are the real, primary effects of the affections, whereas the other phenomena,—motor abnormalities, sensation paralysis, subjective sensations, disturbances of secretion and intelligence—are only secondary disturbances, which have their cause in anomalies of vascular innervation? (1922 translation, 58)

After reviewing the evidence, Lange proposed the following:

We owe all the emotional side of our mental life, our joys and sorrows, our happy and unhappy hours, to our vasomotor system. If the impressions which fall upon our senses did not possess the power of stimulating it, we would wander through life unsympathetic and passionless, all impressions of the outer world would only enrich our experience, increase our knowledge, but would arouse neither joy nor anger, would give us neither care nor fear. (1922 translation, 80)

Neither James nor Lange constructed his "peripheral" theory on anything resembling controlled experimentation. Instead, both theories were based largely on introspection and correlative data. In particular, both James and Lange argued that as much as one could imagine an emotional state, that experience would be flat unless the myriad of bodily changes felt during a real emotional experience could be triggered by the mind. In fact, they argued, it was impossible to imagine real fear, anger, and related emotions on demand. To quote James (1884):

If we fancy some strong emotion, and then try to abstract from our consciousness of it all the feelings of its characteristic bodily symptoms, we find that we have nothing left behind, no "mind stuff" out of which the emotions can be constituted, and that a cold and neutral state of intellectual perception is all that remains. . . . I say that for *us*, emotion dissociated from all bodily feeling is inconceivable. The more closely I scrutinize my states, the more persuaded I become, that whatever moods, affections, and passions I have, are in very truth constituted by, and made up of, those bodily changes we ordinarily call their expression or consequence. (1969 reprint 253–255)

And to quote Lange on the same issue:

Take away the bodily symptoms from a frightened individual; let his pulse beat calmly, his look be firm, his color normal, his movements quick and sure, his thoughts clear; and what remains of his fear? (1922 translation, 66)

James and Lange cited clinical findings to support their ideas. For instance, Lange discussed observations made by Astley Cooper (1839–1905), a respected English surgeon. Cooper observed a man with a skull defect and found that the vascular flow to this man's brain increased every time he was upset.

For his part, James wrote that he was even willing to admit his theory would be proved wrong if one could find cases who were "anesthetic both without and within," yet who could still experience real emotions. He was aware of a few individuals who were capable of exhibiting normal emotions after seemingly profound sensory losses, but after thinking it over, he concluded that these cases did not really represent a good test of his theory.

James included large sections from the earlier *Mind*

paper in his celebrated *Principles of Psychology*, published in 1893. Nevertheless, his chapter on the emotions differed from his 1884 journal article in a number of significant ways. First, James now compared and contrasted his ideas about emotion with those presented by Lange. Second, James unequivocally stated that his theory required no special centers for emotion in the brain, just the basic sensory and motor cortical centers which scientists had already discovered. Finally, he retracted some of his overstatements and, in his words, "the slapdash brevity" of his first paper.

In particular, James (1893) felt he overstated his case and provided a poor example when he wrote that one sees a bear, runs, and only then becomes afraid. James emphasized that he never meant to tie any given emotion to a specific act. He recognized that running does not have to trigger fear and that crying can occasionally be associated with joy rather than sorrow. Still, the core of his theory—which held that until there are bodily changes there could not be "real" emotions—remained very much intact.

The theory that emotional changes in the body must precede a feeling of emotion became broadly known as the James-Lange theory, even though the thoughts of these two men on the role of the vascular system in emotion were dissimilar. The merged theory attracted considerable attention.[17] For example, John Dewey (1859–1952), the American philosopher and educator, published some papers in which he tried to unite some of the evolutionary ideas of Darwin and Spencer with the James-Lange "peripheral theory" of emotion (e.g., Dewey, 1894).

CRITICISMS OF THE JAMES-LANGE THEORY

Despite the stature of William James as a philosopher and the support given to the theory by such learned men as John Dewey, not everyone was willing to accept the James-Lange theory. Indeed, almost immediately after James (1884) and Lange (1885) presented their ideas, a number of critics came forth to do battle. The situation became so heated that William James felt compelled to write a paper just to rebut some of the negative comments of Wilhelm Wundt and other introspective psychologists and philosophers. This rebuttal and defense appeared in the 1894 issue of *Psychological Review*.

In retrospect, the strongest thrust against the James-Lange theory did not come from the psychologists, who were relatively lightly armed, but from a second flank manned by experimental physiologists and neurologists. Their assault was somewhat slower in developing. When it came, however, it included experiments on dogs that still showed signs of emotion after the cervical cord and vagus nerves were cut, as well as human cases, who exhibited marked sensory losses and muscular paralyses, but still experienced seemingly normal fright and joy.

To quote Charles L. Dana (1852–1935), one of the individuals who published an article highly critical of the James-Lange theory:

---

[17]See Meyers (1986) for a scholarly discussion of the scientific impact of the James-Lange theory.

I had a patient . . . who fell from a horse and broke her neck at the third and fourth cervical level. She was completely quadriplegic and suffered complete loss of cutaneous and deep sensation from the neck down—with abolition of all deep reflexes. . . . She lived for nearly a year, and during that time I saw her showing emotions of grief, joy, displeasure and affection. There was no change in her personality or character. . . . It is difficult to understand, on the peripheral theory, why there should have been no change in her emotionality, with the skeletal system practically eliminated and the sympathetic system entirely so. (1921, 636).

Perhaps more than anyone else, Walter Bradford Cannon (1871–1945) contended that the James-Lange theory could not possibly be substantiated. Like James, Cannon (1927, 1931) hailed from Harvard University. But unlike James the philosopher, Cannon entered the arena by conducting a myriad of surgical experiments, most of which were on cats.

For example, Cannon, Lewis, and Britton (1927) noticed that even when the nervous projections from the viscera to the central nervous system were cut, their cats still showed emotional responses indicative of anger, disgust, fear, and joy. Cannon also recognized that the visceral responses were slower than the conscious emotional states they were supposed to trigger. Further, he noted that the visceral responses accompanying the most intense emotional states were diffuse and nonspecific. This raised serious questions about how seemingly similar accelerations of the heart, inhibition of digestive glands, increased blood sugar, and dilation of bronchioles could signal very different emotional experiences. In fact, Cannon showed that some of the same visceral changes occurred in states not associated with intense emotion (chilliness, fever, hypoglycemia, asphyxia). Another observation Cannon felt was particularly damaging to the James-Lange theory was that changes in the internal organs typically are not experienced in consciousness, largely because the viscera have few sensory nerves.

Cannon summarized his experimental findings and ideas in an influential monograph entitled *Bodily Changes in Pain, Hunger, Fear and Rage.* This book appeared in 1915; a revised and expanded second edition was published in 1929.

The major criticism leveled against Cannon's thinking was that although his animals seemed able to act as if they were angry in the absence of visceral feedback, one could not say for sure that they really felt or experienced "true" anger. In this regard, Cannon's protest against the James-Lange theory was seen as unfair or as an instance of setting up a "straw man." These critics argued that James had proposed a true theory of feeling, whereas Cannon was just studying overt expressive behavior and some of its physiological correlates (see Meyers, 1986).

Cannon countered his critics by showing that studies involving humans who could adequately describe their feelings also contradicted the James-Lange theory. For example, some college students who injected adrenalin showed many of the visceral changes associated with emotional states. Their hearts throbbed, their hands trembled, and their bodies acted as if chilled. But, noted Cannon (1927), there were absolutely no emotional feelings accompanying these bodily changes.

In addition, Cannon turned to the literature on pathological laughing and crying. The diseases that can cause these exaggerated displays suggestive of emotion include pseudobulbar paralysis, disseminated sclerosis, and double hemiplegia. Some patients with these disorders had recently been described by Wilson (1924, 1929), who observed that these emotional outbursts correlated poorly with his patients' real feelings. Wilson was surprised to find that some patients were genuinely disturbed by bodily responses at variance with how they really felt. To quote:

> From what has been said it will be understood that the apparent, visible emotion does not necessarily correspond to the patient's real feelings at the time—an observation which has often been made. . . . I have endeavoured to ascertain from intelligent patients whether when thus overcome with laughter against their will and in opposition to their real feelings they do not, in spite of the latter, end by experiencing the emotional state commonly associated with laughter, and I am satisfied it is not so, in some instances at least. (Wilson, 1924, 308–309)

Finally, Cannon cited a number of case studies of totally or nearly completely paralyzed patients who still felt emotions comparable to those of normal individuals. He pointed out that the projections from the viscera were also severely damaged in some of these cases.

Together, the findings and arguments presented by Cannon and others in the 1920s caused many people to abandon the James-Lange theory. As stated by Philip Bard (1898–1977) in his review of the evidence against the James-Lange theory: "The most that can be said for it is that bodily reverberations may add slightly to, and so reinforce, the central processes which underlie emotional consciousness" (Bard, 1929, p. 483).

REFERENCES

Ackerknecht, E. H. 1974. The history of the discovery of the vegetative (autonomic) nervous system. *Medical History, 18,* 1–8.

Ackerly, S. 1935. Instinctive, emotional and mental changes following prefrontal lobe extirpation. *American Journal of Psychiatry, 92,* 717–728.

Aristotle. 1908. *Parva naturalia.* (J. I. Beare, Trans.). Oxford, U.K.: Clarendon Press.

Aristotle. 1966. "On the motion of animals." (A. S. L. Farquharson, Trans.). In M. R. Cohen and I. E. Drabkin (Eds.), *A Source Book in Greek Science.* Cambridge, MA: Harvard University Press. Pp. 554–555.

Aristotle. 1975. *Nichomachean Ethics.* Boston, MA: D. Reidel and Company.

Bard, P. 1929. Emotion. I. The neuro-humoral basis of emotional reactions. In C. Murchison (Ed.), *The Foundations of Experimental Psychology.* Worcester, MA: Clark University Press. Pp. 449–487.

Bell, C. 1806. *Essays on the Anatomy of Expression in Painting.* London: Longman.

Bell, C. 1823. On the motions of the eye, in illustration of the use of the muscles and nerves of the orbit. *Philosophical Transactions of the Royal Society of London, 113,* 166–186.

Bell, C. 1827. Partial paralysis of the face. *London Medical Gazette, 1,* 747–750.

Benton, A. L. 1991. The prefrontal region: Its early history. In H. Levin, H. Eisenberg, and A. L. Benton (Eds.), *Frontal Lobe Function and Dysfunction.* Pp. 3–32.

Bernhardt, M. 1881. *Beiträge zur Symptomatologie und Diagnostik der Hirngeschwülste.* Berlin: Hirschwald.

Bernhardt, M., and Borchardt, M. 1909. Zur Klinik der Stirnhirntumoren nebst Bemerkungen über Hirnpunktion. *Berliner klinische Wochenschrift, 46,* 1341–1347.

Bianchi, L. 1895. The functions of the frontal lobes. (A. de Watteville, Trans.). *Brain, 18,* 497–530.

Bianchi, L. 1920. *La Meccanica del Cervello e la Funzione dei Lobi Frontalii.* Torino: Bocca.

Bianchi, L. 1922. *The Mechanism of the Brain.* (J. H. MacDonald, Trans.). Edinburgh: E. & S. Livingstone.

Bichat, F.-X. 1805. *Recherches Physiologiques sur la Vie et la Mort* (3rd ed). Paris: Brosson/Gabon. Translated by F. Gold as, *Physiological Researches on Life and Death*. Boston: Richardson and Lord, 1827. Reprinted in *Significant Contributions to the History of Psychology, Ser. E, Physiological Psychology: Vol. II, X. Bichat, J. G. Spurzheim, P. Flourens*. D. N. Robinson (Ed.). Washington DC: University Publications of America, 1978. Pp. i–334.

Bigelow, H. J. 1850. Dr. Harlow's case of recovery from the passage of an iron bar through the head. *American Journal of the Medical Sciences, 19*, 13–22.

Boerhaave, H. 1708. *Institutiones medicae in usus annu exercitationis domestic*. Lugduni Batavorum: Petrum van der Eyk.

Brickner, R. M. 1932. An interpretation of frontal lobe function based upon the study of a case of partial bilateral frontal lobotomy. *Research Publications—Association for Research for Nervous and Mental Disease, 13*, 259–351.

Brickner, R. M. 1936. *The Intellectual Functions of the Frontal Lobes*. New York: Macmillan.

Broca, P. 1861. Remarques sur le siège de la faculté du langage articulé; suivies d'une observation d'aphémie (perte de la parole). *Bulletins de la Société Anatomique (Paris), 6*, 330–357, 398–407. In G. von Bonin, *Some Papers on the Cerebral Cortex*. Translated as, "Remarks on the seat of the faculty of articulate language, followed by an observation of aphemia." Springfield, IL: Charles C Thomas, 1960. Pp. 49–72.

Cannon, W. B. 1915. *Bodily Changes in Pain, Hunger, Fear and Rage*. New York: Appleton.

Cannon, W. B. 1927. The James-Lange theory of emotion: A critical examination and an alternative theory. *American Journal of Psychology, 39*, 10–124.

Cannon, W. B. 1929. *Bodily Changes in Pain, Hunger, Fear and Rage*. (2nd ed). New York: Appleton.

Cannon, W. B. 1931. Again the James-Lange and the thalamic theories of emotion. *Psychological Review, 38*, 281–295.

Cannon, W. B., Lewis, J. T., and Britton, S. W. 1927. The dispensability of the sympathetic division of the autonomic nervous system. *Boston Medical and Surgical Journal, 197*, 514–515.

Clark, R. W. 1984. *The Survival of Charles Darwin: A Biography of a Man and an Idea*. New York: Random House.

Cuthbertson, R. A. 1978. The first published clinical photographs? *Practitioner, 221*, 276–278.

Cuthbertson, R. A. 1979. Duchenne de Boulogne. *Australian and New Zealand Journal of Surgery, 49*, 275–278.

Cuthbertson, R. A. 1985. Duchenne de Boulogne and human facial expression. *Clinical and Experimental Neurology, 21*, 55–67.

Cuthbertson, R. A., and Hueston, J. T. 1979. Duchenne de Boulogne and clinical photography, *Annals of Plastic Surgery, 2*, 332–337.

Dana, C. L. 1921. The anatomic seat of the emotions: A discussion of the James-Lange theory. *Archives of Neurology and Psychiatry, 6*, 634–639.

Darwin, C. 1839. *A Journal of the Voyage of the Beagle*. London.

Darwin, C. 1859. *On the Origin of the Species by Means of Natural Selection, or the Preservation of Favoured Races in the Struggle for Life*. London: John Murray.

Darwin, C. 1871. *The Descent of Man and Selections in Relation to Sex*. London: John Murray.

Darwin, C. 1872. *The Expression of the Emotions in Man and Animals*. London: John Murray.

Darwin, C. 1877. A biographical sketch of an infant. *Mind: Quarterly Review of Psychology and Philosophy, 2*, 285–294.

Descartes, R. 1650. *Les Passions de l'Ame*. Amsterdam: L. Elzevir.

Dewey, J. 1894. The theory of emotion: 1. Emotional attitudes. *Psychological Review, 1*, 568–569.

Duchenne, G.-B.-A. 1850. Recherches électro-physiologiques sur les fonctions des muscles de la face. *Bulletin de l'Académie de Médecine, 15*, 491, 533, 634.

Duchenne, G.-B.-A. 1862a. *L'Electrisation Localisée et son Application à la Physiologie, à la Pathologie et à la Thérapeutique* (2nd ed). Paris: J. B. Ballière.

Duchenne, G.-B.-A. 1862b. *Mécanisme de la Physionomie Humaine, ou Analyse Electro-Physiologique de l'Expression des Passions Applicable à la Practique des Arts Plastiques*. Paris: J. B. Ballière.

Duchenne, G.-B.-A. 1867. *Physiologie des Mouvements*. Paris: J. B. Ballière.

Estienne, C. 1545. *De dissectione partium corporis humani*. Paris: S. Colins.

Eulenburg, A., and Guttmann, P. 1863. *Die Pathologie des Sympathicus auf physiologischer Grundlage*. Berlin: A. Hirschwald.

Eustachio, B. 1563. *Opuscula anatomica*. Venetiis: Vincentius Luchinus.

Eustachio, B. 1714. *Tabulae anatomicae*. Roma: F. Gonzaga.

Flechsig, P. E. 1905. Gehirnphysiologie und Willenstheorien. *Fifth International Psychology Congress*, Rome, 73–89. In G. von Bonin, *Some Papers on the Cerebral Cortex*. Translated as, "Brain physiology and theories of volition." Springfield, IL: Charles C Thomas, 1960. Pp. 181–200.

Freud, S. 1920. *A General Introduction to Psychoanalysis*. (J. Riviere, Trans.). New York: Boni and Liveright.

Freud, S. 1927. *The Ego and the Id*. London: Hogarth Press.

Gall, F. J., and Spurzheim, J. 1810–1819. *Anatomie et Physiologie du Système Nerveux en Général, et du Cerveau en Particulier*. Paris: F. Schoell. (Gall was the sole author of the first two volumes of the four in this series.)

German, W. J., and Fox, J. C., Jr. 1934. Observations following unilateral lobectomies. *Research Publications—Association for Research for Nervous and Mental Disease, 13*, 378–434.

Greenblatt, S. H. 1965. The major influences on the early life and work of John Hughlings Jackson. *Bulletin of the History of Medicine, 39*, 346–376.

Guilly, P. 1936. *Duchenne de Boulogne*. Paris: J. B. Ballière.

Harlow, J. M. 1848. Passage of an iron rod through the head. *Boston Medical and Surgical Journal, 39*, 389–393.

Harlow, J. M. 1849. Letter in "Medical Miscellany." *Boston Medical and Surgical Journal, 39*, 506–507.

Harlow, J. M. 1868. Recovery from the passage of an iron bar through the head. *Bulletin of the Massachusetts Medical Society, 2*, 3–20.

Head, H. 1918. Some principles of neurology. *Brain, 41*, 344–354.

Hebb, D. O. 1939. Intelligence in man after large removals of cerebral tissue: Report on four left frontal lobe cases. *Journal of General Psychology, 21*, 73–87.

Hebb, D. O. 1945. Man's frontal lobes: A critical review. *Archives of Neurology and Psychiatry, 54*, 10–24.

Hebb, D. O., and Penfield, W. 1940. Human behavior after extensive bilateral removal from the frontal lobes. *Archives of Neurology and Psychiatry, 44*, 421–438.

Heyfelder, J. F. M. 1839. Gehirntubercle. *Medizniche Annalen, 5*, 119.

Hueston, J. T., and Cuthbertson, R. A. 1978. Duchenne de Boulogne and facial expression. *Annals of Plastic Surgery, 1*, 411–420.

Hun, H. 1887. A clinical study of cerebral localization, illustrated by seven cases. *American Journal of Medical Sciences, 93*, 140–168.

Huschke, E. 1854. *Schaedel, Hirn und Seele des Menschen und der Thiere nach Alter, Geschlecht und Race*. Jena: F. Mauke.

Jackson, J. H. 1881. Remarks on dissolution of the nervous system as exemplified by certain post-epileptic conditions. *Medical Press and Circular*, April, 329–332ff.

Jackson, J. H. 1884. Evolution and dissolution of the nervous system. *British Medical Journal, 1*, 591, 660, 703. (Also in *Popular Science Monthly, 25*, 171–180, 1884.)

Jackson, J. H. 1887–1888. Remarks on evolution and dissolution of the nervous system. *Journal of Mental Science, 33*, 25–48.

Jackson, J. H. 1894. The factors of the insanities. *Medical Press and Circular*, June, 615–619.

James, W. 1884. What is an emotion? *Mind, 9*, 188–205. Reprinted in W. James, *Collected Essays and Reviews*. New York: Russell and Russell, 1969. Pp. 244–275.

James, W. 1893. *Principles of Psychology*. New York: Henry Holt.

James, W. 1894. The physical basis of emotion. *Psychological Review, 1*, 516–529.

James, W., and Lange, C. G. 1922. *The Emotions*. Baltimore, MD: Williams & Wilkins.

Jastrowitz, M. 1888. Beiträge zur Localisation im Grosshirn und über deren praktische Verwerthung. *Deutsche medizinische Wochenschrift, 14*, 81–83, 108–112.

Knörlein, 1865. Krankengeschichte und Sectionsbefund eines basalen Hirntumors. *Allgemeine Wiener medizinische Zentral-Zeitung*, 250–252.

Landis, C. 1929. Emotion: II. The expression of emotion. In C. Murchison (Ed.), *The Foundations of Experimental Psychology*. Worcester, MA: Clark University Press. Pp. 488–523.

Lange, C. S. 1887. *Ueber Gemüthsbewegungen*. Translated from the Danish by H. Kurella. Leipzig. Translated into English in W. James and C. G. Lange (Eds.), *The Emotions*. Baltimore, MD: Williams & Wilkins, 1922. Pp. 33–90.

Lidz, T. 1939. A study of the effect of right frontal lobectomy on intelligence and temperament. *Journal of Neurology and Psychiatry, 2*, 211–222.

Lobstein, J. G. 1823. *De nervi sympathici humani fabrica*. Paris: F. G. Lerault.

Marks, R. L. 1991. *Three Men of the Beagle*. New York: Knopf.

Meyers, G. E. 1986. *William James: His Life and Thought*. New Haven, CT: Yale University Press.

Mohr, C. F. 1840. Hypertrophie . . . der Hypophysis cerebri und dadurch bedingter Druck auf die Hirngrundfläche. . . . *Wochenschrift für die gesamte Heilkunde*, 565.

Müller, E. 1902. Ueber psychische Störungen bei Geschwülsten und Verletzungen des Stirnhirns. *Deutsche Zeitschrift für Nervenheilkunde, 21*, 177–208.

Oppenheim, H. 1889. Zur Pathologie der Grosshirngeschwülste. *Archiv für Psychiatrie und Nervenkrankheiten, 21*, 560–587.

Oppenheim, H. 1890. Zur Pathologie der Grosshirngeschwülste. *Archiv für Psychiatrie und Nervenkrankheiten, 22*, 27–72.

Oppenheim, H. 1902. *Die Geschwülste des Gehirns*. (2nd ed.). Wien: Hölder.

Penfield, W., and Evans, J. P. 1934. Functional deficits produced by cerebral lobectomies. *Research Publications—Association for Research for Nervous and Mental Disease, 13*, 352–377.

Penfield, W., and Evans, J. P. 1935. The frontal lobe in man: A clinical study of maximum removals. *Brain, 58,* 115–133.

Plato. 1952. *Timaeus.* (B. Jowett, Trans.). In *Great Books of the Western World: Vol. 7. Plato.* Chicago: Encyclopaedia Britannica. Pp. 442–477.

Riese, W. 1956. The sources of Jacksonian neurology. *Journal of Nervous and Mental Disease, 124,* 125–134.

Robinson, D. N. 1989. *Aristotle's Psychology.* New York: Columbia University Press.

Ruckmick, C. A. 1936. *The Psychology of Feeling and Emotion.* New York: McGraw-Hill.

Rylander, G. 1940. *Personality Changes after Operations on the Frontal Lobes.* Copenhagen: Munksgaard.

Rylander, G. 1941. Brain surgery and psychiatry. *Acta Chirugica Scandinavica, 85,* 213–234.

Rylander, G. 1948. Personality analysis before and after frontal lobotomy. *Research Publications—Association for Research in Nervous and Mental Disease, 27,* 691–705.

Schuster, P. 1902. *Psychische Störungen bei Hirntumoren.* Stuttgart: Enke.

Spillane, J. D. 1981. *The Doctrine of the Nerves.* Oxford, U.K.: Oxford University Press.

Spurling, R. G. 1934. Notes upon the functional activity of the prefrontal lobes. *Southern Medical Journal, 27,* 4–9.

Starr, M. A. 1884a. Cortical lesions of the brain. A collection and analysis of the American cases of localized cerebral disease. *American Journal of Medical Sciences, 88,* 114–141.

Starr, M. A. 1884b. Cortical lesions of the brain. A collection and analysis of the American cases of localized cerebral disease. *American Journal of Medical Sciences, 87,* 366–391.

Stone, I. 1980. *The Origin.* New York: Signet Books.

Stookey, B., Scarff, J., and Teitelbaum, M. 1941. Frontal lobectomy in the treatment of brain tumors. *Annals of Surgery, 113,* 161–169.

Telfer, W. 1955. *Cyril of Jerusalem and Nemesius of Emesa.* Philadelphia, PA: Westminster Press.

Welt, L. 1888. Ueber Charakterveränderungen des Menschen infolge von Läsionen des Stirnhirns. *Deutsches Archiv für klinische Medicin, 42,* 339–390.

Willis, T. 1664. *Cerebri anatome: cui accessit nervorum descriptio et usus.* London: J. Flesher.

Wilson, S. A. K. 1924. Pathological laughing and crying. *Journal of Neurology and Psychopathology, 4,* 299–333.

Wilson, S. A. K. 1929. *Modern Problems in Neurology.* New York: Wood.

Young, R. M. 1985. *Darwin's Metaphor.* Cambridge, U.K.: Cambridge University Press.

Chapter 20
Defining and Controlling the Circuits of Emotion

The mantle of the hemisphere is composed of one part that is brute—represented by the
great limbic lobe—and another that is intelligent—represented by the rest.

Paul Broca, 1878

Chapter 19 reviewed a number of early, speculative theories about the emotions, beginning with Ancient Greek ideas and culminating with the highly controversial James-Lange theory of the nineteenth century. It also presented certain contributions that proved to be more lasting. Foremost among these was the notion that the circuitry of emotion could best be understood from an evolutionary perspective. This idea stemmed from Charles Darwin's and Herbert Spencer's theories of evolution and natural selection and was applied to neurological thinking by John Hughlings Jackson. He postulated that phylogenetically newer parts of the brain keep tight reins on more primitive parts and that damage to the former can result in a "release" of control and subsequent expression of more animal-like propensities. Jackson suggested that the cerebral cortex, and the frontal lobes in particular, had the job of controlling the beast within. But precisely where that beast resided and how its actions were mediated proved to be much less clear.

This chapter examines four subjects that were only hinted at in the previous chapter. The first concerns how the autonomic nervous system emerged from early anatomical observations and the physiological concept of sympathy. The second entails how investigators came to believe that the hypothalamus exercised control over the sympathetic nervous system. The third subject is the development of the limbic system model of emotional awareness and expression. Lastly, this chapter looks at the idea that so-called emotional disorders can be treated with "psychosurgery."

THE EARLY CONCEPT OF SYMPATHY AND THE AUTONOMIC NERVOUS SYSTEM

During the Roman period, Galen, working primarily on apes and pigs, described the ganglionated (sympathetic) chain that ran alongside the spinal cord and innervated the viscera. He also described the sympathetic ganglia and the white rami communicantes (Sheehan, 1936; Smith, 1971).

Galen mistakenly believed that the ganglionated chain came from the brain.[1] From this error, he deduced that these nerves had to be involved in sensory functions. This was because the brain was soft and therefore capable of receiving sensory impressions. In contrast, the harder projections from the spinal cord presumably gave the viscera their motor power. Galen felt that extensive interconnections among the ganglionated nerves permitted the animal spirits to travel from one organ to another. From this premise, he concluded that there was considerable functional unity, or physiological "sympathy," among the internal organs served by these nerves.

Knowledge about the autonomic nervous system hardly advanced from the time of Galen until the Renaissance. It was not until the sixteenth century that Charles Estienne (1552) and Bartholomeo Eustachio (1545) split the vagus and the sympathetic nerves apart (Eustachio's notes were not published until 1563, and his copper plates were lost and not printed until 1714). Although this was a major advance, Eustachio still showed the ganglionated chain going to the brain (see Figure 20.1).

In 1664, Thomas Willis called the ganglionated nerves the "intercostal" nerves because they were located near the ribs. He believed that these nerves descended from the posterior part of the brain (his broadly defined "cerebellum"), as can be seen in Figure 20.2. This was the part of the brain he associated with involuntary or automatic motion. Hence, Willis associated the sympathetic nerves with motions of the heart and blood, and the process of respiration.

Willis also associated the separate vagus nerves ("the eighth pair" or the "wandering" nerves) with the posterior part of the brain. To his credit, he maintained that the

---

[1]Galen's statements about the origin of the sympathetic nerve were fuzzy and confusing. In *De anatomicis administrationibus* (XIV, 9; S. 1, 273) he wrote that "the thought quickly arises that this nerve is a part of the sixth pair." His sixth pair included the vagus, glossopharyngeal, and accessory nerves. He then signified that it was really a part of his fourth pair, even though it intermingled with the branches of the sixth (especially the vagus). In *De nervorum dissectione*, which was written soon after *De anatomicis administrationibus*, the sympathetic nerve was attributed to the third pair of nerves. See Smith (1971) for a thorough discussion of this issue.

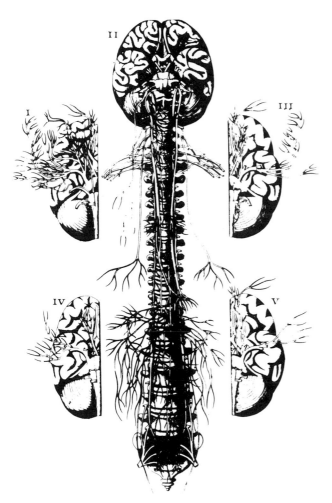

Figure 20.1. A copper plate commissioned by Bartolomeo Eustachio for his *Tabulae anatomicae,* published in 1563. This plate shows the sympathetic nerves arising from the cranium. Eustachio's plates were lost until 1714.

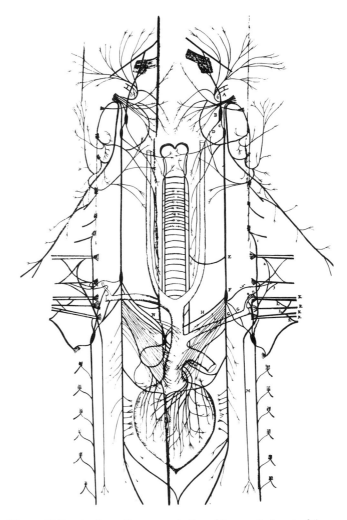

Figure 20.2. A schematic representation of the cervical part of the sympathetic nervous system and the vagus nerve in humans. This figure, from Thomas Willis' (1664) *Cerebri anatome,* erroneously shows the sympathetic trunk originating from the cranium.

vagus branched off to the arch of the aorta "so that it may react to changes in the pulse." He observed that cutting the vagus nerves in a dog resulted in "great trembling" of the heart.

Willis did not disregard the cerebrum in his model of control of the visceral functions. He maintained that especially strong spirits from the viscera could travel all the way to the cerebrum to be consciously appreciated. He also postulated that spirits from the cerebrum had the potential to pass through the cerebellum, allowing them to affect largely involuntary acts, such as the depth and rate of breathing.

To the extent that Willis postulated both higher and lower levels of control over the viscera and the potential for interaction, his theory represented an advance over earlier thinking. Nevertheless, Willis still accepted the Galenic idea that sympathy (communication) could be maintained by the flow of animal spirits among the nerves connecting the internal body parts. And, in a traditional way, he continued to view the sympathetic ganglia as small "storehouses" for the animal spirits.

In 1727, François Pourfour du Petit cut the cervical sympathetic nerves in a dog and noted that this affected pupil size, the nictitating membrane, and secretions from the eye. This meant that some branches of the intercostal nerves actually projected to the face. His work cast doubt on the old idea that the sympathetic nerves were sensory nerves to the brain.

Jacobus ("Jacques") Winslow (1669–1760), a Danish-born professor working in Paris, agreed with Willis that the ganglia of the intercostal nerves acted like "small brains." He also accepted Pourfour du Petit's belief that the intercostal nerves sent projections to the face and head. But because Winslow did not like the terms "intercostal nerves" or "ganglionated nerves," he proposed yet another term. Thus, the name "great sympathetic nerve" was introduced in 1732 to replace the older terminology.

Winslow (1732) actually described three "sympathetic" nerves. In addition to the great sympathetic nerve, he wrote about a "small sympathetic nerve" and a "middle sympathetic nerve." His small sympathetic was the portio dura of the facial nerve, and his middle sympathetic was, in fact, the

Figure 20.3. Walter H. Gaskell (1847–1914). Working at Cambridge University, Gaskell conducted many studies on the autonomic nervous system late in the nineteenth century. He concluded that the "involuntary system" was, in fact, composed of two antagonistic subsystems.

Figure 20.4. John Newport Langley (1852–1925), the Cambridge University physiologist who in 1898 gave the "autonomic nervous system" its name. Langley later called Walter Gaskell's two opposing divisions of the autonomic nervous system "sympathetic" and "parasympathetic."

vagus nerve. Winslow's association of facial and vagus nerves with the great sympathetic nerve received little support (these two structures would later be identified as components of the parasympathetic nervous system; Langley, 1916). Nevertheless, as a result of Winslow's strong belief that the sympathetic nerves were concerned with both the control of the viscera and the production of sympathy, the word "sympathetic," which was initially presented as a physiological construct, became a part of the anatomical vocabulary (Ackerknecht, 1974; Sheehan, 1936).

SYMPATHETIC AND PARASYMPATHETIC SYSTEMS

The nervous control over the viscera was the subject of a number of important physiological experiments during the nineteenth century. In 1845, the brothers Ernst Heinrich Weber and Eduard Weber (1806–1871) stopped the heart by stimulating the vagus nerve.[2] The results of their experiment were soon confirmed by others, who also noted that stimulation of sympathetic nerves could accelerate the heart's action (see Ackerknecht, 1974).

These contrasting findings led Walter H. Gaskell (1847–1914; shown in Figure 20.3) to conduct a series of experiments on heart physiology. From his results, he concluded that the "involuntary system" was, in fact, composed of two antagonistic subsystems. In 1886, he wrote:

From what has been said we may with safety draw the conclusion that the muscular tissue of both the visceral and vas-

[2]The experiment by Weber and Weber, which took place in Leipzig, was especially important because it led to the notion of inhibition in neurophysiology. The work was described in French in 1846.

cular systems is supplied by two sets of nerves, of which the one sets that tissue in activity and causes contraction, the other inhibits its action and causes relaxation . . . the evidence is becoming daily stronger that every tissue is innervated by two sets of nerve fibres of opposite characters so that I look forward hopefully to the time when the whole nervous system shall be mapped out into two great districts of which the function of the one is katabolic, of the other anabolic, to the peripheral tissues. (Gaskell, 1886, 40–41, 50)

Gaskell's colleague at Cambridge University, John Newport Langley (1852–1925), introduced the terms "preganglionic" and "postganglionic" in 1893 and gave the autonomic nervous system its name in 1898 (Windle, 1970). "Autonomic" replaced François-Xavier Bichat's "ganglionic nervous system" and the "vegetative nervous system" of Johann Christian Reil. Langley, shown in Figure 20.4, further maintained that the autonomic nervous system has both peripheral and central parts. Nevertheless, he confined his own descriptions and discussions to the peripheral nerves and organs, primarily because he felt too little was known about the central connections.

Langley (1900, 1903, 1916, 1921) agreed with Gaskell (1886, 1920) that the autonomic nervous system was composed of two antagonistic subsystems. For the division that has its cell bodies in the lateral horns of the thoracic and lumbar spinal cord, and which seemed to be concerned with activation and intense muscular action, he retained Winslow's (1732) term, "sympathetic." Langley then termed the cranial and sacral divisions of the autonomic nervous system, which seemed to be involved with restoration and the conservation of bodily resources, "parasympathetic."

The research of Gaskell, Langley, and others at the turn of the century showed that most preganglionic sympathetic axons project via the ventral spinal roots and white rami communicantes to the sympathetic trunk, where they end in the vertebral ("chain") ganglia. Some sympathetic projections, however, were found to pass through the sympathetic chain to terminate in irregular aggregations of cells called "collateral" or "prevertebral" ganglia (i.e., mesenteric neural plexuses surrounding the aorta and larger visceral arteries).

One of Langley's most important observations was that the preganglionic fibers never end on a single ganglion. Instead, those from each spinal root go to several ganglia, assuring a diffuse autonomic discharge when activated. The postganglionic sympathetic fibers, which are usually unmyelinated, terminate on blood vessels, heart muscles ("accelerator nerves"), eye muscles, sweat glands, abdominal viscera, and other smooth muscles.

The sympathetic nervous system was also found to innervate the adrenal medulla. This innervation seemed atypical because it came directly from preganglionic fibers (splanchnic nerves). This led to the theory that the adrenalin-secreting cells of the adrenal medulla and the postganglionic sympathetic neurons have not only common features but a common parentage (the neural crest). Thomas R. Elliott (1877–1961), who had been a physiology student with Langley at Cambridge, is often cited for advancing this explanation (Elliott, 1904, 1913).

Elliot (1904) provided the first good evidence for the sympathetic nervous system liberating an adrenalin-like substance on the smooth muscles. Walter Cannon called this substance "sympathin" and at first believed it was nothing more than adrenalin itself (see Cannon, 1934). In 1933, Cannon and his Mexican colleague Arturo Rosenblueth (1900–1970) realized there were subtle physiological differences between the nerve and glandular secretions. This eventually led to the use of the term "noradrenalin" (norepinephrine) for the nerve substance.

The parasympathetic nervous system, which arises from the brainstem and the sacral portion of the spinal cord, was also the object of considerable attention by these investigators. As defined, the brainstem projections of the parasympathetic system included the vagus nerves from the medulla to the throat, thoracic, and abdominal viscera and the oculomotor nerves from the midbrain to the eyes. The cranial outflow also involved certain facial nerve and glossopharyngeal nerve projections. The sacral nerves in this system were found to innervate the pelvic colon, rectum, bladder, uterus, and accessory generative organs (e.g., the prostate gland).

In contrast to the preganglionic sympathetic nerves (excluding those to the adrenal glands), the preganglionic parasympathetic nerves from the cranium and sacrum were found to project to individual ganglia close to, or within, specific visceral organs. These ganglia were called either "terminal" or "peripheral" ganglia. The projections to and from these ganglia appeared to be much more specific than those of the sympathetic nervous system. In short, the more restorative (anabolic) parasympathetic (or "craniosacral") system did not appear to be organized to ensure the simultaneous activation of its various component parts.

In the opening decade of the twentieth century, Walter E. Dixon (1871–1931) noted a similarity in the actions of muscarine and extracts taken from activated vagus nerves. Dixon (1907) noticed that both substances inhibited the heart and were antagonized by atropine. Further experimentation by

Figure 20.5. Otto Loewi (1873–1961). In 1921, Loewi demonstrated that the heart could be inhibited by a substance liberated by vagus nerve stimulation. Loewi shared the Nobel Prize in 1936 with Henry Dale for research on chemical transmission at the synapse.

Sir Henry Dale (1875–1968) revealed that certain esters of choline resembled muscarine and could also inhibit heart rate, cause secretion of saliva, and lead to contractions of the esophagus, stomach, and intestines (Dale, 1914).

In 1921, Otto Loewi (1873–1961; shown in Figure 20.5) published his first paper on *Vagusstoff*. He showed that a heart could be inhibited by submerging it in a solution which had previously bathed another heart inhibited by vagus nerve stimulation. Cannon related how Loewi came to conduct his critical experiment:

One night, having fallen asleep while reading a light novel, he awoke suddenly and completely, with the idea fully formed that if the vagus nerves inhibit the heart by liberating a muscarin-like substance, the substance might diffuse out into a salt solution left in contact with a heart while it was subjected to vagal inhibition, and that then the presence of this substance might be demonstrated by inhibiting another heart through the influence of the altered solution. He scribbled the plan of the experiment on a scrap of paper and went to sleep again. Next morning, however, he could not decipher what he had written! Yet he felt that it was important. All day he went about in a distracted manner, looking occasionally at the paper, but wholly mystified as to its meaning. That night he again awoke, with vivid revival of the incidents of the previous illumination, and after this experience he remembered in his waking state both occasions. He set up a frog's heart filled with Ringer's fluid, and after inhibiting the heart by stimulating the vagus nerve, found that the fluid had acquired the new property—that of being able to induce in another frog heart typical vagal

Figure 20.6. Walter Bradford Cannon (1871–1945) in his laboratory. Cannon maintained a long-standing interest in the physiology of the autonomic nervous system and published numerous papers on both central and peripheral correlates of emotion. (Courtesy of Bradford Cannon.)

effects. Furthermore, he found also that when the sympathetic nerves were stimulated, to make the heart beat more rapidly, the Ringer solution in contact with it became endowed with cardio-accelerator power, i.e., an agent was added to it, which, like adrenalin, had sympathomimetic influence. (1934, 149)

Acetylcholine, the neurotransmitter in the parasympathetic nervous system, was isolated in Dale's laboratory eight years later.[3] This led Dale to apply the terms "adrenergic" and "cholinergic" to postganglionic sympathetic and parasympathetic fibers, respectively.[4] In 1936, Dale and Loewi, two old friends with common interests, shared the Nobel Prize for their research on chemical transmission at the synapse.[5]

THE HYPOTHALAMUS AND EMOTIONAL EXPRESSION

Walter Cannon, shown in Figure 20.6, became interested in the autonomic nervous system after he noticed that movements of the stomach and intestines decreased when animals were aroused (Bard, 1970). Beginning in 1911, he conducted many experiments on the autonomic nervous system. In particular, Cannon devoted considerable time and energy to the problem of central regulation of the sympathetic nervous system.

Cannon (1915) noticed early on that emotional excitement was associated with increased diffuse activity of the

sympathetic system. There appeared to be a release of adrenalin into the bloodstream and a decrease in blood going to the viscera, whereas more blood went to the skeletal muscles and the brain. Other changes included increases in blood sugar and blood pressure. Taken as a whole, these widespread sympathetic changes seemed to prepare the organism to fight or to escape from a predator. In Cannon's words, these were vigorous responses directed toward mobilizing bodily resources and preserving life under challenging and stressful conditions.

Cannon also recognized that animals could show uncontrollable rage reactions, even after the cerebral cortex had been removed. This behavior had been observed in 1892 by Friedrich Leopold Goltz. He removed the left hemisphere of a dog and then the right hemisphere in sequential operations. Only a small part of the base of the temporal lobe was left in order to prevent damage to the optic tract. Goltz noted that his decerebrate dogs reacted to stimuli with uncharacteristic fear, attack, and rage, even when taken to be fed and petted.

> The stupidest, unoperated animal would soon have learned that feeding followed lifting from the cage, and it would have been pleased when it was gently led to the feeding place. However, our decerebrate dog, as soon as he was seized and lifted out of the cage, behaved on the last day of his life in the same enraged fashion as he had months before. This fit of temper, expressed by kicking, barking, and biting, only subsided when he was put on the table. (Translated in Clarke and O'Malley, 1968, 564)

This finding was confirmed with cats by Robert S. Woodworth (1869–1962) and Charles Sherrington in 1904. They emphasized that these "pseudo-affective" states were brief in duration. In 1919, the unusual rage response was described again, this time by Dusser de Barenne. His cats spit, growled, scratched, and raised their fur when the experimenter touched them (also see Rothmann, 1923).

---

[3]Histamine was isolated in the same study (Dale and Dudley, 1929).

[4]Later work showed that Dale's biochemical association was not absolute. For example, the sympathetic nerves to the sweat glands are cholinergic.

[5]Because he was Jewish, Loewi was arrested by the Nazis two years later. Most likely because of his fame, he was released after only a few months in captivity, enabling him to emigrate to the United States (Browne, 1970).

Figure 20.7. Philip Bard (1898–1977), who conducted many experiments on the role of the hypothalamus in the expression of the basic emotions.

Together, the data suggested an important role for the upper brainstem in emotional expression.

Cannon also knew that stimulation of the diencephalon could result in vocalizations, changes in respiration and circulation, piloerection, and coordinated movements. This was previously shown by Vladimir Bekhterev in 1887. In addition, two highly respected German investigators, Johann Paul Karplus (1866–?) and Alois Kreidl, conducted experiments demonstrating that electrical stimulation of the hypothalamus could elicit "sympathetic" responses (e.g., Karplus and Kreidl, 1909, 1910, 1911).[6]

Armed with this knowledge from lesion and stimulation experiments, Cannon and his associates began to conduct their own experiments on the role of the brainstem in emotion. They found that cats whose cerebral cortices were disconnected from the brainstem showed rage responses upon coming out of the anesthetic. They also noted that although the rage responses seemed to be coordinated, they never amounted to an effective mechanism of escape (Cannon and Britton, 1925). Moreover, they did not seem to be associated with real anger and were not always directed toward the triggering stimulus. Accordingly, these reactions were called "sham rage."

Cannon and his coworkers noted that sham rage was associated with a rise in blood sugar to five times its normal level and with increased secretions of the adrenal medulla

(Bulatao and Cannon, 1925). In contrast, cats with extensive sympathetic nervous system damage did not show these visceral responses, even when confronted with a dog (Cannon, Lewis, and Britton, 1927). From all indications, sham rage involved both the brain and the sympathetic nervous system.

Cannon's initial findings encouraged Philip Bard to try to discover the specific parts of the brain responsible for sham rage. Bard (1928, 1929a, 1929b, 1930), shown in Figure 20.7, first decorticated his cats to produce sham rage and then made various transections in the brainstem. Working caudally, he found that sham rage was unaffected until he cut through the posterior hypothalamus.

Other investigators soon confirmed that lesions of the posterior hypothalamus eliminated sham rage and caused the animals to become emotionally apathetic (Ingram, Barris, and Ranson, 1936; Masserman, 1938). Such lesions reduced heart rate, lowered blood pressure, and led to constriction of the pupils. In contrast, mild electrical stimulation of the posterior hypothalamus was found to have the opposite effects, confirming the earlier reports by Karplus and Kreidl, and suggesting that the posterior hypothalamus played an important role in emotional expression (Beattie, Brow, and Long, 1930a, 1930b; Fulton, 1932; Ranson, 1937; Ranson, Kabat, and Magoun, 1935).

Some of these observations were reported by Walter Hess, who began to stimulate brainstem structures in free-moving cats with electricity in the 1920s. Working in Zurich, Hess found that electrical stimulation of the posterior hypothalamus made his formerly good-natured cats spit and hiss, while their fur stood erect, their pupils dilated, and their ears moved back. Hess found that these

[6]Karplus and Kreidl utilized a new device for placing their electrodes deep into the brain with accuracy, the stereotaxic instrument. This instrument was first constructed in 1908 by Victor Horsley and Robert Henry Clarke (1850–1921).

aggressive displays ended abruptly when the hypothalamic stimulation was stopped.[7,8]

The importance of the hypothalamus for anger and rage responses seemed clear enough by the 1930s, but what about other basic emotions? Were they also dependent upon the hypothalamus, or at least on the upper brainstem, for integrated expression?

The answer seemed to be yes. This was suggested in part by some neurological patients whose emotional displays no longer appeared to be under the control of the cortex. For example, in 1924 and again in 1929, Wilson described some brain-damaged patients who lost the ability to move their facial muscles voluntarily (presumably a cortical function). But when these patients were very happy or sad, they still exhibited appropriate facial expressions. In addition, Wilson noted that pseudobulbar palsy patients (individuals who are unable to close their eyes or make mouth movements voluntarily) may show "pathological" bouts of uncontrollable laughter or crying, even though these expressive acts may not complement their true feelings. Further, Bard (1929a) noted that patients under anesthetics ("functional decortication") may sing, groan, weep, and laugh without conscious awareness. As far as Wilson, Bard, and Cannon were concerned, the expressions of joy, fear, and grief also involved the brainstem.

The accumulating experimental and neurological data continued to offer strong support for the hypothesis that the cerebral cortex, and the frontal lobes in particular, exerted some inhibitory control over the brainstem centers (e.g., Fulton and Ingraham, 1929). These observations were in accord with John Hughlings Jackson's (1881, 1884, 1887–1888, 1894) concept of levels of functioning and his notion of "release" after damage to higher levels of the system (also see Head, 1918). Such findings suggested a clear dissociation between conscious experience of emotion, which seemed to be cortical, and primitive aspects of emotional expression, which appeared to be controlled by the hypothalamus.

As the 1920s drew to a close, scientists still had not delineated a direct anatomical pathway from the hypothalamus to the sympathetic nervous system. At best, the evidence for such a pathway was suggestive but not conclusive. In 1930, however, a pathway that ran from the hypothalamus to the preganglionic fibers of the sympathetic nervous system was identified (Beattie, Brow, and Long, 1930a, 1930c). Hypothalamic activation of this pathway accelerated the heart rate, raised the core temperature, and mobilized the energy reserves of the body.

Walter Cannon, who was more involved with studying the central control of the autonomic nervous system than any other investigator, tried to combine all of these facts into a unified theory of emotion. In papers published in 1927 and 1931, Cannon suggested that neurons in the diencephalon

are responsible for primitive (unlearned) emotional expression. The diencephalon was also viewed as essential for simple sensation and for activating the cortex so that there could be full, conscious appreciation of the emotions. Ideas generated in the cortex, Cannon reasoned, could also activate the diencephalon, thus triggering autonomic changes.

These ideas attracted many followers during the 1920s and 1930s and were soon used to account for some symptoms of mental illness (Alpers, 1940a, 1940b). With the passage of time, however, Cannon's theory became less influential. This was not because Cannon's or Bard's ideas failed to make sense in the light of new data; rather, it was because their basic thoughts were incorporated into a new, more anatomically sophisticated theory. The newer concept held that emotion was mediated by a "limbic system," a larger collection of related cortical and subcortical centers, and their interconnecting parts.

THE GREAT LIMBIC LOBE OF BROCA

In 1878, just two years before he died, Paul Broca published a lengthy paper on the "great limbic lobe" *(le grand lobe limbique).* He had given the "limbic convolution" *(la circonvolution limbique)* its name a year earlier at the meetings of the *Société d'Anthropologie* after recognizing it completely surrounded the lower threshold of the hemisphere, somewhat like a circular edge or border (Latin *limbus*).

Broca gave credit to anatomists before him (e.g., Gerdy, Foville, and Gratiolet) who also noticed this large arc when dissecting the brain. He then divided his great limbic lobe, which is shown in Figure 20.8, into two parts. One part, the callosal gyrus (or cingulate gyrus), was above and anterior to the second part, the hippocampal gyrus, which fed into it.

Broca intended to associate his great limbic lobe with the sense of smell. With regard to the hippocampal gyrus at least, he was not the first to propose such a relationship. Around 1820, Gottfried Reinhold Treviranus (1776–1837), another comparative anatomist, noted that the hippocampal gyrus seemed to vary in size with the olfactory nerve. Treviranus (1816–1821) even hypothesized that the hippocampal

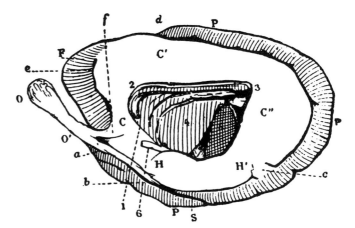

Figure 20.8. Paul Broca's drawing of the inferior-medial face of the right hemisphere of an otter. This drawing, which appeared in Broca's 1878 paper on the great limbic lobe, shows this lobe unshaded.

---

[7]For reviews and summaries of these experiments, see Hess (1948, 1957, 1962).

[8]Hess was not the first scientist to stimulate the brain of an unrestrained animal with electricity. This type of work may have begun with Julius R. Ewald, who put platinum "button" electrodes on the cortex of the dog in 1896. Ewald (1898) walked his dog on a leash and stimulated its brain by connecting the wires to a battery. Ewald did not write up his work in any detail, but a young American who visited Germany extended Ewald's work and then published a more complete report of these experiments (Talbert, 1900; also see Valenstein, 1973).

gyrus was involved with memory, a function easily aroused by olfactory sensations (see Meyer, 1971).

As for Broca, he saw that his great limbic lobe could be found in all mammals, although it varied considerably in size and development across species. He felt that mammals could be divided into two groups on the basis of the comparative anatomy of the limbic lobe and the degree to which they relied on the sense of smell. "Osmatic" animals, such as rabbits, horses, deer, and pigs, rely heavily on olfaction and have well-developed limbic lobes. In contrast, "anosmatic" mammals, such as humans, other primates, and the great sea mammals (e.g., dolphins and seals), have small limbic lobes and are not as dependent upon the sense of smell.

Broca concluded that the two basic parts of his limbic lobe, the hippocampal gyrus and the callosal gyrus, were different anatomically and therefore must also be functionally different. True, Broca reasoned, there appeared to be a strong association between the hippocampal gyrus, which received projections from the lateral olfactory pathway, and whether an animal is osmatic or anosmatic. This was the lower part of the limbic ring, found to be small in primates and smaller still in aquatic mammals that have little, if any, sense of smell ("an olfactory desert"). If anything, the hippocampal gyrus could be the part of the limbic lobe involved in signaling the presence of an odor and mediating certain reflexive or primitive responses to it.

The anterior callosal lobe, however, was another matter. On the one hand, it seemed to receive some projections from the medial olfactory tract. On the other, it was still found to be robust in anosmatic sea mammals and primates who had little in the way of hippocampal lobes. Broca noted that the only remaining link between the anterior callosal lobe and the reduced hippocampal gyri of these anosmatic mammals was a small tract, *la bandelette diagonale* (diagonal band of Broca).

Broca was confident that the anterior callosal lobe functioned in a more primitive way than the remaining convoluted mass of the cerebral hemispheres. The limbic lobe in general represented the beast or brute within. Therefore, it made sense to think that it was associated with primitive functions basic for survival. The remaining part of the cerebrum, which was still growing relative to this stationary and evolutionarily primitive limbic mass, served as the source of true intelligence. It also made sense to conclude from both anatomical and behavioral perspectives that, as one ascended the phylogenetic scale, the "intellectual" cortex (especially the frontal lobes) exerted more and more control over the limbic lobe and its brute functions.

After pondering these ideas, Broca speculated that the anterior callosal lobe may appreciate odors "from the animal's point of view with respect to pleasure and pain." Thus, while the hippocampal lobe might signal the presence of an odor, the anterior callosal lobe could be involved in determining whether an odor is agreeable or disagreeable, or perhaps even in determining drives and emotions in a more general way (see Schiller, 1979).

THE PAPEZ LIMBIC SYSTEM

In 1937, James Papez, the highly respected comparative neurologist at Cornell University, shown in Figure 20.9,

Figure 20.9. James Papez (1883–1958), the anatomist who conceptualized the limbic system and associated it with emotion in 1937. (From Livingston and Hornykiewicz, 1978; courtesy of Plenum Press.)

published a journal article in which he outlined a "new" circuit to account for emotion.[9] He hypothesized that the hippocampus, the cingulate gyrus (Broca's callosal lobe), the hypothalamus, the anterior thalamic nuclei, and the interconnections among these structures (e.g., the fornix) "constitute a harmonious mechanism which may elaborate the functions of central emotion, as well as participate in emotional expression" (p. 743). Although Papez never mentioned Broca's limbic lobe in his paper, others noted that his circuit had more than a few things in common with Broca's great limbic lobe (Bonin, 1950).

Papez, like others before him, differentiated between those structures involved in primitive emotional expression and those involved in subjective emotional experience. Here he returned to the work of Cannon and Bard and to the evolutionary neurology of John Hughlings Jackson (e.g., 1887–1888), all of whom discussed different levels of functioning in the nervous system. Papez wrote that physiological experiments suggested that the hypothalamus mediated emotive expression. He specifically noted that Bard had found that this aspect of emotion depended on the integrative contribution of the hypothalamus rather than on the dorsal thalamus or cortex. Papez (1937, p. 726) then stated: "For subjective emotional experience, however, the participation of the cortex is essential." Here he cited the theoretical reviews of Walter Cannon from 1927 and 1931, as well as

[9]According to his biographers, Papez had been thinking about this topic for some time, even though his paper was written on the spur of the moment (MacLean, 1978). The importance of what Papez had to say was not immediately appreciated by other scientists (Yakovlev, 1978).

the paper written by Charles L. Dana in 1921 on the importance of the cortex for emotional consciousness.

Papez (1937, 1958) assigned a central role in emotion to the hippocampus. He reasoned that the hippocampus sends messages via the fornix to the posterior region of the hypothalamus, the part of the brain most closely associated with emotional expression. This area gives rise to the mamillothalamic tract, which ends in the anterior nucleus of the thalamus. The cells of this nucleus then send the information to the cingulate gyrus, where the conscious experience of emotion can take place. To quote Papez:

> The central emotional process of cortical origin may then be conceived as being built up in the hippocampal formation and as being transferred to the mamillary body and thence through the anterior thalamic nuclei to the cortex of the gyrus cinguli. The cortex of the cingular gyrus may be looked upon as the receptive region of the experiencing of emotion as the result of impulses coming from the hypothalamic region, in the same way as the area striata is considered the receptive cortex for photic excitations coming from the retina. Radiation of the emotive process from the gyrus cinguli to other regions of the cerebral cortex would add emotional coloring to psychic processes occurring elsewhere. This circuit would explain how emotion may arise in two ways: as a result of psychic activity and as a consequence of hypothalamic activity. (1937, 728)

Papez cited pathological data in support of his theory. For example, he noted that rabies victims show rage, terror, anxiety attacks, and apprehensiveness and that rabies has a predilection for hippocampal neurons. He also mentioned that epileptic seizures associated with visceral auras and emotional manifestations (fear and terror) have their foci in the inferior temporal cortex. Additionally, Wernicke-Korsakoff patients, known to have lesions of the mamillary bodies, seemed "affectless."

Papez wrote that tumors impinging upon the cingulum cause marked changes in personality, including a loss of affect. Although he did not cite it, one of the most interesting early cases suggesting a link between the cingulum and the loss of emotion was observed before 1670 by Felix Platter (see Ward, 1948). The case involved a knight, Caspar Bonecurtius, who became affectless and unresponsive to his environment. He spent much of each day at a table with his head in his hands. In the months before he died, he showed little speech, and what he did say made little sense. After he died, a tumor was found in the anterior cingulate gyrus.

Papez recognized that many of the structures comprising his functional system had been loosely linked to olfaction. In fact, many of these structures were classified as "rhinencephalon," meaning "nose brain." Papez argued, however, that there was little evidence to support this old assertion. For example, data linking the hippocampal lobe with olfaction were weak at best. There were movements of the nostrils to temporal lobe stimulation in monkeys and olfactory auras in some cases of temporal lobe epilepsy (Ferrier, 1876; Jackson and Beevor, 1890). Yet in opposition to this, there was a growing literature showing few if any direct connections between the olfactory pathways and the limbic system, and increasing recognition of the fact that lesions of these structures do not cause olfactory

deficits.[10] In short, Papez felt that the time had come to overthrow the old olfactory theory and to link these structures more closely to another function.

THE KLÜVER-BUCY SYNDROME

In 1937, Heinrich Klüver (b. 1897) and Paul Bucy, scientists at the University of Chicago, published the first of a series of reports in which they described three unusual effects of bilateral temporal lobectomies in the rhesus monkey. Their intent was to study the effects of hallucinogens, but to their amazement they found that an adult female monkey behaved as if she were "psychically blind" after sequential right and left temporal lobe surgeries. That is, she seemed able to see things adequately, but appeared unable to recognize the meaning of familiar objects, such as the difference between food, a stick, and a pocketknife. A second feature of this monkey was that she seemed to put everything into her mouth. Her third distinguishing characteristic, and the one most pertinent to the Papez theory of emotion, was that she acted as if she had lost her normal fear and anger responses.

The findings obtained with this monkey were discussed in greater detail by Klüver and Bucy one year later. With regard to the monkey's emotional changes, the authors wrote:

> The picture she presented after the second operation was one of complete loss of all emotional reactions. In spite of being active and exhibiting great interest and curiosity in her surroundings at all times, she appeared never to discover anything causing resentment, anger, fear or pleasure. . . . When on the 25th day a 4-foot long bull snake was held within reach she licked the tongue of the hissing snake. . . . At the end of the four months, the rudiments of affective expressions generally associated with aggressive responses could be said to be occasionally present. (1938, 50–51)

A third paper in the series, published in 1939, showed that these findings generalized to other monkeys with bilateral temporal lobe lesions. Klüver and Bucy again described extreme emotional blunting, even though they deliberately used very wild monkeys in this study. Not only did the animals fail to show signs of fear or anger but their vocalizations and facial expressions also seemed unemotional. This lasted for a minimum of a few months and, in some cases, for the duration of the study.[11]

Klüver and Bucy noted that some features of this syndrome had been observed years earlier by Sanger Brown and Edward Albert Schäfer. In a report to the Royal Society in 1887 (published in 1888), Brown and Schäfer described one monkey with very deep lesions of both superior temporal gyri and a second monkey with complete removals of both temporal lobes. They found that following surgery these animals became surprisingly tame and less fearful, while displaying

---

[10]See Brodal (1947) for evidence against the limbic structures being basically olfactory in nature.

[11]Klüver and Bucy also observed that the monkeys exhibited markedly increased sex drives. This included atypical exploration and manipulation of their own genitalia and heterosexual and homosexual acts toward other monkeys.

problems in visually recognizing familiar objects. These experimenters, however, did not know what to make of these unusual changes in memory and emotion (see Chapter 24).

Klüver and Bucy (1939) did not think that this syndrome could be reduced to a single fundamental defect. They simply noted that none of these symptoms was observed in animals with the hippocampal gyrus left intact. And, in terms of theory, they cited the work of Papez (1937), which associated the temporal lobes, including the hippocampus, with emotional functions.

EXPANSION OF THE LIMBIC SYSTEM CONCEPT

The strongest supporter of the ideas presented by Papez was Paul MacLean (b. 1913), who worked at Harvard Medical School and then Yale University. MacLean had been studying patients who had temporal lobe epilepsy when he came across the 1937 Papez paper. It provided him with a theoretical explanation for the visceral symptoms and unusual emotional feelings that characterized the auras of some of his patients.

MacLean visited Papez in Ithaca, New York, and in 1949 wrote his most important paper on the central control of emotion (Durant, 1985). He maintained that the so-called rhinencephalon could well be a "visceral brain." He associated this part of the brain (shown in Figures 20.10 and 20.11) with both emotional behavior and the basic drives of eating, drinking, and reproduction. As MacLean saw it, the visceral brain was common to all mammals, and its functions were critical for the preservation of the self and the species. MacLean (1949, p. 347) also suggested that some structures in this system may function in opposite ways to each other: "Since the amygdala seems to project predominantly to the parasympathetic centers of the hypothalamus, and the hippocampus to the sympathetic, is it possible that these respective parts of the visceral brain are mutually antagonistic?"

MacLean (1949), like Papez, at first did not recognize Broca's (1878) contribution. But, in 1952, he adopted Broca's (1878) terminology. MacLean now called the system that he had associated with "early neural development involved in the elaboration of emotional experiences and expression" the "limbic system." Anatomically, he defined the limbic system as "the cortex contained in the great limbic lobe of Broca together with its subcortical cell stations" (MacLean, 1952, p. 407).

In subsequent years, MacLean (e.g., 1954, 1978) continued to expand the anatomical foundation of the Papez theory. The higher centers now included the hippocampal gyrus, the cingulate gyrus, and the subcallosal gyri. The nuclear structures placed under the limbic umbrella included the amygdala, the septal nuclei, the epithalamus, the anterior thalamic nuclei, parts of the basal ganglia, and several nuclei in the hypothalamus.

Like others who followed in the footsteps of John Hughlings Jackson, MacLean envisioned the human brain as a conglomerate of three brains. He called these the reptilian brain, the old mammalian brain, and the new mammalian brain. The reptilian brain, localized in the brainstem, was deemed responsible for instincts and ritualistic or stereotypical behaviors. The limbic system made up the middle level, or the old mammalian brain, and was the part presumed most responsible for simple feelings, emotional expression, and basic reproductive behavior. The neocortex, in contrast, constituted the new mammalian brain and represented the problem solver, the source of higher thoughts, and the voice of reason. Unlike the other parts of the brain, only this newer part was viewed as verbal and analytic.

MacLean (1952) argued that the three brains within a single shell are unable to communicate effectively with one another. One reason for this is that only the newer brain is able to use symbolic language. Another reason for ineffective communication is the absence of major pathways between component brains. MacLean postulated that this often put the new brain in conflict with the reflexive or

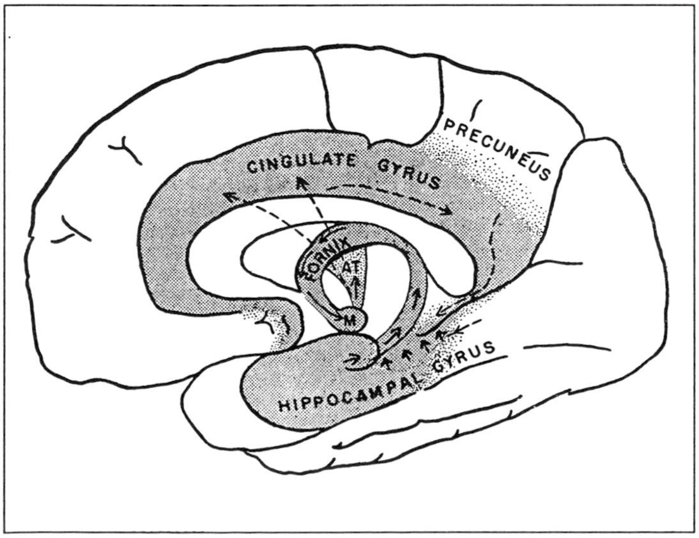

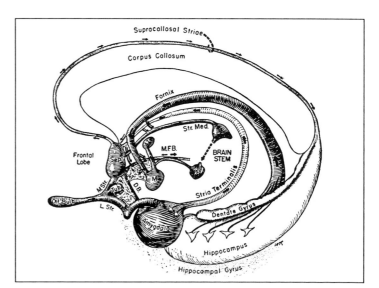

Figure 20.10. Diagram of the brain by Paul MacLean (1949), shaded to show the parts designated as "visceral brain." M = mammillary body, AT = anterior thalamic nucleus. (Courtesy of the American Psychosomatic Society.)

Figure 20.11. Schematic representation of the subcortical structures comprising the rhinencephalon as seen through the medial aspect of the right hemisphere. (From MacLean, 1949; courtesy of the American Psychosomatic Society.)

instinctive features of the more primitive parts of the brain. Earlier he had suggested that this relative absence of communication could even account for certain neuroses:

> Therefore if the visceral brain were the kind of brain that could tie up symbolically a number of unrelated phenomena, and at the same time lack the analyzing ability of the word brain to make a nice discrimination of differences, it is possible to conceive how it might become foolishly involved in a variety of ridiculous correlations leading to phobias, obsessive-compulsive behaviour, etc. (1949, 348)

This was hardly a new theme. As noted in Chapter 19, the battle between the passions and reason was a major concern of Victorian physicians. It affected not just the neurological thinking of men like Jackson but the emergence of psychoanalytic thought. Freud was "fixated" on the struggle of the id to break through the defenses of the more rational ego (e.g, Freud, 1920, 1927). But unlike Freud or even Jackson, MacLean sought to explain why this should be the case on the basis of "real" neuroanatomy.

EARLY PSYCHOSURGICAL INTERVENTIONS FOR "EMOTIONAL" DISORDERS

The brain lesion literature had long supported the idea that "mental" illness could be due to pathological changes in higher structures, especially those that might control brainstem areas and give the basic emotions personal meaning. On a theoretical level, it was easy to hypothesize that it might be better to destroy certain circuits than to allow them to function in an aberrant way.

One of the first individuals to attempt to damage the brain in order to help the mentally ill was Gottlieb Burckhardt (1836–?), the director of a small asylum in Préfargier, Switzerland (see Bach, 1907; Schiller, 1985). Burckhardt was influenced by the work of Friedrich Goltz (1874), who noted that although large frontal lobe ablations made dogs both hyperactive and hyperemotional, posterior cortical ablations made other dogs more placid than normal. To quote Goltz (1888) when later summarizing his observations:

> I have mentioned that dogs with a large lesion in the anterior part of the brain generally show a change in character in the sense that they become excited and quite apt to become irate. Dogs with large lesions of the occipital lobe on the other hand become sweet and harmless, even when they were quite nasty before. (1960 translation, 157)

Burckhardt began to perform his therapeutic brain operations in 1890. He attempted to remove or disconnect Wernicke's area and even Broca's speech area from the rest of the brain in six mental patients who showed agitated behavior and auditory hallucinations. In 1891, Burckhardt reported partial improvement in Case 2 and wrote: "Cases 1, 3, and 5 I take to be results that could not have been achieved by any other method." As for his other two patients, Case 4 showed some improvement, but then committed suicide, whereas Case 6 experienced convulsions and died near the end of the first week.

Although Burckhardt was impressed with the possibilities of psychosurgery, he stopped operating at this time. This was probably because he realized on his own that his results, while somewhat gratifying, were not good enough

to justify the surgical risk. It may also have been because the death of his sixth patient triggered strong professional and community opposition to human brain surgery as an experimental procedure (Fulton, 1952; Ramsey, 1952).

Nevertheless, the idea that more had to be done for emotionally unstable people was prevalent. It was driven by the belief that institutionalized patients were not any more likely to recover now than at the beginning of the century, when Philippe Pinel and Benjamin Rush (1745–1813) were actively crusading for better care of the insane. Many patients continued to be restrained and treated poorly, and many attendants and supervisors still viewed their institutions as final destinations rather than as therapeutic centers. It was partly for these reasons that others opted to continue what Burckhardt had started.

Early in 1900, Ludvig Puusepp (1875–1942), from Estonia, attempted to sever the association fibers between the frontal and parietal lobes in three patients with manic-depressive psychosis. He made lesions on only one side of the brain. Although he regarded his initial results as unsuccessful, he returned to the operating room again in 1910. This time he made four or five holes over the frontal lobes in some patients suffering from general paresis and then inserted chemical agents (a cyanide-mercury mixture, or Neosalvarsan) under the dura. This led to some improvement, but like Burckhardt, Puusepp (1914) concluded that his patients did not benefit enough to justify subjecting others to such dangerous procedures.

MONIZ AND PREFRONTAL LEUCOTOMY

During the 1930s a number of new "somatic" therapies were introduced and quickly put into practice with the mentally ill. These included three types of "shock" treatment. The first, insulin coma therapy, was described in 1933 by Manfred Sakel (1900–1957), a graduate of the University of Vienna who worked on drug dependency in Berlin (Wortis, 1958). Sakel discovered that his diabetic addicts were less anxious during opiate withdrawal when he induced light comas by accidently giving them overdoses of insulin. Upon returning to Vienna, Sakel (1937, 1938) began to treat his schizophrenic patients with insulin and claimed high rates of remission from the psychoses.

Metrazol convulsion therapy was introduced in 1934 by Ladislas von Meduna (1896–1964), a physician from Budapest. Meduna (1935) believed there was a biological antagonism between epilepsy and schizophrenia and thus decided to try to induce "therapeutic" convulsions in his schizophrenic patients. After trying various drugs and then intramuscular camphor injections, Meduna turned to pentylenetetrazol (metrazol, cardiazol), a more reliable synthetic substance.[12] His first metrazol patient was a catatonic schizophrenic who showed dramatic improvement after five convulsive treatments. Encouraged by these and other results, he treated other psychotic patients, claiming good recovery in most instances (see Meduna, 1938).

---

[12]Meduna first tried strychnine, thebaine, nikethamide, caffeine, and absinthe before he turned to camphor. Camphor had been used to treat insanity at least as far back as the eighteenth century (Fink, 1984).

Figure 20.12. Carlyle F. Jacobsen (1902–1974), whose experiments on monkeys and apes with John Fulton stimulated Egas Moniz to perform prefrontal lobotomies on humans.

Electroconvulsive shock was initiated in Italy by Ugo Cerletti (1877–?) in collaboration with Lucio Bini (b. 1908) (see Cerletti, 1956). Cerletti, believing that the body produced a "vitalizing substance" when convulsed, first sought to collect this elusive substance by shocking animals on the side of the head. In 1938, Cerletti and Bini tried to "revitalize" a 39-year-old schizophrenic who had received eight cardiazol treatments a year earlier. Mild current seemed to have no effect on this homeless man who had been picked up by the police in Rome. But with enough current to induce loss of consciousness, their patient became considerably more coherent and a year later was back at work. In this way, therapeutic electroconvulsive shock, the most popular shock treatment, was born.

Prefrontal lobotomies emerged in the same decade as Sakel, Meduna, Cerletti, and Bini were claiming that somatic interventions were the wave the future. Lobotomies for the insane were ushered in by Egas Moniz (1874–1955), a professor of neurology in Lisbon.[13]

Moniz's work prior to prefrontal lobotomy had established his reputation as a gifted scientist. In 1918 and 1919, Walter Dandy replaced cerebrospinal fluid with air or gas injected into the brain (ventriculography) or spinal canal (pneumoencephalography or pneumography) and then x-rayed the brain. Moniz, assisted by the surgeon Pedro Almeida Lima (b. 1903), applied Dandy's reasoning

to the blood vessels of the brain. On June 28, 1927 an opaque substance was injected into the carotid artery of a patient with a pituitary tumor. An x-ray showed displacement of the blood vessels by the tumor. In a second patient, a temporal lobe tumor that had been diagnosed by angiography was confirmed at autopsy. Moniz was elated and over the next seven years published 112 articles and two books on angiographic procedures!

Before 1935, the year in which the first prefrontal lobotomy was conducted, Moniz had thought about damaging the frontal lobes in some of the more severe mental patients in his wards. The idea was precipitated by reviewing the literature on patients with frontal lobe tumors. Moniz did nothing, however, until he attended a meeting of the International Neurological Congress in August 1935. There he exhibited his work on cerebral angiography and attended a popular all-day symposium on the frontal lobes.

A heralded speaker at this popular symposium was Carlyle F. Jacobsen, shown in Figure 20.12. He described some experiments that he and John Fulton had conducted on two chimpanzees. Most of the talk centered on the impairment that the chimpanzees had in learning (see Chapter 23). Significantly, Jacobsen also told his audience about a personality change in one of their animals, a volatile, compulsive chimpanzee named "Becky." This chimpanzee became very upset when the experimenter even lowered the screen between her and the baited food cup. She flew into tantrums, defecating and urinating when she made mistakes on the delayed response problem. "Becky's" "experimental neurosis" eventually became so bad that she had to be dragged from her home cage just to be tested (Crawford et al, 1948; Fulton, 1948; Jacobsen, 1936).

Extirpation of just one frontal lobe did not relieve this neurosis. A profound change occurred, however, when the remaining frontal association area was ablated.

> The chimpanzee . . . went promptly to the experimental cage. The usual procedure of baiting the cup and lowering the opaque screen was followed. . . . If the animal made a mistake, it showed no evidence of emotional disturbance but quietly awaited the loading of the cups for the next trial. It was as if the animal had joined the "happiness cult of the Elder Micheaux," and had placed its burdens on the Lord! (Jacobsen, Wolf, and Jackson, 1935, 10)

After listening to Jacobsen's presentation, Moniz, shown in Figure 20.13, asked whether such a procedure should now be attempted to reduce severe anxiety and delusional states in humans (Fulton, 1948, 1952). This question seemed to alarm both Jacobsen and Fulton, but Moniz walked away from the symposium convinced that this was something worth trying. His first patient was operated upon three months later.[14] On a theoretical level, he reasoned that mental disorders are caused by "fixed" ideas whose circuits can be found in the frontal lobes. His theory was vague, but Moniz was a risk-taker who, at the age of 61, was driven by the desire for still greater fame.

The first frontal lobotomy took place on November 12 in a Lisbon hospital. Two holes were drilled into the skull of a

---

[13]Moniz belonged to an aristocratic Portuguese family and was trained in neurology in Bordeaux and Paris. For biographies, see Damásio (1975), Lima (1974), and Valenstein (1986).

[14]Moniz subsequently tried to convey the impression that he was not influenced by Jacobsen's talk. One reason for this might have been that he wanted to receive more of the credit for the idea itself.

Figure 20.13. Egas Moniz (1875–1942), the neurologist who advocated the use of frontal lobotomies for treating humans with mental disorders. Moniz worked in Lisbon and published his first highly influential findings on prefrontal lobotomy in 1936.

female mental patient who was suffering from involutional melancholia with anxiety and paranoid ideas. Absolute alcohol was then injected into the subcortical white matter of the prefrontal area in an attempt to destroy the association tracts. Because Moniz's hands were badly distorted by gout, the 30-minute operation was performed by Lima, who later stated, "Moniz was my chief and he needed my hands" (see Valenstein, 1986, p. 102). The patient seemed to be less anxious and less paranoid after the operation. Moniz therefore classified the case as "a clinical cure."

Six more operations were performed within the next five weeks, each somewhat different from the one before. The results were mixed. The affective cases showed the most signs of improvement, whereas the chronic schizophrenics seemed to benefit least. For their eighth patient, the surgeon turned to a special needle that was put into the brain. They manipulated it to open a steel loop at its end, and then twisted it to cut cores of tissue (like an apple) in each hemisphere. At first Lima cut one or two cores on each side, but this was soon expanded to four cores on each side. Moniz called the new tool that he had made for this surgery a "leucotome" (Greek for "white matter" and "knife"), and the procedure itself was soon called the "core operation." Moniz also coined the terms "prefrontal leucotomy" and "psychosurgery."

Within four months of the first surgery, Moniz reported the results of his first group of 20 patients (Moniz, 1936; Moniz and Lima, 1936). He stated that prefrontal leucotomies led to recovery in 7 patients and substantially improved the conditions of 7 others from the initial group of 20 patients. The remaining 6 patients, who had more severe delusions and hallucinations, were not as improved as the previously agitated and depressed patients.

Moniz continued to modify his procedure. Thus, for his

next series of 18 patients, he had Lima use a newly designed leucotome to cut six cores more posteriorly in each frontal lobe. Moniz (1937) again reported very favorable results with his anxious, depressed, manic, or schizophrenic patients, with the exception of those most deteriorated.

A major problem with these early cases was that Moniz followed his patients for only a few days after surgery. Although there seemed to be a reduction in agitation, there was nothing concrete to indicate that the patients were able to take care of themselves outside the hospital or that they were cured. In fact, a 12-year follow-up study by another Lisbon neurologist suggested that there were many relapses, seizures, and a high number of deaths (Furtado, 1949).

After this time, Moniz supervised only a limited number of prefrontal leucotomies. Always the diplomat, he explained that he wanted to follow his patients for a number of years before embarking on a full-scale program. He was forced to retire from his professorship when he turned 70 in 1943, but he stayed fairly active in the field until 1949 when a paranoid patient (who did not have a leucotomy) shot him four times. Although he survived the assault, one bullet lodged in his spine and could not be removed.

Moniz was awarded the Nobel Prize in 1949 "for his discovery of the therapeutic value of prefrontal leucotomy in certain psychoses". Gösta Rylander (b. 1903), who performed frontal lobe surgeries in Sweden, was on the selection committee, which also honored Walter Hess for his pioneering work on brainstem stimulation in unrestrained animals.

THE RISE AND FALL OF PREFRONTAL LOBOTOMY

Soon after 1935, prefrontal leucotomies were conducted in many other countries, including Italy, Cuba, Brazil, and Rumania. In the United States, Walter J. Freeman (1895–1972; shown in Figure 20.14) and James W. Watts (b. 1904) carefully followed the scientific papers and news releases coming from Portugal. Freeman, like Moniz, had attended the symposium on the frontal lobes in London. He was a Yale graduate who corresponded with Fulton about the monkey and ape experiments. Watts studied neurology in Boston and was trained in neurophysiology by Fulton. Soon after the London conference, he joined Freeman on the staff of the George Washington Hospital (Washington, DC). But unlike Freeman, who was assertive and demanding, Watts was a quiet, gentle man who shunned controversy (Valenstein, 1986).

On September 14, 1936, nearly one year after the Portuguese investigators performed their first operations, Freeman and Watts began to operate. Their first patient was a woman from Kansas who suffered from agitated depression. After six "cores" were cut, the woman reported that she was no longer fearful and was able to feel more relaxed. Encouraged, they operated upon 19 more patients before the end of the year. These therapeutic interventions were viewed as successful, even though Freeman and Watts increased the number of cores cut and operated at least a second time on eight of their patients.

Freeman and Watts opted for the term "lobotomy," as opposed to "leucotomy," for the operation because lobotomy

Figure 20.14. Walter J. Freeman (1895–1972), the American physician who popularized prefrontal lobotomy.

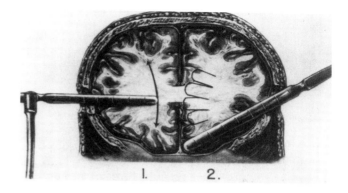

Figure 20.15. A sketch of the procedure used by Walter Freeman and James Watts for their standard lobotomy. (From Freeman and Watts, 1950; courtesy of Charles C Thomas, Springfield, IL)

implied destruction of tracts and cell bodies, whereas leucotomy signified just destruction of the white matter. During 1937, they also changed their technique to one called "the precision method" ("Freeman-Watts standard lobotomy"). This involved opening the skull from the side, inserting a blunt spatula and cutting frontal fibers by arcing the spatula up and down.

Freeman and Watts believed that their frontal lobe surgeries reduced the emotional sting (bleached the affect) of their patients. The rationale offered for their choice of surgical sites showed that they were influenced by the Cannon-Bard and later the Papez theories of emotion. The idea was to "break" the circuit of emotion by cutting the pathway from the dorsomedial thalamus to the prefrontal cortex (see Figure 20.15). This can be appreciated from the following lines, which appeared in their book *Psychosurgery: Intelligence, Emotion and Social Behavior Following Prefrontal Lobotomy for Mental Disorders:*[15]

> The dorsal medial nucleus of the thalamus by its connections with the cerebral cortex may well supply the affective tone to a great number of intellectual experiences and through its connections with the hypothalamus may well be the means by which emotion and imagination are linked. . . . This fasciculus of fibers connecting the prefrontal region with the thalamus is unquestionably of great significance, and there is little doubt that its interruption is of major importance in the alteration of the personality seen after frontal lobotomy. (1942, 27)

The methods for isolating the frontal association areas continued to undergo further modification (Fiamberti, 1937, 1939; Lyerly, 1938; Ody, 1938). One popular alternative to the precision method was to sever the thalam-

ofrontal radiations by entering the frontal lobes through the eye socket (roof of the orbit). During the late 1930s, a "transorbital" procedure was used in Italy, but it involved injecting alcohol or formalin into the front of the brain (Fiamberti, 1937, 1939). This idea attracted little attention outside Italy until Walter Freeman modified and popularized the operation in 1946. Freeman chose to cut pathways via the eye sockets. He initially used a pick from the Uline Ice Company as his leucotome; electroconvulsive shock was his anesthetic.

Freeman (1948) came away convinced that transorbital lobotomy was best for disorders whose severity fell between those that required shock therapy alone and those that required standard frontal lobotomy. He wrote that transorbital lobotomy substantially reduced the emotional distress and moodiness of his patients. Freeman emphasized that this was a procedure which could be used early in the course of the disease, when interventions were thought to be most effective.

In part because fewer fibers were cut, the transorbital patients did not show as many undesirable side effects or require the recovery times associated with the precision method. Freeman (1949) reported only four mortalities after conducting transorbital lobotomies on more than 350 cases, stating that many patients were able to leave the hospital in a day or two.

Watts, a neurosurgeon, strongly opposed Freeman's suggestion that psychiatrists should be taught this rapid "office procedure." This led the two men to go their own ways in 1947. Freeman, now on his own, crusaded back and forth across the United States promoting transorbital lobotomy and grabbing newspaper headlines with an operation that could be performed in as few as 10 minutes. But unlike Moniz, who followed his patients only for days, Freeman kept in touch with his patients for as long as he could.[16] In one long-term review of how they fared, and with special emphasis on the transorbital patients, Freeman wrote:

> High-level performance is possible after major prefrontal lobotomy, and especially after transorbital lobotomy. This is indicated by the competence of lobotomized patients in such

---

[15]This book, first published in 1942, was revised and expanded in 1950.

[16]Walter Freeman visited and corresponded with many of his patients and their families for more than 20 years (see Freeman, 1957, 1958).

professional fields as medicine, law, teaching, nursing, etc. One patient is a member of a first-rate symphony orchestra; another is functioning well as a missionary in the Far East; a psychiatrist was promoted to chief of service in a large mental hospital; a fourth is confidential secretary in a sensitive spot in the Federal Government. (1957, 877)

Lobotomized patients running hospitals and governments! Could there be a stronger endorsement?

The number of lobotomies increased tenfold between 1946 and 1949, the high-water mark for psychosurgery, not just in the United States but around the world (Fulton, 1952; Valenstein, 1973). Freeman himself performed or supervised approximately 3,500 operations before he retired in 1970.

The popularity of psychosurgery can best be understood in the light of three facts. First, it was not difficult to perform. Second, it was a way of reducing overcrowded institutional populations where the budgets were tight. And third, many enthusiastic believers clearly overstated their results.

After the enthusiasm generated by the initial reports abated, physicians began to question the overall effectiveness of these interventions. A major problem was not that the surgeries were producing armies of zombies but that people were finding some conditions (e.g., severe schizophrenia) refractory to lobotomy. Another significant problem was that the degree of recovery suggested by Moniz and other enthusiasts was not often seen.

Commissions were formed in different countries to evaluate just how successful the operations really were and which groups of patients benefited the most. In England, a large study published by the Board of Control in 1947 revealed that severing the pathways between the prefrontal lobes and the nucleus medialis dorsalis of the thalamus was helpful, but that the success rate varied with the severity of the disorder. It was found that 50 percent of the manic-depressive cases were discharged, but the schizophrenics did not do nearly as well. In the United States, Freeman himself concluded that frontal lobotomy was of limited value with chronic schizophrenia or the severe psychoses of childhood (Freeman, 1971; Williams and Freeman, 1953).

One of the most comprehensive analyses of the effectiveness of psychosurgery was the Columbia-Greystone Project, organized in 1947. Forty-eight American psychotic patients who had been hospitalized for at least two years and who were not showing any appreciable improvement served as subjects. They were arranged in matched pairs. One member of each pair underwent a topectomy, which involved removing a relatively small amount of prefrontal cortex, whereas the other member did not undergo brain surgery. Four months later, the patients were examined by psychiatrists who were not told whether or not the subjects had surgery.

The data suggested that the surgery did not affect intelligence, learning, or general neurological status. This was a good start. Even better, various inventories, interviews, and indexes of social adjustment revealed that the operated patients exhibited decreased anxiety and improved mental attitudes. Of the 24 experimental subjects, 20 were recommended for parole, whereas only 4 of the 24 control subjects were recommended (Columbia-Greystone Associates, 1949).

Encouraging psychiatric findings were also reported in other countries, at least to the extent that many agitated and anxious individuals experienced some normalization of their aberrant behaviors and required less intensive care. Nevertheless, investigators such as Gösta Rylander (1940, 1941, 1948) were already realizing that abstract thinking, planning, social judgment, insight, attention, motivation, and spontaneity were often impaired by these operations. Rylander acknowledged that some psychiatric patients were not benefiting enough from psychosurgery to justify the danger of the operation or its potential side effects. The message seemed to be that not all so-called mental disorders were so severe that it was worth exchanging them for an organic brain syndrome.

In the 1950s, the world saw the introduction of new, more effective tranquilizers, led by chlorpromazine. As physicians recognized the clinical effectiveness of the phenothiazine derivatives on their psychiatric populations, lobotomies were less frequently performed.[17] The era of radical surgery for severe "emotional" disorders, a movement based in part on theoretical notions about the circuitry of emotion, had come to an end.

---

[17]For early clinical reports on the use of phenothiazine derivatives with psychiatric patients, see Bower (1954), Delay, Deniker, and Harl (1952), Lehmann and Hanrahan (1954), and Wortis (1953).

REFERENCES

Ackerknecht, E. H. 1974. The history of the discovery of the vegetative (autonomic) nervous system. *Medical History, 18*, 1–8.

Alpers, B. J. 1940a. Personality and emotional disorders associated with hypothalamic lesions. *Research Publications—Association for Research in Nervous and Mental Disease, 20*, 725–752.

Alpers, B. J. 1940b. Personality and emotional disorders associated with hypothalamic lesions. *Psychosomatic Medicine 2*, 286–303.

Bach, C. 1907. Dr. Gottlieb Burckhardt. *Allgemeine Zeitschrift für Psychiatrie und ihre Grenzgebiete, 64*, 529–534.

Bard, P. 1928. A diencephalic mechanism for the expression of rage with special reference to the sympathetic nervous system. *American Journal of Physiology, 84*, 490–515.

Bard, P. 1929a. Emotion. I. The neuro-humoral basis of emotional reactions. In C. Murchison (Ed.), *The Foundations of Experimental Psychology*. Worcester, MA: Clark University Press. Pp. 449–487.

Bard, P. 1929b. The central representation of the sympathetic nervous system as indicated by certain physiologic observations. *Archives of Neurology and Psychiatry, 22*, 230–246.

Bard, P. 1930. The central representation of the sympathetic nervous system as indicated by certain physiologic findings. *Research Publications—Association for Research in Nervous and Mental Disease, 9*, 67–91.

Bard, P. 1970. Walter Bradford Cannon (1871–1945). In W. Haymaker and F. Schiller (Eds.), *The Founders of Neurology* (2nd ed.). Springfield, IL: Charles C Thomas. Pp. 279–281.

Beattie, J., Brow, G. R., and Long, C. N. H. 1930a. Physiological and anatomical evidence for the existence of nerve tracts connecting the hypothalamus with spinal sympathetic centres. *Proceedings of the Royal Society, 106B*, 253–275.

Beattie, J., Brow, G. R., and Long, C. N. H. 1930b. The hypothalamus and the sympathetic nervous system. Part 1. The dependence of the extrasystolic arrhythmia of the heart produced by chloroform, upon the integrity of sympathetic nervous system; and the use of this arrhythmia as an indicator of sympathetic activity. *Research Publications—Association for Research in Nervous and Mental Disease, 9*, 249–294.

Beattie, J., Brow, G. R., and Long, C. N. H. 1930c. The hypothalamus and the sympathetic nervous system. Part 2. The higher connections of the sympathetic nervous system as studied by experimental lesions of the hypothalamus. *Research Publications—Association for Research in Nervous and Mental Disease, 9*, 295–316.

Bekhterev, V. 1887. Die Bedeutung der Sehhügel auf Grund von experimentellen und pathologischen Daten. *Virchow's Archiv für pathologische Anatomie, 110*, 322–365.

Bichat, X. 1805. *Recherches Physiologiques sur la Vie et la Mort* (3rd ed). Paris: Brosson/Gabon. Translated by F. Gold as, *Physiological Researches on Life and Death*. Boston, MA: Richardson and Lord, 1827.

Reprinted in D. N. Robinson (Ed.), *Significant Contributions to the History of Psychology, Ser. E, Physiological Psychology: Vol. II, X. Bichat, J. G. Spurzheim, P. Flourens.* Washington DC: University Publications of America, 1978. Pp. i–334.

Board of Control. 1947. Prefrontal lobotomy: Report on 1000 cases. *Lancet, 1,* 265–266.

Boerhaave, H. 1708. *Institutiones medic in usus annuae exercitationis domesticae.* Lugduni Batavorum: Petrum van der Eyk.

Bonin, G. von. 1950. *Essay on the Cerebral Cortex.* Springfield, IL: Charles C Thomas.

Bower, W. H. 1954. Chlorpromazine in psychiatric illness. *New England Journal of Medicine, 251,* 689–692.

Broca, P. 1877. Sur la circonvolution limbique et la scissure limbique. *Bulletins de la Société d'Anthropologie, 12,* 646–657.

Broca, P. 1878. Anatomie comparée des circonvolutions cérébrales. Le grand lobe limbique et la scissure limbique dans la série des mammifères. *Revue d'Anthropologie, Ser. 2, 1,* 385–498.

Brodal, A. 1947. The hippocampus and the sense of smell. *Brain, 70,* 179–222.

Brown, S., and Schäfer, E. A. 1888. An investigation into the functions of the occipital and temporal lobes of the monkey's brain. *Philosophical Transactions of the Royal Society of London (Biology), 179,* 303–327. (Paper read December 15, 1887)

Browne, J. S. L. 1970. Otto Loewi (1873–1961). In W. Haymaker and F. Schiller (Eds.), *The Founders of Neurology* (2nd ed.). Springfield, IL: Charles C Thomas. Pp. 293–296.

Bulatao, E., and Cannon, W. B. 1925. The role of the adrenal medulla in pseudoaffective hyperglycemia. *American Journal of Physiology, 72,* 295–313.

Burckhardt, G. 1891. Über Rindenexcisionen, als Beitrag zur operativen Therapie der Psychosen. *Allgemeine Zeitschrift für Psychiatrie, 47,* 463–548.

Cannon, W. B. 1915. *Bodily Changes in Pain, Hunger, Fear and Rage.* New York: Appleton.

Cannon, W. B. 1927. The James-Lange theory of emotion: A critical examination and an alternative theory. *American Journal of Psychology, 39,* 10–124.

Cannon, W. B. 1931. Again the James-Lange and the thalamic theories of emotion. *Psychological Review, 38,* 281–295.

Cannon, W. B. 1934. The story of the development of our ideas of chemical mediation of nerve impulses. *American Journal of the Medical Sciences, 188,* 145–159.

Cannon, W. B., and Britton, S. W. 1925. Pseudoaffective medulliadrenal secretion. *American Journal of Physiology, 72,* 283–294.

Cannon, W. B., Lewis, J. T., and Britton, S. W. 1927. The dispensability of the sympathetic division of the autonomic nervous system. *Boston Medical and Surgical Journal, 197,* 514–515.

Cannon, W. B., and Rosenblueth, A. 1933. Studies on conditions of activity in endocrine organs. XXIX. Sympathin E and sympathin I. *American Journal of Physiology, 104,* 557–574.

Cerletti, U. 1956. Electroshock therapy. In F. Martí-Ibáñez, A. A. Sackler, M. D. Sackler, and R. R. Sackler (Eds.), *The Psychodynamic Therapies in Psychiatry.* New York: Hoeber-Harper. Pp. 258–270.

Clarke, E., and O'Malley, C. D. 1968. *The Human Brain and Spinal Cord.* Berkeley: University of California Press.

Columbia-Greystone Associates. 1949. *Selective Partial Ablation of the Frontal Cortex.* (F. A. Mettler, Ed.). New York: P. Hoeber.

Crawford, M. P., Fulton, J. F., Jacobsen, C. F., and Wolfe, J. B. 1948. Frontal lobe ablation in chimpanzee: A résumé of "Becky" and "Lucy." *Research Publications—Association for Research in Nervous and Mental Disease, 27,* 3–58.

Dale, H. H. 1914. The action of certain esters and ethers of choline, and their relation to muscarine. *Journal of Pharmacology and Experimental Therapeutics, 6,* 147–190.

Dale, H. H., and Dudley, H. W. 1929. The presence of histamine and acetylcholine in the spleen of the ox and the horse. *Journal of Physiology, 68,* 97–123.

Damásio, A. 1975. Egas Moniz, pioneer of angiography and leucotomy. *Mt. Sinai Journal of Medicine, 42,* 502–513.

Dana, C. L. 1921. The anatomic seat of the emotions: A discussion of the James-Lange theory. *Archives of Neurology and Psychiatry, 6,* 634–639.

Delay, J., Deniker, P., and Harl, J. M. 1952. Traitement des états d'excitation et d'agitation par une méthode médicamenteuse dérivée de l'hibernothérapie. *Annuals Médico-Psychologique, 110,* 267–273.

Dixon, W. E. 1907. The mode of action of drugs. *Medical Magazine, 16,* 454–457.

Durant, J. R. 1985. The science of sentiment: The problem of the cerebral localization of emotion. In P. P. G. Bateson and P. H. Klopfer (Eds.), *Perspectives in Ethology: Vol. 6: Mechanisms.* New York: Plenum Press. Pp. 1–31.

Dusser de Barenne, J. G. 1919. Recherches expérimentales sur les fonctions du système nerveux central, faites en particulier sur ceux dont le néopallium avait été enlevé. *Archives Neerlandaises de Physiologie, 4,* 31–123.

Elliott, T. R. 1904. On the action of adrenalin. *Journal of Physiology, 31,* xx–xxi.

Elliott, T. R. 1913. The innervation of the adrenal glands. *Journal of Physiology, 46,* 285–290.

Estienne, C. 1545. *De dissectione partium corporis humani.* Paris: S. Colins.

Eustachio, B. 1563. *Opuscula anatomica.* Venetiis: Vincentius Luchinus.

Eustachio, B. 1714. *Tabulae anatomicae.* Roma: F. Gonzaga.

Ewald, J. R. 1898. Über künstlich erzeugte Epilepsie. *Berliner klinische Wochenschrift, 35,* 689.

Ferrier, D. 1876. *The Functions of the Brain.* London: Smith, Elder and Company.

Fiamberti, A. M. 1937. Proposta di una tecnica operatoria modificata e semplificata per gli interventi alla Moniz sui lobi frontali in malati di mente. *Rassegna di Studi Psichiatria, 26,* 797.

Fiamberti, A. M. 1939. Considerazioni sulla leucotomia prefrontale con il metodo transorbitario. *Giornale di Psichiatrie e di Neuropatologia, 67,* 291–295.

Fink, M. 1984. Meduna and the origins of convulsive therapy. *American Journal of Psychiatry, 141,* 1034–1041.

Freeman, W. 1948. Transorbital lobotomy: Preliminary report of ten cases. *Medical Annals of the District of Columbia, 17,* 257–261.

Freeman, W. 1949. Transorbital leucotomy: The deep frontal cut. *Proceedings of the Royal Society of Medicine, 42,* 8–12.

Freeman, W. 1957. Frontal lobotomy 1936–1956. A follow-up study of 3000 patients from one to twenty years. *American Journal of Psychiatry, 113,* 877–886.

Freeman, W. 1958. Prefrontal lobotomy: Final report of 500 Freeman and Watts patients followed for 10 to 20 years. *Southern Medical Journal, 51,* 739–745.

Freeman, W. 1971. Frontal lobotomy in early schizophrenia: Long follow-up in 415 cases. *British Journal of Psychiatry, 119,* 621–624.

Freeman, W., and Watts, J. W. 1936. Prefrontal lobotomy in agitated depression: Report of a case. *Medical Annals of the District of Columbia, 5,* 326–329.

Freeman, W., and Watts, J. W. 1942. *Psychosurgery: Intelligence, Emotion and Social Behavior Following Prefrontal Lobotomy for Mental Disorders.* Springfield, IL: Charles C Thomas.

Freeman, W., and Watts, J. W. 1950. *Psychosurgery in the Treatment of Mental Disorders and Intractable Pain* (2nd ed.). Springfield, IL: Charles C Thomas.

Freud, S. 1920. *A General Introduction to Psychoanalysis.* (J. Riviere, Trans). New York: Boni and Liveright.

Freud, S. 1927. *The Ego and the Id.* London: Hogarth Press.

Fulton, J. F. 1932. New horizons in physiology and medicine: The hypothalamus and visceral mechanisms. *New England Journal of Medicine, 207,* 60–68.

Fulton, J. F. 1948. The surgical approach to mental disorder. *McGill Medical Journal, 17,* 133–145.

Fulton, J. F. 1952. *The Frontal Lobes and Human Behavior.* Springfield, IL: Charles C Thomas.

Fulton, J. F., and Ingraham, F. D. 1929. Emotional disturbances following experimental lesions of the base of the brain (prechiasmal). *Journal of Physiology, 67,* xxvii–xxviii.

Furtado, D. 1949. Results of leucotomy. A twelve-year follow-up. *Psychosurgery, First International Congress.* London. Pp. 171–172.

Gaskell, W. H. 1886. On the structure, distribution, and function of the nerves which innervate the visceral and vascular systems. *Journal of Physiology, 7,* 1–80.

Gaskell, W. H. 1920. *The Involuntary Nervous System.* London: Longmans, Green.

Goltz, F. L. 1874. Über die Functionen des Lendenmarks des Hundes. *Pflüger's Archiv für die gesammte Physiologie, 8,* 460–498.

Goltz, F. 1888. Über die Verrichtungen des Grosshirns. *Pflüg er's Archiv für die gesammte Physiologie, 42,* 419–467. In G. von Bonin (Ed.), *Some Papers on the Cerebral Cortex.* Translated as, "On the functions of the hemispheres." Springfield, IL: Charles C Thomas, 1960. Pp. 118–158.

Goltz, F. L. 1892. Der Hund ohne Grosshirn. *Archiv für die gesammte Physiologie, 51,* 570–614.

Head, H. 1918. Some principles of neurology. *Brain, 41,* 344–354.

Hess, W. R. 1948. *Die funktionelle Organisation des vegetativen Nervensystems.* Basel: Benno Schwabe. Translated as, *Diencephalon: Autonomic and Extrapyramidal Functions.* New York: Grune & Stratton, 1954.

Hess, W. R. 1957. The Functional Organization of the Diencephalon. New York: Grune & Stratton.

Hess, W. R. 1962. *Psychologie in biologischer Sicht,* Stuttgart: Georg Thieme. Translated as, *The Biology of Mind.* Chicago: University of Chicago Press, 1964.

Horsley, V. A. H., and Clarke, R. H. 1908. The structure and functions of the cerebellum examined by a new method. *Brain, 31,* 45–124.

Ingram, W. R., Barris, R. W., and Ranson, S. W. 1936. Catalepsy: An experimental study. *Archives of Neurology and Psychiatry, 35,* 1175–1197.

Jackson, J. H. 1881. Remarks on dissolution of the nervous system as exemplified by certain post-epileptic conditions. *Medical Press and Circular,* April, 329–332ff.

Jackson, J. H. 1884. Evolution and dissolution of the nervous system. *British Medical Journal, 1,* 591, 660, 703. (Also in *Popular Science Monthly, 25,* 171–180, 1884.)

Jackson, J. H. 1887–1888. Remarks on evolution and dissolution of the nervous system. *Journal of Mental Science, 33,* 25–48.

Jackson, J. H. 1894. The factors of the insanities. *Medical Press and Circular,* June, 615–619.

Jackson, J. H., and Beevor, C. E. 1890. Case of tumour of the right tempero-sphenoidal lobe, bearing on the localisation of the sense of smell and on the interpretation of a particular variety of epilepsy. *Brain, 12,* 346–357.

Jacobsen, C. F. 1936. Studies on cerebral function in primates. *Comparative Psychological Monographs, 13,* 1–60.

Jacobsen, C. F., Wolf, J. B., and Jackson, T. A. 1935. An experimental analysis of the functions of the frontal association areas in primates. *Journal of Nervous and Mental Disease, 82,* 1–14.

Karplus, J. P., and Kreidl, A. 1909. Gehirn und Sympathicus. I. Zwischenhirnbasis und Halssympathicus. *Archiv für die gesammte Physiologie, 129,* 138–144.

Karplus, J. P., and Kreidl, A. 1910. Gehirn und Sympathicus. II. Ein Sympathicuszentrum im Zwischenhirn. *Pflüger's Archiv für die gesammte Physiologie, 135,* 401–416.

Karplus, J. P., and Kreidl, A. 1911. Gehirn und Sympathicus. III. Sympathicusleitung im Gehirn und Halsmark. *Pflüger's Archiv für die gesammte Physiologie, 143,* 109–127.

Klüver, H., and Bucy, P. C. 1937. "Psychic blindness" and other symptoms following bilateral temporal lobectomy in rhesus monkeys. *American Journal of Physiology, 119,* 352–353.

Klüver, H., and Bucy P. C. 1938. An analysis of certain effects of bilateral temporal lobectomy in the rhesus monkey, with special reference to "psychic blindness." *Journal of Psychology, 5,* 33–54.

Klüver, H., and Bucy, P. C. 1939. Preliminary analysis of functions of the temporal lobes in monkeys. *Archives of Neurology and Psychiatry, 42,* 979–1000.

Langley, J. N. 1900. The sympathetic and other related systems of nerves. In E. A. Schäfer (Ed.), *Text-book of Physiology* (Vol. 2). London: Pentland. Pp. 616–696.

Langley, J. N. 1903. The autonomic nervous system. *Brain, 26,* 1–26.

Langley, J. N. 1916. Sketch of the progress of discovery in the eighteenth century as regards the autonomic nervous system. *Journal of Physiology, 50,* 225–258.

Langley, J. N. 1921. *The Autonomic Nervous System.* Cambridge, U.K.: Heffer.

Lehmann, H. D., and Hanrahan, G. E. 1954. Chlorpromazine: New inhibiting agent for psychomotor excitement and manic states. *Archives of Neurology and Psychiatry, 71,* 227–237.

Lima, A. 1974. Egas Moniz: 1874–1955. *Journal für Neurologie, 207,* 167–170.

Livingston, K. E., and Hornykiewicz, O. (Eds.). 1978. *Limbic Mechanisms: The Continuing Evolution of the Limbic System Concept.* New York: Plenum Press.

Loewi, O. 1921. Über humorale Übertragbarkeit der Herznervenwirkung. *Pflüger's Archiv für die gesamte Physiologie, 189,* 201–213.

Lyerly, J. G. 1938. Prefrontal lobotomy in involutional melancholia. *Journal of the Florida Medical Association, 25,* 225–229.

MacLean, P. D. 1949. Psychosomatic disease and the "visceral brain." Recent developments bearing on the Papez theory of emotion. *Psychosomatic Medicine, 11,* 338–353.

MacLean, P. D. 1952. Some psychiatric implications of physiological studies on fronto-temporal portions of the limbic system (visceral brain). *Electroencephalography and Clinical Neurophysiology, 4,* 407–418.

MacLean, P. D. 1954. The limbic system and its hippocampal formation. *Journal of Neurosurgery, 11,* 29–44.

MacLean, P. D. 1978. Challenges of the Papez heritage. In K. E. Livingston and O. Hornykiewicz (Eds.), *Limbic Mechanisms: The Continuing Evolution of the Limbic System Concept.* New York: Plenum Press. Pp. 1–15.

Masserman, J. H. 1938. Destruction of the hypothalamus in cats. *Archives of Neurology and Psychiatry, 39,* 1250–1271.

Meduna, L. 1935. Versuche über die biologische Beeinflussung des Ablaufes der Schizophrenia. I. Camphour und Cardiazol Krampfe. *Zeitschrift für die gesamte Neurologie und Psychiatrie, 1935,* 235–262.

Meduna, L. 1938. General discussion of the cardiazol therapy. *American Journal of Psychiatry, Suppl., 94,* 40–50.

Meyer, A. 1971. *Historical Aspects of Cerebral Anatomy.* London: Oxford University Press.

Moniz, E. 1936. *Tentatives Opératoires dans le Traitement de Certaines Psychoses.* Paris: Masson et Cie.

Moniz, E. 1937. Prefrontal leucotomy in the treatment of mental disorders. *American Journal of Psychiatry, 93,* 1379–1385.

Moniz, E., and Lima, A. 1936. Premiers essais de psycho-chirurgie: Technique et résultats. *Lisboa Medicina, 13,* 152–161.

Ody, F. 1938. Le traitement de la démence-précoce par résection du lobe préfrontal. *Archivo Italiani di Chirurgia, 53,* 321–330.

Papez, J. 1937. A proposed mechanism of emotion. *Archives of Neurology and Psychiatry, 38,* 725–743.

Papez, J. W. 1958. Visceral brain, its component parts and their connections. *Journal of Nervous and Mental Disease, 126,* 40–55.

Platter, F. 1614. *Observationum . . . libri tres.* Basel: Koenig.

Pourfour du Petit, F. 1727. Mémoire dans lequel il est démontré que les nerfs intercostaux fournissent des rameaux qui portent des esprits dans les yeux. *Mémoires de Mathématique et de Physiologie. Histoire de l'Académie Royale des Sciences,* 1–19.

Puusepp, L. 1914. Etat actuel et problèmes prochains sur la question du traitement des maladies mentales. *Congrès International, Moscou.*

Ramsey, G. V. 1952. A short history of psychosurgery. *American Journal of Psychiatry, 108,* 813–816.

Ranson, S. W. 1937. Some functions of the hypothalamus. *Bulletin of the New York Academy of Medicine, 13,* 241–271.

Ranson, S. W., Kabat, H., and Magoun, H. W. 1935. Autonomic responses to electrical stimulation of the hypothalamus, preoptic region and septum. *Archives of Neurology and Psychiatry, 33,* 469–477.

Rothmann, H. 1923. Zusammenfassender Bericht über den Rothmannschen grosshirnlosen Hund nach klinischer und anatomischer Untersuchung. *Zeitschrift für die gesamte Neurologie und Psychiatrie, 87,* 247–313.

Rylander, G. 1940. *Personality Changes after Operations on the Frontal Lobes.* Copenhagen: Munksgaard.

Rylander, G. 1941. Brain surgery and psychiatry. *Acta Chirurgica Scandinavica, 85,* 213–234.

Rylander, G. 1948. Personality analysis before and after frontal lobotomy. *Research Publications—Association for Research in Nervous and Mental Disease, 27,* 691–705.

Sakel, M. 1935. *Neue Behandlung der Schizophrenie.* Wien: Moritz Perles.

Sakel, M. 1937. Origin and nature of hypoglycemic therapy of the psychoses. *Archives of Neurology and Psychiatry, 38,* 188–190.

M. Sakel. 1938. The pharmacological shock treatment of schizophrenia. (J. Wortis, Trans., with Foreword by O. Potzl). *Nervous and Mental Disease Monographs, No. 62.*

Schiller, F. 1979. *Paul Broca: Founder of French Anthropology, Explorer of the Brain.* Berkeley: University of California Press.

Schiller, F. 1985. The mystique of the frontal lobes. *Gesnerus, 42,* 415–424.

Sheehan, D. 1936. Discovery of the autonomic nervous system. *Archives of Neurology and Psychiatry, 35,* 1081–1115.

Singer, C. 1952. *Vesalius on the Human Brain.* London: Oxford University Press.

Smith, E. S. 1971. Galen's account of the cranial nerves and the autonomic nervous system. *Clio Medica, 6,* 77–98, 173–194.

Talbert, G. A. 1900. Ueber Rindenreizung am freilaufenden Hunde nach J. R. Ewald. *Archiv für Anatomie und Physiologie, Physiologische Abteilung, 24,* 195–208.

Treviranus, G. R. 1816–1821. *Vermischte Schriften anatomischen und physiologischen Inhalts.* Göttingen: J. F. Röwer.

Valenstein, E. S. 1973. *Brain Control.* New York: John Wiley & Sons.

Valenstein, E. S. 1986. *Great and Desperate Cures.* New York: Basic Books.

Vesalius, A. 1543. *De humani corporis fabrica.* Basel: J. Oporinus.

Ward, A. A. 1948. The anterior cingulate gyrus and personality. *Research Publications—Association for Research in Nervous and Mental Disease, 27,* 438–445.

Weber, E. F., and Weber, E. H. 1846. Expériences qui prouvent que les nerfs vagues peuvent retarder le mouvement. *Archives Génerales de Médecine, Suppl. 12.*

Williams, J. M., and Freeman, W. 1953. Evaluation of lobotomy with special reference to children. *Research Publications—Association for Research in Nervous and Mental Disease, 31,* 311–318.

Willis, T. 1664. *Cerebri anatome: cui accessit nervorum descriptio et usus.* London: J. Flesher.

Wilson, S. A. K. 1924. Pathological laughing and crying. *Journal of Neurology and Psychopathology, 4,* 299–333.

Wilson, S. A. K. 1929. *Modern Problems in Neurology.* New York: Wood.

Windle, W. F. 1970. John Newport Langley (1852–1925). In W. Haymaker and F. Schiller (Eds.), *The Founders of Neurology* (2nd ed.). Springfield, IL: Charles C Thomas. Pp. 289–292.

Winslow, J. B. 1732. *Exposition Anatomique de la Structure du Corps Humain.* Paris: G. Desprez. Translated by G. Douglas as, *An Anatomical Exposition of the Structure of the Human Body.* London: A. Bettsworth and C. Hitch, 1734.

Woodworth, R. S., and Sherrington, C. S. 1904. A pseudoaffective reflex and its spinal path. *Journal of Physiology, 31,* 234–243.

Wortis, J. 1953. Physiological treatment. *American Journal of Psychiatry, 110,* 507–511.

Wortis, J. 1958. Manfred Sakel, M.D., 1900–1957. *American Journal of Psychiatry, 115,* 287–288.

Yakovlev, P. I. 1978. Recollections of James Papez and comments on the evolution of the limbic system concept. In K. E. Livingston and O. Hornykiewicz (Eds.), *Limbic Mechanisms: The Continuing Evolution of the Limbic System Concept.* New York: Plenum Press. Pp. 351–354.

Part V
Intellect and Memory

Chapter 21
Intellect and the Brain

> The brain of a first-class genius like Friedrich Gauss is as far removed from that of the savage bushman as that of the latter from the brain of the nearest related ape.
>
> Edward A. Spitzka, 1907

This chapter examines the relationship between intellectual functions and the brain, with particular emphasis on brain size. Although "intellectual functions" can be defined in many ways, the intent here is to look at functions that relate to the concept of general intelligence, problem solving, and the ability to adapt to the ever-changing challenges of life. For example, the idea that some individuals or races are better equipped than others to master the problems of civilized life because of greater brain development is among the topics considered in this chapter.

The study of intellectual functions has not been restricted to humans, and for this reason experiments on animals, such as dogs and monkeys, are also examined. The use of animals served as a way of testing hypotheses drawn from clinical case material and led to new ways of viewing different populations of people, including those with brain injuries.

It should be noted that some topics which might fit under the general heading of intellect are not addressed in detail in this chapter. The emphasis here is largely on the brain and the cerebral hemispheres as a whole and then on the basic concept of "association" cortex. Chapter 22 examines the large literature on the frontal association areas and higher cognitive acts. Memory and language, which can be viewed as two specific intellectual functions, are examined in subsequent chapters.

BRAIN SIZE AND INTELLECT

In the third century B.C., Erasistratus, who worked in Alexandria, suggested that the superior intellect of humans was due to the human cerebral hemispheres being the most complex. Whether this was his own idea or one that really originated with Herophilus, his associate, is not known with certainty. Nevertheless, it represented one of the first attempts to associate intellect with the size or development of the cerebrum.

The theory proposed by Erasistratus was challenged in Rome some four centuries later. Galen attacked the idea after examining the brain of a donkey. Although donkeys possess highly convoluted brains, these animals had to be—at least in Galen's opinion—among the stupidest ani-

mals on the face of the earth. To Galen, "donkey intelligence" was an oxymoron—a true contradiction in terms! This being the case, he concluded that the "temperament" of the thinking body was more important than its size or complexity.

In the hands of the Church Fathers of the fourth century, attention was drawn to the ventricles of the brain as the most likely location for the rational soul—the soul responsible for perceiving, thinking, and memory (Chapter 2). It was not until the sixteenth century that this theory was challenged.

In Book VII of *De humani corporis fabrica*, Andreas Vesalius drew attention from the ventricles back to the brain substance itself. He emphasized that the human brain was not unique when it came to structures. Differences between humans and animals related more to size and organization of the brain. To quote Vesalius:

> Certainly in the brain of sheep, goat, ox, cat, ape, dog, and of such birds as I have dissected, there is a shaping of the parts corresponding to that of the human brain, and specifically is this so of the ventricles. There is hardly any difference that we have detected except in bulk, (though) the brains do vary according to the intelligence with which the animals are endowed. For to man has been given the largest brain; next after him the ape, the dog, and so on, according to the order that we have learned of the power of reason in animals. And to man's lot falls a brain not only bigger in proportion to the bulk of his body, but actually bigger than the brain of any other animal. (Translated in Singer, 1952, 6–7)

Understanding the functional organization of the brain was more problematic to Vesalius than issues that could be solved with a newer, more realistic anatomy. Indeed, in the same book from which the aforementioned quotation was taken, he lamented:

> I can in some degree follow the brain's functions in dissections of living animals, with sufficient probability and truth, but I am unable to understand how the brain can perform its office of imagining, meditating, thinking, and remembering, or, following various doctrines, however you may wish to divide or enumerate the powers of the Reigning Soul. (Translated in Singer, 1952, 7)

In 1664, Thomas Willis specifically proposed that "the origine and fountain of all motions and conceptions" can be found in the cerebrum (p. 91). Willis also thought that the cerebrum served in the production of animal spirits and played a role in memory (the gyri) and imagination (he actually assigned imagination to the corpus callosum, which he broadly defined as the white matter of the hemispheres). He came to these conclusions in part because the cerebrum varied more across species than did the cerebellum or the corpus striatum. He associated the cerebellum with involuntary movement and considered the striatum a "mart" for the dispensation of the spirits (see Chapters 2 and 15).

By the end of the seventeenth century, it was fairly well established that the size of the cerebral hemispheres set humans apart from other animals and that the cerebrum was the substrate of intellect as well as of the will. But until 1861, when Paul Broca made his case for a center for articulate language in the frontal cortex, the feeling among most scientists was that the cerebrum was equipotential—that is, it was not divisible into functional units or specific centers. The equipotentiality idea was embraced and championed by Flourens, a leading member of the French scientific elite. More than any other person, Flourens (1824, 1842, 1846) was responsible for thrusting the "pseudoscience" of phrenology, along with all notions of distinct cortical faculties, into disrepute (see Chapter 3).

Because the great mass of the cerebral hemispheres accounted for much of the overall size of the human brain and served higher functions, some of Flourens' contemporaries reasoned that there should be differences in overall brain size not only across species but within the human family. These scientists further reasoned that there should be a strong positive correlation between brain size and intellectual accomplishments. Armed with these biases, some investigators set forth to collect data to test these hypotheses. The first goal was to compare brain sizes among individuals from different cultures, races, and social strata. This was accomplished either by measuring brains themselves or by measuring cranial vaults and using them as an index of brain size.

EUROPEAN SCIENCE AND THE FAMILY OF MAN

The biological study of races may have begun in 1684 when François Bernier (1620–1688), a French physician and world traveler, published his "New Division of the Earth by the Different Species or Races of Its Inhabitants" (see Slotkin, 1965). Bernier associated race with skin color. Although he related color to the four known continents, he also presented other physical features of his various races (e.g., Asians have "pigs eyes, long and deep set").

In 1735, Carolus Linnaeus (Linné) expressed the unpopular idea that on morphological grounds, disregarding size, man could not be distinguished from the apes (e.g., the "Orang-Outang," or "Man of the Woods"). Thus, *Simia* and *Homo* were placed together in the same class. Later, in the 1758–1759 edition of his *Systema naturae,* he coined the term "primates" and subdivided the first order *(Homo sapiens)* of his new first class into *Europaeus, Americanus, Asiaticus,* and *Afer.* Linnaeus wrote that *Europaeus* wore clothes and was gentle, inventive, and governed by laws. In contrast, *Afer* was described as crafty, negligent, indolent,

and shameless. Of course, this was a classification more easily accepted by the majority of white Europeans.

At this time, about the only dark-skinned people living in northern European countries were the servants or "ornaments" of a few wealthy aristocrats. In this context, the Landgrave of Hesse brought a small group of black Africans to his residence in Kassel. When some of these individuals died, their bodies were examined by members of the German scientific community. One scientist interested in them was Samuel Thomas von Sömmerring.

In 1784, Sömmerring examined the skulls of two black African males who died at 14 and 20 years of age, respectively. He found that their skulls were considerably smaller than those usually associated with European males. In fact, Sömmerring felt that these skulls, with their corresponding small brains, had more in common with the orangutan than with European specimens (see Dougherty, 1985).

Sömmerring's reports (1785, 1788) were followed by Astley Cooper's study on a large black African man (cited by Peacock, 1865a). This African's brain weight was comparable to that of white Europeans.

In 1836 and 1837, Friedrich Tiedemann (1781–1861; shown in Figure 21.1), another German anatomist, presented his measurements of a very small African man who had a very low brain weight. In spite of this specimen's small size, Tiedemann wrote it was wrong to conclude that the average African's brain is necessarily smaller than the average European brain. He based this conclusion, which he related to mental capacity, on additional measurements of skulls, not brains, and on his belief that the Africans under investigation were the poor remnants of subjugated populations.

Tiedemann was severely criticized for drawing conclusions that were "not consistent" with his own measurements, as well as for the imprecise nature of his measurements. One of his critics was Thomas B. Peacock (1812–1882). In his own

Figure 21.1. Friedrich Tiedemann (1781–1861), who compared the brains of black Africans and Caucasians and concluded they were not necessarily different.

reports, Peacock (1865a, 1865b) described seven brains from black Africans, two of which came from other investigators.[1] The five male specimens were all below the average for Europeans. But to his surprise, Peacock found that the weights of the brains from the two females were slightly above the European average. Even so, the skull measurements of his specimens led Peacock to conclude that, overall at least, European crania were larger than others. Yet, unlike most of his contemporaries, he did not speculate on the intellectual capabilities of black Africans, or on the supposed relation between brain size and intellect.

Another of Tiedemann's critics was Paul Broca (1873a), who reasoned that Tiedemann was too influenced by the preconceived notion that all human races must have the same cranial capacity. Broca also questioned the reliability of using seed poured into the foramen magnum to obtain a measure of brain weight. The weight of the seed was affected by humidity, and findings also varied with how compressed the seeds were within the skull cavity. Broca preferred lead shot instead of seeds, the procedure also recommended by Samuel George Morton (1799–1851) of Philadelphia (see the following section).

By this time, Broca was no stranger to the field of craniometry or to the study of brains. Twelve years earlier, he had written an important paper on intelligence and brain volume. Broca (1861b) had also founded the first anthropological society, the *Société d'Anthropologie*. Between February and July 1861, most of the meetings of this society included discussions about the relation between brain size and intelligence, the "perfectibility" of the races, and the future of the not so noble savage.

Broca (1861b) maintained that educated people have bigger brains than those not educated. He stated that "all other things being equal, there is a remarkable relationship between the development of intelligence and the volume of the brain" (p. 188). But he also stressed that "an enlightened person cannot begin to get the idea of measuring intelligence by measuring the brain" (Broca, 1861b, p. 302).[2]

Broca's (1861b, 1862a, 1862b) early ideas about intellect and the brain were widely quoted. John Thurnam (1810–1873), among others, cited his work and made the following statement:

> The average weight of the brain of the educated, and of those who occupy a superior social position, is no doubt greater than that of the uneducated and lower class. . . . Mr. Broca's researches on the dimensions of the heads of students of medicine, as compared with those of servants in the large hospital of the Bicêtre, show a decided preponderance in favour of the students. He considers it as "certain that, other things being equal, whether the result of education or whether hereditary, the volume of the skull, and consequently of the brain, is greater in the superior than in the inferior classes." (1866, 17–18)

Paul Broca also entertained the idea that the races might not be equal intellectually. This was based on what he knew about the achievements of the different races and his belief that the different races came from separate cre-

ations (polygenism; Tiedemann was a monogenist who believed all humanity derived from Adam). His own studies showed that Europeans had larger brains than individuals from other races.

Although not a believer in racial equality (along with virtually everyone else in the eighteenth century), Broca rejected the notion of ideal racial types and the view that only pure races could endure. He was also a humanitarian, a man repulsed by the concept of slavery and opposed to racial suppression.[3] Thus, Broca (1862b, 1867) went on to point out that black Africans were not below Caucasians on all measures commonly associated with "apishness" and that some of the favorite measures of his contemporaries were weak at best. For example, although he found that black Africans had longer forearms than Caucasians, he also noted they were decidedly less apelike than Caucasians when upper limb/lower limb ratios were compared. To cite but one more example, although Caucasians have more anteriorly located foramen magnums, Broca noted that this difference disappeared when measurements were corrected for the protrusion of the maxillary bone (prognathism).

Broca's (1873b) papers show that as his data amassed he continued to retreat from his emphasis on large brain size as a sure marker of higher intelligence. One possible reason was that some "yellow" subgroups (Eskimos, Lapps, Malays, Tartars) seemed to show even greater brain weights than Europeans. Another was that the archaic Cro-Magnon specimens unearthed in his own country had crania that exceeded those of modern Frenchmen.

But while Broca gave up on the idea that large brains were the exclusive property of the Caucasian race, he did not retreat as much from the notion that very small brains were a sign of inferiority. The groups singled out for having notably small brains included West African blacks, Caffirs, Nubians, Tasmanians, Hottentots, and native Australians, all of whom, he recognized, had smaller body sizes than Europeans.

The so-called intellectually deficient races were also the subject of other studies at this time. In one such investigation, John Marshall (1818–1891), a respected British surgeon, found that the brain of an African Bushwoman was small, even when adjusted for her body size. Marshall (1864) estimated her brain to body weight ratio at 1:45, whereas comparable Europeans averaged 1:17. Her cerebrum was singled out for further inspection. It was found to be long and to have all of the primary convolutions, but it was "defective" in both width and height and was "less complicated." Marshall noted that the frontal area was especially very narrow and shallow.

Marshall wrote that the Bushwoman's brain was comparable to another brain that had been studied, the "Hottentot Venus." In the same year, Carl Christoph Vogt compared the brain of the Venus to those of an anthropoid ape and a fellow German. He wrote:

> I find a remarkable resemblance between the ape and the lower human type, especially with reference to the development of the temporal lobe. The simplicity of the parallel fissure, the arrangement of the gyri, accord so much with those of the orang, that the brain of this bushwoman would

[1]Unlike Tiedemann, Peacock used fresh brains because he realized brain weight can be influenced not only by the nature of the disease producing death but by fixation.

[2]Interestingly, Broca's brain weighed in at just slightly above average at 1,484 g. Tiedemann's brain was a very light 1,254 g (Spitzka, 1907).

[3]Broca's personality and beliefs are described in detail by Francis Schiller (1979). A very different view of Broca, one that portrays him more as a racist, appears in the writings of Stephen J. Gould (1981).

certainly be rather placed by the side of the ape than of the
white man, were there not a decided difference in the form
of the posterior lobe and the operculum at its end. The
frontal, parietal, and temporal lobes are, by their coarse
simple gyri, decidedly simious, still the brain of this woman
belongs to the human type by the size of the hemispheres
and the character of the posterior lobe. (1864, 183–184)

Carl Vogt classified the Venus as a "Negro type" and
without question saw this group as inferior to Caucasians.
He theorized that the two groups were intellectual equals
early in life, but that by the time of puberty, the intellectual faculties of the Africans no longer grew as their skulls
closed. The belief was that from this time on, black
Africans were incapable of further intellectual progress.
Thus, in Vogt's (1864, p. 192) opinion, adult black Africans
resemble white children, females, and senile males in their
capacity for imitation, in caring little for the future, and in
never contributing "anything tending to the progress of
humanity or worthy of preservation."

The message was clear: People at the bottom rung of the
human ladder were made of intrinsically inferior material,
particularly when it came to the brain. August D. Waller
summarized the progression in his 1891 textbook, *Human
Physiology*. Using the ratio of brain weight to body weight,
there was a decrease in the ratio from fish (1:5,000) to reptiles (1:1,500) to birds (1:220) and to mammals (1:180). It
was then shown that the brain to body weight ratio continued to decrease with primates. For the orangutan it was
1:120 and for humans it was 1:50. Within the human family,
Waller wrote that the brain weight of the average European
topped the charts at 1,390 g, whereas that of the average
black African was just 1,250 g.

BRAIN SIZE, RACE, AND INTELLECT IN AMERICA

On the opposite side of the Atlantic, Samuel George Morton,
President of the Academy of Natural Sciences in Philadelphia, amassed one of the largest collections of skulls in the
world (Gould, 1978, 1981). Louis Agassiz (1807–1873), the
Swiss-born naturalist who was one of America's leading
opponents of Darwinian evolution, wrote that seeing this
physician's collection of more than 1,000 skulls was by itself
worth the rigors and stress of a trip across the North
Atlantic. Morton won fame, however, not just for the skull
collection he started in 1820, but for the books, tables, and
conclusions that he based on his collection and for his interest in associated artifacts.

Morton believed he found evidence to suggest that
blacks and Caucasians were already distinct groups in
Ancient Egypt. He concluded that the major races did not
have time to split since humanity began with the flood that
took Noah to Mount Ararat, about 1,000 years before the
height of the Egyptian empire. This was consistent with
his polygenist beliefs about the races and their separate
creations. Much assured by this, Morton set forth to rank
the various races. The underlying assumption was that
cranial capacity was a valid correlate of innate differences
in intelligence.

Morton found that Caucasian skulls, especially those of
Western Europeans, as opposed to Jews and "Hindoos,"
were the largest among the races. In contrast, black
Africans had notably smaller skulls. Even smaller were the
skulls that came from Ancient Peruvians and Native Aus-

Figure 21.2. Josiah Clark Nott (1804–1873), the American physician
who argued for white superiority and for the institution of slavery. Nott
and George Robins Gliddon presented drawings in their books that falsely
gave the impression that black Africans had more in common with the
apes than with Caucasians.

tralians. Such findings led Morton to write that even in
ancient times black Africans could have worked only as
servants and slaves because of their limited intellectual
capacities. This, he believed, was consistent with the way
they were depicted in Ancient Egyptian art.

Morton's attempts to associate intellect with brain size
appeared in three major works: his *Crania Americana* of
1839, his "Crania Aegyptiaca" of 1844, and a more general
work from 1849 in which he described his measurements of
623 human crania. Morton's reputation as one of America's
greatest scholars soared with each of these publications.
Important questions about the validity of his findings did
not surface at this time.[4]

As might be expected, Morton, "the objectivist," was not
the only American to find racial differences in cranial capacity and to relate them to intellect. Comparable results were
obtained by many others. One series of studies was conducted by Josiah Clark Nott (1804–1873) and George Robins
Gliddon (1809–1857). Nott, shown in Figure 21.2, was a
physician from Alabama and one of Morton's most able

[4]More than a century after Morton died, Gould (1978, 1981) showed
that Morton bolstered his Caucasian values by omitting the Hindu skulls
that were small relative to the others in the Caucasian group. At the same
time, he included larger subsamples of South American Indian skulls to
pull down the mean of the Native American group. This manipulation
allowed him to claim that American Indians could not be educated and
could not reason abstractly. In addition, Morton used the skulls of small
female Hottentots to argue more strongly for the inferiority of blacks,
while relying on all-male samples of Englishmen to demonstrate the superiority of Caucasians. When Morton's data were "corrected" for gender,
stature, and all of his slips, omissions, miscalculations, and inconsistencies, the average cranial differences across the races turned out to be
much smaller than he reported. In what would have been an even worse
shock to Morton, Native Americans, and not Caucasians, now came out on
the top of the racial ladder.

Figure 21.3. Drawings from Nott and Glidden (1868) showing black Africans with distorted features to make them look more like apes than Caucasians or Ancient Egyptians (bottom of Row 2).

Figure 21.4. Louis Agassiz (1807–1873), the Swiss-born naturalist who emigrated to America where he argued that black Africans represented a distinct, lower form of the human species than Caucasians.

pupils. Gliddon, a polygenist like Nott, had served the U.S. government in Cairo and was a recognized Egyptologist. Nott and Gliddon (1854, 1868) argued for the inferiority of nonwhites and for the continuation of slavery. Their books are especially noteworthy because the drawings falsely extend the jaws of the black Africans while changing the skulls and faces of apes to give the impression that these people had more in common with the apes than with Caucasians.[5] One of their plates appears as Figure 21.3.

The thoughts conveyed by Nott and Gliddon were shared by Louis Agassiz, Baron Georges Cuvier's Swiss disciple who had become a professor at Harvard University. Agassiz, who appears in Figure 21.4, had accepted the doctrine of human unity before coming to the United States in 1840. It was only after he met the black servants at the Philadelphia hotel where he stayed upon his arrival that he was led by his internal feelings to conclude that this race represented a lower form of the species (Gould, 1981).

Agassiz (1850), like Morton, challenged the idea that all humans originated from a single pair. He wrote: "We would particularly insist upon the propriety of considering Genesis as chiefly relating to the history of the Caucasian race, with special reference to the history of the Jews" (p. 138). He then argued that the different races had distinct, stable features.

[5]See Gould (1981) for a discussion of these drawings.

And it seems to us to be mock-philanthropy and mock-philosophy to assume that all races have the same abilities, enjoy the same powers, and show the same natural dispositions, and that in consequence of this equality they are entitled to the same position in human society. History speaks here for itself. Ages have gone by, and the social developments which have arisen among the different races have at all times been different; and not only different from those of other races, but particularly characteristic in themselves, evincing peculiar dispositions, peculiar tendencies, peculiar adaptations in the different races. (1850, 142)

Agassiz concluded that history has always shown black Africans to be apathetic and indifferent to the advantages of civilized society. As far as he was concerned, they never have, and never could, achieve a state resembling a true civilization. In his mind, the Egyptians, the Phoenicians, the Arabs, and even the American Indians had done considerably better than the submissive, imitative blacks.

Agassiz urged recognition and acceptance of these "facts" of inequality, especially by individuals who developed educational and social policies. He taught that blacks, having limited intellectual faculties, should be trained solely as laborers, whereas Caucasians should be educated differently to make better use of their more highly developed minds.

MENTAL DEFICIENCY AS AN ATAVISTIC FEATURE

Franz Joseph Gall, the leading figure in phrenology, was among the early nineteenth-century scientists who associated mental power with brain size. Gall (1822–1826) considered it to be a fact without exception that idiocy was always associated with heads less than 13–14 inches in circumference. In his mind, complete intelligence was impossible in such a small brain. In contrast, individuals with large brains may not always be very bright, although his expectation was that they would show at least isolated talents corresponding to the superior development of some brain parts.

Gall's assertion was "verified" by Félix Voisin (1794–1872), the medical superintendent of the Bicêtre Hospital in Paris. He wrote that in the lowest class of idiots, the skull's horizontal circumference just above the orbit varied from 11 to 13 inches (see Hollander, 1920). Individuals whose skulls were within the 14- to 17-inch horizontal circumference range showed little attention, but some rudimentary signs of intellect. People with 18- to 19-inch circumferences fared somewhat better, whereas the norm for a healthy, intelligent adult was found to be about 22 inches.

Many of the scientists who were reporting differences in brain size across racial and cultural groups looked upon mental deficiency as an atavistic condition, a throwback to a more primitive state. In particular, idiocy was viewed as a condition of arrested brain development that created a form intermediate between the great apes and the higher races of mankind (see Vogt, 1864).

John Marshall (1864), who studied the Bushwoman's brain, also examined the brains of two Europeans classified as idiots. The first brain came from a microcephalic woman who had little speech, could not feed or dress herself, and was unable to walk normally. The second brain came from a young male with a very small head, an individual even more impaired. The woman died at age 42, and the boy died at age 12. Marshall stated that their brains weighed 10½ and 8½ ounces, respectively. The normal means for the two age/gender groups were estimated at 42 and 44 ounces.

Two years later, John Langdon Haydon Down (1828–1896), a man described as a brilliant student of medicine from a distinguished English family, published a short paper on congenital "idiots" (Down, 1866).[6] Down had decided to devote himself to the humane treatment of the mentally retarded and had become the director of an asylum for the severely retarded in Surrey. He made a startling discovery at his asylum. He wrote that although all of his patients had English fathers and mothers, "a very large number of congenital idiots are typical Mongols" (p. 260). In addition to being intellectually deficient, they had obliquely placed eyes, yellowish skin, sparse straight hair, a small nose, and thick lips.

Down also recognized the feeble circulation and the poor life expectancy in his "Mongolian" children. Nevertheless, the emphasis in his short paper was not on these features but on the atavistic changes that had occurred. These children seemed to exhibit features of "degeneration" from a higher race (Caucasian) to a lower one (Mongol). In Down's words:

> The boy's aspect is such that it is difficult to realize that he is the child of Europeans, but so frequently are these characters presented, that there can be no doubt that these ethnic features are the result of degeneration. (1866, 261)

Although Down showed biases that were very much in line with common beliefs in this era, he remained the

humanist. He did not give up working with his "Mongolian" children and was able to show that they could exhibit improvement with good teaching and proper care.[7]

GENDER DIFFERENCES

Postulated differences between the brains of men and women, and the idea that women must have inferior brains to men, have been discussed for thousands of years. In his treatise, *De partibus animalium,* Aristotle wrote: "Man has the largest brain for his size; and men have a larger brain than women." Aristotle also wrote that humans have "more sutures in the skull than any other animal, and males have more than females" for the sake of ventilation.

Throughout the centuries the belief that females were intellectually inferior to males persisted, and in the nineteenth century this was tied to differences in brain size. Paul Broca (1861b, p. 153), for example, thought that even after corrections were made for body size, females had smaller brains:

> We might ask if the small size of the female brain depends exclusively upon the small size of her body. Tiedemann has proposed this explanation. But we must not forget that women are, on the average, a little less intelligent than men, a difference which we should not exaggerate but which is, nonetheless, real. We are therefore permitted to suppose that the relatively small size of the female brain depends in part upon her physical inferiority and in part upon her intellectual inferiority. (Translated in Gould, 1981, 104)

In 1854, Emil Huschke (1797–1858), a German anthropologist, wrote that the frontal lobe weight as a percentage of brain was 1 percent greater in a sample of 15 males than in a sample of 7 females. Huschke believed the same held true for surface measurements. These findings were endorsed and supported by many nineteenth-century scientists.[8]

The beliefs that women are intellectually inferior to men and therefore should have fewer opportunities for higher education were popular in this era. Gustave Le Bon (1841–1931), for example, did everything he could to compare the intellectual capacities of women to those of the so-called inferior races.[9] In 1879, he even compared the female brain to that of a gorilla. Le Bon was anything but tactful and discreet when he proclaimed:

> All psychologists who have studied the intelligence of women, as well as poets and novelists, recognize today that they represent the most inferior forms of human evolution and that they are closer to children and savages than to an adult, civilized man. They excel in fickleness, inconstancy, absence of thought and logic, and incapacity to reason. Without doubt there exist some distinguished women, very superior to the average man, but they are as exceptional as

---

[7]Many years would pass before it would be recognized that what had been called Mongolian idiocy (Down's syndrome) was due to a genetic mutation causing an extra chromosome in pair 21. The degree of retardation appears to be dependent upon how early in cell division the trisomy first appears.

[8]See the review by Mall (1909).

[9]Born of wealth, Le Bon earned a medical degree. But rather than practice medicine, he first traveled and wrote ethnographic books about Europe, Africa, and Asia. These books showed his strong racial biases and depicted Europeans well on the top of the intellectual scale.

[6]Down's paper has been reprinted with a brief commentary (see Wilkins and Brody, 1971).

the birth of any monstrosity, as, for example, of a gorilla with two heads; consequently, we may neglect them entirely. (pp. 60–61; translated in Gould, 1981, 104–105)

Le Bon only later became interested in group behavior. In 1895, he wrote his most famous work, *Psychologie des Foules* ("The Crowd" in the English edition of 1896). Being a staunch political conservative, he tried to depict the masses in the most negative ways. Thus, he gave crowds the characteristics of children, the inferior races, and women, all of whom tended to be grouped together at this time (e.g., Vogt, 1864).

The idea that there were differences in the gross anatomy of male and female brains did not hold up well in the twentieth century, especially when adjustments were made for overall differences in body size. In particular, the idea that female brains were more fetal, simpler, or more simian proved to be without solid scientific foundation. As one scientist at the end of the first decade of the twentieth century put it:

It is often said that the brains of women are of a simple type, but if their weight is not considered it is questioned whether a collection of brains could be assorted according to sex with any degree of certainty. (Mall, 1909, 24)

CRIMINALITY AND THE APISH BRAIN

Some anthropological studies of the crania and brains of criminals seemed to reveal that their brains were also smaller than normal. This helped to lead to the conclusion that "criminals might be considered as an anthropological variation of the species or at least of the cultured races" (Benedikt, 1879, p. 110).

This notion was developed by the Italian criminologist Cesare Lombroso (1836–1909). His best-known work, *Uomo Delinquente (Criminal Man),* first appeared in 1876 and was soon translated into several languages. Lombroso's (1876, 1887) basic thesis was that born criminals and the ancestors of civilized man had much more in common than had previously been recognized.

Lombroso reasoned that if this were true for criminals' behavior, it might also be true for their physical features. Thus, he postulated that criminals should show physical features that would betray their simian propensities. He turned to the skull in particular and hypothesized that the shape and structure of the cranium might reliably betray the born criminal's more apish brain and propensities. Lombroso hoped that craniometry would provide an objective correlate of the psychic peculiarities of the born criminal: instability of character, deranged passions, and overdeveloped vanity. In later years, while trying to expand his concept of physical correlates of criminality, Lombroso even remarked that just about every born criminal suffers from epilepsy (see Gould, 1981).

Lombroso knew his thesis would run into difficulty if his critics found that man's ancestors and animals lower on the phylogenetic scale acted with civility toward one another. If these more primitive species were compassionate and followed rules based on higher principles, being apish or animal-like would not support his hypothesis that criminal activity is atavistic. With this in mind, in the first part of his *Uomo Delinquente,* Lombroso presented case after case of animal cruelty, including murder. His examples included everything from ants killing aphids to insects eating certain plants.

Once this was done, Lombroso was free to compare criminal behavior in civilized societies to "normal" behavior in more primitive societies. Within the human family, the primitive groups were those whose members had thicker skulls, crania with simpler sutures, smaller brains, longer arms, darker skin, and so on. In short, these groups were decidedly more apish than the civilized, cultured Europeans with whom he identified.

Do the brains of criminals always show signs of degeneration, and are they usually smaller than average? The answer to these questions proved to be no. In fact, the large size of many criminal brains bothered many criminologists and anthropologists who wanted to accept Lombroso's assertion that among criminals, "the small capacities dominate and the very great are rare" (1887, p. 144).

Lombroso himself felt it necessary to divide would-be criminals into two groups. The first group, to whom the atavistic statements applied, were the born criminals—men and women who were guided largely by their poor heredity. For society, these were the real offenders and the evolutionary throwbacks with smaller brains. The second group consisted of individuals who acted against their good hereditary traits. These were the fine, upstanding men who committed crimes of passion, such as killing a stranger for having an affair with one's wife. Lombroso and Enrico Ferri (1856–1929) maintained that the distinction between the two types of criminals was all the more reason it was necessary to judge people on the basis of their minds and not just on the basis of the crime (see Lombroso, 1911; Zimmern, 1898).

Some anthropologists, who thought that the brains of even the most hardened criminals were not necessarily smaller than those of good men, pointed to the cause of death for the failure to find group differences. They reasoned that executing healthy criminals precluded the loss of brain weight due to fatal diseases, biasing the statistics in favor of heavier criminal brains over those good, moral citizens who died from natural causes. It was also argued that death by hanging filled the brain with blood, adding to its weight.

In some circles it was thought that not only did some criminals have heavier brains but that this itself could be a correlate of criminality. This idea was expressed in 1888, when Paul Topinard (1830–1911) wrote that a heavier brain might be correlated with a great exuberance of disordered cerebral activity, leading the individual to act against society.

INTERPRETIVE DILEMMAS

Attempts to measure brain size and to correlate it with intellect played an important role in how different groups of people were perceived and treated. Yet the findings were often variable, differing not just from study to study but sometimes within a single study.

A number of reasons have been given for the inconsistency of the findings. Without question, there were attempts by people like Samuel George Morton (1839, 1844, 1849) and

Figure 21.5. Rudolph Wagner (1805–1864) of the University of Göttingen. Wagner's descriptions of the exceptionally convoluted brain of Carl Friedrich Gauss stimulated considerable research on the brains of other "great men" (see Wagner, 1862, 1864).

Josiah Nott and George Gliddon (1854, 1868) to manipulate their samples and data to conform to their beliefs. In addition, certain unreliable methods, such as using millet or mustard seed to measure cranial capacity, allowed some experimenters to pack skulls more or less tightly, consciously or unconsciously, in accord with their a priori beliefs.

Even when experimenters approached their specimens with the best of intentions, and despite improved methods of measurement, problems remained. The fact of the matter was that there was little appreciation in the nineteenth century or even early in the twentieth century of the wide range of variables that could affect brain size and weight.[10] Many investigators, especially those in the first half of the nineteenth century, did not recognize the importance of body size, not only in comparisons across racial groups but also when comparing specimens of the two genders. In addition, many did not realize that preserving brains can lead to a considerable decrease in brain weight over time, making direct comparisons between fresh and "stale" brains almost meaningless. Until the end of the nineteenth century, relatively little was said about whether an individual might have died from a disease or another condition that could have caused some wasting of the brain. Further, other factors that could affect brain size, such as early malnutrition, remained unrecognized.

As the perceived importance of the myriad of factors affecting brain measurements gained recognition, earlier interpretations of the data became increasingly suspect. The complexity of the problem was recognized in 1902 by Heinrich Matiegka (1862–1941), a member of the Bohemian Science Society, who listed some 15 factors that could affect brain weight.

The demands of the scientific community were also changing at this time. Rather than just looking at overall brain weight across races and subgroups, scientists also wanted to know more about the morphology of the cerebrum itself. In particular, many wanted to know whether the cerebral convolutions were more complex, or at least different, in geniuses and in individuals who were the acknowledged leaders in their fields; they wanted to know about the top of the continuum.

THE CALL FOR THE BEST BRAINS

The study of the brains of gifted individuals really began with Rudolph Wagner (Figure 21.5), who worked in the German city of Göttingen (Spitzka, 1907). In 1855, Wagner obtained for study the brain of Carl Friedrich Gauss (1777–1855), the great physicist, astronomer, and mathematician. Wagner described this brain in two publications, one in 1862 and the other in 1864. He found that this brain weighed only 1,492 g, or just a bit above average. But the brain of Gauss was truly exceptional in its intricate pattern of fissures. In fact, it was more richly convoluted than any brain previously described.

The problem Wagner faced was that this observation did not hold true for five other brains obtained from professors at Göttingen. Some brains weighed considerably less than average, and not one showed the complex pattern of fissures that characterized the brain of Gauss. Hence, Wagner was forced to conclude that neither the complexity of the convolutions nor brain weight was a perfect correlate of intellectual development.

Wagner's conclusions were not readily accepted by many scientists, especially those who held strong preconceived biases. Paul Broca, for example, rejected Wagner's assertion on the basis of his German counterpart's definition of the intellectual elite. Broca (1861b, p. 166), somewhat tongue in cheek, remarked: "A professorial robe is not necessarily a certificate of genius; there may be, even at Göttingen, some chairs occupied by not very remarkable men."

Wagner's pioneering work had more impact than he might have expected. In its wake, many eminent individuals made arrangements for the preservation of their own brains in a kind of surrealistic competition to see who had been endowed with the best brains during life. Eight years before he died, George Grote (1796–1871), a classical scholar and historian, expressed his wish to have his brain examined:

> I desire that after my decease my cranium shall be opened by the Professor of Anatomy in University College, London, or by some other competent Anatomist.
>
> I desire that my brain shall be carefully weighed and examined, and that the weight thereof shall be communicated to Professor Bain, together with any other peculiarities which may be found, especially whether the cerebellum is deficient as compared with the cerebrum. (Cited in Marshall, 1892–1893, 21)

Professor Grote's wish was carried out. His brain was found to weigh 3.75 ounces above the average for a man his

---

[10]For references on factors that can affect brain size and weight, see Gould (1981) and Tobias (1970).

age, although it might have been somewhat heavier were it not for the wasting in the last few months of his life. The cerebrum to cerebellar ratio proved to be simply average. But Grote's frontal lobes were considerably more complex than average, and the frontoparietal region was also above average in weight.

The study of the brains of "great" men became even more fashionable when the Mutual Autopsy Society of Paris was founded in 1881. Its expressed purpose was to secure the brains of truly elite people. This led to the birth of similar associations in other parts of the world (Spitzka, 1905, 1907). Among these were the Cornell Brain Association and the American Anthropometric Society.[11]

Nevertheless, not all calls for brains from exceptional people were successful. In Stockholm, Gustaf Magnus Retzius, the celebrated anatomist, and Robert Tigerstedt (1853–1923), a highly respected physiologist, asked their esteemed colleagues to bequeath their brains to them. This appeal brought forth only two signatures—one from Retzius and the other from Tigerstedt (see Spitzka, 1907).

THE BRAINS OF THE ELITE

Although Retzius did not get a rapid, overwhelming response from his Swedish colleagues when he asked them to bequeath their brains to science, over the years he was able to secure the brains of a number of prominent individuals. This allowed him to describe the brains of an outstanding astronomer, a female mathematician, a noted physicist, a leading statesman, and an important biologist. Retzius (1898, 1900a, 1900b, 1904, 1905) wrote that some of these brains were characterized by unusual "secondary" gyri in the frontal cortex and by exceptional growth near the back of the Sylvian fissure (supramarginal and angular gyri). Nevertheless, the picture was anything but consistent.

Edward Anthony Spitzka (1876–1922), who studied medicine in New York and then embryology, morphology, and psychiatry in Berlin, Leipzig, and Vienna, conducted some of the most extensive surveys of the brains of exceptional people. Spitzka, who was described by Charles Dana in 1928 as "a red headed, hot-headed man, but finely trained in anatomy under Meynert," described a few brains in great detail, but added short descriptions of more than 130 others from greatly admired individuals. These appeared in Spitzka's papers of 1901, 1903, 1905, and 1907. The data on these brains came from various authorities around the world. They allowed Spitzka to include information on the brains of Ludwig von Beethoven (1770–1827), William Thackeray (1811–1863), Abraham Lincoln, and even the late Louis Agassiz.

The highly regarded Spitzka, who became president of the American Neurological Association, graphed brain weights as a function of age at the time of death. For every age category (35 to 85 years), the brains of the eminent people were well above the norm. The average difference was a full 100 g, or about 7 percent. This is shown in Figure 21.6.

Spitzka also showed that eminent people from "the exact

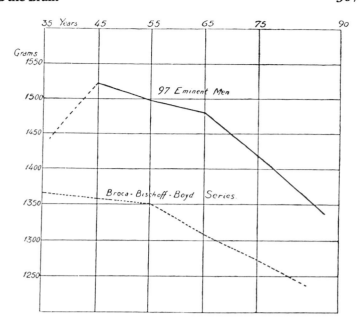

Figure 21.6. Curves of the brain weights of 97 "eminent men" compared to a large group of brains from "normal" individuals (Broca-Bischoff-Boyd series). The eminent men in each age group exhibited heavier average brain weights than the control subjects. (From Spitzka, 1907.)

sciences" had, on average, the heaviest brains. Nevertheless, the variability across subjects was high, and the heaviest brain in his sample came from Ivan Sergeyevitch Turgenev (1818–1883), the Russian poet and novelist. Turgenev's brain weighed in at 2,012 g, or about 50 percent more than the average European brain. It was almost 68 percent heavier than the brain of Gall, which weighed 1,198 g, well below normal for individuals of just average intelligence.

Spitzka dramatized his material by comparing the brains of some exceptional people with those of individuals representing the "lower" races. For some comparisons, he described the brains of Bushwomen and Papuans. The fact that these individuals were very small in stature relative to his eminent Caucasians did not seem to matter to him. Spitzka added some brains from the great apes (gorilla, orangutan) to complete the picture, both literally and figuratively. The results can be seen in Figures 21.7 and 21.8.

The conclusion was inescapable. The assertion reached by some scientists that brain size and complexity were unreliable correlates of intellectual powers or faculties could not possibly be correct. The brain of Friedrich Gauss was as different from the Bushwoman as the latter was from the brain of an ape (Spitzka, 1907). Moreover, by looking at the brains of three brothers just executed in New York State, Spitzka (1904) found anatomical "peculiarities," which he believed conclusively demonstrated the role of heredity in brain morphology.

At about the time Spitzka was presenting his material, Vladimir Bechterev was examining the brain of Dmitri Ivanovich Mendeléyev (Mendelev or Mendelew in German; 1834–1907), the chemist responsible for arranging the elements into a periodic table according to their atomic weights. Bechterev saw not only that the left parietal region was dilated but also that the chemist's frontal lobes showed exceptional development (Bechterew and Weinberg, 1909).

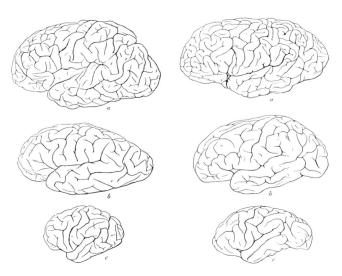

Figure 21.7. Edward Spitzka's (1907) drawings contrasting brains from different people and two great apes. Left column: Hermann Helmholtz, a Papuan, and a chimpanzee. Right column: Karl Friedrich Gauss, a Bush-woman, and a gorilla.

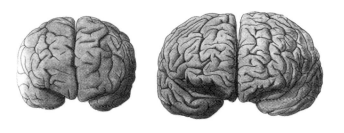

Figure 21.8. Frontal view of the brains of a Papuan and George Francis Train. Edward Spitzka (1907) wrote that Train, an American merchant and traveler, had "one of the best developed brains on record."

Korbinian Brodmann (1908, 1909), Theodor Kaes (1905, 1909a, 1909b), and Oscar Vogt (1929), are among the scientists who presented detailed case studies of brains comparable to those sampled by Spitzka and Bechterev. For example, Vogt studied the brain of Vladimir Lenin (1870–1924), the intellectual leader of the 1917 Russian revolution. Vogt was so impressed by the neuronal arrangements in Lenin's brain that he referred to him as an "association athlete."

As the twentieth century progressed, the variability noted within and between studies was given greater recognition. The failure of many investigators, such as Ludwig Stieda (1837–1918) in 1908, to confirm earlier findings or to find consistent correlates of intellectual development led to other interpretations of the data and less faith in the earlier reports. As stated by the American anatomist Franklin Paine Mall (1862–1917):

> It certainly would be important if it could be shown that the complexity of the gyri and sulci of the brain varied with the intelligence of the individual, that of genius being the most complex, but the facts do not bear this out, and such statements are only misleading. I may be permitted to add that brains rich in gyri and sulci, of the Gauss type, are by no means rare in the American negro. (1909, 24)

Rudolph Wagner, who began this sort of investigation in the 1850s, might have been correct after all. With the brains of Gauss and five other intelligent men on the table in front of him, he had warned his colleagues that it would be dangerous to make broad generalizations from the gross morphology of the highly complex cerebrum of Gauss. Wagner, as noted, was especially worried because the brains of five other notable scholars of Göttingen did not show any special or unusual features.

THE GROWING CONCEPT OF ASSOCIATION CORTEX

The search for brain correlates of intellect followed a logical progression. At first, the search concentrated on the entire brain, and then it focused on just the cerebral hemispheres. In the closing years of the nineteenth century, the emphasis shifted again. With the motor and most sensory areas of the cerebral cortex fairly well delineated by the localizationists, attention was drawn to the "silent areas" of the cortex.

The person most responsible for the anatomical studies on the silent areas was Paul Flechsig, the anatomist who later developed an interest in psychiatry (Schröder, 1930). Flechsig, shown in Figure 21.9, attempted to delineate a number of different functional zones in the cerebral cortex by determining when the different parts of the cortex became myelinated. He began his work on myelogenesis in 1893 and continued it into the twentieth century (see Chapters 3 and 6). Flechsig (1895, 1896, 1898, 1900, 1901, 1905) was guided by the theory that temporal differences in myelination must signify functional differences and that phylogenetically newer areas matured later than more primitive areas.

In 1898, Flechsig reported that his studies of myelination in the human brain revealed 40 distinct cortical areas. By 1901, however, he had combined several of these areas to reduce the total to 36. Three types of functional areas were recognized. In Group 1, the primordial zones, Flech-

Figure 21.9. Paul Flechsig (1847–1929), best known for his developmental studies on myelination of tracts in the central nervous system. Flechsig's anatomical investigations helped to delineate the association areas of the cortex (From Schröder, 1930.)

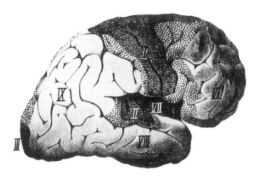

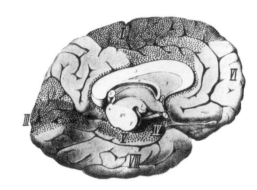

Figure 21.10. Paul Flechsig's (1896) map of the human brain, based on myelination. Area VI is the frontal association area, Area III represents the "visual sphere," and area II defines the "auditory sphere."

sig placed the sensory areas for vision, hearing, and so on. In humans, these areas showed signs of myelination at birth. They were numbered 1–10. Group 2, which received numbers 11–31, was comprised of regions of intermediate development. The third group, having numbers 32–36, was characterized by myelination that became visible only at least one month after birth. One of his plates appears as Figure 21.10.

From the start, Flechsig was not hesitant to ascribe intellectual functions to the slowest-developing areas. He called these areas "association cortex" and noted that their emergence correlated with the development of intellect. As far as he was concerned, this relationship was confirmed by comparative anatomy and clinical neurology. To quote Flechsig (1900):

> The central regions of the association areas are centers which are in more or less direct relation, each with several sensory areas, but some with all of them; they probably combine the activities in themselves (association). Following their bilateral destruction, the intellect appears to be diminished; the association of ideas is especially disturbed. Therefore, the central areas, based upon their appearance, are of the utmost importance for the exercise of intellectual activities, for the formation of mental images composed of several sensory qualities, for the performance of acts such as the naming of objects, reading, etc. (Translated in Clarke and O'Malley, 1968, 551–552)

Flechsig initially distinguished four association areas in humans, but later combined his parietal and temporal areas to form a single posterior center of association. Clinical studies, especially with patients having general paresis and soft-

enings of the cortex as a result of vascular disease, linked Flechsig's frontal and posterior association areas with intellectual functions. Nevertheless, there was some question about which was the more important of the two areas.

Flechsig (1905) thought his posterior association center was the most important part of the brain for intellectual functions. He wrote that the posterior association center was responsible for connecting words with their content, understanding notations, forming intelligent conceptions of the external world, and grasping complex situations. In contrast, Flechsig emphasized the role of the frontal cortex in emotion and consciousness, but concluded that the frontal lobes were not centers for abstract thinking, as Eduard Hitzig had proposed (see Chapter 22).

But what did the studies of the brains of exceptional people suggest? Some case studies, including one on the brain of Hermann von Helmholtz, did, in fact, show an overdevelopment of Flechsig's posterior association area (Hansemann, 1899). For this reason, when Frederick W. Mott reviewed the literature, he accepted the idea that the posterior association cortex was the most important center in the brain for intellect. Mott wrote:

> Rudinger, who studied the brains of quite a number of distinguished men, including Bischoff, Döllinger, Tiedemann, and Liebig, asserted that the higher the mental endowment of an individual the greater is the relative extent of the upper part of the parietal lobe. Other observers, however, have fixed upon the lower part of the parietal lobe as that in which there is a correlation between a high order of intelligence and a relative increase of surface. Thus, Retzius has stated that such was the case in the brains of the astronomer, Hugo Gylden, and the mathematician, Sophie Kovalesky. Hansemann described a similar condition in the brain of Helmholtz. Moreover, the wide parietal development of the cranium in the skull of Beethoven, Sebastian Bach, and of Kant, indicate that an increased size of the parietal lobe is especially related to a high order of intellect, and, we may assume, to genius and constructive imagination. (1907, 29)

In many cases from his sample of "great" men, Edward Spitzka (1901, 1903, 1907) specifically pointed to the superior development of the left temporal region. With regard to a highly respected anthropologist who had just died, he eulogized:

> Keenness of observation, therefore, with a superior ability of forming concrete concepts, profound insight into the interrelations of what he saw or heard, great capacity for associating and generalizing his thoughts and giving them expression in words; all these with the musical and poetical faculties characterized Major Powell's mind. And since these mental qualities, taken collectively, are known to reside, without doubt, in this great "posterior association area" of Flechsig, which we have seen so extraordinarily developed in this brain (particularly on the right side), we may feel justified in saying that here we have found a somatic expression of mental ability of a pronounced kind in the anatomical appearances of a distinguished man's brain. (1903, 642)

Spitzka noted, however, that different talents and strategies may underlie success in fields as diverse as biology, statesmanship, and theoretical physics. Thus, while a scientist with unusual powers of observation would be expected to show greater development of the posterior cortical areas,

Spitzka did not doubt that a scholar known for his powers of abstract thought could have superior frontal lobes.[12]

ANTHROPOLOGY AND THE FRONTAL ASSOCIATION AREAS

In 1854, Emil Huschke wrote that "the frontal lobe is the brain of intelligence, the parietal lobe that of temperament" (p. 180). Whereas this was not truly revolutionary, he further claimed that the parietal lobes were more voluminous in females and the frontal lobes more voluminous in males.

With his acceptance of a center for articulate language in the frontal lobes, Paul Broca (1861a) also suggested that the frontal lobes served the highest functions (speaking, judgment, reflection, abstraction) and the other lobes served the feelings and passions. This was important for Broca, who had been wrestling with the role the Celts may have played in the formation of his own nation and the belief of many of his countrymen that "longheadedness" (dolichocephaly), as opposed to having a broad, short head (brachycephaly), was a physical criterion of national superiority. The new issue involved the Basques in the South of France—an inferior group to many erudite Frenchmen who looked upon the dolichocephalic peoples of northern France as more noble.

In 1862, Broca and one of his friends presented some 60 Basque skulls to the Société d'Anthropologie. Broca found that the Basque skulls, collected from a graveyard on the Gulf of Biscay, were less brachycephalic and had larger cranial capacities (1,486 vs. 1,428 cc) than a group of medieval (twelfth century) skulls from the graves of aristocrats in Paris. Moreover, the Basque average was above the 1,403-cc average for skulls obtained from newer pauper graves in Paris (e.g., Broca, 1861c). If anything, the Basque skulls were more comparable to skulls from newer, private graves (1,484 cc) and were eclipsed only by a series of 17 skulls (1,517 cc), primarily made up of suicide victims from a Paris morgue.

In attempting to account for these surprising findings, Broca (1865) reasoned that intelligence depended not just on the size of the brain but on the growth of certain parts of the brain. In this regard, he pointed to data showing that the anterior part of the average Basque skull was somewhat smaller than the same part of the average Parisian skull. The large size of the Basque skull seemed due to the extensive development of the occipital region. And, as Broca (1861c) had suggested earlier, the front of the brain housed the most important parts for intellect.

Broca (1873b) presented a similar argument to account for the fact that many of the Cro-Magnon specimens that began to appear in France at this time had cranial capacities that far exceeded the best Parisian skulls. Broca concluded that the anterior sutures of the Cro-Magnon skull closed early, limiting the growth of the forebrain. The great size of the Cro-Magnon cranium was again due to the excessive development of the non-intellectual posterior brain.

These arguments were soon embraced by individuals who were absolutely convinced that Caucasians had to be superior to other races. For example, Sir Humphry Davy Rolleston (1862–1944) asked: "What material differences are

there between the brain of an educated moral man and that of a sensual animal-like savage?" Rolleston then answered his rhetorical question with reference to the brain of an Australian aborigine. It was simple:

> If the convolutions of this Australian brain be compared with those of an average European brain the simplicity of the former is at once thrown into relief. The convolutions of the frontal lobe, which is connected with intellectual processes, are seen to have a marked antero-posterior arrangement, to be four instead of three in number, and to be separate, not to join each other at every turn and twist, as is so notably the case in the described brains of many eminent men, and generally in the more civilized nations. The simplicity of the frontal region is a point of importance, and may be considered as characteristic of a primitive brain. (1888, 33)[13]

In the opening decade of the twentieth century, when the concept of association cortex was well accepted and when much more was known about the frontal lobes, the brains of 103 black and 49 white Americans were compared by Robert Bennett Bean (1874–1944). This physician from Virginia reported that the brains of the Caucasians were heavier. He argued that the difference was largely due to differences in the front of the brain. In blacks, Bean stated, the whole posterior part of the brain is large, and the whole anterior portion is small. With regard to relative sizes of the frontal areas, he placed this group firmly between whites and the orangutan.

Bean (1906) also examined the corpus callosum, having become interested in it after Edward Spitzka (1905) concluded it was larger in great men than in ordinary men. This observation was based on poorly conducted research in which overall brain size and weight were not taken into account. Nevertheless, Bean accepted Spitzka's conclusions and reported that the part of the commissure associated with the front of the brain (the genu of the corpus callosum) was absolutely and relatively largest in the white male, smaller in the white female, smaller still in the black male, and smallest in the black female.

These findings, Bean thought, were impressive because the genu not only connected the parts of the brain involved with intelligence, but because it also contained many olfactory fibers. He proudly announced that since blacks have fewer fibers in the front of the corpus callosum than whites, and yet have a much better sense of smell, one could only guess how little frontal cortex they really possessed to handle intellectual functions. Bean was especially intrigued with his data because he believed only lower-class Caucasians (e.g., prostitutes and the depraved) and the very best black specimens would be left for dissection at a Baltimore medical school.

Bean felt his data could explain why black people show instability of character, poor social consciousness, and poor cultural taste. In his own words, "One would naturally expect some such character for the Negro, because the whole posterior part of the brain is large, and the whole anterior portion is small, this being especially true in regard to the anterior and posterior association centers" (p. 379). He went on to say that the association between anatomy and intellect was meaningful because it meant that giving blacks the opportunity to study difficult subjects could not possibly be

---

[12]It is worth noting that Albert Einstein (1879–1955), considered one of the greatest abstract thinkers of all time, showed normal frontal lobes, but some exceptional features in the posterior parietal lobe. These unusual features appeared only on the left side of his brain (Diamond et al, 1985).

[13]The Australian aborigine's brain showed the same percentage of brain mass in front of the Rolandic fissure as did the average European brain (also see Karplus, 1902).

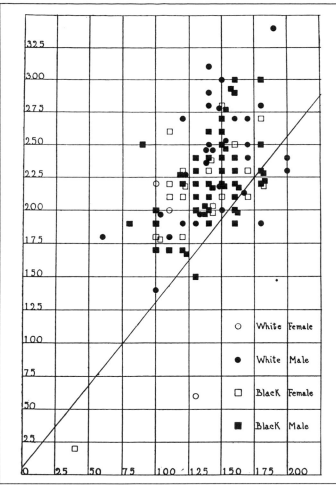

Figure 21.11. Franklin Mall's (1909) figure plotting the area of the genu of the corpus callosum (ordinate) against that of the splenium (abscissa). The diagonal line is positioned to show Robert Bean's (1906) previous separation of Caucasians from blacks.

Figure 21.12. Francis Galton (1822–1911), a pioneer in the attempt to measure intelligence and a staunch supporter of racial and gender differences in intellect.

successful. In the mind of Robert Bennett Bean, they could comprehend abstract concepts and multiple ideas about as well as a horse could understand "the rule of three" (see Chase, 1977).

Bean's "facts," however, did not hold up. An attempt by his mentor at Johns Hopkins, Franklin Mall, to confirm his findings provided no evidence for racial differences related to the corpus callosum, the frontal lobes, or the brain in general (see Figure 21.11). Worse still for Bean, Mall's study of 106 brains included the 10 brains from Caucasians and the 8 brains from black Americans that Bean had used. Even with just this sample under scrutiny, there was no evidence to support Bean's contentions (also see Wilder, 1909).

One major difference between Mall's newer study and the original work by Bean stood out. Unlike Bean, Mall conducted his measurements without advance knowledge about which brains were from Caucasians.

THE NEED FOR OBJECTIVE MEASURES OF INTELLECT

To many scientists and physicians, the studies on the frontal association areas that merited the most attention were not these anthropological investigations but those investigations dealing with brain injuries and diseases in humans, as well

as focal lesions in laboratory animals. This literature is examined in detail in Chapter 22. But before closing this chapter, mention must be made of concurrent efforts to improve the measurement of intellect. To say the least, there was a clear need to go well beyond whether a person was famous, or, as Broca had remarked, on the faculty at Göttingen.

This point was recognized early on by Francis Galton (1822–1911; shown in Figure 21.12), a cousin of Charles Darwin.[14] When Darwin's *The Origin of the Species* appeared in 1859, Galton was quick to grasp its meaning. The most distinctive characteristics of humans were intellectual, and Galton hypothesized that such abilities must follow the same rules of inheritance as the physical traits of height and eye color. These thoughts inspired Galton to begin studying the families of known geniuses.

A few years after Galton's (1869) *Hereditary Genius* was published, he began to labor on a related project, one for improving the human species through selective breeding. He coined the word "eugenics" for this idea, conceived in 1865. Galton recognized that if he wanted to encourage the brightest men to marry the smartest women, he would have to identify these "prize specimens" relatively early in life. He could not wait for them to accomplish great goals, like the individuals described in *Hereditary Genius*, because most such people achieve success only after marriage. Galton's eugenic ideas demanded a test for detecting geniuses at a relatively early age.

[14]Galton spent two years as a medical student at Birmingham General Hospital before enrolling at Cambridge University with the specific intent of winning highest honors in mathematics. A severe nervous breakdown prevented him from achieving his goal, but Galton still received his college degree in 1844. His father died one year later, and he was left with enough money to enable him to spend his life among the idle rich. Galton did, in fact, play the role of a man of leisure for a few years. Then, in need of intellectual stimulation, he spent time exploring parts of Africa and planning expeditions for the Royal Geographic Society (1850–1852). Only later did Galton turn to mental measurement (see Fancher, 1990).

In 1884, Galton established the Anthropometric Laboratory at London's International Health Exhibition. Here he tried to correlate a wide assortment of test results with real-life accomplishments. In line with the times, he took measurements of the skull, which he accepted as indirect measures of brain size. He assumed that the brightest people had to have the biggest brains and, hence, the biggest skulls. Galton also measured sensory abilities, reaction times, and a variety of other functions.

There was logic behind each of his measures. Galton included reaction time measures because he believed bright people could process information faster than average people. And he tested for sensory abilities because the literature suggested that the mentally retarded were not as responsive to pain or temperature as normal people. Further, he added that women, being less intelligent than men, do not possess the sensory abilities to be professional wine testers or great musicians. With such arguments in his Victorian male arsenal, Francis Galton seemed like Julius Caesar (100–44 B.C.) preparing to cross the Rubicon. He soon had people busy bisecting lines, trying to detect high-pitched sounds, and attempting to discriminate colors, weights, and other sensory stimuli.

The International Exhibition of 1884 was a great success. Galton's laboratory drew large crowds of people, who paid three pence each for the opportunity to be evaluated and to see how they stood relative to the averages (Fancher, 1990). But on a practical level, Galton's idea that tests of this sort would correlate with great achievements in real life fizzled. Still, the idea that many small tests could be used to construct an instrument which could measure intelligence had been put into play.

FROM CRANIOMETRY TO IQ

As a young Frenchman, Alfred Binet (1857–1911; shown in Figure 21.13) did not know what he wanted to do with his life.[15] He earned a degree in law, but decided not to practice. He then tried medical school, but suffered a nervous breakdown and withdrew. At age 22, he was a man without direction, passing his time reading in the *Bibliothèque Nationale* in Paris. Quite by accident he seemed to find some books on the new experimental psychology, and with them came his calling.

Binet, being independently wealthy, volunteered as an unpaid trainee and assistant to Jean-Martin Charcot, who was studying the hysterias at this time. He remained with Charcot for eight years, but was greatly embarrassed on a number of occasions when his own experiments were found to be flawed.[16] The humiliation was too much for him to handle and, in 1891, he left the Salpêtrière despondent.

Later that year, Binet had a chance encounter with Henri Beaunis (1831–1921), the physiologist who was in charge of the physiological psychology laboratory at the Sorbonne in

Figure 21.13. Alfred Binet (1857–1911), the psychologist who attempted to measure intellect objectively. Binet constructed an objective, practical test of intellect for children "at risk" in French schools.

Paris. Binet asked if he could work without pay in the laboratory, much as he had done at the Salpêtrière. By 1894, Binet was running the laboratory.[17]

An often overlooked fact in the history of psychology is that Alfred Binet, who had always had a strong interest in individual differences, began his work at the Sorbonne as a staunch believer in craniometry as a way of assessing intelligence. Following in the footsteps of Galton and Broca, Binet (1898) wrote that there was a clear and certain relation between the size of the skull and intelligence and that this had been confirmed on hundreds of subjects. The ideas of craniometry so fascinated Binet that he published a series of papers on cranial measurements in *L'Année Psychologique,* the journal he launched in 1895.

Nevertheless, not all of his studies demonstrated the correlations he had expected to find. In fact, Binet became less sure of himself as he continued to measure the heads of students identified by their teachers as the brightest or the least talented. Binet increased his sample to over 200 pupils, but the craniological differences that emerged seemed too small to be of any consequence.[18] Most disheartening to Binet was that no differences appeared in the front of the skull, where most French scientists were now localizing intellect. By 1900, Binet was forced to write: "The idea of measuring intelligence by measuring heads seemed ridiculous" (p. 403).

As frustrated as Binet had become with his craniometry, he did not give up on the idea of developing an objective measure of intellect. Thus, in 1904, he accepted a commission from the ministry of public education to develop a test

---

[15]Many biographies describe how Binet worked his way into intelligence testing (e.g., Fancher, 1990; Wolf, 1973).

[16]Binet's harshest critic was Joseph Delboeuf (1831–1896). Among other things, Delboeuf criticized Binet's claim that he could use a magnet to move the hysterical symptoms of deeply hypnotized patients from one side of the body to the other. Delboeuf concluded that this was due to nothing more than suggestion. This was a major victory for France's less flamboyant Nancy school over the Salpêtrière, and it contributed to Binet's departure from Charcot.

[17]Binet never looked for another position, nor did he ever accept a salary for his work at the Sorbonne.

[18]In this respect, Binet's findings were not very different from those of Galton's famous pupil Karl Pearson (1857–1936). In 1902, Pearson found only very slight positive correlations between head size and intelligence among soldiers and among children (also see Eyerich and Löwenfeld, 1905; Pearl, 1906).

that would identify children who were not likely to do well in traditional classroom settings. The request for a new instrument, one which would take the place of lengthy interviews and case reports, stemmed from the fact that France had just passed stricter education laws. These laws mandated special schools and curricula for mentally deficient children, who almost certainly would have dropped out in the past.

Binet looked upon the early diagnosis of learning problems as one of the most pressing issues in education. Being a pragmatic man, he now turned to situations young people encounter daily, such as making change in a monetary transaction, understanding the meaning of words, visual coordination, object identification, and so on, and decided to construct a test around these practical skills. In particular, he wanted a test with many short questions related to the practical problems of everyday life and one that would not draw too heavily on reading and writing. This approach was a radical departure not only from his old interest in craniometry but from the reaction time and sensory discrimination tests previously attempted by Francis Galton.

Binet was joined on this project by Théodore Simon (1873–1961), a young physician. The two men delivered their test in 1905. It consisted of 30 items, arranged in order of difficulty. The first questions, for example, asked children to follow a lighted match with their eyes, to grasp an object put into their hands, to eat candy, and to shake hands with the examiner. These were basic skills shown by normal children at age 2. More difficult problems, typically accomplished by 5–6-year-olds, included telling the differences between pairs of objects (e.g., paper and cardboard) and repeating sentences. The most difficult items required the schoolchildren to make rhymes with certain words and to put words into meaningful sentences. These tasks were normally achieved by children 11–12 years of age.

Binet and Simon did not stop here. They continued to perfect their test so that by 1908 it almost doubled in size. Each of the 58 items was now associated with the age at which 50 percent of the normal children could pass it. By tabulating the place on the test where a child missed a sequence of questions, Binet and Simon calculated the child's cutoff, or his or her "mental age."

A third version of the Binet-Simon test appeared in 1911. It now contained five questions for each age from 5 to 15, plus five adult questions. The notion of dividing mental age by chronological age came a year later, but this was not a contribution from Binet or from the French. The IQ ratio was proposed by William Stern (1871–1938), a German psychologist.[19]

Binet and Simon (1911, 1916) were careful to point out that their test was not a device for ranking all pupils. What they had constructed was really a tool for screening pupils at risk. They also backed away from the notion that their test was a pure measure of innate ability. The reason was simple. Binet (1909) had set up special education classrooms and had seen that students at risk could make great intellectual strides in the program he called "mental orthopedics."

The importance of measuring intellect objectively was sorely needed and never really debated. Nevertheless,

Figure 21.14. Louis Terman (1877–1956), the psychologist who popularized standardized intelligence testing in America. Terman extended the Binet-Simon test and modified the questions for North Americans in 1916. Because Terman worked at Stanford University, his revision became known as "the Stanford-Binet."

Binet's idea that intelligence was a loose collection of different abilities (reasoning, memory, attention, etc.) tied together by judgment or good sense, brought out many critics. In many ways, the situation paralleled the neurological debate over whether the cortex functioned as a whole in intellect or whether there were specialized centers for different facets of intelligence (see Chapter 4).

In England, Charles Spearman (1863–1945) found high correlations among the different types of questions on intelligence tests and concluded that there must be a single general intelligence ("g") in addition to any specific faculties ("s") (see Spearman, 1927). And, in the United States, Henry H. Goddard (1866–1957), who translated Binet's articles into English, looked upon intelligence as a unitary mental process.[20]

Louis Terman (1877–1956; shown in Figure 21.14) went on to popularize intelligence testing in the United States. In 1916, Terman modified the questions for North Americans and extended the Binet-Simon test from the midteen years to include superior adults. Because Terman worked at Stanford University, his Stanford Revision of the Binet-Simon Scale simply became known as the Stanford-Binet. Terman accepted Stern's ratio for intelligence, but suggested multiplying the ratio by 100 to eliminate the decimal point. Both the IQ measure and the Stanford-Binet soon gained wide acceptance not only in educational circles but in medical fields as well.

---

[19]Binet died of a stroke in 1911, just before Stern (1912) published his IQ ratio.

[20]Goddard was the director of the Training School for Feeble-Minded Girls and Boys in Vineland, New Jersey. Unlike Binet, Goddard (1914) argued that intellect was largely, if not entirely, inborn.

REFERENCES

Agassiz, L. 1850. The diversity of origin of the human races. *Christian Examiner, 49,* 110–145.

Aristotle. 1937–1955. *De partibus animalium.* (A. L. Pick, Trans.). Cambridge, MA: Harvard University Press.

Bean, R. B. 1906. Some racial peculiarities of the Negro brain. *American Journal of Anatomy, 5, Ser. 3,* 353–432.

Bechterew, W. von, and Weinberg, R. 1909. Das Gehirn des Chemikers D. J. Mendelew. In W. Roux (Ed.), *Anatomische und entwicklungsgeschichtliche Monographien, Heft 1.* Leipzig: Engelmann. Pp. 1–22.

Benedikt, M. 1879. *Anatomische Studien an Verbrecher-Gehirnen.* Wien: Wilhelm Braumüller.

Bernier, F. 1684. Nouvelle division de la terre, par les différentes espèces ou races d'homme qui l'inhabitant. *Journal des Savants, 12,* 148–155.

Binet, A. 1898. Historique des recherches sur les rapports de l'intelligence avec la grandeur et la forme de la tête. *L'Année Psychologique, 5,* 245–298.

Binet, A. 1900. Recherches sur la technique de la mensuration de la tête vivante. *L'Année Psychologique, 7,* 314–429.

Binet, A. 1909. *Les Idées Modernes sur les Enfants.* Paris: Flammarion. (Reprinted 1973 with preface by J. Piaget)

Binet, A., and Simon, T. 1911. *A Method of Measuring the Development of the Intelligence of Young Children.* Lincoln, IL: Courier Company.

Binet, A., and Simon, T. 1916. *The Development of Intelligence in Children.* Translated from articles in *L'Année Psychologique* (1905, 1908, 1911) by E. S. Kite. Baltimore, MD: Williams & Wilkins.

Broca, P. 1861a. Remarques sur le siège de la faculté du langage articulé; suivies d'une observation d'aphémie (perte de la parole). *Bulletins de la Société Anatomique (Paris), 6,* 330–357, 398–407. In G. von Bonin, *Some Papers on the Cerebral Cortex.* Translated as, "Remarks on the seat of the faculty of articulate language, followed by an observation of aphemia." Springfield, IL: Charles C Thomas, 1960. Pp. 49–72.

Broca, P. 1861b. Sur le volume et la forme du cerveau suivant les individus et suivant les races. *Bulletins de la Société d'Anthropologie, 2,* 139–207, 301–321.

Broca, P. 1861c. Sur des crânes d'un cimetière de la Cité antérieur au treizième siècle. *Bulletins de la Société d'Anthropologie, 2,* 139–207, 501–513.

Broca, P. 1862a. Sur les proportions relatives du bras, de l'avant-bras, et de la clavicule chez les Nègres et les Européens. *Bulletins et Mémoires de la Société d'Anthropologie, 3,* 162–172.

Broca, P. 1862b. Sur la capacité des crânes Parisiens des diverses époques. *Bulletins et Mémoires de la Société d'Anthropologie, 3,* 102–116.

Broca, P. 1865. Histoire des travaux de la Société d'Anthropologie de 1859 à 1863, lue dans la séance solennelle du 4 juin 1863. *Mémoires de la Société d'Anthropologie, 2,* VII–LI.

Broca, P. 1867. Sur les proportions relatives des membres supérieurs et des membres inférieurs chez les Nègres et les Européens. *Bulletins de la Société d'Anthropologie, Ser. 2, 2,* 641–653.

Broca, P. 1873a. Sur la mensuration de la capacité du crâne. *Memoires de la Société d'Anthropologie, Ser. 2,* 1–92.

Broca, P. 1873b. Sur les crânes de la caverne de l'Homme Mort (Lozère). *Revue d'Anthropologie, 2,* 1–53.

Brodmann, K. 1908. Beiträge zur histologischen Lokalisation der Grosshirnrinde. VII. Mitteilung: Die cytoarchitektonische Cortexgliederung der Halbaffen (Lemuriden). *Journal für Psychologie und Neurologie, 10,* 287–334.

Brodmann, K. 1909. Antwort an Herrn Dr. Th. Kaes. *Neurologisches Centralblatt, 18,* 635–639.

Chase, A. 1977. *The Legacy of Malthus.* New York: Knopf.

Clarke, E., and O'Malley, C. D. 1968. *The Human Brain and Spinal Cord.* Berkeley: University of California Press.

Dana, C. L. 1928. Early neurology in the United States. *Journal of the American Medical Association, 90,* 1421–1424.

Diamond, M. C., Scheibel, A. B., Murphy, G. M., and Harvey, T. 1985. On the brain of a scientist: Albert Einstein. *Experimental Neurology, 88,* 198–204.

Dougherty, F. W. P. 1985. Johann Friedrich Blumenbach und Samuel Thomas Soemmerring: Eine Auseinandersetzung in anthropologischer Hinsicht. In G. Mann and F. Dumont (Eds.), *Samuel Thomas Soemmerring und die Gelehrten der Goethezeit.* Stuttgart: Fischer. Pp. 35–56.

Down, J. L. H. 1866. Observations on an ethnic classification of idiots. *London Hospital Reports,* 259–262.

Eyerich, G., and Löwenfeld, 1905. L. *Über die Beziehungen des Kopfumfangs zur Körperlänge und zur Geistigen Entwicklung.* Wiesbaden: J. F. Bergmann.

Fancher, R. E. 1990. *Pioneers of Psychology* (2nd ed.). New York: W. W. Norton.

Flechsig, P. E. 1895. Weitere Mittheilungen über die Sinnes- und Associationscentren des menschlichen Gehirns. *Neurologisches Centralblatt, 14,* 1118–1124, 1177–1179.

Flechsig, P. E. 1896. *Gehirn und Steele.* Leipzig: Veit.

Flechsig, P. E. 1898. Neue Untersuchungen über die Markbildung in den menschlichen Grosshirnlappen. *Neurologisches Centralblatt, 17,* 977–996.

Flechsig, P. E. 1900. Les centres de projection et d'association du cerveau humain. *XIIIe Congrès International de Médecine, Paris. Section de Neurologie.* Pp. 115–121.

Flechsig, P. E. 1901. Developmental (myelogenetic) localisation of the cerebral cortex in the human subject. *Lancet, ii,* 1027–1029.

Flechsig, P. E. 1905. Gehirnphysiologie und Willenstheorien. *Fifth International Psychology Congress,* Rome, 73–89. In G. von Bonin, *Some Papers on the Cerebral Cortex.* Translated as, "Brain physiology and theories of volition." Springfield, IL: Charles C Thomas. Pp. 181–200.

Flourens, M.-J.-P. 1824. *Recherches Expérimentales sur les Propriétés et les Fonctions du Système Nerveux dans les Animaux Vertébrés.* Paris: J. B. Ballière.

Flourens, M.-J.-P. 1842. *Recherches Expérimentales sur les Propriétés et les Fonctions du Système Nerveux dans les Animaux Vertébrés* (2nd ed.). Paris: J. B. Ballière.

Flourens, M.-J.-P. 1846. *Phrenology Examined* (2nd ed.). (C. de L. Meigs, Trans.). Philadelphia, PA: Hogan and Thompson. Reprinted in D. N. Robinson (Ed.), *Significant Contributions to the History of Psychology, Ser. E, Physiological Psychology: Vol. II, X. Bichat, J. G. Spurzheim, P. Flourens.* Washington DC: University Publications of America. Pp. iv–144.

Galen. 1968. *De usu partium.* Translated by M. T. May as, *On the Usefulness of the Parts of the Body.* Ithaca, NY: Cornell University Press.

Gall, F. 1822–1826. *Sur les Fonctions du Cerveau* (6 vols.). Paris. J. B. Ballière.

Galton, F. 1869. *Hereditary Genius.* London: Macmillan.

Goddard, H. H. 1914. *Feeble-mindedness: Its Causes and Consequences.* New York: Macmillan.

Gould, S. J. 1978. Morton's ranking of races by cranial capacity. *Science, 200,* 503–509.

Gould, S. J. 1981. *The Mismeasure of Man.* New York: W. W. Norton.

Hansemann, D. 1899. Ueber das Gehirn von Hermann v. Helmholtz. *Zeitschrift für Psychologie und Physiologie der Sinnesorgane, 20,* 1–12.

Hitzig, E. 1884. Zur Physiologie des Grosshirns. *Archiv für Psychiatrie und Nervenkrankheiten, 15,* 270–275.

Hollander, B. 1920. *In Search of the Soul.* London: Kegan Paul.

Karplus, J. P. 1902. Ueber ein Australiergehirn nebst Bemerkungen über einige Negergehirne. *Arbeiten aus dem Neurologisches Institut an der Wiener Universität, 9,* 26–27.

Le Bon, G. 1879. Recherches anatomiques et mathématiques sur les lois des variations du volume du cerveau et sur leurs relations avec l'intelligence. *Revue d'Anthropologie, Ser. 2, 2,* 27–104.

Le Bon, G. 1895. *Psychologie des Foules.* Paris: F. Alcan.

Le Bon, G. 1896. *The Crowd.* London: Ernest Benn.

Linné, von. 1735. *Systema naturae.* Lugduni Batavorum: Ex typographia, J. W. de Groot.

Linné, C. von. 1735. *Systema naturae.* Holmi: L. Salvii.

Lombroso, C. 1876. *Uomo Delinquente.* Milano: Heopi.

Lombroso, C. 1887. *L'Homme Criminel.* Paris: F. Alcan.

Lombroso, C. 1911. *Crime: Its Causes and Remedies.* Boston, MA: Little, Brown.

Mall, F. P. 1909. On several anatomical characters of the human brain, said to vary according to race and sex, with especial reference to the weight of the frontal lobe. *American Journal of Anatomy, 9,* 1–32.

Marshall, J. 1864. On the brain of a bushwoman; and on the brains of two idiots of European descent. *Philosophical Transactions of the Royal Society of London, 154,* 501–558.

Marshall, J. 1892–1893. On the brain of the late George Grote, F.R.S., with comments and observations on the human brain and its parts generally. *Journal of Anatomy and Physiology, 7, New Ser.,* 21–65.

Matiegka, H. 1902. Ueber das Hirngewicht, die Schädelkapazität und die Kopfform, sowie deren Beziehungen zur psychischen Thätigkeit des Menschen. *Sitzungsbericht der königlich böhmischen Gesellschaft der Wissenschaften. Math.-natur. Classe.* 1–75.

Morton, S. G. 1839. *Crania Americana or, A Comparative View of the Skulls of Various Aboriginal Nations of North and South America.* Philadelphia, PA: Remington.

Morton, S. G. 1844. Crania Aegyptiaca: Observations on Egyptian ethnography, derived from anatomy, history, and the monuments. *Transactions of the American Philosophical Society, 9,* 93–159.

Morton, S. G. 1849. Observations on the size of the brain in various races and families of man. *Proceedings of the Academy of Natural Sciences of Philadelphia, 4,* 221–224.

Mott, F. W. 1907. The progressive evolution of the structure and functions of the visual cortex in mammalia. *Archives of Neurology, 3,* 1–48.

Nott, J. C., and Gliddon, G. R. 1854. *Types of Mankind.* Philadelphia, PA: Lippincott.

Nott, J. C., and Gliddon, G. R. 1868. *Indigenous Races of the Earth.* Philadelphia, PA: J. B. Lippincott.

Peacock, T. B. 1865a. On the weight of the brain in the Negro. *Memoirs of the Anthropological Society of London, 1,* 65–71.

Peacock, T. B. 1865b. On the weight of the brain and capacity of the cranial cavity of a Negro. *Memoirs of the Anthropological Society of London, 1,* 520–524.

Pearl, R. 1906. On the correlation between intelligence and the size of the head. *Journal of Comparative Neurology, 16,* 189–199.

Pearson, K. 1902. On the correlation of intellectual ability with the size and shape of the head. *Proceedings of the Royal Society of London, 69,* 333–342.

Retzius, G. 1898. Das Gehirn des Astronomen Hugo Gyldens. *Biologische Untersuchungen, neue Folge, 8,* 1–22.

Retzius, G. 1900a. Das Gehirn des Mathematikers Sonja Kowalewski. *Biologische Untersuchungen neue Folge, 9,* 1–16.

Retzius, G. 1900b. Das Gehirn des Physikers und Pädagogen Per Adam Siljeström. *Biologische Untersuchungen, neue Folge, 10,* 1–14.

Retzius, G. 1904. Das Gehirn eines Staatsmannes. *Biologische Untersuchungen, neue Folge, 11,* 89–102.

Retzius, G. 1905. Das Gehirn des Histologen und Physiologen Christian Lovén. *Biologische Untersuchungen, neue Folge, 12,* 33–49.

Rolleston, H. D. 1888. Description of the cerebral hemispheres of an adult Australian male. *Journal of the Anthropological Institute of Great Britain and Ireland, 17,* 32–42.

Schiller, F. 1979. *Paul Broca: Founder of French Anthropology, Explorer of the Brain.* Berkeley: University of California Press.

Schröder, P. 1930. Paul Flechsig. *Archiv für Psychiatrie und Nervenkrankheiten, 91,* 1–8.

Singer, C. 1952. *Vesalius on the Human Brain.* London: Oxford University Press.

Slotkin, J. S. 1965. *Readings in Early Anthropology.* Chicago: Aldine Publishing Company.

Sömmerring, S. T. von. 1785. *Ueber die körperliche Verschiedenheit des Negers vom Europäer.* Frankfurt und Mainz: Varrentrapp Sohn und Wenner.

Sömmerring, S. T. von. 1788. *Vom Hirn und Rückenmark.* Mainz: P. A. Winkop.

Spearman, C. 1927. *The Abilities of Man, Their Nature and Measurement.* New York: Macmillan.

Spitzka, E. A. 1901. The redundancy of the preinsula in the brains of distinguished educated men. *Medical Record, 59,* 940–943.

Spitzka, E. A. 1903. A study of the brain of the late Major J. W. Powell. *American Anthropologist, New Ser., 5,* 585–643.

Spitzka, E. A. 1904. The brains of three brothers. *American Journal of Anatomy, 3,* IV–V.

Spitzka, E. A. 1905. Report of a study of the brains of six eminent scientists and scholars belonging to the American Anthropometric Society; together with a brief description of the skull of one of them. *American Journal of Anatomy, 4,* III–IV.

Spitzka, E. A. 1907. A study of the brains of six eminent scientists and scholars belonging to the American Anthropometric Society, together with a description of the skull of Professor E. D. Cope. *Transactions of the American Philosophical Society, New Ser., 21,* 175–308.

Stern, W. 1912. *Die psychologischen Methoden der Intelligenzprüfung und deren Anwendung an Schulkindern.* Leipzig: Barth.

Stieda, L. 1908. Das Gehirn eines Sprachkundigen. *Zeitschrift für Morphologie und Anthropologie, 11,* 83–138.

Terman, L. 1916. *The Measurement of Intelligence.* Boston, MA: Houghton Mifflin.

Terman, L. 1926. *Genetic Studies of Genius: Vol. 1. Mental and Physical Traits of a Thousand Gifted Children.* Stanford, CA: Stanford University Press.

Thurnam, J. 1866. On the weight of the brain, and on the circumstances affecting it. *Journal of Mental Science, 12,* 1–43.

Tiedemann, F. 1836. On the brain of the Negro compared with that of the European and the orang-outang. *Philosophical Transactions of the Royal Society of London, 126,* 497–527.

Tiedemann, F. 1837. *Das Hirn des Negers mit dem des Europäers und Orang-Outangs Verglichen.* Heidelberg: K. Winter.

Tobias, P. V. 1970. Brain-size, grey matter and race—Fact or fiction? *American Journal of Physical Anthropology, 32,* 3–26.

Topinard, P. 1888. Le poids de l'encéphale d'après les registres de Paul Broca. *Mémoires de la Société d'Anthropologie, Ser. 2, 3,* 1–41.

Vesalius, A. 1543. *De humani corporis fabrica.* Basel: J. Oporinus.

Vogt, C. C. 1864. *Lectures on Man: His Place in Creation, and in the History of the Earth.* London: Longman, Green, Longman, and Roberts.

Vogt, O. 1929. Bericht über die Arbeiten des Moskauer Staatsinstituts für Hirnforschung. *Journal für Psychiatrie und Neurologie, 40,* 108–118.

Wagner, R. 1862–1864. *Vorstudien zu einer wissenschaftlichen Morphologie und Physiology des menschlichen Gehirns als Seelenorgan.* Göttingen: Dieterichsche Buchhandlung.

Waller, A. D. 1891. *Human Physiology.* London: Longmans, Green.

Wilder, B. G. 1909. *The Brain of the American Negro.* New York: National Negro Committee.

Wilkins, R. H., and Brody, A. 1971. Down's syndrome. *Archives of Neurology, 25,* 88–90.

Willis, T. 1664. *Cerebri anatome: cui accessit nervorum descriptio et usus.* London: J. Martyn and J. Allestry. Tercentenary ed., 1664–1964, *Thomas Willis, The Anatomy of the Brain and Nerves.* Montreal: McGill University Press, 1965.

Wolf, T. H. 1973. *Alfred Binet.* Chicago: University of Chicago Press.

Zimmern, H. 1898. Criminal anthropology in Italy. *Popular Science Monthly, 52,* 743–760.

Chapter 22
The Frontal Lobes and Intellect

> I believe . . . that intelligence, or more correctly, the treasury of ideas, is to be sought for in all parts of the cortex, or rather, in all parts of the brain; but I hold that abstract thought must of necessity require particular organs and those I find in the frontal brain.
> Eduard Hitzig, 1884

Although many of the scientists mentioned in the last chapter searched for broad anatomical correlates of intellect, it is clear that if most scientists had been asked to choose just one part of the brain to associate with intellectual functions, they would have pointed to the frontal lobes. To some extent, this was seen in the anthropological and anatomical studies of the frontal cortex described in Chapter 21. Without question, as the twentieth century approached, and as cortical localization continued to gain favor, the relation between intellect and the frontal association areas became better accepted.

This chapter looks at the development of various ideas on intellect and the frontal lobes, excluding speech, which is treated in detail in Chapter 25. It begins with some early attempts to define the boundaries and the parts of the frontal lobes and then turns to some early clinical cases involving tumors and injuries to the frontal areas anterior to the motor zones. The special status granted to the most anterior parts of the brain by the phrenologists is then described.

Interest in the frontal lobes grew after Paul Broca ignited the cannons of the localizationist revolution in 1861 with his discovery of a frontal center for articulate language, and after 1870, when Gustav Fritsch and Eduard Hitzig identified the motor cortex in the dog. In this era, the major thrust toward understanding the different parts of the frontal lobes came from scientists who studied the effects of brain lesions in laboratory animals.

Nonetheless, the clinicians, who were largely reporting tumor cases in the closing decades of the nineteenth century, saw many of the same symptoms being observed in brain-damaged dogs and monkeys. Two additional sources of information about the frontal association areas in humans arose early in the twentieth century. These were the large numbers of soldiers who suffered head wounds during World War I (1914–1918) and some relatively rare patients with Pick's disease, a degenerative disorder that usually damages the frontal association areas more than other parts of the cortex.

THE FRONTAL CORTEX DEFINED AND DIVIDED

The history of the anatomy of the frontal lobes can be traced to Constanzo Varolio. In 1573, this Italian anatomist wrote

Figure 22.1. François Chaussier (1746–1828). In 1807, Chaussier delineated the four lobes of the cerebral hemispheres, including the *lobus frontalis.*

that three *prominentia* (prominences) made up each hemisphere. Varolio called these the anterior, medial, and posterior prominences. He believed they corresponded to the first, second, and third ventricles, respectively.

In 1664, Thomas Willis concluded that there were only two such prominences. An anterior and a posterior lobe made up each hemisphere in his schema. The frontal and parietal lobes made up his anterior lobe, and the temporal and occipital lobes comprised his posterior lobe. These two lobes were separated by a branch of the carotid artery (the middle cerebral artery) that coursed along the Sylvian fissure.[1]

It was François Chaussier (1746–1828; shown in Figure 22.1) who, in 1807, first adequately delineated the currently

---

[1]The central fissure was named after Franciscus de le Boë ("Sylvius"), who described it in 1663. He may have identified it in 1641, if not earlier (Baker, 1909; also see Chapter 2).

Figure 22.2. Richard Owen (1804–1892). In his book from 1868, Owen presented maps of the cortex in different species and subdivided the frontal cortex in higher mammals. He called one of his areas the "prefrontal" cortex.

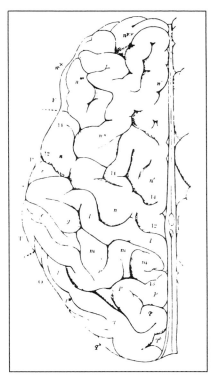

Figure 22.3. Richard Owen's (1868) subdivision of the cortex anterior to the motor region into the "prefrontal," "suprafrontal," "midfrontal," "subfrontal," and "ectofrontal" cortical regions in 1868. His "prefrontal" area has the symbol n[xx].

accepted four lobes of the cerebral hemispheres. The term *lobus frontalis,* or "frontal lobe," was coined by him (see Arnold, 1938; Schiller, 1985). Among German writers, including Oskar Eberstaller (1886–1939), the frontal lobe was called *Stirnhirn.*

The term "prefrontal cortex" appears to have been introduced in 1868 in a book by written by Richard Owen (1804–1892; shown in Figure 22.2) (Divac, 1988). This work contained a section with many detailed drawings on the comparative anatomy of the brain surface. In brains with highly complex sulci, Owen described the frontal fissure at the anterior limits of the area that would eventually become known as the motor cortex. He then subdivided the region in front of the frontal fissure into other areas: "suprafrontal," "midfrontal," "subfrontal," and "ectofrontal" cortex. In his drawing of the human brain, presented as Figure 22.3, the prefrontal region was shown as the most anterior of these regions.

The term "prefrontal" was used by Luigi Luciani and Arturo Tamburini (1878). It was also used by David Ferrier (1878) to define the entire lobe (anterofrontal region) in front of the coronal suture. Ferrier found this region not electrically excitable and not associated with motor deficits in his experiments on monkeys.

In the twentieth century, the term "prefrontal" was frequently used to define all of the cortex anterior to the motor zones. For example, in a journal article of 1935, Carlyle F. Jacobsen stated in his introduction that the terms "frontal association area" and "prefrontal area" would be "used interchangeably to designate that portion of the frontal lobe that lies anterior to the motor and premotor areas" (p. 558). His definition of the prefrontal areas is used here.

THE FRONTAL LOBES IN THE PRESCIENTIFIC ERA

Examining ancient skulls from Europe and the New World to see where they might have been trepanned has provided no support for the idea that the frontal parts of the brain were perceived as special in these early cultures (Walker, 1958). The trepan openings were not placed in locations suggesting that the front of the brain had special powers. If anything, the bias was slightly in favor of the parietal lobes, but this could have been due to the fact that this part of the skull was most accessible for this surgery.

The idea that the frontal lobes could play a leading role in higher mental functions might have begun with the Greeks. Some indirect evidence for the special role of the frontal region has come from the study of art. Gods and demigods from classical Greece and Rome were sometimes sculpted with overdeveloped skulls and especially large foreheads (Cuvier, 1805). This is shown at the top of Figure 22.4, which is taken from Nott and Gliddon (1868). The Greeks also tended to paint and sculpt their most revered poets and artists with projecting foreheads. In contrast, laborers, athletes, gladiators, and women (e.g., Venus de Milo) were given short, broad heads or retreating foreheads.[2]

Nevertheless, questions can be raised about whether the ancients considered the front of the skull or the front of the brain special, at least for higher functions. Galen associated the front of the brain with sensation. With the rise of ventricular theory (cell doctrine) in the hands of the

[2]In some ancient cultures, newborn infants had their heads tightly bandaged in order to give them abnormally high foreheads. This practice, which was probably followed to confer higher status on these individuals, was maintained in some "primitive" cultures well into the twentieth century (Schiller, 1985).

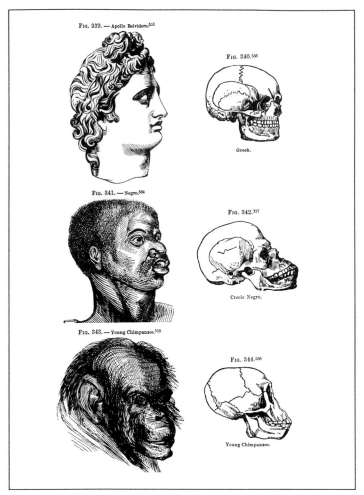

Figure 22.4. A drawing from Nott and Gliddon (1868) showing the Greek god Apollo with an extensively high forehead. The forehead is shown shortened in a black African to make him look more like the chimpanzee at the bottom of the first column than Apollo. Differences in facial angles were also distorted in the second column of this figure to convey the impression of Caucasian superiority.

Church Fathers in the fourth and fifth centuries, the anterior ventricles were typically assigned roles in sensation and perception (Clarke and Dewhurst, 1972; Clarke and O'Malley, 1968; Pagel, 1958). This was probably because the anterior ventricles were closest to the sense organs of vision, taste, and smell, and to the hands, basic for touch. The higher functions of ideation, judgment, and memory typically were allocated to the ventricles located more caudally in the brain. Thinking along these lines would suggest that the ancients could have portrayed their gods and artists with high foreheads simply because this characteristic signified the power to see the world better.

EARLY DESCRIPTIONS OF FRONTAL LOBE INJURIES AND TUMORS

Clinical observations suggestive of a special role for the frontal lobes in intellect began to emerge during the Renaissance. These reports did not, however, have enough impact to suggest that the cerebral cortex could be divided into functional parts. At best they contributed to the more basic idea that the cerebrum proper, and not the ventricles,

mediated behavior. Yet even this advance was slow in coming.

An interesting description of frontal lobe injury was provided by Guido Lanfranchi (Lanfranc, Lanfrancus; d. 1315). He examined two soldiers who sustained frontal lobe damage. Although both survived partial loss of the anterior cortex, they were described as dull-witted with poor memory (Lanfranc, 1565; also see Bakay, 1985).

Additional observations suggesting that the frontal lobe may be important for higher functions came from physicians studying brain tumor patients. In 1614, the case of a patient with a large encapsulated tumor in the front of the brain (above the corpus callosum) was described by Felix Platter in Basel. This tumor caused apathy and dementia during the two-year period preceding the patient's death (see Chapter 20).

FROM SWEDENBORG TO PHRENOLOGY

During the 1740s, Emanuel Swedenborg wrote about the organization of the brain and maintained that the cerebrum had to be composed of distinct functional units. He called these centers *cerebellula* and related those at the front of the brain with thought, imagination, and memory. He believed that injury to the front of the brain could weaken the will and the power of determination by affecting "the highest court of the cerebrum." Unfortunately, his manuscripts on the brain were not made public in the eighteenth century (see Chapter 2).

In contrast to Swedenborg, the physiognomists attempted to associate mental characteristics with physical features, especially those of the face. Baron Georges Cuvier, a geologist, paleontologist, and comparative anatomist, was a leading physiognomist. Cuvier was especially interested in the study of Petrus Camper's (1722–1789; shown in Figure 22.5) facial angle, a measure of how the face sloped relative to the

Figure 22.5. Petrus Camper (1722–1789), whose measure of facial angle became one of the first anthropometrical statistics.

Figure 22.6. Baron Georges Cuvier (1769–1832). In the opening decades of the nineteenth century, Cuvier reported that the facial angle increases as one ascends from lower apes to the great apes to black Africans and on up to white Europeans.

body. The facial angle was the first of many anthropometric measures. Dating from 1760, it was obtained by drawing a straight line from the hole in the ear to the nostril and a second line from the nostril to the forehead (see Camper, 1791).

Cuvier (1805, 1812; shown in Figure 22.6) found that Camper's facial angle increased as one ascended from lower apes to the great apes to humans. He also reported that white Europeans—the group he viewed as the most intellectually advanced of the races—had larger facial angles and more fully developed foreheads than other races and groups. Cuvier (1912, p. 105) called black Africans "the most degraded of the human races, whose form approaches that of the beast and whose intelligence is nowhere great enough to arrive at regular government." They were contrasted with images of the gods from classical times who possessed not only superhuman powers but enormous skulls and the largest facial angles of all.

Franz Gall and the phrenologists were the first to launch a real case for the idea that the cerebrum was composed of distinct functional units (Chapter 3). Between 1810 and 1819, Gall and Johann Spurzheim described the various organs of the mind and their locations. Of these, the highest functions, man's greatest attainments and social characteristics, were placed in specific parts of the frontal lobes. Lower functions, including instincts, were assigned to more posterior parts of the brain. Animals, such as dogs and monkeys, were more developed posteriorly than anteriorly. The phrenologists reasoned, however, that because these animals have at least small frontal lobes, they must be capable of rudimentary forms of intellect (also see Gall, 1822–1826).

In 1822, John Bell, a leading supporter of phrenology in the United States, presented a seminal paper to the Central Phrenological Society, the organization he cofounded

with Charles Caldwell earlier the same year. Bell, a physician, pointed to the direct relation he saw between brain size and intellect. This correlation seemed especially noticeable when the brains of mentally deficient individuals were contrasted with those of great men. But true to his phrenological beliefs, Bell then argued that this association represented only a good first step for the science of the mind. The ultimate goal of the phrenologists had to be learning about those parts of the brain most responsible for intellect.

Bell went on to divide the brain into two basic parts. He associated the anterior division with intellect and wrote:

> Here then in the healthy state we have two grand divisions of faculties and corresponding organic structure marked out. In those unfortunate creatures who have never displayed the usual signs of intelligence, and whom we term idiots, there is not simply imperfect development, but absolute deficiency of organization—the anterior part of the brain seems wanting, and its height is much below the common standard—yet many of these beings display the lower propensities strongly, while the higher feelings and intellect are wanting. If the brain were a unit, ought we not in these idiots, as in children, to see a proportionate activity or a proportionate feeble display of passion and intellect, of feeling and reason . . . ? (1822, 83–84)

The studies of the physiognomists on facial angle and the theories of the phrenologists contributed to Emil Huschke's (1854; shown in Figure 22.7) declaration that the Caucasian "race" was the "frontal race" (also see Gould, 1981; Rylander, 1940). Black Africans, seen as more apelike, belonged to the "occipital race," whereas the Mongolians made up the "parietal race" in Huschke's schema. Some writers took this even a step further by calling these respective races "men of the day," "men of the night," and

Figure 22.7. Emil Huschke (1797–1858), the German anthropologist who studied brain weights of different groups of people. Huschke specifically associated the frontal lobes with intelligence.

"men of the twilight," with the statement that the forehead corresponded to the day, in contrast to the occipit, which corresponded to the "shady or night side" of nature (see Vogt, 1864).

Within the Caucasian sample described by Huschke, the ratio of frontal lobe weight to the rest of the brain was greater by 1 percent for males than for females.[3] And, as expressed by Pierre Gratiolet (1854), it was the great development of the frontal lobes that distinguished people from animals, and it was with the frontal lobes that one encountered the majesty of the human brain.

COMPARATIVE ANATOMY, TUMORS, AND INJURIES BEFORE 1861

Although Gall's phrenological theory was severely criticized in some quarters, some of his critics agreed that the frontal lobes were of paramount importance for intellectual functions. These scientists recognized that the frontal lobes were the least developed lobes in lower animals and also pointed to the fact that animals lacked the tremendous reasoning power of human beings. Indeed, whereas the frontal lobes constituted only about 7 percent of the brain of the dog and some 17 percent of the brain of higher apes, they accounted for almost 30 percent of the human brain.

Karl Friedrich Burdach (1819–1826) was among the anatomists who maintained that because the frontal region was less developed in lower mammals than in humans, it had to be particularly involved in intellectual functions. He called it the special workshop for thinking. Burdach, who attributed some importance to the form of the forehead, noted feelings of tension in the front of his own head when he was thinking intensely. He added that this part of the brain was also close to the olfactory bulb and olfactory tract. Although smell might be thought of as a primitive sense, he explained that the sense of smell, "in contrast to the other senses, it is turned toward the future" (see Meyer, 1970).

Another believer in the high status of the frontal lobes was Dominique-Jean Larrey, the surgeon Napoleon Bonaparte called "the most virtuous man I have ever known." Larrey saw many head injuries in his role as a chief surgeon in the Napoleonic Wars and kept remarkably detailed case histories of wounded French soldiers.[4] Larrey (1829) wrote that frontal injuries often resulted in loss of memory, confusion, and lack of understanding (also see Bakay, 1985). He even sent some of his aphasic frontal lobe cases to Franz Gall, who used them to argue that the center for language, man's most remarkable achievement, was located in the front of the brain (see Chapters 3 and 25).

Jean-Baptiste Bouillaud (1825a, 1825b, 1830, 1848), who associated himself with the phrenologists early in his career, placed language firmly in the frontal lobes. He based this association on pathological material (tumors, injuries, etc.). Moreover, in 1827 Bouillaud "pierced with a gimlet the anterior part of the brain of a dog" in the homologue of Broca's region in humans (see Bateman, 1870). He reported that not only did the dog lose the power of barking but it was also "less intelligent" than before the operation.

One of the most cited early case studies suggesting changes in intellect after acute damage to the frontal lobes was that of Phineas Gage (see Chapter 19).[5] In 1848, at the age of 25, Gage suffered severe brain damage when he turned and accidently dropped his tamping iron on a rock and ignited some blasting powder. The explosion caused the huge iron to shoot point first through the left side of his jaw and then through his cranium.

Gage's recovery over the next few months, while unsteady, was far better than anyone expected. Nevertheless, it was clear that Gage's intellect had been altered. He now exhibited poor judgment, impulsiveness, and lack of restraint. Because "the equilibrium or balance . . . between his intellectual faculties and animal propensities, seems to have been destroyed," he was unable to return to his job as a foreman for the railroad, even though he had previously been regarded as the most efficient and capable man for the job (Harlow, 1868, p. 13). Gage had become so "childish" and his mind so altered that friends and acquaintances said he was no longer the same man.

Gage's limitations were described by John Harlow:

Mentally the recovery was only partial, his intellectual faculties being decidedly impaired, but not totally lost; nothing like dementia, but they were enfeebled in their manifestations, his mental operations being perfect in kind, but not in degree or quantity. This may perhaps be satisfactorily accounted for in the fact that while the left cerebrum must have been destroyed as to function, its functions suspended, its fellow was left intact, and conducted its operations singly and feebly. (1868, 19)

In addition to the Gage case, a number of interesting frontal tumor cases were reported at about this time. One was presented by Jean Cruveilhier, who had an extensive clinical practice at both the Salpêtrière and the Charité in Paris (Bakay, 1989). Between 1829 and 1842, Cruveilhier published a monumental work, *L'Anatomie Pathologique du Corps Humain,* in which he summarized what he knew about brain tumors, especially the meningiomas. In it, he described a 45-year-old schoolteacher with the classic symptoms of a tumor. She experienced headaches, loss of balance, weakness of the left lower limb, and occasional urinary incontinence. Further, she became unusually apathetic and uninterested in what was going on around her. She replied slowly to all questions and asked nothing and refused nothing. Cruveilhier called her unusual mental status, which was like that seen in Platter's (1614) patient, "turpor." An autopsy confirmed that a huge meningioma had pressed upon her right frontal lobe. This is depicted in Figure 22.8. Cruveilhier suggested that the apathy resulted from compression of the frontal convolutions.

Cruveilhier also described a girl of 15 years who was an idiot from the time of birth. The anterior two thirds of her frontal lobes were found to be, in the words of David Ferrier (1878), "completely wanting." After mentioning this early case, Ferrier (p. 447) said: "But, indeed, the frequent association of idiocy with such defect of the frontal lobes is a generally recognized fact." Ferrier also discussed the Gage case in his paper.

---

[3]See Mall (1909) for a critical review of this finding.
[4]The Napoleonic Wars began in 1800 and ended with the Battle of Waterloo in 1815.

[5]The Gage case was described by Henry Bigelow (1850) and by John Harlow (1848, 1849, 1868). Bigelow called it the "crowbar" case, although it was a tamping iron that shot through Gage's skull.

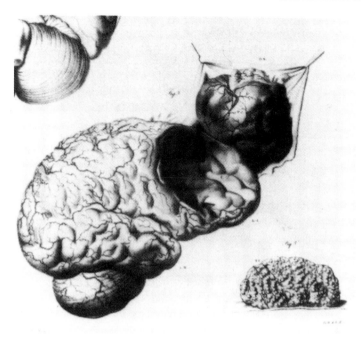

Figure 22.8. A huge meningioma of the frontal lobe of a 45-year-old woman. Jean Cruveilhier, who described this case, wrote that a distinguishing feature of this patient was that she became unusually apathetic. (From Cruveilhier, 1829–1842; Livraison 8, Planche 2.)

The basic clinical findings reported by men like Cruveilhier, Harlow, and Larrey in the first half of the nineteenth century set the stage for the destruction of the cortical equipotentiality theory championed by Marie-Jean-Pierre Flourens (1824, 1842). With the birth of localization theory in the 1860s and 1870s, the experimentalists added to the literature on intellect and the prefrontal lobes. The advantages the experimentalists had over the clinicians was their ability to control lesion placements and to verify the damage when animals were used as subjects.

HITZIG'S EXPERIMENTS

In his 1861 presentation before the *Société d'Anatomie* in Paris, Paul Broca stated that he considered the frontal lobes important for speaking and for higher intellectual functions, including judgment, reflection, comparison, and abstraction. The other lobes of the cerebrum were thought to be more involved with feelings, penchants, and passions. Broca's acceptance of a frontal localization for articulate language and his broad interpretation of the data had a tremendous impact not only on the clinicians but also on the experimentalists. Laboratory scientists wanted to learn more about the frontal lobes, especially about the region anterior to the precentral gyrus, once the motor strip was identified in 1870 by Fritsch and Hitzig.

The scientist most responsible for initiating lesion experiments on the frontal association cortex in laboratory animals was Eduard Hitzig. Even Leonardo Bianchi, who devoted a lifetime to studying the frontal lobes, stated that Hitzig should be given full credit for first bringing frontal lobe research into the laboratory where various ideas gleaned from comparative anatomy could be tested (Bianchi, 1920, 1922).

Hitzig's work on the frontal association cortex began in 1870 when he and Gustav Fritsch made two important dis-

coveries about this region in the dog. The first was that electrical stimulation of the area did not elicit movements; the second was that damage to it did not cause paralysis. The idea that these areas must be contributing to higher mental functions seemed clear to Hitzig (1874a, 1874b) early on.

At an 1884 meeting of German neurologists and psychiatrists, Hitzig summarized more than a decade of experiments on the anterior frontal areas. He stated that dogs that knew how to find food on a table before the surgery seemed unable to do so after bilateral frontal cortex lesions. The animals even appeared to forget about pieces of food they had just been shown. But they were healthy, as evidenced by the fact that they ate normally when the food was placed directly in front of them. He added that this unusual deterioration in intellect was not observed after other cortical lesions.

Hitzig did not state, however, that only the frontal lobes were involved in intellectual functions. In fact, he did not believe intellect was limited to any one part of the brain. Instead, he made the case that the frontal lobes were special for just one aspect of intelligence—abstract thought. He believed this function required its own cerebral organ, and the growing body of evidence suggested that this organ had to be located in the massive frontal lobes (see the quotation at the beginning of this chapter).

FERRIER'S MONKEYS

By 1876, David Ferrier had three monkeys with near complete ablations of the anterior frontal areas. These animals showed nothing noteworthy in the domain of sensory or motor symptoms, essentially confirming Hitzig's observations. Nevertheless, the monkeys seemed to be abnormal. Ferrier wrote:

> Notwithstanding this apparent absence of physiological symptoms, I could perceive a very decided alteration in the animal's character and behaviour, though it is difficult to state in precise terms the nature of the change. The animals operated on were selected on account of their intelligent character. . . . Instead of, as before, being actively interested in their surroundings, and curiously prying into all that came within the field of their observation, they remained apathetic, or dull, and dozed off to sleep, responding only to the sensations or impressions of the moment, or varying their listlessness with restless and purposeless wanderings to and fro. While not actually deprived of intelligence, they had lost, to all appearance, the faculty of attentive and intelligent observation. (1876, 231–232)

Ferrier also conducted experiments on dogs. He noted that when he threw a bone to a normal dog, it ran to it with great speed. In contrast, his dogs with frontal lesions took off but did not always stop at the right spot. Rather than reorienting themselves, these dogs seemed to forget why they were running in the first place. Sometimes they would keep running until the experimenter showed them another bone to attract their attention.

Ferrier was not sure what his findings meant. His electrical stimulation experiments showed only that applying current to the cortex anterior to the classical motor areas would occasionally elicit irregular eye movements.[6] He first

---

[6]Ferrier described eye movements elicited by frontal lobe stimulation in several of his papers and books (see Ferrier, 1875, 1876 and 1886).

theorized that the prefrontal cortex was not a motor center per se, but that it might serve some sort of inhibitory function. He viewed the selection of one idea or sensation and the inhibition of competing mental operations as the essential factor of attention. Further, he not only associated inhibition with attention, but correlated attention with intellectual power. In Ferrier's own words:

> In proportion to the development of the faculty of attention are the intellectual and reflective powers manifested. This is in accordance with the anatomical development of the frontal lobes of the brain, and we have various experimental and pathological data for localizing in these centres of inhibition, the physiological substrata of this psychological faculty. (1876, 287)

In the 1886 edition of *The Functions of the Brain*, Ferrier did not even hint at the relation between inhibition and attention when he discussed the prefrontal areas. The inhibition idea may have been abandoned soon after 1878, when he delivered his Gulstonian Lectures to the Royal Society. The primary role of the frontal lobes in attention, however, remained intact. Thus, Ferrier still theorized that it was because of attentional processes, and how they contributed to the intellect, that his animals with frontal lobe lesions showed a pathological absence of curiosity, apathetic behaviors, restlessness, and impulsiveness. As for intellect itself, Ferrier added:

> It would be absurd to speak of a special seat of intelligence in the brain. Intelligence and will have no local habitation distinct from the sensory and motor substrata of the cortex generally. (1886, 467)

Friedrich Goltz (1877), Eduard Hitzig (1874a, 1874b), Hermann Munk (1882), and Leonardo Bianchi (1895) were of the opinion that Ferrier's work left much to be desired, primarily because he examined his animals too soon after they had come out of surgery. All four shared the belief that Ferrier should have given his animals more recovery time. Munk seemed the most caustic about the problems with Ferrier's experiments when he wrote:

> The question of just how Mr. Ferrier was able to ascribe the apathy, sleepiness, nervousness, etc. in the monkeys to the loss of function in the frontal lobe (instead of considering them the obvious results of only the mechanical lesioning and of the encephalomeningites, commensurate with what occurs after other larger brain injuries in luckless experiments) would not be understandable at all except by realizing that Mr. Ferrier interprets whatever he observes after his extirpation experiments without criticism for his own set opinion. (1882, 773)

Nevertheless, the idea that frontal lobe damage could affect attentional processes was put just as firmly into play as was Hitzig's theory, which emphasized the role of the prefrontal areas in abstract thought. Both ideas received considerable experimental and clinical validation after this time.

BIANCHI'S OBSERVATIONS AND THEORIES

As noted, the most dedicated student of the frontal lobes at the end of the nineteenth century and into the opening decades of the twentieth century was Leonardo Bianchi; shown in Figure 22.9. In 1888, this Italian psychiatrist and

Figure 22.9. Leonardo Bianchi (1848–1927), a lifetime student of the frontal lobes.

neurologist began to work on the prefrontal areas of monkeys and dogs. In 1920, Bianchi summarized his own findings and those of previous investigators in an influential book on frontal lobe functions. This book was translated into English two years later.

In his book and in several other publications (e.g., 1894, 1895), Bianchi stated he confirmed the finding that the areas in front of the motor zones were electrically nonexcitable. He further agreed with Hitzig and Ferrier that lesions in these areas did not result in paralyses or sensory disturbances in his dogs or monkeys, at least once the diffuse effects of the surgery had subsided. Instead, Bianchi noted that intellect appeared to be correlated with the growth of the prefrontal area. The progression was clear from lower vertebrates to cats to dogs to monkeys and to man, just as Gratiolet (1854), Hitzig (1884), and others had noted.

Convinced that the "silent" areas had to be involved in higher mental functions, Bianchi began by making unilateral lesions in the anterior frontal regions. He found that this did not have any noteworthy effects on mental functions. In contrast, bilateral lesions markedly changed the personalities of his animals. Bianchi wrote the following about one of his monkeys:

> Her behaviour is altered, her physiognomy stupid, less mobile; the expression of the eyes is as if uncertain and cruel, devoid of any flashes of intelligence, curiosity, or sociability. Shows terror, even by shrieks and gnashing of teeth, when threatened or hurt, but never reacts aggressively. She is in a state of unrest; when placed in a large closed room, it walks aimlessly around it, always in the same direction, without stopping near any object or person. Any action done with apparent purpose remains incomplete, unfinished; if she runs towards a door she stops near it, goes back, runs to the door again, and so on several times. . . . She is unsociable with the other monkeys, does not play; cannot overcome the least difficulties in her way by new adaptations, nor learn

anything new, nor recover what she has forgotten. (1895, 517)

Similar findings were obtained when Bianchi studied dogs with bilateral prefrontal damage, which he later described as follows:

The author's first experiments on dogs warranted the opinion . . . (that) bilateral destruction brought into evidence a distinct alteration of character, marked especially by weakening of all the psychic manifestations: defect in perceptive judgment; exaggerated fear, resulting from defective critical capacity and inability to make use of the physical powers although these were well preserved; amnesia and, generally speaking, a psychically blind behavior; defect in initiative and resource; lack of finality in the complex movements, as betrayed by incoherent conduct and diminished vivacity (lowering of the psychic tone). In fact, any ordinary individual would have classed these dogs as weak-minded. (1922, 71)

What did all of this mean? Was the deficit one of abstract thought or of inattention, or was it something else? In contrast to Hitzig and Ferrier, Bianchi came to the conclusion that the bilateral prefrontal lesions had affected the "psychical tone" of his animals:

The frontal lobes would thus sum up into series the products of the sensori-motor regions, as well as the emotive states which accompany all the perceptions, the fusion of which constitutes what has been called the psychical tone of the individual. (1895, 521)

Bianchi argued that the "dissolution of the psychical personality" can manifest itself in many ways, including the loss of careful planning and assessment, intellectual blunting, and restlessness. Still, he shared Hitzig's belief that the frontal lobes were not *the* organ of intellect. Instead, he viewed this newer part of the brain as *an* organ of intellect. He theorized that there had to be a coordinated division of labor among the different organs of mind, but certainly nothing like the equipotentiality envisioned by Flourens (1824, 1842).

Another of Bianchi's conclusions was that frontal cortex damage served to exaggerate some of the normal characteristics of his subjects. Animals that were affectionate seemed to become more lovable and dependent on social contact after their lesions, whereas timid animals became more withdrawn.[7]

Bianchi's ability to recognize such changes may have stemmed from the fact that he used every conceivable opportunity to observe his animals. Not only did he study them in planned experiments but he observed them interacting in social settings, often with himself as a participant.

He referred to the character changes as alterations in the social sense, and understandably so, for he lived with his monkeys, ate with them—allowing them even to choose morsels from his plate—permitted them to run about in his office while he was writing, and made every effort to gain their confidence and friendship. Thus he was able to detect after bilateral destruction of the frontal lobes that they were lacking in perceptive judgment, that their memory was poor and unreliable, that associative power was reduced, that initiative and resourcefulness were lost and that sociability disappeared. (Ferraro, 1953, 112)

OTHER OPINIONS FROM THE EXPERIMENTALISTS

Although the frontal lobe studies of Hitzig, Ferrier, and Bianchi are usually cited as being the most important from the experimentalist camp, a number of other investigators in this era conducted lesion experiments on laboratory animals. Some of these researchers reached conclusions different from those mentioned above. Four such individuals were Hermann Munk, Jacques Loeb, Edward A. Schäfer, and Luigi Luciani.

Hermann Munk (1877, 1881) did not deny that the frontal areas played a role in higher cognitive processes, but emphasized that intelligence was not confined to any one part of the cortex. In a way, Munk viewed intellectual functions as everywhere in the cortex and yet nowhere in particular. Thus, he thought that the intellectual deficit was strictly proportionate to the size of the cortical lesion.[8]

Jacques Loeb (1886, 1902) was also convinced that the cerebral hemispheres acted as a whole and not as a mosaic of independent parts, at least in processes of association and intellect. As he summarized it:

It has been claimed that the intellect is the function of specialized parts of the brain. Hitzig and others assumed that the frontal lobes of the cerebral hemispheres are the organs of attention. I have repeatedly removed both frontal lobes in dogs. It was impossible to notice the slightest difference in the mental functions of the dogs. There is perhaps no operation which is so harmless for a dog as removal of the frontal lobes. (Loeb, 1902, 275).

Hitzig (1887) formulated a sharply worded attack in response to Loeb's 1886 report. In fact, Hitzig was so incensed by Loeb's conclusion that removal of the frontal lobes was harmless, that he called him an "Apostle of Lawlessness" in the field of brain physiology. Loeb did not weaken under this barrage. He even found a supporter in Edward A. Schäfer, who would later write:

I have in several cases completely removed in the monkey the whole of the inexcitable area of both frontal lobes without producing the slightest sign of the mental and intellectual dullness and alterations of character which has been regarded as pathognomonic of a lesion of this region. (1900, 772)

THE TUMOR LITERATURE IN THE ERA OF LOCALIZATION

In the final quarter of the nineteenth century, there were many case reports on the effects of frontal lobe tumors. In some of these reports, the investigators were as stymied as the experimentalists by the absence of intellectual deterioration (e.g., Hadden, 1888). In others, the situation was different, although hardly constant from patient to patient. Among the symptoms most frequently described were the loss of attention, apathetic behaviors, and intellectual deterioration

---

[7]See Rylander (1940) for clinical confirmation of this finding.

[8]A similar position was strongly advocated in the first half of the twentieth century by Karl Lashley (see Chapters 4 and 23).

or "mental dullness."[9] These symptoms were comparable to the changes described by Hitzig, Ferrier, and Bianchi in their experiments on laboratory animals.

In the mid-1880s, Moses Allen Starr published two papers in which he attempted to review the American literature on the behavioral effects of tumors, abscesses, diseases, and acute injuries in different parts of the brain. Starr (1884a, 1884b) noted that frontal lobe damage, confirmed by autopsy, frequently affected attention and intellectual processes, as well as temperament and personality (e.g., "deficiency of self control" and loss of inhibition). He acknowledged, however, that frontal lobe damage may not necessarily give rise to symptoms in all cases. With regard to the symptoms that were likely to appear, he stated:

> In respect of judgment and reason the power of man surpasses that of the lower animals. The brain of man differs from that of the lower animals and of idiots chiefly in the greater development of the frontal lobes. It seems possible, therefore, that the processes involved in judgment and reason have as their physical basis the frontal lobes. If so, the total destruction of these lobes would reduce man to the grade of an idiot. Their partial destruction would be manifested by errors of judgment and reason of a striking character. One of the first manifestations would be a lack of that self-control which is the constant accompaniment of mental action, and which would be shown by an inability to fix the attention, to follow a continuous train of thought, or to conduct intellectual processes. It is this very symptom which was present in one-half of the cases cited here. . . . It did not occur in lesions of other parts of the brain here cited. (1884b, 380)

In addition to symptoms like these, Moritz Jastrowitz (1888) and Hermann Oppenheim (1889, 1890, 1902) noted that some of their patients with frontal tumors became loquacious, jocular, and childlike (see Chapter 19).

Poor attention, defective judgment, changes in emotion, and apathetic behaviors continued to be presented into the twentieth century as characteristic symptoms of frontal lobe tumors.[10] The following three examples from the literature are representative of what was being written about frontal lobe patients at this time:

1. He could not sustain his attention for any length of time; any subject which was brought before him, although it might attract his attention for a moment, seemed to pass from his mind very soon. As regards the affection of his memory, it appeared to be not so much a blotting-out of his past impressions as a want of power of associating memories, and comparing and contrasting them. Lack of judgment, therefore, was a marked feature of his mental condition. (Elder and Miles, 1902, 364)

2. The most prominent and constant changes are inattention and inability to keep the mind fixed on the subject at issue, associated with a tendency to irrelevance and incoherence in conversation. . . . The patient is casual and irresponsible in word and deed, and may attempt jokes and make remarks which are quite irrelevant and often rude; this is apparently due to a faulty or too free association of ideas which is not held in check by a rational judgment and not normal power of concentration on the subject. (Stewart, 1906, 1209–1210)

3. Upon one occasion he walked into his place of business and in various ways betrayed either gross inattention to his surroundings or marked failure of vision. It was noticed also that his capacity in business had suffered decidedly. . . . He smiles readily and seems quietly pleased. He manifests no anxiety as to his condition. (Dercum, 1910, 473)

One of the individuals most interested in the effects of frontal lobe tumors was Paul Schuster (1867–?). In 1902, he wrote a monograph based on 785 cerebral tumor cases, which included a well-balanced summary of what was known about the frontal lobes at the time. He then divided the psychical changes seen in these patients into categories. Schuster wrote that the most frequent symptoms were general "mental paralyzation," irritability, hypomania, and *Witzelsucht* (sarcastic joking). In agreement with other authors of the day, he acknowledged that unilateral prefrontal cortex damage was not always associated with these mental and emotional symptoms. Yet he added that when symptoms were seen with unilateral tumors, left frontal damage had more of an effect on intellect, whereas right frontal damage had more of an effect on emotion and temperament.

The literature on frontal tumors was also summarized by Gordon Holmes (1931). He concluded that tumors of no other part of the brain, with the possible exception of the corpus callosum, produced mental symptoms as frequently as those of the frontal lobes. He stated that the most frequent symptom is apathy or indifference to the state of one's own health and to things ordinarily of interest. Holmes went on to say that delayed answers to questions and lack of initiative can make some frontal patients look more retarded than they really are. Attention, Holmes noted, is usually poor and could affect memory for recent events.

Holmes also discussed the variability in symptomatology found across frontal lobe patients. He contrasted the most frequently encountered apathetic patient (described above) with a more jovial and euphoric type, who is less inhibited and more childlike.[11]

THE BRICKNER CASE

One of the most widely cited frontal tumor cases was described by Richard M. Brickner in an article in 1932 and in more detail in his book *The Intellectual Functions of the Frontal Lobes*, which appeared in 1936. The case was looked upon as unique at the time because it involved the intensive study of a man, Joe, who had undergone surgery on both prefrontal lobes. Walter Dandy performed the first of the two operations to remove Joe's meningioma in 1930, and the extent of both amputations was known.

Before the onset of his tumor, Joe was described as a kind, popular, hard-working man who achieved financial success on the New York Stock Exchange. He was probably above aver-

---

[9]For reports describing aberrant behaviors, see Beevor (1898), Clarke (1898), McBurney and Starr (1893), Sharkey (1898), and Williamson (1891, 1896). For a more recent review of this literature, see Benton (1991).

[10]The early-twentieth-century clinical literature on patients with frontal lobe tumors is large. Some representative papers are Dercum (1910), Donath (1912), Elder and Miles (1902), Herzfeld (1901), Höniger (1901), Pfeifer (1910), and Stewart (1906).

[11]Phineas Gage's personality after his accident, or Oppenheim's cases of *Witzelsucht* (Chapter 19), would best describe this more childish type of frontal lobe patient.

age in intelligence, but his tumor led to absent-minded behavior, which interfered with his ability to conduct business.

Following the removal of the tumor, one of the most pronounced mental changes was an increase in distractibility, which showed itself as an inability to maintain fixed attention. Brickner expressed the belief that Joe's basic ability to learn and remember things remained intact and attributed his "memory" problems to distractibility and/or lack of initiative.

Joe showed a lack of insight into the gravity of his illness and treated his situation with surprising indifference and apathy. Yet his emotional constitution had changed considerably from the pre-morbid period. This previously quiet, kind, hard-working man now displayed boasting, hostility, and self-aggrandizement. His behavior was often infantile, and Brickner thought Joe had lost his good moral sense, social judgment, and love of family.

Brickner hypothesized that the primary deficit displayed by his patient was the loss of synthesis. In Brickner's (1932, p. 328) eyes, "all of the interpretable changes may be explained by a diminution in the associative function of synthesizing simple mental engrams into more complex ones." This, he thought, accounted not only for Joe's various intellectual problems but for his emotional symptoms. It also accounted for Joe's ability to look normal under some circumstances. In fact, when some visiting neurologists first saw Joe in the hospital, they had no idea there was anything wrong with him.

The reduction in the ability to synthesize, Brickner thought, may not be critical for very simple thoughts or ideas, but as more and more new elements had to be put into place, the significance of the loss would be felt. This placed limits on the complexity of the thought process and led to secondary emotional disturbances.

Brickner expressed the belief that whereas the frontal cortex received inputs from other cortical areas and played a leading role in intellectual processes, there was no evidence to prove it functioned alone in this capacity. Instead, he expressed the opinion that the frontal lobes were only quantitatively (and not qualitatively) more special than the rest of the brain. Brickner argued this point from the fact that Joe showed various reductions of function, not complete losses of function.

> From the point of view of the intellect, then, the frontal cortex may be thought of as tissue that enriches the intellect immensely, by virtue of multiplication of the associative possibilities. The frontal lobes would then be, from a biological standpoint, luxuries; intellectual activity of a fairly high order can proceed without them. (1932, 336)

And to quote Brickner a few years later:

> It is difficult to think of localization of function in the frontal cortex, in the sense in which the term is generally used. There is nothing to suggest frontal localization comparable to that, for instance, of the calcarine cortex. (1936, 305)

ACUTE FRONTAL LOBE DAMAGE IN HUMANS

The chance to study large numbers of otherwise healthy people with acute frontal lobe injuries came with World War I. The reports on soldiers wounded in the Great War provided clear indications of attentional and intellectual deficits as well as emotional and personality changes.

In 1918, Walther Poppelreuter (1886–1939), who had studied large numbers of wounded men, wrote that frontal injuries affected thinking and reasoning. In 1920, Hans Berger, best known for his work on the electroencephalogram (EEG), also described intellectual changes after prefrontal cortex injuries. Later, Berger (1923) presented one of the first good descriptions of perseveration after damage to the frontal lobes.

One of the most thorough studies on war victims was conducted by Ernst Feuchtwanger (1889–?) and was summarized in his monograph of 1923. Working in Munich, he examined over 400 soldiers with brain injuries and showed that some symptoms were more characteristic of frontal lobe injuries than others. The frontal symptoms included reduced attentional capabilities, loss of drive or initiative, and emotional changes (both depressive and euphoric changes). He also saw intellectual deterioration, but thought this was a secondary reaction to the loss of drive and emotional changes. Feuchtwanger proposed a theory similar to the one entertained by Bianchi (1920, 1922). Feuchtwanger wrote the role of the frontal lobes was to assure the synthesis of volition and emotion in the construction of the personality. In contrast, he thought that intellect itself was not as well localized.

By this time, it had become clear that frontal lobe damage could be followed by a myriad of deficits. Most scientists now gave up trying to put the aberrant behaviors under a single heading as had been frequently tried by experimentalists and clinicians prior to World War I.[12] There were, however, some exceptions. One individual who still envisioned a single, all-encompassing deficit was Kurt Goldstein.

Goldstein followed up on Hitzig's notion that dogs and monkeys with frontal lesions seem to be especially blunted in the domain of abstract thought. Over a span of three decades, he argued that the loss of the ability to abstract was *the* key feature of frontal lobe damage (e.g., Gelb and Goldstein, 1917; Goldstein, 1934, 1936, 1939, 1944; Goldstein and Katz, 1937; Goldstein and Scheerer, 1941).

The underlying loss of the power of abstraction can manifest itself in many ways. To quote Kurt Goldstein and Martin Scheerer (1900–1961):

> This dependence upon immediate claims can take on the characteristic of rigidity and "lack of shifting" well known in abnormal psychology. But it can also take on the characteristic of fluidity which manifests itself in an extreme susceptibility to the varying stimuli in the surroundings. The stimuli are followed as ever newly arising; the person is delivered to their momentary valences. This may appear to be distractibility or continual spontaneous shifting of attention whereas, in reality, the individual is being shunted passively from one stimulus to the next. (1941, 3)

Goldstein added that the loss of the ability to think abstractly after frontal lobe lesions may underlie maladaptive emotional responses and those defects that seem to suggest problems with memory. He maintained that problems with abstraction may not be apparent in those everyday situations where things are familiar and routine. In contrast, when changes are demanded of the individual, or if things become unfamiliar and challenging, a gross defect in the abstract thinking process may emerge. Under such

---

[12]Wundt's (1908) theory of "apperception" is representative of a turn-of-the-century theory stressing a single broad deficit.

circumstances, the patient is likely to act as if he or she had defective initiative, foresight, activity, and choice.

Kurt Goldstein and his associates developed a number of tests for measuring abstract and concrete behaviors and corresponding "attitudes" or capacity levels. These included a variety of sorting tests (colors, objects, forms, sticks) in which war-wounded and other subjects were required to shift strategies at various points to achieve high scores (Goldstein and Scheerer, 1941).

A number of individuals shared Goldstein's interest in abstraction after brain damage. One was Gösta Rylander, who used fables and proverbs to study abstraction after frontal lobotomies. The concrete nature of his patients' interpretations of proverbs can be seen in the following three examples, all of which were taken from his 1948 paper (p. 700):

1. After frontal lobe surgery, one patient interpreted "All is not gold that glitters" as "Other substances also glitter." Before surgery the same patient had responded, "One should not be taken in by any old thing; for instance, by fine talk."

2. In response to "People in glass houses should not throw stones," the response changed from "One should not criticize others: One can make the same mistake oneself" to "Otherwise they will break the walls around them."

3. And to the proverb "Too many cooks spoil the broth," the interpretation changed from "If too many people are working with something it won't be good" to "If too many people do the cooking, the food will not be as good." (Rylander, 1948, p. 700)

But while more and more researchers were agreeing on the cardinal symptoms following frontal lobe damage, the idea that all frontal lobe behavior can be explained by a single common factor continued to wane in popularity. As far as most investigators were concerned, attempts to conjure catch phrases to account for all facets of frontal lobe dysfunction were doomed to failure. Although impressed with the loss of abstracting abilities, Rylander (1940) himself was of this opinion.

PICK'S DISEASE

Arnold Pick (1851–1924), born in Moravia, studied medicine in Vienna. He went on to direct a psychiatric hospital and became a professor at the University of Prague. Pick was a cultured, modest man who conducted extensive pathological studies on patients with neuropsychiatric disorders. Describing his various findings and ideas in more than 350 publications and a major textbook on neuropathology, he was also a most prolific investigator.

In 1892, Pick provided the first of many detailed descriptions of the disorder that now bears his name (e.g., Pick, 1898, 1901, 1904, 1906, 1908). The most salient early clinical features of the disease described by Pick were aphasia (typically beginning as an amnesic aphasia but not confined to this type), followed by a progressive dementia with a dramatic loss of initiative. His patients progressively weakened and eventually died from pneumonia or another infection.

Pick believed he was dealing with a senile dementia resulting from focal lesions of the frontal association centers and, to a somewhat lesser extent, the temporal associational areas. Subsequent observers who conducted careful histological studies on the brains of patients with Pick's disease, and who looked more carefully at the symptomatology and the ages of their patients, were not entirely in agreement. Some came away convinced that this was not a senile dementia at all. Rather, they saw the disorder as a separate, distinct disorder, and they recognized Pick's efforts to understand it by calling it Pick's disease.[13]

The clinical course of Pick's disease was divided into three stages, each about equal in length (Schneider, 1929). Symptoms suggesting frontal cortex damage tend to appear even in the earliest stage. For this reason, and because the symptoms were found to evolve fairly slowly and to vary across subjects, the early signs of Pick's disease tended to be confused with those due to frontal lobe tumors. The "frontal symptoms" included indifference to work, poor insight and judgment, careless dressing, apathy, reduction in creative and associative thinking, motor restlessness, emotional lability, antisocial behavior, and personal isolation.

> The first symptoms are usually noted by the family and consist of peculiar changes in personality. The patient becomes aimlessly over-active, roams for hours in the streets and is irritable, angry and difficult to manage. Professional interests and household duties are neglected. A frequent observation is moral deterioration and loss of the sense of propriety. Women occasionally become sexually promiscuous. Other patients steal, prevaricate, contract debts, collect trash and act thoughtlessly and carelessly. . . . Disturbance of speech is one of the most important symptoms and consists of either diminution of voluntary speech (sensory or motor aphasia) or pressure of speech. (Löwenberg, 1936)

At first it was thought that recent memory was also directly affected early in this disorder. Subsequent investigators have concluded, however, that the problem is not one of memory per se (Kahn and Thompson, 1934; Nichols and Weigner, 1938). The fact that memory itself can be remarkably well preserved throughout most of the course of this dementia is consistent with much of the frontal lobe literature and stands in sharp contrast to Alzheimer's disease, where memory problems are among the first symptoms encountered (Davison, 1938; Gordon, 1935; Nichols and Weigner, 1938; also see Chapter 24). As stated in an early review of Pick's disease, "One may get the impression that the patient does not find access to memory material that may be there, whereas, in Alzheimer's disease the memory itself is destroyed" (Kahn and Thompson, 1934, p. 944).[14]

The so-called frontal lobe symptoms slowly worsen in the second stage of the disease. There is nearly a complete loss of initiative and emotional expression, and behaviors become stereotyped. The motor symptoms gradually increase as patients cannot resist tapping the floor, rocking back and

---

[13]For examples of Pick's disease in the German literature, excluding the cases described by Pick himself, see Gans (1922), Richter (1917), Rosenfeld (1917), Schneider (1929), Spatz (1937), and Stertz (1926). Some of the first case studies in English were written by Benedek and Lehoczky (1939), Ferraro and Jervis (1936), Gordon (1935), Löwenberg (1936), and Thorpe (1932).

[14]Psychotic symptoms such as hallucinations, delusions, and confabulation, which present themselves in Alzheimer's disease, are encountered only rarely in Pick's disease (Thorpe, 1932).

forth, making noises, rubbing their thighs, and other types of "restless," hyperactive behavior. In addition, patients in this stage show more marked changes in their speaking habits (Goldstein and Katz, 1937). Answers to questions are usually given with a single word or by using the same words spoken by the questioner in a declarative sentence.

> She giggled freely and her emotional responses in general were superficial. On a few occasions, when things were not exactly to her liking, she rose in fleeting childish indignation. Restlessness was present and of a diffuse, purposeless type, consisting of pacing about the ward and her room, fumbling with newspapers, magazines, flowers and other articles. At times she was noisy, tapping her feet loudly, clapping her hands and slapping her thighs, and no time giving any reason for this behaviour. Her spontaneous speech consisted of superficial, isolated remarks concerning her husband, her home and her telephone number. When questioned, she invariably repeated the question. . . . For example, when asked, "How do you feel?" she replied: "How do I feel? I feel fine." In her conversation, patient showed a marked restriction of the ideational field. She was unable to give reasoned responses or to elaborate thought material, and, when pressed, made some superficial evasion. (Nichols and Weigner, 1938, 242)

In the final stage, which resembles other dementias and ends with the death of the patient, the childish stereotyped and hyperactive behaviors give way to extreme lethargy with the patient unable to leave the bed. This stage of inertia, incontinence, and emaciation is often reached in three to five years. With disease onset typically between 50 and 60 years of age, most patients die before age 70.

Anatomically, Pick's disease is characterized by fairly symmetrical areas of marked atrophy in the frontal and temporal association areas, in contrast to the more general atrophy of Alzheimer's disease (Davison, 1938; Horn and Stengel, 1930; Onari and Spatz, 1926; Thorpe, 1932). Most patients with Pick's disease show combined frontotemporal involvement, but the degree of involvement of these two areas, and the possibility of some posterior parietal involvement, may vary from patient to patient (Löwenberg, 1936; Malamud and Boyd, 1940).[15] The brain in Figure 22.10, for example, shows extreme frontal involvement.

The characteristic anatomical features are degenerative. Typically, the damage is asymmetrical, usually more severe on the left than on the right side of the brain (Kahn and Thompson, 1934). Vascular changes, originally thought by Pick to be important, do not seem to play a significant role in the clinical syndrome.

Because the damage is much more severe in the frontal and temporal association areas than in other parts of the cortex, some authors have referred to Pick's disease as "circumscribed cortical atrophy," "circumscribed senile atrophy," "localized atrophy of the cerebral cortex," or "involutional focal atrophy" (Davison, 1938; Thorpe, 1932).

Although the basis of Pick's disease was not clear in 1892, Pick recognized that it ran in families. Others confirmed his hunch that the disorder has a genetic basis. In

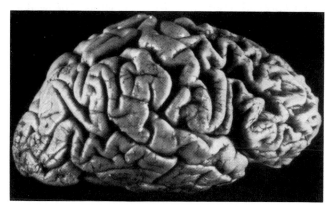

Figure 22.10. A brain with Pick's disease showing extensive atrophy in the frontal areas. (From Spatz, 1937.)

the 1930s, a number of clinical investigations describing siblings and parents with the disorder appeared. At the same time it was recognized that females were twice as likely as males to get the disease.[16]

To the localizationists, Pick's disease was viewed as a further confirmation of frontal lobe function. Although the lesions were never really restricted to just the frontal lobes, the information gleaned from these patients was often in agreement with the growing body of knowledge gained from lesion experiments on laboratory animals and from clinical research on human patients with tumors and acute lesions of the prefrontal lobes.

THE CRITICS SPEAK OUT

As noted earlier in this chapter, some experimentalists (Munk, Loeb, Luciani, Schäfer) were unwilling to assign special, higher functions to just the frontal lobes (also see Chapter 4). As might be expected, a parallel situation also emerged among some neurologists and neurosurgeons in the twentieth century. Although they were in the minority, most of these dissidents agreed with Constantin von Monakow, who stressed that intellect is not localizable.

> Intelligence is a complicated inference of an enormous sum of the singular processes which are acquired successively over the years by the action of the senses and of the muscular apparatus. But by now the most primary psychic processes which follow the acts of each singular sense are already such complicated things occurring under the collaboration of the total cortex that such processes can be considered only rather reluctantly under the viewpoints of localization with respect to gyri and cortical islands. (Monakow, 1904, 106)

Monakow (1904) bolstered his argument against specialization in the prefrontal cortex by pointing to the large size of the frontal lobes in ungulates, where they add some 30 percent to cortical volume. Not having a very high opinion

---

[15]These gross and microscopic differences, and the association of some cases of Pick's disease with neuronal swellings and cytoplasmic (argentophilic) inclusions, were noted by Alois Alzheimer in 1911 and were later confirmed by others (e.g., Ferraro and Jervis, 1936).

[16]Gender differences have been reported in many studies of Pick's disease (e.g., Benedek and Lehoczky, 1939; Grünthal, 1930; Nichols and Weigner, 1938; Thorpe, 1932). For more recent analyses, see Constantinidis, Richard, and Tissot (1974) and Heston (1978).

about the intellectual capabilities of horses or their close relatives, he thought this directly challenged the theory of a causal association between the evolution of the frontal lobes and intellectual activity.

Monakow (1910, 1914) further argued that investigators dealing with lesion material were not giving sufficient recognition to the effects of inflammatory processes, the lack of aseptic conditions during surgery, or the distant and time-dependent effects of local damage. Even if there were localization of higher functions in the brain, he argued, factors such as these would make it very difficult to formulate definitive structure-function associations. The presumed role of the prefrontal cortex in intellectual or executive processes was not exempt from these difficulties, and the data in favor of a special role of the frontal lobes were simply not reliable.

The ideas that higher functions were not the exclusive domain of the frontal lobes and that the cortex may be somewhat equipotential for intellectual activities were put forth to explain some cases of frontal lobe damage without noticeable deficits. A number of these cases were due to attempted suicides or accidents with firearms, whereas others involved surgical interventions (e.g., Leszynsky, 1909; Lindsay, 1904; Wyeth, 1904). For example, after examining one of his patients with both frontal lobes "completely extirpated" because of a brain tumor, Walter Dandy wrote:

> There has been no appreciable disturbance of mentality. The patient is perfectly oriented as to time, place and person; the memory is unimpaired; he reads, writes and conducts mathematical tests accurately; his conversation is seemingly perfectly normal. (1930, 643)

Dandy's findings coincided with those reported by several other investigators and with his own earlier observations involving unilateral frontal cortex removals.[17,18] In some cases, the patients even seemed to improve after surgeons successfully removed large pieces of frontal cortex containing tumorous cells (Spurling, 1934; also see Hebb, 1945). This is not to say that these patients did not show any symptoms immediately after surgery. Indeed, Spurling's patient, who had a large piece of left prefrontal cortex surgically removed because of a slow-growing neoplasm, showed classic "frontal" signs for about four months after her surgery. But after this time, she was described as being able to plan and execute duties around the house as well as she did before she had her tumor, even though she showed some personality changes (Ackerly, 1935).

These findings fostered the hypothesis that the so-called classic effects of frontal lobe tumors and accidental lesions might be due to pathological tissue and foreign agents causing dysfunction in the remaining areas, perhaps in a way comparable to epileptic discharges. As Spurling stated:

> Every neurosurgeon is familiar with cases of frontal lobe tumors presenting preoperatively, profound mental aberra-

tions which disappear promptly when the neoplasm is removed, even when in the process of removal there had occurred more extensive destruction of cortical tissue than that produced by the tumor itself. The prompt regaining of the higher psychical activity in such circumstances invalidates the assumption that the mental aberrations resulted from deficiency or destruction of cortical tissue. It does point to the assumption that some discharge mechanism—which I shall call irritation for want of a better term—is at work upon the pathways and centers of psychical activity. The source of the irritation must be the tumor since, after its removal, aberrations disappear promptly. (1934, 9)

Similar arguments were presented by Clovis Vincent (1879–1947) in 1936 and by Byron Stookey (1887–1966) and his associates five years later. Stookey, Scarff, and Teitelbaum made the following statement about glioma patients whose tumors were successfully excised:

> If distorted impulses from one lobe reach the opposite side, disturbance of function becomes evident. If, however, the disturbing lobe is removed, whether this be dominant or nondominant, as has been done in the cases recorded here, the remaining lobe is able to carry on so well that the loss is hardly to be detected. The "static," so to speak, produced in one frontal lobe interferes with the function of the opposite lobe as well as with its own. . . . A critical search for frontal lobe signs following lobectomy has yielded little of importance. . . . They showed little impairment of general intelligence, they planned and executed household duties competently. . . . In some instances a lack of distractibility was notable. (1941, 164)

The idea that individuals with "clean" surgical removals of prefrontal tissue may behave normally and show no decrement in intellectual ability was also the subject of a number of papers by Donald Hebb (1904–1985). Hebb addressed this topic in 1939 and in numerous papers written in the 1940s. Hebb's statements were based on work he did between 1937 and 1939 at the Montreal Neurological Institute, where his duties included evaluating Wilder Penfield's neurological patients after surgery.

These dissenting opinions, however, did not effectively shake the idea that the prefrontal areas are involved in higher cognitive functions. If anything, the defenders of the faith thought that many of the negative findings resulted from inadequate observations. For example, in 1934 and again in 1935, Wilder Penfield and Joseph Evans (b. 1904) wrote about two patients with excisions of the frontal association cortex. One, who had a part of the left frontal region removed, was given intelligence tests. Although these tests showed nothing very different about the man, in real life he had lost some of his initiative, was poor at calculating, and was no longer able to play a game of bridge.

The second case was Penfield's 43-year-old sister, a housewife who had suffered from Jacksonian seizures for some 23 years. A calcified tumor was found and removed along with about half of her right frontal cortex. She showed an immediate improvement, due to decreased compression, which allowed her to resume her role as a mother of six children. Penfield noted, however, that she had lost her initiative and ability to organize, and that she, like Phineas Gage a century earlier, was no longer the same person.

In addition to the defense based on insensitive measures, many physicians and scientists thought that it was

---

[17]There were many cases of "clean" removals of frontal tumors and surrounding tissue in the literature of the 1930s and early 1940s. Very often the patients seemed to act surprisingly normal in non-challenging situations (e.g., Ackerly, 1935; German and Fox, 1934; Hebb, 1939; Hebb and Penfield, 1940; Jefferson, 1937; Lidz, 1939; Spurling, 1934; Stookey, Scarff, and Teitelbaum, 1941).

[18]Before presenting this case of bilateral frontal cortical resection, Dandy (1922) reported that the entire right or left frontal lobes could be removed without observable mental effects.

unrealistic to expect really bizarre symptoms to appear after limited, unilateral lesions. This was especially true for patients with slow-growing tumors who underwent "modern" aseptic surgery.

The arguments for the frontal lobes playing a role in intellect were reasonably strong. Moreover, the concept of a frontal lobe syndrome was by now too well established to be effectively challenged. The real demand was not for the violent overthrow of the association between the frontal cortex and higher functions; the institution itself was safe. The call was just for more fine tuning and a better understanding of conditions and situations that would bring out the underlying deficits.

It is clear that in the period between the two world wars most scientists had clearly rejected the opinion expressed by Louis Francisque Lélut (1804–1877) a hundred years earlier. This French scientist was pessimistic about the ability of the scientists of his day to answer some of the most important questions about the brain. He wrote: "The question of the relations to be established between the brain and higher acts of thought is very obviously one of those questions condemned by their nature to perpetual indetermination."[19] As far as most twentieth-century scientists were concerned, Lélut was wrong, and the best evidence for this was the wealth of rapidly accumulating clinical and experimental material on behavioral changes after damage to the frontal lobes.

[19]Lélut was quoted by Bianchi (1922, p. 69).

REFERENCES

Ackerly, S. 1935. Instinctive, emotional and mental changes following prefrontal lobe extirpation. *American Journal of Psychiatry, 92,* 717–728.

Alzheimer, A. 1911. Über eigenartige Krankheitsfälle des späteren Alters. *Zeitschrift für die gesamte Neurologie und Psychiatrie, 4,* 356–385.

Arnold, F. 1938. *Bemerkungen über den Bau des Hirns und Rückenmarks nebst Beiträgen zur Physiologie des zehnten und elften Hirnnerven, mehrern kritischen Mittheilungen so wie verschiedenen pathologischen und anatomischen Beobachtungen.* Zürich: S. Höhr.

Bakay, L. 1985. *An Early History of Craniotomy.* Springfield, IL: Charles C Thomas.

Bakay, L. 1989. Cruveilhier on meningiomas (1829–1842). *Surgical Neurology, 32,* 159–164.

Baker, F. 1909. The two Sylviuses. An historical study. *Bulletin of the Johns Hopkins Hospital, 20,* 329–339.

Bateman, F. 1870. *On Aphasia.* London: Churchill.

Beevor, C. E. 1898. The accurate localisation of intra-cranial tumours, excluding tumours of the motor cortex, motor tract, pons and medulla. *Brain, 21,* 291–305.

Bell, J. 1822. On phrenology, or the study of the intellectual and moral nature of man. *Philadelphia Journal of Medical and Physical Sciences, 4,* 72–113.

Benedek, L., and Lehoczky, T. 1939. The clinical recognition of Pick's disease: Report of three cases. *Brain, 62,* 104–122.

Benton, A. L. 1991. The prefrontal region: Its early history. In H. Levin, H. Eisenberg, and A. L. Benton (Eds.), *Frontal Lobe Function and Dysfunction.* New York: Oxford University Press. Pp. 3–32.

Berger, H. 1920. Vorstellung eines Falles von Stirnhirntumor(?). *Münchener medizinische Wochenschrift, 67,* 201.

Berger, H. 1923. Klinische Beiträge zur Pathologie des Grosshirns. I. Mitteilung: Herderkrankungen der Präfrontalregion. *Archiv für Psychiatrie und Nervenkrankheiten, 69,* 1–46.

Bianchi, L. 1894. Ueber die Function der Stirnlappen. *Berliner klinische Wochenschrift, 31,* 309–310.

Bianchi, L. 1895. The functions of the frontal lobes. (A. de Watteville, Trans.). *Brain, 18,* 497–530.

Bianchi, L. 1920. *La Meccanica del Cervello e la Funzione dei Lobi Frontali.* Torino: Bocca.

Bianchi, L. 1922. *The Mechanism of the Brain.* (J. H. MacDonald, Trans.). Edinburgh: E. & S. Livingstone.

Bigelow, H. J. 1850. Dr. Harlow's case of recovery from the passage of an iron bar through the head. *American Journal of the Medical Sciences, 19,* 13–22.

Bouillaud, J.-B. 1825a. Recherches cliniques propres à démontrer que la perte de la parole correspond à la lésion des lobules antérieurs du cerveau et à confirmer l'opinion de M. Gall sur le siège de l'organe du langage articulé. *Archives Générale de Médecine (Paris), 8,* 25–45.

Bouillaud, J.-B. 1825b. *Traité Clinique et Physiologique de l'Encéphalite ou Inflammation du Cerveau.* Paris: J. B. Ballière.

Bouillaud, J.-B. 1830. Recherches expérimentales sur les fonctions du cerveau (lobes cérébraux) en général, et sur celles de sa portion antérieure en particulier. *Journal Hebdomadaire de Médecine (Paris), 6,* 527–570.

Bouillaud, J.-B. 1848. *Recherches Cliniques Propres à Démontrer que le Sens du Langage Articulé et le Principe Coordinateur des Mouvements de la Parole Résident dans les Lobules Antérieurs du Cerveau.* Paris: J. B. Ballière.

Brickner, R. M. 1932. An interpretation of frontal lobe function based upon the study of a case of partial bilateral frontal lobotomy. *Research Publications—Association for Research for Nervous and Mental Disease, 13,* 259–351.

Brickner, R. M. 1936. *The Intellectual Functions of the Frontal Lobes.* New York: Macmillan.

Broca, P. 1861. Remarques sur le siège de la faculté du langage articulé; suivies d'une observation d'aphémie (perte de la parole). *Bulletins de la Société Anatomique (Paris), 6,* 330–357, 398–407. In G. von Bonin, *Some Papers on the Cerebral Cortex.* Translated as, "Remarks on the seat of the faculty of articulate language, followed by an observation of aphemia." Springfield, IL: Charles C Thomas, 1960. Pp. 49–72.

Burdach, K. F. 1819–1826. *Von Baue und Leben des Gehirns* (3 vols.). Leipzig: Dyk.

Camper, P. 1791. *Dissertation Physique sur les Différences Réelles que Présentent les Traits du Visage des Hommes des Différents Pays et de Différents Ages . . .* Utrecht: B. Wild and J. Altheer.

Chaussier, F. B. 1807. *Exposition Sommaire de la Structure et des Différentes Parties de l'Encéphale du Cerveau.* Paris: Barrois.

Clarke, E., and Dewhurst, K. 1972. *An Illustrated History of Brain Function.* Berkeley: University of California Press.

Clarke, E., and O'Malley, C. D. 1968. *The Human Brain and Spinal Cord.* Berkeley: University of California Press.

Clarke, J. M. 1898. Localisation of intra-cranial tumours. *Brain, 21,* 305–332.

Constantinidis, J., Richard, J., and Tissot, R. 1974. Pick's disease: Histological and clinical correlations. *European Neurology, 11,* 208–217.

Cruveilhier, J. 1829–1842. *L'Anatomie Pathologique du Corps Humain.* (2 vols.). Paris: J. B. Ballière.

Cuvier, G. B. 1805. *Leçons d'Anatomie Comparée.* Paris.

Cuvier, G. B. 1812. *Recherches sur les Ossements Fossiles* (Vol. 1). Paris: Deterville.

Dandy, W. E. 1922. Treatment of non-encapsulated brain tumors by extensive resection of contiguous brain tissue. *Bulletin of the Johns Hopkins Hospital, 33,* 188.

Dandy, W. E. 1930. Changes in our conceptions of localization of certain functions in the brain. *American Journal of Physiology, 93,* 643.

Davison, C. 1938. Circumscribed cortical atrophy in the presenile psychoses—Pick's disease. *American Journal of Psychiatry, 94,* 801–818.

Dercum, F. X. 1910. A report of three pre-frontal tumors. *Journal of Nervous and Mental Disease, 37,* 465–480.

Divac, I. 1988. A note on the history of the term "prefrontal." *IBRO News, 16,* 2.

Donath, J. 1912. Gliom des linken Stirnlappens. Operation; Besserung. Gleichzeitig ein Beitrag zur Bedeutung des Stirnhirns. *Zeitschrift für die gesamte Neurologie und Psychiatrie, 13,* 205–216.

Eberstaller, O. 1890. *Das Stirnhirn. Ein Beitrag zur Anatomie der Oberfläche des Grosshirns.* Wein und Leipzig: Urban und Schwarzenberg.

Elder, W., and Miles, A. 1902. A case of tumour of the left prefrontal lobe removed by operation. *Lancet, 1,* 363–366.

Ferraro, A. 1953. Leonardo Bianchi (1848–1927). In W. Haymaker and F. Schiller (Eds.), *The Founders of Neurology.* Springfield, IL: Charles C Thomas. Pp. 110–114.

Ferraro, A., and Jervis, G. A. 1936. Pick's disease. *Archives of Neurology and Psychiatry, 36,* 739–767.

Ferrier, D. 1875. Experiments on the brain of monkeys (Second ser.). *Philosophical Transactions of the Royal Society, 165,* 433–488.

Ferrier, D. 1876. *The Functions of the Brain.* London: Smith, Elder and Company.

Ferrier, D. 1878. The Gulstonian Lectures of the localization of cerebral disease. *British Medical Journal, 1,* 397–402, 443–447.

Ferrier, D. 1886. *The Functions of the Brain.* New York: G. P. Putnam's Sons. (Reprinted 1978; Washington, DC: University Publications of America)

Feuchtwanger, E. 1923. *Die Funktionen des Stirnhirns. Ihre Pathologie und Psychologie.* Berlin: Springer.

Flourens, M.-J.-P. 1824. *Recherches Expérimentales sur les Propriétés et les Fonctions du Système Nerveux dans les Animaux Vertébrés.* Paris: J. B. Ballière.

Flourens, M.-J.-P. 1842. *Recherches Expérimentales sur les Propriétés et les Fonctions du Système Nerveux dans les Animaux Vertébrés* (2nd ed.). Paris: J. B. Ballière.

Fritsch, G., and Hitzig, E. 1870. Über die elektrische Erregbarkeit des Grosshirns. *Archiv für Anatomie und Physiologie*, 300–332. In G. von Bonin, *Some Papers on the Cerebral Cortex*. Translated as, "On the electrical excitability of the cerebrum." Springfield, IL: Charles C Thomas, 1960. Pp. 73–96.

Gall, F. J. 1822–1826. *Sur les Fonctions du Cerveau* (6 vols.). Paris. J. B. Ballière.

Gall, F. J., and Spurzheim, J. 1810–1819. *Anatomie et Physiologie du Système Nerveux en Général, et du Cerveau en Particulier*. Paris: F. Schoell. (Gall was the sole author of the first two volumes of the four in this series.)

Gans, A. 1922. Betrachtungen über Art und Ausbreitung des krankhaften Prozesses in einem Fall von Pickscher Atrophie des Stirnhirns. *Zeitschrift für die gesamte Neurologie und Psychiatrie*, 80, 10–28.

Gelb, A., and Goldstein, K. 1920. *Psychologische Analysen hirnpathologischer Faelle*. Leipzig: Barth.

German, W. J., and Fox, J. C., Jr. 1934. Observations following unilateral lobectomies. *Research Publications—Association for Research for Nervous and Mental Disease*, 13, 378–434.

Goldstein, K. 1934. *Der Aufbau des Organismus*. The Hague: Martinus Nijhoff.

Goldstein, K. 1936. The significance of the frontal lobes for mental performance. *Journal of Neurology and Psychopathology*, 17, 27–40.

Goldstein, K. 1939. *The Organism*. New York: American Book Company.

Goldstein, K. 1944. The mental changes due to frontal lobe damage. *Journal of Psychology*, 17, 187–208.

Goldstein, K., and Katz, S. E. 1937. The psychopathology of Pick's disease. *Archives of Neurology and Psychiatry*, 38, 473–490.

Goldstein, K., and Scheerer, M. 1941. Abstract and concrete behavior: An experimental study with special tests. *Psychological Monographs*, 53, 1–151.

Goltz, F. 1877. Ueber die Verrichtungen des Grosshirns. III. Abhandlung. *Pflüger's Archiv für die gesammte Physiologie des Menschen under der Tiere*, 20, 1–54.

Gordon, A. 1935. Pick's disease: With report of a case and some differential points. *Medical Times and Long Island Medical Journal*, 63, 80–81, 94.

Gould, S. J. 1981. *The Mismeasure of Man*. New York: W. W. Norton.

Gratiolet, P. 1854. *Mémoire sur les Plis Cérébraux de l'Homme et des Primates*. Paris: Bertrand.

Grünthal, E. 1930. Über ein Brüderpaar mit Pickscher Krankheit. *Zeitschrift für die gesamte Neurologie und Psychiatrie*, 129, 350–375.

Hadden, W. B. 1888. A case of tumour of the brain with a long history and with few symptoms. *Brain*, 11, 523–527.

Harlow, J. M. 1848. Passage of an iron rod through the head. *Boston Medical and Surgical Journal*, 39, 389–393.

Harlow, J. M. 1849. Letter in "Medical Miscellany." *Boston Medical and Surgical Journal*, 39, 506–507.

Harlow, J. M. 1868. Recovery from the passage of an iron bar though the head. *Bulletin of the Massachusetts Medical Society*, 2, 3–20.

Hebb, D. O. 1939. Intelligence in man after large removals of cerebral tissue: Report on four left frontal lobe cases. *Journal of General Psychology*, 21, 73–87.

Hebb, D. O. 1945. Man's frontal lobes: A critical review. *Archives of Neurology and Psychiatry*, 54, 10–24.

Hebb, D. O., and Penfield, W. 1940. Human behavior after extensive bilateral removal from the frontal lobes. *Archives of Neurology and Psychiatry*, 44, 421–438.

Herzfeld, J. 1901. Rhinogener Stirnlappenabscess, durch Operation geheilt. *Berliner klinische Wochenschrift*, 47, 1180–1183.

Heston, L. L. 1978. The clinical genetics of Pick's disease. *Acta Psychiatrica Scandinavica*, 57, 202–206.

Hitzig, E. 1874a. *Untersuchungen über das Gehirn*. Berlin: A. Hirschwald.

Hitzig, E. 1874b. Ueber Localisation psychischer Centren in der Hirnrinde. *Zeitschrift für Ethnologie*, 6, 42–47.

Hitzig, E. 1884. Zur Physiologie des Grosshirns. *Archiv für Psychiatrie und Nervenkrankheiten*, 15, 270–275.

Hitzig, E. 1887. Erwiderung dem Herrn Professor Zuntz. *Pflüger's Archiv für die gesammte Physiologie des Menschen und der Tiere*, 40, 129–136.

Hitzig, E. 1904. *Untersuchungen über das Gehirn*. Berlin: A. Hirschwald.

Holmes, G. 1931. Discussion on the mental symptoms associated with cerebral tumours. *Proceedings of the Royal Society of Medicine*, 24, 997–1000.

Höniger. 1901. Zur diagnose der Geschwülste des Stirnhirns. *Münchener medizinische Wochenschrift*, 19, 740–743.

Horn, L., and Stengel, E. 1930. Zur Klinik und Pathologie der Pickschen Atrophie. *Zeitschrift für die gesamte Neurologie und Psychiatrie*, 128, 673–701.

Huschke, E. 1854. *Schaedel, Hirn und Seele des Menschen und der Thiere nach Alter, Geschlecht und Race*. Jena: F. Mauke.

Jacobsen, C. F. 1935. Functions of frontal association area in primates. *Archives of Neurology and Psychiatry*, 33, 558–569.

Jastrowitz, M. 1888. Beiträge zur Localisation im Grosshirn und über deren praktische Verwerthung. *Deutsche medizinische Wochenschrift*, 14, 81–83, 108–112.

Jefferson, G. 1937. Removal of right or left frontal lobes in man. *British Medical Journal*, 2, 199–206.

Kahn, E., and Thompson, L. J. 1934. Concerning Pick's disease. *American Journal of Psychiatry*, 13, 937–946.

Lanfranc, G. 1565. *Chirurgia magna*. London: Marshe.

Larrey, D. J. 1829. *Clinique Chirurgicale, Exercée Particulièrement dans les Camps et les Hôpitaux Militaires, Depuis 1792 jusqu'en 1829*. Paris: Gabon.

Leszynsky, W. M. 1909. A case of gunshot wound of the brain without focal symptoms. *Journal of Nervous and Mental Disease*, 36, 714–715.

Lidz, T. 1939. A study of the effect of right frontal lobectomy on intelligence and temperament. *Journal of Neurology and Psychiatry*, 2, 211–222.

Lindsay, W. S. 1904. Perforating wound of both cerebral hemispheres. *Medical Record*, 65, 186–187.

Loeb, J. 1886. Beiträge zur Physiologie des Grosshirns. *Pflüger's Archiv für die gesammte Physiologie des Menschen under der Tiere*, 39, 265–346.

Loeb, J. 1902. *Comparative Physiology of the Brain and Comparative Psychology*. New York: G. P. Putnam's Sons.

Löwenberg, K. 1936. Pick's disease: A clinicopathologic contribution. *Archives of Neurology and Psychiatry*, 36, 768–790.

Luciani, L., and Tamburini, A. 1878. Ricerche sperimentali sulle funzioni del cervello. *Rivista di Freniatria e Medicina Legale*, 4, 69–89, 225–280.

Malamud, N., and Boyd, D. A. 1940. Pick's disease with atrophy of the temporal lobes. *Archives of Neurology and Psychiatry*, 43, 210–222.

Mall, F. P. 1909. On several anatomical characters of the human brain, said to vary according to race and sex, with especial reference to the weight of the frontal lobe. *American Journal of Anatomy*, 9, 1–32.

McBurney, C., and Starr, M. A. 1893. A contribution to cerebral surgery. Diagnosis, localization, and operation for removal of three tumors of the brain: With some comments upon the surgical treatment of brain tumors. *American Journal of the Medical Sciences*, 105, 361–387.

Meyer, A. 1970. Karl Friedrich Burdach and his place in the history of neuroanatomy. *Journal of Neurology, Neurosurgery and Psychiatry*, 33, 553–561.

Monakow, C. von. 1904. Ueber den gegenwärtigen Stand der Frage nach der Lokalisation im Grosshirn. *Ergebnisse der Physiologie. II Abteilung: Biophysik und Psychophysik*, 3–122.

Monakow, C. von. 1910. Neue Gesichtspunkte in der Frage nach der Lokalisation im Grosshirn. *Zeitschrift für Physiologie der Sinnesorgane. I Abteilung: Zeitschrift für Psychologie*, 54, 161–182.

Monakow, C. von. 1914. *Die Lokalisation im Grosshirn und der Abbau der Funktion durch Kortikale Herde*. Wiesbaden: J. F. Bergmann. In K. H. Pribram (Ed.), *Mood, States and Mind*. (G. Harris, Trans./Excerpter). London: Penguin Books, 1969. Pp. 27–37.

Munk, H. 1877. Zur Physiologie der Grosshirnrinde. *Berliner klinische Wochenschrift*, 14, 505–506.

Munk, H. 1881. *Ueber die Functionen der Grosshirnrinde. Gesammelte Mittheilungen aus den Jahren 1877–80*. Berlin: Hirschwald.

Munk, H. 1882. Ueber die Stirnlappen des Grosshirns. *Sitzungsberichte der königlich preussischen Akademie der Wissenschaften*, 36, 753–789.

Nichols, I. C., and Weigner, W. C. 1938. Pick's disease—A specific type of dementia. *Brain*, 61, 237–249.

Nott, J. C., and Gliddon, G. R. 1854. *Types of Mankind*. Philadelphia, PA: Lippincott.

Nott, J. C., and Gliddon, G. R. 1868. *Indigenous Races of the Earth*. Philadelphia, PA: Lippincott.

Onari, K., and Spatz, H. 1926. Anatomische Beiträge zur Lehre von der Pickschen Atrophie. *Zeitschrift für die gesamte Neurologie und Psychiatrie*, 101, 470.

Oppenheim, H. 1889. Zur Pathologie der Grosshirngeschwülste. *Archiv für Psychiatrie und Nervenkrankheiten*, 21, 560–587.

Oppenheim, H. 1890. Zur Pathologie der Grosshirngeschwülste. *Archiv für Psychiatrie und Nervenkrankheiten*, 22, 27–72.

Oppenheim, H. 1902. *Die Geschwülste des Gehirns* (2nd ed.). Wien: Hölder.

Owen, R. 1868. *On the Anatomy of Vertebrates: Vol. 3. Mammals*. London: Longmans, Green.

Pagel, W. 1958. Medieval and Renaissance contributions to knowledge of the brain and its functions. In M. W. Perrin (Chairman), *The Brain and Its Functions*. Wellcome Foundation. Oxford, U.K.: Blackwell. Pp. 95–114.

Penfield, W., and Evans, J. P. 1934. Functional deficits produced by cerebral lobectomies. *Research Publications—Association for Research for Nervous and Mental Disease*, 13, 352–377.

Penfield, W., and Evans, J. P. 1935. The frontal lobe in man: A clinical study of maximum removals. *Brain*, 58, 115–133.

Pfeifer, B. 1910. Psychiatrische Störungen bei Hirntumoren. *Archiv für Psychologie und Nervenkrankheiten*, 47, 558–569.

Pick, A. 1892. Ueber die Beziehungen der senilen Hirnatrophie zur Aphasie. *Prager medizinische Wochenschrift*, 17, 165–167.

Pick, A. 1898. *Beiträge zur Pathologie und pathologischen Anatomie des Centralnervensystems mit Bemerkungen zur normalen Anatomie desselben*. Berlin: S. Karger.

Pick, A. 1901. Senile Hirnatrophie als Grundlage von Herderscheinungen. *Wiener klinische Wochenschrift*, 14, 403–404.

Pick, A. 1904. Zur Symptomatologie der Linksseitigen Schläfenlappenatrophie. *Monatschrift für Psychiatrie, 16,* 378–388.

Pick, A. 1906. Ueber einen weiteren Symptomenkomplex in Rahmen der Dementia senilis, bedingt durch umschriebene stärkere Hirnatrophie. *Monatschrift für Psychiatrie, 19,* 97–108.

Pick, A. 1908. *Studien zur Hirnpathologie und -Psychologie.* Berlin: S. Karger.

Platter, F. 1614. *Observationum . . . libri tres.* Basel: Koenig.

Poppelreuter, W. 1918. *Die psychischen Schädigungen durch Kopfschuss im Kriege 1914–1916: Vol. 2. Die Herabsetzung der körperlichen Leistungsfähigkeit im Vergleich zu Normalen und Psychogenen.* Leipzig: Voss.

Richter, H. 1917. Eine besondere Art von Stirnhirnschwund mit Verblödung. *Zeitschrift für die gesamte Neurologie und Psychiatrie, 38,* 127–161.

Rosenfeld, M. 1917. Ueber psychische Störungen bei Schussverletzung beider Frontallappen. *Archiv für Psychiatrie, und Nervenkrankheiten 57,* 84–90.

Rylander, G. 1940. *Personality Changes after Operations on the Frontal Lobes.* Copenhagen: Munksgaard.

Rylander, G. 1941. Brain surgery and psychiatry. *Acta Chirurgica Scandinavica, 85,* 213–234.

Rylander, G. 1948. Personality analysis before and after frontal lobotomy. *Research Publications—Association for Research in Nervous and Mental Disease, 27,* 691–705.

Schäfer, E. A. 1900. The cerebral cortex. In E. A. Schäfer (Ed.), *Text-book of Physiology.* Edinburgh: Young J. Pentland. Pp. 697–782.

Schiller, F. 1979. *Paul Broca: Founder of French Anthropology, Explorer of the Brain.* Berkeley: University of California Press.

Schiller, F. 1985. The mystique of the frontal lobes. *Gesnerus, 42,* 415–424.

Schneider, C. 1929. Weitere Beiträge zur Lehre von der Pickschen Krankheit. *Zeitschrift für die gesamte Neurologie und Psychiatrie, 120,* 340–384.

Schuster, P. 1902. *Psychische Störungen bei Hirntumoren.* Stuttgart: Enke.

Sharkey, S. J. 1898. Localization of intra-cranial tumours. *Brain, 21,* 319–329.

Spatz, H. 1937. Über die Bedeutung der basalen Rinde. Auf Grund von Beobachtungen von Pickscher Krankheit und bei gedeckten Hirnverletzungen. *Zeitschrift für die gesamte Neurologie und Psychiatrie, 158,* 208–231.

Spurling, R. G. 1934. Notes upon the functional activity of the prefrontal lobes. *Southern Medical Journal, 27,* 4–9.

Starr, M. A. 1884a. Cortical lesions of the brain. A collection and analysis of the American cases of localized cerebral disease. *American Journal of Medical Sciences, 88,* 114–141.

Starr, M. A. 1884b. Cortical lesions of the brain. A collection and analysis of the American cases of localized cerebral disease. *American Journal of Medical Sciences, 87,* 366–391.

Stertz, G. 1926. Über die Pick'sche Atrophie. *Zeitschrift für die gesamte Neurologie und Psychiatrie, 101,* 729–747.

Stewart, T. G. 1906. The diagnosis and localisation of tumours of the frontal regions of the brain. *Lancet, 2,* 1209–1211.

Stookey, B., Scarff, J., and Teitelbaum, M. 1941. Frontal lobectomy in the treatment of brain tumors. *Annals of Surgery, 113,* 161–169.

Swedenborg, E. 1882–1887. *The Brain, Considered Anatomically, Physiologically, and Philosophically.* (R. L. Tafel, Trans./Ed./Ann.). London: Speirs.

Sylvius, F. 1663. *Disputationes medicarum . . .* Amstelodami: van den Bergh.

Thorpe, F. T. 1932. Pick's disease (circumscribed senile atrophy) and Alzheimer's disease. *Journal of Mental Science, 78,* 302–314.

Varolius, C. 1573. *Anatomia sive de resolutione corporis humani.* Patavii: P. and A. Meiettos.

Vincent, C. 1936. Neurochirurgische Betrachtungen über die Funktionen des Frontallappens. *Deutsche medizinische Wochenschrift, 62,* 41–45.

Vogt, C. C. 1864. *Lectures on Man: His Place in Creation, and in the History of the Earth.* London: Longman, Green, Longman, and Roberts.

Walker, A. E. 1958. The dawn of neurosurgery. *Clinical Neurosurgery, 6,* 1–38.

Williamson, R. T. 1891. A case of abscess in right frontal lobe of the brain. *Medical Chronicle, 13,* 423–427.

Williamson, R. T. 1896. On the symptomatology of gross lesions (tumours and abscesses) involving the prfrontal region of the brain. *Brain, 19,* 346–365.

Willis, T. 1664. *Cerebri anatome: cui accessit nervorum descriptio et usus.* London: J. Martyn and J. Allestry. Tercentenary ed., 1664–1964, *Thomas Willis, The Anatomy of the Brain and Nerves.* Montreal: McGill University Press, 1965.

Wundt, W. 1908. *Grundzüge der Physiologischen Psychologie* (Vol. 1, 6th ed.). Leipzig: Engelmann.

Wyeth, J. A. 1904. A case of gunshot wound of the brain. *Medical Record, 65,* 195.

Chapter 23
The Nature of the Memory Trace

> I would have you imagine, then, that there exists in the mind of man a block of wax. . . .
> Let us say that this tablet is a gift of Memory, the mother of the Muses; and that when
> we wish to remember anything . . . we hold the wax to the perceptions and thoughts,
> and in that material receive the impression of them as from the seal of a ring; and that
> we remember and know what is imprinted as long as the image lasts; but when the
> image is effaced, or cannot be taken, then we forget and do not know.
>
> Socrates, in Plato's *Theaetetus*, fourth century B.C.

BIRDHOUSES AND WAX TABLETS IN THE GRECO-ROMAN PERIOD

Today, people think of creativity, intellect, and imagination as our highest intellectual achievements. Individuals who "simply memorize" things rarely receive our highest accolades, and some may even be scorned. Yet rote memory was revered in the ancient world. Poets, scholars, and theologians who could memorize long passages were honored, admired, and held in highest awe by their peers and the populace in general.

Citizens in Greco-Roman times were formally trained in *memoria,* one of the divisions of rhetoric. In part, this was because books were not readily available as a means of communication. Indeed, as Hesiod (c. 800 B.C.) put it, Mnemosyne (Memory) was not just the wife of Zeus but the spiritual source of the Arts, the mother of the Muses.

But memory was also something more. Memory was a virtue of prudence; it built character, good judgment, citizenship, and piety. Good memory was a sign of moral perfection (Caruthers, 1990).

Memory was deemed to be such an important function by the Greeks and Romans that it was the subject of considerable philosophical and quasi-scientific speculation. One often used metaphor held that the mind was like a thesaurus or storeroom. This metaphor is found in Plato's *Theaetetus,* where Socrates likened memory to a pigeon coop with many compartments and birds flying everywhere. Plato wrote that Socrates compared learning and knowing to filling the empty roosts with birds. In this ancient model, memory was viewed as the process of taking specific birds out of the aviary when needed and then being able to return them to their correct compartments for future use.[1]

The idea of an aviary was popular in part because it used a design that allowed categorization and prevented chaos. It rested on the assumption that one must file memory in a specific way to prevent bits of knowledge from being heaped together and confused. The basic idea was that large blocks of information must be divided into smaller units which could be remembered more easily. This process required ordering the units systematically, in this case in distinct loci.

Aviaries and storerooms were not the only memory metaphors used in ancient times. The belief that the storage of ideas was analogous to putting a hard seal into warm wax is implicit in the writings of the Greek epic poet Homer. The wax tablet analogy was considered quite old by Plato, who had Socrates present it along with the storehouse model in his *Theaetetus.* It was in this work that Socrates made the statement presented at the beginning of this chapter.

In his dialogues, Socrates stated that wax impressions may be very long lasting, but this demanded wax that was abundant, deep, smooth, and of the proper consistency. This type of wax characterized individuals who learn and remember rapidly and are not easily confused. Socrates suggested, however, that problems could arise if the wax were too soft or too hard. Very soft wax would lead to fast learning but rapid forgetting. In contrast, very hard wax would make for good retention but very difficult learning. Additional problems might arise if there were only a small wax tablet, or if the wax had impurities. The latter was associated with indistinct impressions on the tablet, meaning especially unreliable memory.

The metaphor of a wax tablet inscribed with images or stamped with a signet in specific localities was also accepted by Zeno. This Stoic writer of the fourth century B.C. also used the concept of a *thesaurismos phantasion* (storehouse of the mental images) to describe memory. He thought that the *thesaurismos phantasion* began as a *tabula rasa* (blank tablet) at birth.

The wax tablet metaphor is also found in Cicero's *De oratore,* a work that provides a good discussion of the art of memory training. Among other things, this Roman statesman, philosopher, and orator suggested that one should memorize the order of words and phrases by associating them with objects in specific locations.

Near the end of the first century, Marcus Fabius Quintil-

---

[1]During the Middle Ages, pigeons were sometimes replaced by bees in honeycombs, but the meaning remained the same.

ian (c. 35–100), an admirer of Cicero and a teacher of rhetoric, retold the well-known story about how Simonides Melicus of Ceos (c. 556–470 B.C.), the Greek lyric poet, discovered the "art of memory."[2] Simonides had been invited to a banquet to honor a victorious boxer, but had to leave early when he was unexpectedly summoned by some youths. Just as he left the banquet hall, the roof caved in, killing all the guests and crushing their bodies beyond recognition. As the sole survivor of this great tragedy, Simonides was asked to reconstruct the guest list. He found he was able to do this by visualizing where the guests had been placed at the banquet table.

Quintilian taught his students the mnemonic technique of placing images in an orderly architectural display to take full advantage of the sense of sight, the most acute of the senses.[3] Six centuries earlier, Heraclitus of Ephesus had stated that "The eyes are more exact witnesses than the ears" (see Patrick, 1888). This idea was widely accepted by later Greeks and then the Romans. Quintilian knew that when a person returned to a place previously visited, he or she often remembered the events and the people who had been there, even after a considerable amount of time had elapsed. Quintilian therefore reasoned that memories should be tied to localities which could be visualized sharply and in sequence. It is thought that the mnemonic strategy of associating memories with various physical settings may have given rise to the phrases "in the first place," "in the second place," and so on. (Burnham, 1888).

Quintilian established the first public school in Rome. His theories and techniques—and his version of the story of Simonides—can be found in his most important work, the *Institutio oratoria*, which he finished around A.D. 95. His message was that division and logical organization can improve memory, but that exercise and labor are still the most important elements of memory training. This remarkable multivolume work served as a comprehensive course for becoming a respected orator.

ARISTOTLE AND THE LAWS OF ASSOCIATION

Aristotle, who spent 20 years in Plato's Academy before starting his own academy in the Lyceum, went well beyond his mentor when he wrote a special treatise on memory. In *De memoria et reminiscentia (On Memory and Reminiscences)*, memory was not tied to questions about the nature of the eternal soul. Instead, Aristotle, more the naturalist and empiricist than Plato, linked memory with reason. He wrote that both declined with old age and perished with the body. He also recommended the systematic cultivation of memory.

Aristotle believed that external events generated movements in the sensory organs. These movements were then propagated to the heart by spirits or pneuma in the blood. The movements of the pneuma had the ability to continue on a decreased scale after the stimulus was removed. The images within us, he taught, were the subjective side of these persisting movements. If they could be anchored to the past, they were memory. If not, they were imagination or fantasy. In this regard, simple memory was viewed as a

function of the faculty of sense perception, the same faculty responsible for the perception of time.

Aristotle dissociated this simple type of sensory memory—which can be observed in animals—from voluntary recollection. The latter, he believed, distinguished humans from animals. As stated in his *De memoria et reminiscentia*, "many of the other animals (as well as man) have memory, but of all that we are acquainted with, none, we venture to say, except man, shares in the faculty of recollection" (1931 translation, p. 453a). He reasoned that recalling past events at will requires the association and proper ordering of ideas. It also requires searching for images impressed or painted on the mind, as well as deliberation, a form of inference. Aristotle did not, however, associate these or other mental functions directly with the brain.

Aristotle's *De memoria et reminiscentia* is best known to psychologists and philosophers as the tract in which he presented three laws of association to account for voluntary recollection (Burnham, 1888). These were the laws of contiguity, similarity, and contrast. The most important of these, the law of contiguity, was described as follows:

> Acts of recollection, as they occur in experience, are due to the fact that one movement has by nature another that succeeds it in regular order. If this order be necessary, whenever a subject experiences the former of two movements thus connected, it will (invariably) experience the latter. (1931 translation, 451b)

VENTRICULAR LOCALIZATION OF MEMORY

Galen believed that the vital spirits, produced in the heart, were carried to the brain by the carotid arteries where they were transformed into animal spirits and stored in the ventricles. The animal spirits could then pass through the nerves to force the muscles into action or to mediate sensation. Although these spirits formed the instrument of the soul, Galen considered the brain itself the seat of the soul. This was based on the observation that damage to the brain could affect the mind and even life itself (see Chapter 1; also Pagel, 1958).

The perceived role of the ventricles in mental processes was elevated by the Church Fathers. These men began with Aristotle's contention that there were three types of internal activity: (1) the formation of mental images, (2) establishing opinions about them, and (3) recollection of the mental images. Galen recognized these activities as brain functions that could be affected independently, but stopped short of placing them in separate loci.

By the fourth century, the Church Fathers had localized these three activities firmly in the ventricles. Most often, the most posterior ventricle was associated with recollection. The paired anterior ventricles received the sensory information and were treated as a single cavity. And the middle ventricle usually was assigned the functions of ideation, cognition, and reasoning. Thus, the progression of receiving images, forming impressions, and recording the impressions proceeded from front to back.

Nemesius, the influential Syrian Bishop of Emesa, who maintained a strong interest in medicine, championed the aforementioned theory of ventricular localization. Writing around 400, he maintained:

---

[2]The story was also told by Cicero.
[3]See Yates (1966) for a history of mnemonic techniques.

So, then, the faculty of imagination hands on to the faculty of intellect things that the senses have perceived, while the faculty of intellect receives them, passes judgment on them, and hands them on to the faculty of memory. The organ of this faculty is in the hinder part of the brain and the vital spirit there contained. (Translated in Telfer, 1955, 342)

Recollection was also placed in the back of the posterior ventricle by Posidonius of Byzantium (fl. 370). Supposedly, he or one of his sources noted that injury in the front part of the head affected only imagination, that damage to the middle ventricle affected understanding, and that injury to the back of the head destroyed memory (see Sudhoff, 1914).

This posterior location for memory continued to be favored some 800 years later by Albertus Magnus (1193–1280). It was also accepted by his most famous pupil, Tommaso d'Aquino (Saint Thomas Aquinas, 1225–1274), the Christian metaphysical philosopher, shown with Albertus Magnus in Figure 23.1. Not all writers who accepted "cell doctrine," however, favored the posterior locus for memory. For example, in the fifth century, Saint Augustine assigned recollection to the middle cell, movement to the most posterior ventricle, and sensation to the anterior cavities.

Saint Augustine followed Aristotle in maintaining that animals do not have memory for ideas. He also agreed with Aristotle that even fish have a type of sense memory, which allows them to recognize images that had previously been impressed upon the sense organs. Saint Augustine was convinced that memory undergoes a gradual decay. He maintained that it begins to slip away from the moment it is engraved. Yet he stood in awe of the power of memory for ideas.

In his *Confessions,* Saint Augustine related memory (for ideas) to a large and boundless inner hall and to a massive storehouse with hidden recesses. He also mentioned a large cave, another fitting analogy to the importance of the brain's hollow ventricles. To quote:

> The vast cave of memory, with its numerous and mysterious recesses, receives all these things and stores them up, to be recalled and brought forth when required. Each experience enters by its own door, and is stored up in the memory. And yet the things themselves do not enter it, but only the images of the things perceived are there for thought to remember. (1955 translation, 115)

Hand in hand with localization of memory in the ventricles went the belief that the cold and moist brain had to be protected against overheating. One was taught to do things that would maintain the proper balance among the humors and to avoid things that would disrupt this balance. In this context, alcohol was looked upon as being especially bad for cognition and memory. It also was thought that onions, vinegar, fatty meats, and beans should be avoided. Some writers even suggested drinking sugared water and anointing one's head with special concoctions to preserve and improve memory.

One might think that the very old association of memory with the heart would have left the vocabulary by this time. Yet the use of the word "heart" in the context of memory did not fade as the Greek and the Roman empires crumbled, nor did it fade during the Middle Ages. *Recordari,* meaning "to recollect," has *cor* as its root, a word meaning "heart." Middle English dictionaries give the word *herte* ("heart") for memory, and even in the modern languages one is still told to memorize *par coeur,* or "by heart."

THE CULTIVATION OF MEMORY IN THE MIDDLE AGES

The "physiology" of memory most accepted during the Middle Ages held that the mind scanned the posterior ventricle for images when the time came for recall. The stored images, however, were not sterile, isolated entities. Along with them came other aspects of the original experience, including emotion, cognition, and judgment. Individuals in monasteries were taught that the best way to remember something was to form images very rich in scenery (e.g., imaginary towns), in emotion, and in personal associations. Indeed, during the Middle Ages failure to remember typically was not attributed to a loss of recall or recollection. Instead, it was due to a failure to imprint the material properly in the first place (Caruthers, 1990).

Physical prescriptions, such as avoiding alcohol, were viewed as secondary to forming rich images and repeatedly practicing the material to be remembered. This was the pre-

Figure 23.1. Albertus Magnus (1193–1280), shown teaching, and his famous pupil, Saint Thomas Aquinas, localized recollection in the posterior ventricle.

scription advocated by Quintilian. Only in these ways could one hope to emulate one of Petrarch's (Francesco Petrarca, 1304–1374) revered friends. As Petrarch described him in *Rerum memorandarum libri:*

> It was enough for him to have seen or heard something once, he never forgot; nor did he recollect only according to *res,* but by means of the words and time and place where he had first learned it. . . . Even after the passage of many years, if the same things were spoken of, and if I were to say much more or less or say something different, at once he would gently remind me of this and moreover correct the word in question; and when I wondered and asked just how he could have remembered it, he recalled not only the time during which he would have heard it from me, but under what shady tree, by what river-bank, along what sea shore, on the top of what hill. (Book 2, 13)

CARTESIAN MECHANICAL MODELS OF MEMORY

During the fifteenth and sixteenth centuries, a number of new models were proposed to account for learning and memory. One of the most notable post-Renaissance theories was developed by René Descartes. In *De homine,* published posthumously in 1662, Descartes theorized that filaments within each nerve tube operated tiny "valvules" that controlled the flow of animal spirits into the nerves.[4] For example, heat from a fire would move the skin and pull on these filaments to open a valvule in the ventricles of the brain. This action would allow the animal spirits to be released into the nerves, which in turn would make the muscles move a limb away from the flame.[5]

Descartes presented this reflexive theory only to account for involuntary, machine-like behavior. He believed that voluntary behavior was quite another matter. It demanded the interaction of the rational soul with the automaton. This interaction, he proposed, occurred through the pineal gland, a small central body suspended between the cavities (ventricles). Thus, the specific job of the pineal gland was to control the flow of spirits through the system of pipes and valves (see Chapter 2).

In *Les Passions de l'Ame,* published in 1650, the pineal theory was applied to the memory of ideas:

> When the soul desires to recollect something, this desire causes the gland, by inclining successively to different sides, to thrust the spirits toward different parts of the brain until they come across that part where the traces left there by the object which we wish to recollect are found; for these traces are none other than the fact that the pores of the brain, by which the spirits have formerly followed their course because of the presence of this object, have by that means acquired a greater facility than the others in being once more opened by the animal spirits which come towards them in the same way. (Translated in Herrnstein and Boring, 1965, 209–210)

[4]*L'Homme,* the French edition of *De homine,* appeared in 1664.
[5]Descartes developed this mechanistic theory after seeing some mechanical dolls and examining the hydraulic machines in the Royal Gardens in France (Fancher, 1990).

A number of closely related ideas about memory were introduced after Descartes argued that experience can open certain pores in the brain and facilitate the future opening of those pores (see Diamond, 1969). These ideas involved such things as the "smoothing" of pathways traversed by the animal spirits. For example, in 1666, Louis de La Forge (fl. 1661–1677), who had edited Descartes' writings, wrote the following about associations in the brain:

> When two objects have traced their species together, and the two streams of spirits composing them have been joined somewhere in the brain . . . it often suffices that one of them should appear again on the gland (even though this may be the weaker one, which was not capable of producing any considerable effect by itself, or which might even have produced a contrary effect) to reproduce all the same action which formerly came from both together. (Translated in Diamond, 1969, 4)

A comparable passage was written by Pierre Sylvain Régis (1632–1707) in 1691:

> A word makes us think of the thing which it signified, and conversely, the thing makes us think of the word, because of the ease with which the spirits move from one to the trace of the other; for by this means the spirits involved in the latter movement form a little path between the two traces, by means of which the spirits always flow from one to the other. (Translated in Diamond, 1969, 7)

This sort of thinking was used by Régis to explain why a young boy who once loved a crossed-eyed girl could now feel an attraction to other cross-eyed girls.

Nicolas de Malebranche (1638–1715) was another French philosopher strongly influenced by Descartes. In fact, Malebranche declared that Descartes discovered more truths in 30 years than all of the other philosophers put together. Malebranche (1674–78) agreed that traces impressed on the brain simultaneously are revived together because animal spirits can pass more easily along these paths than others. He differentiated memory from habit by the consciousness or awareness that accompanies memory.

Cartesian models of human memory were maintained into the eighteenth century, but they too underwent an evolution. One of the most important modifications was the elimination of the metaphysical soul from the mechanical plan. In 1748, Jules de La Mettrie (1709–1751) developed the idea that human action was unique not because of a soul, but because of its complexity. To this French materialist, humans and animals were both machine-like.

WILLIS ON THE CEREBRUM

Although the perceived role of the ventricles declined during the Renaissance, some scientists continued to maintain that memory resided in the back of the head. In particular, the cerebellum was singled out for a special role in memory. This was largely because the cerebellum was close to the posterior ventricle, the usual seat of memory in "cell" doctrine. The idea that the cerebellum functioned in memory was briefly mentioned by Andreas Vesalius in *De humani corporis fabrica* (1543). Johann Vesling and Nicolaas Tulp, contemporaries of Vesalius, also considered the cerebellum the likely seat of memory functions (Neuberger, 1981).

In England, Thomas Willis entertained the idea that the cerebellum (which he broadly defined to include the pons and colliculi) must have something to do with memory. But after pondering this idea, he rejected it in favor of the notion that memory for ideas most likely resided in the gyri of the cerebral hemispheres. The cerebellum, in contrast, was given the role of regulating involuntary motions, such as breathing.

The logic behind Willis' transferring memory functions to the cerebrum was not especially solid, as can be seen in this passage from *Cerebri anatome* (1664):

> But it is far more probable, that this faculty resides in the *cortical* spires of the Brain, as we have elsewhere shewn. For as often as we endeavour to remember objects long since past, we rub the Temples and the fore part of the Head, we erect the Brain, and stir up or awaken the Spirits dwelling in that place, as if endeavouring to find out something lurking there. . . . Besides, we have shewn, that the Phantasie and Imagination are performed in the Brain; but the Memory depends so upon the Imagination, that it seems only a reflected or inverse act of this: wherefore that it should be placed with it in the same Cloister, to wit, in the Brain, is but necessary. (1965 reprint, 111)

In *Cerebri anatome,* Willis also discussed the autopsies of two mentally deficient patients. In one, he noted that he could find no brain defect "less that its substance or bulk was very small" (p. 162). In the legend for Figure IV, he wrote the following about the other:

> The Effigies of an humane Brain of a certain Youth that was foolish from his birth, and of that sort which are commonly called Changelings; the bulk of whose Brain, as it was thinner and less than is usual, its border could be farther lifted up and turned back, that all the more interior parts might be more deeply beheld together.

In his *De anima brutorum* (1672), Willis described four degrees of "stupidity" and stated that deficiency could be hereditary or acquired.[6] He commented that "Fools beget Fools," but that "Parents being too much given to study, reading, and meditation" can also have mentally deficient children! As for acquired mental deficiency, Willis pointed out that this can be caused by alcohol, opiates, and head injuries.

HARTLEY AND THE ASSOCIATIONISTS

The Aristotelian principle of association by contiguity was kept alive by Thomas Hobbes (1588–1679), the philosopher who, in 1651, wrote that all knowledge is acquired by experience. Association by contiguity was later accepted by another associationist, John Locke (1632–1704), a pupil of Thomas Willis at Oxford and afterward an apprentice to Thomas Sydenham in London. Locke (1690) treated the mind as a storehouse for the experiences of the senses (the

[6]This work was translated from the Latin as *Two Discourses on the Soul of Brutes* in 1683. See Cranefield (1961) for sections of this book dealing with memory and for some commentary.

Figure 23.2. David Hartley (1705–1757), who developed a physiological memory theory by combining the associationism of John Locke with the vibrational theories of Isaac Newton.

ancient concept of the "repository of memory") and wrote that when ideas are not renewed, color fades from the pictures and "the print wears out." He maintained that the mind has neither innate nor unconscious ideas and that experiences are combined and united into more complex ideas by association. Locke introduced the popular phrase "association of ideas."

David Hartley, shown in Figure 23.2, was the associationist who made the first significant foray into neurophysiological theory. His conceptions were primarily based on the ideas of Sir Isaac Newton, who theorized that all bodies contain a hidden, vibrating subtle spirit ("ether"), which caused particles to attract and repel each other over distances. In his *Philosophiae naturalis principia* (1687) and again in his *Opticks* (1704), Newton hypothesized that the vibrating spirit traveled at the command of the will along the solid filaments ("Capillaments") of the nerves from the sensory organs to the brain and from the brain to the muscles

Hartley (1749) blended Newton's vibration theory with Locke's associationism to explain memory. He wrote that vibrations in the nerves and brain were the very cause of associations. Specifically, he postulated that external stimuli caused small backward and forward motions, or oscillations, like the "trembling of sounding bodies," within the sensory nerves. The vibrating particles in the nerves were postulated to lead to diminutive vibrations ("vibratiuncles") in the soft, "medullary substance" of the brain, which he believed was the seat of the rational soul.

Hartley thought the vibrations ordinarily died away or decayed after a short period of time. Nevertheless, he maintained that with repeated presentation of a stimulus, the medullary substance, which is always in motion, may become more permanently affected by the impressed wave form. Thus, repeated exposure to a particular vibration could change the medullary substance so that it would be more likely to vibrate in a specific way when disturbed again and again.

Hartley assigned the role of memory to the medullary substance because he observed that sensation, voluntary motion, memory, and intellect were seriously affected by deep lesions of the brain. In addition, the medullary tissue seemed uniform and thus more suited to such a function than the highly irregular cortex. Automatic reflexes, he reasoned, are different and do not require a rational soul.

Hartley's theory of association followed from these premises. He was now able to maintain that if two distinct vibrations repeatedly overlapped, the modifications in the medullary substance would be such that if just one vibrated, it would still be capable of triggering the vibratory changes of the other. Hartley further argued that a given sensation could bring forth many ideas, not just as a result of being synchronous but because of other associationistic principles and the physics of resonance. Simple ideas could be turned into very complex ideas by such mechanisms.

Hartley, who did not have a medical degree but still practiced medicine in England until he died at the age of 52, first presented his vibration theory in 1746 as a "trial balloon" (Buckingham, 1991). The theory was included in the second edition of a small Latin book about a medical solvent. It reappeared virtually unaltered three years later, in 1749, in his most famous work, *Observations of Man, His Fame, His Duty, and His Expectations*.

In 1775, Joseph Priestley, the esteemed chemist and a devoted Hartley disciple, edited a new edition of Hartley's *Observations* under the title *Hartley's Theory of the Human Mind*. Priestley chose to eliminate much of the "quite dispensable" religious dogma from Hartley's work. But along with this, Priestley also dropped the vibration theory, largely because he thought it was too difficult for the average reader to comprehend (Smith, 1987).

The physiological aspects of Hartley's theory suffered not only because Priestley left them out of his book but also because Herman Boerhaave (1743), the influential Leyden physiologist, pronounced that there was no truth to the idea that soft, flaccid nerves could perform their actions by vibrations in a way comparable to a piano wire or tense cord. Interestingly, Hartley never proposed such an analogy or anything like it. Still, Boerhaave's statements had a damaging effect.

In contrast to the rather cold reception that awaited Hartley's theory of vibrations, his emphasis on the association of ideas proved considerably more durable (Oberg, 1976). His philosophy influenced the later associationists, including Thomas Reid (1710–1796), Dugald Stewart (1753–1828), Thomas Brown (1778–1820), James Mill (1773–1836), John Stuart Mill (1806–1873), and Alexander Bain (1818–1903). It also had a significant impact on the thinking of Herbert Spencer, William James, and Wilhelm Wundt. The greater enduring success of Hartley's mental associationism, as opposed to his physiological postulates, would have come as no surprise to Hartley himself. He had assumed from the beginning that even if his theory of vibrations could not

withstand the test of time, his doctrine of associations was so solid that it would remain in place to guide future philosophers and natural scientists interested in memory and the association of ideas.

LATER VIBRATION THEORIES

The vibration theories, which began with the Greeks (e.g., Aristotle), did not end with Hartley in the 1740s. Charles Bonnet (1720–1792; shown in Figure 23.3) suggested a vibration theory fairly similar to the one proposed by Hartley. Bonnet (1755, 1760, 1764) also began with the idea that sensations caused nerves to vibrate, each with its own specific character. He postulated that this feature allowed the fibers to preserve a tendency to vibrate in the same way again and that the sensory fibers transmitted this information to the brain. Memory once more was represented physically in residual or persisting low-intensity vibrations.

Debates over vibration theories continued during the second half of the nineteenth century. Jules Luys (1828–1895), for example, argued for vibrations, but borrowed his terminology from chemistry to give the theory a more modern flavor. Nevertheless, Luys (1865) did not provide new physiological or scientific evidence for his ideas. Thus, he maintained that vibrations persisted after stimulation, but simply called this phenomenon "phosphorescence," to suggest a chemical glowing after the stimulation ended. Luys further wrote that neural vibrations were analogous to vibrating particles on a photographic plate, not knowing that the salts used in photography do not exhibit Brownian movements (Gomulicki, 1953).

Belief in the ancient theory of vibrations finally came to a halt early in the twentieth century. The last staunch

Figure 23.3. Charles Bonnet (1720–1792), who agreed with David Hartley when it came to the idea that memory was represented physically by low-intensity vibrations in the brain.

defender of a vibration type of theory was Jacques Loeb, who, in 1902, hypothesized that the integrated nature of the nervous system had much to do with the periodicity of the discharges from its cells. Loeb did not believe, however, in strict localization of memory functions. In Loeb's (1902, p. 262) words, "in processes of association the cerebral hemispheres act as a whole, and not as a mosaic of a number of independent parts."

ORGANIC MEMORY: ANOTHER QUESTIONABLE THEORY

Another highly questionable theory which was defended into the opening years of the twentieth century was that all organic matter has memory. In 1870, Ewald Hering outlined his ideas about "organic memory." Speaking before the Imperial Academy of Science in Vienna, Hering suggested that memory was not just a function of specialized parts of the higher nervous system. He viewed heredity as memory that crossed generations via the germ cells, and he accepted the Lamarckian notion of inheritance of acquired characteristics.

The organic memory theory was received with enthusiasm by a number of scientists. Samuel Butler (1835–1902), for one, claimed: "Life is that property of matter whereby it can remember—matter which can remember is living" (Butler, 1880, p. 175). The theory was also entertained in 1881 by Théodule Ribot (1839–1916), who added that organic memory differed from "psychological memory" only by the consciousness associated with the latter.[7]

Another individual influenced by Hering was Richard Semon (1859–1918), a scholar who worked in relative isolation in Munich.[8] Semon's ideas about memory appeared in two books, *Die Mneme* and *Die mnemischen Empfindungen*, published in 1904 and 1909, respectively. He postulated that memories were represented by physical changes in the brain; that the traces were unified complexes not confined to just one cell; and that each repetition left a fresh trace, as opposed to strengthening an existing trace. The memory traces themselves were not looked upon as perfect copies of experiences but as less vivid than the original, subject to distortion, and tied to other memory complexes by their overlapping chronologies.

In contrast to his relatively fresh ideas, Semon followed Hering in believing that memory and hereditary mechanisms were one and the same. Not only did he adhere to Lamarkian doctrine, he did so at a time when the concept of inheritance of acquired characteristics was already being discredited (e.g., Weismann, 1906). Other problems with his Gestalt-like theory were its failure to account for forgetting, and its inability to account for the selective nature of memory.

Semon believed that his new biologically oriented ideas required a new vocabulary, and he invented a number of words to fit his theories. He coined the word "engraphy" to describe how stimuli leave behind connected, unified trace complexes (i.e., how memory is stored); added the word "ecphory" to describe the awakening of the trace from its latent state to a state of activity (i.e., engram retrieval); and used the term "homophony" to describe how different sensations and ideas could become associated and recognized as memories and not new experiences. These three terms fared poorly, but another of his words, "engram," did far better. This term has been popular in the sciences since Semon's time as a synonym for "memory trace."

THE BIRTH OF MEMORY SCIENCE

The doors to the scientific measurement of memory were opened in 1885 when Hermann Ebbinghaus (1850–1909) published his monograph *Über das Gedächtnis: Untersuchungen zur experimentellen Psychologie* (translated in 1913 as *Memory, A Contribution to Experimental Psychology*). Ebbinghaus had been interested in "mental measurement" for some time, but unlike his predecessors, he wanted to apply measurement techniques to "higher mental processes." In contrast to sensation, memory was something his contemporaries thought too subjective and fleeting to be studied scientifically. He felt differently, however, and began to design and conduct memory experiments. In fact, Ebbinghaus, shown in Figure 23.4, was both the subject and the experimenter in the investigations presented in his monumental work of 1885.

Ebbinghaus' book contained new stimuli and paradigms for studying memory. The most famous of these were his "nonsense syllables." These were meaningless combinations of letters (usually two consonants with a vowel between them), which he viewed as "neutral" (not meaning-

Figure 23.4. Hermann Ebbinghaus (1850–1909) revolutionized the scientific measurement of human memory in 1885 with the publication of his monograph *Über das Gedächtnis: Untersuchungen zur experimentellen Psychologie*. This work later appeared in English as *Memory, A Contribution to Experimental Psychology*.

---

[7]Ribot (1883) is much better remembered for his "law of regression." This "law" holds that new, unstable memories succumb before well-organized, long-established ones. Ribot derived this "law" by studying patients. It is discussed in Chapter 24.

[8]Semon's life and work have been the subject of several books and articles (see Schacter, 1982; Schacter, Eich, and Tulving, 1978).

ful) stimuli. With over 2,300 such stimuli to draw at random, Ebbinghaus measured acquisition, recall, and recognition of learned information, while varying such factors as the amount of material to be learned, the number of repetitions, and the time intervals between exposure and recall.

Ebbinghaus (1885, 1908) effectively took the study of memory out of the hands of the philosophers and turned it over to the experimentalists for quantitative analysis. A few years after his monograph appeared, quantitative experiments on learning and memory in laboratory animals also began to be published (e.g., Thorndike, 1898). These studies led to controlled experiments on the effects of specific brain lesions on memory.

CHANGES AT THE SYNAPSE

Ebbinghaus did not speculate on the changes in the brain that might account for memory, nor did he comment on where the engrams were stored. In contrast, the emergence of neuron doctrine at the same time led to a new set of theories about the memory trace.[9]

One idea that grew out of neuron doctrine was that dendritic or axonal branches could grow to enhance or block communication at the synapse. Interestingly, this notion may have been anticipated by Alexander Bain (1872), one of the later associationists. He philosophized that "for every act of memory, every exercise of bodily aptitude, every habit, recollection, train of ideas, there is a specific grouping or coordination of sensations and movements, by virtue of specific growths in the cell junctions" (p. 91).

Especially during the 1890s, when scientists could do no more than speculate about neuronal movements, and when the synaptic concept was itself embryonic, neural growth was envisioned as extremely rapid. A number of theorists thus likened the hypothesized growth of axons and dendrites to the movements of the pseudopodia of the simple ameba.

One such person was Hermann Rabl-Rückhard. As early as 1890, just as Santiago Ramón y Cajal was completing an important study on the growth cone, Rabl-Rückhard proposed that some psychic acts could be explained by a type of "ameboidism," by which neurons could move their processes with ease to different locations.

This basic idea was embraced by Eugenio Tanzi (1856–1934) in 1893. Tanzi theorized that the passage of a nervous impulse could cause a neuron to become longer by elongating its protoplasmic filaments. He postulated that with additional activating stimuli, the filaments could grow even closer to the postsynaptic cells, making conduction across the synapse, and an associative bond, even more likely.

Neural ameboidism was also accepted by Jacques Raphaël Lépine (1894, 1895) and Mathias Duval (1895). Duval, shown in Figure 23.5, applied his ameboid theory to sleep and then to learning and memory. He wrote the following:

> In man when sleeping, the cerebral arborizations of the (central) sensory neurons are retracted like the pseudopods of an anesthetized leukocyte. The weak excitations pro-

Figure 23.5. Mathias Duval (1844–1907). In 1895, Duval wrote that memory may involve neurons stretching their arborizations to bring them closer to each other. Duval also hypothesized that neural "ameboidism" may also play a role in sleep and wakefulness.

duced in the sensory nerves . . . do not reach the cerebral cortex; stronger excitations cause the stretching or relaxation of the cerebral arborizations of the sensory neuron; the impulse, on passing to the cortical cells, leads to awakening, the successive phases of which clearly suggest the reestablishment of a series of (connections) previously interrupted by retraction and withdrawal of the pseudopodial arborizations. . . . Anesthesias and hysterical paralyses could be explained in the same way, as well as the increase in energy of the imagination, in memory, in the association of ideas, under the influence of various agents, such as tea and coffee, that could excite the ameboidism of nerve extremities that are in contiguity, in order to bring their arborizations closer and facilitate the passage of impulses. (1895a; translated by DeFelipe and Jones, 1988, 480)

Duval presented a number of observations, not all of which were his own, to support this hypothesis. They included the idea that the cones in the fish retina seemed to shrink when exposed to light; investigations supposedly showing dendrites retracting under the influence of various anesthetics; experiments on fatigued mice revealing possible alterations in their dendrites; changes in dendrites during hibernation; and the discovery that olfactory nerve cilia can move. Duval (1895b) also talked about Lépine's ideas in relation to his own.

By 1895, Ramón y Cajal was convinced that cell growth was governed at least in part by chemical secretions from target cells.[10] But he added that the most likely candidate for the type of ameboid action which could underlie functions such as learning and sleep was not the neuron but the glial cell. Ramón y Cajal (1895) argued that glial cells

---

[9]See Chapter 3 for a history of neuron doctrine (also Shepherd, 1991).

[10]For a history of the chemoaffinity hypothesis, see Hunt and Cowan (1990).

were more capable of lengthening their appendages to cut off the activity between neurons. And, upon retraction of these glial processes, neurons could once again return to activity. While intrigued with this idea, Ramón y Cajal did not seem to have great faith in his own version of the ameboid theory. In his mind, it was simply an idea worthy of at least some consideration.

The hypotheses of neural and glial ameboidism had their share of critics. In particular, the neural ameboid hypothesis was criticized by Ramón y Cajal's German friend Rudolph Albert von Kölliker (1896). Kölliker argued that axons simply do not contract to any kind of excitation, that they are solid, and that ameboid movements are typically related to thermal and nutritive stimuli, as opposed to psychic functions. As far as Kölliker was concerned, the glial variation on this theme was no better.

Later scientists tended to look upon the growth at the synapse as a somewhat slower and less dynamic affair than Duval envisioned. One such person was Jacques Loeb, who concluded:

> The amoeboid changes in the ganglion-cells have been utilized to account for the phenomena of association. As far as normal processes of association are concerned, these amoeboid changes cannot play any role, as they are much too slow. (1902, 257)

An alternative idea was that increased currents within brain regions stimulate the growth of axon collaterals and dendrites, but not the kind of rapid ameboid growth postulated earlier by Duval. This type of growth theory was proposed by C. U. Ariëns Kappers (1877–1946), who called it "neurobiotaxis" (e.g., Ariëns Kappers, 1917; Ariëns Kappers, Huber, and Crosby, 1936).

Ariëns Kappers and his associates (1936) cited Ramón y Cajal (1893) as the first investigator to hypothesize that trophic ("growth") factors play a role in the direction of the outgrowth of dendrites (and cell migration) during development. Then they presented a number of studies to show that stimulation-induced changes in the electrical potentials of neurons could be trophic. These changes were hypothesized to attract the growth of functionally related cells.

Charles Manning Child (1869–1954) and Edwin B. Holt (1873–1946) proposed other theories involving the growth of neural fibrils toward one another to narrow the synaptic gap (Child, 1921, 1924; Holt, 1931). Yet when the literature on memory mechanisms was reviewed three decades later, Karl Lashley argued that there was still no evidence to support any of these "growth" theories. Lashley specifically criticized these ideas on the grounds that (1) cell growth seemed too slow to account for the rapidity with which some learning can take place and (2) that he was unable to localize the memory trace.

Lashley (1929b) also felt there was no real support for the hypothesis that learning could be attributed to even more subtle changes at the synapse. Scientists such as Hulsey Cason (1893–1950) and Harry Miles Johnson (1885–1953) had proposed that reduced synaptic resistance following a nervous impulse and changes in the permeability of the synaptic membrane might account for learning (e.g., Cason, 1925; Johnson, 1927). Lashley looked upon these ideas as fanciful, wishful thinking. Such notions were not backed by solid scientific evidence and contradicted the facts as he saw them.

Figure 23.6. Ivan Pavlov (1849–1936), the Russian physiologist best known for his classical conditioning experiments.

DRAINAGE AND IRRADIATION THEORIES

Even in the face of Lashley's repeated criticisms, the theories that emphasized specific synaptic changes drew more attention than other memory theories. These theories were given recognition in part because the concept of the synapse was attractive to just about all neurophysiologists and because most other theories had even weaker scientific foundations.

One class of theories that did not last long in this new *Zeitgeist* involved the concepts of drainage and irradiation. A theory of this sort was developed by Ivan Pavlov, the Russian physiologist who won the Nobel Prize in 1904 for his work on digestive secretions. Pavlov, shown in Figure 23.6, became even more famous for his studies of "classical conditioning," especially after his work on "psychic secretions" was described and translated into English (1917, 1927–28).[11] These studies showed that by pairing a "conditional" (conditioned) "neutral" stimulus (a bell) with an "unconditional" (unconditioned) stimulus (food), dogs would learn to salivate to the sound of the bell alone.

Pavlov first talked about classical conditioning in 1903 at the 14th International Medical Congress in Madrid. But, as he stated in his 1904 Nobel lecture, he was not the first to observe psychic stimulation of both the salivary and the gastric glands. Although Pavlov did not mention names at this time, one of the first to experiment on psychic secretions may have been William Beaumont (1785–1853), a U.S. Army surgeon assigned to Fort Mackinac (see Myer, 1912, for his biography). Beaumont, shown in Figure 23.7, gave an early account of psychic secretions of gastric juices in Alexis St. Martin (1803–1886), the Canadian trapper shown in Figure

[11]So little was known about Pavlov before the 1920s that when the surgeon E. V. Pavlov died in St. Petersburg, Russia, in 1916, obituaries were written for Ivan Pavlov in *Lancet*, the *British Medical Journal*, and the *Journal of the American Medical Association* (see Fulton, 1959).

Figure 23.7. A daguerreotype of William Beaumont (1785–1853), the physician who attended and studied the trapper, Alexis St. Martin, after he was shot in the stomach.

Figure 23.8. A portrait of Alexis St. Martin (1803–1886) taken when he was 67 years old. After St. Martin was shot in the stomach (1822), he allowed his gastric secretions to be studied by William Beaumont.

23.8, who had been shot in the stomach in 1822 (see also Figures 23.9 and 23.10). Beaumont wrote that just the smell of cooked steak, which St. Martin adored, led to a flooding of gastric secretions in his "open" stomach.[12]

Pavlov (1904) realized that the phenomenon of psychic secretions demanded a physiological explanation. In this context, he proposed that conditioning took place at the cortical level, even though the stimuli for conditioning might also excite subcortical areas. The notion that the cerebral cortex might function in a reflexive way was not a novel one. This idea had been proposed by the British neurologist John Hughlings Jackson and the Russian physiologist Ivan Mikhailovich Sechenov (1829–1905). Their orientations, which influenced Pavlov's thinking, are worth noting.

Jackson believed that behavior could be understood in terms of sensory-motor connections often involving the cerebral cortex. His ideas about cortical reflexes were influenced by his old friend Thomas Laycock (1812–1876), who had written: "the brain, although the organ of consciousness, is subject to the laws of reflex action . . . it in this respect does not differ from the ganglia of the nervous system" (1845, p. 298). Laycock added: "Like the association of movements, the true explanation of ideas is to be found in the doctrine of the reflex functions of the brain" (p. 311).

In a number of papers, including his 1873 paper on the localization of movements in the brain, Jackson developed the notion that nervous arrangements at higher levels differ only in degree from those at lower levels. Importantly, he believed that relative to the lower centers, the cortex could be more easily modified by experience.

Sechenov's ideas were presented in his classic *Reflexes of the Brain,* which appeared in 1863. He wrote that the most complex association is nothing more than an interrupted series of reflexes. He also proposed that some reflexes could be excitatory whereas others could be inhibitory. To the extent that Sechenov, as well as Jackson, theorized that psychic life was based on reflex action, Pavlov thought it could be analyzed with his new behavioral methods.

Pavlov went on to develop the idea that afferent stimulation set up excitatory processes. These "irradiated" in all directions from the stimulated part of the cortex. The hypothesized irradiation diminished in intensity with distance and, after a few seconds to a few minutes, it supposedly contracted. He postulated that the stronger areas of cortical excitation could attract contiguous but weaker impulses to their foci (Pavlov, 1927; also see Babkin, 1949). Thus, in conditioning, the excitation produced by a weak conditioned stimulus could be drawn to the area of greater excitation produced by the unconditioned stimulus. If this were repeated, new connections would be formed allowing the previously "neutral" sound of the bell to produce new impulses in the same cortical regions as the taste of food.[13]

Pavlov developed these ideas in part by studying how well conditioned stimuli applied to one part of the body

---

[12]St. Martin suffered an accidental shotgun wound that created a large cavity in his abdomen and stomach. He survived, but was left with a hole providing direct access to his stomach. Before he left Fort Mackinac, and then while at Fort Niagara and in other locations, William Beaumont studied St. Martin's digestive functions (e.g., Beaumont, 1826; also see Myer, 1912). He removed the dressing over the hole and conducted a variety of experiments. Some involved placing different kinds of food on a string directly into St. Martin's stomach. In 1832, 10 years after the accident, Beaumont even drew up a legal contract to give him the right to conduct scientific experiments on St. Martin's stomach secretions for one year at a cost of $147. Beaumont's many findings were published in 1833 in his classic *Experiments and Observations on the Gastric Juice and the Physiology of Digestion.*

[13]Pavlov further postulated that unpaired stimuli could somehow trigger inhibitory impulses across the cortex, which could suppress conditioned reflexes.

Figure 23.9. The small frame hospital at Fort Mackinac where Alexis St. Martin stayed after his gunshot wound.

would generalize to other body parts. He noticed that conditioning was best with stimuli close to the point of original stimulation, but was difficult when it involved body parts very far away from each other in the periphery, as well as in terms of their cortical representations.

Pavlov may also have been influenced by prevailing, more philosophical theories. In 1890, William James had written that "when two elementary brain processes have been active together or in immediate succession, one of them on recurring, tends to propagate its excitement to the other" (p. 566). James also postulated that new associations could be attributed to "drainage," the idea that two or more areas of excitation could somehow "drain" into each other to form new functional pathways between unrelated parts of the brain.[14]

Pavlov's theory drew attention largely because behavioral aspects of his work on conditioning were considered important. Nevertheless, his physiological thoughts were challenged by experiments showing that some conditioning could take place in decerebrate animals or in fetuses without functional cortices (e.g., Culler and Mettler, 1934; see Hilgard and Marquis, 1961). The theory fared worse as more was learned about the electrical activity of the brain and as controlled studies on brain-damaged animals began to cast a shadow over theories that placed engrams in specific cortical locations. The most influential lesion experiments arguing against invariant, strict localization of the memory trace were conducted in the United States by Franz and Lashley.

FRANZ AND LASHLEY

Shepherd Ivory Franz was the first to use the new behavioral methods from experimental psychology to assess the effects of brain lesions on learning and memory functions. In his autobiographical statement, Franz wrote the following:

[14]A comparable idea was promoted in 1905 by another psychologist, William McDougall.

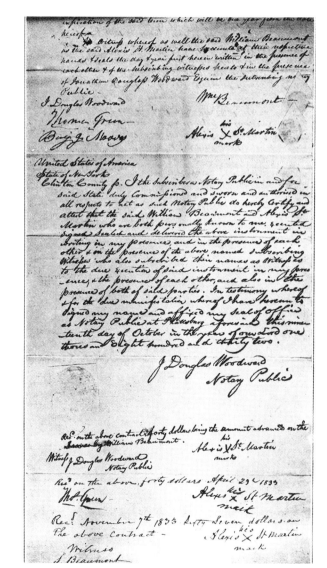

Figure 23.10. Part of the legal contract that allowed William Beaumont to study Alexis St. Martin's (1803–1886) exposed stomach. The contract was for one year and required St. Martin to travel with Beaumont for his $147 fee. This was in 1832, 10 years after the accident. (From Myer, 1912.)

I had read the account of experiments of the frontal lobes by Professor Bianchi in which he cited the fact that a baboon which had acquired habits of saluting lost those habits following removal of the frontal areas. Corresponding observations were, I think, reported also by Sir Edward Sharpey-Schäfer. It was these suggestions or stimuli which immediately gave rise to the idea of the use of the training method of investigating the functions of the frontal lobes. At the time Bianchi and Schäfer reported their findings there were not available the data on methods of animal learning which had developed in the hands of Lloyd Morgan and especially of Thorndike, but it was a simple step from the latter's results to try the combination of animal training and extirpation as a method whereby some additional cerebral problems might be solved. (1961, 97)

In the autumn of 1900, Franz, who had just received his doctorate from Columbia University, began to work with the puzzle box (Figure 23.11) just developed by Edward L. Thorndike (1874–1949). Thorndike (1898) had put unoperated cats into the enclosed box and then measured how long it took them to pull a string or a loop, or press a button, to open the door and escape. Thorndike believed the cats opened the door accidentally at first, while they were flailing and clawing at anything loose and shaky, but that with additional trials they mastered the problem, and the time it took them to escape decreased.

Franz published his findings on the effects of frontal lobe lesions on puzzle box performance in 1902. He reported that cats given frontal cortical lesions showed a loss of memory for solving the previously mastered puzzle box. Franz explained:

After the operation it was noticed that when put in the boxes the cat would put its nose or its paws through the slats, and it would scratch around just as it had when learning the habit. To all who observed the cat at this time it seemed that no memory of the habit was present. . . . This case is typical of all the other experiments upon the frontal lobes. (1902, 11)

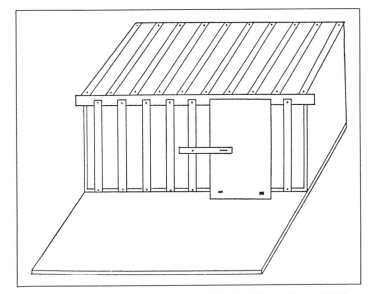

Figure 23.11. The puzzle box used by Shepherd Ivory Franz (1902) to test memory in his cats. Franz measured how long it took a cat to pull a string to open the door and escape. He reported that frontal lobe lesions were most likely to affect this habit, but that even his frontal cats could still relearn it. The puzzle box itself was developed by Edward L. Thorndike in 1898.

Franz found that cortical lesions which left the prefrontal areas intact did not affect the learned habit. He further reported that although the recently formed habit of pulling the string to escape the box had been lost, the cats with frontal lobe lesions still retained the impulses to escape from the box and still came when called; they were motivated and not at all sick. These ancillary findings suggested that the frontal lobes played a role in memory and that the loss was not due to surgical shock. Nevertheless, Franz observed something else very important: Lost habits could be relearned, often quite rapidly. This meant that whereas the frontal lobes might ordinarily contribute to memory, they were not essential for learning or memory.

Franz (1906, 1907) confirmed his findings in experiments using primates. He first taught monkeys to open a food box and to complete a hurdle task to obtain food. Then, after the monkeys had mastered the problems, he damaged their frontal lobes and tested them for retention. Again, the newly formed habits were lost, whereas some older, stable habits, such as eating out of his hand or jumping on his shoulder, remained intact.

In discussing these as well as later findings, Franz emphasized the dynamic interplay he believed had to exist among brain areas. He doubted whether specific learning or memory centers existed at all, but suggested that if they did, they had to be more diffuse and loosely organized than most of his contemporaries were willing to admit. This idea, he maintained, was the best one to account for the fact that cats could still learn an escape response even if their frontal lobes had been removed before training. It could also account for why monkeys that were preoperatively trained until a habit was second nature to them ("overtrained") did not show deficits on those problems after their frontal lobes were removed (Franz, 1906). With obvious pessimism, Franz concluded that investigators had come no farther from where they had been 50 years earlier when it came to localizing the memory trace.

Franz continued to reflect upon this state of affairs many years later. In his autobiography he wrote:

Everything tended to show that there are not the definite and exact functions for parts of the cerebrum which were posited, but that there is rather a possibility of substitution. This does not mean that there is no localization, but that there is not a localization in the sense in which such localizations were reported. (1961, 103)

Karl Lashley, who had just finished graduate school at the time, collaborated with Franz on five projects. Two of their experiments were published in 1917. Franz was largely responsible for the surgery and the histology, and Lashley did the behavioral testing (Franz and Lashley, 1917; Lashley and Franz, 1917; also see Chapter 4). They found that rats trained on brightness or on simple maze problems did not lose their learned habits when large parts of their frontal lobes were damaged. They also noted that lesions of other parts of the cortex had no effects on these behaviors.

Lashley summarized his studies with Franz and many other earlier experiments in *Brain Mechanisms and Intelligence*, published in 1929.[15] In this book, he reported that retention could be affected by large cortical lesions, but that it did not seem to matter where the cortical lesions were placed,

[15]Lashley's book was critically reviewed by James Papez in 1931.

so long as a critical percentage of the cortex was destroyed. Furthermore, the animals still showed a remarkable ability to relearn mazes like those shown in Figure 23.12.

Thus, Lashley (1929a) hypothesized that learning and remembering a maze did not depend upon any one sensory function or upon engrams confined to the cortical motor areas. In fact, he concluded that conceptions of learning and memory based on restricted reflexes of the brain, or on fixed cortical pathways involving the association areas, could not possibly be correct.

> The original program of research looked toward the tracing of conditioned-reflex arcs though the cortex, as the spinal paths of simple reflexes seemed to have been traced through the cord. The experimental findings have never fitted into such a scheme. Rather, they have emphasized the unitary character of every habit, the impossibility of stating any learning as a concentration of reflexes, and the participation of large masses of nervous tissue in the functions rather than the development of restricted conduction-paths. (1963 reprint, 14)

Lashley knew that his book on the brain and intelligence raised more questions than it answered. Recognizing this, he continued to work on problems associated with the locus of memory. His later findings served as the basis for a major review article in 1950, a paper given the provocative title "In Search of the Engram." Among other things,

this paper described (1) studies in which Lashley isolated the visual and motor regions of the cortex and found no impairments in rats trained to jump to a specific visual target; (2) experiments on monkeys that remembered how to open latch boxes after motor cortex damage; (3) other primate studies showing that damage to the visual "association" areas did not affect color, brightness, or visual form discriminations; and (4) experiments in which multiple knife cuts were made throughout the cortex in an unsuccessful attempt to disrupt specific point-to-point connections needed for maze learning.

These experiments crystallized Lashley's belief that the memory trace was not confined to well-defined cortical circuits. In fact, they led him to suggest that the memory trace was either diffused throughout the cortex or was passed to subcortical centers.

> Memory traces, at least of simple sensori-motor associations, are not laid down and stored within associative areas; at least not within the restricted associative area supposedly concerned with each sense modality. Memory disturbances of simple sensory habits follow only upon very extensive experimental destruction, including almost the entire associative cortex. Even combined destruction of the prefrontal, parietal, occipital and temporal areas, exclusive of the primary sensory cortex, does not prevent the animal from forming such habits, although pre-existing habits are lost and their reformation is greatly retarded. (Lashley, 1950, 464)

Lashley added that whereas his findings were obtained largely on rats and other nonprimates, he believed that the same held for humans. In particular, he noted that bilateral removal of the entire prefrontal cortex in five chimpanzees did not result in memory defects. Thus, he theorized that amnesia after a brain injury is rarely due to the loss of engrams. Amnesia, he reasoned, can better be attributed to the inability to activate traces, to a problem in organization, or to some other higher-level deficit (also see Lashley, 1938).

After decades of searching for the engram, Lashley shrugged his shoulders and made the following statement:

> This series of experiments has yielded a good bit of information about what and where the memory trace is not. It has discovered nothing directly of the real nature of the engram. I sometimes feel, in reviewing the evidence on the localization of the memory trace, that the necessary conclusion is that learning is just not possible. (1950, 477–478)

Lashley could well have quoted his intellectual mentor, Franz, who years earlier had written: "What memory means physiologically we do not know; where memories are stored we do not know; and how they are stored we do not know" (1912 p. 328). The American researchers only felt sure that memory was not stored in circumscribed areas and did not involve single memory cells, as Theodor Ziehen (1862–1950) had postulated in 1891.

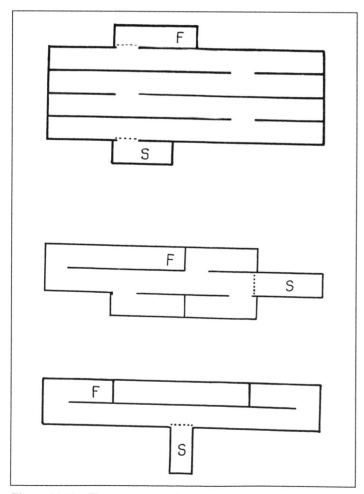

Figure 23.12. Three mazes used by Karl Lashley to study learning and memory in rats. The level of difficulty was determined by the number of blind alleys. S = Startbox; F = Finish area. (After Lashley, 1929.)

PATTERN THEORIES

The work of Franz and Lashley not only challenged specific "connectionistic" models of memory but contributed to the rise of pattern theories of memory (see Gomulicki, 1953). The basic idea was that diffuse neural groupings mediated memory. This view was popularized by the founders of the Gestalt school of psychology, especially Wolfgang Köhler and Kurt Koffka.

Koffka (1935) and Köhler (1929) agreed that there was a close association between memory and the principles of perception. They argued that contiguity alone was not adequate to explain memory and that it was necessary to consider how complex neural patterns might be organized by the "rules" of closure, similarity, and contiguity. They felt this was the only way there could be any hope of explaining both normal memory and common errors of memory.

The Gestalt psychologists hypothesized that multiple memory traces were formed in the cortex. They believed that visual memories involved successive and simultaneous stimuli in different parts of the visual field. But these traces could not be static or rigid point-to-point representations. Rather, they were multidimensional symbolic representations with central figures superimposed on halo-like background fields. They felt that the traces did not even have to be confined to the visual areas. The neural substrate depended upon the complexity of the whole sensory experience, including associated factors.

Koffka and Köhler also postulated that new records might be inscribed on top of old patterns. These new memories could be affected by older memories and could, in turn, affect older memories. If two patterns became associated, the result was the formation of a newly organized "group-unit."

This orientation toward memory was further developed by the neurologist Kurt Goldstein (e.g., 1942). In contrast to Köhler and Koffka, who worked with normal apes and humans, Goldstein dealt extensively with brain-damaged patients, many of whom had memory disorders. Their patterns of behavior, more than anything else, led him to postulate that memory involved engram patterns spread throughout the cortex, each of which involved a central "figure" and a peripheral "ground."

From this premise, Goldstein argued that brain lesions can easily disrupt complex engram patterns, although lesions are unlikely to destroy all aspects of a memory trace. Goldstein maintained that when the figure-ground relationship of a memory pattern is disrupted, memories show a loss of integration, complexity, and meaning. Answers to questions seem funny or odd, and memory may appear faulty, especially if these patients also lose their ability to think abstractly (see Chapter 22).

THE DELAYED RESPONSE PROBLEM

Gestalt concepts of memory seemed to fit well with the lesion data generated by Franz and Lashley, but some experimenters strongly disagreed with the idea that the cortex does not possess specialized parts for storing of engrams. One important debate centered on the frontal lobe lesion studies of Carlyle Jacobsen. In 1935, Jacobsen and his associates described two chimpanzees, "Becky" and "Lucy," who had been trained on some behavioral tasks prior to sequential ablations of the left and right frontal association cortices.[16] After bilateral lesions, both chimps

[16]This study also played a major role in the study of emotions and the history of psychosurgery. For this part of the story, see Chapter 20.

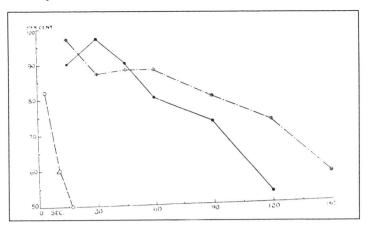

Figure 23.13. Diagram showing performance of a baboon on delayed-response tasks. The ordinate is percentage correct and the abscissa is delay time (seconds). The rapidly declining broken line (hollow circles) on the left of the graph shows performance after bilateral lesions of the frontal association areas. The solid black line shows performance after a unilateral lesion, and the dashed line with crosses within circles shows performance prior to surgery. Although the baboon did well on 90-second delayed-response problems prior to bilateral prefrontal cortex lesions, performance fell to chance after surgery, even with delays as short as just a few seconds. (From Jacobsen, 1935; courtesy of *Archives of Neurology and Psychiatry*.)

could still use a stick to retrieve a piece of food just beyond their reach and could even use one stick to draw a slightly longer stick (up to five sticks) to get to the food (also see Jacobsen, 1931).

Jacobsen's findings differed dramatically on tasks that had a time element between the presentation of the stimulus and the opportunity to make a response. In particular, his chimpanzees did poorly on delayed-response problems. These tasks involved placing a piece of food under one of two cups in full view of an animal and then lowering a screen between the animal and the food cup. After several seconds to a few minutes, the screen was raised and the chimp had to select the cup that contained the previously placed morsel.

After bilateral removal of the frontal areas, the animals showed a continuing eagerness to work on the delay problems they had solved before surgery. Yet even with delays of just five seconds, the animals now performed poorly. This effect seemed specific to the frontal lobes, since no other cortical lesion affected delayed-response performance.

As can be seen in Figures 23.13 and 23.14, Jacobsen (1935) obtained essentially the same findings when he tested baboons with frontal association areas ablated. He also noted that monkeys with frontal lesions that were taught to alternate between two doors (remember the door previously opened and open the opposite one) showed impaired performance on this task, even after more than 1,000 trials (for details of these findings, see Crawford et al, 1948). These observations led Jacobsen to conclude:

After an injury to the prefrontal areas those activities of the organism which in their very nature demand integration over a period of time cannot be carried out effectively. Temporal patterning fails because the subject can no longer remember a single experience for even a few seconds in the face of new incoming sensory data. (1935, 565)

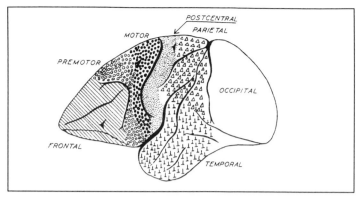

Figure 23.14. A diagram of the monkey brain showing the frontal areas (striped) associated with the deficit on the delayed-response task. (From Jacobsen, 1935; courtesy of *Archives of Neurology and Psychiatry*.)

To say the least, not everybody was ready to agree with Jacobsen's theory of a loss of "immediate" (short-term) memory. The criticisms came from both the clinical and experimental camps. For example, Richard Brickner (1934, 1936), whose human case study had been cited by Jacobsen as being consistent with his own work on primates, did not believe that frontal lobe patients show deficits in immediate memory per se. He argued that the apparent forgetfulness is due to loss of interest in the subject matter and to a decline in the ability to synthesize and organize new material.

Among the experimentalists, Karl Lashley also did not interpret Jacobsen's data as showing a loss of immediate memory. Lashley wrote:

> Loss of the delayed reaction after removal of the prefrontal lobes of the monkey has been interpreted as a loss of immediate memory. However, this task and others, which are affected by prefrontal injury, all involve a series of conflicting actions. Difficulty in maintaining a constant set or attitude is the real basis of the loss. (1950, 463)

Some of the specific experiments that followed Jacobsen's reports showed that the impairment on the delayed-response problem could diminish if the cues were made more salient, if distractions were limited, or if the excessive activity that sometimes follows frontal lobe damage could be controlled (e.g., Malmo, 1942). Other studies showed that the deficit was less specific than originally believed (Fuster, 1980). In general, competing behavioral tendencies had to be suppressed; the more complex the problem, the worse animals with frontal cortical lesions seemed to perform.

It was clear that the deficits observed after frontal cortical lesions could be due to problems with sustained attention, distractibility, hyperactivity, inhibition, or higher-order functions, such as analysis or synthesis. Realizing this, Jacobsen's confidence in his own theory of immediate memory waned. In 1948, Jacobsen and his coworkers reexamined their years of work on "Becky" and "Lucy." They now wrote that the enduring poor performance on the delayed-response problem, and the erratic and variable character of performance in general, were probably not due to a failure of recent memory per se, and that their earlier hypothesis had not withstood the test of time (Crawford et al, 1948).

PROGRESS?

By the middle of the twentieth century, the field of learning was characterized by a plethora of studies aimed at defining the conditions that can make learning successful. This period was also marked by a continued appreciation of various mnemonic devices for enhancing memory, such as the rhyme for learning the names of the cranial nerves that physiology and medical students memorize today.[17] Nevertheless, there was little of real substance in the literature about the locus or nature of those changes that, just about everyone agreed, must underlie learning.

The following passage almost certainly reflects the frustration many of these scientists must have felt in their efforts to learn more about the physical basis of memory:

> But for my part I wonder at memory in a still greater degree. For what is it that enables us to remember, what character has it, or what is its origin? . . . Do we think there is . . . a sort of roominess into which the things we remember can be poured as if into a kind of vessel? . . . Or do we think that . . . memory consists of the traces of things registered in the mind? What can be the traces of words, of actual objects, what further could be the enormous space adequate to the representation of such a mass of material? (Translated by King, 1927, 80)

This passage was not written by a late-nineteenth- or by a twentieth-century investigator. It is found in the *Tusculan Disputations,* written by Cicero in Ancient Rome. In spite of the best efforts of some of the brightest scientists, the nature and locus of the engram have remained as elusive and mysterious to twentieth-century investigators as they were to Cicero and other philosophers and naturalists who pondered the characteristics of the memory trace long ago.

---

[17]The original rhyme came from Oliver Wendell Holmes, a professor of anatomy at Harvard University who was also known for his poetry and humor (see Dana, 1928). Holmes taught:

On old Monadnock's treeless tops
A Finn and Frenchman
Picked some hops.

In this and in later versions of the rhyme, the first letter of each word stands for the first letter of each cranial nerve in correct numerical sequence. When Holmes wrote the rhyme, the nerves were known as (I) olfactory; (II) optic; (III) motor oculi; (IV) trochlearis; (V) trigeminus; (VI) abducens; (VII) facial; (VIII) auditory; (IX) pharyngeal; (X) pneumogastric; (XI) spinal accessory; and (XII) hypoglossal.

REFERENCES

Ariëns Kappers, C. U. 1917. Further contributions on neurobiotaxis. IX. An attempt to compare the phenomena of neurobiotaxis with other phenomena of taxis and tropism. *Journal of Comparative Neurology,* 27, 261–298.

Ariëns Kappers, C. U., Huber, G. C., and Crosby, E. C. 1936. *The Comparative Anatomy of the Nervous System of Vertebrates Including Man* (Vol. 1). New York: Hafner.

Aristotle. 1931. *De memoria et reminiscentia.* (J. I. Beare, Trans.). In W. D. Ross (Ed.), *The Works of Aristotle,* Book 3. Oxford, U.K.: Clarendon Press.

Augustine. 1955. *Confessions and Enchiridion.* (A. C. Outler, Ed./Trans.). Library of the Christian Classics, Vol. VII. Philadelphia, PA: Westminster Press.

Babkin, B. P. 1949. *Pavlov: A Biography.* Chicago: University of Chicago Press.

Bain, A. 1872. *Mind and Body: The Theories of Their Relation.* London: Henry S. King.

Beaumont, W. 1826. Further experiments on the case of Alexis San. Martin, who was wounded in the stomach by a load of duck-shot, detailed in the Recorder for Jan. 1825. *Medical Recorder, 9,* 94–97.

Beaumont, W. 1833. *Experiments and Observations on the Gastric Juice and the Physiology of Digestion.* Plattsburgh, NY: F. P. Allen.

Boerhaave, H. 1743. *Academical Lectures on the Theory of Physic, Being a Genuine Translation of His Institutes and Explanatory Comment* (Vol. 2). London: W. Innys.

Bonnet, C. 1755. *Essai de Psychologie. . . .* Londres.

Bonnet, C. 1760. *Essai Analytique sur les facultés de l'âme.* Copenhague: C. & A. Philibert.

Bonnet, C. 1764. *Contemplation de la Nature.* Amsterdam: Chez Marc-Michel Rey.

Brickner, R. M. 1934. An interpretation of frontal lobe function based upon the study of a case of partial bilateral frontal lobotomy. *Research Publications—Association for Research for Nervous and Mental Disease, 13,* 259–351.

Brickner, R. M. 1936. *The Intellectual Functions of the Frontal Lobes.* New York: Macmillan.

Buckingham, H. 1991. David Hartley, 1746: Various conjectures on the perception, motion, and generation of ideas. Papers presented at the Conference on the History of Brain Function, Ft. Myers, FL.

Burnham, W. H. 1888. Memory, historically and experimentally considered. I. An historical sketch of the older conceptions of memory. *Journal of Psychology, 2,* 39–90.

Burnham, W. H. 1889. Memory, historically and experimentally considered. II.-IV. *Journal of Psychology, 2,* 225–270, 431–464, 568–622.

Butler, S. 1880. *Unconscious Memory.* London: David Bogue.

Caruthers, M. J. 1990. *The Book of Memory.* Cambridge, U.K.: Cambridge University Press.

Cason, H. 1925. The physical basis of the conditioned response. *American Journal of Psychology, 36,* 371–393.

Child, C. M. 1921. *The Origin and Development of the Nervous System.* Chicago: University of Chicago Press.

Child, C. M. 1924. *Physiological Foundations of Behavior.* New York: Henry Holt.

Cicero, M. T. 1942–1948. *De oratore, Partitiones oratoriae.* (E. W. Sutton and H. Rackham, Eds./Trans.). Loeb Classical Library. London: Heinemann.

Cicero, M. T. 1927. *Tusculan Disputations.* (J. E. King, Trans.). Loeb Classical Library. London: Heinemann.

Cranefield, P. F. 1961. A seventeenth century view of mental deficiency and schizophrenia: Thomas Willis on "stupidity or foolishness." *Bulletin of the History of Medicine, 35,* 291–316.

Crawford, M. P., Fulton, J. F., Jacobsen, C. F., and Wolfe, J. B. 1948. Frontal lobe ablation in chimpanzee: A résume of "Becky" and "Lucy." *Research Publications—Association for Research in Nervous and Mental Disease, 27,* 3–58.

Culler, E., and Mettler, F. A. 1934. Conditioned behavior in a decorticate dog, *Journal of Comparative Psychology, 18,* 291–303.

Dana, C. L. 1928. Early neurology in the United States. *Journal of the American Medical Association, 90,* 1421–1424.

DeFelipe, J., and Jones, E. G. 1988. *Cajal on the Cerebral Cortex.* New York: Oxford University Press.

Descartes, R. 1650. *Les Passions de l'Ame.* Amsterdam: L. Elzevir.

Descartes, R. 1662. *De homine figuris et latinitate donatus a Florentio Schuyl.* Leyden: Franciscum Moyardum and Petrum Leffen.

Diamond, S. 1969. Seventeenth century French "connectionism": La Forge, Dilly, and Regis. *Journal of the History of the Behavioral Sciences, 5,* 3–9.

Duval, M. 1895a. Hypothèse sur la physiologie des centres nerveux; théorie histologique du sommeil. *Comptes Rendus Hebdomadaires des Séances et Mémoires de la Société de Biologie, 47,* 74–77.

Duval, M. 1895b. Remarques à propos de la communication de M. Lépine. *Comptes Rendus Hebdomadaires des Séances et Mémoires de la Société de Biologie, 47,* 86–87.

Ebbinghaus, H. 1885. *Über das Gedächtnis: Untersuchungen zur experimentellan Psychologie.* Leipzig: Dunker und Humbolt. Translated by H. A. Ruger and C. E. Bussenius as, *Memory, A Contribution to Experimental Psychology.* New York: Teachers College, Columbia University, 1913.

Ebbinghaus, H. 1908. *Abriss der Psychologie.* Leipzig: Dunker und Humbolt. Translated by M. Meyer as, *Psychology: An Elementary Textbook.* Boston, MA: Heath, 1908.

Ehrenberg, C. G. 1833. Nothwendigkeit einer feineren mechanischen Zerlegung des Gehirns und der Nerven vor der chemischen, dargestellt aus Beobachtungen von C. G. Ehrenberg. *Annalen der Physik und Chemie, 28,* 449–465, 471–473.

Fancher, R. E. 1990. *Pioneers of Psychology* (2nd ed.). New York: W. W. Norton.

Franz, S. I. 1902. On the functions of the cerebrum: I. The frontal lobes in relation to the production and retention of simple sensory-motor habits. *American Journal of Physiology, 8,* 1–22.

Franz, S. I. 1906. Observations on the functions of the association areas (cerebrum) in monkeys. *Journal of the American Medical Association, 47,* 1464–1467.

Franz, S. I. 1907. On the functions of the cerebrum: The frontal lobes. *Archives of Psychology, 2,* 1–64.

Franz, S. I. 1912. New phrenology. *Science, 35,* 321–328.

Franz, S. I. 1961. Shepherd Ivory Franz. In C. Murchison (Ed.), *A History of Psychology in Autobiography.* New York: Russell and Russell. Pp. 89–113.

Franz, S. I., and Lashley, K. S. 1917. The retention of habits by the rat after destruction of the frontal portion of the cerebrum. *Psychobiology, 1,* 3–18.

Fulton, J. F. 1959. Historical reflections on the backgrounds of neurophysiology. In C. McC. Brooks and P. F. Cranefield (Eds.), *The Historical Development of Physiological Thought.* New York: Hafner. Pp. 67–79.

Fuster, J. M. 1980. *The Prefrontal Cortex.* New York: Raven Press.

Goldstein, K. 1942. *Aftereffects of Brain Injuries in War: Their Evaluation and Treatment.* London: Heinemann.

Gomulicki, B. R. 1953. *The Development and Present Status of the Trace Theory of Memory.* Cambridge, U.K.: Cambridge University Press.

Hartley, D. 1834. *Observations on Man, His Fame, His Duty, His Expectations* (6th ed.). London: Thomas Tegg. (First published 1749)

Hartley, D. 1846. *Various Conjectures on the Perception, Motion, and Generation of Ideas.* Translated from the Latin by R. E. A. Palmer with notes by M. Kallick. (Reprinted 1959; Los Angeles: Augustan Reprint Society)

Hering, E. 1870. Über das Gedächtniss als eine allgemeine Function der organisirten Materie. *Almanach der kaiserlicheten Akademie der Wissenschaften, Wien, 20,* 253–278.

Herrnstein, R. J., and Boring, E. G. 1965. *A Source Book in the History of Psychology.* Cambridge, MA: Harvard University Press.

Hilgard, E. R., and Marquis, D. G. 1961. *Conditioning and Learning.* (G. Kimble, Ed.). New York: Appleton-Century-Crofts.

Hobbes, T. 1651. *Leviathan, or The Matter, Forme and Power of a Commonwealth Ecclesiasticall and Civill.* London: Printed for Andrew Crooke at the Green Dragon in St. Paul's Churchyard.

Holt, E. B. 1931. *Animal Drives and the Learning Process.* New York: Henry Holt.

Hunt, R. K., and Cowan, W. M. 1990. The chemoaffinity hypothesis: An appreciation of Roger W. Sperry's contributions to developmental biology. In C. Trevarthen (Ed.), *Brain Circuits and Functions of Mind.* Cambridge, U.K.: Cambridge University Press. Pp. 19–74.

Jackson, J. H. 1873. On the anatomical and physiological localisation of movements in the brain. *Lancet, 1,* 84–85, 162–164.

Jacobsen, C. F. 1931. A study of cerebral function in learning. The frontal lobes. *Journal of Comparative Neurology, 52,* 271–340.

Jacobsen, C. F. 1935. Functions of frontal association area in primates. *Archives of Neurology and Psychiatry, 33,* 558–569.

Jacobsen, C. F., Wolf, J. B., and Jackson, T. A. 1935. An experimental analysis of the functions of the frontal association areas in primates. *Journal of Nervous and Mental Disease, 82,* 1–14.

James, W. 1890. *Principles of Psychology* (Vol. 1). New York: Henry Holt.

Johnson, H. M. 1927. A simpler principle of explanation of imaginative and ideational behavior, and of learning. *Journal of Comparative Psychology, 7,* 187–235.

Koffka, K. 1935. *Principles of Gestalt Psychology.* New York: Harcourt, Brace.

Köhler, W. 1929. *Gestalt Psychology.* London: Bell and Sons.

Kölliker, A. 1896. *Handbuch der Gewebelehre des Menschen* (6th ed.). Leipzig: Engelmann.

Korsakoff, S. S. 1887. Disturbance of psychic functions in alcoholic paralysis and its relation to the disturbance of the psychic sphere in multiple neuritis of non-alcoholic origin. *Vestnik Psichiatrii, 4,* Fasc. 2.

La Forge, L. de 1666. *Traité de l'Esprit de l'Homme de ses Facultés et Fonctions.* Paris.

La Mettrie, J. O. de. 1748. *L'Homme Machine.* Leyde: E. Luzac, fils.

Lashley, K. S. 1929a. *Brain Mechanisms and Intelligence.* Chicago: University of Chicago Press. (Reprinted 1963; New York: Dover Publications).

Lashley, K. S. 1929b. Learning I: Nervous mechanisms in learning. In C. Murchison (Ed.), *The Foundations of Experimental Psychology.* Worcester, MA: Clark University Press.

Lashley, K. S. 1938. Factors limiting recovery after central nervous system lesions. *Journal of Nervous and Mental Disease, 88,* 733–755.

Lashley, K. S. 1950. In search of the engram. *Symposia of the Society for Experimental Biology, 4,* 454–482.

Lashley, K. S., and Franz, S. I. 1917. The effects of cerebral destruction upon habit-formation and retention in the albino rat. *Psychobiology, 1,* 71–139.

Laycock, T. 1845. On the reflex function of the brain. *British-Foreign Medical Review, 19,* 298–311.

Lépine, R. 1894. Cas d'hystérie à forme particulière. *Revue de Médecine, 14,* 727–728.

Lépine, R. 1895. Théorie mécanique de la paralysie hystérique, du somnambulisme, du sommeil naturel et de la distraction. *Comptes Rendus Hebdomadaires des Séances et Mémoires de la Société de Biologie, 47,* 85–86.

Locke, J. 1690. *An Essay Concerning Human Understanding.* London: T. Basset.

Loeb, J. 1902. *Comparative Physiology of the Brain and Comparative Psychology.* New York: G. P. Putnam's Sons.

Luys, J. 1865. *Recherches sur le Système Nerveux Cérébrospinal: Sa Structure, ses Fonctions et ses Maladies.* Paris: J. B. Ballière.

Malebranche, N. de. 1674–1678. *De la Recherche de la Vérité, où l'on Traite de la Nature de l'Espirit de l'Homme et de l'Usage qu'il en Doit Faire pour Eviter l'Erreur dans les Sciences.* Paris: A. Pralard.

Malmo, R. B. 1942. Interference factors in delayed response in monkeys after removal of frontal lobes. *Journal of Neurophysiology, 5,* 295–308.

McDougall, W. 1905. *Physiological Psychology.* London: Dent.

Myer, J. S. 1912. *Life and Letters of Dr. William Beaumont.* St. Louis, MO: C. V. Mosby.

Neuburger, M. 1981. *The Historical Development of Experimental Brain and Spinal Cord Physiology before Flourens.* Translated and edited with additional material by E. Clarke. Baltimore, MD: Johns Hopkins University Press.

Newton, I. 1687. *Philosophiae naturalis principia mathematica.* Londini: Jussi Societatis Regiae ac typis Josephi Streater.

Newton, I. 1704. *Opticks.* London: Smith and Walford. (Reprinted 1952: New York: Dover Publications)

Oberg, B. B. 1976. David Hartley and the association of ideas. *Journal of the History of Ideas, 37,* 441–454.

Pagel, W. 1958. Medieval and Renaissance contributions to knowledge of the brain and its functions. In M. W. Perrin (Chairman), *The Brain and Its Functions.* Wellcome Foundation. Oxford, U.K.: Blackwell. Pp. 95–114.

Papez, J. W. 1931. Book review of K. S. Lashley's *Brain Mechanisms and Intelligence. American Journal of Psychology, 43,* 527–529.

Patrick, G. T. W. 1888. A further study of Heraclitus. *American Journal of Psychology, 1,* 557–690.

Pavlov, I. P. 1904. Physiology of digestion. In *Nobel Lectures Physiology or Medicine 1901–1921.* New York. Elsevier, 1967. Pp. 141–155.

Pavlov, I. P. 1927. *Conditioned Reflexes. An Investigation of the Physiological Activity of the Cerebral Cortex.* New York: Oxford University Press.

Petrarca, F. 1945. *Rerum memorandarum libri:* (G. Billanovich, Ed.). Florence: Sansoni.

Plato. 1921. *Theaetetus and Sophist.* (H. N. Fowler, Ed./Trans.). Loeb Classical Library. London: Heinemann.

Priestley, J. 1775. *Hartley's Theory of the Human Mind on the Principle of the Association of Ideas; with Essays Relating to the Subject of It.* London: Johnson.

Quintilian, M. 1921. *The Institutio Oratoria of Quintilian.* (H. E. Butler, Trans.). Loeb Classical Library. London: Heinemann.

Rabl-Rückhard, H. 1890. Sind die Ganglienzellen amöboid? Eine Hypothese zur Mechanik psychischer Vorgägne. *Neurologisches Centralblatt, 9,* 199.

Ramón y Cajal, S. 1890. A quelle époque apparaissent les expansions des cellules nerveuses de la moëlle épinière du poulet? *Anatomischer Anzeiger, 5,* 631–639.

Ramón y Cajal, S. 1893. La rétine des vertébrés. *La Cellule, 9,* 119–255.

Ramón y Cajal, S. 1895. *Algunas conjecturas sobre el mechanismo anatómico de la ideación, asociación y atención. Revista de Medicina y Cirugia Practica.* Madrid: Moya. Pp. 1–14. Also published in German as, "Einige Hypothesen über den anatomischen Mechanismus der Ideenbildung, der Association und der Aufmerksamkeit." *Archiv für Anatomie und Physiologie,* 367–378.

Régis, P. S. 1690. *Système de la Philosophie, Contenant la Logique, Métaphysique, & Morale.* Lyon: Imprimé de D. Thierry, aux dépens d'Anisson, Poseul et Rigaud.

Ribot, T. 1881. *Les Maladies de la Mémoire.* Paris: J. B. Ballière. Translated by J. Fitzgerald as, *The Diseases of Memory.* New York: *Humboldt Library of Popular Science Literature, 46,* 453–500, 1883.

Schacter, D. L. 1982. *Stranger behind the Engram.* Hillsdale, NJ: Lawrence Erlbaum.

Schacter, D. L., Eich, J. E., and Tulving, E. 1978. Richard Semon's theory of memory. *Journal of Verbal Learning and Verbal Behavior, 17,* 721–743.

Sechenov, I. M. 1863. *Reflexes of the Brain.* Translated from the Russian by S. Belsky, edited by G. Gibbons, with notes by S. Gellerstein. Cambridge, MA: M.I.T. Press, 1965.

Semon, R. W. 1904. *Die Mneme als erhaltendes Prinzip im Wechsel des organischen Geschehens.* Leipzig: Engelmann.

Semon, R. W. 1909. *Die mnemischen Empfindungen.* Leipzig: Engelmann.

Shepherd, G. 1991. *Foundations of the Neuron Doctrine.* New York: Oxford University Press.

Smith, C. U. M. 1987. David Hartley's Newtonian neuropsychology. *Journal of the History of Behavioral Sciences, 23,* 123–136.

Sudhoff, W. 1914. Die Lehre von den Hirnventrikeln im textlicher und graphischer Tradition des Altertums und Mittelalters. *Archiv für Geschichte der Medizin, 7,* 149–205.

Tanzi, E. 1893. I fatti e la induzione nell'odierna istologia del sistema nervoso. *Revista Sperimentale di Freniatria e Medicina Legale delle Alienazioni Mentali, 19,* 149.

Telfer, W. 1955. *Cyril of Jerusalem and Nemesius of Emesa.* Philadelphia, PA: Westminster Press.

Thorndike, E. L. 1898. Animal intelligence: An experimental study of the associative process in animals. *Psychological Monographs,* No. 2.

Vesalius, A. 1543. *De humani corporis fabrica.* Basel: J. Oporinus.

Weismann, A. 1906. Semon's "Mneme" und die Vererbung erworbener Eigenschaften. *Archiv für Rassen- und Gesellschafts Biologie, 38,* 1–27.

Willis, T. 1664. *Cerebri anatome: cui accessit nervorum descriptio et usus.* London: J. Martyn and J. Allestry. Tercentenary ed., 1664–1964, *Thomas Willis, The Anatomy of the Brain and Nerves.* Montreal: McGill University Press, 1965.

Willis, T. 1672. *De anima brutorum.* Oxonii: R. Davis.

Willis, T. 1683. *Two Discourses Concerning the Soul of Brutes. . . .* (S. Pordage, Trans.). London: Dring, Harper and Leigh.

Yates, F. 1966. *The Art of Memory.* London: Routledge and Kegan Paul.

Ziehen, T. 1891. *Leitfaden der physiologischen Psychologie.* Jena: Fischer.

Chapter 24
The Neuropathology of Memory

As to memory . . . no other human faculty is equally fragile: injuries from, and even apprehensions of, diseases and accident may affect in some cases a single field of memory and in others the whole.

Pliny the Elder, first century

The pathological side of memory was occasionally mentioned by the Greeks and Romans, who seemed especially impressed with the vulnerability of memory. They observed that only some aspects of memory were affected after certain injuries. For example, in the first century Pliny the Elder said that he knew of a man who had forgotten only the names of the letters after he received a blow from a stone; another who, after falling from a roof, could not even remember his mother, his neighbors, or his friends; and yet another whose disease left him unable to remember his servants.

The history of speech and language disorders, which the ancients considered memory disorders ("loss of memory for words"), is examined in Chapter 25. The present chapter deals with the broader losses of memory associated with disease and injury, as exemplified by Pliny's patients who could no longer remember their friends and acquaintances.

Unlike Chapter 23, which gives a chronological history of the development of the memory trace from Ancient Greece through the Renaissance and into relatively modern times, this chapter examines some neurological disorders associated with memory impairments. It begins with two dementias, Alzheimer's disease and Creutzfeldt-Jakob disease, and then turns to Korsakoff's psychosis, a more treatable disorder characterized by memory problems. Diseases and injuries affecting the temporal lobes and hippocampus are discussed in the final sections of this chapter.

EARLY DESCRIPTIONS OF THE DEMENTIAS

The Greeks and Romans associated dementia with the aging process, but did not feel that dementia necessarily had to be a part of aging. For example, Cicero wrote an essay (De senectute) in the first century B.C. in which he stated:

As wantonness and licentiousness are faults of the young rather than of the old, yet not of all young men but only the depraved, so the senile folly called dotage is characteristic not of all old men but only of the frivolous. (Translated in Tibbitts, 1957, 126)

Galen, who practiced medicine in Rome two centuries later, used the word "morosis" rather than "dotage" in a way that suggested dementia. In De symptomatum differ-

entiis liber, he spoke of "morosis" as a characteristic of those

in whom the knowledge of letters and other arts are totally obliterated; indeed they can't even remember their own names. . . . Even now it is seen, that on account of extreme debility in old age, some are afflicted with similar symptoms. (1821–1833 translation, 200–201)

In "On Anatomical Procedures," Galen noted that "in very old animals the brain is much too small to fill the cavity of the skull," thus giving one of the earliest descriptions of senile atrophy of the brain.

After the Renaissance, descriptions of dementia stressed not just the loss of memory but the fact that demented people acted childish. In 1599, André Du Laurens (1558–1609), a French court physician, wrote about the body and mind becoming feeble and "the memorie lost and the judgement failing so that they become as they were in their infancie" (1938 translation, pp. 174–175).

The idea that dementia may take one into a "second childhood" was entertained in a number of plays by William Shakespeare in this same period.[1] In *The Winter's Tale* (c. 1611) Polixenes states:

Is not your father grown incapable
Of reasonable affairs? is he not stupid
With age and alt'ring rheums? can he speak? hear?
Know man from man? dispute his own estate?
Lies he not bed-rid? and again does nothing
But what he did being childish?
(Act IV, scene III)

In *As You Like It* (c. 1599), one find's Shakespeare's famous portrayal of the "seven ages" as seen through the eyes of Jacques. He begins with the famous lines:

All the world's a stage,
And all the men and women are merely players.
And one man in his time plays many parts,
His acts being seven ages.

---

[1]Shakespeare displayed considerable knowledge of the medicine of his day. He may have been helped by his son-in-law, who was a physician (Edgar, 1934).

After having described life from infancy through the sixth stage, in which man is depicted wearing spectacles and having a whistling voice and a changed physique, Jacques went on:

> Last scene of all,
> What ends this strange eventful history,
> Is second childishness and mere oblivion,
> Sans teeth, sans eyes, sans taste, sans every thing.
> (Act II, scene VII)

Nowhere in Shakespeare, however, is there a better overall description of senile dementia and a return to a second childhood than in *King Lear* (c. 1605). Lear, the aging monarch, exhibits disorientation for time and place, poor judgment, amnesia, and an inability to remember recent events, such as how he came to wear the clothes he had on. Recognizing her father's demented state, Goneril lamented:

> Idle old man,
> That still would manage those authorities
> That he hath given away!—Now, by my life,
> Old fools are babes again . . .
> (Act I, scene II)

Throughout the Renaissance, the cause of dementia was thought to be a cooling of the brain. This idea, which was developed and became widely accepted in Ancient Greece and Rome, was even maintained late in the seventeenth century by Théophile Bonet (see Boneti, 1700).

The term "dementia" *(démence)* was itself coined by Philippe Pinel in 1797, in order to describe some patients at the Bicêtre (Torack, 1983). Pinel was surprised to find that autopsies on the insane rarely revealed any pathology to the naked eye or even under the microscope. In fact, he finally concluded that insanity probably has its origin in the stomach, affecting the brain only by "sympathy" (see Chapter 20 for a discussion of the concept of sympathy).

Pinel's most outstanding student, Jean Esquirol (1782–1840; shown in Figure 24.1), divided the dementias into three types: acute, chronic, and senile. Esquirol (1838) provided an especially good description of the senile variety. He wrote that senile dementia is a slow, progressive disease that begins with "enfeeblement of memory, particularly the memory of recent impressions." He saw that sensations and movements were also affected and that attention deteriorated and eventually became impossible. The "will" itself was described as "uncertain and without direction."

One year earlier, James Prichard (1786–1848) had defined four landmarks in the relentless progression of the dementias. These were (1) loss of recent memory, (2) loss of reason, (3) loss of comprehension, and (4) loss of "instinctive action."

Surprisingly, few physicians before the second half of the nineteenth century discussed brain atrophy in cases of dementia. One notable exception was Matthew Baillie (1761–1823; shown in Figure 24.2), an English pathologist. In 1795, Baillie wrote that the "ventricles are sometimes found enlarged and full of water."

By the middle of the nineteenth century, Samuel Wilks (1824–1911) and Wilhelm Griesinger, who appear in Figures 24.3 and 24.4, clearly recognized the dementias as brain diseases. In 1864, Wilks discussed brain atrophy in a number of dementing conditions, including senile dementia. He described the enlarged ventricles and "the wasted convolutions" and added that these signs of atrophy are always associated with severe behavioral disability.

> On the discovery of such a brain, I have always been well assured that the possessor of it has been enfeebled both in mind and body. If the case has been treated as mental, it

Figure 24.1. Jean Esquirol (1782–1840), who divided the dementias into acute, chronic, and senile varieties in 1838. He provided an especially good description of senile dementia.

Figure 24.2. In 1795, Matthew Baillie (1761–1823) discussed brain atrophy and ventricular enlargement in cases of dementia.

Figure 24.3. Samuel Wilks (1824–1911) studied brain atrophy and noted that it was always associated with marked behavioral deterioration.

Figure 24.4. Wilhelm Griesinger (1817–1868), a psychiatrist, viewed "mental" diseases as brain diseases. Greisinger looked for gross brain atrophy to distinguish the dementias from other diseases of the brain.

has been found to be one of dementia or general paralysis, where both physical and mental powers have long failed. . . . Should it be the body of an old man where such wasted brain is found, I know that he has long been decrepit; that he has tottered in his walk and that he has been sinking mentally into the stage of second childhood. (1864, 385)

As for Griesinger, he was a psychiatrist who traveled to Paris to study with François Magendie. In the first edition of his *Die Pathologie und Therapie der psychischen Krankheiten,* which appeared in 1845, he made it clear that he believed mental diseases are brain diseases. A subsequent edition of this book was translated into English in 1882 as *Mental Pathology and Therapeutics.* In it, Griesinger described how events of the present are quickly forgotten by the demented individual. He then said that whereas some patients may have reasonable memory for the past, others may have problems remembering the past. Still others, Griesinger pointed out, may show the past completely erased from the "tablets of the memory." Griesinger used gross brain atrophy to distinguish dementia from syphilis and arterial diseases of the brain.

Another individual interested in the dementias was Théodule Ribot, whose most important work, *Les Maladies de la Mémoire,* appeared in 1881 and was translated into English as *The Diseases of Memory* two years later. At this time, the dementias were still classified as senile, paralytic, epileptic, and so on. Ribot astutely recognized that the amnesias accompanying the dementias follow a typical path. They begin with problems remembering recent events, then encompass ideas and "intellectual acquisitions" overall, eventually extending to the older memories. After this, feelings and affections fail to be recalled and, finally, when the dementia is even more severe, the individual is unable to remember even the automatic daily acts and routines to which he or she "has long been addicted." Ribot theorized

that the cortex played a role in the higher forms of memory and that "the inferior centers" of the brain had more to do with the last memories to succumb in the dementias.

The pioneering findings of Wilks, Griesinger, and Ribot were generally accepted, but did not trigger a wealth of further investigative work. The study of the dementias had to wait another 40 years to become a more fashionable endeavor, one capable of attracting the brightest scientific minds. This shift was stimulated by the development of better histological techniques and by Alzheimer's publications on an unusually early type of dementia.

ALZHEIMER'S DISEASE

In 1906, Alois Alzheimer, sometimes depicted as a scientist with a cigar in one hand and a microscope in the other, presented a case report on one of his female patients. The report was given at a meeting of the South-West German Society of Alienists.[2] This meeting took place in Tübingen, the city in which both Griesinger and Alzheimer began their medical studies.

Alzheimer's female patient had been institutionalized with symptoms of dementia when she was still middle-aged and was then observed over the course of her progressive illness. Her first symptoms included changes in personality with strong feelings of jealousy toward her husband. Soon after, she showed a memory impairment that increased to the point at which she could no longer make her way around her own house. She was institutionalized after she became increasingly paranoid about people trying to kill her.

Alzheimer, shown in Figure 24.5, described how her

[2]This was an organization whose members believed in the physical basis of "mental" illness.

Figure 24.5. Alois Alzheimer (1864–1915), who devoted a lifetime to studying the symptoms and anatomical correlates of diseases of the nervous system. His first description of the dementia now bearing his name appeared in 1906.

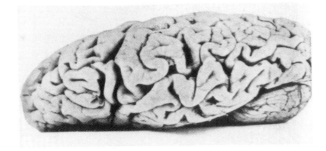

CASE 1.—Showing some atrophy of the gyri and widening of the sulci in the region of the Sylvian fissure.

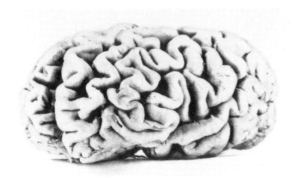

CASE 2.—Showing atrophy of gyri and widening of the sulci.

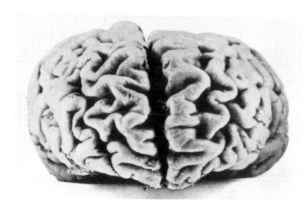

Figure 24.6. Brains from two Alzheimer's disease patients showing excessive cortical atrophy leading to very wide sulci and fissures. (From Henderson and Maclachlan, 1930.)

condition continued to deteriorate and how she became increasingly confused, disoriented, and delirious. He noted that after she was shown some objects, which she could identify, she immediately forgot about them, much as if the exercise had never taken place. Four and one-half years after her first clinical signs appeared, she became incontinent, apathetic, and bedridden. She died in a fetal position at the relatively young age of 51.

Behaviorally, Alzheimer's first case was fairly typical of the disorder that would soon become attached to his name. In the beginning, these patients typically show mild memory lapses and attentional problems. Personal affairs and interest in the surroundings also begin to show slight deterioration at this time. These symptoms increase as the disease progresses, with memory problems being more severe for recent events than for events of the distant past, just as Ribot had described for the dementias in general. Disorientation and hesitancy in speech continue to grow in severity, and the loss of memory steadily increases until there is an inability to remember anything said even a few minutes earlier. Patients in this advanced stage of dementia often wander off, hopelessly disoriented as to person, place, and time. Communication and ambulation also fail as the disease progresses. Eventually the weakened victim dies from pneumonia or a related infection.

The autopsy conducted on Alzheimer's first "presenile dementia" patient revealed gross atrophy of the cerebrum, as opposed to the localized foci of degeneration, which would be more characteristic of syphilis or a vascular disease. (Figure 24.6 shows the atrophied cerebral hemispheres of typical Alzheimer's cases.)

The application of the silver stain developed in 1902 by Max Bielschowsky (1869–1940; shown in Figure 24.7), a pupil of Franz Nissl (1860–1919), revealed that many (25–33 percent) of the cortical neurons were reduced to dense, intricately tangled, thick bundles of neurofibrils. These neurofibrillary tangles looked like skeletons of degenerated cells whose nuclei and cytoplasm were no longer present.

In addition to the neurofibrillary tangles, sclerotic plaques were scattered over the entire cortex, especially in the upper layers. Alzheimer noted that the plaques could even be recognized without staining. He thought that they were due to

deposits of chemical substances *(Aufbau Produkte)* resulting from pathological changes in the neurons.

Alzheimer's report was only noted in the *Neurologisches Centralblatt* in 1906, but a full accounting of the case appeared in two other German journals in 1907. Alzheimer concluded one of these papers by stating he believed he had described a little-known disease process, one that should not be forced into an existing group of disease patterns. He was convinced that his new early form of dementia had unique clinical and anatomical characteristics.

In contrast to the neurofibrillary tangles, the plaques described by Alzheimer had been seen before. In 1892, Georges Marinesco (shown in Figure 24.8) and Paul Blocq saw plaques in the brains of two demented elderly patients. The plaques were further described in 1898 by Emil Redlich (1866–1930), who saw them in the atrophied cortices of two senile individuals with gross memory defects, mental confusion, aphasia, and apraxia. Redlich

Figure 24.7. Max Bielschowsky (1869–1940), a pupil of Franz Nissl and the developer of a stain that allowed investigators to detect neurofibrillary tangles.

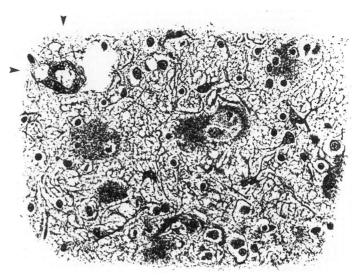

Figure 24.9. The cortex of a senile person showing extensive plaque formation. One seemingly hollow plaque in the upper left corner is identified by two border arrows. This picture was published by Emil Redlich in 1898.

called this "miliary sclerosis" and attributed it to glial proliferation after the ganglion cells degenerated. One of his photographs appears as Figure 24.9.

In 1904, Alzheimer himself noted plaques in patients with senile dementia. These plaques, also called glial rosettes, were observed by several other investigators in demented senile patients at about this time (e.g., Mijake, 1906). But prior to Alzheimer's 1906 case study, they had been found only in aged, typically senile patients and not in association with the neurofibrillary tangles in middle-aged demented patients.

Figure 24.8. Georges Marinesco (1863–1938), who with Paul Blocq modified the staining techniques for glia (previously developed by Santiago Ramón y Cajal) and observed plaques in the brains of two senile patients. The findings of Blocq and Marinesco were reported in 1892.

The plaques and tangles received more attention after this. In 1907, Oskar Fischer (1885–?) observed diffuse plaques in 12 of 16 senile brains. Fischer noted that normal brains and specimens from individuals with paralyses or psychotic disturbances did not show these plaques. Further, Fischer made the first strong case for the plaques correlating with the severity of the memory loss and the senility. His four aged patients who did not have plaques showed a simple form of "senile dementia," characterized by an overall diminution of mental capacity. In contrast, those with plaques exhibited the more bizarre, highly exaggerated symptoms of "presbyophrenia" ("old-mindedness").

Fischer (1907, 1910), who also found neurofibrillary tangles in many of his dementia cases, argued that the plaques and tangles were largely neural. He thought they represented the outcome of necrotic processes, but entertained the idea that there might be an unidentified foreign substance in the central plaque. These ideas received a mixed response (Bonfiglio, 1908).

Shortly after Fischer's 1907 paper appeared, detailed histological descriptions were published of the brains of four individuals who had Alzheimer's "atypical" or "presenile" dementia (Perusini, 1910).[3] The individuals, aged 40 to 65 years, exhibited progressive intellectual deterioration, and their brains contained both neurofibrillary tangles and plaques. It was suggested that the plaques and tangles, which occurred side by side, had to be related. The basic theory was that the odd-shaped plaques resulted from degenerating neural elements and were made of an inner ring of variable chemical composition encapsulated by an outer ring of reactive glial "felt." Later studies confirmed that the plaques associated with Alzheimer's disease contain degenerating neural processes, reactive cells (microglia, phagocytes), and a central core of amyloid. The latter was first noted by Bielschowsky in 1911 (see Wisniewski, 1983, 1989).

Many studies on the symptomatology and pathology of Alzheimer's disease followed these reports. The most noticeable anatomical changes were in the association

[3]One of the brains was, in fact, Alzheimer's first case.

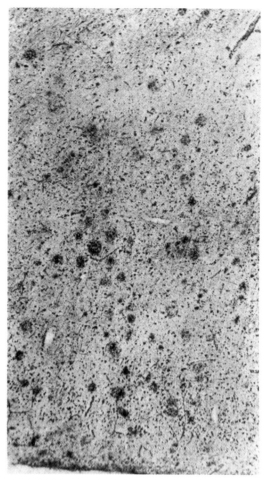

Figure 24.10. Photomicrograph from an Alzheimer's patient showing a rich deposit of plaques in the frontal association cortex. (From Fuller, 1912.)

areas of the cortex, but atrophy, plaques, and tangles were also found in other cortical areas and in the hippocampus, amygdala, basal ganglia, and nucleus basalis of Meynert (Perusini, 1911). For example, in 1911 and 1912, Solomon Fuller (1872–?) presented a series of unusually fine photographs of stained brain sections of

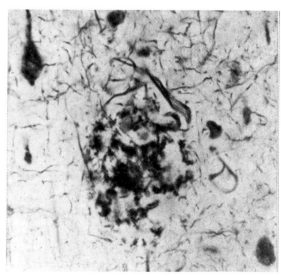

Figure 24.11. A typical plaque from the brain of an Alzheimer's disease patient. (From Fuller, 1912.)

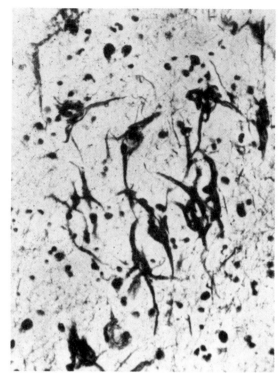

Figure 24.12. Neurofibrillary tangles in the hippocampal gyrus of an Alzheimer patient. (From Fuller, 1912.)

Alzheimer patients. They showed extensive degeneration and many plaques and tangles in the hippocampus. Three of Fuller's photographs appear as Figures 24.10, 24.11, and 24.12.

In other reports, there were statements accompanying the photographs, such as:

> In the hippocampal region the greater proportion of nerve-cells were affected by this degeneration. . . . The cornu ammonis showed typical and abundant Alzheimer's cells—the most prominent form being that of thickened agglutinated fibrils, like bundles of wire. Senile plaques were very numerous in this area. (Henderson and Maclachlan, 1930, 651)

In 1910, the dementia described by Alzheimer was given the name "Alzheimer's disease" by his friend Emil Kraepelin (1856–1926; shown in Figure 24.13). In 1902, Kraepelin called Alzheimer to Heidelberg to help him study the anatomical correlates of psychiatric diseases. Kraepelin was so impressed with Alzheimer that when he and Franz Nissl moved to Munich a year later, he asked Alzheimer to join them and to head the pathology department of the new psychiatric institute.[4]

Kraepelin, the "Linnaeus of psychiatry," devoted most of his time to the classification of mental diseases (Braceland, 1957). He specifically described "Alzheimer's disease" in the eighth edition of his *Psychiatrie.* Nevertheless, he looked upon the disorder as very closely related to the senile dementias. In fact, Kraepelin (1910) seemed only barely convinced by Alzheimer that the early-onset dementia was, in fact, a distinct disease entity.

[4]This institute was later called the Max Planck Institute. Alzheimer remained in Munich until 1912. He then left to assume the chair at the University of Breslau. He died of rheumatic fever just three years later (Torack, 1978).

Figure 24.13. Emil Kraepelin (1856–1926), the psychiatrist best known for his categorization and classification of mental diseases.

The clinical significance of Alzheimer's disease is at present still unclear. While the anatomic findings suggest that this condition deals with an especially severe form of senile dementia, some circumstances speak to a certain extent against this, namely the fact that the disease may arise even at the end of the 5th decade. One would describe such cases at least in terms of *Senium precox,* if not more preferably that this disease is more or less independent of age. (Translated in Torack, 1978, 11)

Although Kraepelin expressed his doubts about Alzheimer's disorder of middle age as a distinct disease entity, his authoritative work nevertheless had the effect of establishing Alzheimer's "presenile dementia" as a unique disorder. Yet by the end of 1911, 5 of the 13 generally agreed-upon Alzheimer's cases were outside the prescribed limit of 40 to 60 years of age for "presenile" dementia; only 3 showed the triad of apraxia, aphasia, and agnosia, the three symptoms investigators thought might distinguish Alzheimer's disease from senile dementia; and not every case showed both plaques and tangles.[5] Further, while all 13 patients exhibited severe memory disorders and dementia, comparable behavioral changes, and plaques and tangles, were also observed in varying degrees in cases of late-onset dementia (e.g., Barrett, 1911; Betts, 1911; Fuller, 1911).

By the end of 1912, more than 45 articles had appeared describing plaques and tangles in various forms of dementia. The mushrooming literature included descriptions of the brains of over 500 individuals. Slowly, positions shifted away from Alzheimer's idea of a presenile dementia qualitatively distinct from senile dementia (Torack, 1978, 1983). These investigations led to broader definitions of Alzheimer's dis-

ease and a corresponding increase in the number of patients classified as having the disorder.

Nevertheless, whether the more broadly defined Alzheimer's disorder is a single disease entity with only one cause has remained unclear.[6] Should different forms of Alzheimer's disease eventually be recognized, one might again see more restricted definitions of this disorder, definitions like those that characterized thinking when Alzheimer first began his dementia studies.

CREUTZFELDT-JAKOB DISEASE

Alfons Jakob (1884–1931) was a pathologist interested in multiple sclerosis and movement disorders. He had worked with Nissl and Alzheimer in Munich before he moved to Hamburg in 1911 (Jacob, 1970). Jakob, who returned to Munich after service in World War I, described a new type of dementia at the Leipzig meetings of the German Neurological Society in 1920 and in three publications one year later. His three patients seemed to show a clinical picture mildly suggestive of multiple sclerosis, but with a deranged mental state.

Jakob's first case involved a 52-year-old housewife whose initial signs of the disease were ataxia and exaggerated tendon reflexes. Her posture then became stiff. Abnormal jerky movements appeared, and she soon found herself unable to walk. In addition to these motor signs, this woman exhibited a growing apathy early in the disease. Her possible depression seemed to progress into fright and delirious confusion and then into disorientation and severe dementia. The woman's illness lasted one year, after which she died of pneumonia. The autopsy revealed diffuse neuronal degeneration and glial cell (astrocyte) proliferation, especially in the cortex and the basal ganglia.

A second patient observed by Jakob, a 34-year-old female factory worker, also showed progressive mental disturbances, including problems with memory, loss of emotional control, disorientation, stupor, and confusion. Her speech became slow and monotonous, her extremities rigid, and her gait spastic. As with the first patient, involuntary jerky movements were noted. Her illness also progressed rapidly, and she died six months after her clinical signs first appeared. The autopsy revealed the same changes in the cerebral cortex and basal ganglia that had been seen in the first patient.

Jakob's third patient was a 42-year-old salesman who early in the disease exhibited memory problems, progressive weakness in his legs, and handwriting and speech deterioration. His symptoms rapidly progressed, and he soon showed severe memory loss, disorientation, twitching movements, ataxia, and aberrant emotional behaviors. He died nine months after his first signs of illness. As with the other cases, an autopsy on this man's brain revealed extensive cortical and basal ganglia degeneration.

Jakob (1923) found a fourth case of this type described in his notes from 1912, and then saw a 38-year-old woman with the disorder in 1922. In searching the literature, he

---

[5]Schnitzler presented a questionable 34-year-old "Alzheimer's" case with tangles, but not plaques, in 1911. See Torack (1978) for a table of the key features shown by the 13 cases classified as Alzheimer's disease between 1907 and 1911.

[6]In newer studies, it was noted that almost all Down's syndrome patients who live into middle age develop features of Alzheimer's disease. In part because Down's syndrome involves a trisomy of chromosome 21, it was theorized that the gene responsible for at least a genetic form of Alzheimer's disease may be on chromosome 21, the same chromosome that also seems to control the production of amyloid (Goldgaber et al, 1987; St. George-Hyslop et al, 1987; Tanzi et al, 1987).

also noted that another German, Hans-Gerhard Creutzfeldt (1885–1964), had described a seemingly related case in 1920 and 1921. Creutzfeldt's patient showed symptoms comparable to those found by Jakob, and the histopathological picture looked fairly similar. His case, however, was not nearly as "clean" as the cases Jakob reported. There was a family history of feeble-mindedness and institutionalization, and the woman, who died at age 22, exhibited an onset of clumsiness five years before the "pseudosclerosis" manifested itself in its more typical way (see Kirschbaum, 1968).

Jakob felt that these six patients displayed a syndrome distinct from all other known clinical entities—a syndrome of middle-age characterized largely by cortical, pyramidal, and extrapyramidal degeneration, and one with mental and physical correlates. He classified this new disorder as "spastic pseudosclerosis with disseminated encephalo-myelopathy." Nevertheless, it was soon called Creutzfeldt-Jakob disease, Jakob-Creutzfeldt disease, or Jakob's syndrome (Spielmeyer, 1922). Whereas there was some disagreement about whether Jakob's or Creutzfeldt's name should appear first, physicians agreed that the disorder represented a rare form of presenile dementia (May, 1968).

The literature on Creutzfeldt-Jakob disease grew during the 1920s and 1930s. The emerging studies showed that males and females were equally likely to be affected; that about 90 percent of the cases appeared in people between 35 and 65 years of age; and that Jakob (1923) was correct when he stated there was spongiform degeneration ("vacuolization") of the brain. The variations in the clinical picture suggested that there might even be distinct subgroups of the disorder (e.g., Davison, 1932; Heidenhain, 1929).[7]

One major question remained unanswered, however: What was the cause of Creutzfeldt-Jakob disease? Was it an inherited disease, as Creutzfeldt (1920, 1921) had hypothesized, or was it something quite different?

The discovery that Creutzfeldt-Jakob disease is caused by a virus resulted from the veterinary work of Björn Sigurdsson (1913–1959; shown in Figure 24.14) of the Institute for Experimental Pathology in Iceland. Sigurdsson was interested in a number of slowly developing progressive diseases of sheep. One such disorder was mædi, a lethal pneumonia (Sigurdsson, Grímsson, and Pálsson, 1952; Sigurdsson, Pálsson, and Tryggvadóttir, 1953). Two other disorders directly involved the central nervous system and were called rida, an Icelandic word for "ataxia" or "tremor", and visna, meaning "wasting." At first rida and visna were thought to be the same disease, but further study showed that although both disorders involved ataxia and tremor, they differed in their clinical manifestations and underlying pathological markers. In contrast to rida, visna was found to be primarily a disease of the glial cells (Sigurdsson and Pálsson, 1958; Sigurdsson, Pálsson, and van Bogaert, 1982).

The first signs of rida were stumbling, trailing from the flock, and head tremor. These symptoms progressed into a paralysis of the hindlegs and eventually of the whole body. The disorder ran its course in just a few months. Autopsies revealed widely distributed degenerative changes in the central nervous system, with the cerebellum, frontal lobes, and brainstem especially involved.

Figure 24.14. Björn Sigurdsson (1913–1959), the investigator who first identified a number of slow-growing viruses. (Courtesy of Johannes Björnsson, University of Iceland.)

Sigurdsson thought that rida, which was epidemic in northern Iceland, had something to do with sheep introduced from Germany in 1933. He did not believe the disorder was hereditary, but efforts to find parasites or bacterial agents in the diseased sheep proved unsuccessful. This investigation led him to consider the possibility that the sheep were suffering from a virus, although all known viruses at this time appeared to cause acute inflammatory diseases, not slowly developing degenerative diseases.

Sigurdsson (1954) eventually demonstrated that rida was, in fact, due to a "filter-passing virus," which could be transmitted by inoculating healthy sheep with brain extracts from sick sheep. This unusual virus was found to have a one- to two-year preclinical incubation period. Sigurdsson concluded that rida was probably introduced into the Icelandic flocks in 1933 by one or possibly just a few infected animals from Germany.[8]

Rida is now considered an Icelandic form of scrapie, a disorder well known in England and Scotland. Scrapie gets its name from the tendency of sick sheep to scrape themselves on trees, rocks, and fences, resulting in patchy losses of wool. Sheep with scrapie also lose their sense of balance, become uncoordinated, and show tremors as they begin to waste away. Histologically, scrapie is characterized by widespread neuronal degeneration, including shrinkage of the cell body and "soap-bubble" vacuolization (spongiform degeneration), especially of the brainstem. The scrapie virus, which is spread naturally through feces dropped on the grass that the sheep eat, can also be transmitted by injecting sheep intracerebrally with central nervous system extracts from sick animals (Gajdusek, 1981; Hadlow, Kennedy, and Race, 1982; Wilson, Anderson, and Smith, 1950).

Sigurdsson (1954) used the term "slow viruses" to describe progressive diseases with long "silent" periods (sev-

---

[7]These subgroups were sometimes based on the clinical course of the disease and sometimes on the histological picture (May, 1968).

[8]Visna and mædi were also found to be due to slow viruses that could be transmitted by experimental inoculation.

eral months to years) and then highly predictable clinical courses (e.g., *rida*, *visna*, *mædi*, and scrapie). His third criterion for a slow virus, its limitation to a single species, would soon be challenged.

In the Okapa subprovince of the Eastern Highlands of New Guinea, where the Fore tribe still practiced cannibalism in the first half of the twentieth century, the leading cause of death was a disease the natives called *kuru*. *Kuru* literally means "to shiver," and the disease begins with fine tremors of the head and limbs and slight problems with balance. Within a few months the disorder progresses to the point where the afflicted person is unable to speak clearly, to walk, or to stop involuntary movements during activity.[9] Victims also exhibit a progressive dementia with emotional changes (Hornabrook, 1978). Death from *kuru* typically occurs within 18 months after the onset of clinical signs and often within a half year. Brains from the afflicted show spongiform degeneration of the cerebrum and basal ganglia, and especially extensive atrophy of the cerebellum.

William J. Hadlow (b. 1921) recognized that human *kuru* looked very much like scrapie in sheep, both clinically and pathologically. Hadlow (1959, p. 290) suggested that "it might be profitable, in view of veterinary experience with scrapie, to examine the possibility of the experimental induction of *kuru* in a laboratory primate." This suggestion led to studies in which chimpanzees were inoculated with brain extracts from human *kuru* victims. The apes showed signs of illness one and one-half to four years later, and the clinical and pathological features were just like those seen among the New Guinea natives.

The slow virus causing *kuru* was identified in 1966 by Daniel Carleton Gajdusek (b. 1923), who began his work on *kuru* while on a visit to a remote part of New Guinea in 1957. Gajdusek subsequently won the Nobel Prize for his research.[10] *Kuru* thus became the first chronic neural degenerative disorder of humans found to be due to a virus, and the first to be transmitted to a laboratory animal, and then from animal to animal.

Kuru was attributed to sorcery by the Fore people. Once it was found to be due to a slow virus, it became clear that its transmission was linked to cannibalization at mourning rites (Glasse, 1967; Hornabrook, 1978; Klitzman, Alpers, and Gajdusek, 1984). In addition to family members eating flesh carrying the virus, the individuals handling the deceased could have acquired *kuru* through open cuts and sores.[11] With the eventual elimination of cannibalism in New Guinea, the frequency of new *kuru* cases has dropped dramatically.

These observations and discoveries led to a much better understanding of Creutzfeldt-Jakob disease, which, unlike epidemic *kuru*, typically appears in rare isolated cases. Gajdusek and his associates saw the clinical and pathological resemblance between *kuru* and Creutzfeldt-Jakob disease. They even showed that "classic" Creutzfeldt-Jakob disease—with its dementia and involuntary movements—could be transmitted to chimpanzees and monkeys by intracerebral inoculation of human brain biopsy material (e.g., Gibbs et al, 1968).[12]

These research findings, and interviews with older members of the Fore tribe, lent support to the theory that *kuru* began in New Guinea early in the twentieth century (Gajdusek, 1981; Gajdusek and Zigas, 1957; Hornabrook, 1978). The most plausible hypothesis is that the virus emerged as a rare, isolated case of Creutzfeldt-Jakob disease, which was then spread by cannibalism. This one form of Creutzfeldt-Jakob disease (a more cerebellar type) reached epidemic proportions only because of the unusual mourning rites of the natives in one well-defined, isolated part of the world.

KORSAKOFF'S PSYCHOSIS

Alzheimer's disorder and Creutzfeldt-Jakob disease are progressive dementias that eventually claim the life of the afflicted. Korsakoff's disease, in contrast, is a disorder that more selectively affects memory and can be reversed, at least within limits. Unlike the progressive dementias, it is a disorder associated with high rates of survival when properly treated.

Sergei Korsakoff (Korsakov, 1853–1900), after whom the disorder was named, was a highly respected Russian psychiatrist whose credo was "help others" (Katzenelbogen, 1970). Korsakoff's disorder was the subject of a number of articles that he published soon after writing his thesis on alcoholic paralysis in 1887. The most striking "psychical" features of his patients were their memory impairments and confabulation, a word used to designate fabricated answers.

In his first report, Korsakoff (1887) described a group of patients, mostly alcoholics, with polyneuritis (paralyses, contractures, pains, muscular atrophies, weakness, staggering) and faulty memories. In some patients, the mental symptoms, which included memory loss, irritability, anxiety, fear, and depression, were dominant. In others, the peripheral symptoms dominated. Still, because both central and peripheral symptoms always seemed present, Korsakoff called the disorder *psychosis polyneuritica*, meaning "psychosis associated with polyneuritis." He hypothesized that whatever was affecting the peripheral nervous system must also be affecting the central nervous system.[13]

Korsakoff (1889a) presented several additional cases of the disorder in a paper whose title translates as, "A Few Cases of Peculiar Cerebropathy in the Course of Multiple Neuritis." Korsakoff now stated that the multiple neuritis can be so slight that it can be hard to detect in some patients showing the memory disorder. Korsakoff also made it even clearer that the symptoms of this disorder might be associated not just with alcoholism (the most frequent cause) but with a variety of illnesses, including typhoid fever, puerperal sepsis and intestinal obstructions. He further suggested that the cause of the disorder was

---

[9]These involuntary movements subside with rest and disappear during sleep.

[10]Gajdusek authored many papers on *kuru*. Some are Farquhar and Gajdusek (1981), Gajdusek, Gibbs, and Alpers (1966, 1967), Gajdusek and Zigas (1957), and Klitzman, Alpers, and Gajdusek (1984).

[11]The fact that adult females were more involved with these rites than males in the Fore tribe was suggested to account for the much higher percentage of females who came down with *kuru*.

[12]Later work showed that individuals carrying this rare "sticky" virus can transmit it to others via human growth hormone extracts, corneal and dura transplants, and contaminated surgical instruments (Brown, 1988a, 1988b; Brown et al, 1985; Rappaport, 1987). The real surprise to most investigators was that the disease sometimes failed to express itself until 15 to 20 years had passed.

[13]See Victor and Yakovlev (1955) for an English translation of Korsakoff's paper.

likely to be a toxin and, for this reason, now chose to call the disorder *cerebropathia psychica toxaemica*.

In a second article published in 1889, the young Russian psychiatrist wrote that the mental symptoms of *cerebropathia psychica toxaemica* usually began with slight irritability and listlessness, along with the peripheral neuritis (Korsakoff, 1889b). He added that these mild behavioral symptoms then developed in one of three directions: (1) delirium tremens (vivid hallucinations and abnormal perceptions most likely due to alcohol withdrawal) with extreme restlessness, insomnia, agitation, terrible fears, and hallucinations; (2) global confusion with memory problems, apathy, indifference, and depression, but sometimes with mania (the confusion making the assessment of memory impossible); and (3) a disorder in which the patients are alert and in which the amnesia stands well in front of other symptoms. Korsakoff clearly defined the amnesia in these cases as a disturbance of memory for recent events and new material. Thus, he recognized both anterograde and retrograde components of the amnesia (Clarke, 1912–13). He thought it remarkable that his patients remembered events that had occurred long before their illnesses.

Korsakoff (1889b) noted that his patients who showed this characteristic amnesia had a tendency to repeat themselves. He wrote:

> On occasion, the patient forgets what happened to him just an instant ago: you came in, conversed with him, and stepped out for one minute; then you come in again and the patient has absolutely no recollection that you had already been with him. Patients of this type may read the same page over and over again sometimes for hours, because they are absolutely unable to remember what they have read. In conversation, they may repeat the same thing 20 times. . . . It often happens that the patient is unable to remember those persons whom he met only during the illness . . . so that each time he sees them, even though seeing them constantly, he swears that he sees them for the first time. (Translated by Victor and Yakovlev, 1955, 398)

The memory disturbance may lead a patient to believe that he is in the same setting he had been in long ago. It may also lead a patient to think that people who have been dead for some time are still living, such as a long-deceased relative. Nevertheless, the amnesia is anything but constant. Rather, it is a symptom that can wax and wane over the course of the disease and vary with degrees of alertness or fatigue (Turner, 1910).

Korsakoff (1889b) described the confabulation that accompanied the amnesia in some detail. This involved fabricating answers and experiences, and stringing facts together imperfectly. His patients showed little insight into what they were doing, and seemed unaware that they were suffering from a memory disorder. As he described it:

> Thus, when asked to tell how he has been spending his time, the patient would very frequently relate a story altogether different from that which actually occurred, for example, he would tell that yesterday he took a ride in town, whereas in fact he has been in bed for two months, or he would tell of conversations which have never occurred, and so forth. On occasion, such patients invent some fiction and constantly repeat it, so that a peculiar delirium develops, rooted in false

recollections (pseudo-reminiscences). (Translated by Victor and Yakovlev, 1955, 399)

Near the end of his second 1889 paper, Korsakoff added modestly that his descriptions of the psychic manifestations of alcoholism were nothing new. He gave priority to Magnus Huss (1807–1890), who in the middle of the nineteenth century drew attention to the derangement of memory in some forms of chronic alcoholic disease.

Without question, Magnus Huss was not the only early writer to comment on memory impairments in alcoholics (see Starr, 1887). In 1822, James Jackson (1777–1867; shown in Figure 24.15) described alcoholic multiple neuritis and stated that "the mind is weakened, but is free from delirium" in those cases characterized by pains and numbness in the limbs. This Boston physician also recognized that "the functions of the stomach are almost always impaired. . . . The appetite is lost, or is morbid, . . . the food is often rejected, and constipation or diarrhoea take place" (pp. 353–353). Jackson did not, however, see the causal relationship between nutrition and this neuropathological condition.

Further, in 1879, Robert Lawson (1815–1894) clearly preempted Korsakoff when he wrote that excessive drinking can be associated with "cases of dementia of which the principal feature is almost absolute loss of memory for recent events" (p. 183). Lawson mentioned that such patients can be alert, cheerful, and understanding, but that when a medical officer makes his third visit of the day and asks a patient if he or she had seem him before, the answer will probably be "no."

Additional cases of memory loss associated with chronic alcoholism were mentioned by a number of other early

Figure 24.15. Dr. James Jackson (1777–1867) of Boston. In 1822, Jackson described multiple neuritis associated with alcoholism and noted that "the mind is weakened" in this condition.

investigators, but statements about recent memory impairments were typically not as clear as the one made by Lawson (e.g., Charcot, 1884; Lancereaux, 1864; Moeli, 1883; Robinson, 1877; Strümpell, 1883).

As for the better-recognized peripheral neuritis associated with alcoholism, one of the earliest descriptions of it can be traced to John Coakley Lettsom (1744–1815). Before the nineteenth century began, Lettsom wrote the following about the "dyspepsia" of alcoholism:

> The legs become as smooth as polished ivory, and the soles of the feet even glassy and shining, and at the same time so tender, that the weight of the finger excites shrieks and moaning. . . . The legs, and the whole lower extremities, lose all power of action; wherever they are placed, there they remain till moved again by the attendant; the arms and hands acquire the same paralysis, and render patients incapable of feeding themselves. (1787, 159)

Lettsom also described the mental symptoms that can accompany this suffering:

> Their minds appear idiotish. . . . They talk freely in the intervals of mitigation, but of things that do not exist; they describe the presence of their friends, as if they saw realities, and reason tolerably clearly upon false premises. (1787, 160)

Cases like these notwithstanding, Sergei Korsakoff concluded that others before him had not looked upon the mental symptoms and the peripheral neuritis as manifestations of a single disease entity. He also concluded that his predecessors had by no means carefully examined and analyzed the mental symptoms in these patients. This, he maintained, was his unique contribution. Nevertheless, Korsakoff did not provide a description of the anatomical correlates of his disorder. He mentioned only casually that there were signs in these patients, such as visual disturbances, to suggest that the brainstem had to be involved.

Soon after Korsakoff's (1887, 1889a, 1889b, 1890) papers appeared, controversy swelled over whether the peripheral neuritis was as firmly linked to the amnestic disorder as Korsakoff believed (see Soukhanoff and Boutenko, 1903). Partly because many investigators did not accept the polyneuritis as a necessary correlate of the amnesia, Friedrich Jolly (1844–1904), a professor from Berlin, coined the terms "Korsakoff's psychosis" or "Korsakoff's syndrome" to describe just the memory and confabulatory aspects of the disorder. These new terms, introduced at the Moscow Medical Congress of 1897, led to the abandonment of Korsakoff's own term, *cerebropathia psychica toxaemica*. In addition, they led to the belief that different traumas, diseases, toxins, and perhaps even diet, might determine whether primarily peripheral, largely central, or mixed pathological changes will be exhibited (Cole, 1902).

By 1903, there were at least 192 recorded cases of Korsakoff's disease in the literature, 38 of which Korsakoff studied himself.[14] The number grew even more dramatically after this time (e.g., Butts, 1913; Clarke, 1912–1913; Stanley, 1909–1910). The case studies, however, were remarkably similar. For example, the following story was told about a 33-year-old patient who had been an excessive drinker. This man was admitted to Boston City Hospital in 1903 with pains in his feet:

> It was noticed also that his memory was becoming poor. Many things of recent happening he was unable to recall. At one time he was consulted and agreed to a change of policy on the part of the company which he represented. A few weeks later when this change went into effect he was very bitter because he had not been consulted, denying all knowledge of the event. During the last six months he had frequently contracted bills and borrowed money, but later proved unable to recall the occurrence. This memory weakness increased until at the time of admission to the hospital he was unable to give a satisfactory account of the last months of his life. (Sims, 1905, 165)

In some circles, there were attempts to classify the amnesic fabrications. One such scheme listed (1) random fabrications invented ad hoc; (2) distorted recollections of real facts; and (3) true recollections of real facts that were wrongly oriented in time and place (Moll, 1915). The confabulations were most marked during the acute stage of the disorder; they were rarely encountered in patients who had endured Korsakoff's disease for more than a few years (Moll, 1915; Turner, 1910).

WERNICKE'S ENCEPHALOPATHY AND KORSAKOFF'S PSYCHOSIS

From all indications, Korsakoff (e.g., 1887, 1889a, 1889b, 1890) was unaware of the relationship of the disorder he was describing and the encephalopathy with its confusional state, ataxia, and ophthalmoplegia (gaze palsies, nystagmus) described by Carl Wernicke in 1881. Additionally, there was no indication that Wernicke saw the relationship between the disorder he was describing and the disorder presented by Korsakoff.[15] Wernicke, who by nature was a much more difficult and taciturn man than Korsakoff, believed he too had discovered a new disease entity (Goldstein, 1970). He called his syndrome *Polioencephalitis haemorrhagica superior*, but it soon went by the name Wernicke's encephalopathy.

Wernicke first saw his encephalopathy in two alcoholic men and in a 20-year-old seamstress. The latter was admitted to the Charité in Berlin after she had swallowed sulfuric acid in an attempted suicide. The woman left the hospital after a few days, but protracted vomiting set in, and her condition deteriorated. The seamstress showed bilateral optic neuritis, ataxia, apathy, disorientation, fright, somnolence, and restlessness before she passed away several weeks later. Many of the same symptoms also characterized the two alcoholic men who had been hospitalized in a delirious state and died soon afterward.

Wernicke noted that the disorder characterized by visual problems, abnormal gait, and mental confusion had been seen by others before him. But, like Korsakoff, he claimed that the earlier descriptions of "his" disease were at best

---

[14]See Soukhanoff and Boutenko (1903) for a review of the early studies, most of which were in Russian and German.

[15]This state of affairs may seem even more unusual when it is realized that Wernicke gave a good clinical account of four patients with Korsakoff's psychosis in 1900.

fragmentary (see Victor, Adams, and Collins, 1989). Of particular interest was his citation of a case by Samuel Annan (1797–1868) from 1840. This brief case study involved a 60-year-old male, "intemperate and partially deranged," who refused food and who at autopsy showed softening of the thalamus, septum, and fornix. Other early suggestions of the disorder came in papers describing alcoholics with diplopia (double vision), who showed symptoms ranging from stupor to delirium tremens (e.g., Canton, 1860; Dardel, 1868; Robinson, 1877).

In the opening years of the twentieth century, Karl Bonhoeffer (1868–1943; shown in Figure 24.16) noted that the majority of alcoholic and nonalcoholic patients who contracted Wernicke's syndrome later developed Korsakoff's amnesia, and that most patients with Korsakoff's amnesia first showed signs of having Wernicke's encephalopathy. As a result of Bonhoeffer's (1901, 1904) observations and authoritative writings, the possible association between the two disorders became more accepted (Coriat, 1906; Moll, 1915; Turner, 1910). According to Bonhoeffer, the major initial features of the combined Wernicke-Korsakoff syndrome are multiple neuritis, ophthalmoplegia, ataxia, and confusion, followed by an inability to remember new facts and poor memory for the recent past.

It was soon suggested that if a patient with Wernicke's disorder did not go on to show Korsakoff's psychosis, it was only because he or she recovered or died before the amnesic state had a chance to emerge from under the cover of the delirium or confusion (Bumke and Foerster, 1936). Still, others were not sure about this and emphasized that in addition to the combined syndrome, the Wernicke and Korsakoff disorders could appear independently (e.g., Coriat, 1906).

At the same time, there was growing appreciation of the

Figure 24.16. Karl Bonhoeffer (1868–1943). As the twentieth century opened, Bonhoeffer noted that many patients who contracted Wernicke's syndrome later developed Korsakoff's amnesia, whereas most with Korsakoff's amnesia first showed signs of having Wernicke's encephalopathy.

fact that the combined syndrome could emerge with any of a variety of diseases and insults. Although chronic alcoholism seemed to be the major culprit, the list also included brain tumors, acute infections, senile dementias, lead intoxication, syphilis of the nervous system, depression, pneumonia, protracted diarrhea, carbon dioxide poisoning, and cancers of the gastrointestinal tract (Coriat, 1906; Patterson, 1905).

THE LESIONS IN KORSAKOFF'S SYNDROME

During the 1880s and 1890s, when Korsakoff and Wernicke were presenting their cases, there was good agreement that changes in the peripheral nervous system were largely if not entirely responsible for the polyneuritic symptoms. The real controversy concerned the sites in the central nervous system thought to be responsible for the mental changes, such as the confusion and the memory problems.

In this context, both Korsakoff and Wernicke postulated that brainstem lesions were behind many of the central symptoms. Although Korsakoff did not provide a description of the anatomical correlates of this disorder, he noted that there were some signs suggesting that the brainstem was involved. In contrast to Korsakoff, Wernicke (1881) carefully examined the brains of his patients and found numerous small lesions (petechial hemorrhages) in the gray matter of the brainstem near the third ventricle, as well as near the floor of the fourth ventricle. In particular, he thought that some of the focal clinical signs were due to acute inflammation of the ocular motor nuclei.

By the beginning of the twentieth century, there was fairly good evidence to support the idea that pathology in a region near the third ventricle was responsible for many of the symptoms in these cases (Gudden, 1896). Notably, this region encompassed the mammillary bodies *(corpora mamillaria)*. The mammillary bodies drew more sustained attention when the brains of 16 patients who showed Korsakoff's psychosis with symptoms of Wernicke's encephalitis were examined in an extensive study during the 1920s (Gamper, 1928). The mammillary bodies were severely affected in all 16 cases. These patients also showed some involvement of the central gray of the midbrain, the oculomotor nuclei, the walls of the third ventricle, the inferior colliculi, and certain other parts of the brainstem.

Although there was good evidence for mammillary body lesions in Korsakoff cases, the exact role of the mammillary bodies remained controversial. One reason was that many patients also showed damage to the dorsomedial thalamus. Another was the belief that the cerebral cortex had to be involved in higher functions. In this context, some early investigators argued that cortical changes could account for the symptoms of memory loss, delirium, and confusion. Not only did this idea seem logical, but it was based on autopsies on chronic alcoholics that tended to show cortical atrophy with widening of the sulci (e.g., Carmichael and Stern, 1931; Cole, 1902). In addition, brains of patients with Wernicke-Korsakoff disease weighed considerably less than would be expected if the degeneration were limited to just the mammillary bodies or even to the mammillary bodies and the dorsomedial thalamic nuclei. The precise roles of these structures and other parts of the brain in the various central symptoms of this disorder have remained matters of controversy long after Korsakoff and Wernicke first described the features of this syndrome.[16]

AVITAMINOSIS

Neither Wernicke nor Korsakoff understood the cause of the disease they were studying. Korsakoff attributed his disorder to a toxin, and Wernicke regarded his encephalopathy as an acute inflammation of the brain similar to a poliomyelitis. Of these two ideas, the toxin hypothesis seemed to attract more supporters, in part because these disorders appeared to be complications of a variety of other diseases (Hurd, 1905; Jolly, 1897). The growing idea that the lesions were most noticeable in parts of the central nervous system closest to the ventricles, such as the mammillary bodies, also suggested a noxious agent, one that might be identified in the cerebrospinal fluid if not in the blood itself (Bender and Schilder, 1933).

One thought shared by many early-twentieth-century researchers was that the toxin might be alcohol itself. But the fact that similar lesions were seen in patients who had not been alcoholics argued against this specific idea. In addition, the observation that the lesions can continue to grow in alcoholics after they stop drinking argued in favor of a "secondary" toxin (Bender and Schilder, 1933). As stated by one investigator:

Neither polyneuritis, delirium tremens, nor the confusional disorders can be regarded as the expression of a mere alcohol poisoning. They have a more complex etiology, and present the characters of a severe general disease, of the nature of a toxaemia, depending probably on an auto-intoxication from disordered metabolism. They may, most of them, result from other poisons besides alcohol, though their appearance is evidently favored in a very special way by chronic alcoholism. (Cole, 1902, 359)

The autointoxication idea found many supporters, some of whom thought the poisoning could be brought about by a variety of conditions that lowered the body's resistance (e.g., Bonhoeffer, 1904; Clarke, 1912–1913; Moll, 1915; Stanley, 1909–1910).[17]

At the same time, the belief that the disorders could be treated by the withdrawal of alcohol and by a return to good nutrition gained slow and steady acceptance (Butts, 1913; Clarke, 1912–1913; Pershing, 1892). "Absence from spirituous liquors of every kind" and proper nutrition had been advocated by James Jackson in 1822 (p. 353).

The concept of a vitamin deficiency can be traced to seventeenth-century observations on individuals who developed beriberi in the Far East. Nicolaas Tulp (1593–1674), who was painted by his friend Rembrandt Harmensz van Rijn (1606–1669; see Figure 24.17), was one of the first Europeans to describe this peripheral neuritis.[18] A century later, John Blackall (1771–1860), an Oxford-trained physician, noted that sailors off Canton, China, who developed beriberi improved when given fermented bread (see Strauss, 1956). Later in the eighteenth century, it was found that fowl kept on a diet of polished rice not only developed polyneuritic symptoms, but that this and the associated peripheral nerve degeneration could be reversed by returning to a diet with "crude" rice (Eijkman, 1897).

In 1911, Casimir Funk (1884–1967), a scientist associated with the Lister Institute in London, isolated a water-soluble crystalline substance from rice polishings that cured the polyneuritis induced in pigeons fed polished rice. He named the substance "vitamine" because he believed he had discovered an amine necessary for *vita,* or life (see Funk, 1911, 1912).

Funk's research led to the isolation of the pure vitamin from rice polishings in the 1920s and to the synthesis of the vitamin a decade later (Jansen, 1956; Williams, 1936). Nevertheless, not until the mid-1930s was it thought that the agent causing Wernicke-Korsakoff disease might be a specific vitamin deficiency (Bender and Schilder, 1933; Campbell and Biggart, 1939; Ironside, 1939; Tanaka, 1934; Wagener and Weir, 1937).

The vitamin hypothesis took hold when it was demonstrated that rats with vitamin B_1 deficiencies showed hemorrhagic changes in the brainstem just like those of Wernicke-Korsakoff patients (Prickett, 1934). In addition, an important study conducted by Leo Alexander (b. 1905) and his associates demonstrated that central "Wernicke symptoms" appeared as a complication of beriberi only when pigeons deprived of vitamin B_1 were given sufficient amounts of vitamins A, B_2, C, and D (Alexander, Pijoan, and Myerson, 1938). In contrast, pigeons kept on a totally vitamin-free diet did not show central symptoms, although they still developed beriberi. Further work by Alexander (1940) also demonstrated that the central lesions produced by depriving pigeons of vitamin B_1 (while administering ample supplies of other vitamins) were similar histologically to those of Wernicke-Korsakoff disease (also see Zimmerman, 1939).

The idea of treating Wernicke-Korsakoff patients with large doses of thiamine (vitamin B_1 in synthesized form) made good sense in this context. Thiamine supplements soon were found to reduce the danger of fatal midbrain hemorrhages in these patients. The supplements also reversed the ophthalmoplegia, disorientation, and clouding of consciousness.

The effects of large dosages of vitamin B_1 on the amnesic part of the disorder, however, were not as impressive. The memory disorder improved more slowly, and the recovery was often less complete than that witnessed for the other symptoms (Bowman, Goodhart, and Jolliffe, 1939; Jolliffe,

---

[16]More recent work has suggested that (1) approximately 74–96 percent of Wernicke-Korsakoff patients show gross neuropathological changes in the mammillary bodies (see Charness and De La Paz, 1987; Malamud and Skillicorn, 1956; Victor, Adams, and Collins, 1989); (2) changes in the dorsomedial thalamic nucleus may play an important role in the amnesia (Mair et al, 1988; Victor, 1969; Victor, Adams, and Collins, 1989); (3) pathology of the mammillary bodies and the dorsomedial nuclei may represent different disease processes (Martin, Eckhardt, and Linnoila, 1989); and (4) many alcoholics show much lower brain weights than would be expected if they suffered from only small, highly localized mammillary body or thalamic lesions (Torvik, Lindbloe, and Rodge, 1982). In particular, the latter finding has led to renewed interest in the role of the cerebral cortex, especially the prefrontal areas, in the mental changes seen in Korsakoff patients.

[17]To cite one theory, in 1931 Karl Neubürger (1890–?) speculated that the poisoning was the result of disorders which depressed the functions of the liver, thus allowing endogenous toxins from the gut to escape detoxification and eventually reach the brain.

[18]Beriberi had also been recognized by Jacobus Bontius (1592–1631), a physician employed by the Dutch East India Company. Sections of the early Bontius and Tulp descriptions of beriberi have been translated from the Dutch into English (see Major, 1978). For historical reviews of beriberi, see Funk (1912), Jansen (1956), and Shattuck (1928).

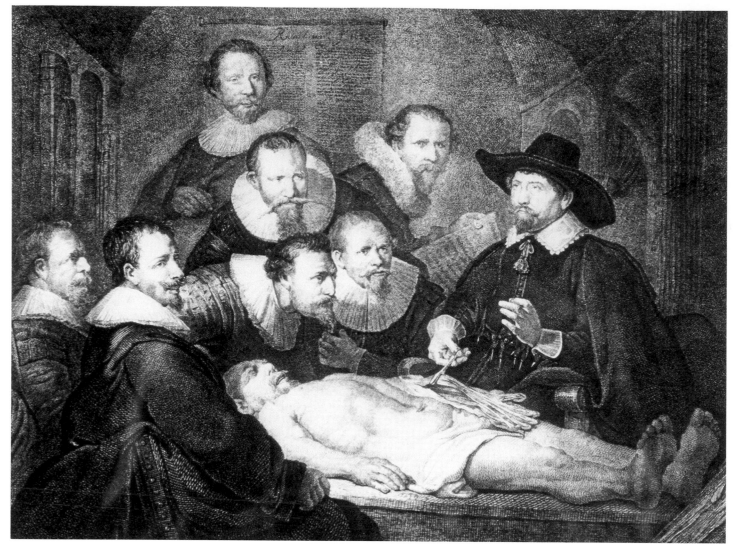

Figure 24.17.   Nicolaas Tulp (1593–1674), a Dutch physician, is shown teaching anatomy in this print based on the famous Rembrandt Harmensz van Rijn (1606–1669) painting of 1662. As a doctor employed by the Dutch East India Company, Tulp gave one of the earliest descriptions of beriberi.

Wortis, and Fein, 1941). These findings were in accord with studies of "recovered" patients prior to the advent of vitamin therapy (e.g., Butts, 1913; Clarke, 1912–13; Moll, 1915).[19]

In part because the amnesic condition usually did not improve as nicely as the other symptoms, some researchers speculated that the complete syndrome could be the result of several agents (Campbell and Biggart, 1939; Campbell and Russell, 1941; Jolliffe, Wortis, and Fein, 1941). They questioned whether all aspects of the Wernicke-Korsakoff syndrome could be accounted for by the deficiency of just vitamin B_1 or whether a toxin or other vitamins might also be involved.[20] Despite sustained efforts to achieve a better understanding of this disease, this issue—like the related

issue of the lesion responsible for each of the symptoms of the combined syndrome—has continued to generate debate.

THE TEMPORAL LOBES AND THE HIPPOCAMPUS

The belief that damage to the temporal lobes and the underlying hippocampus may affect memory can be traced to 1887, when Sanger Brown and Edward Albert Schäfer presented a report on monkeys with brain lesions. Speaking before the Royal Society of London, these men stated that they had ablated the temporal or occipital lobes partially or completely in an effort to localize sensory functions. One of their monkeys had very deep lesions of the superior temporal gyri, and a second monkey had complete removals of both temporal lobes. Brown and Schäfer (1888) noted that these animals were deficient in memory. In fact, they wrote that the intellectual faculties of the two monkeys had been reduced to such a degree that they "resembled idiots." This, they emphasized, was not seen after lesions of the occipital lobes.

---

[19]In 1908, Smith Ely Jelliffe (1866–1945), who strongly recommended nutritional therapy along with massage, electricity, and strychnine for the neuritis, advocated special training for the memory impairment. His initial efforts with pedagogical techniques, however, were not very successful.

[20]The idea that the amnesic part of the syndrome might be due to multiple vitamin deficiencies or to a single vitamin deficiency plus a neurotoxic agent is discussed by Butters (1984).

The description provided of the rhesus monkey with complete temporal lobe ablations is especially noteworthy. Brown and Schäfer wrote:

> Every object with which he comes in contact, even those with which he was previously most familiar, appears strange and is investigated with curiosity. . . . This is the case not only with inanimate objects, but also with persons and his fellow Monkeys. And even after having examined an object in this way with the utmost care and deliberation, he will, on again coming across the same object accidently even a few minutes afterward, go through exactly the same process, as if he had forgotten his previous experiments. (1888, 311)

The experimenters went on to describe how a wild monkey was put into the cage with this animal. The experimental monkey went over to investigate the newcomer and was savagely attacked. After breaking loose, and without any signs of fear, this animal went right back to investigate the stranger as if it had completely forgotten what had just transpired.[21]

Brown and Schäfer did not seem to know what to make of these findings or of the partial recovery displayed by the monkeys. They postulated only that the temporal lobe lesions may have disrupted the vasculature to a large part of the cerebrum, which could have led to a general depression of intellectual functions, including memory.

Twelve years after Brown and Schäfer made their presentation to the Royal Society, Vladimir Bekhterev ("Bechterew" in German) attended a scientific meeting in St. Petersburg, where he presented a clinical case report. He described a 40-year-old patient who had exhibited a distinct and profound memory disturbance. An autopsy of this patient's brain showed bilateral softening of the uncinate gyrus, of Ammon's horn (hippocampus), and of the adjoining temporal cortex (see Bechterew, 1900).

Neither the temporal lobes nor the hippocampus attracted much attention in the memory literature early in the twentieth century. This may seem a little surprising because these two reports were followed by clinical studies demonstrating memory disturbances along with hippocampal atrophy in many Alzheimer patients. Furthermore, the Brown and Schäfer and Bekhterev reports appeared at about the same time that Pick's disease was beginning to attract interest (see Chapter 22). The latter was first described by Arnold Pick in 1892, and reduced associative memory was thought by some early investigators to be one of the symptoms of the disease. Autopsies on Pick's cases often revealed selective temporal cortex atrophy, in addition to degeneration of the prefrontal lobes.

This neglect of the temporal lobes began to change in the 1930s when Wilder Penfield observed that his conscious epileptic patients would occasionally report "flashbacks" while the superior or upper lateral surfaces of their temporal cortices were electrically stimulated. A typical response to stimulation of the right temporal cortex was hearing a particular song. These "psychical experiences" stopped abruptly when the stimulating electrode was withdrawn and did not appear with stimulation of other parts of the brain.

Penfield first thought he had tapped "the stream of consciousness" or the experiential record in the gray matter of "the interpretative cortex."[22] He looked upon the temporal lobes as special, making statements such as:

> The records of an individual's thinking lie dormant in the patterns of his temporal cortex until he activates them in some normal process of recall or until they come into spontaneous, and perhaps distorted, existence in his dreams. Whenever a normal person is paying conscious attention to something, he is simultaneously recording it in the temporal cortex of each hemisphere (Penfield, 1952, 185)

Thirty years after making his initial observations with temporal lobe patients, Penfield (1974) altered his position to suggest that stimulation of the hippocampus was responsible for the temporal lobe memory phenomena. His new proposal held that the hippocampus played the leading role of selectively stimulating the cortex in order to bring certain material from the experiential record to the conscious mind. Penfield's change in position came as a result of some neurosurgical cases that involved destruction of the hippocampus.

Oddly, the hippocampus had been studied surprisingly little by neurologists and behavioral scientists before the second half of the twentieth century. Even the derivation of its name has been shrouded in mystery. The individual most responsible for the term was Guilio Cesare Aranzio (Arantius, 1587), a "dull pupil" of Andreas Vesalius. Aranzio provided a description that left one wondering whether he was referring to a seahorse, a white silkworm, a horse caterpillar, or a bottlenose dolphin.[23]

The hippocampus was not even mentioned by Vesalius (1543), Thomas Willis (1664), or Raymond Vieussens (1684), the real illuminati of this time period. Bartholomeo Eustachio finally showed the structure on some plates, but his illustrations were not published until 1714.

In the beginning of the nineteenth century, Gottfried Reinhold Treviranus (1816–1821), a German comparative anatomist, noted that the hippocampus was associated with many different structures, including the corpus striatum, the fornix, the corpus callosum, and the olfactory nerves. Because of its multiple connections and the fact that it varied in size with the olfactory nerve, he wrote: "Therefore, the hippocampus is probably involved with a higher mental function, perhaps that of memory, which is so greatly aroused by olfactory sensations" (translated in Meyer, 1971, p. 87). Paul Broca (1878) also linked the hippocampus with smell on the basis of comparative anatomy. Later anatomists, however, questioned the idea of direct connections between the olfactory tracts and the hippocampus (e.g., Ramón y Cajal, 1901–2).

In the closing decade of the nineteenth century, the fine anatomy of the cell structure of the hippocampus was worked out by many individuals, including Theodor Meynert, Camillo Golgi, and Santiago Ramón y Cajal.[24] But while significant advances were made on the structural front, the functions of the hippocampus remained a matter of loose speculation.

Several striking new cases of amnesia accompanied by

---

[21]See Chapter 20, which describes Brown and Schäfer's work in the context of the Klüver-Bucy syndrome.

[22]This theory was also described in many of Penfield's other writings (e.g., 1952, 1958, 1968).

[23]See Lewis (1923) for a history of Aranzio and the word "hippocampus," and a translation of Aranzio's Latin text.

[24]See Meyer (1971) for a review of the early history of hippocampal anatomy.

Figure 24.18. Dr. William Scoville (1906–1984), the neurosurgeon who operated on Case H. M. in 1953.

hippocampal pathology were reported in the mid-twentieth century.[25] The descriptions of these individuals seemed to confirm the long-forgotten descriptions of monkeys and humans with deep temporal lobe lesions by Brown and Schäfer (1888) and Bekhterev (Bechterew, 1900), respectively. Typically, these patients could not remember people's names, whether they had a spouse, or if their marital partner had died. But unlike Korsakoff patients, their mammillary bodies seemed normal upon autopsy, and there was no history of peripheral neuropathy.

A still better appreciation of the importance of the hippocampus followed a few "cleaner" surgical cases, some of which involved bilateral resection of the temporal lobes in order to control severe epilepsy or psychotic behaviors. One case involved a 27-year-old man who suffered from such severe epilepsy that he underwent surgery to remove a part of each temporal lobe. This man was Case H. M. (b. 1926). He was operated upon in 1953 by an accomplished Connecticut neurosurgeon, Dr. William Scoville (1906–1984; shown in Figure 24.18). H. M. sustained lesions that destroyed the anterior two thirds of the hippocampus, in addition to the parahippocampal gyrus, anterior temporal cortex, uncus, and amygdala. Scoville (1954) initially described his patient as being unable to remember the rooms in which he lived, the names of his associates, or even his way to the toilet. H. M. had no idea of the year in which he was living, when he had last eaten a meal, or who was caring for him.

The striking memory loss exhibited by H. M. caught everyone by surprise. It led to more formal testing of him, of a psychotic woman who underwent comparable surgery, and of numerous other patients (mostly schizophrenics)

given smaller temporal cortex lesions by Scoville. The testing, much of which was directed by Brenda Milner (b. 1918), revealed extremely severe memory deficits in H. M. and in Scoville's psychotic patient with very large hippocampal lesions. These two patients showed an inability to retain new factual information and a retrograde amnesia that extended back for years prior to the surgeries (Scoville and Milner, 1957).

Patients with more anteriorly placed bilateral lesions that spared the hippocampus did not show the amnesic disorder. In addition, only 2 of more than 90 unilateral temporal lobectomy cases seen by Wilder Penfield and Brenda Milner showed severe memory deficits. Both "unilateral" patients were later found to have suffered natural lesions of the opposite temporal lobe early in life (Milner and Penfield, 1955; Penfield and Milner, 1958; see Penfield, 1974, for autopsy reports).

Follow-up studies on H. M. showed that his profound anterograde amnesia continued relatively unabated for years after his surgery.[26] At best he showed slight signs of remembering a few of the constant features of his environment, such as the arrangement of the rooms and the furniture in his home. Occasionally he seemed to recall that his father had died, the name of an astronaut, or the profile of President John F. Kennedy (1917–1963) on a coin, but little more in the way of new factual information.[27]

In contrast to this anterograde amnesia and a retrograde amnesia that extended back many years before the surgery, H. M. could acquire relatively simple motor and perceptual skills (e.g., mirror writing, puzzles, stylus mazes). This indicated that not all types of memory were equally devastated by his lesions. "Declarative" (factual) memory seemed to depend more upon the integrity of the hippocampus than did "procedural" (ways of doing things) memory.

The distinction between declarative and procedural memory was recognized well before Case H. M. It was discussed by Théodule Ribot in 1881. Ribot cited a number of cases from the literature to illustrate the occasional separation of procedural from declarative memory, not just in the dementias but after trauma or with other types of insult. To quote Ribot on the vulnerability of declarative memory:

> In cases belonging to this morbid group neither the habits nor skill in any handicraft, as sewing or embroidery, . . . disappears. The destruction of memory in these cases affects only its highest and most instable forms, those personal in character and which, being accompanied by consciousness and localization in time, constitute that which . . . we called psychic memory. (1883 translation, 473)

After remarking that the amnesia for declarative memory was most severe for new events, Ribot added:

> This law, however universal it may be with regard to memory, is but a particular expression of a still more general law—a biological law. It is a fact well known in biology that the structures that are latest formed are the first to degenerate. . . . Hughlings Jackson was the first to prove in detail

---

[25]Some of these neurological and neurosurgical cases were described by Glees and Griffith (1952), Grünthal (1947), and Hegglin (1953). See De Jong, Itabashi, and Olson (1969), and Victor et al (1961) for newer cases.

[26]There have been many long-term follow-up studies on Case H. M. Descriptions appear in Corkin (1968, 1984); Drachman and Arbit (1966); Milner (1972); Milner, Corkin, and Teuber (1968); and Scoville (1968).

[27]After many years, H.M. still was unable to recognize the people who had been visiting him regularly. Formal testing showed that complex stimuli which could not be rehearsed were still retained less than one minute.

that the higher, complex, voluntary functions of the nervous system disappear first, and that the lower, simple, general and automatic functions disappear latest. We have seen both these facts verified in the dissolution of the memory: what is new dies out earlier than what is old, what is complex earlier than what is simple. The law we have formulated is therefore only the psychological expression of a law of life. (1883 translation, 482)

Ribot hypothesized that the highest forms of memory were dependent upon the cortex and that automatic, procedural memory was essentially a brainstem function. Scoville, Milner, and many others involved with H. M. and related clinical cases concluded that damage to the hippocampus was largely responsible for the declarative memory disorder.[28]

These striking clinical findings motivated scientists to conduct a plethora of lesion experiments on laboratory animals. Impairments in retention of previously learned material and difficulties in acquiring new information were often noted.[29] Nevertheless, performance varied with the species, the nature of the task, and the characteristics of the hippocampal damage. As the data continued to accumulate, investigators realized that whereas the hippocampus might play an important role in at least some types of human and animal memory, much remained to be learned about its precise role and its functional relationships with other parts of the brain.

---

[28]Scoville and Milner (1957) viewed the hippocampus and the mammillary bodies as functionally related structures. They noted that damage to either one could cause severe recent memory losses and anterograde amnesia (see Barbizet, 1963; Sweet, Talland, and Ervin, 1959). The fornix runs from the hippocampus to the mammillary bodies (Papez, 1958). The main pathway from the mammillary bodies, the mamillothalamic tract, goes to the anterior nucleus of the thalamus. Although the idea of functional unity is attractive, the literature on lesions of the fornix in humans has not revealed consistent memory defects, and researchers believe that lesions in these structures do not produce precisely the same deficits (Squire, 1987).

[29]For descriptions of some of the research on animals with hippocampal lesions, see Drachman and Ommaya (1964) and the reviews by Brierley (1966), Squire (1986), and Squire and Zola-Morgan (1983)

REFERENCES

Alexander, L. 1940 Wernicke's disease. Identity of lesions produced experimentally by B₁ avitaminosis in pigeons with hemorrhagic polioencephalitis occurring in chronic alcoholism in man. *American Journal of Pathology, 16,* 61–69.

Alexander, L., Pijoan, M., and Myerson, A. 1938. Beriberi and scurvy. *Transactions of the American Neurological Association, 64,* 135–139.

Alzheimer, A. 1904. *Histologische Studien zur Differentialdiagnose der progressiven Paralyse. Histologische und histopathologische Arbeiten über die Grosshirnrinde* (Band 1). Jena: Fischer.

Alzheimer, A. 1906. Über einen eigenartigen schweren Krankheitsprozess der Hirnrinde. *Neurologisches Centralblatt, 25,* 1134.

Alzheimer, A. 1907a. Über einen eigenartigen schweren Krankheitsprozess der Hirnrinde. *Centralblatt für Nervenheilkunde und Psychiatrie, 30,* 177–179.

Alzheimer, A. 1907b. Über eine eigenartige Erkrankung der Hirnrinde. *Allgemeine Zeitschrift für Psychiatrie und psychisch-gerichtliche Medizin, 64,* 146–148. Reproduced and translated by L. Jarvik and H. Greenson as, "About a peculiar disease of the cerebral cortex." In *Alzheimer Disease and Associated Disorders, 1,* 5–8, 1987. (Also translated in Bick, Amaducci, and Pepeu, 1987)

Alzheimer, A. 1911. Über eigenartige Krankheitsfälle des späteren Alters. *Zeitschrift für die gesamte Neurologie und Psychiatrie, 4,* 356–385.

Annan, S. 1840. Idiocy from alcoholism. *Maryland Medical Surgery Journal, 1,* 333–334.

Arantius. G. C. 1587. *De humano foetu . . .* Venetiis: Apud Jacobum Brechtanum.

Baillie, M. 1795. *The Morbid Anatomy of Some of the Most Important Parts of the Human Body* (first Amer. ed.). Albany, NY: Barber and Southwick. (Reprinted 1977; Oceanside, NY: Dabor)

Barbizet, J. 1963. Defect of memorizing of hippocampal-mammillary origin: A review. *Journal of Neurology, Neurosurgery and Psychiatry, 26,* 127–135.

Barrett, A. M. 1911. Degenerations of intracellular neurofibrils with miliary gliosis in psychoses of the senile period. *American Journal of Insanity, 67,* 503–516.

Bechterew, V. 1900. Demonstration eines Gehirns mit Zerstörung der vorderen und inneren Theile der Hirnrinde beider Schläfenlappen. *Neurologisches Centralblatt, 19,* 990–991.

Bender, L., and Schilder, P. 1933. Encephalopathia alcoholica: Polioencephalitis haemorrhagica superior of Wernicke. *Archives of Neurology and Psychiatry, 29,* 990–1053.

Betts, J. B. 1911. On the occurrence of nodular necroses (Drusen) in the cerebral cortex. A report of twenty positive cases. *American Journal of Insanity, 68,* 43–56.

Bick, K., Amaducci, L., and Pepeu, G. 1987. *The Early Story of Alzheimer's Disease.* Padova: Liviana Editrice.

Blackall, J. 1820. *Observations on the Nature and Cure of Dropsies, and Particularly on the Presence of the Coagulable Part of the Blood in Dropsical Urine . . .* Philadelphia, PA: J. Webster.

Blocq, P., and Marinesco, G. 1892. Sur les lésions et la pathogénie de l'épilepsie dite essentielle. *Semaine Médicale, 12,* 445–446.

Boneti, T. 1700. *Sepulchretum sive anatomia practica.* Lugduni: Cramer & Perachon.

Bonfiglio, F. 1908. Di speciali reperti in un case di probabile sifilide cerebrale. *Rivista Sperimentale di Freniatria e Medicina Legale delle Alienazioni Mentali, 34,* 196–206. (Translated in Bick, Amaducci, and Pepeu, 1987)

Bonhoeffer, K. 1901. *Die akuten Geisteskrankheiten der Gewohnheitstrinker. Eine klinische Studie. VIII.* Jena: G. Fischer.

Bonhoeffer, K. 1904. Der Korsakowsche Symptomenkomplex in seinen Beziehungen zu den verschiedenen Krankheitsformen. *Allgemeine Zeitschrift für Psychiatrie, 61,* 744–752.

Bowman, K. M., Goodhart, R., and Jolliffe, N. 1939. Observations on role of vitamin B₁ in the etiology and treatment of Korsakoff psychosis. *Journal of Nervous and Mental Disease, 90,* 569–575.

Braceland, F. J. 1957. Kraepelin, his system and his influence. *American Journal of Psychiatry, 113,* 871–876.

Brierley, J. B. 1966. The neuropathology of amnesic states. In C. W. M. Whitty and O. L. Zangwill (Eds.), *Amnesia.* London: Butterworths. Pp. 150–180.

Broca, P. 1878. Anatomie comparée des circonvolutions cérébrales. Le grand lobe limbique et la scissure limbique dans la série des mammifères. *Revue d'Anthropologie, Ser. 2, 1,* 385–498.

Brown, P. 1988a. Human growth hormone therapy and Creutzfeldt-Jakob disease: A drama in three acts. *Pediatrics, 81,* 85–99.

Brown, P. 1988b. The decline and fall of Creutzfeldt-Jakob disease associated with human growth hormone therapy. *Neurology, 38,* 1135–1136.

Brown, P., Gajdusek, D. C., Gibbs, C. J., Jr., and Asher, D. M. 1985. Potential epidemic of Creutzfeldt-Jakob disease from human growth hormone therapy. *New England Journal of Medicine, 313,* 728–731.

Brown, S., and Schäfer, E. A. 1888. An investigation into the functions of the occipital and temporal lobes of the monkey's brain. *Philosophical Transactions of the Royal Society of London (Biology), 179,* 303–327. (Paper read Dec. 15, 1887)

Bumke, O., and Foerster, O. 1936. *Handbuch de Neurologie* (Vol. 13). Berlin: Springer.

Butters, N. 1984. Alcoholic Korsakoff's syndrome: An update. *Seminars in Neurology, 4,* 226–244.

Butts, H. 1913. Korsakow's psychosis, with report of a case. *United States Naval Medical Bulletin, 7,* 113–121.

Campbell, A. C. P., and Biggart, J. H. 1939. Wernicke's encephalopathy (Polioencephalitis haemorrhagica superior): Its alcoholic and non-alcoholic incidence. *Journal of Pathology and Bacteriology, 48,* 245–262.

Campbell, A. C. P., and Russell, W. R. 1941. Wernicke's encephalopathy: The clinical features and their probable relationship to vitamin B deficiency. *Quarterly Journal of Medicine, 10,* 41–64.

Canton, 1860. Chronic alcoholism with impending delirium tremens, treated by suspension of the stimulus. *Lancet, 2,* 237–238.

Carmichael, E. A., and Stern, R. O. 1931. Korsakoff's syndrome: Its histopathology. *Brain, 54,* 189–213.

Charcot, P. 1884. Les paralyses alcooliques. *Gazette des Hôpitaux, 57,* 785–787.

Charness, M. E., and De La Paz, R. L. 1987. Mamillary body atrophy in Wernicke's encephalopathy: Antemordem identification using magnetic resonance imaging. *Annals of Neurology, 22,* 595–600.

Clarke, F. B. 1912–1913. Korsakoff's syndrome with synopsis of cases. *American Journal of Insanity, 69,* 323–336.

Cole, S. J. 1902. On changes in the central nervous system in the neuritic disorders of chronic alcoholism. *Brain, 25,* 326–363.

Coriat, I. H. 1906. The mental disturbances of alcoholic neuritis. *American Journal of Insanity, 62,* 571–613.

Corkin, S. 1968. Acquisition of motor skill after bilateral medial temporal-lobe excision. *Neuropsychologia, 6,* 255–265.

Corkin, S. 1984. Lasting consequences of bilateral medial temporal lobectomy: Clinical course and experimental findings in H. M. *Seminars in Neurology, 4,* 249–259.

Creutzfeldt, H.-G. 1920. Über eine eigenartige herdförmige Erkrankung des Zentralnervensystems. *Zeitschrift für die gessamte Neurologie und Psychiatrie, 57,* 1–18.

Creutzfeldt, H.-G. 1921. Über eine eigenartige herdförmige Erkrankung des Zentralnervensystems. In F. Nissl and A. Alzheimer (Eds.), *Histologische und histopathologische Arbeiten über die Grosshirnrinde.* Jena: G. Fischer. Pp. 1–21.

Dardel. 1868. Alcoholism with delirium tremens complicated by transient diplopia. *Mémoires de la Société de Sciences et Médecine de Lyon, 7,* 392–396.

Davison, C. 1932. Spastic pseudosclerosis (cortico-pallido-spinal degeneration). *Brain, 55,* 247–264.

De Jong, R. N., Itabashi, H. H., and Olson, J. R. 1969. Memory loss due to hippocampal lesions: Report of a case. *Archives of Neurology, 20,* 339–349.

Drachman, D. A., and Arbit, J. 1966. Memory and the hippocampal complex. *Archives of Neurology, 15,* 52–61.

Drachman, D. A., and Ommaya, A. K. 1964. Memory and the hippocampal complex. *Archives of Neurology, 10,* 411–425.

Du Laurens, A. 1599. *A Discourse of the Preservation of Sight: Of Melancholike Diseases; of Rheumes and of Old Age.* (R. Surphlet, Trans.). London: Oxford University Press, 1938.

Edgar, I. I. 1934. Shakespeare's medical knowledge with particular reference to his delineation of madness. *Annals of Medical History, 6,* 150–168.

Eijkman, C. 1897. Eine beriberi-ähnliche Krankheit der Hühner. *Virchow's Archiv für pathologische Anatomie und Physiologie und für klinische Medizin, 148,* 523–532.

Esquirol, J. E. D. 1838 *Des Maladies Mentales* (Vol. 2). Bruxelles: J. B. Tircher. (E. K. Hunt, Trans., 1845). (Reprinted 1938; New York: Arno)

Eustachius, B. 1714. *Tabulae anatomicae.* (J. M. Lancisius, Ed.). Romae: ex officina typographica Francisci Gonzagnae.

Farquhar, J., and Gajdusek, D. C. 1981. *Early Letters and Field Notes from the Collection of D. Carleton Gajdusek.* New York: Raven Press.

Fischer, O. 1907. Miliare Nekrosen mit drusigen Wucherungen der Neurofibrillen, eine regelmässige Veränderung der Hirnrinde bei seniler Demenz. *Monatsschrift für Psychiatrie und Neurologie, 22,* 361–372. Translated as, "Miliary necrosis with nodular proliferation of the neurofibrils, a common change of the cerebral cortex in senile dementia," in Bick, Amaducci, and Pepeu, 1987.

Fischer, O. 1910. Die presbyophrene Demenz, deren anatomische Grundlagen und klinische Abgrenzung. *Zeitschrift für Neurologie und Psychiatrie, 3,* 371–471.

Fuller, S. C. 1911. A study of the miliary plaques found in brains of the aged. *Transactions of the American Medico-Psychological Association, 18,* 109–181.

Fuller, S. C. 1912. Alzheimer's disease (senium prcox): The report of a case and review of published cases. *Journal of Nervous and Mental Disease, 39,* 440–455, 536–557.

Funk, C. 1911. On the chemical nature of the substance which cures polyneuritis in birds induced by a diet of polished rice. *Journal of Physiology, 43,* 395–400.

Funk, C. 1912. The etiology of the deficiency diseases: Beriberi, polyneuritis in birds, epidemic dropsy, scurvy, experimental scurvy in animals, infantile scurvy, ship beriberi, pellagra. *Journal of State Medicine, 20,* 341–368.

Gajdusek, D. C. 1981. Introduction. In J. Farquhar and D. C. Gajdusek (Eds.), *Early Letters and Field Notes from the Collection of D. Carleton Gajdusek.* New York: Raven Press. Pp. xxi–xxviii.

Gajdusek, D. C., Gibbs, C. J., Jr., and Alpers, M. 1966. Experimental transmission of a kuru-like syndrome to chimpanzees. *Nature, 209,* 794–796.

Gajdusek, D. C., Gibbs, C. J., Jr., and Alpers, M. 1967. Transmission and passage of experimental "kuru" to chimpanzees. *Science, 155,* 212–214.

Gajdusek, D. C., and Zigas, V. 1957. Degenerative disease of the central nervous system in New Guinea: The endemic occurrence of "kuru" in the native population. *New England Journal of Medicine, 257,* 974–978.

Galen, C. 1821–1833. *De symptomatum differentiis liber.* Translated in K. Kuhn, *Opera omnia* (Vol. 7). Leipzig: Knobloch.

Galen, C. 1962. On anatomical procedures. In M. C. Lyons and B. Towers (Eds.), *Galen on Anatomical Procedures.* (W. H. L. Duckworth, Trans.). Cambridge, U.K.: Cambridge University Press.

Gamper, E. 1928. Zur Frage der Polioencephalitis haemorrhagica der chronischen Alkoholiker: Anatomische Befund beim alkoholischen Korsakow und ihre Beziehungen zum klinischen Bild. *Deutsche Zeitschrift für Nervenheilkunde, 102,* 122–129.

Gibbs, C. J., Jr., Gajdusek, D. C., Asher, D. M., Alpers, M. P., Beck, E., Daniel, P. M., and Matthews, W. B. 1968. Creutzfeldt-Jakob disease (spongiform encephalopathy): Transmission to the chimpanzee. *Science, 161,* 388–389.

Glasse, R. B. 1967. Cannibalism in the kuru region of New Guinea. *Transactions of the New York Academy of Sciences, Ser. 2, 29,* 748–754.

Glees, P., and Griffith, H. B. 1952. Bilateral destruction of the hippocampus (cornu ammonis) in a case of dementia. *Monatsschrift für Psychiatrie und Neurologie, 123,* 193–204.

Goldgaber, D., Lerman, M. I., McBride, W., Saffiotti, U., and Gajdusek, D. C. 1987. Characterization and chromosomal localization of cDNA encoding brain amyloid of Alzheimer's disease. *Science, 235,* 877–880.

Goldstein, 1970. Carl Wernicke (1848–1904). In W. Haymaker and F. Schiller (Eds.), *The Founders of Neurology.* Springfield, IL: Charles C Thomas. Pp. 531–535.

Griesinger, W. 1867. *Die Pathologie und Therapie der psychischen Krankheiten* (2nd ed.). Translated by C. L. Robinson and J. Rutherford as, *Mental Pathology and Therapeutics* (1st ed. 1845). New York: Wood, 1882. (Reprinted 1965; New York: Hafner)

Grünthal, E. 1939. Ueber das Corpus mamillare und den Korsakowschen Symptomenkomplex. *Confinia Neurologica, 2,* 64–95.

Grünthal, E. 1947. Über das Bild nach umschriebenem beiderseitigen Ausfall der Ammonshornrinde. Ein Beitrag zur Kenntnis der Funktion der Ammonshorns. *Monatsschrift für Psychiatrie und Neurologie, 113,* 1–16.

Gudden, H. 1896. Klinische und anatomische Beiträge zur Kenntniss der multiplen Alkoholneuritis nebst Bemerkungen über die Regenerations vorgänge im peripheren Nervensystem. *Archiv für Psychiatrie, 28,* 643–741.

Hadlow, W. J. 1959. Scrapie and kuru. *Lancet, 2,* 289–290.

Hadlow, W. J., Kennedy, R. C., and Race, R. E. 1982. Natural infection of Suffolk sheep with scrapie virus. *Journal of Infectious Diseases, 146,* 657–664.

Hegglin, K. 1953. Über einen Fall von isolierter, linksseitiger Ammonshornerweichung bei präseniler Demenz. *Monatsschrift für Psychiatrie und Neurologie, 125,* 170–186.

Heidenhain, A. 1929. Klinische und anatomische Untersuchungen über eine eigenartige Erkrankung des Zentralnervensystems im Präsenium. *Zeitschrift für die gesamte Neurology und Psychiatrie, 118,* 49–114.

Henderson, D. K., and Maclachlan, S. H. 1930. Alzheimer's disease. *Journal of Mental Science, 76,* 646–661.

Hornabrook, R. W. 1978. Slow virus infections in the central nervous system. In P. J. Vinken and G. H. Bruyn, in collaboration with H. L. Klawans, *Handbook of Clinical Neurology: Vol. 34. Infections of the Nervous System, Pt. II.* Amsterdam: North Holland Publishing Company. Pp. 275–291.

Hurd, A. W. 1905. Korsakoff's psychosis—Report of cases. *American Journal of Insanity, 62,* 63–76.

Ironside, R. 1939. Neuritis complicating pregnancy. *Proceedings of the Royal Society of Medicine, 32,* 588–595.

Jackson, J. 1822. On a peculiar disease resulting from the use of ardent spirits. *New England Journal of Medicine and Surgery, 11,* 351–353.

Jacob, H. 1970. Alfons Jakob (1884–1931). In W. Haymaker and F. Schiller (Eds.), *The Founders of Neurology* (2nd ed.). Springfield, IL: Charles C Thomas. Pp. 338–342.

Jakob, A. M. 1921a. Über eigenartige Erkrankungen des Zentralnervensystems mit bemerkenswerten anatomischen Befunden (spastische Pseudosclerose-encephalomyelopathie mit disseminierten Degenerationsherden). *Deutsche Zeitschrift für Nervenheilkunde, 70,* 132–146.

Jakob, A. M. 1921b. Über eigenartige Erkrankungen des Zentralnervensystems mit bemerkenswerten anatomischen Befunden (spastische Pseudosclerose-encephalomyelopathie mit disseminierten Degenerationsherden). *Zeitschrift für die gessamte Neurologie und Psychiatrie, 64,* 147–228.

Jakob, A. M. 1921c. Über eigenartige Erkrankungen des Zentralnervensystems mit bemerkenswerten anatomischen Befunden (spastische Pseudosclerose-encephalomyelopathie mit disseminierten Degenerationsherden). *Medizinische Klinik, Berlin 17,* 372–376.

Jakob, A. M. 1923. *Die extrapyramidalen Erkrankungen . . .* Berlin: Springer.

Jansen, B. C. P. 1956. Early nutritional researches on beriberi leading to the discovery of vitamin B_1. *Nutrition Abstracts and Reviews, 26,* 1–14.

Jelliffe, S. E. 1908. The alcoholic psychoses. Chronic alcoholic delirium (Korsakoff's psychosis). *New York Medical Journal, 88,* 769–777.

Jolliffe, N., Wortis, H., and Fein, H. D. 1941. The Wernicke syndrome. *Archives of Neurology and Psychiatry, 46,* 569–597.

Jolly, F. 1897. Über die psychischen Störungen bei Polyneuritis. *Charité Annalen, 22,* 579–612.

Kant, F. 1932. Die Pseudoencephalitis Wernicke der Alkoholiker (polioencephalitis haemorrhagica superior acuta). Ein Beitrag zur Klinik. *Archiv für Psychiatrie und Nervenkrankheiten, 98,* 702–768.

Katzenelbogen, S. 1970. Sergei Korsakov (1853–1900). In W. Haymaker and F. Schiller (Eds.), *The Founders of Neurology* (2nd ed.). Springfield, IL: Charles C Thomas. Pp. 460–463.

Kirschbaum, W. R. 1968. *Jakob-Creutzfeldt Disease.* New York: Elsevier.

Klitzman, R. L., Alpers, M. P., and Gajdusek, D. C. 1984. The natural incubation period of kuru and the episodes of transmission in three clusters of patients. *Neuroepistemology, 3,* 3–20.

Korsakoff, S. S. 1887. Disturbance of psychic functions in alcoholic paralysis and its relation to the disturbance of the psychic sphere in multiple neuritis of non-alcoholic origin. *Vestnik Psichiatrii, 4,* Fascicle 2.

Korsakoff, S. S. 1889a. A few cases of peculiar cerebropathy in the course of multiple neuritis. *Ejenedelnaja Klinischeskaja Gazeta*, Nos. 5–7.

Korsakoff, S. S. 1889b. Psychic disorder in conjunction with multiple neuritis (Psychosis polyneuritica s. Cerebropathia psychica toxaemica). *Medizinskoje Obozrenije, 31*, No. 13.

Korsakoff, S. S. 1890. Ueber eine besondere psychische Störung, combinirt mit multipler Neuritis. *Archiv für Psychiatrie und Nervenkrankheiten, 21*, 669–704.

Kraepelin, E. 1910. *Psychiatrie, Vol. 2* (8th Ed.). Leipzig: J. Barth.

Lancereaux, E. 1864. *Alcoolisme. Dictionaire Encyclopédique des Sciences Médicales*. Paris: Masson.

Lawson, R. 1879. On the symptomatology of alcoholic brain disorders. *Brain, 1*, 182–194.

Lettsom, J. C. 1787. Some remarks on the effects of lignum quassiae amarae. *Memoires of the Medical Society of London, 1*, 128–165.

Lewis, F. T. 1923. The significance of the term *hippocampus*. *Journal of Comparative Neurology, 35*, 213–230.

Mair, R. G., Anderson, C. D., Langlais, P. J., and McEntee, W. J. 1988. Behavioral impairments, brain lesions and monoaminergic activity in the rat following a bout of thiamine deficiency. *Behavioural Brain Research, 27*, 223–239.

Major, R. H. 1978. *Classic Descriptions of Disease* (3rd ed.). Springfield, IL: Charles C Thomas.

Malamud, N., and Skillicorn, S. A. 1956. Relationship between the Wernicke and the Korsakoff syndrome. *Archives of Neurology and Psychiatry, 76*, 585–596.

Martin, P. R., Eckhardt, M. J., and Linnoila, M. 1989. Treatment of chronic organic mental disorders associated with alcoholism. In M. Galanter (Ed.), *Recent Developments in Alcoholism* (Vol. 7). New York: Plenum Press. Pp. 329–350.

May, W. W. 1968. Creutzfeldt-Jakob disease. 1. A survey of the literature and clinical diagnosis. *Acta Neurologica Scandinavica, 44*, 1–32.

Meyer, A. 1971. *Historical Aspects of Cerebral Anatomy*. London: Oxford University Press.

Mijake, K. 1906. Beiträge zur Kenntnis der Altersveränderungen der menschlichen Hirnrinde. *Arbeiten aus dem Obersteinerschen neurologischen Institute und der Wiener Universität. 13*, 212–259.

Milner, B. 1972. Disorders of learning and memory after temporal lobe lesions in man. *Clinical Neurosurgery, 19*, 421–446.

Milner, B., Corkin, S., and Teuber, H.-L. 1968. Further analysis of the hippocampal amnesic syndrome: 14-year follow-up of H. M. *Neuropsychologia, 6*, 215–234.

Milner, B., and Penfield, W. 1955. The effect of hippocampal lesions on recent memory. *Transactions of the American Neurological Association, 80*, 42–48.

Minot, G. R., Strauss, M. B., and Cobb, S. 1933. "Alcoholic" polyneuritis: Dietary deficiency as a factor in its production. *New England Journal of Medicine, 208*, 1244–1249.

Moeli, C. 1883. Statistisches und Klinisches über Alcoholismus. *Charité Annalen, 9*, 524.

Moll, J. M. 1915. "Amnestic" or "Korsakoff's" syndrome with alcoholic etiology: An analysis of 30 cases. *Journal of Mental Science, 61*, 424–443.

Neubürger, K. 1931. Über Hirnveränderung nach Alkoholmissbrauch. *Zeitschrift für die gesamte Neurologie und Psychiatrie, 135*, 159–209.

Papez, J. W. 1958. Visceral brain, its component parts and their connections. *Journal of Nervous and Mental Disease, 126*, 40–55.

Patterson, R. V. 1905. Two cases of Korsakoff's psychosis. One with necropsy. *Journal of Nervous and Mental Disease, 32*, 798–800.

Penfield, W. 1952. Memory mechanisms. *Archives of Neurology and Psychiatry, 67*, 178–198.

Penfield, W. 1958. *The Excitable Cortex in Conscious Man*. Liverpool, U.K: Liverpool University Press.

Penfield, W. 1968. Engrams in the human brain. *Proceedings of the Royal Society of Medicine, 61*, 831–840.

Penfield, W. 1974. Memory. Autopsy findings and comments on the role of the hippocampus in experiential recall. *Archives of Neurology, 31*, 145–154.

Penfield, W., and Milner, B. 1958. Memory deficit produced by bilateral lesions in the hippocampal zone. *Archives of Neurology and Psychiatry, 79*, 475–497.

Pershing, H. T. 1892. Alcoholic multiple neuritis with characteristic mental derangement. *International Medical Magazine, 1*, 803–808.

Perusini, G. 1910. Über klinische und histologisch eigenartige psychische Erkrankungen des späteren Lebensalters. *Histologische und histopathologische Arbeiten, 3*, 297–351.

Perusini, G. 1911. Sul valore nosografico di alcuni reperti istopatologici caratteristici per la senilità. *Revista Italiana di Neuropatologia, Psichiatria ed Elettroterapia, 4*, 193–213.

Pick, A. 1892. Ueber die Beziehungen der senilen Hirnatrophie zur Aphasie. *Prager medizinische Wochenschrift, 17*, 165–167.

Pliny the Elder. 1952. *Natural History* (Vol. II, Libri III–VII). (H. Rackham, Trans.). Loeb Classical Library. London: Heinemann.

Prichard, J. C. 1937. *A Treatise on Insanity and Other Disorders Affecting the Mind*. Philadelphia, PA: Carey and Hart.

Prickett, C. O. 1934. The effect of a deficiency of vitamin B₁ upon the central and peripheral nervous systems of the rat. *American Journal of Physiology, 107*, 459–470.

Ramón y Cajal, S. 1901–1902. Estudios sobre la corteza cerebral humana. *Trabajos del Laboratorio de Investigaciónes Biologías, 1*, 1–227.

Rappaport, E. B. 1987. Iatrogenic Creutzfeldt-Jakob disease. *Neurology, 37*, 1520–1522.

Redlich, E. 1898. Über miliare Sklerose der Hindrinde bei seniler Atrophie. *Jahrbücher für Psychiatrie und Neurologie, 17*, 208–216.

Ribot, T. 1881. *Les Maladies de la Mémoire*. Paris: J. B. Ballière. Translated by J. Fitzgerald as, *The Diseases of Memory*. New York: Humboldt Library of Popular Science Literature (Vol. 46). 453–500, 1883.

Robinson, C. H. 1877. Cirrhosis of the liver: Alcoholic paralysis. *British Medical Journal, 1*, 352–353.

St. George-Hyslop, P., Tanzi, R. E., Polinski, R. J., Haines, J. L., Nee, L., Watkins, P. C., Meyers, R. H., Feldman, R. G., Pollen, D., Drachman, D., Growdon, J., Bruni, A., Fonsin, J. F., Salmon, D., Frommelt, P., Amaducci, L., Sorbi, S., Piacentini, S., Stewart, G. D., Hobbs, W. J., Conneally, P. M., and Gusella, J. F. 1987. The genetic defect causing familial Alzheimer's disease maps on chromosome 21. *Science, 235*, 885–889.

Sandy, W. C. 1913. Polyneuritic delirium—Korsakoff's psychosis. *American Journal of Insanity, 69*, 739–754.

Schnitzler, J. G. 1911. Zur Abgrenzung der sog. Alzheimerschen Krankheit. *Zeitschrift für die gesamte Neurologie und Psychiatrie, 7*, 34–57.

Scoville, W. B. 1954. The limbic lobe in man. *Journal of Neurosurgery, 11*, 64–66.

Scoville, W. B. 1968. Amnesia after bilateral mesial temporal-lobe excision: Introduction to Case H. M. *Neuropsychologia, 6*, 211–213.

Scoville, W. B., and Milner, B. 1957. Loss of recent memory after bilateral hippocampal lesions. *Journal of Neurology, Neurosurgery, and Psychiatry, 20*, 11–21.

Shakespeare, W. 1961. *The Works of William Shakespeare Gathered into One Volume*. New York: Oxford University Press.

Shattuck, G. C. 1928. Relation of beri-beri to polyneuritis from other causes. *American Journal of Tropical Medicine, 8*, 539–543.

Sigurdsson, B. 1954. Rida, a chronic encephalitis in sheep. With general remarks on infections which develop slowly and some of their special characteristics. *British Veterinary Journal, 110*, 341–354.

Sigurdsson, B., Grímsson, H., and Pálsson, P. A. 1952. Maedi, a chronic progressive infection of sheep's lungs. *Journal of Infectious Diseases, 90*, 233–241.

Sigurdsson, B., and Pálsson, P. A. 1958. Visna of sheep. A slow, demyelinating infection. *British Journal of Experimental Pathology, 39*, 519–528.

Sigurdsson, B., Pálsson, P. A., and van Bogaert, L. 1962. Pathology of Visna. *Acta Neuropathologica, 1*, 343–362.

Sigurdsson, B., Pálsson, P. A., and Tryggvadóttir, A. 1953. Transmission experiments with Maedi. *Journal of Infectious Diseases, 93*, 166–175.

Sims, F. R. 1905. Anatomical findings in two cases of Korsakoff's symptomcomplex. *Journal of Nervous and Mental Disease, 32*, 160–171.

Soukhanoff, S., and Boutenko, J. 1903. A study of Korsakoff's disease. *Journal of Mental Pathology, 4*, 1–33.

Spielmeyer, W. 1922, Die histopathologische Forschung in der Psychiatrie. *Klinische Wochenschrift, 1*, 1817–1819.

Squire, L. R. 1986. Mechanisms of memory. *Science, 232*, 1612–1619.

Squire, L. R. 1987. *Memory and the Brain*. New York: Oxford University Press.

Squire, L. R., and Zola-Morgan, S. 1983. The neurology of memory: The case for correspondence between the findings for human and nonhuman primates. In J. A. Deutsch (Ed.), *The Physiological Basis of Memory*. San Diego, CA: Academic Press. Pp. 199–267.

Stanley, C. E. 1909–1910. A report of three cases of Korssakow's psychosis. *American Journal of Insanity, 66*, 613–622.

Starr, M. A. 1887. Multiple neuritis and its relation to certain peripheral neuroses. *Medical News, 141*–154.

Strauss, M. B. 1956. John Blackall's observations on the relation between nutritional deprivation and "wet" beri-beri. *New England Journal of Medicine, 255*, 181.

Strümpell, A. 1883. Zur Kenntniss der multiplen degenerativen Neuritis. *Archiv für Psychiatrie und Nervenkrankheiten, 14*, 339–358.

Sweet, W. H., Talland, G. A., and Ervin, F. R. 1959. Loss of recent memory following section of fornix. *Transactions of the American Neurological Association, 84*, 76–82.

Tanaka, T. 1934. So-called breast milk intoxication. *American Journal of Diseases of Children, 37*, 1286–1298.

Tanzi, R. E., Gusella, J. F., Watkins, P. C., Bruns, G. A. P., St. George-Hyslop, P., van Keuren, M. L., Patterson, K., Pegan, S., Kurnit, D. M., and Neve, R. L. 1987. Amyloid beta protein gene: cDNA, mRNA distribution and genetic linkage near the Alzheimer locus. *Science, 1987, 235*, 880–884.

Tibbitts, E. (Ed.). 1957. *Aging in the Modern World*. Ann Arbor: University of Michigan Press.

Torack, R. M. 1978. *The Pathologic Physiology of Dementia*. Berlin: Springer.

Torack, R. M. 1983. The early history of senile dementia. In B. Reisberg (Ed.), *Alzheimer's Disease*. New York: Free Press. Pp. 23–28.

Torvik, A., Lindboe, C. F., Rodge, S. 1982. Brain lesions in alcoholics: A neuropathological study with clinical correlations. *Journal of the Neurological Sciences, 56,* 233–248.

Treviranus, G. R. 1816–1821. *Vermischte Schriften anatomischen und physiologischen Inhalts.* Göttingen: J. F. Röwer.

Turner, J. 1910. Alcoholic insanity (Korsakow's polyneuritic psychosis): Its symptomatology and pathology. *Journal of Mental Science, 56,* 25–63.

Vesalius, A. 1543. *De humani corporis fabrica* (Bk. VII). Basel: J. Oporinus.

Victor, M. 1969. The amnesic syndrome and its anatomical basis. *Canadian Medical Association Journal, 100,* 1115–1124.

Victor, M., Adams, R. D., and Collins, G. H. 1989. *The Wernicke-Korsakoff Syndrome and Related Neurologic Disorders Due to Alcoholism and Malnutrition.* Philadelphia, PA: F. A. Davis Company.

Victor, M., Angevine, J. B., Mancall, E. L., and Fisher, C. M. 1961. Memory loss with lesions of hippocampal formation. *Archives of Neurology, 5,* 244–263.

Victor, M., and Yakovlev, P. I. 1955. S. S. Korsakoff's psychic disorder in conjunction with peripheral neuritis. A translation of Korsakoff's original article with brief comments on the author and his contributions to clinical medicine. *Neurology, 5,* 394–406.

Vieussens, R. 1684. *Neurographia universalis.* Lyon: J. Certe.

Wagener, H. P., and Weir, J. F. 1937. Ocular lesions associated with postoperative and gestational nutritional deficiency. *American Journal of Ophthalmology, 20,* 253–259.

Wernicke, C. 1881. *Lehrbuch der Gehirnkrankheiten* (Vol. 2). Kassel und Berlin: G. Fischer. Pp. 229–242.

Wernicke, C. 1900. *Grundriss der Psychiatrie.* Leipzig: Thieme.

Wilks, S. 1864. Clinical notes on atrophy of the brain. *Journal of Mental Science, 10,* 381–392.

Williams, R. R. 1936. Structure of vitamin B_1. *Journal of the American Chemistry Society, 58,* 1063–1064.

Willis, T. 1664. *Cerebri anatome: cui accessit nervorum descriptio et usus.* London: J. Martyn and J. Allestry. Tercentenary ed., 1664–1964, *Thomas Willis, The Anatomy of the Brain and Nerves.* Montreal: McGill University Press, 1965.

Wilson, D. R., Anderson, R. D., and Smith, W. 1950. Studies in scrapie. *Journal of Comparative Pathology, 60,* 267–282.

Wisniewski, H. M. 1983. Neuritic (senile) and amyloid plaques. In B. Reisberg (Ed.), *Alzheimer's Disease.* New York: Free Press. Pp. 57–61.

Wisniewski, H. M. 1989. Milestones in the history of Alzheimer disease research. In I. Khalid, H. M. Wisniewski, and B. Winblad (Eds.), *Alzheimer's Disease and Related Disorders.* New York: Alan R. Liss. Pp. 1–11.

Zimmerman, H. M. 1939. The pathology of the nervous system in vitamin deficiencies. *Yale Journal of Biology and Medicine, 12,* 23–29.

Part VI
Speech and Cerebral Dominance

Chapter 25
Speech and Language

I sat down and wrote the first two words that were required; but in that moment I was not able either to find the following words in my mind or to carry out the necessary strokes with my pen. I strained my attention to the utmost; I tried to print very slowly one letter after the other. . . . I said to myself that these were not the letters which I wanted; not in the least could I imagine what really was wrong with them. I tried to speak, to train myself, as it were, and to try to test whether I could say something in a connected order; but much as I forced my attention and my thoughts and proceeded with the utmost slowness I became aware very soon of shapeless monstrous words that were absolutely different from those I intended; my immortal soul was at present as little master of the inner tools of language as it had been before my writing.

Johann Joachim Spalding, 1783

EARLY DESCRIPTIONS OF LOSS OF SPEECH

As noted in Chapter 1, the *Edwin Smith Surgical Papyrus,* written approximately 3,600 years ago, was composed of much older records (Breasted, 1930; Woollam, 1958). It has been suggested that the author of the original papyrus, upon which this document was based, might have been Imhotep, a contemporary of King Zoser (Djoser, twenty-eighth to late twenty-seventh century B.C.) and the first physician-like figure to emerge from antiquity.[1] Two cases described by the author of this papyrus (Numbers 20, 22) specifically associated head injury with loss of speech.

James Breasted, who translated the papyrus in 1930, found Case 20 described as follows:

If thou examinest a man having a wound in his temple penetrating to the bone (and) perforating his temple bone . . . if thou puttest thy fingers on the mouth of that wound and he shudder exceedingly; if thou ask of him concerning his malady *and he speak not to thee* while copious tears fall from both his eyes . . . (this is) an ailment not to be treated.

The following was said about Case 22:

If thou examinest a man having a smash in his temple . . . Cleanse (it) for him with a swab of linen until thou seest its fragments (of bone) in the interior of his ear. If thou callest to him (and) *he is speechless (and) cannot speak,* thou shouldst say concerning him: One having a smash in his temple; he discharges blood from his two nostrils and from his ear; *he is speechless;* (and) he suffers with stiffness in his neck. An ailment not to be treated.

Three points about these two short descriptions merit clarification. The first is that "speechless" does not appear very often in the papyrus. Breasted took it from "silent in

sadness, without speaking, like one suffering with (feebleness) because of something which has entered from outside." Second, the term "speechless" does not always imply "an ailment not to be treated." The early Egyptians had extensive pharmacies and rituals for treating disorders like loss of speech (see Chapter 28). The third point is that the *Edwin Smith Surgical Papyrus* is not the only ancient document in which loss of speech is described.

The writers of the *Samhitas,* such as the *Caraka Samhita,* a first-century text of classical Indian medicine, also described loss of speech. Like the *Edwin Smith Surgical Papyrus,* the *Samhitas* were based upon ancient traditions, some dating as far back as the second millennium B.C. (Banjeri, 1981; Filliozat, 1964; Sharma, 1983). Although not as old as the *Edwin Smith Surgical Papyrus,* the *Samhitas* also make it clear that disorders of speech have always been of great interest to physicians.

THE GRECO-ROMAN PERIOD

Distinguishing between central, peripheral, or even psychogenic causes of loss of voice can be difficult when attempting to understand older writings. This applies not just to the Egyptian papyruses but also to documents from the Greco-Roman period. *Aphonos,* for example, is a word that appears in *Epidemics* and other writings associated with the Hippocratic school of the fifth and fourth centuries B.C. *Aphonos* has been translated in many different ways. Some translators define it as "speechless," "loss of speech," "loss of power of speech," "aphonia," "loss of voice," "loss of articulation," "voiceless," and "aphasia" (see Bateman, 1870; Benton and Joynt, 1960).

In Volume VII of *Epidemics,* there is a description of Anechelus, a boy who had considerable difficulty with his pronunciation and stumbled over the beginnings of words. In the *Coan prognosis,* transient loss of speech *(anaudie)*

---

[1]See Hurry (1926) for a biography of Imhotep, an individual who was a physician, as well as an architect, statesman, priest, and astronomer.

was associated with "paralysis of the tongue, or of the arm and the right side of the body," strongly suggesting a lesion of the central nervous system (Chadwick and Mann, 1950).

Pliny the Elder, Celsus, Soranus of Ephesus (A.D. 98–135), and Galen were among the writers during the Roman Empire to describe speech and other language disorders. Although Pliny's descriptions in his *Natural History* were fragmentary, he did write: "A man has been known when struck by a stone to forget how to read and write but nothing else" (xxiv.88; translated by Rackham).

Celsus, like most of his predecessors in Greece and Rome, believed that the tongue, and not the brain, was the source of most speech disorders. He proposed massages and gargles to loosen the tongue from paralysis. In contrast, Soranus of Ephesus (whose thoughts are known through the writings of Caelius Aurelianus, a fifth-century writer) distinguished between paralysis of the tongue and other diseases that may cause faulty articulation or loss of speech (Drabkin, 1950). In what might be considered a fairly accurate description of aphasia, he noted that in some types of speech disturbances the tongue does not change its color or shape, or lose sensation or motility.

Galen, who followed Soranus, emphasized that word memory could be lost not just by disease but as a consequence of head injury.

FROM THE DARK AGES THROUGH THE RENAISSANCE

Historians seem to agree that during the Dark Ages and the Middle Ages little was added to our knowledge about the relationship between speech disturbances and the brain (Critchley, 1970). Even Avicenna, the most famous of the Islamic writers, avoided references to speechlessness. This was a time when classical doctrines were not challenged and when teaching was based more on philosophical and religious theory than on hard clinical facts. The dominant belief that seemed to have emerged with the Church Fathers by the fourth century was that the fourth ventricle of the brain served memory (see Chapters 2 and 23). In accordance with this theory, aphasia, which was looked upon as a loss of memory for words, was associated with problems of the fourth ventricle.

Somewhat better descriptions of speech disturbances began to appear during the Renaissance. For example, in his *Opera medica* (1481), Antonio Guainerio described two old men with speech disorders. One could not recall the correct name of anyone (possibly amnestic aphasia), and the other could not remember more than three words (possibly motor aphasia). Guainerio followed traditional thinking by ascribing the faulty memories of these two men to excessive phlegm in the fourth ventricle.

Baverius de Baveriis (1543) described cases of paralyses associated with aphasia, but tended to think that these disorders were due to problems in the periphery. In contrast, Paracelsus, who lived in the same century, concluded that centrally caused speech disturbances could occur with or without paralysis. Johann Schenck a Grafenberg (1585; 1530–1598) might have been the clearest of all the writers of this period when it came to stating that loss of memory

for words could occur without paralysis of the tongue (see Benton and Joynt, 1960).

It can be argued that the association between speech and the brain would have been recognized more readily if greater attention had been paid to the surgeons who attempted to remove pieces of broken skull or foreign objects from the brain. A good case study was provided by Francisco Arceo (c. 1493–1573), a Spanish surgeon. He described a workman who was hit on the head by a falling stone. The workman was unable to move or speak until the bone pressing upon the brain was resected several days later. Arceo (1588) wrote that the man then showed excellent recovery, although he at first spoke "imperfecklie, as men of troubled minde are wonte to doe." Another good sixteenth-century description of a successful surgery was provided by Niccolò Massa (see Chapter 26).

THE SEVENTEENTH CENTURY

Before the seventeenth century, descriptions of speech disorders continued to be brief and deficient in essential details. The descriptions provided by Johann Schmidt (1624–1690) and Peter Rommel (1643–1708) late in the seventeenth century, however, represented a marked improvement over those of their predecessors.

Schmidt (1676) described a patient who exhibited disorderly arrangements of spoken words and right-side paralysis after suffering a stroke. Schmidt's case provided the first unmistakable description of a paraphasic disorder. In addition, Schmidt provided one of the first good descriptions of alexia. He wrote: "He did not know a single letter nor could he distinguish one from another."

In contrast to Schmidt's case study, Rommel's (1683) patient suffered from a severe motor aphasia. As can be seen in Rommel's description, she exhibited a marked deficit in conversational speech, but could still recite some verses as well as understand oral and written language:

> After a fairly strenuous walk which she took after dinner, she suffered a mild delirium and apoplexy with paralysis of the right side. She lost all speech with the exception of the words "yes" and "and." She could say no other word, not even a syllable, with these exceptions; the Lord's Prayer, the Apostles' Creed, some Biblical verses and other prayers, which she could recite verbatim and without hesitation but somewhat precipitously. . . . Nevertheless, her memory was excellent. She grasped and understood everything that she saw and heard and she answered questions, even about events in the remote past, by affirmative or negative nods of the head. (Translated in Benton and Joynt, 1960, 210)

Thomas Willis, a contemporary of Rommel, also discussed aphasia. He described a man who suffered a stroke that left him unconscious. When this man regained his consciousness, he recognized his friends, but could not utter their names. This individual was also paralyzed on the right side. Willis then described a second case, very similar to the first, in which there was word-finding difficulty and right-side paralysis (see Critchley, 1970).

Because the basis of speech loss remained poorly understood, seventeenth-century physicians really did not know what to do for their aphasic patients. When William Harvey, famous for his understanding of the circulatory system, lost his power of articulation in 1657, his apothecary made a cut in the frenulum of the tongue to loosen it.[2] Bleeding, leeching, and cupping to draw blood were also accepted treatments for the aphasias, and remained so into the nineteenth century.

THE EIGHTEENTH CENTURY

One of the most interesting eighteenth-century descriptions of aphasia came from Linné (Carolus Linnaeus), a Swede better known for introducing binary nomenclature and for classifying over 6,000 plants (Synek, 1985). Linné was also a physician, and his case of "forgetfulness for Substantives and Names in particular" appeared in 1745. The case is often cited as one of a man who lost his memory for nouns ("amnestic aphasia"), but at the height of his deficit he could not even produce repetitive speech.[3] Another notable aspect of the case was the patient's agraphia for names. Not only could he not write the names of his wife and children but he could not even write his own name. This stood in marked contrast to his ability to read and understand names.

Within a few years of the publication of the aforementioned case, the King of Sweden also became aphasic. In 1748, at the age of 72, King Frederick I (1676–1751) suffered a stroke that caused a paralysis of the right side of his body and a "loss of memory" for the names of his associates. The single exception was the State Councillor, Count Tessin, whose name the King was able to mention clearly. His Majesty called everyone else "doctor," whether male or female.

Eighteenth-century medical science was characterized by more attention being devoted to the cause and nature of the underlying deficits. This included increased interest in the bases of speech and language disorders. Giovanni Battista Morgagni (1761), the renowned Italian pathologist, shown in Figure 25.1 wrote his five-volume *De sedibus et causis morborum per anatomen indagatis (The Seats and Causes of Diseases Investigated by Anatomy)* when he was nearly 80. His huge collection of cases was put together as a sort of correspondence course for a young man whose status might be compared to that of a present-day graduate student (Schiller, 1970).

In his volume devoted to the head and the nervous system, Morgagni emphasized that loss of speech was frequently associated with stroke. He distinguished among (1) loss of voice, (2) uttering sounds devoid of meaning, and (3) speech impairments associated with abnormal appearances of the tongue. Throughout his massive work, Morgagni stressed the importance of looking at the morbid anatomy in conjunction with the clinical symptoms displayed.[4]

Johann Augustin Phillip Gesner (1738–1801) also wrote

Figure 25.1. Giovanni Battista Morgagni (1682–1771), author of *De sedibus,* written when he was 79 years old. Morgagni's five-volume work emphasized the importance of autopsies in understanding neurological disorders.

a five-volume opus of "observations on medical and natural science." It appeared between 1769 and 1776, at about the same time Morgagni was working on his celebrated books. In the section dealing with loss of speech *(Die Sprachamnesie),* Gesner theorized that language deficits were neither due to an all-encompassing disorder of thinking nor to a loss of memory in general, but to very specific impairments in verbal memory. He specifically characterized the aphasic deficit as "the inability to associate images or abstract ideas with their expressive verbal symbols." Inappropriate words and terms, and even new words (neologisms), thus were thought to be the result of inaccurate associations between thoughts and words. Gesner believed this breakdown was caused by diseases of the brain that weakened certain parts. His associationistic model of brain function has been hailed as the first truly modern theory of aphasia (Benton, 1965; Benton and Joynt, 1963).

Associationistic thinking also permeated the writings of Alexander Crichton (1763–1846), a Scottish physician who spent much of his professional life in Russia, where he attended the tzar's family. Crichton's book was published in 1798, before he left the United Kingdom. It contains many fascinating cases of aphasia, although not all were his own.

A good illustration of Crichton's flare for exceptional material and his flowery style can be seen in the case of an aged, married attorney who appeared to have had a stroke after visiting his mistress. Crichton wrote:

---

[2]For a critique of Harvey's own neurological ideas, see Hunter and MacAlpine (1957).

[3]Linné's patient was probably Arvid Arrhenius, a professor at the University of Uppsala (Viets, 1943).

[4]Oddly, Morgagni completely avoided the microscope over his nearly 90 years of life. He never utilized the tool that had already led to advances in the hands of other workers, such as Marcello Malpighi, who worked a century earlier.

The arms of Venus are not wielded with impunity at the age of 70. He was suddenly seized with a great proliferation of strength, giddiness, forgetfulness. . . . When he wished to ask for any thing, he constantly made use of some inappropriate term. Instead of asking for a piece of bread, he would probably ask for his boots; but if these were brought, he knew they did not correspond with the idea he had of the thing he wished to have, and was therefore angry; yet he would still demand some of his boots, meaning bread. If he wanted a tumbler to drink out of, it was a thousand to one he did not call for a certain chamber utensil; and if it was the said utensil he wanted, he would call it a tumbler, or a dish. He evidently was conscious that he pronounced the wrong words, for when the proper expression were spoken by another person, and he was asked if it was not such a thing he wanted, he always seemed aware of his mistake. (1798, 372–373)

Much like Gesner, Crichton (p. 371) looked upon this man's problem as an inability to associate thoughts with words. As he put it: "The person, although he has a distinct notion of what he means to say, cannot pronounce the words which ought to characterize his thoughts."

The eighteenth century was also a time of self-descriptions of aphasic attacks, not just by physicians and scientists but also by famous men from other walks of life. Johann Joachim Spalding (1714–1804), for example, whose own aphasic episode was described at the beginning of this chapter, was an eminent theologian. Other interesting self-reports came from Jean-Paul Grandjean de Fouchy (1707–1788), an astronomer and physicist, and from Samuel Johnson, the great man of letters who seemed to have suffered a stroke in 1783.

The notes provided by Johnson are especially fascinating. They describe his condition immediately after his illness and through his recovery. Johnson showed not only an aphasic disturbance but his penned letters reveal that he also suffered from dysgraphia, a disturbance in the ability to write. For example, the usually fluent Johnson wrote: "I am sorry that your pretty Letter has been so long without being answered; but when I am not pretty well, I do not write plain enough for young ladies." The many peculiarities of Johnson's case history have been discussed in a number of articles, some of which compare his writing style before and after his possible stroke.[5]

In this context, it is interesting to note that John William Ogle (1824–1905), who described the aphasia and dysgraphia of Samuel Johnson in 1874, was the person who first introduced the term "agraphia" to signify the loss of written communication. In 1867, he distinguished between two forms of agraphia, which paralleled and often accompanied the two best-known types of aphasia. He wrote that in "ataxic agraphia," which parallels Broca's (1861) motor aphasia, the ability to write is lost, and there may be only a succession of up and down strokes, bearing little or no resemblance to letters. In "amnemonic agraphia," which is more like the "sensory aphasia" described by Wernicke (1874), patients write fluently, but substitute incorrect words and letters for appropriate ones.[6]

---

[5]Johnson's possible stroke and its effects were described in 1874 by Ogle and in 1970 by Critchley. Johnson, however, showed behavioral abnormalities even from the time of birth. This subject is discussed in Chapter 16.

[6]See Bianchi (1883) and Marcie and Hécaen (1879) for papers on brain pathology and changes in handwriting.

At this time, disorders of communication were of sufficient interest to be incorporated into the fictional literature. An excellent example of aphasia, for instance, can be found in Johann Wolfgang von Goethe's early novel *Wilhelm Meister's Lehrjahre (William Meister's Apprenticeship)*, published in 1795. The case involved Johann Trexler, who had something important he wanted to tell his daughter, Theresa. Unfortunately, Trexler suffered a stroke just before he had a chance to speak to her. It was frustrating for him to realize that he could no longer use spoken language effectively.

APHASIA AND PHRENOLOGY

Franz Joseph Gall (1757–1828) attracted the attention of the scientific community early in the 1800s when he claimed that the cerebral cortex could be subdivided into functional units (see Chapter 3). Gall's theory began with some observations concerning language. He stated that the exceptional "word memory" of a classmate with large bulging eyes "gave the first impulse to my researches, and was the occasion of all my discoveries." Thinking that an abundance of underlying brain tissue was pushing this student's eyes out from behind, he localized the "faculty of attending to and distinguishing words, recollection of words, or verbal memory" in the frontal lobes.

Many of Gall's ideas about speech and language appeared in his *Anatomie et Physiologie du Système Nerveux* (Vol. IV), published in 1819. Yet he did not rely entirely on skull features or bulging eyes when promoting centers for speech in the anterior lobes. He also made his case from comparative anatomy. Even beyond this, Gall cited several cases of "accidental mutilations" to support his phrenological ideas. For

Figure 25.2. Dominique-Jean Larrey (1766–1842), surgeon to Napoleon and the French Army, provided Franz Joseph Gall with several interesting cases of aphasia due to wounds of the frontal lobes. (From Balance, 1922, with permission of Macmillan, London.)

example, he described two men who, after being injured above the eye by the point of a sword, could not remember the names of friends or family members.

Baron Larrey (shown in Figure 25.2), Napoleon Bonaparte's leading battlefield surgeon, sent Gall a third case, Edouard de Rampan, who sustained a fencing injury. The foil entered near the left canine and penetrated the nasal fossa and cribiform plate to protrude into the anterior lobe. The victim showed a loss of memory for words, but not for images and places, and exhibited a paralysis on the right side of his body. De Rampan showed excellent memory for Baron Larrey when his physician entered the room, but he found himself unable to give Baron Larrey's name.[7]

Gall also described a man who suffered from a vascular lesion of the brain. This man could move his tongue, but his speech was so poor that he could communicate only with gestures.

Although Gall might have had the opportunity to recognize the special status of the left hemisphere for speech, he failed to do so. Instead, he followed François-Xavier Bichat in believing that each side of the brain could serve as a complete organ of mind, just as one could conceive that each eye could serve as a complete organ for seeing (Harrington, 1985). Gall thought that unilateral lesions, like those seen in some of his cases, caused deficits by upsetting the balance between the two perfectly equal sides.

Gall's followers in the United States also failed to recognize the importance of the side of the lesion. They agreed that when loss of speech occurred in relative isolation from other mental defects, the condition could best be understood by assuming a focal lesion in the cortical organ responsible for "the faculty of conveying ideas by words" (Freemon, 1991).

An exemplary case was described in 1828 by Samuel Jackson (1787–1872), who was on the medical faculty of the University of Pennsylvania. Jackson's patient had full control of his senses and his intellect, but supposedly could not utter an intelligible word. When Jackson saw that his patient could still move the muscles of his tongue and mouth, he realized that this was not a simple case of paralysis. In fact, when furnished with pen and paper, it became clear that his writing was also affected. His patient wrote: "Didoes doe the doe." Obviously, this was a case of damage that was centered in the organ specifically responsible for word memory. As Jackson stated it:

> The inferences to be drawn from these facts, are, 1st, that as the cerebral irritation produced no general affection or disturbance of the functions of the brain, it was local or limited; and 2nd, as loss of language was the only function derangement of the intellectual faculties, that faculty must have been connected with the portion of the brain, the seat of the irritation; and 3rd, that an organ of language exists in the brain. This case lends strong confirmation to the general truth of the doctrines of Phrenology. (1828, 274)

Jackson's case study was followed by the case of a 55-year-old man who would speak only occasionally and seemed "to have lost the conventional connexion between *an idea and the word* denoting it." S. Henry Dickson (1798–1872), who presented this case in 1830—and who also noted loss of

comprehension in reading and writing—chose not to interpret the deficits in the context of phrenology. Nathaniel Chapman (1780–1853), the editor of the *American Journal of Medical Sciences,* however, wrote a follow-up to Dickson's paper. Chapman (1835) interpreted Dickson's case phrenologically and compared the case to another patient described in the *British Phrenological Journal.*

Examples of phrenology being applied to explain problems with proper nouns and paraphasic speech can also be found in the American literature of this decade. One such report was published by Daniel Drake (1785–1852) in 1834. His patient made word substitutions, such as "Kentucky" for "Louisville," after being shot with a musket ball in the left temple.

BOUILLAUD AND LOCALIZATION THEORY

Although Gall's phrenological beliefs were soon rejected by the majority of people trained in science and medicine, the idea that speech may be localized in the cortex was not ready to die. The most ardent supporter of cortical speech localization was Jean-Baptiste Bouillaud, a man who began within the phrenological camp but then moved away from phrenology, being unable to accept cranioscopy as a viable method for studying brain function (see Chapter 3).

Although he was a founding member of the *Société Phrénologique,* Bouillaud, shown in Figure 25.3, turned to clinical examinations and autopsy material to argue for cortical localization. Unlike Gall, who based his conclusions on small numbers of exceptional cases, Bouillaud used large samples of subjects for his studies. By 1825 he had amassed information on many individuals who showed speech pathology, and his collection of cases grew to over 500 in his lifetime. In fact, Bouillaud was probably the first brain scientist to analyze large samples.

Figure 25.3. Jean-Baptiste Bouillaud (1796–1881), who began arguing for localization of speech functions in the frontal cortex long before Broca presented his celebrated case of "Tan" in 1861.

---

[7]Some writers have cited the case of Edouard de Rampan as the first really good description of speech loss associated with a discrete frontal lobe injury (Fancher, 1979; Riese, 1947).

The following passage illustrates Bouillaud's logic as applied to the localization of speech functions:

> It is evident that the movements of the organs of speech must have a special centre in the brain, because speech can be completely lost in individuals who present no other signs of paralysis, whilst on the contrary other patients have the free use of speech coincident with paralysis of the limbs. But it is not sufficient to know that there exists in the brain a particular centre destined to produce and coordinate the marvellous movements by which man communicates his thoughts and feelings, but it is above all important to determine the exact situation of this coordinating centre. From the observations I have collected, and from the large number I have read in the literature, I believe I am justified in advancing the view that the principle lawgiver of speech is to be found in the anterior lobes of the brain. (1825b; Translated in Head, 1926, 15)

It is less well known that Bouillaud conducted experiments on animals. In particular, he used chickens and dogs to study the effects of brain lesions. In one experiment, dated 1827, he damaged the anterior part of the brain of a dog. He stated that the dog could utter cries of pain, showing it still had control over the muscles for vocalizations, although it lost the power to bark. As retold by Ogle:

> M. Bouillaud pierced with a gimlet the head of a living dog from side to side. The instrument passed through the brain at the junction of the anterior and middle lobes—that is, about the part which corresponds to Broca's region. After death it was found that there was a canal as big as one's finger in this part, and that all the rest of the brain was perfectly healthy. The dog recovered completely from this fearful operation, though with the loss of much of its former intelligence. The point, however, of chief interest is this: the dog entirely lost its power of barking. It could still make sounds, uttering sharp cries when in pain. But during the two months that it lived in perfect health after the operation it was never heard to bark, as it had been used to do, so as to express its emotions. (1867, 87)

Loss of speech as revealed by the clinico-pathological method continued to be a central theme in many of Bouillaud's later publications (e.g., 1830, 1839–40). Yet because the scientific community was reacting so negatively to Gall and to those who had been associated with him even in spirit, most scientists also rejected or ignored Bouillaud's ideas about localization. Jean Cruveilhier, for example, responded to Bouillaud in the following way:

> If it is pathologically demonstrated that a lesion of the anterior lobes is constantly accompanied by a corresponding alteration of speech; on the other hand, if lesions of all parts of the brain other than the anterior lobes never entail speech alterations, the question is solved, and I immediately become a phrenologist. . . . In fact, the loss of the faculty of sound articulation is not always the result of a lesion of the anterior lobes of the brain; moreover, I am able to prove that the loss of the faculty of sound articulation can accompany a lesion of any other part of the brain. (Translated by Riese, 1977, 59)

As one might expect from statements like the above, there were many examples in the literature from 1800 to 1860 of frontal lobe damage without loss of speech (see Bateman, 1870; Sondhaus and Finger, 1988). In addition, lesions in other lobes sometimes affected speech, especially soon after acute insult.

Figure 25.4.   Gabriel Andral (1797–1876). In 1840, Andral wrote that "loss of speech is not a necessary result of lesions in the anterior lobe."

For instance, Gabriel Andral (1840), shown in Figure 25.4, discussed 14 cases of aphasia with damage outside the anterior lobe, as well as 37 cases of anterior lobe injuries, of which only 21 showed serious disruption of speech. Andral (p. 368) was forced to conclude that "loss of speech is not a necessary result of lesions in the anterior lobe and furthermore it can occur in cases in which anatomical investigation shows no changes in these lobes." Oliver Wendell Holmes (1883, p. 420), who studied medicine in Paris from 1833 to 1835, spoke of Andral as a man "whose natural eloquence made it delightful to listen to him." More important, Andral's thoughts and opinions, which were not favorable to localization at the time, carried considerable weight.

In 1843, Jacques Lordat (1773–1870), one of the principle figures in aphasia research and a faculty member at the Montpellier School of Medicine, described his own stroke and subsequent loss of speech.[8] Although Lordat's aphasia ("verbal amnesia") was most severe in 1825, he showed enough recovery to return to his university position several months later, and then survived his aphasic episode by 45 years (Lecours, Nespoulous, and Pioger, 1987). Lordat did not advocate cortical localization in his "Montpellier lessons" or in other self-reports and case studies. His contributions served to polarize members of the French scientific societies even more. By the middle of the nineteenth century, the meetings of the academies were everything but calm.

This was the *Zeitgeist* into which M. Belhomme presented some material in support of Bouillaud's ideas in 1848. This stimulated Bouillaud (1848) to present even

---

[8]Jacques Lordat trained at a military hospital and received the title *chirurgien militaire*. He joined the Montpellier faculty as *Professeur libre* in 1797, served on the faculty for years, and even became dean. Lordat's aphasiology was mentioned by Marc Dax, another Montpellieran, in Dax's 1836 manuscript on aphasia (translated by Joynt and Benton, 1964). Unlike Dax, who argued for cortical specialization, Lordat (1843) viewed the cortex as essential for intellectual and voluntary activities, but not as a functionally divisible structure.

more cases and to make one of the most famous bets in the history of the brain sciences: "Herewith I offer 500 francs to anyone who will provide me with an example of a deep lesion of the anterior lobules of the brain without a lesion of speech."

Bouillaud's challenge did not go unanswered, but the prize was not awarded until 1865 (Joynt, 1961; Schiller, 1979). The money was claimed by Alfred Velpeau (1795–1867) for a case he had presented to the *Académie de Médecine* in 1843. His patient had a massive cancerous tumor, which Velpeau believed "had taken the place of the two anterior lobes." The individual not only spoke fluently but proved to be loquacious ("a greater chatterer never existed"). His fluent speech was not a matter of contention. Rather, it was the idea that a fair amount of frontal tissue had been spared (see Kussmaul, 1887).[9]

BROCA AND THE REVOLUTION OF 1861

Early in 1861, at the *Société d'Anthropologie* in Paris, Pierre Gratiolet described the large skull of a Central American Totonac Indian. A debate erupted about whether the size of the brain was related to intelligence or whether the critical factor was the development of individual parts of the brain, especially the frontal lobes. The latter position was taken by Simon Alexandre Ernest Aubertin, Bouillaud's son-in-law. Aubertin was convinced that it would take only one localization to destroy the idea of an equipotential cortex and that speech was precisely the function to prove the point.

At a continuation of the meeting, Aubertin described the extraordinary case of a patient at the Hôpital St. Louis who attempted suicide and shot away his frontal bone, thereby exposing the anterior lobes. Aubertin (1861) stated:

During the interrogation the blade of a large spatula was placed on the anterior lobes; by means of light pressure speech was suddenly stopped; a word that had been commenced was cut in two. The faculty of speech reappeared as soon as the compression ceased. It has been claimed that this observation proved nothing because pressure could be transmitted to other parts of the brain; but this pressure was directed in such a way that only the anterior lobes were affected, and, besides, it produced neither paralysis nor a loss of consciousness. Again it has been objected that similar results have been obtained in individuals with defects in other parts of the cranial vault, and in particular in that celebrated beggar whose skull cap has been entirely lost to necrosis, and who revealed his skull to passersby in order to secure alms. In these individuals, in fact, compression exerted on the middle part of the brain suddenly suppressed speech, but at the same time it suppressed all other functions of the brain and produced complete loss of consciousness. In the wounded patient in the Hospital St. Louis, on the contrary, compression applied with moderation and discretion did not affect the general functions of the brain; lim-

ited to the anterior lobes it suspended only the faculty of speech. (Translated by Clarke and O'Malley, 1968, 492)

Aubertin acknowledged that some people with frontal lobe injuries may have speech preserved because the precise location of the faculty for articulate speech was not yet established within this lobe, and because one side could compensate for the other in cases of unilateral injury. He felt that it would take a large bilateral lesion to prove his hypothesis wrong. He even added that he would renounce his ideas if an aphasic patient named Bache, whom he had studied for a long time and was close to death, did not show anterior lobe disease. Paul Broca seemed interested, but quiet and noncommittal, during these heated sessions. With regard to localization, Broca stated only that much might be learned from the study of anatomy, embryology, and related material.

On April 12, 1861, Monsieur Leborgne, a 51-year-old man, was transferred to Broca's surgical service after a long hospitalization. Other patients called him "Tan" because this sound and a few obscenities were his only utterances. He had suffered from epilepsy since his youth, had lost his power to speak in 1840, and had lost his ability to use his right arm 10 years after this. Broca invited Aubertin to examine Leborgne as a test case.

In view of the fact that M. Aubertin only a few days before had declared that he would renounce the idea of cerebral localization if a single case in which the loss of the faculty of articulate language were demonstrated to him without a lesion in the anterior lobes, I invited him to see my patient, to know above all what would be his diagnosis and if this observation would be one of those whose findings he would accept as conclusive. In spite of the complications which had supervened during the last 11 years, my colleague found the present condition sufficiently clear to conclude without hesitation that the lesion had its origin in the anterior lobes. (Translated in Stookey, 1954, 572)

Leborgne died six days later. Broca (1861a) presented the brain of his patient to the *Société d'Anthropologie* the next day and issued only a brief statement. At the meetings of the *Société d'Anatomie* four months later, however, Broca (1861b) gave a full report of his case and presented a strong argument for a frontal lobe localization of the faculty for articulate language and in a circumscribed area at that. Broca congratulated Aubertin and Bouillaud for having amassed considerable clinical evidence for the role of the anterior lobes in speech. In fact, in his opening line, Broca stated: "The observations which I present to the *Société d'Anatomie* support the ideas of M. Bouillaud on the seat of the faculty for language." His presentation was met by an enthusiastic audience, one that viewed "Tan" as the landmark case which would revolutionize ideas about brain function.

Why did the Broca paper have so much impact on clinical thought? And why did his predecessors, notably Aubertin and Bouillaud, have such little impact, given their plethora of cases? There is no simple answer to this question, in part because a host of important factors converged with Broca and the case of "Tan" (Sondhaus and Finger, 1988). First, Broca provided greater detail about his case than could be found in previous reports. There was detail in the case history, in emphasizing articulate speech as opposed to all speech defects, and in the search for a precise locus for this

---

[9]Advances in aphasiology did not just come from France in the early and middle 1800s. Many case studies were also presented in other parts of Europe and in the United States at the same time (for reviews, see Bateman, 1870; Winslow, 1868). Among these was an excellent description of jargon aphasia by Jonathan Osborne (1795–1864), who practiced in Great Britain. In his 1833 paper, Osborne also introduced a formal program for speech therapy.

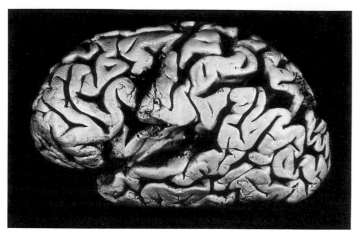

Figure 25.5. A photograph of the brain of Lelong, Broca's second aphasic case. Photographs of the brain of Leborgne, his first case, appear in Chapter 3. (From the Musée Dupuytren. Courtesy of Assistance Publique, Hôpitaux de Paris.)

faculty. Second, Broca went out of his way to show that his speech area was different from the frontal lobe localization proposed by the phrenologists, whose ghosts were still haunting the halls of science. Third, the *Zeitgeist* had changed. The scientific community, perhaps after listening to Bouillaud and others, was now more willing to entertain the idea of cortical localization (see Head, 1926; Young, 1970). And fourth, it was Paul Broca, the highly respected scientist, distinguished physician, head of learned societies, and founder and secretary of the Société d'Anthropologie who was now ready to stand up and fight for cortical localization of function.[10]

BROCA'S LATER OBSERVATIONS

In 1861, Paul Broca presented another case study of a patient with aphasia, an 84-year-old laborer named Lelong. He was able to say only a few simple words, such as *oui, non, trois, toujours* ("yes," "no," "three," "always"), and could not write, after he collapsed at age 83. Because he used gestures, such as holding up eight fingers to indicate he had been at the Bicêtre for eight years, and because he seemed to show good comprehension, Broca concluded that his intelligence was not seriously affected and that a paralysis did not underlie the speech deficit.

When Lelong died, Broca found a depression in the posterior third of the second and third frontal convolutions of the left hemisphere. In a case thought to be considerably less complex and less fraught with uncertainty than his previous one, Broca deduced he had confirmed his previous localization of a center for articulate language. The brain of Lelong is shown in Figure 25.5.[11]

In 1863, Broca collected more cases of aphasia. All involved the left hemisphere, and all but one showed damage in the third frontal convolution. Broca still had little to say

about the role of the left hemisphere in speech. He simply presented the numbers and stated it was remarkable that all of his cases had lesions which involved the left hemisphere.

Prior to this time, Marc Dax (1770–1837) and his son Gustav Dax (1815–1874) might have been the only ones to believe that the left hemisphere was special for speech functions (Critchley, 1964; Joynt and Benton, 1964). In 1836, Marc Dax associated aphasia with left hemispheric injury after examining over 40 cases of aphasia. Yet for reasons that are still unclear, Dax did not present his revolutionary findings at the regional medical congress at which he was supposed to speak. His findings were published only in 1865, after they were sent to Paris by his son Gustav, an established physician who accepted and defended his father's ideas about cerebral dominance for speech (see Chapter 26).

As for Broca, he now had to face the fact that not all patients showing loss of articulate language exhibited lesions of the third frontal convolution. A case described by Charcot in 1861 was the first serious exception. The patient had very little speech and a right hemiplegia, with memory and intelligence supposedly unaffected. The autopsy revealed a lesion of the left hemisphere posterior to the frontal lobe. Broca first considered the possibility that the speech area was more extensive than he had originally thought, but then fell back on the fact that his other 19 cases showed damage anterior to this area in the frontal lobes. This led him to emphasize caution and to call for more data.

In his celebrated paper of 1865, Broca did not mince words when he localized speech in the left hemisphere, at least in right-handed people. To account for some instances of aphasia after right hemispheric lesions, he postulated that speech in left-handed people is controlled by the frontal region of the right hemisphere. Broca further proposed that the right hemisphere could become the speech hemisphere if the left hemisphere were injured, especially if the insult occurred early in life. He felt that this allowed one to account for individuals who were missing the speech area of the left frontal lobe and who were not severely impaired in articulate speech. These ideas seemed logical to his contemporaries, as well as to many scientists late in the nineteenth century (e.g., Gowers, 1885).

In 1868, Broca visited Norwich to read a paper to the British Association for the Advancement of Science. He was invited by Frederic Bateman (1824–1904), a physician at the Norfolk and Norwich Hospital. Bateman had already written in the *Lancet* that the record favored speech localization in the left hemisphere, although he added that more work was required to substantiate Broca's claim that the posterior part of the third frontal convolution was the critical area for this function.[12] John Hughlings Jackson was invited to speak at the same session, which was billed as an important

---

[10]Broca's stature and the respect he received from his colleagues and associates in many walks of life are described by Schiller (1979) in his scholarly biography of Broca as a scientist and as a man.

[11]Actually, Lelong's hemispheres showed severe atrophy, similar to the type characterizing brains of individuals with senile dementia.

[12]Broca's discovery was not embraced by all scientists and physicians, or even by all Frenchmen. There were many questions raised. One concerned Phineas Gage, the New England railroad foreman whose left frontal lobe was shattered by an explosion that sent a heavy tamping iron through his skull in 1848 (See Chapters 19 and 22; also Bigelow, 1850; Harlow, 1848, 1849, 1868). Even in the period immediately after his accident, Gage was never really aphasic nor paralytic, although his doctors said that the anterior and middle left lobes of his cerebrum had been disintegrated. This questionable finding was seized upon by E. Dupuy (1873, 1877), who emphasized Gage's absence of aphasia and who assumed that the destruction of his left frontal lobe was absolute. David Ferrier wrote a strong rebuttal to Dupuy's criticisms of localization theory. After Ferrier's Gulstonian Lecture of 1878, the Gage case was not used again as evidence against Broca's thesis.

"debate" involving the best of the French and British medical minds.

In his presentation, which was published in 1869, Broca defined four different forms of speech afflictions. One type was due to loss of intelligence and ideation, and a second type involved mechanical problems of the speech apparatus, such as peripheral nerve injury. His remaining two categories proved to be of greater theoretical interest. One pertained to the loss of the associations between ideas and words. Here, fluent pronunciation and intellect were preserved *(amnésie verbale)*. The last variety was the loss of articulate speech with the possible preservation of gestures and occasional swear words. This was his original *aphémie*, which later became *aphasie*, or "Broca's aphasia."[13] This was the disorder he had localized in 1861 in the left frontal lobe.

Broca's last statement on aphasia was made in 1877, at which time he reiterated some of the ideas he had expressed in England nine years earlier. Between 1861, when Broca presented his case study of "Tan" and the time of his final words on aphasia, the scientific world had changed dramatically. One notable difference was that the French were no longer completely dominating the race to understand disorders of speech and language.

THE BRITISH NEUROLOGISTS

Frederic Bateman, who had invited Broca to Norwich to present his findings and to meet Great Britain's leading brain scientists, collected 75 observations of aphasia with 27 autopsies. Eighteen failed to confirm Broca's localization but, as noted in a summary of the famous Norwich meetings, Broca showed great ability in replying to this and other challenges to his theory. Bateman summarized his own findings and the existing literature in his book *On Aphasia*, published in 1870.

As for John Hughlings Jackson, he distinguished between the intellectual and the emotional aspects of speech and discussed the possibility that one function could be affected but not the other. For instance, he observed that a man who could not find the word for a simple object could still swear vigorously when provoked.

Jackson was not the first to draw attention to this phenomenon. He credited François Baillarger with seeing fluent emotional speech in aphasic patients, possibly in an address that Baillarger gave to the *Académie de Médecine* in 1865. Further, Broca had described exactly the same sort of thing in his 1861 report on "Tan" to the *Société d'Anatomie*. Broca added that Aubertin had witnessed this phenomenon in one of his patients whose "responses begin with a bizarre word of six syllables and end with this supreme invocation, *Sacré nom de Dieu*" (p. 52). In addition, Lordat (1843) described a parish priest who had a stroke that left him able to utter only a few simple words and "the most forceful oath of the tongue, which begins with an 'f' and which our Dictionaries have never dared to print" (translated in Lecours, Nespoulous, and Pioger, 1987, p. 10).

Jackson viewed speech as a very complicated process and felt that neurologists could not readily distinguish

motor from mental symptoms in aphasia because both were likely to be affected. In the summary of the Norwich meetings, it was stated that Jackson proved "that aphasia is by no means so rigorously separated off from other varieties of cerebral disease, affecting the intellect on one side, and muscular movements on the other, as had been carelessly supposed by some." Jackson thus approached the deficit as a quasi-mental defect in which the motor and mental contributions differed only in degree.

At least on a clinical level, some of Jackson's views may not have been that different from Broca's later thinking. In 1865, Broca wrote that most strokes associated with aphasia also affected the intellect, which is why relearning by an adult aphasic patient can be difficult. But unlike Jackson, Broca seemed to place greater theoretical emphasis on the fact that the lesions which affected speech often were large and went well beyond the boundaries of the faculty for articulate language. Jackson, in contrast, seemed to imply that there would be intellectual blunting even with lesions confined to Broca's area.

Actually, Jackson was never completely convinced that a discrete faculty for language really existed. In 1874, he argued that to locate the damage that destroys speech and to locate speech are two different things. He further maintained that to discuss the functions of the cortex as if they were based on abrupt geographical locations was logically absurd. Nevertheless, Jackson, who was by nature a shy and quiet man, remained extremely polite and tried as best he could to show humility and high regard for Broca's discoveries, especially when the two leaders were together.

Although Henry Charlton Bastian also wrote on aphasia, his ideas on this subject differed considerably from those of Jackson. In 1869, he anticipated the thinking of Carl Wernicke when he concluded that "we think in words, in fact, and these words are received as sound impressions in the auditory receptive centres of the cerebral hemispheres." Bastian also provided one of the earliest wiring diagrams of the brain to show how speech in the auditory area, and writing in the visual area, could be linked with motor pathways to the tongue and hands, respectively. Aphasia, he felt, could be classified according to the brain areas affected. Bastian also may have been one of the first scientists to describe word deafness and word blindness in detail. Unfortunately, many of his observations and theories were not made public until he published his monograph in 1898, some 19 years after he first became interested in aphasia.

WERNICKE AND THE CIRCUITRY OF LANGUAGE

Carl Wernicke, shown in Figure 25.6, was 26 years old when he wrote *Der aphasischer Symptomenkomplex (The Aphasic Symptom Complex)* in 1874. Like Theodor Meynert, with whom he worked for six months, he believed that all sensory nerves ended in the occipital and temporal lobes. From this, he deduced that a lesion of the auditory center in the first temporal convolution would leave people fluent, but unable to understand speech or use words properly. In other words, such a lesion would abolish sound images *(Klangbilder)*.

Wernicke also spoke about other types of aphasia. One was motor aphasia, due to damage in Broca's area in the frontal

---

[13]The term "aphasia" can be traced to Armand Trousseau, who was told by a Greek physician that "aphemia" *(aphémie)* meant "infamous" and that a better word for the disorder should be found. Hence, *aphasie*, or "aphasia," was coined in 1864.

Figure 25.6. Carl Wernicke (1848–1904), author of *Der aphasische Symptomenkomplex* (1874). Wernicke attempted to explain different disorders of communication by referring to functionally distinct parts of the brain and their presumed connections. (From *Monatschrift für Psychiatrie und Neurologie,* 1905.)

cortex, the supposed seat of movement images, necessary for articulate speech. This type of patient can understand all that is said orally and in print, but cannot speak. Another type was "conduction aphasia," a disorder characterized by misapplied words and an inability to read and possibly write, but one in which there is fairly good comprehension and fluent output. Conduction aphasia was postulated to be due to dam-

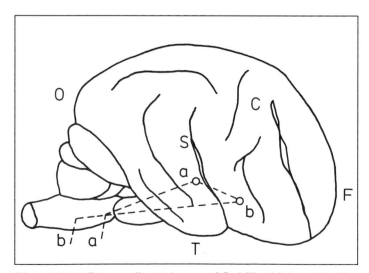

Figure 25.7. Diagram illustrating one of Carl Wernicke's models. The central terminal of the acoustic nerve is given the letter **a**, while **b** represents the center for motor imagery in the frontal cortex. These components, their associated pathways to and from lower levels, and the theorized projection between cortical centers **a** and **b** could account for speech production, according to Wernicke (1874).

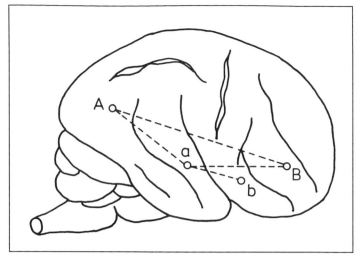

Figure 25.8. Diagram illustrating another of Carl Wernicke's models. This diagram shows a center for acoustic speech imagery (**a**) and motor speech imagery (**b**), as well as centers for graphic motor imagery (**B**) and visual letter imagery (**A**). This diagram was used to explain reading and writing. (Based on Wernicke, 1874.)

age to the pathways connecting the center responsible for acoustic word images and Broca's region.

In his 1881 book, Wernicke defined "total aphasia" as a lack of both expression and comprehension of speech, and attributed this to a lesion of both Broca's area and his center for acoustic word images. Wernicke then constructed a diagram to show how different disorders of communication could result from damage to the connections between his various processing centers in well-defined parts of the cortex. Two of his diagrams are shown as Figures 25.7 and 25.8.

Nevertheless, Wernicke (1874) did not believe that writing disorders could be independent of spoken language. Thus, he did not propose a special writing center in the brain. In contrast, the opposing view, namely that "motor graphic images" essential for writing have their own specialized cortical centers, was proposed by another German investigator, Sigmund Exner (1846–1926), in 1881.[14] Although the issue of a cortical center for writing was subsequently debated, there was agreement over the finding that agraphia or dysgraphia (a milder writing disorder) almost always accompanied the aphasias (Marcie and Hécaen, 1979).

It is notable that Wernicke's (1874) most important work was based on just 10 cases. Of these, only 4 were backed by autopsy material. Henry Head (1926) openly chastised Wernicke for his fixed preconceptions when analyzing his limited number of clinical cases and for his attempts to deduce clinical manifestations from "hypothetical lesions." In contrast, Head praised the efforts of some of Wernicke's countrymen, individuals who emphasized the need to achieve a better understanding of the symptoms of the various language disturbances before attempting to associate symptoms with functions and structures.

GERMAN AND AUSTRIAN FUNCTIONAL APPROACHES

Relatively little is known about Finkelnburg (1832–1896), one of the individuals praised by Head (1926) in his book on

[14]A writing center was also postulated by Henry Charlton Bastian in 1898.

aphasia. What is known is that he was a professor of psychiatry from Bonn who addressed the Society of the Lower Rhine in 1870. There he presented the idea that aphasia had to be much more than simply a disorder of speech. On the basis of five case studies (two with autopsies), he concluded that speech difficulties were typically associated with impairments in verbal comprehension, reading, and writing. In fact, reasoned Finkelnburg, aphasia was associated with a wide range of deficits involving the use of symbolic images. Indeed, it was accompanied by the poor use of common signs, with problems involving musical and monetary notations, and even with the inability to recognize the symbols of rituals and accepted social conventions.

Having come to this realization, Finkelnburg proposed that "asymbolia" would be a good term to signify the underlying deficit behind the "aphasic" disorder, which he believed always manifested itself in both receptive and expressive modes.

> The specific loss of the ability to correctly receive and utter symbols—thus impairing reception and transmission of meaning through symbols—which is so characteristic of this clinical disorder, cannot possibly be precisely and completely encompassed by the term "aphasia" since this term signifies only the disruption of word formation. Thus, in addition to "aphasia," English scientists already describe an "agraphia"; and to these two disorders several others (with the privative suffix "a") would have to be added if one wanted to capture exhaustively the syndrome in all its aspects. It seems much more simple and correct to speak of a disruption of the symbolic function of the brain or to choose the comprehensive term, "asymbolia." Thus, asymbolia would be that disturbance of function in which there is a partial or complete loss of the ability to comprehend and express concepts by means of acquired signs. (Translated in Duffy and Liles, 1979, 164)

Finkelnburg thus chose to place his emphasis on the nature of the symptom complex itself, rather than on its underlying anatomy. With regard to the anatomy, he believed that scientists were unable to support Broca's precise anatomical localization for a faculty for fluent speech and cited the case reports of many other investigators in this context. He urged his audience to return to the task of carefully describing the full range of symptoms of brain damage, to look for fundamental deficits, and to avoid being distracted by issues of localization of function.

Adolf Kussmaul (1887) followed Finkelnburg's lead in attacking the linguistic-localizationist model. He believed that Finkelnburg's theory was basically correct. Kussmaul felt that the notion of asymbolia was incompatible with a special seat in the brain and contended that no part of the brain could serve speech alone. To quote:

> Finally, it has come to this at the present day, that under aphasia we no longer understand merely the disturbances of speech alone, but also the collective symptomatic phenomena, whether abundant or scanty, under which the execution or comprehension of any given signs by which it is sought to communicate conceptions or feelings is impaired. . . . Finkelnburg is certainly right in proposing instead of the term aphasia, which only has reference to phonetic symbols, a more comprehensive one, asymbolia, which would embrace all the numerous as well as different clini-

Figure 25.9. Sigmund Freud (1856–1939), author of *Zur Auffassung der Aphasien* in 1891. Freud stressed the need to learn more about the symptoms of speech disorders and was critical of the early "diagram makers" in this essay. (From Kolle, *Grosse Nervenärzte,* 1970; with permission of George Thieme Verlag.)

cal forms of disturbance in the formation and comprehension of symbols. With Steinthal, however, we should prefer the term "asemia," the idea contained in "symbol" being more restricted than that contained in "sign." (1887, 609)[15]

In 1891, Sigmund Freud, who like Wernicke had been a student of Meynert's in Vienna (at least for five months), published *Zur Auffassung der Aphasien,* a book later translated as *On Aphasia.* In 1885, before he started to work on this book, Freud had visited Adolf Baginsky (1843–1918), an early "diagram maker" (Schiller, 1979). Nevertheless, Freud, shown in Figure 25.9, did not feel comfortable with the approaches of Baginsky and Wernicke. Like Finkelnburg and Kussmaul, he chose to emphasize the functional aspects of language disorders. He stated his philosophy very clearly when he wrote: "It appears to us, however, that the significance of the factor of localization for aphasia has been overrated, and that we should be well advised once again to concern ourselves with the functional states of the apparatus of speech" (p. 105).[16]

---

[15]John Hughlings Jackson read Kussmaul's book and through it he also became familiar with Finkelnburg's theory. Jackson joined those who did not like the term "aphasia," since he had been emphasizing that many of his patients also showed nonverbal defects, such as an inability to pantomime well. In 1879, Jackson stated that he preferred Hamilton's term "asemasia," which meant without signs or language. Jackson added, however, that he feared it was already too late to displace the word "aphasia" from the common vocabulary.

[16]Actually, Freud was less than perfectly consistent with this position. That is, he divided the theoretical language area into two basic functional parts, a center and a periphery. Freud thought that the speech area was a large, continuous cortical region in which associations and transmissions took place with "a complexity beyond comprehension." But he went on to state that lesions in the periphery of this region would affect images from nearby sensory areas which play a role in effective word formation. Lesions in the center of the language zone, however, would result in a more complicated disorder affecting several aspects of language function.

Freud subtitled his short book *A Critical Essay,* and aimed his criticisms especially at Carl Wernicke. Freud's often biting remarks stemmed largely from the fact that he was unable to discern specific structure-function associations from his analyses of cases in the existing literature. In contrast to his treatment of Wernicke and the diagram makers, Freud praised the work of John Hughlings Jackson. He found Jackson's notion of dissolution (the opposite of evolution) very appealing for approaching various speech and language phenomena.

> In assessing the functions of the speech apparatus under pathological conditions we are adopting as a guiding principle Hughlings Jackson's doctrine that all these modes of reaction represent instances of functional retrogression (dis-involution) of a highly organized apparatus, and therefore correspond to earlier states of its functional development. This means that under all circumstances an arrangement of associations which, having been acquired later, belongs to a higher level of functioning, will be lost, while an earlier and simpler one will be preserved. From this point of view, a great number of aphasic phenomena can be explained. (For example) The loss, through damage to the speech apparatus, of new languages as super-associations, while the mother tongue is preserved. Next, the nature of the speech remnants in motor aphasia which are so frequently only "yes" and "no" and other words in use since the beginning of speech development. (1891, 87)

Freud showed a strong interest in paraphasia, a condition in which the appropriate word is replaced by a less appropriate word, but one still retaining some relation to the correct word (e.g., similar sounds or meanings). Freud's statements about paraphasic "slips" have attracted considerable attention, in part because he spoke about them in aphasia cases and in otherwise healthy people, and because they exemplified his strong emphasis on understanding how symptoms might change under different conditions. To quote:

> We only want to mention that the paraphasia observed in aphasic patients does not differ from the incorrect use and the distortion of words which the healthy person can observe in himself in states of fatigue or divided attention or under the influence of disturbing affects,—the kind of thing that frequently happens to our lecturers and causes the listener painful embarrassment. It is tempting to regard paraphasia in the widest sense as a purely functional symptom, a sign of reduced efficiency of the apparatus of speech associations. This does not exclude that they may occur in most typical form as organic focal symptoms. Allen Starr is the only author of distinction who has taken the trouble of searching for the anatomical causes of paraphasia. He arrived at the conclusion that this symptom could be produced by lesions of a great variety of localization; he found it impossible to establish a consistent difference in the pathology of cases of sensory aphasia with or without paraphasia. (1891, 13)

Starr, who was cited favorably by Freud, was an American neurologist. He had written:

> It was impossible to ascertain any constant pathological difference between the cases of sensory aphasia with and without paraphasia. Nor did the power to repeat words one after another seem to depend upon the relative situation of the lesion, as might be supposed from Wernicke's assertion that this defect appears with paraphasia when the tem-

poro-frontal cortex is involved. . . . The analysis of pathological lesions therefore, does not bring out as clear a differentiation of the different forms of aphasia as might be desired. (1889, 87)

THE SEARCH FOR THE TRUE REALITY

Carl Wernicke, and the map makers in general, were also criticized by some scientists in France and England. In France, Pierre Marie claimed that by 1906 he had examined almost 100 patients with aphasia and had done autopsies on over 50 (see Bailey, 1970). From these efforts, he had become convinced that all such patients have difficulty with spoken language and intellectual problems. From this premise, Marie went on to reject Wernicke's concept of pure "sensory aphasia."

Marie (1906a) also pointed out that he found symptoms of motor (Broca's) aphasia in only about half of the stroke patients at the Bicêtre who had lesions of the third left frontal convolution. He theorized that the third frontal convolution was just one part of the large region damaged by these strokes, and perhaps "an unimportant factor, since the most well-characterized aphasia can be observed without any lesion of this convolution" (p. 59). He concluded that Broca's aphasia was probably also fictional:

> I use aphasia in the singular intentionally, for as one can realize from the preceding discussion, aphasia is one. We have seen, as a matter of fact, that the only notable difference between the aphasia of Wernicke and the aphasia of Broca is that in the first, patients speak more or less badly, while in the second they do not speak at all. But in both forms one finds that intellectual deficit which had as its immediate consequence difficulty in understanding sentences which are a little bit complicated and difficulty with or the total loss of reading or writing. To characterize it in two words, the aphasia of Broca is none other than aphasia complicated by anarthria, or if one prefers, according to the case, anarthria complicated by aphasia. But since aphasia is one, its localization ought to be one.—And so it is in reality. The only cerebral territory whose lesion produces aphasia is the so-called territory of Wernicke. (1906a, 63)

Marie presented these challenging thoughts in a paper with the provocative title *La Troisième Circonvolution Frontale Gauche ne Joue aucun Role Spécial dans la Fonction du Langage* ("The Third Frontal Convolution Plays No Special Role Whatever in the Function of Language"). In addition, Marie (with Moutier) published a second paper in 1906, which described in detail the case of a right-handed man whose third left frontal convolution was noticeably atrophied, but who showed excellent preservation of language functions.

This manuscript was followed by a third in the same year, this time criticizing Broca for never sectioning the two brains (Leborgne and Lelong) that had led him to believe it was the third left frontal convolution which was mediating articulate language. Marie (1906b) argued that Leborgne's lesion of "Broca's area" extended well into the temporal lobe and that Lelong was not a classic aphasic at all, but a senile old man who did not even have a focal lesion.

Perhaps partly because he pulled out all the stops and was so brash, Marie (1922) found that the reactions to his ideas were sometimes hostile. Nevertheless, he did seem to have some impact on the establishment. His arguments demanded that simply looking at the surface of the brain was no longer going to be an acceptable procedure for describing the lesions and that more attention would have to be paid to the basic nature of the disorder. In this regard, he was successful.

What Marie had to say was appreciated by Henry Head. This leading British neurologist conducted a lengthy review of the literature on aphasia, as well as many studies of his own. He was vigorously opposed to the simplistic logic of the map makers. In this context, he made the following statement:

> Most of the observers mentioned . . . failed to contribute anything of permanent value to the solution of the problems of aphasia, because they were dominated by a philosophical fallacy of their day, which can still count its victims amongst writers on the subject. They imagined that all vital processes could be explained by some simple formula. With the help of a few carefully selected assumptions, they deduced the mechanism of speech and embodied it in a schematic form. For every mental act there was a neural element, either identical with it or in exact correspondence. From diagrams, based on a priori principles, they deduced in turn the defects in function which must follow destruction of each "centre" or internuncial path. They never doubted the validity of their postulates, based as they were on rules of human reason. (1926, 65)

Head's emphasis was clearly on observable symptoms, and he went on to devise specific tests to measure verbal, syntactical, nominal, and semantic functions. Following in the footsteps of Finkelnburg and Kussmaul, he thought that aphasic disorders were at least partly defects of symbol formation and expression, and he avoided localizing each of these functions in different discrete areas of the cerebral cortex.

Although his tests and functional analyses attracted followers, Head's (1926, p. 66) charge that "the time was ripe for a ruthless destruction of false gods and a return to systematic empirical observation" was not embraced as the path to follow by many scientists. In fact, scientists differed just as dramatically in their orientations to aphasia after 1926—the year Head published his classic *Aphasia and Kindred Disorders of Speech*—as they did in 1891, when Freud launched his attack on the map makers in his book on aphasia. To this day, the field of speech and language has remained divided with regard to the identification of the false gods and the pathways to truth and reality.[17]

---

[17]Head's words were warmly received by Walter Dandy, an American neurosurgeon who had written: "Three times it has been necessary to remove Broca's area, apparently completely and to a depth of 2–3 cm. Although a complete motor aphasia appeared at once, the speech began to come back in a week or 10 days and gradually returned to normal" (1922, p. 188). In contrast, others accepted and expanded the connectionistic models and viewed them as representative of truly modern science (see Geschwind, 1965).

REFERENCES

Arceo, F. 1588. *A most Excellent and Compendious Method of Curing Woundes in the Head.* London: Thomas East for Thomas Cadman.

Andral, G. 1840. *Clinique Médicale. . . .* (4th ed.). Paris: Fortin, Massonet Cie.

Aubertin, E. 1861. Discussion of Gratiolet's paper, Sur le volume et la forme du cerveau. *Bulletins de la Société d'Anthropologie (Paris), 2,* 71–72, 74–75, 81, 209–220, 275–276, 278–279, 421–446.

Bailey, P. 1970. Pierre Marie (1853–1940). In W. Haymaker and F. Schiller (Eds.), *The Founders of Neurology* (2nd ed.). Springfield, IL: Charles C Thomas. Pp. 476–479.

Banjeri, D. C. 1981. Historical and socio-cultural foundations of health service systems. In G. R. Gupta (Ed.), *The Social and Cultural Context of Medicine in India.* New Delhi: Vikas Publishing House.

Bastian, H. C. 1869. On the various forms of loss of speech in cerebral disease. *British and Foreign Medical-Chirurgical Review, 43,* 209–236, 470–492.

Bastian, H. C. 1898. *A Treatise on Aphasia and Other Speech Defects.* London: H. C. Lewis.

Bateman, F. 1870. *On Aphasia.* London: Churchill.

Baverius de Baveriis. 1543. *Conciliorum de re medica sive morborum curationibus liber.* Argentinae: Libaria Balthasari Pistoris.

Belhomme, J.-E. 1848. De la localisation de la parole ou plutôt de la mémoire des mots dans les lobes antérieurs du cerveau. (Summarized by J. Cloquet and J. Ferrus). *Bulletin de l'Académie Royale de Médecine, 13,* 527–536.

Benton, A. L. 1965. Johann A. P. Gesner on aphasia. *Medical History, 1,* 54–60.

Benton, A. L., and Joynt, R. J. 1960. Early descriptions of aphasia. *Archives of Neurology, 3,* 205–221.

Benton, A. L., and Joynt, R. J. 1963. Three pioneers in the study of aphasia. *Journal of the History of Medical Science, 18,* 381–383.

Berker, E. A., Berker, A. H., and Smith, A. 1986. Translation of Broca's 1865 Report: Localization of speech in the third left frontal convolution. *Archives of Neurology, 43,* 1065–1072.

Bianchi, A. 1883. Changes in handwriting in relation to pathology. *The Alienist and Neurology, 4,* 566–590.

Bigelow, H. J. 1850. Dr. Harlow's case of recovery from the passage of an iron bar through the head. *American Journal of the Medical Sciences, 19,* 13–22.

Bouillaud, J.-B. 1825a. Recherches cliniques propres à démontrer que la perte de la parole correspond à la lésion des lobules antérieurs du cerveau et à confirmer l'opinion de M. Gall sur le siège de l'organe du langage articulé. *Archives Générale de Médecine (Paris), 8,* 25–45.

Bouillaud, J.-B. 1825b. *Traité Clinique et Physiologique de l'Encéphalite ou Inflammation du Cerveau.* Paris: J. B. Ballière.

Bouillaud, J.-B. 1830. Recherches expérimentales sur les fonctions du cerveau (lobes cérébraux) en général, et sur celles de sa portion antérieure en particulier. *Journal Hebdomadaire de Médecine (Paris), 6,* 527–570.

Bouillaud, J.-B. 1839–1840. Exposition de nouveaux faits à l'appui de l'opinion qui localise dans les lobules antérieurs du cerveau le principe législateur de la parole; examen préliminaire des objections dont cette opinion à été sujet. *Bulletin de l'Académie Royale de Médecine (Paris), 4,* 282–328, 333–349, 353–369.

Bouillaud, J.-B. 1848. *Recherches Cliniques Propres à Démontrer que le Sens du Langage Articulé et le Principe Coordinateur des Mouvements de la Parole Résident dans les Lobules Antérieurs du Cerveau.* Paris: J. B. Ballière.

Breasted, J. H. 1930. *The Edwin Smith Surgical Papyrus.* Chicago: University of Chicago Press.

Broca, P. 1861a. Perte de la parole, ramollissement chronique et destruction partielle du lobe antérieur gauche du cerveau. *Bulletins de la Société d'Anthropologie, 2,* 235–238.

Broca, P. 1861b. Remarques sur le siège de la faculté du langage articulé; suivies d'une observation d'aphémie (perte de la parole). *Bulletins de la Société Anatomique (Paris), 6,* 330–357, 398–407. In G. von Bonin, *Some Papers on the Cerebral Cortex.* Translated as, "Remarks on the seat of the faculty of articulate language, followed by an observation of aphemia." Springfield, IL: Charles C Thomas, 1960. Pp. 49–72.

Broca, P. 1863. Localisation des fonctions cérébrales. Siège du langage articulé. *Bulletins de la Société d'Anthropologie, 4,* 200–203.

Broca, P. 1865. Sur le siège de la faculté du langage articulé. *Bulletins de la Société d'Anthropologie, 6,* 337–393.

Broca, P. 1869. Sur le siège de la faculté du langage articulé. *Tribune Médicale, 74,* 254–256; 75, 265–269.

Celsus. 1938. *De medicina* (Book. VII). (W. G. Spencer, Trans.). London: Heinemann.

Chadwick, J., and Mann, W. N. 1950. *The Medical Works of Hippocrates.* Oxford, U.K.: Blackwell.

Chapman, N. 1835. Note to J. Grattan, "Case of Amnesia." *American Journal of Medical Sciences, 17,* 467–469.

Clarke, E., and O'Malley, C. D. 1968. *The Human Brain and Spinal Cord.* Berkeley: University of California Press.

Cole, M. F., and Cole, M. 1971. *Pierre Marie's Papers on Speech Disorders.* New York: Hafner.

Crichton, A. 1798. *An Inquiry into the Nature and Origin of Mental Derangement, Comprehending a Concise System of the Physiology and Pathology of the Human Mind and a History of the Passions and Their Effects.* London: T. Cadell, Jr. and W. Davies.

Critchley, M. 1962. Dr. Samuel Johnson's aphasia. *Medical History, 6,* 27–44.

Critchley, M. 1964. Dax's law. *International Journal of Neurology, 4,* 199–206.

Critchley, M. 1970. *Aphasiology and Other Aspects of Language.* London: E. Arnold.

Dandy, W. E. 1922. Treatment of non-encapsulated brain tumors by extensive resection of contiguous brain tissue. *Bulletin of the Johns Hopkins Hospital, 33,* 188.

Dax, M. 1865. Lésion de la moitié gauche de l'encéphale coïncidant avec l'oubli des signes de la pensée (lu a Montpellier en 1836). *Gazette Hebdomidaire de Médecine et de Chirurgie, 2 (2nd ser.),* 259–260.

Dickson, H. 1830. Case of amnesia. *American Journal of Medical Sciences, 7,* 359–360.

Drabkin, I. E. 1950. *Caelius Aurelianus: On Acute Diseases and on Chronic Diseases.* Chicago: University of Chicago Press.

Drake, D. 1834. Case of partial amnesia, in which the memory for proper names was lost. *American Journal of Medical Sciences, 15,* 551–553.

Duffy, R. K., and Liles, B. Z. 1979. A translation of Finkelnburg's (1870) lecture on aphasia as "asymbolia" with commentary. *Journal of Speech and Hearing Disorders, 44,* 156–168.

Dupuy, E. 1873. *Examen de Quelques Points de la Physiologie du Cerveau.* Paris: Delahaye.

Dupuy, E. 1877. A critical review of the prevailing theories concerning the physiology and pathology of the brain. *Medical Times and Gazette, 2,* 11–13, 32–34, 84–87, 356–358, 474–475, 488–490.

Eliasberg, W. 1950. A contribution to the prehistory of aphasia. *Journal of the History of Medicine, 5,* 96–101.

Exner, S. 1881. *Untersuchungen über die Localisation der Functionen in der Grosshirnrinde des Menschen.* Wien: Wilhelm Braumüller.

Fancher, R. E. 1979. *Pioneers of Psychology.* New York: Norton.

Ferrier, D. 1878. The Goulstonian Lectures of the localization of cerebral disease. *British Medical Journal, 1,* 397–402, 443–447.

Filliozat, J. 1964. *The Classical Doctrine of Indian Medicine.* Delhi: Mushi Ram Manohar Lal, Oriental Publishers.

Finkelnburg, F. C. 1870. Asymbolie und Aphasie. *Berliner klinische Wochenschrift, 7,* 449–450, 460–462.

Freemon, F. R. 1991. American medicine and phrenology. Paper presented at Neurohistory Conference, Ft. Myers, FL.

Freud, S. 1891. *Zur Auffassung der Aphasien.* Vienna: Deuticke. Translated by E. Stengel as, *On Aphasia.* New York: International Universities Press, 1953.

Gall, F. J. 1819. *Anatomie et Physiologie du Système Nerveux en Général, et du Cerveau en Particulier* (Vol. 4). Paris: F. Schoell.

Geschwind, N. 1965. Disconnection syndromes in animals and man. *Brain, 88,* 237–294.

Gesner, J. A. P. 1769–1776. *Samlung von Beobachtungen aus der Arzneygelahrtheit und Naturkunde* (5 vols.). Nördlingen: C. G. Beck.

Goethe, J. W. 1795. *Wilhelm Meister's Lehrjahre.* Translated by R. D. Boylan as, *Wilhelm Meister's Apprenticeship.* London: Bell and Daldy, 1871.

Gowers, W. R. 1885. *Epilepsy.* London: William Woods and Company. (Reprinted 1964; New York: Dover Publications)

Grandjean de Fouchy, J.-P. 1784. Observation anatomique. *Histoire de l'Académie Royale des Sciences, Mémoires,* 399–401.

Guainerio, A. 1481. *Opera medica.* Pavia: Antonius de Carcano.

Harlow, J. M. 1848. Passage of an iron rod through the head. *Boston Medical and Surgical Journal, 39,* 389–393.

Harlow, J. M. 1849. Letter in "Medical Miscellany." *Boston Medical and Surgical Journal, 39,* 506–507.

Harlow, J. M. 1868. Recovery from the passage of an iron bar through the head. *Bulletin of the Massachusetts Medical Society, 2,* 3–20.

Harrington, A. 1985. Nineteenth-century ideas on hemisphere differences and "duality of mind." *Behavioral and Brain Sciences, 8,* 617–660.

Head, H. 1926. *Aphasia and Kindred Disorders of Speech.* New York: Macmillan.

Holmes, O. W. 1883. *Medical Essays 1842–1882.* Boston: Houghton Mifflin.

Hunter, R. A., and MacAlpine, I. 1957. William Harvey: His neurological and psychiatric observations. *Journal of the History of Medicine and Allied Sciences, 12,* 126–139.

Hurry, J. B. 1926. *Imhotep.* Milford, U.K.: Oxford University Press.

Jackson, J. H. 1874. On the nature of the duality of the brain. *Medical Press and Circular, 1,* 63ff. Reprinted in *Brain, 38,* 1915, 80–103.

Jackson, J. H. 1879. On affections of speech from disease of the brain. *Brain, 1,* 304–330.

Jackson, S. 1828. A case of amnesia. *American Journal of Medical Science,* 272–274.

Joynt, R. J. 1961. Centenary of patient "Tan." *Archives of Internal Medicine, 108,* 197–200.

Joynt, R. J., and Benton, A. L. 1964. The memoir of Marc Dax on aphasia. *Neurology, 14,* 851–854.

Kolle, K. 1970. *Grosse Nervenäzrte.* Stuttgart: Georg Thieme.

Kussmaul, A. 1887. Disturbances of speech: An attempt in the pathology of speech. In H. von Ziemssen (Ed.), *Cyclopedia of the Practice of Medicine* (Vol. 14). New York: Wood. (Translated from the German, 1877.)

Lecours, A. R., Nespoulous, J.-L., and Pioger, D. 1987. Jacques Lordat or the birth of cognitive neuropsychology. In E. Keller and M. Gopnik (Eds.), *Motor and Sensory Processes and Language.* Hillsdale, NJ: Lawrence Erlbaum. Pp. 1–16.

Linnaeus, C. 1745. Glomska af alla substantiva och isynnerhet namn. *Kunglig Svenska Vetenskapsakademien en Handlingar, Stockholm, 6,* 116–117.

Lordat, J. 1843. *Analyse de la Parole pour Servir à la Théorie de Divers cas d'Alalie et de Paralalie (du Mutisme et d'Imperfection du Parler) que les Nosologistes on Mal Connus.* Paris: J.B. Ballière, Germer-Ballière, Fortin Masson et Cie.

Marcie, P., and Hécaen, H. 1979. Agraphia: Writing disorders associated with unilateral cortical lesions. In K. M. Heilman and E. Valenstein (Eds.), *Clinical Neuropsychology.* New York: Oxford University Press. Pp. 24–31.

Marie, P. 1906a. La troisième circonvolution frontale gauche ne joue aucun role spécial dans la fonction du langage. *Semaine Médicale, 26,* 241–247. In M. F. Cole and M. Cole (Eds./Trans.), *Pierre Marie's Papers on Speech Disorders.* New York: Hafner, 1971. Pp. 51–71.

Marie, P. 1906b. Aphasia from 1861 to 1866: Essay of historical criticism on the genesis of the doctrine of aphasia. *Semaine Médicale, 26,* 565–571. In M. F. Cole and M. Cole (Eds./Trans.), *Pierre Marie's Papers on Speech Disorders.* New York: Hafner, 1971. Pp. 111–133.

Marie, P. 1922. Do preformed or inborn centers of language exist in man? Conference given to the Faculté de Médecine de Paris, recorded by A.-P. Marie, and published in *Questions Neurologiques d'Actualité.* Paris: Masson et Cie. In M. F. Cole and M. Cole (Eds./Trans.), *Pierre Marie's Papers on Speech Disorders.* New York: Hafner, 1971. Pp. 241–260.

Marie, P., and Moutier, F. 1906. New case of cortical lesion of the foot of the third left frontal in a right-handed man without language disorders. *Bulletins et Mémoires de la Société Médicale des Hôpitaux de Paris, Ser. 3, 23,* 1295–1298. In M. F. Cole and M. Cole (Eds./Trans.), *Pierre Marie's Papers on Speech Disorders.* New York: Hafner, 1971. Pp. 141–143.

Marx, O. 1966. Aphasia studies and language theory in the 19th century. *Bulletin of the History of Medicine, 40,* 328–349.

Marx, O. 1967. Freud and aphasia: An historical analysis. *American Journal of Psychiatry, 124,* 815–825.

Massa, N. 1558. *Epistolarum medicinalium, tomus primus,* Venetiis: ex Officina Stellae Iordani Zilleti.

Morgagni, G. B. 1761. *The Seats and Causes of Diseases Investigated by Anatomy.* (Translated by B. Alexander, 1769, from *De sedibus et causis morborum per anatomen indagatis*).

Ogle, J. W. 1867. Aphasia and agraphia. *St. George's Hospital Reports, 2,* 83–122.

Ogle, J. W. 1874. Part of a clinical lecture on aphasia. *British Medical Journal, 2,* 163–165.

Osborne, J. 1833. On the loss of the faculty of speech depending on forgetfulness of the art of using the vocal organs. *Dublin Journal of Medical and Chemical Sciences, 4,* 157–170.

Pliny the Elder. 1952. *Natural History* (Vol. II, Libri III-VII). (H. Rackham, Trans.). Loeb Classical Library. London: Heinemann.

Riese, W. 1947. The early history of aphasia. *Bulletin of the History of Medicine, 21,* 322–334.

Riese, W. 1977. Dynamic in brain lesions. In K. Hoops, Y. Lebrun, and E. Buyssens (Eds.), *Selected Papers on the History of Aphasia: Neurolinguistics* (Vol. 1). Amsterdam: Swets and Zeitling. Pp. 70–85.

Rommel, P. 1683. De aphonia rara. *Miscellanea Curiosa Medico-Physica Academiae Naturae Curiosorum, 2 (Ser. 2),* 222–227.

Schenck a Grafenberg, J. 1585. *Observationes medicae de capite humano.* Lugduni.

Schiller, F. 1970. Concepts of stroke before and after Virchow. *Medical History, 14,* 115–131.

Schiller, F. 1979. *Paul Broca: Founder of French Anthropology, Explorer of the Brain.* Berkeley: University of California Press.

Schmidt, J. 1676. De oblivione lectionis ex apoplexia salva scriptione. *Miscellanae Curiosa Medico-Physica Academiae Naturae Curiosorum, 4,* 195–197.

Sharma, P. 1983. *Caraka-Samhita.* Varnasi: Chaukhambha Orientalia.

Sondhaus, E., and Finger, S. 1988. Aphasia and the C.N.S. from Imhotep to Broca. *Neuropsychology, 2,* 87–110.

Spalding, J. J. 1783. Ein Brief an Sulzern über eine an sich selbst gemachte Erfahrung. *Magazin für Erfahrungsseelenkunde, 1 (Pt. 1),* 38–43.

Starr, M. A. 1899. The pathology of sensory aphasia with an analysis of fifty cases in which Broca's centre was not diseased. *Brain, 12,* 82–99.

Stookey, B. 1954. A note on the early history of cerebral localization. *Bulletin of the New York Academy of Medicine, 30,* 559–578.

Synek, V. M. 1985. Linnaeus and aphasia. *Neurology, 35,* 1555.

Viets, H. R. 1943. Aphasia as described by Linnaeus and as painted by Ribera. *Bulletin of the History of Medicine, 13,* 328–329.

Wernicke, C. 1874. *Der aphasische Symptomenkomplex: eine psychologische Studie auf anatomischer Basis.* Breslau: Cohn and Weigert. (G. H.

Eggert, Trans.). In, *Wernicke's Works on Aphasia: A Sourcebook and Review*. The Hague: Mouton, 1977.

Wernicke, C. 1881. *Lehrbuch der Gehirnkrankheiten*. Kassel: Fischer.

Winslow, F. 1868. *On the Obscure Diseases of the Brain and Disorders of the Mind* (4th ed). Ch. XIX. Cerebral morbid phenomena of speech, London: Churchill. Pp. 355–393.

Woollam, D. H. M. 1958. Concepts of the brain and its functions in classical antiquity. In M. W. Perrin (Chairman), *The Brain and Its Functions*. Wellcome Foundation, Oxford, U.K.: Blackwell.

Young, R. M. 1970. *Mind, Brain and Adaptation in the 19th Century*. Oxford, U.K.: Clarendon Press.

Chapter 26
The Emergence of the Concept of Cerebral Dominance

> I believe it possible to conclude not that all diseases of the left hemisphere necessarily impair verbal memory but that, when this form of memory is impaired by disease of the brain, it is necessary to look for the cause of the disorder in the left hemisphere, and to look for it there even if both hemispheres are diseased.
>
> Marc Dax, 1865 (written in 1836)

The current concept of cerebral dominance can be thought of as having evolved in a number of stages. The first stage involved occasional speculations about the two hemispheres being functionally different in the face of a strong majority, who maintained that the hemispheres were structurally and functionally identical. These ideas of equivalence were based more on philosophical notions than on careful observations, and this type of thinking extended from the time of the early Greeks well into the 1800s.

The second phase began in the 1860s, when the clinical discoveries of Marc Dax and Paul Broca on the special role of the left hemisphere in articulate language were made public, and also when John Hughlings Jackson began to build his case for right hemispheric specialization for spatial functions. More than any others, the publications of these three men rocked the longstanding theory of hemispheric equality.

The third phase in the evolution of the concept of cerebral dominance was its rapid expansion to encompass a wealth of new data and ideas. During the late nineteenth century and into the twentieth century, a number of "new" syndromes, such as the apraxias, disorders of reading (alexia and dyslexia), denial of illness, and some types of neglect, were associated with either the "dominant" (left) or the "nondominant" (right) hemisphere.

The first two stages of the emergence of the concept of cerebral dominance are examined in this chapter. The growth of the concept of dominance, especially from the end of the nineteenth century through the first quarter of the twentieth century, is the subject of Chapter 27.

AN ANCIENT GREEK THEORY
OF BRAIN LATERALITY

The earliest known theory of cerebral dominance may be about 2,400 years old. It is known from a twelfth-century medical treatise whose author has not been identified (Lokhorst, 1982). This work was probably based on the ideas of Vindicianus, who lived in the fourth century A.D. His own source could well have been Soranus of Ephesus, the Greek physician who lived two centuries earlier. It cannot be stated with certainty whose doctrine Soranus was expounding, but it is likely to have been that of Diocles of Carystus, a fourth-century B.C. Greek physician whom Soranus held in high esteem (Lokhorst, 1982, 1982–1983).

The subject matter of the section concerned with asymmetry was phrenitis, a disorder then characterized by delirium, frenzy, raving, and fever. Diocles thought that phrenitis was due to an inflammation of the diaphragm which spread to the heart. He believed this inflammation eventually affected both sensory perception and understanding by the brain. This belief is significant because the author of the manuscript proposed that these two basic functions, sensation and cognition, were controlled by different hemispheres in the brain.

> Accordingly, there are two brains in the head, one which gives understanding, and another which provides sense perception. That is to say, the one which is lying on the right side is the one that perceives; with the left one, however, we understand. (Translated in Lokhorst, 1982–1983, 34–35)

Although this passage may sound surprisingly modern, the author then went on to say that the heart also participated in these functions.

> Because of this, this is also being done by the heart, which is lying under this (i.e., the left) side (of the head), and which is also continually vigilant, listening and understanding, because it also has ears to hear.

The ancient idea that the heart can play a role in hearing certainly reduces the impact of what may be the first theory of cerebral dominance. In addition, the fact remains that no clinical cases were cited to support this early notion of hemispheric specialization. Perhaps for these reasons, this ancient philosophical theory of hemispheric specialization soon fell into oblivion.

MAINSTREAM GREEK AND ROMAN SCIENCE

Diocles lived at a time when the Hippocratic physicians were observing that unilateral head wounds resulted in convulsions and paralyses on the opposite side of the body. Passages from the Hippocratic canon state that "an incised wound in one temple produces a spasm in the opposite side of the body" (Chadwick and Mann, 1950, p. 263). It was also recognized that temporary loss of speech can occur with "paralysis of the tongue or of the arm and the right side of the body" (p. 248). From these two brief quotations, one would think that Greek physicians would have put these two observations together to associate loss of speech (with right hemiplegia) and damage to the left side of the brain with each other. But either this did not occur, or if it did, it did not become a widely accepted position.

In the first centuries A.D., when the brain was gaining recognition as the organ of mind, there was greater acceptance of the idea that the two cerebral hemispheres controlled the opposite sides of the body. Aretaeus the Cappadocian (c. second to third centuries A.D.), for example, wrote that after damage to one side of the head, paralysis will appear on the opposite side owing to the decussation of the nerves.

The statements made by Aretaeus suggest that aphasia, which is characteristically accompanied by right hemiplegia, should have been associated with damage to the left hemisphere in the first few centuries of the current era, even if this relationship had not been recognized earlier. Yet the association between speech defects and the left hemisphere still was not given serious recognition at this time (Benton, 1976, 1984). The basic idea of decussation proposed by Aretaeus was not universally accepted, and cases of contralateral paralysis were often explained as contrecoup effects or instances of more diffuse lesions.[1]

THE SYMMETRICAL BRAIN AND THE MIND-BODY PROBLEM

The prevailing view of the brain that emerged from the classical period was that each part of the brain was duplicated on the other side. This view remained so entrenched through the Renaissance that it led the French philosopher René Descartes to select the tiny, unitary pineal gland (glandula pinealis) on the top of the brain as the single most logical site where body and mind could interact (see Chapter 2).

Others at this time proposed ideas similar to those of Descartes, but theories that granted the same high honors to different unitary brain structures. The septum was sometimes mentioned, and the corpus callosum was elevated to the loftiest of positions in a number of instances. At this time, however, the corpus callosum often included more than just the great cerebral commissure. For example, in his Cerebri anatome, Thomas Willis (1664) used the term to refer to all cerebral white matter.

Giovanni Maria Lancisi (1654–1720), a papal physician, was among the early writers who proposed that the corpus callosum was the seat of the soul (see Clarke, 1968). Lancisi (1712), however, referred to the corpus callosum in the seemingly modern way when he wrote:

It is quite clear that the part formed by the weaving together of innumerable nerves is both unique and situated in the

[1]It was not until early in the eighteenth century that Domenico Mistichelli and Pourfour du Petit convincingly demonstrated the crossing of the corticospinal tract in the pyramids of the medulla (see Chapter 14).

middle: and so it can be said it is like a common marketplace of senses, in which the external impressions of the nerves meet. But we must not think of it merely as a storehouse for receiving the movements of the structures: we must locate in it the seat of the soul, which imagines, deliberates, and judges. (Translation in Neuburger, 1981, 50)

Perhaps the most ardent supporter of Lancisi's position was the French surgeon François Gigot de La Peyronie (1678–1747), whose thoughts on this subject began to appear early in the eighteenth century (see Neuburger, 1981). In a report to the Académie des Sciences in Paris, La Peyronie (1741) stated that cerebral injuries did not inevitably cause death, even if they involved the pineal body, cerebellum, or thalamus. He argued that fatal cases always involved the corpus callosum, now viewed as essential to life. He added that by examining brain-damaged patients before they died, it also became clear that the corpus callosum was the seat of intellectual faculties. La Peyronie, however, did not back up his statements with cases that exhibited lesions restricted to just the corpus callosum (Neuburger, 1897).

Experimental proof that the corpus callosum was not essential to life came from Albrecht von Haller and his students. One of them, Johann Gottfried Zinn (1749), conducted a number of experiments on dogs and cats in which he damaged the corpus callosum. Because some animals survived for over a day and even looked healthy, Zinn concluded that La Peyronie was wrong; the corpus callosum was not basic to life.

The important point on which all of these philosophers and scientists seemed to agree, in their attempts to understand these unitary midline structures, was that the two sides of the brain were similar. There seemed to be no hint that either hemisphere was special—no idea at all that one was more important than the other for functions such as speech, reasoning, or spatial orientation.

HEMISPHERIC BALANCE IN THE FIRST HALF OF THE NINETEENTH CENTURY

A new and different philosophical idea—namely, that there might be two distinct minds, each associated with a hemisphere—began to emerge late in the 1700s. This concept was proposed by Meinard Simon Du Pui (1754–1834) in 1780. He wrote that man is Homo duplex, meaning that man possessed a double brain. He added that "man's nervous system is just as bipartite as the rest of his body, with the result that one half of it may become affected while the other half continues to carry out its proper functions" (see Lokhorst, 1985, pp. 184–185).

As the 1800s began, the view that the two hemispheres were associated with two minds seemed to grow, although the dominant position was still that the two sides were identical. In this context, Karl Friedrich Burdach (1826) maintained that the corpus callosum served to unite the psyches, although he also looked upon it as an organ of reason. He wrote:

By means of the corpus callosum both hemispheres can act together in perceiving sensations. The activity aroused by impressions on corresponding points in the two hemispheres will therefore be able to produce one and the same perceptive image. (Translation in Neuburger, 1981, 280)

The phrenologists, who were challenging scientific dogma with their revolutionary theories of cortical localization at this time, also viewed the cortex as a dual organ (see Chapter 3). Franz Joseph Gall, the leader of the phrenological movement and an outstanding anatomist, recognized no anatomical or functional differences between the two hemispheres. Even though he occasionally described aphasic cases with cortical damage in his attempt to localize language in the anterior lobes, he did not realize that the left hemisphere was special for speech (see Chapter 25).

The fact that speech and other higher-order deficits frequently appeared after unilateral lesions may seem at odds with the theory of hemispheric duality. One might think that if one side were damaged, the other, being functionally equal (redundant), could still serve to mediate normal behavior. Viewed in this way, there should be no deficits after unilateral lesions once the shock from the trauma has a chance to dissipate.

In *Sur les Fonctions du Cerveau (On the Functions of the Brain),* Gall described a clergyman whose right hemisphere was entirely "disorganized" just before he died, but who still preached in an astonishing manner. Gall (1822, p. 247) added:

> I have proved . . . that the nervous system of the spinal marrow, of the organs of the senses, and of the brain are double, or in pairs. We have two optic nerves and two nerves of hearing, just as we have two eyes and two ears; and the brain is in like manner double, and all its integrant parts are in pairs. Now, just as when one of the optic nerves, or one of the eyes is destroyed, we continue to see with the other eye; so when one of the hemispheres of the brain, or one of the brains, has become incapable of exercising its functions, the other hemisphere, or the other brain, may continue to perform without obstructing those belonging to itself. In other words, the functions may be disturbed or suspended on one side, and remain perfect on the other. (Translation from Anonymous, 1845)

The theory advanced to account for the loss of speech long after damage to just one hemisphere was that some lesions may throw the organ of mind into a state of imbalance. This idea was expressed by the highly influential French vitalist François-Xavier Bichat, even before Gall's phrenological theory became popular. Bichat (1805), shown in Figure 26.1, taught anatomy, physiology, and surgery in Paris and defined life as the sum total of the forces that resisted death. He believed that the organs of the brain had to be symmetrical for "animal life" (as opposed to "organic life") and wrote the following about the symmetry of the nervous system:

> The Nerves, which transmit the impressions received by the senses, are evidently assembled in symmetrical pairs. The brain, the organ (on which the impressions of objects are received) is remarkable also for the regularity of its form. Its double parts are exactly alike, and even those which are single, are all of them symmetrically divided by the median line. (1827 translation, 20–21)

Bichat, who had a remarkably asymmetrical head himself, believed the two sides of the brain had to work in unison. He wrote that any change in this equality or harmony would lead to chaos by not allowing the organism to deal

Figure 26.1. François-Xavier Bichat (1771–1802), the French scientist who emphasized the importance of brain symmetry for animal life.

appropriately with the right and left sides of its world. Damage to the hemispheres could not only affect sensory and motor functions but could also disrupt one's ability to meld two impressions into one.

From this premise, Bichat concluded that under some circumstances it might be better to have bilateral brain damage as opposed to a unilateral lesion. He even reasoned that a blow to the healthy side of the head in a patient with unilateral brain damage could be therapeutic. Bichat expressed this as follows:

> If both hemispheres of the brain, however, be affected equally, the judgment though weaker, will be more exact. Perhaps it is thus, that we should explain those observations so frequently repeated, of an accidental stroke upon one side of the head having restored the intellectual functions, which had long remained dormant in consequence of a blow received upon the other side. (1827 translation, 31–32)

Marie-Jean-Pierre Flourens (1824, 1842), the leading French experimentalist (and a scientist greatly influenced by Bichat), also did not differentiate between the two hemispheres, either structurally or functionally. In this regard, Flourens sided with Charles Bell (1811), who also argued that the two hemispheres were similar. To quote Bell:

> Whatever we observe on one side has a corresponding part on the other; an exact resemblance and symmetry is preserved in all the lateral divisions of the brain. And so, if we take the proof of anatomy, we must admit that as the nerves are double, so is the brain double; and every sensation conveyed to the brain is conveyed to the two lateral parts; and the operations performed must be done in both lateral portions at the same moment. I speak of the lateral divisions of the brain being distinct brains combined in

function. . . . Betwixt the lateral parts there is a strict resemblance in form and substance; each principle part is united by transverse tracts of medullary white matter, and there is every provision for their acting with perfect sympathy. (1936 reprint, 112)

Nevertheless, the notion of perfect *anatomical* symmetry, especially in man, was beginning to be questioned. In 1805, Félix Vicq d'Azyr, another Frenchman, wrote that the convolutional pattern is more variable on the two sides in humans than it is in lower forms. He used this as an explanation for the superiority of humans over lower animals.

Three decades later, a contributor to *The Phrenological Journal and Miscellany* wrote that some inequality between the hemispheres allows them to perform antagonistic actions or to act jointly (Watson, 1836). This author proposed that we have the power of dividing the consciousness of the two hemispheres to some extent, but that the hemispheres still communicate with each other, just as we communicate with other people.

By the mid-nineteenth century, the belief that the two hemispheres may not be functionally equivalent was more readily accepted. Even the author of a feature article in a leading phrenological publication *(Buchanan's Journal of Man)* proposed that the hemispheres were not precisely the same (an illustration from this article is presented as Figure 26.2). This author (presumably Buchanan) thought that one was more "male" and the other more "female" in function. He then asked why we do not have two distinct, competing

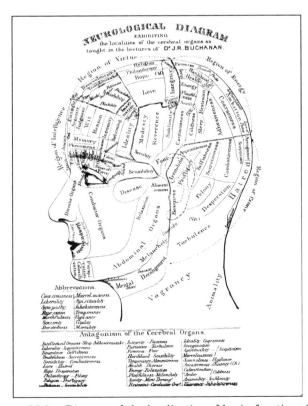

Figure 26.2. Diagram of the localization of brain functions from *Buchanan's Journal of Man,* 1850. The article that followed this picture stated that although the right and left hemispheres were equal in most respects, the left hemisphere was "male" and the right hemisphere was "female." The author proposed that male and female minds were able to work in harmony by communicating through the corpus callosum and other commissures.

minds if there are two distinct brains. He answered his rhetorical question by pointing to the corpus callosum and the other commissures:

> Thus whatever excitement may be manifested by one hemisphere, will be echoed by the other, the secondary excitement being, in all cases, in exact correspondence with the primitive action of the other hemisphere. Thus the two hemispheres of the cerebrum, which I have already shown to have relatively a masculine (left) and a feminine (right) character, sustain to each other in action the same relation which exists in a harmonious family, where a perfect marriage has thoroughly united husband and wife. The will of each is influential upon the other, and generates a corresponding will or desire. (Anonymous, 1850, 528)

Just a few years earlier, Henry Holland (1788–1873) had drawn attention to the role of the corpus callosum in the transmission of information between the two sides of the brain. Holland, who became physician to Queen Victoria (1819–1901), wrote a popular essay in 1840 entitled "On the Brain as a Double Organ." In it, he emphasized that the ability of the two hemispheres to function noncompetitively had to be due to the commissures. He recognized, however, that experiments and accidental lesions of the commissures had not yet provided unequivocal support for his ideas.

MENTAL ILLNESS AND HEMISPHERIC DYSFUNCTION

Two conflicting views of mental illness prevailed in the first half of the nineteenth century. Moral theorists looked upon mental illness as an affliction of the mind that could be treated with kindness and by appeals to the intellect. Philippe Pinel, shown at the Salpêtrière in Figure 26.3, was a leader of the school that emphasized the spiritual causes and treatment of madness. Another was the Reverend John Barlow (1798–1869), who preached that madness was a moral defect that could be cured without recourse to physical or somatic approaches. In contrast, the "physicalists" thought mental illness was due to a defective brain. Consequently, they usually proposed somatic treatments for insanity (Jacyna, 1982).

During the nineteenth century, some physicians who emphasized the physical basis of mental illness fought to put the care of the mentally ill solely under their control. These doctors, sometimes referred to as "alienists," attempted to apply emerging findings about the brain to the treatment and understanding of their patients. The alienists also searched hard for structural changes to associate with madness, deviance, epilepsy, and other behavioral abnormalities. At first, many turned to phrenology, but as the study of skull features fell into disrepute, most started to look more carefully at the brain itself.

One investigator, for example, surveyed the brains of over 100 mentally defective people and said that he found only 8 brains free from morbid signs, especially in the gray matter (Davey, 1853). His conclusion was that in the "grey matter of the brain, then, is located the proximate cause of insanity" (p. 28).

The history of alienist psychiatry and neurology in the first half of the nineteenth century contains many claims for "madness" being due to pathology on just one side of the brain (Harrington, 1985, 1986, 1987). The earlier writers

Figure 26.3. Philippe Pinel (1745–1826) with mental patients in the courtyard of the Salpêtrière.

generally assumed it was the result of the disruption of symmetry between the two sides. The belief that mental illness could result from a loss of hemispheric unity was expressed by Jean Esquirol (1838) in France, Henry Holland (1840) in England, and Benjamin Rush (1812) in the United States (see Rush, 1981).

In Great Britain, Arthur Ladbroke Wigan (d. 1847) wrote that the two hemispheres were, in fact, two separate wholes, each associated with a distinct and entire mind (Wigan, 1844a, 1844b). He initially based this idea on a man who seemed rational, even though one cerebral hemisphere was destroyed. Wigan searched the medical literature and found other cases of patients with severe damage to one hemisphere who exhibited seemingly sane behavior. To Wigan, this meant that there were not only two distinct hemispheres but two minds.[2]

Wigan (1844a, 1844b) presumed that even if the two hemispheres were not perfectly equal in healthy people (as a result of "nutritional" differences due to asymmetric blood supplies), they normally worked without discord because one was superior in power and could dominate the other. If anything, the left hemisphere was seen as superior since most people preferred their right hand for writing and other manual functions.

In Wigan's opinion, the fact that the hemispheres were usually in harmony was attributable to education. That is,

he viewed functional unity as an acquired skill. To quote Wigan:

> That the object and effect of a well-managed education is to give the power of concentrating the energies of both brains on the same subject at the same time; that is, to make both cerebra carry on the same train of thought together; and that the object of moral discipline is to strengthen the power of self-control; not merely to augment and confirm the power of both intellectual organs to govern the animal propensities and passions, but to corroborate the intellectual antagonism of the two brains; each, so to speak, a sentinel and security for the other. (1844b, 40)

Because Wigan was aware of cases of normal emotion and high intellect after the corpus callosum had been damaged, he felt the callosum did not play an especially unique or critical role in maintaining this functional unity.[3] As he put it: "The corpus callosum seems little more than a bond of mechanical union" (1844b, p. 39).

Wigan's position on hemispheric organization was somewhat like that of the phrenologists, who saw one hemisphere as a backup for the other.[4] Wigan's emphasis, however, was more on the possibility that the incongruous actions of the

---

[2]Wigan did not go so far as to claim that there were two souls.

[3]One such case was that of James Cardinal, described by Bright (1831).

[4]See the anonymous review in *The Phrenological Journal, and Magazine of Moral Science* of 1845.

two hemispheres could cause madness. Indeed, Wigan was a physicalist in the sense that he recognized mental illness could have physical causes. Yet he was also a moralist, at least to the extent of accepting the idea that some aberrant behaviors could be curtailed, and possibly even cured, by treatments based on self-control and discipline.

Wigan cited *Man's Power over Himself to Prevent or Control Insanity,* a book written by the Reverend John Barlow in 1843, as the immediate incentive for the development of his own treatment theory. Wigan proposed that diseases rarely affected both hemispheres equally and maintained that the insane could learn to use the healthier hemisphere to overcome maladaptive thoughts, emotions, and antisocial actions coming from the diseased hemisphere.

In the reviews that followed, some of Wigan's ideas were claimed to be unoriginal, whereas others were simply assailed. In one short paper that appeared in *Lancet,* both guns were fired (Davey, 1844). First, it was argued that not only had the phrenologists acknowledged the double brain, but that this idea had been advanced even earlier by Gerard van Swieten (1700–1772). The following scathing statement appeared later in the same review:

> The controlling power of which he speaks is not generally exercised by one hemisphere over the other, but by individual parts of each hemisphere, simultaneously, over other parts, or *organs,* of the brain more or less distant. To speak, therefore, of "presenting motives of encouragement to the *sound* brain, to exercise and strengthen its control over the *unsound* brain" meaning thereby the two hemispheres of the organ, is advice which applies only to the exception and not the *rule.* . . . To suppose that the hemispheres of the brain have each one faculty, and that one is either *good* or *bad* is, indeed, to simplify the labours of the great Gall. (Davey, 1844, 377–378)

Additional criticisms of Wigan's claims appeared in other letters to the *Lancet* (e.g., Ryan, 1844), in the minutes of the Westminster Medical Society (Fisher, 1844), and in the phrenological journals (e.g., Anonymous, 1845).

CLINICAL REPORTS ON APHASIA IN THE PERIOD BEFORE BROCA

Before 1860, a number of clinical reports confirmed the decussation of the motor pathways and suggested that left hemispheric damage, or right hemiplegia, was associated with loss of speech more frequently than right hemispheric damage, or left hemiplegia. The authors of these studies, however, did not recognize possible laterality effects for speech, often being occupied with other concerns or dealing with small samples.

Giovanni Battista Morgagni (1769), for example, described 10 cases of unilateral paralysis in noncomatose or nonstuporous patients. Morgagni found loss of speech in four of five patients with right-side paralysis and in only one of four with left-side paralysis (see Benton, 1984).

Gall, as already noted, presented some clinical cases, but saw them only as a confirmation of what his skull measurements had already suggested. Thus, Gall assumed on a priori grounds that the organs of speech, which he localized in the anterior lobes, were duplicated.

Jean-Baptiste Bouillaud, who had been trained as a phrenologist, agreed with the idea of a frontal lobe localization for speech. But unlike Gall, he presented a large number of cases of people with brain damage to support his contentions. Bouillaud (1825) postulated the existence of two frontal lobe speech areas—one for the creation of ideas and the other for the movements needed for articulation. He maintained that the two could be altered independently, but he failed to see the special significance of the left hemisphere before Paul Broca drew attention to it in the 1860s. By 1825, Bouillaud had, in fact, described 11 cases of left hemispheric lesions and 24 of right hemispheric lesions. In retrospect, it is interesting to note that 73 percent of the people with left hemisphere damage were aphasic, whereas only 29 percent with right hemispheric damage were aphasic (Benton, 1984). To Bouillaud, the important point was only that, after unilateral lesions, the aphasia cases exhibited damage within or close to the anterior lobes.

Gabriel Andral (1840) also examined a number of individuals who appeared to have unilateral lesions. Loss of speech was seen in three of five of his left hemispheric patients and in none of his six right hemispheric patients. These numbers were too small to direct Andral's attention to hemispheric differences, so instead he chose to emphasize the fact that loss of speech is not always associated with frontal lobe insult.

THE MARC DAX MANUSCRIPT OF 1836

When Broca (1861) presented his famous case of "Tan," the lateralization issue was still far from his mind. But in 1863, Gustave Dax sent a paper, handwritten by his father in 1836, to the *Académie de Médecine* in Paris. This paper was deposited on March 24, shortly before Paul Broca described eight aphasia cases, all with lesions on the left side. The paper showed that Marc Dax, a physician from Sommières (a town near Montpellier in the South of France), had already recognized a strong association among lesions of the left hemisphere, aphasia, and paralysis of the right side of the body. This correlation was based on the records of more than 40 patients collected over a 20-year period.

The Marc Dax paper of 1836 on the significance of the left hemisphere was reviewed by Jean-Baptiste Bouillaud, Paul Augustin Béclard (1785–1825), and Louis Francisque Lélut, but was not read before the *Académie* until December 1864. The paper was published in 1865, accompanied by a note from Gustav Dax, who was also a physician interested in aphasia and a dutiful son who was strongly supportive of his father's claims. Although Gustav Dax stated that his father's paper had been read at the *Congrès Méridional de Montpellier* in 1836, no evidence to support this claim has emerged (Critchley, 1979; Joynt and Benton, 1964). Nevertheless, a few copies of the paper appeared to have been distributed to friends and colleagues (Caizergues, 1879). In fact, the handwritten copy sent to Paris came from a former dean at Montpellier, a school whose faculty during the 1830s was at least equal to the faculty of Paris, and a place where the oral transmission of knowledge seemed to count at least as much as the written document (see Bayle, 1939; Lecours, Nespoulous, and Pioger, 1987).

The sequence of events and the possibility that Broca

may have seen or heard of the Dax finding prior to his own
written reports on cerebral dominance have generated considerable interest among historians. It has been pointed out,
however, that Broca's cautious and brief 1863 reports, which
raised the possibility of left hemisphere specialization for
language, were probably written before he knew of the Dax
report. In particular, Broca applied for membership in the
Académie de Médecine and submitted an annotated bibliog-
raphy containing his views on cerebral dominance before the
Dax papers (father and son) were even deposited. Moreover,
Broca's application was mentioned on the same page of the
bulletin of the *Académie* that announced the receipt of the
Marc Dax manuscript. This would suggest that Broca must
have noted that the statistical evidence favored the left side
of the brain for language sometime before March 24, 1863
(Schiller, 1979).

BROCA'S CASE FOR DOMINANCE

In his note in the *Bulletins de la Société d'Anthropologie,*
published in May 1863, Paul Broca described eight cases of
aphasia, all with lesions on the left side. At the time, this
seemed surprising to him. Unlike Marc Dax, Broca could
only say that whereas this caught his attention, he could
not draw firm conclusions until he had accumulated more
facts.

Later that year, Jules Parrot (1829–1883) presented the
case of a patient who had a lesion of the right frontal lobe
without an articulate language disturbance. Broca recognized that this finding also pointed to the special role of the
left hemisphere in speech. Nevertheless, Broca seemed to
realize that given how hard it was for the conservative scientific establishment to accept a frontal lobe localization
for articulate speech, it was going to be even more difficult
to present the novel idea that only the left frontal lobe was
essential for this function.

In 1864, Broca described two more patients who had
traumatic head injuries on the left side (cited in Duval,
1864). Here he referred to the mounting evidence for left-
side involvement.

> Numerous observations gathered during the last three years
> have a tendency to indicate that lesions of the left hemi-
> sphere are solely susceptible for causing aphemie. This
> proposition is no doubt strange, but however perplexing it
> may be for physiology, it must be accepted if subsequent
> findings continue to indicate the same view point. (Trans-
> lated in Berker, Berker, and Smith, 1986, 1066)

In contrast to this relative conservatism, and perhaps
now stirred by his knowledge of the Dax findings, Broca's
1865 paper dealt directly and extensively with the issue of
left hemispheric dominance for language. This extraordinary paper appeared in the June 15 issue of the *Gazette
Hebdomadaire de Médecine et de Chirurgie,* the same journal that published the Marc and Gustave Dax paper less
than two months earlier. Broca now took a firm position
about left hemispheric dominance for articulate speech, but
was careful to add that language was not the exclusive
function of the left hemisphere.

> This does not mean to say that the left hemisphere is the
> exclusive center of the general capacity of language, which

consists of establishing a determined relationship between
an idea and a sign, nor even of the special capacity of articulate speech, which consists of establishing a determined
relationship between an idea and an articulate word. The
right hemisphere is not more a stranger than the left hemisphere to this special faculty, and the proof is that the person rendered speech disabled through a deep and extensive
lesion of the left hemisphere is, in general, deprived only of
the faculty to reproduce the sounds of articulate speech; he
understands perfectly the connection between ideas and
words. In other words, the capacity to conceive these connections belongs to both hemispheres, and these can, in the
case of a malady, reciprocally substitute for each other;
however, the faculty to express them by means of coordinated movements in which the practice requires a very long
period of training, appears to belong to but one hemisphere,
which is almost always the left hemisphere. (Translated in
Berker, Berker, and Smith, 1986, 1068)

Armand Trousseau (1864), like Bouillaud, felt that the
statistics offered good support for the statements made by
Marc Dax and Paul Broca about the significance of the left
side of the brain for speech. He thought the data were also
relevant to the question of the seat of the intellect. This view
stemmed from his observations of many aphasic patients
who showed impairments in general intelligence that went
well beyond their language difficulties. Trousseau, shown in
Figure 26.4, speculated that the cause of these functional
asymmetries might be different blood supplies and other
anatomical differences between the hemispheres.

These ideas were acknowledged positively by Jules
Gabriel François Baillarger, who postulated that the more
rapid growth of the left hemisphere could also explain handedness. In addition, Pierre Gratiolet and François Leuret
(1797–1851) had claimed some years earlier that there were

Figure 26.4. Armand Trousseau (1801–1867), the French neurologist
who associated the left hemisphere with both aphasia and intellect.
Trousseau also coined the word "aphasia" (*aphasie*) and studied movement
disorders. A portrait of Trousseau as a young man appears in Chapter 16.

differences of a few grams between the two hemispheres during development, with the left hemisphere being the heavier (Gratiolet and Leuret, 1839–1857).

Broca himself accepted these ideas. Hence, in 1865, he made a strong case for the two hemispheres not being innately different from each other, but for the left being more precocious. The implication was that the left hemisphere would be educated first and would take the lead in the acquisition of speech and handedness.

In his 1865 paper, Broca wrote that he believed speech was mediated by the right hemisphere in left-handed people. The concept of reversed dominance was readily accepted, as seen in this quotation from William Gowers 20 years later:

> In left-handed persons the "speech-centre" is usually on the right, and not on the left side of the brain, and the association just mentioned was well exemplified by a left-handed man, who, at the age of thirty-one, became liable to fits which commenced by spasms in the left side of the face, spreading thence to the left arm, without loss of consciousness. Inability to speak preceded each attack for ten minutes, and persisted afterwards for the same time. (1885, 38)

Only in the twentieth century did scientists begin to wonder whether Broca might have been wrong about dominance in left-handed people. In his 1914 review of cerebral functions, Shepherd Ivory Franz stated that the general belief that aphasias in left-handed people are produced by lesions in the right hemisphere was not borne out by case studies. Franz specifically discussed one case in which the author found that a left-handed person developed an aphasia after a lesion of the left frontal lobe.[5] With more time, it was recognized that a majority of left-handed people still have speech localized on the left side of the brain or exhibit more of a "mixed" dominance.

WERNICKE'S IMPACT

The idea that the left hemisphere is special received a boost in 1874 when Carl Wernicke demonstrated that left hemispheric injuries were also associated with a more "sensory" sort of aphasia—a variety in which speech is fluent but not meaningful. Although cases of "Wernicke's aphasia" could be found in the older literature, this disorder had not been linked in the past with a lesion of just one hemisphere.

Since many patients with Wernicke's aphasia also showed diverse cognitive defects and inappropriate conduct, the possibility was further entertained that these patients were exhibiting broader defects in conceptual thinking. For example, pantomime, a form of symbolic communication, is often difficult for aphasic patients to understand. This observation was made by Finkelnburg in 1870 and by John Hughlings Jackson in 1878, and suggested an even broader role for the "dominant" hemisphere.

JACKSON'S CASE FOR THE RIGHT HEMISPHERE

John Hughlings Jackson's first reference to cerebral dominance appeared in 1864 when he said he thought Paul Broca was correct in suggesting that aphasia was associated with lesions on the left side of the brain.[6] Jackson (1864a) had examined some 70 patients with loss or defect of speech, and in all but one the hemiplegia was on the right side, indicating left hemispheric injury.

Nevertheless, Jackson had trouble with Broca's concept of a cerebral center for language, as opposed to a sensitivity for disrupting articulate language with a particular lesion. He wrote:

> Whilst I believe that the hinder part of the left third frontal convolution is the part most often damaged, I do not localise speech in any such small part of the brain. To locate the damage which destroys speech and to locate speech are two different things. . . . *But the matter of most significance is that damage to but one hemisphere will make a man speechless.* This no physician denies, so far as I know. (1874a, 19)

Jackson, unlike Broca, tended to look at the brain in terms of sensory-motor relationships, even with regard to "higher functions" such as language. Yet he always remained the gentleman, never criticizing Broca and always speaking of the Frenchman's important discoveries with praise and admiration.

Jackson is often associated with the idea that even patients with severe "motor" (Broca's) aphasia can curse and blurt out emotional exclamations. Actually, Jackson (1866, 1874a) repeatedly credited Jules Baillarger for drawing his attention to this phenomenon. But even before this, Broca (1861) himself had stated that aphasia patients could still swear and come forth with fluent, crude oaths when provoked.

Just what did this retained ability mean? Jackson (1874a) felt that propositional speech was perhaps the most voluntary activity humans possessed. The importance of this was, he believed, that whereas voluntary activities were mediated more by one hemisphere than the other, involuntary activities were associated with more functionally equivalent neural structures.

> The right hemisphere is the one for the most automatic use of words, and the left the one in which automatic use of words merges into voluntary use of words—into speech. Otherwise stated, the right is the half of the brain for the automatic use of words, the left the half for both the automatic and the voluntary use. The expression I have formerly used is that the left is the "leading" half for words (speech). (1915 reprint, 81–82)

Swearing and cursing were involuntary emotional functions to Jackson. This meant they had to have more equal representation in the two hemispheres than ordinary expressive speech, which was more firmly rooted in the left hemisphere. Jackson emphasized this difference by teaching

---

[5]Franz cited the case of Long (1913).

[6]Jackson may have learned of Broca's discovery though Charles-Edouard Brown-Séquard, who traveled extensively between Paris and London. It is also possible that he saw or was told about the July 1863 issue of the *Bulletins de la Société Anatomique*, which mentioned Broca's lateralization findings (Greenblatt, 1970; Harrington, 1987).

that although a patient with motor aphasia could swear when aroused, he or she could not "say," or repeat, the same oath that had just been shouted in a fit of emotion. Moreover, he felt that left hemispheric lesions, which affected propositional speech, could "heighten" the activity of the opposite hemisphere, where largely involuntary functions would be represented.

Jackson (1873) also believed comprehension was a bilateral function and perhaps the most involuntary language function of all. He theorized the right hemisphere could still learn the meanings of words as a result of associative laws (hearing the word "horse" while also seeing a horse) and that it possessed a consciousness for places and things. Yet, it seemed clear to him that only the left hemisphere could truly become "conscious in words."

Jackson began to express himself on these matters before Broca even published his 1865 paper on laterality.[7] For example, he raised the following point in 1864:

> If, then, it should be proved by wider evidence that the faculty of expression resides in one hemisphere, there is no absurdity in raising the question as to whether perception, its corresponding opposite—may not be seated in the other. (From *National Hospital for Epilepsy and Paralysis Reports*, 604)

In 1868, Jackson elaborated further on cerebral dominance. One of his major points was that left hemispheric patients with aphasia were not poor in their perceptual abilities.

But this was "negative" evidence, and Jackson recognized that positive instances were required to strengthen his case for right hemispheric specialization, especially for perceptual abilities. These cases began to come forth in the 1870s. For example, in 1872 Jackson described a man with a left hemiplegia who could not recognize people, including his wife, or places and things. He presented a comparable case four years later. Elisa P., 59 years old, had trouble dressing herself and did odd things. Her most noticeable symptom was the loss of directional sense (topographical memory loss). Jackson wrote:

> She was going from her own house to Victoria Park, a short distance and over roads that she knows quite well, as she has lived in the same house for 30 years, and has had frequent occasion to go to the park; on this occasion, however, she could not find her way there, and after making several mistakes she had to ask her way, although the park gates were just in front of her. When she wished to return she was utterly unable to find her way, and had to be taken home by a country relation to whom she was showing the Park for the first time. (1876, 438)

Elisa P. deteriorated and died three weeks after the onset of her illness. Her brain was examined by William Gowers, who found a large glioma in the posterior part of the right temporal lobe.[8]

Jackson (1874b, 1876) coined the term "imperception" to describe the "loss or defect of memory for persons, objects, and places." He associated imperception with right hemi-

spheric lesions. In his mind, imperception was every bit as special as the language disturbances associated with left hemispheric lesions.

Jackson (1868) accepted the belief that the left frontal lobe grew faster than the right frontal lobe and took the lead in voluntary speech. He also believed the right posterior lobe grew in advance of that on the left, and from this he deduced that the posterior right hemisphere was the leading part for perception and imagination. To use Jackson's (1874a) own words: "The 'important side' of the left hemisphere is the anterior lobe, and the important side of the right, the posterior lobe."

THE SEARCH FOR ANATOMICAL CORRELATES OF CEREBRAL DOMINANCE

By 1869, Paul Broca was writing that asymmetry was a hallmark of the human brain and discussing the increase in symmetry as one went down the phylogenetic scale. This position was fairly widely accepted at the time, as can be seen in the following lines from one of Broca's contemporaries in England, John William Ogle:

> The secondary convolutions are never arranged in the same way in the two hemispheres. And it is a notable fact, that this want of symmetry is peculiar to man, or at any rate much more strongly marked in man than in any other animal. . . . It would thus appear that man, distinguished from other animals by his faculty of speech and by his higher intellectual capacity, is also distinguished from them by the want of symmetry in his two hemispheres. We may go even further than this. At that period of life when the intelligence is undeveloped, and the faculty of speech not yet acquired, the brain even in man presents perfect symmetry on its two sides. "In the imperfectly developed brains of the infant or young child" (I again quote from Todd) "the convolutions are quite symmetrical." . . . In this case we may suppose with Dr. Moxon the difference between the two sides to arise in the course of education. (1867, 88–89)

Broca believed that asymmetry was greater in the brains of Caucasians than in other people. This idea, accepted throughout Europe, corresponded well with the belief that the brains of white Europeans were also notably larger than those of "inferior races" (Thurnam, 1866). For example, Henry Charlton Bastian (1880) placed "Negroids" between apes and Caucasians on a scale of brain symmetry. It was also thought that female heads and brains were more symmetrical than those of males, and that children showed more symmetry than adults. Cesare Lombroso even believed "born" criminals were more likely than upstanding citizens to have symmetrical nervous systems and to be ambidextrous (see Lombroso, 1903; Marro and Lombroso, 1883; also see Chapter 21). Lombroso is shown in Figure 26.5.

Along with this sort of thinking went a search not just to show that the hemispheres were more asymmetrical as one went up the phylogenetic scale, but to find reliable structural correlates of dominance. In 1864, Hans Carl Barkow (1798–1873) reported that he had made an extensive study of human skulls and had found the left frontal region more pronounced than its corresponding member on the right side. He also observed a trend for the right posterior region to be larger than the left posterior region. Broca

---

[7]Jackson mentioned functional differences between the hemispheres in two of his own papers from 1864 (1864a, 1864b). The *National Hospital for Epilepsy and Paralysis Reports* also cited this finding.

[8]For another striking case of topographical memory loss due to damage to the right hemisphere, see Case 1 in Meyer (1900).

Figure 26.5. Cesare Lombroso (1836–1909), the Italian criminologist who believed "born" criminals suffer from nervous systems that are too symmetrical and therefore primitive.

(1866) viewed this as consistent with the findings of Gratiolet and said he had even seen such a trend himself.

Nevertheless, these measurements were on skulls, not brains. Hence, even more attention was drawn to reports showing that there was more gray matter in the left frontal region than in the right frontal region (Roques, 1869).

In 1890, Oskar Eberstaller, a Viennese anatomist, examined 170 human brains and found the left Sylvian fissure more than 12 percent longer than the right. Two years later, D. J. Cunningham (1850–1909) confirmed Eberstaller's findings using the ratio of Sylvian fissure length to hemisphere length. He further described differences in the angle of the Sylvian fissure and extended his laterality findings to the insula. Cunningham showed that a longer left Sylvian fissure could also be seen in 7½ to 8½ month-old fetuses and newborns, although the difference was not as marked as it was in adults. He added that a small difference between the Sylvian fissures also existed in chimps and macaque monkeys, but not in orangutans, mangabys, or cebus monkeys. But because monkeys and apes do not show dominance or consistent side preferences, Cunningham downplayed the significance of his findings with infrahuman primates.

Scientists also compared the weights of the two hemispheres and the individual lobes of the brain. In 1861, Robert Boyd (1808–1883), an Englishman, reported a right versus left hemispheric difference. He did not look at specific lobes when he concluded that the left side was slightly heavier in his sample of 528 individuals from an asylum. This finding, however, failed to be confirmed in follow-up studies (e.g., Crichton-Browne, 1878; Thurnam, 1866). Paul Broca, for one, made a more careful analysis and found no meaningful differences between the hemispheres as a whole, although in 1875 he noted that the left frontal region was four grams heavier than its counterpart on the right side.

Some researchers turned to the study of prehistoric hominids to see if there might be important trends in the evolution of the two hemispheres. The more noteworthy anthropological studies, however, did not appear in the nineteenth century. In the 1920s, it was reported that the left frontal lobe was slightly larger than the right in Java man, whereas the region of Broca's speech area was more "complex" on the left side of Neanderthal man relative to the more ancient Java specimens (Tilney, 1927; see also Smith, 1924).

THEORIES OF PERSONALITY, EMOTION, AND INSANITY

Many theories about personality, behavioral deviance, and hemispheric specialization emerged in the last quarter of the nineteenth century. For example, Jules Bernard Luys wrote that he had observed entirely different personality changes after left and right hemisphere lesions.

> Whereas . . . right hemiplegics, are more or less apathetic, more or less silent, passive and stricken with hebetude;— left, emotional hemiplegics, are more or less afflicted with an abnormal impressionability. They respond, when one questions them, with limping tones broken up with a type of sobbing. . . . In other circumstances . . . they are boisterous and loquacious. (I)t is not unusual to see some of them in the grips of a veritable fit of excitation, maniacal, having false conceptions, delusions of persecution, and even attempting suicide. (1881, 380)

Luys' ideas were consistent with the notion that the hysterias were more likely to affect the left side of the body than the right side (and thus the right hemisphere more than the left hemisphere). The ratio was put at 3:1 by Paul Briquet (1796–1881), who worked in Paris at the Charité. Brown-Séquard cited Briquet on this subject and came to similar conclusions after examining a large number of cases.

Luys (1879) argued that in a state of madness the weight difference between the two hemispheres was reversed. The brute in civilized man predominated over reason when the right hemisphere was bigger. The supposed weight differences favoring the right side of the brain in lunatics was, to say the least, controversial. One need only contrast the 1861 measurements of Robert Boyd with those of John Thurnam (1866) and James Crichton-Browne (1878) to appreciate the variability.

Nevertheless, the general concept of an "uncivilized" right hemisphere drew additional support from studies showing a greater frequency of auditory hallucinations associated with the left ear. It also drew support from some questionable experiments on magnetism and hypnotism—these supposedly demonstrated that the left side of the brain was moral and intellectual, whereas the right side was melancholic and irrational (see Harrington, 1985, 1987).

The data on which these ideas rested could be interpreted in at least two ways. One was that there were innate differences between the hemispheres. The other was that two fundamentally equivalent hemispheres were differentially affected by education and/or by disease. From the latter perspective, it was not so much that the right hemisphere naturally grew larger or faster than the left, but the idea that the educated, intellectual left hemisphere had atrophied more than the right hemisphere in cases of madness and degeneracy (Montyel, 1884).

As would be expected, there were also growing literary allusions to people showing two opposing personalities—one reflecting the emerging concept of the right hemisphere and the other exemplifying the growing concept of the left hemisphere. One famous case of opposing minds formed the basis of Robert Louis Stevenson's (1850–1894) *Strange Case of Dr. Jekyll and Mr. Hyde*, the title page of which appears as Figure 26.6.

The *Zeitgeist* could not have been better in 1886 when Stevenson's classic about two distinct and opposing personalities housed in a single body appeared. This work followed the 1880 publication of a play entitled *Deacon Brodie; or the Double Life; a Melodrama Founded on the Facts, in Four Acts and Ten Tableaux* by Stevenson and William Ernest Henley (1849–1903). The subject of the play, William Brodie (1741–1788), was a real person who had lived in Edinburgh. He was a respected cabinetmaker during the day and a nefarious burglar by night. The dual nature of Brodie so intrigued Stevenson and Henley that they continued to modify their drafts of the melodrama throughout the 1880s.

The idea to carry the theme of dual minds even further came to Stevenson in a terrifying nightmare. In fact, he admonished his wife, Fanny, for breaking off such "a fine bogy-tale" by waking him. In addition, the hypnotic influence of Edgar Allan Poe's (1809–1849) morbid psychology, and especially his short story "William Wilson," may have influenced Stevenson to write his horror story in just a few weeks and then, within just three days, to recast it as an allegory (Daiches, 1973; Hammond, 1984). This was the context in which Stevenson, ever conscious of the battle between good and evil, wrote his *Strange Case of Dr. Jekyll and Mr. Hyde*.

The novel itself deals with a respected citizen, Dr. Jekyll, who no longer wanted to contend with the war between the two parts of his mind. By taking a potion, he found that one part of himself was able to break away in the guise of the smaller, younger Mr. Hyde. This evil-looking figure was "more express and single, than the imperfect and divided countenance" of the good Doctor, who remained a mix of two opposing minds. As stated by Henry Jekyll:

> And it chanced that the direction of my scientific studies, which led wholly towards the mystic and the transcendental, reacted and shed a light on this consciousness of the perennial war among my members. With every day, and from both sides of my intelligence, the moral and the intellectual, I thus drew steadily nearer to that truth, by whose partial discovery I have been doomed to such a dreadful shipwreck: that man is not truly one, but truly two. . . . It was on the moral side, and in my own person, that I learned to recognise the thorough and primitive duality of man. I saw that, of the two natures that contended in the field of my consciousness, even if I could rightly be said to be either; it was only because I was radically both. . . . It was the curse of mankind that these incongruous faggots thus were bound together—that in the agonised womb of consciousness, these polar twins should be continually struggling. (Stevenson, 1986 edition, 113–114)

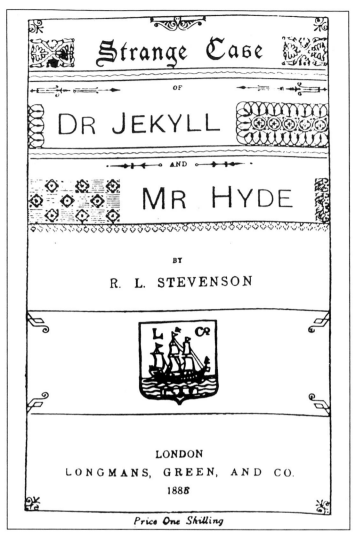

Figure 26.6. Original title page of *Strange Case of Dr. Jekyll and Mr. Hyde* by Robert Louis Stevenson (1886).

Hyde did not represent violence or lust in the pure sense but rather an unrestrained, primitive soul who could not abide by the notion that certain acts, including murder, were simply not permissible.

The Jekyll and Hyde story, which first sold as "a shilling shocker," was an immediate success. Within six months, over 40,000 copies of the little book had been sold in Great Britain alone. It is fascinating to think about how Stevenson could have been affected by the science of his day, and how the Jekyll and Hyde story could have affected the ways some scientists subsequently viewed their case material. Was the respected Dr. Jekyll not largely the personification of cultivated left hemispheric function? And was Mr. Hyde, the doctor's morally defective stunted self, not the personification of the primitive right hemisphere that simply had to be suppressed?

In 1895, nine years after Stevenson's book appeared, Lewis Campbell Bruce (1866–1949) published a scientific paper on an individual who seemed to have "two distinct consciousnesses." One consciousness was demented and spoke in Welsh. It could not comprehend spoken language, made jabbering sounds, and was shy and suspicious. When it predominated, the man's left hand was used for writing. The second consciousness showed fluency in English, but was "restless, destructive and thievish." It drew pictures of ships and incidents related to its past life, but had no memory for anything that had occurred during the other consciousness.

Bruce stated explicitly that the two distinct personalities or consciousnesses, named the Welsh and the English, were due to the actions of the two different hemispheres. The Welsh consciousness was thought to reflect the functioning of the right hemisphere, and the English consciousness, which had to be superior to the Welsh, was believed to be due to the action of the left hemisphere. But unlike others before him, Bruce (1897) went on to provide a mechanism that he thought could account for the switching back and forth. As far as Bruce was concerned, the switching could well be due to epilepsy. When the seizures "paralyzed" the civilized left hemisphere, the right hemisphere was allowed to carry out its separate mental existence.

Bruce used this explanation to account for similar behavior in a second subject. He wrote:

In her normal state the left cerebral hemisphere controlled the functions of the right, and was also in a higher state of education than the right, the mental development of which must have been peculiarly low, either due to atavism—throwing back to some remote and degraded ancestor—or through congenital deficiency, not having the capability to develop. After a right-sided grand mal attack the left cerebral hemisphere was temporarily disorganized by the violence of the brain storm, thus leaving the uneducated right hemisphere alone capable of receiving and assorting external stimuli. Her mental state, never very high, was then practically that of an uneducated savage, suddenly brought face to face with strange surroundings, with only hereditary or animal instincts to guide her. . . . After attacks of petit mal her mental condition was similar to that occurring after right-sided grand mal attacks. The only difference was that she was not left-handed. (1897, 117)

This sort of thinking had legal ramifications. Could a "right mind" (i.e., hemisphere) be found guilty if it is really behaving only in a way that is natural for it? Indeed, no one would want to say that a wild animal is guilty for killing its prey. And, among civilized men, can the left hemisphere be brought to trial for not being its brother's keeper if there are signs of epilepsy or disease? However one may choose to answer these questions, one thing seemed clear in Victorian circles. Scientists had to find ways to suppress or even change the basic nature of the uneducated, amoral, and even dangerous right hemisphere.

MOVEMENTS TO EDUCATE THE TWO HEMISPHERES

Among some individuals, there was a growing belief that the inferior right hemisphere could be educated, perhaps even as much as the "intelligent" left hemisphere. In particular, this was presumed to be true early in life. Such thinking naturally led to the presumption that early "bilateral training" could produce a new kind of superior intellectual being.

Brown-Séquard (1874a, 1874b) was among those who felt that the differences seen between the two hemispheres

reflected the failure of existing school systems to educate the right hemisphere. At this time, he believed the left hemisphere was the larger of the two and that this was partly a function of use (see also Moxon, 1866). He wrote that "if we make use of only one for most of our actions, . . . we leave useless one-half of the most important of our organs as regards manifestations of intelligence, will, and perception or sensation" (1874a, p. 2).

Brown-Séquard, never one to shy from a controversial theory, proposed that if the left side of the body could be used more, the right hemisphere would become better developed and more like the left in size and function. Thus, he recommended special bilateral training as a means of stimulating both hemispheres.

I think, therefore, the important point should be to try to make every child, as early as possible, exercise the two sides of the body equally—to make use of them alternately. One day or one week it would be one arm which would be employed for certain things, such as writing, cutting meat, or putting a fork or spoon in the mouth, or in any of the other various duties in which both the hands and the feet are employed. (1874a, 20)

Brown-Séquard felt this made perfect sense. He assumed the two hemispheres were not fundamentally different, and he had data that suggested both hemispheres were involved with the sensory and motor functions of the two sides of the body. He wrote that he knew of a number of people who sustained destruction of one half of the brain without loss of voluntary movement or sensation on either side of the body. With training, the right hemisphere, which he associated with the emotions in its natural state, could be taught to play a more active role in higher mental functions.

Brown-Séquard continued to express these revolutionary ideas well after he had introduced them.

(We) have a great many motor elements in our brain and our spinal cord which we neglect absolutely to educate. Such is the case particularly with the elements which serve for the movements of the left hand. Perhaps fathers and mothers will be more ready to develop the natural powers of the left hand of a child, giving it thereby two powerful hands, if they believe, as I do, that the condition of the brain and spinal cord would improve if all their motor and sensitive elements were fully exercised. (1890, 643)

These notions were put into practice in a few schools in the United States and Great Britain. Further, "ambidextrous culture societies" were spawned to train people to do two things at once, such as playing a piano with one hand while writing a letter with the other (Jackson, 1905). It was even thought that training the two hands could prevent permanent aphasia and speed recovery after unilateral brain lesions. Training supposedly would do this by strengthening the language areas of the weaker hemisphere (see Harris, 1985).

The movement to educate the two hemispheres also had its share of sharp, vocal critics. One of these individuals was James Crichton-Browne, who wrote:

In this present movement in favour of ambidexterity I fancy I detect the old taint of faddism. Some of those who promote it are addicted to vegetarianism, hatlessness, or anti-vaccination, and other aberrant forms of belief; but it must be allowed that beyond that it has the support of a large number of intelligent and reasonable people, and of some men of light and leading. . . . An eminent admiral gives his blessing to ambidexterity, or either-handedness, and evidently thinks it will conduce to our naval supremacy; an eminent soldier assures us that it is of the utmost value from a military point of view; an eminent artist opines that it will contribute to personal beauty; an eminent surgeon testifies that it will increase brain power, and therefore, working and intellectual ability by 25 per cent; and an eminent physician predicts that it will to a large extent ward off paralysis. (1907, 624)

Crichton-Browne then presented his belief that increased right-handedness was positively correlated with the emergence of modern man and the growth of civilization. He stated he was sure it would be an absolute disaster to try to alter such a wonderfully successful evolutionary trend. As he put it:

To "raze out the written troubles of the brain" is no easy matter; to delete its deeply engraven records is a task impossible. . . . Pushed towards that consummation which its ardent apostles tell us is so devoutly to be wished for, when the two hands will be able to write on two different subjects at the same time, it must involve the enormous enlargement of our already overgrown lunatic asylums. Right-handedness is woven into the brain. . . . My own conviction is that, as regards right-handedness, our best policy is to let well alone and to stick to dexterity and the bend sinister. (1907, 652)

Data on the effects of learning to use the nonpreferred hand and on equal use of both hands were notably missing from the heated debates taking place at this time. This changed when Philip Boswood Ballard (1865–?) presented the results of his empirical research in 1912. He found that left-handed students who were forced to become right-handed ("dextro-sinistrals") had rates of stammering some four times higher than normal. In contrast, left-handed students not forced to use their right hands in school had a low frequency of stammering. Ballard thus concluded that forced training of the nonpreferred hand might actually disorganize the admirable functional organization of the brain. He felt compelled to end his paper by stating that, in his opinion, no kind of interference with a child's natural preference would be educationally justifiable. As might be expected, not everyone was ready and willing to agree with him.

REFERENCES

Andral, G. 1840. *Clinique Médicale* (4th ed.). Paris: Fortin, Masson.

Anonymous. 1845. Review of *The Duality of the Mind* by A. L. Wigan. *Phrenological Journal, and Magazine of Moral Science*, 17, 168–181.

Anonymous. 1850. Duality and decussation. *Buchanan's Journal of Man*, 1, 513–528.

Aretaeus, The Cappadocian. 1856. On epilepsy; on paralysis. In F. Adams (Ed.), *The Extant Works of Aretaeus, the Cappadocian*. London: Sydenham Society.

Ballard, P. B. 1912. Sinistrality and speech. *Journal of Experimental Pedagogy*, 1, 298–310.

Barkow, H. C. L. 1864. *Bemerkungen zur Pathologischen Osteologie*. Breslau: Ferdinand Hirt's Königliche Universitäts -Buchhandlung.

Barlow, J. 1843. *Man's Power over Himself to Prevent or Control Insanity*. London: William Pickering.

Bastian, H. C. 1880. *The Brain as an Organ of Mind*. London: Kegan Paul.

Bayle, J. M. J. 1939. *Les Fondateurs de la Doctrine Française de l'Aphasie*. Bordeaux, France: Masson.

Bell, C. 1811. *Idea of a New Anatomy of the Brain*. London: Strahan and Preston. (Reprinted in 1936 in *Medical Classics*, 1, 105–120.)

Benton, A. 1976. Historical development of the concept of hemispheric cerebral dominance. In S. F. Spicker and H. T. Engelhardt (Eds.), *Philosophical Dimensions of the Neuro-Medical Sciences*. Dordrecht, Holland: D. Reidel Publishing Company. Pp. 35–57.

Benton, A. 1984. Hemispheric dominance before Broca. *Neuropsychologia*, 22, 807–811.

Berker, E. A., Berker, A. H., and Smith, A. 1986. Translation of Broca's 1865 Report: Localization of speech in the third left frontal convolution. *Archives of Neurology*, 43, 1065–1072.

Bichat, F.-X. 1805. *Recherches Physiologiques sur la Vie et la Mort* (3rd ed.). Paris: Brosson/Gabon. Translated by F. Gold as, *Physiological Researches on Life and Death*. Boston, MA: Richardson and Lord, 1827. Reprinted in D. N. Robinson (Ed.), *Significant Contributions to the History of Psychology*, Ser. E, Physiological Psychology: Vol. II, X. Bichat, J. G. Spurzheim, P. Flourens. Washington DC: University Publications of America, 1978. Pp. i–334.

Bouillaud, J.-B. 1825. *Traité Clinique et Physiologique de l'Encéphalite ou Inflammation du Cerveau*. Paris: J. B. Ballière.

Boyd, R. 1861. Tables of the weights of the human body and internal organs in the sane and insane of both sexes and of various ages, arranged from 2614 post-mortem examinations. *Philosophical Transactions of the Royal Society of London*, 151, 241–262.

Bright, R. 1831. *Reports of Medical Cases: Vol. 2. Diseases of the Brain and Nervous System*. London: Longman, Brown, Green and Longmans.

Broca, P. 1861. Remarques sur le siège de la faculté du langage articulé, suivies d'une observation d'aphémie (perte de la parole). *Bulletins de la Société Anatomique*, 36, 330–357.

Broca, P. 1863. Localisations des fonctions cérébrales.—Siège du langage articulé. *Bulletins de la Société d'Anthropologie*, 4, 200–204.

Broca, P. 1865. Sur le siège de la faculté du langage articulé. *Bulletins de la Société d'Anthropologie*, 6, 337–393.

Broca, P. 1866. Poids comparé des lobes frontaux et occipitaux, et des deux hémisphères. In "Correspondance." *Bulletins de la Société d'Anthropologie*, Ser. 2, 1, 195–196.

Broca, P. 1869. L'ordre des primates. Parallèle anatomique de l'homme et des singes. XI. Le cerveau. *Bulletins de la Société d'Anthropologie*, Ser. 2, 7, 879–896.

Broca, P. 1875. Sur les poids relatifs des deux hémisphères cérébraux et de leurs lobes frontaux. *Bulletins de la Société d'Anthropologie*, 10, 534.

Brown-Séquard, C.-E. 1874a. Dual character of the brain. *Smithsonian Miscellaneous Collections*, 15, 1–21.

Brown-Séquard, C.-E. 1874b. The brain power of man: Has he two brains or has he only one? *Cincinnati Lancet and Observer*, 17, 330–333.

Brown-Séquard, C.-E. 1890. Have we two brains or one? *Forum*, 9, 627–643.

Bruce, L. C. 1895. Notes of a case of dual brain action. *Brain*, 18, 54–65.

Bruce, L. C. 1897. Dual brain action and its relation to certain epileptic states. *Medico-Chirurgical Society of Edinburgh*, 16, 114–119.

Burdach, K. F. 1826. *Vom Baue und Leben des Gehirns* (Vol. 3). Leipzig: Dyk.

Caizergues, R. 1879. Notes pour servir à l'histoire de l'aphasie. *Montpellier Médecine*, 42, 178–180.

Chadwick, J., and Mann, W. N. 1950. *The Medical Works of Hippocrates*. London: Blackwell.

Clarke, E. 1968. Brain anatomy before Steno. In G. Scherz (Ed.), *Steno and Brain Research in the Seventeenth Century*. Analecta Medico-Historica (Vol. 3). Oxford, U.K.: Pergamon Press. Pp. 27–34.

Crichton-Browne, J. 1878. On the weight of the brain and its component parts in the insane. *Brain*, 1, 504–518.

Crichton-Browne, J. 1907. Dexterity and the bend sinister. *Proceedings of the Royal Institution of Great Britain*, 18, 623–652.

Critchley, M. 1979. The Broca-Dax controversy. In M. Critchley (Ed.), *The Devine Banquet of the Brain*. New York: Raven Press. Pp. 72–82.

Cunningham, D. J. 1892. *Contribution to the Surface Anatomy of the Cerebral Hemispheres*. Dublin: Royal Irish Academy.

Daiches, D. 1973. *Robert Louis Stevenson and His World*. London: Thames and Hudson.

Davey, J. G. 1844. The duality of the mind known to the early writers on medicine. *Lancet*, 1, 377–378.

Davey, J. G. 1853. *On the Nature, and Proximate Causes of Insanity*. London: Churchill.

Dax, G. 1865. Notes sur le même sujet. *Gazette Hebdomadaire de Médecine et de Chirurgie*, 2, 262.

Dax, M. 1865. Lésion de la moitié gauche de l'encéphale coïncidant avec l'oubli des signes de la pensée (lu à Montpellier en 1836). *Gazette Hebdomidaire de Médecine et de Chirurgie*, 2, 259–262.

Descartes, R. 1649. *Les Passions de l'Ame*. Paris: Henry Le gras.

Du Pui, M. S. 1780. *Dissertatio medica inauguralis de homine dextro et sinestro*. Leiden: Apud Fratres Murray.

Duval, A. 1864. Deux cas d'aphémie traumatique, produite par des lésions de la troisième circonvolution frontale gauche: Diagnostic chirurgical. *Société de Chirurgerie de Paris Bulletin*, 5, 51–63.

Eberstaller, O. 1890. *Das Stirnhirn*. Wien: Urban und Schwartzenberg.

Esquirol, J. E. D. 1838. *Mental Maladies*. (E. K. Hunt, Trans., 1845). (Facsimile ed.; New York: Hafner, 1865)

Finkelnburg, F. C. 1870. Asymbolie und Aphasie. *Berliner klinische Wochenschrift*, 7, 449–450, 460–462.

Fisher. 1844. Minutes of the Westminster Medical Society, Saturday, March 16, 1844. *Lancet, 1,* 85.

Flourens, M.-J.-P. 1824. *Recherches Expérimentales sur les Propriétés et les Fonctions du Système Nerveux dans les Animaux Vertébrés.* Paris: J. B. Ballière.

Flourens, M.-J.-P. 1842. *Recherches Expérimentales sur les Propriétés et les Fonctions du Système Nerveux dans les Animaux Vertébrés* (2nd ed.). Paris: J. B. Ballière.

Franz, S. I. 1914. The functions of the cerebrum. *Psychological Bulletin, 11,* 131–140.

Gall, F. 1822. *Sur les Fonctions du Cerveau et sur Celles de Chacune de ses Partes* (Vol. 2). Paris: J. B. Ballière.

Gowers, W. R. 1885. *Epilepsy.* London: William Woods and Company. (Reprinted 1964; New York: Dover Publications)

Gratiolet, P., and Leuret, F. 1839–1857. *Anatomie Comparée du Système Nerveux, Considérée dans ses Rapports avec l'Intelligence* (2 vols.; Vol. 2 by Gratiolet alone). Paris: J. B. Ballière.

Greenblatt, S. H. 1970. Hughlings Jackson's first encounter with the work of Paul Broca: The physiological and philosophical background. *Bulletin of the History of Medicine, 44,* 555–570.

Hammond, J. R. 1984. *A Robert Louis Stevenson Companion.* New York: Macmillan.

Harrington, A. 1985. Nineteenth-century ideas on hemisphere differences and "duality of mind." *Behavioral and Brain Sciences, 8,* 617–660.

Harrington, A. 1986. Models of mind and the double brain: Some historical and contemporary reflections. *Cognitive Neuropsychology, 3,* 411–427.

Harrington, A. 1987. *Medicine, Mind and the Double Brain.* Princeton, NJ: Princeton University Press.

Harris, L. J. 1985. The ambidextral culture society and the "duality of mind." *Behavioral and Brain Sciences, 8,* 639–640.

Holland, H. 1840. On the brain as a double organ. In *Chapters on Mental Physiology.* London: Longman, Brown, Green and Longmans, 1852.

Jackson, J. 1905. *Ambidexterity or Two-Handedness and Two-Brainedness: An argument for Natural Development and Rational Education.* London: Kegan, Paul.

Jackson, J. H. 1864a. Hemiplegia on the right side, with loss of speech. *British Medical Journal, 1,* 572–573.

Jackson, J. H. 1864b. Clinical remarks on cases of defects of expression (by words, writing, signs, etc.) in diseases of the nervous system. *Lancet, 2,* 604–605.

Jackson, J. H. 1866. Remarks on those cases of diseases of the nervous system, in which defect of expression is the most striking symptom. *Medical Times and Gazette, 1,* 659–662.

Jackson, J. H. 1868. Hemispheral coordination. *Medical Times and Gazette, 2,* 208–209.

Jackson, J. H. 1872. Case of disease of the brain—left hemiplegia—mental affection. *Medical Times and Gazette, 1,* 513–514.

Jackson, J. H. 1873. On the anatomical and physiological localisation of movements in the brain. *Lancet, 1,* 84–85, 162–164.

Jackson, J. H. 1874a. On the nature of the duality of the brain. *Medical Press and Circular, New Ser., 17,* 19–21, 41–44, 63–66. (Reprinted in *Brain, 38,* 1915, 80–103.)

Jackson, J. H. 1874b. Remarks on systematic sensations in epilepsies. *British Medical Journal, 1,* 174.

Jackson, J. H. 1876. Case of large cerebral tumour without optic neuritis and with left hemiplegia and imperception. *Ophthalmic Hospital Reports, 8,* 434–444.

Jackson, J. H. 1878. On affections of speech from disease of the brain. *Brain, 1,* 304–330.

Jacyna, L. S. 1982. Somatic theories of mind and the interests of medicine in Britain. *Medical History, 26,* 233–258.

Joynt, R. J., and Benton, A. L. 1964. The memoir of Marc Dax on aphasia. *Neurology, 14,* 851–854.

Lancisi, G. M. 1712. *Dissertatio Altera de Sede Cogitantis Animae.* Rome: D. Joannem Fantonum. In *Opera quae Hactenùs Prodierunt; Omnia . . .* (Vol. 2, 1718). Geneva: Cramer and Perachon. Pp. 302–318.

La Peyronie, F. 1741. Observations par lesquelles on tâche de découvrir la partie du cerveau ou l'âme exerce ses fonctions. *Mémoires de l'Académie Royale des Sciences à Paris,* 119–218.

Lecours, A. R., Nespoulous, J.-L., and Pioger, D. 1987. Jacques Lordat or the birth of cognitive neuropsychology. In E. Keller and M. Gopnik (Eds.), *Motor and Sensory Processes and Language.* Hillsdale, NJ: Lawrence Erlbaum. Pp. 1–16.

Lokhorst, G.-J. C. 1982. The oldest printed text on hemispheric specialization. *Neurology, 21,* 762.

Lokhorst, G.-J. C. 1982–1983. An ancient Greek theory of hemispheric specialization. *Clio Medica, 17,* 33–38.

Lokhorst, G.-J. C. 1985. Hemisphere differences before 1800. *Behavioral and Brain Sciences, 8,* 642.

Lombroso, C. 1903. Left-handedness and left-sidedness. *North American Review, 177,* 440–444.

Long, E. 1913. Un cas d'aphasie par lésion de l'hémisphère gauche chez un gaucher. *L'Encéphale, 8,* 520–536.

Luys, J. B. 1879. Etudes sur le dédoublement des opérations cérébrales et sur le rôle isolé de chaque hémisphère dans les phénomènes de la pathologie mentale. *Bulletin de l'Académie de Médecine, 2nd. Ser., 8,* 516–534, 547–565.

Luys, J. B. 1881. Recherches nouvelles sur les hémiplégies émotives. *Encéphale, 1,* 378–398.

Marro. A., and Lombroso, C. 1883. Ambidestrismo nei pazzi e nei criminali. *Archivio di Psichiatria, Antropologia Criminale e Scienze Penali, 4,* 229–230.

Meyer, I. O. 1900. Ein und Doppelseitige homonyme Hemianopsie mit orientierungs Störungen. *Monatsschrift für Psychiatrie und Neurologie, 8,* 440–456.

Mistichelli, D. 1709. *Trattato dell'apoplessia.* Roma: A de Rossi alla Piazza di Ceri.

Montyel, M. de. 1884. Contribution à l'étude de l'inégalité de poids des hémisphères cérébraux dans la folie nervosique et la démence paralytique. *Encéphale, 4,* 574–589.

Morgagni, G. 1769. *The Seats and Causes of Diseases Investigated by Anatomy.* (B. A. Alexander, Trans.). London: A. Millar and T. Cadell.

Moxon, W. 1866. On the connexion between loss of speech and paralysis of the right side. *British and Foreign Medico-Chirurgical Review, 37,* 481–489.

National Hospital for Epilepsy and Paralysis. 1864. Clinical remarks on cases of defects of expression (by words, writing, signs, etc.) in diseases of the nervous system. (Under the care of Dr. Hughlings Jackson). *Lancet, 2,* 604–606.

Neuburger, M. 1981. *The Historical Development of Experimental Brain and Spinal Cord Physiology before Flourens* (translated and edited with additional material by E. Clarke). Baltimore, MD: Johns Hopkins University Press.

Ogle, J. W. 1867. Aphasia and agraphia. *St. George's Hospital Reports, 2,* 83–122.

Pourfour du Petit, F. 1710. *Lettres d'un Médecin des Hôpitaux du Roi, à un Autre Médecin de Ses Amis.* Naumur: Charles Gerhard Albert.

Roques, F. 1869. Sur un cas d'asymétrie de l'encéphale, de la moëlle, du sternum et des ovaires. *Bulletins de la Société d'Anthropologie de Paris, Ser. 2, 4,* 727–732.

Rush, B. 1981. *Lectures on the Mind.* (E. T. Carlson, J. L. Wollock, and P. S. Noel, (Eds.). Philadelphia, PA: American Philosophical Society.

Ryan, M. 1844. the non-duality of the brain. *Lancet, 1,* 154.

Schiller, F. 1979. *Paul Broca: Founder of French Anthropology, Explorer of the Brain.* Berkeley: University of California Press.

Smith, G. E. 1924. *The Evolution of Man.* Oxford, U.K.: Oxford University Press.

Stevenson, R. L. 1886. *Strange Case of Dr. Jekyll and Mr. Hyde.* London: Longmans, Green. Centenary ed., 1986; Edinburgh: Cannongate Publishing.

Stevenson, R. L., and Henley, W. E. 1880. *Deacon Brodie; or the Double Life; a Melodrama, Founded on the Facts, in Four Acts and Ten Tableaux.* London.

Thurnam, J. 1866. On the weight of the brain, and on the circumstances affecting it. *Journal of Mental Science, 12,* 1–43.

Tilney, F. 1927. The brain of prehistoric man. A study of the psychologic foundations of human progress. *Archives of Neurology and Psychiatry, 17,* 723–769.

Trousseau, A. 1864. De l'aphasie, maladie décrite récemment sous le nom impropre d'aphémie. *Gazette des Hôpitaux, Paris, 37,* 13–14, 25–26, 37–39, 48–50.

Vicq d'Azyr. F. 1805. *Oeuvres de Vicq d'Azyr.* (J. L. Moreau, Ed.). Paris: L. Duprat-Duverger.

Watson, H. 1836. What is the use of the double brain? *Phrenological Journal and Miscellany, 9,* 608–611.

Wernicke, C. 1874. *Der aphasische Symptomenkomplex: eine psychologische Studie auf anatomischer Basis.* Breslau: Cohn and Weigert. (G. H. Eggert, Trans.) In *Wernicke's Works on Aphasia: A Sourcebook and Review.* The Hague: Mouton, 1977.

Wigan, A. L. 1844a. *A New View of Insanity: The Duality of the Mind.* London: Longman, Brown, Green and Longmans.

Wigan, A. L. 1844b. Duality of the mind, proved by the structure, functions, and diseases of the brain. *Lancet, 1,* 39–41.

Willis, T. 1664. *Cerebri anatome: cui accessit nervorum descriptio et usus.* London: J. Martyn and J. Allestry. Tercentenary ed., 1664–1964, *Thomas Willis, The Anatomy of the Brain and Nerves.* Montreal: McGill University Press, 1965.

Zinn, J. G. 1749. In A. von Haller, *Experimenta quaedam circa corpus callosum, cerebellum, duram meningem . . . Inaugural dissertation, J. G. Zinn, respondent and author.* Göttingae: Apud A. Vanderhoek.

Chapter 27
Expansion of the Concept of Cerebral Dominance

> Apraxia may be termed an *aphasia of the extremities*. (The patient) is unable to awaken the motor associations of waving, clenching the fists, etc. . . . Thus, an essential component is missing from his conception of fist-clenching, waving, etc; the words "make a fist" awaken in him something less than in us.
>
> Hugo Liepmann, 1905

Chapter 26 reviewed how the concept of cerebral dominance evolved from the widespread belief that the two cerebral hemispheres were structurally and functionally equal. Although Marc Dax almost certainly appeared to recognize the importance of the left hemisphere for language in 1836, his ideas were not made public at that time. The scientific break with the past had to wait for Paul Broca to associate aphasia with left hemispheric damage and for the Dax manuscript to be published. These events took place between 1863 and 1865. Although the left hemisphere dominated attention at this time, John Hughlings Jackson soon raised the possibility that the right hemisphere may be the lead hemisphere for perceptual functions.

After these pioneering studies appeared, the concept of cerebral dominance grew to encompass a wealth of additional clinical findings. Alexia, dyslexia, amnesia for faces, ideomotor and constructional apraxias, denial of illness, and unilateral spatial neglect are among the many subjects addressed in this chapter. Sections on the perceived functions of the corpus callosum late in nineteenth century and how the hypothesized role of the callosum changed in the mid-twentieth century conclude the present chapter.

ACQUIRED ALEXIA

In approximately A.D. 31, Valerius Maximus, a Roman best known for his collection of historical anecdotes, described a man from Athens who was struck on the head by a stone. This educated man lost his memory for letters, but did not show other memory deficits (Benton and Joynt, 1960; Kempf, 1888). This very early case report of "acquired alexia," a loss of the ability to read words or letters not attributable to poor eyesight, may also have been cited by Pliny the Elder. The relevant passage from Pliny has been translated as follows: "One with the stroke of a stone, fell presently to forget his letters only, and could read no more; otherwise his memory served him well ynough" (Holland, 1601; also see Benton and Joynt, 1960).

A number of other interesting case reports on alexia appeared well before the modern period of localization of function (Benson and Geschwind, 1969; Benton and Joynt, 1960). Girolamo Mercuriale, for example, had noted "a truly astonishing thing; this man would write but could not read what he had written" after a seizure (see Albert, 1979).

Mercuriale's description of alexia, which appeared in 1588, was followed a century later by an excellent case report by Johann Schmidt. Schmidt's patient was a leading 65-year-old citizen who had a stroke that affected his left hemisphere. This man remained alexic after his motor aphasia receded. Schmidt wrote:

> He could not read characters, much less combine them in any way. He did not know a single letter nor could he distinguish one from another. But it is remarkable that, if some name were given to him to be written, he could write it readily, spelling it correctly. However, he could not read what he had written even though it was in his own hand. Nor could he distinguish or identify the characters. For if he were asked what letter this or that was or how the letters had been combined, he could answer only by chance or through his habit of writing. It appeared that he wrote without deliberation. No teaching or guidance was successful in inculcating recognition of letters in him. (Translated in Benton and Joynt, 1960, 209)

Schmidt described a second case of alexia in the next section of his report. This man was a stonecutter who showed eventual recovery of the ability to read. Another individual who showed recovery of reading (more for Latin than for German) was reported by Johann Gesner in the 1770s. Gesner's patient, however, also exhibited receptive speech and cognitive deficits.

At this time, alexia typically was interpreted as a partial loss of memory, a memory defect more or less restricted to reading. The fact that the lesion causing alexia almost always involved the left hemisphere was not considered important until Paul Broca's reports on hemispheric dominance appeared in the 1860s (e.g., Broca, 1865).

In 1872, William Henry Broadbent (1835–1907) described a patient with alexia, mild aphasia, and agraphia.[1] Broad-

---

[1] For early descriptions of writing difficulties, see Chapter 26 and the 1867 paper by Ogle, the individual who first used the term "agraphia."

bent's patient suffered two lesions of the left hemisphere, one of which was an infarction beneath the parietal-occipital junction. Broadbent postulated that this lesion may have broken the connections between the visual and verbal areas of the brain.

Five years after Broadbent's material appeared, Adolf Kussmaul described another case of alexia. He hypothesized that the lesion damaged a special center for "visual-verbal images" in the left hemisphere. Both Broadbent and Kussmaul thus implicated the left hemisphere in cases of alexia.

In the last decade of the nineteenth century, Joseph Jules Dejerine published a series of papers in which he carefully distinguished among many different types of acquired reading disorders. He thought this distinction had to be made since one could interfere with the reading process at different stages and with lesions in different places (Dejerine, 1891, 1892; also see Dejerine, 1914). In particular, Dejerine distinguished between "pure word blindness" and "alexia with agraphia."

Dejerine's pure word blindness was not associated with speech or writing defects. He proposed that this kind of alexia was caused by a lesion in the left occipital cortex that disconnected the calcarine region (basic for vision) from the angular gyrus. He presented an alexic patient with this type of lesion. Although this patient also had a lesion of the splenium of the corpus callosum, Dejerine did not consider pathology of the corpus callosum critical for pure alexia early in the 1890s. Nevertheless, the role of posterior callosal lesions in preventing visual images in the healthy, right hemisphere from crossing to the left hemisphere was soon pointed out (Brissaud, 1900; Redlich, 1895).

Dejerine thought pure word blindness should be viewed as a specialized form of aphasia. Another approach to this disorder was to consider it a specialized type of visual agnosia. In 1891, Sigmund Freud coined the term "agnosia" to describe defects in recognition, as opposed to defects in naming. "Agnosia" was soon applied to a number of different disorders, replacing older terms such as "mind-blindness". The argument that pure alexia was an agnosia came in part from cases showing a close association between pure alexia and agnosias for colors, objects, and pictures (Lissauer, 1889; Nodet, 1889; Poetzl, 1928).

Dejerine's second major type of alexia was alexia with agraphia. This disorder was characterized as an inability to read and understand long words or sentences, accompanied by severe difficulties in spelling and writing. Nevertheless, individual letters typically can be read, and disorders of spoken language are absent or mild. Dejerine attributed this to a lesion in the left angular gyrus, the parietal lobe center for the visual memory of letters. He considered this variety of alexia to be a type of sensory (Wernicke's) aphasia with more severe word blindness than word deafness.[2]

DEVELOPMENTAL DYSLEXIA

Whereas most disorders of hemispheric specialization have been discovered by studying the effects of acute brain lesions in adults, one defect in particular has been thought to represent a congenital defect of the left hemisphere. This is congenital word blindness, or developmental dyslexia, a problem with reading usually seen in school-age children (Benton, 1980).

The syndrome of congenital word blindness went virtually unrecognized until 1896. In that year, William Pringle Morgan (1854–?) published his case report of an intelligent 14-year-old boy, Percy F., who had great difficulty reading and writing. Percy F. could read only individual letters and struggled through short words, despite having had tutors since he was 7 years old. He seemed utterly unable to appreciate how lengthy words should be read or spelled. Morgan wrote:

> I then asked him to read me a sentence out of an easy child's book without spelling the words. The result was curious. He did not read a single letter correctly, with the exception of "and," "the," "of," "that," etc.; the other words seemed to be quite unknown to him, and he could not even make an attempt to pronounce them. I next tried his ability to read figures, and found that he could do so easily. . . . He says that he is fond of arithmetic, and finds no difficulty with it, but that printed or written words "have no meaning to him." (1896, 1378)

Morgan contended that Percy was "word-blind." On the basis of conversations with the boy's father, he added that he believed Percy had previously been "letter-blind," but that extensive and exhaustive training had allowed him to overcome this more severe reading problem. Since adult "acquired" alexics were thought to have lesions involving the left angular gyrus, Morgan theorized that congenital reading problems, like those displayed by Percy F., stemmed from defective development of this cortical area.

Two years after Morgan's report appeared, Henry Charlton Bastian described another case of congenital word blindness. Like Morgan, he attributed the disability to congenital weakness of, or early damage to, the "visual word centre" in the left angular gyrus.

A great many papers on congenital word blindness followed these pioneering reports. In 1900 and 1902, James Hinshelwood (1859–1919) presented some interesting cases of intelligent boys who simply could not learn to read or spell words, but who had visual memory for objects and reasonable skills at reading numbers. Hinshelwood accepted the idea that the disorder was due to a defect in the left angular and supramarginal gyri, which he viewed as critical for the memory of words.

Because most of these children could work well with numbers, Hinshelwood postulated a separate, independent cortical center for the memory of numbers. He described this as follows:

> In the four cases of congenital word-blindness quoted in my first paper it was particularly observed that none of these patients had any difficulty in learning to read or manipulate figures. In Case I of the present paper it will be observed that the child experienced no difficulty in learning figures. . . . In Case 2, the child read figures much better than letters, but his training in arithmetic had not really begun. These cases afford further proof of my contention that the visual memories of letters, words and figures, are deposited in different areas of the cerebral cortex. . . . This very fact would at once suggest to the careful observer that

---

[2] Alexia with agraphia has been called parietal alexia by some authors (e.g., Hermann and Poetzl, 1926).

the difficulty of such a patient in learning to read was not due to an ocular but to a cerebral defect. (1902, 97–98)

Hinshelwood concluded that children with "congenital word-blindness" should not be taught to read with the same methods used with unimpaired students. He suggested that one might try to train the corresponding parts of the right hemisphere to become involved with reading. Thus, he turned to methods that utilized touch-sight associations, such as the use of blocks with raised letters which could be both seen and felt.

Not everyone agreed with this mixed-modality teaching strategy or with the idea that words should be taught one letter at a time. Another suggested approach was learning to recognize whole words rather than piecing elements together (Fisher, 1905). The word "cat," for example, would be taught as a whole, or as a symbol, and not as c–a–t = cat. This "look and say" method was compared to the Chinese system of using pictures and symbols to convey words and ideas (Harman, 1915).

In 1906, Edward Jackson (1856–1942), an ophthalmologist from Colorado, theorized that "developmental alexia" was a more appropriate term for this disorder than congenital word blindness. This stemmed from his belief that the disorder was one of delayed development.

Estimates of the frequency of developmental dyslexia ranged from 1 in 100 to 1 in 2,000 at this time (Fisher, 1905; Warburg, 1911). It was also noticed early on that the great majority of children with congenital word blindness were boys (Nettleship, 1901).

The period before World War I (1914–18) was also characterized by further studies of the functional capabilities of these children (see Waites, 1982). J. Herbert Claiborne (1828–1905) presented two cases with this in mind. He then suggested that dyslexic children should learn to become left-handed to stimulate the right side of the brain to take over for the defective left hemisphere. Claiborne explained:

When symbol or auditory amblyopia exists in such children, it is a reasonable idea that they may be taught to be left-handed. It is improbable that the corresponding cells on the right side are similarly affected, and thus the speech center and the centers for symbols and sounds may be transferred entirely to that side, or the right side be so educated that it takes command. (1906, 1816)

On the same page, Claiborne added, however, that "at present no postmortem corroboration of the assumed congenital lesions is extant."

In 1912, Phillip Boswood Ballard recognized that about one third of the nonreaders were, in fact, left-handed.[3] It was pointed out that dyslexics often begin to read from the right instead of the left, so that the words "saw" and "was" are likely to be confused. It was wondered whether some early transfer of handedness could have caused a reading difficulty comparable to that of stammering.

Between 1925 and 1929, Samuel T. Orton (1879–1948) of the University of Iowa found that dyslexics could read mirror image text about as well as regular text. This led him to theorize that dyslexia, characterized by reversals of

letters ("static reversals") and words ("kinetic reversals"), was due to the lack of dominance by one hemisphere and the resulting competition between the two hemispheres. Orton wrote:

In the process of early visual education, the storage of memory images of letters and words occurs in both hemispheres, and that with the first efforts at learning to read the external visual stimuli irradiate equally into the associative cortices of both hemispheres, and are there recorded in both dextrad and sinistrad orientation. . . . This suggests the hypothesis that the process of learning to read entails the elision from the focus of attention of the confusing memory images of the nondominant hemisphere which are in reversed form and order, and the selection of those which are correctly oriented and in correct sequence. . . . The frequency, in these cases of reading disability, of reversals of letter pairs, such as we see in M.P.'s gary for gray, of whole syllables, as tar-shin for tar-nish, or of the major parts of words, as tworrom for tomorrow, strongly suggests that there has been an incomplete elision of the memory patterns in the nondominant hemisphere, and that therefore either right or left sequence may be followed in attempting to compare presented stimuli with memory images, and that this leads to confusion or to delay in selection. (1925, 608–609)

Orton (1925) thought that training should involve repetitive drilling of phonic associations with letter forms, both seen and written, until the correct associations were built up and the incorrect ones suppressed (also see Orton, 1966).

PROSOPAGNOSIA

Facial amnesia, or prosopagnosia (a term introduced by Bodamer in 1947), is a higher-order disorder in which individuals with adequate vision are unable to recognize faces. Nevertheless, they typically retain the ability to recognize that a face is a face and can make out specific features. Thus, while normal people may have some difficulty remembering faces, a person with prosopagnosia may not even be able to tell if the face is that of a family member or a close friend unless there is a mole, eyeglasses, or some other telling feature associated with the particular person. In the absence of such distinguishing features, the person may not be recognized. For example, Jean-Martin Charcot once had a patient who tried to shake hands with another person who bumped into him. It turned out that the "other" person was nothing more than his own reflection in the mirror (De Romanis and Benfatto, 1973).

An especially good description of prosopagnosia was given by Arthur Wigan in 1844. Although he did not recognize the specific disorder or the causal lesion, Wigan described the problem as follows:

A gentleman of middle age, or a little past that period, lamented to me his utter inability to remember faces. He would converse with a person for an hour, but after an interval of a day could not recognize him again. Even friends, with whom he had engaged in business transactions, he was unconscious of ever having seen. . . . When I inquired more fully into the matter, I found that there was no defect in vision. (1844, 128–129)

---

[3]The relationship between dyslexia and left-handedness was further expanded by Ballard's students (e.g., Hincks, 1926).

Antonio Quaglino (1817–1894) and Giambattista Borelli (1813–1891) went beyond Wigan by describing prosopagnosia more fully and by attempting to localize the causal lesion. In 1867, these two ophthalmologists described a 54-year-old man who had suffered a right hemispheric stroke one year earlier and showed defective recognition of faces. Although this patient also exhibited a loss of color vision (achromatopsia), some deficits in spatial orientation, and a left visual field defect, he was able to read and showed good central vision. A notable feature of the Quaglino and Borelli case was that their patient not only had trouble identifying faces but was also poor at remembering the façades of houses.

Quaglino and Borelli (1867) maintained that their patient was demonstrating specific perceptual disabilities rather than a general visual agnosia. They believed that some basic functions, such as color, might be localized in discrete parts of the cortex and felt that their case confirmed the newly emerging idea of discrete cortical localization of functions. They held that their patient had suffered a cerebral hemorrhage of his right hemisphere, but wondered whether his left hemisphere could also have suffered some damage.

It is not known whether John Hughlings Jackson was aware of Quaglino and Borelli's (1867) description of prosopagnosia in the 1870s, when he described his own patients with serious disturbances of spatial organization (Benton, 1990). The male patient described by Jackson in 1872 could not remember people in addition to places and things, and his female patient, who was described in 1876, often mistook her niece for her daughter. Jackson classified the inability to recognize people as a part of his more general disorder of "imperception," which he associated with lesions of the right hemisphere.

The question of whether lesions of the right hemisphere alone are able to cause prosopagnosia, as Jackson suggested, remained an open question. Quaglino and Borelli (1867), while stressing the role of the right hemisphere, were unsure that this was the sole cause of the prosopagnosia in their patient. As noted, they entertained the possibility that damage to the left hemisphere may have contributed to the disorder. The fact that the disorder is rare suggests that a stroke of only the right posterior cerebral artery, which is not a rare occurrence, may not be sufficient to cause prosopagnosia.

Unfortunately, only a few cases of prosopagnosia were autopsied before the second half of the twentieth century. One such case involved a 63-year-old woman who lost her ability to recognize her friends, her relatives, and the street on which she lived (Wilbrand, 1892). She had visual field defects that were more severe on the left than on the right. When she died, bilateral lesions were found in the temporo-occipital region, although the lesion on the right side was more severe. In another case, bilateral lesions of the occipital cortex were also encountered (Heidenhain, 1927).

An analysis of 11 confirmed cases of prosopagnosia with autopsies, beginning with these two, has provided some evidence for the idea that bilateral damage to visual association cortex may underlie prosopagnosia.[4] In most cases, the lesions were found to encroach on the part of the lin-

[4]The anatomical correlates of prosopagnosia have been discussed by several more contemporary investigators (see Damasio and Damasio, 1983; Meadows, 1974).

gual and fusiform gyri that had been implicated in achromatopsia (see Chapter 7; also Zeki, 1990). This is probably why patients with prosopagnosia often show a loss of color vision.

Another disorder often seen in patients with prosopagnosia is simultaneous agnosia (Walsh, 1978). In 1924, Ilja E. Wolpert (1891–1967) characterized this disorder as an inability to integrate units into a coherent, meaningful whole. Individuals with simultaneous agnosia are able to see single aspects of a complex stimulus or a stimulus array, but are unable to appreciate more than one aspect of a stimulus at a time. A pair of reading glasses may look like two glass circles and some sticks, whereas the composite may remain a mystery.

Prosopagnosia, acquired achromatopsia, and simultaneous agnosia are relatively rare disorders. They are usually not associated with paralysis or other "hard" neurological signs. For these reasons, it seems possible, and even likely, that in the past most physicians not used to such bizarre complaints would have treated isolated cases of these disorders as hysterias and "functional" problems (Benton, 1990).

ONE OR MANY SPATIAL DISORDERS?

One of the most important issues addressed in the closing years of the nineteenth century was whether disorders of spatial thinking cut across sensory systems or were confined to just one system. In 1888, Jules Badal (1840–1929), an ophthalmologist from Bordeaux, France, described a 31-year-old woman, Valerie Clem, who probably suffered a diffuse cerebrovascular insult (Benton and Meyers, 1956). Even though her central visual acuity was good, this woman could not find her way around her neighborhood or her own apartment. Significantly, she also showed deficits in localizing the source of sounds, in recognizing the right and left sides of her body, and in dressing herself. Badal believed Valerie Clem showed errors of spatial localization and orientation that went beyond vision into the auditory and somesthetic spheres.

Following Badal's lead, other investigators soon described additional individuals with multimodal spatial deficits. Nevertheless, there were conflicting reports showing that not only could spatial deficits be restricted to a single modality, but that they could be limited to just certain types of spatial discriminations within that modality.

One such report came from Thomas D. Dunn, a Philadelphia physician. Dunn described a 67-year-old man who could not remember streets or even rooms in his house long after his right hemisphere became diseased. One month later:

> There was no improvement in the sense of locality. He could form no conception of the geography of his own house or of any place he had ever been. He could recollect that he lived on the corner of two streets and their names, but their relation to each other, or to other streets, was completely lost. (1895, 47)

In contrast, this patient was able to recognize faces and objects soon after the attack that paralyzed his left side.

Dunn believed he had demonstrated that even within the visual modality, some visuospatial functions (e.g., facial recognition) may be spared, whereas others (e.g., geography) may be lost after focal brain lesions. He postulated a center in the right hemisphere for the sense of visual location, which was functionally independent from the centers for object or facial recognition.

Dunn's case study helped to show that there are many facets to disorders of visual perception and spatial thinking. Scientists who followed him continued to debate which symptoms are likely to occur together and which would really prove to be independent of other higher-order perceptual deficits and basic sensory defects. They also continued to argue about the critical sites responsible for the various deficits.

LIEPMANN'S APRAXIAS

John Hughlings Jackson (1866) noted that patients of normal strength sometimes have severe difficulties performing voluntary skilled movements. In particular, his aphasia patients often showed such disturbances, especially protruding the tongue on demand. This observation was consistent with his belief that voluntary acts are lost before involuntary ones.

In 1887, Carl Wilhelm Hermann Nothnagal followed Jackson by describing a "paralysis of intent." He speculated that this kind of "paralysis" was due to impaired memories for motor patterns. Theodor Meynert (1890) referred to this as a loss of motor symbols. These disturbances of purposeful, skilled movements, or "apraxias," were also related to intellect. Carl Wernicke, for example, thought they were due to a separation of the motor cortex from the intellectual centers (see Kertesz, 1979).

Hugo Liepmann (1863–1925), a Berlin physician who had served as an assistant to Carl Wernicke in Breslau, presented the apraxias as distinct neurological entities. In the first decade of the twentieth century, Liepmann defined apraxia as a higher-order disorder in the execution of skilled movements, but one that could not be attributed to sensory defects, tremor, posture, paralysis, impaired intellect, or poor verbal comprehension.[5]

Liepmann actually described several apraxias, the most notable being his "ideomotor apraxia." He characterized this type of apraxia by the inability of the individual to perform acts in response to verbal commands. The same acts, however, might be performed automatically or when the person is aroused emotionally. For example, when asked to show the examiner how he would comb his hair or use a hammer, a person with this type of apraxia either would fail to make a response or would make an incorrect response. Yet the individual could perform these acts under other circumstances and would not experience difficulty comprehending verbal instructions.

Liepmann provided descriptions of a number of individuals with this disorder (Liepmann, 1900; Liepmann and Maas, 1907). One was a man who suffered a left hemispheric stroke and who tried to use his left hand to perform responses to verbal commands. This patient was unable to carry out verbal commands and could not imitate skilled movements with this hand, even though it was not paralyzed. Liepmann was confident that the man was not intellectually blunted and that he did not have a language comprehension deficit.

Liepmann explained this kind of apraxia with reference to two mechanisms. First, he theorized that the left hemisphere was "the organizer" for skilled movements. He found that right-handed patients with right hemispheric brain damage and paralysis of the left hand were not apraxic with their good right hand. Patients with lesions of the left hemisphere, however, showed apraxias with what should have been their good left hand. Liepmann therefore became convinced that the left hemisphere played a dominant role in the will to initiate and perform complex voluntary acts. He suggested that the supramarginal gyrus of the left parietal lobe, which could be damaged by middle cerebral artery disease, housed these critical "engrams."

Second, Liepmann (1906) suggested that a break in the callosal pathways between the motor-organizing (and language) areas on the left side of the brain and the motor areas for the left hand on the right side of the brain could cause ideomotor apraxia. This could prevent the left hemisphere, which is dominant for planning skilled movements, from communicating the necessary information to the right hemisphere, which controls the left hand. A patient with only a lesion of the critical part of the corpus callosum should show an apraxia for the left, but not the right, side of the body.

The newly proposed role of the corpus callosum did not receive the initial wide support granted to Liepmann's concept of a left hemispheric center for skilled movements. Nevertheless, the callosal idea grew in importance with the passage of time (see Geschwind, 1975).

Liepmann distinguished ideomotor apraxia from two other types of apraxias. One was "limb-kinetic apraxia" ("innervatory apraxia"). This type was characterized by a loss of precision and very clumsy movements and was thought to be caused by destruction of the center housing the kinesthetic engrams (proprioceptive memory). The other was "ideational apraxia." Here, the patient achieves the wrong objective due to poor conceptualization. For example, a person might put a match in his mouth to light a cigarette (Kertesz, 1979). Liepmann thought this behavior was due to the loss of the goal object. Ideational apraxia has been associated with large and diffuse lesions, and some investigators have looked upon it as nothing more than a very severe form of ideomotor apraxia (e.g., Sittig, 1931).

CONSTRUCTIONAL APRAXIA

The types of apraxias described by Liepmann were not the only ones to receive renewed attention early in the twentieth century. An apraxia characterized by an inability to put parts together into meaningful wholes was also gaining recognition.

In 1917, Walther Poppelreuter classified difficulties in building, drawing, and assembling as "optic ataxias." He also developed a number of tests (e.g., overlapping drawings) to illustrate the optic ataxias of some of his head-injured men. Unlike patients with ideomotor apraxia, who cannot respond appropriately to verbal commands and do better spontaneously or when shown models, the patients described by Poppelreuter generally had a difficult time

---

[5]See Liepmann (1900, 1905, 1906) and Liepmann and Maas (1907).

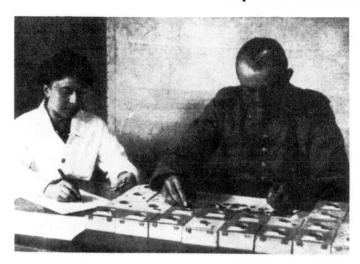

Figure 27.1. A neurological patient working on a constructional test. This photograph appeared in Walther Poppelreuter's (1917) book on soldiers wounded during World War I.

building or drawing under all conditions. (See Figures 27.1 and 27.2 for a photograph and a drawing from Poppelreuter's 1917 book.)

At this time, Karl Kleist (1879–1960) also classified constructional problems as optic apraxias. By 1934, however, Kleist was using the term "constructive apraxia." Kleist (1912, 1923, 1934) characterized the problem as a disturbance in carrying out appropriate action patterns for drawing, building, and so on, in people who have adequate visual perception and the ability to make single move-

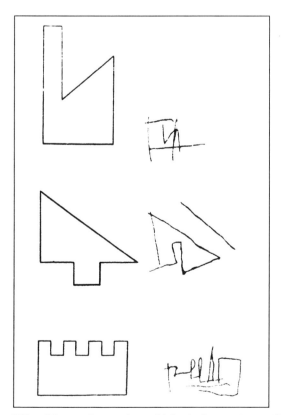

Figure 27.2. Attempts by a patient with brain damage to copy the three geometric figures in the left column. (From Poppelreuter, 1917.)

ments. Nevertheless, he did not take credit for first seeing this, but instead cited Conrad Rieger (1855–1939) as probably the first to describe this form of apraxia. Rieger wrote about it in 1909. Unlike his predecessors, however, Kleist forcefully argued that constructional apraxia was a separate and distinct disorder.

Kleist emphasized that his disorder of purposeful constructional activities was neither sensory nor motor, although he recognized that other types of constructional defects could be caused by poor perception or defective motor functions. Simply stated, he viewed constructional apraxia as a problem of organizing and sequencing activities to achieve a synthesis. That is, Kleist saw it as an inability to translate adequate visual impressions into action patterns.

The tests used to study constructional apraxia included freehand drawing, making stick models, assembling puzzles, making block designs, and building with blocks. Some of these tests, such as the block design test, had been used as measures of general intelligence in the past (Kohs, 1923).

Kleist looked upon constructional apraxia as a "disconnection" syndrome. It was attributed to a break in the connections between the visual and proprioceptive centers in the occipital and Rolandic areas, respectively. Thus, he maintained that this defect of "executive" function was most likely associated with lesions situated in the left posterior parietal lobe.

Kleist was very successful in attracting attention to constructional apraxia as a disorder of higher function associated with brain disease. His rather restrictive definition, however, did not fare well. The term "constructional apraxia" was soon used to designate almost any visuoconstructive disability, even those associated with perceptual impairments.

Using modified definitions of constructional apraxia, so as to allow other higher-order visuospatial difficulties to appear with it, led to some changes in thinking. Investigators defining constructional apraxia broadly showed that defects in building, drawing, and related tasks may appear more frequently after injuries of the posterior right hemisphere than after left hemispheric damage (Benton, 1967, 1972). This finding was in accord with John Hughlings Jackson's notion of overall right hemispheric superiority for visuospatial functions, but on both conceptual and anatomical grounds, it represented a challenge to Kleist's original conception of constructional apraxia.

Should constructional apraxia be treated as a single entity, or might there be more than one kind of constructional apraxia? "Lumpers" opted for just one disorder called constructional apraxia, whereas "splitters" suggested there may be at least two basic types. One was the type Kleist defined and associated with damage to the left hemisphere. The other was a right hemispheric variety, a type that was associated with other defects of spatial thinking. This very important distinction, with its broad ramifications, has continued to elicit strong opinions long after Rieger, Poppelreuter, and Kleist first described their cases of constructional apraxia (Benton, 1967).

DENIAL OF ILLNESS

The first good descriptions of denial of illness came in the 1890s (Anton, 1893, 1898, 1899; Dejerine and Vialet, 1893;

Figure 27.3. Joseph Babinski (1857–1932), the neurologist who coined the word "anosognosia" and described this disorder in 1914. (Courtesy of Arthur Benton.)

Monakow, 1897). Gabriel Anton noted that patients with severe, bilateral visual cortex lesions tended to deny that they were blind and seemed unaware of their hemiplegias. He attributed this lack of recognition to a disturbance of intellect, which accompanied the blindness.

In 1914, Joseph Babinski provided several examples of denial of hemiplegia following focal, unilateral cortical lesions, and showed that this disorder could not be explained by primary sensory or motor deficiencies. Babinski, shown in Figure 27.3, called this disorder "anosognosia."

Anosognosia is almost always associated with the rupturing of the middle cerebral artery in the posterior parietal region of the right hemisphere. Patients with the disorder tend to rationalize their failure to use the paralyzed left limb and sometimes have the delusion that the left limb is not their own. Hemianesthesia and hemianopia may accompany the paralysis. This disorder is seen much less frequently after left hemispheric injuries.

UNILATERAL INATTENTION AND NEGLECT

In the 1880s, Jacques Loeb, who worked on dogs, and Hermann Oppenheim, who applied Loeb's technique to patients, showed that some stimuli, which could be appreciated when applied to just one visual field or side of the body, might not be experienced when both visual fields or the two sides of the body were stimulated simultaneously.[6] Specifically, they noted that the stimulus contralateral to the lesion was likely to be appreciated less than the ipsilat-

eral stimulus ("obscuration"), if it were even experienced at all ("extinction"), on those trials which called for double simultaneous stimulation. Neither Loeb (1884, 1885) nor Oppenheim (1885, 1898) nor their contemporaries, seemed to realize that neglect or inattention might be greater after right hemispheric lesions in humans, even under conditions of just unilateral stimulation.

Unilateral visual "inattention" or "spatial agnosia" began to receive attention in the opening decades of the twentieth century. It was described by Walther Poppelreuter (1917, 1923) and Gordon Holmes (1918) in brain-damaged soldiers. And, in 1928, Achille Alexandre Souques (1860–1944) presented a case report of a person who seemed to neglect or forget about his left side after a lesion of the right hemisphere.

Surprisingly, it was not until 1941 that sustained attention was devoted to unilateral neglect and its association with the right hemisphere. In that year, W. Russell Brain (1895–1966) described three patients who showed left hemispatial inattention, behaving as if the left half of space did not exist. This was linked to negligence for the left half of the body. All three had large lesions on the right side in the parietal lobe or the parieto-occipital junction. Brain (1945) showed that neglect for the left half of space, and the left side of the body, could be found much more frequently than neglect for the right side. For the first time, the disorder was firmly associated with damage to the right hemisphere.

Brain's empirical findings were well accepted by most investigators, but some wondered whether the proposed functional asymmetry had something to do with the fact that patients with large left hemispheric lesions were often severely aphasic and "untestable." While this objection had to be taken seriously, the mounting evidence continued to show that neglect after posterior right hemispheric lesions was both more common and more severe than neglect after lesions of the left hemisphere, and that this could not be attributed to "untestable" subjects.[7]

Neglect of one half of visual space can easily be tested and measured with drawing and writing tasks. These patients tend to draw clocks with most or all of the numbers compressed on the right half of the clock's face, to draw people, houses, and other objects on the right side of the page, and to leave much wider margins on the left side of a piece of writing paper. In cancellation tasks, on which they must cross out specific letters of the alphabet scattered among others, they typically perform well on the right side of the page but poorly on the left. Individuals who do poorly on these paper and pencil tests tend to swing one limb more than the other when walking. In severe cases, they may attend only to their teeth and hair on the right side and may button only the right sleeve of a shirt.

The underlying basis of this unusual disorder has been a source of controversy. Among the many theories proposed to account for unilateral neglect is one that maintains it reflects a defect in sensory integration. Other theorists have argued the problem is more of an attentional deficit. A third approach is one that emphasizes low levels of arousal.

[6]The history of double simultaneous stimulation has been traced by Benton (1956).

[7]The literature on unilateral neglect was reviewed by Critchley in 1953, and by Weintraub and Mesulam in 1989.

THE GERSTMANN "SYNDROME"

"Finger agnosia," or the inability to name, select, differentiate, or indicate the individual fingers of either hand, was described by Josef Gerstmann (1887–1969) in 1924, in an intellectually blunted patient who was not aphasic. This defect existed not only when the patient's own hands were tested but also when the tester pointed to the hands of another person. His patient also showed agraphia and acalculia (difficulties in mathematics) and some problems in right-left orientation. At this time, however, Gerstmann did not link these other problems with finger recognition.

In 1927, Gerstmann described two more patients with finger agnosia and agraphia and now concluded that these two deficits were, in fact, associated. In 1930, he combined finger agnosia with agraphia, right-left disorientation, and acalculia into the single four-symptom syndrome that still bears his name. He did not feel this syndrome reflected an underlying psychiatric disorder, a critically weakened intellect, or sensory blunting; nor did he believe it was due to a general agnosia, apraxia, or aphasia.

Gerstmann thought his four-component syndrome was not especially rare and that it reflected a higher-order disturbance of body schema. He believed it was caused by a "cerebral lesion in the transitional area of the lower parietal and middle occipital convolution" of the left hemisphere (1940, p. 398). As far as he was concerned, the localizing value of the syndrome was proved.

It is interesting to note that Jules Badal (1888) saw all four symptoms of Gerstmann's syndrome in Valerie Clem, the patient described earlier in this chapter whose clinical features led Badal to suggest that disorders of spatial thinking could transcend modalities. Regarding the key feature, finger agnosia, Badal wrote:

> It happened frequently that the patient was not able to say which of her fingers had been touched or pricked, and for example, would report the index finger when the ring finger had been pricked. . . . Sometimes the same errors were made even when the eyes were allowed to remain open so that they could guide the tactile impressions; moreover, if one asked Valerie to name the five fingers of the hand, one after the other and in the order in which they were placed, it was rare for her not to commit some error, either in naming them or in classifying them from 1 to 5. (Translated by Benton and Meyers, 1956, 840)

Badal also stated that Valerie Clem showed a moderate but clear dyscalculia and right-left disorganization. Dysgraphia, the fourth part of the Gerstmann syndrome, showed itself in her inability to write from dictation or from a model whose letters could be visualized and named, and to draw even simple designs from memory or a model.[8]

Many questions have been raised about the Gerstmann syndrome since its conception. One is whether it is really a syndrome. Controlled studies have shown that all four components of the Gerstmann syndrome only rarely occur together (Benton, 1961, 1977).

In addition, researchers have shown that these patients almost always show other perceptual and cognitive deficits, including problems with spatial and temporal orientation, constructional apraxia, and difficulties with configurational thinking. Seen in this light, the disorder appears to be more than a problem of body schema. In fact, the case has been made that other symptom combinations could be even more closely associated than Gerstmann's four symptoms.

The importance of the left angular gyrus, suggested by Gerstmann as the cause of the disorder, has also been questioned. It was found that extensive cerebral damage in the region of the left posterior perisylvian territory was even more likely than a limited lesion of the angular gyrus to produce the four Gerstmann symptoms, and that the four symptoms have different localizing values.[9]

Given these findings, whether finger agnosia, acalculia, agraphia, and right-left disorientation will continue to be thought of as a syndrome is uncertain. Investigators, however, have tended to agree with Gerstmann that each of these symptoms is a fairly good sign of left posterior parietal lobe damage.

THE CORPUS CALLOSUM REDISCOVERED

The clinical literature on the possible role of the corpus callosum was anything but consistent from the time of Henry Holland's (e.g., 1840) publications in the mid-1800s into the middle of the twentieth century. Eduard Hitzig found that atrophy of the corpus callosum was not associated with mental defects or with sensory and motor changes. William Ireland (1832–1909), who cited Hitzig in 1886 and 1891, also emphasized that complete absence of the corpus callosum could occur without notable mental defects (speech, intelligence, sensation, reflexes), at least if the rest of the brain were normal. He wrote that there are

> at least three instances where the corpus callosum was found to be entirely wanting, without any mental derangement or deficiency of intellect being observed during life, and without any manifestation of a double personality. It seems, therefore, impossible to avoid the conclusion that the two hemispheres of the brain can perform their usual functions without this structure, which serves to bind them together, but whose other functions are unknown. (1886, 318)

Not everybody agreed with Ireland's assessment. Some individuals adhered to the opposing position, one holding that callosal lesions are associated with weaker intellects. The evidence for this position came in part from case studies of people with highly abnormal commissures or callosal agenesis. One case involved a man who died at age 48, after having spent 25 years in an asylum. He was described as "never sharp," and irritable and violent. He showed, among other things, no corpus callosum, possibly no anterior commissure, and abnormal gyri crossing the top of the hemispheres (Turner, 1877–1878). A drawing of his brain appears as Figure 27.4.

As stated by another author after presenting his own case study and a review of the literature:

---

[8]Valerie Clem showed a variety of other signs of diffuse cerebral damage. Nevertheless, she was notably free of color agnosia, the symptom that seems to accompany the four classic Gerstmann signs more than any other.

[9]For data and discussions pertaining to the lesions in the Gerstmann "syndrome," see Benton (1961, 1977), Critchley (1953), and Heimburger, Demeyer, and Reitan (1964).

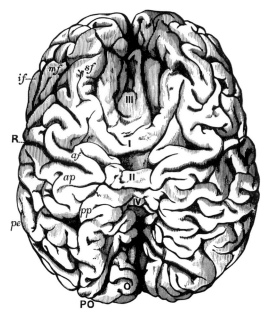

Vertex view of Cerebrum. R, fissure of Rolando ; PO, parieto-occipital fissure ;
O, occipital lobe ; *pe*, convolution of parietal eminence ; *pp*, postero-parietal
convolution ; *ap*, ascending parietal convolution ; *af*, ascending frontal convolution ; *sf, mf, if*, superior, middle, and inferior frontal convolutions ; I.,
anterior, and II., posterior transverse convolutions ; III., grey matter in
front of anterior transverse convolutions ; IV., mesial convolutions behind
posterior transverse convolution.

Figure 27.4.   Diagram of a human brain showing gyri connecting the
two hemispheres across the dorsal midline. This individual did not have a
corpus callosum and was irritable, violent, and "never sharp." He showed
seizures and dementia before dying at age 48. (From Turner, 1877–78.)

The conclusions to which I have come from a careful study
of all the cases are these—First, in those cases in which the
commissure system is wholly awanting or very rudimentary, idiocy or imbecility prevails; and second, in those in
which the corpus callosum is only partially defective, while
the other commissures are present, intelligence but slightly
below the average, together with dullness or levity, melancholy, or perhaps childishness are the usual mental characteristics. (Knox, 1875, 237)

By the end of the nineteenth century, there were many
cases in the literature describing individuals whose corpora callosa were severely damaged or absent. A summary
of many of these cases by Alexander Bruce (1854–1911) in
the 1889–90 volume of *Brain* showed that most of these
individuals exhibited aberrant behaviors and were not of
normal intelligence ("idiotic," "weak-minded," "retarded"),
although there was much variability across patients. There
were also considerable differences with regard to other
structures missing or damaged.

One thing seemed fairly clear at this time. Even with
callosal lesions or congenital abnormalities, these patients
did not show a doubling of consciousness or personality.
Rather, there was still a unified being as far as the clinicians or family members could tell (Descourtis, 1890).
William McDougall (1911) thought about how this related
to the theory that postulated two distinct minds in a single
skull. He once told Sir Charles Sherrington

that if ever he should be smitten with an incurable disease,
Sherrington should cut through his corpus callosum. "If the

physiologists are right the result should be a split personality. If I am right, my consciousness will remain a unitary
consciousness." (Cited by Zangwill, 1974, 265)

Some tongue-in-cheek remarks that were made about the
functions of the corpus callosum by later investigators are
also worth noting. One was Warren S. McCullough's
(1898–1969) quip that the only demonstrated function of the
corpus callosum was to permit the transmission of epileptic
seizures from one hemisphere to the other. Another was that
its only known function was to prevent the hemispheres
from sagging. The latter was stated by Karl Lashley (see
Bogen, 1979).

These negative conclusions about the corpus callosum
were based not just on the growing literature on callosal
agenesis, but on studies of epileptic patients whose corpora
callosa had been surgically sectioned. Walter Dandy
(1930), the combative American neurosurgeon shown in
Figure 27.5, was one of those who argued that no symptoms follow splitting of the corpus callosum. Dandy wrote
that since the corpus callosum may be split without any
disturbance in function, "this structure is, therefore, eliminated from participation in the important functions which
hitherto have been ascribed to it" (p. 643).

Dandy continued his attack a few years later:

The corpus callosum is split longitudinally, from its posterior extremity to a point anteriorly where the third or the
lateral ventricle comes into view; this incision is bloodless.
Usually this incision takes most, and sometimes all, of this
structure to its downward bend. No symptoms follow its
division. This simple experiment at once disposes of the
extravagant claims to function of the corpus callosum.
(1936, 40)

In addition to Dandy's work, a large number of "split-
brain" patients were described by Andrew Akelaitis (1904–
1955) in the early 1940s.[10] Akelaitis and his coworkers concluded that grasping, handwriting, speech, "skin writing,"
praxis, stereognosis, orientation, and other higher-order
functions were not disrupted by this surgery, provided the
hemispheres themselves were free from serious, chronic
damage. This was amazing. A structure with two million
nerve fibers had been cut, and the split-brain patients
showed no deficits on tests of higher brain functions. Moreover, they experienced a clinical benefit—their seizures were
fewer and more controllable.

Akelaitis thought his negative results could be due to
three factors: (1) the representation of some functions in
both hemispheres, (2) the utilization of subcortical commissural pathways, and (3) bilateral projections to and from
the hemispheres. For some functions, such as praxis, Akelaitis (1942a) believed that all three possibilities had to be
given serious consideration. For others, such as language,
he emphasized that subcortical commissural systems were
probably used for communication between the nondominant and the dominant hemisphere (Akelaitis, 1943).

It was not until Roger Sperry (b. 1913) and Ronald Myers
(b. 1929) did a series of experiments on split-brain cats that
the role of the corpus callosum became clearer. After the

---

[10]A sampling of Akelaitis' papers includes Akelaitis (1940, 1941a,
1941b, 1942a, 1942b, 1943), Akelaitis et al (1941), and Akelaitis, Risteen,
and Van Wagenen (1943).

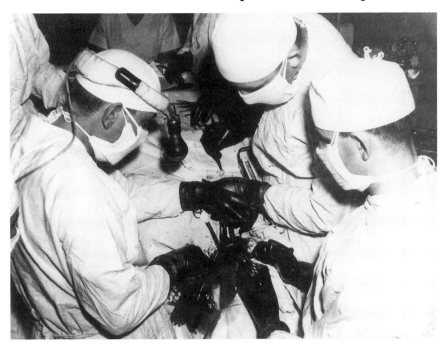

Figure 27.5. Walter Dandy (1886–1946), left, in an operating room. In 1936, Dandy argued that splitting the corpus callosum did not cause any cognitive changes. (Courtesy of the Chesney Medical Archives, Johns Hopkins University, Baltimore, MD.)

hemispheres and the optic chiasma had been surgically divided, the cats were tested one eye (hemisphere) at a time under highly controlled conditions. The results showed that the two hemispheres were not able to communicate with each other and that each learned the required visual task in about the same amount of time (Myers and Sperry, 1953).

This finding was confirmed with monkeys. As summarized by Sperry:

> Callosum-sectioned cats and monkeys are virtually indistinguishable from their normal cagemates under most tests and training conditions. (But) if one studies such a "split-brain" monkey more carefully, under special training and testing conditions . . . one finds that each of the divided hemispheres now has its own mental sphere or cognitive system—that is, its own independent perceptual, learning, memory, and other mental processes . . . as if the animals had two separate brains. (Sperry, 1961, 1749)

Sperry and his associates followed up their laboratory experiments with an important set of behavioral studies on humans who suffered from severe epilepsy (Bogen, 1979). Cutting the corpus callosum was again successful in reducing the seizures, and to the casual observer the patients looked normal. They showed no radical changes in personality, temperament, or intelligence. The differences from earlier findings with split-brain humans emerged when better tests and evaluative procedures were used to study these patients (Sperry, 1962, 1968). Under highly controlled testing conditions, the left hemisphere showed that it was much more verbal than the right, as well as more analytic and more rational. The right hemisphere, in turn, came out being more holistic, emotional, impulsive, and artistic. In the laboratory, the separated hemispheres functioned as two distinct minds, although in everyday life these patients continued to exhibit a unified psyche.

Joseph Bogen (1969), the neurosurgeon who worked with Roger Sperry, used the term "neowiganism" to describe these findings. He did this to honor Arthur Wigan, who, in 1844, had written philosophically about each hemisphere housing its own distinct mind (see Chapter 26). Wigan wrote that the two minds ordinarily worked in a unified, highly adaptive manner, but also possessed the ability to function independently. More than 100 years after he wrote his book, Wigan's theory was embraced by the experimentalists as well as the clinicians.

REFERENCES

Akelaitis, A. J. 1940. A study of gnosis, praxis and language following partial and complete section of the corpus callosum. *Transactions of the American Neurological Association, 66,* 182–185.

Akelaitis, A. J. 1941a. Psychobiological studies following section of the corpus callosum. *American Journal of Psychiatry, 97,* 1147–1157.

Akelaitis, A. J. 1941b. Studies on the corpus callosum. II. The higher visual functions in each homonymous field following complete section of the corpus callosum. *Archives of Neurology and Psychiatry, 45,* 788–796.

Akelaitis, A. J. 1942a. Studies on the corpus callosum. III. A contribution to the study of dyspraxia and apraxia following partial and complete section of the corpus callosum. *Archives of Neurology and Psychiatry, 47,* 971–1008.

Akelaitis, A. J. 1942b. Studies on the corpus callosum. VI. Orientation (temporal-spatial gnosis) following section of the corpus callosum. *Archives of Neurology and Psychiatry, 48,* 914–937.

Akelaitis, A. J. 1943. Studies on the corpus callosum. VII. Study of language functions (tactile and visual lexia and graphia) unilaterally following section of the corpus callosum. *Journal of Neuropathology and Experimental Neurology, 2,* 226–262.

Akelaitis, A. J., Risteen, W. A., Herren, R. Y., and Van Wagenen. 1941. A contribution to the study of dyspraxia and apraxia following partial and complete section of the corpus callosum. *Transactions of the American Neurological Association, 67,* 75–78.

Akelaitis, A. J., Risteen, W. A., and Van Wagenen, W. P. 1943. Studies on the corpus callosum. IX. Relationship of the grasp reflex to section of the corpus callosum. *Archives of Neurology and Psychiatry, 49,* 820–825.

Albert, M. L. 1979. Alexia. In K. M. Heilman and E. Valenstein (Eds.), *Clinical Neuropsychology.* New York: Oxford University Press. Pp. 59–91.

Anton, G. 1893. Beiträge zur klinischen Beurteilung zur Localisation der Muskelsinnstörungen im Grosshirn. *Zeitschrift für Heilkunde, 14,* 313–348.

Anton, G. 1898. Über Herderkrankungen des Gehirns, welche vom Patienten selbst nicht wahrgenommen werden. *Wiener klinische Wochenschrift, 11,* 227–229.

Anton, G. 1899. Über die Selbstwahrnehmung der Herderkrankungen des Gehirns durch den Kranken bei Rindenblindheit und Rindentaubheit. *Archiv für Psychiatrie, 32*, 86–127.

Babinski, J. 1914. Définition de l'hystérie. *Revue Neurologique, 9*, 1074–1080.

Badal, J. 1888. Contribution à l'étude des cécités psychiques: alexie, agraphie, hémianopsie inférieure, trouble du sens de l'espace. *Archives d'Opthalmologie, 8*, 97–117.

Ballard, P. B. 1912. Sinistrality and speech. *Journal of Experimental Pedagogy, 1*, 298–310.

Bastian, H. C. 1898. *A Treatise on Aphasia and Other Speech Defects.* London: H. K. Lewis.

Benson, D. F., and Geschwind, N. 1969. The alexias. In P. J. Vincken and G. W. Bruyn (Eds.), *Handbook of Clinical Neurology: Disorders of Speech, Perception, and Symbolic Behavior.* New York: Elsevier. Pp. 112–140.

Benton, A. L. 1956. Jacques Loeb and the method of double stimulation. *Journal of the History of Medicine and Allied Sciences, 11*, 47–53.

Benton, A. L. 1961. The fiction of the "Gerstmann syndrome." *Journal of Neurology, Neurosurgery and Psychiatry, 24*, 176–181.

Benton, A. L. 1967. Constructional apraxia and the minor hemisphere. *Confinia Neurologica, 29*, 1–16.

Benton, A. L. 1972. The "minor" hemisphere. *Journal of the History of Medicine, 27*, 5–14.

Benton, A. L. 1977. Reflections on the Gerstmann syndrome. *Brain and Language, 4*, 45–62.

Benton, A. L. 1979. Visuoperceptive, visuospatial, and visuoconstructive disorders. In K. M. Heilman and E. Valenstein (Eds.), *Clinical Neuropsychology.* New York: Oxford University Press. Pp. 186–232.

Benton, A. L. 1980. Dyslexia: Evolution of a concept. *Bulletin of the Orton Society, 30*, 10–26.

Benton, A. L. 1982. Spatial thinking in neurological patients: Historical aspects. Chapter 11 in *Spatial Abilities: Developmental and Physiological Foundations.* New York: Academic Press. Pp. 253–275.

Benton, A. L. 1990. The fate of some neuropsychological concepts: An historical inquiry. In E. Goldberg (Ed.), *Contemporary Neuropsychology and the Legacy of Luria.* Hillsdale, NJ: Lawrence Erlbaum. Pp. 171–179.

Benton, A. L., and Joynt, R. J. 1960. Early descriptions of aphasia. *Archives of Neurology, 3*, 205–221.

Benton, A. L., and Meyers, R. 1956. An early description of the Gerstmann syndrome. *Neurology, 6*, 838–842.

Bodamer, J. 1947. Die prosop-Agnosie. *Archiv für Psychiatrie und Nervenkrankheiten Vereinigt mit Zeitschrift für die gesamte Neurologie und Psychiatrie, 179*, 6–53.

Bogen, J. 1969. The other side of the brain. II. An appositional mind. *Bulletin of Los Angeles Neurological Societies, 34*, 135–162.

Bogen, J. 1979. The callosal syndrome. In K. M. Heilman and E. Valenstein (Eds.), *Clinical Neuropsychology.* New York: Oxford University Press. Pp. 308–357.

Brain, R. 1941. Visual disorientation with special reference to lesions of the right hemisphere. *Brain, 64*, 144–172.

Brain, W. R. 1945. Speech and handedness. *Lancet, 2*, 837–842.

Brissaud, E. 1900. Cécité verbale sans aphasie ni agraphie. *Revue Neurologique, 8*, 757.

Broadbent, W. H. 1872. Cerebral mechanism of speech and thought. *Royal Medical and Chirurgical Society of London, Ser. 2*, 55, 145–194.

Broca, P. 1861. Remarques sur le siège de la faculté du langage articulé, suivies d'une observation d'aphémie (perte de la parole). *Bulletins de la Société Anatomique, 36*, 330–357.

Broca, P. 1865. Sur le siège de la faculté du langage articulé. *Bulletins de la Société d'Anthropologie, 6*, 337–393.

Bruce, A. 1889–1890. On the absence of the corpus callosum in the human brain, with the description of a new case. *Brain, 12*, 171–190.

Claiborne, J. H. 1906. Types of congenital symbol amblyopia. *Journal of the American Medical Association, 47*, 1813–1816.

Critchley, M. 1953. *The Parietal Lobes.* London: Arnold.

Damasio, A. R., and Damasio, H. 1983. Localization of lesions in achromatopsia and prosopagnosia. In A. Kertesz (Ed.), *Localization in Neuropsychology.* New York: Academic Press. Pp. 417–428.

Dandy, W. E. 1930. Changes in our conceptions of localization of certain functions in the brain. *American Journal of Physiology, 93*, 643.

Dandy, W. E. 1936. Operative experience in cases of pineal tumor. *Archives of Surgery, 33*, 19–46.

Dejerine, J. J. 1891. Sur un cas de cécité verbale avec agraphie, suivi d'autopsie. *Comptes Rendus Hebdomadaire des Séances et Mémoires de la Société de Biologie, 43*, 197–201.

Dejerine, J. J. 1892. Contribution à l'étude anatomopathologique et clinique des différentes variétés de cécité verbale. *Comptes Rendus Hebdomadaire des Séances et Mémoires de la Société de Biologie, 44*, 61–90.

Dejerine, J. J. 1914. *Sémiologie des Affections du Système Nerveux* (2nd ed.). Paris: Masson et Cie.

Dejerine, J. J., and Vialet, N. 1893. Sur un cas de cécité corticale. *Comptes Rendus Hebdomadaire des Séances et Mémoires de la Société de Biologie, 11*, 983.

De Romanis, F., and Benfatto, B. 1973. Presentazione e discussione di quattro casi di prosopagnosia. *Rivista di Neurologia, 43*, 111–132.

Descourtis, G. 1890. Les deux cerveaux de l'homme. *Revue d'Hypnologie Théorique et Pratique, 1*, 97–106.

Dunn, T. D. 1895. Double hemiplegia with double hemianopsia and loss of geographical centre. *Transactions of the College of Physicians of Philadelphia, 17*, 45–56.

Fisher, J. H. 1905. Case of congenital word-blindness. (Inability to learn to read). *Ophthalmic Review, 24*, 315–318.

Freud, S. 1891. *Zur Auffassung der Aphasien.* Vienna: Deuticke. Translated by E. Stengel as, *On Aphasia.* New York: International Universities Press, 1953.

Gerstmann, J. 1924. Fingeragnosie: eine umschriebene Störung der Orienterung am eigenen Körper. *Wiener klinische Wochenschrift, 37*, 1010–1012.

Gerstmann, J. 1927. Fingeragnosie und isolierte Agraphie—ein neues Syndrom. *Zeitschrift für die gesamte Neurologie und Psychiatrie, 108*, 152–177.

Gerstmann, J. 1930. Zur Symptomatologie der Hirnlaesionen im Uebergangsgebiet der unteren Parietal-und mittleren Occipitalwindung. *Nervenarzt, 3*, 691–695.

Gerstmann, J. 1940. Syndrome of finger agnosia, disorientation for right and left, agraphia and acalculia. *Archives of Neurology and Psychiatry, 44*, 398–408.

Geschwind, N. 1975. The apraxias: Neural mechanisms of disorders of learned movements. *American Scientist, 63*, 188–195.

Gesner, J. A. P. 1769–1776. *Samlung von Beobachtungen aus der Arzneygelahrtheit und Naturkunde* (5 vols.). Nördlingen: C. G. Beck.

Harman, U. B. 1915. *Kelynack's Defective Children.* New York: William Woods and Company.

Heidenhain, A. 1927. Beitrag zur Kenntnis der Seelenblindheit. *Monatsschrift für Psychiatrie und Neurologie, 66*, 61–116.

Heimburger, R. F., Demyer, W., and Reitan, R. M. 1964. Implications of Gerstmann's syndrome. *Journal of Neurology, Neurosurgery and Psychiatry, 27*, 52–57.

Hermann, G., and Poetzl, O. 1926. *Über die Agraphie und ihre lokaldiagnostischen Beziehungen.* Berlin: Karger.

Hincks, E. M. 1926. Disability in reading and its relation to personality. *Harvard Monographs on Education, Ser. 1, 2*, 1–92.

Hinshelwood, J. 1900. Congenital word-blindness. *Lancet, 1*, 1506–1508.

Hinshelwood, J. 1902. Congenital word-blindness, with reports of two cases. *Ophthalmic Review, 21*, 91–99.

Holland, H. 1840. On the brain as a double organ. In *Chapters on Mental Physiology.* London: Longman, Brown, Green and Longmans, 1852.

Holland, P. 1601. *The Historie of the World, Commonly Called the Natural Historie of C. Plinius Secundus.* London.

Holmes, G. 1918. Disturbances of vision of cerebral lesions. *British Journal of Ophthalmology, 2*, 353–384.

Ireland, W. W. 1886. *The Blot upon the Brain.* New York: G. P. Putnam's Sons.

Ireland, W. W. 1891. On the discordant action of the double brain. *British Medical Journal, 1*, 1167–1169.

Jackson, E. 1906. Developmental alexia (congenital word-blindness). *American Journal of Medical Science, 131*, 843–849.

Jackson, J. H. 1866. Remarks on those cases of diseases of the nervous system, in which defect of expression is the most striking symptom. *Medical Times and Gazette, 1*, 659–662.

Jackson, J. H. 1872. Case of disease of the brain—left hemiplegia—mental affection. *Medical Times and Gazette, 1*, 513–514.

Jackson, J. H. 1876. Case of large cerebral tumour without optic neuritis and with left hemiplegia and imperception. *Royal London Hospital Reports, 8*, 434–444.

Kempf, K. 1888. *Valerii Maximi Factorum et Dictorum Memorabilium Libri Novem.* Leipzig: B. G. Teubner.

Kertesz, A. 1979. *Aphasia and Associated Disorders: Taxonomy, Localization, and Recovery.* New York: Grune & Stratton.

Kleist, K. 1912. Der Gang und der gegenwärtige Stand der Apraxieforschung. *Ergebnisse der Neurologie und Psychiatrie, 1*, 342–452.

Kleist, K. 1923. Kriegsverletzungen des Gehirns in ihrer Bedeutung für die Hirnlokalisation und Hirnpathologie. In O. von Schjerning (Ed.), *Handbuch der Arztlichen Erfahrung im Weltkriege, 1914/1918* (Vol. 4). Leipzig: J. A. Barth.

Kleist, K. 1934. *Gehirnpathologie.* Leipzig: J. A. Barth.

Knox, D. N. 1875. Description of a case of defective corpus callosum. *Glasgow Medical Journal, New Ser., 7*, 227–237.

Kohs, S. C. 1923. *Intelligence Measurement.* New York: Macmillan.

Kussmaul, A. 1877. *Die Störungen der Sprache.* Leipzig: Vogel.

Liepmann, H. 1900. Das Krankheitsbild der Apraxia (motorischen Asymbolie) auf Grund eines Falles von einseitiger Apraxie. *Monatsschrift für Psychiatrie und Neurologie, 8*, 15–44, 102–132, 182–197.

Liepmann, H. 1905. Die linke Hemisphäre und das Handeln. Reprinted in *Drei Aufsätze aus dem Apraxiegebiet (neu durchgesehen und mit Zusatzen versehen).* Berlin: Von Karger, 1908. Pp. 17–50.

Liepmann, H. 1906. Der weitere Krankheitsverlauf bei dem einseitig

Apraktischen und der Gehirnbefund auf Grund von Serienschnitten. *Monatsschrift für Psychiatrie und Neurologie, 19,* 217–243.

Liepmann, H., and Maas, O. 1907. Fall von linksseitiger Agraphie und Apraxie bei rechtsseitiger Lähmung. *Zeitschrift für Psychiatrie und Neurologie, 10,* 214–227.

Loeb, J. 1884. Die Sehstörungen nach Verletzung der Grosshirnrinde. *Pflüger's Archiv für die gesammte Physiologie, 34,* 67–172.

Loeb, J. 1885. Die elementaren Störungen Functionen nach oberflächlicher, umschriebener Verletzung des Grosshirns. *Pflüger's Archiv für die gesammte Physiologie, 37,* 51–56.

McDougall, W. 1911. *Body and Mind: A History and Defense of Animism.* New York: Macmillan.

Meadows, J. C. 1974. The anatomical basis of prosopagnosia. *Journal of Neurology, Neurosurgery and Psychiatry, 37,* 489–501.

Mercuriale, G. 1588. *Variae lectiones.* Venetiis: Apud Juntas.

Meynert, T. 1890. *Klinische Vorlesungen über Psychiatrie.* Wien: W. Braumüller.

Monakow, C. von. 1897. *Gehirnpathologie.* Wien: Nothnagel.

Morgan, W. P. 1896. A case of congenital word-blindness. *British Medical Journal, 2,* 1378.

Myers, R., and Sperry, R. 1953. Interocular transfer of a visual form discrimination habit in cats after section of the optic chiasma and corpus callosum. *Anatomical Record, 115,* 351–352.

Nettleship, E. 1901. Cases of congenital word-blindness (inability to learn to read). *Ophthalmic Review, 20,* 61–67.

Nodet, V. 1889. *Les Agnosies.* Paris: Fayard.

Nothnagal, C. W. H. 1887. *Über die Lokalisation der Gehirnkrankheiten.* Weisbaden: J. F. Bergmann.

Ogle, W. 1867. Aphasia and agraphia. *St. George's Hospital Reports, 2,* 83–122.

Oppenheim, H. 1885. Ueber eine durch klinisch bisher nicht verwerthete Untersuchungsmethode ermittelte Form der Sensibilitätsstörung bei einseitigen Erkrankungen des Grosshirns. *Neurologisches Centralblatt, 4,* 529–533.

Oppenheim, H. 1898. *Lehrbuch der Nervenkrankheiten für Aerzte und Studirende, II Aufl.* Berlin: Karger.

Orton, S. T. 1925. "Word-blindness" in school children. *Archives of Neurology and Psychiatry, 14,* 581–615.

Orton, S. T. 1966. *"Word-blindness" in School Children and Other Papers on Strephosymbolia (Specific Language Disability-Dyslexia) 1925–1946.* Towson, MD: Orton Society.

Poetzl, O. 1928. *Die Optisch-Agnosischen Störungen.* Vienna: Deuticke.

Poppelreuter, W. 1917. *Die psychischen Schädigungen durch Kopfschuss im Kriege 1914–1916: die Störungen der niederen und höheren Sehleistungen durch Verletzungen des Okzipitalhirns.* Leipzig: Voss. Translated by J. Zihl with the assistance of L. Weiskranz as, *Disturbances of Lower and Higher Visual Capacities Caused by Occipital Damage.* Oxford, U.K.: Clarendon Press, 1990.

Poppelreuter, W. 1923. Zur Psychologie und Pathologie der optischen Wahrnehmung. *Zeitschrift für die gesamte Neurologie und Psychiatrie, 83,* 26–152.

Quaglino, A., and Borelli, G. 1867. Emiplegia sinistra con amaurosi; guarigione; perdita totale della percezione dei colori e della memoria della configurazione degli oggetti. *Giornale d'Oftalmologia Italiano, 10,* 106–117.

Redlich, E. 1895. Über die sogenannte subcoricale Alexie. *Jahrbucher für Psychiatrie und Neurologie, 13,* 1–60.

Rieger, C. 1909. *Über Apparate in dem Hirn.* Jena: G. Fischer.

Schmidt, J. 1676. De oblivione lectionis ex apoplexia salva scriptione. *Miscellanae Curiosa Medico-Physica Academiae Naturae Curiosorum, 4,* 195–197.

Sittig, O. 1931. *Über Apraxie.* Berlin: Karger.

Souques, A. 1928. Quelques cases d'anarthrie de Pierre Marie: Aperçu historique sur la localisation du langage. *Revue Neurologique, 2,* 319–368.

Sperry, R. 1961. Cerebral organization and behavior. *Science, 133,* 1749–1757.

Sperry, R. 1962. Some general aspects of interhemispheric integration. In V. B. Mountcastle (Ed.), *Interhemispheric Relations and Cerebral Dominance.* Baltimore, MD: Johns Hopkins University Press. Pp. 43–49.

Sperry, R. 1968. Hemispheric disconnection and unity in conscious awareness. *American Psychologist, 23,* 723–733.

Turner, J. 1877–1878. A human cerebrum imperfectly divided into two hemispheres. *Journal of Anatomy and Physiology, 12,* 241–253.

Waites, L. 1982. American pioneers in the topic of word blindness. In F. C. Rose and W. F. Bynum (Eds.), *Historical Aspects of the Neurosciences.* New York: Raven Press. Pp. 115–116.

Walsh, K. W. 1978. *Neuropsychology.* Edinburgh: Churchill Livingstone.

Warburg, F. 1911. Ueber die angeborene Wortblindheit und die Bedeutung ihrer Kenntnis für den Unterricht. *Zeitschrift für Kinderforschung, 16,* 97–113.

Weintraub, S., and Mesulam, M.-M. 1989. Neglect: Hemispheric specialization, behavioral components and anatomical correlates. In F. Boller and J. Graffman (Eds.), *Handbook of Neuropsychology* (Vol. 2). Amsterdam: Elsevier.

Wigan, A. L. 1844. *A New View of Insanity: The Duality of the Mind.* London: Longman, Brown, Green and Longmans.

Wilbrand, H. 1892. Ein Fall von Seelenblindheit und Hemianopsie mit Sectionsbefund. *Nervenheilkunde, 2,* 361–387.

Wolpert, I. 1924. Die Simultanagnosie: Störung der Gesamtauffassung. *Zeitschrift für die gesamte Neurologie und Psychiatrie, 93,* 397–415.

Zangwill, O. 1974. Consciousness and the cerebral hemispheres. In S. Dimond and J. G. Beaumont (Eds.), *Hemisphere Function in the Human Brain.* New York: John Wiley and Sons. Pp. 264–278.

Zeki, S. 1990. A century of cerebral achromatopsia. *Brain, 113,* 1721–1777.

Part VII
Treatments and Therapies

Chapter 28
Treatments and Therapies: From Antiquity through the Seventeenth Century

> Now everyone can see whence disease arises. There are four natures out of which the body is compacted, earth and fire and water and air, and the unnatural excess or defect of these, or the change in any of them from its own natural place into another, or . . . the assumption by any of these of a wrong kind, or any similar irregularity, produces disorders and diseases.
>
> Plato, fourth century B.C.

This chapter examines ways in which individuals with head and brain damage have been cared for from the time of the earliest known Egyptian medical papyruses through the 1600s. The number of different treatments is enormous, and it would take volumes to describe and illustrate the history or justification of each known drug or procedure. Thus, of necessity, this chapter and the next are selective. The emphases are on some of the major interventions, such as bleeding and craniotomies, as well as on showing how slowly some treatments of limited or questionable value gave way to more successful methods.

ANCIENT EGYPT

Diseases in Ancient Egypt, including those now associated with disorders of the nervous system, were first thought to be caused by sorcery, witchcraft, and evil spirits. From this premise, it was surmised that the earliest physicians probably were priest-magicians. They were interveners who treated their patients largely by calling upon the powers of omnipotent deities and by employing "magic."[1]

The early Egyptian medical papyruses contain descriptions of many spells and incantations. The simplest spells merely commanded the demon or "poison" to leave the body. These orders repeatedly included the patient's name, so that the evil spirits would know what was being demanded of them. They also included the names of those gods who were thought to exert great power. In particular, the goddess Sachmet was believed to be a cause of illness, whereas Re, the sun god, was viewed as a very powerful physician and apothecary.

Isis was the patroness of magic and the wife of Osiris. In legend, Osiris was murdered by his evil brother Set (Setekh). Isis found the fragments of her husband's dismembered body scattered throughout Egypt. By her divine pow-

ers, Isis reunited his body and brought him back to life. In later forms of the story, she did this while giving birth to a son, Horus.[2] Set, in another fit of rage, ripped the eyes out of Isis' son. This time Thoth, the ibis-god, came forth to heal Horus. The significance of these myths is that sick and injured people in Egypt identified with Osiris and Horus and called upon Isis, Thoth, and other powers associated with healing to make them well again.

In many spells, the spoken word was accompanied by gestures and the use of figures, amulets, and symbols. Special incantations were often said over these magical objects to give them even more power. In addition, the patient might have been given something special to eat or drink. Mixed in water, wine, beer, or honey, the medicinal substances could come from animal, mineral, or vegetable sources. To say the least, some of the medicines mentioned in the Egyptian papyruses were foul or offensive. One idea behind this sort of treatment seemed to be that fetid substances would make the body disgusting and uninhabitable to a spirit with any self-respect.

The Ancient Egyptians also used narcotics. Henbane (*Hyoscyamus*), for example, produced sleep accompanied by fantasies and hallucinations. This drug was also used by early European physicians to reduce pain. Its active ingredient was later found to be scopolamine (Thorwald, 1963). There is also evidence of the use of opium by the Egyptians (Majno, 1975).

One might expect that there would have been numerous cases in which chants, amulets, and some medicines would not have been successful in promoting healing. The Egyptian healers recognized the effect that repeated failures could have on their credibility. Thus, the papyri show that these men often used alternative procedures if their preferred methods failed. In addition, a priest-physician could always claim the patient's sins were so serious that the deities were not willing to listen to pleas for help or requests for forgiveness.

The title physician *(swnw* or *sunu)* first appeared at the

---

[1]Indeed, in Ancient Egypt, existence was itself thought to be the result of the breath of life entering through the right ear, whereas death was associated with the left ear (*Ebers Papyrus*, c. 1500 B.C.).

[2]Motherhood became one of the functions of goddesses in the New Kingdom.

time the great pyramids were built. One of the earliest known medical practitioners was Hesy Re, referred to as Chief of Dentists and Physicians to the pyramid builders of the Third Dynasty, approximately 2600 B.C. (Majno, 1975). During the time Hesy Re was practicing medicine, fractures were already being set with splints and bandages.

From surprisingly early times, the Egyptians appeared to rely on medical specialists. A hieroglyph of a man named Iry, from the Fourth Dynasty (c. twenty-seventh to twenty-sixth centuries B.C.), describes this individual as an "eye doctor of the palace, doctor of the abdomen and guardian of the royal bowel movement." Based on a monument erected in his honor, Sekhet-n-ankh, who lived around 2550 B.C., was identified as the "nose doctor" to King Sahure. Surgery, however, was not one of his specialties.

Herodotus of Halicarnassus, the Greek father of history who traveled to Egypt more than a thousand years after the Fourth Dynasty ended, wrote that medicine was still practiced on a plan of specialization. Each physician was thought to treat only a single organ or disorder and no more. Some dealt with the eyes, some with the teeth, some with the head, and so on.

Medical historians believe that with the rise of medical schools in the New Kingdom (1567–1085 B.C.), the roles of magic and religion in Egyptian medicine slowly began to diminish (e.g., Dawson, 1929). Yet some procedures, rituals, and beliefs that seemed to have little if any scientific justification survived over the centuries. A good example is the use of charms. The early physicians sometimes put charms on the site of the injury, a painful spot, or the area thought to be most affected by a disease. In addition, some charms were simply worn around the neck, whereas others were tied to a hand or a foot, even for headaches *(Leiden Papyrus)*.

During the New Kingdom, every part of the body continued to be associated with a different god, and diseases were still thought to be caused by evil intruders. Thus, the prescriptions given by the "new" physicians were not really intended to "cure" but to terrify, banish, and drive out the forces of evil. The malevolent spirits could leave by feces, urine, sweat, gas, or breath. New openings could also be made in the body to help them out, although trepanning was rarely if ever practiced in Ancient Egypt (Ghalioungui, 1963; Majno, 1975).

This orientation was consistent with the belief that there were 36 channels to the heart, which were called *metu.* These channels went to the nose, bladder, rectum, ears, and other organs. There is no evidence that the Ancient Egyptians understood the functions of the nerves. They seemed to believe that nerves simply formed one part of this system of channels.

Some problems were thought to be due to blockages in the *metu,* and one job of the physician was to keep the channels open. Bloodletting was practiced in this context. A wall painting in the tomb of Userhat, a scribe from the Eighteenth Dynasty (1567–1308 B.C.), shows a leech being taken from a bowl for application to a patient's head.

In general, observable wounds were dealt with more rationally than diseases, probably because diseases were easier to attribute to the supernatural. One common treatment for a head wound involved covering the injury with a piece of meat on the first day and after this with a linen cloth dipped in grease or honey. A patient with a more severe injury would probably be immobilized in a sitting position, with bricks placed on both sides of the head for support.

The prescriptions for wounds of the head were often very precise. The *Ebers Papyrus* refers to a mixture that called for the fats from a lion, hippopotamus, crocodile, snake, and bird (the ibex). The instructions were: "Mix into one mass; anoint the head therewith." This papyrus, which contains over 900 prescriptions, is estimated to have been written around 1555 B.C. Because it was based on older material, this prescription might have been used for centuries before this time.

One learns from the *Edwin Smith Surgical Papyrus* that Ancient Egyptians who suffered head injuries during military campaigns had their wounds classified as (1) an ailment to be treated, (2) an ailment that may be treated, or (3) an ailment not to be treated (see Chapter 1). These categories most likely corresponded to, respectively, injuries where recovery could be expected; injuries where recovery was possible but not assured; and cases in which there was little or no chance for a successful outcome. With severe head injuries that resulted in loss of speech, for instance, the prognosis was likely to be poor, and the classification usually was "an ailment not to be treated."

Case 8 in the papyrus involved a head injury that could be treated. Here the writer simply stated: "His treatment is sitting, until he (gains color), (and) until thou knowest he has reached the decisive point."

More interesting is Case 9, which involved an individual with a wound of the forehead that produced a compound fracture of the skull. James Breasted (1930) translated this case as follows:

> Thou shouldst prepare for him an egg of an ostrich, triturated with grease (and) placed in the mouth of his wound. Now afterward thou shouldst prepare for him the egg of an ostrich, triturated and made into poultices for drying up that wound. Thou shouldst apply to it for him a covering for physician's use; thou shouldst uncover it the third day, (and) find it knitting together the shell, the color being like the egg of an ostrich.

The following, which was said as a charm over this recipe, shows that the physician's intent was to drive out the evil causing the mischief. Its use also shows that rational medicine and religious beliefs had not yet gone completely separate ways, even in the treatment of a wound suffered in battle:

> Repelled is the enemy that is in the wound!
> Cast out the (evil) that is in the blood.
> The adversary of Horus, (on every) side of the
>   mouth of Isis.
> This temple does not fall down;
> There is no enemy of the vessel therein.
> I am under the protection of Isis;
> My rescue is the son of Osiris.
> (Breasted, 1930, 220)

ASSYRIA

The clay tablets of the Royal Library of King Ashurbanipal (668–626 B.C.) at Nineveh show that the Assyrians treated diseases and injuries in ways fairly similar to those of the Ancient Egyptians (Gordon, 1939). Medicine was based primarily on ritual and magic. Although drugs were used, some seemed to have had a rational basis, whereas others did not. Relative to those of the Egyptians, the Assyrian records provide less information about the quantity of drugs to be taken (Thompson, 1923).

The following is an incantation used to drive a foreign body from the eye. It was said in a ceremony involving a charm made from "magical" red wool with seven knots:

> O failing eyes, O painful eyes, O eyes sundered by a dam of blood! Why do ye fail, why do ye hurt? Why hath the dust of the river come nigh you, (or) the spathe of the date-palm whereof ye have chanced to catch the pollen which the fertilizer hath been shaking? Have I invited you, Come to me? I have not invited you, come not to me, or ever the first wind, the second wind, the third wind, the fourth wind cometh to you! (Translated in Thompson, 1924, 31)

ANCIENT GREECE

At the height of the Ancient Greek empire, medicine was tied to the four humors: yellow bile, blood, phlegm, and black bile (for a brief history of humoral theory, see Chapter 1). These were products of fire, air, water, and earth, respectively. Each in turn was associated with two qualities: hot or cold and moist or dry. The basic theory underlying Greek medicine in the fourth century B.C. was that symptoms arose when the balance among the humors was disrupted. This belief is illustrated in the quotation at the beginning of this chapter from Plato's *Timaeus*.

The job of the physician was to try to reestablish the equilibrium when it was disrupted. This process was often accomplished by treating the patient with agents thought to affect one of the specific humors. When the problem was assumed to lie in too much blood, the physicians considered bloodletting a reasonable way of restoring the balance. Two signs of excessive blood in the head were a red face and bloodshot eyes. The Greeks also turned to bleeding individuals who complained about headaches. By the time of Hippocrates, they were letting blood in a variety of ways. One method involved opening a vein (venesection), and another involved cupping. The latter approach was often selected when more topical bleeding was demanded.

Hippocratic physicians may have learned how to cup from the Ancient Egyptians. They used cupping for painful disorders such as headaches, pleurisy, and throat infections, as well as for delirium, stupor, insanity, and depression. The placement of the cups varied with the disorder, but in general the logic was to draw blood close to the affected organ, which sometimes meant cupping the head. Writers of this period described different types of cups and specific cupping procedures (see Mapleson, 1813).

Typically, superficial cuts were made on the skin (scarification) to promote local bleeding in a procedure known as wet cupping (dry cupping was cupping without scarification). The Greeks often used a small gourd with a large opening that fit over the scarified skin. A second, smaller opening allowed the physician to suck out the air to create a partial vacuum, which was then closed with a bit of wax. This vacuum caused the superficial cuts to open and blood to flow.

The Greeks also relied on a cup (often of brass) with only one aperture. With this type of cup, the scarified skin was lightly greased and a piece of flax, linen, or tow (coarse broken fiber), was lighted and put into the cup. After the material was burned, the heated cup with its expanded air was attached to the skin. As the air cooled, it contracted, pulling the skin into the cup and suctioning blood vessels. The cups containing blood were then taken off or simply allowed to fall off. This procedure remained popular for more than 2,000 years.

Leeching was used as an alternative procedure for blood letting, especially among the lower classes. Nicander, a Greek who died in Athens around 135 B.C., wrote a treatise on how and where to apply leeches.

The word "apoplexy," indicating brain hemorrhage or stroke, can be found in a number of writings of the Hippocratic school (Clarke, 1963). The expectation for recovery depended to a large extent on whether a fever accompanied the stroke. This can be gleaned from the following passage from *On Diseases II:*

> The healthy subject is taken with sudden pain; he immediately loses his speech and rattles in his throat. His mouth gapes and if one calls him or stirs him he only groans but understands nothing. He urinates copiously without being aware of it. If fever does not supervene, he succumbs in seven days, but if it does he usually recovers. (Translated in Clarke, 1963, 307)

Physicians at this time attributed apoplexy to an accumulation of black bile that was allowed to cool. The fever was thought to be a good sign, because it warmed the blood and affected the cool black bile. Most victims of apoplexy died three to five days after the stroke, but if they survived past the first week, chances for survival were seen as improved.

The Hippocratic physicians relied heavily on Nature, but also intervened to increase the likelihood of recovery after brain damage. As can be seen in the following passage, they had many different procedures in their arsenals:

> First the patient should be washed with plenty of hot water and hot applications placed on his head; or his head should be placed in a vapour bath. He must then be dried well and a luke-warm mixture of honey and vinegar placed in his mouth. If he recovers he should be fed, and when he seems to have gained strength, a substance such as myrrh mixed with flowers of sulphur is put in his nose to promote sneezing. Sternutories are particularly necessary if the headache persists, for as the nasal and intracranial cavities were considered to be directly connected, they were thought to purge the head. Soup, such as a decoction of barley, is given and water to drink, but alcoholic beverages are strictly forbidden. After an interval of a few days a purgative is to be administered, and this is an important part of the treatment, because without purging the patient may fail to recover. Incision of the scalp is likewise obligatory if all other measures have failed; if it is not carried out in these cases, the patient will probably die on the eighteenth or twentieth day. The physician is directed to incise the scalp in the region of the obliterated anterior fontanelle, and when the blood has ceased to flow, the edges of the wound are brought together and a bandage is applied. (Clarke, 1963, 313)

Craniotomies, seemingly not performed by earlier Egyptian physicians, were conducted in Ancient Greece and were made with a number of different instruments (Phillips, 1973; Thompson, 1938). Greek physicians used a drill, called a *terebra,* spun by a strap wrapped around the center. They also used a trepanon, which could be operated by a bow (Horne, 1894). These tools allowed them to make a ring of holes around the piece of bone they wished to resect.

Another instrument, the *terebra serrata,* or *prion charactos,* consisted of a conical piece of metal with a circular serrated edge at the bottom. A center pin kept the saw in place while it cut, and there was a large handle that was rolled rapidly between the palms or rotated with a bow. The center pin was removed after the drill made a starting channel in the bone.

Not all parts of the skull were trepanned. Greek physicians seemed to stay away from areas with major blood vessels. They also tried to make sure the dura was not broken when they operated.

Craniotomies were performed primarily because the physicians believed this procedure could restore the balance among the four humors. If the skull were opened by the injury, the theory held that the humors probably could escape and the trepan was not needed. The demand for the trepan was much greater for cases of closed head injuries where the humors could not escape and might form harmful dark pus. This can be seen in the following passage from the Hippocratic work *On Injuries of the Head:*

> Of these modes of fracture, the following require trepanning: the contusion, whether the bone be laid bare or not; and the fissure, whether apparent or not. And if, when an indentation *(hedra)* by a weapon takes place in a bone it be attended with fracture and contusion, and even if contusion alone, without fracture, be combined with the indentation, it requires trepanning. A bone depressed from its natural position rarely requires trepanning; and those which are most pressed and broken require trepanning the least; neither does an indentation *(hedra)* without fracture and contusion require trepanning; nor does a notch, provided it is large and wide; for a notch and a hedra are the same. (1952 translation, 65)

Craniotomies were helpful because opening the skull was a way of reducing the rising intracranial pressure, which could kill or debilitate a patient after a closed head injury. The successes achieved when the meninges were not damaged can help to explain the rise in the popularity of this sort of surgery, tied to humoral theory in Hippocratic Greece and most likely to demonology in neolithic cultures (see Chapter 1).

ANCIENT ROME

Galen combined the doctrine of the four humors with his astrological beliefs, especially the idea that the positions of the planets and the moon may signify the nature of an illness, when a crisis will occur, and if the patient would survive (Stahl, 1937).

Like other physicians in the Roman Empire, he continued to cup, bleed, and leech. Galen suggested clipping the tail of the leech to allow more blood to be drawn by each leech. This recommendation, however, resulted in increased mortality among the leeches and was not rigorously followed. The Roman physicians also used cups of brass or horn to draw blood and generally relied on procedures very much like those of the Hippocratic writers.

These physicians also performed trepanations and used a wide assortment of drugs to treat injuries and diseases. As might be expected, some of their treatments bordered on the bizarre. Individuals with obstinate cases of epilepsy, for example, were given the warm blood of slain gladiators to drink (Bakay, 1985).

Celsus, one of the most learned individuals in the first century, pointed to the tongue as the source of most speech disorders. In *De medicina,* he described the following treatment for "paralysis of the tongue."

> When the tongue is paralyzed, either from a vice of the organ or as a consequence of another disease, and when the patient cannot articulate, gargles should be administered of a decoction of thyme, hyssop, pennyroyal; he should drink water, and the head and neck and mouth and the parts below the chin be well rubbed. The tongue should be rubbed with lazerwort, and he should chew pungent substances such as mustard, garlic, onions, and make every effort to articulate. He must exercise himself to retain his breath, wash his head with cold water and then vomit.

Celsus recommended dressing head wounds with wool soaked with oil and vinegar and stated that "in process of time flesh grows from the bone itself and fills up the cavity left after the operation" (Thompson, 1938). He also relied on the cautery, believing that cauterization allowed the excess humors to escape through the burns. He thought cauterizing was especially helpful for depression and epilepsy. In classic Hippocratic fashion, the latter was attributed to an accumulation of phlegm and thick bile.

In the first century A.D., Scribonius Largus used electricity to treat headache and gout. The medical writings of this Roman physician were discovered early in the sixteenth century and were published in 1529. For headache, he placed a large Mediterranean electric ray, *Torpedo mamorata,* across the brow of his patients and allowed the fish to discharge its electricity "until the patient's senses were benumbed." For gout, the live fish was placed under the sufferer's feet.

The torpedo continued to be used by later Romans, some of whom kept several torpedoes for multiple shock treatments. Although it is occasionally stated that the torpedo was used by Pliny the Younger (c. 61–113) for the pains of childbirth, Pliny did not utilize the shock from the fish but rather the edible flesh of dead fish (Plinius Secundus, 1601). In contrast, Galen questioned the pharmacological use of the dead torpedo and concluded it did not have healing power. Thus, he turned to the shock of the live torpedo as a remedy for headache and other pains. He was enthusiastic about the medical properties of the live fish and stated it was the best remedy known for epilepsy.

The powers of electric fish were probably known well before Roman times, but whether the shock was used for medical purposes is unclear. Fishing scenes on the walls of Egyptian tombs, such as the one shown in Figure 28.1 from the burial place of the architect Ti (middle of the third century B.C.) at Sakkara, show the common Nile catfish, *Malopterurus electricus,* very realistically (Kellaway, 1946). Nevertheless, the electric catfish was not mentioned in early Egyptian works, and all that is certain is that the early Egyptians used the electric fish for food.

As for the Greeks, in the second book of *On regimen,* which is a part of the Hippocratic corpus, 18 fish are described. One fish had the name *narke* or *narce,* the Greek

Figure 28.1. Bas-relief from the Tomb of Ti in Sakkara, dated the middle of the third century B.C. This relief shows two realistically drawn electric Nile catfish *(Malopterurus electricus)* in the center and bottom of the middle column of fish.

word for "numbness" (as in "narcosis"). It is not possible to tell which fish the Hippocratic writer was describing, but it seems clear that both the Nile catfish and the torpedo (electric ray) were known to the Greeks. Figure 28.2 shows the electric torpedo decorating a Greek plate.

The effects of the torpedo's electricity were described by Plato in the *Meno* (Section 80a). Plato wrote that the electric ray leaves anyone who comes into contact with it feeling numb. A more naturalistic description of the torpedo can be found in Aristotle's *Historia animalium,* written later in the fourth century B.C.:

> The torpedo narcotizes the creatures that it wants to catch, overpowering them by the power of shock that is resident in its body, and feeds upon them; it also hides in the sand and mud, and catches all the creatures that swim in its way and come under its narcotizing influence. (1956 translation, 620b)

Aristotle further described how the torpedo breeds, the nature of its young, and the features of some of its organs (gills, gall bladder).

It was left for the Romans, however, to speak not just about the tender flesh of the electric fish and its wondrous powers of capturing prey and defending itself, but also about its use in the healing arts (Kellaway, 1946). In fact, there is even a mosaic of the torpedo in the front of an ancient herb shop in Pompeii, indicating this was a place where these miraculous fish could be bought for therapeutic purposes.

MITHRIDATES AND UNIVERSAL CURE-ALLS

The idea of a universal antidote is usually associated with the historical figure Mithridates VI (120–63 B.C.), the last king of the small land of Pontus in what is now northeastern Turkey. Mithridates VI Eupator, called the Great, skillfully fought three long wars with the Romans. Nevertheless, he became known not just for his military campaigns but for his preoccupation with poisons and attempts to prevent them from working.

Mithridates believed that some of his guardians were trying to poison him slowly, even before he took the throne. Gradually poisoning a person's food in order to make the death look natural was relatively common at this time. Indeed, the strategy had worked many times before in Pontus, as well as in Rome. In order to guard against this, Mithridates took preventatives daily. He then gave himself small amount of poison to ensure the preventatives were working.[3]

After becoming King, Mithridates continued to study poisons and to use them against his enemies. Galen wrote in *De antidotis* that Mithridates often used condemned criminals to test his antidotes for poisons from different insects, animals, minerals, and plants. Among the poisons that most concerned him were those from spiders, scorpions, sea slugs, and the plant aconite.

Realizing that he had been protecting himself against

Figure 28.2. An ancient Greek ceramic plate (Campanian ware) showing three fish. The skatelike fish on the right is the electric *Torpedo ocellata.* Perhaps because the torpedo was considered good eating, it was often shown on Greek plates. (From Blumenbach, 1810.)

[3]Alexipharmacy, or the study of preventative medicines, did not begin with Mithridates VI. The Greeks studied preventative medicines, and Middle Eastern and Oriental preventative drugs were known when Mithridates VI lived.

only some of these poisons, Mithridates set out to create a single medication, with the hope that it would give him universal protection. This concoction became known as a mithridate (Jarcho, 1972). The number of ingredients in the mithridate was put at 36 by Celsus and at 54 by Pliny the Elder. The latter stated in his *Natural History* that no two ingredients had the same weight.

The mithridate proved so effective that the King of Pontus lived to be a relatively old man—and even took poisons in public to show his immunity. In fact, when his army was finally defeated by Pompey the Great's (106–48 B.C.) Roman legions, Mithridates could not even poison himself. On his last day, he turned to some poison hidden in his sword sheath. His two daughters, who insisted on taking the poison first—and who had not been taking any preventatives—died immediately. The King of Pontus, however, was unaffected. He even tried to increase his metabolism by running, hoping to hasten the drug's action, but it did not help. As a result, he finally ordered one of his chieftains to slay him with his own sword, not a first choice, but still more desirable than falling into the hands of the advancing Roman legions.

The formula for the mithridate was changed in the years that followed, always with the hope that it could be made even more effective and universal, not just as a preventative, but as a cure. Andromachus (fl. A.D. 54–68), for example, created "theriac" by eliminating some ingredients from the original formula and adding others. He emphasized the importance of including the flesh of vipers in the formula for theriac (see Andromachus 1607, 1668). Galen, who wrote a treatise entitled *De theriaca ad pisonem,* agreed that this ingredient made theriac a more powerful concoction than mithridate, especially for snake bites (see Jarcho, 1972).

Over the centuries, theriac was recommended not just for poisons and bites, but for a wide variety of injuries and diseases. Although of questionable medicinal value at best, theriac remained remarkably popular into and through the Renaissance, and could still be found on the shelves of many pharmacies in Western Europe in the nineteenth century.

ARABIC MEDICINE BEFORE THE RENAISSANCE

The Islamic and Nestorian physicians represented the most notable successors of Galen and other writers from classical antiquity and early Byzantine times (see Chapters 1 and 5). Because dissections on humans were not permitted, Yuhanna Ibn-Masawayh (d. 857) dissected apes at a special institute created on the banks of the Tigris River. Here, the authorities supplied him with ample material for his studies (Shanks and Al-Kalai, 1984).

Rhazes, who died in 925, was one of the most influential figures in the golden period of Arabian medicine. Among other things, he provided many descriptions of charlatans and their tricks. A common deception involved making a circular opening in the skull and then pretending to remove an object, often a hidden stone, with some sleight of hand maneuver. This procedure was advertised as a treatment for epilepsy. In 931, perhaps because of the widespread charlatanism, Abbasid Caliph Al-Muqtadir required physicians to take a qualifying examination before they could practice medicine. As a result, physicians rose in power and esteem, although they still had to compete with astrologers and geomancers.

Rhazes practiced in Baghdad, and it is said that when he was asked to select a site for a new hospital, he hung meat in different quarters of the city and chose the place where it showed the least decomposition (Shanks and Al-Kalai, 1984). His book *Kitab Al-hawi fi-altibb* combined his own experiences with those of others who studied diseases, including Greeks, Persians, and Indians (Hai and Ahmed, 1981). Indeed, it is said that Rhazes once remarked, "when Galen and Aristotle agree, it is obvious that their opinion is correct, but that when they disagree, judgment is indeed difficult" (see Mettler, 1947, p. 32).

The Caliphs appointed physicians to hospitals and paid their salaries. Records indicate that each major city had its own physician and ophthalmologist. The names of the chief surgeons were not often mentioned, but surgery gained in respectability with Albucasis, also known as Abul-Quasim Khalif ibn Abbasmal-Zahrawi. Albucasis was born in Moorish Spain in Al-Zahra, near Cordoba, in 936 and died in 1013 (Al-Rodhan and Fox, 1986; Izquierdo et al., 1981). His thirteenth book of a 30-volume treatise on medicine *(Kitab-al-Tasfif)* was concerned with surgery. It had three parts: (1) cauterization, (2) wounds and bloodletting, and (3) fractures and dislocations. Although Albucasis wrote in Arabic, his works were soon translated into Latin by Gerard of Cremona and then into Hebrew and Provençal.

Albucasis used Paul of Aegina, a Byzantine figure who lived in the seventh century, as one of his most valued sources (Bakay, 1985).[4] For example, he included a chapter dealing with hydrocephalus that was copied from Paul, who had made an earlier survey of the literature. He also relied on some of Paul's techniques for trepanation. Albucasis described how to drill holes in the skull with a "nonsinking drill" that prevented penetration into the brain and how to cut away the bone. He stated that the operation was so simple that "you cut through the bone in the confident knowledge that nothing inward can happen to the membrane even though the operator be the most ignorant and cowardly of men; yes even if he be sleepy" (translated in Bakay, 1987a, p. 11). The wound was dressed with lily roots, vetch flour, frankincense, and birthwort. If the dura turned black, however, "you may know that he/the patient/ is doomed."

In cases of severe headaches from the temporal region, Albucasis recommended cutting the superficial temporal artery and ligating it with silk or lute string. He also described how to cauterize the arteries for paralysis, epilepsy, tremors, and numbness. For unilateral facial palsy, he used the cautery (possibly on the facial nerve) to make the other side paretic. Restoring symmetry was intended to make the face appear less distorted (Al-Rodhan and Fox, 1986).

Albucasis used alcohol to clean the wound and opium as

---

[4]Only Paul's seven-volume *Epitome* remains. Although this work contains some original thoughts, it is largely a compilation of earlier classics. Paul devoted some of his text to skull fractures and head injuries. He advised removing loose pieces of bone when there was a depressed fracture, but added that this should be done "before the 14th day in winter and before the 7th day in summer." He recognized that injury to the brain could cause paralysis, loss of speech, convulsions, headache, and intellectual problems. *Epitome* was republished in 1528 in Venice.

an anesthetic. He followed the Greeks and Romans, who also used mandrake root seeped in wine as an anesthetic beverage prior to surgery. His texts are notable for his descriptions of more than 150 different early surgical instruments.

CURES DURING THE PRE-RENAISSANCE PERIOD IN EUROPE

A medical practitioner in Anglo-Saxon England, defined as the period between the departure of the Romans in 596 to the invasion of William the Conqueror (1028–1087) in 1066, was sometimes called a leech (Wellcome, 1912). This term obviously came from the extensive use of leeches, although this was not the only method for drawing blood.

Before the thirteenth century, most physicians were associated with the Church in one way or another. Consequently, treatments for injuries and diseases tended to mix religious dogma with rational medicine. Under the Church, disease was associated with sin and the wrath of God, and injuries could also be ascribed, at least indirectly, to sinning and pacts with Satan.

The edict *Ecclesia abhorret a sanguine* ("the Church abhors the shedding of blood") was issued at the Council of Tours in 1163 and in many other places in the twelfth and thirteenth centuries. It was never really clear whether this pertained to all levels of the clergy or to just the highest levels (see Amundsen, 1972). As the Renaissance approached, these edicts effectively turned surgery over to the barbers, and occasionally to executioners and bathkeepers (see Chapter 2).

THE RENAISSANCE

When the Renaissance began in the mid-fourteenth century, injuries that penetrated the meninges were believed to be fatal because they allowed the animal spirits to escape. This was an ancient idea, even mentioned by Pliny the Elder in the eleventh book of his *Natural History:* "If the tunic (pia mater) be pierced and wounded, there is no way but present death." As stated more than 1,300 years later by John of Mirfield (?–1407): "If the pia mater has been injured, cure is impossible and . . . when the brain is injured humors will rush into the brain, either the brain will become abscessed, or the patient will die suddenly because of seizures" (see Bakay, 1987a, p. 21). The paradox is that Galen, who was often cited as the authority on such matters, had himself described a Nubian who survived a serious head injury that penetrated his meninges and damaged his ventricles.

There were some excellent depictions of head injuries, surgical tools, and pharmacological remedies late in the fifteenth century (see Figures 28.3, 28.4, and 28.5), but it remained for Jacopo Berengario da Carpi, the son of a barber surgeon, to make the case that one could recover from injuries that penetrated the dura (Bakay, 1986). In *De fractura calve sive cranei,*[5] published in Bologna in 1518, he described six patients who survived severe brain damage

[5]*De fractura calve sive cranei* was the first comprehensive book dealing entirely with head injuries.

(see Figure 28.6 for a glimpse of some of the weapons associated with head injuries at this time). Some of Berengario da Carpi's patients were important people, which would be expected from his reputation for charging very high fees. The most notable was Lorenzo de Medici (1492–1519), Duke of Urbino, who had been shot in the head.

Another of Berengario da Carpi's patients was his nephew, who had been wounded with a halberd, a weapon with a long shaft topped with an axe and a spike. The wound broke the skull and went into a ventricle. Berengario da Carpi used a forceps to remove fragments of bone lodged deep in the brain and conveyed that "a notable portion" of brain escaped from the wound. He drained the wound with a cannula, but later the brain showed signs of infection and he had to put a stylus into it to remove milk-colored, watery material. Although his patient experienced some seizures, he regained his health and even "lived a long life."

The year in which Berengario da Carpi's book appeared was the birth year for charters and licenses for physicians in England. Medical training, however, was only loosely based on anatomy and physiology at this time, and dissection played only a small role in the curriculum (Cooperman, 1960).

The first well-known physician of the Tudor period in England was Thomas Boorde (1490–1519), a Carthusian monk educated at Oxford. Like other physicians before him, Boorde was taught Greco-Roman medicine. But unlike most of his predecessors, he had serious doubts about the value of what he was taught. In his own words: "If docteurs of Physicke should at all times follow their bookes they should do more harme than goode. . . . Such practicing doeth kill many men" (cited in Cooperman, 1960, p. 8).

John Caius (1510–1573) introduced the new Vesalian anatomy to the English in 1546. Still, for all practical purposes, the physiology and medicine remained that of Hippocrates and Galen. It was based on the four humors; the animal, natural, and vital spirits; and the delicate balance among these factors. The logic that heat expels moist, moist expels dryness, and so on, dictated most treatments, just as they had in Greece and Rome.

Bleeding and purging remained frequent ways for reducing the buildup of harmful humors, and the sites and times for each remained entrenched in astrological theory. Leeches were sometimes used to bleed patients, especially those of the lower classes "who were afeared of the lancet." Since each leech could consume about its own weight in blood before falling off, one leech was needed for each ounce of blood to be drawn. In many cases 20 to 30 leeches would be applied. The leeches were usually rented from the apothecaries, although some physicians proudly maintained their own prized collections.

Botany was still important, as it was in ancient times through the period preceding the Renaissance, when Albertus Magnus conducted his studies. The writings of Paracelsus in the first half of the sixteenth century added more remedies to the shelves of the apothecaries. Paracelsus, shown in Figure 28.7, was an alchemist early in his unusual career. He later taught that the first aim of science was not to transmute gold but to prepare new medicines for the benefit of mankind. During the Renaissance, his approach led to a decrease in many of the ineffective medicines from the Galenic pharmacy. Nevertheless, some truly exotic

Figure 28.3. Cover plate of Hieronymous Brunschwig's *Buch der Cirurgia,* published in Strassburg in 1497.

Figure 28.5. Shelves filled with jars of medicines. (Brunschwig, 1497.)

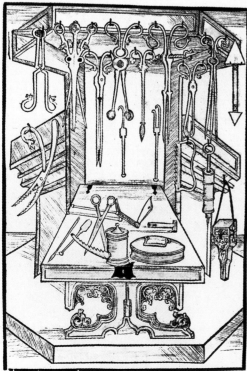

Figure 28.4. Illustration from Hieronymous Brunschwig's (1497) *Buch der Cirurgia* showing surgical and medical instruments.

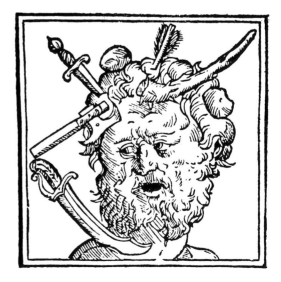

Figure 28.6. Cover page of Berengario da Carpi's (1535) *Tractatus de fractura cranei,* the first comprehensive book dealing exclusively with head injuries. The author was an innovative surgeon who developed many surgical tools.

Figure 28.7. Paracelsus (1493–1541). Although Paracelsus taught that the first aim of science was to prepare new medicines for the benefit of man, he still allowed astrology to play a role in medicine. The Latin under his name in this plate states that he was a "rumor monger."

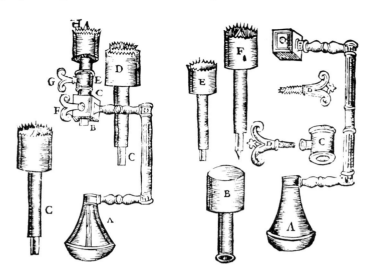

Figure 28.8. Trepan instruments used by Ambroise Paré in the mid-sixteenth century.

medicines, like "unicorn's horn" and Egyptian mummy powder, would not be withdrawn from the jars on the shelves.

Typical of the "older" remedies from this period was a potion used to treat epilepsy. It called for a mixture of mistletoe, powdered skull from a man, and peony roots and seeds gathered in the wane of the moon. Because peony was noted for its dryness, it was chosen to neutralize the cold, moist brain.

Although Paracelsus was revolutionary in his search for new medicines, he, like many others at this time, remained faithful to the ancient astrological beliefs that the moon ruled the brain and the sun ruled the heart (Garrison, 1929). Additional parts of this archaic theory maintained that Saturn controlled the spleen, Mercury the lungs, Mars the bile, and Venus the kidneys. Moreover, the planets remained associated with different metals. These fanciful ideas had more than a passing influence on treatments during the sixteenth century.

To say the least, the sixteenth century was a period of strange contrasts. This can be appreciated by looking at the writings of Ambroise Paré, the French physician who began his remarkable medical career as a barber-surgeon, without a university education or knowledge of Latin. To his credit, Paré made many notable advances in the development of artificial limbs. He also made many contributions to surgery, including a variety of new instruments (see Figure 28.8) and one of the first treatises on gunshot wounds.[6] He directed surgeons to find and remove bone fragments that might have been driven into the brain.

In contrast to these achievements, Paré believed in venesection and in a variety of questionable concoctions. He mentioned a medicine made by boiling newly whelped puppies in oil of lilies and then mixing in earthworms prepared in turpentine of Venice (Bakay, 1985). He was introduced to this incredible prescription by an Italian surgeon. Paré also believed in theriac, the ancient cure-all from the Roman period.

One of Paré's (1564) most interesting cases dealt with a man who asked for a brain transplant.[7]

> A gentleman otherwise well, had the idea his brain was rotten. He went to the King, begging him to command M. le Grand, Physician, M. Pigray, King's Surgeon-in-Ordinary, and myself, to open his head, remove his diseased brain and replace it with another. (Translated in Hamby, 1960, 8)

Paré provided few clues to suggest that anything like a transplant was attempted (see Finger, 1990). He wrote: "We did many things to him, but it was impossible for us to restore his brain."[8] Ironically, a portrait of Paré from this time shows him next to a bottle containing a fetus (Figure 28.9). More than 400 years would pass before it would be recognized that the use of fetal brain tissue could markedly increase the chances of brain graft survival.

Like Paré, Wilhelm Fabry von Hilden (1560–1634), who settled in Bern as official wound surgeon, also devised a number of new tools for operating on the brain. They included a device for raising depressed bone with a screw, hook, and lever. He also used screws and pins to prevent the broken bone from again collapsing, and devised a number of techniques for use with head-injured children.[9] Some

---

[6]Another early treatise on gunshot wounds was written in 1552 by Maggius in Italy.

[7]Investigators began to try to graft brain tissue in 1900. The aim was to see the general viability of transplanted brain tissue and to determine whether the grafts could stimulate regeneration (Forssman, 1900; Saltykow, 1905; Thompson, 1890). For historical reviews, see Björklund and Stenevi (1985) and Gash (1984). Relative to these "early" experiments on rabbits, dogs, and cats, the idea that brain tissue transplantation might serve as a therapeutic intervention for brain disorders, such as Parkinson's disease, was a much newer concept (Perlow et al, 1979).

[8]Among Paré's usual treatments for brain disease and injury were drawing palettes of blood, cauterizing with the seton, applying poultices, and administering drugs.

[9]Although Fabry von Hilden had to deal with bullet wounds, he did not attempt to remove lead from the brain.

Figure 28.9.   Ambroise Paré (1510–1590) next to a jar appearing to contain a human fetus.

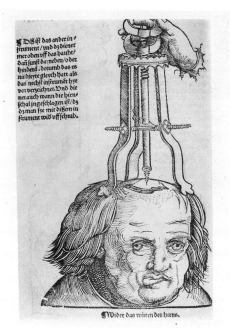

Figure 28.10.   Elevating a depressed skull fracture. (From Hans von Gersssdorff, *Feldbuch der Wundartzney,* published in 1517.)

and extracted the bone from the wound, whereupon the patient began to speak at once saying, "Praise God, I am cured." This drew much applause from the doctors, nobles, and attendants who were present. (Translated in Benton and Joynt, 1960, 208)

It should be noted that Andreas Vesalius, who figured prominently in the history of neuroanatomy, also performed craniotomies. In 1562, when he was a physician to the Spanish court, Don Carlos (1545–1568), the son of Philip II (1527–1598), fell and hit his head while indiscretely trying to corner a kitchen maid. Don Carlos went into coma, and Vesalius

of these tools, which come from Fabry von Hilden and other sources, are shown in Figures 28.10 through 28.13. Figure 28.14 shows a craniotomy in progress

In the mid-1600s, the brace and drill stock, to which a circular saw or sharp perforator was screwed, was one type of trepan used to open the skull. The mechanical cogwheel trepan, invented in Antwerp in 1575, was also available (Thompson, 1938; Figure 28.15). The drill was held in one hand and cranked with the other, much like a modern hand drill. But because the cogwheel trepan was very heavy, its popularity was not great among the surgeons of this period.

A number of good examples of successful brain surgeries can be found in sixteenth-century books. One of the better descriptions was provided by Nicolo Massa in 1558. He described the successful effects of an operation to remove pieces of bone from the skull:

> Also restored to health by my efforts was a handsome young man, Marcus Goro, who was wounded by a sharp point of a spear. . . . There was a fracture not only of the cranial bone but also of the meninges and the brain substance extending to the basal bone. Because this was protruding, a silver tube, which extended to the basal bone and exerted pressure on it, was placed in the wound. In addition to all his other misfortunes, the young man was speechless for 8 days. . . . I concluded that the reason for the loss of voice was that part of the bone was lodged in the brain. I took an instrument . . .

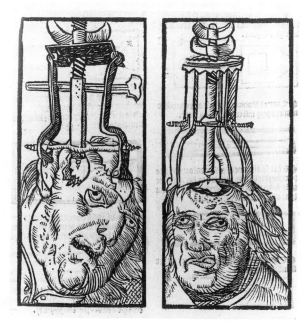

Figure 28.11.   Woodcut of two heads with drilling instruments applied to them. (From Brunschwig, 1525.)

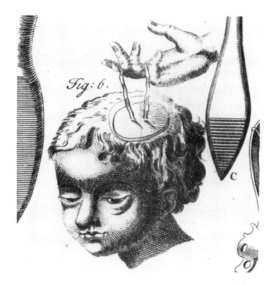

Figure 28.12. Elevating bone in a child with a cranial fracture. The method shown used plaster with a loop of string. The technique is that of Fabry von Hilden, who lived from 1560 to 1624, although this illustration is from Lorenz Heister, 1739.

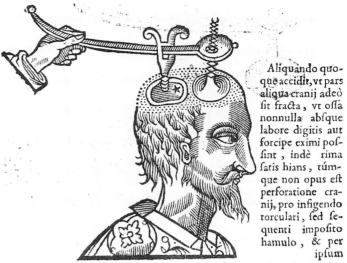

Aliquando quoque accidit, vt pars aliqua cranij adeò sit fracta, vt ossa nonnulla absque labore digitis aut forcipe eximi possint, indè rima satis hians, túmque non opus est perforatione cranij, pro infigendo torculari, sed sequenti imposito hamulo, & per ipsum

Figure 28.13. Device using a lever to correct a depressed skull fracture. (From Fabry von Hilden, 1641.)

Figure 28.14. Woodcut showing a craniotomy in progress. Two people assist the surgeon, while a man warms a cloth, a woman prays, and two others watch. (From Della Croce, 1573.)

Figure 28.15. A mechanical drill introduced by Matthia Narvatio in 1575.

recommended opening his skull to try to remove what might have been an epidural blood clot. The young man eventually recovered and attributed his salvation not to the bevy of physicians who had attended him, but to the cadaver of San Diego. The dead friar's body had been placed next to his bed by the people of the town, with the hope that it would stimulate a divine miracle (Bakay, 1985).

SEVENTEENTH-CENTURY TREATMENTS

Surgery to remove bone and objects pressing on or entering the brain continued to grow in popularity after the Renaissance. This can be seen in the wonderful description by Robert Boyle (1691) of a melancholy knight who underwent surgery to remove a flake of bone lodged in his brain (see Chapter 2). Nevertheless, medicine was still not on very solid footing, especially with regard to pharmacologic treatments of injuries and diseases. For example, Boyle also recommended some strange prescriptions to his patients. To cite one: "Take about one dram of *Album Graecum* or white dog's turd, burnt to a perfect whiteness, and with about one ounce of Honey of Roses, or clarified honey, make thereof a linctus to be very slowly let down the throat" (see Wright, 1898, p. 136).

Boyle was not alone when it came to the use of strange and unusual medicines. Matthias Purmann (1648– 1721), one of his contemporaries, recommended salves made of earthworms, hog brains, mummy powders, and the "moss of a man's skull that was either killed or hanged, and gathered when the star of Venus predominates." Purmann's call for mummy powders can be traced to the early Egyptian belief in the rejuvenating powers and magical abilities of mummification. These powders had been put on wounds, or even served in beverages, for centuries. Of course, to be effective, real Egyptian mummies had to be used. In fact, very specific parts of the mummy were often called for. But as prices for real mummy powders increased, a flourishing trade in fakes came into existence (Bakay, 1985).

"Dragon's blood" was another pharmacological ingredient associated with fakery, or at best, with half truths. This was a fairly common ingredient in prescriptions of the time. Actually, it had nothing to do with animals of any sort. It came from *Dracona draco,* the dragon tree. Arab merchants, who profited handsomely selling "dragon's blood," did nothing to dispel the European belief that the "blood" really came from fierce Eastern dragons.

Many if not most medicines during the seventeenth century were still tied to the Greco-Roman theory of the four humors. François du Port (c. 1540–1624), a doctor who lived during the reign of Louis XIII (1601–1643) of France, was representative of the conservative physicians who were not about to veer from the belief that every disease could be attributed to the excess or deficiency of one of the four humors. In a book written in 1694, du Port declared:

> Etna's heat is nullified by glacial cold, dryness and moisture interact, fire absorbs fluid and renders dry, the thin sunders the thick, hard iron resists what is soft, the rough opposes the smooth, the porous the solid, what is closed or joined is contrary to what is open, the sticky to the polished. Thus, lettuce overcomes the bile: calamint and thyme digest and subtilize thick phlegm: sweet almond and ointments soften whatever is rough that they touch: mel roseum removes the slough from a foul ulcer: sticky willow, elm, sanicle and bugle compress an open wound and restrain haemorrhage. Whenever purple humour distends the vessels, opening a vein is helpful: and any residues adherent in the passages are purged by vomiting, urine and sweat, or by the bowel, the nasal openings or the uterus, ridding the body of its humoural content. And if the fire devours and flourishes within the marrow of the bones, a healing draught will cool and water give relief, countering the body's drought: since anything whatever is expelled by its enemy. So that all that penetrates the human body and acts upon it is driven out by an ill that is contrary. (84–85)

Du Port used this logic to treat syncope, apoplexy, and paralysis. He recommended massaging the weakened part of the body with balsams and olive oil, bleeding from the head and tongue, and evacuating the bowels with senna, turbith, and agaric. The list of additional agents he used to treat paralyses was extensive. For example, du Port recommended that one should follow these directives:

> Sweeten the water drunk with sugar and cinnamon, or infuse with hydromel, sugar and iris, for wine alone is bad for the nerves. Roasts are useful food. Sage, marjoram, calendula, the true primula, wild thyme, origanum, laurel, dwarf elder, garden thyme and juniper, boiled together with a fox in water, make a bath to immerse the paralytic: or you can make a fermentation by soaking towels in the hot liquor, and let the vapours given off be absorbed by the skin and mouth and nose. . . . The so-called sarsaparilla, macerated with guaiacum in warm water, makes a drink which dislodges the coarse humors of the body and expels the sweat through the fine pores of the skin. (109)

One of the most enlightened surgeons of the seventeenth century was Joannis Scultetus (1595–1645), who practiced during the Thirty Years' War (1618–1648). Scultetus believed in "informed consent," as can be seen in the following quotation:

> The learned and well disposed surgeon should tell the patient and point out the danger of the operation. . . . The

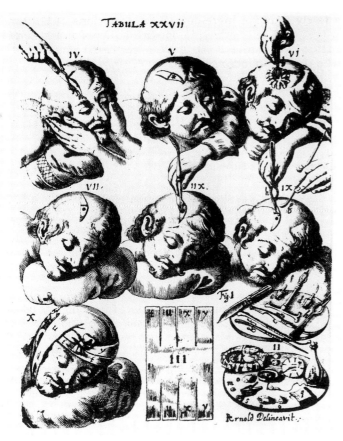

Figure 28.16. Illustration from Scultetus (1653) showing a craniotomy.

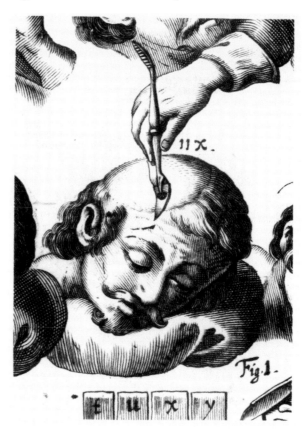

Figure 28.17. Detail of the center figure in the illustration from Scultetus (1653) depicting a craniotomy.

surgeon in such a case (demanding trephination) should immediately make a statement in which he predicts the danger to the patient's life, to avoid stupid gossip by laymen. (Cited in Bakay, 1985, 90)

Scultetus (1653) described craniotomies and provided beautiful illustrations of how they were performed, such as Figure 28.16, a detail of which is shown in Figure 28.17. Scultetus performed craniotomies only after making sure that his patient's condition had stabilized. He did not believe in the old superstition that tied the operation to the size of the moon. He described a *trioploides,* a device with three legs and a screw for raising depressed fractures. He also used a trepan, which had an attachment that prevented the drill from slipping into the brain. Calling the drill with a center pin the "male" and the one without it a "female," he wrote:

Before we use females we must make some print upon the skull with the male so that the females may stand faster upon it. Now for to trepan the skull, the Chyrurgian must have at hand at least three trepans exactly equal to the other; one male and two females, so that he may oft-times change them. (Translated in Thompson, 1938, 732)

Scultetus packed his patients' wounds with flax or hemp that had been dipped in egg white and rose oil. If done properly, the belief was that *pus laudable* would replace the ominous dark pus that would drain out. For contusions, Scultetus recommended local applications of hides of animals that were just killed, such as dogs or mice. They were thought to create heat, which would decrease swelling within a day. Scultetus (1653) wrote that "since it is occasionally difficult to get hold of a dog or a mouse skin, I personally prefer the skin of a lamb which can be easily obtained at any hour and has the same properties."

One other surgeon from this period who deserves mention is Richard Wiseman (c. 1621–1676), who served the Dutch, English, and Spanish (Bakay, 1987b). Wiseman dealt mostly with battle wounds. In *Severall Chirurgicall Treatises,* published in 1676, he described over 600 wound cases. Some were from swords, halberds, axes, and related weapons, but he also had to contend with the marked increase in the use of gunshot.

Wiseman had less regard than many of his contemporaries for some of the widely prescribed drugs: "For neither Crabs-eyes, nor any Medicine inwardly prescribed, or Wine outwardly applied with Lint will help." He recommended removing epidural hematomas and opening the dura to get to subdural hematomas.

Still, old traditions would not die. Wiseman accepted that part of classical doctrine which related the appropriate timing of trepanation to astrological conditions. Thus, he wrote that he delayed surgery on one of his patients after he was stabbed in the head because the moon was not right. Further, he considered most large brain injuries hopeless, explaining that they permitted the air to "overcool" the brain. In addition, he did not believe one could survive gunshot wounds that penetrated the meninges. Thus, although Wiseman spoke about a new beginning, he still kept one foot anchored in the waters of the past.

REFERENCES

Al-Rodhan, N. R. F., and Fox, J. L. 1986. Al-Zahrawi and Arabian neurosurgery. *Surgical Neurology, 26,* 92–95.

Amundsen, D. W. 1972. Medieval canon law on medical and surgical practice by the clergy. *Bulletin of the History of Medicine, 52,* 22–44.

Andromachus. 1607. *De theriaca.* . . . In F. Tidicaeo, *Andromachi Senioris.* . . . Thorvnii: Ex officina typ. A Cotenij.

Andromachus. 1668. *Theriaque d'Andromachus.* . . . Paris: Chez O. de Varennes.

Aristotle. 1956. *Historia animalium.* (D. W. Thompson, Trans.). In J. A. Smith and W. D. Ross (Eds.), *The Works of Aristotle* (Vol. IV). London: Oxford University Press.

Bakay, L. 1971. *The Treatment of Head Injuries in the Thirty Years' War (1618–1648), Joannis Scultetus and His Age.* Springfield, IL: Charles C Thomas.

Bakay, L. 1985. *An Early History of Craniotomy.* Springfield, IL: Charles C Thomas.

Bakay, L. 1986. Surgical treatment of brain laceration in 1513. *Surgical Neurology, 25,* 568–570.

Bakay, L. 1987a. *Neurosurgeons of the Past.* Springfield, IL: Charles C Thomas.

Bakay, L. 1987b. Richard Wiseman, a Royalist surgeon of the English Civil War. *Surgical Neurology, 27,* 415–418.

Benton, A. L., and Joynt, R. 1960. Early descriptions of aphasia. *Archives of Neurology, 43,* 1065–1072.

Berengario da Carpi, J. 1535. *Tractatus de fractura calve sive cranei* (3rd ed.). Bologna: de Benedictis.

Björklund, A., and Stenevi, U. 1985. Intracerebral neural grafting: A historical perspective. In A. Björklund and U. Stenevi (Eds.), *Neural Grafting in the Mammalian CNS.* Amsterdam: Elsevier. Pp. 2–14.

Blumenbach, J. F. 1810. *Abbildungen naturhistorischen Gegenstand.* Göttingen: H. Dieterich.

Boyle, R. 1691. *Experimenta et observationes physicae.* London: Printed for John Taylor at the Ship and John Wyat in St. Paul's Church-Yard.

Breasted, J. H. 1930. *The Edwin Smith Surgical Papyrus.* Chicago: University of Chicago Press.

Brunschwig, H. 1497. *Buch der Cirurgia.* Strasbourg: Grüninger.

Brunschwig, H. 1525. *The Noble Experyence of the Vertuous Handy Warke of Surgeri.* London: P. Trevaris.

Celsus, A. C. 1935–1938. *De medicina* (3 vols.). Cambridge, MA: Loeb Classical Library.

Clarke, E. 1963. Apoplexy in the Hippocratic writings. *Bulletin of the History of Medicine, 37,* 301–314.

Cooperman, W. S. C. 1960. *Doctors and Disease in Tudor Times.* London: Dawson's of Pall Mall.

Dawson, W. R. 1929. *Magician and Leech.* London: Methuen.

du Port, F. 1694. *La Décade de Médecine ou Le Médecin des Riches et des Pauvres.* Paris: Laurent d'Houry. Translated by M. Diehl as, *The Decade of Medicine: The Physician of the Rich and the Poor.* Berlin-Heidelberg: Springer, 1988.

Finger, S. 1990. A 16th century request for brain tissue transplantation. *Restorative Neurology and Neuroscience, 1,* 367–368.

Forssman, J. 1900. Zur Kenntniss des Neurotropismus. *Zeigler's Beiträge zur pathologische Anatomie, 27,* 407–430.

Garrison, F. H. 1929. *An Introduction to the History of Medicine.* Philadelphia, PA: W. B. Saunders.

Gash, D. M. 1984. Neural transplants in mammals: A historical overview. In J. R. Sladek and D. M. Gash (Eds.), *Neural Transplants: Development and Function.* New York: Plenum Press. Pp. 1–12.

Gersdorff, H. 1517. *Feldbuch der Wundartzney.* Strassburg: J. Schott.

Ghalioungui, P. 1963. *Magic and Medical Science in Ancient Egypt.* London: Hodder and Stoughton.

Gordon, B. L. 1939. Oculists and occultists. Demonology and the eye. *Archives of Ophthalmology, 22,* 25–65.

Hai, A. A., and Ahmed, S. W. 1981. Islamic legacy to modern surgery. In *Proceedings of the First International Conference on Islamic Medicine* (Vol. 1, 2nd ed.). Kuwait: Kuwait Ministry of Public Health. Pp. 290–298.

Hamby, W. B. 1960. *The Case Reports and Autopsy Records of Ambroise Paré.* Springfield, IL: Charles C Thomas.

Herodotus. 1972. *The Histories.* (A. de Sélincourt, Trans., with notes by A. R. Burn). London: Penguin Books.

Hildanus, F. 1641. *Observationum & curatonium chirurgicarum centuriae, nunc primum simul in unum opus congestae.* Leyden: Huguetan.

Hippocrates. 1952. *On Injuries of the Head.* In *Hippocrates and Galen. Great Books of the Western World* (Vol. 10). Chicago: William Benton. Pp. 63–70.

Horne, J. F. 1894. *Trephining in Its Ancient and Modern Aspect.* London: John Bale and Sons.

Izquierdo, J. M., Coca, J. M., Diaz de Tuesta, J., San Emeterio, F., and Tejada, J. 1981. Contribution of the Islamic-Spanish surgeon Albucasis to neurosurgery. In *Proceedings of the First International Conference on Islamic Medicine* (Vol. 1, 2nd ed.). Kuwait: Kuwait Ministry of Public Health. Pp. 322–324.

Jarcho, S. 1972. Medical numismatic notes, VII: Mithridates IV. *Bulletin of the New York Academy of Medicine, 48,* 1059–1064.

Kellaway, P. 1946. The part played by electric fish in the early history of bioelectricity and electrotherapy. *Bulletin of the History of Medicine, 20,* 112–137.

Maggius. 1552. *De vulnerum a bombardorum et sclopetorum globulis illatorum et de eorum symptomatorum curatione tractus.* Bologna: Bonardo.

Majno, G. 1975. *The Healing Hand.* Cambridge, MA: Harvard University Press.

Mapleson, T. 1813. *Treatise on the Art of Cupping.* London: Printed for the author by G. Sidney.

Massa, N. 1558. *Epistolarum medicinalium tomus primus.* Venetiis: ex Officina Stellae Iordani Zilleti.

Mettler, C. C. 1947. *History of Medicine.* Philadelphia, PA: Blakiston.

Mirfield, J. 1969. *Surgery. A Translation of His Breviarium Bartholomei.* (J. B. Colton, II, Trans.). New York: Hafner.

Paré, A. 1564. *Ten Books of Surgery.* (R. W. Linker and N. Womack, Trans.). Athens, GA: University of Georgia Press, 1969.

Perlow, M. J., Freed, W. J., Hoffer, B. J., Seiger, A., Olson, L., and Wyatt, R. J. 1979. Brain grafts reduce motor abnormalities produced by destruction of nigrostriatal dopamine system. *Science, 204,* 643–647.

Phillips, E. D. 1973. *Aspects of Greek Medicine.* New York: St. Martin's Press.

Plato. 1985. *Meno.* (Trans. with notes by R. W. Sharpless). Warminster U. K.: Aris and Phillips Ltd.

Plinius Secundus. 1601. *The History of the World.* London: Islip.

Pliny the Elder. 1963. *Natural History.* (W. H. S. Jones, Trans.). Loeb Classical Library. Cambridge, MA: Harvard University Press.

Purmann, M. G. 1706. *Chirurgia curiosa.* (J. Sprengell, Trans.). London: Browne, Smith and Brown.

Saltykow, S. 1905. Versuche über Gehirnreplantation, sugleich ein Beitrag sur Kenntniss der Vorgänge an den zelligen Gehirnelementen. *Archiv für Psychiatrie und Nervenkrankheit, 40,* 329–388.

Scribonius Largus. 1529. *De compositionibus medicamentorum. Liber unus.* (J. Ruel, Ed.). Paris: Wechel.

Scultetus, J. 1653. *Armamentarium chirurgicum.* Ulm: Kühnen. (Reprinted for *Editions Medicina Rara Ltd.* by Agathon Press, Baiersbroon, Germany).

Shanks, N. J., and Al-Kalai, D. 1984. Arabian medicine in the Middle Ages. *Journal of the Royal Society of Medicine, 77,* 60–65.

Sterling, W. 1902. *Some Apostles of Physiology.* London: Waterlow and Sons.

Thompson, C. J. S. 1938. The evolution and development of surgical instruments. *British Journal of Surgery, 25,* 726–734.

Thompson, R. C. 1923. *Assyrian Medical Texts.* Milford, U.K.: Oxford University Press.

Thompson, R. C. 1924. Assyrian medical texts. *Proceedings of the Royal Society of Medicine, 17,* 1–34.

Thompson, W. G. 1890. Successful brain grafting. *New York Journal of Medicine, 51,* 701–702.

Thorwald, J. 1963. *Science and the Secrets of Early Medicine.* New York: Harcourt, Brace and World.

Wellcome, H. S. 1912. *Anglo-Saxon Leechcraft.* London: Burroughs Wellcome and Company.

Wiseman, R. 1676. *Severall Chirurgicall Treatises.* London: Royston.

Wright, J. 1898. *The Nose and Throat in Medical History.* St. Louis, MO: Matthews.

Chapter 29
Treatments and Therapies: From 1700 to World War I

> At first I permitted him (the patient, Mr. P., who could speak only in an imperfect whisper) to receive several shocks from a Leyden jar . . . causing the fluid to pass from one hand to the other . . . I then gave the jar very nearly its full charge, and passed it two or three times through the breast. He spoke in a very audible manner, saying "my voice is coming." His voice then resembled that of a man somewhat hoarse with a cold. He added, "I feel a little of it (meaning the numbness) yet about the lower part of my breast." I then passed two or three similar charges through the part in which the numbness was felt. By this time, Mr. P. spoke with his natural voice.
>
> Parker Cleaveland, 1813

HEAD AND BRAIN INJURIES IN THE EIGHTEENTH CENTURY

François Quesnay (1694–1774), an eighteenth-century apprentice to a barber-surgeon, graduated from the College of Surgery in 1718. He advanced in his profession and became physician to Louis XV (1710–1774) and his mistress, Madame de Pompadour (1721–1764).[1] Quesnay, who saved the King's life, lived during the Age of Enlightenment and was decidedly more optimistic about treating brain injuries than most of his predecessors and contemporaries. In 1743, he wrote:

> The brain is formed of such delicate substance, and its functions are generally so important to life, that it would seem as if the smallest shock, or the slightest wound, would cause irreparable injury, and attack life at its fountainhead. . . . However, we found that wounds of this organ, especially those of the cortical and medullary substances, heal almost as readily as wounds of other viscera. (Cited in Bakay, 1987, 53)

Quesnay provided experimental support for his assessment by putting nails into the brains of dogs and showing they could survive. Their wounds were kept clean by pouring Rhine wine or arquebusade water on them. The latter was made of vinegar, alcohol, sugar, dilute sulfuric acid, and lead salts. As far as Quesnay was concerned, wounds of the cortical and medullary substances were not that different from wounds of the viscera.

Quesnay was one of the first physicians to suggest that one should remove not just surface meningiomas but cancers of the brain itself.[2] This practice was probably based on his successes in removing growths from the meninges, as well as abscesses and objects from the brain substance. In 1743, he wrote: "We may attempt certain operations of the brain itself in desperate cases . . . we may, for instance, open an abscess in the substance of the brain, or, when the circumstances require it, seek for a foreign body." He advised going beyond a simple trepanation and exploring for possible tumors and abscesses in cases where the patient would probably die miserably if left alone. He noted that the brain itself was insensible to pain and that the surgery, "should it fall on an abscess, it might save the patient's life." Still, whether Quesnay ever operated upon intracerebral tumors is not certain (Bakay, 1985).

A similar position was taken by Antoine Louis (1723–1792), who wrote an essay in 1774 on the surgical removal of tumors and who clearly operated to remove cancers under the meninges. Nevertheless, Louis dealt mostly with meningiomas. He diagnosed these tumors primarily by skull deformations, but also relied on symptoms such as visual defects, headaches, weakness of the limbs, and convulsions. It was generally believed that these tumors, which were often called fungus growths, were caused by head injuries. At the time, attempts to destroy tumors by applying caustic solutions (e.g., potash) often were catastrophic, whereas wine, herbs, rose honey, and related medicines were ineffective. Still, not everybody accepted Louis' surgical recommendations. Most patients refused to undergo surgery, and many physicians did not think operating would save the patient's life.

Louis was also interested in contrecoup injuries and organized a session on this subject at the April 10, 1766 meeting of the *Académie Royale de Chirurgie*. Twenty-six papers were received, but no prize was awarded.[3] The contest for the *Prix de l'Académie Royale de Chirurgie* was rescheduled for 1768. This time, Louis Sebastian Saucerotte wrote the

---

[1]Later in his life, Quesnay's interest in medicine waned, and he devoted his attention to economics.

[2]Surgery to remove tumors may have begun in 927, when two brothers removed a tumor from the King of Dhar. The surgeons used one drug, called *Samohini*, to anesthetize the King and another agent to revive him (O'Connor and Walker, 1967; Thorwald, 1963).

[3]The author of the paper chosen to receive the 1766 *Prix de l'Académie Royale de Chirurgie* refused to make the recommended changes and was not given the coveted award (Bakay, 1987).

prize-winning paper on contrecoup injuries. In his 47-page essay, he described 27 patients, some of whom he had operated upon. He also described a series of experiments on dogs.

Saucerotte suggested that patients with contrecoup injuries should undergo bloodletting from the veins of the upper body, dietary changes, and warm infusions of *incisives et céphaliques*. He also advocated rubbing their shaved heads with certain herbs. Saucerotte said that these patients should not lie on the paralyzed side because a hematoma could migrate to the opposite side. If a hematoma formed and did not dissipate rapidly, the prize-winning author felt that surgery was justified.

As the eighteenth century progressed, surgeons found they had to deal more and more with gunshot wounds. At first, it was thought that gunshot was poisonous or diabolical. After all, explosives were made of saltpeter, sulfur, and charcoal. This opinion gradually changed and carried little weight after Benjamin Bell (1749–1806) and others made the case that gunshot was not in itself poisonous (see Bell, 1782–87).

CUPPING AND LEECHING IN THE NINETEENTH CENTURY

During the first half of the nineteenth century, leeching remained a common treatment for brain diseases (Pfeiffer, 1985). In fact, leeches represented the remedy of choice for "cerebral congestion" and were applied to the temples for relief of seizures. Leeches were also used to treat injuries and inflammations of the nerves, such as those arising from amputated limb stumps.

Marc Dax made a reference to leeching in his 1836 paper on the role of the left hemisphere in speech (see Chapter 26). Dax wrote:

> A lady fainted and fell from the chair in which she had been sitting. Although I left promptly to see her, when I appeared, she had already recovered. Had she suffered a fainting spell or a syncope? Did the short duration of the attack permit one to suspect an apoplectic affection? I did not think so at first but the patient, in describing to me what she had experienced, said: "After regaining consciousness, I was unable to speak for a moment." These words were a ray of light to me, and, two days later, called in great haste to the same patient (who had just suffered a similar accident, but one much more intense than the first, because I found her completely mute this time), I had no need for reflection to know the nature, the site or the treatment of this illness. I promptly applied a large number of leeches to the left temple and in a few minutes, as the blood flowed, her speech was gradually restored. A half hour later, the patient had recovered and, taking some precautions, she has continued to enjoy good health for several years. (Translated by Joynt and Benton, 1964, 853)

Léon Rostan, a mid-nineteenth-century physician with a strong interest in aphasiology also described the beneficial effects of leeching. He was not just praising the effects of leeching patients under his care, however. He spoke about the role of the leech in treating his own aphasic disorder. After he became ill, he was unable to speak, write, or comprehend written material. Rostan stated he began to regain his speech soon after he was bled and that within 12 hours his speech had returned to normal. Armand Trousseau (1864), who described Rostan's case, cited additional instances in which leeching was effective for restoring lost speech.

Leeches were used so frequently in the British Isles in the first few decades of the nineteenth century that they became a scarce commodity. Imports from Lisbon, Bordeaux, and other places had to be purchased to meet the demand. During this period, one commentator even remarked that "100 foreign leeches were employed for every English one" (see Pfeiffer, 1985, p. 47). The enhanced demand caused prices to rise sharply, and many people who would have undergone leeching reluctantly turned to other remedies because they could not pay the high prices. The rise in the cost of leeches also led to articles in the medical and scientific journals telling apothecaries and physicians how to start and maintain their own leech colonies.

Why was the leech so popular? The answer probably lies in two basic observations. First, drawing approximately 20 ounces of blood did not seem to make patients any worse. And second, leeching was occasionally correlated with improvement.

As for their beneficial effects, the following points may be made. First, leeches can draw blood from hematomas. Second, although nineteenth-century scientists did not realize it, leeches secrete hirudin (an anticoagulant) when they suck blood. The hirudin could have reduced some blood clots and obstructions and thinned the blood, thus increasing cerebral blood flow. Third, a certain percentage of cases might have been showing coincidental spontaneous recovery. And fourth, there were probably placebo effects.

The two main types of cupping used in Ancient Greece and Rome also remained popular in the first half of the nineteenth century (Bayfield, 1823). One was dry cupping, which involved heating a cup, now likely to be made of glass, and then applying it to the skin to create a vacuum as the air cooled. The other was wet cupping, the procedure which called for lightly scarring the skin with a knife or an instrument called a scarificator, so that blood could be drawn more easily into the cup.

At this time, Thomas Mapleson was the official cupper to the English King, the Westminister Hospital, and the St. Pancras Parochial Infirmary. In 1813, he wrote *A Treatise on the Art of Cupping*. Mapleson stated that he was distressed by the fact that cupping—which was performed by regular surgeons until just a century earlier—was now in the hands of unskilled practitioners. He especially disapproved of the growing trend of bathhouses to advertise cupping services in cheap popular magazines, including *The Tattler* and *The Spectator*.

Mapleson looked upon cupping somewhat like a sacred gift and advocated it for a wide range of disorders. His list included headaches, apoplexy, phrenitis ("inflammation of the brain"), insanity, and hydrocephalus. He was especially optimistic about cupping as a treatment for epilepsy.

> In EPILEPSY, taking blood from the immediate vicinity of the brain is often very beneficial. In a case of epilepsy, which recently occurred at the Westminster Hospital . . . I applied cupping glasses, under the direction of Dr. Buchan; and so sensible was the patient (a young man) of relief,

that he frequently solicited to be cupped, even when it was not ordered by his physician, and by this means, I believe, alone, he obtained a perfect cure. (1813, 38)

Mapleson added that the small loss of blood did not shorten a patient's life. He cited another physician who was cupped twice a year and who lived "upwards of ninety." In his professional opinion, drawing 6 to 20 ounces of blood properly and regularly was all that was needed to assure good health and longevity.

The drawing of blood as a therapeutic procedure, and the ancient theory of the four humors, diminished considerably in Western Europe after the first quarter of the nineteenth century. Cupping and related archaic procedures, however, were slower to succumb in Eastern Europe and in many undeveloped countries (Lennon and Davenport, 1987). For example, both wet and dry cupping remained popular in North Africa (e.g., Tunisia and Nigeria) as a means of treating aches, pains, inflammations, and congestive states well into the twentieth century.

ANIMAL ELECTRICITY AND ELECTROTHERAPY BEFORE GALVANI

As noted in Chapter 28, the Romans used shock from fish with electric organs to treat illnesses. The fish included electric eels and the black torpedo. Typically, the patient would stand barefoot on one of these creatures until all of its electrical power was spent. Among the disorders the Roman physicians attempted to treat in this way were paralysis, headaches, gout, and arthritis.

The shocks imparted by the fish may have contributed to the notion of vitalism, which holds that animate objects possess a "force of life," or a spirit, that distinguishes them from inanimate objects. Both Aristotle and Galen believed in such a force, and the idea that the nervous system worked by animal spirits or "juices" that flowed through nerves continued to be accepted centuries later by Marcello Malpighi, Jan Swammerdam (1637–1680), René Descartes, Giovanni Borelli (1608–1679), and Herman Boerhaave.

Stephen Hales (1677–1761) and Alexander Monro (1697–1762) were among the first researchers to hint that electricity—a term recently introduced into the vocabulary—could be the fluid of the nervous system.[4] Felice Fontana (1730–1805) took this observation a step further when he suggested that the brain acted like an electrical generator and that the nerves were its conducting wires.

The first true scientific studies of electric fish were conducted by John Walsh (1725–1795). His experiments took place on the Atlantic coast of France. They were reported in 1772 to a scientific academy in La Rochelle and were published one year later. Walsh thought that the shock from the fish was due to the release of compressed electrical fluid. But because this shock was not accompanied by light or sound, many scientists continued to doubt whether living things could really produce electricity. Finally, in 1847, Carlo Matteuci (1811–1862) demonstrated convinc-

ingly that the torpedo fish could, in fact, generate a spark, and the debates abated.

Special machines for creating electricity appeared in the middle of the seventeenth century. The early friction machines were made of globes of sulphur, porcelain, or glass and later from revolving disks.[5] As early as 1672, Otto von Guericke (1602–1686) conducted experiments in Magdebourg on attraction and repulsion with his "globe of sulphur."

On the basis of early experiments with manmade electricity, seventeenth-century physicians suggested that friction machines could be used to treat paralysis.

But what is the usefulness of electricity, for all things must have a usefulness, that is certain. We have seen it cannot be looked for either in theology or in jurisprudence, and therefore nothing is left but medicine. The best effect would be found in paralyzed limbs to restore sensation and reestablish the power of motion. (Kruger, 1744; cited in Petrofsky, 1988, 273)

In this context, Robert Whytt (1714–1766), one of England's most brilliant physicians, wrote:

A man aged 25, who, from a palsy of twelve years continuance, had lost all power of motion in his left arm, after trying other remedies in vain, at last had recourse to electricity; by every shock of which the muscles of this arm were made to contract; and the member itself, which was very much withered, after having been electrified for some weeks, became plumper. (1768, 14)

As major advances were made in the development of machines that could produce electrical charges, scientists became increasingly willing to try out their new devices in therapeutic settings. By the middle of the eighteenth century, a convent hospital for electrical therapy was established in Montpellier by François Boissier de Sauvages de la Croix.

Therapeutic applications were made easier with the invention and dissemination of the Leiden (Leyden) jar in 1745. The inventor was Petrus (Pieter) van Musschenbroek (1692–1761) of the University of Leiden, a pupil of Herman Boerhaave. The Leiden jar allowed electricity to be built up and conserved until released, a marked advance over machines that simply produced sparks by friction. The sparks produced by the Leiden jar and variations of this apparatus led to well over 100 publications on electrotherapy in the eighteenth century (Licht, 1967). Nevertheless, the first really "conclusive" demonstrations of contraction of skeletal muscles from the Leiden jar had to wait for Giambattista Beccaria (1716–1781) in 1753 and Leopoldo Caldani in 1756 (Clarke and Jacyna, 1987).

John Wesley (1703–1791), who founded the Methodist Church, was not a physician by training. This, however, did not dampen his enthusiasm for the new technology (Schiller, 1981). In *Primitive Physick,* which appeared in 1747, he listed 288 medical conditions he thought could be prevented or healed by electricity. Wesley wrote that he was "firmly persuaded there is no remedy in nature for nervous disorders of every kind, comparable to the proper and consistent use of the electrical machine" (1973 reprinting, p. 92). He

---

[4]The word "electricity" was introduced by William Gilbert (1546–1603), President of the Royal College of Physicians. He derived the word from *elektron,* the Greek word for "amber," which he used to produce frictional electricity.

[5]The role of friction machines in medicine has been examined by Brazier (1958), Pfeiffer (1985), and Schiller (1982).

suggested that 50 to 100 gentle shocks should be used for most conditions. For paralysis, he specifically recommended that "the palsey be electrified, daily, for three months, from the places wherein the nerves sprang." He added that he had not known a single instance in which this practice had caused harm. Wesley had been using his "electrical fire" for many years when he published *Electricity Made Plain and Useful by a Lover of Mankind and of Common Sense* in 1759.

Other notable individuals who were not physicians by training but who were enamored of electricity at this time were Benjamin Franklin (1706–1790) and Jean-Paul Marat (1743–1793). Franklin worked on electrical science from 1746 to 1752. In 1751, he published *Experiments and Observations on Electricity,* a pamphlet that made him famous not just in North America but in Europe. In fact, early believers in the electrical fire were sometimes called Franklinists. Nevertheless, on December 21, 1757, when the use of electricity to treat paralysis was already widespread, Franklin penned a letter to John Pringle (1707–1782) in which he expressed doubts about electricity as a treatment for this particular disorder. He wrote:

> I never knew any advantage from electricity in palsies that was permanent. And how far the apparent temporary advantage might arise from the exercise in the patients' journey, and coming daily to my house, or from the spirits given by the hope of success, enabling them to exert more strength in moving their limbs, I will not pretend to say. (Cited in Smith, 1905–7, 426)

As for Marat, he conducted many experiments on animals and humans. He reported success after using electricity for a variety of ailments, including paralysis and pain, as well as with "underdeveloped" children (Schiller, 1981). He published important works on electricity in 1782 *(Recherches Physiques sur l'Electricité)* and 1784 *(Mémoires sur l'Electricité Médicale).* The latter received first prize in a competition held by the Royal Academy of Rouen. Marat's work with medical electricity was overshadowed by his role in the French Revolution of 1779 and afterwards by his dramatic assassination in a bathtub at the hands of Charlotte Corday (1768–1793).

Small wonder that static machines soon found their way into many major hospitals. In England, for example, they were introduced into Middlesex Hospital in 1767 and St. Bartholomew's in 1777 (Garrison, 1929). Early in the nineteenth century, Thomas Addison (1793–1860) and Marshall Hall practiced galvanotherapy in Guy's Hospital.

THE ROLES OF GALVANI AND VOLTA

Luigi Galvani (1737–1798; shown in Figure 29.1) used frictional machines, primitive condensers, and the Leiden jar for his experiments with electricity, which began in the 1780s, if not earlier (Clarke and Jacyna, 1987). He studied with Leopoldo Caldani and was familiar with Benjamin Franklin's writings on friction, contact, and electrostatic conduction.

Although Galvani was a professor of anatomy at the University of Bologna, much of his work on the excitability of the nervous system (intrinsic animal electricity) was carried out at his house, where he kept some electrostatic

Figure 29.1. Luigi Galvani (1737–1798), a pioneer in the study of electrical phenomena in animals.

machines. On September 20, 1786, when he hung some fresh frog legs on brass hooks from the iron railings of his garden, he unexpectedly discovered the powers of "bimetallic electricity." This date has been hailed as the "birthday of electrophysiology" (Brazier, 1973).

Another of Galvani's famous encounters with animal electricity occurred in a room in which some of his pupils were amusing themselves with one of his electrical machines. One machine threw a spark at the same time that Galvani touched a nerves of a frog with a metal scalpel. The electrostatic effect caused the frog's body to be thrown into a violent convulsion.

In addition to his work on the peripheral nerves and musculature, Galvani tried to stimulate the brain electrically. He was unsuccessful in his endeavor to cause muscular contractions this way. So were Caldani and Fontana, who both tried similar experiments in the 1750s (see Brazier, 1958).

Galvani published his experiments in 1791. They had a major impact on the field, even though all were not entirely original. For example, his work on the generation of electricity with dissimilar metals was preceded by experiments by Johann Sulzer, who showed that two different metals applied to the tongue could produce "electrical tastes" (see Luciani, 1917). In addition, Galvani used the same frog leg preparation as Jan Swammerdam had used in 1666–67. But while Swammerdam induced contractions only by mechanical stimulation, Galvani was able to do this electrically.

Galvani looked upon the nerve as a conductor of electricity, much like a wire from a generator. His position was not radically different from that of the ancients, who viewed the brain as a secreting gland. Galvani, however, substituted electricity for animal spirits in his formulation. He believed the muscles contained animal electricity secreted by the brain and distributed by the nerves. The

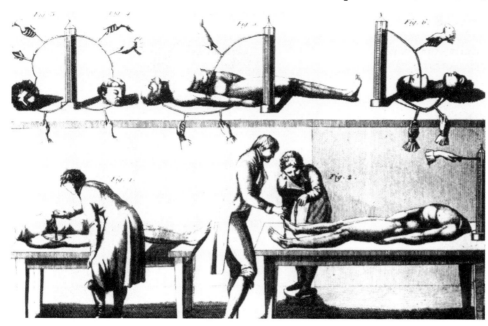

Figure 29.2. Plate from Aldini (1804) showing experiments with electricity being performed on humans after decapitation.

nerves were hypothesized to have oily coverings, which prevented leakage of this electricity.

Johann Friedrich Blumenbach, Galvani's contemporary, praised Galvani in the strongest possible terms:

> By the combined labours of experimental physiologists in different parts of the world, this branch of science was at length matured for giving birth to another discovery, which will probably be found of equal importance, in explaining the phenomena, and in removing the diseases of the animal system, with that which consigned to immortality the name of the illustrious Harvey. The discovery to which I wish at present to direct the attention of the reader is that of, what is usually called "animal electricity," or, the existence and operation of a fluid extremely similar to electricity in the living animal system. For the fortunate Galvani, professor of anatomy at Bologna, was referred the honour of lighting by accident on this beautiful and divine discovery—a discovery which entitles its author to be ranked with the great promoters (of) science and the essential benefactors of man. (1795, 217–218)

Early in the nineteenth century, Giovanni Aldini (1762–1834), a nephew of Galvani who also lived in Bologna, conducted experiments on the exposed brains of oxen. Unlike his uncle, Aldini (1804) showed this could produce movements of the eyelids, lips, and eyes. He and many others went on to experiment on decapitated criminals, individuals dying from natural causes, and women who had died in childbirth (Pfeiffer, 1985). Some of these experiments are depicted in Figure 29.2.

Aldini vigorously defended his uncle's theories against Alessandro Volta (1745–1827; shown in Figure 29.3), who had no problem with the concept of bimetallic electricity, but who had difficulty with Galvani's concept of animal electricity. The debate between Volta and Galvani was protracted and bordered on the irrational (see Brazier, 1958; Clarke and Jacyna, 1987). It was Alexander von Humboldt who eventually proved that animal electricity did, in fact, exist (Licht, 1967).

Although Volta was shown to be wrong on this issue, he too made a major contribution to the study of electricity. In 1800, he invented the wet cell battery, or "pile," which allowed whole streams of current to be produced—a marked improvement over just single sparks.

It is interesting to consider the broader ramifications of these developments. On the one hand, there was a rise in interest in electrotherapy, which played a major role in the development of neurology (Schiller, 1982). On the other, the

Figure 29.3. Alessandro Volta (1745–1826), the Italian physicist who invented the electric battery (voltaic cell), which served as a source of continuous current. Volta did not accept Galvani's concept of a vital force unique to animals ("animal electricity").

more basic scientific work relating to the electrical conduction of the nervous impulse formed the basis of modern neurophysiology (Brazier, 1958). In particular, the work of Emil Du Bois-Reymond in the 1840s on nerve transmission gave neurophysiology a solid scientific start.

THE EARLY 1800S: FROM ELECTROTHERAPY TO GOTHIC HORROR

Electrotherapy advanced rapidly in therapeutic settings, enhanced by the experiments from the early 1800s, a myriad of new devices (see Figures 29.4 and 29.5), and claims of successful clinical trials. The logic was simple. Nerves were electrically excitable, and nervous energy was electrical. From this premise scientists and physicians deduced that nervous diseases could be explained as electrical breakdowns. It followed that electrical stimulation of the nerves might help to correct these defects.

Some disorders of the nervous system treated with electrotherapy in the 1800s were paralysis, asphyxia, insanity, and hydrophobia (rabies). Galvanism was also used to treat blindness and deafness. In the case of hearing loss, one electrode was applied to the affected ear, and another was held in the hand. The current was then turned on and off rapidly. In addition, electricity was used to treat pain, convulsive disorders, and aphasia. The quotation at the beginning of this chapter exemplifies this dimension of galvanism. It was written in 1813 by Parker Cleaveland (1780–1858), a professor at Bowdoin College, Maine.

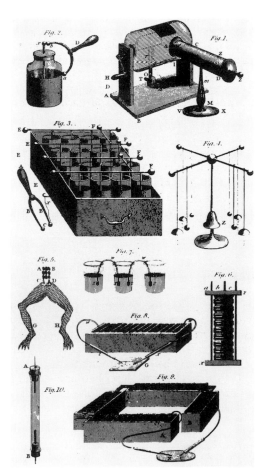

Figure 29.4. Some of the batteries used at the beginning of the nineteenth century. (From Joyce, 1810.)

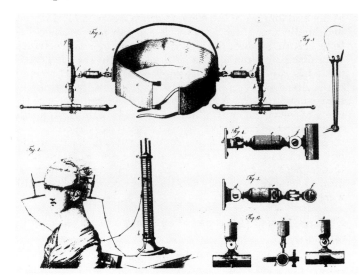

Figure 29.5. Galvanic devices from the early nineteenth century. (From Bischoff, 1802.)

As can be surmised, the clinicians made many extravagant claims for electrotherapy. This enthusiasm was somewhat predictable, given what investigators working with laboratory animals were telling their clinical colleagues and friends about the powers of electricity. An extreme example of this exuberance may be found in the work of Karl August Weinhold (1782–1829) of Halle (Clarke and Jacyna, 1987). Weinhold was among the experimenters who believed the central nervous system was like a galvanic battery, and he set out to prove this hypothesis. In 1817, he reported he had removed the cerebrum, the cerebellum, and the spinal cord of a kitten. He then filled the cranium and spinal canal with an amalgam of zinc and silver. Weinhold proudly stated that the kitten subsequently revived! His argument was that bimetallic energy could, in fact, replace the natural electricity of the nervous system.

Weinhold's story could easily have come from an imaginative novelist. In fact, his tale was not that different in concept, or in time of dissemination, from the better-known Gothic horror story told by Mary Wollstonecraft Shelley (1797–1851). Shelley began her narrative in 1816 when she was in Geneva. She, Percy Bysshe Shelley (1792–1822), George Gordon Byron (1788–1824), and John William Polidori (1795–1821) each agreed to make up a ghost story. Their nightly conversations covered terror, mythology, theories of the origin of life, and of course galvanism.

Mary, who was not quite 19 years old at the time, was especially intrigued by the Prometheus concept, as were the others. Indeed, Byron wrote his celebrated poem "Prometheus" in 1816. In the original Greek epic *Prometheus Bound* by Aeschylus (c. 525–456 B.C.), Prometheus *pyrphoros* gave mankind fire from the sun, for which Zeus severely punished him. In a second version, which was more popular in Rome, Prometheus *plasticator* recreated mankind by giving life to a clay figure. The two ideas were fused around the second to third century A.D. (Joseph, 1990). The result was the frequently told story about Prometheus stealing fire from the sun and using it to give life to inanimate human forms.

Mary Shelley published her story in 1818. She called it *Frankenstein, or the Modern Prometheus*. In 1831, in the revised version of her novel, she wrote:

Many and long were the conversations between Lord Byron and Shelley, to which I was a devout but nearly silent listener. During one of these, various philosophical doctrines were discussed, and among others the nature of the principle of life, and whether there was any probability of its ever being discovered and communicated. They talked of the experiments of Dr. (Erasmus) Darwin . . . who preserved a piece of vermicelli in a glass case, till by some extraordinary means it began to move with voluntary motion. Not thus, after all, would life be given. Perhaps a corpse would be re-animated; galvanism had given a token of such things: perhaps the component parts of a creature might be manufactured, brought together, and endued with vital warmth.

Chapter 5 of the text itself opened with the following paragraph:

It was on a dreary night of November, that I beheld the accomplishment of my toils. With an anxiety that almost amounted to agony, I collected the instruments of life around me, that I might infuse the spark of being into the lifeless thing that lay at my feet. It was already one in the morning; the rain pattered dismally against the panes, and my candle was nearly burnt out, when, by the glimmer of the half-extinguished light, I saw the dull yellow eye of the creature open; it breathed hard, and a convulsive motion agitated its limbs.

One difference between Mary Shelley and Karl August Weinhold was that the former seemed well aware of the fact that she was making up a story involving the ability of electricity to resurrect a long-deceased corpse. Weinhold, in contrast, seemed to think he really could replace the central nervous system of a dead cat with metals that could generate currents.

GALVANISM AFTER 1850

Claims for the healing powers of galvanism continued to be made in the second half of the nineteenth century. In 1855, Guillaume-Benjamin-Amand Duchenne de Boulogne wrote *De l'Electrisation Localisée, et de son Application à la Pathologie et à la Thérapeutique (On Localized Electrification, and on its Application to Pathology and Therapy)*. Jean-Martin Charcot was so impressed with Duchenne and his electrotherapy that he invited him to join his staff in 1862 (see Chapter 19).

Nevertheless, the concept of electrotherapy began to be questioned by some of the most respected authorities of the time. John Hughlings Jackson made it clear that he had little use for galvanism. Although a special "electrical room" had been built in 1867 at the National Hospital, it was soon left for the general staff to manage. In 1888 and again in 1893, William Gowers stated emphatically that electrotherapy was ineffective in treating softening of the brain, ocular paralysis, migraine headaches, *paralysis agitans, tabes dorsalis,* and muscular atrophy:

Electricity in all forms is useless. Static electricity was suggested long ago by Reynolds, and later by Charcot. I have tried it in several cases (of Parkinson's disease) very thoroughly, but could not observe improvement from its use. So too, with voltaic electricity. I can entirely confirm the conclusion of Berger, who treated twenty cases sedulously in

this manner without any of them being in the least improved by it. (1893, 657)

Wilhelm Erb, a foremost German neurologist, disagreed. He continued to talk about electrotherapy in glowing terms. In his influential textbook, Erb wrote:

We may state, unhesitatingly, that electricity is an extremely powerful and many sided remedy, and that more evident and undoubted curative effects may be attributed to it in diseases of the nervous system than to almost any other remedy. The experience of the last thirty years leaves not the least doubt that electricity is valuable in the treatment of neuralgia, anesthesia, spasms, and paralysis, in diseases of the peripheral nerves as in those of the central nervous system, and that its introduction into therapeutics has caused a more favorable prognosis in many forms of disease. I am not guilty of exaggeration when I say that the curative effects not infrequently astonish even the most experienced physician by their magical rapidity and completeness. (1883, 103)

Erb's praise of electrotherapy had a profound effect on Sigmund Freud. But after being drawn to it, the father of psychoanalysis soon soured on its powers. Freud (1924) later wrote:

My therapeutic arsenal contained only two weapons, electrotherapy and hypnotism. . . . My knowledge of electrotherapy was derived from W. Erb's text-book (1882), which provided detailed instructions for the treatment of all the symptoms of nervous diseases. Unluckily I was soon driven to see that following these instructions was of no help whatever and that what I had taken for an epitome of exact observations was merely the construction of phantasy. The realization that the work of the greatest name in German neuropathology had no more relation to reality than some "Egyptian" dream-book, such as is sold in cheap bookshops, was painful, but it helped to rid me of another shred of the innocent faith in authority from which I was not yet free. So I put my electrical apparatus aside. (see Freud, 1959, 16.)

Freud cited Paul Julius Möbius (1853–1907), a Leipzig physician, who maintained that electrotherapy was for the most part just psychotherapy and suggestion. Like Freud, Möbius was critical of many of the claims made for electrotherapy. The views of these two eminent men served to temper enthusiasm for electrotherapy in German-speaking countries.[6]

RE-EDUCATION

One of the most interesting questions raised in the 1800s was whether patients who suffered brain damage could be "re-educated" to perform better than they would have if they had been treated in traditional ways (e.g, drugs and massage) or if they had been left untreated.

In 1833 Jonathan Osborne described one of the first systematic attempts at re-education (see Chapter 25). His patient had jargon aphasia. When the man was first seen, he

---

[6]Although electrotherapy reached its peak well before the nineteenth century ended, it emerged again in the 1930s, this time as a treatment for psychoses (see Chapter 20). After this application abated, electrical stimulation evolved as a means of preventing muscular atrophy after paralyses caused by central nervous system damage (Petrofsky, 1988).

was asked to repeat the following passage from the bylaws of the College of Physicians: "It shall be in the power of the College to examine or not to examine any Licentiate, previous to his admission to a Fellowship, as they shall think fit." This passage was said aloud as: "An the be what in temother of the trothotodoo to majorum or that emidrate ein einkrastrai mestreit to ketra totombreidei to ra fromtreido asthat kekritest."

After presenting the salient clinical characteristics of his case, Osborne suggested that speech therapy, based on the way a child learns to speak, could be of help.

> Having explained to the patient my view of the particular nature of his case, and having produced a complete conviction in his mind that the defect lay in his having lost, not the power, but the art, of using the vocal organs, I advised him to commence learning to speak like a child, repeating the first letters of the alphabet, and subsequently words after another person. The result has been most satisfying. (1833, p. 169)

Paul Broca (1865) also attempted this type of therapy on a patient with motor aphasia (see Berker, Berker, and Smith, 1986). Broca thought he might help his patient learn to speak more effectively after an injury to the critical part of the left hemisphere. His hospitalized patient was an adult who was successful in relearning the alphabet and in working with syllables. Nevertheless, he did not do well in constructing higher units of language. Broca concluded that retraining an adult to speak can be more difficult than teaching a child to talk. Not only do adult aphasia patients have to overcome previous habits but they usually have intellectual deficiencies.

Amand Trousseau (1864) also attempted to re-educate aphasic subjects, but he was not as impressed with the idea as Osborne or Broca had been. One of his patients was a woman with aphasia and right hemiplegia. She was given speech therapy for a year, but her speech improved very little despite intensive practice and repetition. Trousseau wrote the following about another aphasic patient: "In spite of months of lessons and effort, poor Marcou never remembers the word 'hair,' and of his nightcap *(bonnet de coton)* can only say the last part, *de coton,* which he produced with real satisfaction" (p. 680; translated by Howard and Hatfield, 1987).

Hermann Gutzmann (1865–1922) followed Adolf Kussmaul's (1876) dictum that speech movements should be seen and memorized by motor aphasia patients as a part of their retraining. He also thought that use of the left hand might stimulate the development of the right side of the brain. Gutzmann's procedures (1896) differed somewhat for sensory aphasia patients, but, in general, the materials used were also those designed for children, as well as for the deaf and dumb.

In England, Henry Charlton Bastian (1898) accepted and exemplified the growing position that the right hemisphere housed latent centers for speech and that re-education could stimulate the development of these speech centers.[7] He also mentioned trying to teach aphasia patients to speak by beginning with the simplest sounds and slowly working into words, phrases, and sentences, and he emphasized the need for extensive practice and repetition. Reading and writing disorders were treated in the same general way as speech defects. Bastian cited Osborne's 1833 paper and John Bristowe's (1827–1895) monograph from 1880 in his chapter entitled "The Treatment of Speech Defects."

In 1904, Charles Mills published a review of the treatment of aphasia by different methods of training. He provided two of his own cases, in which the motivation of the subject and self-initiative for retraining seemed to be key ingredients in the successful recovery from left hemisphere strokes. Although he talked about relearning being somewhat analogous to learning in childhood, Mills also emphasized important differences.[8] He noted that an aphasic adult may have an advantage over a child because the adult has an organized and educated brain. In contrast, the normal child may have an advantage over the aphasic adult by not having a speech mechanism affected by disease.

As for techniques, Mills (p. 1942) listed the following pedagogic methods for treating aphasia patients:

1. The method of repetition after others which later becomes that of spontaneous recall as the patient improves; and allied or assisting methods like reading aloud, copying and writing from dictation.

2. Phonetic methods such as the method of the physiologic alphabet suggested by Wyllie and the use of phonetic readers.

3. The employment of vision to assist in the training as when the aphasic initiates the movements of articulation, enunciation and vocalization as made by others or by himself, in the latter case observing these in a mirror.

4. The retraining of the patients in the grammar of language when the aphasic is educated and when not, reorganizing so far as possible such language as he originally had.

5. Various special methods suggested by different authorities, as, for instance, that of Goldscheider of training the patient to repeat meaningless syllables.

In some circles, it did not seem to matter whether the patient suffered a traumatic injury, a vascular accident, or a disease. The expectation was that re-education would work. On a theoretical level, the dominant idea was that new brain parts were trained to take over lost functions. In this *Zeitgeist*, tautologies were common, and the amazing plasticity of the brain was heralded. The sharp boundaries between theories and actual facts often seemed to evaporate into thin air in the enthusiasm of the moment.

---

[7]Many other physicians around the turn of the century accepted these basic ideas. Alfred Goldscheider in 1902 and J. Wyllie (1844–1916) in 1894, for example, both believed that the minor hemisphere, or undamaged parts of the healthy left hemisphere, could be stimulated into action by elementary verbal exercises.

[8]The association emerging between recovery and developmental processes had an interesting experimental connection. Otto Soltmann (1844–1912) applied electrical stimulation to the motor cortex of very young animals and found that the responsiveness of this area changed dramatically with increasing age and possibly with functional exercise. There were no motor responses to stimulation in earliest infancy. Among animals just a few days older, however, he observed responses of the anterior extremities to electrical stimulation of the cortex. Proportionally, the responsive area was larger than in mature animals. The posterior extremities came in next, also taking up a larger than expected territory, and then the face. In Paris, Charcot learned of Soltmann's experimental work and asked: "When the centres are destroyed by lesions, may not a similar process of development be the means by which such centres are replaced under the influence of functional excitement?" (Dodds, 1877–1878, p. 345).

MEASURING THE EFFECTS OF THERAPY

In the opening decades of the twentieth century, Shepherd Ivory Franz took it upon himself to measure, graph, and quantify the effects of different therapies in experimental situations. These characteristics of Franz's research—and how the data were used to test various theories of recovery—can be found in a study he published in 1906. It involved a stroke victim who was virtually unable to speak, had poor written and auditory comprehension, and suffered from a right-side paralysis. The paralysis subsided in the first few weeks, as did the motor aphasia, although "a great sensory speech defect" remained.

Franz retaught this 57-year-old man a list of 10 familiar colors, the numbers 1–9, a stanza of poetry, "The Lord's Prayer," and a collection of German words. His technique involved going through lists, item by item, giving the subject the opportunity to respond and providing the correct answer in the case of an error. Franz compared his patient's scores with data for acquiring a new habit and for relearning an old, forgotten habit. In the former instance, the habit is acquired gradually, whereas in the latter there is a quick return to a high level of efficiency.

> In the case of this patient we find that this process, i.e., sudden return to former efficiency after little practice, has not gone on, but that the gradual betterment is more like the acquirement of a new habit. This would indicate, it seems to me, that not all the old brain paths were being reopened, but that in all probability new connections were being made. In harmony with this view we have the fact of "vicarious" function of the cerebrum, that one part or one hemisphere may take up the lost functions of another part or side. It seems probable that this is what happens always in cases of aphasia, in which there is any amount of reeducation. (1906, 593)

In another experiment, Franz (1924) used syphilitic subjects. Syphilis was a common cause of aphasia at the time, and Franz's patients had longstanding aphasias before they started his program. He began with extensive testing to assess word-finding skills, repetition capabilities, and related functions. He then began therapy, which concentrated at first on relearning the names of objects.

Sixty objects were selected, from which Franz first chose 20. He placed them in front of the subject one at a time and named each one. The subject had to repeat the name and was told the correct answer if unable to provide it. The objects were then presented in a different order. As the criterion was reached for each object, it was replaced by a new one from the remaining collection of 40 objects until the names of all were spoken. Franz did equivalent testing with colors, reading the alphabet, and matching printed names with pictures of objects. He also retested his subjects several months later.

When the recovery curves for each item were plotted, Franz found that the subjects learned more rapidly as testing continued. Objects introduced late in testing generally took fewer trials to master than those introduced early in testing. Perhaps most important, Franz demonstrated that the effects of therapy could be measured with considerable precision under controlled conditions (also see Franz, Scheetz, and Wilson, 1915).

Franz emphasized that others had also found beneficial effects of re-education with neurological patients. He cited a study in which the experimenters worked with a patient who had a gunshot wound of the left postcentral region (Brown and Stewart, 1916). When training began, the patient was unable to localize stimuli on the right hand. For just one finger, the experimenter provided feedback to show the patient where he had been touched. The results showed that localization improved more on the finger for which feedback was provided than on the other fingers. Franz (1921) asserted that these findings were similar to those obtained with his own patients.

In addition to studies of this sort, and parallel studies on his own monkeys, Franz cited many examples to illustrate the importance of motivating patients outside the usual office or hospital setting.[9] He mentioned a number of cases in this context. One involved a man who used a cane and became so excited while playing baseball that he forgot to use it when running the bases. Another case concerned a wounded soldier who walked without his cane when ordered to do so by an officer (see Garrett, 1951). Franz was not beyond bribing patients to do certain things for candy, food, and other rewards, often with good results.

Franz's observations about motivation led him to stress that motivation should play a major role in all rehabilitation programs. This point was not lost on Karl Lashley, who had worked with Franz (see Chapter 23). After citing Franz's work on the importance of motivation in "reëducation," Lashley told the following story:

> With the patient of Dr. Franz, who had failed to learn the alphabet after 900 repetitions, I tried betting 100 cigarettes that he could not learn it in a week. After 10 trials he was letter perfect and remembered it until the debt was paid. Especially with older patients it is often only under pressure of this sort that rapid progress is made. (1938, 751)

Interestingly, Lashley was not as optimistic about restitution at this time as he had been in 1929 when he wrote *Brain Mechanisms and Intelligence*. He now carefully distinguished between simple sensory and motor abilities and functions such as attention and abstraction. The higher functions, he thought, posed more serious and intractable problems when it came to recovery, even with special rehabilitation programs. Lashley concluded his paper with the following statement:

> It is true that such activities as new combinations of movement, maze habits, visual memories, vocabulary, can be restored by persistent training with strong motivation after almost any degree of central nervous destruction. But the fundamental and important effects of cerebral injury are not the simple sensorimotor disorders or the amnesias. They are the defects in the capacity for organization, the lowered level of abstraction, the slowing of learning and reduced retentiveness, the loss of interest and spontaneous motivation, and there is little evidence that these improve with time or training. (1938, 752)

Although Lashley was more cautious than Franz when it came to attentional deficits and problems with abstraction,

---

[9] Franz and his coworkers also conducted a number of studies on re-education in monkeys. In one case, they compared the effects of different types of therapies (e.g., massage of a limb and forced usage of a paralyzed limb) on monkeys with motor cortex damage (Ogden and Franz, 1917). They found little recovery of motor function when the animals were left to their own devices. In contrast, recovery was rapid and extensive when the animals were forced to use their paralyzed limbs, especially when combined with massage.

Franz's initial work did much to promote the development of "data-based" special therapies. These efforts also stimulated the growth of multifaceted approaches to treating brain-damaged patients (Eldridge, 1968; Shewan, 1986). Such programs emphasized not only traditional medical treatments, such as surgery and drugs, but also counseling to enhance motivation and understanding, and the application of behavioral techniques from the field of experimental psychology.

THE BIRTH OF "MODERN" NEUROSURGERY

Along with several other aspects of treatment, neurosurgery underwent a major transformation in the closing decades of the nineteenth century. This was when the localizationists began to publish "functional maps" of the cortex (see Chapter 3). For the first time, surgeons were able to localize where tumors, infections, and foreign objects were most likely to be found solely on the basis of symptoms, such as a paralysis of one of the body parts, or problems recognizing sounds. This advance was especially important when there were no clues on the skull to suggest the location of a tumor, an abscess, or a comparable problem.

William Macewen (1848–1924), a Glasgow surgeon who read Ferrier's (1876) *The Functions of the Brain,* is often cited as the first neurosurgeon to rely on functional maps of the cortex for human surgery. In 1876, Macewen came close to operating upon a boy with a brain abscess. He was prevented from doing so, however, by the family physician. After the boy died, an abscess was found in the location Macewen predicted, strengthening his belief in the importance of timely surgical interventions based on the new physiology.

The first of Macewen's brain operations came three years later. It involved a teenage patient with a meningioma above the left frontal area. This patient exhibited seizures with twitching of the arm and face. Macewen (1879) realized that this implied a tumor in the motor region of the cortex. Surgery for removal of the tumor was successfully accomplished in 1879, and this patient lived for eight more years. Macewen was so impressed with the advantages afforded by the new maps of the cortex that he then performed brain operations on five more cases (Walker, 1967).

One of the most celebrated early surgical cases was that of London surgeon Rickman J. Godlee (1849–1925). It was first mentioned in a letter defending animal experimentation in *The Times* on December 16, 1884. The letter was sent by James Crichton-Browne. Its opening paragraph read as follows:

> Sir,—While the Bishop of Oxford and Professor Ruskin were, on somewhat intangible grounds, denouncing vivisection at Oxford last Tuesday afternoon, there sat at one of the windows of the Hospital for Epilepsy and Paralysis, in Regent's Park, in an invalid chair, propped up with pillows, pale and careworn, but with a hopeful smile on his face, a man who could have spoken a really pertinent word upon the subject, and told the rev. prelate and great art critic that he owed his life, and his wife and children their rescue from bereavement and penury, to some of these experiments on living animals which they so roundly condemned. The case of this man has

been watched with intense interest by the medical profession, for it is of a unique description, and inaugurates a new era in cerebral surgery.

Four days later, a full description of the case appeared in *Lancet.* A. Hughes Bennett (1850–?), who was on the staff of the Hospital for Epilepsy and Paralysis in Regent's Park, diagnosed the tumor and persuaded Godlee, also affiliated with the hospital, to operate (Trotter, 1934).

The patient seen by Bennett was a farmer who had suffered from Jacksonian epilepsy for three years and showed a useless left arm, weakness in the left leg, double optic neuritis, severe headaches, and vomiting. The diagnosis was a tumor "probably of limited size involving the cortex of the brain and situated in the middle part of the Fissure of Rolando." This was told to Henderson, the 25-year-old Scottish patient, who agreed to the surgery. The operation took place on November 25, 1884, with Ferrier, Jackson, and Victor Horsley present. Godlee found a hard glioma, "about the size of a walnut," just below the cortical surface, where it was predicted to be.

Gordon Holmes told an interesting story about John Hughlings Jackson—a droll man not often cited for anything resembling a colorful wit—after the operation was over. Jackson supposedly said to Ferrier, who stood next to him, "awful, awful." "Awful?" questioned Ferrier, "the operation was performed perfectly." "Yes," said Jackson in his typical slow drawl, "but he opened a Scotsman's head and failed to put a joke in it!" (Holmes, 1954, p. 72).

Henderson showed improvement in the days following the operation. He stopped vomiting, his headache ceased, and his seizures ended. This was reported in *Lancet* before the year ended (Bennett and Godlee, 1884). Unfortunately, meningitis set in and he died 28 days after surgery. Although Bennett and Godlee (1885) dejectedly reported Henderson's death to the public, the importance of what they had accomplished was immediately recognized (Trotter, 1934).

Before the decade was over, symptoms were also being used to diagnose tumors of the spinal cord. In 1887, William Gowers diagnosed a nonmalignant tumor in the dorsal region of the spinal cord in a 42-year-old well-traveled businessman. "Captain G," as he was called, first experienced signs of nerve irritation at the level of the lesion (scapula) and then a gradual paralysis with one leg affected before the other. By the time Gowers saw him, the patient was paraplegic and showed a corresponding loss of cutaneous sensitivity. Captain G told him he had tried many cures, including Turkish baths, digitalis, iodide, blisters, and morphia, but none had given him lasting relief from the pain.

Gowers was certain that Captain G's problems were due to compression of the spinal cord and called upon Victor Horsley to confirm his suspicion. Horsley, shown in Figure 29.6, saw the patient early in the afternoon of June 9, 1887 and with the Captain's blessing decided to operate at the National Hospital, Queens Square. A few hours later, Horsley successfully removed the oval-shaped encapsulated tumor (a fibromyxoma) that had pressed upon the lateral surface of the spinal cord (see Figure 29.7). Within two weeks, Captain G started to regain power in his legs and by the end of the year was again able to walk.

Figure 29.6. Victor Horsley (1857–1916), a pioneer in the field of neurosurgery.

On June, 6, 1888, Gowers and Horsley received a letter from Captain G. He wrote that he was in excellent health and had just worked a 16-hour day, which involved considerable walking and standing (Gowers and Horsley, 1888).

Surgical interventions for tumors, abscesses, and related neurological problems rose rapidly in popularity as the curtain closed on the nineteenth century and the twentieth century began.

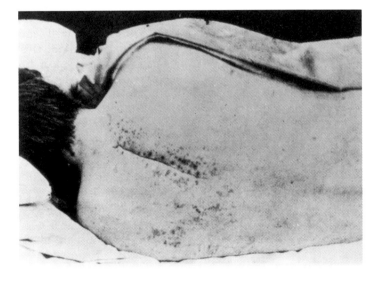

Figure 29.7. The incision over the dorsal spinal cord of Captain G, the man who had a spinal tumor removed by William Gowers and Victor Horsley in 1887. (From Gowers and Horsley, 1888.)

REFERENCES

Aldini, J. 1804. *Essai Théorique et Expérimental sur le Galvanisme.* Paris: Fournier.

Bakay, L. 1985. François Quesnay and the birth of brain surgery. *Neurosurgery, 17,* 518–521.

Bakay, L. 1987. *Neurosurgeons of the Past.* Springfield, IL: Charles C Thomas.

Bastian, H. C. 1898. *A Treatise on Aphasia and Other Speech Defects.* London: Lewis.

Bayfield, S. 1823. *A Treatise on Practical Cupping.* London: Cox.

Bell, B. 1782–1787. *A System of Surgery.* Edinburgh: Elliott.

Bennett, A. H., and Godlee, R. 1884. Excision of a tumour from the brain. *Lancet, ii,* 1090–1091.

Bennett, A. H., and Godlee, R. 1885. Sequel to the case of excision of a tumour from the brain. *Lancet, i,* 13.

Berker, E. A., Berker, A. H., and Smith, A. 1986. Translation of Broca's 1865 report: Localization of speech in the third left frontal convolution. *Archives of Neurology, 43,* 1065–1072.

Blumenbach, J. F. 1795. *Elements of Physiology.* Philadelphia, PA: Thomas Dobson at the Stone-House.

Brazier, M. 1958. The evolution of concepts relating to the electrical activity of the nervous system, 1600–1800. In M. W. Perrin (Chair), *The Brain and Its Functions.* Wellcome Foundation. Oxford, U.K.: Blackwell.

Brazier, M. 1973. The electrophysiology of muscle and nerve in man from the pre-galvanic era to about 1930. In J. A. Simpson (Ed.), *Handbook of Electroencephalography and Clinical Neurophysiology: Vol. 16. Neuromuscular Diseases.* Amsterdam: Elsevier. Pp. 16B-5.

Bristowe, J. S. 1880. *The Physiological and Pathological Relations of the Voice and Speech.* London: Bogue.

Broca, P. 1865. Sur le siège de la faculté du langage articulé. *Bulletins de la Société d'Anthropologie, 6,* 337–393.

Brown, T. G., and Stewart, R. M. 1916. On disturbances of the localization and discriminations of sensations in cases of cerebral lesions, and on the possibility of recovery of these functions after a process of training. *Brain, 39,* 348–454.

Clarke, E., and Jacyna, L. S. 1987. *Nineteenth Century Origins of Neuroscientific Concepts.* Berkeley: University of California Press.

Cleaveland, P. 1813. Account of the effects of electricity. *New England Journal of Medicine and Surgery, 2,* 26–29.

Dodds, W. J. 1877–1878. On the localization of the functions of the brain. *Journal of Anatomy, 12,* 340–363, 454–494, 636–660.

Duchenne, G.-B.-A. 1855. *De l'Electrisation Localisée, et de son Application à la Pathologie et à la Thérapeutique.* Paris: J. B. Ballière.

Eldridge, M. 1968. *The History of the Treatment of Speech Disorders.* Edinburgh: E. & S. Livingstone Ltd.

Erb, W. H. 1883. *Handbook of Electrotherapeutics.* (L. Pützel, Trans.). New York: William Woods.

Ferrier, D. 1876. *The Functions of the Brain.* London: Smith, Elder and Company.

Franz, S. I. 1906. The reeducation of an aphasic. *Journal of Philosophy, Psychology and Scientific Method, 2,* 589–597.

Franz, S. I. 1921. Cerebral-mental relations. *Psychological Review, 28,* 81–95.

Franz, S. I. 1924. Studies in re-education: The aphasias. *Journal of Comparative Psychology, 4,* 349–429.

Franz, S. I., Scheetz, M. E., and Wilson, A. A. 1915. The possibility of recovery of motor function in long-standing hemiplegia. *Journal of the American Medical Association, 65,* 2150–2154.

Freud, S. 1959. *An Autobiographical Study, The Standard Edition of the Complete Psychological Works* (Vol. 20). London: Hogarth Press.

Galvani, L. 1791. *De viribus electricitatis in motu muscalari commentarius.* Translated by M. G. Foley as, *Luigi Galvani: Commentary on the Effects of Electricity on Muscular Motion.* Norwalk, CT: Burndy Library, 1953.

Garrett, H. E. 1951. *Great Experiments in Psychology* (3rd ed.). New York: Appleton-Century-Crofts.

Garrison, F. H. 1929. *An Introduction to the History of Medicine.* Philadelphia, PA: W. B. Saunders.

Gibson, W. C. 1962. Pioneers in localization of function in the brain. *Journal of the American Medical Association, 180,* 944–951.

Goldscheider, A. 1902. *Handbuch der Physikalischen Therapie.* Leipzig.

Gowers, W. R. 1888. *A Manual of Diseases of the Nervous System.* Philadelphia, PA: Blakiston.

Gowers, W. R. 1893. *A Manual of Diseases of the Nervous System* (2nd ed., Vol. 2). Philadelphia, PA: Blakiston.

Gowers, W. R., and Horsley, V. 1888. A case of tumour of the spinal cord. Removal: recovery. *Medical and Chirurgical Transactions, 71,* 377–428.

Guericke, O. van. 1672. *Experimenta nova (ut vocantur) Magdeburgica de vacuo spatio. . . .* Amsterdam: J. Jansson-Waesberg.

Gutzmann, H. 1896. Heilungsversuche bei centromotorischer und centrosensorischer Aphasie. *Archiv für Psychiatrie und Nervenkrankheiten, 28,* 354–378.

Holmes, G. 1954. *The National Hospital, Queens Square; 1860–1948.* Edinburgh: Livingstone.

Howard, D., and Hatfield, F. M. 1987. *Aphasia Therapy: Historical and Contemporary Issues*. Hillsdale, NJ: Lawrence Erlbaum.

Joseph, M. K. 1990. *Mary Shelley: Frankenstein*. New York: Oxford University Press.

Joynt, R. J., and Benton, A. L. 1964. The memoir of Marc Dax on aphasia. *Neurology, 14*, 851–854.

Kussmaul, A. 1876. *Die Stoerungen der Sprache* (4th ed.). (H. Gutzmann, Ed.). Leipzig: Vogel, 1910.

Lashley, K. S. 1929. *Brain Mechanisms and Intelligence*. Chicago: University of Chicago Press.

Lashley, K. S. 1938. Factors limiting recovery after central nervous lesions. *Journal of Nervous and Mental Disease, 88*, 733–755.

Lennon, S., and Davenport, S. 1987. A long-discontinued practice. *Lancet, 2*, 1514, 1534.

Licht, S. 1967. *Therapeutic Electricity and Ultraviolet Radiation* (2nd ed.). Baltimore, MD: Waverly Press.

Louis, A. 1774. *Mémoire sur les tumeurs fongueuses de la dure-mère*. *Mémoire de l'Académie Royale de Chirurgerie, 5*, 1.

Louis, A. 1776. *Recueil d'Observations d'Anatomie et de Chirurgerie, pour Servir de Base à la Théorie des Lésions de la Tête, par Contre-coup*. Paris: Cavelier.

Luciani, L. 1917. *Human Physiology*. (F. Welby, Trans.). London: Macmillan.

Macewen, W. 1879. Tumour of the dura mater removed during life in a person affected with epilepsy. *Glasgow Medical Journal, 12*, 210–213.

Mapleson, T. 1813. *Treatise on the Art of Cupping*. London: Printed for the author by G. Sidney.

Marat, J. P. 1782. *Recherches Physiques sur l'Electricité*. Paris: Clousier.

Marat, J. P. 1784. *Mémoires sur l'Electricité Médicale*. Paris: Méquignon.

Matteuci, C. 1847. *Leçons sur les Phénomènes Physiques des Corps Vivants*. Paris: Masson.

Mills, C. K. 1904. Treatment of aphasia by training. *Journal of the American Medical Association, 43*, 1940–1949.

O'Connor, D. C., and Walker, A. E. 1967. Prologue. In A. E. Walker (Ed.), *A History of Neurological Surgery*. New York: Hafner. Pp. 1–22.

Ogden, R., and Franz, S. I. 1917. On cerebral motor control: The recovery from experimentally produced hemiplegia. *Psychobiology, 1*, 33–49.

Osborne, J. 1833. On the loss of the faculty of speech depending on forgetfulness of the art of using the vocal organs. *Dublin Journal of Medical and Chemical Sciences, 4*, 157–170.

Petrofsky, J. S. 1988. Functional electrical stimulation and its application in the rehabilitation of neurologically injured individuals. In S. Finger,

T. E. LeVere, C. R. Almli, and D. G. Stein (Eds.), *Brain Injury and Recovery: Theoretical and Controversial Issues*. New York: Plenum Press. Pp. 273–305.

Pfeiffer, C. J. 1985. *The Art and Practice of Western Medicine in the Early Nineteenth Century*. Jefferson, NC: McFarland and Company.

Quesnay, F. 1743. Remarques sur les playes du cerveau. . . . *Mémoires de l'Académie Royale de Chirurgie, 1*, 310–336.

Schiller, F. 1981. Reverend Wesley, Doctor Marat and their electrical fire. *Clio Medica, 15*, 159–176.

Schiller, F. 1982. Neurology: The electrical root. In F. C. Rose and W. F. Bynum (Eds.), *Historical Aspects of the Neurosciences*. New York: Raven Press. Pp. 1–11.

Shelley, M. 1818. *Frankenstein, or the Modern Prometheus*. London. Printed for Lackington, Hughes, Harding, Mavor and Jones.

Shelley, M. 1831. *Frankenstein, or the Modern Prometheus* (3rd ed.). London: Colburn and Bentley.

Shewan, C. M. 1986. The history and efficacy of aphasia treatment. In R. Chapey (Ed.), *Language Intervention Strategies in Adult Aphasia* (2nd ed.). Baltimore, MD: Williams & Wilkins. Pp. 28–43.

Smith, A. H. 1905–1907. *The Writings of Benjamin Franklin* (Vol. 3). New York: Macmillan.

Thorwald, J. 1963. *Science and the Secrets of Early Medicine*. New York: Harcourt, Brace and World.

Trotter, W. 1934. A landmark in modern neurology. *Lancet, 2*, 1207–1210.

Trousseau, A. 1864. De l'aphasie, maladie décrite récemment sous le nom impropre d'aphémie. *Gazette des Hôpitaux, 37*, 13–14, 25–26, 37–39, 49–50.

Walker, A. E. 1967. Sir William Macewen. In A. E. Walker (Ed.), *A History of Neurological Surgery*. New York: Hafner. 178–179.

Walsh, J. 1773. On the electric property of the torpedo. *Philosophical Transactions, 63*, 461–480.

Weinhold, K. A. 1817. *Versuche über das Leben und seine, Grundkräfte, auf dem Wege der experimental-Physiologie*. Magdeburg: Creutz.

Wesley, J. 1747. *Primitive Physick*. (Reprinted 1973 as *Primitive Remedies*; Santa Barbara, CA: Woodridge)

Wesley, J. 1759. *The Desideratum: or, Electricity Made Plain and Useful. By a Lover of Mankind and of Common Sense* (2nd ed.). London: Ballière, Tindall and Cox, 1871.

Whytt, R. 1768. *The Works of Robert Whytt, M.D.* (3rd ed.; published by his son). Edinburgh: Balfour, Auld and Smellie.

Wyllie, J. 1894. *The Disorders of Speech*. Edinburgh: Oliver and Boyd.

Epilogue

What's past is prologue.
William Shakespeare, 1611

The present volume was written as a survey of the history of the neurosciences, with emphasis on how specific functions were eventually associated with particular brain structures. It began with the realization that the practice of opening skulls in living human beings is thousands of years old, suggesting that the brain was perceived as important even by our Neolithic forefathers. After reviewing what the Ancient Egyptians, Greeks, and Romans had to say about the brain, those landmark developments that firmly established the brain as the organ of mind were examined. The central chapters examined specific brain functions, such as sensory and motor functions, sleep, emotion, intellect, learning, and speech. The last two chapters were devoted to the treatment of brain injuries and diseases over the millennia.

The purpose of this work was to provide a broad introduction to the infancy and formative years of the neurosciences in its basic and applied forms. Thus, little was said about developments in the field after the opening decades of the twentieth century. In part, it made sense to end in the period between the two World Wars because most contemporary books in neurology, neuropsychology, neurophysiology, neuroanatomy, and related disciplines tend to cover these newer developments in some detail. Less important, but on the practical side, this volume was already so large that it was hard to justify adding additional pages to it.

When the twentieth century began, most neuroscientists wanted to know more about how human behavior was affected by specific neurological diseases and brain injuries in certain locations, such as the frontal association cortex. Although formal tests were beginning to be developed, the approach was basically the same as that utilized by previous generations of scientists and physicians. Systematic behavioral research on brain-damaged laboratory animals, stimulated by the work of individuals like Eduard Hitzig and David Ferrier, was still in its toddlerhood and was carried out in only a small number of laboratories in Europe and North America. The same can be said about electrophysiological recordings. When animals were used at the turn of the century, more likely than not the intent was to trace tracts and to define the anatomical boundaries of specific areas of the brain.

The theory of cortical (and subcortical) localization of function, which was firmly in place by this time, changed little during the middle decades of the twentieth century. Nevertheless, after World War I, greater emphasis was placed on data generated by electrophysiological recordings and on quantitative behavioral research using laboratory animals. Armed with more sensitive tests and with a greater number of procedures, many investigators set forth to determine if various "functional areas" could be divided into smaller and smaller specialized parts, despite the best efforts of the Gestalt psychologists in Germany, and Shepherd Ivory Franz and Karl Lashley in the United States, to question the dissection, fragmentation, and compartmentalization of intellectual functions.

Perhaps the greatest fundamental change to take place in the neurosciences in the twentieth century was the shift toward "molecularization." When the century began, neuron doctrine was relatively new and scientists were turning to synaptic events, real and imagined, to explain functions as varied as sleep and memory. As the years passed, biochemistry ascended in importance, rising to become an indispensable tool in many areas of contemporary neuroscience. The list of possible neurotransmitters is still increasing, and there is now a rapidly growing appreciation of how these substances are released, what their actions are at the receptor level, and what roles they play in certain diseases of the nervous system. The ability to modulate the actions of different neurotransmitters with pharmacological agents, as well as the hope of identifying and controlling neurotrophic factors, rank among the most exciting developments in the molecularized neurosciences of today. The perceived importance of this sort of work has triggered a virtual explosion of research in university and medical school settings, as well as by drug companies hoping to find the mithridates and panaceas of tomorrow.

An equally dramatic change has been the development of new technologies and probes. Some of these methodological advances could not have been envisioned when the twentieth century opened. They range from the construction of powerful electron microscopes to the development of a myriad of new stains for sophisticated anatomical and physiological analyses. Perhaps most remarkable has been the advent of noninvasive probes, such as the CAT and PET scans, for better determining the status of the brain and shedding new light on the functional properties of its circuitry.

As awe inspiring as our latest intellectual and technological achievements in the brain sciences may seem, it would be a mistake to think that they magically sprang out of thin air. Many of the advancements in the neurosciences can be considered revolutionary. But in a very real sense, all such achievements are also evolutionary. William Shakespeare had it right in 1611, when he wrote in *The Tempest* that the past is prologue. The new discoveries and technologies that draw more and more attention to the neurosciences can be traced to ideas and conceptions of earlier pioneers, many of whose names—hopefully not to be forgotten—can be found in this book.

Johann Wolfgang von Goethe (1810) could not have expressed it better when he wrote: "We cannot clearly be aware of what we possess till we have the means of knowing what others possessed before us." He was equally perceptive when he added: "We cannot really and honestly rejoice in the advantages of our own time if we know not how to appreciate the advantages of former periods." Goethe died in Germany in 1832, and Shakespeare in England in 1616. Time has passed and memories have faded, but the significance of what these intellectual giants had to say about history should not be forgotten, and should serve well as a point of departure for future generations of explorers of the brain.

REFERENCES

Goethe, J. W. von. 1810. *Zur Farbenlehre*. Weimar. Translated by C. L. Eastlake in 1840 as, "Goethe's Theory of Colours." Frank Cass & Company, 1967.

Shakespeare, W. 1811. *The Tempest*. (1670 ed.; London: J. M. for Henry Herringham.)

Appendix
Dates of Birth and Death

Abbasmal-Zahrawc, Abul-Quasim Khalif ibn (*see* Albucasis)
Abu `Ali al-Husain ibn `Abdullah ibn Sina (*see* Avicenna)
Abu-`Ali al-Hasan ibn al-Haitham al-Haytham (*see* Alhazen)
Abu-l-Walid Muhammad ibn Rushd (*see* Averroes)
Ackermann, Jacob Fidelis   1765–1815
Addison, Thomas   1793–1860
Adie, W. J.   1886–1935
Adrian, Edgar D.   1889–1977
Aegidus   1247–1316
Aeschylus   c. 525–456 B.C.
Aesculpius of Epidaurus   c. 1100 B.C.
Agassiz, Louis   1807–1873
Akelaitis, Andrew   1904–1955
al-Hakim, Fatimid Khalif   996–1021
Al-Muqtadir, Abbasid Caliph   fl. 931
Albertus Magnus   1193–1280
Albucasis   936–1013
Alcmaeon   c. fifth century B.C.
Aldini, Giovanni   1762–1834
Alexander, Leo   b. 1905
Alhazen   c. 965–1039
Allen, Harrison   1841–1897
Alzheimer, Alois   1864–1915
Amman, Johann Conrad   1663–1730
Ammar   c. 1000
Anaxagoras   500–428 B.C.
Anaximander   638–547 B.C.
Anaximenes   d. 504 B.C.
Andral, Gabriel   1797–1846
Andromachus   fl. 54–68
Annan, Samuel   1797–1868
Anton, Gabriel   1858–1933
Antoninus, Marcus Aurelius   121–180
Aquapendente, Fabricius ab   c. 1533–1619
Arantius, Julius Caesar   1530–1589
Aranzi, Guilio Cesare (*see* Arantius)
Aranzio, Guilio Cesare (*see* Arantius)
Arbiter, Petronius   c. A.D. 66
Arceo, Francisco   1493–1573?
Archigenes of Apamea   second century
Aretaeus the Cappadocian   second–third centuries
Ariëns Kappers, C. U.   1877–1946
Aristotle   384–322 B.C.
Arnemann, Justus   1763–1806
Artemidorus of Ephesus   c. A.D. 140
Aserinski, Eugene   b. 1921
Ashurbanipal (King)   668–626 B.C.
Asoka   274–232 B.C.
Aubert, Rudolph   1826–1892
Aubertin, Simon Alexandre Ernest   1825–1865

Augustine (*see* Saint Augustine)
Aurelianus, Caelius   fifth century
Averroes   1126–1198
Avicenna, Abu Ali Hussain   980–1037

Baader, Joseph Lambert   1723–1773
Babinski, Joseph   1857–1932
Bacon, Francis   1561–1626
Bacon, Roger   c. 1220–1292
Badal, Jules   1840–1929
Baginsky, Adolf   1843–1918
Baillard, Edme   1648–1677
Baillarger, Jules Gabriel François   1806–1891
Baillie, Matthew   1761–1823
Bain, Alexander   1818–1903
Ballard, Philip Boswood   1865–?
Bard, Philip   1898–1977
Barkow, Hans Carl   1798–1873
Barlow, John   1798–1869
Bartholow, Roberts   1831–1904
Basil the Great   330–379
Bastian, Henry Charlton   1837–1915
Bateman, Frederick   1824–1904
Bauhin, Kaspar   1560–1624
Bean, Robert Bennett   1874–1944
Beard, George   1839–1883
Beaumont, William   1785–1853
Beaunis, Henri   1831–1921
Beccaria, Giambattista   1716–1781
Beck, Adolf   1863–1942
Béclard, Paul Augustin   1785–1825
Beddoes, Thomas   1754–1805
Beethoven, Ludwig von   1770–1827
Beevor, Charles   1854–1908
Békésy, Georg von   1889–1972
Bekhterev, Vladimir   1857–1927
Bell, Alexander Graham   1847–1922
Bell, Benjamin   1749–1806
Bell, Charles   1774–1842
Bell, John   1796–1872
Bellingeri, Carlo Francesco   1789–1848
Bennett, A. Hughes   1850–?
Berengario da Carpi, Jacopo   c. 1460–1530
Berger, Hans   1873–1941
Bernard, Claude   1838–1878
Bernhardt, M.   1844–1915
Bernier, François   1620–1688
Betz, Vladimir Alexandrovich   1834–1894
Bianchi, Leonardo   1848–1927
Bichat, François-Xavier   1771–1802
Bidder, Friedrich Heinrich   1810–1894

Bielschowsky, Max   1869–1940
Bigelow, Henry   1818–1890
Binet, Alfred   1857–1911
Bini, Lucio   b. 1908
Binns, Edward   d. 1851
Bismarck, Otto von   1815–1898
Blackall, John   1771–1860
Blix, Magnus   1849–1904
Blocq, Paul   1860–1896
Blumenbach, Johann Friedrich   1752–1840
Börnstein, Walter   1890–?
Boerhaave, Herman   1668–1738
Boethius, Anicius   c. 480–524
Bohn, Johannes   1640–1718
Bolk, Louis   1866–1930
Boll, Franz Christian   1849–1879
Bonaparte, Napoleon   1769–1821
Bonet, Juan Pablo   1579–1633
Bonet, Théophile   1620–1689
Bonhoeffer, Karl   1868–1943
Bonnet, Charles   1720–1792
Bontekoe, Cornelius van   1640–1685
Bontius, Jacobus   1592–1631
Boorde, Thomas   1490–1519
Borelli, Giambattista   1813–1891
Borelli, Giovanni   1608–1679
Boring, Edwin   1886–1968
Borogognoni, Theodorico   1205–1296
Boswell, James   1740–1795
Bouchard, Charles Jacques   1837–1915
Bouillaud, Jean-Baptiste   1796–1881
Bouteille, Etienne Michel   1732–1816
Boyd, Robert   1808–1883
Boyle, Robert   1627–1691
Brain, W. Russell   1895–1966
Bray, Charles W.   b. 1904
Breasted, James   1865–1935
Bremer, Frédéric   1892–1959
Brickner, Richard M.   1896–1959
Bright, Richard   1789–1858
Brill, Nathan E.   1860–1925
Briquet, Paul   1796–1881
Brissaud, Edouard   1852–1909
Bristowe, John   1827–1895
Broadbent, William Henry   1835–1907
Broca, Paul   1824–1880
Brodie, William   1741–1788
Brodmann, Korbinian   1868–1918
Brown, James Crichton   1840–1938
Brown, Sanger   1852–1928
Brown, Thomas   1778–1820
Brown-Séquard, Charles Edouard   1817–1894
Bruce, Alexander   1854–1911
Bruce, Lewis Campbell   1866–1949
Brunschwig, Hieronymous   c. 1450–1533
Bucy, Paul   b. 1904
Buddha (see Gautama)
Budge, Julius Ludwig   1811–1884
Bunyon, John   1628–1688
Burckhardt, Gottlieb   1836–?
Burdach, Karl Friedrich   1776–1847
Burney, Fanny   1752–1840

Butler, Samuel   1835–1902
Buzzard, Edward Farquhar   1871–1945
Buzzard, Thomas   1831–1919
Buzzi, Francesco   1751–1805
Byron, George Gordon (Lord)   1788–1824

Caesar, Julius   100–44 B.C.
Caesar, Nero Claudius   37–68
Caius, John   1510–1573
Cajal (see Ramón y Cajal)
Calcar, Jan Stephan van   1499–1546
Caldani, Leopoldo   1725–1813
Caldwell, Charles   1772–1853
Caligula   12–41
Calkins, Mary   1863–1930
Camerarius, Joachim   1534–1598
Campbell, Alfred Walter   1868–1937
Camper, Petrus   1722–1789
Cannon, Walter Bradford   1871–1945
Capivaccio   1523–1589
Cardano, Girolamo   1501–1576
Carlos, (Don) (Prince)   1527–1598
Carpi (see Berengario da Carpi)
Carus, Carl Gustav   1789–1869
Cason, Hulsey   1893–1950
Casserio, Giulio   c. 1545–1616
Cassius   c. 450
Caton, Richard   1842–1926
Celsus   25 B.C.–A.D. 50
Cerletti, Ugo   1877–?
Chaldi, Ernst   1756–1827
Chapman, Nathaniel   1780–1853
Charcot, Jean-Martin   1825–1893
Charles I of England   1600–1649
Charles II of England, Scotland, and Ireland   1630–1685
Charles IX of France   1550–1574
Charpentier, Augustin   1852–1916
Chauliac, Guy de   c. 1298–1368
Chaussier, François   1746–1828
Child, Charles Manning   1869–1954
Chirac, Pierre   1650–1732
Chrysostom, St. John   347–407
Cicero, Marcus Tullius   104–43 B.C.
Claiborne, J. Herbert   1828–1905
Clarke, Robert Henry   1850–1921
Cleaveland, Parker   1780–1858
Cloquet, Hippolyte   1787–1840
Coiter, Volcher   1534–1600
Colton, Gardner Quincy   1814–1898
Combe, George   1788–1858
Commodus, Lucius Aelius Aurelius   161–192
Constantinus Africanus   c. 1020–1087
Cooke, John   1756–1838
Cooper, Astley   1839–1905
Corday, Charlotte   1768–1793
Cordus, Valerius   1515–1544
Corti, Alphonso Giocomo   1822–1876
Cotugno, Domenico   1736–1822
Creutzfeldt, Hans-Gerhard   1885–1964
Crichton, Alexander   1763–1846
Crichton-Browne, James   1840–1938
Croce (see Della Croce)

Cruveilhier, Jean   1791–1874
Cunningham, D. J.   1850–1909
Cushing, Harvey   1869–1939
Cuvier, Georges   1769–1832

Dale, Henry   1875–1968
Dalton, John   1766–1844
Dalton, John Call   1825–1889
Dana, Charles L.   1852–1935
Dandy, Walter   1886–1946
Dart, Raymond   b. 1893
da Rivalto, Giordano   1260–1311
Darwin, Charles   1809–1882
Darwin, Erasmus   1731–1802
da Vinci, Leonardo   1472–1519
Davis, Hallowell   1896–1992
Davy, Humphry   1778–1829
Dax, Gustav   1815–1874
Dax, Marc   1770–1837
de Baveriis, Baverius   c. 1480
Deen, Izaak van   1804–1869
Deiters, Otto Friedrich Karl   1834–1863
Dejerine, Joseph Jules   1849–1917
de la Böe, Franciscus (*see* Sylvius, F.)
de la Forge, Louis   fl. 1661–1677
de la Mettrie, Jules   1709–1751
de la Peyronie, François Gigot   1678–1747
Delboeuf, Joseph   1831–1896
Della Croce, Giovan Andrea   1514–1575
Democritus   c. 460–370 B.C.
Dercum, Francis X.   1856–1931
Descartes, René   1596–1650
Dewey, John   1859–1952
Dickson, S. Henry   1798–1872
Diocles of Carystus   fourth century B.C.
Dixon, Walter E.   1871–1931
Djoser (King)   c. twenty-eighth–late twenty-seventh
    century B.C.
Dogiel, Alexander   1852–1922
Dogliotti, Achille Mario   1897–1966
Donaldson, Henry   1857–1938
Donders, Frans Cornelius   1818–1889
Down, John Langdon Haydon   1828–1896
Drake, Daniel   1785–1852
du Bocage, Barbie   1797–1834
du Bois-Reymond, Emil   1818–1896
Du Laurens, André   1558–1609
du Port, François   c. 1540–1624
Du Pui, Meinard Simon   1754–1834
Duchenne de Boulogne, Guillaume Benjamin Amand
    1806–1870
Dugès, Antoine Louis   1796–1838
Dunglison, Robley   1798–1869
Dupuytren, Guillaume   1777–1835
Durham, Arthur E.   1833–1895
Dusser de Barenne, Johannes   1885–1940
Duval, Mathias   1844–1907
Duverney, Joseph Guichard   1648–1730

Earle, Pliny   1809–1892
Ebbinghaus, Hermann   1850–1909
Ebers, Georg   1837–1898

Eberstaller, Oskar   1886–1939
Eckhard, Conrad   1822–1905
Economo, Constantin von   1876–1931
Edinger, Ludwig   1855–1918
Ehrenberg, Christian Gottfried   1795–1876
Ehrenfels, Christian von   1859–1932
Einstein, Albert   1879–1955
Elliotson, John   1791–1868
Elliott, Thomas R.   1877–1961
Empedocles   490–430 B.C.
Epicurus   c. 341–270 B.C.
Erasistratus   c. 310–250 B.C.
Erb, William   1840–1921
Erlanger, Joseph   1874–1965
Errera, Leo   1858–1905
Eschenmeyer, Adam Karl August   1768–1852
Esquirol, Jean   1782–1840
Estienne, Charles   1503–1564
Euclid   fl. 300 B.C.
Eulenberg, Albert   1840–1917
Eustachio, Bartolomeo   1524–1574
Evans, Joseph   b. 1904
Ewald, Carl Anton   1845–1915
Ewald, Julius R.   1855–1921

Fabritius, Harald   1877–1946
Fabry von Hilden, Wilhelm   1560–1634
Fallopius, Gabriel   1523–1562
Fernel, Jean   1497–1558
Ferri, Enrico   1856–1929
Ferrier, David   1843–1928
Feuchtersleben, Ernst von   1806–1849
Feuchtwanger, Ernst   1889–?
Finkelnburg   1832–1896
Fischer, Oskar   1885–?
FitzRoy, Robert   1805–1865
Flechsig, Paul Emil   1847–1929
Flourens, Marie-Jean-Pierre   1794–1867
Foderà, Michele   1793–1848
Foerster, Otfrid   1873–1941
Fontana, Felice Gaspar Ferdinand   1730–1805
Forel, August   1848–1937
Fourier, Jean-Baptiste   1768–1830
Fournie, Edouard   1833–1886
Foville, Achille Louis   1799–1878
Franklin, Benjamin   1706–1790
Franz, Shepherd Ivory   1874–1933
Frederick I, King of Sweden   1676–1751
Freeman, Walter J.   1895–1972
Freud, Sigmund   1856–1939
Frey, Max von   1852–1932
Friedlander, Carl   1847–1887
Fritsch, Gustav   1838–1927
Fukala, Vincenz   1847–1911
Fuller, Solomon   1872–?
Fulton, John   1899–1960
Funk, Casimir   1884–1967

Gage, Phineas   1823–1860
Gajdusek, Daniel Carleton   b. 1923
Galen   130–200
Galilei, Galileo   1564–1642

Gall, Franz   1757–1828
Galton, Francis   1822–1911
Galvani, Luigi   1737–1798
Gaskell, Walter H.   1847–1914
Gasser, Herbert   1888–1963
Gauss, Friedrich   1777–1855
Gautama (Buddha)   c. 560–480 B.C.
Gehuchten, Arthur van   1861–1914
Gélineau, Edouard   1828–1906
Gennari, Francesco   1750–1797
Geoffroy-Saint-Hilaire, Etienne   1772–1844
Gerard of Cremona   c. 1114–1187
Gerlach, Joseph von   1820–1896
Gerssdorff, Hans von   c. 1456–1517
Gerstmann, Josef   1887–1969
Gesner, Johann   1738–1801
Gilbert, William   1546–1603
Gilles de la Tourette, Georges   1857–1904
Glidden, George Robins   1809–1857
Goddard, Henry H.   1866–1957
Godlee, Rickman   1849–1925
Goethe, Johann Wolfgang von   1749–1832
Goldscheider, Alfred   1858–1935
Goldstein, Kurt   1878–1965
Golgi, Camillo   1843–1926
Goltz, Friedrich   1834–1902
Gowers, William   1845–1915
Goya y Lucientes, Francisco Josè de   1746–1828
Graefe, Albrecht von   1828–1870
Grandjean de Fouchy, Jean-Paul   1707–1788
Gratiolet, Pierre   1815–1865
Greisinger, William   1817–1868
Grohmann, Johann   1769–1847
Grote, George   1796–1871
Grünbaum, Sidney Frankau   1861–1921
Gryllus, Laurentius   1484–1560
Guainerio, Antonio   d. 1440
Guericke, Otto von   1602–1686
Guttmann, Paul   1834–1893
Gutzmann, Hermann   1865–1922

H. M. (Case)   b. 1926
Hadlow, William J.   b. 1921
Hales, Stephen   1677–1761
Hall, G. Stanley   1844–1924
Hall, Marshall   1790–1857
Haller, Albrecht von   1707–1777
Halstead, William   1852–1922
Hammond, William   1828–1900
Hammurabi   c. 1792–1750 B.C.
Hannover, Adolph   1814–1894
Harlow, John   1819–1907
Harrison, John P.   1796–1849
Hartley, David   1705–1757
Hartmann, Karl Eduard   1842–1906
Harvey, William   1578–1657
Hasse, Carl   1841–1922
Haycraft, John   1857–1922
Head, Henry   1861–1940
Hebb, Donald   1904–1985
Heister, Lorenz   1683–1758
Held, Hans   1866–?

Helmholtz, Hermann   1821–1894
Henle, Jakob   1809–1885
Henley, William Ernest   1849–1903
Henning, Hans   1885–?
Hensen, Victor   1835–1924
Henschen, Salomon E.   1847–1930
Heraclitus of Ephesus   556–460 B.C.
Hering, Ewald   1834–1918
Hermann, Ludimar   1838–1914
Herodotus of Halicarnassus   c. 490–425 B.C.
Herophilus   c. 335–280 B.C.
Herz, Ernst   b. 1900
Heschl, Richard L.   1824–1881
Hess, Walter   1881–1973
Hesy Re   c. 2600 B.C.
Heyfelder, Johann F.   1798–1869
Hildanus (see Fabry)
Hilden (see Fabry)
Hinshelwood, James   1859–1919
Hippocrates   460–370 B.C.
His, William   1831–1944
Hitzig, Eduard   1838–1907
Hobbes, Thomas   1588–1679
Hoffmann, H.   1819–1891
Holland, Henry   1788–1873
Holmes, Gordon   1876–1966
Holmes, Oliver Wendell   1809–1894
Holt, Edwin B.   1873–1946
Homer   c. 700 B.C.
Hooke, Robert   1635–1703
Hooker, Joseph   1817–1911
Horsley, Victor   1857–1916
Howell, William   1860–1945
Hua To   190–265
Huddart, Joseph   1741–1816
Hugh of Lucca   c. 1250
Humboldt, Alexander von   1769–1859
Hun, Henry   1854–1924
Huntington, George   1850–1916
Hurst, C. Herbert   1856–1898
Huschke, Emil   1797–1858
Huss, Magnus   1807–1890
Hutchinson, Johnathan   1828–1913
Huygens, Christiaan   1629–1695

Imhotep   c. twenty-eighth–late twenty-seventh
   century B.C.
Ingrassia, Giovanni Filippo   1510–1580
Ireland, William   1832–1909
Is-Hâq, Hunain ibn   809–873
Itard, Jean   1775–1838

Jackson, Charles T.   1805–1880
Jackson, Edward   1856–1942
Jackson, James   1777–1867
Jackson, John Hughlings   1835–1911
Jackson, Samuel   1787–1872
Jacobsen, Carlyle F.   1902–1974
Jakob, Alfons   1884–1931
James, William   1842–1910
Jastrow, Joseph   1863–1944
Jastrowitz, Moritz   1839–1912

Jelliffe, Smith Ely 1866–1945
Johnson, Harry Miles 1885–1953
Johnson, Samuel 1709–1784
Jolly, Friedrich 1844–1904

Kant, Immanuel 1694–1778
Karplus, Johann Paul 1866–?
Keen, William Williams 1837–1932
Kennard, Margaret 1899–1972
Kennedy, John F. 1917–1963
Kepler, Johannes 1571–1630
Kiesow, Friedrich 1858–1940
Kleist, Karl 1879–1960
Kleitman, Nathaniel b. 1895
Klüver, Heinrich b. 1897
Köhler, Wolfgang 1887–1967
Kölliker, Rudolph Albert von 1817–1905
König, Arthur 1856–1901
Koffka, Kurt 1886–1941
Korsakoff, Sergi 1853–1900
Koyter (see Coiter)
Kraemer, Hans d. 1508
Kraepelin, Emil 1856–1926
Krause, Fedor 1857–1937
Kries, Johannes von 1853–1928
Kühne, Willy 1837–1900
Kussmaul, Adolf 1822–1902

Ladd, George T. 1842–1921
Ladd-Franklin, Christine 1847–1930
Lallemand, Claude-François 1790–1854
Lamarck, Jean 1744–1829
Lancisi, Giovanni Maria 1654–1720
Lanfranchi, Guido d. 1315
Lange, Carl G. 1834–1900
Langley, John Newport 1852–1925
Larrey, Dominique Jean 1776–1842
Larsell, Olaf 1886–1964
Lashley, Karl 1890–1958
Lavater, Johann Kaspar 1741–1801
Lawson, Robert 1815–1894
Layard, Austin Henry 1817–1894
Laycock, Thomas 1812–1876
Le Bon, Gustave 1841–1931
Le Cat, Claude 1700–1768
Leeuwenhoek, Antony van 1632–1723
Legallois, Julien-Jean-César 1770–1840
Leibniz, Gottfried Wilhelm 1646–1716
Leidy, Joseph 1823–1891
Lélut, Louis Françisque 1804–1877
Lenin, Vladimir 1870–1924
Lépine, Jacques Raphaël 1840–1919
Lettson, John Coakley 1744–1815
Leuret, François 1797–1851
Lewandowsky, Max 1876–1912
Lewis, W. Bevan 1847–1929
Leyton (see Grünbaum)
Lhermitte, Pierre Marie 1877–1959
Liepmann, Hugo 1863–1925
Lima, Pedro Almeida b. 1903
Lincoln, Abraham 1809–1865
Linnaeus, Carolus 1707–1778

Lobstein, Johann G. 1777–1835
Locke, John 1632–1704
Loeb, Jacques 1859–1924
Löwenthal, Max Solly 1867–1960
Loewi, Otto 1873–1961
Lombroso, Cesare 1836–1909
Lomonosow, Mikhail Vasilevich 1711–1765
Long, Crawford 1815–1878
Longet, François Achille 1811–1871
Lonitzer, Adam 1528–1586
Loomis, Alfred 1887–1975
Lordat, Jacques 1773–1870
Lorente de Nó, Rafael b. 1902
Lorry, Antoine Charles de 1726–1783
Lort, Michael 1725–1790
Louis XIII of France 1601–1643
Louis XIV of France 1638–1715
Louis XV of France 1710–1774
Louis, Antoine 1723–1792
Lovén, Otto Christian 1835–1904
Luciani, Luigi 1840–1919
Lucretius c. 98–55 B.C.
Luys, Jules Bernard 1828–1895
Lyell, Charles 1797–1875
Lyon, Irving 1840–1896

Macewen, William 1848–1924
Mach, Ernst 1838–1916
MacKenzie, Stephen 1853–1925
MacLean, Paul b. 1913
Magendie, François 1783–1855
Magnus, Albertus (see Albertus Magnus)
Magnus, Rudolf 1873–1927
Magoun, Horace 1907–1991
Malacarne, Michele Vincenzo 1744–1816
Malebranche, Nicolas 1638–1715
Mall, Franklin Paine 1862–1917
Malpighi, Marcello 1628–1694
Mancinius, Thomas 1550–1620
Marat, Jean-Paul 1743–1793
Mareschal, Georges 1658–1736
Marie, Pierre 1853–1940
Marinesco, Georges 1864–1938
Mariotte, Edme 1620–1684
Marshall, Henry Rutgers 1852–1927
Marshall, John 1818–1891
Martin, Edward 1859–1938
Marxov, Ernst Fleischl von 1846–1892
Masawayh, Yuhanna lbn d. 857
Massa, Nicolo 1489–1569
Matiegka, Heinrich 1862–1941
Matteuci, Carlo 1811–1862
Maury, Louis 1817–1892
Mauthner, Ludwig 1840–1894
McCullough, Warren Sturgish 1898–1969
McDougall, William 1871–1938
Medici, Lorenzo de, Duke of Urbino 1492–1519
Meduna, Ladislas von 1896–1964
Mella, Hugo 1888–1969
Mendelev, Vladimir 1834–1907
Mercuriale, Girolamo 1530–1606
Meyer, Adolf 1866–1950

Meyer, Max Friedrich   1873–?
Meynert, Theodor   1833–1892
Mill, James   1773–1836
Mill, John Stewart   1806–1873
Mills, Charles   1845–1931
Milner, Brenda   b. 1918
Minkowski, Oskar   1858–1923
Mistichelli, Domenico   1675–1715
Mitchell, John Kearsley   1869–1927
Mitchell, Silas Weir   1829–1914
Mithridates VI of Pontus   120–63 B.C.
Möbius, Paul Julius   1853–1907
Mohammed   c. 570–632
Mohr, Carl Friedrich   1806–1879
Molinelli, Pietro Paolo   1702–1764
Monakow, Constantin von   1853–1930
Moniz, Egas   1874–1955
Monroe, Alexander   1697–1762
Monroe, Alexander   1737–1817
Montfauchon, Bernard de   1655–1741
Morehouse, George Reed   1829–1905
Morgagni, Giovanni Battista   1682–1771
Morton, Samuel George   1799–1851
Morton, William Pringle   1854–?
Morton, William T. G.   1819–1868
Moruzzi, Guiseppe   1910–1986
Mosso, Angelo   1846–1910
Mott, Frederick W.   1853–1926
Moyer, Harold   1858–1923
Müller, Heinrich   1820–1864
Müller, Johannes   1801–1858
Munk, Hermann   1839–1912
Musschenbroek, Petrus van   1692–1761
Muybridge, Eadweard   1830–1904
Myers, Ronald   b. 1929

Nansen, Fridtjof   1861–1930
Nemesius   fl. A.D. 390
Neubürger, Karl   1890–?
Newton, Isaac   1642–1727
Nicander   d. 135 B.C.
Niemann, Albert   1831–1917
Nissl, Franz   1860–1919
Nothnagal, Carl Wilhelm Hermann   1841–1905
Nott, Josiah Clark   1804–1873

Obersteiner, Heinrich   1847–1922
Ogle, John William   1824–1905
Ogle, William   1827–1912
Ohm, Georg Simon   1787–1854
Öhrwall, Hjalmar   1851–1929
Oppenheim, Hermann   1858–1919
Oppolzer, Johannes von   1808–1871
Orton, Samuel T.   1879–1948
Osborne, Johnathan   1795–1864
Osler, William   1849–1919
Ovid   43 B.C.–A.D. 18
Owen, Richard   1804–1892

Pacini, Filippo   1812–1883
Panizza, Bartolomeo   1785–1867
Papez, James   1883–1958
Paracelsus   1493–1541
Paré, Ambroise   1510–1590

Parinaud, Henri   1844–1905
Parkinson, James   1755–1824
Parrot, Jules   1829–1883
Pasteur, Louis   1822–1895
Paul of Aegina   seventh century
Pavlov, Ivan   1849–1936
Pawlow (see Pavlov)
Peacock, Thomas B.   1812–1882
Pearson, Karl   1857–1936
Pecham, John   d. 1292
Penfield, Wilder   1891–1976
Pepper, William   1843–1898
Perrault, Claude   1613–1688
Pertinax, Publius Helvius   126–193
Peterson, Frederick   1859–1938
Petrarch, Francesco   1304–1374
Peyligk, Johannes   1474–1522
Pflüger, Eduard Friedrich Wilhelm   1829–1900
Phillip II (King)   1527–1598
Pick, Arnold   1851–1924
Piéron, Henri   1881–1964
Pinel, Philippe   1745–1826
Plato   c. 429–348 B.C.
Platter, Felix   1536–1614
Pliny the Elder   c. 23–79
Pliny the Younger   c. 61–113
Plutarch   c. 46–119
Poe, Edgar Allan   1809–1849
Poliak, Stephen   1889–1955
Polidori, John William   1795–1821
Politzer, Adam   1835–1920
Poljak (see Poliak)
Polyak (see Poliak)
Pompadour, Madame   1721–1764
Pompey the Great   106–48 B.C.
Ponce de Leon, Pedro   1520–1584
Poppelreuter, Walther   1886–1939
Posidonius of Byzantium   fl. 370
Pourfour du Petit, François   1664–1741
Preyer, Thierry Wilhelm   1841–1877
Prichard, James   1786–1848
Priestley, Joseph   1733–1804
Prince, Morton   1854–1929
Pringle, John   1707–1782
Procháska, Jirí   1749–1820
Pruner, Franz   1808–1882
Pruner-Bey (see Pruner)
Ptolemaeus, Claudius   second century
Ptolemy I   367–283 B.C.
Puchelt, Friedrich A.   1784–1856
Purkinje (see Purkyně)
Purkyně, Jan   1787–1869
Purmann, Matthias   1648–1721
Puusepp, Ludvig   1875–1942
Pyl, Johann Theodor   1749–1794
Pythagoras   580–489 B.C.

Quaglino, Antonio   1817–1894
Quesnay, François   1694–1774
Quintilian, Marcus Fabius   c. 35–100

Rabl-Rückhard, Hermann   1839–1905
Ramón y Cajal, Santiago   1852–1934
Ramsay, William   1852–1916

Ramses II   1304–1237 B.C.
Ranson, Stephen   1880–1942
Ranvier, Louis Antoine   1835–1922
Ray, John   1627–1705
Redlich, Emil   1866–1930
Régis, Pierre Sylvain   1632–1707
Reid, Thomas   1710–1796
Reil, Johann Christian   1759–1813
Reissner, Ernst   1824–1878
Remak, Robert   1815–1865
Rembrandt (see van Rijn)
Retzius, Gustaf Magnus   1842–1919
Reynolds, Joshua   1723–1792
Rhazes   865–925
Ribot, Théodule   1839–1916
Rieger, Conrad   1855–1939
Rinne, Heinrich Adolf   1819–1868
Rivers, William   1864–1922
Roger, Henri   1809–1891
Rolando, Luigi   1773–1831
Rolleston, Humphry Davy   1862–1944
Rommel, Peter   1643–1708
Rosenblueth, Arturo   1900–1970
Rostan, Léon Louis   1790–1866
Roussy, Gustave   1874–1948
Rowntree, Leonard   1883–1959
Rudio, Eustachio   d. 1611
Rufus of Ephesus   fl. A.D. 98–117
Rush, Benjamin   1745–1813
Rutherford, William   1839–1899
Rylander, Gösta   b. 1903

Sagar, Johann   1732–1813
Sakel, Manfred   1900–1957
Saliceto, Guglielmo de   1210–1280
Saint Anthony   c. 250–355
Saint Augustine   354–430
Saint Martin, Alexis   1803–1886
Saint Thomas Aquinas   1225–1274
Santorini, Giovanni Domenico   1681–1737
Saucerotte, Louis Sebastian   1741–1814
Saul   1020–1000 B.C.
Saul, Leon J.   b. 1901
Sauvages de la Croix, François Boissier de   1706–1767
Savart, Félix   1791–1841
Schäfer, Edward Albert   1850–1935
Scheerer, Martin   1900–1961
Scheiner, Christopher   1571–1650
Schelhammer, Günther Christoph   1649–1716
Schenck a Grafenberg, Johann   1530–1598
Schiff, Moritz   1823–1896
Schmidt, Johann   1624–1690
Schneider, Conrad Victor   1614–1680
Schüller, Arthur   1874–1957
Schultze, Max   1825–1874
Schuster, Paul   1867–?
Schwalbe, Gustav   1844–1911
Schwann, Theodor   1810–1882
Scoville, William   1906–1984
Scribonius Largus   first century
Scultetus, Joannis   1595–1645
Sechenov, Ivan Mikhailovich   1829–1905
Sée, Germain   1818–1896
Seebeck, August   1805–1849

Sekhet-n-ankh   c. 2550 B.C.
Semon, Richard   1859–1918
Septimus, Lucius Severus   146–211
Seqenenre (King)   c. 1580 B.C.
Serres, Antoine   1776–1868
Sewall, Thomas   1786–1845
Shakespeare, William   1564–1616
Shaw, John Cargyll   1845–1900
Shelley, Mary Wollstonecraft   1797–1851
Shelley, Percy Bysshe   1792–1822
Sherrington, Charles Scott   1857–1952
Sigurdsson, Björn   1913–1959
Simon, Théodore   1873–1961
Simonides Melicus of Ceos   c. 556–470 B.C.
Smith, Edwin   1822–1906
Smith, Grafton Elliot   1871–1937
Socrates   c. 470–399 B.C.
Soltmann, Otto   1844–1912
Sömmerring, Samuel Thomas von   1755–1830
Sophocles   c. 497–406 B.C.
Soranus of Ephesus   98–135
Souques, Achille Alexandre   1860–1944
Spearman, Charles   1863–1945
Spencer, Herbert   1820–1903
Sperry, Roger   b. 1913
Spiller, William Gibson   1863–1940
Spina, Alessandro   d. 1313
Spitzka, Edward Anthony   1876–1922
Sprenger, Jakob   1436–1495
Spurling, Roy Glenn   b. 1894
Spurzheim, Johann   1776–1832
Squier, Ephraim George   1821–1888
Starr, Moses Allen   1854–1932
Steno (see Stensen)
Stensen, Niels   1638–1686
Stephanus (see Estienne)
Stern, William   1871–1938
Stevenson, Robert Louis   1850–1894
Stewart, Dugald   1753–1828
Stewart, T. Granger   1837–1900
Stieda, Ludwig   1837–1918
Stookey, Byron   1887–1966
Strabo   63 B.C.–A.D. 24
Strato d.   c. 270 B.C.
Strughold, H.   b. 1898
Sulzer, Johann   1720–1779
Swammerdam, Jan   1637–1680
Swanzy, Henry R.   1843–1913
Swedenborg, Emanuel   1688–1772
Sydenham, Thomas   1624–1689
Sylvius, François   1614–1672
Sylvius, Jacobinus   1478–1555

Tamburini, Augusto   1848–1919
Tanzi, Eugenio   1856–1934
Taylor, John   1703–1772
Terman, Louis   1877–1956
Thackery, William   1811–1863
Thales   652–548 B.C.
Theophrastus   c. 371–288 B.C.
Thomas Aquinas, Saint   1225–1274
Thorndike, Edward L.   1874–1949
Thurman, John   1810–1873
Ti   c. mid-third century B.C.

Tiberius   42 B.C.–A.D. 37
Tiedemann, Friedrich   1781–1861
Tigerstedt, Robert   1853–1923
Titchener, Edward Bradford   1867–1927
Titian   c. 1488–1576
Todd, Robert Bentley   1809–1859
Topinard, Paul   1830–1911
Tourtual, Caspar   1802–1865
Train, George Francis   1829–1904
Trajan   53–117
Treviranus, Gottfried Reinhold   1776–1837
Trousseau, Armand   1801–1867
Tulp, Nicolaas   1593–1674
Türck, Ludwig   1810–1868
Turgenev, Ivan Sergeyevitch   1818–1883
Turner, William   1832–1916
Tyndall, John   1820–1893

Uexküll, Jacob von   1864–1944

Valentin, Gabriel   1813–1883
Valsalva, Antonio Maria   1666–1738
van der Kolk, Jacob Ludwig Conrad Schroeder
   1797–1862
van Rijn, Rembrandt Harmensz   1606–1669
van Swieten, Gerard   1700–1772
Varolio, Constanzo   1543–1575
Vater, Abraham   1684–1751
Velpeau, Alfred Armand Louis Marie   1795–1867
Venturi, Giovanni   1746–1822
Verger, Henri   1873–1930
Vesalius, Andreas   1514–1564
Vesling, Johann   1598–1649
Vetruvius   first century
Vicq d'Azyr, Félix   1748–1794
Victoria, Queen   1819–1901
Vierordt, Karl von   1818–1884
Vieussens, Raymond   1641–1716
Vincent, Clovis   1879–1947
Vintschgau, Maximilian Ritter von   1832–1902
Virgil   70–19 B.C.
Vitello (*see* Witelo)
Vogt, Carl Christoph   1817–1895
Vogt, Cécile   1875–1962
Vogt, Oscar   1870–1950
Voisin, Félix   1784–1872
Volkmann, Alfred Wilhelm   1800–1877
Volta, Alessandro   1745–1827
Voltolini, Friedrich E. R.   1819–1889
Vulpian, Edme Félix   1826–1887

Wagner, Rudolph   1805–1864
Waldeyer, Wilhelm von   1836–1921
Walker, Alexander   1779–1852
Wallace, Alfred Russel   1823–1913
Waller, Augustus D.   1856–1922
Walsh, John   1725–1795
Walther, Phillip Franz von   1782–1849
Warren, John   1842–1927
Warren, John C.   1778–1856
Waters, Charles O.   1816–1892
Watson, John B.   1878–1958
Watt, Henry Jackson   1879–1925
Watts, James W.   b. 1904
Weber, Eduard   1806–1871
Weber, Ernst Heinrich   1795–1878
Weinhold, Karl August   1782–1829
Wells, Horace   1815–1848
Wernicke, Carl   1848–1904
Wertheimer, Max   1880–1943
Wesley, John   1703–1791
Wesphal, F. C. O.   1833–1890
Wever, Ernest Glenn   b. 1902
Whytt, Robert   1714–1766
Wigan, Arthur Ladbroke   ?–1847
Wilbrand, Hermann   1851–1935
Wilks, Samuel   1824–1911
Willis, Thomas   1621–1675
Wilson, S. A. Kinnier   1848–1937
Winslow, Forbes   1810–1874
Winslow, Jacobus ("Jacques")   1669–1760
Winterbottom, Thomas   1766–1859
Winthrop, John   1588–1649
Wiseman, Richard   c. 1621–1676
Witelo   c. 1220–1275
Wollaston, William Hyde   1766–1828
Wolpert, Ilja E.   1891–1967
Woodworth, Robert S.   1869–1962
Wren, Christopher   1632–1723
Wundt, William   1832–1920
Wyllie, J.   1844–1916

Yeo, Gerald   1845–1909
Young, Thomas   1773–1829

Zeno   356–264 B.C.
Ziehen, Theodor   1862–1950
Zimmermann, Johann Georg   1728–1795
Zinn, Johann Gottfried   1727–1759
Zoser (King) (*see* Djoser)
Zwaardemaker, Hendrik   1857–1930

Index